FOURTH EDITION

Intermediate Algebra

ARNOLD R. STEFFENSEN
L. MURPHY JOHNSON
Northern Arizona University

HarperCollins*Publishers*

To Barbara, Barbara, Becky, Cindy, and Pam

TO THE STUDENT

A *Student's Solutions Manual,* by Joseph Mutter, provides complete, worked-out solutions to all of the exercises in set A, the chapter review exercises, and the chapter tests. Your college bookstore either has this book or can order it for you.

Sponsoring Editor:	Bill Poole
Development Editor:	Terry McGinnis
Project Editor:	Sarah Joseph
Art Direction:	Julie Anderson
Text and Cover Design:	Lucy Lesiak Design/Lucy Lesiak
Cover Photo:	Simeone Huber/TSW/Click/Chicago Ltd.
Photo Research:	Judy Ladendorf
Director of Production:	Jeanie Berke
Production Assistant:	Linda Murray
Compositor:	York Graphic Services
Printer and Binder:	Courier Corporation

Intermediate Algebra, Fourth Edition

ISBN 0-673-46283-8

91 92 93 9 8 7 6 5 4 3 2

PREFACE

Intermediate Algebra, Fourth Edition, is designed for college students who have had a course in introductory algebra or who require a review before taking further courses in mathematics, science, business, or computer science. Informal yet carefully worded explanations, detailed examples with accompanying practice exercises, pedagogical second color, abundant exercises, and comprehensive chapter reviews are hallmarks of the book. The text has been written for maximum instructor flexibility. Both core and peripheral topics can be selected to fit individual course needs. An annotated instructor's edition, testing manual, solutions manual, and test generator are provided for the instructor. A solutions manual is available for student purchase. Interactive tutorial software and a set of instructional videotapes are also available.

FEATURES

STUDENT GUIDEPOSTS 1 Designed to help students locate important concepts as they study or review, student guideposts specify the major topics, rules, and procedures in each section. The guideposts are listed at the beginning of the section, then each is repeated as the corresponding material is discussed in the section.

EXAMPLES More than 700 carefully selected examples include detailed, step-by-step solutions and side annotations in color. Each example is headed by a brief descriptive title to help students focus on the concept being developed and to aid in review.

PRACTICE EXERCISES These parallel each example and keep students involved with the presentation by allowing them to check immediately their understanding of ideas. Answers immediately follow these exercises.

CAUTIONS This feature calls students' attention to common mistakes and special problems to avoid.

COLOR Pedagogical color highlights important information throughout the book. Key definitions, rules, and procedures are set off in colored boxes for increased emphasis. Figures and graphs utilize color to clarify the concepts presented. Examples present important steps and helpful side comments in color.

EXERCISES As a key feature of the text, more than 7000 exercises, including 2000 review exercises, are provided. Two parallel exercise sets (A and B) and a collection of extension exercises (C) follow each section and offer a wealth of practice for students and flexibility for instructors. Exercises ranging from the routine to the more challenging, including application and calculator problems, are provided.

Exercises A This set of exercises includes space for working the problems, with answers immediately following the exercises. Some of these problems, identified by a colored circle 21, have their solutions worked out at the back of the book. Many of these solutions are to exercises that students frequently have difficulty solving.

Exercises B This set matches the exercises in set A problem for problem but is presented without work space or answers.

Exercises C This set is designed to give students an extra challenge. These problems extend the concepts of the section or demand more thought than exercises in sets A and B. Answers or hints are given for selected exercises in this set.

REVIEW EXERCISES To provide ample opportunities for review, the text features a variety of review exercises.

For Review exercises are located at the end of most A and B exercise sets. They not only encourage continuous review of previously covered material, but also often provide special review preparation for topics covered in the upcoming section.

Chapter Review Exercises and a practice **Chapter Test** conclude each chapter. The Chapter Review Exercises are divided into two parts: the problems in Part I are ordered and marked by section. Those in Part II are not referenced to the source section, and are thoroughly mixed to help students prepare for examinations. Answers to all review and text exercises are provided in the text.

Final Review Exercises, referenced to each chapter and with answers supplied, are located at the back of the book.

CHAPTER REVIEWS In addition to the Chapter Review Exercises and Chapter Tests, comprehensive chapter reviews also include **Key Words** and **Key Concepts.** Key Words, listed by section, have brief definitions. Key Concepts summarize the major points of each section.

CALCULATORS It is assumed that most students have a hand-held scientific calculator. Therefore, use of a calculator, while not emphasized throughout the text, is integrated with such topics as exponents, radicals, and logarithms. Step-by-step key entries for algebraic calculators are shown in those sections. Some exercises, many in Exercises C, are marked for calculator use with the symbol ▦

INSTRUCTIONAL FLEXIBILITY

Intermediate Algebra, Fourth Edition, offers proven flexibility for a variety of teaching situations such as individualized instruction, lab instruction, lecture classes, or a combination of methods.

Material in each section of the book is presented in a well-paced, easy-to-follow sequence. Students in a tutorial or lab instruction setting, aided by the student guideposts, can work through a section completely by reading the explanation, following the detailed steps in the examples, working the practice exercises, and then doing the exercises in set A.

The book can also serve as the basis for, or as a supplement to, classroom lectures. The straightforward presentation of material, numerous examples, practice exercises, and three sets of exercises offer the traditional lecture class an alternative approach within the convenient workbook format.

NEW IN THIS EDITION

In a continuing effort to make this text even better suited to the needs of instructors and students, the following are some of the enhanced features of the new edition.

- The student guideposts are more visible.
- Explanations have been polished, reworded, and streamlined where appropriate. More figures and illustrations are used in discussions.
- Exercise sets have been reviewed for grading the balance of coverage. The number and variety of practical applications are increased. Additional challenging exercises and a greater emphasis on calculators have been incorporated into Exercises C.
- Additional For Review exercises review topics in preparation for the next section.
- Chapter Review Exercises are presented in two parts, one with sectional references and one without, to help students recognize problem types and better prepare for tests.
- Key Words given in the Chapter Review are expanded to include brief definitions.
- Greater emphasis has been given to the use of figures and illustrations in examples and exercises, particularly applications.
- Instructions for using a calculator have been expanded.
- Interval notation is introduced in Chapter 2 to help prepare students for later courses.

- Where appropriate, sections from the previous edition were recombined to form new sections more easily manageable in one class period.
- Graphing in the coordinate plane remains early in the text; graphing by plotting points from a table of values is given greater emphasis.
- The Pythagorean theorem and the distance and midpoint formulas are now discussed in Chapter 7.
- Complex numbers are introduced in Chapter 7, allowing for a more thorough discussion of quadratic equations in Chapter 8.
- An expanded treatment of functions and their properties has been added in new Chapter 9.
- The use of a calculator for computing logarithms replaces the previous dual presentation of both the calculator and logarithmic tables. The use of logarithmic tables and interpolation are still presented in the Appendix.
- Chapter 11 is expanded to include a more complete treatment of the conics, especially the circle and the parabola, with an increased emphasis on standard forms and graphing.

SUPPLEMENTS

An expanded supplemental package is available for use with *Intermediate Algebra, Fourth Edition.*

For the Instructor

The ANNOTATED INSTRUCTOR'S EDITION provides instructors with immediate access to the answers to every exercise in the text; each answer is printed next to the corresponding text exercise.

The INSTRUCTOR'S TESTING MANUAL contains a series of **ready-to-duplicate tests,** including a Placement Test, six different but equivalent tests for each chapter, and two final exams, all with answers supplied. More than 1000 **additional problems,** grouped by section, are also included for use as in-class examples or on quizzes and tests. Section-by-section **teaching tips** provide suggestions for content implementation that an instructor, tutor, or teaching assistant might find helpful.

The INSTRUCTOR'S SOLUTIONS MANUAL contains complete, worked-out solutions to every exercise in sets B and C of the text.

HARPERCOLLINS TEST GENERATOR FOR MATHEMATICS Available in Apple, IBM, and Macintosh versions, the test generator enables instructors to select questions by objective, section, or chapter, or to use a ready-made test for each chapter. Instructors may generate tests in multiple-choice or open-response formats, scramble the order of questions while printing, and produce multiple versions of each test (up to 9 with Apple, up to 25 with IBM and Macintosh). The system features printed graphics and accurate mathematics symbols. It also features a preview option that allows instructors to view questions before printing, to regenerate variables, and to replace or skip questions if desired. The IBM version includes an editor that allows instructors to add their own problems to existing data disks.

VIDEOTAPES A new videotape series, *Algebra Connection: The Intermediate Algebra Course* has been developed to accompany *Intermediate Algebra, Fourth Edition.* Produced by an Emmy Award-winning team in consultation with a task force of academicians from both two-year and four-year colleges, the tapes cover all objectives, topics, and problem-solving techniques within the text. In addition, each lesson is preceded by motivational ''launchers'' that connect classroom activity to real-world applications.

For the Student

The STUDENT'S SOLUTIONS MANUAL contains complete, worked-out solutions to each practice exercise in the text, to every exercise in set A, and to all Chapter Review Exercises and Chapter Tests.

INTERACTIVE TUTORIAL SOFTWARE This innovative package is also available in Apple, IBM, and Macintosh versions. It offers interactive modular units, specifically linked to the text, for reinforcement of selected topics. The tutorial is self-paced and provides unlimited opportunities to review lessons and to practice problem solving. When students give a wrong answer, they can request to see the problem worked out. The program is menu-driven for ease of use, and on-screen help can be obtained at any time with a single keystroke. Students' scores are automatically recorded and can be printed for a permanent record.

ACKNOWLEDGMENTS

We extend our sincere gratitude to the students and instructors who used the previous editions of this book and offered many suggestions for improvement. Special thanks go to the instructors at Northern Arizona University and Yavapai Community College. In particular, the assistance given over the years by James Kirk and Michael Ratliff is most appreciated. Also, we sincerely appreciate the support and encouragement of the Northern Arizona University administration, especially President Eugene M. Hughes. It is a pleasure and privilege to serve on the faculty of a university that recognizes quality teaching as its primary role.

We also express our thanks to the following instructors who responded to a questionnaire sent out by HarperCollins: Janet Brougher, University of Oregon; Georgia Hart, Arizona Western College; Richard Hilliard, Central Washington University; Susan Hollar, Kalamazoo Valley Community College; Mary F. King, Waycross College; Ann Thorne, College of DuPage; and Susan Townsend, Kalamazoo Valley Community College.

We are also indebted to the following reviewers for their countless beneficial suggestions at various stages of the book's revision:

Jane Edgar, *Brevard Community College*
Paul Eenigenburg, *Western Michigan University*
Ronald D. Faulstich, *Moraine Valley Community College*
Al Giambrone, *Sinclair Community College*
Linda Holden, *Indiana University, Bloomington*
Phyllis H. Jore, *Valencia Community College*
Maureen Kelley, *Northern Essex Community College*
Rebecca R. Kitto, *Antelope Valley College*
Susan C. Meyers, *Sinclair Community College*
Lloyd C. Neitling, *University of Nevada, Las Vegas*
Marilyn P. Persson, *University of Kansas*
Donna Fields Rochon, *Western Washington University*
Mark Serebransky, *Camden County College*
Mark L. Sigfrids, *Kalamazoo Valley Community College*
Anthony Soychak, *University of Southern Maine*
Barbara W. Worley, *Des Moines Area Community College*

We extend special appreciation to Joseph Mutter for writing the *Student's Solutions Manual* and for his countless suggestions and support since the text's first edition. Thanks go to Diana Denlinger for typing this edition and the *Instructor's Solutions Manual*.

We thank our editors, Jack Pritchard, Bill Poole, Terry McGinnis, and Sarah Joseph, whose support has been most appreciated.

Finally, we are indebted to our families and in particular our wives, Barbara and Barbara, whose encouragement over the years cannot be measured.

Arnold R. Steffensen
L. Murphy Johnson

CONTENTS

TO THE STUDENT

During the past several years we have taught intermediate algebra to more than 1500 students and have heard the following comments many times. "I've always been afraid of math and have avoided it as much as possible. Now my major requires algebra and I'm petrified." "I don't like math, but it's required to graduate." "I can't do word problems!" If you have ever made a similar statement, now is the time to think positively and start down the path toward success in mathematics. Don't worry about this course as a whole. The material in the text is presented in a way that lets you take one small step at a time. As you begin, keep in mind these guidelines that are both necessary and helpful.

GENERAL GUIDELINES

1. Mastering algebra requires motivation and dedication. Just as an athlete does not improve without commitment to his or her goal, an algebra student must be prepared to work hard and spend time studying. Make it a point to keep up with your assignments. If you need extra help, *ask* for it right away.
2. Algebra is not learned simply by watching, listening, or reading; *it is learned by doing*. Use your pencil and practice. When your thoughts are organized and written in a neat and orderly way, you have taken a giant step toward success. Be complete and write out all details. The following are samples of two students' work on a word problem. Can you tell which one was more successful in the course?

   STUDENT A

   Let t = time the woman walks
   $7-t$ = time she rides
   $4.5t$ = distance she walks
   $2.5(7-t)$ = distance she rides
   $4.5t = 2.5(7-t)$
   $4.5t = 17.5-2.5t$
   $7t = 17.5$
   $t = 2.5$
   She walks for 2.5 hr.

   STUDENT F

   $7(4.5) = 2.5t$
   t = time
   ~~$4.5+t = 7.5-t$~~
   ~~$4.5t = 2.5t$~~

3. If you are allowed to use a calculator in your course, consult your owner's manual to become familiar with its features. When you are computing with decimals or approximating irrational numbers, a calculator can be a time-saving device. On the other hand, you should not be so dependent on a calculator that you use it for simple calculations that can be done mentally. For example, it would be ridiculous to use a calculator to solve an equation such as $2x = 8$, while it would be helpful to solve $1.12x = 7343.84$. It is important that you learn when to use and when not to use your calculator; some problems for which a calculator may be especially helpful are marked with a calculator symbol ▦.

SPECIFIC GUIDELINES

1. As you begin to study each section, look through the material for a preview of what is coming.
2. Return to the beginning of the section and start reading slowly. The STUDENT GUIDEPOSTS will help you find important concepts as you progress or as you review.
3. Read through each EXAMPLE and make sure you understand every step. The side comments in color will help you if something is not clear.
4. After reading each example, work the parallel PRACTICE EXERCISE. This will reinforce what you have just read and start the process of practice. (*Note:* Complete solutions to the Practice Exercises are available in the *Student's Solutions Manual.*)
5. Now and then you will see a CAUTION. These warn you of common mistakes and special problems to avoid.
6. After you finish the material in the body of the section, you need to practice, practice, practice! Begin with the exercises in set A. Answers to all of these problems are at the end of the set for easy reference. Some of the problems, identified by a blue circle on the number, have step-by-step solutions at the back of the book. After trying these problems, refer to the step-by-step solutions if you have difficulty. To practice more, do the exercises in set B; to challenge yourself, try the exercises in set C. (*Note:* Solutions to all problems in set A are available in the *Student's Solutions Manual.*)
7. After you have completed all of the sections in the chapter, read the CHAPTER REVIEW that contains Key Words and Key Concepts. The Review Exercises provide more practice before you take the Chapter Test. Answers to the tests are at the back of the book. Solutions are provided in the *Student's Solutions Manual.*
8. To help you study for your final examination, we have concluded the book with a comprehensive set of FINAL REVIEW EXERCISES.

If you follow these steps and work closely with your instructor, you will greatly improve your chances for success in algebra.

Best of luck, and remember that you can do it!

CHAPTER 1

Review of Fundamental Concepts

1.1 REAL NUMBERS

STUDENT GUIDEPOSTS

1 Sets and Elements	4 Equality
2 Sets of Numbers	5 Inequality
3 The Number Line	6 Graphing Inequalities

1 SETS AND ELEMENTS

In this chapter we develop the basic terminology used throughout our study of algebra. Two important terms are *set* and *element*. A **set** is a collection of objects. These objects, the **elements** of the set, are listed within braces, { }, and the symbol $\in$ means "is an element of." For example,

$$c \in \{a, b, c, d\}$$

means that c is an element of the set $\{a, b, c, d\}$. The symbol $\notin$ means "is *not* an element of."

$$f \notin \{a, b, c, d\}.$$

Capital letters are used to denote sets. In

$$A = \{\text{Troy, Scott, Jay, Shawn, Andy}\},$$

the letter A represents the set of five people. Another way of describing the set A is to give the common characteristic of its elements.

$$A = \{\text{the starting five on the State University basketball team}\}$$

2 SETS OF NUMBERS

Most sets we deal with are sets of *numbers,* concepts named by symbols called **numerals.** The most basic set of numbers is the set of **natural** or **counting** numbers, N.

$N = \{1, 2, 3, 4, 5, \ldots\}$ Natural (counting) numbers

The three dots mean that the pattern continues on in the same manner. Sets that have no limit or no last element, such as N, are called **infinite sets,** while sets whose elements can be counted, such as A above, are **finite sets.** One special set is the **empty** or **null set** which has no elements. The empty set is denoted by the symbol $\varnothing$.

When the number 0 is included with the set of natural numbers, we have the set of **whole numbers.**

$$W = \{0, 1, 2, 3, 4, \ldots\} \qquad \text{Whole numbers}$$

When the negatives of all the natural numbers are included with the whole numbers, we have the set of **integers.**

$$I = \{\ldots -3, -2, -1, 0, 1, 2, 3, \ldots\} \qquad \text{Integers}$$

We often talk about the **negative integers,** $\{-1, -2, -3, \ldots\}$, and the **positive integers,** $\{1, 2, 3, \ldots\}$.

A **variable** is a symbol such as x that represents any element of a set containing more than one number. For example, if n represents any positive integer, then n is a variable. A **constant** is a symbol denoting a specific number. For example, 4 is a constant.

Using variables, we can describe sets with **set-builder notation.** For example,

$$\{x | x \text{ is a natural number between 2 and 7}\},$$

is read "the set of all x such that x is a natural number between 2 and 7," and describes in set-builder notation the set

$$\{3, 4, 5, 6\}.$$

The set of numbers formed by taking quotients of integers (division by zero is excluded) is the set of **rational numbers.** In set-builder notation, the rational numbers, denoted by Q, are given by

$$Q = \left\{ \frac{a}{b} \,\middle|\, a \text{ and } b \text{ are integers and } b \neq 0 \right\} \qquad \text{Rational numbers}$$

This is read "the set of all $\frac{a}{b}$ such that a and b are integers and b is not zero." The vertical bar represents the words "such that."

Some examples of rational numbers are

$$\frac{2}{3}, \quad \frac{3}{1}, \quad \frac{-1}{5}, \quad \frac{12}{7}, \quad \frac{-8}{1}, \quad \frac{28}{-3}, \quad \text{and} \quad \frac{0}{1}.$$

Since every integer can be expressed as a quotient of itself and 1, such as 3, -8, and 0 above, integers are also rational numbers.

Every rational number can be expressed as a *fraction* or a *decimal*. For example, $\frac{3}{8}$ can be written 0.375, the result obtained by dividing 3 by 8. Likewise, $\frac{3}{11}$ can be expressed as 0.2727 . . . . The decimal 0.375 is a **terminating decimal** since the sequence of digits comes to an end. The decimal 0.2727 . . .

is a **repeating decimal** since a block of digits, in this case 27, repeats indefinitely. We can write such numerals with a bar over the repeating block.

$$\frac{3}{11} = 0.\overline{27} \quad \text{and} \quad \frac{1}{3} = 0.\overline{3}$$

In general, every rational number can be written as either a terminating or an infinitely repeating decimal.

Numbers that are not rational, that *cannot* be expressed as a quotient of two integers, are **irrational numbers.** An irrational number cannot be written as a terminating or repeating decimal. But irrational numbers can be approximated by rational numbers; the irrational number π, the ratio of the circumference of a circle to its diameter, is often approximated by 3.14. Most square roots of natural numbers, such as $\sqrt{2}$, $\sqrt{3}$, $\sqrt{5}$, and $\sqrt{7}$, are irrational numbers, but some, such as $\sqrt{1} = 1$, $\sqrt{4} = 2$, $\sqrt{9} = 3$, and $\sqrt{16} = 4$, are rational.

The set of **real numbers,** denoted by R, consists of the rational numbers together with the irrational numbers. The relationships among the types of numbers we have discussed are displayed in Figure 1.1.

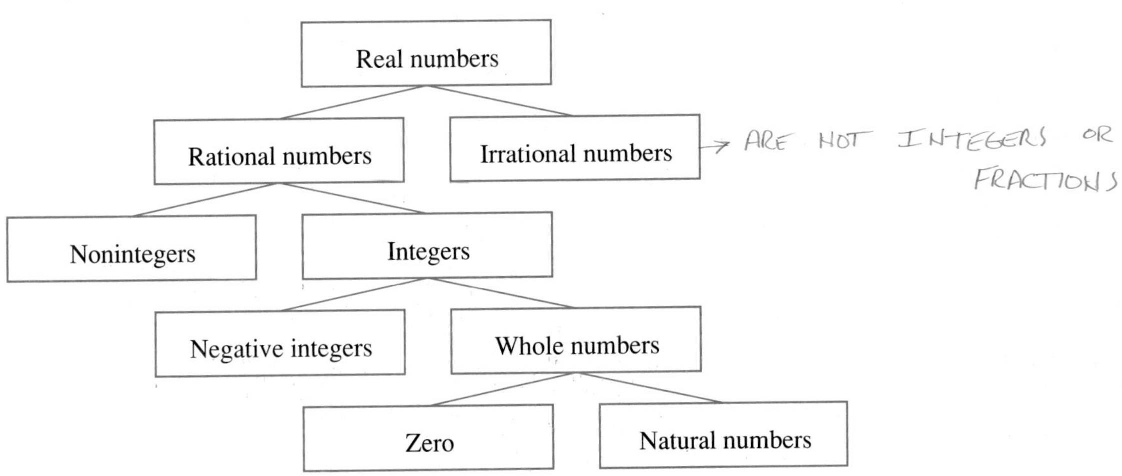

Figure 1.1 Sets of Numbers

EXAMPLE 1

List the numbers in the set $\left\{-8, \sqrt{11}, 2\pi, 0, 7, \frac{2}{3}, -5.6\right\}$ that belong to the specified set.

(a) Natural numbers: 7

(b) Integers: -8, 0, 7

(c) Rational numbers: -8, 0, 7, $\dfrac{2}{3}$, -5.6

(d) Irrational numbers: $\sqrt{11}$, 2π

PRACTICE EXERCISE 1

Using the sets (a)–(d) in Example 1, list the sets to which each number belongs.

(a) 22

(b) 7π

(c) -10

(d) $\dfrac{8}{3}$

Answers: (a) natural numbers, integers, rational numbers
(b) irrational numbers
(c) integers, rational numbers
(d) rational numbers

3 THE NUMBER LINE

An excellent means of displaying numbers and showing some of their properties is a **number line,** as shown in Figure 1.2.

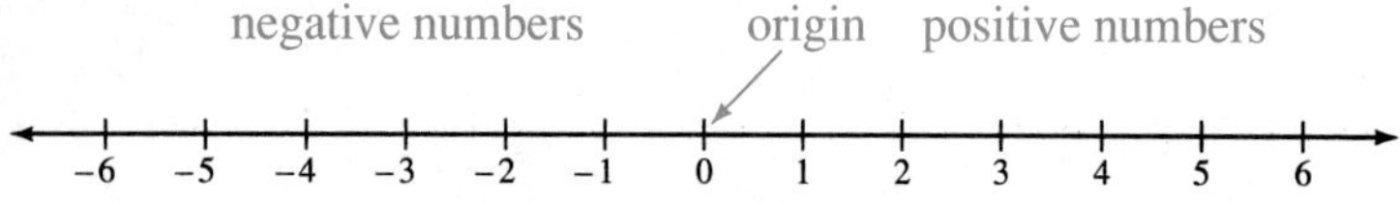

Figure 1.2 Number Line

Every real number corresponds to exactly one point on a number line, and every point on a number line corresponds to exactly one real number. The number line in Figure 1.3 has several numbers plotted on it; each point is the **graph** of the number.

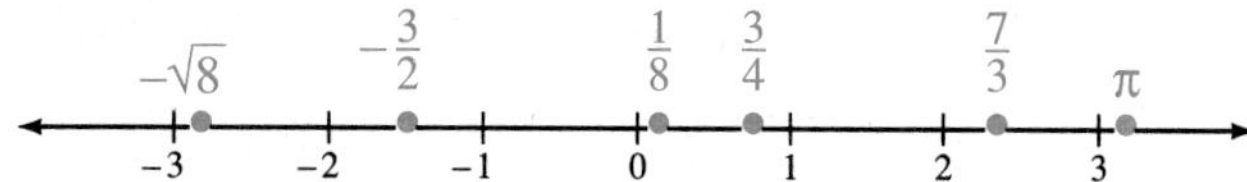

Figure 1.3 Points on a Number Line

4 EQUALITY

Two numbers a and b that correspond to the same point on a number line are **equal,** written $a = b$. For example,

$$\frac{4}{2} = 2, \quad 1.5 = \frac{3}{2}, \quad \text{and} \quad \frac{3}{11} = 0.\overline{27}.$$

Numbers that are not equal (written $a \neq b$) correspond to different points.

The **properties of equality,** assumed to be true without verification and listed below, are used for solving equations and simplifying algebraic expressions.

Properties of Equality

Let a, b, and c be real numbers.

Reflexive Law	$a = a$
Symmetric Law	If $a = b$, then $b = a$.
Transitive Law	If $a = b$ and $b = c$, then $a = c$.
Substitution Law	If $a = b$, then either may replace the other in any statement without affecting the truth of the statement.

EXAMPLE 2 RECOGNIZING PROPERTIES OF EQUALITY

State the property of equality illustrated by each statement.

(a) $4 = 4, \quad -3 = -3, \quad \frac{1}{2} = \frac{1}{2}$ Reflexive law

(b) If $x = 8$, then $8 = x$. Symmetric law

PRACTICE EXERCISE 2

State the property of equality illustrated by each statement.

(a) If $y = 7$ and $7 = w$, then $y = w$.

(b) If $x = 2$ and $z + x = 8$, then $z + 2 = 8$.

Answers: (a) transitive law (b) substitution law

5 INEQUALITY

When one number a is to the *left* of another number b on a number line, we say that a is **less than** b and write $\boldsymbol{a < b}$. We also say that b is **greater than** a and write $\boldsymbol{b > a}$. For example, 1 is to the left of 3 on a number line in Figure 1.3 and we write $1 < 3$. Also, we can see in Figure 1.3 that

$$-3 < -1, \quad -2 < 0, \quad -2 < \frac{3}{4}, \quad \frac{1}{8} < \pi.$$

We also say that 3 is greater than 1 and write $3 > 1$. Similarly,

$$-1 > -3, \quad 0 > -2, \quad \frac{3}{4} > -2, \quad \pi > \frac{1}{8}.$$

The symbol $\leq$ means *less than or equal to*. Similarly, $\geq$ represents the phrase *greater than or equal to*. Thus,

$$-3 \leq -3, \quad -2 \leq 0, \quad 1 \geq 0, \quad -\frac{3}{2} \leq \frac{3}{2}, \quad \frac{7}{4} \geq \frac{7}{4}.$$

The symbols $<$, $>$, $\leq$, and $\geq$ are **inequality symbols.** A more precise definition of these relationships will be given in Section 1.3.

6 GRAPHING INEQUALITIES

An inequality involving a variable such as

$$x > 1$$

means $\{x | x \text{ is greater than } 1\}$. This can be graphed on a number line as in Figure 1.4. The open dot at the point 1 indicates that 1 is not included in the graph. Since the inequality $x \leq 1$ includes the point 1, it has a solid dot at 1 in Figure 1.5.

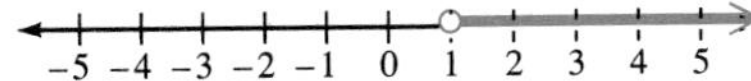

Figure 1.4 $x > 1$

Figure 1.5 $x \leq 1$

Inequalities such as $-2 < x < 1$ indicate the points between -2 and 1; the graph is shown in Figure 1.6. The graph of $-3 \leq x < 2$ is in Figure 1.7.

Figure 1.6 $-2 < x < 1$

Figure 1.7 $-3 \leq x < 2$

1.1 EXERCISES A

Decide whether each statement is true *or* false. *If the statement is false, explain why.*

1. A collection of objects is called an element.

2. The symbol used to name a number is a numeral.

3. $\{1, 2, 3, 4, \ldots\}$ is the set of counting or natural numbers.

4. Sets that have no limit or no last element are finite sets.

5. $\{\ldots, -3, -2, -1, 0, 1, 2, 3, \ldots\}$ is the set of integers.

6. The set of rational numbers is formed by taking all quotients of integers, division by zero excluded.

7. When we identify a point on a number line associated with a number, we are plotting the number.

8. If one number a is to the left of another number b on a number line, we say that a is greater than b.

Write each set by listing its elements within braces.

9. $\{x|x$ is a letter in the word DALLAS$\}$

10. $\{x|x$ is a natural number between 2 and 8$\}$

11. $\{x|x$ is one of the first four whole numbers$\}$

12. $\{x|x$ is a natural-number multiple of 2$\}$

Write each set using set-builder notation. [*Note: More than one description may exist.*]

13. $\{1, 2, 3\}$

14. $\{\pi\}$

15. $\{1, 3, 5, 7, 9, \ldots\}$

16. $\left\{1, \frac{1}{2}, \frac{1}{3}, \frac{1}{4}, \ldots\right\}$

Is the given set finite *or* infinite*?*

17. $\{x|x$ is a whole number$\}$

18. $\{x|x$ is a natural number less than 1000$\}$

19. $\{x|x$ is an integer greater than $-10\}$

20. $\{x|x$ is a natural number less than 2$\}$

Express each rational number as a decimal.

21. $\frac{7}{8}$

22. $\frac{1}{7}$

23. $\frac{12}{5}$

24. $\frac{1}{3}$

Answer true *or* false.

25. Every natural number is a whole number.

26. Every integer is a whole number.

27. Some integers are whole numbers.

28. All rational numbers are real numbers.

29. Given $X = \left\{15, -3, \frac{4}{3}, 1.75, -2.\overline{4}, \pi, 0, \sqrt{3}\right\}$, list the elements of X that are

(a) natural numbers

(b) integers

(c) rational numbers

(d) irrational numbers

(e) real numbers.

30. Using the sets (a)–(e) in Exercise 29, list all the sets to which each of the following numbers belong.

(a) 5

(b) $-\sqrt{2}$

(c) $\frac{1}{5}$

Graph each set of numbers on the given number line.

31. $\{4, -2, 1, 3\}$

−6 −5 −4 −3 −2 −1 0 1 2 3 4 5 6

32. $\left\{1.5, -\frac{5}{3}, \frac{7}{2}, -3\frac{1}{2}\right\}$

−6 −5 −4 −3 −2 −1 0 1 2 3 4 5 6

Decide whether each statement is true *or* false.

33. The fact that $7 = 7$ illustrates the symmetric law of equality.

34. If $a = -3$ then $-3 = a$ by the symmetric law of equality.

35. If $w = 2$ and $2 = c$, then we know that $w = c$ by the transitive law of equality.

36. If $x = 4$ and $a = 1 + x$, then $a = 5$ by the substitution law of equality.

Write the correct symbol (=, <, or >) for each pair of numbers.

37. 4 1

38. −6 0

39. 0 2

40. $\frac{1}{2}$ $-\frac{3}{4}$

41. $-\frac{7}{2}$ $-3\frac{1}{2}$

42. $-\frac{5}{8}$ -2

Graph each inequality on the given number line.

43. $x < 3$

−5 −4 −3 −2 −1 0 1 2 3 4 5

44. $x \geq -\frac{1}{2}$

−5 −4 −3 −2 −1 0 1 2 3 4 5

45. $-2 < x < 3$

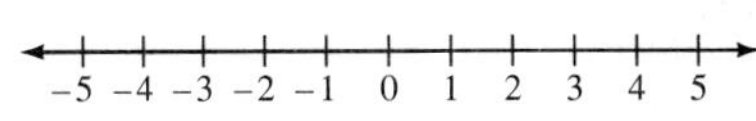

Give the inequality that is graphed on the given number line.

46. −5 −4 −3 −2 −1 0 1 2 3 4 5

47. −5 −4 −3 −2 −1 0 1 2 3 4 5

ANSWERS: **1.** false; it's called a set **2.** true **3.** true **4.** false; infinite sets **5.** true **6.** true **7.** true **8.** false; *a* is less than *b* **9.** {D, A, L, S} **10.** {3, 4, 5, 6, 7} **11.** {0, 1, 2, 3} **12.** {2, 4, 6, 8, . . .} **13.** $\{x|x$ is a whole number between 0 and 4$\}$ **14.** $\{x|x$ is the ratio of the circumference of any circle to its diameter$\}$ **15.** $\{x|x$ is an odd counting number$\}$ **16.** $\{x|x$ is a rational number formed by dividing 1 by a counting number$\}$ **17.** infinite **18.** finite **19.** infinite **20.** finite **21.** 0.875 **22.** $0.\overline{142857}$ **23.** 2.4 **24.** $0.\overline{3}$ **25.** true **26.** false; for example, −1 **27.** true **28.** true **29.** (a) 15 (b) 15, −3, 0 (c) 15, −3, $\frac{4}{3}$, 1.75, $-2.\overline{4}$, 0 (d) π, $\sqrt{3}$ (e) All numbers in *X* are real numbers. **30.** (a) natural numbers, integers, rational numbers, real numbers (b) irrational numbers, real numbers (c) rational numbers, real numbers **31.** −2 1 3 4; −6 −5 −4 −3 −2 −1 0 1 2 3 4 5 6

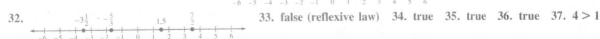

32. $-3\frac{1}{2}$ $-\frac{5}{3}$ 1.5 $\frac{7}{2}$; −6 −5 −4 −3 −2 −1 0 1 2 3 4 5 6 **33.** false (reflexive law) **34.** true **35.** true **36.** true **37.** $4 > 1$

38. $-6 < 0$ **39.** $0 < 2$ **40.** $\frac{1}{2} > -\frac{3}{4}$ **41.** $-\frac{7}{2} = -3\frac{1}{2}$ **42.** $-\frac{5}{8} > -2$ **43.** −5 −4 −3 −2 −1 0 1 2 3 4 5

44. −5 −4 −3 −2 −1 0 1 2 3 4 5 **45.** −5 −4 −3 −2 −1 0 1 2 3 4 5 **46.** $x \leq 0$ **47.** $-3 < x \leq 5$

1.1 EXERCISES B

Decide whether each statement is true *or* false. *If the statement is false, explain why.*

1. The objects that belong to a set are elements of the set.

2. Sets with a fixed or limited number of elements are infinite sets.

3. $\{1, 2, 3, 4, \ldots\}$ is the set of natural or counting numbers.

4. $\{0, 1, 2, 3, \ldots\}$ is the set of whole numbers.

5. $\{\ldots, -3, -2, -1\}$ is the set of positive integers.

6. The set of irrational numbers consists of all numbers that in decimal form either terminate or have a repeating block of digits.

7. When we plot a number on a number line, the point is called the graph of the number.

8. If one number a is to the right of another number b on a number line, we say that a is less than b.

Write each set by listing its elements within braces.

9. $\{x|x$ is a letter in the word ARIZONA$\}$

10. $\{x|x$ is an integer between -1 and $1\}$

11. $\{x|x$ is one of the first three natural numbers$\}$

12. $\{x|x$ is a natural-number multiple of $5\}$

Write each set using set-builder notation. [*Note: More than one description may exist.*]

13. $\{0, 1, 2, 3, 4\}$

14. $\{4\}$

15. $\{2, 4, 6, 8, 10, \ldots\}$

16. $\left\{-1, -\frac{1}{2}, -\frac{1}{3}, -\frac{1}{4}, \ldots\right\}$

Decide whether each set is finite *or* infinite.

17. $\{x|x$ is an integer$\}$

18. $\{x|x$ is a whole number less than $10{,}000\}$

19. $\{x|x$ is an integer less than $10\}$

20. $\{x|x$ is a whole number between 2 and $4\}$

Express each rational number as a decimal.

21. $\frac{3}{4}$

22. $\frac{7}{11}$

23. $\frac{15}{4}$

24. $\frac{2}{3}$

Write true *or* false *for each statement.*

25. Every counting number is an integer.

26. Every integer is a counting number.

27. Some rational numbers are integers.

28. All real numbers are rational numbers.

29. Given $X = \left\{-10, 4, \frac{5}{4}, -2.5, 3.\overline{25}, -\pi, -\sqrt{3}\right\}$, list the elements of X that are

(a) natural numbers

(b) integers

(c) rational numbers

(d) irrational numbers

(e) real numbers.

30. Using the sets (a)–(e) in Exercise 29, list all the sets to which each of the following numbers belong.

(a) -2

(b) $\sqrt{11}$

(c) $-\frac{2}{3}$

Graph each set of numbers on a number line.

31. $\{5, 2, -3, 6\}$

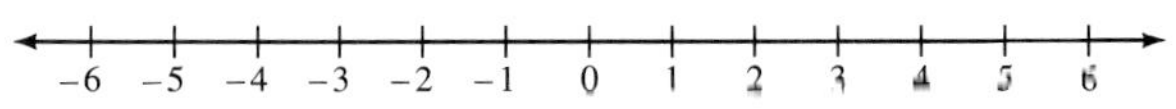

32. 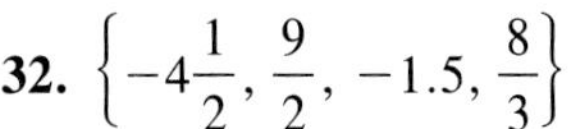 $\left\{-4\frac{1}{2}, \frac{9}{2}, -1.5, \frac{8}{3}\right\}$

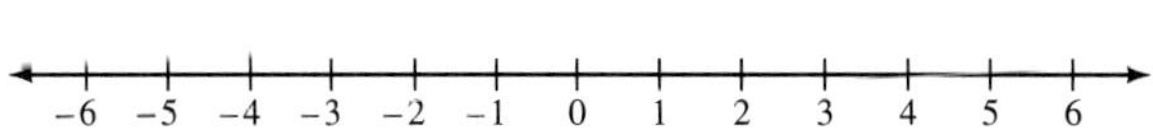

Decide whether each statement is true *or* false.

33. If $x = 2$ then $2 = x$ by the transitive law of equality.

34. If $a = 3$ and $y = a + 3$, then $y = 6$ by the substitution law of equality.

35. The fact that $\sqrt{2} = \sqrt{2}$ illustrates the reflexive law of equality.

36. If $y = -3$ and $-3 = w$ then $y = w$ by the transitive law of equality.

Write the correct symbol (=, <, or >) for each pair of numbers.

37. 2 5

38. 0 -4

39. 4 0

40. $-\frac{7}{4}$ $\frac{5}{2}$

41. $-\frac{9}{2}$ $-4\frac{1}{2}$

42. $-4\frac{1}{2}$ -1.5

Graph each inequality on a number line.

43. $x > -3$

−5 −4 −3 −2 −1 0 1 2 3 4 5

44. $x \le \frac{1}{2}$

−5 −4 −3 −2 −1 0 1 2 3 4 5

45. $-1 < x \le 5$

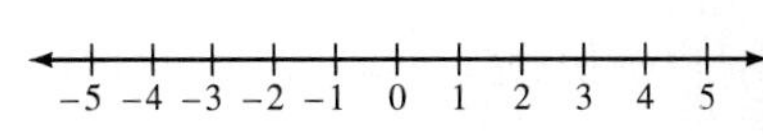

Give the inequality that is graphed on the given number line.

46. −5 −4 −3 −2 −1 0 1 2 3 4 5

47. −5 −4 −3 −2 −1 0 1 2 3 4 5

1.1 EXERCISES C

**Write the correct symbol (=, <, or >) for each pair of numbers.*

1. $\frac{2}{9}$ 0.223

2. $-5\frac{5}{9}$ $-\frac{28}{5}$

3. Express $\frac{93{,}824}{211{,}104}$ as a decimal. [Answer: $0.\overline{4}$]

4. List the elements in the set $\{x | x = 2n - 1, n \in N\}$.

---

*Although not required, a calculator may be helpful for solving some of these problems. Such exercises are marked by a symbol throughout the book.

5. What is the greatest integer that is less than $\frac{32}{7}$?

6. What is the smallest integer that is greater than $\frac{32}{7}$?

1.2 OPERATIONS ON REAL NUMBERS

STUDENT GUIDEPOSTS

1 Negatives	3 Operations on Real Numbers
2 Absolute Value	4 Reciprocals

In this section we consider the operations of addition, subtraction, multiplication, and division on the real numbers. Two concepts necessary for understanding these operations are *negatives* and *absolute value*.

1 NEGATIVES

Numbers on a number line that are on opposite sides of 0 but the same distance from 0, are called **negatives** (or **additive inverses**) of each other. For example, 2 and -2 are negatives, as are $-\frac{3}{4}$ and $\frac{3}{4}$. In general, if a is any number, the negative of a is $-a$, found by changing the sign of a. This means that if a is 2, then $-a = -2$ is the negative of 2. If a is $-\frac{3}{4}$, then

$$-a = -\left(-\frac{3}{4}\right) = \frac{3}{4}$$

is the negative of $-\frac{3}{4}$. Thus, $-a$ is positive if a is negative. This is an example of the following property.

Double-Negation Property

For any real number a,

$$-(-a) = a.$$

EXAMPLE 1 USING THE DOUBLE-NEGATION PROPERTY

Evaluate.

(a) $-(-4) = 4$ 4 is the negative of -4

(b) $-\left(-\frac{3}{2}\right) = \frac{3}{2}$

(c) $-[-(-2)] = -[2]$
$= -2$

PRACTICE EXERCISE 1

Evaluate.

(a) $-(-11)$

(b) $-(-6.5)$

(c) $-[-(-4)]$

Answers: (a) 11 (b) 6.5 (c) -4

Notice that the symbol "$-$" can be used in three ways:

1. As an operational sign for subtraction: $5 - 2 = 3$
2. As part of a number symbol indicating a negative number: -4 and $-\frac{2}{3}$
3. As an operational sign for negation: $-a$, $-(-2)$, $-0 = 0$.

2 ABSOLUTE VALUE

The **absolute value** of a number is its distance from zero on a number line. Numbers that are negatives of each other are located the same distance from zero, and as a result, have the same absolute value. The absolute value of a number a is denoted by $|a|$, read "the absolute value of a." Hence, $|3| = 3$, $|-3| = 3$, $|4| = 4$, and $\left|-\frac{5}{2}\right| = \frac{5}{2}$, as shown in Figure 1.8. *The absolute value of any number except zero is always a positive number, and the absolute value of zero is zero.*

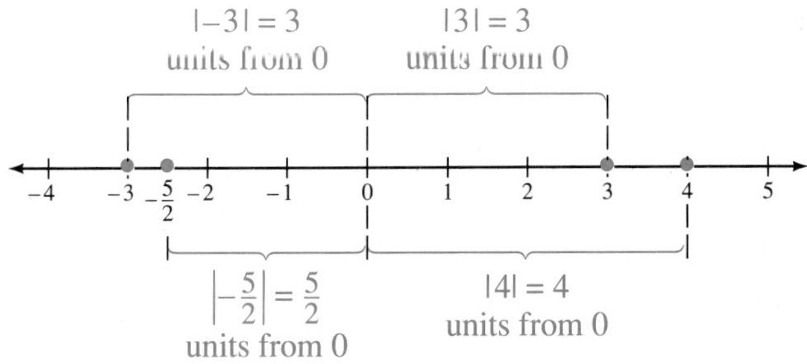

Figure 1.8 Absolute Value

More formally, we define absolute value as follows:

> **Absolute Value**
>
> Let a be a real number. The **absolute value** of a is denoted by $|a|$ and
>
> $$|a| = \begin{cases} a \text{ if } a \geq 0 \\ -a \text{ if } a < 0. \end{cases}$$

If $a = -3$, then

$$|a| = |-3| = -(-3) \qquad a = -3 < 0, \text{ so } |a| = -a$$
$$= 3. \qquad \text{Double-negation property}$$

The double-negation property ensures that the absolute value of a number is *never* a negative number.

3 OPERATIONS ON REAL NUMBERS

To help understand addition of real numbers, look at a number line. Suppose we add -2 and -3. Start at 0, as shown in Figure 1.9, and move 2 units to the left (-2 is negative). This corresponds to the number -2. From this point we move 3 more units to the left, corresponding to the number -3. The net result is a move of 5 units to the left of zero, or -5. Thus, $(-2) + (-3) = -5$.

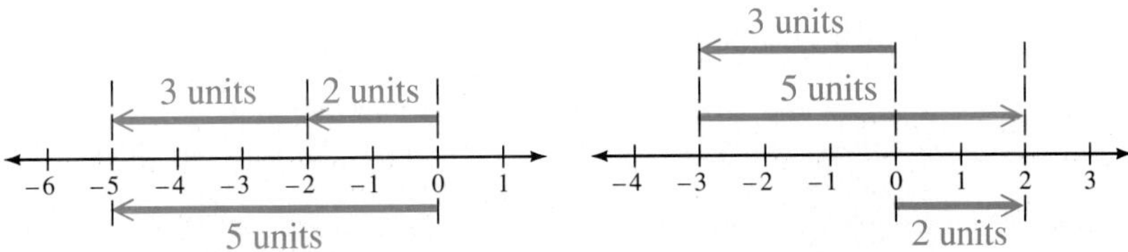

Figure 1.9 **Figure 1.10**

Similarly, to add -3 and $+5$, move 3 units left of zero, then 5 units to the right, as in Figure 1.10. The result is a move of 2 units to the right of zero. Thus, we have $(-3) + (+5) = +2$.

Adding Real Numbers

1. To add two numbers having the same signs, add their absolute values. The sum has the same sign as the numbers being added.
2. To add two numbers having different signs, subtract the smaller absolute value from the larger absolute value. The result has the same sign as the number with the larger absolute value. If the absolute values are the same, the sum is zero.

EXAMPLE 2 ADDING REAL NUMBERS

Add.

(a) $5 + 11 = 16$

(b) $(-5) + (-11) = -(5 + 11) = -16$

(c) $5 + (-7) = -(7 - 5) = -2$ $\quad |-7| > |5|$

(d) $(-5) + 7 = +(7 - 5) = 2$ $\quad |7| > |-5|$

(e) $8 + (-8) = 0$ $\quad |8| = |-8|$

(f) $-4 + 0 = -(4 - 0) = -4$ $\quad |-4| > |0|$

(g) $\frac{3}{5} + \left(-\frac{1}{2}\right) = +\left(\frac{3}{5} - \frac{1}{2}\right)$

$= +\left(\frac{6}{10} - \frac{5}{10}\right) = \frac{1}{10}$ $\quad \left|\frac{3}{5}\right| > \left|-\frac{1}{2}\right|$

PRACTICE EXERCISE 2

Add.

(a) $13 + 7$

(b) $(-13) + (-7)$

(c) $13 + (-7)$

(d) $(-13) + 7$

(e) $11 + (-11)$

(f) $-9 + 0$

(g) $\frac{2}{7} + \left(-\frac{1}{3}\right)$

Answers: (a) 20 (b) -20 (c) 6 (d) -6 (e) 0 (f) -9 (g) $-\frac{1}{21}$

To subtract a real number b from a real number a we *add* the negative or additive inverse of b to a.

Subtracting Real Numbers

To subtract one number b from another number a, change the sign of the number being subtracted, b, and add. That is,

$$a - b = a + (-b).$$

EXAMPLE 3 SUBTRACTING REAL NUMBERS

Subtract.

(a) $13 - 7 = 13 + (-7) = +(13 - 7) = 6$ Change sign and add

(b) $13 - (-7) = 13 + 7 = 20$ Change sign and add

(c) $(-13) - 7 = (-13) + (-7)$ Change sign and add

$= -(13 + 7) = -20$

(d) $(-13) - (-7) = (-13) + 7$

$= -(13 - 7) = -6$

(e) $13 - 13 = 13 + (-13) = 0$

(f) $13 - (-13) = 13 + 13 = 26$

PRACTICE EXERCISE 3

Subtract.

(a) $12 - 5$

(b) $12 - (-5)$

(c) $(-12) - 5$

(d) $(-12) - (-5)$

(e) $12 - 12$

(f) $12 - (-12)$

(g) $\left(-\frac{5}{2}\right) - \left(\frac{3}{4}\right) = \left(-\frac{5}{2}\right) + \left(-\frac{3}{4}\right)$

$= -\left(\frac{5}{2} + \frac{3}{4}\right) = -\left(\frac{10}{4} + \frac{3}{4}\right) = -\frac{13}{4}$

(g) $\left(-\frac{2}{9}\right) - \left(-\frac{1}{3}\right)$

Answers: (a) 7 (b) 17 (c) −17 (d) −7 (e) 0 (f) 24 (g) $\frac{1}{9}$

When two real numbers are multiplied or divided, the answer is positive when the numbers have the same sign. The answer is negative when the numbers have different signs.

Multiplying and Dividing Real Numbers

To multiply or divide two numbers, carry out the operation on their absolute values, and determine the sign by the following rules.

1. The product or quotient of two numbers with the same sign is positive.
2. The product or quotient of two numbers with different signs is negative.
3. If one or both numbers is zero, the product is zero (zero-product rule).
4. Zero divided by any number (except zero) is zero, and division of any number by zero is undefined.

EXAMPLE 4 Multiplying Real Numbers

Multiply.

(a) $4 \cdot 2 = 8$ Same signs; positive product

(b) $4 \cdot (-2) = -(4 \cdot 2) = -8$ Different signs; negative product

(c) $(-4) \cdot 2 = -(4 \cdot 2) = -8$ Different signs

(d) $(-4) \cdot (-2) = 8$ Same signs

(e) $0 \cdot (-2) = 0$ Zero-product rule

(f) $\left(\frac{2}{3}\right)\left(-\frac{5}{4}\right) = -\left(\frac{2}{3} \cdot \frac{5}{4}\right)$

$= -\left(\frac{\cancel{2} \cdot 5}{3 \cdot \cancel{2} \cdot 2}\right) = -\frac{5}{6}$

Practice Exercise 4

Multiply.

(a) $3 \cdot 6$

(b) $3 \cdot (-6)$

(c) $(-3) \cdot 6$

(d) $(-3) \cdot (-6)$

(e) $(-6) \cdot 0$

(f) $\left(-\frac{4}{7}\right)\left(\frac{14}{3}\right)$

Answers: (a) 18 (b) −18 (c) −18 (d) 18 (e) 0 (f) $-\frac{8}{3}$

EXAMPLE 5 Dividing Real Numbers

Divide.

(a) $9 \div 3 = \frac{9}{3} = 3$ Same signs; positive quotient

(b) $9 \div (-3) = -\frac{9}{3} = -3$ Different signs; negative quotient

(c) $(-9) \div 3 = -\frac{9}{3} = -3$ Different signs

(d) $(-9) \div (-3) = \frac{-9}{-3} = 3$ Same signs

Practice Exercise 5

Divide.

(a) $14 \div 7$

(b) $14 \div (-7)$

(c) $(-14) \div 7$

(d) $(-14) \div (-7)$

(e) $9 \div 0$ or $\dfrac{9}{0}$ is undefined.

(f) $0 \div 9 = \dfrac{0}{9} = 0$

(e) $(-14) \div 0$

(f) $0 \div (-7)$

Answers: (a) 2 (b) −2 (c) −2 (d) 2 (e) undefined (f) 0

4 RECIPROCALS

Remember that to divide fractions we need to find the *reciprocal* of the divisor and multiply. The **reciprocal** or **multiplicative inverse** of the fraction $\frac{a}{b}$ is $\frac{b}{a}$, provided $a \neq 0$ and $b \neq 0$. Thus the reciprocal of $\frac{4}{5}$ is $\frac{5}{4}$, the reciprocal of $\frac{1}{3}$ is $\frac{3}{1} = 3$, and the reciprocal of 7 is $\frac{1}{7}$ $\left(7 = \frac{7}{1}\right)$. Of course, 0 has no reciprocal since $\frac{1}{0}$ is undefined.

EXAMPLE 6 DIVIDING FRACTIONS

Divide.

(a) $\dfrac{1}{9} \div \left(-\dfrac{4}{7}\right) = \dfrac{1}{9} \cdot \left(-\dfrac{7}{4}\right) = -\left(\dfrac{1}{9} \cdot \dfrac{7}{4}\right) = -\dfrac{7}{36}$

Multiply by the reciprocal of $-\frac{4}{7}$, $-\frac{7}{4}$

(b) $\left(-\dfrac{3}{4}\right) \div \left(-\dfrac{27}{2}\right) = \left(-\dfrac{3}{4}\right) \cdot \left(-\dfrac{2}{27}\right)$

$= \dfrac{\cancel{3} \cdot \cancel{2}}{\cancel{2} \cdot 2 \cdot \cancel{3} \cdot 9} = \dfrac{1}{18}$

PRACTICE EXERCISE 6

Divide.

(a) $\dfrac{2}{5} \div \left(-\dfrac{7}{10}\right)$

(b) $\left(-\dfrac{4}{15}\right) \div \left(-\dfrac{2}{5}\right)$

Answers: (a) $-\frac{4}{7}$ (b) $\frac{2}{3}$

Absolute value expressions often involve operations on real numbers. Perform all operations within the absolute value bars before taking the absolute value of the result.

EXAMPLE 7 PERFORMING OPERATIONS WITH ABSOLUTE VALUE

Simplify.

(a) $|5 - 7| = |-2| = 2$

(b) $|(-2)(-5)| = |10| = 10$

(c) $-|(-3)(4) - (-2)| = -|-12 - (-2)|$

$= -|-12 + 2| = -|-10| = -10$

PRACTICE EXERCISE 7

Simplify.

(a) $|3 - 14|$

(b) $-|(-8)(3)|$

(c) $|(-2)(-7) - (-9)|$

Answers: (a) 11 (b) −24 (c) 23

1.2 EXERCISES A

Simplify using the double-negation property.

1. $-(-1)$ **2.** $-[-(-2)]$ **3.** $-[-(-0)]$ **4.** $-(-x)$

Rewrite without using absolute value notation.

5. $|-5|$ **6.** $\left|-\frac{3}{4}\right|$ **7.** $-|4|$ **8.** $-|-4|$

9. $-\left|-\frac{2}{3}\right|$ **10.** $|10-3|$ **11.** $|10|-|3|$ **12.** $|10|+|3|$

Perform the indicated operations.

13. $2+(-5)$ **14.** $(-2)+(-5)$ **15.** $2-(-5)$ **16.** $(-2)-5$

17. $(-2)-(-5)$ **18.** $(-2)-(-2)$ **19.** $(-4)(7)$ **20.** $(-4)(-7)$

21. $15\div(-3)$ **22.** $(-15)\div(-3)$ **23.** $\frac{0}{-15}$ **24.** $\frac{-15}{-15}$

25. $\frac{0}{0}$ **26.** $2.1+(-3.2)$ **27.** $(-2.1)+(-3.2)$ **28.** $(-2.1)-(-3.2)$

29. $(-2.1)(-3.2)$ **30.** $2.1\div(-0.7)$ **31.** $\left(-\frac{1}{9}\right)+\left(-\frac{1}{3}\right)$ **32.** $\left(-\frac{1}{9}\right)-\left(-\frac{1}{3}\right)$

33. $\left(-\frac{2}{3}\right)\left(\frac{3}{4}\right)$ **34.** $\left(-\frac{2}{3}\right)\left(-\frac{3}{4}\right)$ **35.** $\frac{2}{3}\div\left(-\frac{3}{4}\right)$ **36.** $\left(-\frac{2}{3}\right)\div\left(-\frac{3}{4}\right)$

37. $|3-8|$ **38.** $|3|-|8|$ **39.** $|(3)(-8)|$ **40.** $|3||-8|$

41. $\left|\frac{8}{-2}\right|$ **42.** $\frac{|8|}{|-2|}$ **43.** $|(-2)+1|$ **44.** $|-2|+|1|$

Answer each of the following.

45. For what real number a will $|a|=|-a|$?

46. Give an example to show that $|a+b|\neq|a|+|b|$.

47. If $a<0$, is $-a$ positive or negative?

48. If $a<0$ is $-(-a)$ positive or negative?

FOR REVIEW

From now on, each section will end with a set of For Review exercises. Sometimes they cover topics that will help you prepare for the presentation in the next section. Unless otherwise identified, For Review exercises are from the preceding section.

49. Give the set of counting numbers.

50. How do you know if a decimal is a rational number?

51. What is the set of real numbers?

52. Express $\frac{4}{9}$ as a decimal.

ANSWERS: 1. 1 2. −2 3. 0 4. x 5. 5 6. $\frac{3}{4}$ 7. −4 8. −4 9. $-\frac{2}{3}$ 10. 7 11. 7 12. 13 13. −3 14. −7 15. 7 16. −7 17. 3 18. 0 19. −28 20. 28 21. −5 22. 5 23. 0 24. 1 25. undefined 26. −1.1 27. −5.3 28. 1.1 29. 6.72 30. −3 31. $-\frac{4}{9}$ 32. $\frac{2}{9}$ 33. $-\frac{1}{2}$ 34. $\frac{1}{2}$ 35. $-\frac{8}{9}$ 36. $\frac{8}{9}$ 37. 5 38. −5 39. 24 40. 24 41. 4 42. 4 43. 1 44. 3 45. every real number 46. Answers will vary; one set that will work is $a = 1$ and $b = -1$. 47. positive 48. negative 49. {1, 2, 3, . . .} 50. The decimal form either terminates or has a repeating block of digits. 51. the set of rational numbers together with the set of irrational numbers 52. $0.\overline{4}$

1.2 EXERCISES B

Simplify using the double-negation property.

1. $-(-7)$ **2.** $-(-0)$ **3.** $-[-(-8)]$ **4.** $-(-c)$

Rewrite without using absolute value notation.

5. $|-1|$ **6.** $\left|-\frac{1}{2}\right|$ **7.** $-|2|$ **8.** $-|-2|$

9. $-\left|-\frac{7}{2}\right|$ **10.** $|7-4|$ **11.** $|7|-|4|$ **12.** $|7|+|4|$

Perform the indicated operations.

13. $3+(-7)$ **14.** $(-3)+(-7)$ **15.** $3-(-7)$ **16.** $(-3)-7$

17. $(-3)-(-7)$ **18.** $(-3)-(-3)$ **19.** $(-5)(11)$ **20.** $(-5)(-11)$

21. $12\div(-4)$ **22.** $(-12)\div(-4)$ **23.** $\frac{-12}{0}$ **24.** $\frac{0}{-12}$

25. $\frac{-12}{-12}$ **26.** $1.5+(-2.6)$ **27.** $(-1.5)+(-2.6)$ **28.** $(-1.5)-(-2.6)$

29. $(-1.5)(-2.6)$ **30.** $(-1.8)\div(-0.9)$ **31.** $\frac{2}{9}+\left(-\frac{4}{3}\right)$ **32.** $\left(-\frac{2}{9}\right)-\left(-\frac{4}{3}\right)$

33. $\left(-\frac{1}{3}\right)\left(\frac{1}{4}\right)$ **34.** $\left(-\frac{1}{3}\right)\left(-\frac{1}{4}\right)$ **35.** $\frac{1}{3}\div\left(-\frac{1}{4}\right)$ **36.** $\left(-\frac{1}{3}\right)\div\left(-\frac{1}{4}\right)$

Properties of 0, 1, and −1

Let a be a real number.	
Properties of 0 under subtraction	$a - 0 = a$ $0 - a = -a$
Properties of 0 under multiplication	$a \cdot 0 = 0$ $0 \cdot a = 0$
Properties of 0 under division	$0 \div a = 0 \quad (a \neq 0)$ $a \div 0$ is undefined
Properties of 1 under division	$1 \div a = \frac{1}{a} \quad (a \neq 0)$ $a \div 1 = a$
Property of -1 under multiplication	$(-1)a = -a$

EXAMPLE 5 RECOGNIZING PROPERTIES OF 0, 1, AND −1

State the property that verifies the truth of each statement.

(a) $0 \cdot (4.5) = 0$ — Property of 0 under multiplication

(b) $5 \div 1 = 5$ — Property of 1 under division

(c) $0 \div \frac{3}{4} = 0$ — Property of 0 under division

(d) $(-1) \cdot (-3) = -(-3) = 3$ — Property of −1 under multiplication

PRACTICE EXERCISE 5

State the property that verifies the truth of each statement.

(a) $8 - 0 = 8$

(b) $2 \div 1 = 2$

(c) $4.8 \div 0$ is undefined.

(d) $\frac{1}{7}(0) = 0$

Answers: (a) property of 0 under subtraction (b) property of 1 under division (c) property of 0 under division (d) property of 0 under multiplication

3 NEGATIVE FRACTION PROPERTY

A property that will be useful when we study algebraic fractions is the **negative fraction property.**

$$-\frac{a}{b} = \frac{-a}{b} = \frac{a}{-b}$$

Thus, we may associate a minus sign with the fraction as a whole, with the numerator, or with the denominator. For example,

$$-\frac{8}{11} = \frac{-8}{11} = \frac{8}{-11}.$$

4 PROPERTIES OF INEQUALITIES

In Section 1.1, we discussed the order relationships ''less than'' and ''greater than'' from an intuitive standpoint using the terms *left* and *right* when related to points on a number line. On the following page we present a formal definition of these relationships.

Inequalities

Let a and b be real numbers.

1. We say that a **is less than** b, denoted $a < b$, if there is a positive number c such that $a + c = b$.
2. We say that a **is greater than** b, denoted $a > b$, if there is a positive number c such that $a - c = b$.

For example,

$$5 < 12 \text{ since } 7 \text{ is a positive number such that } 5 + 7 = 12,$$

and

$$7 > 3 \text{ since } 4 \text{ is a positive number such that } 7 - 4 = 3.$$

Two more inequality properties can now be added to the list of properties of the set of real numbers.

Properties of Inequality

Let a, b, and c be real numbers.

Trichotomy Law	Exactly one of the following holds: $a > b$, $a = b$, $a < b$.
Transitive Law	If $a < b$ and $b < c$, then $a < c$.

By the trichotomy law, either

$$6 < z \text{ or } 6 = z \text{ or } 6 > z.$$

By the transitive law,

$$\text{if } x < 9 \text{ and } 9 < y, \text{ then } x < y.$$

1.3 EXERCISES A

Name the property that justifies each statement. Assume that all letters represent real numbers.

1. $3 \cdot x = x \cdot 3$

2. $x + 0 = x$

3. $8(xy) = (8x)y$

4. $2 + y = y + 2$

5. $1 \cdot 5 = 5$

6. $7 + (-7) = 0$

7. $\left(-\frac{1}{4}\right)(-4) = 1$

8. $(4 + x) + 1 = 4 + (x + 1)$

9. $3 + 9$ is a real number

10. $3(a + b) = 3a + 3b$

11. $(2 + 3) + 9 = 9 + (2 + 3)$

12. $0 + (a + b) = (0 + a) + b$

13. $3(a + 7) = (a + 7)3$

14. $-(-3) = 3$

15. $7 \div 0$ is undefined

16. $0 \cdot (-8) = 0$

17. $-\frac{2}{3} = \frac{-2}{3} = \frac{2}{-3}$

18. $(-1)(10) = -10$

19. If $a < b$ and $b < 7$ then $a < 7$.

20. $2 < 5$ since $3 > 0$ and $2 + 3 = 5$.

Complete each statement using the specified law.

21. Commutative law of addition: $8 + a =$ ________

22. Associative law of multiplication: $4(y \cdot 7) =$ ________

23. Existence of a multiplicative identity: $1 \cdot 28 =$ ________

24. Distributive law: $2(a + y) =$ ________

25. Associative law of addition: $x + (3 + 7) =$ ________

26. Distributive law: $3a + 3w =$ ________

27. Property of 0 under division: $3 \div 0$ is ________

28. Existence of reciprocals: 6(________) = 1

29. Property of 0 under multiplication: $5 \cdot 0 =$ ________

30. Double-negation property: $-(-15) =$ ________

Perform the indicated operations.

31. $(-2) + (-3) + 5 + 3 + (-5) + 2 + (-7) + 4$

32. $(-2)(-2) + (-3)(3) + (-1)(4) + (0)(5)$

33. $(-1.7) + 2.3 - 4.1 - (-3.2) + (-2.5) - (-1.1)$

34. $(3)(-2)(-2)(-1)(-1)(2)(-1)(1)(-1)$

35. $(-1)(-1)(-1)(-1)(-1)(-1)(-1)(-1)$

36. $\left(-\frac{1}{2}\right)\left(\frac{2}{3}\right)\left(-\frac{3}{7}\right)\left(\frac{7}{4}\right)\left(-\frac{4}{5}\right)\left(-\frac{5}{3}\right)$

37. Henry Sheppard had \$531.61 in his bank account on the first of the month. He deposited a scholarship check for \$1200.00 and a paycheck for \$385.16 during the month. What was his balance at the end of the month if he wrote checks for \$314.50, \$617.85, \$491.25, and \$62.42 during the month?

Give **(a)** *the negative (additive inverse) and* **(b)** *the reciprocal (multiplicative inverse) of each number.*

38. 5

39. -2

40. 0

41. $\sqrt{2}$

42. Give an example to show that subtraction does not satisfy the commutative law. That is, find real numbers a and b such that $a - b \neq b - a$.

43. Give an example to show that division does not satisfy the associative law. That is, find real numbers a, b, and c such that $(a \div b) \div c \neq a \div (b \div c)$.

FOR REVIEW

Perform the indicated operations.

44. $\left(\frac{4}{9}\right)\left(-\frac{3}{16}\right)$

45. $\left(-\frac{5}{28}\right) \div \left(-\frac{15}{14}\right)$

46. $\left(-\frac{3}{5}\right) + \left(-\frac{1}{15}\right)$

47. $\left(-2\frac{1}{3}\right) \div \left(-1\frac{1}{9}\right)$

Complete each of the following using one of the words positive, negative, *or* zero.

48. The product of two nonzero numbers with opposite signs is always ______________.

49. The quotient of two negative numbers is always ______________.

50. The sum of two negative numbers is always ______________.

51. The product of zero and any negative number is always ______________.

ANSWERS: 1. commutative law of multiplication 2. existence of additive identity 3. associative law of multiplication 4. commutative law of addition 5. existence of multiplicative identity 6. existence of negatives 7. existence of reciprocals 8. associative law of addition 9. closure law of addition 10. distributive law 11. commutative law of addition 12. associative law of addition 13. commutative law of multiplication 14. double-negation property 15. property of zero under division 16. property of zero under multiplication 17. negative fraction property 18. property of -1 under multiplication 19. transitive law of inequality 20. definition of *less than* 21. $a + 8$ 22. $(4y) \cdot 7$ 23. 28 24. $2a + 2y$ 25. $(x + 3) + 7$ 26. $3(a + w)$ 27. undefined 28. $\frac{1}{6}$ 29. 0 30. 15 31. -3 32. -9 33. -1.7 34. 24 35. 1 36. $\frac{1}{3}$ 37. \$630.75 38. (a) -5 (b) $\frac{1}{5}$ 39. (a) 2 (b) $-\frac{1}{2}$ 40. (a) 0 (b) 0 has no reciprocal 41. (a) $-\sqrt{2}$ (b) $\frac{1}{\sqrt{2}}$ 42. $2 - 3 = -1 \neq 1 = 3 - 2$ 43. $(8 \div 4) \div 2 = 2 \div 2 = 1 \neq 4 = 8 \div 2 = 8 \div (4 \div 2)$ 44. $\frac{-1}{12}$ 45. $\frac{1}{6}$ 46. $\frac{-2}{3}$ 47. $\frac{21}{10}$ 48. negative 49. positive 50. negative 51. zero

1.3 EXERCISES B

Name the property that justifies each statement. Assume that all letters represent real numbers.

1. $3(y + 2) = 3y + 3 \cdot 2$

2. $5 \cdot c = c \cdot 5$

3. $0 \div (-13) = 0$

4. $7\left(\frac{1}{7}\right) = 1$

5. $3 + z = z + 3$

6. $(3 + 5)a = 3a + 5a$

7. $1 \div 15 = \frac{1}{15}$

8. $8 \div 0$ is undefined

9. $(2 + a) + 5 = 2 + (a + 5)$

10. $4(y + 2) = (y + 2)4$

11. $3 \cdot 14$ is a real number

12. $(4 + 1) + 8 = 8 + (4 + 1)$

13. $0 + (3 + d) = (0 + 3) + d$

14. $7 \div 1 = 7$

15. $(13)(0) = 0$

16. $(-1)(5) = -5$

17. $0 \div 2 = 0$

18. $-(-2.5) = 2.5$

19. If $x < 2$ and $2 < w$ then $x < w$.

20. $4 > -1$ since $5 > 0$ and $4 - 5 = -1$.

Complete each statement using the specified law.

21. Existence of an additive identity: $6 + 0 =$ ____

22. Closure law of multiplication: $4 \cdot 8$ is ____

23. Distributive law: $3(x + w) =$ ____

24. Commutative law of multiplication: $3x =$ ____

25. Existence of negatives: $(-12) +$ ____ $= 0$

26. Distributive law: $4a + 4m =$ ____

27. Existence of reciprocals: $4($____$) = 1$

28. Property of 0 under division: $0 \div 5 =$ ____

29. Negative fraction property: $-\frac{4}{9} = \frac{-4}{9} =$ ____

30. Double-negation property: $-(-\sqrt{3}) =$ ____

Perform the indicated operations.

31. $(-8) + 3 + (-2) + 5 + (-4) + (-1) + (-5) + (-1)$

32. $(3)(-3) + (5)(-1) + (2)(-2) + (3)(0)$

33. $4.2 - (-1.7) + (-1.7) - 8.2 - (-2.2) - (4.1)$

34. $(-1)(4)(-2)(-1)(-1)(1)(2)(-1)(1)(1)(-1)$

35. $(-1)(-1)(-1)(-1)(-1)(-1)(-1)(-1)(-1)(-1)(-1)$

36. $\left(\frac{2}{3}\right)\left(-\frac{3}{5}\right)\left(-\frac{5}{4}\right)\left(-\frac{1}{4}\right)\left(-\frac{4}{7}\right)\left(\frac{7}{2}\right)\left(-\frac{2}{9}\right)$

37. Lisa Winn had a temperature of 101.2° when she entered the hospital. It dropped 2.5° with medication but rose 3.6° when the medicine was stopped. What was Lisa's temperature after further treatment brought it down 3.1°?

Give **(a)** *the negative and* **(b)** *the reciprocal of each number.*

38. 9

39. -4

40. $-\sqrt{5}$

41. 0

42. Give an example to show that division does not satisfy the commutative law. That is, find real numbers a and b such that $a \div b \neq b \div a$.

43. Give an example to show that subtraction does not satisfy the associative law. That is, find real numbers a, b, and c such that $(a - b) - c \neq a - (b - c)$.

FOR REVIEW

Perform the indicated operations.

44. $\left(\frac{2}{9}\right)\left(-\frac{3}{8}\right)$

45. $\left(-\frac{3}{14}\right) \div \left(-\frac{6}{7}\right)$

46. $\left(-\frac{2}{5}\right) + \left(-\frac{7}{15}\right)$

47. $\left(-3\frac{1}{3}\right) \div \left(-2\frac{2}{9}\right)$

Complete each of the following using one of the words positive, negative, *or* zero.

48. The quotient of two nonzero numbers with opposite signs is always ________.

49. The product of two negative numbers is always ________.

50. The sum of two positive numbers is always ________.

51. The product of zero and any positive number is always ________.

1.3 EXERCISES C

Name the property that justifies each statement.

1. $\left(1.5 + 3\frac{2}{3} + x\right)\left(\frac{1}{7} + 0\right) = \left(\frac{1}{7} + 0\right)\left(1.5 + 3\frac{2}{3} + x\right)$

2. $(x + 0)\left[(3 + y) - \left(\frac{3}{5} + 0.7\right)\right] = x\left[(3 + y) - \left(\frac{3}{5} + 0.7\right)\right]$

Use the definition of inequalities to explain why each statement is true.

3. $-15 < 25$
[Answer: Since $40 > 0$ and $-15 + 40 = 25$, we know that $-15 < 25$.]

4. $-3 < -1$

5. $6 > -4$

1.4 POWERS, ROOTS, AND ORDER OF OPERATIONS

STUDENT GUIDEPOSTS

1. Exponential Notation
2. Squares and Square Roots
3. Radicals
4. Order of Operations
5. Symbols of Grouping

1 EXPONENTIAL NOTATION

To multiply a number or a variable by itself several times, such as

$$3 \cdot 3 \cdot 3 \cdot 3 \quad \text{or} \quad a \cdot a \cdot a,$$

it is convenient to use **exponential notation** to avoid writing long strings of **factors.** For example, we write $3 \cdot 3 \cdot 3 \cdot 3$ as 3^4,

$$\underbrace{3 \cdot 3 \cdot 3 \cdot 3}_{\text{4 factors}} = \underset{\uparrow \text{ base}}{3}^{4 \leftarrow \text{exponent}}$$

where 3 is called the **base,** 4 (the number of factors) is called the **exponent,** and 3^4 the **exponential expression** (read "3 to the fourth power"). In the same way, $a \cdot a \cdot a = a^3$ is the third power of a, or the **cube** of a. The second power of a number y, y^2, is called the **square** of y, and the first power of a number w, w^1, is written simply as w. In general, if a is any number and n is a natural number,

$$a^n = \underbrace{a \cdot a \cdot a \cdot a \ . \ . \ . \ a}_{n \text{ factors}}.$$

EXAMPLE 1 Writing Expressions Using Exponents

Write in exponential notation.

(a) $\underbrace{5 \cdot 5 \cdot 5 \cdot 5 \cdot 5 \cdot 5}_{\text{6 factors}} = 5^6$

(b) $\underbrace{3 \cdot 3 \cdot 3 \cdot 3 \cdot 3}_{\text{5 factors}} \cdot \underbrace{x \cdot x \cdot x \cdot x \cdot x \cdot x \cdot x}_{\text{7 factors}} = 3^5x^7$

PRACTICE EXERCISE 1

Write in exponential notation.

(a) $m \cdot m \cdot m \cdot m \cdot m$

(b) $2 \cdot 2 \cdot 2 \cdot w \cdot w \cdot w \cdot w \cdot w \cdot w$

Answers: (a) m^5 (b) 2^3w^6

EXAMPLE 2 WRITING EXPRESSIONS WITHOUT USING EXPONENTS

Write without using exponents.

(a) $4^3 = 4 \cdot 4 \cdot 4 = 64$

(b) $2^5x^2 = 2 \cdot 2 \cdot 2 \cdot 2 \cdot 2 \cdot x \cdot x$
$= 32 \cdot x \cdot x$

PRACTICE EXERCISE 2

Write without using exponents.

(a) a^5

(b) 5^2n^4

Answers: (a) $a \cdot a \cdot a \cdot a \cdot a$
(b) $5 \cdot 5 \cdot n \cdot n \cdot n \cdot n$

2 SQUARES AND SQUARE ROOTS

If a and x are real numbers and

$$x = a^2,$$

then x is the **square** of a and a is a **square root** of x. If a is an integer, then x is called a **perfect square.**

EXAMPLE 3 USING PERFECT SQUARES AND SQUARE ROOTS

(a) Four is a perfect square since $4 = 2^2$. We call 2 a square root of 4. Also, since $4 = (-2)^2$, -2 is also a square root of 4.

(b) Eighty-one is a perfect square since $81 = 9^2$. The two square roots of 81 are 9 and -9.

(c) Zero is a perfect square since $0 = 0^2$. Unlike other perfect squares, 0 has only one square root, namely itself, 0.

PRACTICE EXERCISE 3

(a) Is 25 a perfect square?

(b) Is 121 a perfect square?

(c) Is 90 a perfect square?

Answers: (a) yes; $25 = 5^2$
(b) yes; $121 = 11^2$ (c) no

CAUTION

Do not confuse the terms *square* and *square root!*

3 is a *square root* of 9 → $3^2 = 9$ ← 9 is the *square* of 3

3 RADICALS

Since all perfect squares other than 0 have two square roots, one positive and one negative, confusion can result when we refer to the square root of a number. Which one do we mean, the positive root or the negative root? Usually we emphasize the nonnegative root, also called the **principal square root.** We use the symbol $\sqrt{}$ as shown below.

radical → $\sqrt{9} = 3$ ← principal square root
↑
radicand

The radical symbol by itself only represents the principal or positive (possibly zero) square root of a number. Negative square roots are denoted by $-\sqrt{}$, for example, $-\sqrt{9} = -3$.

A fraction that has two identical fractional factors is a perfect square, and each factor is a square root of the fraction. For example, $\frac{4}{9}$ is a perfect square and since $\left(\frac{2}{3}\right)^2 = \frac{4}{9}$,

$$\sqrt{\frac{4}{9}} = \frac{2}{3}.$$

Also,

$$\sqrt{\frac{25}{81}} = \frac{5}{9} \quad \text{and} \quad -\sqrt{\frac{25}{81}} = -\frac{5}{9}.$$

Clearly, a fraction is a perfect square if its numerator and denominator are natural number perfect squares. However, a fraction can be a perfect square without this feature. For example, $\frac{8}{18}$ is a perfect square. If we reduce $\frac{8}{18}$ to lowest terms, $\frac{4}{9}$, we see it is a ratio of natural number perfect squares, with principal square root $\frac{2}{3}$.

4 ORDER OF OPERATIONS

When the operations of addition, subtraction, multiplication, division, exponents or powers and roots are combined in a numerical expression, confusion may result unless we specify some order of operations. For example,

	$3 \cdot 2 + 5$	could equal	$3 \cdot 7$ or 21,	Add first, then multiply
or	$3 \cdot 2 + 5$	could equal	$6 + 5$ or 11.	Multiply first, then add

Similarly,

	$3 \cdot 4^2$	could equal	12^2 or 144,	Multiply first, then square
or	$3 \cdot 4^2$	could equal	$3 \cdot 16$ or 48.	Square first, then multiply

The second procedure in each of these examples is the correct one to use, according to the following rule.

To Evaluate a Numerical Expression

1. Evaluate all powers and roots.
2. Perform all multiplications and divisions in order from left to right.
3. Perform all additions and subtractions in order from left to right.

EXAMPLE 4 EVALUATING NUMERICAL EXPRESSIONS

Evaluate each numerical expression.

(a) $3 - 5 \cdot 2 = 3 - 10 = -7$ — Multiply first, then subtract

(b) $5 + 9^2 = 5 + 81 = 86$ — Square first, then add

(c) $\sqrt{4} \cdot 3 - 15 \div 5 = 2 \cdot 3 - 15 \div 5$ — Take square root first

$= 6 - 3 = 3$ — Multiply and divide next, then subtract

(d) $20 \div 4 + 2 \cdot 3^3 = 20 \div 4 + 2 \cdot 27$ — Cube first

$= 5 + 54$ — Divide and multiply second

$= 59$ — Add last

PRACTICE EXERCISE 4

(a) $\sqrt{9} + 12 \div 4$

(b) $5^3 - 100 \div 5$

(c) $10 \div 2 + 8 \cdot 3$

(d) $3 \cdot 2^4 \div 6 + 1$

Answers: (a) 6 (b) 105 (c) 29 (d) 9

5 SYMBOLS OF GROUPING

Suppose we want to evaluate three times the sum of 2 and 5. If we write $3 \cdot 2 + 5$ and use the above rule, we obtain $6 + 5$, or 11. However, the first sentence states that we want 3 times 7, or 21. **Symbols of grouping** (such as parentheses ( ), square brackets [ ], or braces { } help symbolize our problem correctly as $3 \cdot (2 + 5)$. The expression inside the grouping symbols is evaluated first.

$$3 \cdot (2 + 5) = 3 \cdot (7) = 21$$

Using a multiplication dot along with parentheses is unnecessary. Usually one or the other is omitted. The problem above can be written

$$3(2 + 5) = 3(7) = 21.$$

Other methods of grouping involve fraction bars, radicals, and absolute value bars. For example, in

$$\frac{7 + 5}{2 \cdot 3},$$

the fraction bar acts like parentheses since the expression is the same as $(7 + 5) \div (2 \cdot 3)$. We evaluate first above the fraction bar and then below it. Then we divide the results.

$$\frac{7 + 5}{2 \cdot 3} = \frac{12}{6} = 2$$

Similarly, in

$$\sqrt{4^2 + 3^2},$$

we evaluate under the radical first, $4^2 + 3^2 = 16 + 9 = 25$, then take the square root of the result.

$$\sqrt{4^2 + 3^2} = \sqrt{25} = 5$$

Also,

$$|(-2)(3) + 5| = |-6 + 5| = |-1| = 1,$$

since we evaluate inside the absolute value bars first, and then take the absolute value of the result.

EXAMPLE 5 Evaluating Expressions Involving Grouping

Evaluate each numerical expression.

(a) $(5 + 9)7 - 16 = (14)7 - 16$ Work inside parentheses first
$= 98 - 16$ Multiply before subtracting
$= 82$

(b) $[2(7 - 12) + 4]3 = [2(-5) + 4]3$ Innermost parentheses first
$= [-10 + 4]3$ Multiply before adding inside brackets
$= [-6]3$ Combine inside brackets first
$= -18$

(c) $4|-5 + 2(6 + (-6))| = 4|-5 + 2(0)|$
$= 4|-5 + 0|$
$= 4|-5|$
$= 4(5) = 20$

Practice Exercise 5

Evaluate each numerical expression.

(a) $15 - 3(6 - 8)$

(b) $4[3(2 + 1) - 10]$

(c) $-2|3(-2 + 2) - 5|$

(d) $\dfrac{3 + 2^2}{\sqrt{2 \cdot 25 - 1^2}} = \dfrac{3 + 4}{\sqrt{50 - 1}}$

$= \dfrac{7}{\sqrt{49}} = \dfrac{7}{7} = 1$ $\sqrt{49} = 7$, *not* -7

(d) $\dfrac{\sqrt{5^2 - 9}}{|3 \cdot 2 - 8|}$

Answers: (a) 21 (b) −4 (c) −10 (d) 2

The following summarizes the method for simplifying numerical expressions.

Order of Operations

1. Evaluate within grouping symbols first, beginning with the innermost set if more than one set is used.
2. Evaluate all powers and roots.
3. Perform all multiplications and divisions in order from left to right.
4. Perform all additions and subtractions in order from left to right.

EXAMPLE 6 EVALUATING USING ORDER OF OPERATIONS

Evaluate each numerical expression.

(a) $2^2 + 3^2 = 4 + 9 = 13$ Square first, then add

(b) $(2 + 3)^2 = 5^2 = 25$ Add first, then square

(c) $4^3 - 3^3 = 64 - 27 = 37$ Cube first, then subtract

(d) $(4 - 3)^3 = 1^3 = 1$ Subtract first, then cube

(e) $2 \cdot 5^2 = 2 \cdot 25 = 50$ Square first, then multiply

(f) $(2 \cdot 5)^2 = 10^2 = 100$ Multiply first, then square

PRACTICE EXERCISE 6

Evaluate each numerical expression.

(a) $4^2 + 1^2$

(b) $(4 + 1)^2$

(c) $5^2 - 2^2$

(d) $(5 - 2)^2$

(e) $8 \cdot 3^3$

(f) $(8 \cdot 3)^3$

Answers: (a) 17 (b) 25 (c) 21 (d) 9 (e) 216 (f) 13,824

CAUTION

Three common errors were avoided in Example 6. Notice that

$$2^2 + 3^2 \neq (2 + 3)^2, \quad 4^3 - 3^3 \neq (4 - 3)^3, \quad 2 \cdot 5^2 \neq (2 \cdot 5)^2.$$

1.4 EXERCISES A

Write true *or* false *for each statement. If the statement is false, explain why.*

1. Relative to the exponential expression a^3, a is called the base, and 3 is called the exponent.

2. The whole number obtained as the product when an integer is squared is called a square root.

3. The principal square root of a positive number is the positive square root of the number.

4. The radical symbol by itself only designates the principal or nonnegative square root of a number.

5. The number written under the radical symbol is called the index.

Write each expression in exponential notation.

6. $7 \cdot 7 \cdot 7$ **7.** $b \cdot b$ **8.** $3 \cdot 3 \cdot 3 \cdot 3 \cdot w \cdot w \cdot w \cdot w \cdot w$

Write without using exponential notation.

9. 10^3 **10.** 3^2 **11.** $2^2 \cdot a^5$

Evaluate the square roots.

12. $\sqrt{49}$ **13.** $\sqrt{225}$ **14.** $-\sqrt{64}$ **15.** $\sqrt{0}$

16. $-\sqrt{169}$ **17.** $\sqrt{\frac{196}{25}}$ **18.** $-\sqrt{\frac{49}{16}}$ **19.** $\sqrt{\frac{48}{12}}$

20. **(a)** Compute 16^2. **(b)** Compute $\sqrt{256}$. **21.** **(a)** Compute 18^2. **(b)** Compute $\sqrt{324}$.

22. **(a)** What is the square of 25? **(b)** What is the principal square root of 25?

Evaluate the following numerical expressions.

23. $2 + 3 \cdot 5$ **24.** $2 - \sqrt{9} + 5$ **25.** $2 - (3 + 5)$ **26.** $4 \cdot 2 - 6 \div 3$

27. $15 \div 5 \cdot 2 + 3$ **28.** $(4 - 2)3 - 7$ **29.** $-[3 + 2(4 - 6)]$ **30.** $-[-(1 - 3)]$

31. $8\{(-2) - 3[1 + (-5)]\}$ **32.** $-[1 - 2(1 - 2)]$ **33.** $-|-(-[1 - 3])|$ **34.** $2^3 + 1^3$

35. $(2 + 1)^3$ **36.** $(4 - 7)^2$ **37.** $4^2 - 7^2$ **38.** $3 \cdot 2^2$

39. $(3 \cdot 2)^2$ **40.** $\frac{(-2)(-3) + 4}{25 \div 5}$ **41** *$\frac{4 \cdot 5^2}{\sqrt{6^2 + 8^2}}$ **42** $3 - (-2)\frac{4 - (-1)}{2 + (-3)}$

FOR REVIEW

Name the property that justifies each statement. Assume that all variables represent real numbers.

43. $2 + y = y + 2$ **44.** $3(a + 7) = 3a + 3 \cdot 7$ **45.** $0 + (-35) = -35$

46. $3x$ is a real number **47.** $0 \div (-3) = 0$ **48.** $\left(\frac{3}{4}\right)\left(\frac{4}{3}\right) = 1$

*A blue circle on the number indicates that a complete solution is supplied in the back of the text. See *To the Student* for more details.

Evaluate.

49. $(-1)(-1)(-1)(2)(-2)(2)(-1)(1)(1)(-1)(1)(-2)(-2)(1)(-1)$

50. Graph $x \geq -4$ on the number line.

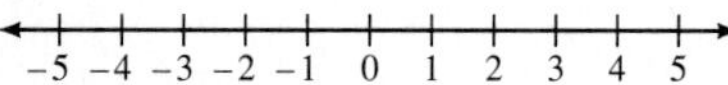

ANSWERS: 1. true 2. false; perfect square 3. true 4. true 5. false; radicand 6. 7^3 7. b^2 8. $3^4 \cdot w^5$ 9. $10 \cdot 10 \cdot 10$ or 1000 10. $3 \cdot 3$ or 9 11. $2 \cdot 2 \cdot a \cdot a \cdot a \cdot a \cdot a$ 12. 7 13. 15 14. −8 15. 0 16. −13 17. $\frac{14}{5}$ 18. $-\frac{7}{4}$ 19. 2 20. (a) 256 (b) 16 21. (a) 324 (b) 18 22. (a) 625 (b) 5 23. 17 24. 4 25. −6 26. 6 27. 9 28. −1 29. 1 30. −2 31. 80 32. −3 33. −2 34. 9 35. 27 36. 9 37. −33 38. 12 39. 36 40. 2 41. 10 42. −7 43. commutative law of addition 44. distributive law 45. existence of additive identity 46. closure law of multiplication 47. property of 0 under division 48. existence of reciprocals 49. −32 50. −5 −4 −3 −2 −1 0 1 2 3 4 5

1.4 EXERCISES B

Write true *or* false *for each statement. If the statement is false, explain why.*

1. Relative to the exponential expression x^2, x is called the exponent and 2 is called the base.

2. Either of the identical factors of a perfect square is called a square root of the number.

3. The negative square root of a number is called its principal square root.

4. The symbol $\sqrt{}$ is called a radical.

5. To represent the negative square root of a number using the radical symbol, we write $-\sqrt{}$.

Write in exponential notation.

6. $4 \cdot 4 \cdot 4 \cdot 4 \cdot 4$

7. $x \cdot x \cdot x$

8. $2 \cdot 2 \cdot w \cdot w \cdot w \cdot w \cdot w \cdot w$

Write without using exponential notation.

9. 10^4

10. 5^2

11. $3^2 \cdot b^3$

Evaluate the square roots.

12. $\sqrt{25}$

13. $\sqrt{100}$

14. $-\sqrt{36}$

15. $-\sqrt{0}$

16. $-\sqrt{196}$

17. $\sqrt{\frac{1}{144}}$

18. $-\sqrt{\frac{36}{4}}$

19. $\sqrt{\frac{50}{32}}$

20. **(a)** Compute 17^2. **(b)** Compute $\sqrt{289}$.

21. **(a)** Compute 19^2. **(b)** Compute $\sqrt{361}$.

22. **(a)** What is the square of 9? **(b)** What is the principal square root of 9?

Evaluate the following numerical expressions.

23. $4 \cdot 2 + 6$

24. $4 - \sqrt{25} + 1$

25. $4 - (5 + 1)$

26. $5 \cdot 3 - 8 \div 4$

27. $26 \div 13 \cdot 2 + 7$

28. $(2 - 8)4 - 1$

29. $-[5 + 7(2 - 3)]$

30. $-[-(4 - 5)]$

31. $3\{(-1) - 2[3 + (-4)]\}$ **32.** $-[4 - 7(4 - 7)]$ **33.** $-|-[-(5 - 9)]|$ **34.** $3^2 + 2^2$

35. $(3 + 2)^2$ **36.** $(2 - 5)^3$ **37.** $2^3 - 5^3$ **38.** $2 \cdot 4^3$

39. $(2 \cdot 4)^3$ **40.** $\dfrac{(-3)(4) + 2}{14 \div 7}$ **41.** $\dfrac{26 \cdot 2^3}{\sqrt{5^2 + 12^2}}$ **42.** $4 - (-3)\dfrac{1 + (-3)}{-5 - (-2)}$

FOR REVIEW

Name the property that justifies each statement. Assume that all variables represent real numbers.

43. $4 \cdot z = z \cdot 4$ **44.** $8\left(\frac{1}{8}\right) = 1$ **45.** $5 + 12$ is a real number

46. $8u + 8w = 8(u + w)$ **47.** $(3 + y) + 2 = 3 + (y + 2)$ **48.** $(-2.1) + 2.1 = 0$

Evaluate.

49. $(-2)(-1)(1)(-1)(-2)(2)(-1)(-1)(-1)(-1)(-1)(1)(-2)(2)(-1)(2)$

50. Graph $x < 3$ on the number line.

−5 −4 −3 −2 −1 0 1 2 3 4 5

1.4 EXERCISES C

Evaluate the numerical expressions.

1. $\sqrt{\dfrac{4^2 \div 2 + 1}{\left(-\frac{1}{3}\right) \div (-3)}}$ [Answer: 9] **2.** $\sqrt{(6 \div 2 \cdot 4 - 5)^2 + 3 \cdot 5}$

Place parentheses in the expression to make the equation true.

3. $8 \div 2 + 3 - 6 \cdot 7 = -17$ **4.** $\sqrt{4 - 2^2 - 8 \div 8 + 6} = 3$

1.5 ALGEBRAIC EXPRESSIONS AND THE DISTRIBUTIVE LAW

STUDENT GUIDEPOSTS

1. Terms and Factors
2. Evaluating Algebraic Expressions
3. Factoring Algebraic Expressions
4. Collecting Like Terms
5. Removing Parentheses

1 TERMS AND FACTORS

We now extend our treatment of the arithmetic of real numbers to the algebra of real numbers. An **algebraic expression** involves sums, differences, products, quotients, powers, or roots of numbers *and* variables. For example,

$$7x, \quad 2a + 5, \quad x^2 + 3y + \sqrt{z}, \quad 2x - 5a + 3 - 4x$$

are algebraic expressions. Portions of an expression that are separated by plus or minus signs are called **terms.** The numbers and letters multiplied in a term are its **factors,** and the numerical factor is the **coefficient** of the term. The expression $2a + 5$ has two terms, $2a$ and 5. The term $2a$ has factors 2 and a, and 2 is the coefficient of the term.

Two terms are **like terms** if they contain the same variables raised to the same power. In $2x - 5a + 3 - 4x$, the terms $2x$ and $-4x$ are like terms. Note that the minus sign goes with the term, and thus the coefficient of $-4x$ is -4. A good way to remember this is to consider terms as separated by plus signs and write, for example,

$$2x - 5a + 3 - 4x = 2x + (-5a) + 3 + (-4x).$$

In this way the sign is part of the coefficient.

2 EVALUATING ALGEBRAIC EXPRESSIONS

Using the basic rules for evaluating numerical expressions, it is a simple step to evaluate algebraic expressions as summarized in the following.

To Evaluate an Algebraic Expression

1. Replace each variable with its specified value.
2. Proceed as in evaluating numerical expressions.

EXAMPLE 1 Evaluating Algebraic Expressions

(a) Evaluate $3(x + y) - w$ when $x = 2$, $y = 4$, and $w = 8$.

$$\begin{aligned} 3(x + y) - w &= 3(2 + 4) - 8 && \text{Replace variables with numerical values} \\ &= 3(6) - 8 && \text{Evaluate inside parentheses first} \\ &= 18 - 8 && \text{Multiply before subtracting} \\ &= 10 \end{aligned}$$

(b) Evaluate $4[1 - 2(a + b) + b] + 2c$ when $a = 1$, $b = 3$, and $c = 5$.

$$\begin{aligned} 4[1 - 2(a + b) + b] + 2c &= 4[1 - 2(1 + 3) + 3] + 2(5) \\ &= 4[1 - 2(4) + 3] + 10 \\ &= 4[1 - 8 + 3] + 10 \\ &= 4[-4] + 10 = -16 + 10 = -6 \end{aligned}$$

Practice Exercise 1

(a) Evaluate $5(a - b) - m$ when $a = 3$, $b = 5$, and $m = 2$.

(b) Evaluate

$$3x - 2[1 - 3(y + x) - z]$$

when $x = 1$, $y = 5$, and $z = 4$.

Answers: (a) -12 (b) 45

When evaluating expressions using negative numbers, start with parentheses around the number as shown in the next example.

EXAMPLE 2 Evaluating Algebraic Expressions

Evaluate $\dfrac{xy + 5}{1 - a}$ for the following values.

(a) $x = -2$, $y = 5$, $a = 1$

$$\frac{xy + 5}{1 - a} = \frac{(-2)(5) + 5}{1 - 1} = \frac{-10 + 5}{0} = \frac{-5}{0} \qquad \text{Substitute } (-2) \text{ for } x$$

Since we cannot divide by 0, the expression is undefined when $a = 1$.

Practice Exercise 2

Evaluate $\dfrac{u - 6v}{1 + w}$ for the following values.

(a) $u = 12$, $v = 2$, $w = 1$

(b) $x = -1$, $y = 5$, $a = 2$

$$\frac{xy + 5}{1 - a} = \frac{(-1)(5) + 5}{1 - 2} = \frac{-5 + 5}{-1} = \frac{0}{-1} = 0$$

Notice that $\frac{0}{-1}$ is 0, but $\frac{-5}{0}$ obtained in (a) is undefined.

(b) $u = 3$, $v = 1$, $w = -1$

Answers: (a) 0 (b) undefined

A **formula** is an algebraic statement that relates two or more quantities. For example, the formula

$$A = lw$$

relates the area A of a rectangle to its length l and width w. Finding the value of a particular variable in a formula involves evaluating an algebraic expression.

EXAMPLE 3 EVALUATING A FORMULA

When P dollars (called the **principal**) is invested at an **interest rate** r, compounded annually for a period of t years, it will grow to an **amount** A given by the formula $A = P(1 + r)^t$. If a principal of $5000 is invested at 10%, compounded annually, how much will be in the account at the end of 2 years?

$A = P(1 + r)^t$ Start with the given formula

$= 5000(1 + 0.10)^2$ Substitute 5000 for P, 0.10 for r, and 2 for t

$= 5000(1.1)^2$

$= 5000(1.21)$

$= 6050$

There will be $6050 in the account at the end of two years. Note that 10% was converted to a decimal, 0.10, to solve the problem.

PRACTICE EXERCISE 3

If $12,000 is invested at 13%, compounded annually, how much will be in the account at the end of 3 years?

Answer: $17,314.76

CAUTION

Using negative numbers to evaluate expressions may produce two signs in front of a number unless parentheses are used. For example, when $x = -3$,

write $1 + x = 1 + (-3)$ instead of $1 + x = 1 + -3$.

EXAMPLE 4 EVALUATING ALGEBRAIC EXPRESSIONS

Evaluate when $a = -2$ and $b = -1$.

(a) $a - b = (-2) - (-1)$ Use parentheses to substitute

$= -2 + 1 = -1$

(b) $4a + 7 = 4(-2) + 7$

$= -8 + 7 = -1$

(c) $b - [-a] = (-1) - [-(-2)]$

$= (-1) - [+2] = -1 - 2 = -3$

PRACTICE EXERCISE 4

Evaluate when $m = -3$ and $n = -1$.

(a) $m - n$

(b) $m - 2n$

(c) $n - (-m)$

(d) $a - 2b = (-2) - 2(-1)$
$= -2 - (-2) = -2 + 2 = 0$

(e) $|-a - b| = |-(-2) - (-1)|$
$= |+2 + 1| = |3| = 3$

(f) $a^2 + b^2 = (-2)^2 + (-1)^2$
$= 4 + 1 = 5$

(g) $(a + b)^2 = ((-2) + (-1))^2$
$= (-3)^2 = 9$

(h) $3a^2 = 3(-2)^2$
$= 3(4) = 12$

(i) $(3a)^2 = [(3)(-2)]^2$
$= (-6)^2 = 36$

Compare (h) and (i). Notice that in $3a^2$, only a is squared (not 3 also). In order to square both 3 and a, we must use parentheses as in (i).

(d) $6n - 2m$

(e) $|-n - m|$

(f) $n^2 + m^2$

(g) $(n + m)^2$

(h) $2m^2$

(i) $(2m)^2$

Answers: (a) -2 (b) -1 (c) -4 (d) 0 (e) 4 (f) 10 (g) 16 (h) 18 (i) 36

3 FACTORING ALGEBRAIC EXPRESSIONS

Remember the distributive law from Section 1.3: for real numbers a, b, and c, $a(b + c) = ab + ac$. If the terms in an algebraic expression have a common numerical or variable factor, the distributive law can be used to remove the common factor by a process called **factoring.**

EXAMPLE 5 Factoring Using the Distributive Law

Use the distributive law to factor the following.

(a) $2x + 2y = 2(x + y)$ The distributive law in reverse order

The two-term expression $2x + 2y$ can be represented as a single-term expression with factors 2 and $(x + y)$.

(b) $5a - 5b = 5(a - b)$ Distributive law: factor out 5

Thus 5 and $(a - b)$ are the factors of $5a - 5b$.

(c) $7a - 7 = 7 \cdot a - 7 \cdot 1$ Express 7 as $7 \cdot 1$
$= 7(a - 1)$ Factor out 7

(d) $6u - 3v + 9 = 3 \cdot 2u - 3 \cdot v + 3 \cdot 3$ 3 is a common factor
$= 3(2u - v + 3)$ Factor out 3

(e) $-3a + 9b = (-3)(a) - (-3)(3b)$
$= (-3)(a - 3b)$ Factor out -3

We could have factored out $+3$ instead.

$$3(-a) + 3(3b) = 3(-a + 3b)$$

Both of these factorizations are correct.

PRACTICE EXERCISE 5

Use the distributive law to factor the following.

(a) $5p + 5q$

(b) $3x - 3w$

(c) $2m - 2$

(d) $4a - 12b - 8$

(e) $-6u + 18v$

Answers: (a) $5(p + q)$ (b) $3(x - w)$ (c) $2(m - 1)$ (d) $4(a - 3b - 2)$ (e) $-6(u - 3v)$

In any factoring problem, we can always check our work by multiplying. For example, since

$$2(x + y) = 2x + 2y, \quad 5(a - b) = 5a - 5b, \quad \text{and} \quad 7(a - 1) = 7a - 7,$$

our factoring in the first three parts of Example 5 is correct.

CAUTION

If you factor $4x + 12$ as $2(2x + 6)$, you are not finished factoring. Always remove the largest common factor, 4 in this case. That is, $4x + 12 = 4(x + 3)$.

4 COLLECTING LIKE TERMS

When an expression contains like terms, the expression can be simplified by **collecting like terms** as in the next example.

EXAMPLE 6 Collecting Like Terms

Use the distributive law to collect like terms.

(a) $2x + 9x = (2 + 9)x$ Use the distributive law to factor out x and then add 2 and 9

$= 11x$

The like terms $2x$ and $9x$ have been collected to form the term $11x$.

(b) $2y + 5y - y + 4 = (2 + 5 - 1)y + 4$ Distributive law

$= 6y + 4$

The terms $6y$ and 4 *cannot* be collected since they are not like terms.

(c) $3a + 5b - a + 4b = 3a + 5b + (-a) + 4b$

$= 3a + (-a) + 5b + 4b$ Commutative law

$= 3a - a + 5b + 4b$

$= 3 \cdot a - 1 \cdot a + 5 \cdot b + 4 \cdot b$

$= (3 - 1)a + (5 + 4)b$ Distributive law

$= 2a + 9b$

With practice, some of these steps can be eliminated.

(d) $0.07x + x = (0.07) \cdot x + 1 \cdot x$ Write x as $1 \cdot x$

$= (0.07 + 1)x$ Distributive law

$= 1.07x$

(e) $3x + 4y - z + 8$ has no like terms to collect.

Practice Exercise 6

Use the distributive law to collect like terms.

(a) $5w + 3w$

(b) $6 + 2a + 7a - 6a$

(c) $7x + 2y - 4x - 3y$

(d) $P + (0.12)P$

(e) $5a + 5b - 3x + 9$

Answers: (a) $8w$ (b) $6 + 3a$ (c) $3x - y$ (d) $1.12P$ (e) no like terms to collect

5 REMOVING PARENTHESES

Using the distributive laws and the fact that $-x = (-1) \cdot x$, we can see that

$$-(a + b) = (-1)(a + b) = (-1)(a) + (-1)(b) = -a - b,$$
$$-(a - b) = (-1)(a - b) = (-1)(a) - (-1)(b) = -a + b,$$
$$-(-a - b) = (-1)(-a - b) = (-1)(-a) - (-1)(b) = a + b.$$

This shows how we can remove parentheses that follow a minus sign.

To Simplify an Expression by Removing Parentheses

1. When a minus sign precedes parentheses, remove the parentheses by changing the sign of every term within the parentheses.
2. When a plus sign precedes parentheses, remove the parentheses without changing any of the signs of the terms.

EXAMPLE 7 Removing Parentheses

Simplify by removing parentheses.

(a) $-(x + 1) = -x - 1$ Change all signs

(b) $-(x - 1) = -x + 1$ Change all signs

(c) $-(-x + 1 + a) = x - 1 - a$ Change all signs

(d) $+(x - 1 + a) = x - 1 + a$ Signs do not change

Practice Exercise 7

Simplify by removing parentheses.

(a) $-(a - 6)$

(b) $-(a + 6)$

(c) $-(-a + 7 - w)$

(d) $+(-a + 7 - w)$

Answers: (a) $-a + 6$ (b) $-a - 6$ (c) $a - 7 + w$ (d) $-a + 7 - w$

EXAMPLE 8 Simplifying Expressions

Simplify by removing parentheses and collecting like terms.

(a) $2x - (6x + 2) = 2x - 6x - 2$ Change all signs within parentheses

$= (2 - 6)x - 2$

$= -4x - 2$ Collect like terms

(b) $y - (3y - 2) + y = y - 3y + 2 + y$ Change all signs within parentheses

$= -y + 2$ Collect like terms

(c) $4 + (z - 3) - 4z = 4 + z - 3 - 4z$ Signs remain the same

$= -3z + 1$

(d) $3a - (-a - 1) + 7 = 3a + a + 1 + 7$ Change signs

$= 4a + 8$

Practice Exercise 8

Simplify.

(a) $4a - (3a + 1)$

(b) $w - (1 - w)$

(c) $7 + (m - 2) - m$

(d) $8 - (-u - 3) + 6u$

Answers: (a) $a - 1$ (b) $2w - 1$ (c) 5 (d) $11 + 7u$

When an expression involves more than one set of parentheses (or brackets or braces), remove the innermost parentheses first. Continue removing from the inside out until all parentheses have been removed.

EXAMPLE 9 Simplifying Expressions

Simplify.

(a) $-4[x - 2(x - 3)] = -4[x - 2x + 6]$ Since $-2(x - 3) = -2x + 6$

$= -4[-x + 6]$

$= 4x - 24$ Since $-4[-x + 6] = (-4)(-x) + (-4)(6) = 4x - 24$

Practice Exercise 9

Simplify.

(a) $-5[a - 3(2 - a)]$

(b) $a - [2a - (1 - 3a)] = a - [2a - 1 + 3a]$
$= a - [5a - 1]$
$= a - 5a + 1$
$= -4a + 1$

(b) $m - [1 - (3m - 2)]$

Answers: (a) $-20a + 30$ (b) $4m - 3$

CAUTION

Is $-(x - 2)$ equal to $-x - 2$? The answer is no.

$$-(x - 2) = -x + 2$$

Change *all* signs when parentheses are removed.

1.5 EXERCISES A

Give the number of terms in each algebraic expression.

1. $2x - 1 + 3b + c$
2. $ax + 7$
3. $5x$

Give the term in each expression that is like the specified term.

4. $3x + 2y + 7;\quad 4y$
5. $ax - 7z + 2;\quad 2z$
6. $-8w + 3 - 12a;\quad -8$

Evaluate when $a = -1$, $b = -2$, and $c = 5$.

7. $a - b$
8. $b - (-a)$
9. $2(a - b + c)$
10. $\frac{a + b}{c}$
11. $\frac{2c - b}{b - 2a}$
12. $|3a + b|$
13. $0 - 2a$
14. $c^2 - b^2$
15 $\frac{\sqrt{c^2 + 11}}{b}$

Evaluate when $x = -3$, $y = -1$, and $z = 5$.

16. $3x + 1$
17. $|-2x - y|$
18. $-|x + y + z|$
19. $|5x + 3z|$
20. $-(-x)$
21. $x \cdot 0$
22. $\frac{x}{0}$
23. $\frac{0}{x}$
24. $0 - x$
25 $-x^2$
26 $(-x)^2$
27 $2x^2$

Evaluate when $u = \frac{1}{4}$, $v = -\frac{2}{3}$, and $w = -\frac{1}{2}$.

28. $u + v$
29. $u - v$
30. $u + vw$
31. $-3v + 4w$

32. $-w$ **33.** $-2w$ **34** $2w^2$ **35** $(2w)^2$

Use the distributive law to factor the following.

36. $2a + 2b$ **37.** $2x - 2y$ **38.** $2u + 4v - 6w$

39. $3b + yb$ **40.** $20x - 10y + 40$ **41.** $36x + 6$

Multiply.

42. $2(a + b)$ **43.** $2(x - y)$ **44.** $2(u + 2v - 3w)$

45. $(3 + y)b$ **46.** $10(2x - y + 4)$ **47.** $6(6x + 1)$

Use the distributive law to collect like terms.

48. $3x + 5x$ **49.** $4z - z$ **50.** $3y - y + 7$

51. $4a - a + 3a$ **52.** $4z - 2x + 3z + 1$ **53.** $2y + a - 2y - a$

54. $\frac{1}{2}x - \frac{1}{4}x + \frac{3}{4}$ **55.** $\frac{2}{3}a + b + \frac{1}{3}a - b$ **56.** $-\frac{3}{4}x + \frac{1}{4} + \frac{3}{4}x - \frac{1}{4}y$

Clear parentheses and simplify.

57. $-(x + 3)$ **58.** $-(a - 3)$ **59.** $-(-a - 1)$

60. $-(x - z - 1)$ **61.** $-(x + 3y - 4)$ **62.** $-(1 + y) + (y + 1)$

63. $-(x + y) - (-x - y)$ **64.** $-2[a - 3(a + 2)]$ **65** $x - [3x - (1 - 2x)]$

66. The perimeter of a rectangle is given by $P = 2l + 2w$ where l is its length and w is its width. What is the perimeter of a rectangle of length 15 cm and width 11 cm?

67. Use $A = P(1 + r)^t$ to find the amount of money in an account at the end of 2 years if a principal of \$1000 is invested at 12% interest, compounded annually.

68. The surface area A of a cylinder with height h and base radius r is given by $A = 2\pi rh + 2\pi r^2$. Use 3.14 for π and find the surface area of a cylinder with radius 3 ft and height 10 ft.

69. The temperature F measured in degrees Fahrenheit can be obtained from degrees Celsius, C, by using $F = \frac{9}{5}C + 32$. Find F when C is 40°.

FOR REVIEW

The following exercises will help you prepare for the next section. Write in exponential notation.

70. $6 \cdot 6 \cdot 6 \cdot 6$

71. $3 \cdot 3 \cdot x \cdot x \cdot x$

Write without using exponential notation.

72. 7^3

73. 2^4x^2

Evaluate.

74. $-\sqrt{\dfrac{25}{121}}$

75. $\sqrt{\dfrac{18}{50}}$

76. $4 \div 2 + 3 \cdot 5$

77. $-|-[-(2-3)]|$

Simplify.

78. $\dfrac{5 \cdot 5 \cdot 5 \cdot 5}{5 \cdot 5}$

79. $\dfrac{a \cdot a \cdot a \cdot a}{a \cdot a \cdot a}$

80. $\dfrac{x \cdot x \cdot x}{x} \cdot \dfrac{x \cdot x}{x}$

ANSWERS: 1. 4 2. 2 3. 1 4. $2y$ 5. $-7z$ 6. 3 7. 1 8. -3 9. 12 10. $-\frac{3}{5}$ 11. undefined 12. 5 13. 2 14. 21 15. -3 16. -8 17. 7 18. -1 19. 0 20. -3 21. 0 22. undefined 23. 0 24. 3 25. -9 26. 9 27. 18 28. $-\frac{5}{12}$ 29. $\frac{11}{12}$ 30. $\frac{7}{12}$ 31. 0 32. $\frac{1}{2}$ 33. 1 34. $\frac{1}{2}$ 35. 1 36–41. Answers given in Exercises 42–47. 42–47. Answers given in Exercises 36–41. 48. $8x$ 49. $3z$ 50. $2y+7$ 51. $6a$ 52. $7z-2x+1$ 53. 0 54. $\frac{1}{4}x+\frac{3}{4}$ 55. a 56. $\frac{1}{4}-\frac{1}{4}y$ 57. $-x-3$ 58. $-a+3$ 59. $a+1$ 60. $-x+z+1$ 61. $-x-3y+4$ 62. 0 63. 0 64. $4a+12$ 65. $-4x+1$ 66. 52 cm 67. \$1254.40 68. 244.92 ft^2 69. 104° 70. 6^4 71. 3^2x^3 72. $7 \cdot 7 \cdot 7 = 343$ 73. $2 \cdot 2 \cdot 2 \cdot 2 \cdot x \cdot x = 16 \cdot x \cdot x$ 74. $-\frac{5}{11}$ 75. $\frac{3}{5}$ 76. 17 77. -1 78. $5 \cdot 5 = 25$ 79. a 80. $x \cdot x \cdot x = x^3$

1.5 EXERCISES B

Give the number of terms in each algebraic expression.

1. $3x + 2 - a$

2. $-7b$

3. $2x + 5$

Give the term in each expression that is like the specified term.

4. $3a + 2b + 11;\quad 5b$

5. $by - 3x + 2;\quad 2x$

6. $-8m + 5 + 6n;\quad -8$

Evaluate when $x = -2$, $y = -1$, and $z = 4$.

7. $x - y$

8. $y - (-x)$

9. $3(x - y + z)$

10. $\dfrac{x+y}{z}$

11. $\dfrac{2x-y}{y-2x}$

12. $|2x + y|$

13. $0 - 3y$

14. $y^2 + z^2$

15. $\dfrac{\sqrt{z^2+9}}{3-x}$

Evaluate when $a = -1$, $b = -3$, and $c = 2$.

16. $2a + 1$

17. $|3a - b|$

18. $-|a + b + c|$

19. $|3c + 2b|$

20. $-(-a)$

21. $a \cdot 0$

22. $\dfrac{a}{0}$

23. $\dfrac{0}{a}$

24. $0 - a$

25. $-a^2$

26. $(-a)^2$

27. $2a^2$

Evaluate when $u = \frac{1}{2}$, $v = -\frac{3}{4}$, and $w = -\frac{1}{4}$.

28. $u + v$

29. $u - v$

30. $u + vw$

31. $-4v + 12w$

32. $-w$

33. $-4w$

34. $4w^2$

35. $(4w)^2$

Use the distributive law to factor the following.

36. $3x + 3z$

37. $3x - 3z$

38. $3u - 6w + 9v$

39. $az + yz$

40. $10a - 30b + 50$

41. $49x + 7$

Multiply.

42. $3(x + z)$

43. $3(x - z)$

44. $3(u - 2w + 3v)$

45. $(a + y)z$

46. $10(a - 3b + 5)$

47. $7(7x + 1)$

Use the distributive law to collect like terms.

48. $4a + 7a$

49. $3x - x$

50. $2y - y + 5$

51. $3w - w + 5w$

52. $2x - 2y + 5x + 1$

53. $4a + b - 2a + 3b$

54. $\dfrac{1}{2}a - \dfrac{3}{4}a + \dfrac{1}{2}$

55. $\dfrac{1}{5}a + b - \dfrac{2}{5}a - b$

56. $-\dfrac{3}{7}y + \dfrac{2}{3} + \dfrac{3}{7}y - \dfrac{2}{3}$

Clear parentheses and simplify.

57. $-(a + 2)$

58. $-(a - 2)$

59. $-(-x - 1)$

60. $-(a - b - 5)$

61. $-(w + 3y - 1)$

62. $-(2 + a) + (a + 2)$

63. $-(a + b) - (-a + b)$

64. $-3[a - 2(a - 1)]$

65. $y - [2y - (1 - y)]$

66. The temperature measured in degrees Celsius, C, can be obtained from degrees Fahrenheit, F, by using $C = \frac{5}{9}(F - 32)$. Find C when F is 5°.

67. Use $A = P(1 + r)^t$ to find the amount of money in an account at the end of 2 years if a principal of \$2000 is invested at 9% interest, compounded annually.

68. The surface area A of a cube is given by $A = 6e^2$ where e is the length of an edge. Find the surface area of a cube with an edge of 1.5 in.

69. The area of a trapezoid is $A = \frac{1}{2}(b_1 + b_2)h$, where b_1 and b_2 are bases and h is the height. Find A when $b_1 = 3$ m, $b_2 = 5$ m, and $h = 4$ m.

FOR REVIEW

The following exercises will help you prepare for the next section. Write in exponential notation.

70. $9 \cdot 9 \cdot 9 \cdot 9$

71. $8 \cdot 8 \cdot 8 \cdot a \cdot a$

Write without using exponential notation.

72. 5^4

73. 3^3y^4

Evaluate.

74. $-\sqrt{\frac{64}{49}}$

75. $\sqrt{\frac{8}{162}}$

76. $9 \div 3 - 2 \cdot 4$

77. $-|-[-(4-9)]|$

Simplify.

78. $\frac{4 \cdot 4 \cdot 4 \cdot 4}{4}$

79. $\frac{y \cdot y \cdot y \cdot y \cdot y}{y \cdot y}$

80. $\frac{w \cdot w}{w \cdot w \cdot w} \cdot \frac{w \cdot w \cdot w}{w}$

1.5 EXERCISES C

1. Use the distributive law to factor $(x + y)a + (x + y)3$.

2. Clear parentheses and simplify. $-(x - [-(x - 3) - (-x - 7)])$ [*Hint:* Start with the innermost parentheses.]

Evaluate when $a = \frac{1}{2}$, $b = -\frac{1}{4}$, *and* $c = -2$.

3. $\frac{ac - b^2}{a^2 - b}$

4. $\frac{ca^2 + cb}{\sqrt{abc}}$ [Answer: 0]

Solve.

5. A type of finishing material is priced at \$5.25 per square foot. How much must be paid for enough of this material to cover an area shaped like a trapezoid with bases 10.5 ft and 12.5 ft and height 8.0 ft?

6. A certain perfume sells for \$52.50 per cubic centimeter. What would be the total cost of a bottle of this perfume which is in the shape of a cylinder with radius 2.1 cm and height 5.6 cm? The volume of a cylinder is given by $V = \pi r^2h$, where r is the radius of the base and h is the height of the cylinder.

1.6 INTEGER EXPONENTS AND SCIENTIFIC NOTATION

STUDENT GUIDEPOSTS

1. Exponential Notation
2. Rules of Exponents
3. Zero as an Exponent
4. Negative Exponents
5. Scientific Notation
6. Significant Digits

1 EXPONENTIAL NOTATION

In Section 1.4 we introduced exponents and exponential notation. On the following page we review the definition of a^n.

Exponential Notation

If a is any number and n is a positive integer,

$$a^n = \underbrace{a \cdot a \cdot a \ldots a}_{n \text{ factors}},$$

where a is the base, n the exponent, and a^n is the exponential expression.

For example, we write

$$2 \cdot y \cdot y \cdot y = 2y^3 \quad \text{and} \quad (2y)(2y)(2y) = (2y)^3 = 8y^3.$$

Also, $$2x^2 = 2 \cdot x \cdot x \quad \text{and} \quad (2x)^2 = (2x)(2x) = 4x^2.$$

Notice the difference between $2x^2$ and $(2x)^2$. An exponent only applies to the factor immediately next to it and does not extend to other factors without using parentheses. Also, $-x^2 \neq (-x)^2$. Thus, $-3^2 = -9$ and $(-3)^2 = 9$.

2 RULES OF EXPONENTS

When we multiply, divide, or take powers of terms containing exponents, our work can be simplified by using the rules of exponents. For example,

$$a^4a^2 = \underbrace{(a \cdot a \cdot a \cdot a)}_{4 \text{ factors}}\underbrace{(a \cdot a)}_{2 \text{ factors}} = \underbrace{a \cdot a \cdot a \cdot a \cdot a \cdot a}_{6 \text{ factors}} = a^6.$$

When two exponential expressions *with the same base* are multiplied, the product is that base raised to the sum of the exponents on the original pair of expressions.

Product Rule

If a is any number, and m and n are positive integers,

$$a^m a^n = a^{m+n}.$$

(To multiply powers with the same base, add exponents.)

To verify the product rule for exponents, consider

$$a^m a^n = \underbrace{a \cdot a \ldots a}_{m \text{ factors}} \cdot \underbrace{a \cdot a \ldots a}_{n \text{ factors}} = \underbrace{a \cdot a \ldots a}_{m+n \text{ factors}} = a^{m+n}.$$

EXAMPLE 1 Using the Product Rule

Use the product rule to simplify each expression.

(a) $a^2a^5 = a^{2+5} = a^7$

(b) $4^3 \cdot 4^7 = 4^{3+7} = 4^{10}$ — *Not* 16^{10}

(c) $2^2 \cdot 2^3 \cdot 2^5 = 2^{2+3+5} = 2^{10}$ — The rule also applies to more than two factors

(d) $2x^2x^9 = 2x^{2+9} = 2x^{11}$ — Note that only x is squared in $2x^2$, not the 2

Practice Exercise 1

Use the product rule to simplify.

(a) x^3x^9

(b) $9^5 \cdot 9^5$

(c) $5^3 \cdot 5^2 \cdot 5^8$

(d) $3w^6w^2$

Answers: (a) x^{12} (b) 9^{10} (c) 5^{13} (d) $3w^8$

When two powers with the same base are divided, for example

$$\frac{a^7}{a^3} = \frac{\overbrace{a \cdot a \cdot a \cdot a \cdot \not{a} \cdot \not{a} \cdot \not{a}}^{7\text{ factors}}}{\underbrace{\not{a} \cdot \not{a} \cdot \not{a}}_{3\text{ factors}}} = \underbrace{a \cdot a \cdot a \cdot a}_{4\text{ factors}} = a^4,$$

the quotient can be found by raising the base to the difference of the exponents $(7 - 3 = 4)$.

Quotient Rule

If a is any number except zero and m, n, and $m - n$ are positive integers, that is, $m > n$, then

$$\frac{a^m}{a^n} = a^{m-n}.$$

(To divide powers with the same base, subtract exponents.)

To verify the quotient rule for exponents, assume m and n are positive integers with $m > n$, and consider

$$\frac{a^m}{a^n} = \frac{\overbrace{a \cdot a \ldots a}^{m\text{ factors}}}{\underbrace{a \cdot a \ldots a}_{n\text{ factors}}} = \frac{\overbrace{\not{a} \cdot \not{a} \ldots \not{a}}^{n\text{ factors}}\overbrace{a \cdot a \ldots a}^{m-n\text{ factors}}}{\underbrace{\not{a} \cdot \not{a} \ldots \not{a}}_{n\text{ factors}}} = \overbrace{a \cdot a \ldots a}^{m-n\text{ factors}} = a^{m-n}.$$

EXAMPLE 2 Using the Quotient Rule

Use the quotient rule to simplify.

(a) $\frac{a^5}{a^2} = a^{5-2} = a^3$

(b) $\frac{4^7}{4^3} = 4^{7-3} = 4^4$

(c) $\frac{3^2}{4^3}$ Cannot be simplified by quotient rule since the bases are different

(d) $\frac{2x^5}{x} = 2x^{5-1}$ $x = x^1$

$= 2x^4$

(e) $\frac{3y^2y^5}{y^3} = \frac{3y^{2+5}}{y^3}$ Use product rule first

$= \frac{3y^7}{y^3} = 3y^{7-3} = 3y^4$

Practice Exercise 2

Use the quotient rule to simplify.

(a) $\frac{m^7}{m^5}$

(b) $\frac{8^9}{8^4}$

(c) $\frac{x^5}{y^3}$

(d) $\frac{c^3}{5c}$

(e) $\frac{7a^7}{a^2a^4}$

Answers: (a) m^2 (b) 8^5 (c) cannot be simplified (d) $\frac{c^2}{5}$ (e) $7a$

When raising a power to a power, for example,

$$(a^2)^3 = \underbrace{(a^2)(a^2)(a^2)}_{3\text{ factors}} = \underbrace{a \cdot a \cdot a \cdot a \cdot a \cdot a}_{6\text{ factors}} = a^6,$$

the resulting exponential expression can be found by raising the base to the product of the exponents $(2 \cdot 3 = 6)$.

Power Rule

If a is any number, and m and n are positive integers,

$$(a^m)^n = a^{mn}.$$

(To raise a power to a power, multiply exponents.)

The power rule is verified in much the same way as the product and quotient rules.

EXAMPLE 3 USING THE POWER RULE

Use the power rule to simplify.

(a) $(a^5)^2 = a^{5 \cdot 2} = a^{10}$

(b) $(4^7)^3 = 4^{7 \cdot 3} = 4^{21}$

PRACTICE EXERCISE 3

Use the power rule to simplify.

(a) $(c^3)^5$

(b) $(8^2)^4$

Answers: (a) c^{15} (b) 8^8

CAUTION

Do not confuse the power rule with the rule for multiplying exponential expressions with the same base. In Example 3(a), $(a^5)^2 = a^{10}$, but $a^5a^2 = a^7$.

Often, a product or quotient of expressions is raised to a power. For example,

$$(2x^2)^3 = \underbrace{(2x^2)(2x^2)(2x^2)}_{3\text{ factors}} = \underbrace{2 \cdot 2 \cdot 2}_{3\text{ factors}} \cdot \underbrace{x^2 \cdot x^2 \cdot x^2}_{3\text{ factors}} = 2^3 \cdot (x^2)^3$$

$$\left(\frac{3}{y^3}\right)^2 = \underbrace{\frac{3}{y^3} \cdot \frac{3}{y^3}}_{2\text{ factors}} = \frac{\overbrace{3 \cdot 3}^{2\text{ factors}}}{\underbrace{y^3 \cdot y^3}_{2\text{ factors}}} = \frac{3^2}{(y^3)^2}.$$

These examples illustrate the following rules.

Powers of Products and Quotients

If a and b are any numbers, and n is a positive integer, then

$$(ab)^n = a^nb^n \quad \text{and} \quad \left(\frac{a}{b}\right)^n = \frac{a^n}{b^n} \quad (b \text{ not zero}).$$

EXAMPLE 4 SIMPLIFYING PRODUCTS AND QUOTIENTS

Simplify.

(a) $(3y)^4 = 3^4 \cdot y^4 = 3^4y^4$

(b) $(2a^2b^3)^5 = 2^5(a^2)^5(b^3)^5$ Raise each factor to the fifth power

$= 2^5a^{10}b^{15}$ Use power rule

(c) $\left(\frac{3z^2}{u^3}\right)^4 = \frac{(3z^2)^4}{(u^3)}$ $\left(\frac{a}{b}\right)^n = \frac{a^n}{b^n}$

$= \frac{3^4(z^2)^4}{(u^3)^4}$ $(a \cdot b)^n = a^nb^n$

$= \frac{3^4z^8}{u^{12}}$ $(a^m)^n = a^{mn}$

(d) $(-2x^2)^3 = (-2)^3(x^2)^3$

$= -8x^6$

(e) a^2b^3 is *not* $(ab)^5$. To see this, substitute 3 for a and 2 for b

Only exponential expressions with the same base can be combined.

PRACTICE EXERCISE 4

Simplify.

(a) $(5a)^3$

(b) $(3u^3v^4)^2$

(c) $\left(\frac{8w^4}{z^2}\right)^3$

(d) $(-7u^4)^2$

(e) u^3v^5

Answers: (a) 5^3a^3 (b) $3^2u^6v^8$ (c) $\frac{8^3w^{12}}{z^6}$ (d) $(-7)^2u^8$ (e) cannot be simplified

3 ZERO AS AN EXPONENT

By the quotient rule we know that if a is not zero and $m > n$,

$$\frac{a^m}{a^n} = a^{m-n}.$$

If we extend the quotient rule to include $m = n$,

$$\frac{a^m}{a^m} = a^{m-m} = a^0 \quad \text{and also} \quad \frac{a^m}{a^m} = 1$$

since any nonzero number divided by itself is 1. This suggests the following definition.

Zero Exponent

If a is any number except zero,

$$a^0 = 1.$$

EXAMPLE 5 USING ZERO AS AN EXPONENT

Simplify.

(a) $5^0 = 1$

(b) $17^0 = 1$

(c) 0^0 is not defined.

(d) $(2x^2y)^0 = 1$, assuming $x \neq 0$ and $y \neq 0$. (Why?)

PRACTICE EXERCISE 5

Simplify.

(a) 25^0

(b) $(ab)^0$

Answers: (a) 1 (b) 1, provided $a \neq 0$ and $b \neq 0$

4 NEGATIVE EXPONENTS

Considering $a^m/a^n = a^{m-n}(a \neq 0)$ again, what happens when $n > m$? For example, if $n = 5$ and $m = 2$, and we extend the quotient rule once more, we have

$$\frac{a^m}{a^n} = \frac{a^2}{a^5} = a^{2-5} = a^{-3}.$$

If we consider this same problem in another way, we have

$$\frac{a^2}{a^5} = \frac{\not{a} \cdot \not{a}}{\not{a} \cdot \not{a} \cdot a \cdot a \cdot a} = \frac{1}{a \cdot a \cdot a} = \frac{1}{a^3}.$$

Thus we conclude that $a^{-3} = \frac{1}{a^3}$. This suggests a way to define exponential expressions involving negative integer exponents.

Negative Exponents

If $a \neq 0$ and n is a positive integer ($-n$ is a negative integer), then

$$a^{-n} = \frac{1}{a^n}.$$

EXAMPLE 6 USING NEGATIVE EXPONENTS

Simplify.

(a) $7^{-3} = \frac{1}{7^3} = \frac{1}{343}$

(b) $\frac{1}{3^{-2}} = \frac{1}{\frac{1}{3^2}} = \frac{1}{\frac{1}{9}} = 1 \cdot \frac{9}{1} = 9 = 3^2$

(c) $(-2)^{-3} = \frac{1}{(-2)^3} = \frac{1}{-8} = -\frac{1}{8}$

PRACTICE EXERCISE 6

Simplify.

(a) 9^{-2}

(b) $\frac{1}{2^{-4}}$

(c) $(-5)^{-2}$

Answers: (a) $\frac{1}{81}$ (b) $2^4 = 16$ (c) $\frac{1}{25}$

We often "remove" negative exponents by simply moving a factor with a negative exponent from denominator to numerator (or numerator to denominator) while simultaneously changing the sign of the exponent. This occurs in (b) above when

$$\frac{1}{3^{-2}} \text{ becomes } \frac{3^2}{1} = 3^2.$$

CAUTION

Is it true that 5^{-2} is the same as -5^2? as $(-2)(5)$? Neither is true since

$$5^{-2} = \frac{1}{25}, \qquad -5^2 = -25, \qquad (-2)(5) = -10.$$

All the rules of exponents developed in terms of positive integer exponents apply to *all* integer exponents: positive, negative, and zero. They are summarized here.

Rules for Exponents

Let a and b be any two numbers, m and n any two integers.

1. $a^m a^n = a^{m+n}$
2. $\dfrac{a^m}{a^n} = a^{m-n} \quad (a \neq 0)$
3. $(a^m)^n = a^{mn}$
4. $(ab)^n = a^n b^n$
5. $\left(\dfrac{a}{b}\right)^n = \dfrac{a^n}{b^n} \quad (b \neq 0)$
6. $a^0 = 1 \quad (a \neq 0)$
7. $a^{-n} = \dfrac{1}{a^n} \quad (a \neq 0)$
8. $\dfrac{1}{a^{-n}} = a^n \quad (a \neq 0)$

EXAMPLE 7 Reviewing Rules of Exponents

Simplify and express without negative exponents.

(a) $(2y)^{-1} = \dfrac{1}{(2y)^1} = \dfrac{1}{2y}$ — $(2y)^{-1}$ is *not* $-2y$

(b) $2y^{-1} = 2 \cdot \dfrac{1}{y^1} = \dfrac{2}{y}$ — Compare with (a). Only y is raised to the -1 power

(c) $x^3x^{-2} = x^{3+(-2)}$ — $a^m a^n = a^{m+n}$

$= x^1 = x$

(d) $\dfrac{x^2y^{-3}}{x^{-1}y^4} = x^{2-(-1)}y^{-3-4}$ — $\dfrac{a^m}{a^n} = a^{m-n}$

$= x^3y^{-7}$

$= x^3 \cdot \dfrac{1}{y^7} = \dfrac{x^3}{y^7}$

(e) $(-2)^{-4} = \dfrac{1}{(-2)^4} = \dfrac{1}{16}$ — $a^{-n} = \dfrac{1}{a^n}$

Practice Exercise 7

Simplify and express without negative exponents.

(a) $(6u)^{-1}$

(b) $6u^{-1}$

(c) w^4w^{-7}

(d) $\dfrac{m^5n^{-2}}{m^{-2}n^3}$

(e) $(-8)^{-2}$

Answers: (a) $\frac{1}{6u}$ (b) $\frac{6}{u}$ (c) $\frac{1}{w^3}$ (d) $\frac{m^7}{n^5}$ (e) $\frac{1}{64}$

EXAMPLE 8 Simplifying Exponential Expressions

Simplify and express without negative exponents.

(a) $\left(\dfrac{a}{b}\right)^{-1} = \dfrac{a^{-1}}{b^{-1}}$ — $\left(\dfrac{a}{b}\right)^n = \dfrac{a^n}{b^n}$

$= \dfrac{\frac{1}{a}}{\frac{1}{b}} = \dfrac{1}{a} \cdot \dfrac{b}{1} = \dfrac{b}{a}$ — Thus $\left(\dfrac{a}{b}\right)^{-1} = \dfrac{b}{a}$

(b) $(2x^2y^{-3})^4 = 2^4(x^2)^4(y^{-3})^4$ — $(ab)^n = a^nb^n$

$= 16x^8y^{-12}$ — $(a^m)^n = a^{mn}$

$= 16x^8\dfrac{1}{y^{12}} = \dfrac{16x^8}{y^{12}}$

Practice Exercise 8

Simplify and express without negative exponents.

(a) $\left(\dfrac{c}{d}\right)^{-2}$

(b) $(5a^3b^{-5})^2$

(c) $6^0(-3x^2)^{-3} = 1 \cdot (-3x^2)^{-3}$ $\quad 6^0 = 1$

$$= (-3)^{-3}(x^2)^{-3} \qquad (ab)^n = a^n b^n$$

$$= \frac{1}{(-3)^3} x^{2(-3)}$$

$$= \frac{1}{-27} x^{-6}$$

$$= -\frac{1}{27} \cdot \frac{1}{x^6} = -\frac{1}{27x^6}$$

(d) $\left(\frac{a^5}{3x^{-3}}\right)^{-2} = \frac{(a^5)^{-2}}{3^{-2}(x^{-3})^{-2}}$

$$= \frac{a^{-10}}{\frac{1}{3^2} \cdot x^6} = \frac{\frac{1}{a^{10}}}{\frac{x^6}{9}} = \frac{1}{a^{10}} \cdot \frac{9}{x^6} = \frac{9}{a^{10}x^6}$$

(c) $7^0(-4w^5)^{-3}$

(d) $\left(\frac{m^{-2}}{2n^3}\right)^{-4}$

Answers: (a) $\frac{d^2}{c^2}$ (b) $\frac{25a^6}{b^{10}}$ (c) $-\frac{1}{64w^{15}}$ (d) $16m^8n^{12}$

5 SCIENTIFIC NOTATION

One important application of integer exponents is scientific notation. For example, chemists use Avogadro's number,

$$602{,}000{,}000{,}000{,}000{,}000{,}000{,}000,$$

but instead of writing all the zeros, they write 6.02×10^{23}. The second notation is easier to use in computations. Likewise, a scientist might use the number

$$0.0000000000000000084,$$

but he or she would write 8.4×10^{-18}.

A number is written in **scientific notation** if it is expressed as the product of a number greater than or equal to 1 and less than 10 with a power of 10. Notice that scientific notation uses the symbol $\times$ instead of the multiplication dot. A number can be converted to scientific notation using the following algorithm.

To Write a Number in Scientific Notation

1. Move the decimal point to the position immediately to the right of the first nonzero digit.
2. Multiply by a power of ten which is equal in absolute value to the number of decimal places moved. The exponent is positive if the original number is greater than or equal to 10 and negative if the number is less than 1.
3. If the number is negative, follow steps 1 and 2, then attach a negative sign to the result.

Thus, the number 78,100 would be written as 7.81×10^4 and the number 0.0000027 as 2.7×10^{-6} (not 2.7^{-6}). To check these notations, simply multiply.

$$7.81 \times 10^4 = 7.81(10{,}000) = 78{,}100$$

$$2.7 \times 10^{-6} = 2.7\left(\frac{1}{10^6}\right) = \frac{2.7}{1{,}000{,}000} = 0.0000027$$

EXAMPLE 9 USING SCIENTIFIC NOTATION

Write in scientific notation.

(a) $5,300,000 = 5.3 \times 10^6$
6 places

(b) $0.0000053 = 5.3 \times 10^{-6}$
6 places

(c) $10 = 1^1 \times 10 = 1 \times 10$
1 place

(d) $0.1 = 1 \times 10^{-1}$
1 place

(e) $-6.2 = -6.2 \times 10^0$

PRACTICE EXERCISE 9

Write in scientific notation.

(a) 42,500

(b) 0.000029

(c) 25

(d) 2.5

(e) −0.00659

Answers: (a) 4.25×10^4 (b) 2.9×10^{-5} (c) 2.5×10^1 (d) 2.5×10^0 (e) -6.59×10^{-3}

Students sometimes feel that scientific notation is only for scientists and does not really concern them. However, any student who owns a hand-held scientific calculator will soon discover that it uses scientific notation extensively. For example, the product of (250,000)(18,000) would appear on a scientific calculator as either display in Figure 1.11 rather than 4,500,000,000. Since most calculators accommodate only eight digits on the display, numbers with more than eight digits must be shown in shortened form using scientific notation.

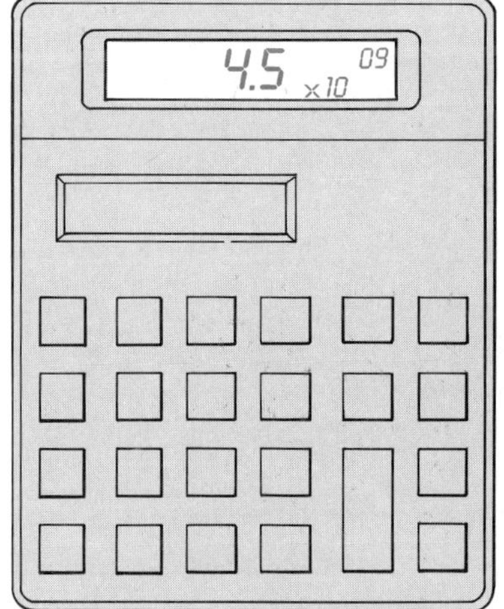

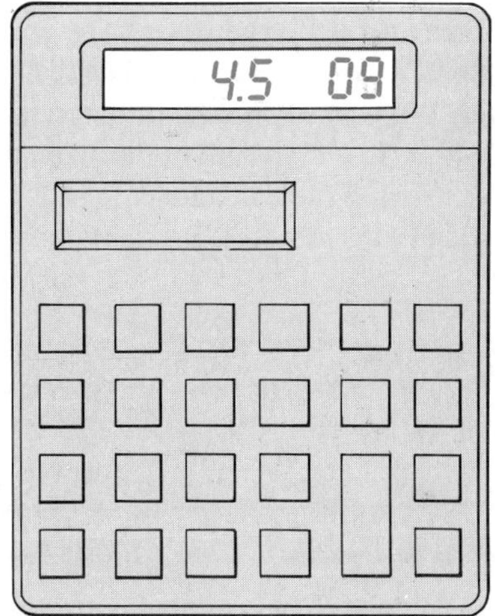

Figure 1.11

Scientific notation not only shortens the notation for certain numbers but also simplifies calculations involving very large or very small numbers.

EXAMPLE 10 PERFORMING OPERATIONS WITH SCIENTIFIC NOTATION

Perform the indicated operation using scientific notation.

(a) $(20,000)(3,000,000) = (2 \times 10^4)(3 \times 10^6)$

$= (2 \cdot 3) \times (10^4 \cdot 10^6)$ Use the commutative law

$= 6 \times 10^{10}$ $10^4 \cdot 10^6 = 10^{4+6} = 10^{10}$

PRACTICE EXERCISE 10

Perform the indicated operation using scientific notation.

(a) (45,000)(6,000,000)

(b) $(2.4 \times 10^{-13})(5.0 \times 10^{6}) = 12 \times 10^{-7}$

$= \mathbf{1.2 \times 10^{1}} \times 10^{-7}$ Obtain a number between 1 and 10

$= 1.2 \times 10^{-6}$

(c) $\dfrac{8.2 \times 10^{-4}}{4.1 \times 10^{8}} = \left(\dfrac{8.2}{4.1}\right) \times \left(\dfrac{10^{-4}}{10^{8}}\right)$

$= 2.0 \times (10^{-4-8})$ $\quad \dfrac{a^m}{a^n} = a^{m-n}$

$= 2.0 \times 10^{-12}$

(d) $\dfrac{(2.65 \times 10^{-3})(4.18 \times 10^{-8})}{3.21 \times 10^{6}}$

We will use a scientific calculator to carry out these operations. If your calculator works in algebraic logic, these are the steps to follow.*

2.65 [EE] 3 [+/−] [×] 4.18 [EE] 8 [+/−] [÷] 3.21 [EE] 6 [=] → [3.4508 − 17]

The result, rounded to two decimal places, is 3.45×10^{-17}.

(b) $(5.6 \times 10^{-8})(3.2 \times 10^{15})$

(c) $\dfrac{9.9 \times 10^{-3}}{1.1 \times 10^{10}}$

(d) $\dfrac{(4.38 \times 10^{4})(6.15 \times 10^{-7})}{1.06 \times 10^{-3}}$

Answers: (a) 2.7×10^{11} (b) 1.792×10^{8} (c) 9.0×10^{-13} (d) 2.54×10^{1}

6 SIGNIFICANT DIGITS

Notice that we rounded the answer in Example 10(d) to the same number of decimal places as the numbers in the original problem. Rounding is necessary when approximate numbers are used in calculations, since it would be inappropriate to give an answer with a higher degree of accuracy than the values used to compute it. The idea of *significant digits* is often used to describe approximate values. The number of **significant digits** is always one more than the number of decimal places given when the number is written in scientific notation. Thus, there are three significant digits in 3.45×10^{-17}, the answer to Example 10(d).

To find the number of significant digits, write the number in scientific notation, count the decimal places, and add 1. The following table gives several examples.

Number	*Scientific Notation*	*Significant Digits*
315.6	3.156×10^{2}	4
0.000 012	1.2×10^{-5}	2
7,860,300	7.8603×10^{6}	5
0.000 920	9.20×10^{-4}	3

The last entry in the table indicates that when one or more zeros *follow* nonzero digits to the *right* of the decimal point, they are significant digits.

1.6 EXERCISES A

Write in exponential notation.

1. $x \cdot x \cdot x$

2. $(3a)(3a)$

3. $3 \cdot a \cdot a$

4. $\dfrac{1}{(6a)(6a)}$

5. $\dfrac{1}{6 \cdot a \cdot a}$

6. $\dfrac{1}{(a+b)(a+b)(a+b)}$

*If you have a calculator that uses Reverse Polish Notation (RPN), consult your operator's manual.

Write without using exponents.

7. x^7

8. $a^2b^3c^4$

9. $3y^3$

10. $(3y)^3$

11. $-x^3$

12. $(-x)^3$

13. $(4a)^{-2}$

14. $4a^{-2}$

15. $\frac{1}{x^2 + y^2}$

Square the following.

16. $5y$

17. $2x^2y$

18. $3a^0 - 3b^0$ (a, b not zero)

Cube the following.

19. $2a$

20. $3a^3$

21. $2x^2y^3z$

Simplify and write without negative exponents.

22. $y^3y^2y^4$

23. $7a^4a^5$

24. $\frac{x^{10}}{x^4}$

25. $\frac{4a^5}{a^3}$

26. $(y^2)^5$

27. $(2ab^4)^3$

28. $\left(\frac{2a^2}{b^5}\right)^2$

29. $\frac{a^4}{b^5}$

30. $3x^0$ ($x \neq 0$)

31. $(3x)^0$ ($x \neq 0$)

32 $(3a)^{-1}$

33 $3a^{-1}$

34 $x^{-1} + a^{-1}$

35 $(x + a)^{-1}$

36. $\frac{2a^2}{a^5}$

37. $(6x^2y^{-3})^{-1}$

38. $\frac{a^2b^{-3}}{a^5b^2}$

39. $\frac{x^{-3}}{y^{-6}}$

40. $\left(\frac{a^2b^3}{2a^{-2}b}\right)^{-2}$

41 $\left(\frac{5^0x^{-2}}{3^{-1}y^{-2}}\right)^2$

42. $\left(\frac{18x^3y^{-3}}{-3x^{-2}y}\right)^{-1}$

43. $\left(\frac{6x^0y}{15x^{-1}y^2}\right)\left(\frac{xy^2}{x^{-1}y^{-1}}\right)$

44 $\left(\frac{x^4y^{-2}}{x^3y^{-4}}\right)^{-2}\left(\frac{x^{-1}y^{-3}}{x^{-4}y^5}\right)^{-3}$

45 $\frac{(x^{-2}y^{-1})^2(xy^{-5})^{-2}}{(3x^2y^{-1})^{-2}(2xy)^{-1}}$

Evaluate.

46. $2^2 + 3^2$

47. $(2 + 3)^2$

48. $(3 \cdot 2)^2$

49. $3^2 \cdot 2^2$

50. $(5 + 2)^{-1}$

51. $5^{-1} + 2^{-1}$

Evaluate when $x = -3$, $y = 2$, and $z = -1$.

52. $2x^2$

53. $(2x)^2$

54. $x^2 - y^2 - z^2$

55. $-z^{-3}$

56. $(-x)^2$

57. $(x - z)^{-3}$

Evaluate when $a = -1$ and $b = -1$.

58. $(-5a)^3$

59. $a^3 + b^3$

60. $(2b + 3a)^{-2}$

61 $(ab)^{-2}$

62 ab^{-2}

63 $a^0 - b^2$

Evaluate when $x = -\frac{2}{3}$.

64. x^{-1}

65. x^{-2}

66. $3x^{-1}$

67. $(3x)^{-1}$

68. $-x^{-1}$

69. $(-x)^{-1}$

70. Given the expression $-y + x^2$, answer the following questions.
(a) What is the exponent on x?
(b) What is the exponent on y?
(c) What is the coefficient of the x^2 term?
(d) What is the coefficient of the y term?

Write in scientific notation.

71. 2,400,000,000

72. −193,000,000

73. −0.000000000000298

Write without scientific notation.

74. 3.6×10^7

75. 3.61×10^{-7}

76. -6×10^{10}

Simplify and give answers in scientific notation.

77. (300,000)(2,000,000)

78. $(2.8 \times 10^6)(5.0 \times 10^{-10})$

79. (0.00000000001)(0.000004)

80. $\dfrac{(4 \times 10^7)(6 \times 10^{-5})}{(8 \times 10^{10})}$

81. $\dfrac{(0.0231)(0.000572)}{866{,}000}$

82. $\dfrac{(3200)(0.000081)}{0.0000012}$

Write the number in each statement using scientific notation.

83. A light-year, the distance that light will travel in one year, is approximately 5,870,000,000,000 mi.

84. The thickness of a particular surface is 0.0000415 cm.

85. The national debt of a country is $86,500,000,000 and the population is 27,300,000. Use scientific notation and a calculator to find the amount of debt per person.

State the number of significant digits in each number.

86. 6805

87. 0.00000802

88. 0.300

FOR REVIEW

How many terms does the given expression have?

89. $2x - y + 7 + w$

90. $7w$

Use the distributive law to factor the following.

91. $5x - 5y + 5$

92. $ax - a$

Simplify.

93. $3x + 2 - (1 - 3x)$

94. $-[2 - (3 - 5x)]$

95. $a - [3a - (-1 - a)]$

96. The surface area A of a sphere with radius r is given by $A = 4\pi r^2$. Use 3.14 for π and find the surface area of a sphere with radius 20 cm.

ANSWERS: 1. x^3 2. $9a^2$ 3. $3a^2$ 4. $\frac{1}{36a^2}$ 5. $\frac{1}{6a^2}$ 6. $\frac{1}{(a+b)^3}$ 7. $xxxxxxx$ 8. $aabbbcccc$ 9. $3yyy$ 10. $(3y)(3y)(3y)$ 11. $-xxx$ 12. $(-x)(-x)(-x)$ 13. $\frac{1}{(4a)(4a)}$ 14. $\frac{4}{aa}$ 15. $\frac{1}{xx + yy}$ 16. $25y^2$ 17. $4x^4y^2$ 18. 0 19. $8a^3$ 20. $27a^9$ 21. $8x^6y^9z^3$ 22. y^9 23. $7a^9$ 24. x^6 25. $4a^2$ 26. y^{10} 27. $8a^3b^{12}$ 28. $\frac{4a^4}{b^{10}}$ 29. $\frac{a^4}{b^5}$ (cannot be simplified further) 30. 3 31. 1 32. $\frac{1}{3a}$ 33. $\frac{3}{a}$ 34. $\frac{1}{x} + \frac{1}{a}$ 35. $\frac{1}{x+a}$ 36. $\frac{2}{a^3}$ 37. $\frac{y^3}{6x^2}$ 38. $\frac{1}{a^3b^5}$ 39. $\frac{y^6}{x^3}$ 40. $\frac{4}{a^8b^4}$ 41. $\frac{9y^4}{x^4}$ 42. $-\frac{y^4}{6x^5}$ 43. $\frac{2x^3y^2}{5}$ 44. $\frac{y^{20}}{x^{11}}$ 45. $\frac{18y^7}{x}$ 46. 13 47. 25 48. 36 49. 36 50. $\frac{1}{7}$ 51. $\frac{1}{5} + \frac{1}{2} = \frac{7}{10}$ 52. 18 53. 36 54. 4 55. 1 56. 9 57. $-\frac{1}{8}$ 58. 125 59. -2 60. $\frac{1}{25}$ 61. 1 62. -1 63. 0 64. $-\frac{3}{2}$ 65. $\frac{9}{4}$ 66. $-\frac{9}{2}$ 67. $-\frac{1}{2}$ 68. $\frac{3}{2}$ 69. $\frac{3}{2}$ 70. (a) 2 (b) 1 (c) 1 (d) -1 71. 2.4×10^9 72. -1.93×10^8 73. -2.98×10^{-13} 74. 36,000,000 75. 0.000000361 76. $-60{,}000{,}000{,}000$ 77. 6×10^{11} 78. 1.4×10^{-3} 79. 4×10^{-17} 80. 3×10^{-8} 81. 1.53×10^{-11} 82. 2.2×10^5 83. 5.87×10^{12} 84. 4.15×10^{-5} 85. \$3170 86. 4 87. 3 88. 3 89. 4 90. 1 91. $5(x - y + 1)$ 92. $a(x - 1)$ 93. $6x + 1$ 94. $1 - 5x$ 95. $-3a - 1$ 96. 5024 cm^2

1.6 EXERCISES B

Write in exponential notation.

1. $x \cdot x \cdot x \cdot x$

2. $(5b)(5b)(5b)$

3. $5 \cdot b \cdot b \cdot b$

4. $\dfrac{1}{(3w)(3w)(3w)}$

5. $\dfrac{1}{3 \cdot w \cdot w \cdot w}$

6. $\dfrac{1}{(x-y)(x-y)(x-y)}$

Write without using exponents.

7. x^8

8. xy^4z^5

9. $2z^4$

10. $(2z)^4$

11. $-y^4$

12. $(-y)^4$

13. $(2w)^{-3}$

14. $2w^{-3}$

15. $\dfrac{1}{w^2 - z^2}$

Square the following.

16. $3w$

17. $3ay^3$

18. $2x^0 - y^0$ (x, y not zero)

Cube the following.

19. $3z$ **20.** $4y^3$ **21.** $3xa^2y$

Simplify and write without negative exponents.

22. $a^2a^3a^5$ **23.** $3y^3y^2$ **24.** $\frac{a^8}{a^5}$ **25.** $\frac{3y^3}{y^2}$

26. $(a^3)^2$ **27.** $(3x^2y^3)^2$ **28.** $\left(\frac{4x^3}{y^2}\right)^3$ **29.** $\frac{x^3}{y^2}$

30. 0^0 **31.** $(4x^2y)^0$ $(x, y \neq 0)$ **32.** $4x^2y^0$ $(y \neq 0)$ **33.** $(5y)^{-1}$

34. $5y^{-1}$ **35.** $a^{-2} + b^{-2}$ **36.** $(a + b)^{-2}$ **37.** $\frac{5y^3}{y^7}$

38. $(2a^3b^{-4})^{-1}$ **39.** $\frac{x^3y^{-5}}{x^7y^3}$ **40.** $\frac{a^{-4}}{b^{-2}}$ **41.** $\left(\frac{x^3y^{-4}}{3x^5y^{-7}}\right)$

42. $\left(\frac{24a^{-2}b^5}{-4a^{-4}b^{-1}}\right)^{-1}$ **43.** $\left(\frac{7a^2b^{-2}}{21ab^{-1}}\right)\left(\frac{a^{-2}b}{ab^{-3}}\right)$ **44.** $\left(\frac{a^3b^2}{a^{-3}b^4}\right)^{-3}\left(\frac{a^{-2}b^{-3}}{ab^{-2}}\right)^{-2}$ **45.** $\frac{(2ab)^{-4}(a^{-5}b)^{-3}}{(3a^{-1}b)^2(4^{-1}a^{-1}b^{-2})^3}$

Evaluate.

46. $2^2 - 3^2$ **47.** $(2 - 3)^2$ **48.** $\frac{3^2}{2^2}$

49. $\left(\frac{3}{2}\right)^{-2}$ **50.** $(2 + 3)^{-1}$ **51.** $2^{-1} + 3^{-1}$

Evaluate when $x = -3$, $y = 2$, *and* $z = -1$.

52. $-2x^2$ **53.** $(-2x)^2$ **54.** $-3z^2 - (x + y)$

55. $-x^2$ **56.** $x^2 - 4yz$ **57.** $x^3 - z^2$

Evaluate when $a = -1$ *and* $b = -1$.

58. $(a + b)^3$ **59.** $(b - a)^{-1}$ **60.** a^{-2}

61. $(ab)^{-1}$ **62.** ab^{-1} **63.** $b^0 - a^2$

Evaluate when $a = -\frac{3}{4}$.

64. a^{-1} **65.** a^{-2} **66.** $4a^{-1}$

67. $(4a)^{-1}$ **68.** $-a^{-1}$ **69.** $(-a)^{-1}$

70. Given the expression $2a - b^2$, answer the following questions.
(a) What is the exponent on a? **(b)** What is the exponent on b?

(c) What is the coefficient of the b^2 term? **(d)** What is the coefficient of the a term?

Write in scientific notation.

71. 6,800,000

72. $-127{,}000{,}000{,}000$

73. -0.00000000000541

Write without scientific notation.

74. 2.8×10^9

75. 5.25×10^{-5}

76. -5×10^{11}

Simplify and give answers in scientific notation.

77. $(3 \times 10^5)(2 \times 10^6)$

78. $(2.8 \times 10^{-6})(5.0 \times 10^{10})$

79. $(2.8 \times 10^7)(5.0 \times 10^0)$

80. $\dfrac{(4 \times 10^5)(6 \times 10^{-3})}{(8 \times 10^{-7})}$

81. $\dfrac{(0.00541)(0.0635)}{1{,}250{,}000}$

82. $\dfrac{(64{,}100)(0.00498)}{0.000000451}$

Write the number in each statement using scientific notation.

83. The earth is approximately 93,000,000 miles from the sun.

84. The measure of one calorie is equal to 0.000000278 kilowatt-hours.

85. The sun is 93,000,000 miles from the earth and light travels at 186,000 miles per second. Use scientific notation and a calculator to find how long it takes light to reach the earth from the sun.

State the number of significant digits in each number.

86. 9,720,300

87. 0.0000003005

88. 0.0005020

FOR REVIEW

Find the number of terms in each expression.

89. $ax - 3$

90. $3y + 2 - z$

Use the distributive law to factor the following.

91. $-3w - 6x + 12$

92. $3y - 3$

Simplify.

93. $2y + 5 - (1 - 5y)$

94. $-[1 - (-2 - 3z)]$

95. $x - [2x - (1 - 2x)]$

96. The surface area A of a cylinder is given by $A = 2\pi rh + 2\pi r^2$ where r is the radius of its base and h is its height. Find A if $r = 20$ in, $h = 15$ in, and $\pi \approx 3.14$.

1.6 EXERCISES C

Simplify each expression and write without negative exponents.

1. $(x^2 - y^2)^{-1}$

2. $\left(\dfrac{-2a^3y^{-2}}{3a^{-4}y^7}\right)^{-3}\left(\dfrac{a^{-1}y^3}{2a^2y}\right)^{-1}\left(\dfrac{-a^3y^{-2}}{4a^{-2}y^3}\right)^2$

3. $\dfrac{(a + b)^{-1}}{a^{-1} + b^{-1}}$ $\left[\text{Answer: } \dfrac{ab}{(a + b)^2}\right]$

4. Evaluate $(x^{x+y} + y^{x+y})(x^{x-y} - y^{x-y})$ when $x = -1$ and $y = 2$. $\left[\text{Answer: } -\frac{9}{8}\right]$

5. Does -2^2 equal $(-2)^2$?

6. For what value of x does $-x^2 = (-x)^2$?

7. $\dfrac{2^{-2} \cdot 5^{-2}}{3^{-1} \cdot 4^{-2}}$

8. $\dfrac{2^{-2} - 5^{-2}}{3^{-1} - 4^{-2}}$ $\left[\text{Answer: } \dfrac{252}{325}\right]$

CHAPTER 1 REVIEW

KEY WORDS

1.1 A **set** is a collection of objects called **elements** of the set.

The **natural numbers** are 1, 2, 3, 4, ...

The **whole numbers** are 0, 1, 2, 3, ...

The **integers** are . . . −3, −2, −1, 0, 1, 2, 3, ...

A **rational number** has the form $\frac{a}{b}$ where a and b are integers, $b \neq 0$.

An **irrational number** has a decimal form that neither terminates nor repeats a block of digits.

The **real numbers** are composed of the rational numbers and the irrational numbers.

A **variable** is a letter that represents a number.

Two numbers that correspond to the same point on a **number line** are **equal.**

1.3 A real number a **is less than** real number b, written $a < b$, if there is a positive number c such that $a + c = b$.

A real number a **is greater than** real number b, written $a > b$, if there is a positive number c such that $a - c = b$.

1.4 An expression a^n is an **exponential expression** with **base** a and **exponent** n.

A number that is the square of an integer is a **perfect square.**

Either of the identical factors of a perfect square number is a **square root** of the number with the nonnegative root called the **principal square root.**

1.5 An **algebraic expression** involves sums, differences, products, quotients, powers, or roots of numbers and variables.

Portions of an algebraic expression separated by plus or minus signs are **terms** of the expression.

Factoring is using the distributive law to remove common factors.

Two terms are **like terms** if they contain the same variables raised to the same power.

We **collect like terms** when we combine the like terms in an expression using the distributive law.

1.6 A number is in **scientific notation** when it is written $a \times 10^m$ where m is an integer and $1 \le a < 10$.

KEY CONCEPTS

1.1 **1.** A rational number can be expressed as a terminating or repeating decimal.

Let a, b, and c be real numbers.

2. Reflexive law: $a = a$

3. Symmetric law: If $a = b$ then $b = a$.

4. Transitive law: If $a = b$ and $b = c$, then $a = c$.

5. Substitution law: If $a = b$, then either may replace the other in any statement without affecting the truth of the statement.

1.2 **1.** The double-negation property: $-(-a) = a$

2. Absolute value of a is defined as

$$|a| = \begin{cases} a \text{ if } a \ge 0 \\ -a \text{ if } a < 0. \end{cases}$$

1.3 Let a, b, and c be real numbers.

1. Closure laws: $a + b$ and ab are real numbers.

2. Commutative laws: $a + b = b + a$ and $ab = ba$

3. Associative laws: $(a + b) + c = a + (b + c)$ and $(ab)c = a(bc)$

4. Existence of identities: $a + 0 = a$, $0 + a = a$, $1 \cdot a = a$, $a \cdot 1 = a$

5. Existence of negatives (additive inverses): $a + (-a) = 0$ and $(-a) + a = 0$

6. Existence of reciprocals (multiplicative inverses): $a\left(\frac{1}{a}\right) = 1$ and $\left(\frac{1}{a}\right)a = 1$, for $a \neq 0$

7. Distributive law: $a(b + c) = ab + ac$

8. $a \div 0$ or $\frac{a}{0}$ is undefined.

9. Trichotomy law: Exactly one of the following holds:

$$a > b,\ a = b,\ a < b.$$

10. Transitive law: If $a < b$ and $b < c$, then $a < c$.

1.4 The radical symbol ($\sqrt{\ }$) by itself designates only the nonnegative or principal square root.

The order of operations:

1. Evaluate within grouping symbols.
2. Evaluate powers and roots.
3. Perform multiplications and divisions from left to right.
4. Perform additions and subtractions from left to right.

1.5 1. Only like terms can be combined. For example, $5y + 2$ is not $7y$ since $5y$ and 2 are not like terms.

2. A minus sign before a set of parentheses changes the sign of every term inside when the parentheses are removed.

1.6 If a and b are any numbers, and m and n are integers, the following statements are true.

$$a^m a^n = a^{m+n}$$

$$\frac{a^m}{a^n} = a^{m-n} \quad \text{if } a \neq 0$$

$$(a^m)^n = a^{mn}$$

$$(ab)^n = a^n b^n$$

$$\left(\frac{a}{b}\right)^n = \frac{a^n}{b^n} \quad \text{if } b \neq 0$$

$$a^0 = 1 \quad \text{if } a \neq 0$$

$$a^{-n} = \frac{1}{a^n} \quad \text{if } a \neq 0$$

$$\frac{1}{a^{-n}} = a^n \quad \text{if } a \neq 0$$

REVIEW EXERCISES

Part I

1.1 1. Write $\{x \mid x \text{ is an integer greater than } -3\}$ using the listing method.

2. Express $\frac{3}{8}$ as a decimal.

Write true *or* false *for each statement. If false, explain why.*

3. Every real number is a rational number.

4. If $3 = x$ then $x = 3$ by the symmetric law.

Insert the correct symbol, =, <, or >, between the given pair of numbers.

5. $-2 \quad 2$

6. $-\frac{3}{2} \quad \frac{1}{2}$

7. $-\frac{3}{4} \quad -\frac{9}{10}$

Graph each inequality on a number line.

8. $x \geq -1$

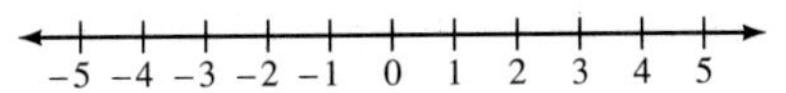

9. $-3 < x \leq 2$

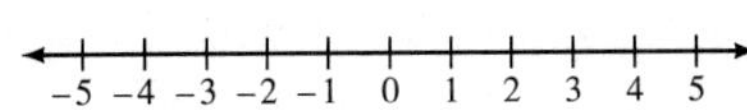

Write true *or* false *for each statement. If the statement is false, explain why.*

1.2 10. For any real number a, $-(-a) = a$.

11. The absolute value of a number is always greater than or equal to zero.

12. The sum of two negative numbers is always a positive number.

13. The quotient of two negative numbers is always a negative number.

14. The product of two nonzero numbers with opposite signs is always a negative number.

Simplify.

15. $-[-(-5)]$

16. $-|-6|$

17. $\left|-\left(-\frac{2}{7}\right)\right|$

18. $|8-3|$

Perform the indicated operations.

19. $-3-(-5)$

20. $(-5)(3)$

21. $(-18) \div (-9)$

22. $\left(-\frac{1}{8}\right)+\left(-\frac{1}{4}\right)$

23. $\frac{3}{4}\cdot\left(-\frac{2}{9}\right)$

24. $|6|-|-10|$

1.3 *Name the property that justifies each statement. Assume that all variables represent real numbers.*

25. $-(-4.5) = 4.5$

26. $5x + 5y = 5(x + y)$

27. $7\left(\frac{1}{7}\right) = 1$

28. $0 + (x + y) = x + y$

29. $9 + (-9) = 0$

30. $x < 3$, $x = 3$, or $x > 3$

31. $2(a + 5) = (a + 5)2$

32. $12 + 7$ is a real number

33. $-\frac{2}{3} = \frac{-2}{3} = \frac{2}{-3}$

Evaluate.

34. $(-3) + 2 + (-5) + (-8)$

35. Give the negative of $-\frac{1}{3}$.

1.4 **36.** Express $2 \cdot 2 \cdot b \cdot b \cdot a \cdot a \cdot a$ using exponential notation.

37. Express $3^2 \cdot a^5$ without exponential notation.

Evaluate the square roots.

38. $\sqrt{49}$

39. $-\sqrt{49}$

40. $-\sqrt{\frac{2}{50}}$

41. What is the square of 16?

42. What is the principal square root of 16?

Evaluate the following.

43. $(8-3)4-1$

44. $|(-2)(-5)+(-3)(3)|$

45. $2-\frac{5-(-3)}{1-(-1)}$

1.5 **46.** Consider the expression $-3x + 2w - 9z + 8$.
(a) Which term is like $-9w$? **(b)** How many terms does the expression have?

Evaluate when $a = -2$, $b = -1$, and $c = 3$.

47. $a - b$ **48.** $b - (-c)$ **49.** $|3a - 6b|$ **50.** $|0 - (-b)|$

51. $\dfrac{2a - b}{c}$ **52.** $5a^2$ **53.** $(5a)^2$ **54.** $-c^2$

Use the distributive law to factor the following.

55. $ab + 4b$ **56.** $-2x - 4y + 8$

Multiply.

57. $-2(4 - 3a)$ **58.** $a(b - 3c)$

Use the distributive law to collect like terms.

59. $3z - z + 5z$ **60.** $\dfrac{1}{2}a - \dfrac{1}{4}a + \dfrac{3}{4}$ **61.** $3a - x + 2a + 4x$

Remove parentheses and simplify.

62. $-(a - 3)$ **63.** $-(-y - 5)$ **64.** $-2[a - 3(a + 1)]$ **65.** $2x - [x - 2(x - 2)]$

1.6 **66.** Given the expression $x - 3y^2$, answer the following questions.
(a) What is the coefficient of the x term? **(b)** What is the coefficient of the y^2 term?
(c) What is the exponent on x? **(d)** What is the exponent on y?

Write true *or* false *for each of the following. If false, explain why.*

67. When multiplying two powers with the same base, we multiply the exponents.

68. When raising a power to a power, we add the exponents.

Simplify and write without negative exponents.

69. $(2x^2y^{-3})^{-2}$ **70.** $\dfrac{a^2b^{-3}}{a^{-3}b}$ **71.** $\dfrac{1}{x^{-3}}$

72. $\left(\dfrac{2x^2}{y^{-1}}\right)^{-3}$ **73.** $\left(\dfrac{2x^2y^{-1}}{xy^3}\right)\left(\dfrac{x^{-3}y^2}{4x^{-1}y^3}\right)^{-2}$ **74.** $\dfrac{x^{-7}}{y^{-3}}$

Evaluate when $x = -2$.

75. $3x^2$

76. $-3x^2$

77. $(3x)^2$

78. $(-3x)^2$

79. $3x^{-2}$

80. $(3x)^{-2}$

81. Write 29,300,000 in scientific notation.

82. Write 2.9×10^{-8} without scientific notation.

83. Use a calculator to simplify and write the following in scientific notation.

$$\frac{(0.000000081)(21{,}000{,}000)}{(0.000055)}$$

84. State the number of significant digits in the number 0.500.

Part II

85. Write $\{0, 2, 4, 6\}$ using set-builder notation.

86. Write 0.000000049 in scientific notation.

Remove parentheses and simplify.

87. $-(y + 2) - (-y - 2)$

88. $3 - 2(y - [y + 1])$

89. Evaluate. $3 + 8 \div 2 - 3 \cdot 5$

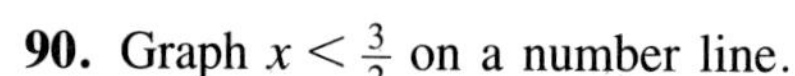

90. Graph $x < \frac{3}{2}$ on a number line.

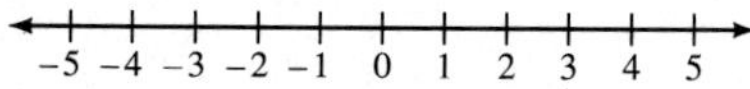

Write true *or* false *for each statement. If the statement is false, explain why.*

91. The product of two negative numbers is always a positive number.

92. The coefficient of the w^2 term in $3x - w^2$ is 2.

93. When dividing two powers with the same base, we subtract the exponents.

94. Every rational number is an integer.

95. By the closure law of multiplication, $x + 2 = 2 + x$.

96. If $x < y$ and $y < 5$, then by the transitive law we know that $x < 5$.

97. The number 423,000,000 can be written in scientific notation as 4.23×10^{-8}.

Simplify. Write without negative exponents when appropriate.

98. $-|-8|\,|-2|$

99. $-[-(-x)]$

100. $\left(\dfrac{3x^{-1}}{y^{-2}}\right)^{-3}$

Evaluate when $x = -3$, $y = 4$.

101. $4x^{-2}$

102. $\sqrt{x^2y}$

103. $-x - y$

104. Give the reciprocal of $-\sqrt{7}$.

105. Use the distributive law and collect like terms.

$$-3x + y - 2x - 5y + 7$$

106. State the number of significant digits in the number 3,205,400.

107. Divide. $\left(-\frac{7}{22}\right) \div \left(-\frac{21}{11}\right)$

108. Use the distributive law to factor $4x - 20y$.

109. Evaluate the square root. $-\sqrt{121}$

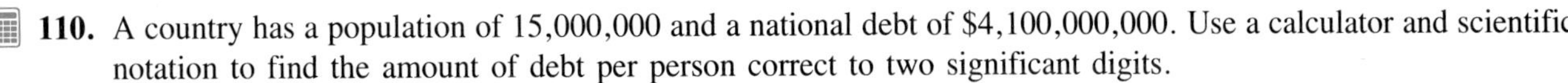

110. A country has a population of 15,000,000 and a national debt of $4,100,000,000. Use a calculator and scientific notation to find the amount of debt per person correct to two significant digits.

ANSWERS: **1.** $\{-2, -1, 0, 1, 2, 3, \ldots\}$ **2.** 0.375 **3.** false; for example π **4.** true **5.** $-2 < 2$ **6.** $-\frac{3}{2} < \frac{1}{2}$ **7.** $-\frac{3}{4} > -\frac{9}{10}$ **8.** **9.** **10.** true **11.** true **12.** false; always negative **13.** false; always positive **14.** true **15.** -5 **16.** -6 **17.** $\frac{2}{7}$ **18.** 5 **19.** 2 **20.** -15 **21.** 2 **22.** $-\frac{3}{8}$ **23.** $-\frac{1}{6}$ **24.** -4 **25.** double-negation property **26.** distributive law **27.** existence of reciprocals **28.** existence of an additive identity **29.** existence of negatives **30.** trichotomy law **31.** commutative law of multiplication **32.** closure law of addition **33.** negative fraction property **34.** -14 **35.** $\frac{1}{3}$ **36.** $4b^2a^3$ **37.** $9aaaaa$ **38.** 7 **39.** -7 **40.** $-\frac{1}{5}$ **41.** 256 **42.** 4 **43.** 19 **44.** 1 **45.** -2 **46.** (a) $2w$ (b) 4 **47.** -1 **48.** 2 **49.** 0 **50.** 1 **51.** -1 **52.** 20 **53.** 100 **54.** -9 **55.** $(a + 4)b$ **56.** $-2(x + 2y - 4)$ **57.** $-8 + 6a$ **58.** $ab - 3ac$ **59.** $7z$ **60.** $\frac{1}{4}a + \frac{3}{4}$ **61.** $5a + 3x$ **62.** $-a + 3$ **63.** $y + 5$ **64.** $4a + 6$ **65.** $3x - 4$ **66.** (a) 1 (b) -3 (c) 1 (d) 2 **67.** false; add the exponents **68.** false; multiply the exponents **69.** $\frac{y^6}{4x^4}$ **70.** $\frac{a^5}{b^4}$ **71.** x^3 **72.** $\frac{1}{8x^6y^3}$ **73.** $\frac{32x^5}{y^2}$ **74.** $\frac{y^3}{x^7}$ **75.** 12 **76.** -12 **77.** 36 **78.** 36 **79.** $\frac{3}{4}$ **80.** $\frac{1}{36}$ **81.** 2.93×10^7 **82.** 0.000000029 **83.** 3.1×10^4 **84.** 3 **85.** $\{x|x$ is an even whole number less than $7\}$ **86.** 4.9×10^{-8} **87.** 0 **88.** 5 **89.** -8 **90.** **91.** true **92.** false; -1 **93.** true **94.** false; for example, $\frac{1}{2}$ **95.** false; commutative law of addition **96.** true **97.** false; 4.23×10^8 **98.** -16 **99.** $-x$ **100.** $\frac{x^3}{27y^6}$ **101.** $\frac{4}{9}$ **102.** 6 **103.** -1 **104.** $-\frac{1}{\sqrt{7}}$ **105.** $-5x - 4y + 7$ **106.** 5 **107.** $\frac{1}{6}$ **108.** $4(x - 5y)$ **109.** -11 **110.** $270

CHAPTER 1 TEST

1. Write $\{x|x$ is a whole number less than $5\}$ using the listing method. 1. ______

2. Express $\frac{7}{8}$ as a decimal. 2. ______

3. True or false: Every integer is a real number. 3. ______

4. Insert $=$, $>$, or $<$ between the pair of numbers: $-\frac{5}{2} \quad \frac{5}{2}$. 4. ______

5. Evaluate: $-|-[-(-4)]|$. 5. ______

6. Give the inequality graphed below. 6. ______

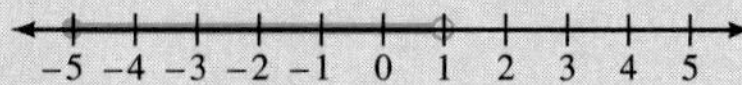

Perform the indicated operations.

7. $(-3) + 8 - 7$ 7. ______

8. $[(-5)(2) - (-3)(-2)]$ 8. ______

9. $\frac{2}{3} - \left(-\frac{1}{3}\right)$ 9. ______

10. $(-3.6) \div (-1.2)$ 10. ______

11. $|-5| - |-2|$ 11. ______

Name the property that justifies each statement.

12. $7 + 5 = 5 + 7$ 12. ______

13. $-(-3.8) = 3.8$ 13. ______

14. If $a = 7$ and $7 = w$ then $a = w$. 14. ______

Evaluate.

15. $\sqrt{36}$ 15. ______

16. $-\sqrt{\frac{3}{75}}$ 16. ______

17. $-3[4 - (2 - 3)]$ 17. ______

18. Use the distributive law to factor $ax + 5x - cx$. 18. ______

19. Multiply: $-3(2 - c)$. 19. ______

Evaluate when $a = 3$ and $b = -3$.

20. $2b^2$ 20. ______

21. $(2b)^2$ 21. ______

22. $-b^2$ 22. ______

23. $(ab)^2$ 23. ______

24. ab^2 24. ______

25. Evaluate $|5 - 8x|$ when $x = -1$. 25. ______

Remove parentheses and simplify.

26. $-(-x + 1)$ 26. ______

27. $2a - [a - 3(a - 1)]$ 27. ______

28. The temperature measured in degrees Fahrenheit (F) can be obtained from degrees Celsius (C) by using $F = \frac{9}{5}C + 32$. Find F when C is 80°. 28. ______

Simplify and write without negative exponents.

29. $2^0x^{-1}x^3x^2$ 29. ______

30. $\left(\dfrac{a^3b^{-3}}{a^2b^2}\right)^3$ 30. ______

31. $\left(\dfrac{x^{-1}y^{-3}}{2xy^{-1}}\right)^{-2}\left(\dfrac{3xy^2}{x^{-4}}\right)$ 31. ______

32. Write 0.0000237 in scientific notation. 32. ______

CHAPTER 2

Linear Equations and Inequalities

2.1 SOLVING LINEAR EQUATIONS

STUDENT GUIDEPOSTS

1. Equations
2. Equivalent Equations
3. Conditional Equations
4. Combining Rules to Solve Equations
5. Collecting Like Terms
6. Grouping Symbols

1 EQUATIONS

A statement that two quantities are equal is an **equation.** The quantities are written with an equal sign (=) between them. Some equations are true, some are false, and for some the truth value cannot be determined. For example,

$3 + 5 = 8$	is true,
$3 + 5 = 3 - 5$	is false,
$x + 5 = 8$	is neither true nor false since the value of x is not known.

Equations in algebra usually involve variables. If the variable in an equation can be replaced by a number that makes the resulting equation true, that number is a **solution** of the equation. The process of **solving the equation** involves finding all its solutions. Some simple equations may be solved by inspection. (For example, it is easy to see that 3 is a solution to the equation $x = 3$.) Other equations require techniques described in the material that follows.

In this chapter we concentrate on **linear equations** in one variable. Since the variable in a linear equation is raised to the first power only, it is often called a **first-degree equation.** Any linear equation in one variable x can be written

$$ax + b = 0$$

where a and b are known real numbers and $a \neq 0$.

2 EQUIVALENT EQUATIONS

Consider the two equations

$$x + 3 = 5 \quad \text{and} \quad x = 2.$$

Both equations have 2 as a solution. While both can be solved by inspection, it is easier to recognize 2 as a solution to the second. **Equivalent equations** are equations that have exactly the same solutions.

To solve an equation we change it into an equivalent equation that can be solved by direct inspection. Usually, the equivalent equation will have the variable isolated on one side, such as in $x = 2$. The first equation-solving rule is the *additive property of equality,* which states that the same number (positive or negative) can be added to both sides of an equation.

Additive Property of Equality

Let a, b, and c be numbers.

$$\text{If } a = b, \quad \text{then } a + c = b + c.$$

That is, $a = b$ and $a + c = b + c$ are equivalent equations.

EXAMPLE 1 Using the Additive Property

Solve.

$$x + 3 = 5$$

$$x + 3 - 3 = 5 - 3 \quad \text{Subtracting 3 is the same as adding } -3$$

$$x + 0 = 2$$

$$x = 2 \quad \text{Equation equivalent to the original } x + 3 = 5$$

Check: $2 + 3 \stackrel{?}{=} 5$ Substitute 2 for x in the original equation

$$5 = 5$$

The solution is 2.

Practice Exercise 1

Solve.

$$x + 7 = 4$$

Answer: -3

3 CONDITIONAL EQUATIONS

With practice the second and third steps in Example 1 can be done mentally. Always check all indicated solutions by replacing the variable with the possible solution. If the resulting numerical equation is true, the number is a solution. An equation, like in Example 1, that is true for some replacements of the variable and false for others is called a **conditional equation.**

EXAMPLE 2 Using the Additive Property

Solve.

$$x - \frac{1}{4} = 3$$

$$x - \frac{1}{4} + \frac{1}{4} = 3 + \frac{1}{4} \quad \text{Add } \tfrac{1}{4} \text{ to both sides to isolate } x$$

$$x + 0 = \frac{3 \cdot 4}{4} + \frac{1}{4}$$

$$= \frac{12}{4} + \frac{1}{4} = \frac{13}{4}$$

$$x = \frac{13}{4}$$

Practice Exercise 2

Solve.

$$x - \frac{2}{3} = 4$$

Check: $\frac{13}{4} - \frac{1}{4} \stackrel{?}{=} 3$

$\frac{12}{4} \stackrel{?}{=} 3$

$3 = 3$

The solution is $\frac{13}{4}$.

Answer: $\frac{14}{3}$

The second equation-solving rule is the *multiplicative property of equality* which states that both sides of an equation can be multiplied (or divided) by the same number.

Multiplicative Property of Equality

Let a, b, and c be numbers, with $c \neq 0$.

$$\text{If } a = b, \quad \text{then} \quad ac = bc.$$

That is, $a = b$ and $ac = bc$ are equivalent equations.

EXAMPLE 3 Using the Multiplicative Property

Solve.

$$3x = 11$$

$$\frac{1}{3} \cdot 3x = \frac{1}{3} \cdot 11 \qquad \text{Multiply both sides by } \tfrac{1}{3} \text{ to isolate } x$$

$$1 \cdot x = \frac{11}{3}$$

$$x = \frac{11}{3}$$

Check: $3 \cdot \frac{11}{3} \stackrel{?}{=} 11$

$11 = 11$

The solution is $\frac{11}{3}$. Equivalently, we may divide both sides of the equation by 3.

$$3x = 11$$

$$\frac{3x}{3} = \frac{11}{3} \qquad \text{Divide both sides by 3}$$

$$\frac{3}{3} \cdot x = \frac{11}{3}$$

$$1 \cdot x = \frac{11}{3}$$

$$x = \frac{11}{3}$$

Practice Exercise 3

Solve.

$$14x = 35$$

Answer: $\frac{5}{2}$

EXAMPLE 4 AN EQUATION WITH A FRACTIONAL COEFFICIENT

Solve.

$$-\frac{2}{5}y = 10$$

$$\left(-\frac{5}{2}\right)\cdot\left(-\frac{2}{5}\right)y = \left(-\frac{5}{2}\right)\cdot 10$$ Multiply both sides by $-\frac{5}{2}$, the reciprocal of $-\frac{2}{5}$

$$1 \cdot y = -25$$

$$y = -25$$

Check: $\left(-\frac{2}{5}\right)\cdot(-25) \stackrel{?}{=} 10$

$$10 = 10$$

The solution is -25.

PRACTICE EXERCISE 4

Solve.

$$\frac{2}{3}y = -50$$

Answer: -75

Equations that have a fraction in a denominator must be solved carefully.

EXAMPLE 5 AN EQUATION WITH A FRACTIONAL DENOMINATOR

Solve.

$$\frac{x}{\frac{1}{2}} = 10$$

In an equation of this type, simplify the left side first.

$$\frac{x}{\frac{1}{2}} = \frac{\frac{x}{1}}{\frac{1}{2}} = \frac{x}{1} \div \frac{1}{2} = \frac{x}{1}\cdot\frac{2}{1} = 2x$$

Thus, we are actually solving

$$2x = 10.$$

$$\frac{1}{2}\cdot 2x = \frac{1}{2}\cdot 10$$ Multiply both sides by $\frac{1}{2}$

$$1 \cdot x = 5$$

$$x = 5$$

The solution is 5. To check, substitute 5 for x in the original equation.

PRACTICE EXERCISE 5

Solve.

$$\frac{x}{\frac{1}{9}} = 36$$

Answer: 4

Study the above examples carefully and observe that in all cases we multiply or divide both sides so that the coefficient of the variable becomes 1.

4 COMBINING RULES TO SOLVE EQUATIONS

Many equations are solved by using a combination of the additive property and the multiplicative property. Generally we do the addition before the multiplication, as in the example on the following page.

EXAMPLE 6 USING A COMBINATION OF RULES

Solve.

$$3x + 7 = 13$$
$$3x + 7 - 7 = 13 - 7 \quad \text{Subtract 7 from both sides}$$
$$3x = 6$$
$$\frac{1}{3} \cdot 3x = \frac{1}{3} \cdot 6 \quad \text{Multiply both sides by } \tfrac{1}{3}$$
$$x = 2$$

The solution is 2. Check by substituting 2 for x in the original equation.

PRACTICE EXERCISE 6

Solve.

$$4x - 2 = 14$$

Answer: 4

EXAMPLE 7 AN EQUATION WITH DECIMAL COEFFICIENTS

Solve.

$$1.2y - 3.6 = 2.4$$
$$1.2y - 3.6 + 3.6 = 2.4 + 3.6 \quad \text{Add 3.6 to both sides}$$
$$1.2y = 6.0$$
$$\frac{1.2y}{1.2} = \frac{6.0}{1.2} \quad \text{Divide both sides by 1.2}$$
$$y = 5$$

The solution is 5. Check in the original equation.

PRACTICE EXERCISE 7

Solve.

$$2.2y + 3.3 = 9.9$$

Answer: 3

We can eliminate decimals by multiplying both sides of the equation by the appropriate power of 10. In the above example, multiplying both sides by 10 would give us

$$12y - 36 = 24.$$

There is less chance of making an arithmetic error solving this equation than the original.

Equations with fractional coefficients can also be simplified. We multiply both sides of the equation by the least common denominator (LCD) of the fractions.

EXAMPLE 8 AN EQUATION WITH FRACTIONAL COEFFICIENTS

Solve.

$$\frac{1}{3}x + \frac{1}{4} = \frac{1}{2}$$
$$12\left(\frac{1}{3}x + \frac{1}{4}\right) = 12\left(\frac{1}{2}\right) \quad \text{Multiply both sides by the LCD 12}$$
$$12 \cdot \frac{1}{3}x + 12 \cdot \frac{1}{4} = 6 \quad \text{Use distributive law}$$

PRACTICE EXERCISE 8

Solve.

$$\frac{1}{5}x - \frac{2}{3} = \frac{1}{2}$$

$$4x + 3 = 6$$
$$4x + 3 - 3 = 6 - 3 \qquad \text{Subtract 3}$$
$$4x = 3$$
$$\frac{1}{4} \cdot 4x = \frac{1}{4} \cdot 3 \qquad \text{Multiply by } \tfrac{1}{4}$$
$$x = \frac{3}{4}$$

Check: $\frac{1}{3}\left(\frac{3}{4}\right) + \frac{1}{4} \stackrel{?}{=} \frac{1}{2}$

$$\frac{1}{4} + \frac{1}{4} \stackrel{?}{=} \frac{1}{2}$$
$$\frac{2}{4} = \frac{1}{2}$$

The solution is $\frac{3}{4}$.

Answer: $\frac{35}{6}$

5 COLLECTING LIKE TERMS

If one side of an equation contains like terms, these terms should be collected before using the additive or multiplicative properties.

EXAMPLE 9 An Equation with Like Terms

Solve.

$$6x + 1 - 4x = 12 - 2x - 11$$
$$2x + 1 = -2x + 1 \qquad \text{Collect like terms on both sides}$$
$$2x + 2x + 1 = -2x + 2x + 1 \qquad \text{Add } 2x \text{ to both sides}$$
$$4x + 1 = 1$$
$$4x + 1 - 1 = 1 - 1 \qquad \text{Subtract 1}$$
$$4x = 0$$
$$\frac{1}{4}(4x) = \frac{1}{4}(0) \qquad \text{Multiply by } \tfrac{1}{4}$$
$$x = 0 \qquad \tfrac{1}{4} \cdot 0 = 0$$

Check: $6(0) + 1 - 4(0) \stackrel{?}{=} 12 - 2(0) - 11$

$$1 \stackrel{?}{=} 12 - 11$$
$$1 = 1$$

The solution is 0.

Practice Exercise 9

Solve.

$$3x + 2 - 8x = 14 + x - 6$$

Answer: -1

The example on the next page shows that not every equation has a solution. Equations with no solution are called **contradictions.**

EXAMPLE 10 Working with a Contradiction

Solve.

$$2x + 3 = 2x + 8$$
$$2x + 3 - 3 = 2x + 8 - 3 \quad \text{Subtract 3}$$
$$2x = 2x + 5$$
$$2x - 2x = 2x - 2x + 5 \quad \text{Subtract } 2x$$
$$0 = 5$$

Whenever a false statement, such as $0 = 5$, is obtained, the original equation has no solution. The equation is a contradiction.

Practice Exercise 10

Solve.

$$x + 4 - 5x = 1 - 4x$$

Answer: no solution (contradiction)

Some equations have many solutions, as illustrated in the next example. When an equation has every real number as a solution, it is called an **identity.**

EXAMPLE 11 Solving an Identity

Solve.

$$3x + 1 = 1 + 3x$$
$$3x + 1 - 1 = 1 - 1 + 3x \quad \text{Subtract 1}$$
$$3x = 3x$$

Whenever a statement such as $3x = 3x$ is obtained, every real number is a solution. The equation is an identity.

Practice Exercise 11

Solve.

$$2x + 1 - x = 1 + x$$

Answer: Every real number is a solution (identity).

6 GROUPING SYMBOLS

When an equation contains grouping symbols, remove all sets of grouping symbols using the distributive law. Then collect like terms and proceed as before.

EXAMPLE 12 An Equation Involving Grouping Symbols

Solve.

$$6(z + 1) - 4(z - 3) = 0$$
$$6z + 6 - 4z + 12 = 0 \quad \text{Remove parentheses and watch signs}$$
$$2z + 18 = 0 \quad \text{Collect like terms}$$
$$2z = -18 \quad \text{Subtract 18 from both sides}$$
$$z = -9 \quad \text{Divide both sides by 2}$$

Check:
$$6(-9 + 1) - 4(-9 - 3) \stackrel{?}{=} 0$$
$$6(-8) - 4(-12) \stackrel{?}{=} 0$$
$$-48 + 48 \stackrel{?}{=} 0$$
$$0 = 0$$

The solution is -9.

Practice Exercise 12

Solve.

$$3(z + 2) - 4(z - 1) = 0$$

Answer: 10

To conclude, we summarize the method for solving linear equations.

To Solve a Linear Equation

1. To simplify both sides, remove grouping symbols and collect like terms.
2. To eliminate fractions or decimals, multiply both sides by an appropriate factor (the LCD of the fractions or a power of ten for the decimals).
3. Use the additive property to isolate all variable terms on one side and all constant terms on the other side. Collect like terms when possible.
4. Use the multiplicative property to give the variable a coefficient of 1.
5. Check the solution by substituting in the original equation.
6. If at any step an identity results, the original equation has every real number as a solution. If a contradiction results, there is no solution.

2.1 EXERCISES A

1. Give an example of a true equation.

2. Give an example of a conditional equation.

3. Give an example of a contradiction.

4. Give an example of a linear equation.

5. Consider the equation $2x - 5 = 17$.

(a) What is the variable?

(b) What is the right side of the equation?

(c) What is the left side of the equation?

(d) Is it an identity?

(e) Is it a contradiction?

(f) Is it a conditional equation?

(g) Is it equivalent to $x = 11$?

(h) What is the solution?

Solve the following equations.

6. $x + 2 = 9$

7. $y - 3 = 8$

8. $z + 2.1 = 3.8$

9. $x - \frac{3}{4} = 4$

10. $-2 + y = 4.2$

11. $-3.1 - z = 1.5$

12. $4x = 16$

13. $-3y = 27$

14. $\frac{1}{5}z = 7$

15. $-\frac{3}{4}x = 9$

16. $\frac{y}{\frac{1}{4}} = 20$

17. $\frac{z}{\frac{1}{3}} = -9$

18. $\dfrac{x}{\frac{2}{3}} = -18$

19. $4y = 12.8$

20. $-1.2z = -72$

21. $2x + 1 = 9$

22. $3y - 2 = 10$

23. $-4z + 1 = 17$

24. $\dfrac{1}{4}x + 2 = 3$

25. $\dfrac{2}{5} - y = \dfrac{3}{5}$

26. $12 = -1.2z + 36$

27. $-\dfrac{1}{10} - \dfrac{3}{5}x = \dfrac{1}{5}$

28. $3y + 2y = 15$

29. $\dfrac{1}{2}z + \dfrac{1}{4}z = -12$

30. $4x + 5 = x - 16$

31. $6 + 5y = 5y - 2$

32. $z + 2z = 8 - 2z + 7$

33. $x - 3 + 4x = 2 + x - 5$

34 $2.1y + 45.2 = 3.2 - 8.4y$

35 $3z + \dfrac{3}{2} + \dfrac{5}{2}z = \dfrac{1}{2}z + \dfrac{5}{2}z$

36. $3(2x + 3) = 15$

37. $20 = 5(y - 1)$

38. $-2(3z - 1) = 10$

39. $2(x + 3) = 3(x - 7)$

40. $4(2y - 1) - 7(y + 3) = 0$

41. $3(2z - 1) = 5 - (3z - 2)$

42. $4(1 - 2z) - 3 = 8z - 1$

43 $2[x - (x + 1)] = -2$

44. $-2[x - 3(x - 5)] = 6 - 5x$

45. $\dfrac{y}{3} - \dfrac{y}{7} = 1$

46. $\dfrac{z}{5} - \dfrac{z}{2} = 3$

47 $\dfrac{x + 1}{3} - \dfrac{x + 7}{2} = 1$

48. $\dfrac{x - 5}{4} + \dfrac{x + 1}{5} = 3$

49. $\dfrac{x + 8}{6} - \dfrac{x + 10}{2} = -5$

50. $x(x - 1) = 2 + x^2$

Find a value of k so that each of the following equations has only the solution −1.

51 $3y + 2 = k - 1$

52. $\frac{x-1}{2} = \frac{k+1}{4}$

Find a value of m so that the equations in each pair are equivalent.

53 $3x + 1 = m$ and $5x - 1 = 9$

54. $4y - m = 12$ and $2y - 3 = 27$

FOR REVIEW

The following exercises combine material from Chapter 1 with solving equations. Evaluate each expression for $a = -2$ and $b = -4$, and then solve for x.

55. $a - [(x + b) - (a - x)] = 3x - (a - b)$

56. $\frac{2a - bx}{a - b} - \frac{x - ab}{ab} = \frac{ax - bx}{a + 4}$

ANSWERS: Answers to Exercises 1–4 will vary; however, one possible answer is given for each. 1. $1 + 3 = 4$ 2. $x + 1 = 5$ 3. $x + 1 = x - 1$ 4. $2x + 5 = 0$ 5. (a) x (b) 17 (c) $2x - 5$ (d) no (e) no (f) yes (g) yes (h) 11 6. 7 7. 11 8. 1.7 9. $\frac{19}{4}$ 10. 6.2 11. −4.6 12. 4 13. −9 14. 35 15. −12 16. 5 17. −3 18. −12 19. 3.2 20. 60 21. 4 22. 4 23. −4 24. 4 25. $-\frac{1}{5}$ 26. 20 27. $-\frac{1}{2}$ 28. 3 29. −16 30. −7 31. no solution 32. 3 33. 0 34. −4 35. $-\frac{3}{5}$ 36. 1 37. 5 38. $-\frac{4}{3}$ 39. 27 40. 25 41. $\frac{10}{9}$ 42. $\frac{1}{8}$ 43. Every real number is a solution (identity). 44. 4 45. $\frac{21}{4}$ 46. −10 47. −25 48. 9 49. 4 50. −2 51. 0 52. −5 53. 7 54. 48 55. $\frac{2}{5}$ 56. $\frac{8}{7}$

2.1 EXERCISES B

1. Give an example of a false equation.

2. Give an example of an equation that is neither true nor false.

3. Give an example of an identity.

4. Give an example of a first-degree equation.

5. Consider the equation $3y - 1 = 20$.

(a) What is the variable?

(b) What is the right side of the equation?

(c) What is the left side of the equation?

(d) Is it an identity?

(e) Is it a contradiction?

(f) Is it a conditional equation?

(g) Is it equivalent to $y = 7$?

(h) What is the solution?

Solve the following equations.

6. $x + 3 = 7$

7. $y - 5 = 4$

8. $z + 1.5 = 2.7$

9. $x - \frac{7}{8} = 2$

10. $-3 + y = 5.4$

11. $-2.7 - z = 1.6$

12. $5x = 20$

13. $-3y = 18$

14. $\frac{1}{5}z = 3$

15. $-\frac{3}{4}x = 12$

16. $\frac{y}{\frac{1}{5}} = 20$

17. $\frac{z}{\frac{1}{3}} = -12$

18. $\frac{x}{\frac{2}{3}} = -15$

19. $4y = 16.4$

20. $-1.2z = -84$

21. $2x + 3 = 7$

22. $3y - 1 = 11$

23. $-4z + 3 = 15$

24. $\frac{1}{5}x + 2 = 5$

25. $\frac{3}{4} - y = \frac{1}{4}$

26. $15 = -1.5z + 45$

27. $-\frac{3}{10} - \frac{2}{5}x = \frac{1}{5}$

28. $5y + 2y = 14$

29. $\frac{1}{2}z + \frac{3}{4}z = -20$

30. $4x + 2 = x - 4$

31. $6 + 5y = 5 + 5y$

32. $2z - 1 + z = 4 + 3z - 5$

33. $x + 2 + 3x = 2 + x - 6$

34. $3.2y + 0.1 = 1.6 - 1.8y$

35. $2z + \frac{1}{4} + \frac{3}{4}z = \frac{1}{2}z$

36. $4(3x + 1) = 28$

37. $35 = 7(y - 2)$

38. $-3(2z - 1) = 9$

39. $4(2y - 1) = 9(y + 3)$

40. $3(x + 1) - 5(x - 7) = 0$

41. $2(3z - 1) = 2 - (4z - 6)$

42. $5(1 - 3y) - 6 = 3 - 11y$

43. $3[x - (2x + 1)] = 4 - 2x$

44. $-2[y - (y + 3)] = 2 - y$

45. $\frac{y}{2} - \frac{y}{7} = 1$

46. $\frac{x}{5} - \frac{x}{3} = 2$

47. $\frac{x + 1}{2} - \frac{x + 5}{3} = 1$

48. $\frac{x - 8}{8} + \frac{x - 4}{4} = -1$

49. $\frac{5 - x}{7} - \frac{x + 4}{5} = 3$

50. $y(y + 1) = y^2 + 4$

Find a value of k so that each of the following equations has only the solution −3.

51. $2y + 8 = k - 5$

52. $\frac{z - 2}{3} = \frac{k + 2}{4}$

Find a value of m so that the equations in each pair are equivalent.

53. $2x - 5 = m$ and $4x - 3 = 9$

54. $6y + m = 15$ and $5y + 1 = 11$

FOR REVIEW

The following exercises combine material from Chapter 1 with solving equations. Evaluate each expression for a = −2 and b = −4, and then solve for x.

55. $a - bx + (a + b)x = a[x - (b - 2x)]$

56. $\frac{ax + b}{2a} + \frac{x + ab}{b} = \frac{ax + bx}{a + 3}$

2.1 EXERCISES C

Solve.

1. $\frac{1}{2}x + 2.5 = 4$

2. $0.000005x = 0.000006$

3. $1.06x + 4.356 = 1.706$

4. $\frac{x}{5} - \frac{1}{3} = x + 1\frac{1}{2}$ [Answer: $-\frac{55}{24}$]

5. $\frac{2}{x^{-1}} + \frac{3}{x^{-1}} = 10$

6. $\frac{1}{(3x)^{-1}} + 2 = \frac{1}{x^{-1}}$

7. Which of the following equations are equivalent to $2x + 1 = x + 1$?
   (a) $3x = 2$ **(b)** $x + 1 = 1$ **(c)** $2x = x$ **(d)** $x = 2$ **(e)** $x = 0$

2.2 TECHNIQUES OF SOLVING APPLICATION PROBLEMS

STUDENT GUIDEPOSTS

1. Translating Words Into Symbols
2. Solving Applied Problems
3. Consecutive Integer Problems
4. Geometry Problems

1 TRANSLATING WORDS INTO SYMBOLS

Solving a word problem with algebra involves two basic steps. First, we translate the words of the problem into an algebraic equation, and second, we solve the equation. We learned how to solve several types of equations in Section 2.1, and now we concentrate on translating words into equations. Some common words and their symbolic translations are presented below.

Symbol	*Stands for*
$+$	and, sum, sum of, added to, increased by, more than
$-$	minus, less, subtracted from, less than, diminished by, difference between, difference, decreased by
$\cdot$	times, product, product of, multiplied by, of
$\div$	divided by, quotient of, ratio
$=$	equals, is equal to, is as much as, is, is the same as, gives, yields

Any letter (we often use x) can stand for the unknown or desired quantity. Some examples of translations of these phrases follow.

A number increased by 5 is 13 .
$$x + 5 = 13$$

Two subtracted from a number is 5 .
$$x - 2 = 5$$

Three percent of a number is 15 .
$$0.03 \cdot x = 15$$

Twice my age in two years will be 20 .
↓ ↓ ↓ ↓
$2 \cdot \quad (x + 2) \quad = \quad 20$

Four times a number, diminished by 12, is as much as twice the number .

$4 \quad \cdot \quad x \quad - \quad 12 \quad = \quad 2x$

It is sometimes helpful to use a letter as the variable that is indicative of the quantity it represents. For example, we might use n for a natural number, t for time, s for salary, V for volume, or T for Ted's age.

EXAMPLE 1 TRANSLATING PHRASES INTO SYMBOLS

Select a variable to represent each quantity and translate the phrase into symbols.

Word phrase	*Symbolic translation*
(a) Twice a natural number	$2n$
(b) The time in 6 hours	$t + 6$
(c) My salary less \$200 for taxes	$S - 200$
(d) The price less 10% of the price	$p - 0.10p$
(e) One-third the volume	$\frac{1}{3}V$

PRACTICE EXERCISE 1

Select a variable and translate each phrase into symbols.

(a) One-half a whole number

(b) The rate, increased by 5 mph

(c) The price less \$15

(d) The population, diminished by 20%

(e) Three times the area

Answers: (a) $\frac{1}{2}w$ (b) $r + 5$ (c) $p - 15$ (d) $p - 0.20p$ (e) $3A$

EXAMPLE 2 TRANSLATING SENTENCES INTO SYMBOLS

Use x for the variable and translate each word sentence into symbols.

Word expression	*Symbolic translation*
(a) The product of a number and 3 is 22.	$3x = 22$
(b) Four times a number is 8.	$4x = 8$
(c) Twice a number, increased by 8, is 11.	$2x + 8 = 11$
(d) Seven is 4 less than three times a number.	$7 = 3x - 4$
(e) Seven is 4 less three times a number.	$7 = 4 - 3x$
(f) One-tenth of a number is 6.	$\frac{1}{10}x = 6$
(g) Twice a number, less five times the number, is the same as 12.	$2x - 5x = 12$
(h) A number is 15% of 28.	$x = (0.15)(28)$
(i) A number subtracted from 2 is 5.	$2 - x = 5$
(j) Two subtracted from a number is 8.	$x - 2 = 8$
(k) Twice the sum of a number and 3 is four times the square of the number.	$2(x + 3) = 4x^2$
(l) Eight more than a number is 12.	$8 + x = 12$
(m) My age 3 years ago was 35. (Let x be my present age.)	$x - 3 = 35$

PRACTICE EXERCISE 2

Use x for the variable and translate into symbols.

(a) Three times a number is 5 less the number.

(b) Thirty is 12% of the principal.

(c) Ann's age in 8 years will be 24.

(d) Five times the price, less \$20, is \$45.

(e) The cube of a number is 24 more than 10 times the number.

Answers: (a) $3x = 5 - x$ (b) $30 = 0.12x$ (c) $x + 8 = 24$ (d) $5x - 20 = 45$ (e) $x^3 = 24 + 10x$

2 SOLVING APPLIED PROBLEMS

Because there is a wide variety of applied problems, no single technique works for all problems. But following these steps will help.

To Solve an Applied Problem

1. Read the problem carefully (perhaps several times) and determine what quantity (or quantities) must be found.
2. Represent the unknown quantity (or quantities) with a letter.
3. Make a sketch or diagram whenever possible.
4. Determine which expressions are equal and write an equation.
5. Solve the equation and state the answer to the problem.
6. Check to see if the answer (or answers) satisfies the conditions of the original problem. This can often be done mentally.

In the remainder of this section we solve several types of applied problems.

EXAMPLE 3 Number Problem

If three times a number is subtracted from 20, the result is twice the number. Find the number.

Let x = the desired number,
$20 - 3x$ = three times the number subtracted from 20.

$$20 - 3x = 2x \quad \text{Symbolic translation of the problem}$$
$$20 = 5x$$
$$4 = x$$

The number is 4. Check: 20 minus $3 \cdot 4$ is indeed twice 4.

Practice Exercise 3

Twenty less a number is 3 times the number. Find the number.

Answer: 5

3 CONSECUTIVE INTEGER PROBLEMS

Consecutive integers are integers that are next to each other in counting order. For example, 3, 4, and 5 are consecutive integers. If x is an integer, the next consecutive integer is $x + 1$, and the next after that is $x + 2$. We can also consider consecutive even or odd integers such as 4, 6, 8, or 9, 11, 13. In these cases, if x is an integer, the next consecutive even (or odd) integer is $x + 2$, and the next after that is $x + 4$.

EXAMPLE 4 Consecutive Integer Problem

The sum of two consecutive integers is 47. Find the integers.

Let x = the first integer,
$x + 1$ = the next consecutive integer.

$$x + (x + 1) = 47$$
$$2x + 1 = 47$$
$$2x = 46$$
$$x = 23 \quad \text{and} \quad x + 1 = 24$$

The integers are 23 and 24. Check: 23 and 24 are indeed consecutive integers that add up to 47.

Practice Exercise 4

Find two consecutive odd integers whose sum is 56.

Answer: 27 and 29

EXAMPLE 5 Number Problem

A 20-ft section of rope is cut into two pieces, one 7 ft longer than the other. How long is each piece?

Let x = the length of the shorter piece,

$x + 7$ = the length of the longer piece. (Why?)

Make a sketch as in Figure 2.1.

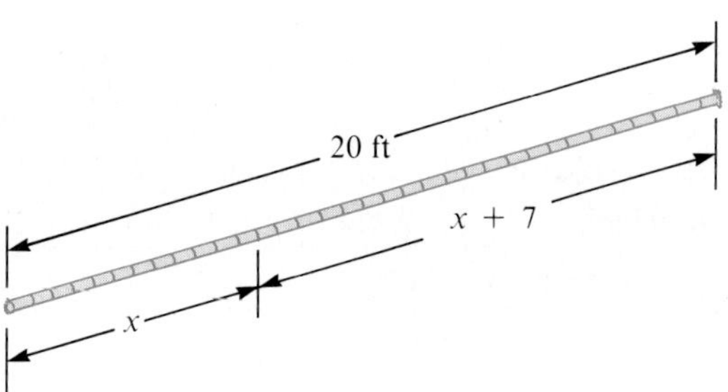

Figure 2.1

$$\begin{aligned} x + (x + 7) &= 20 \\ 2x + 7 &= 13 \\ 2x &= 13 \\ x &= \frac{13}{2} \\ \text{and} \quad x + 7 &= \frac{27}{2} \end{aligned}$$

The pieces are $\frac{13}{2}$ ft and $\frac{27}{2}$ ft. Check: $\frac{27}{2}$ is indeed 7 more than $\frac{13}{2}$, and $\frac{13}{2} + \frac{27}{2} = 20$.

Practice Exercise 5

A 48-in gold chain is cut into two lengths, one 12 in shorter than the other. How long is each piece?

Answer: One is 18 in and the other is 30 in.

4 GEOMETRY PROBLEMS

Many applied problems require certain basic formulas. In particular, problems involving geometric figures might need one of the formulas given inside the back cover. As you read through each problem, try to determine which formula is needed. Remember to make a sketch.

EXAMPLE 6 Geometry Problem

The second angle of a triangle is six times as large as the first angle. If the third angle is 45° more than twice the first, find the measure of each angle.

Recall that the sum of the measures of the angles of any triangle is 180°. Make a sketch as in Figure 2.2.

Let x = the measure of the first angle,

$6x$ = the measure of the second angle,

$2x + 45$ = the measure of the third angle.

$$\begin{aligned} x + 6x + (2x + 45) &= 180 \\ 9x + 45 &= 180 \\ 9x &= 135 \\ x &= 15 \end{aligned}$$

Practice Exercise 6

The second angle of a triangle is four times the first, and the third is 20° less than three times the first. Find the measures of the angles.

$$\text{so} \quad 6x = 90$$
$$\text{and} \quad 2x + 45 = 75$$

Figure 2.2

The measures are 15°, 90°, and 75°. Check: Since 15° + 90° + 75° = 180°, 90° = 6(15°), and 75° = 45° + 2(15°), the measures do check.

Answer: 25°, 100°, and 55°

EXAMPLE 7 CONSTRUCTION PROBLEM

A farmer plans to build a cylindrical storage silo with radius 5.2 m. How high will it need to be to store 724 m^3 of grain? [Use 3.14 for π.]

The volume of a cylinder is given by $V = \pi r^2 h$. See Figure 2.3.

$$V = \pi r^2 h$$
$$724 = (3.14)(5.2)^2 h \qquad \text{Substitution}$$
$$\frac{724}{(3.14)(5.2)^2} = h$$
$$8.53 \approx h \quad \text{(to the nearest hundredth)}$$

Thus, the silo must be about 8.5 m high.

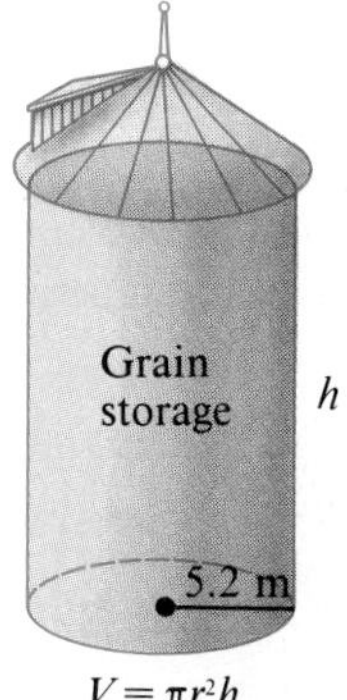

Figure 2.3

PRACTICE EXERCISE 7

Find the height (to the nearest tenth) of a cylindrical tank with surface area 500 ft^2 and radius 3.5 ft. The surface area of a cylinder is $A = 2\pi rh + 2\pi r^2$.

Answer: 19.2 ft

In Example 7 we used the symbol ≈ to represent "*is approximately equal to*." Use ≈ for approximate or rounded numbers.

2.2 EXERCISES A

Select a variable to represent each quantity and translate the phrase into symbols.

1. The sum of a number and 4

2. The time in 2 hours

3. Her salary, increased by $400

4. Twice the volume

5. $4000 less the cost

6. The reciprocal of a number

7. Twice the number of votes, increased by 500

8. 5% of the price

Letting x represent the unknown number, translate each statement into symbols.

9. Eight times a number is 72.

10. A number subtracted from 3 is 10.

11. A number increased by 7 is 21.

12. A number divided by 3 is equal to 4.

13. My age in 8 years will be 26.

14. Twice a number, increased by 3, is 14.

15. Twice the sum of a number and 5 is the same as the number.

16. Seven times a number, less 3, is as much as twice the number, plus 5.

17. A number less 8 is as much as the number squared.

18. Eleven is 5 less than eight times a number.

19. Two-thirds of a number is 36.

20. A number is 12% of 25.

21. When 9 is added to three times a number and the result tripled, the resulting quantity is the same as 13 more than 6 times the number.

22. A number increased by its reciprocal is the same as five times the number, decreased by 7.

Solve each of the following applied problems. (Some problems have been started.)

23. If 3 is added to five times a number, the result is 38. Find the number.

Let x = the number,
$5x$ = five times the number.

24. The sum of two numbers is 24. The larger number is five times the smaller number. Find the two numbers.

Let x = one number,
$5x$ = other number.

25. Sam made $10 more than twice what Pete earned in one month. If together they earned $760, how much did each earn that month?

Let x = amount Pete earned in one month,
$2x + 10$ = amount Sam earned in one month.

26. A woman burns up three times as many calories running as she does when walking the same distance. If she runs 2 mi and walks 5 mi and burns up a total of 770 cal, how many calories does she burn up while running 1 mi?

Let x = calories burned walking 1 mi,
$3x$ = calories burned running 1 mi.

27. The sum of two consecutive odd integers is 76. Find the two integers. [*Hint:* If one integer is x, the other is $x + 2$. Why?]

28. Two-fifths of a man's income each month goes to taxes. If he pays $848 in taxes each month, what is his monthly income?

29. The sum of three consecutive integers is 126. Find the three integers.

30. The sum of Jan's age and Juan's age is 78 years. If Jan is 6 years younger than Juan, how old is each?

31. A board is 11 ft long. It is to be cut into three pieces in such a way that the second piece is twice as long as the first piece and the third piece is 3 ft longer than the first piece. Find the length of each piece.

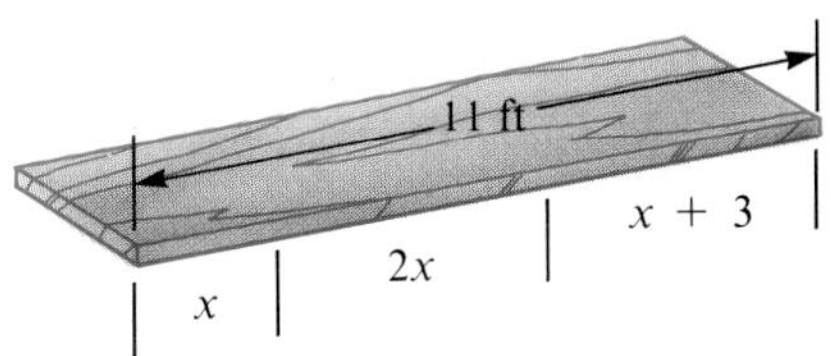

32 A pole is standing in a small lake. If one-sixth of the length of the pole is in the sand at the bottom of the lake, 25 ft are in the water, and two thirds of the total length is in the air above the water, what is the length of the pole?

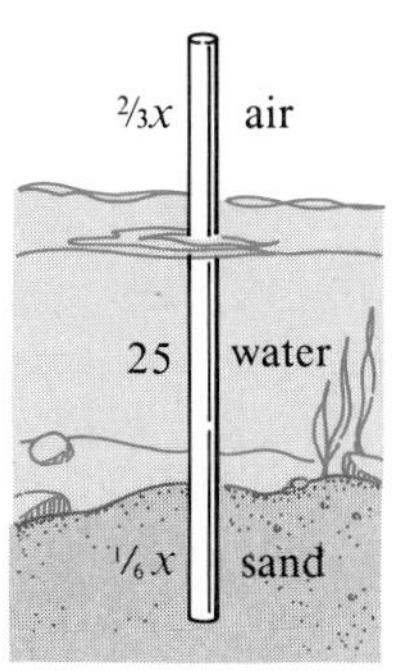

Let x = length of pole.

33 Becky must have an average of 90 on four tests in geology to get an A in the course. What is the lowest score she can get on the fourth test if her first three scores are 96, 78, and 91?

34 International Car Rental will rent a compact car for \$12.50 per day plus 10¢ per mile driven. If Charlene rented a compact car for two days and was billed a total of \$73.00, how many miles did she drive?

35. The perimeter of a rectangle is 86 in. If the length is 19 in more than the width, find the dimensions.

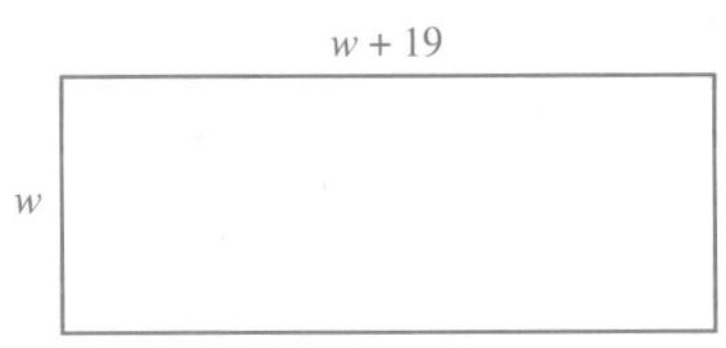

Let w = width,
$w + 19$ = length.

$$P = 2l + 2w$$
$$86 = 2(w + 19) + 2w$$

36. If the second angle of a triangle is 50° more than the first angle, and the third angle is eleven times the first, find the measure of each angle.

Let x = the measure of the first angle.
$x + 50$ =
$11x$ =

37. If two angles are supplementary (they have measures totaling 180°) and the second is 15° less than twice the first, find the measure of each angle.

38. An isosceles triangle has two equal sides called legs and a third side called the base. If each leg of an isosceles triangle is five times the base and the perimeter is 77 m, find the lengths of the legs and base.

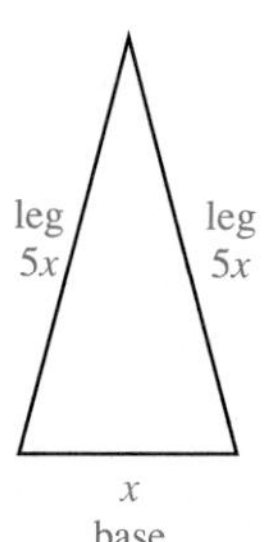

Let x = base.
$5x$ =

39. The circumference of a circular garden is 182 ft. How many feet of pipe are required to reach from the edge of the garden to a fountain in the center of the garden? [Use 3.14 for π.]

40. Find the height of a parallelogram whose base is 19 cm and whose area is 456 cm^2.

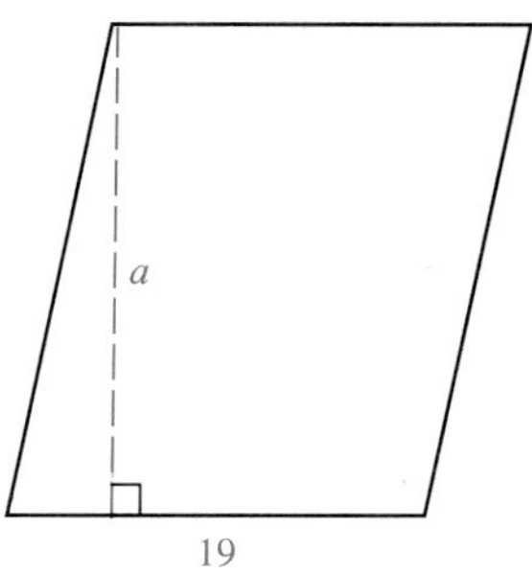

Let a = height.

41 A sphere of radius 2 ft is dropped into a rectangular tank that is 10 ft long and 8 ft wide. To the nearest tenth, how much does this raise the level of water in the tank?

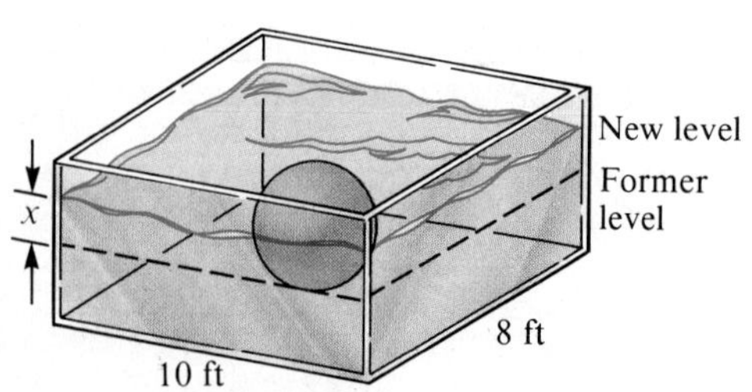

Let x = rise in level.

42. Find the height of a cylindrical tank with volume 3200 in^3 and radius 8 in. [Use 3.14 for π.]

43. A cube with edge 10 in is submerged in a rectangular tank of water. If the tank is 40 in by 50 in, how much does the level of water in the tank rise?

44. A rancher wishes to enclose his north pasture with fence selling for 28¢ per linear foot. If the pasture is rectangular in shape, 3 mi long and 2 mi wide, how much will the project cost?

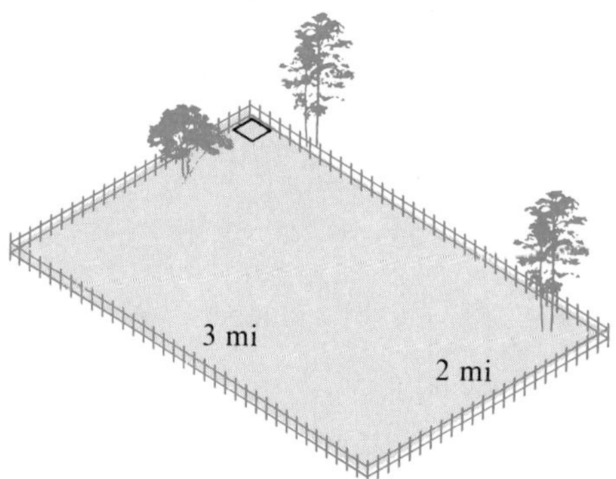

FOR REVIEW

Solve.

45. $\frac{1}{2}y + \frac{3}{4}y = -40$

46. $0.2(y + 1) - 0.3(y - 1) = 0.9$

47. $7(2y - 1) = 3 + 14y$

48. $-3[z - (4z + 1)] = 5z - 1$

49. $z^2 - 3 = (3 + z)z$

50. $\frac{z + 2}{5} - \frac{2z - 1}{2} = 1$

ANSWERS: Some answers are rounded. **1.** $n + 4$ **2.** $t + 2$ **3.** $s + 400$ **4.** $2V$ **5.** $4000 - c$ **6.** $\frac{1}{n}$ **7.** $2v + 500$ **8.** $0.05p$ **9.** $8x = 72$ **10.** $3 - x = 10$ **11.** $x + 7 = 21$ **12.** $\frac{x}{3} = 4$ **13.** $x + 8 = 26$ **14.** $2x + 3 = 14$ **15.** $2(x + 5) = x$ **16.** $7x - 3 = 2x + 5$ **17.** $x - 8 = x^2$ **18.** $11 = 8x - 5$ **19.** $\frac{2}{3}x = 36$ **20.** $x = (0.12)(25)$ **21.** $3(3x + 9) = 6x + 13$ **22.** $x + \frac{1}{x} = 5x - 7$ **23.** 7 **24.** 4, 20 **25.** Pete: $250; Sam: $510 **26.** 210 calories **27.** 37, 39 **28.** $2120 **29.** 41, 42, 43 **30.** Jan is 36 and Juan is 42. **31.** 2 ft, 4 ft, 5 ft **32.** 150 ft **33.** 95 **34.** 480 mi **35.** 12 in, 31 in **36.** 10°, 60°, 110° **37.** 65°, 115° **38.** base: 7 m; legs: 35 m **39.** about 29 ft **40.** 24 cm **41.** 0.4 ft **42.** 15.9 in **43.** 0.5 in **44.** $14,784 **45.** −32 **46.** −4 **47.** no solution **48.** −1 **49.** −1 **50.** $-\frac{1}{8}$

2.2 EXERCISES B

Select a variable to represent each quantity and translate the phrase into symbols.

1. A number subtracted from 20

2. The time 6 hours ago

3. My wages less $60

4. Five times the area

5. Triple the cost

6. A number added to its reciprocal

7. 25% of a number

8. Cost plus 20% of the cost

Letting x represent the unknown number, translate the following into symbols.

9. Seven times a number is 42.

10. A number subtracted from 8 is 1.

11. A number increased by 5 is 37.

12. A number divided by 7 is equal to 3.

13. Jeff's age in 3 years will be 39.

14. Twice a number, increased by 4, is 20.

15. Three times the sum of a number and 2 is the same as the number.

16. Six times a number, less 9, is as much as twice the number, plus 7.

17. A number less 15 is as much as the number squared.

18. The quotient of 4 more than a number and 2 less the number is equal to -2.

19. Three-fourths of a number is 28.

20. A number is 8% of 300.

21. When 11 is added to three times a number and the result doubled, the resulting quantity is the same as 15 more than four times the number.

22. A number decreased by its reciprocal is the same as ten times the number, increased by 5.

Solve each of the following applied problems.

23. If 2 is added to seven times a number, the result is 79. Find the number.

24. The sum of two numbers is 52. The larger is three times the smaller number. Find the two numbers.

25. Carlos made $3000 more than twice what Ernie earned one year. If together they earned $31,500, how much did each earn that year?

26. A man burns up one-third as many calories walking as he does when running the same distance. If he walks 3 km and runs 4 km and burns up a total of 1125 cal, how many calories does he burn up while walking 1 km?

27. The sum of two consecutive even integers is 170. Find the two integers.

28. Three-fourths of the operating budget of a university goes for faculty salaries. If faculty salaries total $6,000,000, what is the operating budget?

29. The sum of three consecutive odd integers is 87. Find the integers.

30. In an election with only two candidates, the winner received 40 votes more than twice the total for the loser. If a total of 1480 votes were cast, how many did each candidate receive?

31. A wire is 40 in long. It is to be cut into three pieces so that the second piece is three times as long as the first piece and the third piece is 5 in longer than the first piece. Find the length of each piece.

32. On a recent trip, Bob Packard traveled one-fifth the total distance by car, 300 miles by boat, and one-half the total distance by air. What was the total distance traveled on the trip?

33. To get a B in history, Bernie must have an average of 80 on four tests. What is the lowest score he can get on the fourth test if his first three scores are 87, 81, and 68?

34. The cost of renting a carpet shampooer is $4.50 per hour plus $3.15 per quart of shampoo. If the Pearsons used 3 quarts of shampoo and were billed $49.95, how many hours did they use the machine?

35. The perimeter of a rectangle is 120 ft. If the length is 22 ft more than the width, find the dimensions.

36. If the second angle of a triangle is 36° more than the first angle, and the third is six times the first, find the measure of each angle.

37. Two angles are supplementary and the second is 30° less than twice the first. Find the measure of each angle.

38. Each leg of an isosceles triangle is four times the base, and the perimeter of the triangle is 99 in. Find the lengths of the legs and base.

39. The circumference of a circular patio is 125.6 yd. How far is it from the edge of the patio to a flagpole in its center? [Use 3.14 for π.]

40. Find the height of a parallelogram whose base is 12.5 cm and whose area is 255 cm^2.

41. A sphere of radius 9 inches is dropped into a rectangular tank 20 in long and 18 in wide. To the nearest tenth, how much does this raise the level of the water in the tank?

42. Find the height of a cylindrical tank with volume 301.44 ft^3 and radius 4 ft. [Use 3.14 for π.]

43. A cube with edge 15 cm is submerged in a rectangular tank of water. If the tank is 30 cm by 40 cm, how much does the level of the water in the tank rise?

44. Mr. Williams wishes to put a fence around his property. If his land is rectangular in shape, 1 mi wide and 2 mi long, and fencing costs 32¢ per linear foot, how much will the project cost?

FOR REVIEW

Solve.

45. $\frac{1}{3}y + \frac{5}{3}y = -9$

46. $0.2(x + 1) - 0.3(x - 2) = 0.5$

47. $6(2z - 1) = 4 + 12z$

48. $-2[x - (2x + 1)] = 5 - x$

49. $y^2 + 2 = y(y - 1)$

50. $\frac{z+1}{3} - \frac{z+2}{2} = 1$

2.2 EXERCISES C

Solve.

1. Three integers are consecutive multiples of 5. If their sum is 90, find the integers.

2. One side of a triangle is 14 cm. Another side is one-fourth the perimeter, and the third side is 4 less than one-third the perimeter. Find the perimeter of the triangle. [Answer: 24 cm]

2.3 APPLICATIONS USING LINEAR EQUATIONS

STUDENT GUIDEPOSTS

1. Converting Decimals and Percents
2. Tax Problems
3. Commission Problems
4. Discount Problems
5. Simple Interest Problems
6. Rate Problems

Applications of percent are common. We may pay a 5% sales tax, read that Mr. Levy won an election with 58% of the vote, or hear that a basketball player shot 82% from the free throw line.

The word *percent* means "per hundred." It refers to the number of parts in one hundred parts and can be written as a fraction or decimal. For example, when the tax rate is 5%, we pay 5¢ tax on each 100¢ purchase.

Percent: 5% Fraction: $\frac{5}{100}$ Decimal: 0.05

The basketball player could have hit 82 of 100 free throws when he shot 82% from the free throw line.

Percent: 82% Fraction: $\frac{82}{100}$ Decimal: 0.82

1 CONVERTING DECIMALS AND PERCENTS

To solve applied problems involving percent, we need to be able to convert from percent to decimal notation, and from decimal to percent. Let's review these techniques.

To Convert a Decimal to a Percent

Multiply the decimal by 100 (move the decimal point two places to the right) and attach the percent symbol, %.

EXAMPLE 1 Converting Decimals to Percents

Convert each decimal to a percent.

(a) $0.67 = (0.67)(100)\% = 67\%$

(b) $3.28 = (3.28)(100)\% = 328\%$

(c) $0.\overline{3} = (0.\overline{3})(100)\%$

$= 33.\overline{3}\% = 33\frac{1}{3}\%$

Practice Exercise 1

Convert each decimal to a percent.

(a) 0.25

(b) 4.31

(c) $2.\overline{6}$

Answers: (a) 25% (b) 431% (c) $266.\overline{6}\%$

To Convert a Percent to a Decimal

Multiply the percent by 0.01 = 1/100 (move the decimal point two places to the left) and remove the percent symbol, %.

EXAMPLE 2 Converting Percents to Decimals

Convert each percent to a decimal.

(a) $35\% = (35)(0.01) = 0.35$

(b) $225.5\% = (225.5)(0.01) = 2.255$

(c) $5\% = (5)(0.01) = 0.05$

(d) $0.5\% = (0.5)(0.01) = 0.005$

(e) $0.05\% = (0.05)(0.01) = 0.0005$

Practice Exercise 2

Convert each percent to a decimal.

(a) 73%

(b) 106.2%

(c) 8%

(d) 0.8%

(e) 0.08%

Answers: (a) 0.73 (b) 1.062 (c) 0.08 (d) 0.008 (e) 0.0008

When working with percent, the word "of" translates to *multiplied by* or *times*. There are three basic questions to answer with percent problems.

(1) What is 30% of 40?

$x = (0.30) \cdot 40$ Thus, $x = 12$

(2) Twelve is 30% of what number?

$12 = (0.30) \cdot x$ Thus, $x = 40$

(3) What percent of 40 is 12?

$x \cdot 40 = 12$ Thus, $x = 0.3 = 30\%$

In each case percent is expressed as a decimal, and the resulting translation is a linear equation.

2 TAX PROBLEMS

Many types of tax problems involve percent. For example, in most states a tax is charged on purchases made in retail stores. The rate of this tax, called a **sales tax,** varies from location to location. In general, however, we have

sales tax = (tax rate) · (selling price)

and **total cost = (selling price) + (sales tax).**

EXAMPLE 3 Sales Tax Problem

The sales tax rate in Maryville, Texas is 4%. How much tax would be charged on a purchase of $42? What is the total cost?

In effect, we are asked: What is 4% of 42?

Let x = the tax. We obtain the following equation.

$$x = (0.04)(42) = 1.68$$

The tax is $1.68, and the total cost is $43.68.

Practice Exercise 3

How much tax would be charged on a purchase totaling $145 if the tax rate is 6%? What is the total cost including tax?

Answer: $8.70; $153.70

3 COMMISSION PROBLEMS

Some salespersons receive all or part of their income as a percent of their sales. This is the **commission,** and the **commission rate** is the percent involved. In general,

commission = (commission rate) · (total sales).

For example, if a person works on a 20% commission and has total sales of $8360, her commission is 20% of $8360, which amounts to

$$(0.20)(\$8360) = \$1672.$$

EXAMPLE 4 Commission Problem

What are the total sales on which Gino received a commission of $1140 if his commission rate is 12%?

We are asked: 12% of what is $1140?

Let x = Gino's total sales. The question then translates to the equation on the following page.

Practice Exercise 4

If Terri earned a commission of $1455 during July and her commission rate is 15%, what were her total sales that month?

$$(0.12) \cdot x = 1140$$

$$x = \frac{1140}{0.12} = 9500$$

Gino's total sales amounted to $9500.

Answer: $9700

4 DISCOUNT PROBLEMS

A **discount** is a reduction in the regular price of an item put on sale. The **discount rate** is the percent of the regular price that the item is reduced, and the **sale price** is the price after the discount is taken. In general,

discount = (discount rate) · (regular price)

and **sale price = (regular price) − (discount).**

EXAMPLE 5 Discount Problem

A sport coat is discounted 30% and the sale price is $87.50. What was the original price?

Let x = the original price of the sport coat,
$(0.30)x$ = the amount of the discount.
Then we need to solve the following equation.

$$x - (0.30)x = 87.50 \quad \text{Regular price − discount = sale price}$$
$$x(1 - 0.30) = 87.50 \quad \text{Distributive law}$$
$$x(0.70) = 87.50$$
$$x = 125$$

Thus, the sport coat originally sold for $125.

Practice Exercise 5

The sale price on a new car is $13,416. If the car has been discounted 14%, what was the original price?

Answer: $15,600

5 SIMPLE INTEREST PROBLEMS

When money is invested or deposited in a bank and earns **simple interest,** the formula for the interest earned, I, is

$$I = Prt,$$

where P is the **principal** or amount invested, r is the **yearly interest rate,** and t is the **time** in years. The amount in the account A at the end of t years is

$$A = P + I = P + Prt.$$

EXAMPLE 6 Simple Interest Problem

In three years, Pat Polinski wishes to have $10,850 available to purchase a car. How much should he invest today at 12% simple interest to have this amount available when he needs it?

Let x = the amount of money Pat should invest today. In three years the interest earned will be

$$I = P\,rt$$
$$I = x \cdot (0.12)(3) = (0.36)x.$$

Practice Exercise 6

In five years the Raffertys wish to have $15,000 available for their daughter's education. How much should they invest today at 11% simple interest to have this amount when needed?

We need to solve the following equation.

$$x + (0.36)x = 10{,}850 \quad A = P + I$$
$$(1 + 0.36)x = 10{,}850 \quad \text{Distributive law}$$
$$x = \frac{10{,}850}{1.36}$$
$$x \approx 7977.94$$

Pat must invest approximately $7977.94.

Answer: approximately $9677.42

6 RATE PROBLEMS

Another important formula relates the distance (d) that an object travels in a given time (t) at a constant or uniform rate (r).

$$d = rt \quad \text{distance} = \text{(rate)(time)}$$

An automobile traveling at a rate of 55 mph for 3 hr will travel a distance of

$$d = rt = 55 \cdot 3 = 165 \text{ miles.}$$

Many rate or distance problems depend in some way on this basic formula.

EXAMPLE 7 Rate-Distance Problem

A hiker crossed a canyon by walking 4 mph the first 2 hr and 3 mph the next 3 hr. What was the total distance that she hiked?

A sketch like Figure 2.4 is helpful.

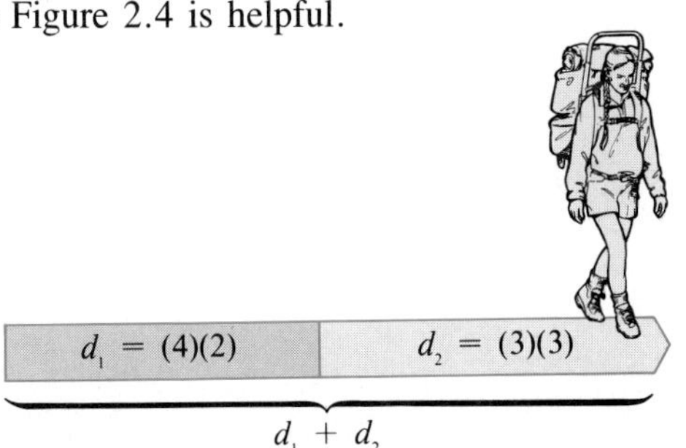

Figure 2.4

Let d_1 = distance hiked the first 2 hr. Then $d_1 = (4)(2) = 8$ mi.
Let d_2 = distance hiked the second 3 hr. Then $d_2 = (3)(3) = 9$ mi.
The total distance hiked is $d_1 + d_2 = 8 + 9 = 17$ mi.

Practice Exercise 7

On a trip a woman drove her car for 6 hr at 55 mph and then rode a boat for 3 hr at 30 mph. What was the total distance traveled?

Answer: 420 mi

EXAMPLE 8 Rate-Distance Problem

Two cars leave the same point traveling in opposite directions. The second car travels 10 mph faster than the first and after 3 hr they are 300 mi apart. How fast is each car traveling?

Make a sketch like Figure 2.5.

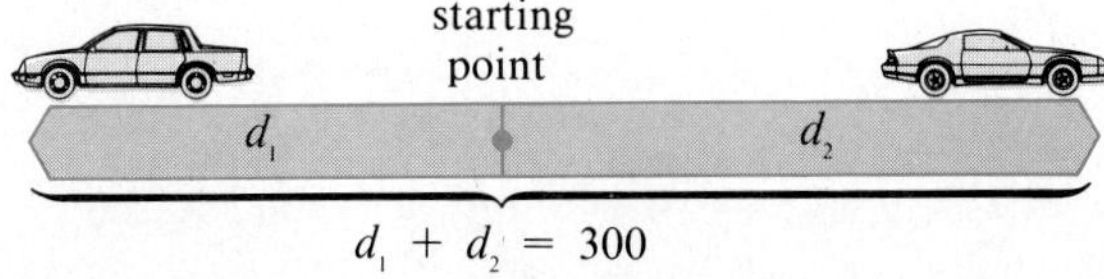

Figure 2.5

Practice Exercise 8

A jet plane leaves Miami and travels toward Chicago at 425 mph. At the same time a small plane leaves Chicago and travels 150 mph toward Miami. If the cities are 1150 mi apart, how long will it take the planes to meet and how far will they be from Miami at that time?

Let d_1 = the distance the first car travels,
r = rate of the first car.
Then $d_1 = 3r$.
Let d_2 = the distance the second car travels,
$r + 10$ = rate of the second car. (Why?)
Then $d_2 = 3(r + 10)$.
We need to solve the following equation.

$$d_1 + d_2 = 300$$
$$3r + 3(r + 10) = 300 \quad \text{Substitution}$$
$$3r + 3r + 30 = 300$$
$$6r = 270$$
$$r = 45$$
$$\text{and} \quad r + 10 = 55$$

The first car is traveling 45 mph and the second, 55 mph.

Answer: 2 hr and 850 mi

Other rate-time problems are solved in a similar manner.

EXAMPLE 9 RATE-TIME PRODUCTION PROBLEM

A machine that was manufactured 15 years ago can produce 20 items per day. A newer machine, manufactured last year, is more efficient and can produce 45 items per day. How long will it take to produce 1105 items if both machines are turned on together?

Let x = number of days required to produce 1105 items when both machines work together,
$n_1 = 20x$ = number of items produced by the older machine in x days,
$n_2 = 45x$ = number of items produced by the newer machine in x days.
We need to solve the following equation.

$$n_1 + n_2 = 1105$$
$$20x + 45x = 1105 \quad \text{Substitution}$$
$$65x = 1105$$
$$x = 17$$

It will take 17 days for the two machines to produce 1105 items.

PRACTICE EXERCISE 9

One printer can print 5 form letters per minute while a second can print 8 per minute. How long will it take both printers together to print 468 form letters?

Answer: 36 min

2.3 EXERCISES A

Convert to a percent.

1. 0.75 **2.** 2.41 **3.** 0.08 **4.** 0.001

5. 1 **6.** 100 **7.** 0.1 **8.** 0.015

Convert to a decimal.

9. 40% **10.** 23.5% **11.** 8.5% **12.** $8\frac{1}{2}\%$

13. $\frac{1}{2}\%$ **14.** 0.3% **15.** 0.03% **16.** 100%

Translate each question into a mathematical statement and solve.

17. 36 is 15% of what? **18.** 12% of 240 is what? **19.** What is $\frac{1}{2}\%$ of 400?

Solve.

20. A family spent \$180 a month for food. This was 15% of their monthly income. What was their monthly income?

Let x = monthly income,
$0.15x$ = amount spent for food.
$0.15x = 180$

21. If \$0.43 sales tax is charged on a coffeepot that sells for \$21.50, what is the tax rate?

Let x = tax rate.
$21.50x =$

22. An automobile salesperson received a 2% commission on the price of each vehicle sold. What total commission will be earned on the sale of a car priced at \$9850 and a truck priced at \$13,500?

23. A retailer discounted all merchandise in her shop by 40%. What are the discount and sale price on a suit regularly priced at \$135.50?

24 A tank contained 985 gal of fuel. An additional quantity of fuel was added making a total of 1182 gal. What was the percent increase?

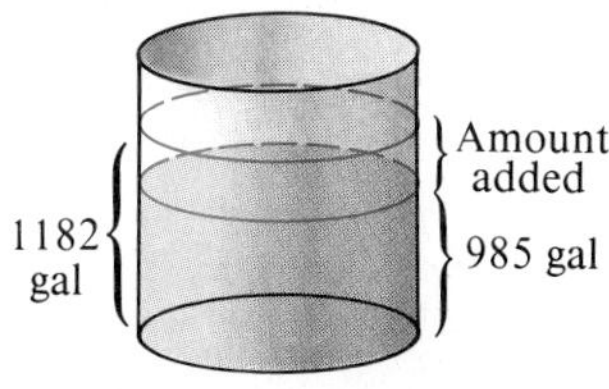

25 The population in Deserted, Arizona was 567 in 1990. This was a 35% increase over the population in 1980. What was the population in 1980?

26. The retail selling price of a sofa is \$806.25. What is the cost to the dealer if she sells at a 25% markup on the cost?

27 Paul Pendergast must have \$11,470 available to pay off a loan in full five years from today. How much should he invest at 11% simple interest in order to obtain this amount?

28 Lucky Lucia won \$10,000 in the New York Lottery. She invested part in a savings account that earned 12% simple interest and the rest in a mutual fund that paid 15% annually. If at the end of one year she earned an income of \$1275 from the two, how much was invested at each rate?

29. Troy Hudson has \$17,000 invested in bonds that pay 13% annually. How much additional money must he invest in a savings account paying 17% so that the average return on the two investments amounts to 14%?

30. A floor diagram of the Perkos' family room is given below. Suppose they purchased enough carpeting to cover the floor wall-to-wall, and the total bill for the job came to \$1719.90, including a 5% sales tax. Assuming there was no waste, how much did the carpet cost per square yard?

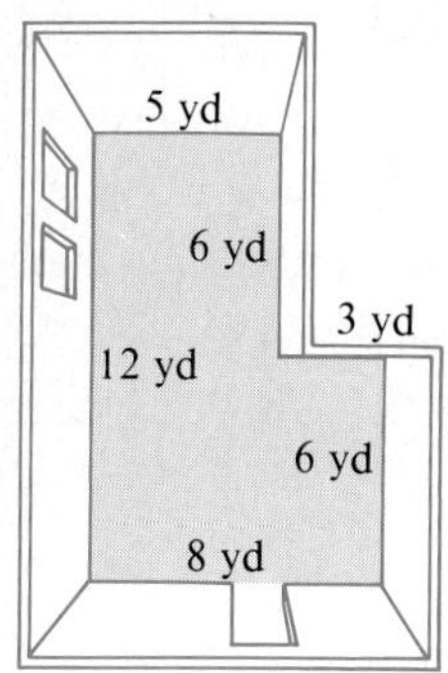

31 The We-Try-Harder Rental Company charges \$13 per day and 8¢ per mile to rent an Oldsmobile Ciera. If Jeff rented a Ciera for 4 days and the total bill was \$199.68, including a 4% sales tax, how many miles did he drive?

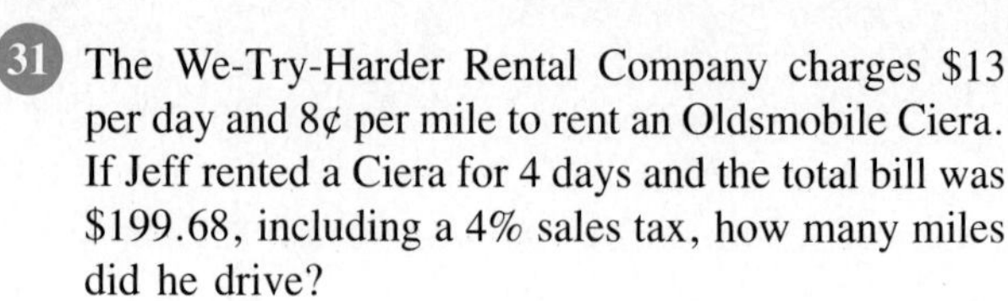

32. Two trains leave the same city, one traveling north and the other traveling south. If the first train is moving 15 mph faster than the second, and if after 3 hr they are 405 mi apart, how fast is each train traveling?

33. Two hikers head towards a mountain 25 mi away. The first travels 4 mph and the second travels 2.5 mph. How far apart will they be after 6 hr?

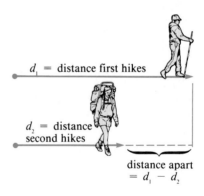

d_1 = distance first hikes

d_2 = distance second hikes

distance apart $= d_1 - d_2$

34 A woman hikes to the bottom of the Grand Canyon at a rate of 4.5 mph and returns to the top by mule traveling at a rate of 2.5 mph. If the total time of the trip is 7 hr, how long did she walk? What was the total length of the trip?

35 Two families leave their homes at 9:30 A.M., planning to meet for a picnic at a point between them. If one travels at a rate of 55 mph, the other travels at 50 mph, and they live 315 mi apart, at what time do they meet?

36. Two machines are turned on at 8:00 A.M. If one can produce 36 items each hour and the other can produce 50 items each hour, at what time will they have produced a total of 645 items?

37. Two distance runners, Bob and Murph, run the same course with Bob running 1.5 mph slower than Murph. If Murph finishes in 4 hr and Bob takes 5 hr to finish, what is the rate of each? How far do they run?

38 Two printing presses print 14,850 fliers. Their rates of production differ by 15 fliers per minute. What is the rate of each if together they complete the job in 150 min?

39 A private boat traveling at 20 knots (a knot is 1 nautical mile per hour) is 10 nautical miles from a harbor when a Coast Guard cutter initiates pursuit on the same course traveling 35 knots. How long will it take for the cutter to overtake the private boat?

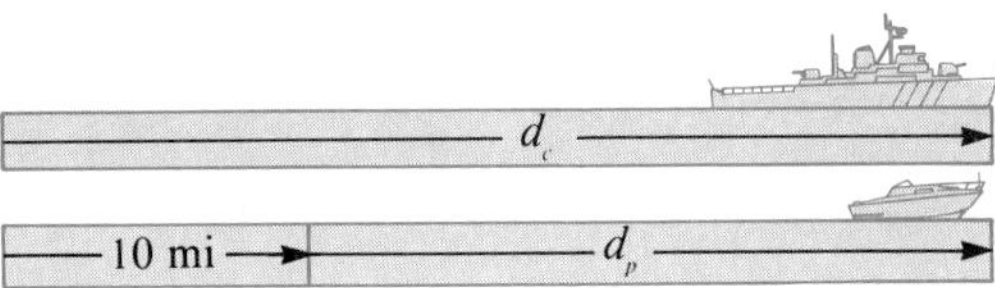

FOR REVIEW

Select a variable to represent each quantity and translate the phrase into symbols.

40. Price plus 6% of the price

41. Twice a number, less its reciprocal

Solve.

42. On a recent football booster trip, it was noted that the bus was completely full and that there were 12 more men than women. If the bus has a capacity of 56 passengers, how many men were on the trip?

43. The sum of two consecutive even integers is 114. Find the integers.

44. Two angles are complementary if their sum is 90°. One of two complementary angles is 30° more than twice the other. Find the measure of each angle.

45. A farmer wishes to enclose a rectangular pen with 168 yd of fencing in such a way that the length is three times the width. What will be the dimensions of the pen?

ANSWERS: 1. 75% 2. 241% 3. 8% 4. 0.1% 5. 100% 6. 10,000% 7. 10% 8. 1.5% 9. 0.4 10. 0.235 11. 0.085 12. 0.085 13. 0.005 14. 0.003 15. 0.0003 16. 1 17. 240 18. 28.8 19. 2 20. $1200 21. 2% 22. $467 23. $54.20, $81.30 24. 20% 25. 420 26. $645 27. $7400 28. $7500 at 12%, $2500 at 15% 29. $5666.67 30. $21.00 31. 1750 miles 32. 60 mph, 75 mph 33. 9 mi 34. 2.5 hr, 22.5 mi 35. 12:30 P.M. 36. 3:30 P.M. 37. Murph: 7.5 mph; Bob: 6 mph; 30 mi 38. 42 fliers/min, 57 fliers/min 39. 40 min 40. $p + 0.06p$ 41. $2n - \frac{1}{n}$ 42. 34 men 43. 56, 58 44. 20°, 70° 45. 21 yd by 63 yd

2.3 EXERCISES B

Convert to a percent.

1. 0.83 **2.** 4.32 **3.** 0.07 **4.** 0.007

5. 3 **6.** 200 **7.** 0.6 **8.** 0.026

Convert each percentage to a decimal.

9. 70%

10. 13.2%

11. 3.7%

12. $3\frac{7}{10}\%$

13. $\frac{3}{4}\%$

14. 0.8%

15. 0.08%

16. 1%

Translate each question into a mathematical statement and solve.

17. 9 is 2% of what?

18. 3% of 180 is what?

19. 8% of what is 68?

Solve.

20. Larry spent \$3600 for a used car. This was 30% of his annual income. What was his annual income?

21. What is the tax rate in Southern, Georgia if a tax of \$6.25 is charged on an item selling for \$250?

22. The Blue Carpet Realty Co. receives a 6% commission on all real estate sales. During August the company sold a house for \$89,500 and a lot for \$22,300. What was the total commission on these sales?

23. Bjorn was making \$24,500 per year. The next year he earned \$27,440. What was the percent increase from one year to the next?

24. Marvin received a 14% raise and now makes \$14,592. What was his former salary?

25. The retail selling price of a video recorder is \$1022. What is the cost to the dealer if he sells at a 40% markup on the cost?

26. What amount of money must be invested today at 14% simple interest to reach \$9672 in four years?

27. A fuel tank contained 125 gal of heating oil. An additional quantity of oil was added, making a total of 165 gal. What was the percent increase?

28. If \$4200 is invested in stocks at 10%, how much additional money must be invested in a savings account that earns 14% to make the total earnings equivalent to an investment of 13%?

29. When his rich aunt passed away, Martin received an inheritance of \$25,000. He put part of the money in an account that paid 14% simple interest, and the rest into a stock that paid a 17% annual dividend. If at the end of the year he had a total income of \$3920 from the two, how much was invested in each?

30. A floor diagram of the Fancher's recreation room is given below. Suppose they purchased enough carpeting to cover the floor wall-to-wall, and the total bill for the job came to \$962.00, including a 4% sales tax. Assuming there is no waste, what was the price per square yard of the carpet?

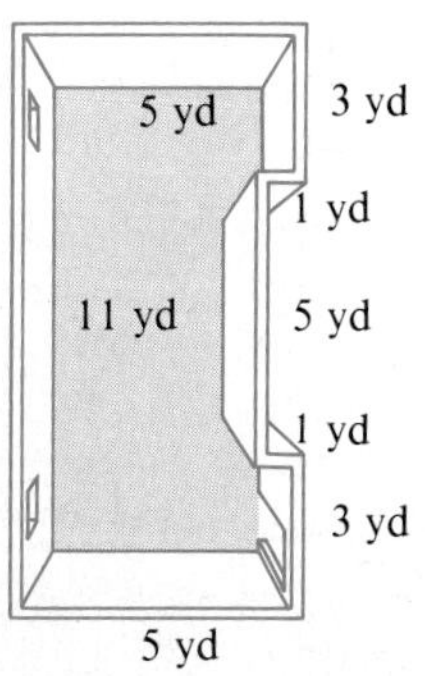

31. Luxury Limo Rentals charges \$50 per day and 21¢ per mile to rent a Continental. If Beth rented a Continental for 3 days and the total bill was \$307.44, including a 5% sales tax, how many miles did she drive?

32. Two cars leave the same park, one traveling east and the other traveling west. If the first is moving 10 mph faster than the second, and if after 4 hr they are 424 mi apart, how fast is each traveling?

33. Two truckers leave San Francisco at the same time heading east. The first is traveling at a rate of 60 mph and the second at 54 mph. How far apart will they be in 7 hr?

34. Jerry hikes to a waterfall at a rate of 3.5 mph and returns at a rate of 3 mph. If the total time of the trip is 6.5 hr, how long did it take to reach the waterfall? What was the total distance traveled?

35. Two families leave their homes at 10:00 A.M., planning to meet for a reunion at a point between them. If one travels at a rate of 60 mph, the other travels at a rate of 54 mph, and they live 285 mi apart, at what time do they meet?

36. Two machines are turned on at 9:00 A.M. If one can produce 40 items each hour and the other can produce 32 items each hour, at what time will they have produced a total of 360 items?

37. Two hikers, Jenny and Christine, hike the same trail with Jenny hiking 1 mph slower than Christine. If Jenny finishes in 4 hr and Christine takes 3 hr to finish, what is the rate of each? How long is the trail?

38. A 1977-model machine, which can produce 30 units an hour, has made 15 units when a 1987-model machine is turned on. If the 1987-model can produce 45 units an hour, how long will it be before it has produced the same number of units as the older model?

39. A fighter plane flying at a speed of 750 mph is to overtake a transport plane flying at a speed of 500 mph. If the transport took off at 8:30 A.M. and the fighter left 1.5 hr later, at what time will it overtake the transport?

FOR REVIEW

Select a variable to represent each quantity and translate the phrase into symbols.

40. My weight before gaining 10 lb

41. Twice the reciprocal of a number

Solve.

42. A pole is standing in a pond. If one-fourth of the height of the pole is in the sand at the bottom of the pond, three-eighths of the total height is in the air above the water, and 9 ft are in the water, what is the length of the pole?

43. Sean, Gary, and Troy were born in consecutive years. If the sum of their ages is 54, how old is each?

44. Marv and Burford leave their campsite and walk in opposite directions around a lake. If the shoreline is 15 miles long, Marv walks 0.5 mph faster than Burford, and they meet in 2 hr, how fast does each walk?

45. A rancher has 700 ft of fencing with which he wishes to enclose a small rectangular pasture with length four times the width. What will be the dimensions of the pasture?

2.3 EXERCISES C

Solve. A calculator may be helpful in some of these exercises.

1. On Tuesday Galen Benning bought ten shares of stock. On Wednesday the value of the shares went up 5%, and on Thursday the value fell 5%. How much did Galen pay for the ten shares on Tuesday if he sold them on Thursday for $99.75? [Answer: $100]

2. The Brinegars plan to sell their home. They want to receive $65,000 after deducting the sales commission of 6% on the selling price. Rounded to the nearest dollar, at what selling price should the house be listed?

3. During a consecutive three-year period, Scott Williams earned a 10%, 20%, and 30% raise in salary, respectively. What was the total percent change from the beginning of the first year to the end of the third year?

4. A container holds 100 qt of a solution that is 12% alcohol and 88% water. How much water must be added to obtain a solution that is 10% alcohol? [Answer: 20 quarts]

5. Aprile Kyle rents a boat at 9:00 A.M. She is told it will travel 20 mph upstream and 25 mph downstream. If she decides to travel as far downstream as possible and be back by 6:00 P.M., at what time should she turn back, and how far downriver will she be at that time?

6. Assume that sound travels at 0.34 km/sec and that a bullet travels 0.68 km/sec. If the gun is fired at a target and 3 sec later the sound of impact is heard, how far away is the target? [Answer: 0.68 km]

7. Six seconds after Kristen fires a gun at a rock cliff she hears the sound of impact. If sound travels at 1095 ft/sec and the bullet at 2190 ft/sec, how far is the cliff from Kristen?

8. A ship located 1 mi from an island fires one of its guns. The sound reaches the island through the water 3.73 sec before it arrives through the air. If sound travels 4.5 times faster through water than through air, approximately how fast does sound travel in water? [Answer: 4954 ft/sec]

9. Distance between ships can be determined when one ship sends signals that travel through the water at different rates of speed and the second ship measures the time between the signals. Two signals traveling at 1.2 mi/sec and 2.5 mi/sec are received 39 seconds apart. How far apart are the ships? [Answer: 90 mi]

10. Repeat Exercise 9 if the signals are received 15 seconds apart. Give answer correct to the nearest tenth of a mile.

11. An object that is thrown downward with an initial velocity of 120 ft/sec satisfies the equation $v = 32t + 120$ where v is the velocity in ft/sec and t is in seconds. What is the velocity of the object after 3 seconds? After how many seconds will the velocity be 280 ft/sec? [Answer: 216 ft/sec; 5 sec]

12. An object that is thrown upward with an initial velocity of 48 ft/sec satisfies the equation $v = -32t + 48$ where v is velocity in ft/sec and t is in seconds. What is the velocity of the object after 2.5 seconds? After how many seconds will the velocity be 0?

A type of problem with many applications in physics is a lever problem. *A* ***lever*** *is a rigid rod or beam rotating about a fixed point, called the* ***fulcrum*** *of the lever. A lever is in* ***equilibrium*** *if the forces on the lever are balanced according to a fundamental law of physics, the* ***law of the lever.*** *Suppose three forces, F_1, F_2, and F_3, are applied to a lever at distances d_1, d_2, and d_3 as shown in the figure. For equilibrium, the law of the lever requires that*

$$F_1d_1 + F_2d_2 = F_3d_3.$$

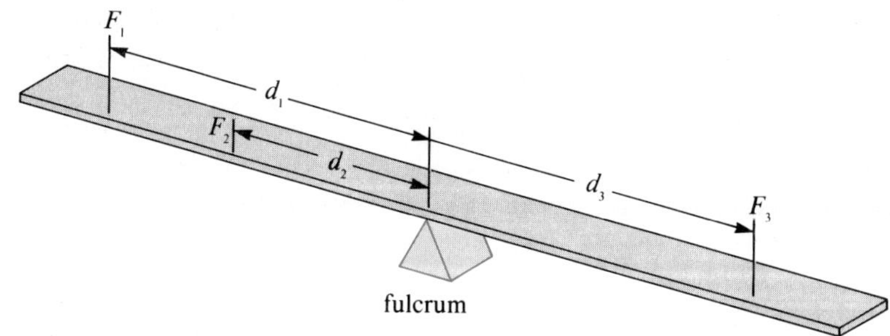

Use this information in the following problems.

13. Jeff, who weighs 160 pounds, and Annie, who weighs 110 pounds, are sitting at opposite ends of a seesaw. Each is 8 feet from the fulcrum. Where should their 80-lb daughter Cindy sit, to balance the system? [Answer: 5 ft from the fulcrum on Annie's side]

14. Repeat Exercise 13 but this time use their son Steve, who weighs 64 pounds.

15. Repeat Exercise 13 but this time use their dog, Chuck, who weighs 32 pounds.

2.4 LITERAL EQUATIONS

STUDENT GUIDEPOSTS

1. Equations With the Unknown In One Term
2. Equations With the Unknown In Several Terms

1 EQUATIONS WITH THE UNKNOWN IN ONE TERM

Thus far the equations we have solved have contained only one variable. A **literal equation** (or **letter equation**) is an equation that has two or more variables. Many formulas familiar to us are literal equations, such as

$$A = lw, \quad P = 2l + 2w, \quad d = rt, \quad A = \pi r^2, \quad I = Prt.$$

At times we want to solve a literal equation for a particular variable. For example, the equation $d = rt$ can be solved for t and for r,

$$t = \frac{d}{r} \quad \text{and} \quad r = \frac{d}{t}.$$

Remember that the letters in a literal equation, that is, the variables, simply represent numbers. Any of the rules for solving equations with numerical values also apply to literal equations. It can be helpful to solve first a numerical equation similar in form to the literal equation. We show this in the examples.

EXAMPLE 1 Solving a Literal Equation

Solve $T = cn$ for n.

Literal equation		*Similar numerical equation*	
$T = cn$		$5 = 3n$	
$\frac{1}{c} \cdot T = \frac{1}{c} \cdot cn$	Multiply by $\frac{1}{c}$	$\frac{1}{3} \cdot 5 = \frac{1}{3} \cdot 3n$	Multiply by $\frac{1}{3}$
$\frac{T}{c} = n$		$\frac{5}{3} = n$	

PRACTICE EXERCISE 1

Solve $P = RT$ for R.

Answer: $R = \frac{P}{T}$

Keep in mind that solving a literal equation (or any other equation) depends on our ability to isolate the variable on one side of the equation.

EXAMPLE 2 Solving a Literal Equation

Solve $F = v + w$ for v.

Literal equation		*Similar numerical equation*	
$F = v + w$		$5 = v + 7$	
$F - w = v + w - w$	Subtract w	$5 - 7 = v + 7 - 7$	Subtract 7
$F - w = v$		$-2 = v$	

PRACTICE EXERCISE 2

Solve $h = u + z$ for z.

Answer: $z = h - u$

Geometry problems often require the solution of a literal equation. In the following example we work with the formula for perimeter of a rectangle.

EXAMPLE 3 GEOMETRY PROBLEM

Solve the formula for the perimeter of a rectangle, $P = 2l + 2w$, for w.

Literal equation	*Similar numerical equation*
$P = 2l + 2w$	$12 = 2 \cdot 3 + 2w$
$P - 2l = 2l - 2l + 2w$	$12 - 2 \cdot 3 = 2 \cdot 3 - 2 \cdot 3 + 2w$
$P - 2l = 2w$	$6 = 2w$
$\frac{1}{2}(P - 2l) = \frac{1}{2} \cdot 2w$	$\frac{1}{2} \cdot 6 = \frac{1}{2} \cdot 2w$
$\frac{P - 2l}{2} = w$	$3 = w$

PRACTICE EXERCISE 3

Solve $m = 3n + 5k$ for k.

Answer: $k = \frac{m - 3n}{5}$

2 EQUATIONS WITH THE UNKNOWN IN SEVERAL TERMS

When the variable for which we wish to solve occurs in more than one term, we collect these terms on one side of the equation first. This is illustrated in the next example.

EXAMPLE 4 SOLVING A LITERAL EQUATION

Solve $By = 5p - by$ for y.

Literal equation		*Similar numerical equation*	
$By = 5p - by$		$3y = 5 \cdot 7 - 2y$	
$By + by = 5p$	Add by	$3y + 2y = 5 \cdot 7$	Add $2y$
$y(B + b) = 5p$	Factor out y using the distributive law	$y(3 + 2) = 35$	Factor out y
		$5y = 35$	Simplify
$y = \frac{5p}{B + b}$	Divide by $(B + b)$	$y = \frac{35}{5} = 7$	Divide by 5

PRACTICE EXERCISE 4

Solve $mx = nx + ky$ for x.

Answer: $x = \frac{ky}{m - n}$

CAUTION

The step at which y is factored out in the numerical equation in Example 4 is usually omitted and performed mentally, resulting in the simplified equation $5y = 35$. In the literal equation, however, we need this step to obtain $y(B + b)$ so we can divide both sides by the coefficient of y, $B + b$. Also, *do not* try to solve $By = 5p - by$ for y by writing

$$y = \frac{5p - by}{B}.$$

Notice that y is still a factor of a term on the right, so the equation has not been solved for y.

Many formulas used by scientists use several variables and require solutions as literal equations. The following is an example from physics.

EXAMPLE 5 PHYSICS PROBLEM

The distance above ground level of an object under the influence of gravity is given by the equation

$$h = \frac{1}{2}gt^2 + vt + s.$$

Solve this equation for v, then find v when $h = 650$ ft, $g = 32$ ft/sec^2, $t = 2$ sec, and $s = 1250$ ft.

To isolate the vt term, subtract $\frac{1}{2}gt^2$ and s from both sides of the equation.

$$h = \frac{1}{2}gt^2 + vt + s$$

$$h - \frac{1}{2}gt^2 - s = \frac{1}{2}gt^2 - \frac{1}{2}gt^2 + vt + s - s$$

$$h - \frac{1}{2}gt^2 - s = vt$$

$$\frac{h - \frac{1}{2}gt^2 - s}{t} = \frac{vt}{t} \qquad \text{Divide by } t$$

$$\frac{h - \frac{1}{2}gt^2 - s}{t} = v$$

Now find v, which is initial velocity, when $h = 650$ ft, $g = 32$ ft/sec^2, $t = 2$ sec, and $s = 1250$ ft.

$$v = \frac{h - \frac{1}{2}gt^2 - s}{t} = \frac{650 - \frac{1}{2}(32)(2)^2 - 1250}{(2)}$$

$$= \frac{650 - 64 - 1250}{2} = -332$$

Thus, the initial velocity is -332 ft/sec, which means 332 ft/sec downward.

PRACTICE EXERCISE 5

Solve $h = \frac{1}{2}gt^2 + vt + s$ for s, then find s when $h = 1000$, $g = 32$ ft/sec^2, $t = 4$ sec, and $v = -50$ ft/sec.

Answer: $s = h - \frac{1}{2}gt^2 - vt$; $s =$ 944 ft

The following summarizes the steps for solving literal equations.

To Solve a Literal Equation

1. Simplify each side of the equation.
2. Using the additive property, change the equation into an equivalent equation having all terms with the specified variable on one side and all other terms on the other side.
3. Factor the specified variable from all terms (if necessary) and divide both sides by the coefficient of the variable, combining all like terms when appropriate.

2.4 EXERCISES A

Solve.

1. $a = b + c$ for c
2. $a = bc$ for b
3. $a = bc + d$ for b
4. $a + b = c + d$ for c
5. $u = 2v + 2w$ for v
6. $x + z + y = w$ for z
7. $abc = w$ for a
8. $aA = a + A$ for a
9. $P = 20a + b$ for a
10. $a = 3(b + c)d$ for d
11. $a = 3(b + c)d$ for b
12. $F = \frac{9}{5}C + 32$ for C
13. $V = lwh$ for w
14. $A = \frac{1}{2}(b_1 + b_2)h$ for b_1
15. $P = a + b + c$ for c
16. $A = P + Prt$ for r
17. $A = P + Prt$ for P
18. $I = Prt$ for r
19. $a_n = a_1 + (n - 1)d$ for d
20. $a_n = a_1 + (n - 1)d$ for n
21. $A = \frac{1}{2}bh$ for b
22. $E = IR + Ir$ for r
23. $S = 2\pi rh + 2\pi r^2$ for h
24. $bx - c = ax$ for x
25. $3by - c = 2ay$ for y
26. $2(x - b) + 5x = 4c$ for x
27. $x(b + 2) = 3b + 4x$ for x

For each of the following, write an appropriate formula for the specified variable, and then find the required value of that variable.

28. If the distance traveled d and the rate of travel r are given, find a formula for the time of travel t. What is the value of t when d is 495 mi and r is 45 mph?

29. If the area A and length l of a rectangle are given, find a formula for the width w. What is the value of w when A is 3.75 cm^2 and l is 2.5 cm?

30. If the rate r, time t, and simple interest I are given, find a formula for the principal P. What is the value of P when r is 15%, t is 4 yr, and I is $1500?

31. If two sides of a triangle, a and b, and the perimeter P are given, find a formula for the third side c. What is the value of c when a is 4.3 in, b is 5.2 in, and P is 17.0 in?

32. If the perimeter P and width w of a rectangle are given, find a formula for the length l. What is the value of l when P is 38 yd and w is 6 yd?

33. If the total cost T of a number of items n is given, find a formula for the cost per item c. What is the value of c when T is $52.50 and n is 35?

34. If the two bases b_1 and b_2 and the area A of a trapezoid are given, find a formula for the height h of the trapezoid. What is the value of h if b_1 is 7 ft, b_2 is 19 ft, and A is 110.5 ft^2?

35. If the volume V and two edges a and b of a rectangular solid are given, find a formula for the third edge c. What is the value of c if V is 31 yd^3, a is 3.1 yd, and b is 2.5 yd?

FOR REVIEW

36. After an 8% raise, Mick's new salary is $14,472. What was his former salary?

37. What amount of money invested at $7\frac{1}{2}$% simple interest will increase to $1612.50 in one year?

38. Jackson Black received a commission of $396 on sales amounting to $5280. What was his commission rate?

39. After receiving a 20% discount on the selling price, Dyane paid $74.76, including 5% sales tax, for a watch. What was the price of the watch before the discount?

40. Becky can ride her bike to school in 10 min, but it takes her an hour if she walks. Her speed walking is 15 mph slower than her speed when riding. Find each rate.

41. One secretary can fold and stuff 12 envelopes per minute while another can only do 9 per minute. How long will it take them to complete the job of folding and stuffing 6000 envelopes if the first works for 45 min before being joined by the other?

In preparation for the next section, the following exercises review material from Section 1.1. Graph each inequality on a number line.

42. $x < 5$

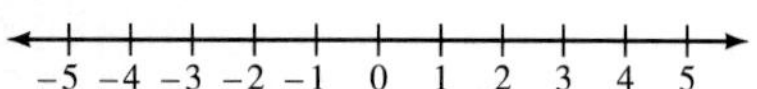

43. $x > -2.5$

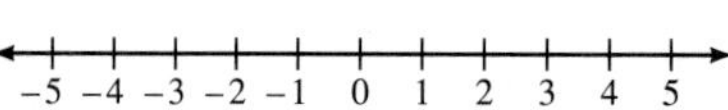

44. $-5 \le x < 2$

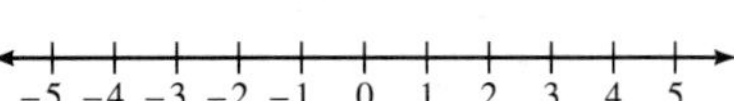

ANSWERS: **1.** $c = a - b$ **2.** $b = \frac{a}{c}$ **3.** $b = \frac{a-d}{c}$ **4.** $c = a + b - d$ **5.** $v = \frac{u-2w}{2}$ **6.** $z = w - y - x$ **7.** $a = \frac{w}{bc}$ **8.** $a = \frac{A}{A-1}$ **9.** $a = \frac{P-b}{20}$ **10.** $d = \frac{a}{3b+3c}$ **11.** $b = \frac{a-3cd}{3d}$ **12.** $C = \frac{5}{9}(F - 32)$ **13.** $w = \frac{V}{lh}$ **14.** $b_1 = \frac{2A - b_2h}{h}$ **15.** $c = P - a - b$ **16.** $r = \frac{A-P}{Pt}$ **17.** $P = \frac{A}{1+rt}$ **18.** $r = \frac{I}{Pt}$ **19.** $d = \frac{a_n - a_1}{n-1}$ **20.** $n = \frac{a_n - a_1 + d}{d}$ **21.** $b = \frac{2A}{h}$ **22.** $r = \frac{E - IR}{I}$ **23.** $h = \frac{S - 2\pi r^2}{2\pi r}$ **24.** $x = \frac{c}{b-a}$ **25.** $y = \frac{c}{3b-2a}$ **26.** $x = \frac{4c+2b}{7}$ **27.** $x = \frac{3b}{b-2}$ **28.** $t = \frac{d}{r}$; 11 hr **29.** $w = \frac{A}{l}$; 1.5 cm **30.** $P = \frac{I}{rt}$; \$2500 **31.** $c = P - a - b$; 7.5 in **32.** $l = \frac{P-2w}{2}$; 13 yd **33.** $c = \frac{T}{n}$; \$1.50 **34.** $h = \frac{2A}{b_1+b_2}$; 8.5 ft **35.** $c = \frac{V}{ab}$; 4.0 yd **36.** \$13,400 **37.** \$1500 **38.** 7.5% **39.** \$89.00 **40.** riding: 18 mph, walking: 3 mph **41.** 4 hr and 20 min **42.** −5 −4 −3 −2 −1 0 1 2 3 4 5 **43.** −5 −4 −3 −2 −1 0 1 2 3 4 5 **44.** −5 −4 −3 −2 −1 0 1 2 3 4 5

2.4 EXERCISES B

Solve.

1. $a = b + c$ for b

2. $a = bc$ for c

3. $a = bc + d$ for c

4. $a + b = c + d$ for d

5. $u = 2v + 2w$ for w

6. $x + z + y = w$ for y

7. $abc = w$ for c

8. $Bb = B + b$ for b

9. $P = 20a + b$ for b

10. $a = 3(b + c)d$ for c

11. $C = \frac{5}{9}(F - 32)$ for F

12. $V = lwh$ for h

13. $A = \frac{1}{2}(b_1 + b_2)h$ for h

14. $A = \frac{1}{2}(b_1 + b_2)h$ for b_2

15. $P = a + b + c$ for c

16. $A = P + Prt$ for t

17. $U = A + Acw$ for A

18. $I = Prt$ for t

19. $a_n = a_1 + (n - 1)d$ for a_1

20. $A = \frac{1}{2}bh$ for h

21. $E = IR + Ir$ for R

22. $C = 2\pi r$ for r

23. $ax = c - 3bx$ for x

24. $4ay = c + 2by$ for y

25. $dx - 3(x - a) = 2b$ for x

26. $a(y + b) + c = dy$ for y

27. $v = k + gt$ for g

For each of the following, write an appropriate formula for the specified variable, and find the required value of that variable.

28. If the distance traveled d and the time of travel t are given, find a formula for the rate of travel r. What is the value of r when d is 350 km and t is 5 hr?

29. If the area A and the width w of a rectangle are given, find a formula for the length l. What is the value of l when A is 18.9 ft^2 and w is 3.5 ft?

30. If the principal P, time t, and simple interest I are given, find a formula for the rate r. What is the value of r when P is \$400, t is 3 yr, and I is \$144?

31. If two sides of a triangle, b and c, and the perimeter P are given, find a formula for the third side a. What is the value of a when b is 2.8 cm, c is 4.3 cm, and P is 15.0 cm?

32. If the perimeter P and length l of a rectangle are given, find a formula for the width w. What is the value of w when P is 40 m and l is 13 m?

33. If the total cost T of a number of items n is given, find a formula for the cost per item c. What is the value of c when T is \$74.00 and n is 40?

34. If the volume V and two edges, b and c, of a rectangular solid are given, find a formula for the third edge a. What is the value of a if V is 10.78 cm^3, b is 3.5 cm, and c is 1.4 cm?

35. If the surface area S and two edges, b and c, of a rectangular solid are given, find a formula for the third edge a. What is the value of a if S is 184 ft^2, b is 4 ft, and c is 8 ft?

FOR REVIEW

36. Paulette received a 6% raise, making her new salary \$11,130. What was her former salary?

37. What amount of money invested at 14% simple interest will increase to \$2125 at the end of five years?

38. Eric Bradley borrowed a sum of money from a savings and loan office at 7% simple interest. At the end of 3 years he repaid the loan in the amount of \$1663.75. How much did he originally borrow?

39. After receiving a 30% discount on the selling price, Verna paid \$29.40, including 5% sales tax, for a pair of shoes. What was the price of the shoes before the discount?

40. Two cars, that are 436 mi apart and whose speeds differ by 5 mph are traveling towards each other. What is the speed of each if they meet in 4 hr?

41. Two airplanes leave Chicago at 12:00 noon and fly in opposite directions. If one flies at 410 mph and the other 120 mph faster, how long will it take them to be 3290 mi apart?

In preparation for the next section, the following exercises review material from Section 1.1. Graph each inequality on a number line.

42. $x > 0$

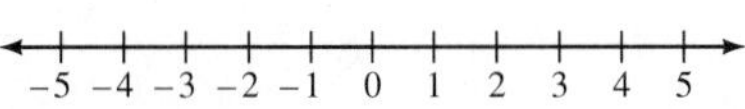

43. $x \leq 3.5$

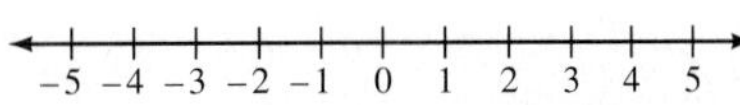

44. $-4 < x \leq 5$

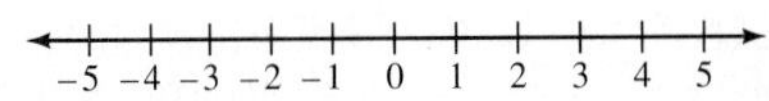

2.4 EXERCISES C

Solve.

1. $S_n = \frac{n}{2}[2a_1 + (n-1)d]$ for d

2. $S_n = \frac{n}{2}[2a_1 + (n-1)d]$ for a_1

3. $b(x+y) - a = Ax$ for x $\left[\text{Answer: } x = \frac{a - by}{b - A}\right]$

4. $y(a+1) = 3a + 2y$ for y

For each of the following, write an appropriate formula for the specified variable, and then find the required value of that variable.

5. If the length of one base a, the height h, and the area A of a trapezoid are given, find a formula for the length of the other base b. What is the value of b when a is 5 in, h is 8 in, and A is 44 in^2?

6. If the surface area S and two edges, a and b, of a rectangular solid are given, find a formula for the third edge c. What is the value of c if S is 62 m^2, a is 3 m, and b is 2 m?
$\left[\text{Answer: } c = \frac{S - 2ab}{2b + 2a}; \text{ 5 m}\right]$

2.5 SOLVING LINEAR INEQUALITIES

STUDENT GUIDEPOSTS

1. Linear Inequalities
2. Interval Notation
3. Solving Linear Inequalities

1 LINEAR INEQUALITIES

In Chapter 1 we graphed simple inequalities such as $x > 1$ and $-2 < x < 1$. In this section, we will solve and graph more complex statements, such as

$$x + 7 < 2, \quad 3x \geq 9, \quad 3x - 1 > 11, \quad \text{and} \quad 2x + 1 \leq 5 - 7x,$$

which are called **linear inequalities.** A linear inequality can be written as

$$ax + b < 0, \quad ax + b > 0, \quad ax + b \leq 0, \quad \text{or} \quad ax + b \geq 0,$$

where a and b are real numbers, $a \neq 0$.

2 INTERVAL NOTATION

Before solving linear inequalities like these, we introduce a new way to write and graph inequalities, *interval notation*. The set described by $-2 < x < 1$ is an **open interval** because it contains neither end point, and is denoted $(-2, 1)$. The set $-2 \leq x \leq 1$ is a **closed interval** because it contains both end points, and is denoted $[-2, 1]$. In general, if $a < b$ we make the following definitions.

$(a, b) = \{x | a < x < b\}$ Open interval

$[a, b] = \{x | a \leq x \leq b\}$ Closed interval

$(a, b] = \{x | a < x \leq b\}$ Half-open interval

$[a, b) = \{x | a \leq x < b\}$ Half-open interval

Note that $(a, b]$ and $[a, b)$ are both called **half-open intervals** because each contains only one end point.

If $x > 1$, we write the **infinite interval,** $(1, \infty)$. The infinity symbol ∞ is not a number but simply indicates that all numbers greater than 1 are included in the set. The symbol $-\infty$, read "minus infinity" or "negative infinity," is also used in the following definitions.

$$(-\infty, a) = \{x|x < a\}$$
$$(-\infty, a] = \{x|x \le a\}$$
$$(a, \infty) = \{x|x > a\}$$
$$[a, \infty) = \{x|x \ge a\}$$
$$(-\infty, \infty) = \{x|x \text{ is any real number}\}$$

EXAMPLE 1 Using Interval Notation

Write each inequality in interval notation and then graph.

(a) $-3 \le x < 0$

In the notation for a half-open interval, $-3 \le x < 0$ becomes $[-3, 0)$. Figure 2.6 shows two methods of graphing the interval, the dot notation we used in Chapter 1, as well as interval notation. Note that [replaces the closed dot on the left and) replaces the open dot on the right.

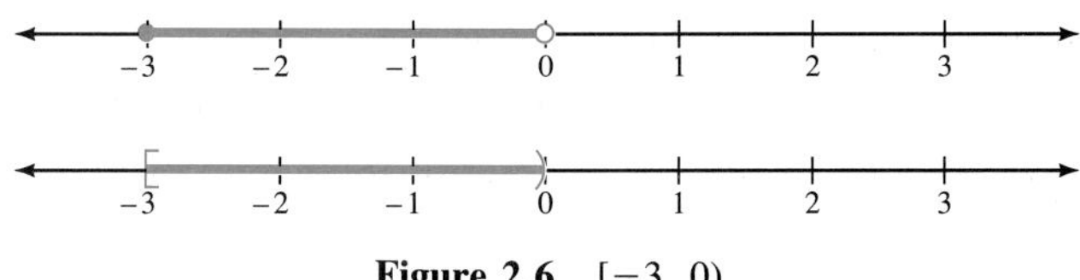

Figure 2.6 $[-3, 0)$

(b) $x < 2$

This infinite interval can be written as $(-\infty, 2)$. Figure 2.7 again shows the interval graphed in both dot notation and interval notation.

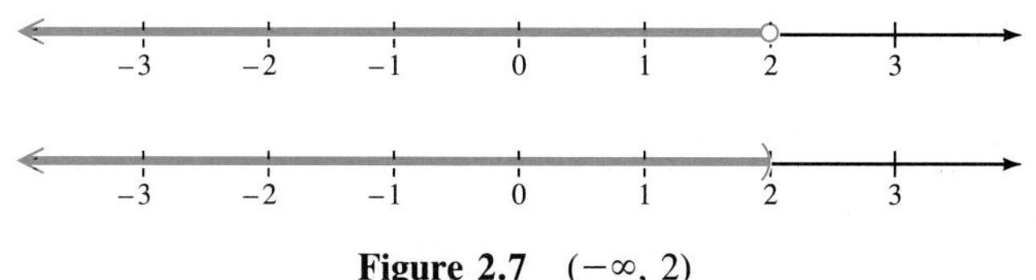

Figure 2.7 $(-\infty, 2)$

Practice Exercise 1

Write each inequality in interval notation and then graph.

(a) $-1 < x \le 2$

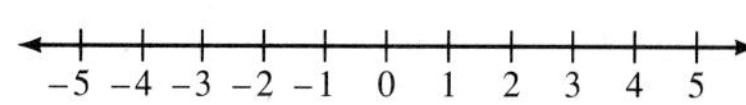

(b) $x \ge -2$

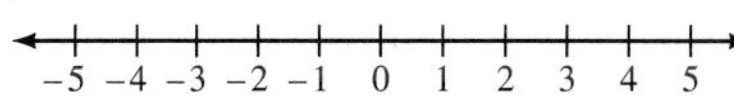

Answers: (a) Graph $(-1, 2]$. (b) Graph $[-2, \infty)$.

3 SOLVING LINEAR INEQUALITIES

We are now ready to solve linear inequalities such as

$$x + 2 < 7, \quad 3x \ge 9, \quad 3x - 1 > 11, \quad \text{and} \quad 2x + 1 \le 5 - 7x.$$

A **solution** to an inequality is a number that, when substituted for the variable, makes the inequality true.

If the inequality symbols ($<$, $\ge$, $>$, and $\le$) in the above statements were replaced with the equality symbol ($=$), we would know how to solve the resulting equations. Fortunately, solving an inequality is much like solving an equation since most of the same basic rules apply. There is one exception that will be discussed shortly. If we add 4 to both sides of the true inequality $5 < 8$, we have another true inequality.

$$5 + 4 < 8 + 4$$
$$9 < 12 \quad \text{True}$$

Similarly, if we subtract 11 from both sides, we again obtain a true inequality.

$$5 - 11 < 8 - 11$$

$$-6 < -3 \quad \text{True}$$

These observations illustrate the property which states that the same number (positive or negative) can be added to both sides of an inequality.

Additive Property of Inequalities

Let a, b, and c be numbers.

If $a < b$, then $a + c < b + c$.

That is, $a < b$ and $a + c < b + c$ are **equivalent inequalities** (they have the same solutions).

This rule is used to solve inequalities in the same way that the corresponding additive property is used to solve equations. The key is to isolate the variable on one side of the inequality. Most of the equations we have solved so far have had only one solution (identities excepted). Inequalities generally have many solutions.

EXAMPLE 2 Using the Additive Property

Solve and graph.

$$x + 7 < 2$$

$$x + 7 - 7 < 2 - 7 \quad \text{Subtract 7 from both sides}$$

$$x < -5$$

The solutions are *all* numbers less than -5, $(-\infty, -5)$, and the graph is given in Figure 2.8.

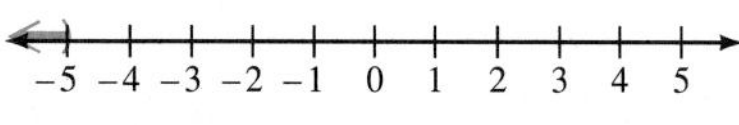

Figure 2.8 $(-\infty, -5)$

Practice Exercise 2

Solve and graph.

$$y + 4 \geq -2$$

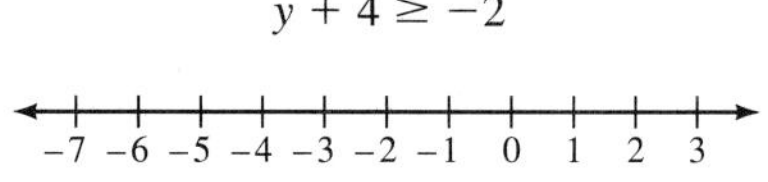

Answer: Graph $y \geq -6$, $[-6, \infty)$

If we start again with the true inequality

$$5 < 8$$

and multiply both sides by 2, we obtain the true inequality

$$10 < 16.$$

However, if we multiply both sides by -2 we obtain the false inequality

$$-10 < -16. \quad -10 \text{ is really greater than } -16$$

To obtain a true inequality when we multiply by the negative number -2, we must reverse the symbol of inequality, that is, change from $<$ to $>$ or from $>$ to $<$. The next property gives the rules for multiplying (or dividing) both sides of an inequality by positive or negative numbers.

Multiplicative Property of Inequalities

Let a, b, and c be numbers.

If $c > 0$ and $a < b$, then $ac < bc$.

That is, if $c > 0$, then $a < b$ and $ac < bc$ are equivalent inequalities.

If $c < 0$ and $a < b$, then $ac > bc$.

That is, if $c < 0$, then $a < b$ and $ac > bc$ are equivalent inequalities.

The only substantial difference between solving an equation and solving an inequality concerns multiplying or dividing both sides by a negative number.

CAUTION

Always reverse the symbol of inequality when multiplying (or dividing) both sides of an inequality by a negative number.

EXAMPLE 3 Using the Multiplicative Property

Solve and graph.

(a) $3x \geq 9$

$$\frac{1}{3}(3x) \geq \frac{1}{3} \cdot 9 \qquad \tfrac{1}{3} \text{ is positive, so the inequality remains the same}$$

$$x \geq 3$$

The solutions are *all* numbers greater than or equal to 3. Generally, we will simply write $x \geq 3$ or $[3, \infty)$. The graph is in Figure 2.9.

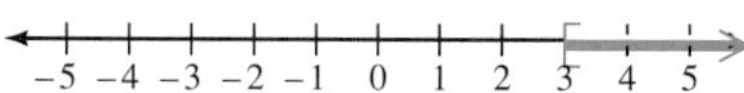

Figure 2.9 $[3, \infty)$

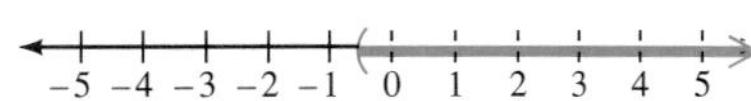

Figure 2.10 $\left(-\frac{1}{2}, \infty\right)$

(b) $$-\frac{1}{2}x < \frac{1}{4}$$

$$(-2)\left(-\frac{1}{2}x\right) > (-2)\left(\frac{1}{4}\right) \qquad -2 \text{ is negative, so the inequality symbol is reversed}$$

$$x > -\frac{1}{2}$$

The solution is $x > -\frac{1}{2}$, or $\left(-\frac{1}{2}, \infty\right)$. The graph is given in Figure 2.10.

Practice Exercise 3

Solve and graph.

(a) $7y < 35$

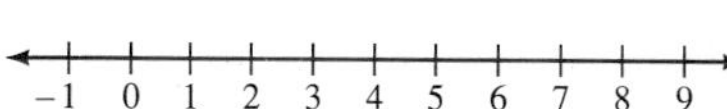

(b) $-3x \geq 4$

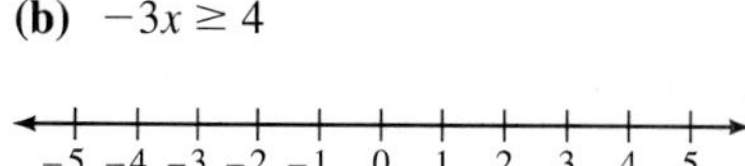

Answers: (a) Graph $y < 5$, $(-\infty, 5)$. (b) Graph $x \leq -\frac{4}{3}$, $\left(-\infty, -\frac{4}{3}\right]$.

As with solving equations, many times we need a combination of the additive and multiplicative properties. We usually use the additive property first, as in the example on the following page.

EXAMPLE 4 Using a Combination of Rules

Solve and graph.

(a) $3x - 1 > 11$

$3x - 1 + 1 > 11 + 1$ Add 1

$3x > 12$

$\frac{1}{3} \cdot 3x > \frac{1}{3} \cdot 12$ Multiply by *positive* $\frac{1}{3}$

$x > 4$

The solution, $x > 4$ or $(4, \infty)$, is graphed in Figure 2.11.

Figure 2.11 $(4, \infty)$

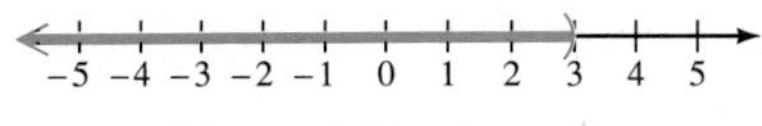

Figure 2.12 $(-\infty, 3)$

(b) $5 - 4x > 2 - 3x$

$5 - 5 - 4x > 2 - 5 - 3x$ Subtract 5

$-4x > -3 - 3x$

$-4x + 3x > -3 - 3x + 3x$ Add $3x$

$-x > -3$

$(-1)(-x) < (-1)(-3)$ Multiply by -1 and *reverse* inequality

$x < 3$

The solution, $x < 3$ or $(-\infty, 3)$, is graphed in Figure 2.12.

Practice Exercise 4

Solve and graph.

(a) $9y + 1 \le 19$

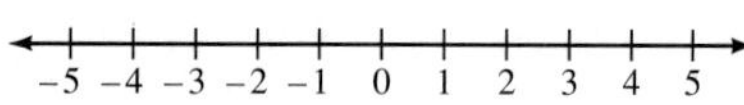

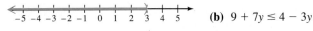

(b) $9 + 7y \le 4 - 3y$

Answers: (a) Graph $y \le 2$, $(-\infty, 2]$. (b) Graph $y \le -\frac{1}{2}$, $\left(-\infty, -\frac{1}{2}\right]$.

When an inequality contains parentheses, remove all parentheses, collect like terms (if such exist) on each side, and proceed as in previous cases.

EXAMPLE 5 Solving an Inequality Containing Parentheses

Solve and graph.

$x - 3(2 + x) > 2(3x - 1)$

$x - 6 - 3x > 6x - 2$ Remove parentheses

$-2x - 6 > 6x - 2$ Collect like terms

$-2x - 6 + 6 > 6x - 2 + 6$ Add 6

$-2x > 6x + 4$

$-2x - 6x > 6x - 6x + 4$ Subtract $6x$

$-8x > 4$

$\left(-\frac{1}{8}\right)(-8x) < \left(-\frac{1}{8}\right)(4)$ Multiply by $-\frac{1}{8}$ and reverse the inequality

$x < -\frac{1}{2}$

The solution, $x < -\frac{1}{2}$ or $\left(-\infty, -\frac{1}{2}\right)$ is graphed in Figure 2.13.

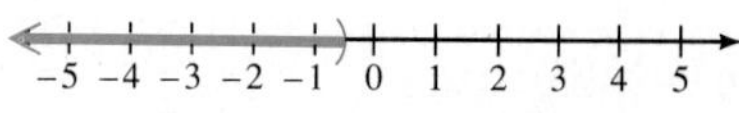

Figure 2.13 $\left(-\infty, -\frac{1}{2}\right)$

Practice Exercise 5

Solve and graph.

$2y - 4(y - 3) \le 3(2y + 1)$

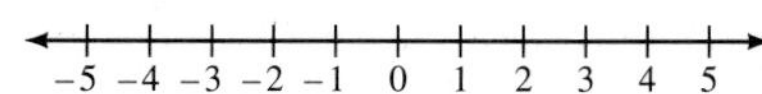

Answer: Graph $y \ge \frac{9}{8}$, $\left[\frac{9}{8}, \infty\right)$.

As is the case with linear equations, it is best to clear all fractions or decimals first.

EXAMPLE 6 Solving an Inequality Involving Fractions

Solve. $\frac{5x-1}{7} \le 2$

$7\left(\frac{5x-1}{7}\right) \le 7(2)$ Multiply both sides by 7 to clear the fraction

$5x - 1 \le 14$

$5x \le 15$ Add 1 to both sides

$x \le 3$ Divide both sides by 5

In interval notation, the solution is $(-\infty, 3]$.

Practice Exercise 6

Solve.

$$-0.2y + 1.8 > 3.2$$

[*Hint:* Multiply both sides by 10 first to clear the decimals.]

Answer: $(-\infty, -7)$

We now give a summary of inequality-solving techniques.

To Solve a Linear Inequality

1. To simplify both sides, remove grouping symbols and combine like terms. To eliminate fractions or decimals, multiply both sides by an appropriate positive number (the LCD of the fractions or a power of ten for decimals).
2. Use the addition-subtraction rule to isolate all variable terms on one side and all constant terms on the other side. Collect like terms when possible.
3. Use the multiplication-division rule to obtain a variable with coefficient 1. Remember to reverse the symbol of the inequality whenever multiplying or dividing both sides by a negative number.

We conclude this section with an application.

EXAMPLE 7 Grade-Computing Problem

Krista must have at least 90% of the points in her math class to get an A. The grade is determined by three 100-point tests and a 150-point final exam. If she has 85, 92, and 96 on her tests, what range of scores on the final will give her an A?

Let x = Krista's total points on final exam,

450 = total points possible,

(0.9)450 = 90% of the total points.

Thus, x must satisfy the following inequality.

$$85 + 92 + 96 + x \ge 0.9(450)$$

$$273 + x \ge 405$$

$$x \ge 132$$

Krista's score must be greater than or equal to 132 to get an A.

Practice Exercise 7

What range of scores on the final will give Krista (see Example 7) an A or B in the course if at least 80% of the points are necessary?

Answer: Krista's score must be greater than or equal to 87 to get an A or B.

2.5 EXERCISES A

Solve the following inequalities. Give answers in interval notation.

1. $x + 3 < 7$

2. $7 < z - 3$

3. $y + 2.1 \geq -3.4$

4. $-2z > 10$

5. $\frac{1}{3}y \leq -5$

6. $-\frac{1}{4}x \geq \frac{3}{2}$

7. $3z - 1 < 8$

8. $1 - 3y < 8$

9. $2z + 3 > 5z - 3$

10. $\frac{2y - 1}{3} < 5$

11. $\frac{x + 3}{-2} < 17$

12. $-y \leq 4$

13. $-1.5x < 4.5$

14. $0.4z + 1.6 < 2.4$

15. $-8y - 21 \geq 7 - 15y$

16. $5(3 - x) - 10 \leq 25$

17. $3(2z + 8) \leq 4(z - 3)$

18. $5(y + 3) + 1 \leq y - 4$

Solve and graph the following inequalities. Give answers in interval notation.

19. $x - 4 > 0$

−5 −4 −3 −2 −1 0 1 2 3 4 5

20. $z + 2 \leq 0$

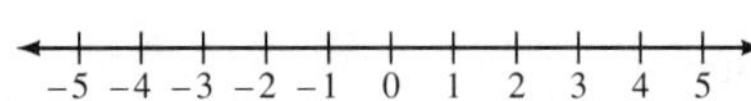

21. $2x - 5 < -1$

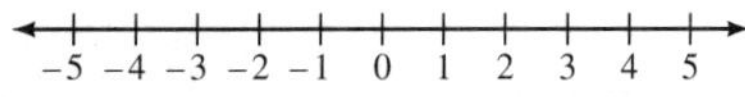

22. $3 - 5x \geq -7$

−5 −4 −3 −2 −1 0 1 2 3 4 5

23. $3(z + 7) \leq 2z + 18$

24. $4(x - 2) + 2 \geq 2(x - 6)$

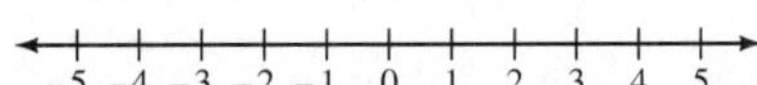

Letting x represent the unknown number, translate the following into symbols and solve.

25. Twice a number is greater than 8.

26. If 3 is subtracted from a number, the result is more than 20.

27. If twice a number is diminished by 7, the result is at least 12.

28 One-third of a number is no more than 15.

Translate to an inequality and solve.

29 Walt has scores of 86, 83, and 97 on three history tests. What range of scores can he make on test four to have an average of 90 or better?

30 To make a profit on the sale of television sets, a store owner knows that the total income on sales, S, must exceed the total of all costs involved, C. If n represents the number of televisions sold, $S = 150n$ and $C = 125n + 350$. How many sets must be sold for the owner to make a profit?

Are the following statements true or false? If false, change the second inequality to make the statement true.

31. If $x > 9$, then $-3x > -27$.

32. If $x < 4$, then $3x < 12$.

33. If $x \geq 8$, then $x - 1 \geq 7$.

34. If $x < -2$, then $-x > 2$.

FOR REVIEW

Solve.

35. $c = \dfrac{wrt}{1000}$ for w

36. $ax + by + c = 0$ for x

37. $e = mc^2$ for m

38. $ax + 2b = 3(x + c)$ for x

39. $3dy = d + 2ay$ for y

40. $3dy = d + 2ay$ for d

ANSWERS: **1.** $x < 4$; $(-\infty, 4)$ **2.** $z > 10$; $(10, \infty)$ **3.** $y \geq -5.5$; $[-5.5, \infty)$ **4.** $z < -5$; $(-\infty, -5)$ **5.** $y \leq -15$; $(-\infty, -15]$ **6.** $x \leq -6$; $(-\infty, -6]$ **7.** $z < 3$; $(-\infty, 3)$ **8.** $y > -\frac{7}{3}$; $\left(-\frac{7}{3}, \infty\right)$ **9.** $z < 2$; $(-\infty, 2)$ **10.** $y < 8$; $(-\infty, 8)$ **11.** $x > -37$; $(-37, \infty)$ **12.** $y \geq -4$; $[-4, \infty)$ **13.** $x > -3$; $(-3, \infty)$ **14.** $z < 2$; $(-\infty, 2)$ **15.** $y \geq 4$; $[4, \infty)$ **16.** $x \geq -4$; $[-4, \infty)$ **17.** $z \leq -18$; $(-\infty, -18]$ **18.** $y \leq -5$; $(-\infty, -5]$ **19.** $(4, \infty)$ **20.** $(-\infty, -2]$ **21.** $(-\infty, 2)$ **22.** $(-\infty, 2]$ **23.** $(-\infty, -3]$ **24.** $[-3, \infty)$ **25.** $2x > 8$; $x > 4$, $(4, \infty)$ **26.** $x - 3 > 20$; $x > 23$, $(23, \infty)$ **27.** $2x - 7 \geq 12$; $x \geq \frac{19}{2}$, $\left[\frac{19}{2}, \infty\right)$ **28.** $\frac{1}{3}x \leq 15$; $x \leq 45$, $(-\infty, 45]$ **29.** Walt's score, s, must satisfy $s \geq 94$. **30.** $n > 14$, that is, 15 or more sets must be sold. **31.** false; $-3x < -27$ **32.** true **33.** true **34.** true **35.** $w = \frac{1000c}{rt}$ **36.** $x = \frac{-by - c}{a}$ **37.** $m = \frac{e}{c^2}$ **38.** $x = \frac{3c - 2b}{a - 3}$ **39.** $y = \frac{d}{3d - 2a}$ **40.** $d = \frac{2ay}{3y - 1}$

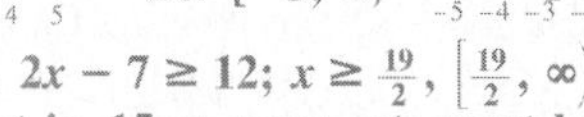

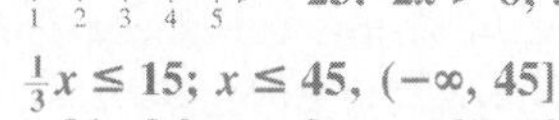

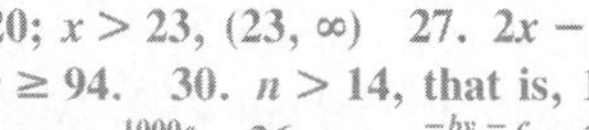

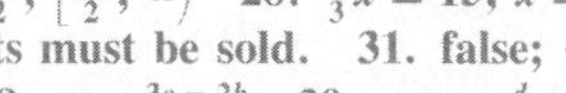

2.5 EXERCISES B

Solve the following inequalities. Give answers in interval notation.

1. $y + 5 \geq 3$

2. $x - 5 > 2$

3. $z - \frac{1}{2} < \frac{3}{4}$

4. $-9 \leq -3x$

5. $-3 \geq \frac{1}{4}z$

6. $\frac{7}{10} < -\frac{1}{5}y$

7. $5 - 2x > 3$

8. $2z - 5 > 3$

9. $4x - 5 \leq 2x + 13$

10. $\frac{3z - 2}{5} \geq 2$

11. $\frac{4 - y}{-3} \geq 13$

12. $-z > 17$

13. $-1.2y \geq -7.2$

14. $3.2 - 0.8x \geq 4.8$

15. $-3z - 39 \leq 13 - 2z$

16. $2(1 - y) - 4 > 6$

17. $2(3x + 1) > 5(x - 4)$

18. $-4z - 3 + z < 2(z + 12)$

Solve and graph the following inequalities. Give answers in interval notation.

19. $y - 3 < 0$

−5 −4 −3 −2 −1 0 1 2 3 4 5

20. $x + 1 \geq 0$

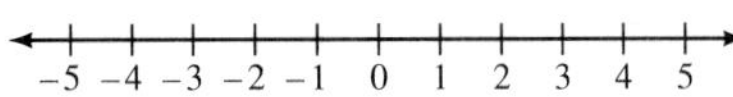

21. $5z - 2 > -7$

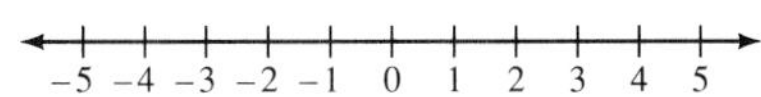

22. $7 - 3y \leq -2$

−5 −4 −3 −2 −1 0 1 2 3 4 5

23. $5(x + 1) < 3x + 1$

−5 −4 −3 −2 −1 0 1 2 3 4 5

24. $4 - (x - 2) < 8 - 2x$

−5 −4 −3 −2 −1 0 1 2 3 4 5

Letting x represent the unknown number, translate the following into symbols and solve.

25. My age in 3 years will be less than 25.

26. If 5 is subtracted from a number, the result is more than 3.

27. If three times a number is increased by 9, the result is at least 21.

28. Three-fourths of a number is no less than 81.

Translate to an inequality and solve.

29. Missy has scores of 80, 86, and 97 on three geology tests. What range of scores can she make on test four to have an average of 90 or better?

30. To make a profit on the sale of clocks, a dealer knows that the total income on sales, S, must exceed the total of all costs involved, C. If n is the number of clocks sold, $S = 40n$ and $C = 25n + 150$. How many clocks must be sold to make a profit?

Are the following statements true or false? If false, change the second inequality to make the statement true.

31. If $y > 5$, then $-2y > -10$.

32. If $y < 3$, then $8y < 24$.

33. If $y \geq 12$, then $y - 1 \geq 11$.

34. If $y < -4$, then $-y > 4$.

FOR REVIEW

Solve.

35. $c = \dfrac{wrt}{1000}$ for t

36. $ax + by + c = 0$ for y

37. $g = mt^2$ for m

38. $ax + 3b = 7(x - c)$ for x

39. $4dz = d + 3bz$ for z

40. $4dz = d + 3bz$ for d

2.5 EXERCISES C

Solve. Give answers in interval notation.

1. $\dfrac{2x}{3} + \dfrac{1}{4}(x - 4) \leq \dfrac{x}{12} - \dfrac{1}{3}(x + 1)$
[*Hint:* Clear all fractions first.]

2. $(x + 1)(x - 2) > (x - 3)(x - 5)$
[Answer: $x > \frac{17}{7}$; $\left(\frac{17}{7}, \infty\right)$]

Are the following statements true or false? If the statement is false, explain why.

3. If $x > -1$, then $3 - 2x < 5$.

4. If $x < 5$, then $x^2 < 5x$.

5. If $x < -2$ and $y < 0$, then $xy > -2y$.

6. If $y < 2$, then $y^2 < 4$. [Answer: false]

2.6 COMPOUND INEQUALITIES

STUDENT GUIDEPOSTS

 *And* Statements

2 *Or* Statements

1 *AND* STATEMENTS

We often encounter problems that combine two linear inequalities. When two inequalities are combined by using the word *and* or *or*, the result is a **compound statement.** Consider the compound statement

$$x + 2 > 0 \quad \textit{and} \quad 2(x - 1) < 4.$$

Solving each inequality gives the equivalent compound statement

$$x > -2 \quad \textit{and} \quad x < 3,$$

which could also be given by

$$-2 < x \quad \textit{and} \quad x < 3,$$

by the **chain of inequalities**

$$-2 < x < 3,$$

or by the open interval $(-2, 3)$. Thus, the numbers that satisfy this chain of inequalities are solutions to both inequalities: numbers that are *both* greater than -2 *and* less than 3. The graphs of $x > -2$, $x < 3$, and $-2 < x < 3$ are all shown in Figure 2.14.

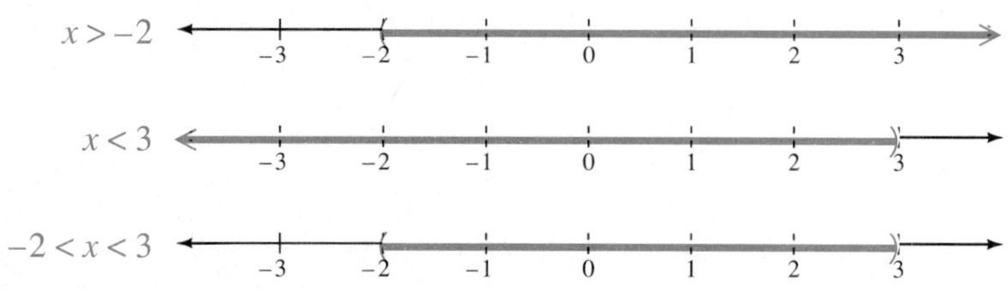

Figure 2.14 $-2 < x < 3$ or $(-2, 3)$

Figure 2.15 $-1 \le x \le 2$ or $[-1, 2]$

Notice that the graph of $-2 < x < 3$ includes all points that $x > -2$ and $x < 3$ have in common. Similarly, the graph of $-1 \le x \le 2$ ($-1 \le x$ and $x \le 2$) is shown in Figure 2.15, the graph of $-3 \le x < 0$ ($-3 \le x$ and $x < 0$) in Figure 2.16, and the graph of $1 < x \le 2$ ($1 < x$ and $x \le 2$) in Figure 2.17.

Figure 2.16 $-3 \le x < 0$ or $[-3, 0)$

Figure 2.17 $1 < x \le 2$ or $(1, 2]$

The compound statement

$$2 < x \quad \text{and} \quad x < -1$$

has no solution, since there are no numbers that are *both* less than -1 *and* greater than 2. If we try to merge the two inequalities into a single chain as before, we obtain

$$2 < x < -1, \qquad \text{This is wrong}$$

which states that 2 is less than -1, a contradiction. Such chains should always be avoided.

A chain of inequalities may be more complex. For example,

$$-1 < 2x + 1 < 3$$

is

$$-1 < 2x + 1 \quad \textit{and} \quad 2x + 1 < 3.$$

Note that -1 is indeed less than 3, so the chain makes sense.

EXAMPLE 1 Solving an *And* Inequality

Solve and graph $-1 < 2x + 1 < 3$.

We need to solve two inequalities.

$-1 < 2x + 1$	and	$2x + 1 < 3$	
$-1 - 1 < 2x + 1 - 1$	and	$2x + 1 - 1 < 3 - 1$	Subtract 1
$-2 < 2x$	and	$2x < 2$	
$-1 < x$	and	$x < 1$	Divide by 2

Practice Exercise 1

Solve and graph $-7 \le 3x - 4 \le 2$.

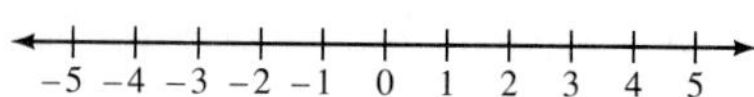

Rewritten as a single chain, the solution is

$$-1 < x < 1,$$

and the graph is shown in Figure 2.18.

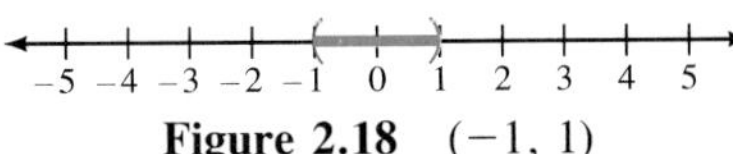

Figure 2.18 $(-1, 1)$

Answer: Graph $-1 \le x \le 2$, $[-1, 2]$.

Alternatively and more compactly, we can solve the original chain by the following method.

$$-1 < 2x + 1 < 3$$

$$-1 - 1 < 2x + 1 - 1 < 3 - 1 \quad \text{Subtract 1 throughout}$$

$$-2 < 2x < 2$$

$$\frac{1}{2}(-2) < \frac{1}{2}(2x) < \frac{1}{2}(2) \quad \text{Multiply throughout by } \tfrac{1}{2}$$

$$-1 < x < 1$$

2 *OR* STATEMENTS

The second type of compound statement uses the connective *or*. For example,

$$x + 1 < 0 \quad or \quad 3(x - 1) > 3$$

simplifies to

$$x < -1 \quad or \quad x > 2.$$

A solution to a compound statement using the word *or* solves *either* (or both) inequalities.

CAUTION

You *cannot* form a single chain when the word *or* is used. Notice that $2 < x < -1$ is a senseless statement (2 is not less than -1) and should *not* be used to describe the compound statement $x < -1$ *or* $x > 2$.

The graph of an *or* combination is usually *two* intervals on a number line, such as the graph of $x < -1$ or $x > 2$ shown in Figure 2.19. Similarly, $x < -2$ or $x \ge 1$ is graphed in Figure 2.20. Also, the graph of $x \le 0$ or $x > 3$ is shown in Figure 2.21, and the graph of $x \le -1$ or $x \ge 0$ is in Figure 2.22. Finally, the graph of $x \le 2$ or $x \ge -1$ in Figure 2.23 is the entire number line.

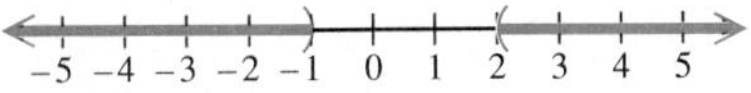

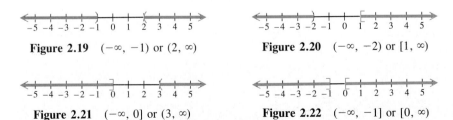

Figure 2.19 $(-\infty, -1)$ or $(2, \infty)$

Figure 2.20 $(-\infty, -2)$ or $[1, \infty)$

Figure 2.21 $(-\infty, 0]$ or $(3, \infty)$

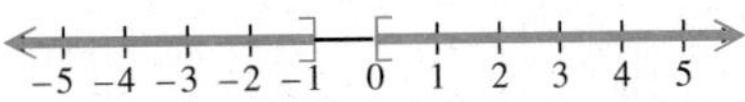

Figure 2.22 $(-\infty, -1]$ or $[0, \infty)$

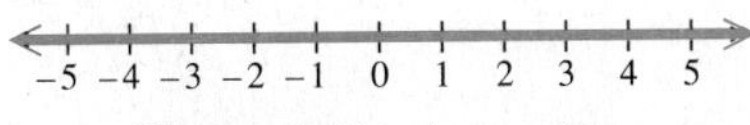

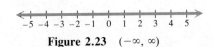

Figure 2.23 $(-\infty, \infty)$

EXAMPLE 2 Solving an *Or* Inequality

Solve and graph the following compound statements.

(a)

$2x + 1 \le -1$	or	$2x + 1 > 3$	
$2x + 1 - 1 \le -1 - 1$	or	$2x + 1 - 1 > 3 - 1$	Subtract 1
$2x \le -2$	or	$2x > 2$	
$x \le -1$	or	$x > 1$	Divide by 2

The solution is $x \le -1$ or $x > 1$,

or, in interval notation, $(-\infty, -1]$ or $(1, \infty)$. The graph is shown in Figure 2.24.

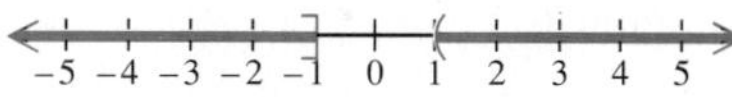

Figure 2.24 $(-\infty, -1]$ or $(1, \infty)$

(b) Solve and graph the following compound statement.

$3 - 4x < -1$	or	$3 - 4x \ge 9$	
$-4x < -4$	or	$-4x \ge 6$	
$x > 1$	or	$x \le -\frac{6}{4}$	Reverse
		$x \le -\frac{3}{2}$	

The solution is $x > 1$ or $x \le -\frac{3}{2}$,

or, in interval notation, $(1, \infty)$ or $\left(-\infty, -\frac{3}{2}\right]$ and the graph is given in Figure 2.25.

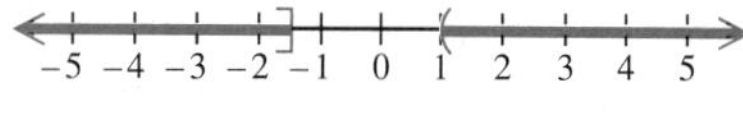

Figure 2.25 $(1, \infty)$ or $\left(-\infty, -\frac{3}{2}\right]$

Practice Exercise 2

Solve and graph.

(a) $4x - 3 < -5$ or $4x - 3 \ge 5$

−5 −4 −3 −2 −1 0 1 2 3 4 5

(b) $2 - 5x \le -3$ or $2 - 5x > 7$

−5 −4 −3 −2 −1 0 1 2 3 4 5

Answers: (a) Graph $x < -\frac{1}{2}$ or $x \ge 2$; $\left(-\infty, -\frac{1}{2}\right)$ or $[2, \infty)$.
(b) Graph $x < -1$ or $x \ge 1$; $(-\infty, -1)$ or $[1, \infty)$.

There are numerous applications of compound inequalities. The following is one example.

EXAMPLE 3 Temperature Problem

Celsius (C) and Fahrenheit (F) temperatures are related by $F = \frac{9}{5}C + 32$. If a hot tub is advertised to have a temperature range of 95°F to 104°F, inclusive, within what range does the Celsius temperature fall?

First write F using a chain of inequalities.

$$95 \le F \le 104$$

Now substitute $F = \frac{9}{5}C + 32$ and solve the chain.

$$95 \le \frac{9}{5}C + 32 \le 104$$

Practice Exercise 3

A student must have an average mark between 70% and 80%, inclusive, to receive a grade of C. If Bobbi has 69%, 82%, and 79% on the first three tests this semester, what range of scores on the fourth test would give her a C?

$$95 - 32 \le \frac{9}{5}C \le 104 - 32 \qquad \text{Subtract 32 throughout}$$

$$63 \le \frac{9}{5}C \le 72$$

$$(63)(5) \le 9C \le (72)(5) \qquad \text{Multiply by 5}$$

$$\frac{(63)(5)}{9} \le C \le \frac{(72)(5)}{9} \qquad \text{Divide by 9}$$

$$35 \le C \le 40$$

The temperature is between 35°C and 40°C, inclusive.

Answer: Bobbi must score between 50 and 90, inclusive.

Remember that an *and* compound inequality (such as $-4 < x$ and $x \le 3$) can be expressed as a single chain of inequalities ($-4 < x \le 3$) and generally has as its graph a single interval on a number line. An *or* compound inequality (such as $x < -1$ or $x > 2$) *cannot* be expressed as a single chain of inequalities (*never write* $2 < x < -1$) and generally has as its graph two separate intervals on a number line.

2.6 EXERCISES A

Solve and graph each compound inequality. Give answer in interval notation.

1. $x + 2 < 2$ and $x + 2 > -1$

−5 −4 −3 −2 −1 0 1 2 3 4 5

2. $z + 3 < 0$ or $z > 0$

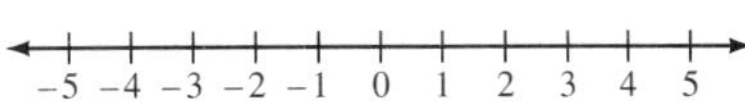

3. $y \ge 0$ and $y - 2 \le 2$

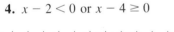

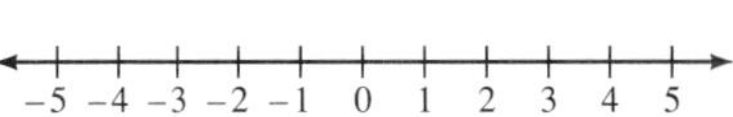

4. $x - 2 < 0$ or $x - 4 \ge 0$

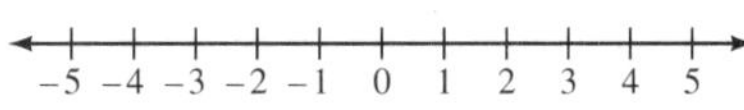

5. $-5 \le 4z + 3 < 5$

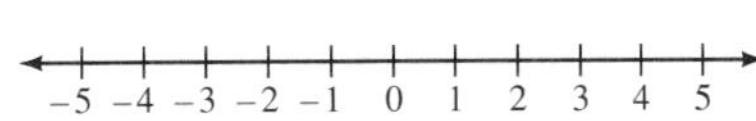

6. $4y + 3 < -5$ or $4y + 3 \ge 5$

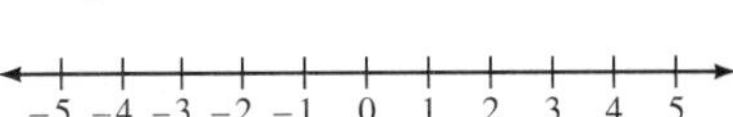

7. $0 > 3 - x$ or $3 - x \ge 1$

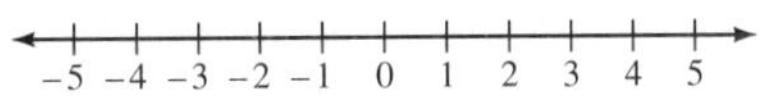

8. $0 < -z < 2$

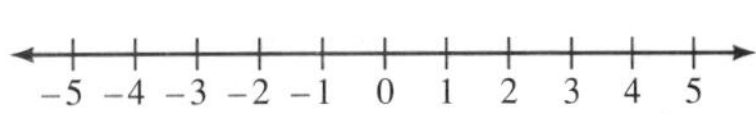

9. $0 \le 3 - x < 1$

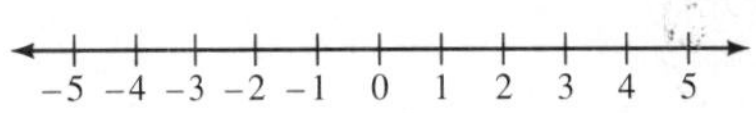

Letting x represent the unknown number, translate the following into symbols and solve.

10. A number is less than 8 and greater than −1.

11. A number is at least 2 and at most 9.

12. When twice a number is increased by 1, the result is between 7 and 13.

13. One less a number is greater than 10 or less than −3.

Solve.

14. Fahrenheit (F) and Celsius (C) temperatures are related by $C = \frac{5}{9}(F - 32)$. If the Celsius temperature is between $-15°$C and 30°C, inclusive, within what range will the Fahrenheit temperature fall?

15 A student must have an average percent between 80% and 90%, inclusive, in order to receive a grade of B. If Jake has 69%, 82%, and 79% on the first three tests this semester, what range of scores on the fourth test will give him a B?

FOR REVIEW

Solve.

16. $\frac{z - 3}{-4} < 3$

17. $3(2 - y) + y > 14$

18. $3(2z - 1) \geq -2(1 - z)$

19. The force in pounds F required to stretch a spring x inches beyond its normal length is given by $F = 5.5x$. What are the corresponding values for x if F is greater than 6.6 pounds?

Exercises 20–22 review material from Section 1.2 to help you prepare for the next section. Evaluate the absolute values.

20. $|(-2) - (-1)(-2)|$

21. $\frac{|(-5)(-1) - 7|}{|-2|}$

22. $\frac{|(-2)(-7) - 20|}{|-18 + (-24)|}$

ANSWERS:

1. $(-3, 0)$;

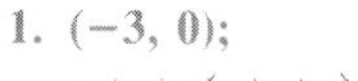

2. $(-\infty, -3)$ or $(0, \infty)$;

3. $[0, 4]$;

4. $(-\infty, 2)$ or $[4, \infty)$;

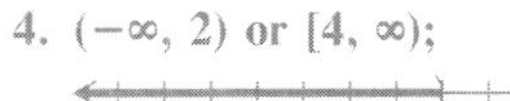

5. $\left[-2, \frac{1}{2}\right)$;

6. $(-\infty, -2)$ or $\left[\frac{1}{2}, \infty\right)$;

7. $(-\infty, 2]$ or $(3, \infty)$;

8. $(-2, 0)$;

9. $(2, 3]$;

10. $-1 < x < 8$; $(-1, 8)$ 11. $2 \leq x \leq 9$; $[2, 9]$ 12. $7 < 2x + 1 < 13$; $(3, 6)$ 13. $1 - x > 10$ or $1 - x < -3$; $(-\infty, -9)$ or $(4, \infty)$ 14. $5° \leq F° \leq 86°$ 15. His score must be between 90% and 100%, inclusive. 16. $z > -9$ 17. $y < -4$ 18. $z \geq \frac{1}{4}$ 19. $x > 1.2$ inches 20. 4 21. 1 22. $\frac{1}{7}$

2.6 EXERCISES B

Solve and graph each compound inequality. Give answer in interval notation.

1. $y + 3 > 2$ and $y + 3 \leq 5$

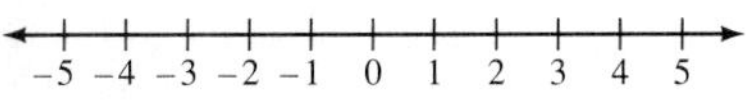

2. $x + 2 < 0$ or $x > 0$

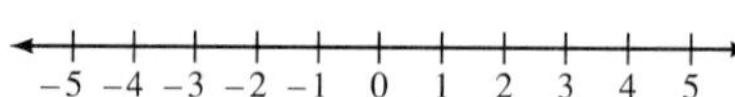

3. $z + 1 \geq 0$ and $z - 1 \leq 0$

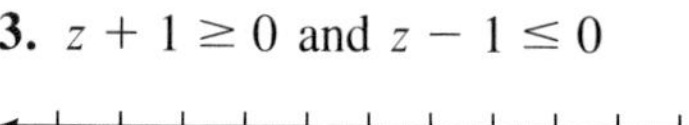

4. $y - 1 < 0$ or $y - 3 \geq 0$

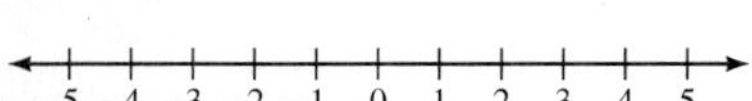

5. $-4 \leq 3x + 2 < 3$

6. $2z + 1 \leq -5$ or $2z + 1 > 5$

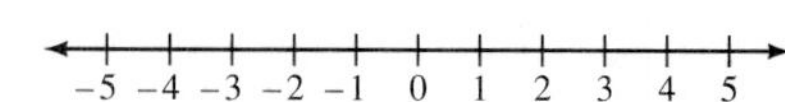

7. $0 \geq 2 - y$ or $2 - y > 1$

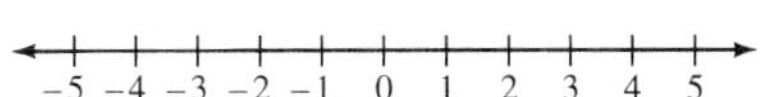

8. $-2 \leq -y < 3$

-5 -4 -3 -2 -1 0 1 2 3 4 5

9. $4 > 1 - y \geq 0$

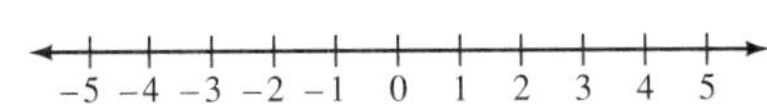

Letting x represent the unknown number, translate the following into symbols and solve.

10. A number is greater than or equal to 3 and less than 7.

11. A number is at most 0 and at least -1.

12. When three times a number is decreased by 5, the result is a number between 4 and 22.

13. Seven less a number is more than 5 or less than -5.

Solve.

14. Fahrenheit (F) and Celsius (C) temperatures are related by $C = \frac{5}{9}(F - 32)$. If the Celsius temperature is between $-5°$C and 30°C, inclusive, within what range will the Fahrenheit temperature fall?

15. A student must have an average percent between 70% and 80%, inclusive, to earn a grade of C. If Syd has 45%, 51%, and 62% on the first three tests this semester, what range of scores on the fourth test will give him a C?

FOR REVIEW

Solve.

16. $\frac{z - 2}{-3} > 5$

17. $2(3 - y) + y < 8$

18. $5(z - 1) < -2(z - 15)$

19. The force in pounds F required to stretch a spring x inches beyond its normal length is given by $F = 6.2x$. What are the corresponding values for x if F is greater than 15.5 pounds?

Exercises 20–22 review material from Section 1.2 to help you prepare for the next section. Evaluate the absolute values.

20. $|(-3) - (4)(-2)|$

21. $\frac{|(-6)(-2) - 12|}{|-5|}$

22. $\frac{|(-8)(4) - (-16)|}{|-20 + (-12)|}$

2.6 EXERCISES C

Solve.

1. The perimeter of a rectangular enclosure must be less than or equal to 100 yd. If one side must be 20 yd long, what is the maximum length for the other side?

2. Mr. James manufactures novelty items. The cost in dollars, c, of producing n items in a day is given by $c = 100 + 3n$. The number of dollars, s, generated by daily sales is given by $s = 5n$. In order to realize a profit, $s > c$. What is the minimum number of items that must be produced each day for Mr. James to realize a profit? [Answer: 51 items]

The formula $W = EI$ is used to describe the state of an electrical circuit, where W represents the power measured in watts, E represents the pressure in volts, and I represents the current in amperes.

3. The range in power demanded in a 110-volt circuit in a house satisfies $220 \le W \le 2200$. What is the range in current in this circuit?

4. Suppose a 20-ampere fuse is used in a 110-volt circuit in a mobile home. What is the maximum total wattage of all items on the circuit before "blowing" the fuse? [Answer: 2200 watts]

2.7 ABSOLUTE VALUE EQUATIONS AND INEQUALITIES

STUDENT GUIDEPOSTS

1. Absolute Value Equations
2. Absolute Value Inequalities

1 ABSOLUTE VALUE EQUATIONS

In Chapter 1 we defined absolute value as

$$|a| = \begin{cases} a & \text{if } a \ge 0 \\ -a & \text{if } a < 0. \end{cases}$$

To solve an absolute value equation, such as

$$|x| = 5,$$

we need to solve two equations, since

$$|x| = \begin{cases} x & \text{if } x \ge 0 \\ -x & \text{if } x < 0. \end{cases}$$

The two equations to solve are

$$x = 5 \quad \text{and} \quad -x = 5.$$

The second equation becomes $x = -5$, giving us the two equations

$$x = 5 \quad \text{and} \quad x = -5.$$

Therefore, $|x| = 5$ implies two solutions, $x = 5$ and $x = -5$.

EXAMPLE 1 SOLVING AN ABSOLUTE VALUE EQUATION

Solve $|x| = 12$.

From the above discussion, $|x| = 12$ gives us the two solutions,

$$x = 12 \quad \text{and} \quad x = -12.$$

Check: $|12| = 12$ and $|-12| = 12$. Thus, the solutions 12 and -12 do check.

PRACTICE EXERCISE 1

Solve.

$$|y| = 9$$

Answer: 9, -9

An absolute value equation in which the expression inside the absolute value bars is more complex, is solved similarly using the following technique.

Note: An equation with absolute value sign will have two solutions.

Absolute Value Equations

To solve an absolute value equation of the form

$$|Expression| = a$$

when $a \geq 0$, solve the two related equations

$$Expression = a \quad \text{and} \quad Expression = -a.$$

If $a < 0$, $|Expression| = a$ has no solution.

EXAMPLE 2 Solving an Absolute Value Equation

Solve $|2x + 1| = 3$.

We need to solve

$$2x + 1 = 3 \qquad \text{and} \qquad 2x + 1 = -3.$$

$$2x + 1 - 1 = 3 - 1 \qquad 2x + 1 - 1 = -3 - 1$$

$$2x = 2 \qquad 2x = -4$$

$$\frac{1}{2}(2x) = \frac{1}{2}(2) \qquad \frac{1}{2}(2x) = \frac{1}{2}(-4)$$

$$x = 1 \qquad x = -2$$

The solutions are 1 and -2. Check by substituting each solution into the original equation.

Practice Exercise 2

Solve.

$$|4x - 2| = 6$$

Answer: -1, 2

EXAMPLE 3 Solving an Absolute Value Equation

Solve $|1 - 3x| = 0$.

We must solve $1 - 3x = 0$ and $1 - 3x = -0$. But since $-0 = 0$, we really have only one equation to solve.

$$1 - 3x = 0$$

$$-3x = -1 \qquad \text{Subtract 1 from both sides}$$

$$x = \frac{1}{3} \qquad \text{Divide both sides by } -3$$

The solution is $\frac{1}{3}$. Check in the original equation.

Practice Exercise 3

Solve.

$$|4 - 5y| = 0$$

Answer: $\frac{4}{5}$

CAUTION

$|5x - 1| = -2$, has no solution. Since the absolute value of a number is always ≥ 0, it cannot equal -2.

Other types of absolute value equations are solved in a similar manner, as illustrated in the next example.

EXAMPLE 4 Solving an Absolute Value Equation

Solve $|3x + 1| = |2 - x|$.

For this equation to be true, $3x + 1$ must equal $2 - x$ or its negative. Thus, we need to solve

Practice Exercise 4

Solve.

$$|2x - 3| = |4 - 3x|$$

$$3x + 1 = 2 - x \quad \text{and} \quad 3x + 1 = -(2 - x).$$

$$4x = 1 \qquad 3x + 1 = -2 + x$$

$$x = \frac{1}{4} \qquad 2x = -3$$

$$x = -\frac{3}{2}$$

The two solutions are $\frac{1}{4}$ and $-\frac{3}{2}$. Check these in the original equation.

Answer: $1, \frac{7}{5}$

2 ABSOLUTE VALUE INEQUALITIES

Solutions to absolute value equations are usually found by solving two related equations. Similarly, solutions to **absolute value inequalities** are solutions to two related inequalities. Consider the inequality

$$|x| < 2.$$

Geometrically, since $|x|$ represents the distance that x is located from the origin, $|x| < 2$ means that x cannot be further from zero than 2 units in either direction; that is, it must be located between -2 and 2. This is the compound statement

$$-2 < x < 2 \qquad \text{Written as a chain of inequalities}$$

or $\quad x > -2 \quad and \quad x < 2 \qquad$ Written using the connective *and*

Thus, to solve $|x| < 2$, we simply give the solution using the related compound statement $-2 < x < 2$ ($x > -2$ *and* $x < 2$).

To solve the inequality

$$|x| > 2$$

we are interested in those numbers x whose distance from zero is more than 2 units in either direction; that is, numbers that are less than -2 or greater than 2. This is the compound statement

$$x < -2 \quad or \quad x > 2.$$

Thus, the solution to $|x| > 2$ is given using the compound statement $x < -2$ *or* $x > 2$.

Absolute value inequalities with more complex expressions within the absolute value bars are solved in a similar manner.

Absolute Value Inequalities

1. To solve an absolute value inequality of the form $|Expression| > a$ where $a > 0$, solve the compound statement

   Expression $< -a$ or *Expression* $> a$.

2. To solve an absolute value inequality of the form $|Expression| < a$ where $a > 0$, solve either the compound statement

   Expression $< a$ and *Expression* $> -a$

   or the chain of inequalities

   $-a <$ *Expression* $< a$.

3. Use similar procedures for $|Expression| \geq a$ and $|Expression| \leq a$.

Since $|Expression|$ represents the distance that *Expression* is located from the origin, $|Expression| < a$ means that *Expression* cannot be farther from zero than

a units in either direction; that is, it must be located between $-a$ and a. Similarly, $|Expression| > a$ means that *Expression* must be more than a units from zero in either direction; that is, it must be less than $-a$ or greater than a. These relationships are shown in Figure 2.26.

Figure 2.26 Absolute Value Inequalities

EXAMPLE 5 Solving an Absolute Value Inequality

Solve $|x| > 3$.

The solution is

$$x < -3 \quad or \quad x > 3, \qquad \text{Expression is } x \text{ in this case}$$

or, in interval notation, $(-\infty, -3)$ or $(3, \infty)$.

Practice Exercise 5

Solve. $|y| \geq 10$

Answer: $y \leq -10$ or $y \geq 10$; $(-\infty, -10]$ or $[10, \infty)$

EXAMPLE 6 Solving an Absolute Value Inequality

Solve $|x| \leq 5$.

The solution is $x \leq 5$ *and* $x \geq -5$, which is equivalent to

$$-5 \leq x \leq 5, \qquad \text{Expression is } x$$

or, in interval notation, $[-5, 5]$.

Practice Exercise 6

Solve. $|y| < 9$

Answer: $-9 < y < 9$; $(-9, 9)$

EXAMPLE 7 Solving an Absolute Value Inequality

Solve $|x - 2| > 7$. Graph the solution.

We need to solve

$$x - 2 < -7 \quad or \quad x - 2 > 7. \qquad \text{Expression is } x - 2$$
$$x - 2 + 2 < -7 + 2 \quad \text{or} \quad x - 2 + 2 > 7 + 2$$
$$x < -5 \quad \text{or} \quad x > 9$$

The solution is $x < -5$ or $x > 9$. (*Do not write* $9 < x < -5$.) In interval notation, the solution is $(-\infty, -5)$ or $(9, \infty)$, which is graphed in Figure 2.27.

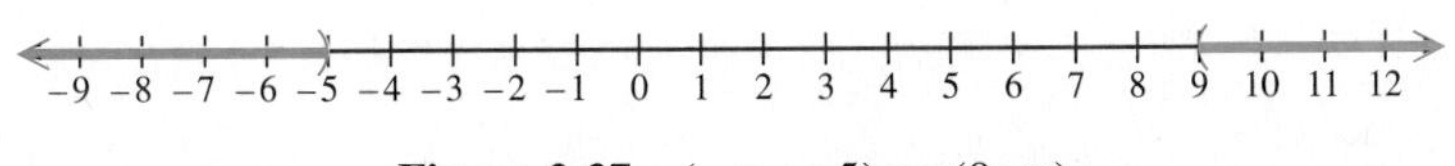

Figure 2.27 $(-\infty, -5)$ or $(9, \infty)$

Practice Exercise 7

Solve and graph.

$$|y + 8| \geq 2$$

−10 −9 −8 −7 −6 −5 −4 −3 −2 −1 0

Answer: Graph $y \leq -10$ or $y \geq -6$; $(-\infty, -10]$ or $[-6, \infty)$.

EXAMPLE 8 Solving Absolute Value Inequalities

(a) Solve $|x + 3| \le 4$. Graph the solution.

We need to solve $x + 3 \le 4$ *and* $x + 3 \ge -4$, which is equivalent to

$$-4 \le x + 3 \le 4. \qquad \textit{Expression is } x + 3$$

$$-4 - 3 \le x + 3 - 3 \le 4 - 3 \qquad \text{Subtract 3 throughout}$$

$$-7 \le x \le 1$$

The solution is $-7 \le x \le 1$. (Equivalently, $x \le 1$ and $x \ge -7$.) In interval notation, the solution is $[-7, 1]$, which is graphed in Figure 2.28.

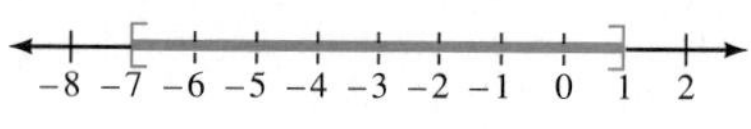

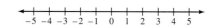

Figure 2.28 $[-7, 1]$

(b) Solve $|1 - 3x| \ge 7$. Graph the solution.

We need to solve

$$1 - 3x \le -7 \quad \textit{or} \quad 1 - 3x \ge 7. \qquad \textit{Expression is } 1 - 3x$$

$$-3x \le -8 \quad \text{or} \quad -3x \ge 6 \qquad \text{Subtract 1}$$

$$x \ge \frac{8}{3} \quad \text{or} \quad x \le -2 \qquad \text{Reverse inequalities}$$

The solution is $x \ge \frac{8}{3}$ or $x \le -2$. (*Do not use* $\frac{8}{3} \le x \le -2$ to describe the solution.) In interval notation, the solution is $(-\infty, -2]$ or $\left[\frac{8}{3}, \infty\right)$, which is graphed in Figure 2.29.

Figure 2.29 $(-\infty, -2]$ or $\left[\dfrac{8}{3}, \infty\right)$

Practice Exercise 8

Solve and graph.

(a) $|y - 5| < 5$

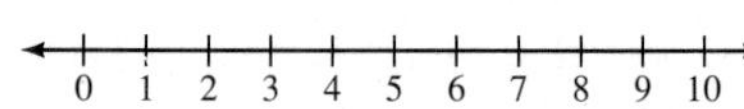

(b) $|2 - 4y| \ge 6$

Answers: (a) Graph $0 < y < 10$; $(0, 10)$. (b) Graph $y \le -1$ or $y \ge 2$; $(-\infty, -1]$ or $[2, \infty)$.

CAUTION

To solve an absolute value inequality, *memorize the rule above,* since the very first step is translating to a compound statement. It is not enough to solve $2x + 1 < 5$ when $|2x + 1| < 5$ is given. This is a common mistake.

Can we solve

$$|1 - x| < -2 \quad ?$$

Notice that -2 is negative. This inequality has no solutions since $|\textit{Expression}|$ is always nonnegative. Similarly, to solve

$$|1 - x| \le 0$$

we would solve $|1 - x| = 0$ or $1 - x = 0$ which has 1 as the only solution. Finally, to solve

$$|1 - x| \ge -2$$

we would observe that since $|1 - x|$ is always greater than or equal to 0, hence greater than or equal to -2, every real number is a solution.

2.7 EXERCISES A

Solve.

1. $|x| = 7$
2. $|z| = 0$
3. $|y| = -7$
4. $|x + 1| = 4$
5. $|z + 1| + 8 = 4$
6. $|y + 1| = 0$
7. $|2x - 5| = 9$
8. $|5 - 2z| - 1 = 8$
9. $|2y - 5| = 0$
10. $|5 - 2y| = 0$
11. $\left|\frac{1}{2}z - \frac{3}{4}\right| = \frac{1}{4}$
12. $|2.2y - 1.1| = 5.5$
13. $|x + 2| = |3x - 4|$
14. $\left|\frac{1}{2}z + 1\right| = \left|\frac{3}{4}z - 1\right|$
15. $|y - 1| = |7 + y|$

Letting x represent the unknown number, translate the following into symbols and solve.

16. The absolute value of a number is 15.
17. If 2 is added to a number, the absolute value of the result is 5.
18. The absolute value of 3 more than one-half of a number is 8.
19. The absolute value of the sum of a number and 1 is equal to the absolute value of the number itself.

Solve. Give answers using interval notation.

20. $|x| < 9$
21. $|x| \geq 10$
22. $|x| < -1$
23. $|x| < 0$
24. $|x| \leq 0$
25. $|x + 1| < 3$
26. $|x - 1| < 3$
27. $|x + 4| > 3$
28. $|x + 4| \geq 3$
29. $|x - 5| \leq 7$
30. $|2x + 1| < 9$
31. $|2x + 1| \geq 9$
32. $|1 - 2x| > 7$
33. $|1 - 2x| \leq 7$
34. $\left|\frac{x + 1}{2}\right| < 3$
35. $\left|\frac{x + 1}{2}\right| > 3$
36. $\left|\frac{x - 3}{-4}\right| \leq 5$
37. $\left|\frac{x - 3}{-4}\right| \geq 5$
38. $|2x + 5| < -1$
39. $|2x + 5| \geq 0$

Letting x represent the unknown number, translate the following into symbols and solve.

40. The absolute value of a number is less than 20.

41 If 3 is added to twice a number, the absolute value of the result is at least 7.

Graph the following absolute value inequalities.

42. $|x - 1| < 2$

43. $|x - 1| \geq 2$

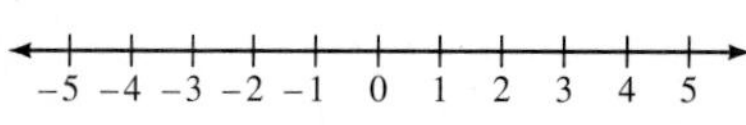

44. $|2x + 3| \geq 3$

45. $|2x + 3| < 3$

46. $|2x + 3| \leq 0$

47. $|2x + 3| < 0$

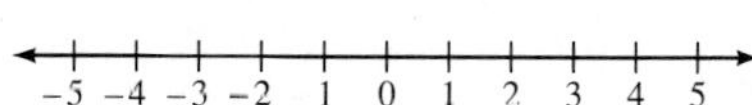

FOR REVIEW

Graph the following on a number line.

48. $(2x + 3) - (4x - 1) \leq 0$

49. $4(2x - 3) > 7x - 10$

50. $0 > 2 - x$ *or* $2 - x \geq 2$

51. $-1 \leq x \leq 3$

52. $x > \dfrac{1}{2}$ *or* $2x < 1$

53. $-3 < x + 1 < 3$

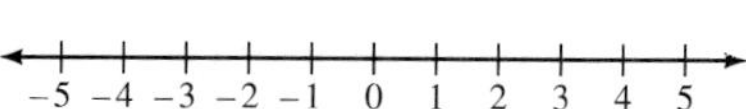

Solve.

54. Three times a number, decreased by 5, is at least -17 and no more than 1. What numbers satisfy this statement?

55. Five less a number is at least 4 or at most -1. What numbers satisfy this statement?

ANSWERS: 1. 7, −7 2. 0 3. no solution 4. 3, −5 5. no solution 6. −1 7. 7, −2 8. 7, −2 9. $\frac{5}{2}$ 10. $\frac{5}{2}$ 11. 2, 1 12. 3, −2 13. 3, $\frac{1}{2}$ 14. 8, 0 15. −3 16. $|x| = 15$; 15, −15 17. $|x + 2| = 5$; 3, −7 18. $|3 + \frac{1}{2}x| = 8$; 10, −22 19. $|x + 1| = |x|$; $-\frac{1}{2}$ 20. $-9 < x < 9$ ($x > -9$ *and* $x < 9$); $(-9, 9)$ 21. $x \leq -10$ or $x \geq 10$ (do not write $10 \leq x \leq -10$); $(-\infty, -10]$ or $[10, \infty)$ 22. no solution 23. no solution 24. 0 25. $-4 < x < 2$; $(-4, 2)$ 26. $-2 < x < 4$; $(-2, 4)$ 27. $x < -7$ or $x > -1$; $(-\infty, -7)$ or $(-1, \infty)$ 28. $x \leq -7$ or $x \geq -1$; $(-\infty, -7]$ or $[-1, \infty)$ 29. $-2 \leq x \leq 12$; $[-2, 12]$ 30. $-5 < x < 4$; $(-5, 4)$ 31. $x \leq -5$ or $x \geq 4$; $(-\infty, -5]$ or $[4, \infty)$ 32. $x > 4$ or $x < -3$; $(-\infty, -3)$ or $(4, \infty)$ 33. $-3 \leq x \leq 4$; $[-3, 4]$ 34. $-7 < x < 5$; $(-7, 5)$ 35. $x < -7$ or $x > 5$; $(-\infty, -7)$ or $(5, \infty)$ 36. $-17 \leq x \leq 23$; $[-17, 23]$ 37. $x \leq -17$ or $x \geq 23$; $(-\infty, -17]$ or $[23, \infty)$ 38. no solution 39. Every real number is a solution. 40. $|x| < 20$; $-20 < x < 20$ 41. $|2x + 3| \geq 7$; $x \leq -5$ or $x \geq 2$

42. 43. 44. 45. 46. $-\frac{3}{2}$ 47. 48. 49. 50. 51. 52. 53.

54. $-4 \leq x \leq 2$; $[-4, 2]$ 55. $x \leq 1$ *or* $x \geq 6$; $(-\infty, 1]$ or $[6, \infty)$

2.7 EXERCISES B

Solve.

1. $|y| = 13$
2. $|x| = 0$
3. $|z| = -2$
4. $|y + 5| = 3$
5. $|3x + 4| + 3 = 1$
6. $|2z + 5| = 0$
7. $|3y - 1| = 14$
8. $|4 - 3x| - 2 = 3$
9. $|7z - 3| = 0$
10. $|8 - 3y| = 0$
11. $\left|\frac{1}{5}x - \frac{3}{10}\right| = \frac{1}{2}$
12. $|1.1z - 4.4| = 7.7$
13. $|3y - 1| = |2y + 5|$
14. $\left|\frac{7}{8}x + \frac{1}{4}\right| = \left|\frac{1}{2}x - \frac{5}{8}\right|$
15. $|2z + 5| = |3 + 2z|$

Letting x represent the unknown number, translate the following into symbols and solve.

16. The absolute value of a number is 0.
17. If 5 is subtracted from a number, the absolute value of the result is 13.
18. The result is 17 when the absolute value of the difference between a number and 4 is obtained.
19. The absolute value of 1 less than a number is equal to the absolute value of the number itself.

Solve. Give answers using interval notation.

20. $|y| < 7$
21. $|x| \geq 3$
22. $|x| < -8$
23. $|z| \geq 0$
24. $|y| > 0$
25. $|x + 2| < 11$
26. $|z - 3| < 7$
27. $|y - 3| > 17$
28. $|x - 3| \leq 7$
29. $|3z + 2| < 11$
30. $|3y + 2| \geq 11$
31. $|2 - x| \leq 3$
32. $|2 - z| > 3$
33. $\left|\frac{y + 5}{3}\right| \leq 5$
34. $\left|\frac{x + 5}{3}\right| \geq 5$
35. $\left|\frac{x - 6}{-2}\right| < 3$
36. $\left|\frac{x - 6}{-2}\right| > 3$
37. $|3z + 2| < -1$
38. $|3z + 2| \leq 0$
39. $|3z + 2| \geq 0$

Letting x represent the unknown number, translate the following into symbols and solve.

40. The absolute value of a number is more than 20.
41. Add -5 to twice a number and take the absolute value of the result. This quantity is no more than 17.

Graph the following absolute value inequalities.

42. $|x - 2| < 3$

−5 −4 −3 −2 −1 0 1 2 3 4 5

43. $|x - 2| \geq 3$

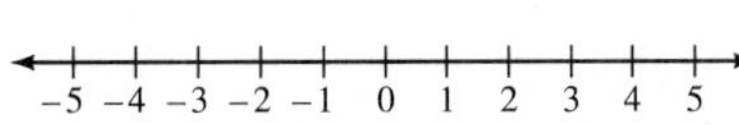

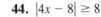

44. $|4x - 8| \geq 8$

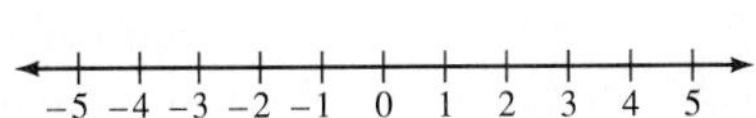

45. $|4x - 8| < 8$

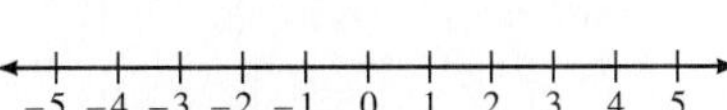

46. $|4x - 8| \geq 0$

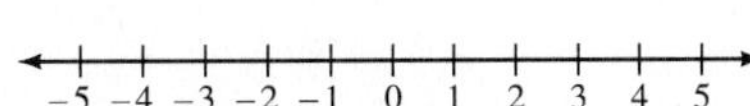

47. $|4x - 8| < 0$

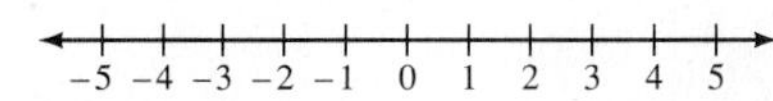

FOR REVIEW

Graph the following on a number line.

48. $(x + 5) - (2x - 1) \geq 0$

−3 −2 −1 0 1 2 3 4 5 6 7

49. $3(2x - 1) < 5x - 2$

−5 −4 −3 −2 −1 0 1 2 3 4 5

50. $0 > 3 - x$ *or* $3 - x \geq 1$

−5 −4 −3 −2 −1 0 1 2 3 4 5

51. $-2 \leq x \leq 1$

−5 −4 −3 −2 −1 0 1 2 3 4 5

52. $x > \frac{1}{3}$ *or* $3x < 1$

−5 −4 −3 −2 −1 0 1 2 3 4 5

53. $-2 < x + 1 < 2$

−5 −4 −3 −2 −1 0 1 2 3 4 5

Solve.

54. Two less a number is at most 17 and at least 3. What numbers satisfy this statement?

55. Three less a number is at most -2 or at least 1. What numbers satisfy this statement?

2.7 EXERCISES C

Solve.

1. $|x + 1| = x + 1$
[*Hint:* When is the absolute value of a number equal to the number itself?]

2. $|x + 1| = -x - 1$

3. $|x - 1| = 1 - x$

4. $|x - 1| = x - 1$

5. $|x + 1| > x$
[Answer: all real numbers]

6. $|x + 1| < x$
[Answer: no solution]

CHAPTER 2 REVIEW

KEY WORDS

2.1 An **equation** is a statement that two quantities are equal.

A **solution** of an equation is a number replacement for the variable that makes the equation true.

A **linear** (or **first-degree**) **equation in one variable** x is an equation of the form $ax + b = 0$.

Two equations are **equivalent equations** if they have exactly the same solution(s).

A **conditional equation** is true for some replacements of the variable and false for others.

A **contradiction** is an equation that has no solution.

An **identity** is an equation that has every real number as a solution.

2.2 **Consecutive integers** are integers that are next to each other in the regular counting order.

2.4 A **literal equation** is an equation that involves two or more variables.

2.5 A **linear inequality** can always be written in one of the forms:

$ax + b < 0$, $ax + b > 0$, $ax + b \leq 0$, or $ax + b \geq 0$.

$(a, b) = \{x | a < x < b\}$ is an **open interval.**

$[a, b] = \{x | a \leq x \leq b\}$ is a **closed interval.**

$(a, b] = \{x | a < x \leq b\}$ and $[a, b) = \{x | a \leq x < b\}$ are **half-open intervals.**

$(-\infty, a) = \{x | x < a\}$, $(-\infty, a] = \{x | x \leq a\}$, $(a, \infty) = \{x | x > a\}$, $[a, \infty) = \{x | x \geq a\}$, and $(-\infty, \infty) = \{x | x \text{ is any real number}\}$ are **infinite intervals.**

A **solution** to an inequality is any number that, when substituted for the variable, makes the inequality true.

Two inequalities are **equivalent inequalities** if they have exactly the same solutions.

2.6 A **compound statement** is an inequality formed by joining two inequalities with the words *and* or *or*.

When an *and* statement such as $a < x$ *and* $x < b$ is written as $a < x < b$, this second form is called a **chain of inequalities.**

KEY CONCEPTS

2.1 1. When using the additive property to solve an equation, be sure to add (or subtract) the same expression on both sides.

2. When using the multiplicative property to solve an equation, be sure to multiply both sides by the same nonzero number.

3. Use the additive property to isolate the variable before using the multiplicative property.

4. When solving an equation involving parentheses, first use the distributive law to remove all parentheses.

5. Multiply both sides of an equation by an appropriate factor to eliminate all fractions and decimals.

2.2 When solving an applied problem, read the problem several times and describe the variable completely. Also, make a sketch showing the information given, if appropriate.

2.3 1. To convert a decimal to a percent, multiply by 100 and attach the % symbol.

2. To convert a percent to a decimal, divide by 100 and remove the % symbol.

3. Sales tax = (tax rate) · (selling price) and total cost = (selling price) + (sales tax)

4. Discount = (discount rate) · (regular price) and sale price = (regular price) − (discount).

5. Commission = (commission rate) · (total sales)

6. The formula for simple interest, I, is $I = Prt$, where P is the principal, r is the yearly interest rate, and t is the time in years. The amount in the account, A, after t years is $A = P + I = P + Prt$.

7. Many distance-rate problems use the formula $d = rt$ [distance = rate · time].

2.4 It may help to solve a similar numerical equation before solving a given literal equation.

2.5 1. If the same number or quantity is added to both sides of an inequality, the result is an equivalent inequality.

2. When both sides of an inequality are multiplied (or divided) by a negative number, be sure to *reverse* the symbol of inequality to obtain an equivalent inequality.

2.6 Inequalities connected by the word *and* can generally be written as a single chain of inequalities; those connected by the word *or* cannot be written as a chain.

2.7 1. If $a > 0$ and $|expression| = a$, then solve

$$expression = a \quad \text{and}$$
$$expression = -a.$$

2. If $a > 0$ and $|expression| > a$, then solve

$$expression < -a \quad \text{or}$$
$$expression > a.$$

3. If $a > 0$ and $|expression| < a$, then solve

$$-a < expression < a.$$

4. An inequality of the form $|expression| > b$, where $b < 0$, has every number as a solution, while $|expression| < b$, where $b < 0$, has no solutions.

5. To solve $|expression| \le 0$, solve the equation

$$expression = 0.$$

6. $|expression| \ge 0$ has every real number as a solution, $|expression| < 0$ has no solution, and $|expression| > 0$ has every real number as a solution except those for which $expression = 0$.

REVIEW EXERCISES

Part I

2.1 *Solve.*

1. $4 - 3x = -5$

2. $\frac{1}{3}y = -2$

3. $3 - 2(x - 1) = x - 10$

4. $4(2y - 4) + 6 = 6y - 2(y - 1)$

5. $(2z + 1) - (3z - 7) = 0$

6. $\frac{x}{2} - \frac{x}{7} = 1$

7. $y(y + 2) = y^2 - 6$

8. $2[x - (1 - 3x)] = 3(x + 1)$

2.2 *Select a variable to represent each quantity and translate the phrase into symbols.*

9. the time 4 hours ago

10. twice the surface area

Solve.

11. If three times a number is increased by 2, the result is the same as 20 less than five times the number. Find the number.

12. The perimeter of a rectangle is 20.8 cm. If the length is 3.5 cm more than twice the width, find the dimensions.

2.3 **13.** 60.5 is 11% of what?

14. 7% of 320 is what?

15. After a 7% raise, Andy Woodring's new salary is $14,445. What was his former salary?

16. What amount of money invested at 15% simple interest will increase to $1140 at the end of six years?

17. A fuel tank contained 620 gal of fuel. An additional quantity of fuel was added, making a total of 775 gal. What was the percent increase?

18. A man leaves Flagstaff at 8:00 A.M., traveling by car to Lake Powell, a distance of 135 miles, at a rate of 45 mph. He then drives a boat to Rainbow Bridge, a distance of 50 miles, at a rate of 20 mph. At what time does he reach the bridge?

2.4 **19.** Solve $M = N + Nat$ for N.

20. Solve $b(x + 2) = x + a$ for x.

2.5 *Solve and graph on a number line.*

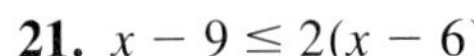

21. $x - 9 \leq 2(x - 6)$

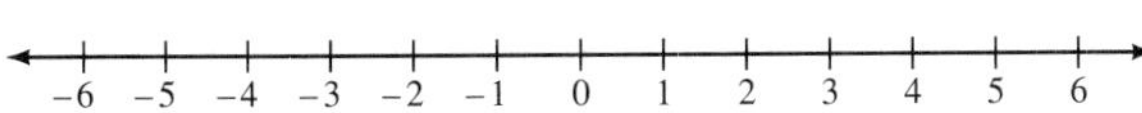

22. $-3y + 1 + 2y < 5(y - 3)$

2.6 *Solve and graph on a number line. Give the answer in interval notation.*

23. $y + 2 \leq -1$ or $y + 2 \geq 1$

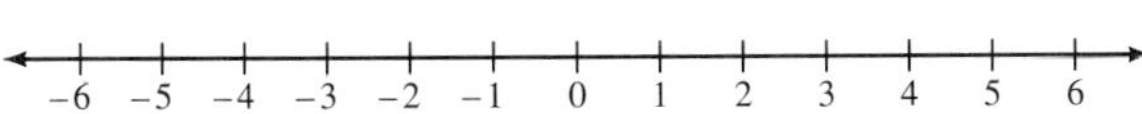

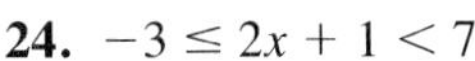

24. $-3 \leq 2x + 1 < 7$

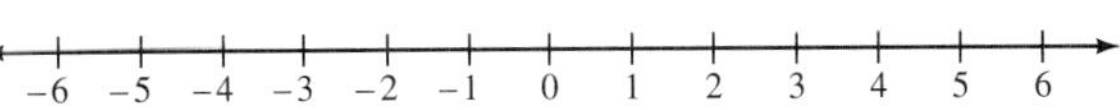

25. A student must have an average percent between 80% and 90%, inclusive, in order to receive a grade of B. Ted has 78%, 64%, and 92% on the first three tests. What range of scores on the fourth test will give him a B?

2.7 *Solve.*

26. $|1 - 7x| = 6$

27. $|1 - 7x| = 0$

28. $|x - 3| = |2 + x|$

Solve. Give answer in interval notation.

29. $|1 - 3x| < 4$

30. $|3y + 4| \geq 28$

31. $|2z + 11| < -3$

Graph each inequality on a number line.

32. $|x - 2| \leq 2$

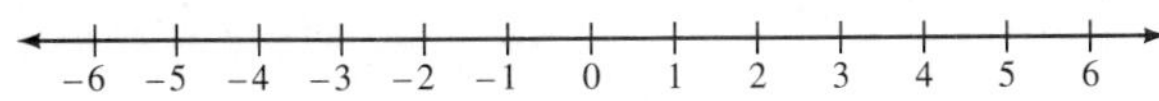

33. $|2 - y| > 1$

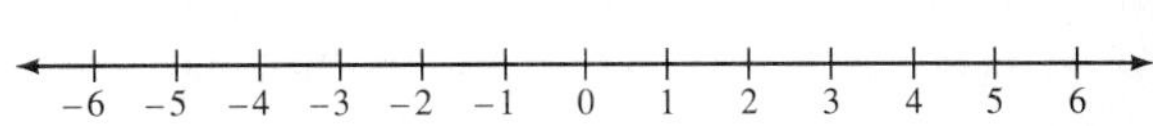

Part II

Solve. When appropriate, give answer in interval notation.

34. Two trains leave the same town, one traveling east and the other west. If one train is moving 10 mph faster than the other, and if after 2 hr they are 260 mi apart, how fast is each traveling?

35. The population of Appleville, Washington was 740 in 1980. This was a 20% decrease in the population of 1970. What was the population in 1970?

36. $|2x - 3| \leq 7$

37. $|1 - 7x| = -3$

38. $\frac{z + 1}{3} - \frac{z - 1}{2} = 1$

39. $3(x - 4) - (x + 7) = 5$

40. A basketball player made 12 shots in 20 attempts. What was her shooting percentage?

41. After receiving a 30% discount on the selling price, Dan Busch paid $35.28, including 5% sales tax, for a running suit. What was the price of the suit before the discount?

42. Peter is 9 years older than Rhoda. If the sum of their ages is 87, how old is each?

43. A steel rod is 17 m long. It is to be cut into two pieces in such a way that one piece is 7 m longer than the other. How long is each piece?

44. $|2x - 5| \leq 15$

45. $|1 - x| > 3$

46. Solve $a = \frac{1}{2}uv + t$ for u.

47. Find the measures of the angles of a triangle if the second is three times the first and the third is 30° more than twice the first.

Solve and graph on a number line. Give answer in interval notation.

48. $x - 1 < -5$ or $x - 1 > 2$

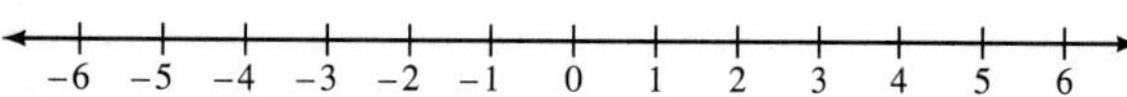

49. $-1 \leq 1 - 2x \leq 3$

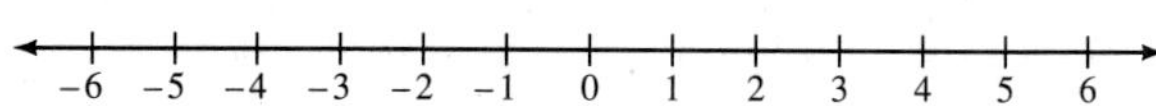

50. $|3x + 2| \leq 5$

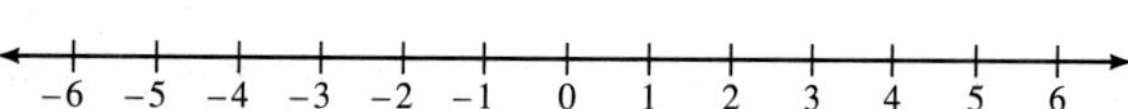

51. $|4 - x| > 1$

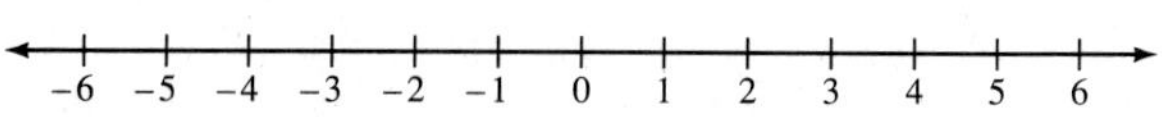

52. Solve. $B = 40a + w$ for a.

53. Solve. $|2y - 3| = |6 - y|$

54. Solve. $3z - (1 - 2z) = 4 - (z - 1)$

55. Solve. $|4x - 5| = 0$

Select a variable to represent each quantity and translate the phrase into symbols.

56. His wages plus 10% of his wages

57. \$1000 less the cost

58. The power in watts W on an electrical circuit is related to the current I in amperes by $W = 110I$. If the power demand on an appliance is between 220 watts and 550 watts, what is the range of current in the circuit?

ANSWERS: 1. 3 2. −6 3. 5 4. 3 5. 8 6. $\frac{14}{5}$ 7. −3 8. 1 9. $t - 4$ 10. $2S$ 11. 11 12. 8.1 cm by 2.3 cm 13. 550 14. 22.4 15. \$13,500 16. \$600 17. 25% 18. 1:30 P.M. 19. $N = \frac{M}{1 + at}$ 20. $x = \frac{a - 2b}{b - 1}$
21. $x \geq 3$

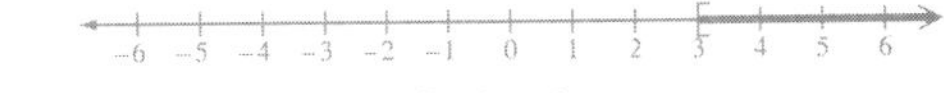

22. $y > \frac{8}{3}$

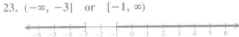

23. $(-\infty, -3]$ or $[-1, \infty)$

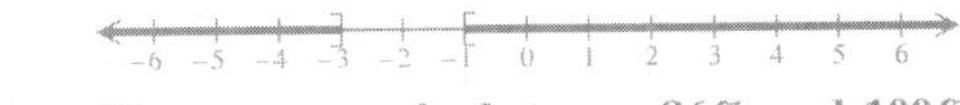

24. $[-2, 3)$

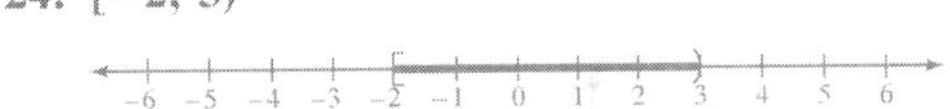

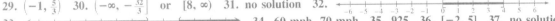

25. His score must be between 86% and 100% (the maximum score possible), inclusive. 26. $-\frac{5}{7}, 1$ 27. $\frac{1}{7}$ 28. $\frac{1}{2}$ 29. $\left(-1, \frac{5}{3}\right)$ 30. $\left(-\infty, -\frac{32}{3}\right]$ or $[8, \infty)$ 31. no solution 32.

33. 34. 60 mph, 70 mph 35. 925 36. $[-2, 5]$ 37. no solution 38. −1 39. 12 40. 60% 41. \$48.00 42. Peter is 48, Rhoda is 39. 43. 5 m, 12 m 44. $[-5, 10]$ 45. $(-\infty, -2)$ or $(4, \infty)$ 46. $u = \frac{2a - 2t}{v}$ 47. 25°, 75°, 80° 48. $(-\infty, -4)$ or $(3, \infty)$

49. $[-1, 1]$ 50. $\left[-\frac{7}{3}, 1\right]$

51. $(-\infty, 3)$ or $(5, \infty)$ 52. $a = \frac{B - W}{40}$ 53. 3, −3 54. 1 55. $\frac{5}{4}$ 56. $w + 0.10w$ 57. $1000 - c$ 58. between 2 amperes and 5 amperes

CHAPTER 2 TEST

Solve.

1. $6 - 5x = 31$ — 1. ______

2. $(4z + 1) - (3z - 5) = 0$ — 2. ______

3. $2(3y - 1) - 4 = 2y - (6 - y)$ — 3. ______

4. $a - (2a - 1) = \frac{2}{3}(5 - a)$ — 4. ______

5. $\frac{x}{6} - \frac{x}{5} = 2$ — 5. ______

6. A wire that is 33 inches long is cut into two pieces in such a way that one piece is 5 inches shorter than the other. How long is each piece? — 6. ______

7. Find the measures of the angles of a triangle if the second angle is three times the first, and the third is 15° more than seven times the first. — 7. ______

8. Two cars leave Denver at 8:00 A.M., one traveling north and the other traveling south. If one is going 55 mph and the other 60 mph, at what time will they be 575 mi apart? — 8. ______

9. 42 is 12% of what? — 9. ______

$42 = (.12)x.$

$\frac{42}{.12} = 350$

10. After receiving an 8% raise, Mary has a new salary of \$16,848. What was her former salary? — 10. ______

11. The population of South Keys, Florida, was 828 in 1980 and 720 in 1970. What was the percent increase in population over this period of time? 11. ______________

12. Solve $W = 30a + c$ for a. 12. ______________

13. Solve $u(v - w) + 2 = vw$ for w. 13. ______________

14. Solve and graph $x - 8 \geq 2(x - 5)$. 14.

−5 −4 −3 −2 −1 0 1 2 3 4 5

15. Solve and graph $y - 1 < -3$ or $y - 1 > 3$. 15.

−5 −4 −3 −2 −1 0 1 2 3 4 5

16. Solve and graph $|z - 3| < 2$. 16.

−5 −4 −3 −2 −1 0 1 2 3 4 5

Solve.

17. $|4 - 3x| = 7$ 17. ______________

18. $|5 - 2y| \leq 9$ (Give answer in interval notation.) 18. ______________

19. $|x - 5| = |2x + 1|$ 19. ______________

20. One-third of a number x is greater than -2. 20. ______________

CHAPTER 3

Graphs, Relations, and Functions

3.1 GRAPHING EQUATIONS IN TWO VARIABLES

STUDENT GUIDEPOSTS

1. The Coordinate System
2. Plotting Points
3. Solutions to Equations in Two Variables
4. Graphing Equations

1 THE COORDINATE SYSTEM

Solutions to inequalities in one variable were graphed on number lines in Chapter 2. We will now be graphing equations and inequalities in two variables. When two number lines are placed together as in Figure 3.1, so that the two origins coincide and the lines are perpendicular, the result is a **Cartesian** or **rectangular coordinate system** (named after French mathematician René Descartes) or a **coordinate plane.**

The horizontal number line is the ***x*-axis,** the vertical number line is the ***y*-axis,** and the point of intersection of the axes is the **origin.**

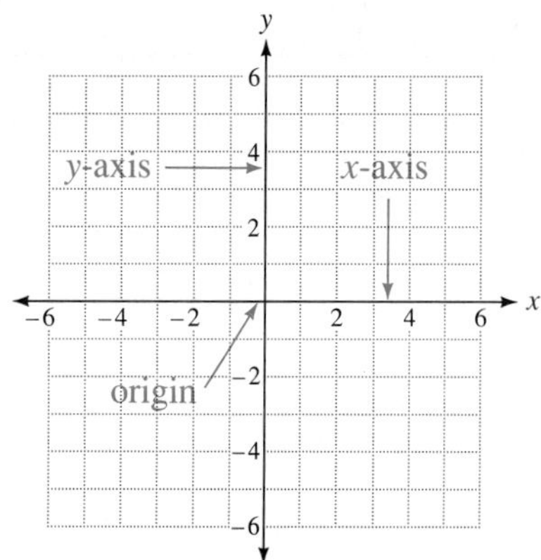

Figure 3.1 Rectangular or Cartesian Coordinate System

2 PLOTTING POINTS

Recall that there is one and only one point on a number line associated with each real number. A similar situation exists for points in a plane and **ordered pairs** of numbers. For example, the ordered pair (2, 3) can be identified with a point in a coordinate plane as follows:

The first number, 2, called the **x-coordinate** or **abscissa** of the point, is associated with a value on the horizontal or x-axis.

The second number, 3, called the **y-coordinate** or **ordinate** of the point, is associated with a value on the vertical or y-axis.

The pair (2, 3) is associated with the point where the vertical line through 2 on the x-axis and the horizontal line through 3 on the y-axis intersect. See Figure 3.2.

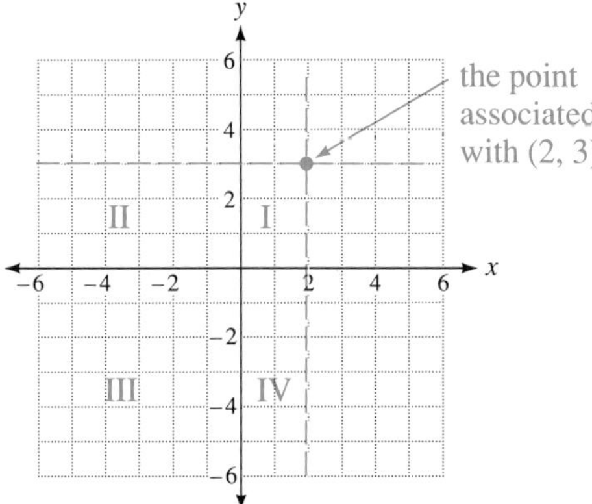

Figure 3.2 Plotting the Point (2, 3)

The axes in a coordinate system separate the plane into four sections called **quadrants.** The first, second, third, and fourth quadrants are identified by the Roman numerals I, II, III, and IV, respectively, in the coordinate plane in Figure 3.2. The signs of the x-coordinate (first) and y-coordinate (second) in the various quadrants are as follows.

I: $(+, +)$ **II:** $(-, +)$ **III:** $(-, -)$ **IV:** $(+, -)$

We often use (x, y) to refer to an ordered pair of numbers. Thus, the point in the plane associated with the pair (x, y) has x as its x-coordinate and y as its y-coordinate. When we identify the point P in a plane corresponding to the ordered pair (x, y), sometimes called "the point (x, y)," we are plotting the point $P(x, y)$.

EXAMPLE 1 Plotting Points

The points associated with the ordered pairs (3, 2), (−1, 3), (−3, −2), and (2, −2) are plotted in the coordinate plane in Figure 3.3.

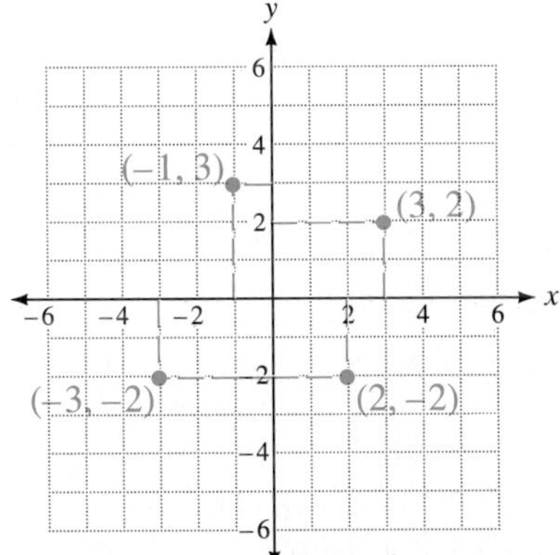

Figure 3.3 Plotting Points

Practice Exercise 1

Give the ordered pair associated with each point and the quadrant in which each is located.

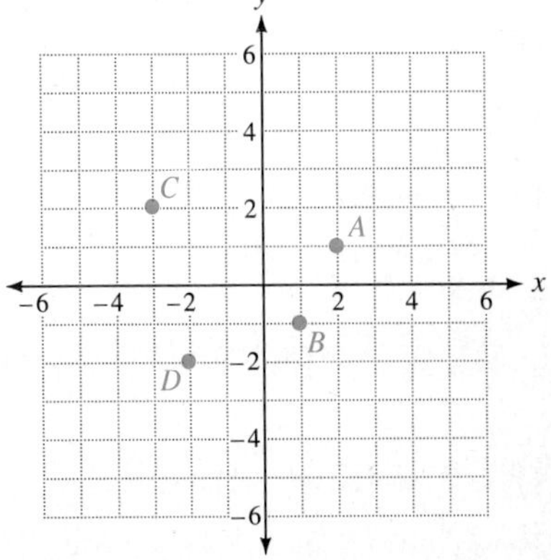

Answers: $A(2, 1)$, I; $B(1, -1)$, IV; $C(-3, 2)$, II; $D(-2, -2)$, III

3 SOLUTIONS TO EQUATIONS IN TWO VARIABLES

Graphing equations with two variables requires a rectangular coordinate system since **solutions** to these equations are ordered pairs of numbers that when substituted for the variables result in a true equation.

For example, the equation

$$y = 3x - 2$$

has infinitely many solutions. One solution is (2, 4), since if x is replaced with 2 and y with 4, the resulting equation is true as shown below.

$$y = 3x - 2$$
$$4 = 3(2) - 2 \quad x = 2 \text{ and } y = 4$$
$$4 = 6 - 2$$
$$4 = 4$$

You can verify that $(0, -2)$ and $(-1, -5)$ are also solutions, while $(-2, 4)$ is not.

EXAMPLE 2 Finding Solutions to an Equation

Given the equation $2x + 3y = 12$, complete the following ordered pairs so that they are solutions to the equation.

$$(0, \ \ ), \quad (\ \ , 0), \quad (3, \ \ ), \quad (\ \ , -2)$$

To complete the ordered pair (0,), substitute 0 for x in $2x + 3y = 12$ and solve for y.

$$2(0) + 3y = 12$$
$$3y = 12$$
$$y = 4$$

The completed ordered pair is (0, 4).

To complete the ordered pair (, 0), substitute 0 for y in $2x + 3y = 12$ and solve for x.

$$2x + 3(0) = 12$$
$$2x = 12$$
$$x = 6$$

The completed ordered pair is (6, 0).

To complete the pair (3,), substitute 3 for x and solve for y.

$$2(3) + 3y = 12$$
$$6 + 3y = 12$$
$$3y = 6$$
$$y = 2$$

The completed pair is (3, 2).

Similarly, to complete (, -2), substitute -2 for y and solve for x, obtaining $(9, -2)$.

Practice Exercise 2

Given $3x - 4y = 24$, complete the following ordered pairs so that they are solutions to the equation.

$$(0, \ \ ), \quad (\ \ , 0), \quad (-4, \ \ ), \quad (\ \ , 1)$$

Answers: $(0, -6)$, $(8, 0)$, $(-4, -9)$, $\left(\frac{28}{3}, 1\right)$

4 GRAPHING EQUATIONS

The **graph** of an equation in two variables is the set of points in a rectangular coordinate system that corresponds to solutions of the equation. Since there are infinitely many solutions, we cannot find and plot each possible pair. Usually we plot enough points to see a pattern and then connect these points with a line or curve, the graph of the equation.

Consider again the equation $y = 3x - 2$. One way to display some of its solutions is to make a **table of values.** We choose several values for x, substitute these values into the equation, and compute the corresponding values for y. We begin by making a table such as the one below. Each y-value we find is placed beside the x-value used to calculate it. (Calculations are usually done mentally or as scratch work.)

Substitution	*Result in* $y = 3x - 2$
$x = 0$	$y = 3(0) - 2 = -2$
$x = 1$	$y = 3(1) - 2 = 1$
$x = -1$	$y = 3(-1) - 2 = -5$
$x = 2$	$y = 3(2) - 2 = 4$
$x = -2$	$y = 3(-2) - 2 = -8$
$x = 3$	$y = 3(3) - 2 = 7$
$x = -3$	$y = 3(-3) - 2 = -11$

x	y	*Solutions*
0	−2	(0, −2)
1	1	(1, 1)
−1	−5	(−1, −5)
2	4	(2, 4)
−2	−8	(−2, −8)
3	7	(3, 7)
−3	−11	(−3, −11)

Now we plot the points that correspond to these ordered-pair solutions in a rectangular coordinate system, as in Figure 3.4. It appears that all seven points lie on a straight line. Thus, it is reasonable to assume that the graph of this equation is the straight line passing through these seven points in Figure 3.5.

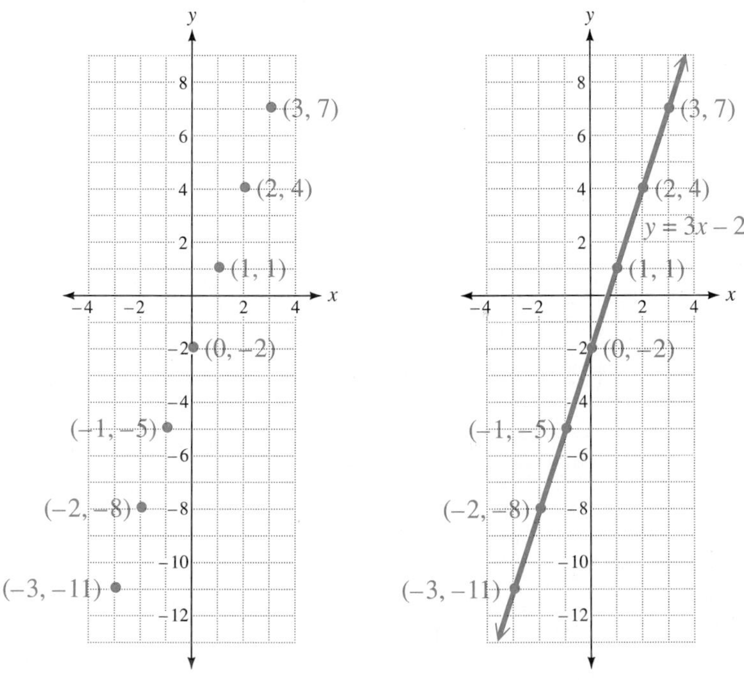

Figure 3.4
Plotting Solutions to $y = 3x - 2$

Figure 3.5
Connecting Points in Straight Line

We are now ready to graph other equations in two variables. In Example 3 on the following page we select values of x which give points that are easily plotted.

EXAMPLE 3 GRAPHING AN EQUATION

Graph $y = x^2$ in a rectangular coordinate system.

We begin by making a table of values.

Substitution	*Result in* $y = x^2$
$x = 0$	$y = (0)^2 = 0$
$x = 1$	$y = (1)^2 = 1$
$x = -1$	$y = (-1)^2 = 1$
$x = 2$	$y = (2)^2 = 4$
$x = -2$	$y = (-2)^2 = 4$
$x = 3$	$y = (3)^2 = 9$
$x = -3$	$y = (-3)^2 = 9$

x	y	*Solutions*
0	0	(0, 0)
1	1	(1, 1)
−1	1	(−1, 1)
2	4	(2, 4)
−2	4	(−2, 4)
3	9	(3, 9)
−3	9	(−3, 9)

Plot the solution points and connect them with a smooth curve to obtain the graph shown in Figure 3.6.

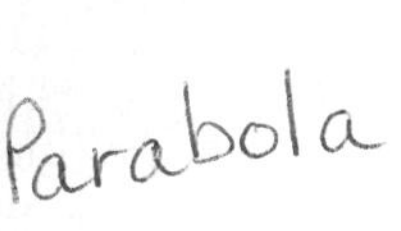

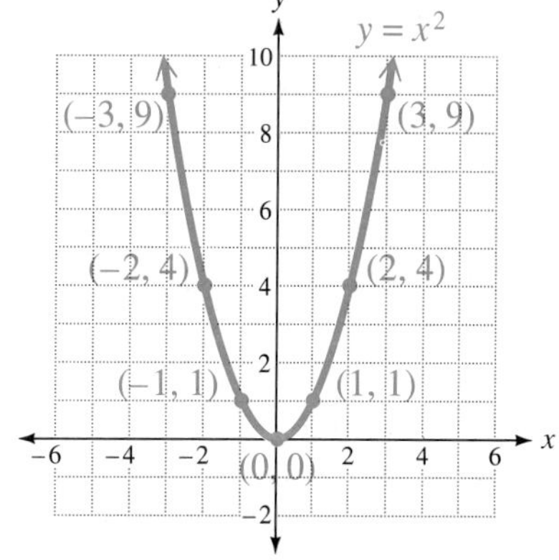

Figure 3.6 Graph Different from a Straight Line

PRACTICE EXERCISE 3

Graph $y = x^2 - 5$ in a rectangular coordinate system.

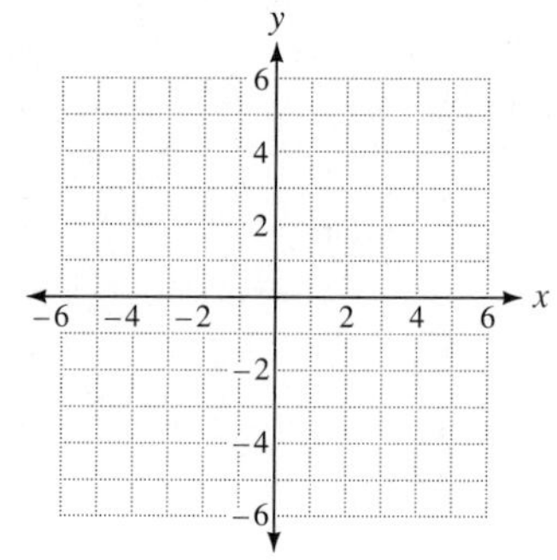

Answer: Plot the points (0, −5), (1, −4), (−1, −4), (2, −1), and (−2, −1). Notice that the graph is similar to $y = x^2$ but moved down 5 units.

Before considering another example of graphing an equation, we present a summary of the method to use.

To Graph an Equation in Two Variables x and y

1. Make a table of values. These values represent the ordered-pair solutions of the equation.
2. Plot the points that correspond to these solutions in a rectangular coordinate system.
3. Connect the points with a line or curve.

EXAMPLE 4 Graphing an Equation

Graph $y = |x|$ in a rectangular coordinate system.

Make a table of values.

Substitution	*Result in* $y = \lvert x \rvert$
$x = 0$	$y = \lvert 0 \rvert = 0$
$x = 1$	$y = \lvert 1 \rvert = 1$
$x = -1$	$y = \lvert -1 \rvert = 1$
$x = 2$	$y = \lvert 2 \rvert = 2$
$x = -2$	$y = \lvert -2 \rvert = 2$
$x = 3$	$y = \lvert 3 \rvert = 3$
$x = -3$	$y = \lvert -3 \rvert = 3$

x	y	*Solutions*
0	0	(0, 0)
1	1	(1, 1)
−1	1	(−1, 1)
2	2	(2, 2)
−2	2	(−2, 2)
3	3	(3, 3)
−3	3	(−3, 3)

Plot the points associated with the ordered-pair solutions in the table to obtain the graph shown in Figure 3.7.

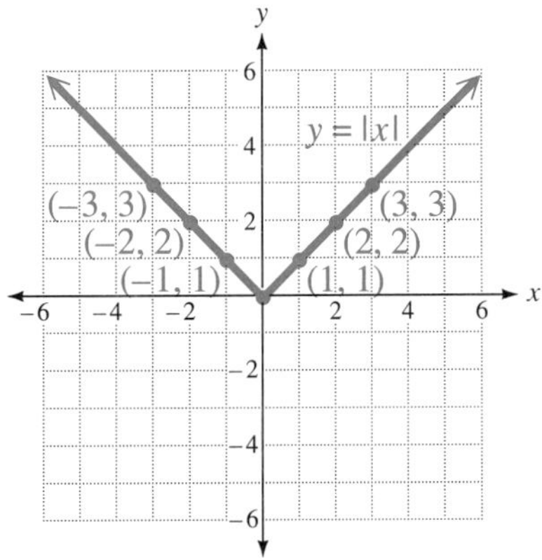

Figure 3.7
Graph with Straight-Line Segments

Practice Exercise 4

Graph $y = |x| + 1$ in a rectangular coordinate system.

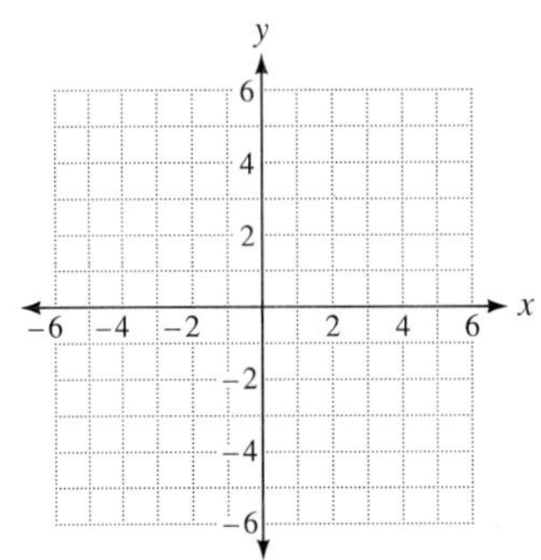

Answer: Plot the points (0, 1), (2, 3), (−2, 3) and notice that the graph is similar to $y = |x|$ but moved up 1 unit.

3.1 EXERCISES A

1. Plot the points associated with the given pairs $A(1, 4)$, $B(4, -2)$, $C(-3, 2)$, $D(-3, 0)$, $E(3, 0)$, $F(0, 0)$, $G(-3, -3)$, and $H(0, -2)$.

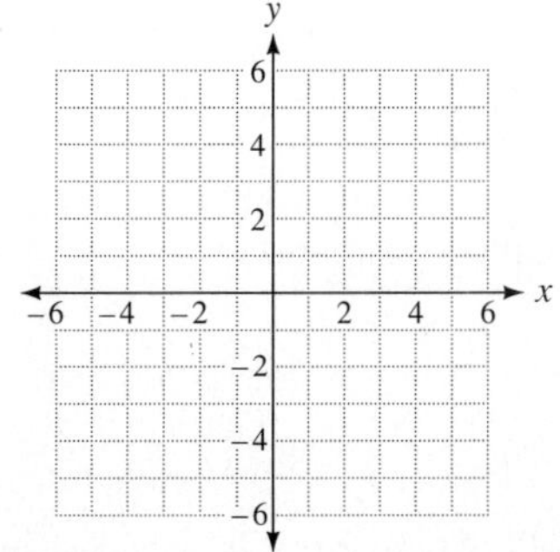

2. Give the coordinates of the points A, B, C, D, E, F, G, and H.

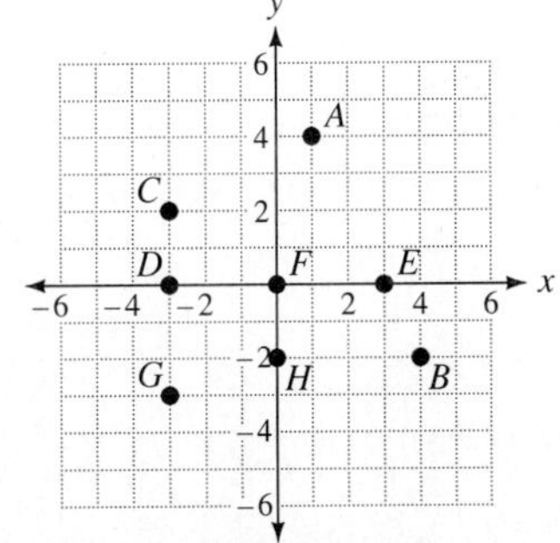

3. Plot the points associated with the pairs $J\left(\frac{1}{2}, 2\right)$, $K\left(-\frac{5}{2}, 1\right)$, $L\left(-2, -\frac{7}{4}\right)$, and $M\left(3, -\frac{3}{4}\right)$.

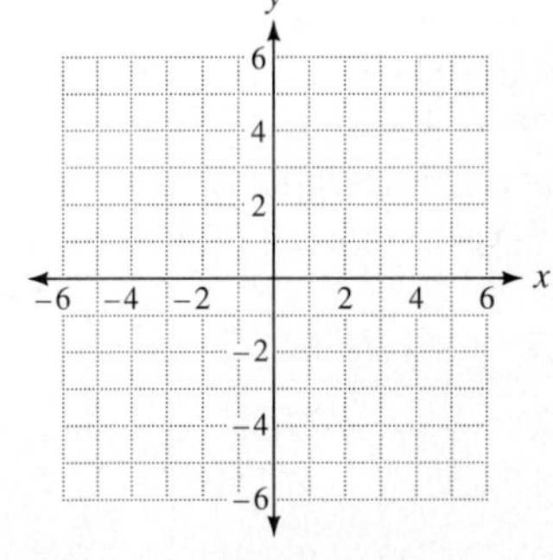

4. In which quadrants are the points J, K, L, and M of Exercise 3 located?

Give the quadrant that contains each point.

5. $(-2, -3)$

6. $(1, 7)$

7. $(-2, 5)$

8. $\left(\sqrt{2}, -\frac{1}{2}\right)$

9. Give the coordinates of each point. What do these points have in common? What do their coordinates have in common?

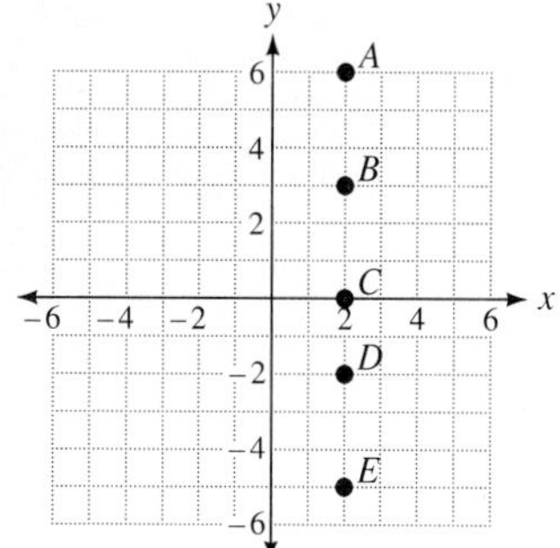

10. Give the coordinates of each point. What do these points have in common? What do their coordinates have in common?

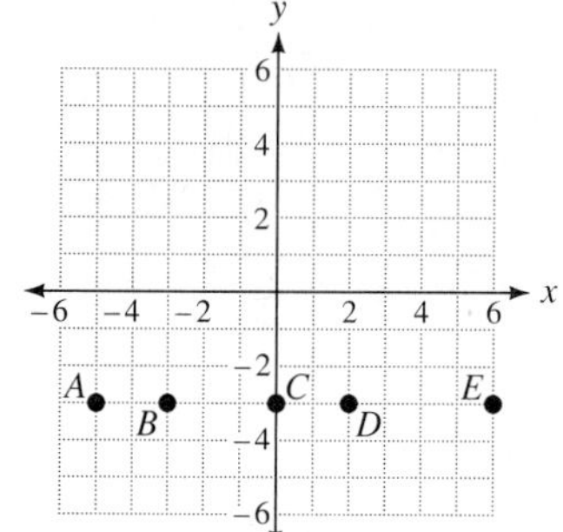

11. Find the coordinates of each point described below.

(a) Point A

(b) Point B, that is 2 units down from A

(c) Point C that is 5 units to the left of A

(d) Point D that is 1 unit up and 2 units to the right of A

(e) Point E that is 4 units down and 1 unit to the left of A

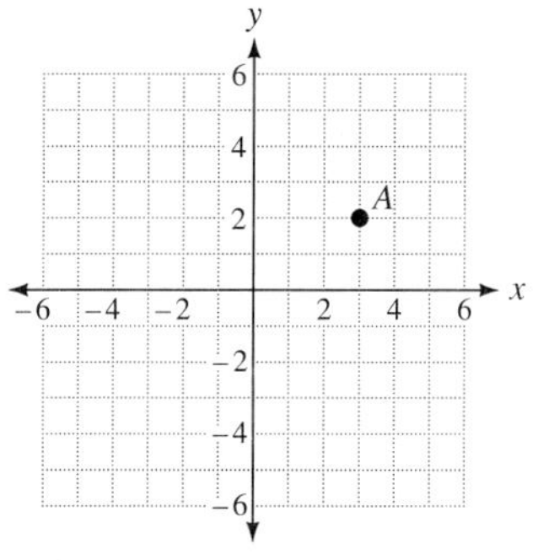

Write true *or* false *for Exercises* 12–21. *If the statement is false, explain why.*

12. A configuration in which points corresponding to ordered pairs of real numbers are plotted in a plane is a rectangular or Cartesian coordinate system.

13. The ordered pair $(3, -4)$ determines a point in quadrant IV with x-coordinate 3 and y-coordinate -4.

14. The ordered pair $(-8, -5)$ determines a point in quadrant III with abscissa -8 and ordinate -5.

15. In quadrant I, the x-coordinate of a point is always negative.

16. In quadrant II, the y-coordinate of a point is always positive.

17. In quadrant III, the x-coordinate of a point is always negative.

18. In quadrant IV, the y-coordinate of a point is always positive.

19. The horizontal axis in a coordinate system is also called the y-axis.

20. The perpendicular lines in a rectangular coordinate system are called the axes.

21. A rectangular coordinate system separates the plane into four regions called quadrants.

Answer each question below.

22. In which quadrants do points have coordinates with opposite signs?

23. The x-coordinate of a point is positive in which quadrants?

24. The y-coordinate of a point is negative in which quadrants?

25. What is the x-coordinate of any point on the y-axis?

26. What is the distance between $A(5, 2)$ and $B(8, 2)$?

27. What is the distance between $C(4, 2)$ and $D(4, -3)$?

In each of the following, find the missing coordinate so that the ordered pair is a solution to the equation.

28. $x + y + 2 = 0$
(a) $(0, \quad)$ **(b)** $(\quad, 0)$
(c) $(1, \quad)$ **(d)** $(\quad, -2)$

29. $x - y = 0$
(a) $(0, \quad)$ **(b)** $(\quad, 0)$
(c) $(5, \quad)$ **(d)** $(\quad, -3)$

Graph the following equations.

30. $x + y + 2 = 0$

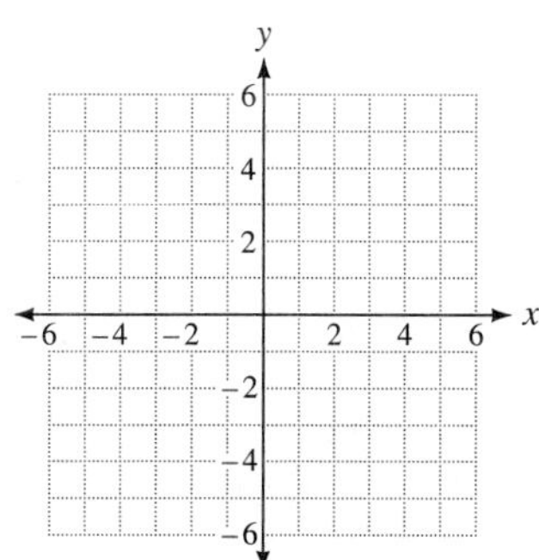

31. $3x + y = 0$

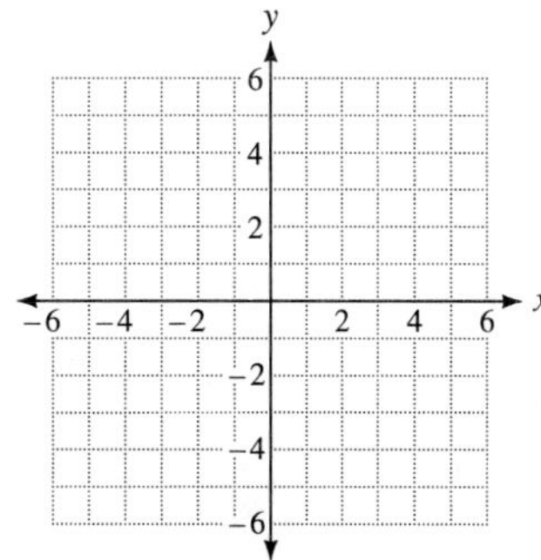

32. $x - y = 2$

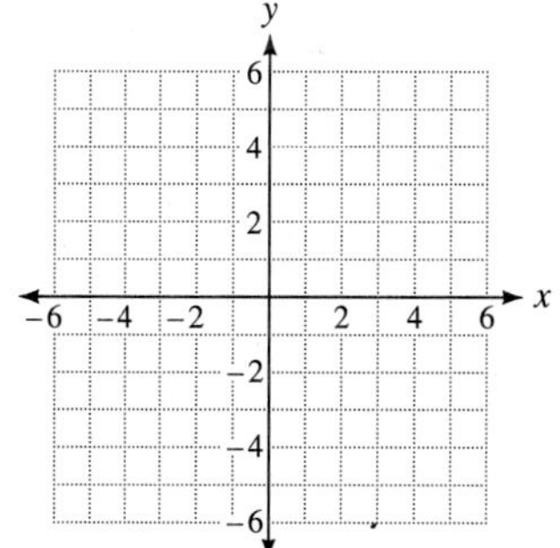

33. $x - y = 0$

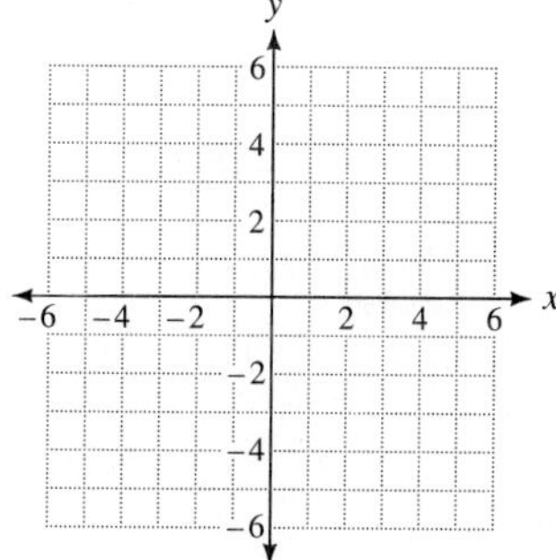

34. $y = x^2 - 1$

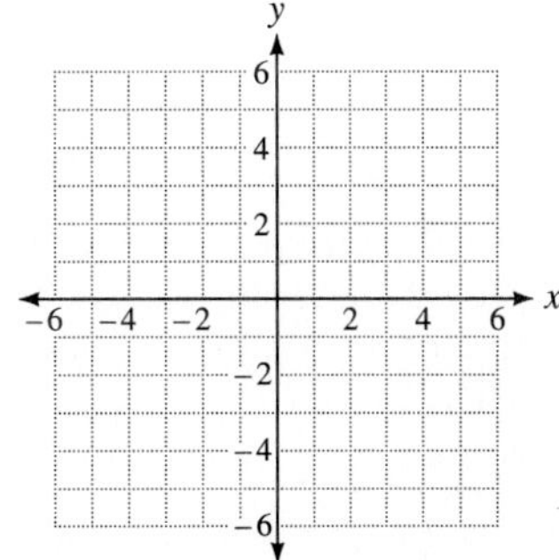

35. $y = x^2 + 1$

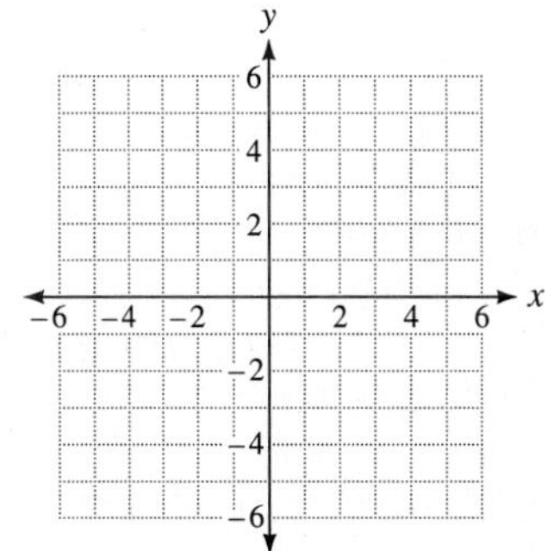

36. $y = |x| + 2$

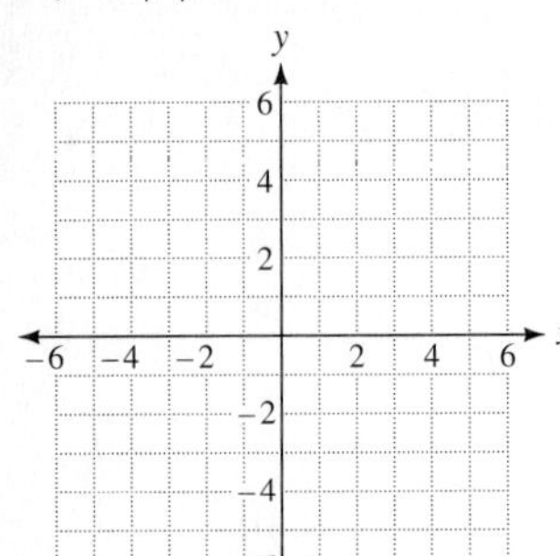

37. $y = |x| - 3$

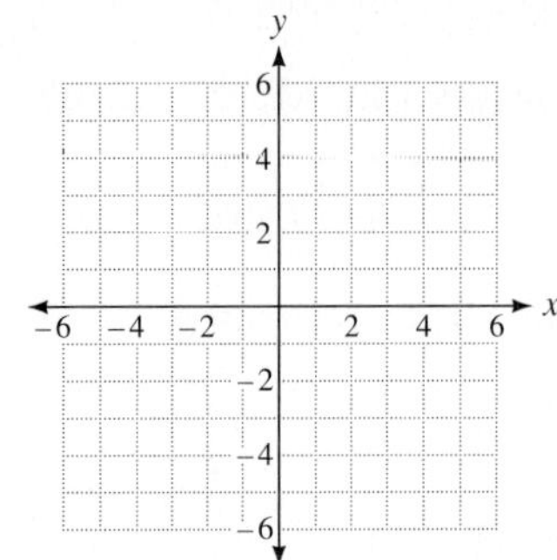

38. $y = |x - 1|$

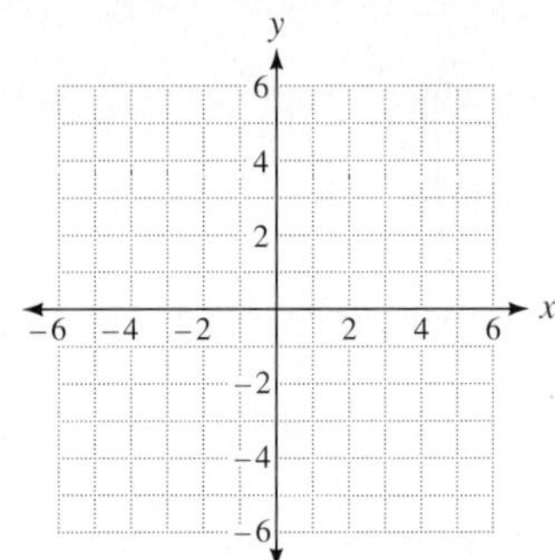

39 Mr. Goldstein has estimated that the cost (in dollars), y, of producing a number, x, of machine parts in his plant is given by the equation $y = 200x + 30$. Complete the following ordered pairs: (1,   ), (5,   ), (10,   ). What is the cost of producing **(a)** 1 part? **(b)** 5 parts? **(c)** 10 parts?

FOR REVIEW

Solve the absolute value equation or inequality.

40. $|2x - 5| = 7$

41. $|3x + 1| < 4$

42. $|2x - 6| \geq 2$

ANSWERS: 1. Answers are given in the coordinate system in Exercise 2. 2. See Exercise 1. 3.

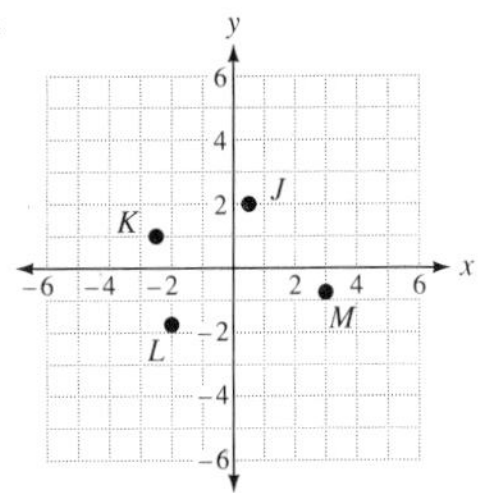

4. J: I; K: II; L: III; M: IV 5. III 6. I 7. II 8. IV 9. $A(2, 6)$, $B(2, 3)$, $C(2, 0)$, $D(2, -2)$, $E(2, -5)$; the points all lie on the same vertical line; all points have x-coordinate 2 10. $A(-5, -3)$, $B(-3, -3)$, $C(0, -3)$, $D(2, -3)$, $E(6, -3)$; the points all lie on the same horizontal line; all points have y-coordinate -3 11. (a) (3, 2) (b) (3, 0) (c) $(-2, 2)$ (d) (5, 3) (e) $(2, -2)$ 12. true 13. true 14. true 15. false; positive 16. true 17. true 18. false; negative 19. false; x-axis 20. true 21. true 22. II and IV 23. I and IV 24. III and IV 25. 0 26. 3 units 27. 5 units 28. (a) -2 (b) -2 (c) -3 (d) 0 29. (a) 0 (b) 0 (c) 5 (d) -3

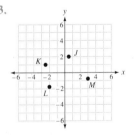

30.

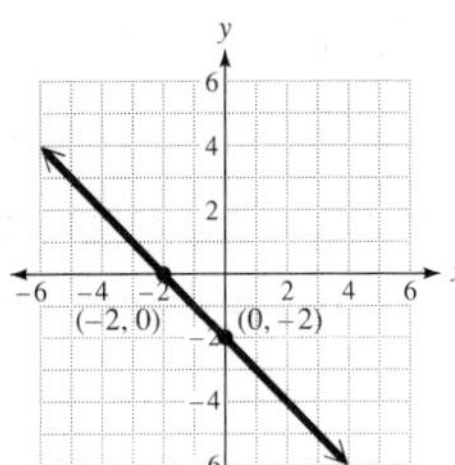

31.

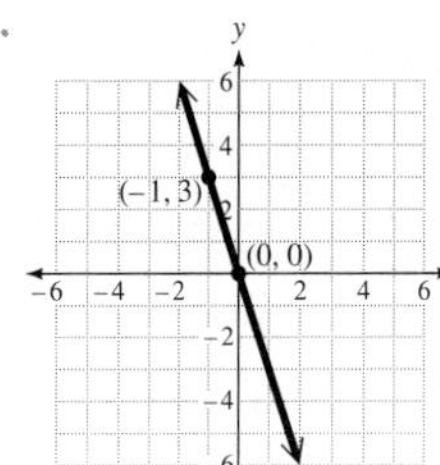

32.

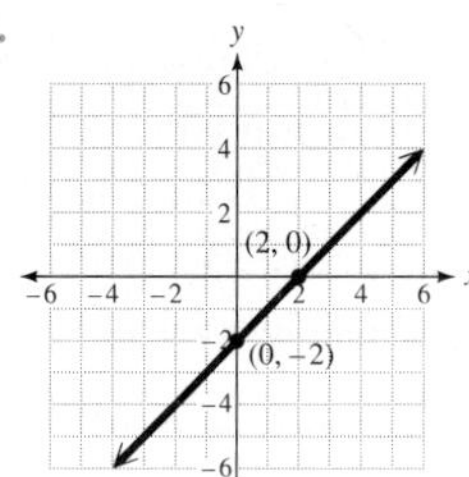

33.

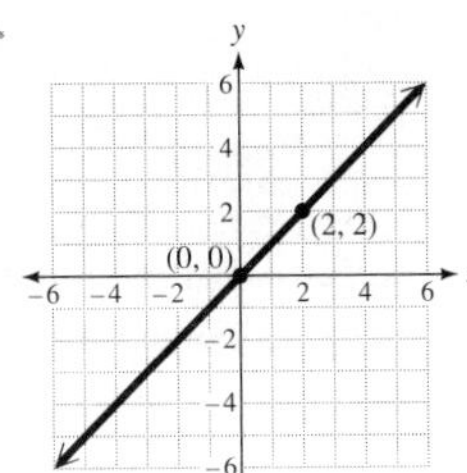

34.

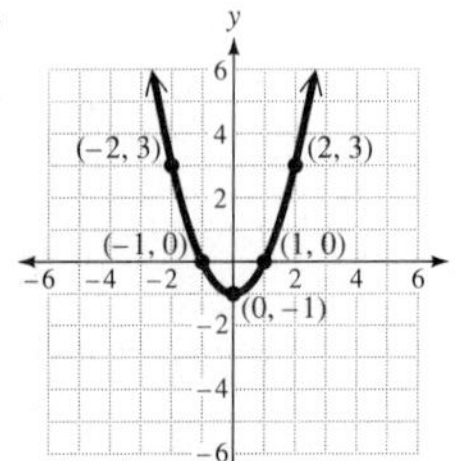

35.

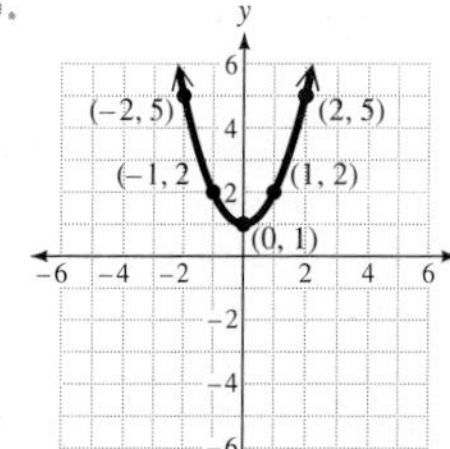

36.

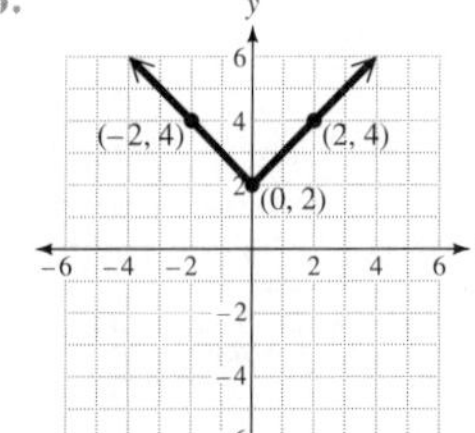

37.

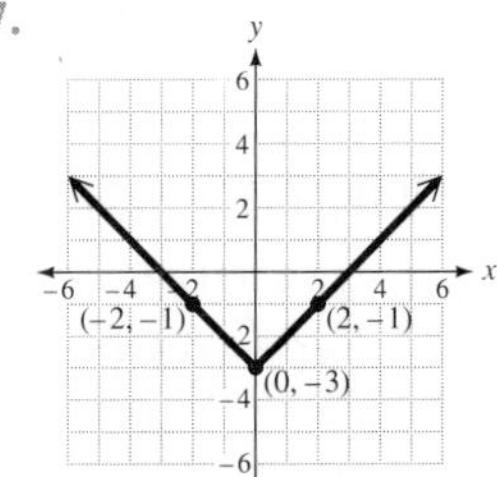

38.

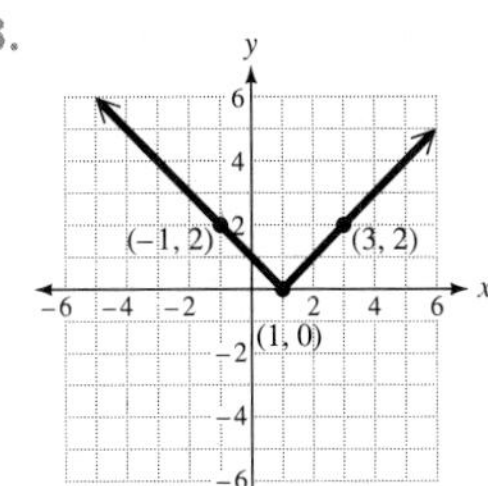

39. (1, 230), (5, 1030), (10, 2030) (a) \$230 (b) \$1030 (c) \$2030 40. 6, −1
41. $-\frac{5}{3} < x < 1$ 42. $x \ge 4$ or $x \le 2$

3.1 EXERCISES B

1. Plot the points corresponding to the ordered pairs $A(1, 5)$, $B(-4, 3)$, $C(5, 2)$, $D(0, 1)$, $E(-6, 0)$, $F(0, -3)$, $G(-6, -6)$, $H(3, -4)$.

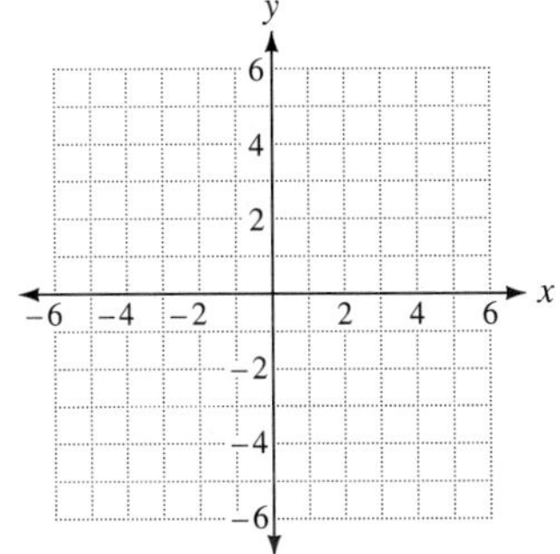

2. Give the coordinates of the points A, B, C, D, E, F, G, and H.

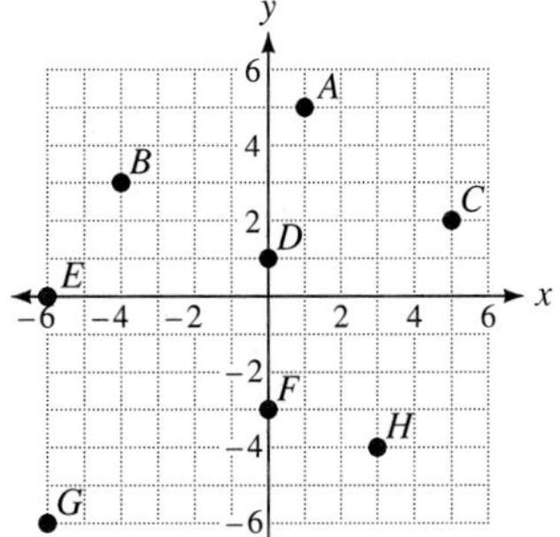

3. Plot the points associated with the ordered pairs $P\left(-\frac{1}{2}, 3\right)$, $Q\left(\frac{7}{2}, 4\right)$, $R\left(-3, -\frac{3}{4}\right)$, and $S\left(2, -\frac{7}{3}\right)$.

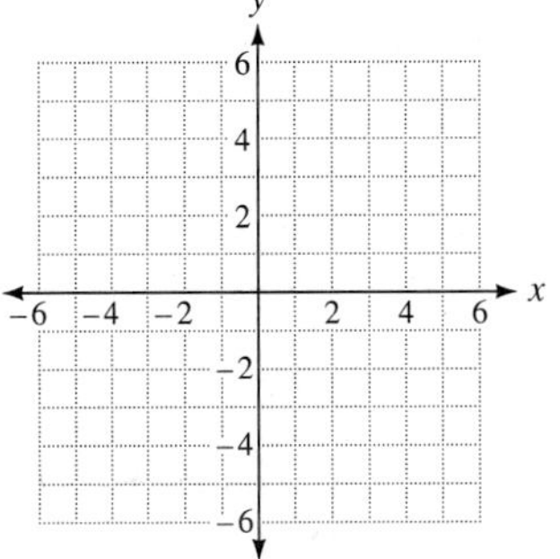

4. In what quadrants are the points P, Q, R, and S of Exercise 3 located?

Give the quadrant that contains each point.

5. $(-4, 5)$

6. $(6, -2)$

7. $(-5, -1)$

8. $\left(\frac{1}{3}, \sqrt{2}\right)$

9. Give the coordinates of each point. What do these points have in common? What do their coordinates have in common?

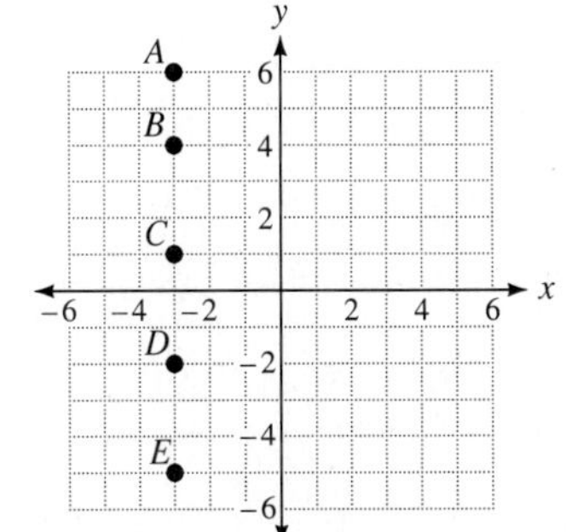

10. Give the coordinates of each point. What do these points have in common? What do their coordinates have in common?

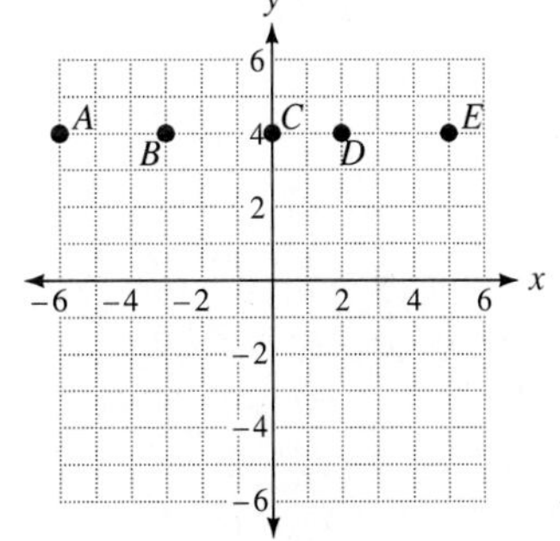

11. Find the coordinates of each point described below.

(a) Point A

(b) Point B that is 3 units up from A

(c) Point C that is 4 units to the right of A

(d) Point D that is 2 units up and 5 units to the left of A

(e) Point E that is 1 unit down and 3 units to the right of A

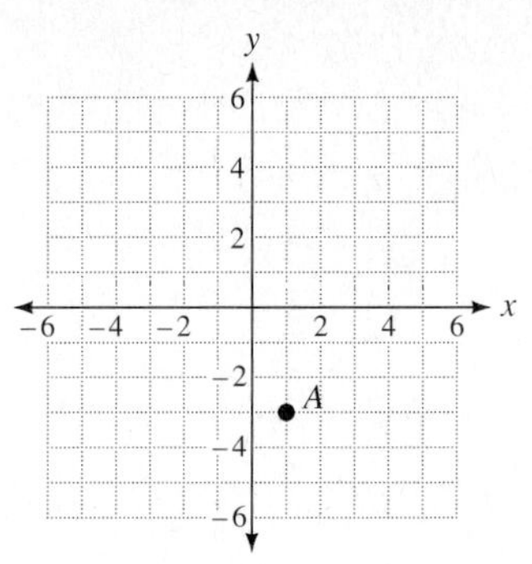

Write true *or* false *for Exercises* 12–21. *If the statement is false, explain why.*

12. Another name for a rectangular or Cartesian coordinate system is a coordinate plane.

13. The ordered pair $(-2, 8)$ determines a point in quadrant I with x-coordinate -2 and y-coordinate 8.

14. The ordered pair $(7, 2)$ determines a point in quadrant I with abscissa 2 and ordinate 7.

15. In quadrant I, the y-coordinate of a point is always positive.

16. In quadrant II, the x-coordinate of a point is always positive.

17. In quadrant III, the y-coordinate of a point is always negative.

18. In quadrant IV, the x-coordinate of a point is always negative.

19. The vertical axis in a coordinate system is also called the y-axis.

20. The coordinates of the origin are $(0, 0)$.

21. When we find the point associated with an ordered pair, we are plotting the point.

Answer each question below.

22. In which quadrants do points have coordinates with the same sign?

23. The x-coordinate of a point is negative in which quadrants?

24. The y-coordinate of a point is positive in which quadrants?

25. What is the y-coordinate of any point on the x-axis?

26. What is the distance between $A(-1, 3)$ and $B(-1, -4)$?

27. What is the distance between $C(6, -2)$ and $D(-2, -2)$?

In each of the following, find the missing coordinates so that the ordered pair is a solution to the equation.

28. $2x - y + 1 = 0$; **(a)** $(0, \quad)$ **(b)** $(\quad , 0)$ **(c)** $(-1, \quad)$ **(d)** $(\quad , 3)$

29. $x + y = 0$; **(a)** $(0, \quad)$ **(b)** $(\quad , 0)$ **(c)** $(4, \quad)$ **(d)** $(\quad , -5)$

Graph the following equations.

30. $x + y - 2 = 0$

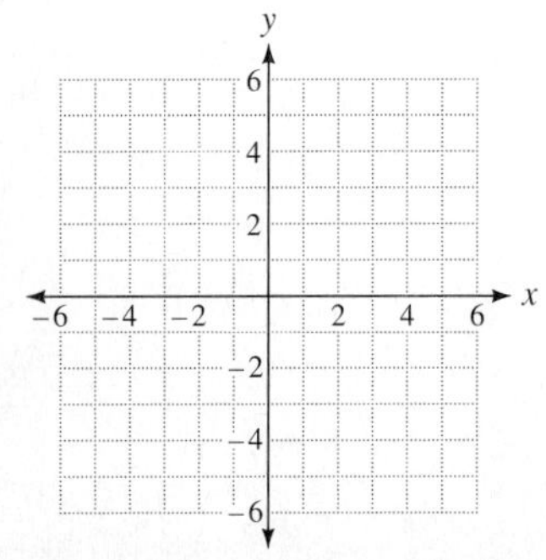

31. $2y - x = 0$

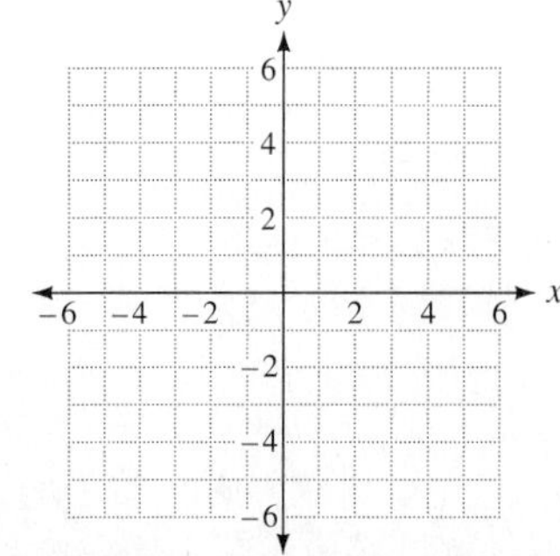

32. $y - x = 1$

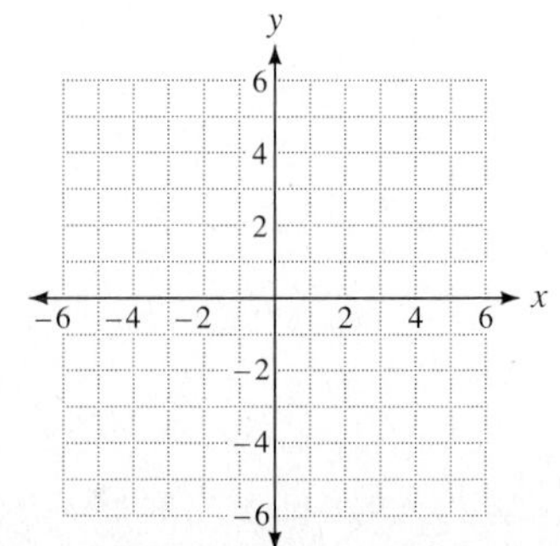

33. $y - x = 0$

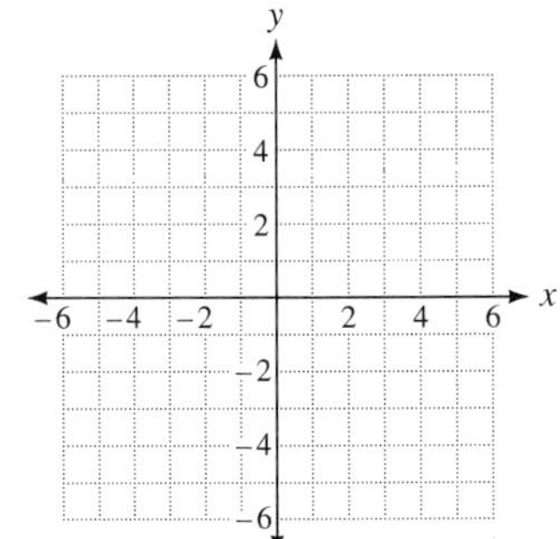

34. $y = x^2 + 2$

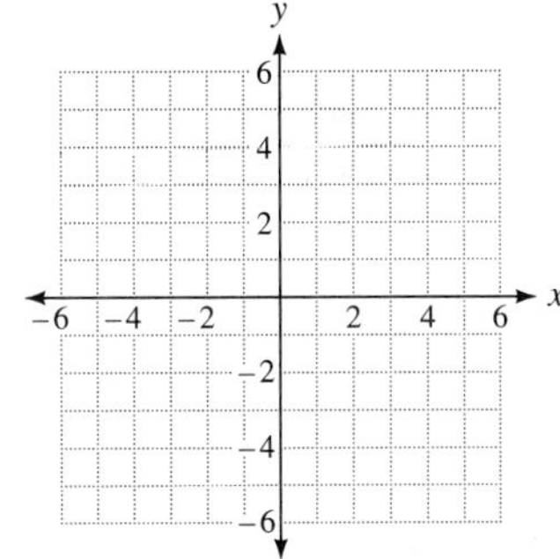

35. $y = x^2 - 3$

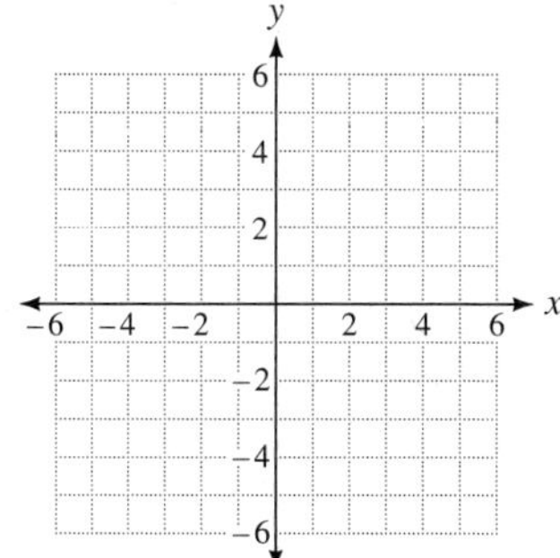

36. $y = |x| - 1$

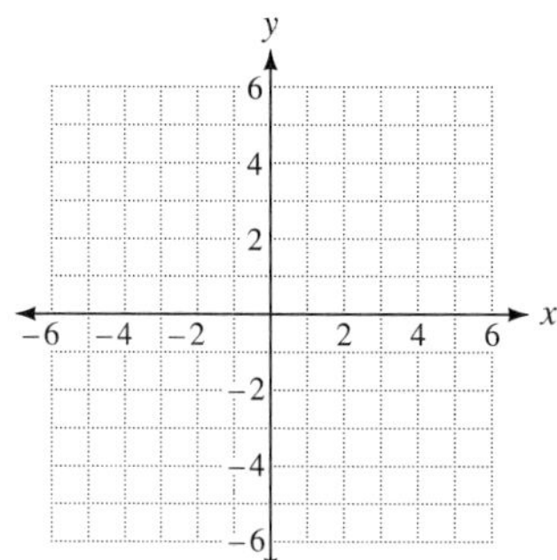

37. $y = |x| + 3$

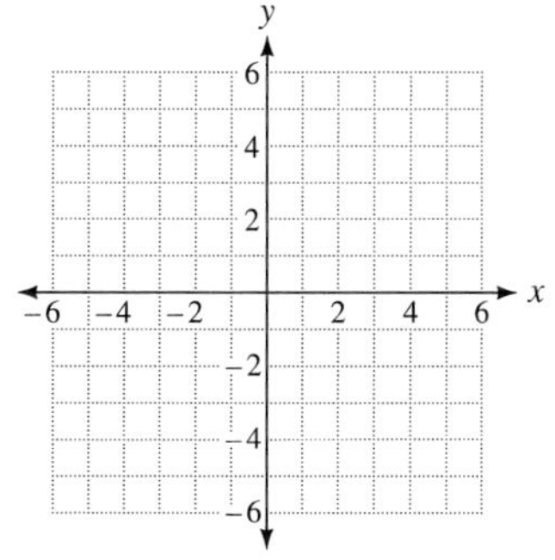

38. $y = |x + 1|$

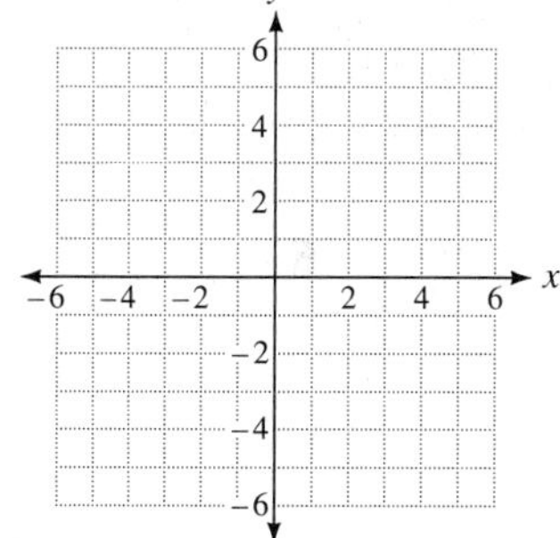

39. The annual salary of Sam Passamante, a salesman, is given in terms of his total sales x by the equation $y = 15{,}000 + (0.2)x$. Complete the ordered pairs (10,000,), (20,000,) and (50,000,). How much will Sam earn if he sells **(a)** 10,000 items? **(b)** 20,000 items? **(c)** 50,000 items?

FOR REVIEW

Solve the absolute value equation or inequality.

40. $|3x + 5| = 4$

41. $|2x - 4| \leq 2$

42. $|2x + 8| > 3$

3.1 EXERCISES C

1. Give the coordinates of the vertices (corners) of a rectangle that is 7 units long and 3 units wide if the rectangle is in the first quadrant, its longest side lies along the x-axis, and one vertex is at the origin. [*Hint:* Make a sketch.]

2. Repeat Exercise 1 for a rectangle that is in the third quadrant with all other conditions the same.

3. Give the equation in two variables x and y in which the value of y is 3 more than twice the value of x. Graph the equation.

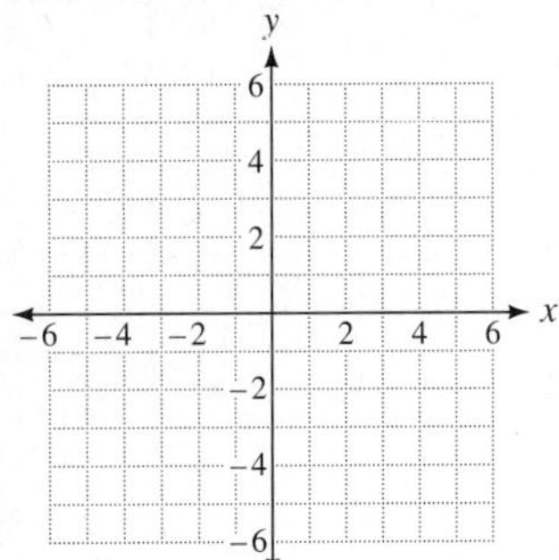

4. Graph $x + y = 2$ and $2x - y = 1$ in the same coordinate system. What are the coordinates of the point of intersection of the two graphs? Is this ordered pair a solution to both equations? [Answer: (1, 1); yes]

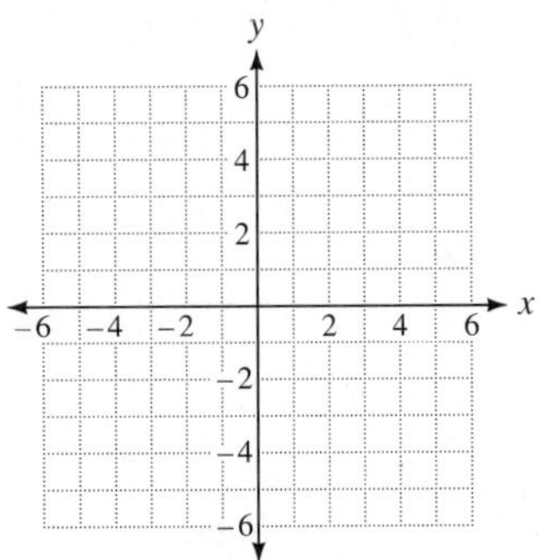

5. A retail store owner estimates that the daily profit in dollars y that she can make on the sale of x dresses is given by $y = 30x - 50$. Find the profit she made on a day when 8 dresses were sold. How many dresses were sold on a day when her profit was \$1000?

Graph the following equations.

6. $y = x^3$

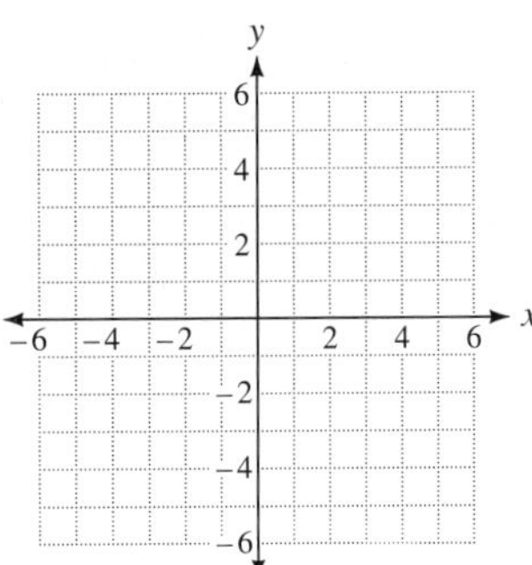

7. $x^2 + y^2 = 4$

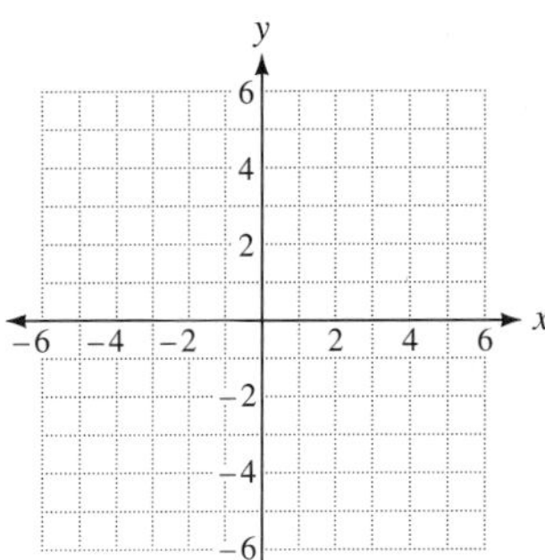

8. $y = -x^2$

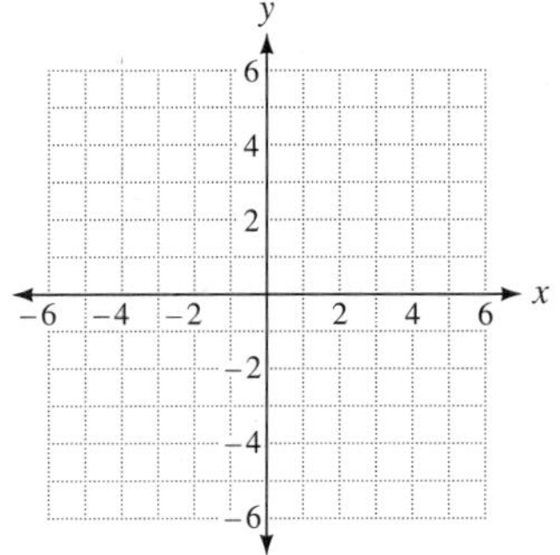

9. Suppose that $(-2, 3)$, $(6, 3)$, and $(4, -2)$ are three of the four vertices of a parallelogram. Find the coordinates of the fourth vertex.

10. What is the area of the parallelogram found in Exercise 9? [Answer: 40 square units]

3.2 LINEAR EQUATIONS

STUDENT GUIDEPOSTS

1 Linear Equations in Two Variables
2 The Intercept Method of Graphing
3 Graphing Special Linear Equations

1 LINEAR EQUATIONS IN TWO VARIABLES

A linear equation in one variable like those we studied in Chapter 2 can always be written in the form

$$ax + b = 0. \quad a, b \text{ constants}$$

A **linear equation in two variables** x and y can be written in the **general form**

$$ax + by + c = 0. \quad a, b, c \text{ constants with } a \text{ and } b \text{ not both zero}$$

Linear equations are also called **first-degree equations** since the variables occur to the first power only. Following are examples of both linear equations and those that are not linear.

EXAMPLE 1 Identifying Linear and Nonlinear Equations

Equation	*Nature*	*General form*	*Constants*
(a) $3x - y = 4$	linear	$3x - y - 4 = 0$	$a = 3,\ b = -1,$ $c = -4$
(b) $x = y + 3$	linear	$x - y - 3 = 0$	$a = 1,\ b = -1,$ $c = -3$
(c) $x + 2 = 0$	linear	$x + 0 \cdot y + 2 = 0$	$a = 1,\ b = 0,$ $c = 2$
(d) $3y - 8 = 0$	linear	$0 \cdot x + 3y - 8 = 0$	$a = 0,\ b = 3,$ $c = -8$
(e) $y = x^2 + 1$	not linear	(x to second power)	
(f) $2x + 3y^3 = 1$	not linear	(y to third power)	
(g) $xy = -1$	not linear	(cannot be written in form $ax + by + c = 0$)	
(h) $y = 3 - \frac{2}{x}$	not linear	(cannot be written in form $ax + by + c = 0$)	

Practice Exercise 1

Decide whether or not each equation is linear.

(a) $8x + 2y = 7$

(b) $y = 2x - 5$

(c) $x^2 - y^2 = 1$

(d) $2 = x - 7y$

(e) $3xy + 2 = 0$

(f) $y = \frac{1}{x}$

(g) $x = y$

(h) $y + 5 = 0$

Answers: (a) yes (b) yes (c) no (d) yes (e) no (f) no (g) yes (h) yes

2 THE INTERCEPT METHOD OF GRAPHING

In Section 3.1 we graphed equations, some of which were linear, by making a table of values of solutions to the equation, plotting these points, and drawing a smooth curve through them. This method can be improved upon if we know ahead of time that an equation has a particular geometric shape as its graph.

Every linear equation in two variables has a straight line for its graph. Since two points determine a line, we need to find only two solutions to graph a linear equation in two variables. In most cases, the two solutions easiest to find correspond to the **intercepts.** The points where a line crosses the x-axis and y-axis, respectively, are the **x-intercept** and the **y-intercept.** Since the y-intercept is a point on the y-axis, it has x-coordinate 0 as shown in the following table. Similarly, the x-intercept is a point on the x-axis, and has y-coordinate 0.

x	y
0	
	0

← **Complete this to find y-intercept** (row 1)

Complete this to find x-intercept → (row 2)

EXAMPLE 2 Graphing by the Intercept Method

(a) Graph $3x - 5y - 15 = 0$.
First find the x- and y-intercepts by completing the following table.

x	y
0	
	0

Practice Exercise 2

(a) Graph $2x - 3y = 6$. Use axes on next page.

Substitute 0 for x and solve for y.

$$3(0) - 5y - 15 = 0 \qquad x = 0 \text{ determines the } y\text{-intercept}$$
$$-5y - 15 = 0$$
$$-5y = 15$$
$$y = -3$$

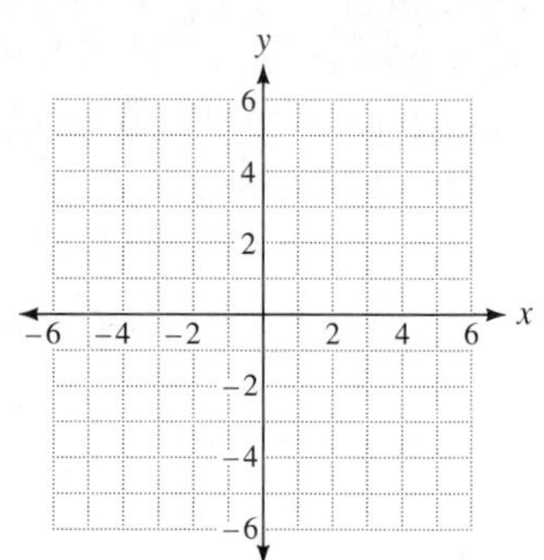

Then substitute 0 for y and solve for x.

$$3x - 5(0) - 15 = 0 \qquad y = 0 \text{ determines the } x\text{-intercept}$$
$$3x - 15 = 0$$
$$3x = 15$$
$$x = 5$$

The completed table

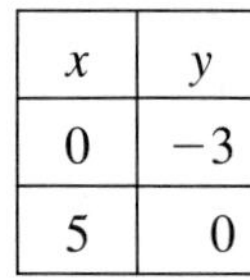

x	y
0	−3
5	0

shows the y-intercept $(0, -3)$ and the x-intercept $(5, 0)$. If we plot these two points and draw the line between them, we obtain the graph of $3x - 5y - 15 = 0$ in Figure 3.8. *Although only two points are necessary, it is wise to calculate a third solution as a check.* For example, if $x = 1$, we calculate y to be $-\frac{12}{5}$ or -2.4, and it certainly appears that $(1, -2.4)$ is on the line $3x - 5y - 15 = 0$.

(b) Graph $x - 3y = 0$.

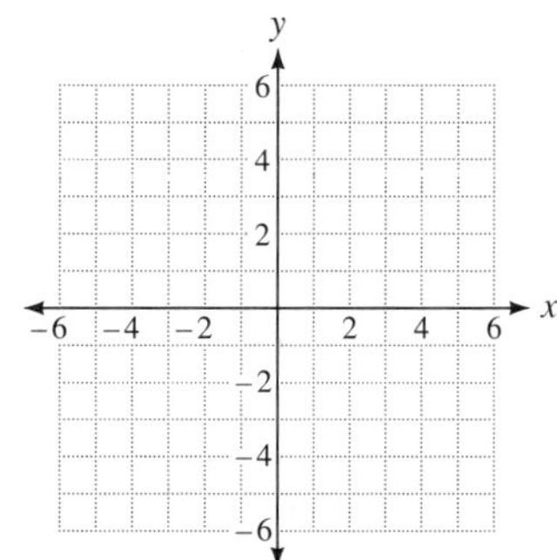

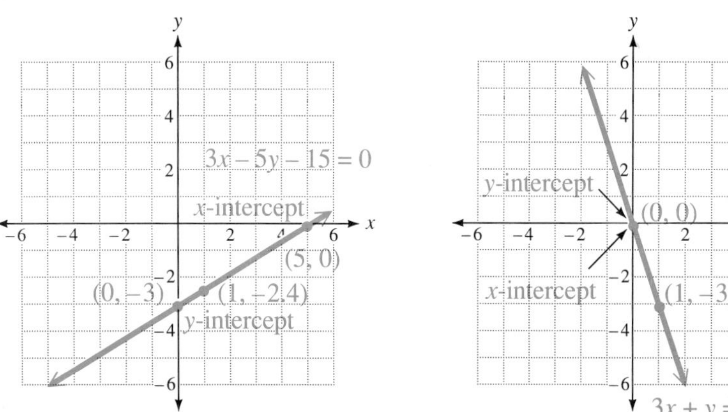

Figure 3.8 Graphing Using Intercepts

Figure 3.9

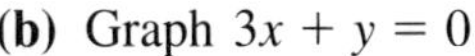

(b) Graph $3x + y = 0$.
Here, the constant term c is 0, and the table of intercepts looks like this.

x	y
0	0
0	0

That is, the x-intercept and y-intercept are the same point, the origin $(0, 0)$. When this occurs, we need to find another point on the line. If $x = 1$, $y = -3$, so $(1, -3)$ is also on the line. Plot $(0, 0)$ and $(1, -3)$ and connect the points to obtain the graph in Figure 3.9.

Answers: (a) line with intercepts $(3, 0)$ and $(0, -2)$ (b) line through $(0, 0)$ and $(3, 1)$

3 GRAPHING SPECIAL LINEAR EQUATIONS

Two special cases of linear equations are presented in the next example which shows what happens to the graph when $a = 0$ or $b = 0$ in an equation $ax + by + c = 0$. When $a = 0$, the graph is a horizontal line, and when $b = 0$, the graph is a vertical line.

EXAMPLE 3 Graphing Horizontal and Vertical Lines

(a) Graph $3y + 5 = 0$.
The x-term is missing so $a = 0$. The equation, $3y + 5 = 0$, can be simplified to

$$y = -\frac{5}{3}$$

so that solutions are of the form

$$\left(x, -\frac{5}{3}\right) \text{ for } x \text{ any real number.}$$

Since y cannot be zero, there is no x-intercept. However, when $x = 0$, $y = -\frac{5}{3}$ so that $\left(0, -\frac{5}{3}\right)$ is the y-intercept. The graph is the horizontal line passing through y-intercept $\left(0, -\frac{5}{3}\right)$ parallel to the x-axis, as shown in Figure 3.10.

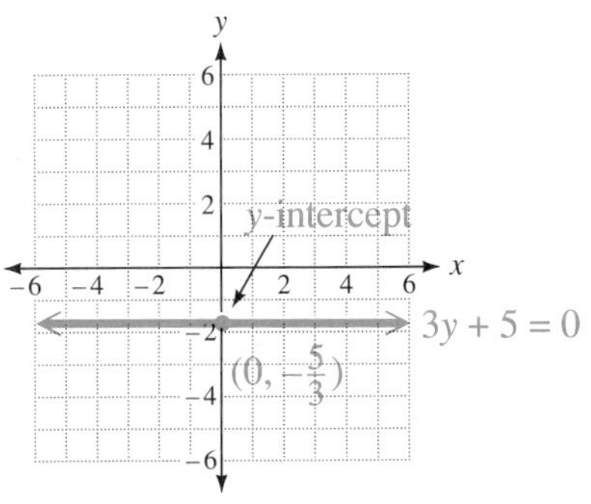

Figure 3.10 Horizontal Line

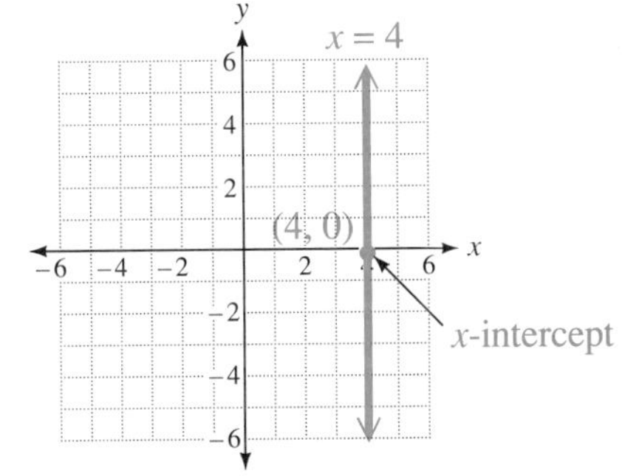

Figure 3.11 Vertical Line

(b) Graph $x = 4$.
This time $b = 0$, so the y-term is missing and the solutions are of the form

$$(4, y) \text{ for } y \text{ any real number.}$$

Since x cannot be zero, there is no y-intercept. However, if $y = 0$, x is 4 so that $(4, 0)$ is the x-intercept. The graph is the vertical line passing through x-intercept $(4, 0)$ parallel to the y-axis, as shown in Figure 3.11.

Practice Exercise 3

(a) Graph $2y - 6 = 0$.

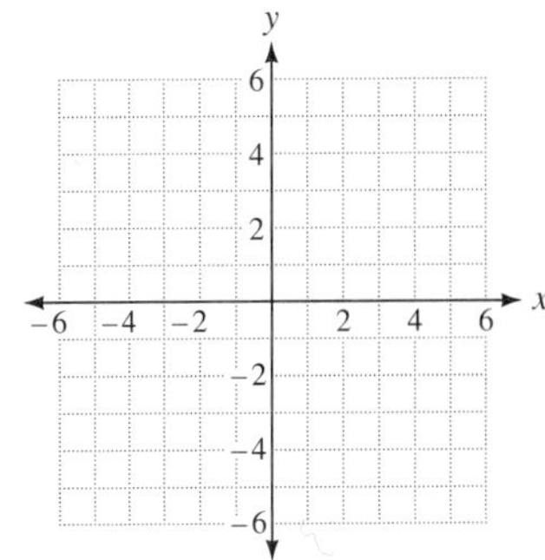

(b) Graph $3x + 9 = 0$.

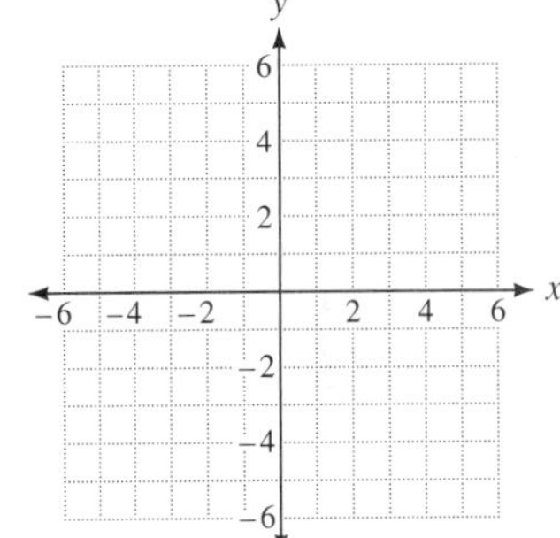

Answers: (a) horizontal line through $(0, 3)$ (b) vertical line through $(-3, 0)$

We conclude this section with a summary of the method for graphing a linear equation using intercepts.

To Graph a Linear Equation $ax + by + c = 0$

1. If $a \neq 0$ and $b \neq 0$: Plot the intercepts. If both intercepts are $(0, 0)$, find an additional point. Draw the line through the two points.
2. If $a = 0$, the equation can be written as $y =$ a constant. The graph is a horizontal line (parallel to the x-axis) through the y-intercept.
3. If $b = 0$, the equation can be written as $x =$ a constant. The graph is a vertical line (parallel to the y-axis) through the x-intercept.

3.2 EXERCISES A

Write true *or* false. *If the statement is false, explain why.*

1. A linear equation has as its graph a straight line.

2. The general form of a linear equation is $ax + by + c = 0$.

3. An equation of the form $ax + c = 0$, $a \neq 0$, has as its graph a straight line parallel to the x-axis.

4. An equation of the form $ax + by = 0$, $a \neq 0$ and $b \neq 0$, has x-intercept and y-intercept $(0, 0)$.

5. A point at which a graph crosses the x-axis is called a y-intercept.

6. Usually, the best two points to use when graphing a linear equation are the intercepts.

7. What are the intercepts of the line $x = 0$? What is its graph?

Is the given equation linear? Explain.

8. $x + y + 1 = 0$

9. $x^2 + xy + y^2 = 7$

10. $x - xy = 0$

11. $\frac{2}{x} + \frac{3}{y} = 1$

12. $x = 5$

13. $3y - 7 = 0$

Find the intercepts and graph the following linear equations using the intercept method.

14. $3x + 4y - 12 = 0$

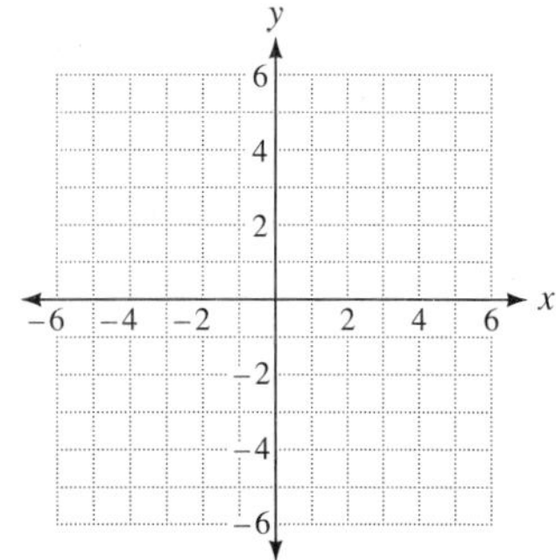

15. $2x - y = 4$

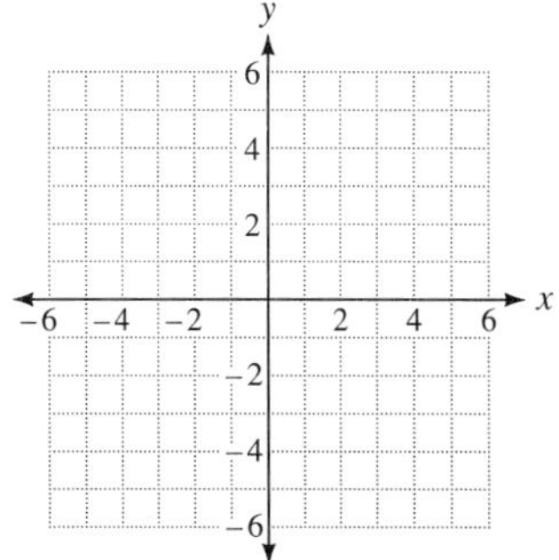

16. $2x + y = 1$

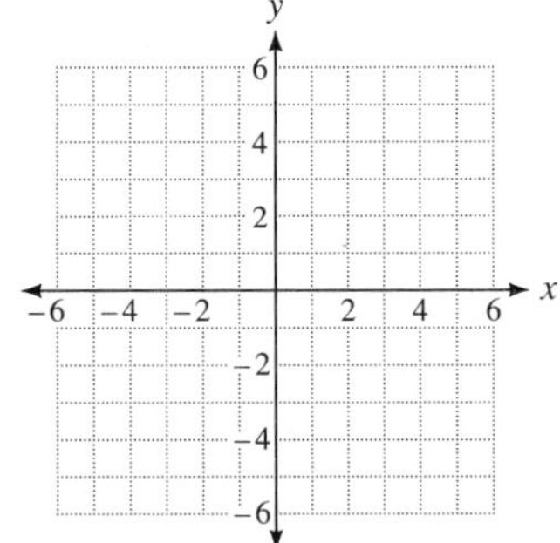

17. $2x - y = 0$

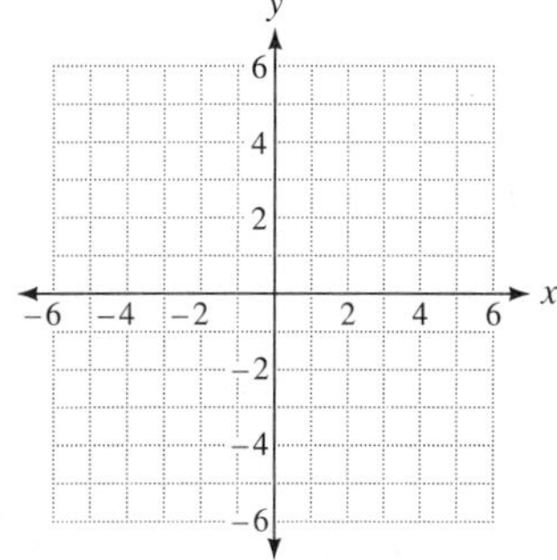

18. $x + 3y = 0$

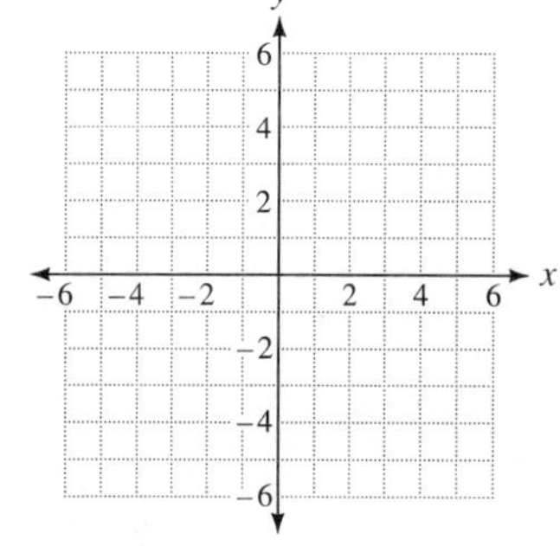

19. $y = -5$

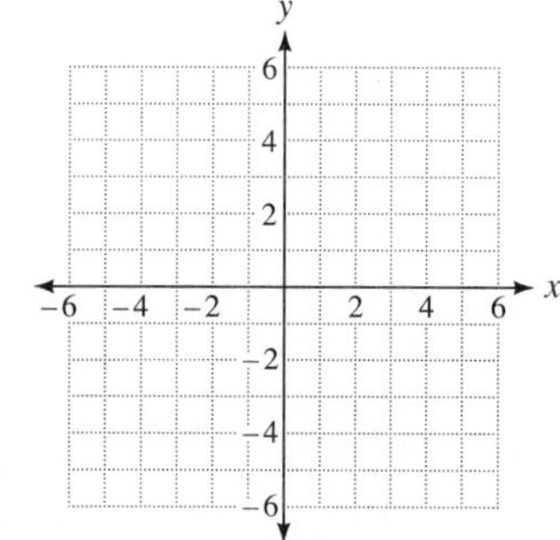

20. $2y - 6 = 0$

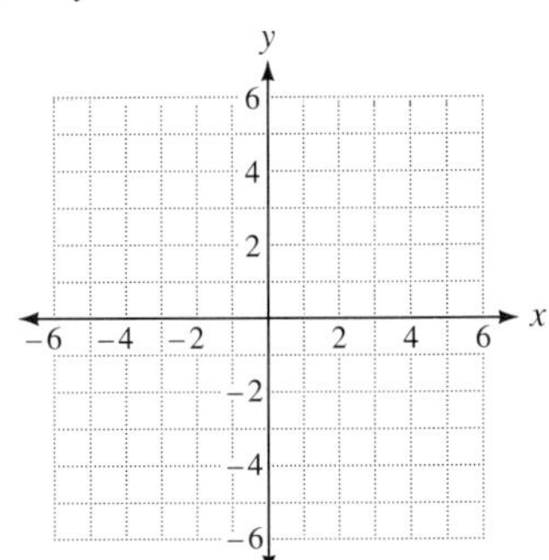

21. $x = 2$

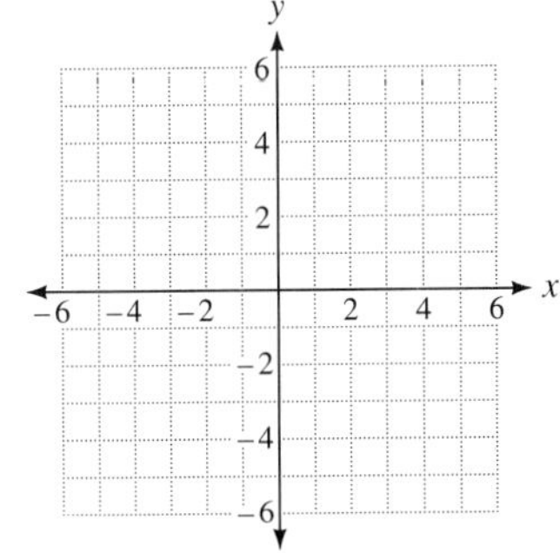

22. $2x + 5 = 0$

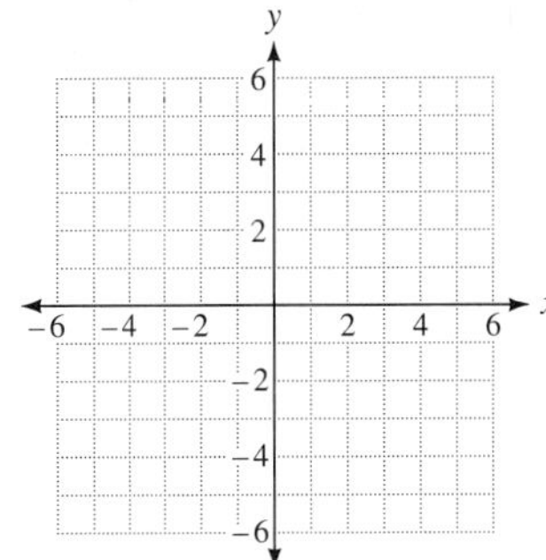

FOR REVIEW

Write true *or* false. *If the statement is false, explain why.*

23. The coordinates of the origin are (0, 0).

24. Every point on the y-axis has y-coordinate equal to zero.

25. Every point above the x-axis has x-coordinate equal to a positive number.

26. Every point to the left of the y-axis has x-coordinate equal to a negative number.

27. Points for which both coordinates are positive lie in quadrant III.

28. Points that have x-coordinate negative and y-coordinate positive lie in quadrant II.

29. In Jefferson City, Missouri, it costs 90¢ plus 10¢ for every tenth of a mile for a taxi ride. The cost in cents, y, of riding a taxi is given by $y = 10x + 90$, where x is the number of tenths of a mile driven. Complete the ordered pairs (5,), (50,), and (100,). How much does it cost to ride **(a)** 5 tenths of a mile? **(b)** 50 tenths of a mile? **(c)** 100 tenths of a mile?

The following exercises review material from Section 1.4 that will help you prepare for the next section. Evaluate each expression.

30. $\dfrac{-2-(-4)}{3-(-1)}$

31. $\dfrac{6-(-5)}{-2-(-2)}$

32. $\dfrac{-1-(-1)}{2-(-3)}$

ANSWERS: 1. true 2. true 3. false; to the y-axis 4. true 5. false; x-intercept 6. true 7. x-intercept (0, 0); every point on the y-axis is a y-intercept; the graph is the y-axis 8. yes 9. no 10. no 11. no 12. yes 13. yes

14.

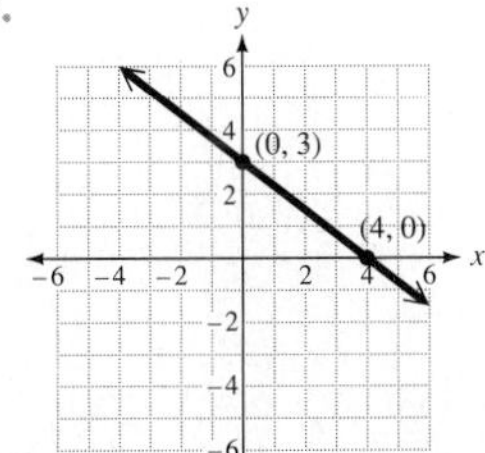

15.

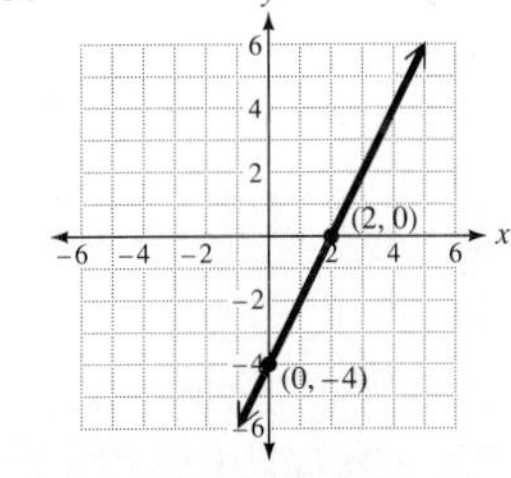

16.

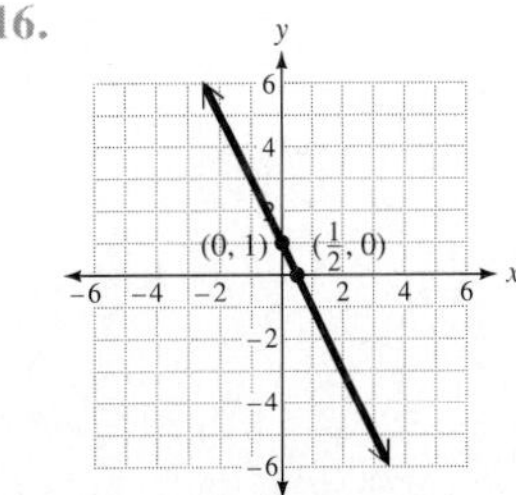

17.

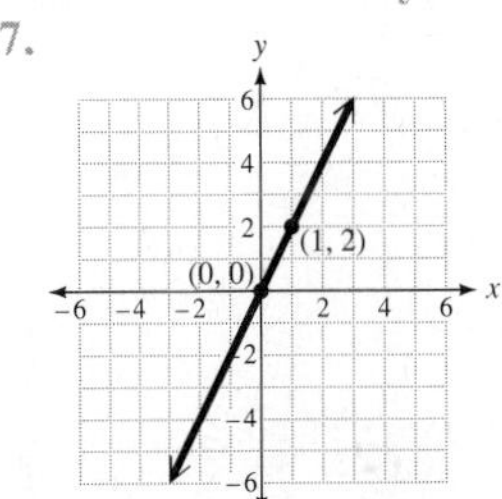

18.

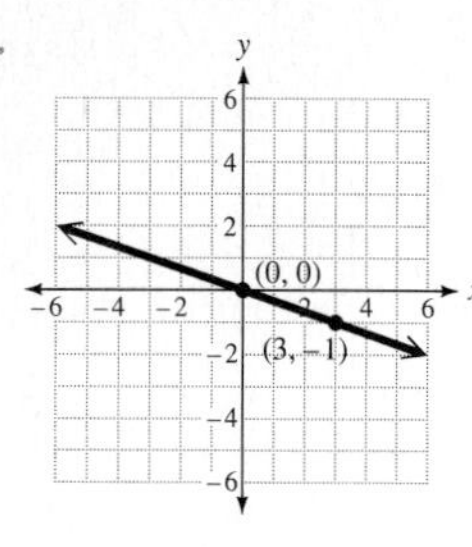

19.

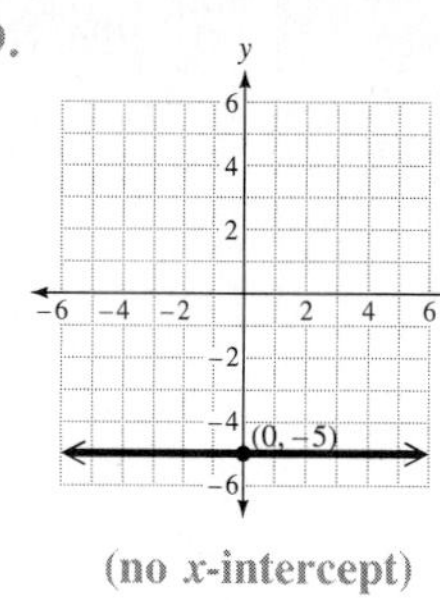

(no x-intercept)

20.

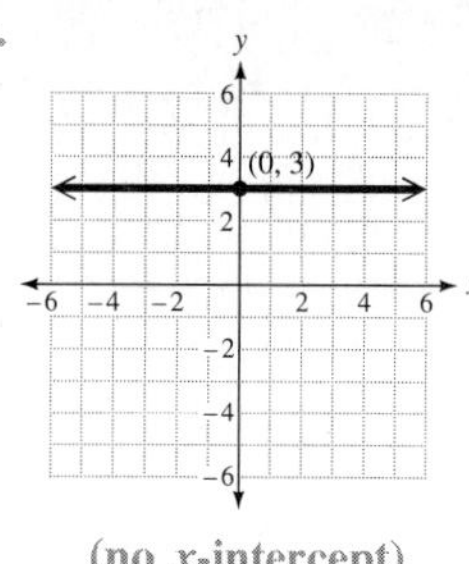

(no x-intercept)

21.

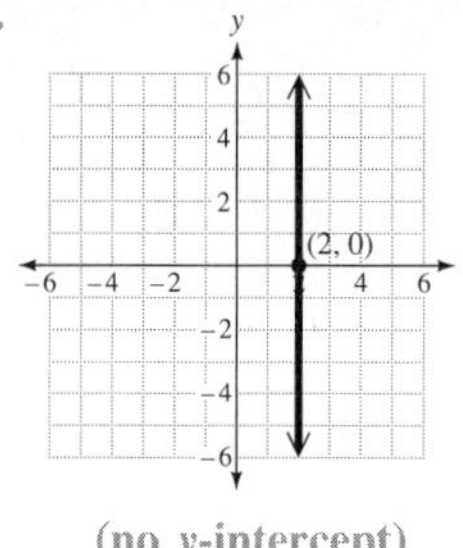

(no y-intercept)

22.

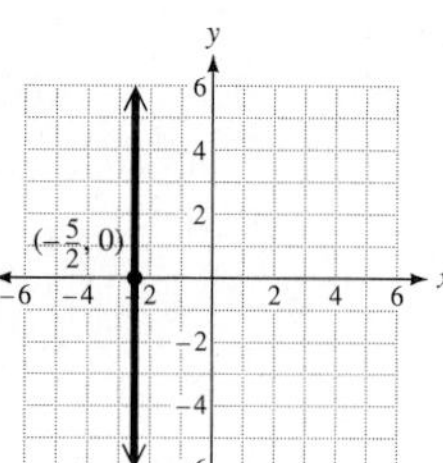

(no y-intercept) 23. true 24. false; x-coordinate is zero 25. false; y-coordinates are positive 26. true 27. false; quadrant I 28. true 29. (5, 140), (50, 590), (100, 1090); (a) \$1.40 (b) \$5.90 (c) \$10.90 30. $\frac{1}{2}$ 31. undefined 32. 0

3.2 EXERCISES B

Write true *or* false. *If the statement is false, explain why.*

1. Linear equations are also called first-degree equations.
2. The equation $ax + by + c = 0$, $a \neq 0$, or $b \neq 0$, is called the general form of a linear equation.
3. An equation of the form $by + c = 0$ has as its graph a straight line parallel to the y-axis.
4. In an equation $ax + by + c = 0$, if $c = 0$, $a \neq 0$, $b \neq 0$, the point (0, 0) is both the x-intercept and the y-intercept.
5. A point at which a graph crosses the y-axis is called an x-intercept.
6. To graph a linear equation, only two points are needed.
7. What are the intercepts of the line $y = 0$? What is its graph?

Is the given equation linear? Explain.

8. $3x - y + 2 = 0$
9. $2x^2 - xy - y^2 = 3$
10. $xy = 3$
11. $x - \dfrac{1}{y} = 8$
12. $3x + 7 = 0$
13. $y = \dfrac{1}{2}$

Find the intercepts and graph the following linear equations using the intercept method.

14. $4x - 3y - 12 = 0$
15. $2x + y = 4$
16. $2x - y = 1$

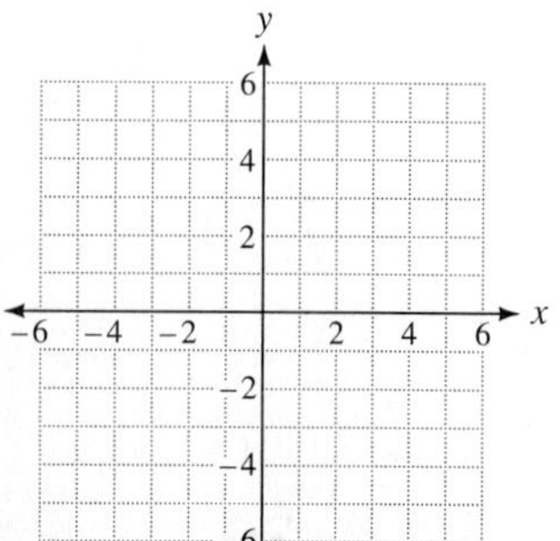

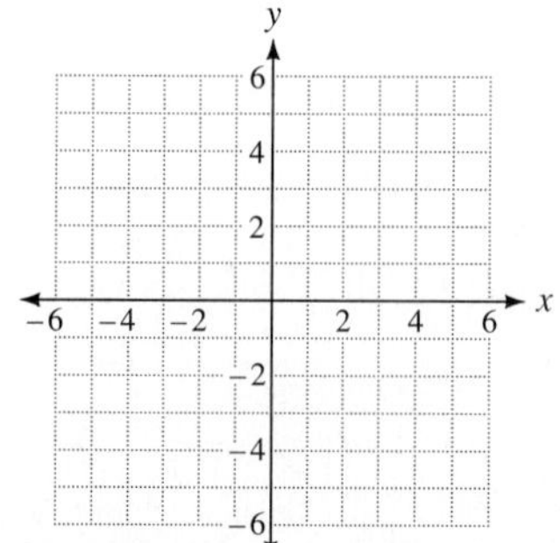

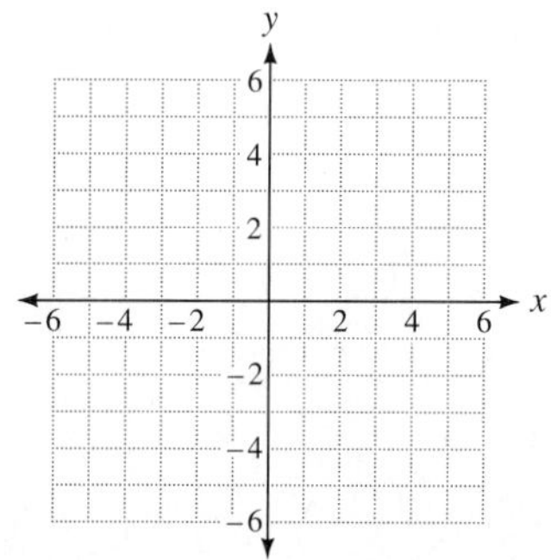

17. $2x + y = 0$

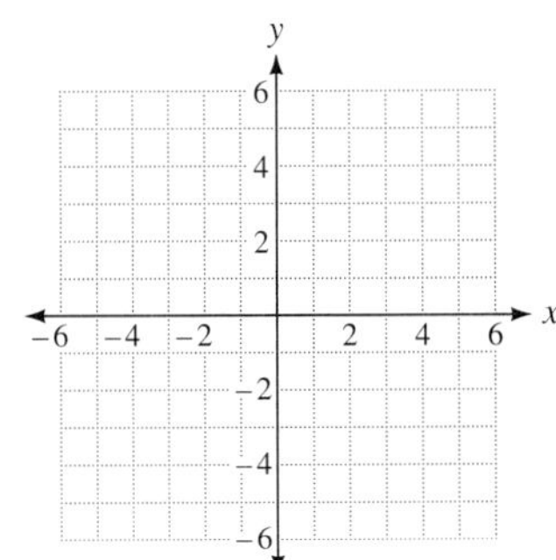

18. $x - 3y = 0$

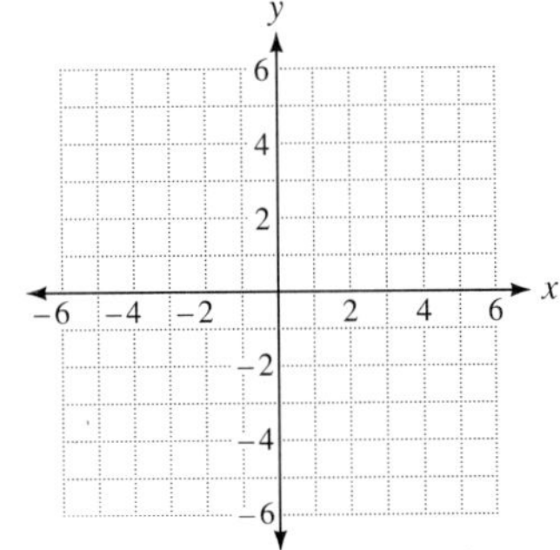

19. $y = 3$

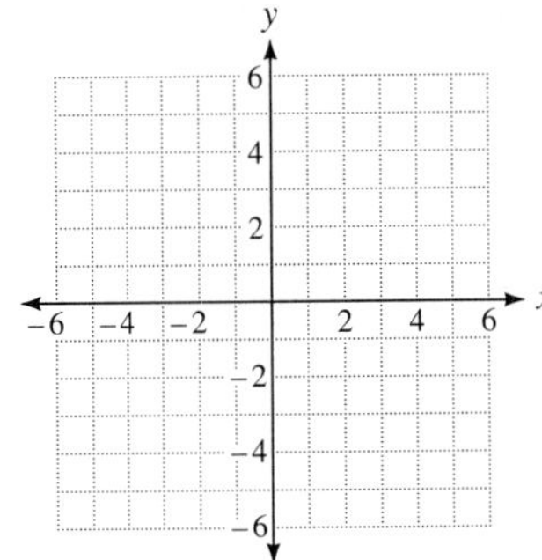

20. $3y + 15 = 0$

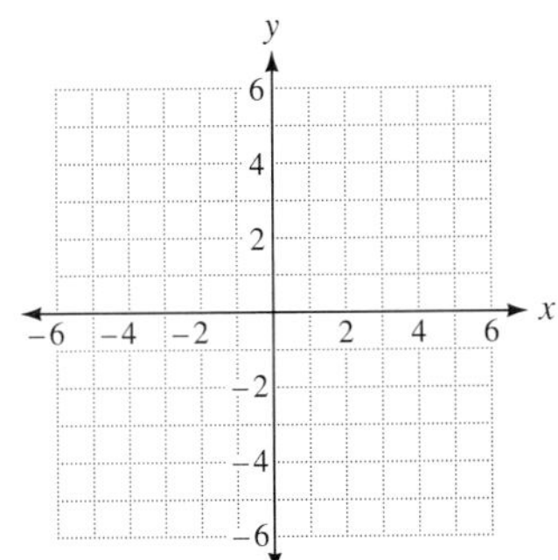

21. $x = -5$

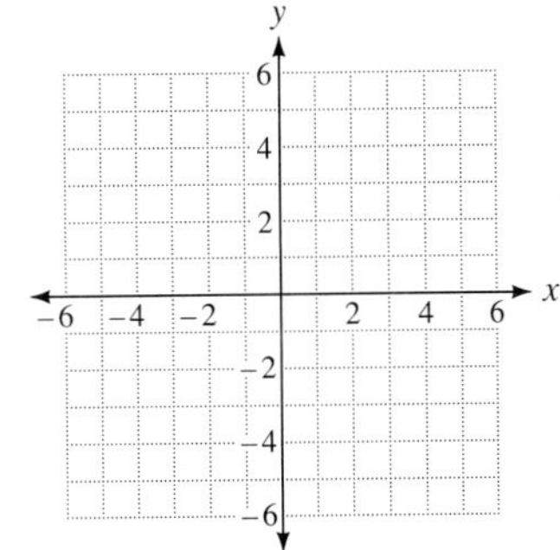

22. $2x - 7 = 0$

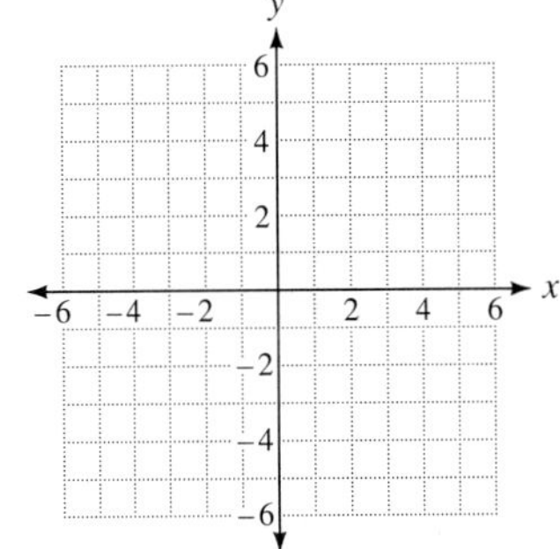

FOR REVIEW

Write true *or* false. *If the statement is false, explain why.*

23. The point with coordinates (0, 0) is called the origin.

24. Every point on the x-axis has x-coordinate equal to zero.

25. Every point below the x-axis has x-coordinate equal to a negative number.

26. Every point to the right of the y-axis has x-coordinate equal to a positive number.

27. Points for which both coordinates are negative lie in quadrant II.

28. Points that have x-coordinate positive and y-coordinate negative lie in quadrant IV.

29. The number, y, of antelope that can live on a prairie is related to the number of acres, x, in the prairie. If this relationship is given by the equation $y = 2x + 10$, complete the ordered pairs (5,), (100,), and (1000,). How many antelope can the prairie sustain if it contains **(a)** 5 acres? **(b)** 100 acres? **(c)** 1000 acres?

The following exercises review material from Section 1.4 that will help you prepare for the next section. Evaluate each expression.

30. $\dfrac{-3-(-5)}{5-(-1)}$

31. $\dfrac{-4-(-4)}{-2-(-3)}$

32. $\dfrac{4-(-7)}{-3-(-3)}$

3.2 EXERCISES C

1. Graph $y = 3x + 2$, $y = \frac{1}{3}x + 2$, $y = 0x + 2$, $y = -\frac{1}{3}x + 2$, and $y = -3x + 2$ in the same coordinate system. What common characteristics do all the lines possess?

2. Graph $y = 2x + 3$, $y = 2x + 1$, $y = 2x + 0$, $y = 2x - 1$, and $y = 2x - 3$ in the same coordinate system. What common characteristics do all the lines possess?

3.3 SLOPE OF A LINE

STUDENT GUIDEPOSTS

1. Definition of Slope
2. Positive, Negative, Zero, and Undefined Slopes
3. Collinear Points
4. Parallel and Perpendicular Lines

1 DEFINITION OF SLOPE

We have seen that the graph of a linear equation may be horizontal (parallel to the x-axis), vertical (parallel to the y-axis), may "slope" upward from lower left to upper right, or may "slope" downward from upper left to lower right. The *slope* of a line (or lack of slope) can be defined precisely.

To define the slope of a line, suppose that $P(x_1, y_1)$ and $Q(x_2, y_2)$ are two points on a line. (The subscript notation distinguishes between the points while identifying the x- and y-coordinates.) The *slope* of the line is the ratio of the vertical change, *rise,* to the horizontal change, *run,* as we move from (x_1, y_1) to (x_2, y_2) along the line. The rise is $y_2 - y_1$, the change in y-coordinates, and the run is $x_2 - x_1$, the change in x-coordinates. Refer to Figure 3.12.

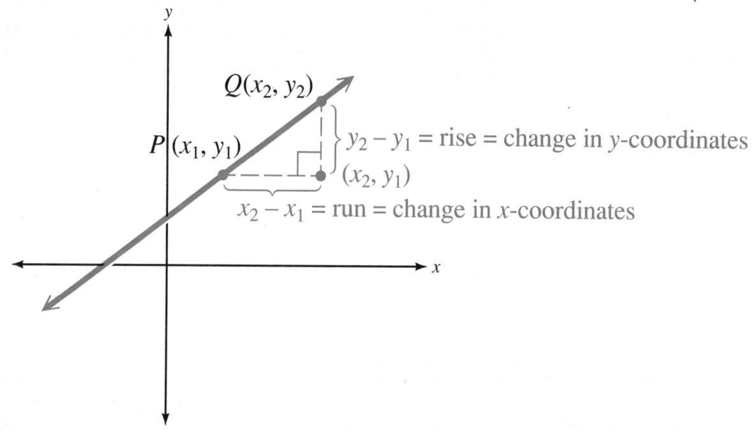

Figure 3.12 Slope $= \dfrac{\text{rise}}{\text{run}} = \dfrac{y_2 - y_1}{x_2 - x_1}$

The Slope of a Line

Let $P(x_1, y_1)$ and $Q(x_2, y_2)$ be two points on a *nonvertical* line. The **slope** m of the line is given by

$$m = \frac{y_2 - y_1}{x_2 - x_1} = \frac{\textbf{rise}}{\textbf{run}} = \frac{\textbf{change in } y\textbf{-coordinates}}{\textbf{change in } x\textbf{-coordinates}}.$$

EXAMPLE 1 FINDING THE SLOPE OF A LINE

Find the slope of the line passing through the given points.

(a) (5, 4) and (2, 1)

Let $P(x_1, y_1)$ be identified with (5, 4) and $Q(x_2, y_2)$ with (2, 1), as shown in Figure 3.13. The slope is given by

$$m = \frac{y_2 - y_1}{x_2 - x_1} = \frac{1 - 4}{2 - 5} = \frac{-3}{-3} = 1.$$

Thus, the line goes up one unit for each unit to the right, and it has *positive slope*. The same slope would have been obtained had we let (2, 1) be $P(x_1, y_1)$ and (5, 4) be $Q(x_2, y_2)$. You should calculate m with the different names of the points to check this. *The slope is the same regardless of how the two points are chosen.*

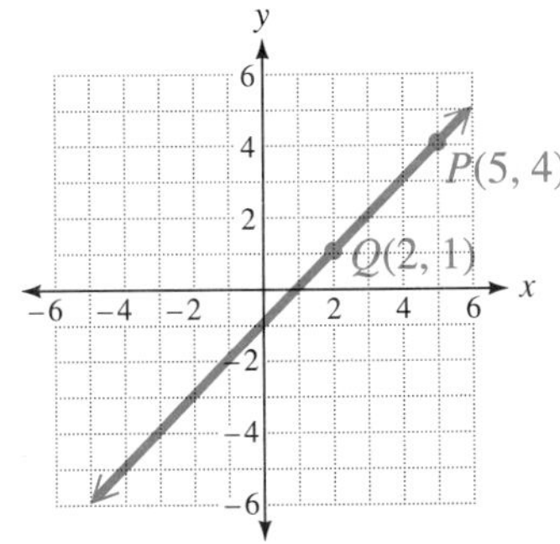

Figure 3.13 Positive Slope

Figure 3.14 Negative Slope

(b) (−1, 3) and (1, −2)

Let $P(x_1, y_1) = P(-1, 3)$ and $Q(x_2, y_2) = Q(1, -2)$, as in Figure 3.14.

$$m = \frac{y_2 - y_1}{x_2 - x_1} = \frac{-2 - 3}{1 - (-1)} = \frac{-5}{1 + 1} = -\frac{5}{2}$$

Thus, the line goes down 5 units for each 2 units to the right, and it has *negative slope*.

PRACTICE EXERCISE 1

Find the slope of the line passing through the given points.

(a) (2, 3) and (4, 6)

(b) (1, −3) and (−2, 4)

Answers: (a) $\frac{3}{2}$ (b) $-\frac{7}{3}$

CAUTION

Are the coordinates subtracted in the same order? To find the slope of the line through $P(1, 2)$ and $Q(4, 3)$, you cannot compute

$$\frac{3 - 2}{1 - 4} = \frac{y_2 - y_1}{x_1 - x_2}.$$

This is wrong

EXAMPLE 2 Finding Zero Slope and Undefined Slope

Find the slope of the line passing through the given points.

(a) $(-2, -2)$ and $(4, -2)$
Identifying $P(x_1, y_1)$ with $(-2, -2)$ and $Q(x_2, y_2)$ with $(4, -2)$ as in Figure 3.15, we obtain

$$m = \frac{y_2 - y_1}{x_2 - x_1} = \frac{-2 - (-2)}{4 - (-2)} = \frac{-2 + 2}{4 + 2} = \frac{0}{6} = 0.$$

The horizontal line has *zero slope*.

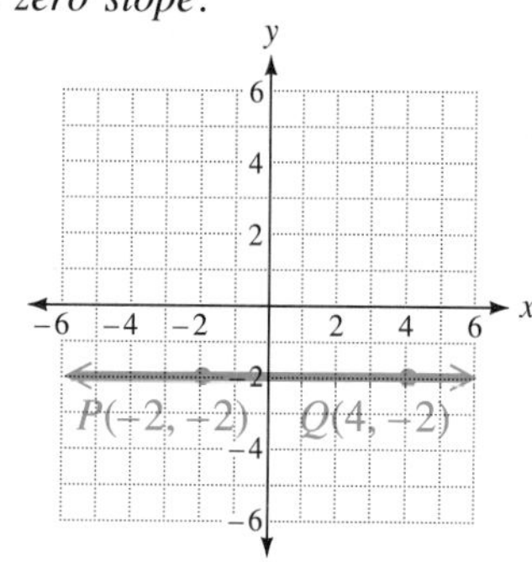

Figure 3.15 Horizontal Line: Zero Slope

(b) $(3, 2)$ and $(3, -1)$
Let $P(x_1, y_1) = (3, 2)$ and $Q(x_2, y_2) = (3, -1)$, as in Figure 3.16.

$$m = \frac{y_2 - y_1}{x_2 - x_1} = \frac{-1 - 2}{3 - 3} = \frac{-3}{0} \quad \text{This is undefined}$$

A vertical line has an *undefined slope*.

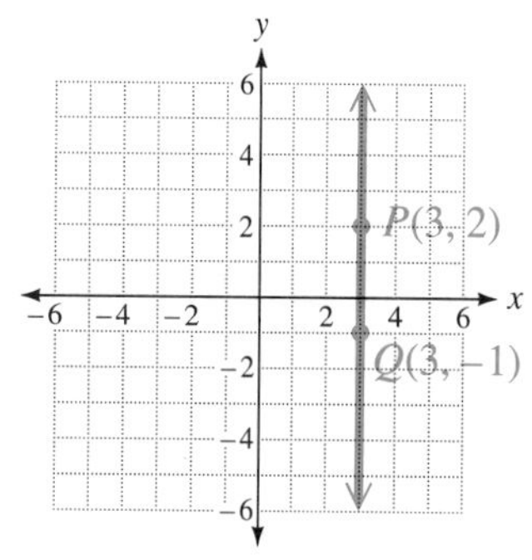

Figure 3.16 Vertical Line: Undefined Slope

Practice Exercise 2

Find the slope of the line passing through the given points.

(a) $(5, 3)$ and $(-4, 3)$

(b) $(-1, 3)$ and $(-1, -2)$

Answers: (a) 0 (b) undefined slope

2 POSITIVE, NEGATIVE, ZERO, AND UNDEFINED SLOPES

Examples 1 and 2 suggest the following summary.

The Slope of a Line

1. A line that "slopes" from lower left to upper right has **positive slope.**
2. A line that "slopes" from upper left to lower right has **negative slope.**
3. A horizontal line (parallel to the x-axis) has **zero slope.**
4. A vertical line (parallel to the y-axis) has an **undefined slope.**

The graphs in Figure 3.17 show the four possibilities listed above.

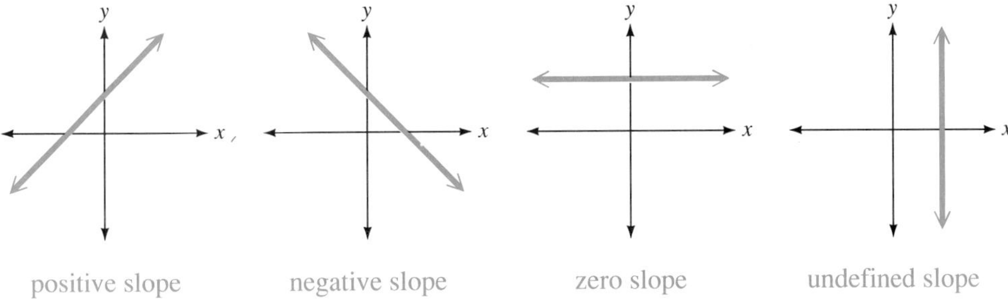

Figure 3.17 Types of Slope

3 COLLINEAR POINTS

Slope can be useful in solving a variety of problems. For example, suppose we are given three points P, Q, and R, and we wish to know if the three points are **collinear** (lie on the same straight line). We find the slope of the line through P and Q, then the slope of the line through Q and R. The points are on the same straight line through P, Q, and R provided the two slopes are the same.

EXAMPLE 3 Showing Points Collinear

Tell whether $P(5, 3)$, $Q(1, 1)$, and $R(-3, -1)$ are collinear.

The slope of the line through P and Q is

$$m_1 = \frac{1-3}{1-5} = \frac{-2}{-4} = \frac{1}{2}.$$

The slope of the line through Q and R is

$$m_2 = \frac{-1-1}{-3-1} = \frac{-2}{-4} = \frac{1}{2}.$$

Since $m_1 = m_2$, the three points lie on the same straight line (that is, they are collinear).

Practice Exercise 3

Tell whether the points $P(2, 3)$, $Q(-1, -2)$, and $R(4, 6)$ are collinear.

Answer: no

4 PARALLEL AND PERPENDICULAR LINES

The slope of a line can also be used to tell when two lines are **parallel** (never intersect) or **perpendicular** (intersect at right angles).

Slopes of Parallel and Perpendicular Lines

Given two distinct nonvertical lines with slopes m_1 and m_2:

1. If $m_1 = m_2$, the lines are parallel (equal slopes determine parallel lines).
2. If $m_1m_2 = -1$, the lines are perpendicular (slopes that are negative reciprocals determine perpendicular lines).

Also, two distinct vertical lines (both with undefined slope) are parallel, and a vertical line and a horizontal line (with slopes undefined and 0, respectively) are perpendicular.

EXAMPLE 4 Showing Lines Parallel

Verify that the line l_1 through $(2, -1)$ and $(-3, 3)$ and the line l_2 through $(4, 1)$ and $(-1, 5)$ are parallel. See Figure 3.18.

$$\text{The slope of } l_1 \text{ is } m_1 = \frac{3-(-1)}{-3-2} = \frac{3+1}{-5} = -\frac{4}{5}.$$

$$\text{The slope of } l_2 \text{ is } m_2 = \frac{5-1}{-1-4} = \frac{4}{-5} = -\frac{4}{5}.$$

Since $m_1 = m_2$, the lines are parallel.

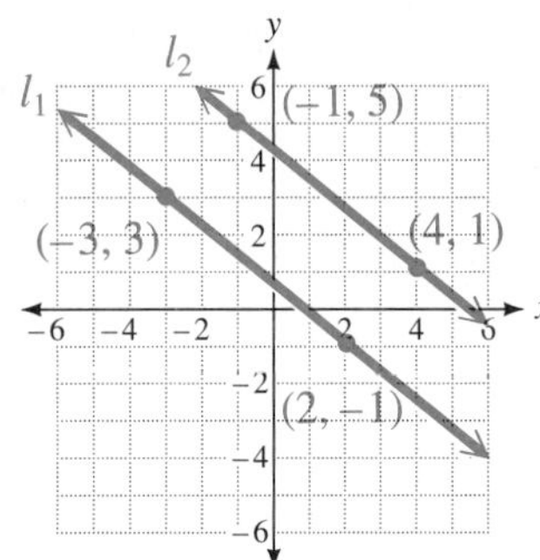

Figure 3.18 Parallel Lines

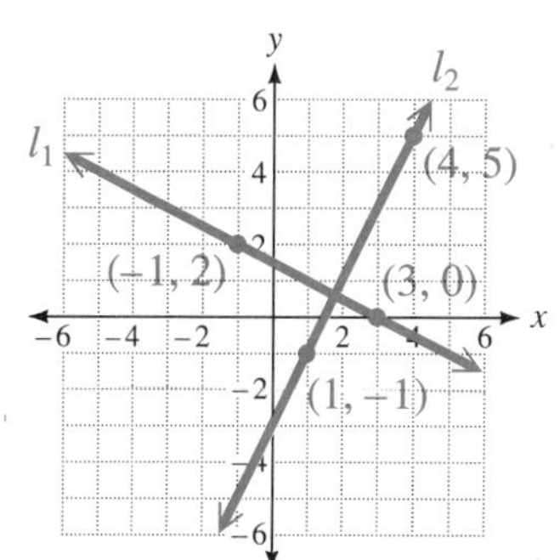

Figure 3.19 Perpendicular Lines

Practice Exercise 4

Is the line l_1 through $(-1, 3)$ and $(2, 4)$ parallel to l_2 through $(5, 2)$ and $(2, 1)$?

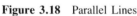

Answer: yes

EXAMPLE 5 Showing Lines Perpendicular

Verify that the line l_1 through $(-1, 2)$ and $(3, 0)$ is perpendicular to the line l_2 through $(4, 5)$ and $(1, -1)$. See Figure 3.19.

$$\text{The slope of } l_1 \text{ is } m_1 = \frac{0-2}{3-(-1)} = \frac{-2}{3+1} = \frac{-2}{4} = -\frac{1}{2}.$$

$$\text{The slope of } l_2 \text{ is } m_2 = \frac{-1-5}{1-4} = \frac{-6}{-3} = 2.$$

Since $m_1 m_2 = \left(-\frac{1}{2}\right)(2) = -1$, the lines are perpendicular. Note that $m_1 = -\frac{1}{2} = -\frac{1}{m_2}$.

Practice Exercise 5

Is l_1 through $(-1, 4)$ and $(6, 3)$ perpendicular to l_2 through $(-1, 2)$ and $(-2, -5)$?

Answer: yes

3.3 EXERCISES A

Answer Exercises 1–5 with one of the following phrases: positive slope, negative slope, zero slope, undefined slope.

1. A line parallel to the y-axis has __________.
2. A line that "slopes" from lower left to upper right has __________.
3. A line that "slopes" from upper left to lower right has __________.
4. A line parallel to the x-axis has __________.
5. The x-axis has __________.

Find the slope of the line passing through the given pair of points.

6. $(7, -1)$ and $(3, 3)$

7. $(-5, 2)$ and $(-1, -6)$

8. $(-4, 1)$ and $(-1, 3)$

9. $(4, 2)$ and $(4, -2)$

10. $(1, 7)$ and $(3, 3)$

11. $(-2, 7)$ and $(-5, 7)$

12. **(a)** What is the slope of a line through points A and B?
(b) A line through A with slope -2 passes through which other marked point?
(c) Does the line through C and E have positive or negative slope?

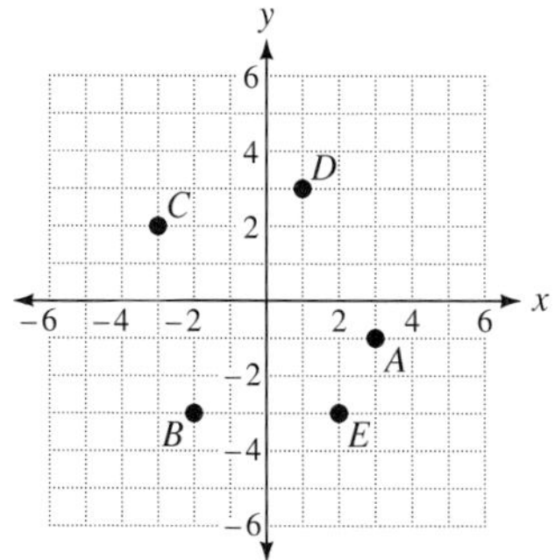

13. **(a)** What is the slope of a line through points H and F?
(b) A line through I with slope $\frac{2}{3}$ passes through which other marked point?
(c) Does the line through G and F have positive or negative slope?

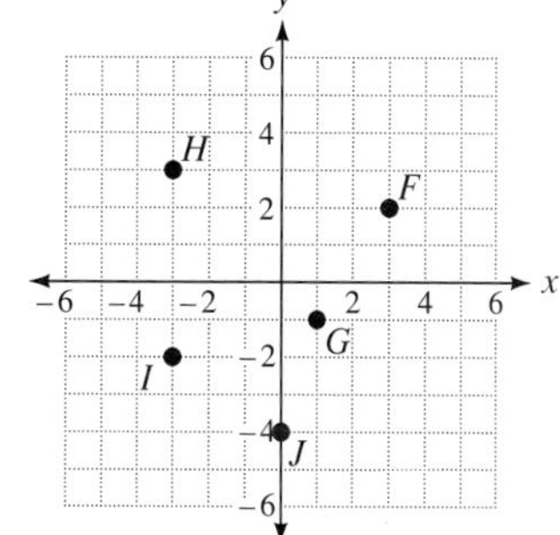

14. Find the slope of a line that is parallel to the line through $(5, -1)$ and $(3, 3)$. Explain.

15. Find the slope of a line that is perpendicular to the line through $(5, -1)$ and $(3, 3)$. Explain.

16. Verify that the line l_1 through $(1, 5)$ and $(-2, -1)$ and the line l_2 through $(-1, -7)$ and $(4, 3)$ are parallel.

17 Verify that the line l_1 through $(0, 2)$ and $(1, -1)$ and the line l_2 through $(0, -1)$ and $(3, 0)$ are perpendicular.

18. Find the slope of the line with equation $x - 3y + 6 = 0$ by first finding two points on the line. Solve the equation for y and compare the coefficient of x with the slope you obtained.

19. Are the points $P(-2, 1)$, $Q(2, 3)$, and $R(0, 2)$ collinear? Explain.

FOR REVIEW

20. Find the intercepts and graph the line with equation $3x - 2y - 6 = 0$.

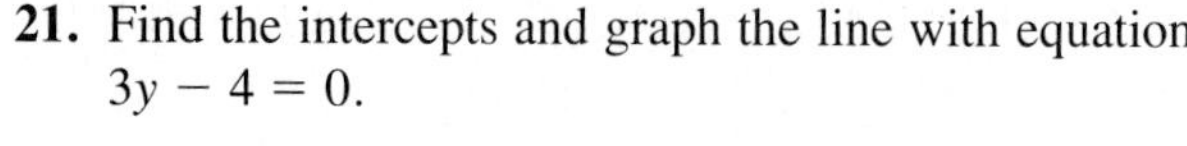

21. Find the intercepts and graph the line with equation $3y - 4 = 0$.

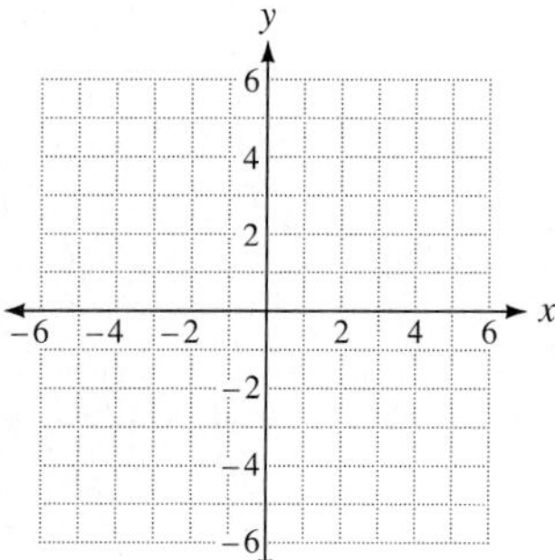

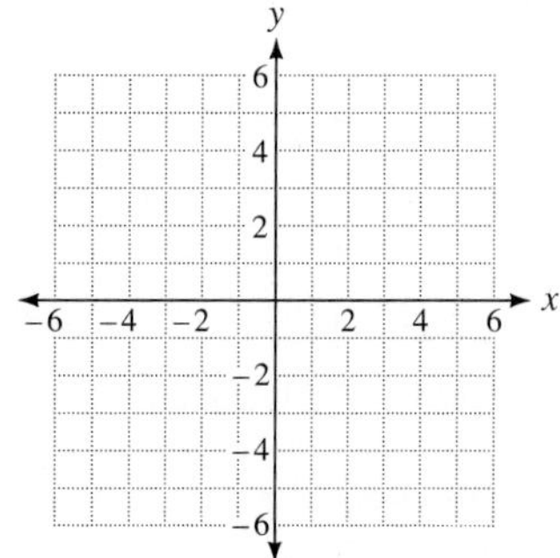

The following exercises will help you prepare for the topics presented in the next section.

22. Remove parentheses and write $y - 2 = \frac{3}{5}(x - (-4))$ in the form $ax + by + c = 0$.

23. Start at point $A(0, -4)$ and give the coordinates of point B if B is 3 units up and 2 units to the right of A.

ANSWERS: **1.** undefined slope **2.** positive slope **3.** negative slope **4.** zero slope **5.** zero slope **6.** -1 **7.** -2 **8.** $\frac{2}{3}$ **9.** undefined slope **10.** -2 **11.** 0 **12.** (a) $\frac{2}{5}$ (b) D (c) negative slope **13.** (a) $-\frac{1}{6}$ (b) F (c) positive slope **14.** -2; parallel lines have the same slope, and the given line has slope -2 **15.** $\frac{1}{2}$; perpendicular lines must have slopes that are negative reciprocals, and the given line has slope -2, the negative reciprocal of which is $\frac{1}{2}$ **16.** Both have slope 2. **17.** $m_1 = -3$ and $m_2 = \frac{1}{3}$ (negative reciprocals) **18.** $\frac{1}{3}$; the coefficient of x is also $\frac{1}{3}$. **19.** Yes. The slope of the line through P and Q is $\frac{1}{2}$, and the slope of the line through Q and R is $\frac{1}{2}$.
20. x-intercept $(2, 0)$; y-intercept $(0, -3)$ **21.** no x-intercept; y-intercept $(0, \frac{4}{3})$ **22.** $3x - 5y + 22 = 0$ **23.** $B(2, -1)$

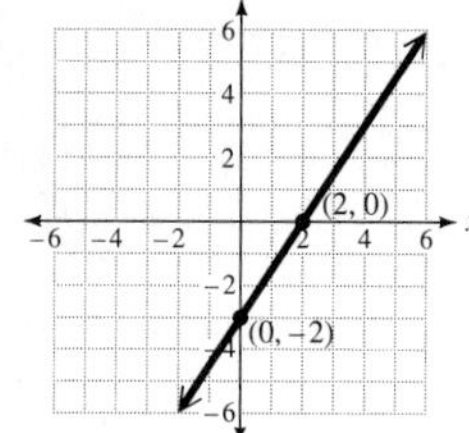

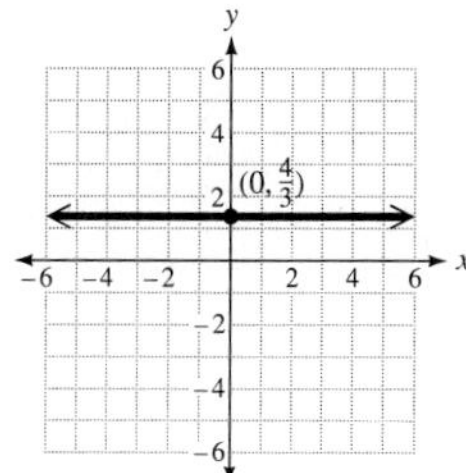

3.3 EXERCISES B

1. A line with zero slope is parallel to which axis?

2. A line with undefined slope is parallel to which axis?

3. Is it true that a line with positive slope ''slopes'' from lower left to upper right?

4. Is it true that a line with negative slope ''slopes'' from upper left to lower right?

5. Which coordinate axis has undefined slope? Which coordinate axis has zero slope?

Find the slope of the line passing through the given pair of points.

6. $(4, -2)$ and $(2, 2)$

7. $(-3, -1)$ and $(1, -5)$

8. $(-5, 2)$ and $(-1, 3)$

9. $(-1, 5)$ and $(9, 5)$

10. $(-2, 5)$ and $(4, -1)$

11. $(-3, 5)$ and $(-3, 4)$

12. **(a)** What is the slope of a line through points B and E?
(b) A line through C with slope 3 passes through which other marked point?
(c) Does the line through A and D have positive or negative slope?

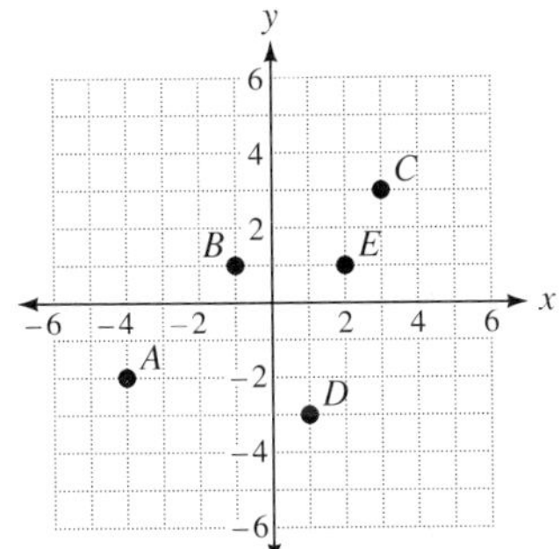

13. **(a)** What is the slope of a line through points F and H?
(b) A line through H with slope $-\frac{1}{2}$ passes through which other marked point?
(c) Does the line through G and F have positive or negative slope?

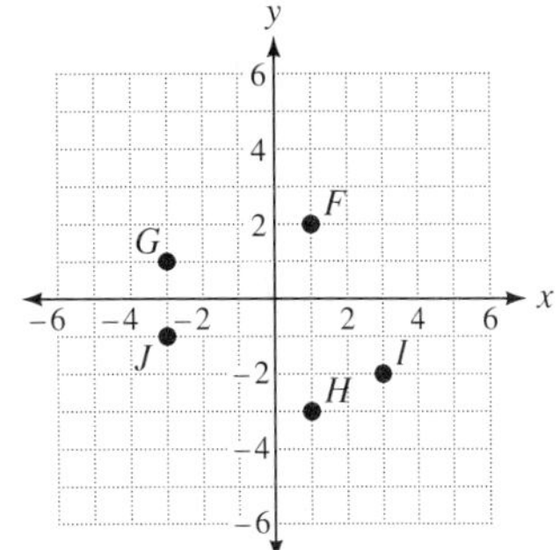

14. What is the slope of a line parallel to the line through $(3, -4)$ and $(2, 5)$?

15. What is the slope of a line perpendicular to the line through $(3, -4)$ and $(2, 5)$?

16. Verify that the line l_1 through $(3, -1)$ and $(4, 2)$ and the line l_2 through $(-2, 1)$ and $(-3, -2)$ are parallel.

17. Verify that the line l_1 through $(4, 0)$ and $(-2, 1)$ and the line l_2 through $(3, 5)$ and $(2, -1)$ are perpendicular.

18. Find the slope of the line with equation $3x + y = 6$ by first finding two points on the line. Solve the equation for y and compare the coefficient of x with the slope you obtained.

19. Are the points $P(2, 2)$, $Q(-1, 0)$, and $R(-3, -1)$ collinear? Explain.

FOR REVIEW

20. Find the intercepts and graph the line with equation $3x + y - 6 = 0$.

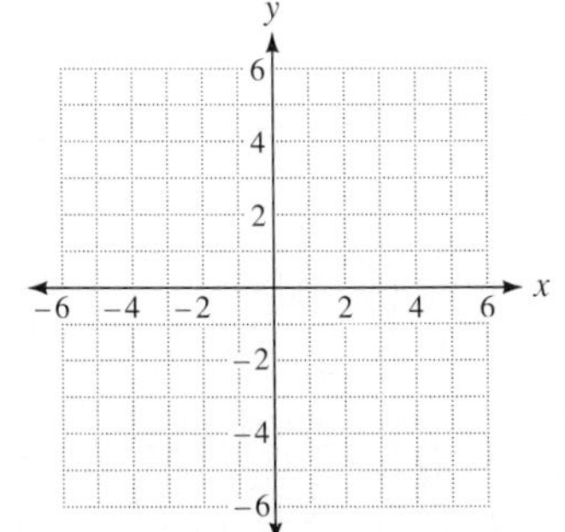

21. Find the intercepts and graph the line with equation $3x + 6 = 0$.

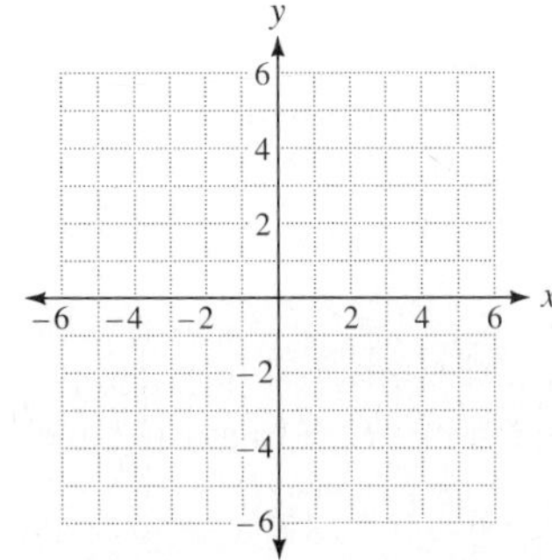

The following exercises will help you prepare for the topics presented in the next section.

22. Remove parentheses and write $y - (-3) = -\frac{4}{3}(x + 2)$ in the form $ax + by + c = 0$.

23. Start at point $A(0, 2)$ and give the coordinates of point B if B is 5 units down and 3 units to the left of A.

3.3 EXERCISES C

1. Is the triangle with vertices $A(-2, 7)$, $B(-3, 3)$, and $C(-6, 8)$ a right triangle? Explain. [*Hint:* Are two adjacent sides perpendicular?]

2. Is the parallelogram with vertices $A(2, 1)$, $B(4, 1)$, $C(-1, -1)$, and $D(1, -1)$ a rectangle? Explain.

3. Find the value of x so that the line passing through $(x, 1)$ and $(3, -x)$ has slope 5. [Answer: 4]

4. Find the slope of the line passing through $(u, 2 - v)$ and $(u + 1, 5 - v)$.

5. A pyramid in Mexico has a square base measuring 400 ft on a side. If the slope of a side of the pyramid is $\frac{6}{5}$, what is the height of the pyramid?

6. A grain storage silo is a cylinder topped by a cone. The height of the cylindrical portion is 20 feet with radius 8 ft. If the slope of the cone is $\frac{1}{4}$, what is the total height of the silo? [Answer: 22 ft]

7. What is the equation of the line through $(2, -1)$ with slope 0? With undefined slope?

8. Find the real number n so that the given three points are collinear: $(6, 3)$, $(n, -1)$, and $(-3, -3)$

3.4 FORMS OF THE EQUATION OF A LINE

STUDENT GUIDEPOSTS

1. Point-Slope Form
2. Slope-Intercept Form
3. Equation of the Line through Two Points
4. Graphing Equations Using Slope

1 POINT-SLOPE FORM

In Section 3.2 we defined the general form of a linear equation in two variables to be

$$ax + by + c = 0 \qquad (a \neq 0 \text{ or } b \neq 0).$$

In this section we will look at two other forms, the *point-slope form* and the *slope-intercept form*. The point-slope form allows us to find the equation of a line when given its slope and any point on the line.

Point-Slope Form of the Equation of a Line

To find the equation of a nonvertical line with slope m and passing through the point (x_1, y_1), substitute these values into the **point-slope form**

$$y - y_1 = m(x - x_1).$$

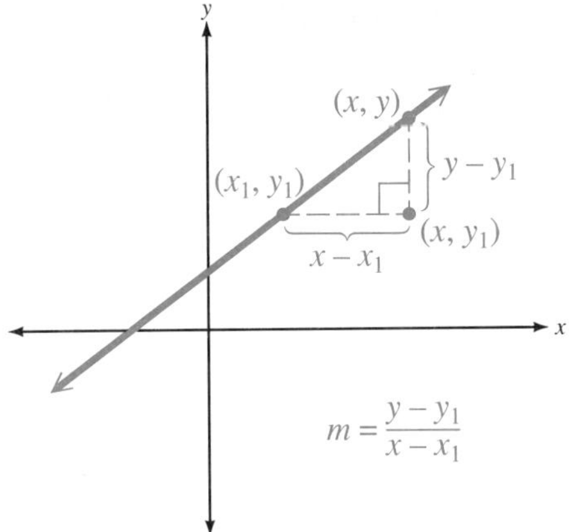

Figure 3.20 Point-Slope Form

To see how this formula is obtained, refer to Figure 3.20 and think of a general point on the line passing through (x_1, y_1) with slope m as having coordinates (x, y). The slope m is given by

$$m = \frac{y - y_1}{x - x_1},$$

where (x, y) plays the role of the point (x_2, y_2) in the formula for m. Multiplying both sides by $(x - x_1)$, we obtain

$$(x - x_1) \cdot m = (x - x_1) \cdot \frac{(y - y_1)}{(x - x_1)},$$

which is equivalent to the point-slope form $y - y_1 = m(x - x_1)$.

EXAMPLE 1 Using Point-Slope Form

Find the point-slope form of the equation of the line with slope $\frac{1}{2}$ passing through the point $(-3, 1)$. Also write the general form of the equation of this line.

We have $(x_1, y_1) = (-3, 1)$ and $m = \frac{1}{2}$, so by substituting we obtain the point-slope form

$$y - y_1 = m(x - x_1)$$

$$y - 1 = \frac{1}{2}[x - (-3)]. \qquad \text{Watch the sign}$$

To obtain the general form of the equation, eliminate fractions, clear parentheses, and then collect all terms on the left side of the equation.

$$y - 1 = \frac{1}{2}(x + 3)$$

$$2(y - 1) = 2 \cdot \frac{1}{2}(x + 3) \qquad \text{Multiply by LCD} = 2$$

$$2y - 2 = x + 3 \qquad \text{Remove parentheses}$$

$$2y - 2 - x - 3 = x + 3 - x - 3 \qquad \text{Subtract } x \text{ and } 3$$

$$-x + 2y - 5 = 0$$

$$x - 2y + 5 = 0 \qquad \text{Multiply by } -1$$

Practice Exercise 1

Give the general form of the equation of a line with slope $\frac{1}{4}$ passing through $(2, -3)$.

Answer: $x - 4y - 14 = 0$

To change $-x + 2y - 5 = 0$ to $x - 2y + 5 = 0$ in Example 1, we multiplied both sides by -1. Either form is acceptable as the general form. However, a positive leading coefficient on x is usually preferred.

EXAMPLE 2 Finding the Equation of a Perpendicular Line

Find the general form of the equation of a line passing through the point $(-1, 2)$ and perpendicular to a line with slope 3.

Since the line is perpendicular to a line with slope 3, it must have slope $-\frac{1}{3}$ (negative reciprocal of 3). Substitute in the point-slope form.

$$y - y_1 = m(x - x_1)$$

$$y - 2 = -\frac{1}{3}(x - (-1))$$

$$y - 2 = -\frac{1}{3}(x + 1) \quad \text{Watch signs}$$

$$3(y - 2) = 3\left(-\frac{1}{3}\right)(x + 1) \quad \text{Eliminate fractions}$$

$$3(y - 2) = -(x + 1)$$

$$3y - 6 = -x - 1 \quad \text{Remove parentheses}$$

$$x + 3y - 5 = 0 \quad \text{Add } x \text{ and 1 to both sides}$$

PRACTICE EXERCISE 2

Find the general form of the equation of a line passing through $(3, -5)$ and perpendicular to a line with slope $\frac{5}{6}$.

Answer: $6x + 5y + 7 = 0$

2 SLOPE-INTERCEPT FORM

The point-slope form lets us find the equation of a line when we are given a point on the line and the slope of the line. A special case occurs when the given point is the y-intercept.

Slope-Intercept Form of the Equation of a Line

If the equation of a nonvertical line is solved for y, the resulting equation is in **slope-intercept** form

$$y = mx + b$$

where m is the slope and $(0, b)$ is the y-intercept.

To see how this formula is obtained, consider the y-intercept $(0, b)$ as a given point on a line with slope m. Substitute in the point-slope form to obtain the slope-intercept form.

$$y - y_1 = m(x - x_1)$$

$$y - b = m(x - 0)$$

$$y - b = mx$$

$$y = mx + b$$

An important use of the slope-intercept form is that it allows us to find quickly the slope (and y-intercept) of a line without first finding two points on the line. To find the slope-intercept form of a line, simply solve the equation for y. The coefficient of x is the slope, and the constant term is the y-coordinate of the y-intercept.

EXAMPLE 3 FINDING SLOPE AND Y-INTERCEPT

Find the slope and y-intercept of the line with the following equation.

$$y = \frac{1}{2}x - 3$$

Since the equation is already solved for y (that is, already in slope-intercept form), the slope is $\frac{1}{2}$ and the y-intercept is $(0, -3)$. (Note that $b = -3$, not 3.) The following diagram might be helpful for remembering these facts.

$$y = \frac{1}{2}x - 3 = \frac{1}{2}x + (-3) = mx + b$$

$m = \frac{1}{2}$ = slope $\qquad (0, -3)$ = y-intercept $[-3 = b]$

PRACTICE EXERCISE 3

Find the slope and y-intercept of the line $y = -4x - 5$.

Answer: -4, $(0, -5)$

EXAMPLE 4 FINDING SLOPE AND Y-INTERCEPT

Find the slope and the y-intercept of the line with equation

$$2x - 3y + 5 = 0.$$

First solve for y to obtain the slope-intercept form.

$$-3y = -2x - 5 \qquad \text{Subtract } 2x \text{ and } 5$$

$$\left(-\frac{1}{3}\right)(-3y) = \left(-\frac{1}{3}\right)(-2x - 5) \qquad \text{Multiply by } -\tfrac{1}{3}$$

$$y = \frac{2}{3}x + \frac{5}{3} \qquad \text{Remove parentheses}$$

$m = \frac{2}{3}$ = slope $\qquad \left(0, \frac{5}{3}\right)$ = y-intercept $\left[\frac{5}{3} = b\right]$

The slope is $\frac{2}{3}$ and the y-intercept is $\left(0, \frac{5}{3}\right)$.

PRACTICE EXERCISE 4

Find the slope and y-intercept of $4x + 3y - 2 = 0$.

Answer: $-\frac{4}{3}$, $\left(0, \frac{2}{3}\right)$

In Exercise 18 of 3.3 Exercises A we found the slope of the line with equation $x - 3y + 6 = 0$ by first finding two points on the line and then using the definition of the slope. We might use the two points $(0, 2)$ and $(3, 3)$, and substitute to obtain

$$m = \frac{y_2 - y_1}{x_2 - x_1} = \frac{3 - 2}{3 - 0} = \frac{1}{3}.$$

With what we have just learned, we can solve the same problem in a much simpler way by solving for y.

$$-3y = -x - 6 \qquad \text{Subtract } x \text{ and } 6$$

$$y = \frac{1}{3}x + 2 \qquad \text{Divide through by } -3$$

m = slope = $\frac{1}{3}$

The ability to find the slope of a line more quickly by solving for y is also helpful when considering parallel and perpendicular lines.

EXAMPLE 5 Finding if Lines Are Parallel or Perpendicular

Tell whether the lines with equations $3x - 2y + 7 = 0$ and $2x + 3y - 6 = 0$ are parallel, perpendicular, or neither.

Solve each equation for y (write each in slope-intercept form) to find the slope.

$$\begin{aligned} 3x - 2y + 7 &= 0 \\ -2y &= -3x - 7 \\ y &= \frac{3}{2}x + \frac{7}{2} \\ &\downarrow \\ m_1 &= \tfrac{3}{2} \end{aligned} \qquad \begin{aligned} 2x + 3y - 6 &= 0 \\ 3y &= -2x + 6 \\ y &= -\frac{2}{3}x + 2 \\ &\downarrow \\ m_2 &= -\tfrac{2}{3} \end{aligned}$$

Since $m_1 = \frac{3}{2}$ and $m_2 = -\frac{2}{3}$ are negative reciprocals, the two lines are perpendicular.

Practice Exercise 5

Tell whether the lines with equations $5x - 2y + 8 = 0$ and $2x + 5y - 17 = 0$ are parallel, perpendicular, or neither.

Answer: perpendicular

EXAMPLE 6 Recognizing Coinciding Lines

Tell whether the lines with equations $x - 4y + 2 = 0$ and $3x - 12y + 6 = 0$ are parallel, perpendicular, or neither.

Solve each equation for y.

$$\begin{aligned} x - 4y + 2 &= 0 \\ -4y &= -x - 2 \\ y &= \frac{1}{4}x + \frac{2}{4} \\ y &= \frac{1}{4}x + \frac{1}{2} \\ &\downarrow \\ m_1 &= \tfrac{1}{4} \end{aligned} \qquad \begin{aligned} 3x - 12y + 6 &= 0 \\ -12y &= -3x - 6 \\ y &= \frac{3}{12}x + \frac{6}{12} \\ y &= \frac{1}{4}x + \frac{1}{2} \\ &\downarrow \\ m_2 &= \tfrac{1}{4} \end{aligned}$$

Since $m_1 = m_2$ we are tempted to conclude that the two lines are parallel. However, the y-intercepts are also equal [both are $(0, \frac{1}{2})$] so the equations determine the same line, that is, the two lines coincide.

Practice Exercise 6

Tell whether the lines with equations $9x + 3y - 6 = 0$ and $3x + y - 2 = 0$ are parallel, perpendicular, or coincide.

Answer: The lines coincide.

Once we have used the slope-intercept form to identify the slope of a line, we can then find the equation of a line parallel or perpendicular to it.

EXAMPLE 7 The Equation of a Line Parallel to a Line

Find the equation of the line passing through the point $(-2, 5)$ parallel to the line with equation $4x + 2y - 9 = 0$.

Since we want a line parallel to the line with equation $4x + 2y - 9 = 0$, we must find the slope of this line. Solving for y,

$$\begin{aligned} 2y &= -4x + 9 \\ y &= -2x + \frac{9}{2}. \end{aligned} \qquad \text{Slope-intercept form}$$

Practice Exercise 7

Find the equation of the line passing through $(4, -2)$ perpendicular to the line with equation $5x - 3y + 10 = 0$.

The slope is -2, and this is also the slope of the desired line (parallel lines have equal slopes) which must pass through $(-2, 5)$. To find the desired line, substitute in the point-slope form.

$$y - y_1 = m(x - x_1)$$
$$y - 5 = -2(x - (-2))$$
$$y - 5 = -2(x + 2)$$
$$y - 5 = -2x - 4$$
$$2x + y - 1 = 0$$

The general form of the equation of the line through $(-2, 5)$ parallel to the line $4x + 2y - 9 = 0$ is $2x + y - 1 = 0$.

Answer: $3x + 5y - 2 = 0$

CAUTION

The point-slope form lets us find the equation of a line when we are given a point and the slope. It does not work, of course, when the line is vertical with undefined slope.

3 EQUATION OF THE LINE THROUGH TWO POINTS

To find the equation of a line passing through two points, use the slope formula with the point-slope form; as in the next example.

EXAMPLE 8 The Equation of the Line through Two Points

Find the general form of the equation of a line passing through the points $(4, -1)$ and $(2, 7)$.

Let $(x_1, y_1) = (4, -1)$ and $(x_2, y_2) = (2, 7)$ and find the slope.

$$m = \frac{y_2 - y_1}{x_2 - x_1} = \frac{7 - (-1)}{2 - 4} = \frac{8}{-2} = -4$$

Now use $m = -4$ and either point, say $(4, -1)$, in the point-slope form.

$$y - y_1 = m(x - x_1)$$
$$y + 1 = -4(x - 4)$$
$$y + 1 = -4x + 16$$
$$4x + y - 15 = 0$$

The general form of the equation of the line passing through $(4, -1)$ and $(2, 7)$ is $4x + y - 15 = 0$.

Practice Exercise 8

Find the general form of the equation of a line passing through $(-1, 5)$ and $(6, -3)$.

Answer: $8x + 7y - 27 = 0$

The name of the form of the equation tells us what we need to know to use that form.

Which Form to Use

1. If we know the *slope* and the *y-intercept,* we use the *slope-intercept* form.
2. If we know the *slope* and *any point,* we use the *point-slope* form.
3. If we know *two points,* we use the *slope formula* followed by the *point-slope* form.

Many applied problems can be described by a linear equation in two variables. The next example illustrates this.

EXAMPLE 9 SALES PROJECTION PROBLEM

In 1982 the total sales of a newly formed manufacturing company amounted to \$55,000. During the year 1989, the total sales were \$111,000. Assuming that the total sales y in year x can be approximated by a linear equation, find this equation and approximate the total sales expected in 1996.

Suppose we identify 1982 as year 1, $x = 1$. Then 1989 corresponds to year 8 and $x = 8$. When $x = 1$, $y = 55{,}000$ and when $x = 8$, $y = 111{,}000$, giving us the two points $(x_1, y_1) = (1, 55{,}000)$ and $(x_2, y_2) = (8, 111{,}000)$ on the line describing total sales. First find the slope of the line.

$$m = \frac{y_2 - y_1}{x_2 - x_1} = \frac{111{,}000 - 55{,}000}{8 - 1} = \frac{56{,}000}{7} = 8000$$

Now use 8000 for m and $(1, 55{,}000)$ for (x_1, y_1) in the point-slope form.

$$y - 55{,}000 = 8000(x - 1)$$
$$y - 55{,}000 = 8000x - 8000$$
$$y = 8000x + 47{,}000 \quad \text{Slope-intercept form}$$

We leave the equation in slope-intercept form since we want the value of y when x is 15 (in year 1996).

$$\begin{aligned} y &= 8000(15) + 47{,}000 \quad x = 15 \\ &= 120{,}000 + 47{,}000 \\ &= 167{,}000 \end{aligned}$$

In 1996, the company can expect total sales of about \$167,000.

PRACTICE EXERCISE 9

Kopy Kat Video Services can duplicate 5 videos at a cost of \$40. With volume purchasing of blank tapes, the owner discovers that 80 videos can be duplicated for \$565. If y is the cost of duplicating x videos, find a linear equation describing this relationship. Use the equation to determine the cost of duplicating 100 videos.

Answer: $y = 7x + 5$; \$705

4 GRAPHING EQUATIONS USING SLOPE

In Section 3.2 we learned how to graph linear equations using the intercepts. We conclude this section by showing how the graph of a linear equation given in slope-intercept form can be easily obtained using the rise-over-run concept of slope. The method is illustrated in the following example.

EXAMPLE 10 GRAPHING USING SLOPE

Graph $y = \frac{4}{3}x - 2$.

Since the y-intercept is $(0, -2)$, we know that the line passes through this point. We must find the one line of the infinitely many through $(0, -2)$ that has slope $\frac{4}{3}$. To do this, we will find a second point using

PRACTICE EXERCISE 10

Graph $y = 2x + 1$. [*Hint:* The slope is $2 = \frac{2}{1}$, so the rise is 2 and the run is 1.]

the rise of 4 units and the run of 3 units from slope $\frac{4}{3}$. Starting at the y-intercept $(0, -2)$, we move up 4 units (a rise of 4) and then move right 3 units (a run of 3), ending up at the point $(3, 2)$. The line passes through $(0, -2)$ and $(3, 2)$ as shown in Figure 3.21.

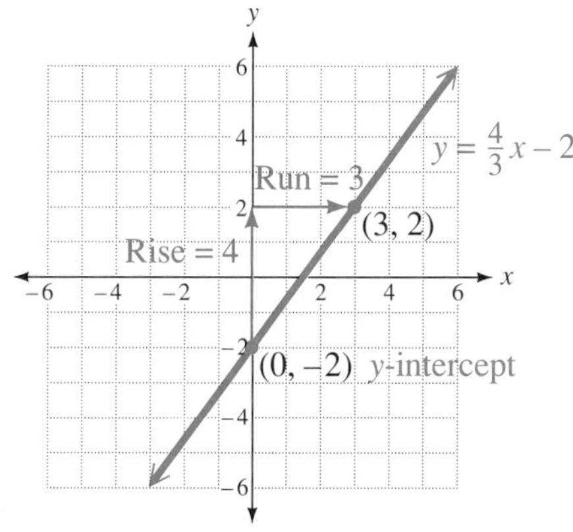

Figure 3.21

Answer: The line passes through $(0, 1)$ and $(1, 3)$.

To graph using slope, we do not need the equation of the line. All that is needed is one point on the line and the slope of the line.

CAUTION

If the slope is $-\frac{2}{3}$, we can say the rise is -2 and the run is 3. A rise of -2 means move *down* two units. Alternatively, if we say the rise is 2 and the run is -3, a run of -3 means move *left* 3 units. Either interpretation will result in the same straight line.

3.4 EXERCISES A

Supply the required information for Exercises 1–21.

1. Find the general form of the equation of the line passing through the point $(3, -1)$ with slope -2. [*Hint:* First find the point-slope form and then transform it to the general form.]

2. Find the general form of the equation of the line passing through the point $(2, -4)$ parallel to a line with slope $\frac{1}{3}$. Also give the x-intercept and y-intercept.

3. Find the general form of the equation of the line passing through the point $(-1, -3)$ perpendicular to a line with slope $\frac{1}{5}$.

4. Find the general form of the equation of the line passing through the points $(7, -1)$ and $(5, 3)$.

5. Find the slope-intercept form of the equation of the line passing through the points $(-2, 3)$ and $(1, -3)$. What is the slope? The y-intercept?

6. Find the slope and y-intercept of the line with equation $5x - 2y + 4 = 0$.

7. Find the slope and y-intercept of the line with equation $5x + 4 = 0$.

8. Find the slope and y-intercept of the line with equation $-2y + 4 = 0$.

9. Tell whether the lines with equations $5x - 3y + 8 = 0$ and $3x + 5y - 7 = 0$ are parallel, perpendicular, or neither. Justify your answer.

10. Tell whether the lines with equations $2x - y + 7 = 0$ and $-6x + 3y - 1 = 0$ are parallel, perpendicular, or neither. Justify your answer.

11. Tell whether the lines with equations $2x + 1 = 0$ and $-3y - 4 = 0$ are parallel, perpendicular, or neither. Justify your answer.

12. Find the general form of the equation of the line passing through $(-1, 4)$ parallel to the line with equation $-3x - y + 4 = 0$.

13 Find the general form of the equation of the line passing through $(-1, 4)$ perpendicular to the line with equation $-3x - y + 4 = 0$.

14. Find the general form of the equation of the line through $(-2, -4)$ with $m = -3$.

$y - y_1 = m(x - x_1)$

15. Find the general form of the equation of the line with y-intercept $(0, 2)$ and slope -5.

16 Find the general form of the equation of the line with x-intercept $(3, 0)$ and slope $\frac{1}{2}$.

17. Find the slope and y-intercept of the line with equation $6x + 2y - 10 = 0$.

18. Find the slope and y-intercept of the line with equation $3y + 9 = 0$.

19. Find the slope and y-intercept of the line with equation $x = 7$.

20 Find the general form of the equation of a line through $(2, 3)$ with undefined slope.

21. Consider the line that passes through the points $(1, 6)$ and $(5, -2)$.
(a) Find the slope of a line parallel to this line. **(b)** Find the slope of a line perpendicular to this line.

Some application problems involve information that can be described by a straight-line graph. If two pieces of information are given, and they are interpreted as points on the graph of a line, an equation describing the situation can be found. Use this idea in the following two exercises.

22 John makes wood-burning stoves. He discovers that 10 stoves can be constructed for \$2800, and 25 stoves can be made for \$6125. If y is the cost required to make x stoves, find a linear equation describing this relationship. Use the equation to calculate the cost of making 31 stoves.

23. In 1987 (call this year 1), the Mutter Manufacturing Company had sales of \$40,000. Three years later, in 1990 (year 4), total sales were \$90,000. If y represents total sales in year x, find a linear equation describing this relationship. Use the equation to approximate total sales for the year 1991 (year 5).

Graph the following equations using the slope method illustrated in Example 10.

24. $y = \frac{2}{3}x - 4$

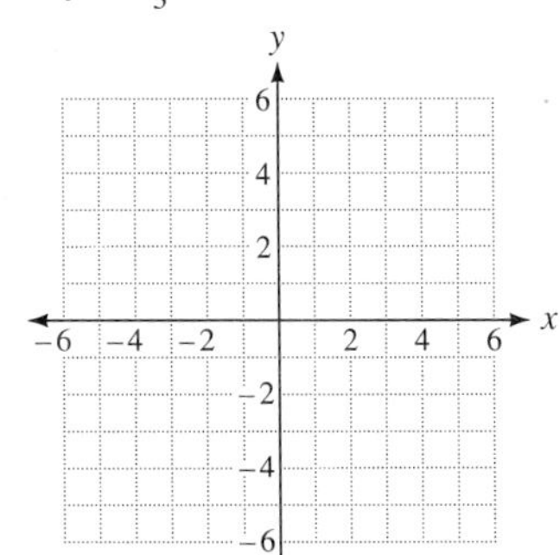

25. $y = 4x + 1$

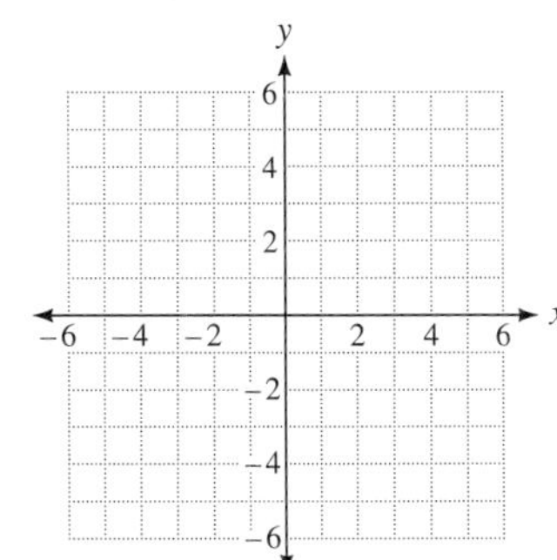

26. $y = -\frac{2}{3}x + 3$

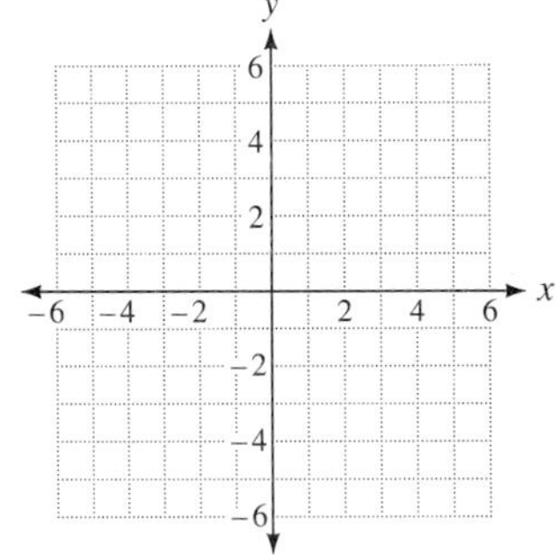

27. $y = -3x + 2$

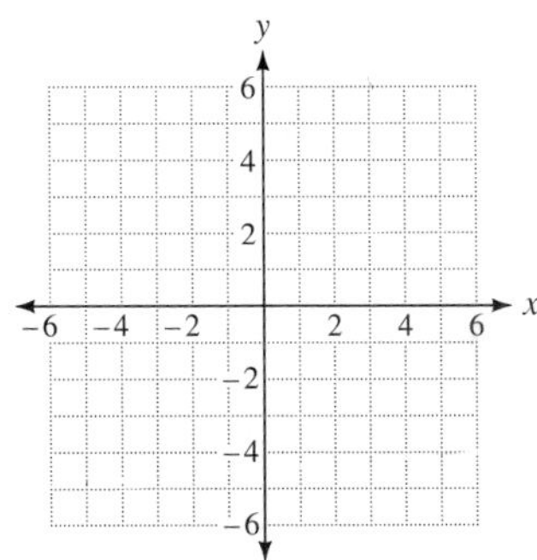

28. $x + 3y - 6 = 0$

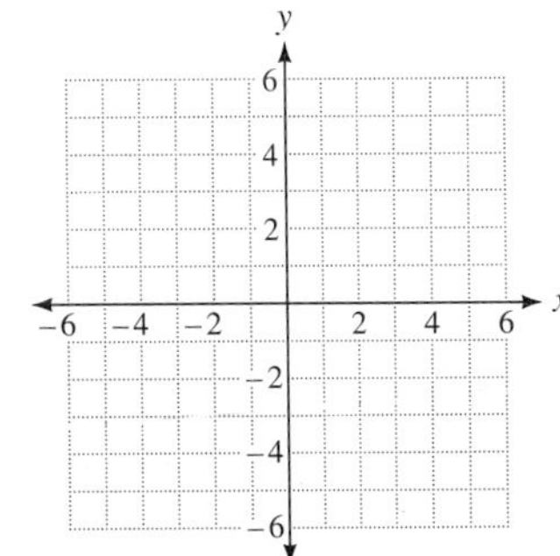

29. $3x - 4y + 8 = 0$

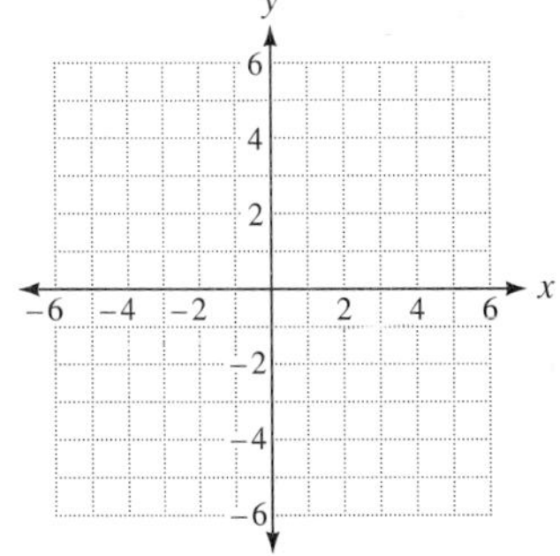

Without finding the equation of the line, graph the line using the given information and the slope method.

30. Through (1, 2) with slope -3

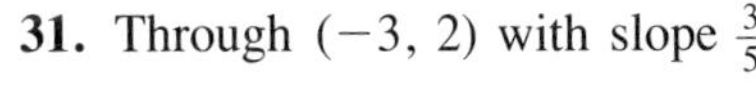
31. Through $(-3, 2)$ with slope $\frac{3}{5}$

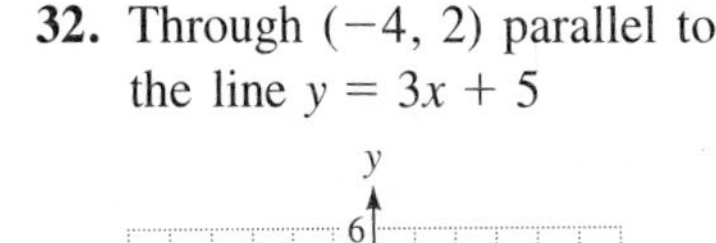
32. Through $(-4, 2)$ parallel to the line $y = 3x + 5$

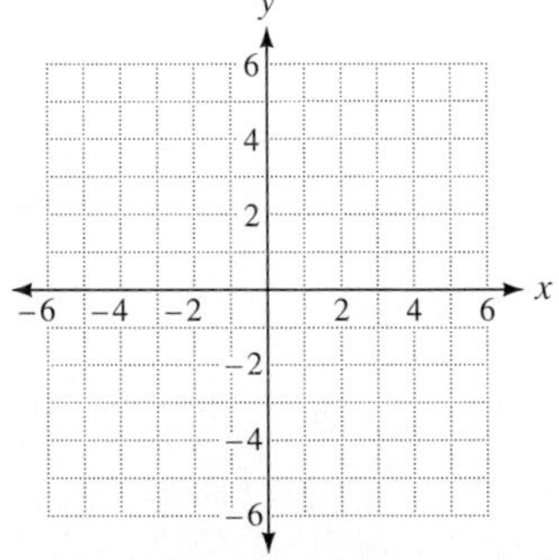

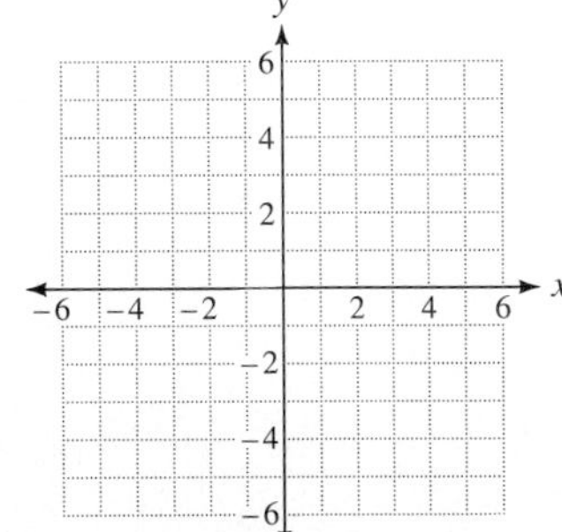

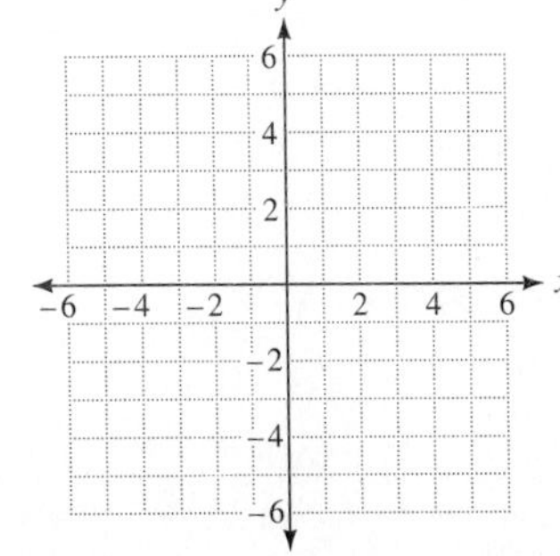

33. The graph shows the relationship between the distance y traveled by a car and the time of travel x. Find the equation that describes this relationship and use it to find the distance traveled after 10 hours.

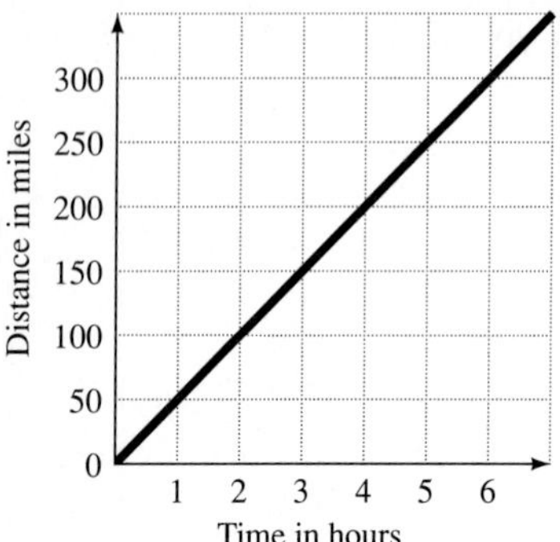

34. The graph shows the relationship between the cost of a rental mobile home y and the depreciation allowed for tax purposes x. Find the equation that describes this relationship and use it to find the value of the home after five and one-half years.

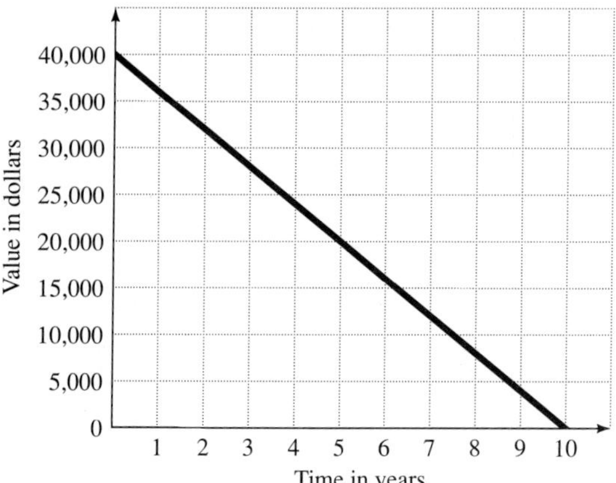

FOR REVIEW

35. Find the slope of the line joining the points with coordinates $(1, 4)$ and $(-7, -12)$. What is the slope of a line parallel to this line? perpendicular to this line?

36. Without finding the actual values, tell whether each line in the figure has positive slope, negative slope, zero slope, or undefined slope.

(a) Line l_1 has ________ slope.
(b) Line l_2 has ________ slope.
(c) Line l_3 has ________ slope.
(d) Line l_4 has ________ slope.
(e) The x-axis has ________ slope.
(f) The y-axis has ________ slope.

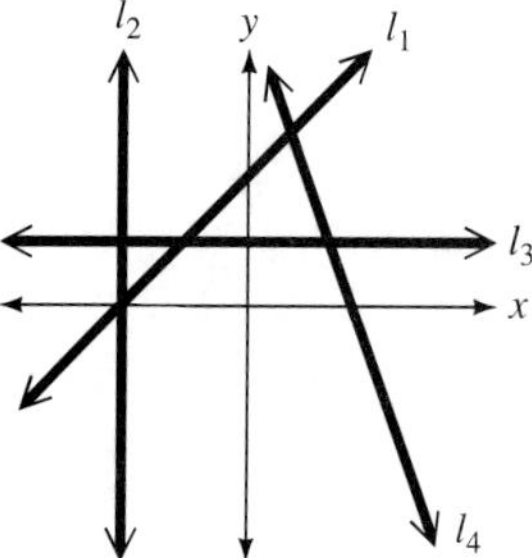

ANSWERS: **1.** $2x + y - 5 = 0$ **2.** $x - 3y - 14 = 0$; x-intercept $(14, 0)$; y-intercept $\left(0, -\frac{14}{3}\right)$ **3.** $5x + y + 8 = 0$ **4.** $2x + y - 13 = 0$ **5.** $y = -2x - 1$; slope $= -2$; y-intercept $(0, -1)$ **6.** slope $= \frac{5}{2}$; y-intercept $(0, 2)$ **7.** undefined slope; no y-intercept **8.** slope $= 0$; y-intercept $(0, 2)$ **9.** perpendicular $\left(\text{slopes are } \frac{5}{3} \text{ and } -\frac{3}{5}\right)$ **10.** parallel (both slopes are 2 and the y-intercepts are different) **11.** perpendicular (each is parallel to one of the axes) **12.** $3x + y - 1 = 0$ **13.** $x - 3y + 13 = 0$ **14.** $3x + y + 10 = 0$ **15.** $5x + y - 2 = 0$ **16.** $x - 2y - 3 = 0$ **17.** slope $= -3$; y-intercept $(0, 5)$ **18.** slope $= 0$; y-intercept $(0, -3)$ **19.** undefined slope; no y-intercept **20.** $x - 2 = 0$ **21.** (a) -2 (b) $\frac{1}{2}$ **22.** $665x - 3y + 1750 = 0$; \$7455 **23.** $50{,}000x - 3y + 70{,}000 = 0$; \$106,666.67

24.
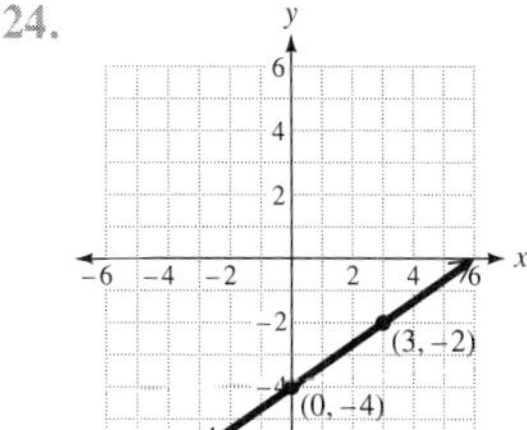

25.
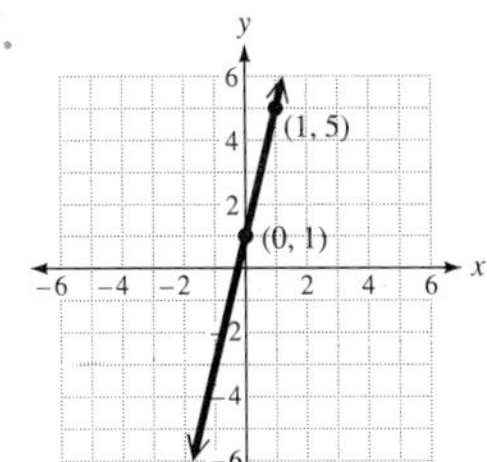

26.
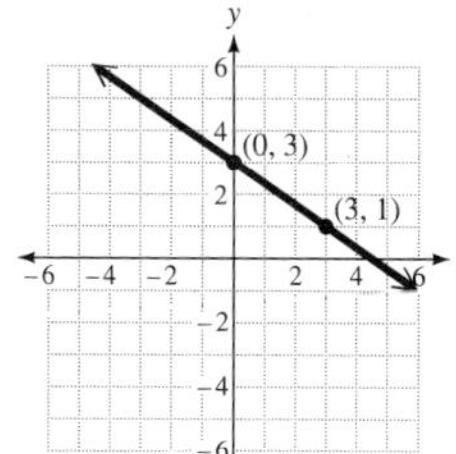

27.
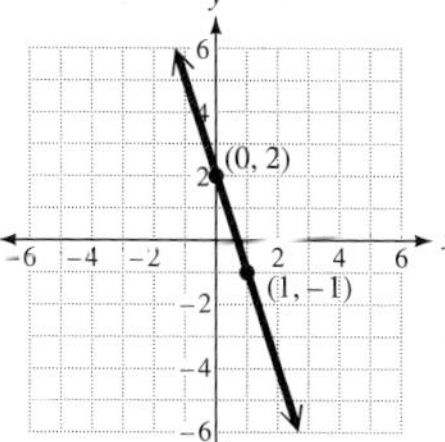

28.
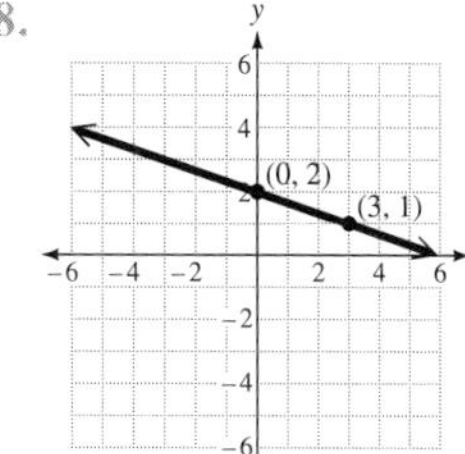

29.
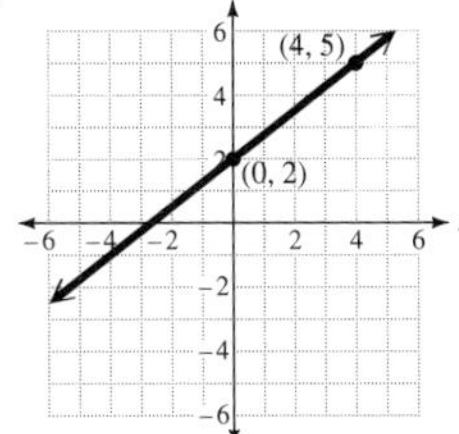

30.
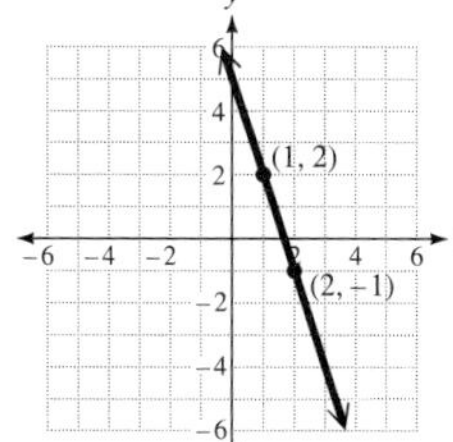

31.
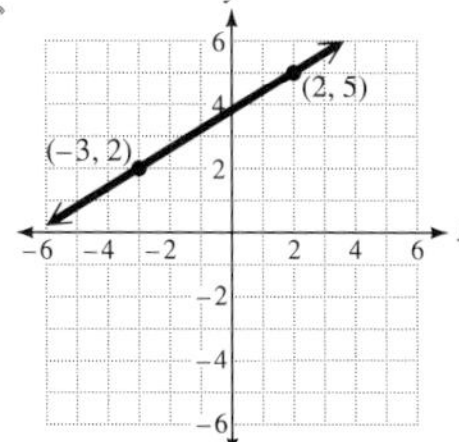

32.
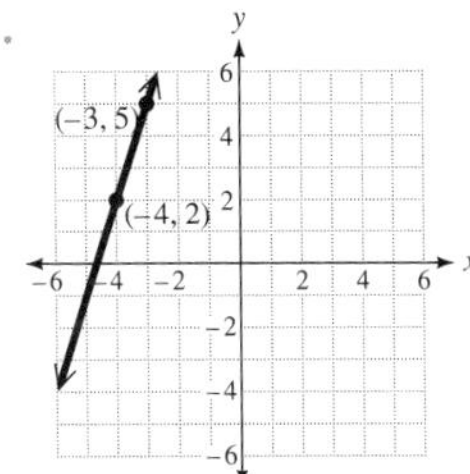

33. $y = 50x$; 500 miles 34. $y = -4000x + 40{,}000$; $18,000 35. 2; 2; $-\frac{1}{2}$
36. (a) positive (b) undefined (c) zero (d) negative (e) zero (f) undefined

3.4 EXERCISES B

Supply the required information for Exercises 1–23.

1. Find the general form of the equation of the line passing through the point $(-2, -4)$, with slope -3.

2. Find the general form of the equation of the line passing through the point $(-3, 5)$ parallel to a line with slope -4. Also give the x-intercept and y-intercept.

3. Find the general form of the equation of the line passing through the point $(3, -5)$ perpendicular to a line with slope 7.

4. Find the general form of the equation of the line passing through the points $(-7, 1)$ and $(3, -5)$.

5. Find the slope-intercept form of the equation of the line passing through the points $(4, -5)$ and $(2, 8)$. What is the slope? The y-intercept?

6. Find the slope and y-intercept of the line with equation $6x + 2y - 10 = 0$.

7. Find the slope and y-intercept of the line with equation $x + 7 = 0$.

8. Find the slope and y-intercept of the line with equation $3y + 9 = 0$.

9. Tell whether the lines with equations $2x - y + 3 = 0$ and $x + 2y - 5 = 0$ are parallel, perpendicular, or neither. Justify your answer.

10. Tell whether the lines with equations $5x - 2y + 3 = 0$ and $-10x + 4y + 3 = 0$ are parallel, perpendicular, or neither. Justify your answer.

11. Tell whether the lines with equations $4x - y + 7 = 0$ and $3x - 1 = 0$ are parallel, perpendicular, or neither. Justify your answer.

12. Find the general form of the equation of the line passing through $(-2, 5)$ parallel to the line with equation $4x + 2y - 9 = 0$.

13. Find the general form of the equation of the line passing through $(4, -7)$ perpendicular to the line with equation $x - 5y + 7 = 0$.

14. Find the general form of the equation of the line through $(3, 0)$ with $m = -\frac{1}{2}$.

15. Find the general form of the equation of the line with y-intercept $(0, 4)$ and slope -3.

16. Find the general form of the equation of the line with x-intercept $(-5, 0)$ and slope 2.

17. Find the slope and y-intercept of the line with equation $3x + 6y - 5 = 0$.

18. Find the slope and y-intercept of the line with equation $2y + 10 = 0$.

19. Find the slope and y-intercept of the line with equation $2x - 5 = 0$.

20. Find the general form of the equation of a line through $(2, 3)$ with slope zero.

21. Consider the line that passes through the points $(2, -1)$ and $(-3, 9)$.
(a) Find the slope of a line parallel to this line. **(b)** Find the slope of a line perpendicular to this line.

22. Willie makes leather bags. He knows that 5 bags can be made for \$45, and 20 bags can be made for \$140. If y is the cost of making x bags, find a linear equation describing this relationship. Use the equation to calculate the cost of making 65 bags.

23. In 1988 (call this year 1), the BRS Construction Company had sales of \$100,000. Two years later, in 1990 (year 3), total sales were \$220,000. If y represents total sales in year x, find a linear equation describing this relationship. Use the equation to approximate total sales for the year 1993 (year 6).

Graph the following equations using the slope method illustrated in Example 10.

24. $y = \dfrac{3}{2}x + 1$

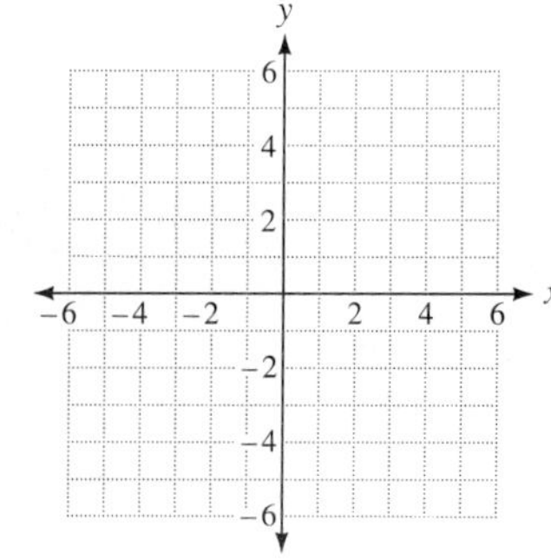

25. $y = 4x - 1$

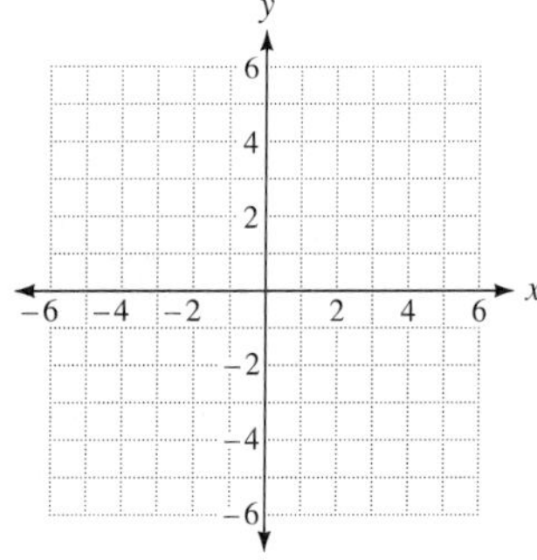

26. $y = -\dfrac{3}{2}x + 5$

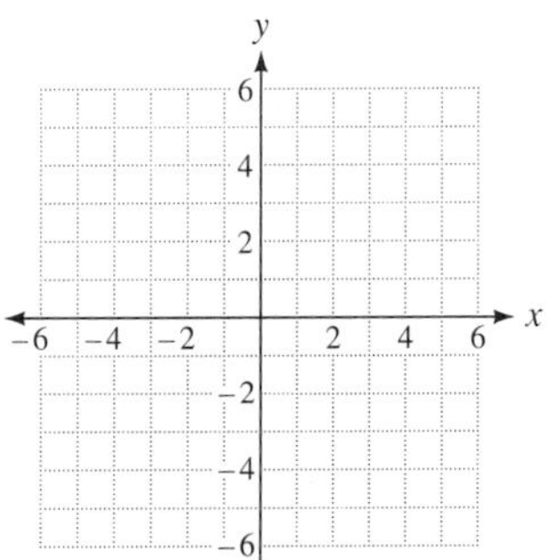

27. $y = -5x + 3$

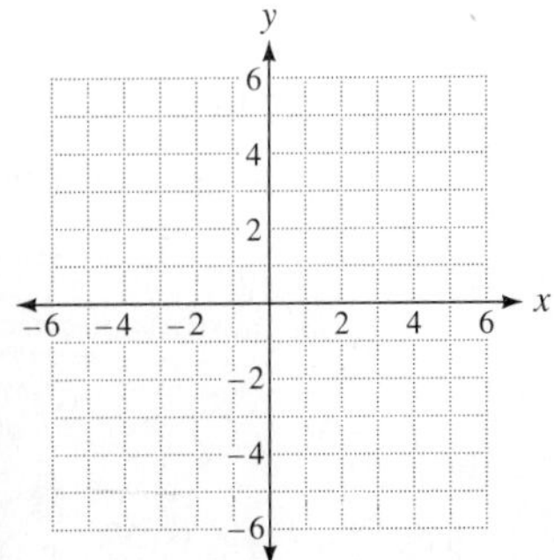

28. $x - 3y + 9 = 0$

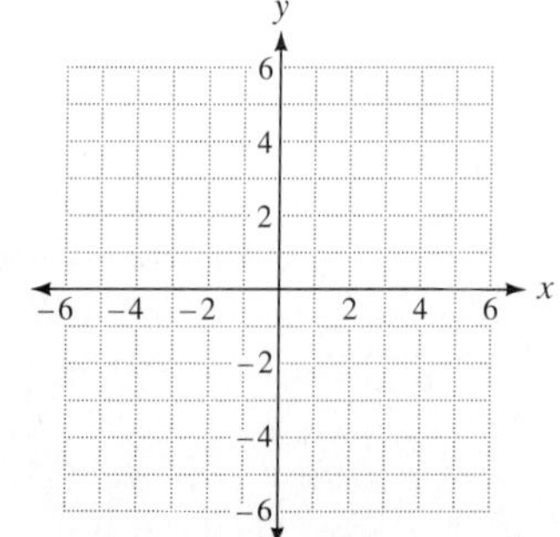

29. $3x + 4y - 8 = 0$

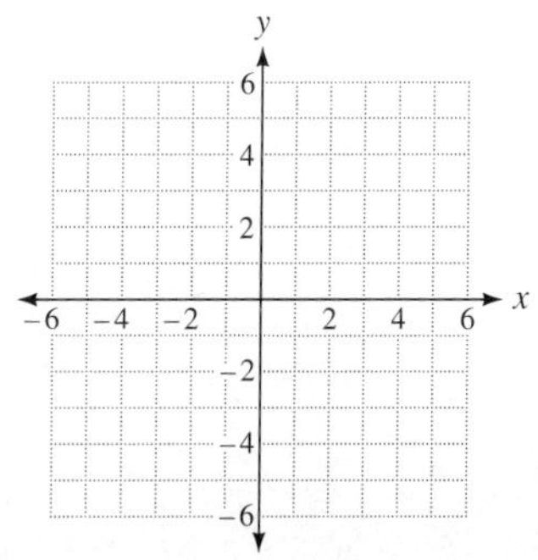

Without finding the equation of the line, graph the line using the given information and the slope method.

30. Through (2, 3) with slope $\frac{1}{2}$

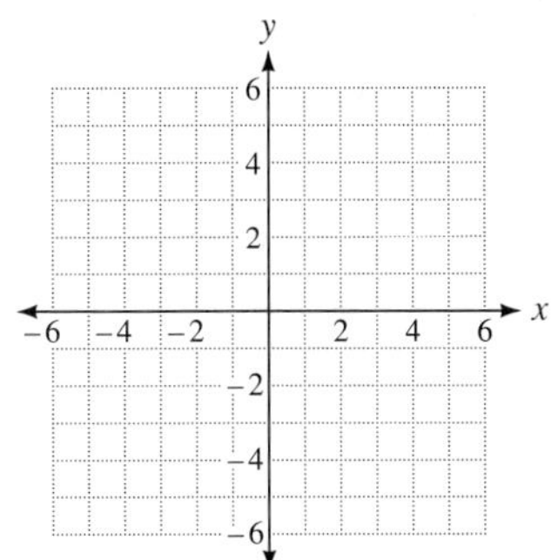

31. Through $(-3, -4)$ with slope $\frac{4}{5}$

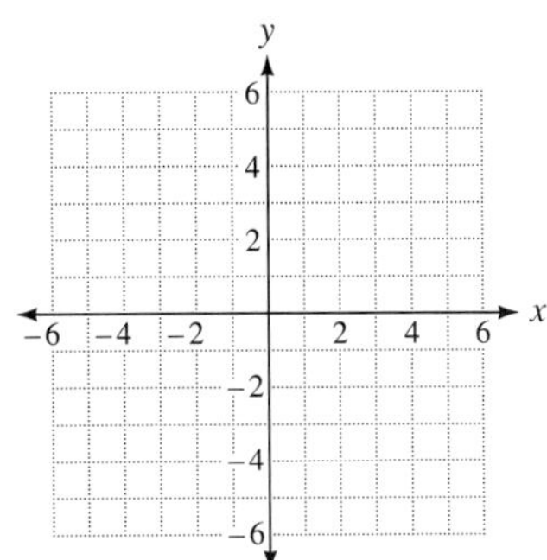

32. Through $(3, -5)$ parallel to the line $y = 4x + 7$

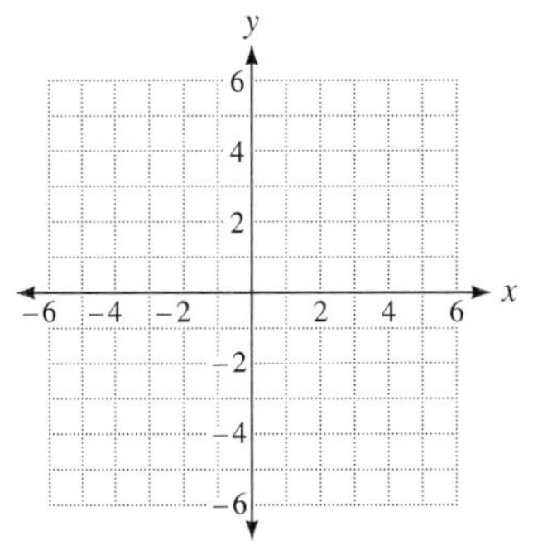

33. The graph shows the relationship between the monthly income y and total sales x of an insurance salesperson. Find the equation that describes this relationship and use it to find the monthly income when monthly sales are $32,000.

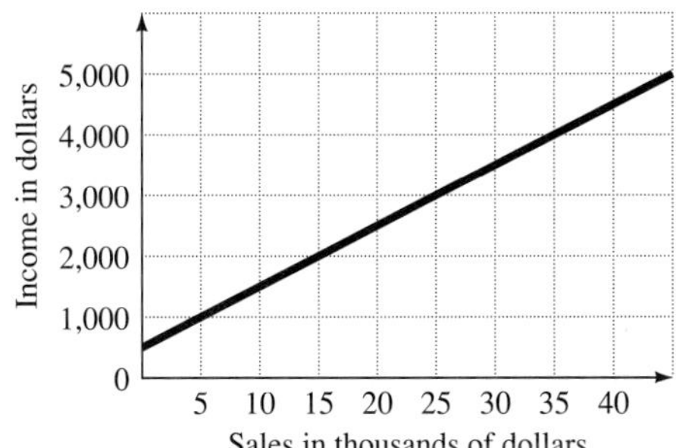

34. The graph shows the relationship between the cost of a delivery truck y and the depreciation allowed for tax purposes x. Find the equation that describes this relationship and use it to find the value of the truck after four and one-half years.

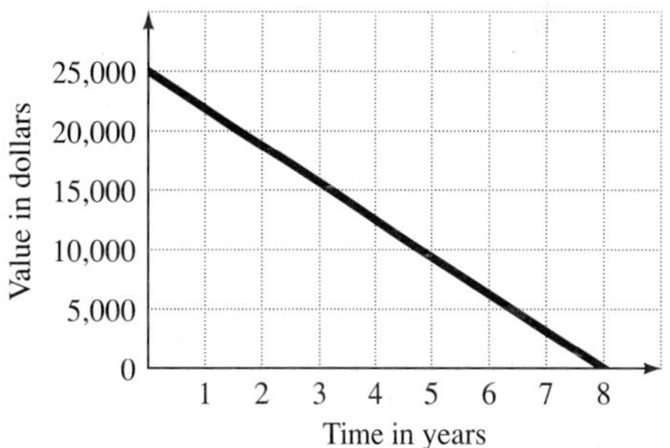

FOR REVIEW

35. Find the slope of the line joining $(2, -5)$ and $(-8, -7)$. What is the slope of a line parallel to this line? perpendicular to this line?

36. Without finding the actual values, tell whether each line in the figure has positive slope, negative slope, zero slope, or undefined slope.

(a) Line l_1 has ________ slope.
(b) Line l_2 has ________ slope.
(c) Line l_3 has ________ slope.
(d) Line l_4 has ________ slope.
(e) The y-axis has ________ slope.
(f) The x-axis has ________ slope.

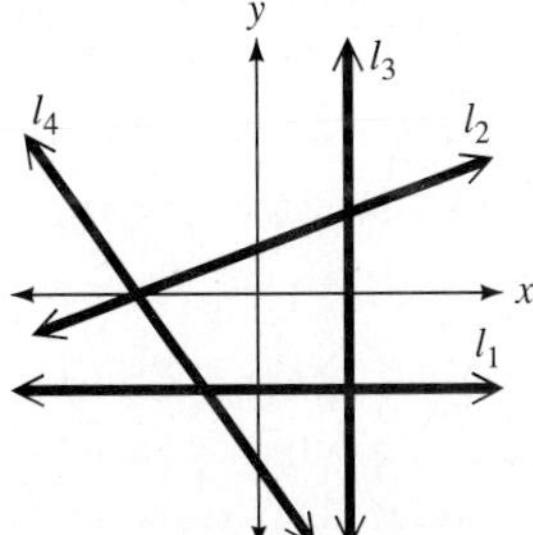

3.4 EXERCISES C

1. Give the general form of the equation of the line passing through the points $\left(\frac{1}{2}, -\frac{3}{4}\right)$ and $\left(-\frac{3}{8}, \frac{5}{12}\right)$.

2. Find the value of a so that the lines with equations $x + 2y = 3$ and $ax + y = 7$ are perpendicular. [Answer: -2]

Find the equations of the (a) vertical and (b) horizontal lines through each point.

3. $(-1, 5)$

4. $(-\sqrt{3}, \sqrt{2})$

5. $(\sqrt{5}, \pi)$

6. Find the slope of a line parallel to the line with equation $ax + by + c = 0$. [*Hint:* The answer will be in terms of a, b, and c.]

7. Find the slope of a line perpendicular to the line with equation $ax + by + c = 0$.

8. An economist uses a demand curve to show the relationship between the price of a product y and the demand by consumers for the product, x. Find the equation of the demand curve graphed here and use it to find the demand for the product when the price is \$2.50 per unit.

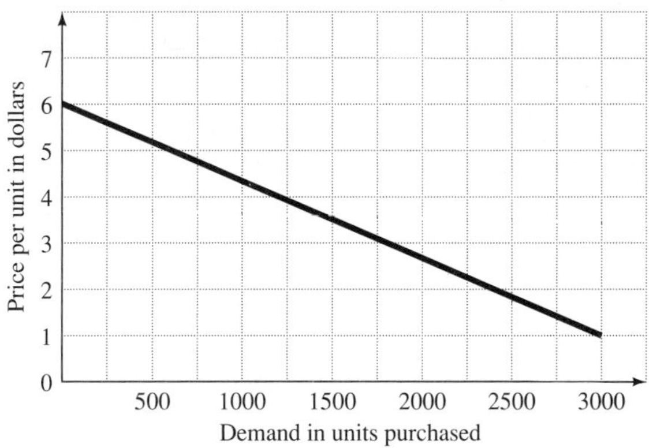

3.5 RELATIONS AND FUNCTIONS

STUDENT GUIDEPOSTS

1. Relations
2. Functions
3. Dependent and Independent Variables
4. Graphing Relations and Functions
5. Vertical Line Test
6. Functions Described by Equations

1 RELATIONS

In this section we introduce two central concepts in mathematics: relations and functions. The idea of a relation is already familiar since relationships or correspondences between two sets of objects play an important role in everyday life. When the terminology of mathematics is applied to these concepts, we discover that many relations can be described precisely using equations. Consider the following *relations:*

1. To each automobile in the state there corresponds a license number.
2. To each person in the United States there corresponds a legal last name.
3. To each triangle there corresponds a perimeter.
4. To each of the starting five on a basketball team in a particular game, there corresponds a number, the number of points each scores in that game.
5. To each parent in the United States there corresponds one or more children.

Definition of a Relation (by Rule)

A correspondence between a first set of objects, the **domain,** and a second set of objects, the **range,** is called a **relation** if to each element of the domain there corresponds *one or more* elements in the range.

2 FUNCTIONS

Each of the above five correspondences is a relation. In fact, the first four share an even more specific property in that they are *functions*.

Definition of a Function (by Rule)

A **function** is a relation with the additional property that to each element in the domain there corresponds *one and only one* element in the range.

CAUTION

By definition, all functions are relations but not all relations are functions. The fifth relation above is not a function since a parent may correspond to more than one child.

EXAMPLE 1 IDENTIFYING RELATIONS AND FUNCTIONS

Which of the following relations are functions?

(a) Relation: item to its price
Domain: items in a discount store
Range: the set of all prices

This relation *is* a function since each item has one and only one selling price. Notice that the same price (an element of the range) may correspond to several different items (elements of the domain). This does not change the fact that the relation of an item to its price is a function.

(b) Relation: number to its square
Domain: the set of integers
Range: the set of whole number perfect squares

This relation *is* a function since each integer has one and only one square.

(c) Relation: number to its square root
Domain: the set of whole number perfect squares
Range: the set of integers

This *is not* a function since each whole number perfect square, other than 0, has two integer square roots. For example, 2 and -2 are both square roots of 4. Thus, two numbers in the range (in this case 2 and -2) correspond to the same number in the domain (in this case 4).

(d) Relation: number to its principal square root
Domain: the set of whole number perfect squares
Range: the set of whole numbers

This relation *is* a function since each whole number perfect square has exactly one principal (nonnegative) square root.

PRACTICE EXERCISE 1

Which of the following relations are functions?

(a) Relation: car to its price
Domain: cars in stock
Range: the set of all prices

(b) Relation: number to its cube root
Domain: integer perfect cubes
Range: integers

(c) Relation: "is the brother of"
Domain: males with siblings
Range: people with brothers

(d) Relation: six times a number
Domain: real numbers
Range: real number multiples of 6

Answers: (a) function (b) function (c) not a function (d) function

One way to represent a function is to use a table. For example, consider the starting five on a basketball team and the points each scored in a given game. The domain of this function is the set of players while the range is the set of scores. We often use arrows to indicate the object in the range (the arrow points *to* it) which corresponds to the object in the domain (the arrow points *from* it).

Domain		*Range*
Troy	⟶	20
Rick	⟶	19
Eric	⟶	14
Jeff	⟶	14
Andy	⟶	9

Notice that one number in the range, 14, is identified with two members in the domain. The definition of a function allows for this (which is possible in reality), but it does not allow a player, an element of the domain, to be identified with two different numbers in the range (which *is impossible* in reality). We usually do not list an object in the range more than once. Thus, the function would take the following form.

Domain		*Range*
Troy	⟶	20
Rick	⟶	19
Eric	⟶	14
Jeff	⟶	14
Andy	⟶	9

EXAMPLE 2 USING TABLES TO DESCRIBE FUNCTIONS

Which of the following relations are functions?

(a) *Domain* *Range*

1 → 4
2 → 3
3 → 5

This relation *is* a function since each object in the domain corresponds to exactly one object in the range.

(b) *Domain* *Range*

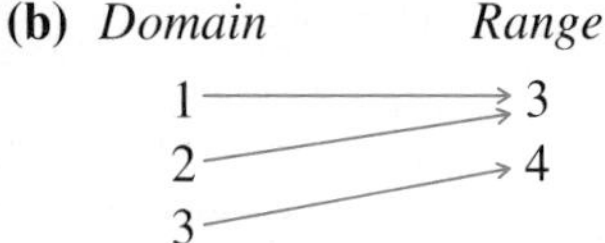

This relation *is* a function since 3 (in the range) can correspond to the two objects 1 and 2 (in the domain).

(c) *Domain* *Range*

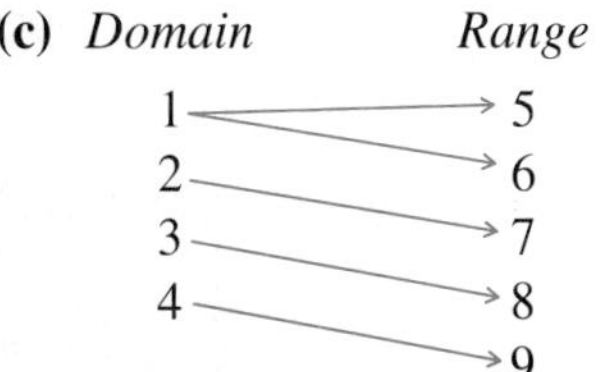

This relation *is not* a function since 1 (in the domain) corresponds to two objects, 5 and 6 (in the range).

PRACTICE EXERCISE 2

Tell whether each relation is a function.

(a) *Domain* *Range*

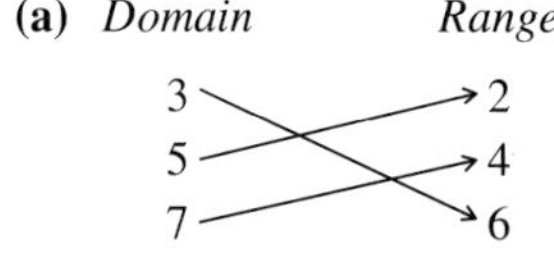

(b) *Domain* *Range*

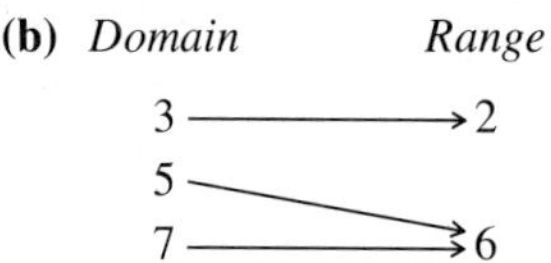

(c) *Domain* *Range*

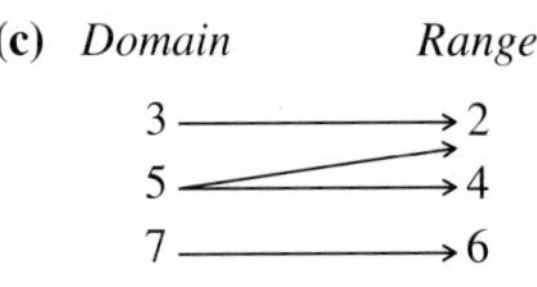

Answers: (a) function (b) function (c) not a function

3 DEPENDENT AND INDEPENDENT VARIABLES

When the objects in the domain and range of a relation are real numbers, a set of ordered pairs of numbers (x, y) can be used to define the relation. If $x \rightarrow y$, that is, if x is related to y, then the pair (x, y) is included in the relation. We think of the value of y as depending on the value of x and call y the **dependent variable** and x the **independent variable.** The set of all values of the independent variable is the domain of the relation, and the set of all values of the dependent variable is the range.

Consider again the relations given in Example 2. Using ordered pairs, the relation in (a) is $\{(1, 4), (2, 3), (3, 5)\}$, the relation in (b) is $\{(1, 3), (2, 3), (3, 4)\}$, and the relation in (c) is $\{(1, 5), (1, 6), (2, 7), (3, 8), (4, 9)\}$.

The observations above are restated in the following alternative definition of relation and function.

Definition of Relation and Function (Using Ordered Pairs)

Any set of ordered pairs of numbers is called a **relation.** A **function** is a relation with the additional restriction that no two ordered pairs can have the same x-coordinate and different y-coordinates.

Notice that the relations in Example 2 (a) and (b) are indeed functions, but the relation in (c) is not since (1, 5) and (1, 6) have the same x-coordinate and different y-coordinates.

4 GRAPHING RELATIONS AND FUNCTIONS

The ordered-pair definition of a relation provides a useful way to graph relations and functions since ordered pairs can be plotted in a rectangular coordinate system.

EXAMPLE 3 GRAPHING RELATIONS AND FUNCTIONS

List the ordered pairs in each relation, plot its graph, and tell whether it is a function.

(a) *Relation*

Domain	*Range*	*Ordered pairs in the relation*
$1 \rightarrow 1$		(1, 1)
$2 \rightarrow 2$		(2, 2)
$3 \rightarrow 2$		(3, 2)
$4 \rightarrow 3$		(4, 3)
$5 \rightarrow 4$		(5, 4)

Graph of the relation

Figure 3.22

Since every number in the domain is paired with exactly one number in the range, this relation is a function.

PRACTICE EXERCISE 3

List the ordered pairs and tell whether the relation is a function.

(a) *Domain* *Range*

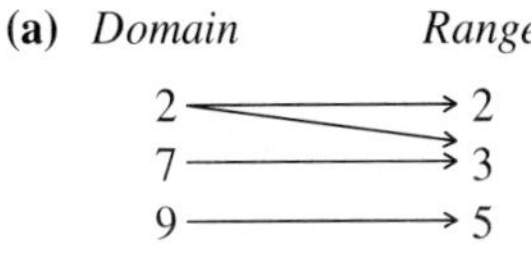

(b) *Relation*

Domain	*Range*
1 → 2	
1 → 3 (from 1)	
2 → 4	
3 → 4	

Ordered pairs in the relation

(1, 2)
(1, 3)
(2, 4)
(3, 4)

Graph of the relation

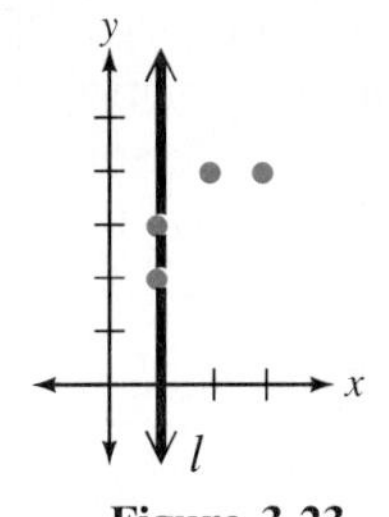

Figure 3.23

Since the number 1 in the domain is paired with two numbers, 2 and 3, in the range, this relation is not a function.

(b) *Domain* *Range*

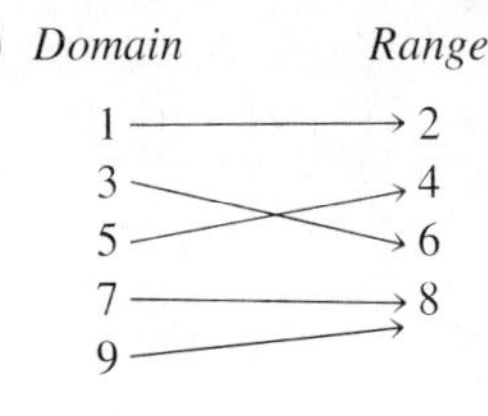

Answers: (a) (2, 2), (2, 3), (7, 3), (9, 5); not a function (b) (1, 2), (3, 6), (5, 4), (7, 8), (9, 8); function

5 VERTICAL LINE TEST

Notice in Figure 3.23 that two points on the graph, (1, 2) and (1, 3), lie on the same vertical line l. This property is true of all relations that are not functions. On the other hand, any collection of points that has the property that no vertical line passes through two or more points, is the graph of a function. This allows us to use the following test.

Vertical Line Test

If each vertical line intersects the graph of a relation in no more than one point, the relation is a function. FROM THE X-AXIS

EXAMPLE 4 Using the Vertical Line Test

Use the vertical line test to tell whether the graph is the graph of a function.

(a) Figure 3.24 *is* the graph of a function since no vertical line crosses the graph twice.

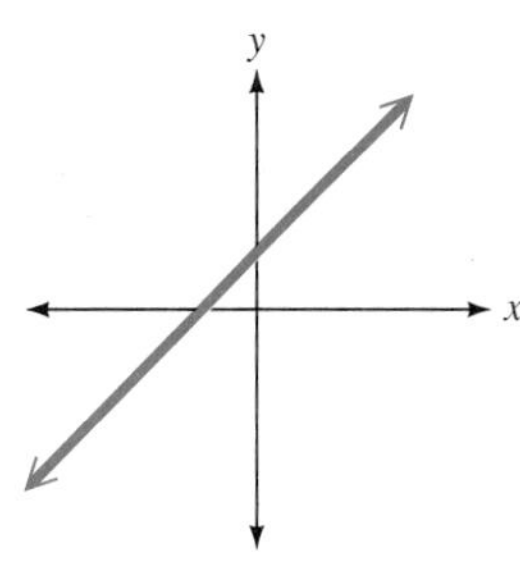

Figure 3.24

(b) Figure 3.25 *is not* the graph of a function since the vertical line l crosses the graph twice. Note that two values of y (y_1 and y_2) correspond to the same value of x (x_1).

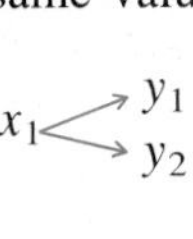

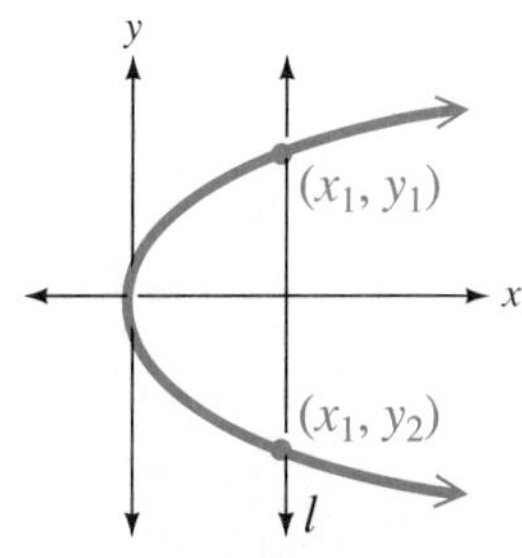

Figure 3.25

Practice Exercise 4

Use the vertical line test to tell whether or not each graph is the graph of a function.

(a)

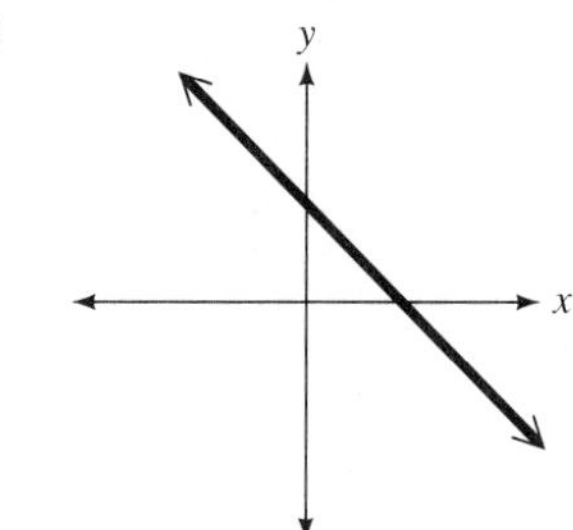

(b)

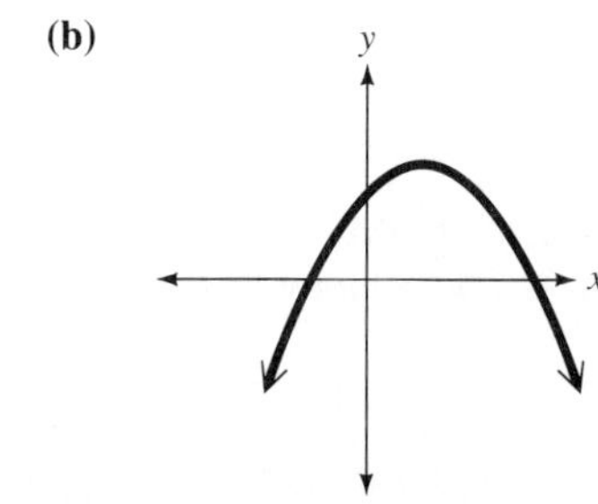

(c) Figure 3.26 *is not* the graph of a function, as shown by the vertical line l. Here a given value of x (x_1) corresponds to three values of y (y_1, y_2, and y_3).

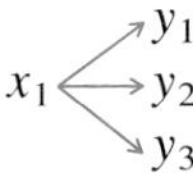

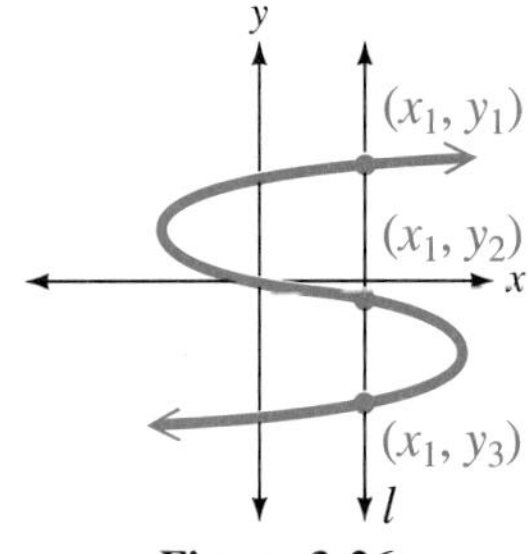

Figure 3.26

(c)

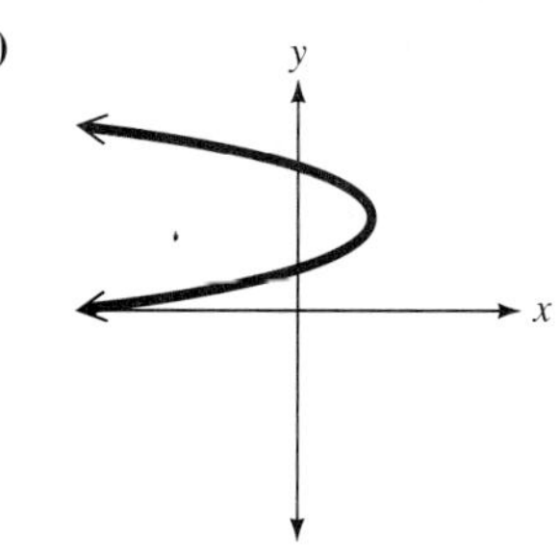

Answers: (a) function (b) function (c) not a function

6 FUNCTIONS DESCRIBED BY EQUATIONS

Up to now we have discussed functions described by diagram, by ordered pairs, and by graphs. Often, an equation is used to describe a relation for which the ordered pair solutions (x, y) fit the definition of a function. In fact, nearly every linear equation

$$ax + by + c = 0$$

describes a function. The only linear equations that do not, have $b = 0$ and a vertical line for a graph. For example, $2x - 3 = 0$ does not describe a function, as is clear from the graph of $2x - 3 = 0$ or $2x = 3$ or $x = \frac{3}{2}$ in Figure 3.27. This cannot be the graph of a function since for one value of x, $\frac{3}{2}$, there are more than one (in fact infinitely many) values of y. However, any other type of line in the plane is the graph of a function. We will consider functions defined by equations in more detail in the following sections.

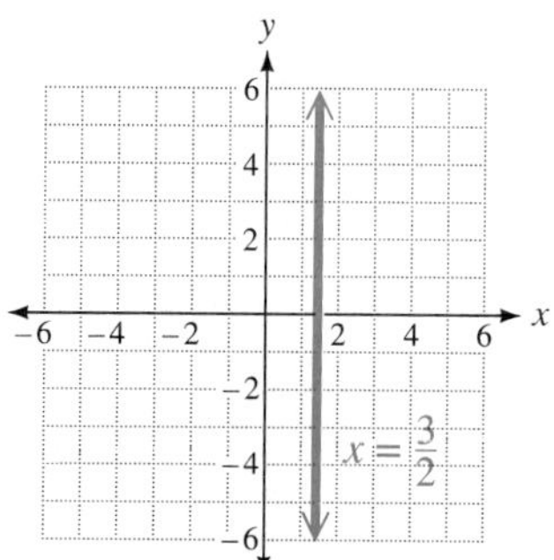

Figure 3.27

3.5 EXERCISES A

Write true *or* false *for Exercises 1 and 2. If the statement is false, explain why.*

1. A relation is a correspondence between a first set called the domain and a second set called the range of the relation.

2. If x corresponds to y ($x \rightarrow y$) in a function, the ordered pair (x, y) belongs to the function, and the point in the plane that corresponds to this ordered pair is a point on the graph of the function.

Tell which of the relations in Exercises 3–4 are functions.

3. Relation: a number to its fourth roots
Domain: the set of whole numbers
Range: the set of fourth roots of whole numbers

4. Relation: person to his/her annual salary
Domain: all salaried people
Range: the set of salaries

Which of the following are functions? Explain.

5. 1 → 3; 2 → 4; 3 → 6; 4 → 5

6. 1 → 4; 2 → 4; 3 → 5; 4 → 5

7. 1 → 3; 2 → 4; 3 → 5; 4 → 6

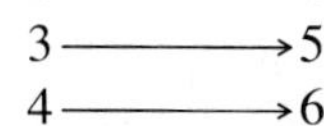

8. 1 → 3; 1 → 5; 2 → 6; 3 → 3; 4 → 4

9. 2 → −1; 2 → 0; 2 → 1; 4; 6; 7

10. 2 → −1; 4 → −1; 6 → 0; 7 → 1

11. 2 → 0; 4 → −1; 6 → 1; 7 → 1

12. 2 → −1; 4 → 1; 6 → 1; 7 → 1

Which of the following are graphs of functions? Explain.

13.

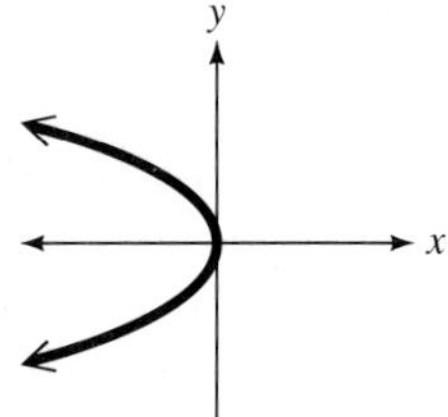

14.

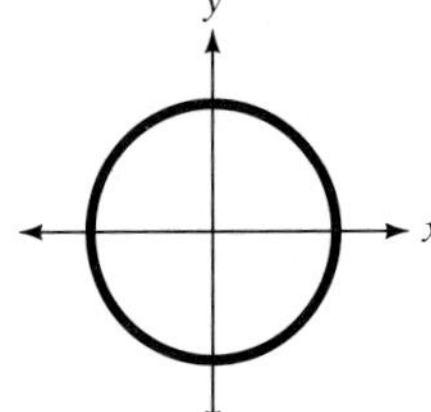

15.

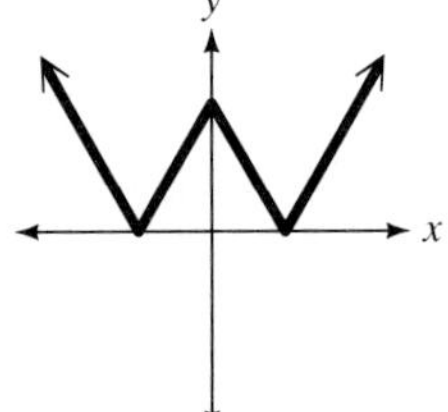

16.

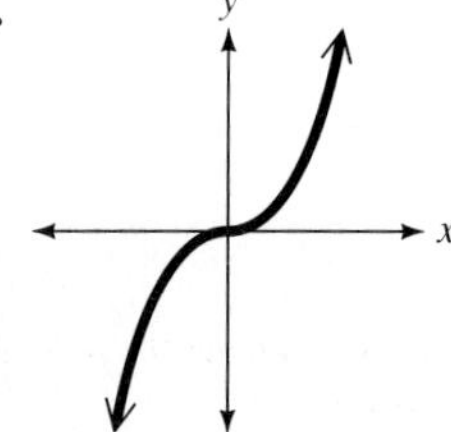

17.

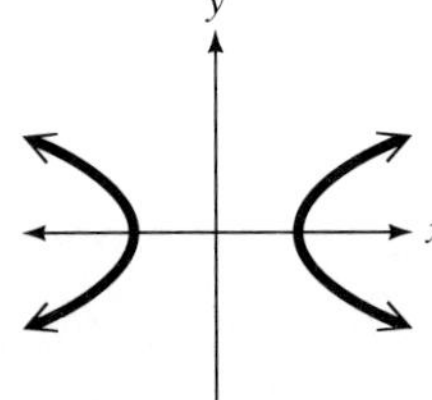

18.

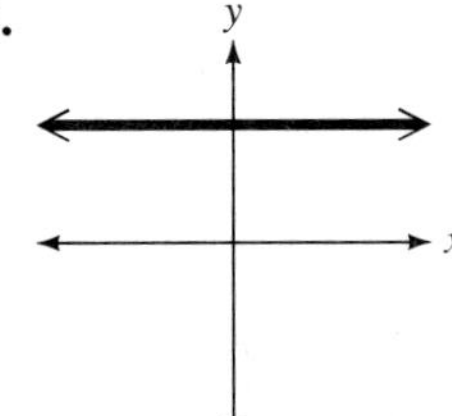

19.

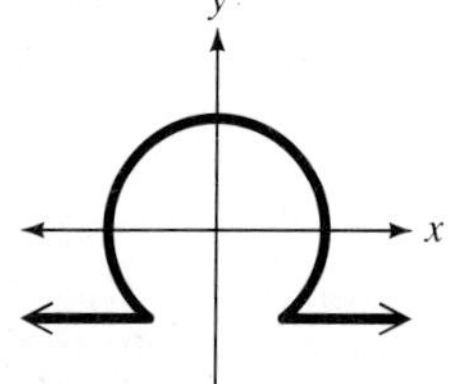

20.

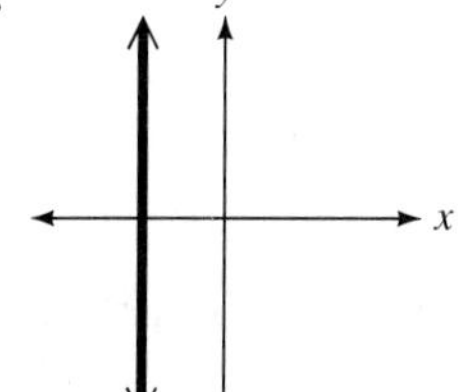

21.

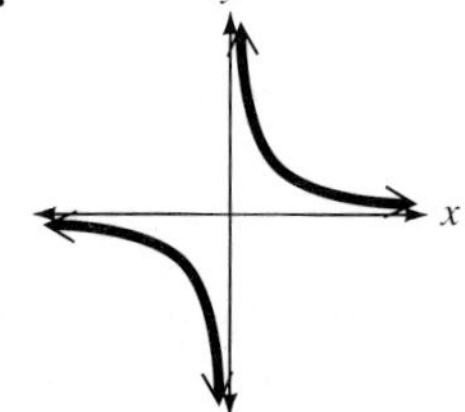

Graph the equation and use your graph to tell whether the equation describes a function.

22. $3x + 5y - 15 = 0$

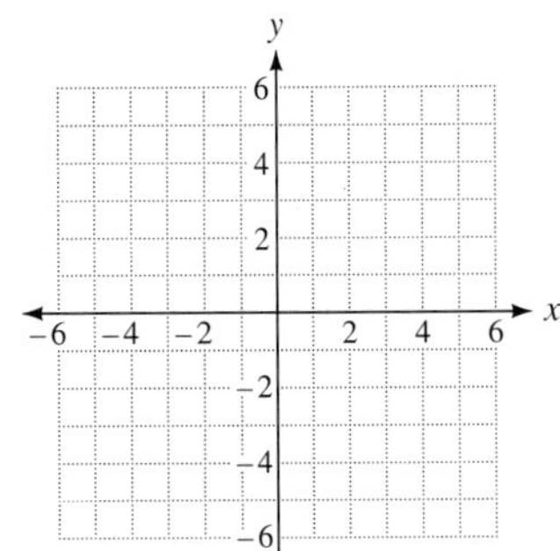

23. $3y - 12 = 0$

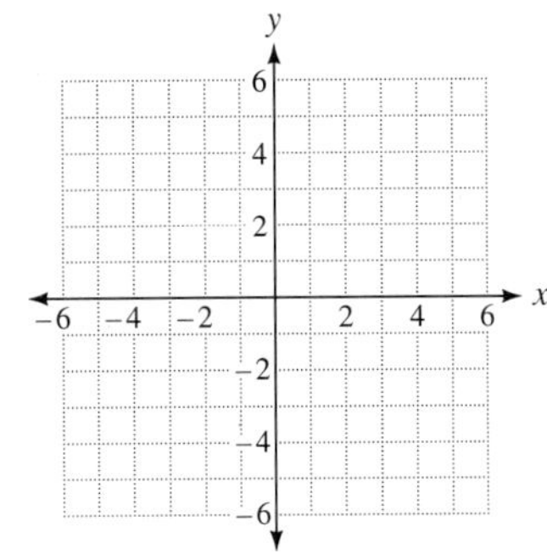

24. $5x - 15 = 0$

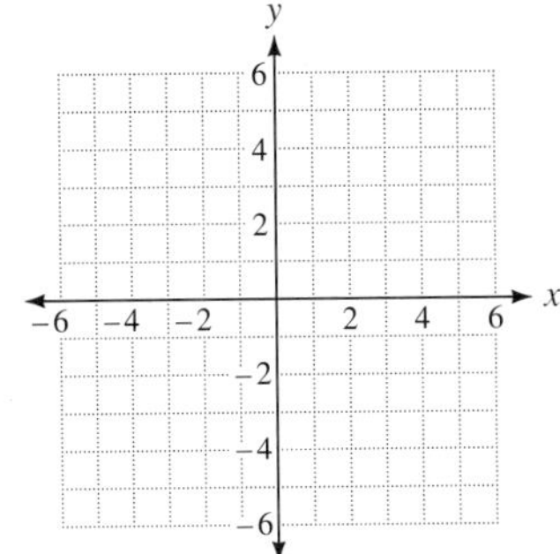

25. Consider the relation described by the set of ordered pairs $\{(1, 3), (2, 5), (3, 4), (4, 2), (5, 2)\}$.

(a) Write the relation using arrows as in Example 3.
(b) Graph the relation in the given rectangular coordinate system.
(c) Is the relation a function? Explain.

(a)

(b)

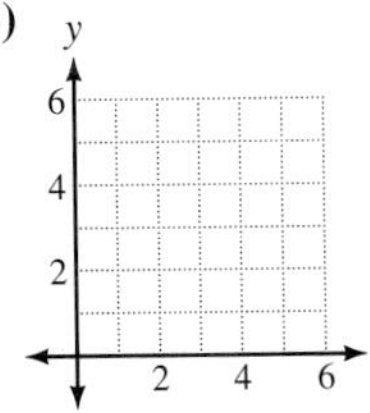

(c)

In Exercises 26–27, list the ordered pairs in each relation. What is the domain of the relation? What is the range of the relation? Is the relation a function?

26. $\{(x, y)$ where $x = y^2$ and $y = 1, 2, 3\}$

27. $\{(x, y)$ where $x^2 + y^2 = 1$ and $x = -1, 0, 1\}$

FOR REVIEW

28. Give the point-slope form of the equation of a line passing through (x_1, y_1) with slope m.

29. Give the slope-intercept form of the equation of a line with slope m and y-intercept $(0, b)$.

30. Give the slope of a line parallel to a line with slope m.

31. Find the slope-intercept form of the line with general equation $6x + 3y - 9 = 0$. What is the slope? The y-intercept?

32. Find the general form of the equation of the line passing through $(-5, 2)$ with slope $m = -3$.

33. Terry Larsen manufactures a special part for farm equipment. He can make 15 parts for \$70 and 60 parts for \$192. If y is the cost of making x parts, find a linear equation describing this relationship and use it to calculate the cost of making 100 parts, to the nearest dollar.

The following exercises review material from Chapter 1 and will help you prepare for the next section.

34. Will the expression $\dfrac{1}{x-2}$ be a real number for every real number x?

35. Evaluate $2x^2 - x + 1$ when $x = -1$.

ANSWERS: 1. true 2. true 3. not a function 4. function 5. function 6. function 7. function 8. not a function 9. not a function 10. function 11. function 12. function 13. not the graph of a function 14. not the graph of a function 15. graph of a function 16. graph of a function 17. not the graph of a function 18. graph of a function 19. not the graph of a function 20. not the graph of a function 21. graph of a function

22. function

23. function

24. not a function

25. (a)

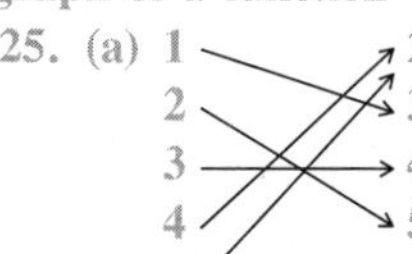

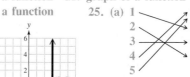

(b)

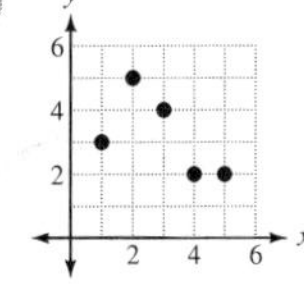

(c) Yes. The graph passes the vertical line test.

26. $\{(1, 1), (4, 2), (9, 3)\}$; $\{1, 4, 9\}$; $\{1, 2, 3\}$; yes 27. $\{(-1, 0), (0, 1), (0, -1), (1, 0)\}$; $\{-1, 0, 1\}$; $\{-1, 0, 1\}$; no 28. $y - y_1 = m(x - x_1)$ 29. $y = mx + b$ 30. m 31. $y = -2x + 3$; $m = -2$; $(0, 3)$ 32. $3x + y + 13 = 0$ 33. $122x - 45y + 1320 = 0$; \$300 34. for every real number except 2 35. 4

3.5 EXERCISES B

Write true *or* false *for Exercises 1 and 2. If the statement is false, explain why.*

1. A relation with the property that every object in its domain corresponds to exactly one object in its range is called a function.

2. If (x, y) belongs to a function, then x corresponds to y.

Tell which of the relations in Exercises 3–4 are functions.

3. Relation: a number to its cube
   Domain: the set of all integers
   Range: the set of all perfect cubes

4. Relation: person to his/her age
   Domain: all living people
   Range: the set of all ages

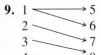

Which of the following are functions? Explain.

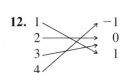

5. 1 → 5; 2, 3 → 6, 7; 4 → 8

6. 1, 2, 3 → 6; 4 → 8

7. 1 → 5; 2, 3, 4 → 7, 8

8. 1, 2, 3, 4 → 5, 6, 7, 8

9. 1, 2, 3, 4 → 5, 6, 7, 8

10. 1, 2, 3, 4 → 0, 1

11. 1, 2, 3, 4 → −1, 0, 1

12. 1, 2, 3, 4 → −1, 0, 1

Which of the following are graphs of functions? Explain.

13.

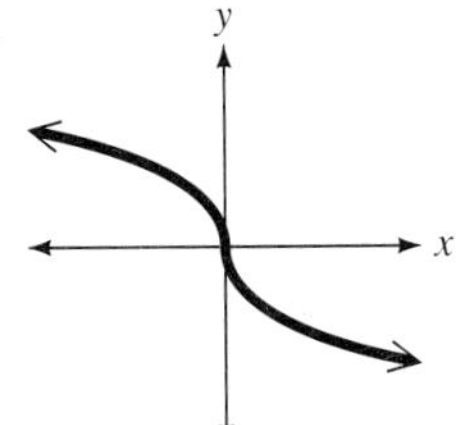

14.

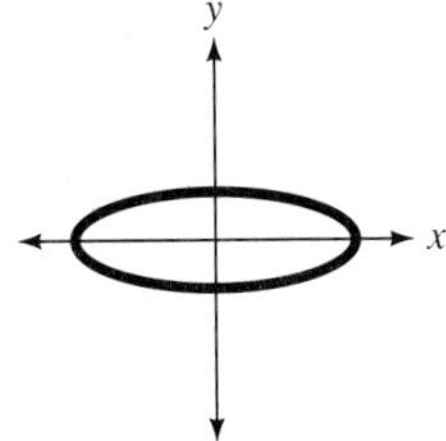

15.

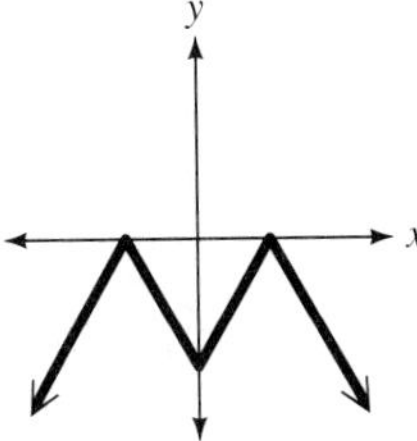

16.

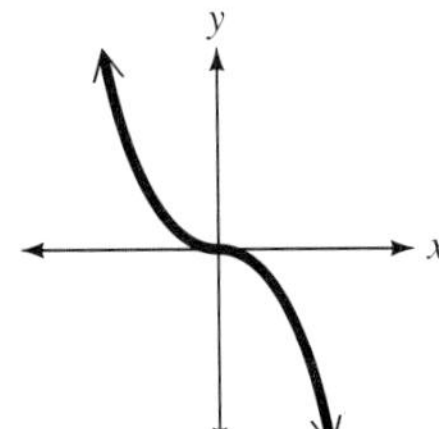

17.

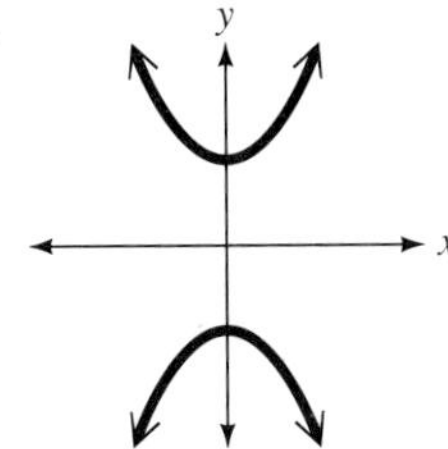

18.

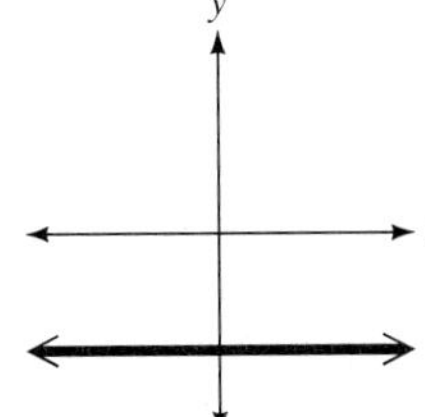

19.

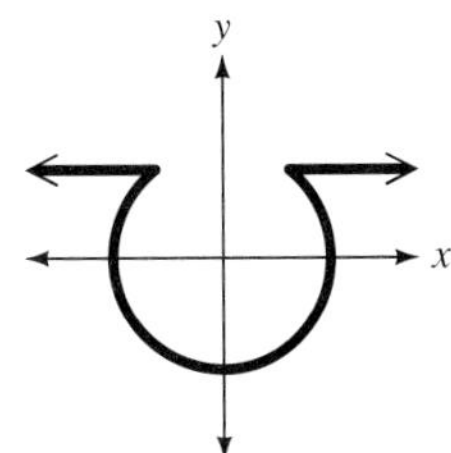

20.

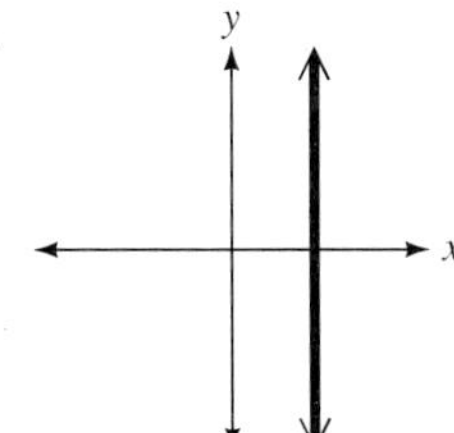

21.

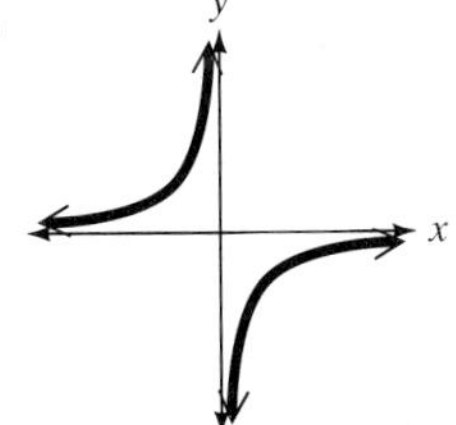

Graph the equation and use your graph to tell whether the equation describes a function.

22. $4y + 12 = 0$

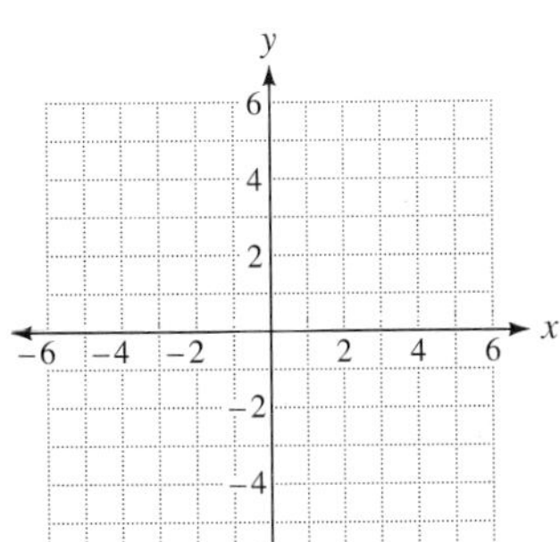

23. $x + y = 0$

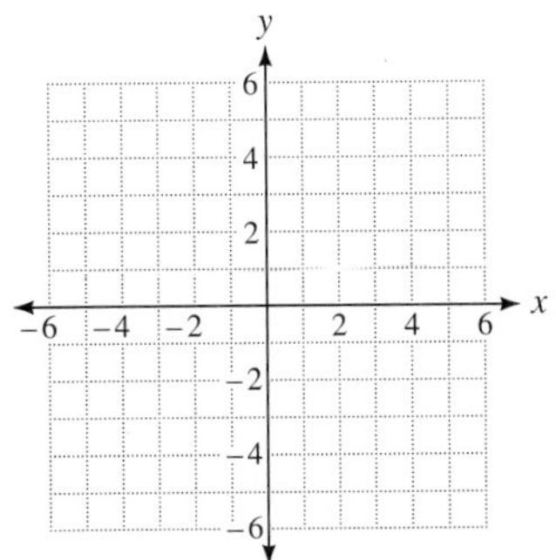

24. $5x + 20 = 0$

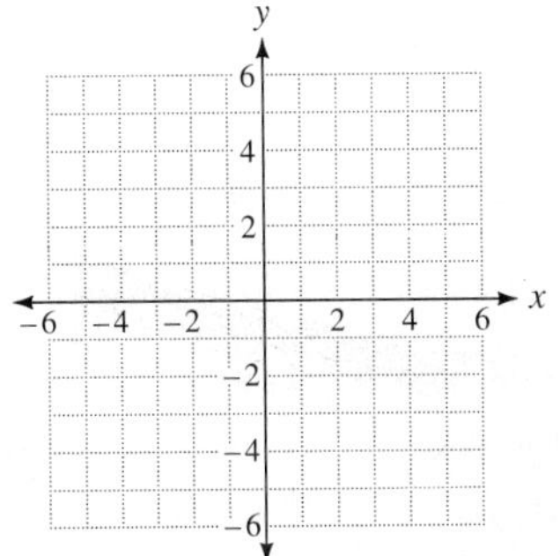

25. Consider the relation described by the set of ordered pairs $\{(1, 2), (1, 3), (2, 1), (3, 5), (4, 5)\}$.
(a) Write the relation using arrows as in Example 3.
(b) Graph the relation in a Cartesian coordinate system.
(c) Is the relation a function? Explain.

(a)

(b)

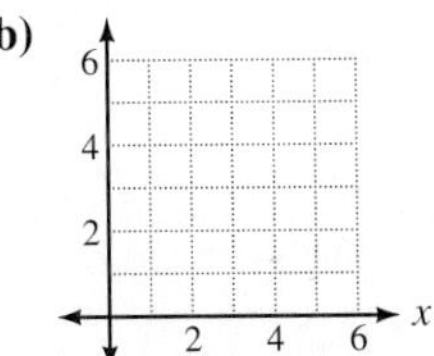

(c)

In Exercises 26–27, list the ordered pairs in each relation. What is the domain of the relation? What is the range of the relation? Is the relation a function?

26. $\{(x, y) \text{ where } y = x^2 \text{ and } y = 0, 1, 4\}$

27. $\{(x, y) \text{ where } x^2 + y^2 = 4 \text{ and } y = -2, 0, 2\}$

FOR REVIEW

28. Give the type of the equation $y - y_1 = m(x - x_1)$.

29. Give the type of the equation $y = mx + b$.

30. Give the slope of a line perpendicular to a line with slope m.

31. Find the slope-intercept form of the line with general form $3x - 6y - 24 = 0$. What is the slope? The y-intercept?

32. Find the general form of the equation of the line passing through $(-3, 7)$ with slope $m = \frac{1}{2}$.

33. Phil Mortensen builds a special type of hitch. He can make 3 hitches for $170, or 7 hitches for $342. If y is the cost of making, x hitches, find a linear equation describing this relationship, and use it to calculate the cost of making 10 hitches.

The following exercises review material from Chapter 1 and will help you prepare for the next section.

34. Will the expression $\dfrac{1}{x^2 + 1}$ be a real number for every real number x?

35. Evaluate $-3x^2 + x - 2$ when $x = -1$.

3.5 EXERCISES C

Which of the following relations are functions?

1. $\{(x, y) | y = \sqrt{x + 1} \quad \text{and} \quad x = 1, 2, 3\}$
[*Hint:* List the ordered pairs.]

2. $\{(x, y) | y^2 = x + 1 \quad \text{and} \quad x = 1, 2, 3\}$

3. $\{(x, y) | y = |x| \quad \text{and} \quad x = -1, 0, 1\}$

4. $\{(x, y) | x = |y| \quad \text{and} \quad x = 0, 1\}$
[Answer: not a function]

3.6 FUNCTIONAL NOTATION AND SPECIAL FUNCTIONS

STUDENT GUIDEPOSTS

1. Specifying the Domain of a Function
2. Functional Notation $f(x)$
3. Linear Functions
4. Constant Functions
5. The Identity Function

1 SPECIFYING THE DOMAIN OF A FUNCTION

When a function consists of a set of ordered pairs (x, y) where x is the independent variable and y is the dependent variable, we think of y *as a function of* x. That is, if a value of x is given, one and only one value of y can be found.

Equations that are solved for y in terms of x often describe a function. For example,

$$y = 5, \quad y = x, \quad y = 2x + 3, \quad y = x^2 + 1, \quad \text{and} \quad y = \frac{1}{x}$$

all describe functions consisting of ordered pairs (x, y). That is, for a given real number x, there is one and only one corresponding value for y making the equation true. Unless we specify otherwise, *the domain of a function described by an equation is the set of all real numbers for which the expression in x is defined.* The domain for each of the functions

$$y = 5, \quad y = x, \quad y = 2x + 3, \quad \text{and} \quad y = x^2 + 1,$$

is the entire set of real numbers. However, the domain of

$$y = \frac{1}{x}$$

is the set of all real numbers except 0. Remember, division by zero is not defined, so no value can be found for y when x is 0. In general, the domain of a function is restricted so that *numbers that make a denominator zero are not permitted.*

EXAMPLE 1 SPECIFYING DOMAINS

Specify the domain of each function.

(a) The function $y = \dfrac{1}{x - 2}$ is defined for all real numbers except $x - 2 = 0$. Thus, the domain is the set of all real numbers except 2 $(x \neq 2)$.

(b) The function $y = \dfrac{1}{x + 1}$ is defined for all real numbers except $x + 1 = 0$. The domain is the set of all real numbers except -1 $(x \neq -1)$.

(c) The function $y = 3x^2 + x + 1$ is defined for all real numbers. The domain is the set of all real numbers.

PRACTICE EXERCISE 1

Specify the domain of each function.

(a) $y = \dfrac{1}{x + 6}$

(b) $y = 2x - 8$

(c) $y = \dfrac{x + 2}{x - 3}$

Answers: (a) $x \neq -6$ (b) all real numbers (c) $x \neq 3$

2 FUNCTIONAL NOTATION $f(x)$

Many times we use a letter to name a function (f and g are common choices). Suppose that a function f is described by the equation

$$y = x^2 + 1.$$

To show that y depends on choices of x, we use $f(x)$ for y and write

$$f(x) = x^2 + 1, \qquad y \text{ is } f(x)$$

where $f(x)$ is read "f of x." To determine the value of the function f for a given value of x, we substitute the value into the formula $f(x) = x^2 + 1$. For example, to compute $f(2)$, we replace each x with 2 and simplify. It is sometimes helpful to place parentheses around the number we are substituting for x.

$$f(2) = (2)^2 + 1 = 4 + 1 = 5$$

Remember that $f(2) = 5$ is the y-value that corresponds to the x-value 2. That is, $2 \rightarrow f(2)$, and the ordered pair $(2, f(2))$ or $(2, 5)$ belongs to the function. Also, when

$$x = -3, \quad y = f(-3) = (-3)^2 + 1 = 9 + 1 = 10;$$
$$x = 0, \quad y = f(0) = (0)^2 + 1 = 0 + 1 = 1.$$

EXAMPLE 2 FUNCTIONAL NOTATION

Consider the function g defined by

$$g(x) = 2x + 3.$$

(a) $g(0) = 2(0) + 3 = 0 + 3 = 3$ $\quad 0 \rightarrow 3$; $(0, 3)$ belongs to g

(b) $g(4) = 2(4) + 3 = 8 + 3 = 11$ $\quad 4 \rightarrow 11$; $(4, 11)$ belongs to g

(c) $g(-2) = 2(-2) + 3 = -4 + 3 = -1$ $\quad -2 \rightarrow -1$; $(-2, -1)$ belongs to g

(d) $g(a) = 2(a) + 3 = 2a + 3$ $\quad a \rightarrow 2a + 3$; $(a, 2a + 3)$ belongs to g

(e) $g(b + 1) = 2(b + 1) + 3 = 2b + 2 + 3 = 2b + 5$

PRACTICE EXERCISE 2

$h(x) = 4x - 5$

(a) $h(0) =$

(b) $h(2) =$

(c) $h(-2) =$

(d) $h(a) =$

(e) $h(b - 1) =$

Answers: (a) -5 (b) 3 (c) -13 (d) $4a - 5$ (e) $4b - 9$

3 LINEAR FUNCTIONS

Since we know from Section 3.4 that a linear equation in slope-intercept form is $y = mx + b$, it follows that a **linear function** can be described by an equation of the form

$$f(x) = mx + b \quad (m \text{ and } b \text{ constants}).$$

The domain of every linear function is the set of real numbers. To graph a linear function, we graph the linear equation that defines it.

EXAMPLE 3 A LINEAR FUNCTION

The linear function f defined by

$$f(x) = 2x + 1$$

has the following values.

$$f(0) = 2(0) + 1 = 1$$
$$f(1) = 2(1) + 1 = 3$$
$$f(2) = 2(2) + 1 = 5$$
$$f(-1) = 2(-1) + 1 = -1$$
$$f(a + 1) = 2(a + 1) + 1$$
$$= 2a + 2 + 1 = 2a + 3$$

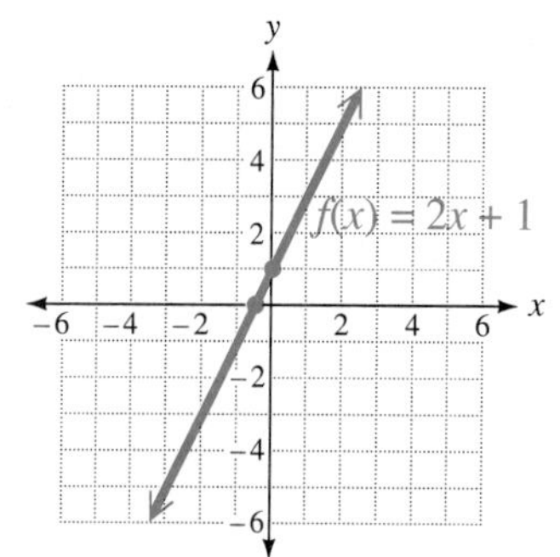

Figure 3.28 Linear Function

To graph the linear function $f(x) = 2x + 1$, graph $y = 2x + 1$ as in Figure 3.28 by finding the intercepts $(0, 1)$ and $\left(-\frac{1}{2}, 0\right)$ or by using the slope method discussed in Section 3.3.

PRACTICE EXERCISE 3

$f(x) = -3x + 2$

(a) $f(0) =$

(b) $f(-2) =$

(c) $f(1) =$

(d) $f(a) =$

(e) $f(b + 1) =$

Answers: (a) 2 (b) 8 (c) -1 (d) $-3a + 2$ (e) $-3b - 1$

4 CONSTANT FUNCTIONS

If $m = 0$ in a linear function $f(x) = mx + b$, we obtain a **constant function**

$$f(x) = b. \qquad (b \text{ a constant})$$

The domain of a constant function is the entire set of real numbers, and no matter what value x assumes, the corresponding y-value, $f(x)$, is the same constant b. Since $f(x)$ is another name for y, and since every equation of the form $y = b$ describes a horizontal line, every constant function has as its graph a straight line parallel to the x-axis.

EXAMPLE 4 A CONSTANT FUNCTION

The constant function f defined by

$$f(x) = 3$$

has the following values and is graphed in Figure 3.29.

$$f(0) = 3$$
$$f(1) = 3$$
$$f(2) = 3$$
$$f(-1) = 3$$
$$f(a + 1) = 3$$

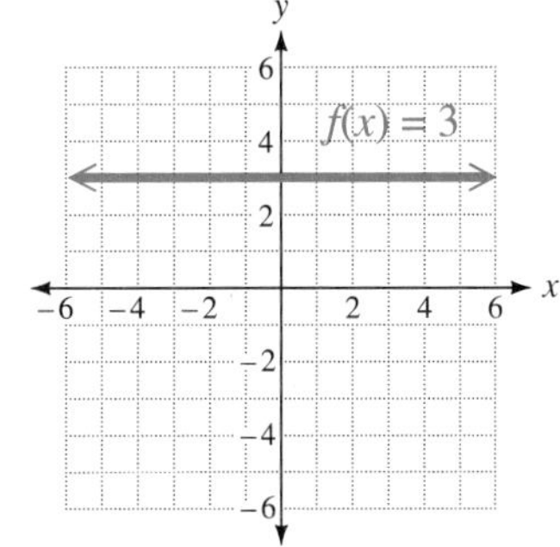

Figure 3.29 Constant Function

PRACTICE EXERCISE 4

$g(x) = -5$

(a) $g(0) =$

(b) $g(3) =$

(c) $g(-6) =$

(d) $g(b) =$

(e) $g(a - 1) =$

Answers: (a) -5 (b) -5 (c) -5 (d) -5 (e) -5

5 THE IDENTITY FUNCTION

Another special linear function results when $b = 0$ and $m = 1$ in $f(x) = mx + b$. The identity function

$$f(x) = x$$

has domain the entire set of real numbers, and identifies a real number with itself. For example, $f(4) = 4$, $f(100) = 100$, and $f(-1) = -1$. The identity function is graphed in Figure 3.30.

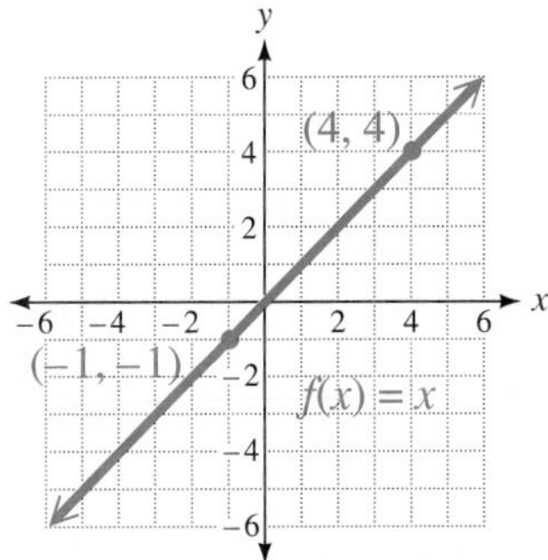

Figure 3.30 Identity Function

The identity function will be considered in more detail in Chapter 9 when additional properties of functions are discussed.

Many applied problems can be simplified using the idea of a function. One quantity or variable may be given in terms of another in an equation that describes a functional relationship. This is illustrated in the next example.

EXAMPLE 5 Sales Problem

A sales representative earns an annual salary of \$12,000 plus a 10% commission on total sales. Express her annual earnings, y, as a function, f, of her total sales, x. What are her annual earnings when her yearly sales total \$85,000?

The function defined by this information is

$$y = f(x) = 12{,}000 + 0.10x.$$

When her yearly sales total \$85,000, her annual earnings amount to $f(85{,}000)$.

$$\begin{aligned} f(85{,}000) &= 12{,}000 + (0.10)(85{,}000) \\ &= 12{,}000 + 8500 = 20{,}500 \end{aligned}$$

Thus, she earned \$20,500 that year.

Practice Exercise 5

A sales representative earns \$9000 plus 8% commission on total sales. Express his earnings as a function of total sales, x, and find the earnings on sales of \$120,000.

Answer: $f(x) = 9000 + 0.08x$; \$18,600

3.6 EXERCISES A

Specify the domain of each function defined by the given equation.

1. $y = x + 1$

2. $y = 5x - 2$

3. $y = \dfrac{5}{x + 5}$

4. $y = \dfrac{1}{4 - x}$

5. $y = x^2 + 3$

6. $y = 9x^2 + 5$

7. $y = \dfrac{x}{x - 1}$

8. $y = \dfrac{5}{x^2 + 5}$

9. $y = 3x^2 + 4x + 1$

Evaluate each of the following.

10. Let $f(x) = 2x + 5$.

(a) $f(0) =$ **(b)** $f(1) =$
(c) $f(3) =$ **(d)** $f(-2) =$
(e) $f(a) =$ **(f)** $f(b - 1) =$

11. Let $g(x) = x^2 - 3$.

(a) $g(0) =$ **(b)** $g(-1) =$
(c) $g(3) =$ **(d)** $g(-2) =$
(e) $g(b) =$ **(f)** $g(-a) =$

12. Let $f(x) = 2x^3 - 1$.

(a) $f(0) =$ **(b)** $f(-1) =$

(c) $f(2) =$ **(d)** $f(-3) =$

(e) $f(a) =$ **(f)** $f(-b) =$

13. Let $g(x) = -7$.

(a) $g(0) =$ **(b)** $g(-1) =$

(c) $g(2) =$ **(d)** $g(-3) =$

(e) $g(a) =$ **(f)** $g(b - 1) =$

14. Let $f(x) = -3x + 1$.

(a) $f(0) =$ **(b)** $f(-1) =$

(c) $f(2) =$ **(d)** $f(-3) =$

(e) $f(b) =$ **(f)** $f(a - 1) =$

15. Let $g(x) = 4x^2 - 3x + 1$.

(a) $g(0) =$ **(b)** $g(-1) =$

(c) $g(2) =$ **(d)** $g(-3) =$

(e) $g(a) =$ **(f)** $g(-b) =$

16. Let $f(x) = 1 - x$.

(a) $f(0) =$ **(b)** $f(-1) =$

(c) $f(2) =$ **(d)** $f\left(-\frac{3}{2}\right) =$

(e) $f(b) =$ **(f)** $f(a - 1) =$

17. Let $f(x) = -x$.

(a) $f(0) =$ **(b)** $f(-1) =$

(c) $f(2) =$ **(d)** $f\left(-\frac{3}{2}\right) =$

(e) $f(b) =$ **(f)** $f(a - 1) =$

State whether the given function is a constant or linear function, and sketch its graph.

18. $f(x) = -2x + 1$

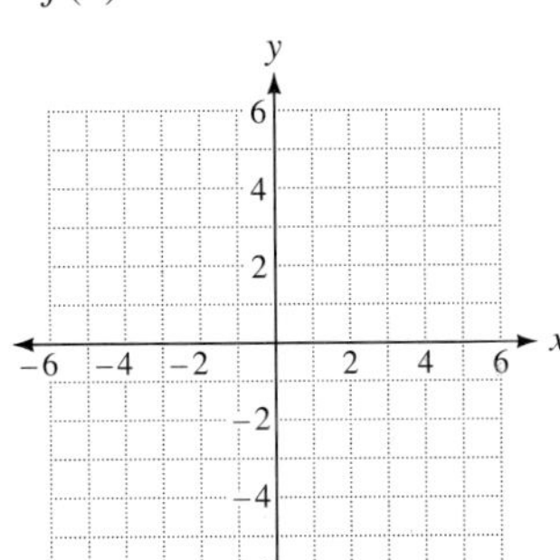

19. $f(x) = 2x + 1$

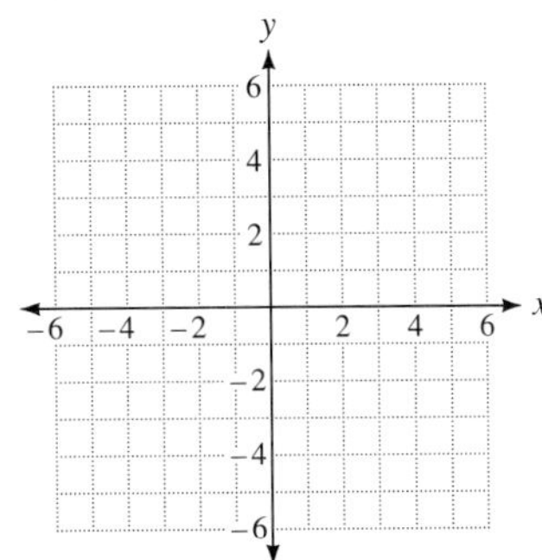

20. $f(x) = -2$

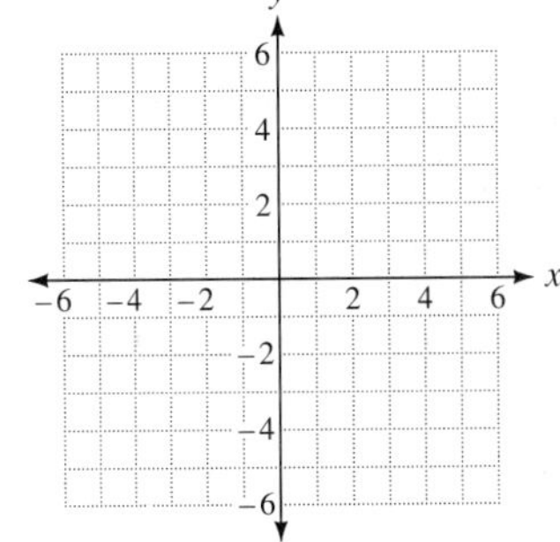

21. Write an equation for the cost function, f, described by selling x books at \$11.95 a book. How much do 15 books cost?

22. The distance that a car travels at 55 mph can be found by multiplying the total number of hours driven, x, by the hourly rate. Express this functional relationship using the letter g, and use it to find the distance traveled in 8 hr.

23 A salesman earns an annual salary of \$15,000 plus a 12% commission on sales. Express his annual earnings as a function, f, of his total sales, x. What are his annual earnings when his yearly sales total \$100,000?

24 An object is thrown upward with initial velocity of 128 ft per sec. If its height after x sec is given by $f(x) = -16x^2 + 128x$, at what height will it be in 2 sec?

FOR REVIEW

Which of the following are functions? Explain.

25. 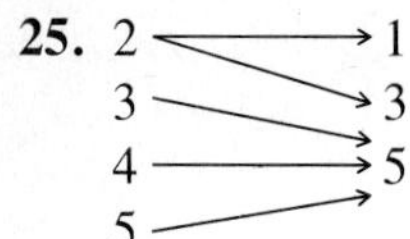

26. 2 → 1, 3 → 3, 4 → 7, 5 → 7

27. 2 → 1, 2 → 3, 2 → 5, 2 → 7; 3, 4, 5

Which of the following are graphs of functions? Explain.

28. 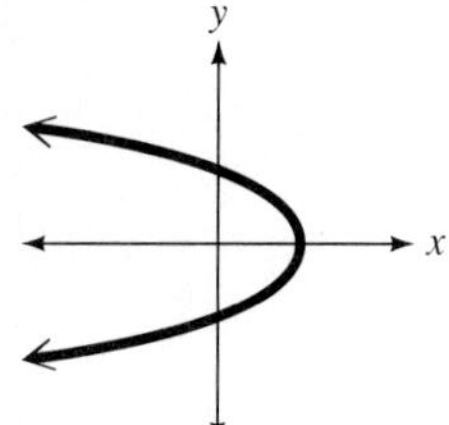

29. 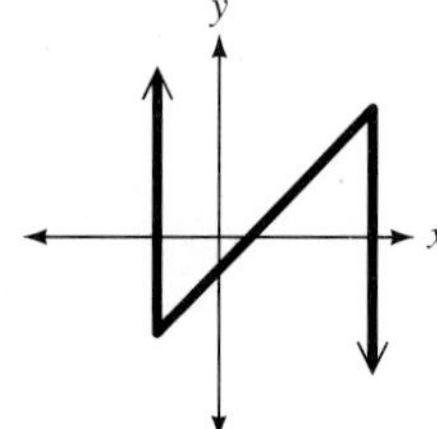

30. 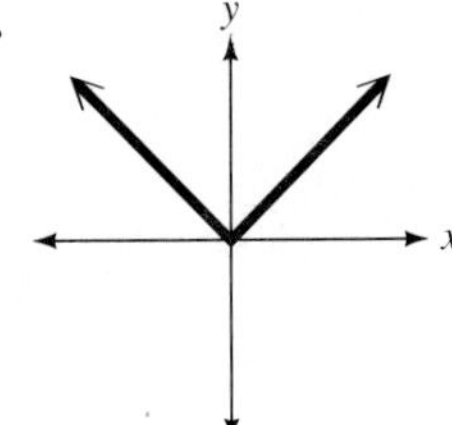

The following exercises, primarily from Section 3.3, will help you prepare for the material in Chapter 4.

31. Find the value of y if x is 2 in $3x - 2y + 8 = 0$.

32. Are the lines with equations $5x - 4y - 8 = 0$ and $5x - 4y + 8 = 0$ parallel? Explain.

33. If a line has slope $-\frac{1}{3}$, what is the slope of a line parallel to it? What is the slope of a line perpendicular to it?

34. If two lines have the same slope are they necessarily parallel lines?

ANSWERS: 1. all real numbers 2. all real numbers 3. all real numbers except -5 4. all real numbers except 4 5. all real numbers 6. all real numbers 7. all real numbers except 1 8. all real numbers 9. all real numbers 10. (a) 5 (b) 7 (c) 11 (d) 1 (e) $2a + 5$ (f) $2b + 3$ 11. (a) -3 (b) -2 (c) 6 (d) 1 (e) $b^2 - 3$ (f) $a^2 - 3$ 12. (a) -1 (b) -3 (c) 15 (d) -55 (e) $2a^3 - 1$ (f) $-2b^3 - 1$ 13. (a) -7 (b) -7 (c) -7 (d) -7 (e) -7 (f) -7 14. (a) 1 (b) 4 (c) -5 (d) 10 (e) $-3b + 1$ (f) $-3a + 4$ 15. (a) 1 (b) 8 (c) 11 (d) 46 (e) $4a^2 - 3a + 1$ (f) $4b^2 + 3b + 1$ 16. (a) 1 (b) 2 (c) -1 (d) $\frac{5}{2}$ (e) $1 - b$ (f) $2 - a$ 17. (a) 0 (b) 1 (c) -2 (d) $\frac{3}{2}$ (e) $-b$ (f) $1 - a$ 18. linear function 19. linear function 20. constant function

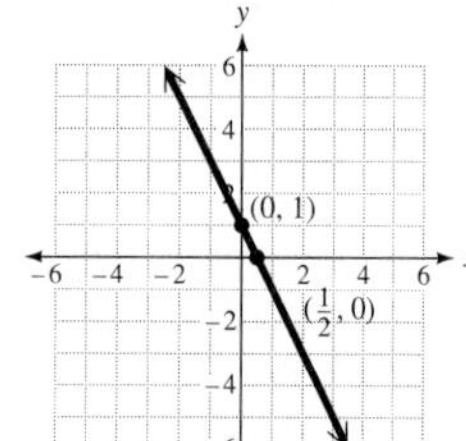

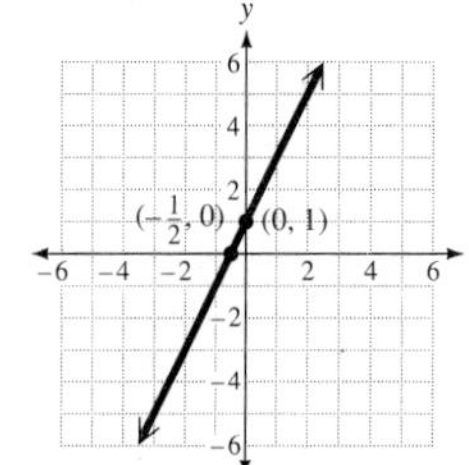

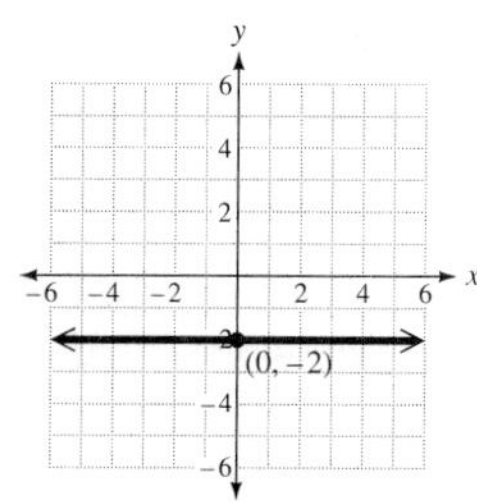

21. $f(x) = 11.95x$; \$179.25 22. $g(x) = 55x$; 440 miles 23. $f(x) = 15{,}000 + 0.12x$; \$27,000 24. 192 feet 25. not a function 26. a function 27. not a function 28. not the graph of a function 29. not the graph of a function 30. graph of a function 31. $y = 7$ 32. Yes. Both have slope $\frac{5}{4}$ but different y-intercepts. 33. $-\frac{1}{3}$; 3 34. No. They could be the same line.

3.6 EXERCISES B

Specify the domain of each function defined by the given equation.

1. $y = x - 1$

2. $y = 9x - 5$

3. $y = \dfrac{2}{x + 2}$

4. $y = \dfrac{1}{3 - x}$

5. $y = x^2 + 1$

6. $y = 8x^2 + 1$

7. $y = \dfrac{x}{1 - x}$

8. $y = \dfrac{1}{x^2 + 1}$

9. $y = 2x^2 + x - 3$

For each function, evaluate ***(a)*** $f(0)$ ***(b)*** $f(1)$ ***(c)*** $f(2)$ ***(d)*** $f(-3)$ ***(e)*** $f(b)$ ***(f)*** $f(-a)$.

10. $f(x) = 3x - 1$
(a) **(b)**
(c) **(d)**
(e) **(f)**

11. $f(x) = x^2 + 2$
(a) **(b)**
(c) **(d)**
(e) **(f)**

12. $f(x) = 3x^3 - 2$
(a) **(b)**
(c) **(d)**
(e) **(f)**

13. $f(x) = 4x^2 - x + 3$
(a) **(b)**
(c) **(d)**
(e) **(f)**

For each function, evaluate ***(a)*** $f(0)$ ***(b)*** $f(-1)$ ***(c)*** $f\left(-\frac{1}{2}\right)$ ***(d)*** $f(3)$ ***(e)*** $f(a)$ ***(f)*** $f(b - 1)$.

14. $f(x) = -3$
(a) **(b)**
(c) **(d)**
(e) **(f)**

15. $f(x) = -2x + 5$
(a) **(b)**
(c) **(d)**
(e) **(f)**

16. $f(x) = 3 - x$
(a) **(b)**
(c) **(d)**
(e) **(f)**

17. $f(x) = x$
(a) **(b)**
(c) **(d)**
(e) **(f)**

State whether the given function is a constant or linear function, and sketch its graph.

18. $f(x) = 2x - 1$

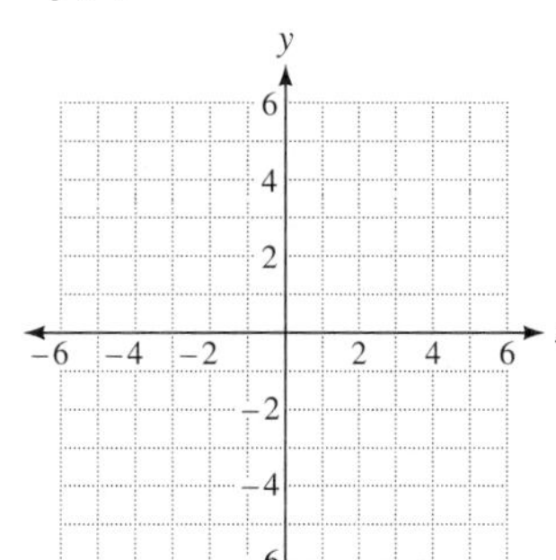

19. $f(x) = -2x - 1$

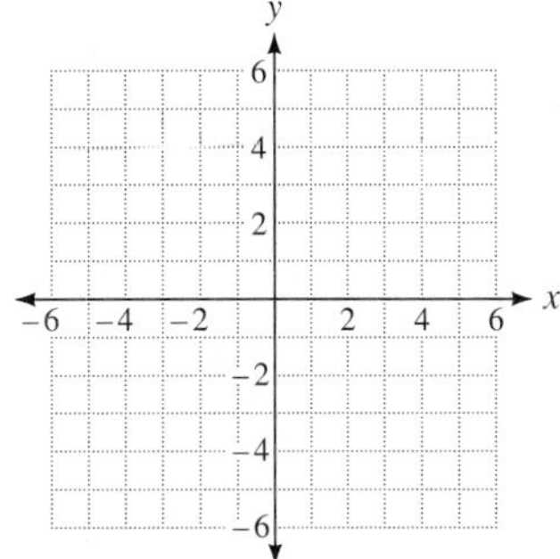

20. $f(x) = 5$

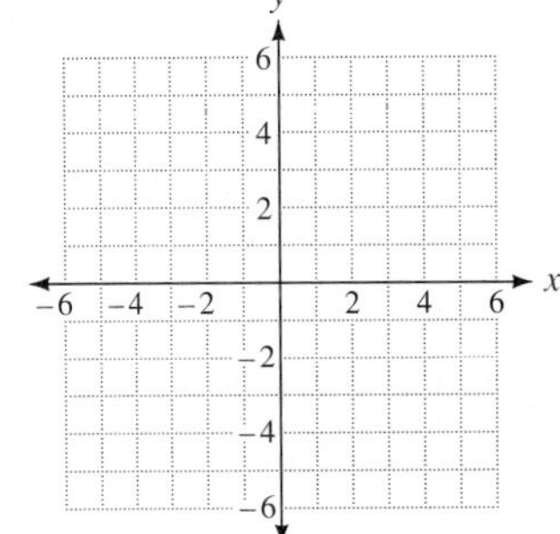

21. Write an equation for the cost function, f, described by selling x tickets at \$8.50 a ticket. How much do 50 tickets cost?

22. The distance that a motor home travels at 50 mph can be found by multiplying the number of hours driven, x, by the hourly rate. Express this functional relationship using the letter g, and find the distance traveled in 14 hours.

23. A realtor earns an annual salary of \$10,000 plus a 6% commission on her sales. Express her annual earnings as a function, f, of her sales, x. What does she earn in a year when she sells \$450,000 worth of real estate?

24. An object is thrown upward with initial velocity of 128 ft per sec. If its height after x sec is given by $f(x) = -16x^2 + 128x$, at what height will it be in 5 sec?

FOR REVIEW

Which of the following are functions? Explain.

25.
$3 \to 1$
$4 \to 2$
$5 \to 4$
$6 \to 4$

26.
$3 \to 2$
$4 \to 2$
$5 \to 2$
$6 \to 4$

27.
$3 \to 1$
$4 \to 1$, $4 \to 2$, $4 \to 3$
$5 \to 3$
$6 \to 4$

Which of the following are graphs of functions? Explain.

28.

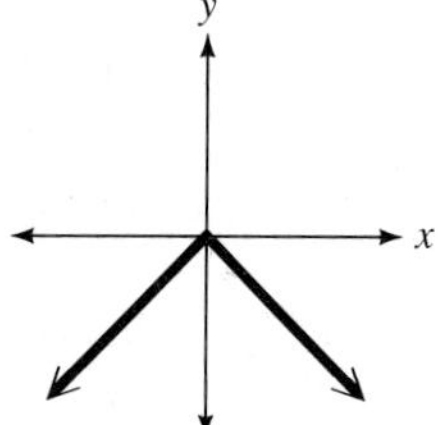

29.

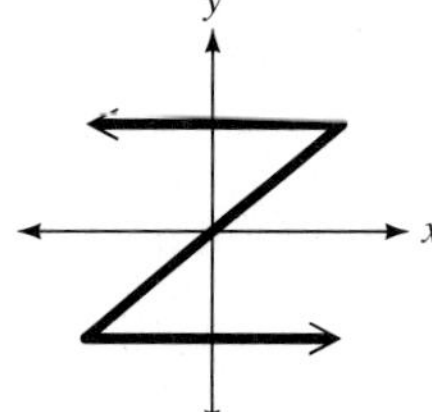

30.

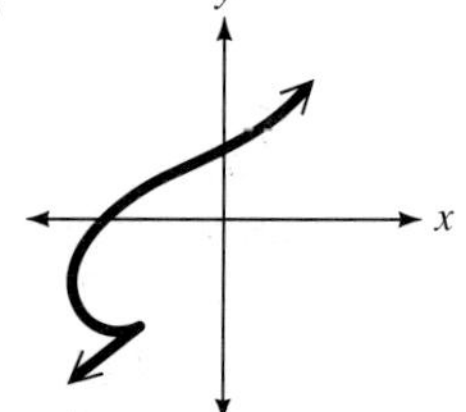

The following exercises, primarily from Section 3.3, will help you prepare for the material in Chapter 4.

31. Find the value of x if y is -2 in $3x - 2y + 8 = 0$.

32. Are the lines with equations $2x - 4y + 10 = 0$ and $x - 2y + 5 = 0$ parallel? Explain.

33. If a line has slope -6, what is the slope of a line parallel to it? What is the slope of a line perpendicular to it?

34. What can be said about the coordinates of a point of intersection of two lines relative to the equations of the lines?

3.6 EXERCISES C

Suppose $f(x) = x + 1$ and $g(x) = x^2 - 5$. Evaluate the following.

1. $f(g(1))$ [Answer: -3]

2. $g(f(1))$

3. $f(g(x))$

4. $f(x + h)$

5. $\dfrac{f(x + h) - f(x)}{h}$
[Answer: 1]

6. $\dfrac{g(x + h) - g(x)}{h}$

CHAPTER 3 REVIEW

KEY WORDS

3.1 A **rectangular (Cartesian) coordinate system** is a configuration used to graph equations in two variables.

The point of intersection of the axes in a rectangular coordinate system is called the **origin.**

A point with coordinates given by the ordered pair (x, y) has ***x*-coordinate (abscissa)** x and ***y*-coordinate (ordinate)** y.

The axes in a coordinate system divide the plane into four regions called **quadrants.**

A **solution** to an equation in two variables is an ordered pair of numbers that when substituted makes the equation true.

The **graph** of an equation in two variables is the set of points in a coordinate system that corresponds to solutions of the equation.

3.2 A **linear** (or **first-degree**) **equation in two variables** can be written in the form $ax + by + c = 0$, a and b not both zero.

Points where the graph of a linear equation (line) crosses the x-axis and y-axis, respectively, are called the ***x*-intercept** and ***y*-intercept.**

3.3 The **slope** of a line is a number which is the ratio of the vertical change (rise) to the horizontal change (run) as we move from one point to another on the line.

Points are **collinear** if they lie on the same line.

Two lines are **parallel** if they never intersect.

Two lines are **perpendicular** if they intersect at right angles.

3.5 A correspondence between a first set of objects, the **domain,** and a second set of objects, the **range,** is a **relation** if to each element in the domain there corresponds one or more elements in the range.

A **function** is a relation with the additional property that to each element in the domain there corresponds one and only one element in the range.

If (x, y) is an element of a function, x is the **independent variable** and y is the **dependent variable.**

KEY CONCEPTS

3.1 To graph an equation, construct a table of values, plot the points corresponding to these solutions in a rectangular coordinate system, and join these points with a smooth curve.

3.2

1. The general form of the equation of a line is $ax + by + c = 0$. If $a = 0$, the line is a horizontal line, parallel to the x-axis, with y-intercept $\left(0, -\frac{c}{b}\right)$. If $b = 0$, the line is a vertical line, parallel to the y-axis, with x-intercept $\left(-\frac{c}{a}, 0\right)$.
2. To graph a linear equation using the intercept method, plot the intercepts and draw the line through them. If both intercepts are (0, 0), plot another solution and draw the line through it and (0, 0).

3.3

1. The slope of a line passing through points with coordinates (x_1, y_1) and (x_2, y_2) is $m = \frac{y_2 - y_1}{x_2 - x_1}$.
2. If m_1 and m_2 are the slopes of two lines and if $m_1 = m_2$, then the lines are parallel; if $m_1 = -\frac{1}{m_2}$, then the lines are perpendicular.

3.4

1. Point-slope form: $y - y_1 = m(x - x_1)$.
2. Slope-intercept form: $y = mx + b$.
3. To find the equation of a line through two points, find the slope of the line first, then use the point-slope form with either of the given points.
4. To graph a line using the slope method, plot the y-intercept and use the slope to determine the rise and run of the graph which will give a second point on the line.

3.5

1. A relation can be thought of as a set of ordered pairs of numbers. If no two pairs have the same x-coordinate and different y-coordinate, the relation is a function.
2. Vertical line test: If every vertical line intersects the graph of a relation in no more than one point, the relation is a function.

3.6

1. Unless specified otherwise, the domain of a function described by an equation is the set of all real numbers for which the expression in x is defined.

2. If a function is written in functional notation, for example, $f(x) = 3x + 5$, $f(x)$ represents y. To find y when x is 2, we calculate $f(2) = 3(2) + 5 = 11$.
3. Linear function: $f(x) = mx + b$.
4. Constant function: $f(x) = c$.
5. Identity function: $f(x) = x$.

REVIEW EXERCISES

Part I

3.1 1. Give the coordinates of the points A, B, C, and D. In which quadrant is each point located?

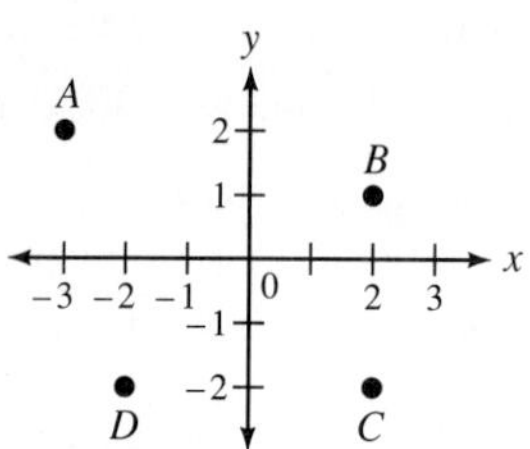

2. Give the coordinates of the vertices of a rectangle 4 units long and 3 units wide if the rectangle is in the first quadrant, its longest side is parallel to the x-axis, and one vertex is at (1, 2).

Graph the following equations.

3. $y = x^2 - 2$

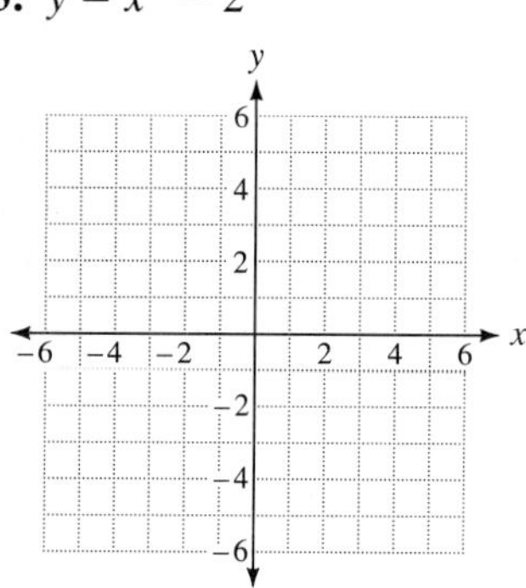

4. $y = |x| - 4$

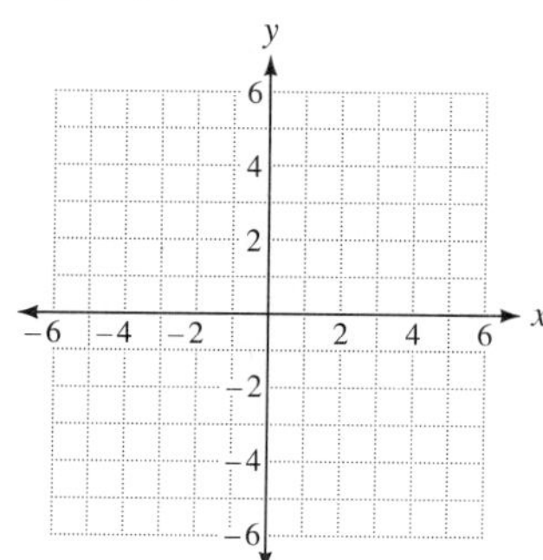

5. Mike Ratliff has estimated that the cost (in dollars), y, of producing a number, x, of computer parts in his plant is given by the equation $y = 3x + 40$. Complete the ordered pairs (1,), (100,), (1000,). What is the cost of producing **(a)** 1 part? **(b)** 100 parts? **(c)** 1000 parts?

3.2–3.3 *Give the intercepts, slope, and graph of the following equations.*

6. $x + 2y - 4 = 0$

x-intercept =

y-intercept =

slope =

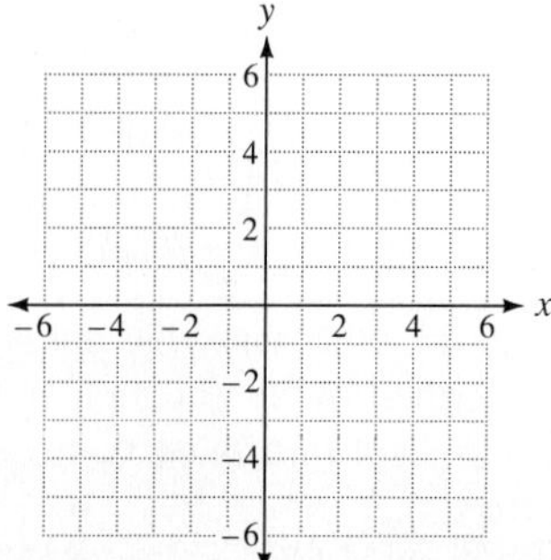

7. $2x + 1 = 0$

x-intercept =

y-intercept =

slope =

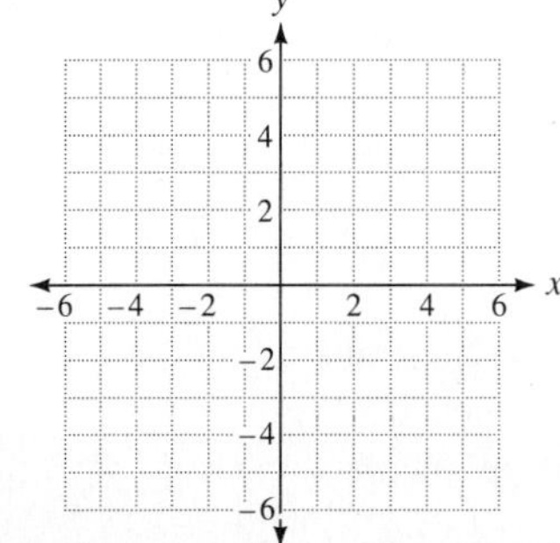

8. $y - 3 = 0$

x-intercept =

y-intercept =

slope =

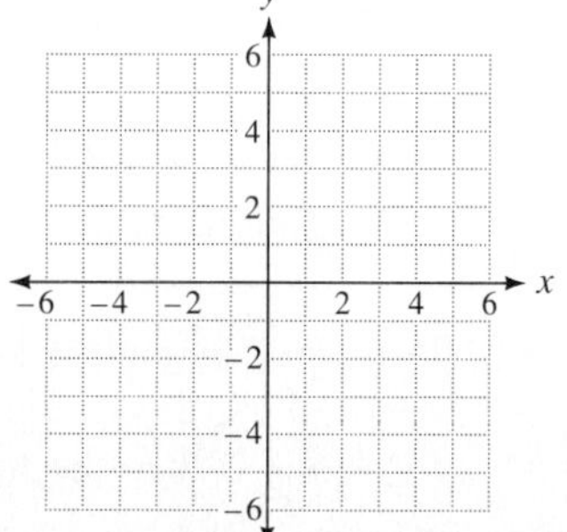

9. Does a line that slopes from upper left to lower right have positive slope or negative slope?

10. Which coordinate axis has undefined slope?

11. What is the slope of the line passing through $(3, -8)$ and $(-1, -2)$? What is the slope of a line that is perpendicular to this line?

12. Are the points $P(2, 5)$, $Q(-4, 1)$, and $R(5, 7)$ collinear? Explain.

3.4 *Are the given pairs of lines* parallel, perpendicular, *or* neither? *Explain.*

13. $3x - y + 2 = 0$
$x + 3y - 7 = 0$

14. $x + 2 = 0$
$y - 2 = 0$

15. $2x - 5y + 2 = 0$
$-4x + 10y - 2 = 0$

16. Find the general form of the equation of the line that has slope -4 and passes through $(-1, 3)$.

17. Find the slope-intercept form of the equation of the line with equation $4x + 2y - 10 = 0$. What is the slope? the y-intercept?

18. Find the general form of the equation of the line passing through the points $(-2, 5)$ and $(6, 9)$.

19. Find the general form of the equation of the line that passes through the point $(-3, 1)$ and is perpendicular to the line $2x + y - 3 = 0$.

20. What are the intercepts and the slope of the line with equation $3y - 15 = 0$?

21. What is the equation of the line passing through $(-4, 5)$ parallel to the x-axis? What is the equation of the line through $(-4, 5)$ parallel to the y-axis?

3.5 *Which of the following are functions?*

22.

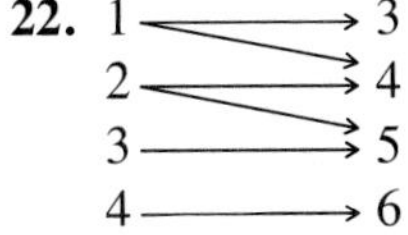

23.

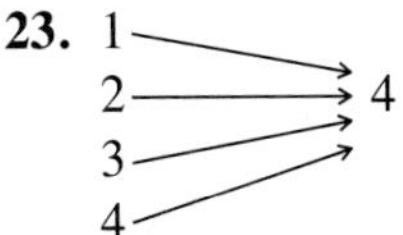

24.

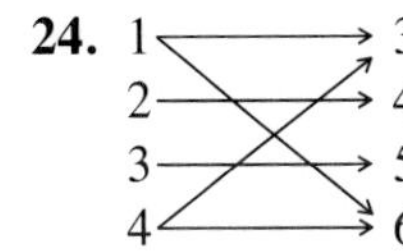

Which of the following are graphs of functions?

25.

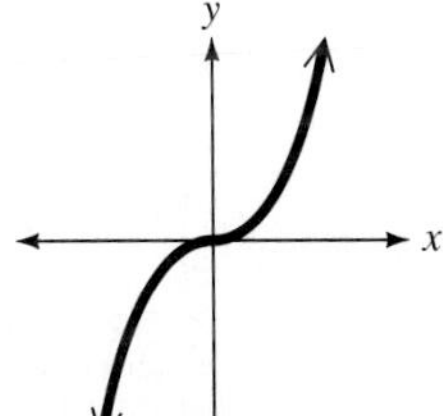

26.

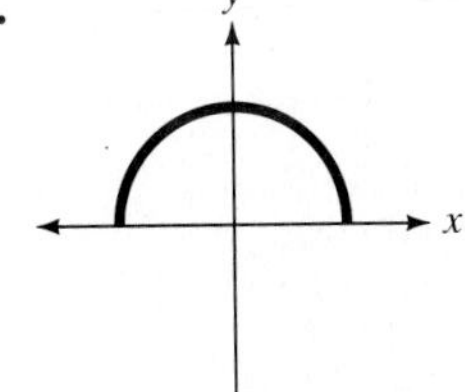

27.

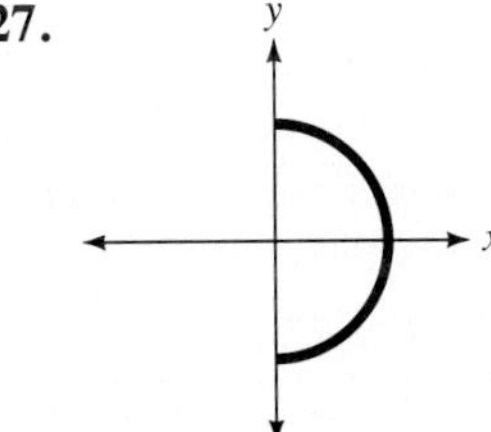

28. Consider the relation described by the set of ordered pairs $\{(2, 3), (1, 3), (3, 3), (4, 4), (4, 1)\}$.
(a) Write this relation using arrows.
(b) Graph the relation in the given Cartesian coordinate system.
(c) Is the relation a function? Explain.

(a) **(b)** **(c)**

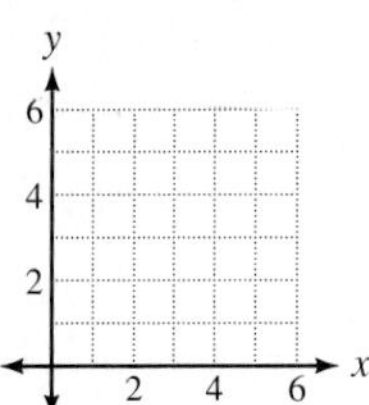

3.6 *For the given function, supply the required information for parts* **(a)** *through* **(f)**. *Graph the equation for part* **(g)**.

29. $f(x) = -2$
(a) $f(0) =$
(b) $f(1) =$
(c) $f(-2) =$
(d) $f(b) =$
(e) $f(-a) =$
(f) Is this a constant or linear function?
(g)

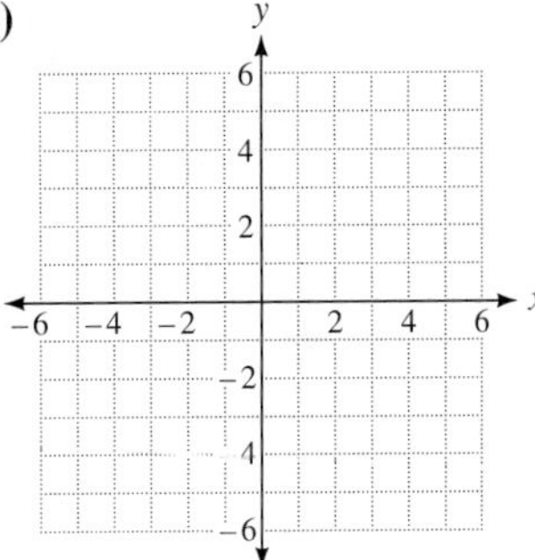

30. $f(x) = -3x + 6$
(a) $f(0) =$
(b) $f(1) =$
(c) $f(-2) =$
(d) $f(b) =$
(e) $f(-a) =$
(f) Is this a constant or linear function?
(g)

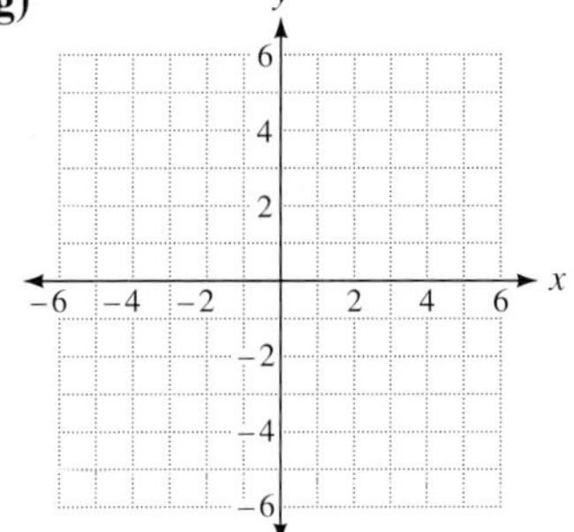

Specify the domain of each function defined by the given equation.

31. $y = 8x + 5$

32. $y = x^2 + 10$

33. $y = \dfrac{7}{7 - x}$

34. $y = \dfrac{1}{x^2 + 10}$

35. Write an equation for the cost function, g, described by selling x shirts at \$16.95 a shirt. How much do 8 shirts cost?

36. What is the equation describing the identity function?

Part II

37. Which coordinate axis has zero slope?

38. Give the slope and y-intercept of the line with equation $3x - 2y + 12 = 0$.

39. Barry Bertrowski earns an annual salary of \$12,000 plus a 15% commission on all of his sales. Express his annual earnings as a function, f, of his total sales, x. What are his annual earnings when his yearly sales total is \$230,000?

40. In which quadrant is the point with the given coordinates located?
(a) $(-2, 8)$ **(b)** $(6, -3)$ **(c)** $(4, \pi)$
(d) $(-\sqrt{2}, -3)$

Graph the following equations.

41. $y = |x - 2|$

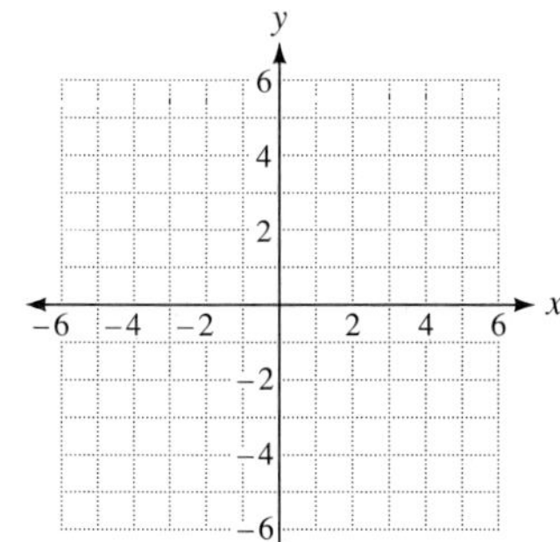

42. $6x + y - 6 = 0$

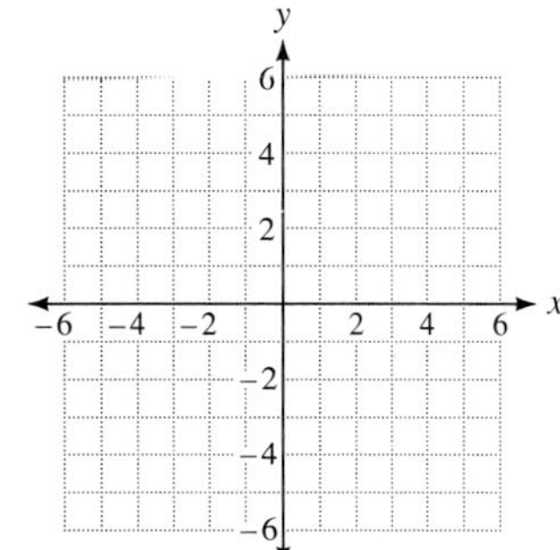

43. What is the slope of the line passing through $(0, -2)$ and $(7, -2)$?

44. Find the general form of the equation of the line that passes through $(-2, 1)$ and is parallel to the line $3x - y + 2 = 0$.

45. $f(x) = 4x - 2$

(a) $f(0) =$ **(b)** $f(1) =$

(c) $f(-2) =$ **(d)** $f(b) =$

(e) $f(-a) =$

(f) Is this a constant or linear function?

(g) Graph the function using the slope method.

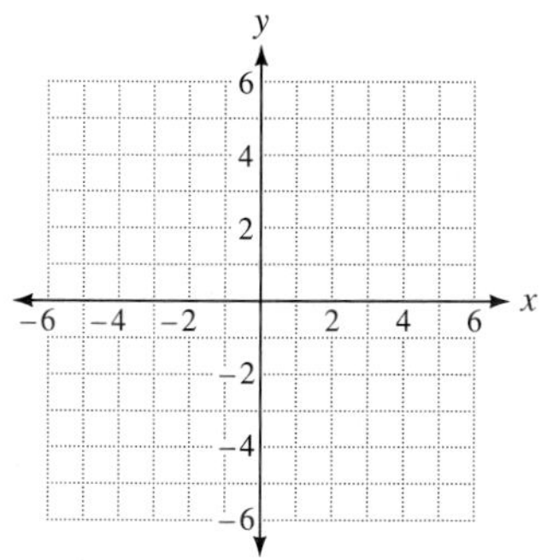

46. Is the following a function?

1 ⟶ 5
2 ⟶ 6
3 ⟶ 6

47. Are the lines with equations

$$2x - y + 3 = 0$$
$$-4x + 2y - 6 = 0$$

parallel, perpendicular, or neither?

48. What are the slope and y-intercept of the line with equation $2x + 10 = 0$?

49. If $g(x) = -x^2 + 6$, find **(a)** $g(-1)$, and **(b)** $g(c)$.

50. Terry McGinnis has found that the cost in dollars, y, of typing x pages of manuscript is given by $y = 8x + 25$. Complete the ordered pair (500,) and use it to find the cost of typing 500 pages.

ANSWERS: 1. $A(-3, 2)$ in II; $B(2, 1)$ in I; $C(2, -2)$ in IV; $D(-2, -2)$ in III 2. (1, 2), (5, 2), (5, 5), (1, 5)

3.

(−2, 2) (2, 2) (−1, −1) (1, −1) (0, −2)

4.

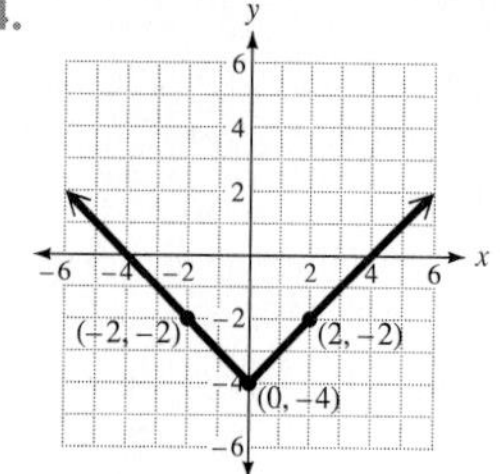

5. (1, 43), (100, 340), (1000, 3040); (a) \$43 (b) \$340 (c) \$3040

6. x-intercept (4, 0); y-intercept (0, 2); slope $-\frac{1}{2}$

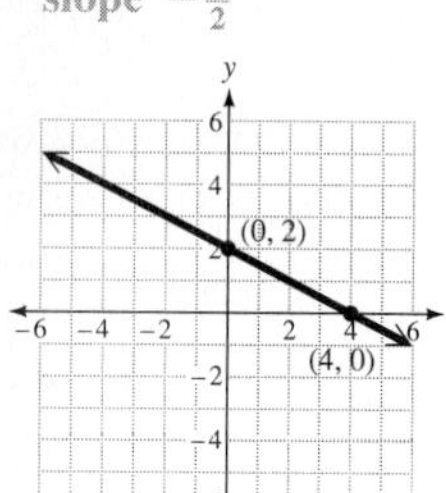

7. x-intercept $\left(-\frac{1}{2}, 0\right)$; no y-intercept; undefined slope

8. no x-intercept; y-intercept (0, 3); slope 0

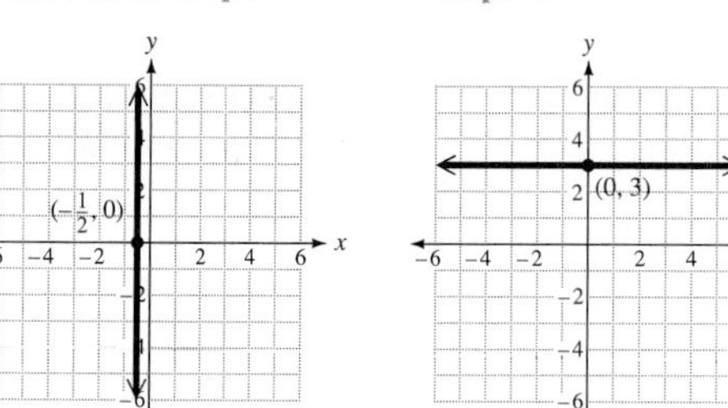

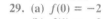

9. negative slope 10. y-axis 11. $-\frac{3}{2}$; $\frac{2}{3}$ 12. Yes. The slope of the line through P and Q is $\frac{2}{3}$, as is the slope of the line through Q and R. 13. perpendicular (slopes are 3 and $-\frac{1}{3}$) 14. perpendicular (each is parallel to an axis) 15. parallel (both have slope $\frac{2}{5}$) 16. $4x + y + 1 = 0$ 17. $y = -2x + 5$; slope -2; (0, 5) 18. $x - 2y + 12 = 0$ 19. $x - 2y + 5 = 0$

20. no x-intercept; y-intercept (0, 5); slope 0 21. $y = 5$; $x = -4$ 22. not a function 23. function 24. not a function 25. function 26. function 27. not a function

28. (a)

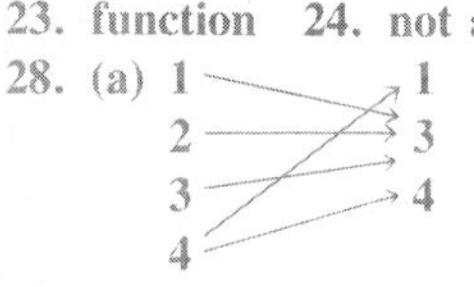

(b)

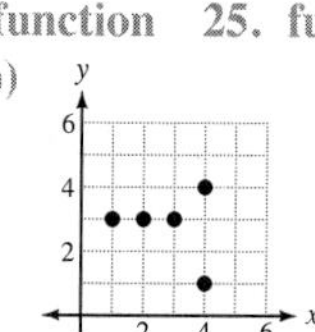

(c) No. The graph does not pass the vertical line test.

29. (a) $f(0) = -2$
(b) $f(1) = -2$
(c) $f(-2) = -2$
(d) $f(b) = -2$
(e) $f(-a) = -2$
(f) constant

(g)

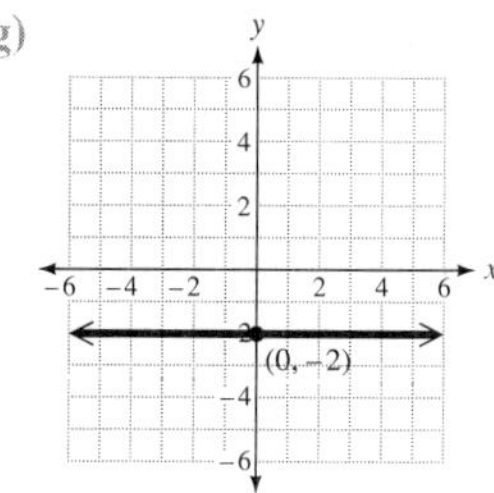

30. (a) $f(0) = 6$
(b) $f(1) = 3$
(c) $f(-2) = 12$
(d) $f(b) = -3b + 6$
(e) $f(-a) = 3a + 6$
(f) linear

(g)

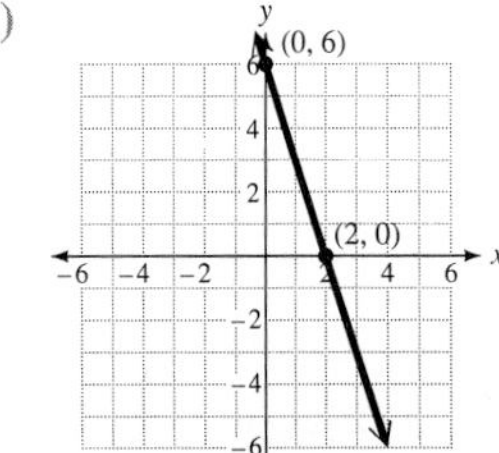

31. all real numbers 32. all real numbers 33. all real numbers except 7 34. all real numbers 35. $g(x) = 16.95x$; \$135.60 36. $f(x) = x$ 37. x-axis 38. $\frac{3}{2}$; (0, 6) 39. $f(x) = 12{,}000 + 0.15x$; \$46,500 40. (a) II (b) IV (c) I (d) III

41.

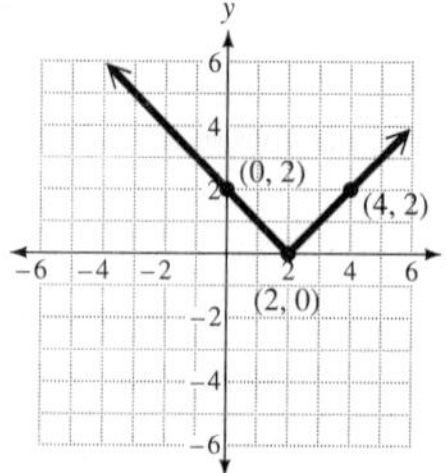

42.

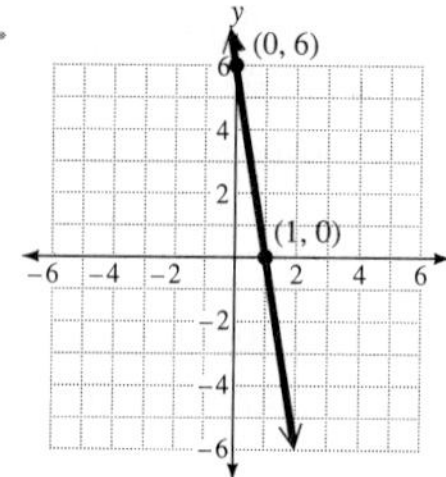

43. 0 44. $3x - y + 7 = 0$

45. (a) $f(0) = -2$
(b) $f(1) = 2$
(c) $f(-2) = -10$
(d) $f(b) = 4b - 2$
(e) $f(-a) = -4a - 2$
(f) linear

(g)

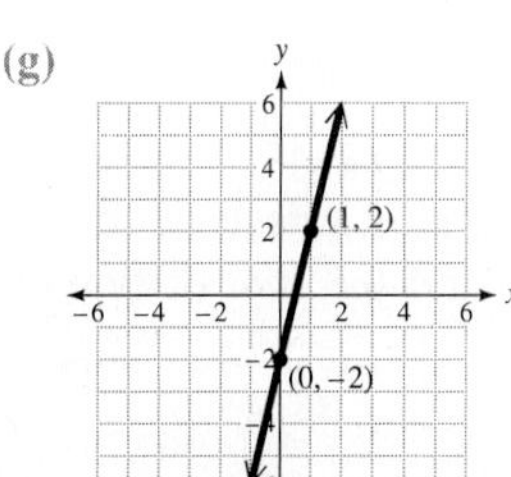

46. yes 47. Neither. The lines are the same. 48. undefined slope; no y-intercept 49. (a) 5 (b) $-c^2 + 6$ 50. (500, 4025); \$4025

CHAPTER 3 TEST

Exercises 1–5 refer to the equation $3x + 2y = 12$.

1. Write the equation in slope-intercept form. 1. ____________

2. What is the x-intercept? 2. ____________

3. What is the y-intercept? 3. ____________

4. What is the slope? 4. ____________

5. Graph the equation. 5.

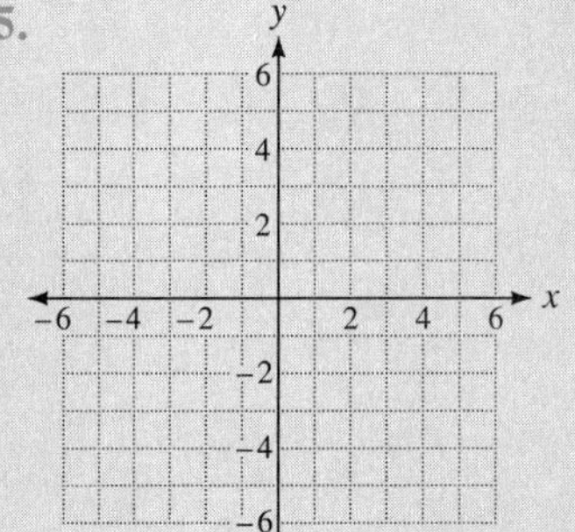

6. Ricardo estimates that the cost (in dollars), y, of repairing a number, x, of electronic detectors is given by the equation $y = 4x + 30$. Complete the ordered pair (10,   ) and use it to find the cost of repairing 10 detectors. 6. ____________

7. What is the slope of the line passing through $(-2, 7)$ and $(4, -5)$? What is the slope of a line perpendicular to this line? 7. ____________

8. Are the lines represented by the given equations parallel, perpendicular, or neither? 8. ____________

$$2x - 5y + 7 = 0$$
$$-2x + 5y + 7 = 0$$

9. Find the general form of the equation of the line that has slope -3 and passes through $(-2, 4)$. 9. ____________

CHAPTER 3 TEST

(CONTINUED)

10. Find the general form of the equation of the line passing through $(-1, 7)$ and $(4, -3)$.

10. ______________

11. What is the equation of the line through $(-2, 8)$ parallel to the x-axis?

11. ______________

12. Is 1 → 4 a function?
2 → 5
3 → 5

12. ______________

13. Is the curve at right the graph of a function?

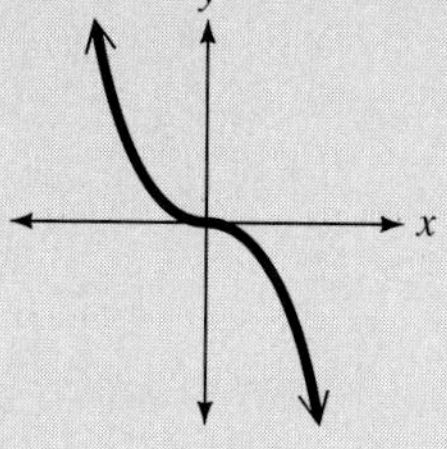

13. ______________

14. Is the relation $\{(2, 1), (1, 3), (1, 4), (3, 5)\}$ a function?

14. ______________

Exercises 15–18 refer to the function $f(x) = 3x + 3$.

15. Find $f(-2)$.

15. ______________

16. Find $f(a - 1)$.

16. ______________

17. Is this a constant or linear function?

17. ______________

18. Graph the function.

18.

19. Specify the domain of the function $y = \dfrac{1}{x + 3}$.

19. ______________

20. Write an equation for the cost function, g, described by selling x pairs of shoes at \$20 a pair.

20. ______________

CHAPTER 4

Systems of Linear Equations and Inequalities

4.1 SYSTEMS OF TWO LINEAR EQUATIONS IN TWO VARIABLES

STUDENT GUIDEPOSTS

1. Systems of Equations
2. The Graphing Method
3. Types of Linear Systems
4. Number of Solutions to a System

In Chapter 3 we saw that a linear equation in the two variables x and y, such as

$$x - y - 1 = 0,$$

has many solutions that are ordered pairs of numbers that make the equation true. Some of the solutions to this equation are

$$(0, -1), \quad (1, 0), \quad (-1, -2), \quad (2, 1), \quad \text{and} \quad (5, 4).$$

The graph of the equation is a straight line, every point of which corresponds to a solution to the equation. Suppose that we have a second linear equation,

$$2x + y - 5 = 0.$$

This equation too has infinitely many solutions, some of which are

$$(0, 5), \quad (1, 3), \quad (-1, 7), \quad (2, 1), \quad \text{and} \quad (-3, 11).$$

1 SYSTEMS OF EQUATIONS

A **system of two linear equations in two variables** is a pair of linear equations such as the following.

$$x - y - 1 = 0$$
$$2x + y - 5 = 0$$

A **solution** to a system of equations is an ordered pair of numbers that is a solution to *both* equations. For our particular example, because (2, 1) is listed as a solution to both of the equations, (2 1) is a solution to the system of equations. We can check this by substituting (2, 1) into each of the equations.

$$2 - 1 - 1 \stackrel{?}{=} 0 \qquad\qquad 2(2) + 1 - 5 \stackrel{?}{=} 0$$
$$0 = 0 \qquad\qquad 4 + 1 - 5 \stackrel{?}{=} 0$$
$$0 = 0$$

2 THE GRAPHING METHOD

If we write both equations above in slope-intercept form,

$$x - y - 1 = 0 \qquad\qquad 2x + y - 5 = 0$$
$$-y = -x + 1 \qquad\qquad y = -2x + 5$$
$$y = x - 1$$

we can see that the slopes are unequal ($m_1 = 1$ and $m_2 = -2$) and that the graphs of the equations are intersecting lines. Graphing both equations in the same rectangular coordinate system in Figure 4.1, we see that the point of intersection corresponds to the solution of the system, (2, 1).

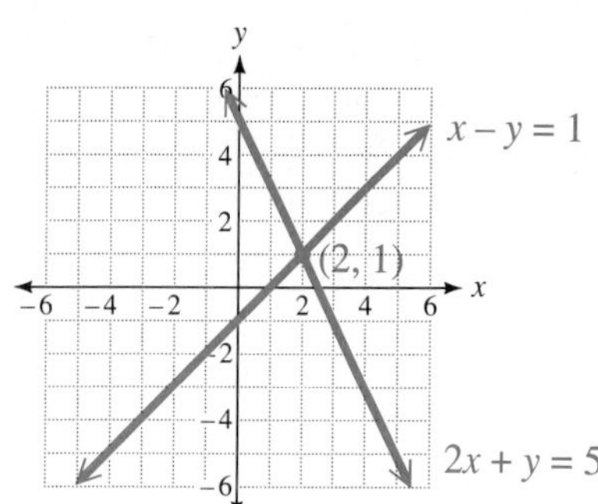

Figure 4.1

This procedure illustrates the **graphing method** for solving a system: We plot the graphs of the equations, estimate the point of intersection (if such a point exists), and verify by substitution that the ordered pair corresponding to the point does indeed solve each equation.

EXAMPLE 1 Solving Systems by Graphing

Solve the system using the graphing method.

$$x + y = 2$$
$$2x - y = -5$$

We graph each equation by finding its intercepts.

$x + y = 2$

x	y
0	2
2	0

$2x - y = -5$

x	y
0	5
$-\frac{5}{2}$	0

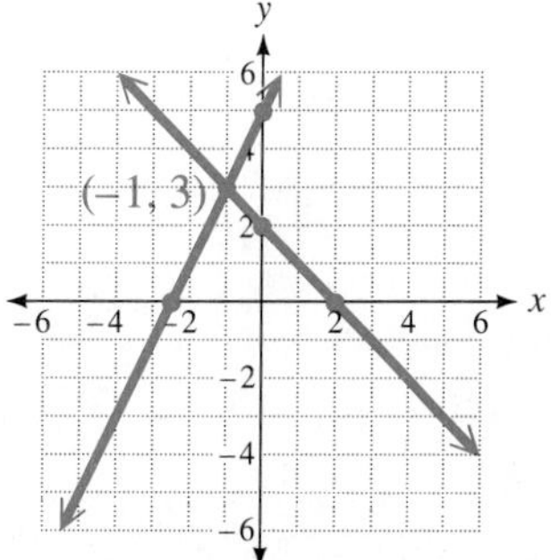

Figure 4.2

Practice Exercise 1

Solve the system using the graphing method.

$$x + y = 0$$
$$2x - y = 6$$

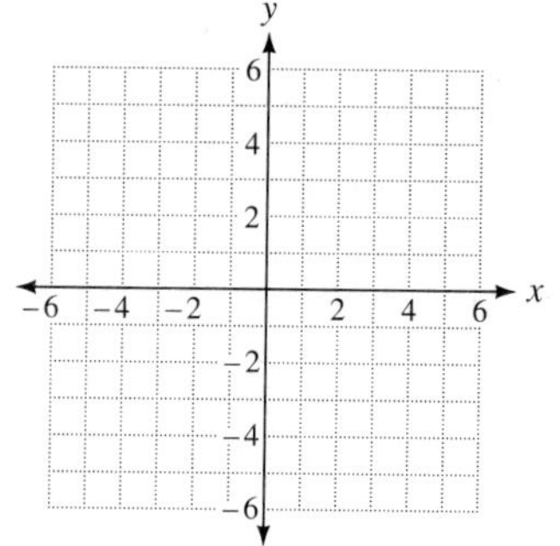

The point of intersection of the two lines, graphed in Figure 4.2, appears to have coordinates $(-1, 3)$. We check by substitution in both equations.

$$(-1) + (3) \stackrel{?}{=} 2 \qquad\qquad 2(-1) - (3) \stackrel{?}{=} -5$$
$$2 = 2 \qquad\qquad -2 - 3 \stackrel{?}{=} -5$$
$$-5 = -5$$

The solution to the system is $(-1, 3)$.

Answer: $(2, -2)$

3 TYPES OF LINEAR SYSTEMS

Before we consider other methods of finding solutions to systems of equations, it is helpful to study the nature of these solutions. When each of two linear equations in a system is graphed in the same rectangular coordinate system, one of three possibilities occurs:

1. The lines coincide, and the system of equations has infinitely many solutions. The system is said to be **dependent.**
2. The lines are parallel, and the system of equations has no solution. The system is said to be **inconsistent.**
3. The lines intersect in exactly one point, and the system of equations has exactly one solution. The system is said to be **independent** and **consistent.**

As examples of these three cases consider the following three systems of equations and their corresponding graphs in Figure 4.3.

(A) $2x - y = -4$ (C) $2x - y = -4$ (E) $2x - y = -4$

(B) $6x - 3y = -12$ (D) $2x - y = 2$ (F) $x + y = 1$

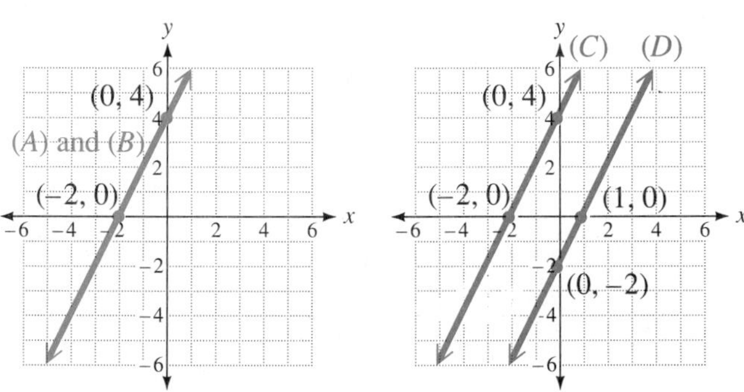

Lines coincide — system dependent

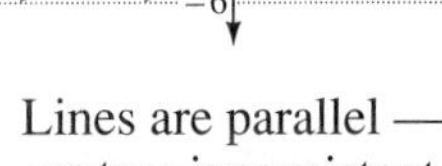

Lines are parallel — system inconsistent

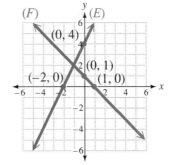

Lines intersect in exactly one point — system independent and consistent

Figure 4.3 Types of Linear Systems

In the first system, lines (A) and (B) coincide; in the second system, lines (C) and (D) are parallel; and in the third system, lines (E) and (F) intersect in one point. Let us write each equation in slope-intercept form (solve each equation for y).

(A) $y = 2x + 4$ (C) $y = 2x + 4$ (E) $y = 2x + 4$

(B) $y = 2x + 4$ (D) $y = 2x - 2$ (F) $y = -x + 1$

same slope, same intercept — same slope, different intercept — different slopes

Remember that in slope-intercept form, the coefficient of x is the slope and the constant term determines the y-intercept. Also, parallel or coinciding lines have the same slope (or both have undefined slope).

4 NUMBER OF SOLUTIONS TO A SYSTEM

Recognizing different types of linear systems allows us to know the number of solutions to a particular system.

To Find the Number of Solutions to a System of Equations

1. Write each equation in slope-intercept form and identify the slope and y-intercept of each.
2. If the slopes are equal and the y-intercepts are equal, the graphs coincide and the system has *infinitely many solutions*.
3. If the slopes are equal and the y-intercepts are unequal, the graphs are parallel and the system has *no solution*.
4. If the slopes are unequal, the graphs intersect in exactly one point and the system has *exactly one solution*.

CAUTION

The rule above does not hold when one or both of the equations in a system is of the form $ax = c$, since such equations cannot be solved for y. However, since the graph of an equation of this form is always a line parallel to the y-axis, we know parallel or coinciding lines result only if both equations in the system are of this type.

EXAMPLE 2 DETERMINING THE NUMBER OF SOLUTIONS

Find the number of solutions to each system.

(a) $3x - 2y + 1 = 0$
$-4y + 6x = -2$

Solve each equation for y.

$$-2y = -3x - 1 \qquad -4y = -6x - 2$$

$$y = \frac{3}{2}x + \frac{1}{2} \qquad y = \frac{6}{4}x + \frac{2}{4} = \frac{3}{2}x + \frac{1}{2}$$

Since the slopes and y-intercepts are equal, the lines coincide, and the system has infinitely many solutions. That is, any solution to one equation also solves the other and is a solution to the system.

(b) $2x - 5y + 1 = 0$
$15y - 6x = 1$

Solve each equation for y.

$$-5y = -2x - 1 \qquad 15y = 6x + 1$$

$$y = \frac{2}{5}x + \frac{1}{5} \qquad y = \frac{6}{15}x + \frac{1}{15} = \frac{2}{5}x + \frac{1}{15}$$

Since the slopes are equal and the y-intercepts are unequal, the lines are parallel. Thus, the system has no solution.

PRACTICE EXERCISE 2

Find the number of solutions to each system.

(a) $4x + y = -1$
$-8x - 2y = 1$

(b) $x - 3y + 5 = 0$
$-2x + 6y = 10$

(c) $x - 5y = 6$
$4 + y = 0$

(c) $2x - 5 = 0$
$2x + y = 5$

Since the first equation has as its graph a line with undefined slope (parallel to the y-axis), and the second equation has as its graph a line with slope -2, the lines intersect and the system has exactly one solution.

Answers: (a) no solution (b) infinitely many solutions (c) exactly one solution

4.1 EXERCISES A

Answer true *or* false *for each statement. If the statement is false, explain why.*

1. In a system of equations, if the lines coincide, the system has infinitely many solutions and is said to be dependent.

2. In a system of equations, if the lines are parallel, the system has no solution and is said to be consistent.

3. When the slopes of the lines in a system of equations are unequal, the graphs are intersecting lines.

4. When the slopes of the lines in a system of equations are unequal, the system is said to be consistent and independent.

5. Is (3, 0) a solution to the given systems?

(a) $3x - 2y - 9 = 0$
$x + 8y - 3 = 0$

(b) $x + y + 1 = 4$
$2x - y + 2 = 4$

(c) $x = 3$
$2x - 4y - 6 = 0$

6. Is $\left(\frac{1}{2}, -5\right)$ a solution to the given systems?

(a) $2x + y + 4 = 0$
$8x - 3y - 11 = 0$

(b) $2x - 1 = 0$
$y + 5 = 0$

(c) $3y + 15 = 0$
$4x - y - 7 = 0$

Without solving, tell the number of solutions to the given system.

7. $5x - 3y + 7 = 0$
$-5x + 3y - 7 = 0$

8. $5x - 3y + 7 = 0$
$3x - 5y - 7 = 0$

9. $2x + 7 = 0$
$2x + 7y = 5$

10. $2x + 7 = 0$
$7x + 2 = 0$

11. $x + y + 1 = 0$
$y + x - 1 = 0$

12. $3x + 3y + 3 = 0$
$5x + 5y + 5 = 0$

13. $y - 6 = 0$
$x + 6 = 0$

14. $x + 2y + 3 = 0$
$3x + 2y + 1 = 0$

15. $3x + y = 0$
$x + 3y = 0$

Solve the following systems using the graphing method.

16. $x + y = 4$
$3x - y = 0$

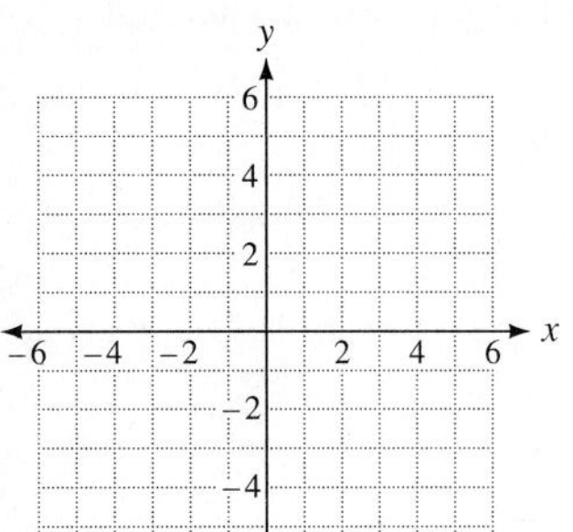

17. $2x - y = 1$
$x + 5y = 6$

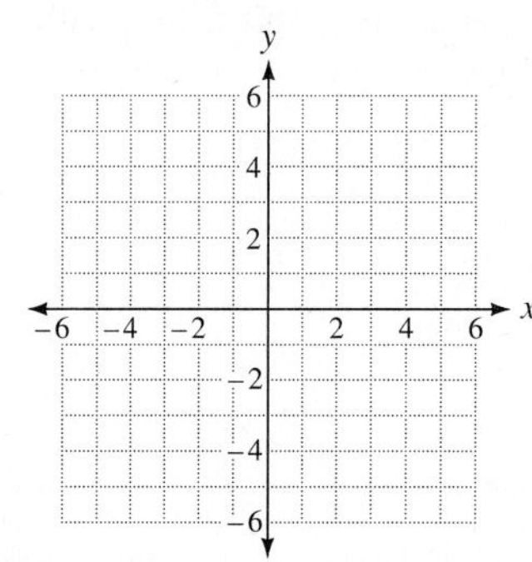

18. $3x + 2y = 5$
$-6x - 4y = 5$

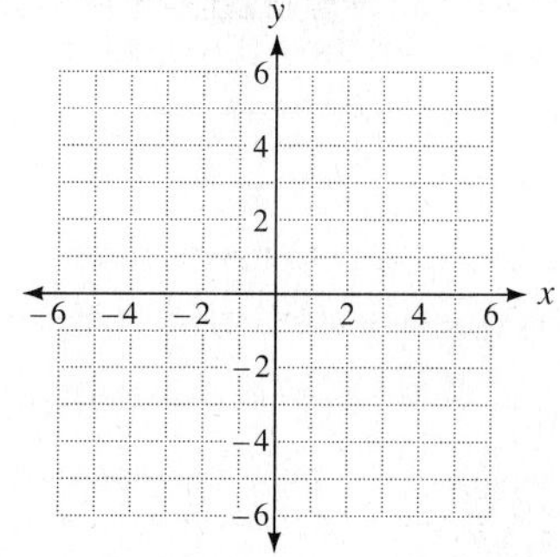

19. $5x - y = 0$
$x + 4y = 0$

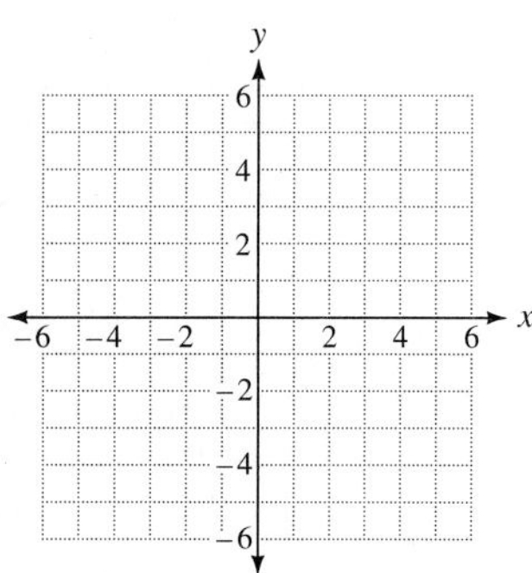

20 $2x + 3y = -2$
$-4x + y = -3$

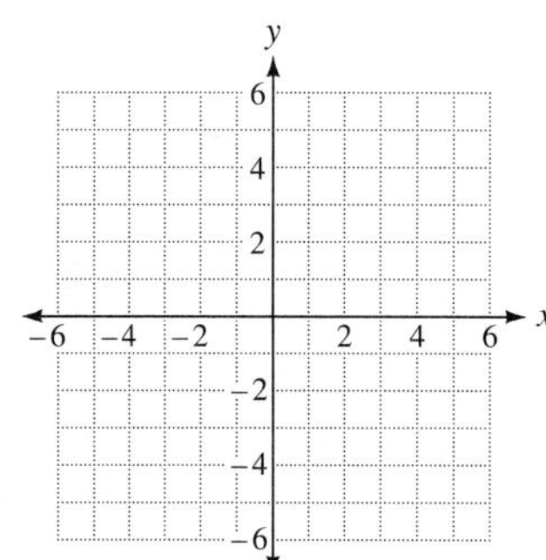

21. $x - 4 = 0$
$2x + y = 7$

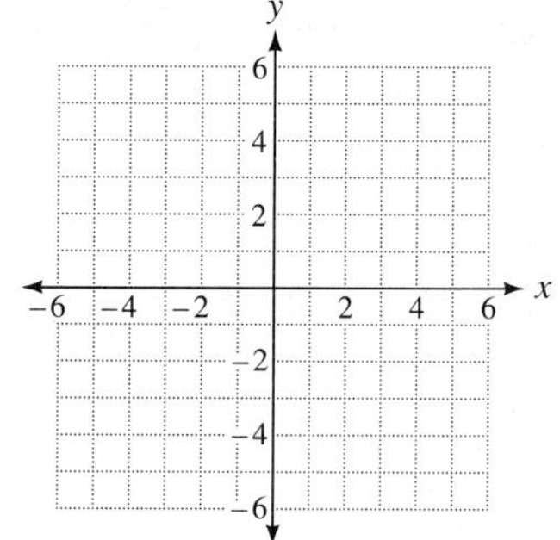

FOR REVIEW

The following exercises will help you prepare for the next section.

22. Solve $6x + 3y - 15 = 0$ for y. Substitute this expression for y into $x - 2y + 5 = 0$. Solve the resulting equation for the numerical value of x.

23. Consider the ordered pairs $(x, x + 2)$, for $x = -2, -1, 0, 1, 2$. Show that each of these ordered pairs is a solution to $3x - 3y = -6$.

24. Find the value of y if $x = 3$ and $y = 2x - 5$.

25. Find the value of x if $y = -2$ and $x = \frac{1}{2}y + 4$.

ANSWERS: **1.** true **2.** false (inconsistent) **3.** true **4.** true **5.** (a) yes (b) no (c) yes **6.** (a) no (b) yes (c) yes **7.** infinitely many **8.** exactly one **9.** exactly one **10.** none **11.** none **12.** infinitely many **13.** exactly one **14.** exactly one **15.** exactly one **16.** (1, 3) **17.** (1, 1) **18.** no solution (graphs are parallel) **19.** (0, 0) **20.** $\left(\frac{1}{2}, -1\right)$ **21.** (4, −1) **22.** $y = -2x + 5$; $x = 1$ **23.** Substituting the values in each pair shows it is a solution. **24.** $y = 1$ **25.** $x = 3$

4.1 EXERCISES B

Answer true *or* false *for each statement. If the statement is false, explain why.*

1. If a system of equations has no solution, the graphs of the system are intersecting lines.

2. If a system of equations has exactly one solution, the graphs of the system are parallel lines.

3. When the slopes of the lines in a system are equal but the y-intercepts are unequal, the graphs of the system are parallel lines.

4. A system has infinitely many solutions whenever the graphs coincide.

5. Is (0, 2) a solution to the given systems?

(a) $3x - y = -2$
$x + 7y = 14$

(b) $y + 2 = 0$
$3x - y = -2$

(c) $4x - 2y = -4$
$x + 6y = 12$

6. Is $\left(-1, \frac{1}{3}\right)$ a solution to the given systems?

(a) $x + 3y = 0$
$2x - 6y = 0$

(b) $3x - 6y = -5$
$x + 9y = 2$

(c) $5x + 9y = -2$
$-x - 3y = 0$

Without solving, tell the number of solutions to the given system.

7. $-5x + 2y + 3 = 0$
$-5x - 2y + 3 = 0$

8. $-5x + 2y + 3 = 0$
$5x - 2y - 3 = 0$

9. $2x + y = 7$
$2x - 3 = 0$

10. $3y + 5 = 0$
$y - 7 = 0$

11. $x + y - 3 = 0$
$x - y + 3 = 0$

12. $2x + 2y - 2 = 0$
$5x + 5y - 5 = 0$

13. $2x = 8$
$8y = 2$

14. $-5x + 2y + 3 = 0$
$10x - 4y + 3 = 0$

15. $x - 7y = 0$
$7x - y = 0$

Solve the following systems using the graphing method.

16. $3x + y = -1$
$x - y = -3$

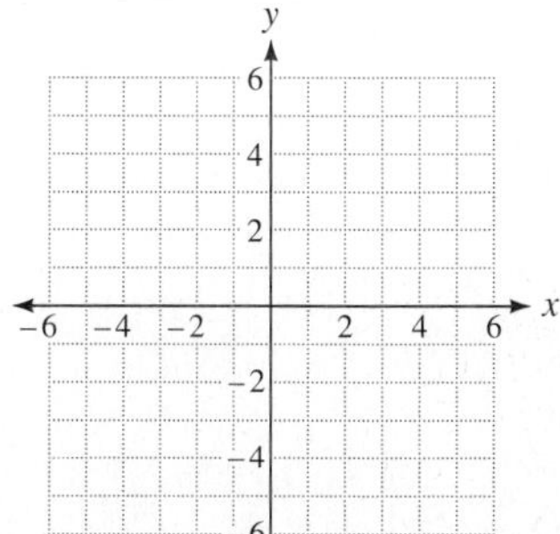

17. $2x - 3y = -2$
$x + y = 4$

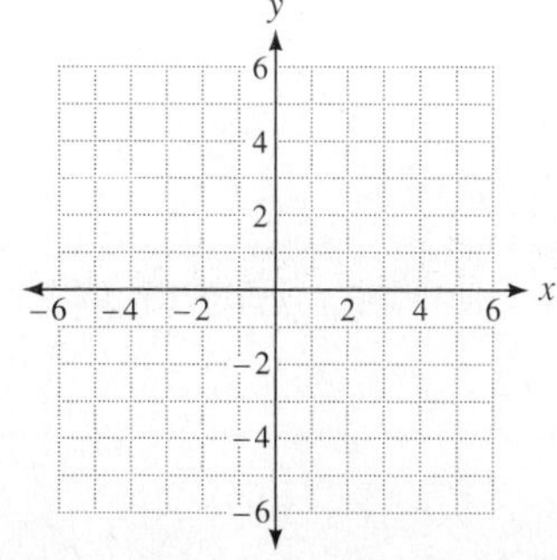

18. $5x - 2y = 7$
$-5x + 2y = 7$

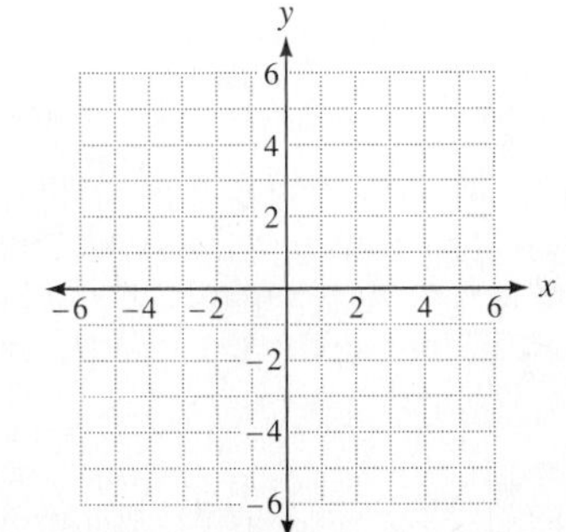

19. $x + 3y = -1$
$3x + 5y = 5$

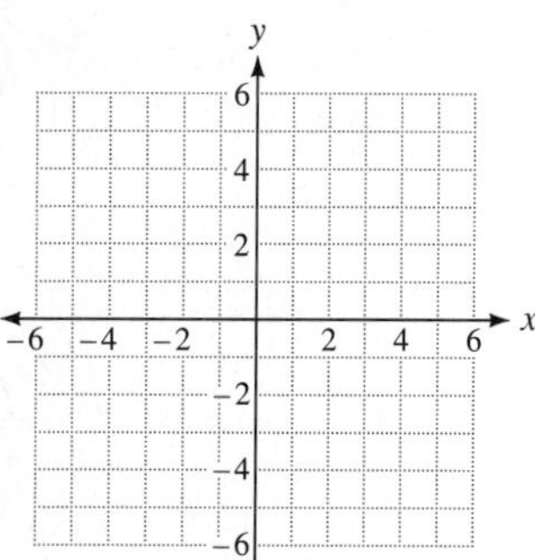

20. $3x - y = -1$
$-6x + 5y = 8$

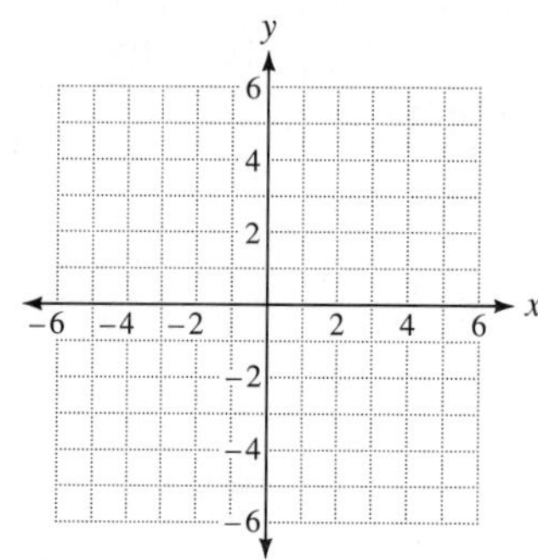

21. $y + 3 = 0$
$3x - 2y = 0$

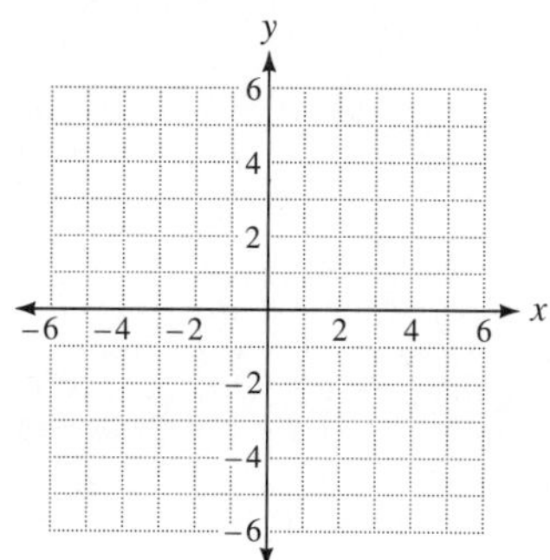

FOR REVIEW

The following exercises will help you prepare for the next section.

22. Solve $x + y + 3 = 0$ for x. Substitute this expression for x into $2x - y - 6 = 0$. Solve the resulting equation for the numerical value of y.

23. Consider the ordered pairs $(x, 2 - x)$, for $x = -2, -1, 0, 1, 2$. Show that each of these pairs is a solution to $2x + 2y - 4 = 0$.

24. Find the value of x if $y = -2$ and $x = 3y + 1$.

25. Find the value of y if $x = 6$ and $y = \frac{1}{3}x - 5$.

4.1 EXERCISES C

Tell whether $(0.25, 1.78)$ *is a solution to the given system.*

1. $0.3x + 1.5y = 2.745$
$2.8x - 0.7y = -0.546$

2. $3.4x + 1.6y - 3.698 = 0$
$-0.7x + 2.1y - 3.913 = 0$

4.2 SOLVING SYSTEMS OF TWO EQUATIONS

STUDENT GUIDEPOSTS

1 Substitution Method
2 Elimination Method

1 SUBSTITUTION METHOD

There are disadvantages to graphing as a method of solving a system. First, it takes too much time; second, the correctness of the answer depends on the accuracy of the graphing techniques. Here we develop a better approach, the **substitution method.** Consider the system

$$x + y = 4$$
$$3x - y = 0.$$

This system can be changed into one equation in one unknown. Solve the first equation for x,

$$x = -y + 4,$$

and substitute this value of x into the second equation.

$$3(-y + 4) - y = 0$$

The result is an equation in the single variable y which can be solved for y.

$$\begin{aligned} -3y + 12 - y &= 0 \\ -4y &= -12 \\ y &= 3 \end{aligned}$$

Substitute this value of y into either of the original equations, for example, the second,

$$3x - (3) = 0,$$

and solve for x.

$$\begin{aligned} 3x &= 3 \\ x &= 1 \end{aligned}$$

The solution to the system is (1, 3). Check by substituting (1, 3) into each of the original equations. Compare your work here with the results of Exercise 16 in the A exercises for Section 4.1. Although we used two different methods to solve the same system, we obtained the same solution each time.

To Solve a System of Equations Using the Substitution Method

1. Solve one of the equations for one of the variables.
2. Substitute that value of the variable in the *remaining* equation.
3. Solve this new equation and substitute the numerical solution into either of the two *original* equations to find the numerical value of the second variable.
4. Check your solution in both original equations.

EXAMPLE 1 Using the Substitution Method

Solve by the substitution method.

$$\begin{aligned} 2y - x - 2 &= 0 \\ 2x + y - 6 &= 0 \end{aligned}$$

We can solve either equation for either variable, but one way to avoid fractions is to solve the second equation for y.

$$y = -2x + 6$$

Substitute $-2x + 6$ for y in the first equation.

$$\begin{aligned} 2(-2x + 6) - x - 2 &= 0 \\ -4x + 12 - x - 2 &= 0 \\ -5x + 10 &= 0 \\ -5x &= -10 \\ x &= 2 \end{aligned}$$

Practice Exercise 1

Solve by the substitution method.

$$\begin{aligned} 2x - y &= -11 \\ x + 5y &= 0 \end{aligned}$$

Now substitute 2 for x in

$$\begin{aligned} y &= -2x + 6. \\ &= -2(2) + 6 \\ &= -4 + 6 \\ &= 2 \end{aligned}$$

The solution is (2, 2). Check in both equations.

Answer: $(-5, 1)$

EXAMPLE 2 A System with Infinitely Many Solutions

Solve by the substitution method.

$$\begin{aligned} 2x - y &= 1 \\ -4x + 2y &= -2 \end{aligned}$$

To avoid fractions, solve the first equation for y.

$$y = 2x - 1$$

Substitute this expression into the second equation.

$$\begin{aligned} -4x + 2(2x - 1) &= -2 \\ -4x + 4x - 2 &= -2 \\ -2 &= -2 \quad \text{An identity} \end{aligned}$$

When we obtain an identity, return to the original system and examine the nature of the solutions by solving each equation for y.

$$\begin{aligned} 2x - y &= 1 \qquad & -4x + 2y &= -2 \\ -y &= -2x + 1 & 2y &= 4x - 2 \\ y &= 2x - 1 & y &= 2x - 1 \end{aligned}$$

The slopes and y-intercepts are equal, so the lines coincide, the system is dependent, and there are infinitely many solutions. In fact, any pair of numbers

$$(x, 2x - 1) \qquad \text{Since } y = 2x - 1$$

for x any real number, is a solution to the system. Several representative solutions are $(0, -1)$, $(1, 1)$, $(2, 3)$, and $(-1, -3)$, obtained by letting $x = 0, 1, 2,$ and -1, respectively.

Practice Exercise 2

Solve by the substitution method.

$$\begin{aligned} 5x + y &= -1 \\ -10x - 2y &= 2 \end{aligned}$$

Answer: $(x, -5x - 1)$ for any real number x

EXAMPLE 3 A System with No Solution

Solve by the substitution method.

$$\begin{aligned} 2x + 4y &= 1 \\ x + 2y &= 8 \end{aligned}$$

To avoid fractions, solve the second equation for x.

$$x = -2y + 8$$

Substitute this value into the first equation.

$$\begin{aligned} 2(-2y + 8) + 4y &= 1 \\ -4y + 16 + 4y &= 1 \\ 16 &= 1 \quad \text{A contradiction} \end{aligned}$$

Practice Exercise 3

Solve by the substitution method.

$$\begin{aligned} x - 3y &= 1 \\ 3x - 9y &= -1 \end{aligned}$$

When we obtain a contradiction, return to the original system and examine the nature of the solutions by solving each equation for y.

$$2x + 4y = 1 \qquad\qquad x + 2y = 8$$
$$4y = -2x + 1 \qquad\qquad 2y = -x + 8$$
$$y = -\frac{1}{2}x + \frac{1}{4} \qquad\qquad y = -\frac{1}{2}x + 4$$

The slopes are equal and the y-intercepts are unequal, so the lines are parallel, the system is inconsistent, and there are no solutions to the system.

Answer: no solution

Examples 2 and 3 lead us to the following result that will apply to the next method we study as well as to the substitution method.

Solving Dependent or Inconsistent Systems

In solving a system of equations,

1. if an **identity** results, the lines coincide, the system is dependent, and there are **infinitely many solutions;**
2. if a **contradiction** results, the lines are parallel, the system is inconsistent, and there is **no solution.**

2 ELIMINATION METHOD

An alternative for solving systems of equations, known as the **elimination method,** is perhaps superior to substitution for systems such as

$$3x + 5y = -2$$
$$5x + 3y = 2.$$

Solving this system by the substitution method would involve fractions regardless of our choice of variable and choice of equation. The elimination method lets us avoid computations with fractions.

To help in our approach to the above system, we first consider a somewhat simpler system.

$$x + y = 4$$
$$2x - y = 5$$

Observe that $2x - y$ and 5 are both names or expressions for the same number. Remember that if the same expression is added to or subtracted from both sides of an equation the results are equal. Therefore, we can add $2x - y$ to the left side of the first equation and 5 to the right to obtain another equation with the same solution.

$$(x + y) + (2x - y) = 4 + 5$$
$$3x = 9$$

Generally, we add vertically and avoid the intermediate steps.

$$\begin{array}{r} x + y = 4 \\ \underline{2x - y = 5} \\ 3x \quad\;\; = 9 \\ x = 3 \end{array}$$

By adding the two equations, we "eliminate" the y term and obtain one equation in one variable. Once we know that x is 3, we substitute this value into either one of the two original equations to find the value of y. Substitute in the first equation.

$$\begin{aligned} 3 + y &= 4 \\ y &= 1 \end{aligned}$$

The solution is (3, 1). Check this by substituting in both original equations.

At times, addition or subtraction alone will not eliminate a variable and yield one equation in only one variable. When this happens, we may have to multiply one or both equations by a number that makes the coefficients of one of the variables negatives of each other. Then by adding the two equations we eliminate a variable and proceed as before. This is illustrated in the next example.

EXAMPLE 4 Using the Elimination Method

Solve by the elimination method.

$$\begin{aligned} 5x - 2y &= -13 \\ 2x + y &= 11 \end{aligned}$$

By adding or subtracting the two equations as given, we do not eliminate a variable. However, if we multiply both sides of the second equation by 2 (using the multiplication rule), the resulting equation is

$$4x + 2y = 22.$$

Add this equation to the first equation.

$$\begin{aligned} 5x - 2y &= -13 \\ 4x + 2y &= 22 \\ \hline 9x \quad &= 9 \\ x &= 1 \end{aligned}$$

Substitute 1 for x in the second equation.

$$\begin{aligned} 2(1) + y &= 11 \\ 2 + y &= 11 \\ y &= 9 \end{aligned}$$

The solution is (1, 9). Check by substitution.

Practice Exercise 4

Solve by the elimination method.

$$\begin{aligned} x - 4y &= -14 \\ 3x + y &= -3 \end{aligned}$$

Answer: $(-2, 3)$

At times the multiplication rule must be applied to both equations before adding or subtracting. Let us solve our original system.

EXAMPLE 5 Using the Elimination Method

Solve by the elimination method.

$$\begin{aligned} 3x + 5y &= -2 \\ 5x + 3y &= 2 \end{aligned}$$

Multiply the first equation by -5 and the second equation by 3, making the coefficients of the x terms -15 and 15, respectively. Add to eliminate x.

Practice Exercise 5

Solve by the elimination method.

$$\begin{aligned} 2x + 7y &= 8 \\ 3x - 5y &= 12 \end{aligned}$$

$$\begin{array}{rl} -15x - 25y = 10 & \text{−5 times first equation} \\ 15x + 9y = 6 & \text{3 times second equation} \\ \hline -16y = 16 & \\ y = -1 & \end{array}$$

Substitute -1 for y in the first equation.

$$3x + 5(-1) = -2$$
$$3x - 5 = -2$$
$$3x = 3$$
$$x = 1$$

The solution is $(1, -1)$. Check by substitution.

Answer: $(4, 0)$

To Solve a System of Equations Using the Elimination Method

1. Apply the multiplication rule to one or both equations (if necessary) to transform them so that addition or subtraction will eliminate a variable.
2. Eliminate the variable by adding or subtracting.
3. Solve the resulting single-variable equation and substitute this value into one of the original equations and solve.
4. Check your answer by substitution in both original equations.

You may ask, "Should I use substitution or elimination for this problem?" If one of the variables has a coefficient of 1 or -1, it is usually easier to solve for it and use substitution. Otherwise, you probably should use elimination.

Suppose we try to solve the following system by elimination.

$$3x - 2y = 5$$
$$-3x + 2y = -5$$

Notice that we can eliminate the variable x by addition.

$$\begin{array}{r} 3x - 2y = 5 \\ -3x + 2y = -5 \\ \hline 0 = 0 \end{array}$$

However, we have actually eliminated both variables and obtained an identity. Just as with the substitution method, this indicates that there are infinitely many solutions. (Verify that the lines do coincide.) If we solve the first equation for y, we obtain

$$-2y = -3x + 5$$
$$y = \frac{3}{2}x - \frac{5}{2}.$$

The solutions to the system take the form $\left(x, \frac{3}{2}x - \frac{5}{2}\right)$ where x is any real number.

We conclude this section with an example of a system that is not linear but that can be solved by the techniques for linear systems, after changing variables.

EXAMPLE 6 Solving a Nonlinear System

Solve the system.

$$\frac{3}{x} - \frac{5}{y} = -2$$
$$\frac{6}{x} + \frac{15}{y} = 16$$

This system is not linear, but if we rewrite it as

$$3\left(\frac{1}{x}\right) - 5\left(\frac{1}{y}\right) = -2$$
$$6\left(\frac{1}{x}\right) + 15\left(\frac{1}{y}\right) = 16$$

and make the substitution $u = \frac{1}{x}$ and $v = \frac{1}{y}$, we obtain

$$3u - 5v = -2$$
$$6u + 15v = 16$$

which is linear in u and v. To solve for u and v, we multiply the first equation by 3 and add.

$$\begin{aligned} 9u - 15v &= -6 \\ 6u + 15v &= 16 \\ \hline 15u \quad &= 10 \\ u &= \frac{2}{3} \end{aligned}$$

Substituting u into one of the equations, we find $v = \frac{4}{5}$. Since $u = \frac{1}{x}$ and $v = \frac{1}{y}$,

$$\frac{2}{3} = \frac{1}{x} \quad \text{and} \quad \frac{4}{5} = \frac{1}{y}$$
$$x = \frac{3}{2} \quad \text{and} \quad y = \frac{5}{4}.$$

Thus, the solution is $\left(\frac{3}{2}, \frac{5}{4}\right)$. Check.

Practice Exercise 6

Solve the system.

$$\frac{2}{x} + \frac{4}{y} = -4$$
$$\frac{3}{x} - \frac{1}{y} = 8$$

Answer: $\left(\frac{1}{2}, -\frac{1}{2}\right)$

CAUTION

When you have found the values of substitute variables such as u and v in Example 6, you're not finished with the problem! Be sure you back-substitute and find the values for the variables x and y in the original problem.

4.2 EXERCISES A

Solve using the substitution method.

1. $x + 2y = -1$
$2x - 3y = 12$

2. $4x - y = -3$
$2x - 3y = 1$

3. $2x + 4y = -12$
$3x + 5y = -14$

4. $2x - 3y = 2$
$-4x + 6y = 2$

5 $5x - 5y = 10$
$x - y = 2$

6. $x - 4 = 0$
$x + y = 1$
[*Hint:* x is always 4 in the first equation so substitute to get y.]

Solve using the elimination method.

7. $3x + 2y = 4$
$3x - 3y = 9$

8. $2x - 5y = 1$
$4x + 2y = 14$

9. $2x - 7y = 24$
$3x + 2y = -14$

10 $5x - 7y = 3$
$-10x + 14y = -1$

11. $5x - 3y = 4$
$2x + 4y = -1$

12. $3x + 3y = 12$
$7x + 7y = 28$

Solve using the substitution or elimination method (whichever seems appropriate).

13. $x + 2y = -9$
$3x + 4y = -17$

14. $x + 2 = 0$
$3x + 2y = 10$

15. $5x - 11y = -6$
$11x - 5y = 6$

16. $3y - 15 = 0$
$3x - 15y = 0$

17. $3x + y = -1$
$-4x + 2y = -7$

18. $7x - 7y = 7$
$-13x + 13y = -13$

19 $\frac{3}{4}x - \frac{2}{3}y = 1$
$\frac{3}{8}x - \frac{1}{6}y = 1$

20 $0.6x - 0.7y = 1.3$
$-1.2x + 1.4y = 7.3$

21 $\frac{4}{x} - \frac{2}{y} = -2$
$\frac{2}{x} + \frac{1}{y} = 3$

22. $\frac{1}{u} + \frac{1}{v} = 5$
$\frac{2}{u} - \frac{1}{v} = 4$

23. Suppose that in the process of solving a system of equations we obtain $0 = 2$. What does this tell us about the system?

24. Suppose that in the process of solving a system of equations we obtain $0 = 0$. What does this tell us about the system?

FOR REVIEW

25. If the two lines in a system coincide, how many solutions does the system have? What is the system called?

26. If the two lines in a system are parallel, how many solutions does the system have? What is the system called?

27. If the two lines in a system intersect, how many solutions does the system have? What is the system called?

28. Solve by the graphing method.
$x + y = 0$
$2x + y = -3$

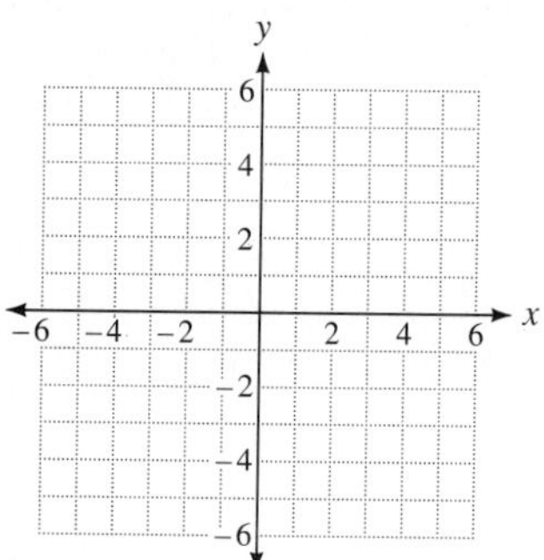

Exercises 29–30 review material from Section 2.2. They will help you prepare for the next section. Let x represent the unknown variable and translate each sentence into symbols. Do not solve.

29. Three times what number, less 8, is 22?

30. Two liters of acid are in how many liters of a 5% acid solution?

ANSWERS: 1. $(3, -2)$ 2. $(-1, -1)$ 3. $(2, -4)$ 4. no solution 5. infinitely many solutions of the form $(x, x-2)$ for x any real number 6. $(4, -3)$ 7. $(2, -1)$ 8. $(3, 1)$ 9. $(-2, -4)$ 10. no solution 11. $\left(\frac{1}{2}, -\frac{1}{2}\right)$ 12. infinitely many solutions of the form $(x, 4-x)$ for x any real number 13. $(1, -5)$ 14. $(-2, 8)$ 15. $(1, 1)$ 16. $(25, 5)$ 17. $\left(\frac{1}{2}, -\frac{5}{2}\right)$ 18. infinitely many solutions of the form $(x, x-1)$ for x any real number 19. $(4, 3)$ 20. no solution 21. $\left(2, \frac{1}{2}\right)$ 22. $\left(\frac{1}{3}, \frac{1}{2}\right)$ 23. The system has no solution. 24. The system has infinitely many solutions. 25. infinitely many; dependent 26. no solution; inconsistent 27. exactly one; consistent and independent 28. $(-3, 3)$ 29. $3x - 8 = 22$ 30. $2 = 0.05x$

4.2 EXERCISES B

Solve using the substitution method.

1. $3x - y = 7$
$x + 4y = -2$

2. $5x - y = 0$
$3x - 2y = -7$

3. $4x + 2y = 3$
$-2x - y = -7$

4. $2x + y = 3$
$-10x - 5y = -15$

5. $4x + y = -1$
$-6x + 5y = -18$

6. $4y - 12 = 0$
$8 - 2x = 0$

Solve using the elimination method.

7. $4x - 2y = -18$
$4x + 2y = 10$

8. $3x - 2y = -4$
$-5x + 4y = 10$

9. $2x - 5y = -14$
$3x + 4y = 2$

10. $4x - 9y = 3$
$-8x + 18y = 3$

11. $3x + y = 3$
$-12x - 4y = -12$

12. $7x - 2y = -1$
$5x + 4y = 2$

Solve using the substitution or elimination method (whichever seems appropriate).

13. $7x + 5y = 2$
$5x + 7y = -2$

14. $4x + 16 = 0$
$2x + y = -1$

15. $x + 5y = -2$
$3x - y = 10$

16. $\begin{aligned} 3x - y &= -9 \\ 2x + 4y &= 8 \end{aligned}$

17. $\begin{aligned} 3x - 4y &= 2 \\ -6x + 4y &= -3 \end{aligned}$

18. $\begin{aligned} 6x - 6y &= 6 \\ -15x + 15y &= -15 \end{aligned}$

19. $\begin{aligned} \frac{5}{2}x + \frac{3}{4}y &= 2 \\ \frac{3}{4}x - \frac{1}{16}y &= -4 \end{aligned}$

20. $\begin{aligned} 1.5x - 2.2y &= 3.8 \\ -4.5x + 6.6y &= -7.6 \end{aligned}$

21. $\begin{aligned} \frac{5}{u} - \frac{3}{v} &= -\frac{1}{2} \\ \frac{10}{u} + \frac{9}{v} &= 4 \end{aligned}$

22. $\begin{aligned} \frac{4}{3u} + \frac{1}{9v} &= 3 \\ \frac{5}{2u} - \frac{2}{3v} &= 3 \end{aligned}$

23. If an identity is obtained when solving a system of equations, what can be said about the solution to the system?

24. If a contradiction is obtained when solving a system of equations, what can be said about the solution to the system?

FOR REVIEW

25. When the two lines in a system have different slopes, how many solutions does the system have? What is the system called?

26. When the two lines in a system have the same slope but different y-intercepts, how many solutions does the system have? What is the system called?

27. When the two lines in a system have the same slope and the same y-intercepts, how many solutions does the system have? What is the system called?

28. Solve by the graphing method.

$2x - y = -3$

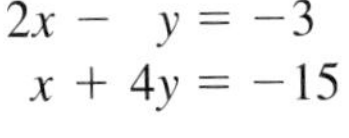

$x + 4y = -15$

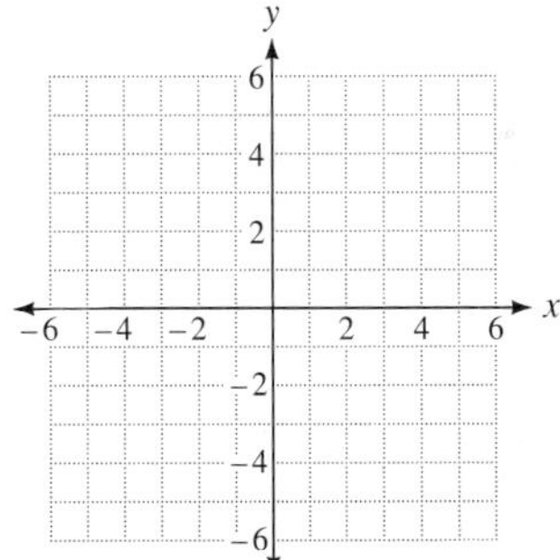

Exercises 29–30 review material from Section 2.2. They will help you prepare for the next section. Let x represent the unknown variable and translate each sentence into symbols. Do not solve.

29. A distance of 150 miles was traveled at what rate of speed if the time of travel was 3 hours?

30. Seven coats were purchased at what price each if the total amount spent was $280?

4.2 EXERCISES C

1. Find the values for a and b so that $(x, y) = (-3, 2)$ is a solution to the following system.

$$\begin{aligned} ax + by &= 4 \\ -2ax - by &= 2 \end{aligned}$$

[*Hint:* Substitute -3 for x and 2 for y in both equations, and solve the resulting system for a and b.]

2. Solve the system for x and y in terms of the nonzero constants a and b.

$$\begin{aligned} ax + y &= b \\ ax - y &= 2b \end{aligned}$$

3. Suppose that $\left(\frac{1}{2}, 3\right)$ and $(-4, 0)$ are two of the solutions to the equation $ax + by = -8$. Find a and b.
[Answer: $a = 2$, $b = -3$]

4. Solve for x and y.

$$\frac{x+y}{2} + \frac{x-y}{4} = 3$$
$$\frac{x+y}{2} - \frac{x-y}{4} = 1$$

5. Find the value(s) of a that will make the following system dependent.

$$2x + 5y = -1$$
$$6x + 15y = a$$

6. Find the value(s) of a that will make the following system inconsistent.

$$2x + 5y = -1$$
$$6x + 15y = a$$

4.3 APPLICATIONS USING SYSTEMS OF TWO EQUATIONS

STUDENT GUIDEPOSTS

 Combination Problems

In previous chapters we solved several kinds of applied problems. Some of those, as well as other problems, can be simplified by using two variables (instead of one) and translating them into a system of equations. Several examples illustrate this idea.

EXAMPLE 1 GEOMETRY PROBLEM

Two angles are **complementary** (that is, their sum is 90°). If one is 10° more than seven times the other, find the angles.

Let x = one angle,
y = the second angle.

Then $x + y = 90$. Why? If x is 10° more than seven times y, then

$$x = 7y + 10.$$

The system

$$x + y = 90$$
$$x - 7y = 10$$

can be solved by substituting $7y + 10$ for x in $x + y = 90$.

$$(7y + 10) + y = 90$$
$$8y + 10 = 90$$
$$8y = 80$$
$$y = 10$$

Now substitute 10 for y in $x = 7y + 10$.

$$x = 7(10) + 10 = 70 + 10 = 80$$

The angles are 10° and 80°. Since 10° + 80° = 90° and 80° = 7(10°) + 10°, the answer checks.

PRACTICE EXERCISE 1

Two angles are **supplementary** (their sum is 180°). If one is 30° less than four times the other, find the angles.

Answer: 138° and 42°

EXAMPLE 2 Distance-Rate Problem

By going 20 mph for one period of time and then 30 mph for another, a boater traveled from Glen Canyon Dam to the end of Lake Powell, a distance of 180 miles. If she had gone 21 mph throughout the same period of time, she would only have reached Hite Marina, a distance of 147 miles. How many hours did she travel at each speed?

Let x = the number of hours that she traveled at 20 mph,
y = the number of hours that she traveled at 30 mph.

Since

$$(\text{distance}) = (\text{rate})(\text{time}),$$

it follows that

$$\text{distance traveled at 20 mph} = 20x$$
$$\text{distance traveled at 30 mph} = 30y.$$

The total distance traveled is the sum of the two distances traveled at the two rates, so the first equation is

$$20x + 30y = 180.$$

She would have traveled 147 miles at a rate of 21 mph for the total time $x + y$. Thus the second equation is

$$21(x + y) = 147 \quad \text{or} \quad 21x + 21y = 147.$$

It is usually wise to simplify equations before solving. Dividing the first equation by 10 and the second by 7, we obtain the following system.

$$2x + 3y = 18$$
$$3x + 3y = 21$$

Subtract the first from the second.

$$x = 3$$

Then

$$2(3) + 3y = 18$$
$$3y = 12$$
$$y = 4.$$

She traveled 3 hours at 20 mph and 4 hours at 30 mph. Check this in the words of the problem.

Practice Exercise 2

A train travels 330 miles by going 60 mph for one period of time and 70 mph for another. Had it gone 62 mph throughout the trip it would have traveled only 310 miles. How many hours did it travel at each speed?

Answer: 2 hr at 60 mph and 3 hr at 70 mph

1 COMBINATION PROBLEMS

In another kind of word problem, a **combination problem,** two quantities are combined in two different ways, resulting in two equations. Many combination problems involve *total value*. The **total value** of something composed of a number of units, each having the same value, is given by

(total value) = (value per unit)(number of units).

For example, the total value of 8 pounds of chicken worth \$1.10 per pound is

$$(\text{total value}) = (\text{value per pound})(\text{number of pounds})$$
$$= (\$1.10)(8)$$
$$= \$8.80.$$

Similarly, the value of a collection of 12 quarters and four dimes is

$$\begin{aligned}(\text{total value}) &= (\text{value per quarter})(\text{number of quarters}) \\ &\quad + (\text{value per dime})(\text{number of dimes}) \\ &= (25¢)(12) + (10¢)(4) \\ &= 340¢ = \$3.40.\end{aligned}$$

The next example applies total value equations to a combination problem.

EXAMPLE 3 COMBINATION PROBLEM

Bill Poole bought 5 shirts (of the same value) and 4 pairs of socks (of the same value) for \$87. He returned to the same store a week later and purchased (at the same prices) 2 more shirts and 6 more pairs of socks for \$48. Find the price of each.

Let x = price of one shirt,
y = price of one pair of socks.
$5x$ = value of 5 shirts, Value = (price per shirt)(number of shirts)
$4y$ = value of 4 pairs of socks.

Thus, the system to solve is

$$5x + 4y = 87$$
$$2x + 6y = 48.$$

Divide the second equation by 2 to simplify.

$$x + 3y = 24$$

Solve for x,

$$x = 24 - 3y,$$

and substitute into the first equation.

$$\begin{aligned}5(24 - 3y) + 4y &= 87 \\ 120 - 15y + 4y &= 87 \\ -11y &= -33 \\ y &= 3\end{aligned}$$

Now substitute 3 for y in $x = 24 - 3y$.

$$\begin{aligned}x &= 24 - 3y \\ &= 24 - 3(3) = 24 - 9 = 15\end{aligned}$$

Each shirt cost \$15 and each pair of socks cost \$3.

PRACTICE EXERCISE 3

Three 5.25-inch computer disks and seven 3.5-inch computer disks cost \$16.25. Ten 5.25-inch disks and four 3.5-inch disks cost \$15.50. Find the price of each type of disk.

Answer: 5.25-inch: \$0.75; 3.5-inch: \$2.00

2 MIXTURE PROBLEMS

In a certain type of combination problem called a **mixture problem,** the mixing together of two quantities results in two equations we call the *quantity equation* and the *value equation*.

EXAMPLE 4 MIXTURE PROBLEM

The owner of a candy store wishes to mix candy selling for \$1.50 per lb with nuts selling for \$1.00 per lb to obtain a party mix to be sold for \$1.20 per lb. How many pounds of each must be used to obtain 50 lb of the mixture?

PRACTICE EXERCISE 4

A shipping crate contains 95 boxes, some weighing 1 lb each and the others 3 lb each. How many of each

Let c = the number of pounds of candy,
n = the number of pounds of nuts.

It sometimes helps to place all given information into a summary table. To avoid decimals, we convert all monetary units to cents.

	Number of pounds	*Value per pound*	*Value*
Candy	c	150	$150c$
Nuts	n	100	$100n$
Total	50	120	120(50)

First we obtain the quantity equation

(number lb of candy) + (number lb of nuts) = (number lb of mixture),

which translates to

$$c + n = 50.$$

Next we obtain the value equation.

(value of candy) + (value of nuts) = (total value of mixture)

That is, the value of a mixture is equal to the sum of the values of its parts. Thus, the value equation is

$$150c + 100n = 120 \cdot 50$$

or $$3c + 2n = 120. \quad \text{Divide through by 50}$$

The system we have to solve is

$$\begin{aligned} c + n &= 50 \\ 3c + 2n &= 120. \end{aligned}$$

Solve the first equation for c.

$$c = 50 - n$$

Then substitute into the second.

$$\begin{aligned} 3(50 - n) + 2n &= 120 \\ 150 - 3n + 2n &= 120 \\ -n &= -30 \\ n &= 30 \end{aligned}$$

Now substitute 30 for n in $c = 50 - n$.

$$c = 50 - 30 = 20$$

The store owner must use 20 lb of candy and 30 lb of nuts to obtain the desired mixture.

type are in the crate if the total weight of the contents of the crate is 135 lb?

Answer: 75 1-lb boxes and 20 3-lb boxes

EXAMPLE 5 Coin Problem

Burford has 30 coins consisting of dimes and quarters. If the total value is \$4.20, find the number of dimes and the number of quarters.

Let d = the number of dimes,
q = the number of quarters.

Once again, a table might be helpful.

Practice Exercise 5

Matt found a coin purse containing 40 coins with a total value of \$2.90. If there are only nickels and dimes in the purse, how many of each are there?

	Number of coins	*Value per coin*	*Value*
Dimes	d	10	$10d$
Quarters	q	25	$25q$
Total	30	—	420

The quantity equation

(number of dimes) + (number of quarters) = (number of coins)

translates to $d + q = 30.$

The value equation

(value of dimes) + (value of quarters) = (value of coins)

becomes

$$10d + 25q = 420, \quad \text{which reduces to} \quad 2d + 5q = 84$$

after dividing both sides by 5. Solve the first equation for d, $d = 30 - q$, and substitute.

$$2(30 - q) + 5q = 84$$
$$60 - 2q + 5q = 84$$
$$3q = 24$$
$$q = 8$$

Now substitute 8 for q in $d = 30 - q$.

$$d = 30 - 8 = 22$$

Thus, Burford has 22 dimes and 8 quarters.

Answer: 18 dimes and 22 nickels

EXAMPLE 6 Mixture Problem in Chemistry

A chemist has two solutions, each containing a certain percentage of acid. If solution A is 5% acid and solution B is 15% acid, how much of each should be mixed to obtain 20 liters of a solution that is 12% acid?

Let x = number of liters of 5% acid solution,
y = number of liters of 15% acid solution.

The quantity equation is

$$x + y = 20.$$

The value equation this time equates the amount of acid in the solutions.

$0.05x$ = amount of acid in x liters of 5% solution
$0.15y$ = amount of acid in y liters of 15% solution
$(0.12)(20)$ = amount of acid in 20 liters of 12% mixture

So we have

$$0.05x + 0.15y = (0.12)(20)$$
$$5x + 15y = 240 \quad \text{Clear all decimals}$$
$$x + 3y = 48. \quad \text{Divide by 5}$$

Solving $x + y = 20$ for x, $x = 20 - y$. Substitute.

$$(20 - y) + 3y = 48$$
$$20 + 2y = 48$$
$$2y = 28$$
$$y = 14$$

Practice Exercise 6

A 15% salt solution is to be mixed with a 40% salt solution to obtain 5 quarts of a 25% salt solution. How many quarts of each should be used?

Now substitute 14 for y in $x = 20 - y$.

$$x = 20 - y = 20 - 14 = 6$$

The chemist should mix 6 liters of 5% solution with 14 liters of 15% solution to get the 12% acid solution.

Answer: 15% solution: 3 qt; 40% solution: 2 qt

4.3 EXERCISES A

Solve using a system of equations.

1. The sum of two numbers is 50 and their difference is 16. Find the numbers.

   Let x = first number,
   y = second number.
   $x + y =$ ____
   $x - y =$ ____

2. Pete is 7 years younger than Jim. Three years from now the sum of their ages will be 33. How old is each now?

   Let x = Jim's age,
   y = Pete's age,
   $x + 3$ = Jim's age in 3 yr,
   $y + 3$ = Pete's age in 3 yr.
   $y = x -$ ____
   $(x + 3) + (y + 3) =$ ____

3. Two angles are supplementary (their sum is 180°), and one is 18° more than five times the other. Find the angles.

   Let x = measure of first angle,
   y = measure of second angle.
   $x + y =$ ____

4. By traveling 30 mph for one period of time and 40 mph for another, Mandy traveled 230 miles. Had she gone 10 mph faster throughout, she would have traveled 300 miles. How many hours did she travel at each rate?

   Let x = time traveling 30 mph,
   y = time traveling 40 mph.

5. Four books (of the same kind) and 6 pens (of the same kind) cost \$9.00. Three books and 9 pens also cost \$9.00. Find the cost of one book and of one pen.

6. A candy mix sells for \$1.10 per lb. If the mix has two kinds of candy, one worth 90¢ per lb and the other worth \$1.50 per lb, how many pounds of each would be in a 30-lb mixture?

7. A collection of dimes and nickels is worth \$3.75. If there are 55 coins in the collection, how many of each are there?

8. If there were 450 people at a play, the total receipts were \$600, and the admission price was \$2.00 for adults and 75¢ for children, how many adults and how many children were in attendance?

9. Mr. Smith is 43 years old. If the sum of Mr. Smith's two daughters' ages is 23, and if three times the age of the younger plus the age of the older daughter is equal to the age of Mr. Smith, how old is each daughter?

10. A 40-ft rope is cut in two pieces. One piece is 1 ft more than twice the other. How long is each piece?

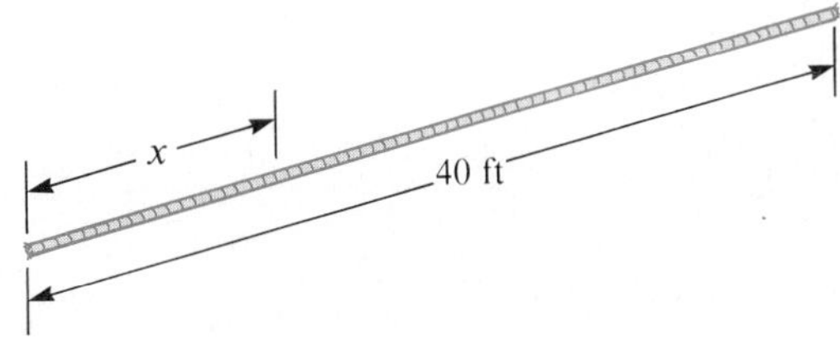

11. In a right triangle, one acute angle is 6° less than three times the other. Find each acute angle.

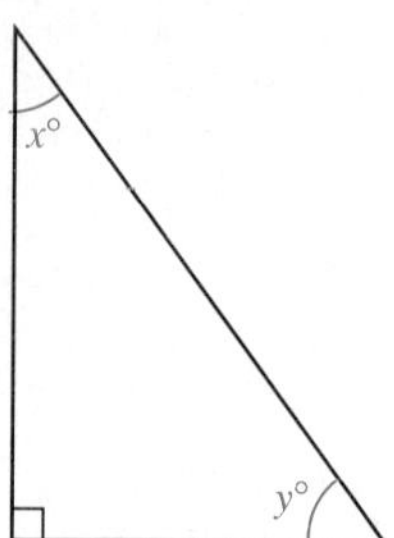

12. A collection of quarters and nickels is worth \$3.50. If there are 30 coins in the collection, how many of each are there?

13. Abby invested \$2000 for one year in two separate savings accounts. One account paid 8% simple interest and the other 9% simple interest. How much was invested in each account if the total interest from both was \$175?

14. The total interest earned on a sum of money invested for one year was \$183. If \$500 more was invested at 10% simple interest than was invested at 9% simple interest, how much was invested at each rate?

15. The owner of Carrie's Coffee House wishes to mix two blends of coffee, one selling for \$3.60 per pound and the other for \$4.80 per pound, to obtain a 20-lb mixture selling for \$4.20 per pound. How many pounds of each must she use?

16. A chemist has one solution which is 50% acid and a second which is 25% acid. How much of each should he use to make 10 L of a 40%-acid solution?

17. The perimeter of a rectangular field is 192 yards. If the length is 8 yards more than the width, find the dimensions.

18. The Small Car Rental Agency charges a daily fee plus a mileage fee. Mr. Green was charged \$143.12 for 3 days and 316 miles, and Mr. Brown was charged \$147.44 for 5 days and 242 miles. What is the company's daily rate, and what is its mileage rate?

19. A houseboat sailed 30 miles downstream in 3 hr and returned to the same marina upstream in 5 hr. Find the rate of the houseboat in still water and the rate of the current in the river.

20. Two trucks leave Kansas City, one traveling east and the other traveling west. After 4 hours they are 492 miles apart. If one truck is traveling 7 mph faster than the other, what is the speed of each?

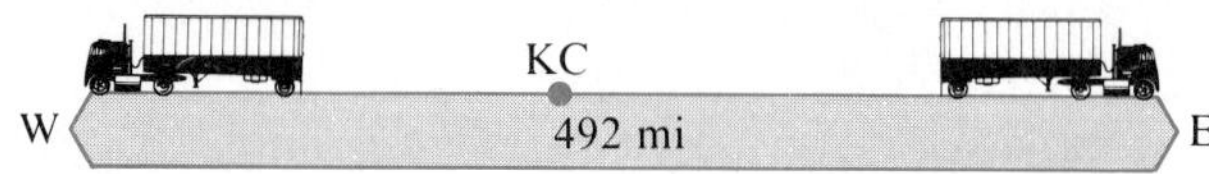

FOR REVIEW

Solve by any method.

21. $3x + 8y = 13$
$2x - 3y = 17$

22. $-x + y = 7$
$12x - 12y = 7$

23. $\frac{2}{3}x + \frac{1}{9}y = 6$
$\frac{1}{4}x + \frac{3}{4}y = 15$

The following exercises from Chapter 1 will help you prepare for the next section. In Exercises 24–26, evaluate each expression when $x = -1$, $y = 3$, and $z = -5$.

24. $2x + 4y - z$

25. $-3x - y + 2z$

26. $-10x + 12y - 3z$

Find the numerical value of y in each equation if $x = 2$ and $z = -4$.

27. $3x + y - 2z = 13$

28. $7x - 2y - z = 4$

29. $-8x + 2y - 7z = 10$

ANSWERS: 1. 17, 33 2. Pete is 10, Jim is 17. 3. 27°, 153° 4. 5 hr at 30 mph, 2 hr at 40 mph 5. book is \$1.50, pen is 50¢ 6. 20 lb of 90¢ candy, 10 lb of \$1.50 candy 7. 20 dimes; 35 nickels 8. 210 adults, 240 children 9. 10 yr, 13 yr 10. 13 ft, 27 ft 11. 24°, 66° 12. 10 quarters, 20 nickels 13. \$500 at 8%, \$1500 at 9% 14. \$1200 at 10%, \$700 at 9% 15. 10 lb of each 16. 4 L of 25% solution, 6 L of 50% solution 17. 52 yd, 44 yd 18. \$14 per day, \$0.32 per mile 19. houseboat: 8 mph; river: 2 mph 20. 58 mph, 65 mph 21. (7, −1) 22. no solution 23. (6, 18) 24. 15 25. −10 26. 61 27. $y = -1$ 28. $y = 7$ 29. $y = -1$

4.3 EXERCISES B

Solve using a system of equations.

1. Twice one number plus a second number is 42. The first minus the second is −6. Find the numbers.

2. The sum of Jan's and Sue's ages is 15. In 3 years, Jan will be twice as old as Sue. What are their present ages?

3. Two angles are complementary (their sum is 90°). If one is 10° more than seven times the other, find the angles.

4. Maria rode her bike at 15 mph for a time but had mechanical problems and had to continue her trip walking at 4 mph. She covered 42 miles. Had she been able to ride all the way, she would have covered 75 miles in the same time. How long did she ride, and how long did she walk?

5. Seven pads of paper and 5 pencils cost \$5.20. Two pads and 18 pencils cost \$4.80. Find the price of each.

6. Nuts worth \$2.25 per lb are mixed with candy worth \$1.50 per lb. The mixture sells for \$1.90 per lb. How many pounds of nuts and how many pounds of candy should be used to make a 60-lb mixture?

7. There are 38 coins in a collection of nickels and quarters. If the total value of the collection is \$5.10, how many of each are there?

8. A total of 8340 people paid to see a college basketball game. Tickets were \$1.50 for children, \$2.50 for adults, and the total receipts were \$19,620. How many children and how many adults paid to see the game?

9. Mari is 26 years older than Larry. In 10 years Mari will be twice as old as Larry is then. How old is each now?

10. Phillip bought 6 pounds of hamburger and 4 pounds of steak for \$22.50. If steak is three times the price of hamburger, what is the price per pound of each?

11. In a basketball game, Troy scored 4 more than twice as many points as Shawn. If one of them had scored 2 more points, their total would have been 60. How many points did each score?

12. Wanda invested \$6000 in bonds, \$7500 in savings certificates, and received \$1515 in interest for the year. If the interest rate on the certificates was 4% higher than the rate on the bonds, what were the respective interest rates?

13. Michele invested \$5000 for one year in two separate savings accounts. One account paid 7% simple interest and the other 9% simple interest. How much was invested in each account if the total interest from both was \$410?

14. The total interest earned on a sum of money invested for one year was \$80. If \$300 more was invested at 8% simple interest than at 6% simple interest, how much was invested at each rate?

15. A shop owner mixes candy worth \$1.40 per lb with candy worth \$1.95 per lb to get 110 lb of a party mix worth \$1.75 per lb. How many pounds of each did she use?

16. A lab technician is diluting a 70% alcohol solution with water. How much water and how much alcohol must be mixed to obtain 21 L of a 30%-alcohol solution?

17. In an endurance race Fred ran 3 times as far as Sam. If their total distance was 12.8 miles, how far did each run?

18. A new machine can produce 120 copies per minute while an older model produces only 30 copies per minute. The new machine had been operating 15 minutes when the older one was started. What was the total time each was operating if together they produced 8850 copies?

19. Grayden Bell sails his boat upstream from a marina 210 miles at constant speed in 7 hr. He returns downstream to the same marina at the same speed in 6 hr. What are the speed of the boat in still water and the speed of the current?

20. A plane flying with the wind flew 600 miles in 4 hr. The return trip against the wind took 5 hr. Find the rate of the plane in calm air and the rate of the wind.

FOR REVIEW

Solve by any method.

21. $3x + y = 10$
$5x + 7y = 6$

22. $-x + y = -7$
$12x - 12y = 84$

23. $\frac{1}{2}x + \frac{3}{4}y = 5$
$\frac{3}{8}x - \frac{3}{2}y = -21$

The following exercises from Chapter 1 will help you prepare for the next section. In Exercises 24–26, evaluate each expression when $x = -2$, $y = 1$, and $z = -3$.

24. $3x - y + 2z$

25. $-2x + y - 4z$

26. $5x - 7y + 8z$

Find the numerical value of x in each equation if $y = -3$ and $z = 0$.

27. $4x - 8y + 7z = 0$

28. $3x + 7y - 9z = 0$

29. $-2x + 8y - 13z = 5$

4.3 EXERCISES C

Solve.

1. The ones digit of a whole number is 1 more than twice the tens digit. When the two digits are reversed, the new number is 4 less than twice the original number. Find the original number. [Answer: 49]

2. A shipping crate contains two types of items, one weighing 3 lb each and the other $\frac{1}{2}$ lb each. Suppose we know there are 90 items in the crate. If the crate weighs 30 lb when empty, and now weighs 200 lb, how many of each type of item are in the crate?

3. The radiator in an automobile holds 14 quarts. How much antifreeze should be mixed with a 20% antifreeze solution to obtain a 40% antifreeze mixture that will fill the radiator?

4. In a laboratory experiment, a biologist must keep a control group of animals on a strict diet. Each animal must receive 19.5 grams of protein and 2.2 grams of fat every day. The biologist has purchased two food blends, HealthChow and GroMeal. If HealthChow contains 30% protein and 4% fat and GroMeal contains 25% protein and 2% fat, how many grams of each should be mixed to provide the correct diet for one animal?

4.4 SYSTEMS OF THREE LINEAR EQUATIONS IN THREE VARIABLES

STUDENT GUIDEPOSTS

1. Linear Equations in Three Variables
2. Systems of Three Linear Equations
3. Inconsistent and Dependent Systems
4. Applications of Systems of Three Equations

1 LINEAR EQUATIONS IN THREE VARIABLES

An equation of the form

$$ax + by + cz = d$$

when a, b, c, and d are constant real numbers and x, y, and z are variables, is called a **linear equation in three variables.** The term *linear equation* in this context is a bit misleading (perhaps *first-degree equation* is better) since the graph of a linear equation in three variables is actually a plane in space, not a line. However, our solutions to linear equations in three variables will be purely algebraic, patterned after similar situations with linear equations in two variables.

A solution to a linear equation such as

$$2x + y - 3z = 3$$

is an **ordered triple** of numbers. For example, $(1, -2, -1)$ is a solution to the above equation since if x, y, and z are replaced with 1, -2, and -1, in that order, the equation is true.

$$\begin{aligned} 2(1) + (-2) - 3(-1) &\stackrel{?}{=} 3 \\ 2 - 2 + 3 &\stackrel{?}{=} 3 \\ 3 &= 3 \end{aligned}$$

The *order* in the ordered triple is always the x-value first, the y-value second, and the z-value third. If we substitute -1 for x, -2 for y, and 1 for z,

$$\begin{aligned} 2(-1) + (-2) - 3(1) &\stackrel{?}{=} 3 \\ -2 - 2 - 3 &\stackrel{?}{=} 3 \\ -7 &\neq 3, \end{aligned}$$

we discover that $(-1, -2, 1)$ is not a solution to the equation.

2 SYSTEMS OF THREE LINEAR EQUATIONS

A **system of three linear equations in three variables** is a trio of linear equations such as the following.

$$\begin{aligned} x + y + z &= 2 \\ -2x - y + z &= 1 \\ x - 2y - z &= 1 \end{aligned}$$

A solution to such a system is an ordered triple of numbers that is a solution to *all three* equations. It is easy to verify by substitution that $(1, -1, 2)$ is a solution to each of the above equations, and hence is a solution to the system.

For a system of three equations in three variables there may be exactly one solution, no solution, or infinitely many solutions. Graphically, these possibilities correspond to three planes, A, B and C, intersecting in exactly one point as in Figure 4.4(a), having no single point found in all three planes as in Figure 4.4(b), and having infinitely many points in common as in Figure 4.4(c).

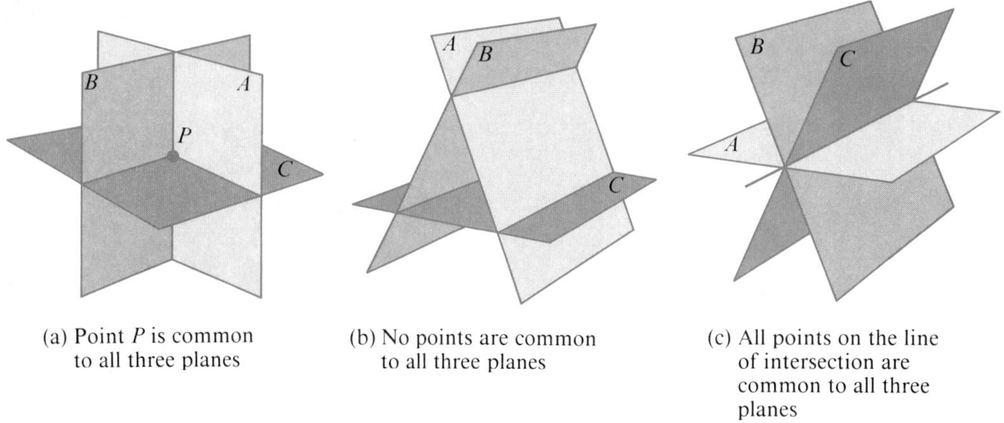

(a) Point P is common to all three planes

(b) No points are common to all three planes

(c) All points on the line of intersection are common to all three planes

Figure 4.4 Number of Solutions

To solve a system we combine methods of elimination and substitution similar to those in Section 4.2 to reduce to a system of two equations in two variables. The basic steps of this technique follow.

To Solve a System of Three Equations in Three Variables

1. Select any two of the three equations and eliminate one of the variables by addition or subtraction.
2. Use the remaining equation from the original system together with either equation used in Step 1, and eliminate the *same* variable as in Step 1.
3. Solve the resulting pair of equations in two variables.
4. Substitute the values obtained in Step 3 into any of the three original equations to obtain the value of the third variable.
5. Check the solution in all three of the original equations.
6. If, at any step, a contradiction is obtained, the system has no solution. If an identity is obtained after using only two equations, the system has either infinitely many solutions or no solution, depending on what happens when the third equation is involved.

In the process of solving a system, letter labels help us keep track of equations and steps. This is shown in the next example.

EXAMPLE 1 Solving a System of Three Equations

Solve the system.

$$\begin{aligned} &\textbf{(A)}\quad 3x + 2y + 3z = 3\\ &\textbf{(B)}\quad 4x - 5y + 7z = 1\\ &\textbf{(C)}\quad 2x + 3y - 2z = 6 \end{aligned}$$

Eliminate x from (A) and (B) by adding 4 times (A) to -3 times (B).

$$\begin{array}{lrl} \textbf{4(A)} & 12x + 8y + 12z = & 12 \\ \textbf{-3(B)} & -12x + 15y - 21z = & -3 \\ \hline \textbf{(D)} & 23y - 9z = & 9 \end{array} \qquad 4(A) - 3(B) = (D)$$

Eliminate x from (B) and (C) by adding (B) to -2 times (C).

$$\begin{array}{lrl} \textbf{(B)} & 4x - 5y + 7z = & 1 \\ \textbf{-2(C)} & -4x - 6y + 4z = & -12 \\ \hline \textbf{(E)} & -11y + 11z = & -11 \end{array} \qquad (B) - 2(C) = (E)$$

Simplify (E) by multiplying through by $-\frac{1}{11}$ (or dividing by -11).

$$\textbf{(F)}\quad y - z = 1 \qquad (F) = -\frac{1}{11}(E)$$

We have the following system.

$$\begin{aligned} &\textbf{(D)}\quad 23y - 9z = 9\\ &\textbf{(F)}\quad y - z = 1 \end{aligned}$$

Solve (F) for y, obtaining $y = 1 + z$, and substitute into (D).

$$\begin{aligned} 23(1 + z) - 9z &= 9\\ 23 + 23z - 9z &= 9\\ 14z &= -14\\ z &= -1 \end{aligned}$$

Substitute -1 for z in (F).

$$\begin{aligned} y - (-1) &= 1\\ y + 1 &= 1\\ y &= 0 \end{aligned}$$

Substitute 0 for y and -1 for z in (A).

$$\begin{aligned} 3x + 2(0) + 3(-1) &= 3\\ 3x - 3 &= 3\\ 3x &= 6\\ x &= 2 \end{aligned}$$

The solution is $(2, 0, -1)$. Check in all three equations.

PRACTICE EXERCISE 1

Solve the system using the steps given.

$$\begin{aligned} &(A)\quad 2x - y + 3z = 5\\ &(B)\quad 3x + 5y - z = 1\\ &(C)\quad -x + 3y + 2z = -5 \end{aligned}$$

Step I: Use (A) and (C) and eliminate x to obtain equation (D).

Step II: Use (B) and (C) and eliminate x to obtain equation (E).

Step III: Solve the system of two equations in the two variables y and z, equations (D) and (E), to obtain the numerical values of y and z.

Step IV: Substitute the values for y and z obtained in III back into (C) to find the value of x.

Answer: $(2, -1, 0)$

EXAMPLE 2 Solving a System with Some Coefficients Zero

Solve the system.

$$\begin{aligned} &\textbf{(A)}\quad x + y + z = 6\\ &\textbf{(B)}\quad x - z = -2\\ &\textbf{(C)}\quad y + 3z = 11 \end{aligned}$$

PRACTICE EXERCISE 2

Solve the system.

$$\begin{aligned} &(A)\quad x + y - 2z = -1\\ &(B)\quad y + z = 5\\ &(C)\quad 2x - y + 4z = -6 \end{aligned}$$

When a variable is missing from one or more equations, we can shorten our work a bit. Since x does not appear in (C), if we subtract (B) from (A) to eliminate x, the resulting equation can be paired with (C) immediately.

$$\begin{array}{lrcl} \textbf{(A)} & x + y + z &=& 6 \\ \textbf{(B)} & x \quad\;\; - z &=& -2 \\ \hline \textbf{(D)} & y + 2z &=& 8 \end{array} \qquad \text{(A) − (B) = (D)}$$

Thus, we have the following system.

$$\textbf{(C)} \quad y + 3z = 11$$
$$\textbf{(D)} \quad y + 2z = 8$$

Subtracting (D) from (C), we eliminate y.

$$\textbf{(E)} \quad z = 3 \qquad \text{(C) − (D) = (E)}$$

Substitute 3 for z in (C).

$$y + 3(3) = 11$$
$$y + 9 = 11$$
$$y = 2$$

Substitute 3 for z in (B).

$$x - 3 = -2$$
$$x = 1$$

The solution is (1, 2, 3). Check in all three equations.

[*Hint:* Use equations (A) and (C) and eliminate x to obtain equation (D) in y and z. Pair (D) with (B) and solve.]

Answer: $(-3, 4, 1)$

INCONSISTENT AND DEPENDENT SYSTEMS

As with systems of two equations, when a linear system of three equations has no solution, it is inconsistent, and a contradiction will result when we try to solve it.

EXAMPLE 3 Working with an Inconsistent System

Solve the system.

$$\begin{array}{lrcl} \textbf{(A)} & 3x - y + 2z &=& 4 \\ \textbf{(B)} & -6x + 2y - 4z &=& 1 \\ \textbf{(C)} & 5x - 3y + 8z &=& 0 \end{array}$$

If we try to eliminate y by adding 2 times (A) to (B)

$$\begin{array}{lrcl} \textbf{2(A)} & 6x - 2y + 4z &=& 8 \\ \textbf{(B)} & -6x + 2y - 4z &=& 1 \\ \hline & 0 &=& 9 \end{array}$$

we obtain a contradiction, so the system is inconsistent and has no solution.

PRACTICE EXERCISE 3

Solve the system.

(A) $x - 3y + z = 2$
(B) $2x + y - 3z = 1$
(C) $3x - 9y + 3z = -1$

Answer: no solution

If we obtain an identity such as $0 = 0$ when solving a linear system of three equations, and only two equations have been used to this point, we know that either the system has infinitely many solutions and is dependent, or else it has no solution and is inconsistent. Then we need to use the third equation to know the nature of the system. Notice this differs from systems of two equations where an identity at *any* step *always* means there are infinitely many solutions.

EXAMPLE 4 WORKING WITH A DEPENDENT SYSTEM

Solve the following system.

$$\begin{aligned} &\textbf{(A)} \quad x - 3y + z = 4 \\ &\textbf{(B)} \quad x + 5y - z = 2 \\ &\textbf{(C)} \quad -2x + 2y - z = -7 \end{aligned}$$

If we add (A) and (B) we obtain

$$\textbf{(D)} \quad 2x + 2y = 6$$

which reduces to (E) by multiplying both sides by $\frac{1}{2}$.

$$\textbf{(E)} \quad x + y = 3$$

If we add (A) and (C) we obtain

$$\textbf{(F)} \quad -x - y = -3.$$

If we add (E) and (F),

$$\begin{aligned} &\textbf{(E)} \quad x + y = 3 \\ &\textbf{(F)} \quad -x - y = -3 \\ &\qquad\quad 0 = 0 \end{aligned}$$

we obtain an identity. Since all three equations have been used in the process of elimination, we can conclude that the system has infinitely many solutions and is dependent.

PRACTICE EXERCISE 4

Solve the system using the steps given.

$$\begin{aligned} &\textbf{(A)} \quad x - y + z = 3 \\ &\textbf{(B)} \quad 2x - 2y + 2z = -1 \\ &\textbf{(C)} \quad -x + y - z = -3 \end{aligned}$$

Step 1: Use (A) and (C) and add to eliminate x. An identity results. Equation (B) must now be used to determine if the system is inconsistent or dependent.

Step 2: Multiply (A) by -2 and add to (B) to again eliminate x. The result is a contradiction, so the system is inconsistent.

Answer: no solution

4 APPLICATIONS OF SYSTEMS OF THREE EQUATIONS

Many applied problems can be translated into a system of three equations in three variables. As with problems in two equations, be careful to write out all details such as complete definitions.

EXAMPLE 5 GEOMETRY PROBLEM

In a triangle, the largest angle is 70° more than the smallest angle, and the remaining angle is 10° more than three times the smallest angle. Find the measure of each angle.

Let x = the measure of the smallest angle,
y = the measure of the middle angle,
z = the measure of the largest angle.

Since z is 70° more than x, we have

$$x + 70 = z$$

or

$$\textbf{(A)} \quad x - z = -70.$$

Since y is 10° more than 3 times x,

$$3x + 10 = y$$

or

$$\textbf{(B)} \quad 3x - y = -10.$$

Finally, the sum of the measures of the angles of a triangle is 180°.

$$\textbf{(C)} \quad x + y + z = 180$$

PRACTICE EXERCISE 5

The smallest angle in a triangle is one-third the measure of the middle-sized angle, and the largest angle is twice the sum of the other two. Find the measure of each angle.

We now have the following system.

$$\begin{aligned} &\textbf{(A)} \quad x \quad\quad - z = -70 \\ &\textbf{(B)} \quad 3x - y \quad = -10 \\ &\textbf{(C)} \quad x + y + z = 180 \end{aligned}$$

y is missing in (A), and if we add (B) and (C), we get another equation with y missing.

$$\textbf{(D)} \quad 4x + z = 170 \qquad \text{(B) + (C) = (D)}$$

We need to solve

$$\begin{aligned} &\textbf{(A)} \quad x - z = -70 \\ &\textbf{(D)} \quad 4x + z = 170. \end{aligned}$$

Add (A) and (D).

$$\begin{aligned} 5x &= 100 \\ x &= 20 \end{aligned}$$

Substitute 20 for x in (A).

$$\begin{aligned} 20 - z &= -70 \\ -z &= -90 \\ z &= 90 \end{aligned}$$

Substitute 20 for x in (B).

$$\begin{aligned} 3(20) - y &= -10 \\ 60 - y &= -10 \\ -y &= -70 \\ y &= 70 \end{aligned}$$

The angles have measures 20°, 70°, and 90°. Check.

Answer: 15°, 45°, 120°

EXAMPLE 6 Interest Problem

Joe has $10,000 divided into three separate investments. Part of the money is invested in certificates at 7%, part in bonds at 8%, and the rest invested in his brother Bob's private business. If the business earns 6%, the total earnings from all the investments will amount to $660. However, if the business slumps, Joe loses 6% on his investment, and the total earnings from the three will amount to only $60. How much is invested in each category?

Let x = the number of dollars invested in certificates,
y = the number of dollars invested in bonds,
z = the number of dollars invested in Bob's business.

Then we must solve

$$\begin{aligned} x + y + z &= 10{,}000 && \text{Total invested is \$10,000} \\ 0.07x + 0.08y + 0.06z &= 660 && \text{Amount earned if business earns 6\%} \\ 0.07x + 0.08y - 0.06z &= 60. && \text{Amount earned if business loses; note the minus } 0.06z \end{aligned}$$

It is best to clear all decimals and then solve.

$$\begin{aligned} &\textbf{(A)} \quad x + y + z = 10{,}000 \\ &\textbf{(B)} \quad 7x + 8y + 6z = 66{,}000 && \text{Multiply by 100} \\ &\textbf{(C)} \quad 7x + 8y - 6z = 6000 && \text{Multiply by 100} \end{aligned}$$

Practice Exercise 6

Vanessa borrowed $80,000 in three different loans to start a small business. She borrowed two amounts, totaling $70,000, at 11% and 10%, respectively. She borrowed the rest at 13%. How much was borrowed at each rate if the total annual simple interest was $8500?

Adding (B) and (C) eliminates z.

(D) $14x + 16y = 72{,}000$ (B) + (C) = (D)

Adding (B) to -6 times (A) will also eliminate z.

(E) $x + 2y = 6000$ (B) − 6(A) = (E)

Solve (E) for x, $x = 6000 - 2y$, and substitute into (D).

$$\begin{aligned} 14(6000 - 2y) + 16y &= 72{,}000 \\ 84{,}000 - 28y + 16y &= 72{,}000 \\ -12y &= -12{,}000 \\ y &= 1000 \end{aligned}$$

Substitute 1000 for y in (E).

$$\begin{aligned} x + 2(1000) &= 6000 \\ x + 2000 &= 6000 \\ x &= 4000 \end{aligned}$$

Substitute 4000 for x and 1000 for y in (A).

$$\begin{aligned} 4000 + 1000 + z &= 10{,}000 \\ 5000 + z &= 10{,}000 \\ z &= 5000 \end{aligned}$$

Joe has invested \$4000 in certificates, \$1000 in bonds, and \$5000 in his brother's business. Check.

Answer: \$20,000 at 11%, \$50,000 at 10%, and \$10,000 at 13%

4.4 EXERCISES A

Solve the systems using the procedures shown in the examples.

1. $\begin{aligned} x + y + z &= 2 \\ -x + y - 2z &= -5 \\ 2x - y + z &= 1 \end{aligned}$

2. $\begin{aligned} x + 2y + 3z &= 6 \\ -2x + y - z &= -2 \\ -x + 3y - 2z &= 0 \end{aligned}$

3. $\begin{aligned} x - 2y - z &= 2 \\ 2x - y + z &= 7 \\ 3x + 2y + z &= 2 \end{aligned}$

4. $\begin{aligned} x + y + z &= 8 \\ x - y - z &= 0 \\ x + 2y + z &= 9 \end{aligned}$

5 $\begin{aligned} 3x + y - z &= 4 \\ y - 2z &= 5 \\ 2x + z &= -1 \end{aligned}$

6. $\begin{aligned} 2x + 3y &= 13 \\ 5x - 2z &= -2 \\ 4y - z &= 6 \end{aligned}$

7. $\begin{aligned} 4x - y + 3z &= 3 \\ -8x + 2y - 6z &= -7 \\ x + 5y - z &= 2 \end{aligned}$

8. $\begin{aligned} 2x - 3y + 4z &= 3 \\ 5x + 4y - 2z &= 7 \\ -3x + 2y - 5z &= -6 \end{aligned}$

9. $\begin{aligned} 2x - 5y + 7z &= 6 \\ x + 4y - 5z &= -2 \\ -6x + 15y - 21z &= 5 \end{aligned}$

10. $\begin{aligned} 3x + y \quad &= -1 \\ 5y - 8z &= 17 \\ -2x \quad + z &= 5 \end{aligned}$

11. $\begin{aligned} 4x \quad &= 8 \\ -5y \quad &= 3 \\ 2z &= 7 \end{aligned}$

12 $\begin{aligned} 0.3x - 2.1y + 1.2z &= -3 \\ 5.7x + 4.2y - 4.8z &= -3 \\ 1.1x - 6.9y + 3.6z &= -19 \end{aligned}$
[*Hint:* Clear decimals first.]

Each system in Exercises 13 and 14 is either inconsistent or dependent.

13. $\begin{aligned} 3x - 2y + 5z &= 8 \\ 6x - 4y + 10z &= 4 \\ 2x + 3y + 4z &= 5 \end{aligned}$

14. $\begin{aligned} 2x - y + z &= 1 \\ -x + 2y - z &= 1 \\ -3x \quad - z &= -3 \end{aligned}$

Solve using a system of three equations.

15. The sum of three numbers is 6. The first, plus twice the second, minus the third is -3. Three times the first, minus the second, plus twice the third is 18. Find the three numbers.

16. A collection of 90 coins consisting of nickels, dimes, and quarters has a value of \$11.50. If the number of quarters is twice the number of nickels, find the number of each type of coin.

17. The smallest angle of a triangle is one-third the middle-sized angle, and the largest angle is 5° more than the middle-sized angle. Find the measure of each.

18 Anita has \$10,000 divided into three separate investments. Part is invested in a mutual fund which earns 8%, part is in time certificates which earn 7%, and the rest is invested in a business. If the business does well, it will earn 10% and her total earnings will amount to \$840. If the business loses 2%, her total earnings will amount to only \$360. How much is invested in each category?

19. The average of Manny's three scores in geometry is 78. If the first is 10 points less than the second and the third is 4 points more than the second, find all three scores. [*Hint:* the first sentence translates to $\frac{x+y+z}{3} = 78$.]

20. Find three numbers such that the sum of the first and second is -1, the sum of the second and third is -4, and the sum of the first and third is 7.

21 The average price of oranges, apples, and limes is 50¢ per pound. Six pounds of oranges, 5 pounds of apples, and 2 pounds of limes cost \$6.00. If limes cost 25¢ per pound more than apples, what is the price per pound of each?

22. At a recent rock concert, a total of 1000 tickets were sold. The prices of the three different types of tickets were \$4, \$5, and \$10, and the total receipts were \$4900. If twice as many \$4 tickets were sold as \$5 tickets, how many tickets at each price were sold?

FOR REVIEW

Solve using a system of equations.

23. By traveling 50 mph for one period of time and then 55 mph for another, Pam traveled 310 miles. If she had gone 5 mph faster throughout, she would have traveled 340 miles. How many hours did she travel at each rate?

24. If 2 lb of nuts and 3 lb of candy cost a total of \$5.40, and 3 lb of nuts and 4 lb of candy cost a total of \$7.50, find the cost of 1 lb of nuts and of 1 lb of candy.

25. A chemist has one solution which is 10% acid and a second which is 5% acid. How much of each should be used to make 20 L of an 8%-acid solution?

26. A boat travels 48 miles downstream in 2 hours and returns the 48 miles upstream in 3 hours. Find the speed of the boat and the speed of the stream.

The following exercises review material from Chapter 1 and will help you prepare for the material in the next section. Find the value of each expression.

27. $(-2)(3) - 4(-1) + (-2)(-3)$

28. $(-4)(-3) - (-2)(5) + (-4)(1)$

ANSWERS: 1. $(-1, 0, 3)$ 2. $(1, 1, 1)$ 3. $(1, -2, 3)$ 4. $(4, 1, 3)$ 5. $(0, 3, -1)$ 6. $(2, 3, 6)$ 7. inconsistent; no solution 8. $(1, 1, 1)$ 9. inconsistent; no solution 10. $(-2, 5, 1)$ 11. $\left(2, -\frac{3}{5}, \frac{7}{2}\right)$ 12. $(10, 20, 30)$ 13. inconsistent; no solution 14. dependent; infinitely many solutions 15. $(3, -1, 4)$ 16. 10 nickels, 60 dimes, 20 quarters 17. 25°, 75°, 80° 18. \$2000 in a mutual fund, \$4000 in certificates, \$4000 in a business 19. 70, 80, 84 20. $(5, -6, 2)$ 21. oranges cost 45¢ per pound, apples cost 40¢ per pound, limes cost 65¢ per pound 22. 600 \$4 tickets, 300 \$5 tickets, and 100 \$10 tickets 23. 4 hours at 50 mph; 2 hours at 55 mph 24. nuts are 90¢ per lb, candy is \$1.20 per lb 25. 8 L of 5% solution, 12 L of 10% solution 26. boat: 20 mph; stream: 4 mph 27. 4 28. 18

4.4 EXERCISES B

Solve the systems using the procedures shown in the examples.

1. $$\begin{aligned} x + y + z &= 1 \\ -x + y - 3z &= -5 \\ 4x - y + 2z &= 0 \end{aligned}$$

2. $$\begin{aligned} x - 2y + z &= 7 \\ 2x + y - z &= 3 \\ -3x + 2y + z &= -9 \end{aligned}$$

3. $$\begin{aligned} 3x - y + 3z &= 2 \\ 2x + 3y + z &= 7 \\ x + y + 2z &= 0 \end{aligned}$$

4. $$\begin{aligned} 2x + y - z &= 2 \\ x - y + z &= 1 \\ x + 2y + 3z &= 6 \end{aligned}$$

5. $$\begin{aligned} -x + 4y + 3z &= 3 \\ 5x + y - z &= 2 \\ 2x - 8y - 6z &= -7 \end{aligned}$$

6. $$\begin{aligned} x + 2y &= -2 \\ 2x + z &= 8 \\ -3y - z &= 9 \end{aligned}$$

7. $$\begin{aligned} 3x - 2y - z &= -2 \\ 5x + y - 5z &= 3 \\ x - 3y + 3z &= -1 \end{aligned}$$

8. $$\begin{aligned} x + y + z &= -3 \\ 4x - 4y + 2z &= -2 \\ -5x + 4y - 2z &= 3 \end{aligned}$$

9. $$\begin{aligned} 2x - 12y + z &= 17 \\ 3x + 12y - 4z &= -2 \\ 4x - 8y - 6z &= -6 \end{aligned}$$

10. $$\begin{aligned} 2x - 3y &= -7 \\ 3y - 2z &= -3 \\ 4x - 4z &= 5 \end{aligned}$$

11. $$\begin{aligned} -3z &= 9 \\ 5x &= 4 \\ 9y &= -12 \end{aligned}$$

12. $$\begin{aligned} \frac{1}{2}x + \frac{1}{3}y - \frac{1}{4}z &= 2 \\ \frac{1}{4}x - \frac{1}{2}y + \frac{1}{2}z &= 2 \\ -\frac{1}{2}x + \frac{1}{2}y - \frac{1}{4}z &= -1 \end{aligned}$$

Each system in Exercises 13 and 14 is either inconsistent or dependent.

13. $$\begin{aligned} 2x - 5y + 7z &= 6 \\ x + 4y - 5z &= -2 \\ -6x + 15y - 21z &= 5 \end{aligned}$$

14. $$\begin{aligned} 2x - y + z &= 5 \\ x - y + 2z &= 1 \\ -3x + y &= -9 \end{aligned}$$

Solve using a system of three equations.

15. The sum of twice a first number, three times a second, and four times a third is 20. The sum of the first and the third is 10 more than the second. The first is the sum of the second and third. Find the numbers.

16. Connie has a total of 27 bills, consisting of ones, fives, and tens, in a cash box. The total value of these bills is \$127. If there is one more ten than there are fives, how many of each are in the box?

17. The largest angle of a triangle is 20° more than twice the middle-sized angle. The sum of the smallest and middle-sized is one-half the largest. Find the measure of each.

18. Howard has a total of \$50,000 invested in stocks, bonds, and certificates. The amount in stocks returned 12% for the year, each of the other accounts returned 9%, and his total income from the investments for the year was \$4860. Had he reversed the amounts invested in stocks and bonds, he would have earned \$5100. How much was invested in each category?

19. Alberta has an average of 88 on three math tests. If her first test score was 13 points higher than the second, and the third was 5 points more than the second, find her three scores.

20. Find three numbers such that the first is five times the third, the second minus the third is -10, and four times the third is 3 less than the first.

21. A chemist has three acid solutions A, B, and C with an average concentration of 25% acid. If he mixes 5 L of A, 4 L of B, and 11 L of C, he obtains 20 L of 24% acid. If the percent acid in A is twice that in C, find the concentration of each solution.

22. A total of 5000 tickets were sold for the conference championship game. The prices of the three different types of tickets were \$2, \$3, and \$5. If three times as many \$2 tickets were sold as \$5 tickets, and the total receipts were \$14,000, how many tickets at each price were sold?

FOR REVIEW

Solve using a system of equations.

23. A 60-ft piece of rope is cut into two pieces. One is 12 ft more than three times the other. How long is each piece?

24. In a right triangle, one acute angle is 15° more than four times the other. Find each acute angle.

25. If there are 26 coins in a collection of dimes and quarters, and the value of the collection is \$4.70, how many dimes and how many quarters are there?

26. A chemist has one solution which is 15% acid and another which is 20% acid. How many gallons of each should be mixed together to obtain 30 gallons of a solution that is 18% acid?

The following exercises review material from Chapter 1 and will help you prepare for the next section. Find the value of each expression.

27. $(-1)(-3) - 5(-2) + (-4)(-2)$

28. $(2)(-5) - (-3)(-2) + (-6)(-1)$

4.4 EXERCISES C

Solve.

1.
$$\begin{aligned} w + x + y + z &= 2 \\ 2w + x + y + z &= 3 \\ w + x + y \quad\quad &= 0 \\ w + x \quad\quad\quad\quad &= 1 \end{aligned}$$

2.
$$\begin{aligned} w + 2z &= 5 \\ 2w - 3y &= -5 \\ 4x + z &= 11 \\ x - 5y &= -3 \end{aligned}$$
[Answer: $w = -1$, $x = 2$, $y = 1$, $z = 3$]

3. $$\begin{aligned} \frac{1}{u} + \frac{1}{v} + \frac{3}{w} &= 4 \\ \frac{2}{u} - \frac{1}{v} + \frac{4}{w} &= -2 \\ \frac{3}{u} - \frac{2}{v} - \frac{1}{w} &= 2 \end{aligned}$$

4. $$\begin{aligned} \frac{1}{u} + \frac{1}{v} \qquad &= -1 \\ \frac{1}{v} + \frac{1}{w} &= 1 \\ \frac{1}{u} \qquad + \frac{1}{w} &= 6 \end{aligned}$$

5. A three-digit whole number has the property that the sum of the three digits is 9, and the ones digit is 3 more than the tens digit. Also, if the hundreds digit and the tens digit are interchanged, the value of the number remains the same. Find the number.

6. The general form of the equation of a circle in a rectangular coordinate system is given by $x^2 + y^2 + Ax + By + C = 0$, where A, B, and C are constants. Find the general form of the equation of a circle passing through $(-1, -2)$, $(1, 0)$, and $(3, -2)$.
[Answer: $x^2 + y^2 - 2x + 4y + 1 = 0$]

7. Ponderosa Products manufactures three models of wood-burning stoves, the Aspen, the Birch, and the Sierra. Each stove is made in three stages requiring the specified times in hours listed in the table below. Due to vacation schedules and equipment maintenance, the cutting, welding, and finishing departments have available a maximum of 75, 47, and 36 workhours per week, respectively. Determine the number of stoves of each model that should be made weekly for Ponderosa Products to be at full capacity operation.

	Aspen	Birch	Sierra
Cutting	4	5	7
Welding	2	3	5
Finishing	2	2	4

[Answer: 3 Aspen stoves, 7 Birch stoves, 4 Sierra stoves]

8. The owner of a chemical plant received an order for a special blend of fertilizer that consists of, among other things, 760 pounds of nitrate, 540 pounds of phosphate, and 106 pounds of iron. He has available three mixes with the compositions shown in the table below. How many pounds of each should he blend to fill the order?

	Nitrate (in %)	Phosphate (in %)	Iron (in %)
GroWell	20	15	4
YieldHi	25	20	2
FastGro	30	18	5

4.5 DETERMINANTS AND CRAMER'S RULE

STUDENT GUIDEPOSTS

1. Second-order Determinants
2. Cramer's Rule for Two Equations
3. Third-order Determinants
4. Cramer's Rule for Three Equations

In this section we define a determinant and use it to solve systems of linear equations. The method we develop is a numerical technique that can easily be programmed in a computer or a hand calculator. However, without a computing device, this method is probably less desirable than either the substitution method or the elimination method.

1 SECOND-ORDER DETERMINANTS

As motivation for our basic definition, suppose we solve a general system of two linear equations in x and y,

$$a_1x + b_1y = c_1$$
$$a_2x + b_2y = c_2,$$

where a_1, b_1, c_1, a_2, b_2, and c_2 are real numbers. Multiply the first equation by b_2 and the second by b_1.

$$a_1 b_2 x + b_1 b_2 y = c_1 b_2 \quad b_2 \text{ times the first equation}$$
$$a_2 b_1 x + b_2 b_1 y = c_2 b_1 \quad b_1 \text{ times the second equation}$$

Subtract the second from the first to eliminate y.

$$a_1b_2 x - a_2b_1 x = c_1b_2 - c_2b_1 \quad b_1b_2y - b_2b_1y = 0 \cdot y = 0$$
$$(a_1b_2 - a_2b_1) x = c_1b_2 - c_2b_1 \quad \text{Factor out } x$$
$$x = \frac{c_1b_2 - c_2b_1}{a_1b_2 - a_2b_1} \quad \text{Divide by } a_1b_2 - a_2b_1, \text{ if } a_1b_2 - a_2b_1 \neq 0$$

Similarly, we may first eliminate x and then solve for y.

$$y = \frac{a_1c_2 - a_2c_1}{a_1b_2 - a_2b_1}$$

We can remember the numbers $a_1b_2 - a_2b_1$, $c_1b_2 - c_2b_1$, and $a_1c_2 - a_2c_1$ if we look at the arrangements of coefficients and constants in the original system, and relate these numbers to the following definition.

Second-Order Determinants

If the four numbers a_1, b_1, a_2, b_2 are written in a square array within vertical bars,

$$\begin{vmatrix} a_1 & b_1 \\ a_2 & b_2 \end{vmatrix}$$

the result is **a second-order determinant** and has **value**

$$\begin{vmatrix} a_1 & b_1 \\ a_2 & b_2 \end{vmatrix} = \mathbf{a_1b_2 - a_2b_1}. \quad \text{Difference of cross-products}$$

EXAMPLE 1 EVALUATING SECOND-ORDER DETERMINANTS

Evaluate the second-order determinants.

(a) $\begin{vmatrix} 1 & 2 \\ 3 & 4 \end{vmatrix} = (1)(4) - (3)(2) = 4 - 6 = -2$

(b) $\begin{vmatrix} c_1 & b_1 \\ c_2 & b_2 \end{vmatrix} = c_1b_2 - c_2b_1$

PRACTICE EXERCISE 1

Evaluate the second-order determinants.

(a) $\begin{vmatrix} 2 & 3 \\ 5 & -1 \end{vmatrix}$

(b) $\begin{vmatrix} a_1 & c_1 \\ a_2 & c_2 \end{vmatrix}$

Answers: (a) -17
(b) $a_1c_2 - a_2c_1$

2 CRAMER'S RULE FOR TWO EQUATIONS

Return to the original system of equations at the beginning of this section with the solutions

$$x = \frac{c_1b_2 - c_2b_1}{a_1b_2 - a_2b_1} \quad \text{and} \quad y = \frac{a_1c_2 - a_2c_1}{a_1b_2 - a_2b_1}.$$

We can use the definition of second-order determinant to rewrite the solutions

$$x = \frac{\begin{vmatrix} c_1 & b_1 \\ c_2 & b_2 \end{vmatrix}}{\begin{vmatrix} a_1 & b_1 \\ a_2 & b_2 \end{vmatrix}} \quad \text{and} \quad y = \frac{\begin{vmatrix} a_1 & c_1 \\ a_2 & c_2 \end{vmatrix}}{\begin{vmatrix} a_1 & b_1 \\ a_2 & b_2 \end{vmatrix}}$$

The determinant

$$A = \begin{vmatrix} a_1 & b_1 \\ a_2 & b_2 \end{vmatrix}$$

which appears in both denominators of the solutions is often called the **determinant of the coefficients of the system of equations.** The determinant in the numerator of the solution for x is denoted by

$$A_x = \begin{vmatrix} c_1 & b_1 \\ c_2 & b_2 \end{vmatrix}$$

and is formed from A by replacing the coefficients of the x terms (the a's) with the constants (the c's). Similarly, the determinant in the numerator of the solution for y is denoted by

$$A_y = \begin{vmatrix} a_1 & c_1 \\ a_2 & c_2 \end{vmatrix}$$

and is formed from A by replacing the coefficients of y (the b's) with the constants (the c's). Using this notation, we have **Cramer's rule for solving a system of two linear equations in two variables.**

Cramer's Rule

The system

$$a_1x + b_1y = c_1$$
$$a_2x + b_2y = c_2$$

has solutions $x = \frac{A_x}{A}$ and $y = \frac{A_y}{A}$, if $A \neq 0$,

where $A = \begin{vmatrix} a_1 & b_1 \\ a_2 & b_2 \end{vmatrix}$, $A_x = \begin{vmatrix} c_1 & b_1 \\ c_2 & b_2 \end{vmatrix}$, $A_y = \begin{vmatrix} a_1 & c_1 \\ a_2 & c_2 \end{vmatrix}$.

EXAMPLE 2 Using Cramer's Rule to Solve Two Equations

Solve using Cramer's rule.

$$3x + 5y = -2$$
$$5x + 3y = 2$$

Practice Exercise 2

Solve using Cramer's rule.

$$3x + 2y = 0$$
$$5x - y = -13$$

The determinant of the coefficients of the system is

$$A = \begin{vmatrix} 3 & 5 \\ 5 & 3 \end{vmatrix},$$

the determinant in the numerator for x is

$$A_x = \begin{vmatrix} -2 & 5 \\ 2 & 3 \end{vmatrix}$$

(we have replaced 3 and 5, the coefficients of x, with the constants -2 and 2), and the determinant in the numerator for y is

$$A_y = \begin{vmatrix} 3 & -2 \\ 5 & 2 \end{vmatrix}$$

(we have replaced 5 and 3, the coefficients of y, with the constants -2 and 2). Then, by Cramer's rule,

$$x = \frac{A_x}{A} = \frac{\begin{vmatrix} -2 & 5 \\ 2 & 3 \end{vmatrix}}{\begin{vmatrix} 3 & 5 \\ 5 & 3 \end{vmatrix}} = \frac{(-2)(3) - (2)(5)}{(3)(3) - (5)(5)} = \frac{-6 - 10}{9 - 25} = \frac{-16}{-16} = 1$$

$$y = \frac{A_y}{A} = \frac{\begin{vmatrix} 3 & -2 \\ 5 & 2 \end{vmatrix}}{\begin{vmatrix} 3 & 5 \\ 5 & 3 \end{vmatrix}} = \frac{(3)(2) - (5)(-2)}{(3)(3) - (5)(5)} = \frac{6 - (-10)}{9 - 25} = \frac{16}{-16} = -1.$$

The solution to the system is $(1, -1)$.

$$A = \begin{vmatrix} 3 & 2 \\ 5 & -1 \end{vmatrix} =$$

$$A_x = \begin{vmatrix} 0 & 2 \\ -13 & -1 \end{vmatrix} =$$

$$A_y = \begin{vmatrix} \quad & \quad \end{vmatrix} =$$

$$x = \frac{A_x}{A} =$$

$$y =$$

Answer: $A = -13$, $A_x = 26$, $A_y = -39$, $x = -2$, $y = 3$; the solution is $(-2, 3)$

If we try to solve a system such as

$$\begin{aligned} 2x - y &= 3 \\ -4x + 2y &= 1, \end{aligned}$$

we have $$x = \frac{A_x}{A} = \frac{7}{0} \quad \text{and} \quad y = \frac{A_y}{A} = \frac{14}{0}.$$

Since the determinant of the coefficients of the system A has value 0, both denominators are zero. Upon closer examination of the system, we see that the lines are parallel, the system is inconsistent, and there is no solution. Likewise, if we solve

$$\begin{aligned} 2x - y &= 3 \\ -4x + 2y &= -6, \end{aligned}$$

we have

$$x = \frac{A_x}{A} = \frac{0}{0} \quad \text{and} \quad y = \frac{A_y}{A} = \frac{0}{0}.$$

Again, the determinant of the coefficients, A, has value 0, resulting in zero denominators, but this time both numerators are also 0. Upon closer examination of the system, we see that the lines coincide, the system is dependent, and there are infinitely many solutions of the form $(x, 2x - 3)$ for any real number x.

Inconsistent and Dependent Systems

When a system is solved using Cramer's rule,

1. if $A = 0$, and either $A_x \neq 0$ or $A_y \neq 0$, then the system is inconsistent and has **no solution;**
2. if $A = 0$, and both $A_x = 0$ and $A_y = 0$, then the system is dependent and has **infinitely many solutions.**

3 THIRD-ORDER DETERMINANTS

Cramer's rule can be applied to systems of three linear equations in three variables (in fact, to systems of n linear equations in n variables, for $n \geq 2$) by suitably defining **third-order determinant** (and ***n*th-order determinant**).

Third-Order Determinants

If the nine numbers $a_1, b_1, c_1, a_2, b_2, c_2, a_3, b_3, c_3$ are written in a square array within vertical bars,

$$\begin{vmatrix} a_1 & b_1 & c_1 \\ a_2 & b_2 & c_2 \\ a_3 & b_3 & c_3 \end{vmatrix}$$

The result is a **third-order determinant** and has **value**

$$\begin{vmatrix} a_1 & b_1 & c_1 \\ a_2 & b_2 & c_2 \\ a_3 & b_3 & c_3 \end{vmatrix} = a_1\begin{vmatrix} b_2 & c_2 \\ b_3 & c_3 \end{vmatrix} - a_2\begin{vmatrix} b_1 & c_1 \\ b_3 & c_3 \end{vmatrix} + a_3\begin{vmatrix} b_1 & c_1 \\ b_2 & c_2 \end{vmatrix}.$$

Notice that the third-order determinant is evaluated by evaluating three second-order determinants. Remember the definition by concentrating on the first column of numbers: each number in the first column multiplies the second-order determinant which remains if the row and column which contain the number are deleted from the third-order determinant.

$$a_1\begin{vmatrix} b_2 & c_2 \\ b_3 & c_3 \end{vmatrix} \qquad -a_2\begin{vmatrix} b_1 & c_1 \\ b_3 & c_3 \end{vmatrix} \qquad + a_3\begin{vmatrix} b_1 & c_1 \\ b_2 & c_2 \end{vmatrix}$$

$$\begin{vmatrix} a_1 & b_1 & c_1 \\ a_2 & b_2 & c_2 \\ a_3 & b_3 & c_3 \end{vmatrix} \qquad \begin{vmatrix} a_1 & b_1 & c_1 \\ a_2 & b_2 & c_2 \\ a_3 & b_3 & c_3 \end{vmatrix} \qquad \begin{vmatrix} a_1 & b_1 & c_1 \\ a_2 & b_2 & c_2 \\ a_3 & b_3 & c_3 \end{vmatrix}$$

Be sure to *subtract* a_2 times the second-order determinant.

EXAMPLE 3 EVALUATING A THIRD-ORDER DETERMINANT

Evaluate the third-order determinant.

$$\begin{vmatrix} 5 & -1 & -2 \\ 4 & -1 & 0 \\ 1 & 2 & 3 \end{vmatrix} = 5\begin{vmatrix} -1 & 0 \\ 2 & 3 \end{vmatrix} - 4\begin{vmatrix} -1 & -2 \\ 2 & 3 \end{vmatrix} + 1\begin{vmatrix} -1 & -2 \\ -1 & 0 \end{vmatrix}$$

$$= 5[(-1)(3) - (2)(0)] - 4[(-1)(3) - (2)(-2)] + 1[(-1)(0) - (-1)(-2)]$$
$$= 5[-3 - 0] - 4[-3 + 4] + 1[0 - 2]$$
$$= 5[-3] - 4[1] + 1[-2] = -15 - 4 - 2 = -21$$

PRACTICE EXERCISE 3

Evaluate the third-order determinant.

$$\begin{vmatrix} 2 & 0 & 5 \\ 3 & 4 & -2 \\ -1 & 1 & 2 \end{vmatrix}$$

Answer: 55

4 CRAMER'S RULE FOR THREE EQUATIONS

The definition we have given for a third-order determinant would seem reasonable if we were to solve a general system of three linear equations.

$$\begin{aligned} a_1x + b_1y + c_1z &= d_1 \\ a_2x + b_2y + c_2z &= d_2 \\ a_3x + b_3y + c_3z &= d_3 \end{aligned}$$

The solutions to this system have the following form.

$$x = \frac{\begin{vmatrix} d_1 & b_1 & c_1 \\ d_2 & b_2 & c_2 \\ d_3 & b_3 & c_3 \end{vmatrix}}{\begin{vmatrix} a_1 & b_1 & c_1 \\ a_2 & b_2 & c_2 \\ a_3 & b_3 & c_3 \end{vmatrix}}, \quad y = \frac{\begin{vmatrix} a_1 & d_1 & c_1 \\ a_2 & d_2 & c_2 \\ a_3 & d_3 & c_3 \end{vmatrix}}{\begin{vmatrix} a_1 & b_1 & c_1 \\ a_2 & b_2 & c_2 \\ a_3 & b_3 & c_3 \end{vmatrix}}, \quad z = \frac{\begin{vmatrix} a_1 & b_1 & d_1 \\ a_2 & b_2 & d_2 \\ a_3 & b_3 & d_3 \end{vmatrix}}{\begin{vmatrix} a_1 & b_1 & c_1 \\ a_2 & b_2 & c_2 \\ a_3 & b_3 & c_3 \end{vmatrix}}$$

This result is also known as **Cramer's rule for a system of three linear equations in three variables.** The determinant of the coefficient appears in the denominator of each fraction, and the determinant in each numerator is formed by replacing the coefficients of the respective variable with the constants. As with systems of two equations in two variables, we use the following notation.

$$A = \begin{vmatrix} a_1 & b_1 & c_1 \\ a_2 & b_2 & c_2 \\ a_3 & b_3 & c_3 \end{vmatrix} \quad A_x = \begin{vmatrix} d_1 & b_1 & c_1 \\ d_2 & b_2 & c_2 \\ d_3 & b_3 & c_3 \end{vmatrix} \quad A_y = \begin{vmatrix} a_1 & d_1 & c_1 \\ a_2 & d_2 & c_2 \\ a_3 & d_3 & c_3 \end{vmatrix} \quad A_z = \begin{vmatrix} a_1 & b_1 & d_1 \\ a_2 & b_2 & d_2 \\ a_3 & b_3 & d_3 \end{vmatrix}$$

The solutions to the system, using Cramer's rule, are

$$x = \frac{A_x}{A}, \quad y = \frac{A_y}{A}, \quad z = \frac{A_z}{A}, \quad \text{if } A \neq 0.$$

EXAMPLE 4 Using Cramer's Rule to Solve Three Equations

Solve using Cramer's rule.

$$\begin{aligned} x + y + z &= 2 \\ -2x - y + z &= 1 \\ x - 2y - z &= 1 \end{aligned}$$

The three solutions are given by

$$x = \frac{A_x}{A} = \frac{\begin{vmatrix} 2 & 1 & 1 \\ 1 & -1 & 1 \\ 1 & -2 & -1 \end{vmatrix}}{\begin{vmatrix} 1 & 1 & 1 \\ -2 & -1 & 1 \\ 1 & -2 & -1 \end{vmatrix}} \qquad y = \frac{A_y}{A} = \frac{\begin{vmatrix} 1 & 2 & 1 \\ -2 & 1 & 1 \\ 1 & 1 & -1 \end{vmatrix}}{\begin{vmatrix} 1 & 1 & 1 \\ -2 & -1 & 1 \\ 1 & -2 & -1 \end{vmatrix}}$$

$$z = \frac{A_z}{A} = \frac{\begin{vmatrix} 1 & 1 & 2 \\ -2 & -1 & 1 \\ 1 & -2 & 1 \end{vmatrix}}{\begin{vmatrix} 1 & 1 & 1 \\ -2 & -1 & 1 \\ 1 & -2 & -1 \end{vmatrix}}$$

Practice Exercise 4

Solve using Cramer's rule.

$$\begin{aligned} x - y + z &= -2 \\ x + 2y - 3z &= -6 \\ 4x + y + 5z &= -7 \end{aligned}$$

In each fraction, the denominator is the determinant of the coefficients, and the numerator has the corresponding column of coefficients of each variable replaced with the column of constants. Next we evaluate each of the four different third-order determinants.

$$A = \begin{vmatrix} 1 & 1 & 1 \\ -2 & -1 & 1 \\ 1 & -2 & -1 \end{vmatrix} = (1)\begin{vmatrix} -1 & 1 \\ -2 & -1 \end{vmatrix} - (-2)\begin{vmatrix} 1 & 1 \\ -2 & -1 \end{vmatrix} + (1)\begin{vmatrix} 1 & 1 \\ -1 & 1 \end{vmatrix}$$

$$= (1)(3) + (2)(1) + (1)(2) = 3 + 2 + 2 = 7$$

$$A_x = \begin{vmatrix} 2 & 1 & 1 \\ 1 & -1 & 1 \\ 1 & -2 & -1 \end{vmatrix} = (2)\begin{vmatrix} -1 & 1 \\ -2 & -1 \end{vmatrix} - (1)\begin{vmatrix} 1 & 1 \\ -2 & -1 \end{vmatrix} + (1)\begin{vmatrix} 1 & 1 \\ -1 & 1 \end{vmatrix}$$

$$= (2)(3) - (1)(1) + (1)(2) = 6 - 1 + 2 = 7$$

$$A_y = \begin{vmatrix} 1 & 2 & 1 \\ -2 & 1 & 1 \\ 1 & 1 & -1 \end{vmatrix} = (1)\begin{vmatrix} 1 & 1 \\ 1 & -1 \end{vmatrix} - (-2)\begin{vmatrix} 2 & 1 \\ 1 & -1 \end{vmatrix} + (1)\begin{vmatrix} 2 & 1 \\ 1 & 1 \end{vmatrix}$$

$$= (1)(-2) + (2)(-3) + (1)(1) = -2 - 6 + 1 = -7$$

$$A_z = \begin{vmatrix} 1 & 1 & 2 \\ -2 & -1 & 1 \\ 1 & -2 & 1 \end{vmatrix} = (1)\begin{vmatrix} -1 & 1 \\ -2 & 1 \end{vmatrix} - (-2)\begin{vmatrix} 1 & 2 \\ -2 & 1 \end{vmatrix} + (1)\begin{vmatrix} 1 & 2 \\ -1 & 1 \end{vmatrix}$$

$$= (1)(1) + (2)(5) + (1)(3) = 1 + 10 + 3 = 14$$

Thus, substituting the values of A, A_x, A_y, and A_z, we have

$$x = \frac{A_x}{A} = \frac{7}{7} = 1, \qquad y = \frac{A_y}{A} = \frac{-7}{7} = -1, \qquad z = \frac{A_z}{A} = \frac{14}{7} = 2,$$

giving the solution $(1, -1, 2)$.

Answer: $A = 23$, $A_x = -69$, $A_y = 0$, and $A_z = 23$, the solution is $(-3, 0, 1)$

As is the case with Cramer's rule for systems of two equations, when the determinant of the coefficients is zero ($A = 0$), there is no solution if at least one of A_x, A_y, or A_z is not zero, and there are infinitely many solutions if all of A_x, A_y, and A_z are zero.

4.5 EXERCISES A

Evaluate the determinants.

1. $\begin{vmatrix} 3 & 5 \\ 2 & 4 \end{vmatrix}$

2. $\begin{vmatrix} 2 & 5 \\ -3 & -1 \end{vmatrix}$

3. $\begin{vmatrix} -3 & -7 \\ -6 & -8 \end{vmatrix}$

4. $\begin{vmatrix} 2 & 5 \\ 0 & 0 \end{vmatrix}$

5. $\begin{vmatrix} 2 & -1 \\ 2 & -1 \end{vmatrix}$

6. $\begin{vmatrix} 10 & 20 \\ 30 & 40 \end{vmatrix}$

7. $\begin{vmatrix} 3 & 1 & 2 \\ 4 & 0 & -2 \\ -2 & 1 & 3 \end{vmatrix}$

8. $\begin{vmatrix} 1 & 1 & 1 \\ -1 & 2 & -2 \\ -3 & 4 & -3 \end{vmatrix}$

9. $\begin{vmatrix} 0 & -4 & -2 \\ 4 & 2 & -1 \\ 0 & 6 & 3 \end{vmatrix}$

10. $\begin{vmatrix} 3 & 0 & 0 \\ 0 & -2 & 0 \\ 0 & 0 & 4 \end{vmatrix}$

Solve the systems using Cramer's rule.

11. $\begin{aligned} x + 2y &= -1 \\ 2x - 3y &= 12 \end{aligned}$

12. $\begin{aligned} 4x - y &= -3 \\ 2x - 3y &= 1 \end{aligned}$

13. $\begin{aligned} 5x + 5y &= 10 \\ x + y &= 2 \end{aligned}$

14 $\begin{aligned} 3x + 2y &= 7 \\ -x + 4y &= 0 \end{aligned}$

15. $\begin{aligned} 3x - y &= -11 \\ 2x + 4z &= -2 \\ -3y + 5z &= -1 \end{aligned}$

16. $\begin{aligned} 2x + 3y &= -1 \\ 3y + z &= 6 \\ 3y + 4z &= 15 \end{aligned}$

17. $\begin{aligned} 2x - y &= 3 \\ 2y - z &= 3 \\ x + y + z &= 9 \end{aligned}$

18 $\begin{aligned} 4x - 3y + 8z &= 12 \\ 2x - \frac{3}{2}y + 4z &= 11 \\ x - 5z &= -10 \end{aligned}$

FOR REVIEW

Solve.

19. $\begin{aligned} 3x - 5y + z &= 4 \\ 5x + 4y &= 5 \\ 2x - 3y - 3z &= -1 \end{aligned}$

20. $\begin{aligned} x + y &= 1 \\ -y + z &= -1 \\ x + z &= 2 \end{aligned}$

21. The average of a student's first three scores on math tests is 84. If the second score was 5 points higher than the first, and the third was 2 points higher than the second, what were the three scores?

22. Jill has \$20,000 invested in stocks, bonds, and securities. She has twice as much in stocks as she does in bonds. The return on her investments was 12% for the stocks, 9% for the bonds, and 8% for the securities. If her income for the year was \$1960, what amount does she have invested in each?

Exercises 23–26, reviewing material covered in Chapter 2, will help you prepare for the next section. Solve each inequality.

23. $2x - 1 \le 3x + 4$

24. $3(x - 4) - (x - 1) > 5$

25. $2(x - 5) < 1 - (x - 1)$

26. $-3x \ge 2 - 4(1 - 2x)$

ANSWERS: 1. 2 2. 13 3. −18 4. 0 5. 0 6. −200 7. 6 8. 7 9. 0 10. −24 11. (3, −2) 12. (−1, −1) 13. infinitely many solutions of the form $(x, 2 - x)$ for x any real number 14. $(2, \frac{1}{2})$ 15. (−3, 2, 1) 16. (−2, 1, 3) 17. (3, 3, 3) 18. no solution 19. (1, 0, 1) 20. inconsistent; no solution 21. 80, 85, 87 22. $8000 in stocks, $4000 in bonds, $8000 in securities 23. $x \geq -5$ 24. $x > 8$ 25. $x < 4$ 26. $x \leq \frac{2}{11}$

4.5 EXERCISES B

Evaluate the determinants.

1. $\begin{vmatrix} -1 & 0 \\ 3 & 4 \end{vmatrix}$ **2.** $\begin{vmatrix} 5 & 0 \\ -3 & 2 \end{vmatrix}$ **3.** $\begin{vmatrix} -1 & -1 \\ -2 & -4 \end{vmatrix}$ **4.** $\begin{vmatrix} 2 & 0 \\ 5 & 0 \end{vmatrix}$ **5.** $\begin{vmatrix} 5 & 5 \\ 6 & 6 \end{vmatrix}$

6. $\begin{vmatrix} -2 & -7 \\ -3 & -4 \end{vmatrix}$ **7.** $\begin{vmatrix} 1 & -2 & 1 \\ 2 & 3 & 2 \\ 3 & 1 & 3 \end{vmatrix}$ **8.** $\begin{vmatrix} -2 & 0 & 6 \\ 1 & 2 & -1 \\ 4 & -2 & -5 \end{vmatrix}$ **9.** $\begin{vmatrix} 3 & -2 & -4 \\ 0 & 0 & 6 \\ 2 & 0 & 8 \end{vmatrix}$ **10.** $\begin{vmatrix} 0 & 0 & 7 \\ 0 & 3 & 0 \\ -1 & 0 & 0 \end{vmatrix}$

Solve the systems using Cramer's rule.

11. $\begin{aligned} 3x + 2y &= -3 \\ 6x - 3y &= 8 \end{aligned}$

12. $\begin{aligned} 2x \phantom{{}+y} &= 4 \\ -3x + y &= 7 \end{aligned}$

13. $\begin{aligned} -3x + 3y &= 5 \\ 2x - 2y &= -1 \end{aligned}$

14. $\begin{aligned} 2x + 2y &= 0 \\ x - 5y &= -18 \end{aligned}$

15. $\begin{aligned} 2x + y - z &= -3 \\ x - y + z &= 0 \\ -3x + 2y - 4z &= -3 \end{aligned}$

16. $\begin{aligned} -x + 5y + z &= 2 \\ 3x - y + 4z &= 3 \\ 6x - 2y + 8z &= 7 \end{aligned}$

17. $\begin{aligned} 4x \phantom{{}+3y} + z &= 5 \\ 3y - 6z &= 0 \\ 5x \phantom{{}+3y} + 5z &= 10 \end{aligned}$

18. $\begin{aligned} 5x + 3y - 2z &= -4 \\ 2y - z &= 8 \\ 3x \phantom{{}+3y} + 4z &= -1 \end{aligned}$

FOR REVIEW

Solve.

19. $\begin{aligned} 3x - 3y + z &= 0 \\ 4x - 2y - z &= 3 \\ 2x + 3y - 2z &= -2 \end{aligned}$

20. $\begin{aligned} x + y \phantom{{}+z} &= 2 \\ -y + z &= -4 \\ x \phantom{{}+y} + z &= 3 \end{aligned}$

21. The smallest angle of a triangle is one-third the middle-sized angle, and the middle-sized angle is one-half the largest. Find the measure of each angle.

22. The average of the weights of the three interior linemen for the Lumberjacks is 257 pounds. If the sum of the first two is 500 pounds, and the third is 21 pounds heavier than the second, what is the weight of each?

Exercises 23–26, reviewing material covered in Chapter 2, will help you prepare for the next section. Solve each inequality.

23. $3x - 4 \geq 4x + 2$

24. $4(x - 1) - 2(x - 3) < 6$

25. $5(x - 3) > 2 - (x - 1)$

26. $-2x \leq 6 - 3(4 - x)$

4.5 EXERCISES C

Solve for x.

1. $\begin{vmatrix} x & 2 \\ 3 & 1 \end{vmatrix} = 3$

2. $\begin{vmatrix} x & 0 & 0 \\ 0 & 1 & 2 \\ 0 & -1 & 3 \end{vmatrix} = 10$

[Answer: $x = 2$]

3. $\begin{vmatrix} x & 1 \\ 1 & x \end{vmatrix} = 3$

4.6 LINEAR INEQUALITIES IN TWO VARIABLES

STUDENT GUIDEPOSTS

1. Linear Inequalities in Two Variables
2. Graphing Linear Inequalities
3. Systems of Linear Inequalities
4. Graphing Systems of Linear Inequalities

In Chapter 2 we solved inequalities in one variable, such as

$$2x + 1 < 3 \quad \text{and} \quad x - 2 \geq 2(x - 4),$$

and graphed their solutions on a number line. For example, if we solve

$$2x + 1 < 3$$

the solution $x < 1$, or $(-\infty, 1)$, is graphed in Figure 4.5.

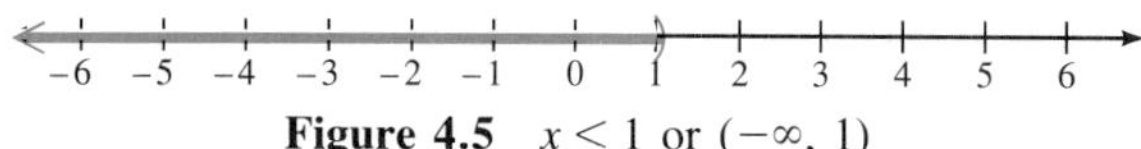

Figure 4.5 $x < 1$ or $(-\infty, 1)$

1 LINEAR INEQUALITIES IN TWO VARIABLES

We now consider inequalities in two variables in which the variables are raised only to the first power. Such inequalities, called **linear inequalities in two variables,** can always be written in one of the forms

$$ax + by > c, \quad ax + by < c, \quad ax + by \geq c, \quad \text{or} \quad ax + by \leq c,$$

where a, b, and c are real number constants, a and b not both zero. For example,

$$2x + 3y > 6$$

is a linear inequality in two variables x and y. As with linear equations, a solution to such an inequality is an ordered pair (x, y) which makes the inequality true. Thus $(-1, 4)$ is a solution to $2x + 3y > 6$ since

$$2(-1) + 3(4) > 6$$
$$10 > 6 \quad \text{is true.}$$

However, $(1, -4)$ is not a solution since

$$2(1) + 3(-4) > 6$$
$$-10 > 6 \quad \text{is false.}$$

Remember that the graph of a linear equation such as $2x + 3y = 6$ is a line in the plane. Notice in Figure 4.6 that the graph of $2x + 3y = 6$ divides the plane into a region above the line, the line itself, and a region below the line.

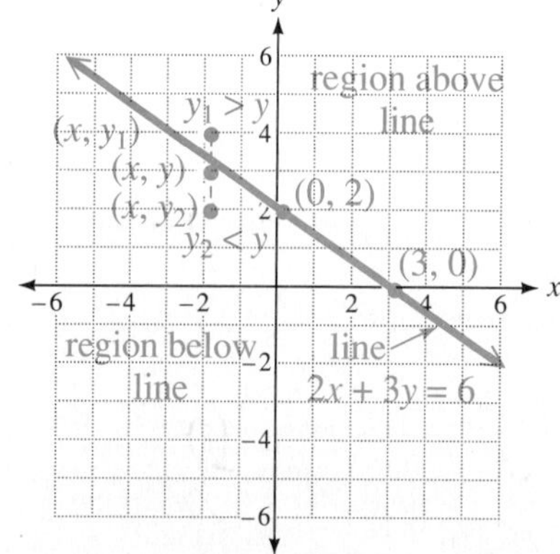

Figure 4.6 Regions Formed by a Line

2 GRAPHING LINEAR INEQUALITIES

The graph of an inequality such as $2x + 3y > 6$ consists of all the points on one side of the boundary line $2x + 3y = 6$, while the graph of $2x + 3y < 6$ is all the points on the other side. To decide which side of the line to graph for an inequality, consider the point (x, y) on the line and the points (x, y_1) and (x, y_2) above and below the line in Figure 4.6. Since $y_1 > y$, the point (x, y_1) would satisfy $2x + 3y > 6$. Similarly, since $y_2 < y$, the point (x, y_2) would satisfy $2x + 3y < 6$.

Suppose we graph $2x + 3y > 6$. We can select any point *not* on the line $2x + 3y = 6$ and use it as a **test point.** Since (0, 0) is not on the line, we choose it for the test point to make the arithmetic easy.

$$\begin{aligned} 2x + 3y &> 6 \\ 2(0) + 3(0) &> 6 \\ 0 &> 6 \quad \text{This is false} \end{aligned}$$

Since this is a false inequality we graph the points on the side of the line not containing (0, 0). If the point (0, 0) is used as a test point for the inequality $2x + 3y < 6$, a true inequality is obtained.

$$\begin{aligned} 2x + 3y &< 6 \\ 2(0) + 3(0) &< 6 \\ 0 &< 6 \quad \text{This is true} \end{aligned}$$

To graph $2x + 3y < 6$ we shade the region containing the test point (0, 0). The graphs of both $2x + 3y > 6$ and $2x + 3y < 6$ are shown in Figure 4.7 with the appropriate region for each inequality shaded.

Note: If the sign is > or <, draw a dotted line.
≥ or ≤ draw solid line.

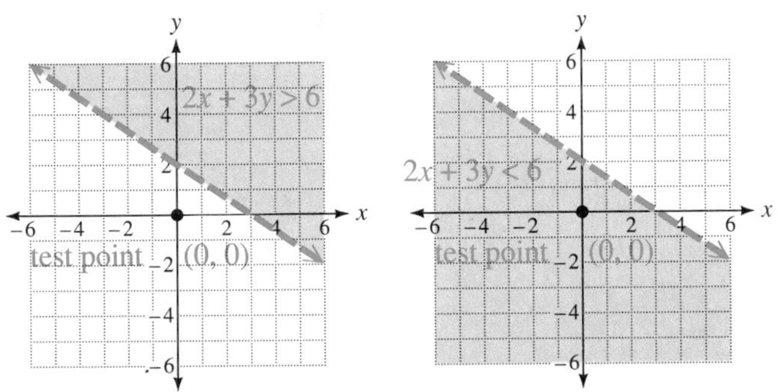

Figure 4.7 Graphing Linear Inequalities

Notice that the lines in Figure 4.7 are dashed. This is done to show that the points on the line are *not* part of the graph. To graph $2x + 3y \geq 6$ and $2x + 3y \leq 6$ we would make the line solid to indicate that points on the line *are* part of the graph.

To Graph a Linear Inequality in Two Variables

1. Graph the boundary line, using a dashed line if the inequality is $<$ or $>$ and a solid line if it is $\leq$ or $\geq$.
2. Choose a test point that is not on the boundary line and substitute it into the inequality.
3. Shade the region that includes the test point if a true inequality is obtained, and shade the region that does not contain the test point if a false inequality results.

EXAMPLE 1 Graphing a Linear Inequality

(a) Graph $2x - 5y \le 10$.

First graph the line $2x - 5y = 10$ using a solid line because the inequality $\le$ includes the points on the line. To graph the line use the intercepts $(0, -2)$ and $(5, 0)$. Now select a test point which is not on the line. The point $(0, 0)$ is again easy to use.

$$2x - 5y \le 10$$
$$2(0) - 5(0) \le 10$$
$$0 \le 10 \quad \text{This is true}$$

Since $0 \le 10$ is true we shade the side of the graph that *contains* the test point $(0, 0)$. See Figure 4.8.

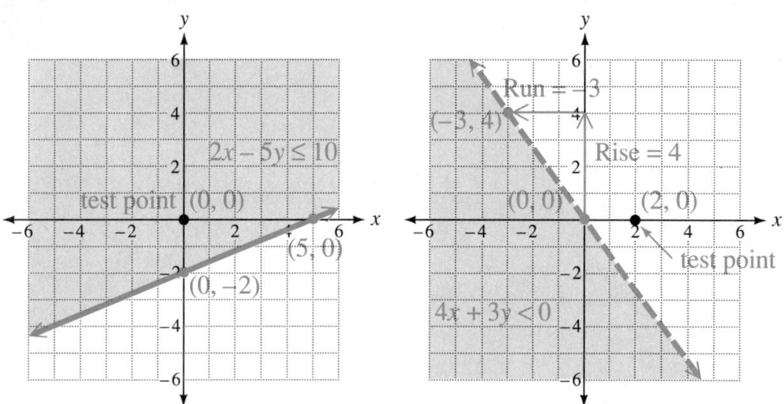

Figure 4.8 Graph Using $\le$ **Figure 4.9** Graph Using $<$

(b) Graph $4x + 3y < 0$.
We graph $4x + 3y = 0$ with a dashed line since the inequality is $<$. This time we graph the line using the slope and y-intercept. Since $3y = -4x$, we have $y = -\frac{4}{3}x + 0$ as the slope-intercept form. Starting at the y-intercept (also the x-intercept) $(0, 0)$, we move up 4 units (a rise of 4) and left 3 units (a run of -3) to obtain a second point $(-3, 4)$ on the line. Since $(0, 0)$ is on the line we choose a different test point such as $(2, 0)$.

$$4x + 3y < 0$$
$$4(2) + 3(0) < 0$$
$$8 < 0 \quad \text{This is false}$$

Since $8 < 0$ is false, we shade the region in Figure 4.9 that does not contain the test point $(2, 0)$.

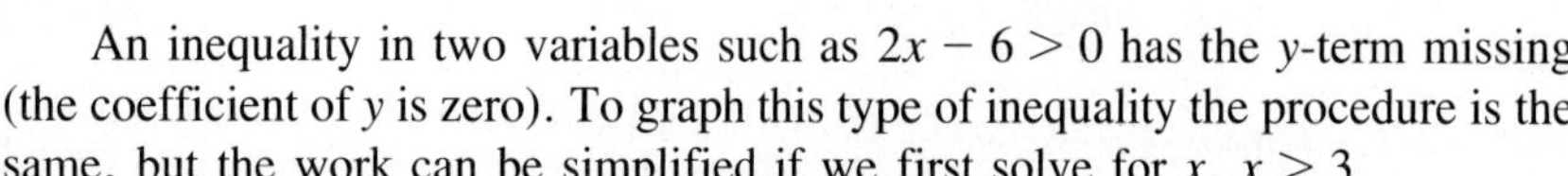

PRACTICE EXERCISE 1

(a) Graph $3x + y \ge 6$.

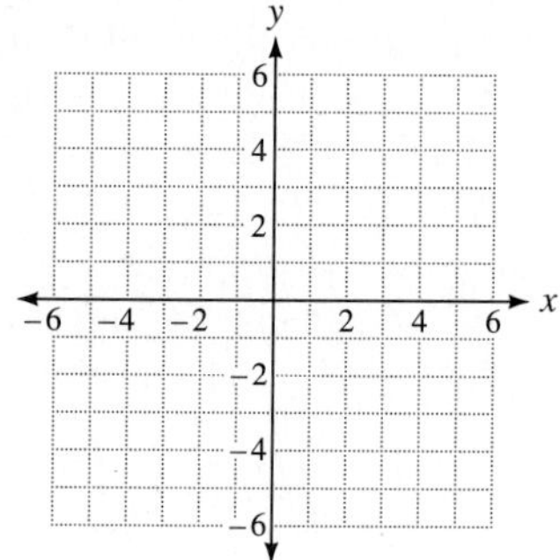

(b) Graph $2x - 5y > 0$.

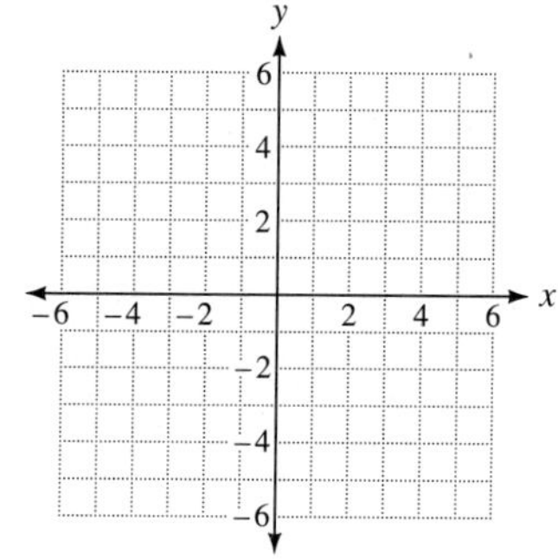

Answers: (a) The graph is the region above and including the line with intercepts (0, 6) and (2, 0). (b) The graph is the region below (not including) the line with intercepts (0, 0) and passing through the point (5, 2).

An inequality in two variables such as $2x - 6 > 0$ has the y-term missing (the coefficient of y is zero). To graph this type of inequality the procedure is the same, but the work can be simplified if we first solve for x, $x > 3$.

We now recognize the graph of the boundary line $x = 3$ as a vertical line through the point $(3, 0)$. Since the inequality is $>$, we dash the line. Next, the test point $(0, 0)$ gives a false inequality.

$$x > 3$$
$$0 > 3 \quad \text{This is false}$$

Shade the region that does not contain (0, 0) as in Figure 4.10.

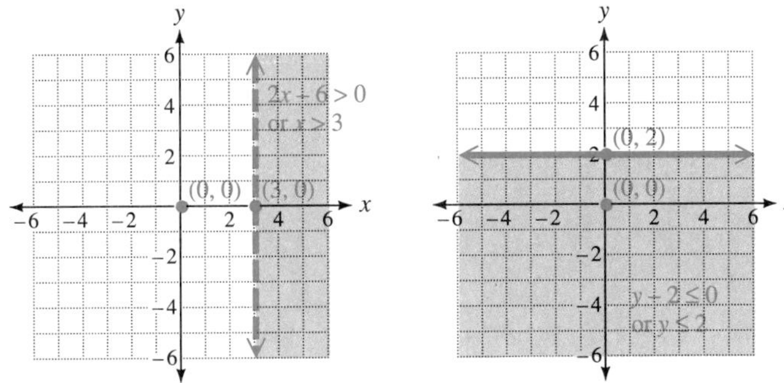

Figure 4.10 Regions Determined by Vertical Line **Figure 4.11** Regions Determined by Horizontal Line

Suppose we graph $y - 2 \leq 0$. This inequality can be written $y \leq 2$; the line to plot is the horizontal line $y = 2$. Use the test point (0, 0) to obtain the true inequality $0 \leq 2$ and shade the region that contains the test point. See Figure 4.11.

3 SYSTEMS OF LINEAR INEQUALITIES

When we graph a linear inequality in two variables, the points on the graph correspond to ordered pairs of numbers that are solutions to the inequality. A **system of linear inequalities in two variables** is simply two or more linear inequalities in two variables. A solution to such a system is an ordered pair of numbers that solves all inequalities in the system.

4 GRAPHING SYSTEMS OF LINEAR INEQUALITIES

Graphing a system of inequalities is outlined in the following.

To Graph a System of Linear Inequalities in Two Variables

1. Draw the graph of each inequality in the system.
2. The graph consists of those points common to the graphs of the individual inequalities.

EXAMPLE 2 Graphing a System of Inequalities

(a) Graph the systems of inequalities.

$$3x + 4y < 12$$
$$y - x \geq 1$$

Graph $3x + 4y = 12$ with a dashed line (the inequality is <) and $y - x = 1$ with a solid line (the inequality is ≥). The test point (0, 0) can be used for each inequality.

$3x + 4y < 12$		$y - x \geq 1$	
$3(0) + 4(0) < 12$		$0 - 0 \geq 1$	
$0 < 12$	True	$0 \geq 1$	False

Thus, the graph of the system includes all points both below the line $3x + 4y = 12$ *and* above and on the line $y - x = 1$. The graph is shaded in Figure 4.12 on the following page.

Practice Exercise 2

(a) Graph the systems of inequalities.

$$3x + 4y > 12$$
$$y - x \leq 1$$

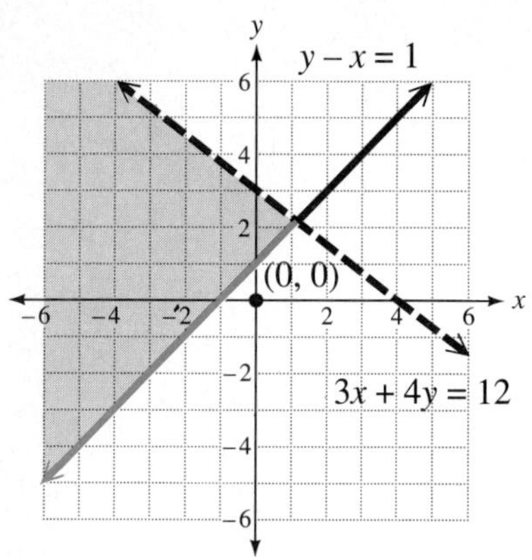

Figure 4.12

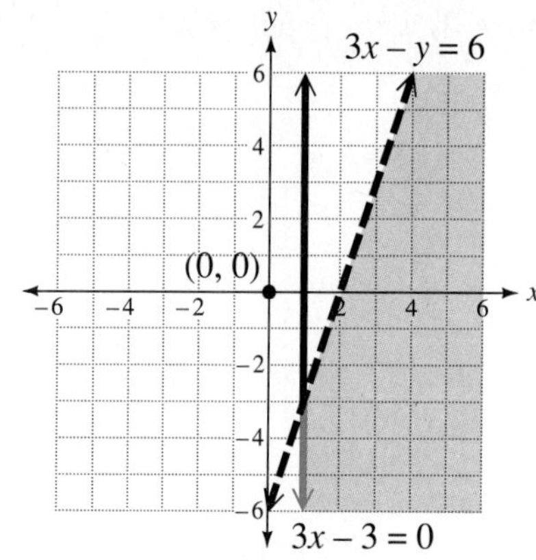

Figure 4.13

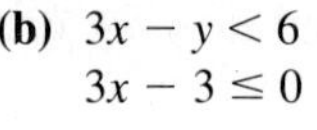

(b) $3x - y < 6$
$3x - 3 \le 0$

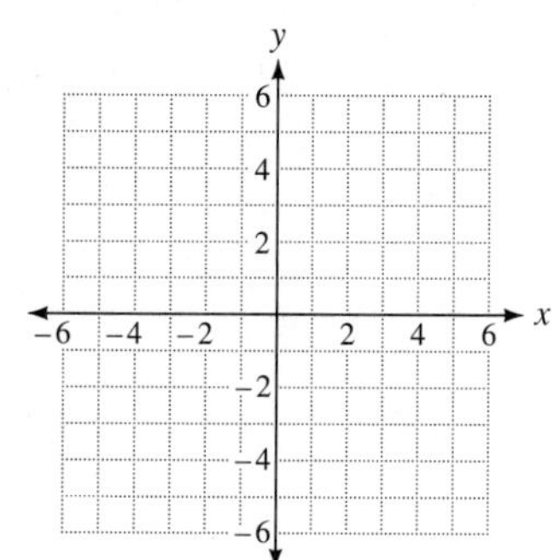

(b) $3x - y > 6$

$3x - 3 \ge 0$

Use the test point (0, 0) after the boundary lines $3x - y = 6$ and $3x - 3 = 0$ are graphed.

$$3x - y > 6 \qquad\qquad 3x - 3 \ge 0$$
$$3(0) - 0 > 6 \qquad\qquad 3(0) - 3 \ge 0$$
$$0 > 6 \quad \text{False} \qquad\qquad -3 \ge 0 \quad \text{False}$$

Thus, the graph consists of all points both below the line $3x - y = 6$ *and* to the right of and on the line $3x - 3 = 0$ ($x = 1$). The graph is shaded in Figure 4.13.

Answers: (a) Your graph should be the region above the line with intercepts (0, 3) and (4, 0) but on and below the line with intercepts (0, 1) and (−1, 0). (b) Your graph should be the region above the line with intercepts (0, −6) and (2, 0) but on and to the left of the vertical line $x = 1$.

Many applied problems, especially in business, can be described and solved using a system of inequalities.

EXAMPLE 3 Production Problem

Bikes Inc. manufactures two types of racing bicycles, the Standard and the Professional models. The table below summarizes the weekly production data. Determine the various combinations of bicycles that can be made each week under the limitations on the workhours available in each stage of production.

	Workhours for Standard model	*Workhours for Professional model*	*Maximum workhours per week*
Construction	8	12	216
Finishing	2	2	48

Let x = number of Standard models made per week,
y = number of Professional models made per week.
The number made of each model must not be negative, so

$$x \ge 0$$
$$y \ge 0.$$

Since the number of weekly workhours for constructing Standard models, $8x$, together with the weekly workhours for constructing Professional

Practice Exercise 3

Graph the system.

$$x \ge 0$$
$$y \ge 0$$
$$x + y \le 4$$
$$x + 2y \le 6$$

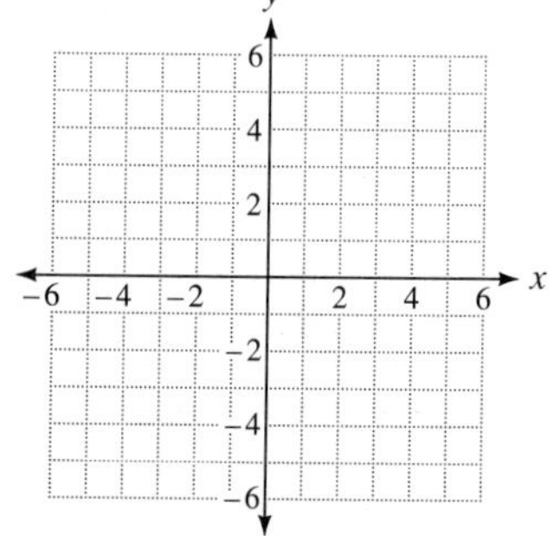

models, $12y$, cannot exceed 216, a third inequality is

$$8x + 12y \leq 216.$$

Similarly, the weekly workhours allocated to finishing define a fourth inequality

$$2x + 2y \leq 48.$$

Any solution to the system

$$\begin{aligned} x &\geq 0 \\ y &\geq 0 \\ 8x + 12y &\leq 216 \\ 2x + 2y &\leq 48 \end{aligned}$$

provides a possible weekly production schedule. The graph of these solutions is shaded in Figure 4.14. Notice that the inequalities $x \geq 0$ and $y \geq 0$ restrict the graph to points in the first quadrant while the other two inequalities restrict the graph to points *below* both boundary lines. For example, one possible production schedule would be 18 Standard models and 6 Professional models, corresponding to the point of intersection of the two lines (18, 6). A point outside the graph, such as (15, 12), corresponding to 15 Standard and 12 Professional models, is impossible.

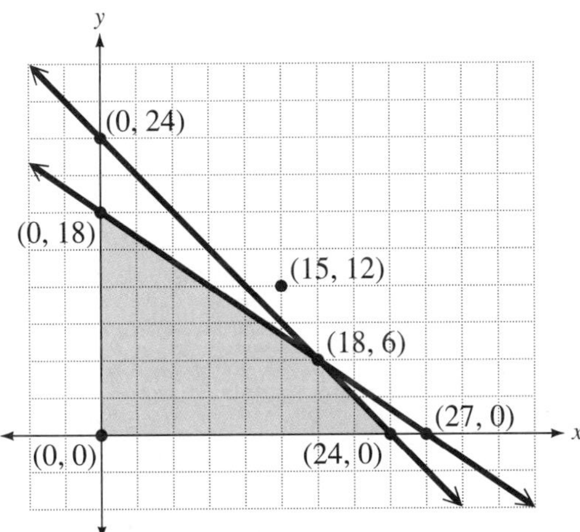

Figure 4.14

Answer: The graph consists of all points in quadrant I below the lines $x + y = 4$ and $x + 2y = 6$ together with points on these boundary lines and the positive axes.

4.6 EXERCISES A

Graph the linear inequalities in two variables.

1. $x + y > 2$

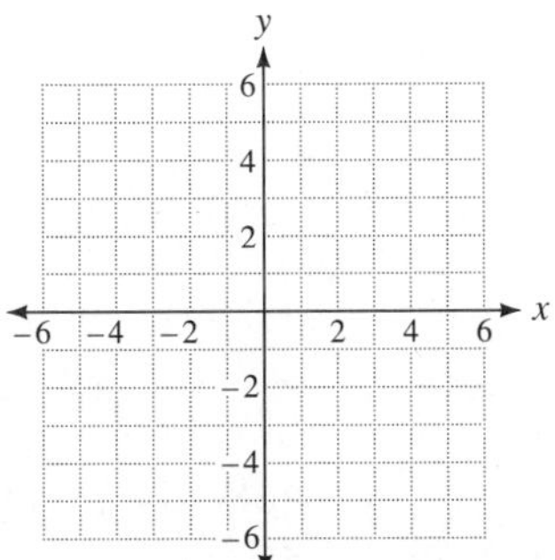

2. $x + y \leq 2$

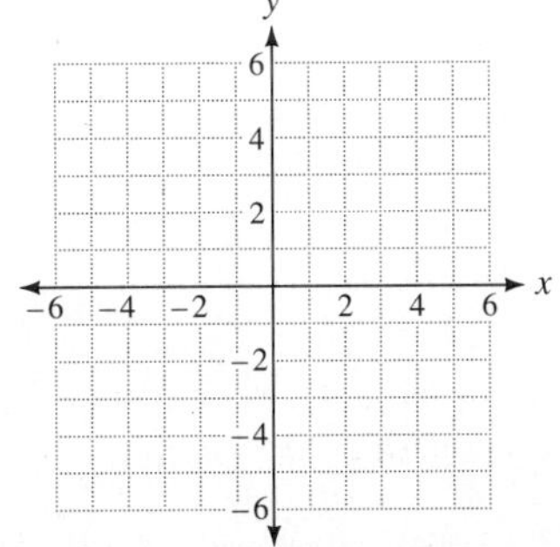

3. $x - y \geq -1$

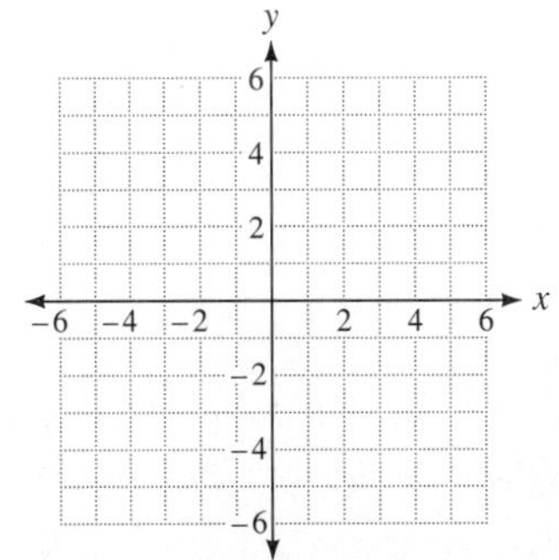

4. $x - y < -1$

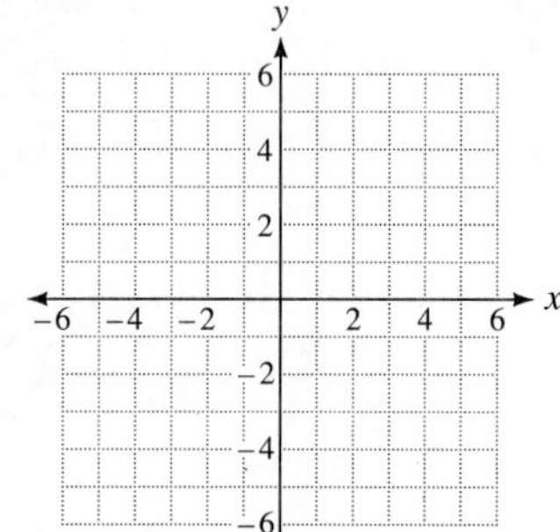

5. $3x + 4y < 12$

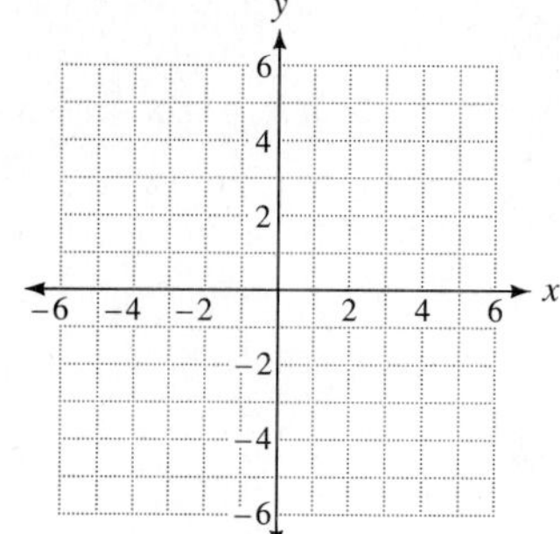

6. $3x + 4y \geq 12$

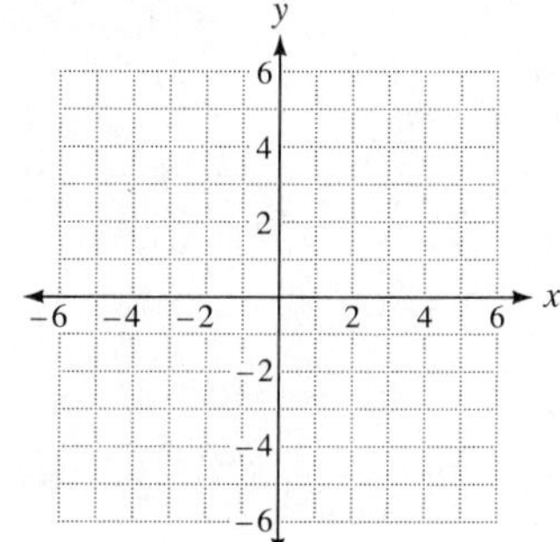

7. $4x + 8 > 0$

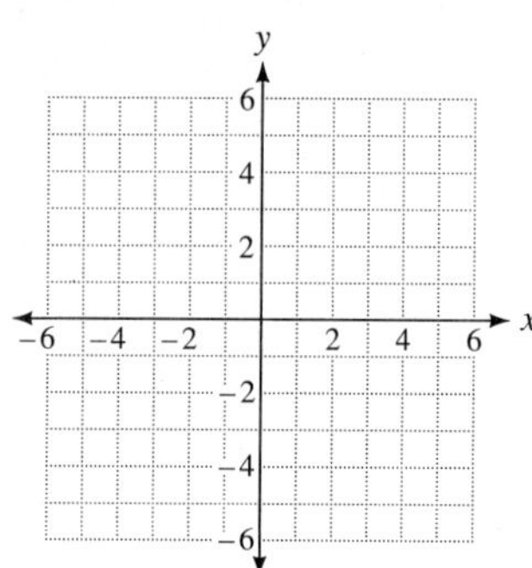

8. $4x + 8 \leq 0$

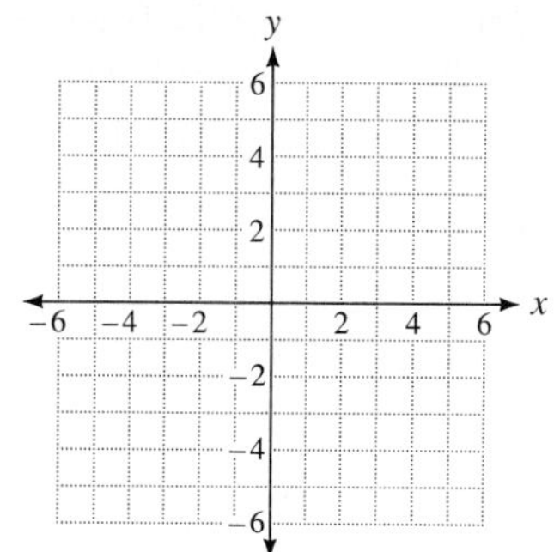

9. $2y - 6 \leq 0$

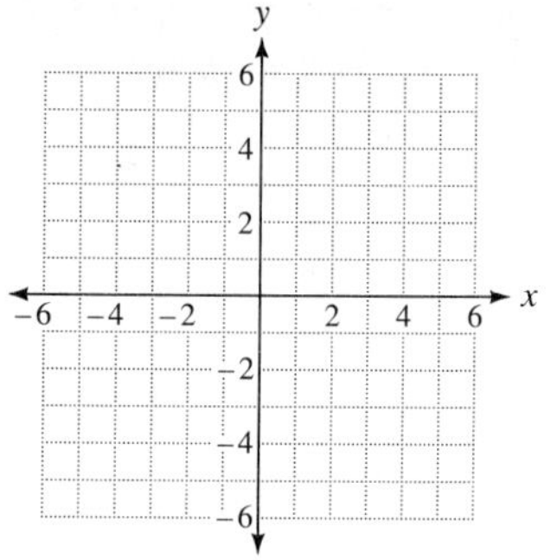

10. $2y - 6 > 0$

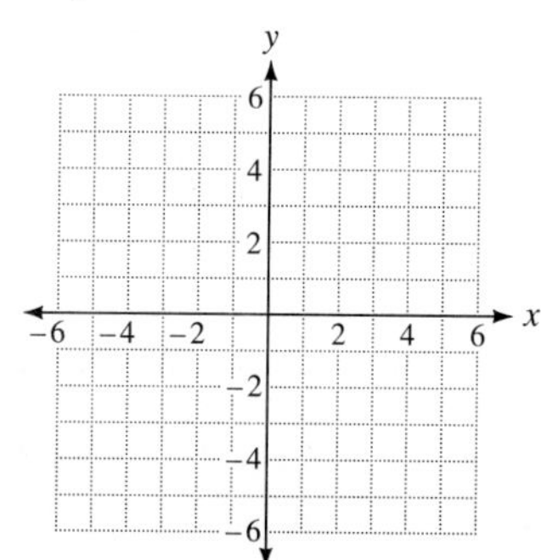

11. $2x + y < 0$

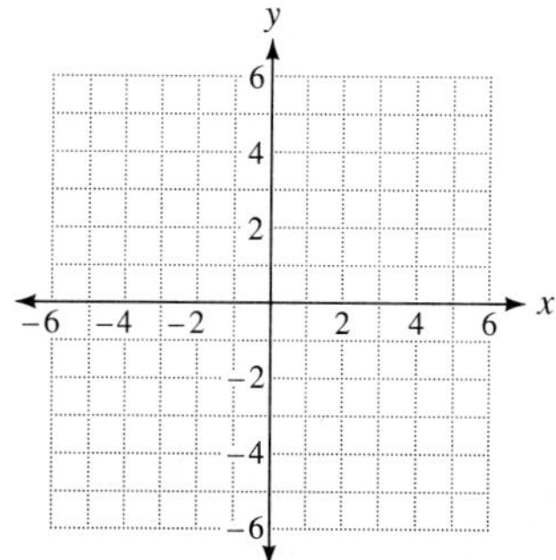

12. $2x + y \geq 0$

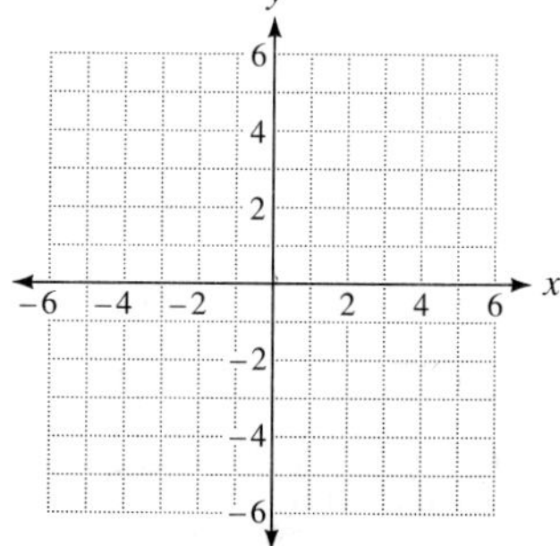

13. $3x + 2 > 5x$

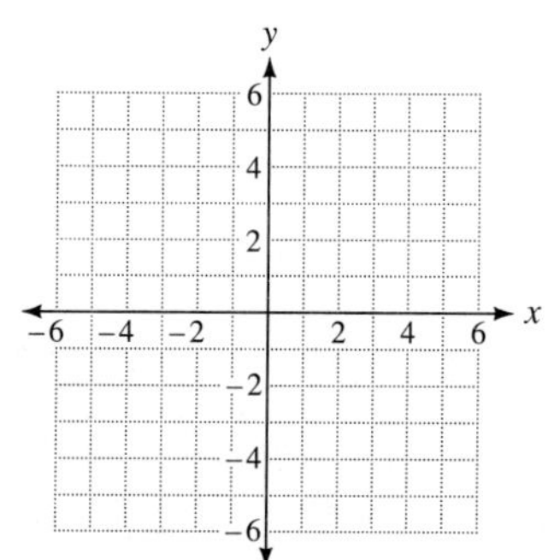

14. $3(x + y) \geq 2(x - y)$

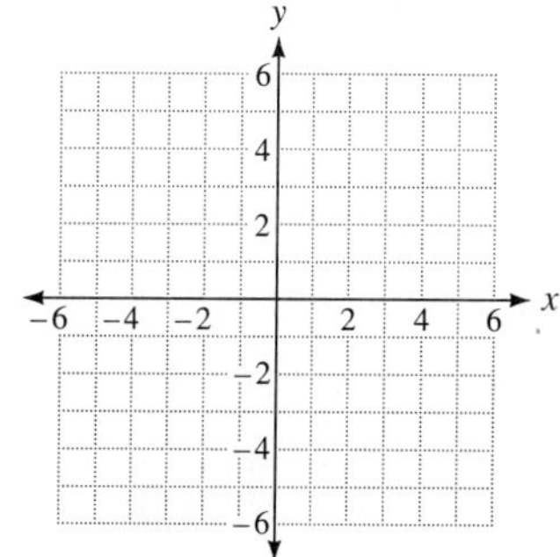

15. $5x + 6 < -3y - 9$

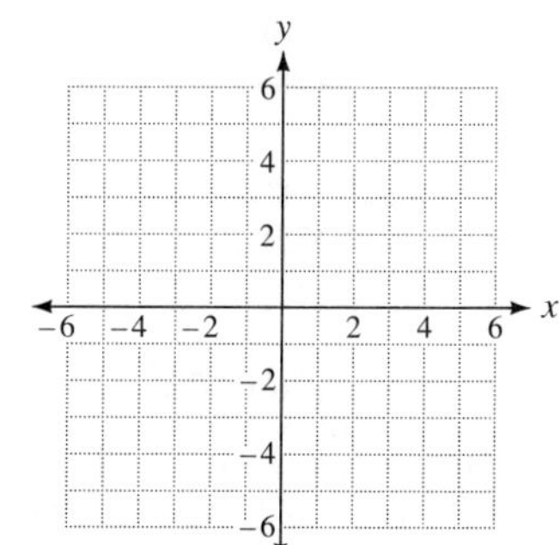

Graph each system of inequalities.

16. $3x - 2y \geq 6$
$x + 2y > -2$

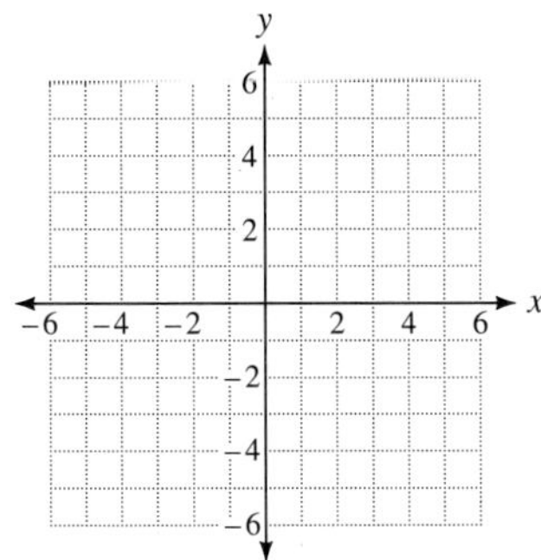

17. $3x - 2y < 6$
$x + 2y \geq -2$

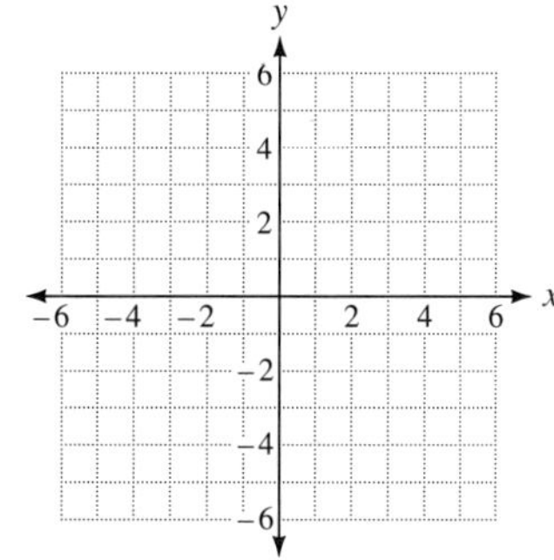

18. $3x - 2y \leq 6$
$x + 2y < -2$

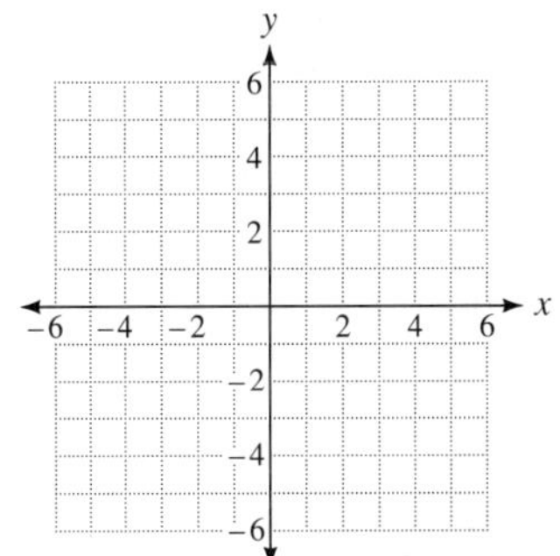

19. $x - y < 1$
$2x - 2y \geq -4$

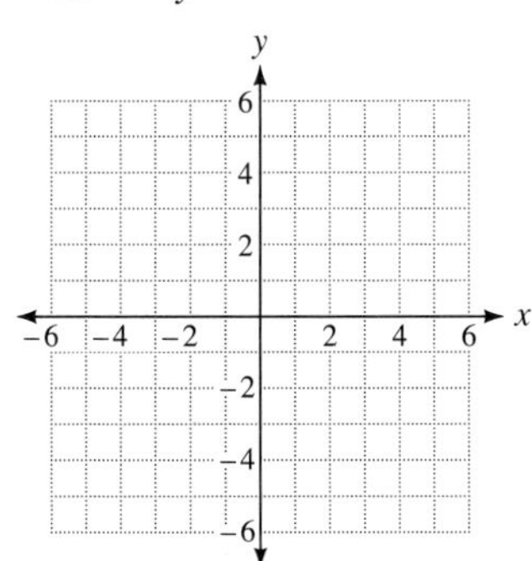

20. $x + 1 \leq 0$
$3y + 6 > 0$

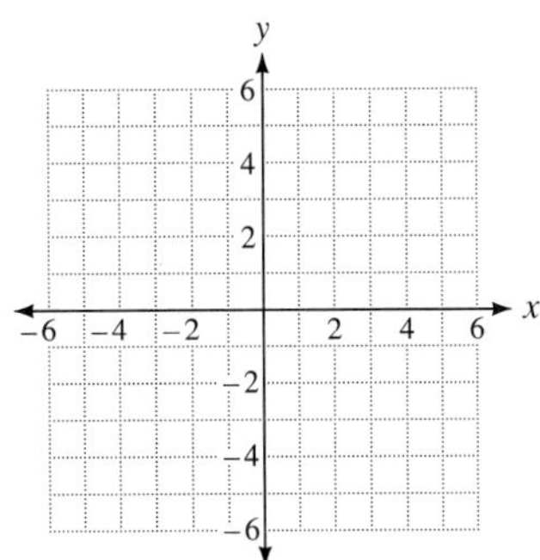

21. $3y - x \leq 3$
$x + 2 > 0$

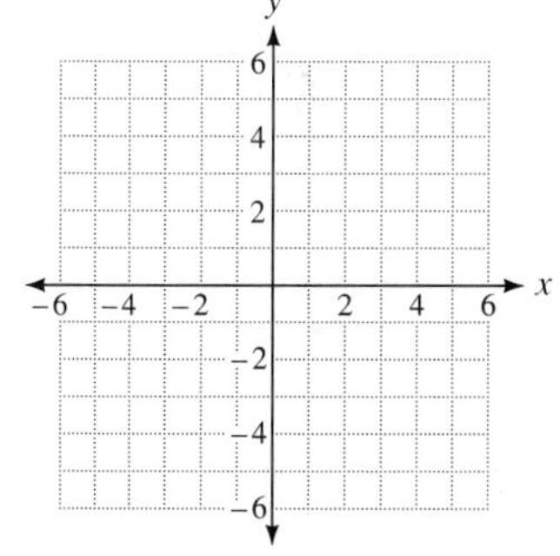

22. $y \geq x$
$y \geq -x$

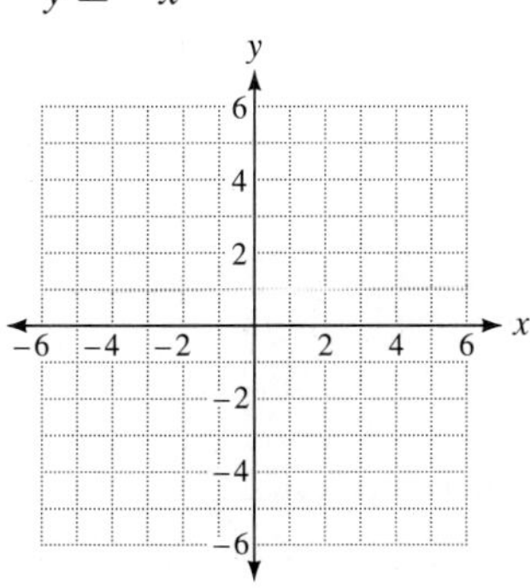

23. $x \geq 0$
$y \geq 0$

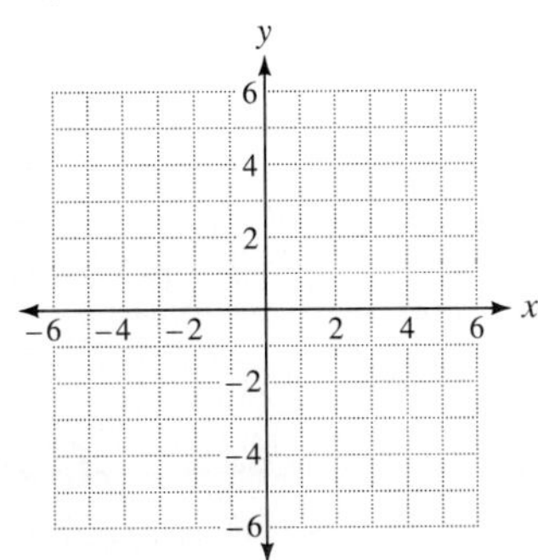

24. $x \geq 0$
$y \geq 0$
$x + y < 1$

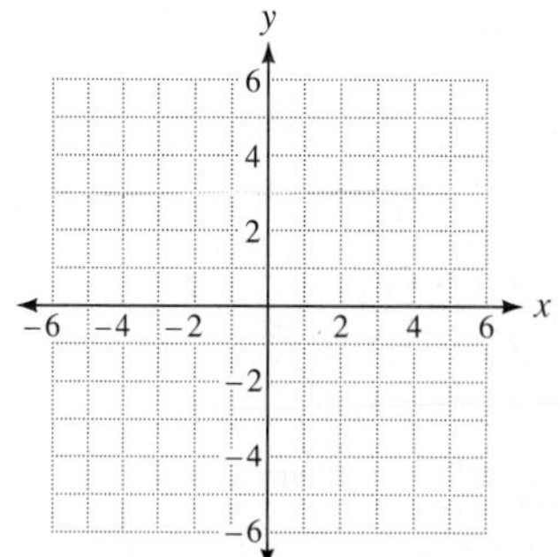

FOR REVIEW

Evaluate the following determinants.

25. $\begin{vmatrix} -3 & 4 \\ 2 & -8 \end{vmatrix}$

26. $\begin{vmatrix} 0 & 0 & 7 \\ 0 & 3 & 0 \\ -1 & 0 & 0 \end{vmatrix}$

Solve using Cramer's rule.

27. $4x + y = -3$
$-3x - 2y = -4$

28. $x + y - z = -6$
$4x + 3y + 2z = 6$
$-2x - 5y + 2z = 12$

ANSWERS:

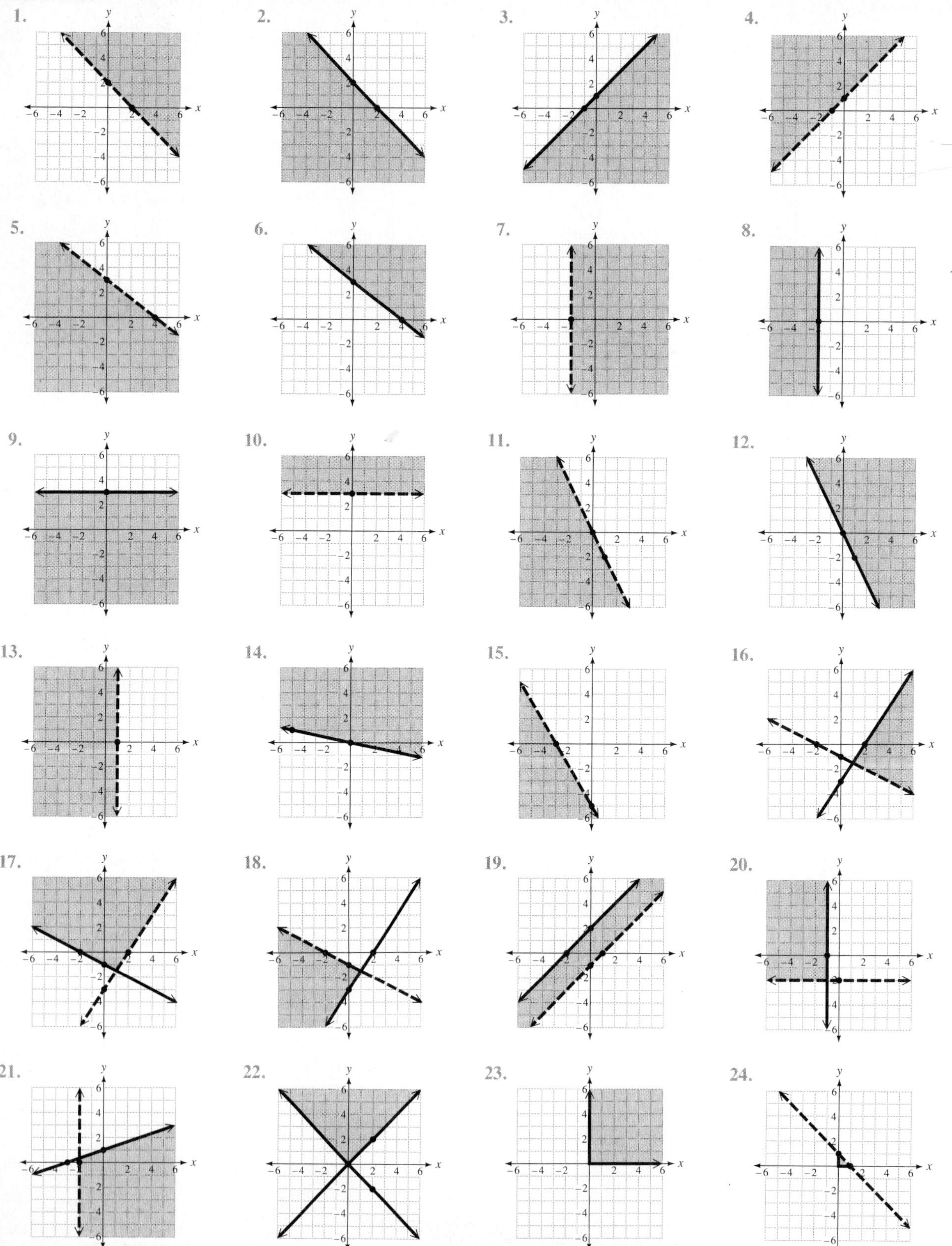

25. 16 26. 21 27. $(-2, 5)$ 28. $(-1, 0, 5)$

4.6 EXERCISES B

Graph the linear inequalities in two variables.

1. $x + y < 1$

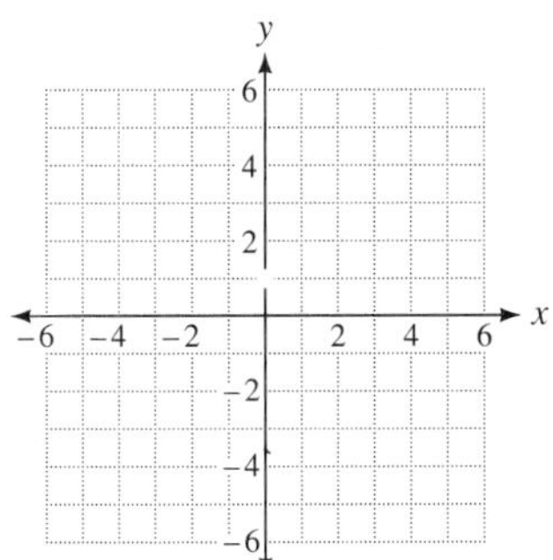

2. $x + y \geq 1$

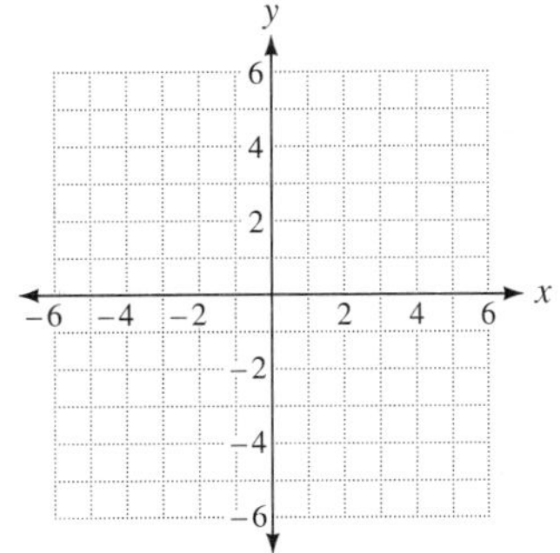

3. $x - y \geq -2$

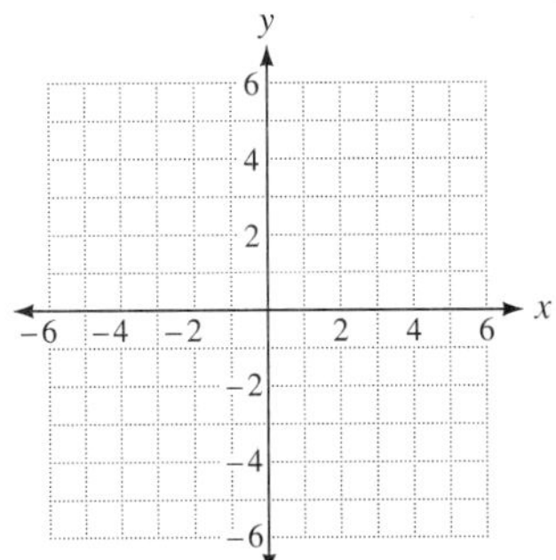

4. $x - y < -2$

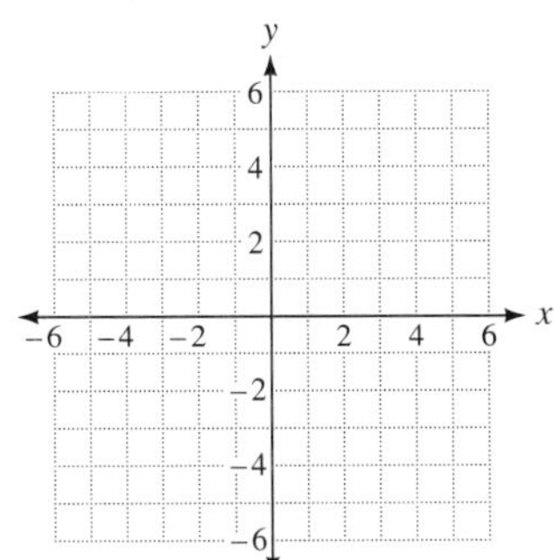

5. $4x - 3y > 12$

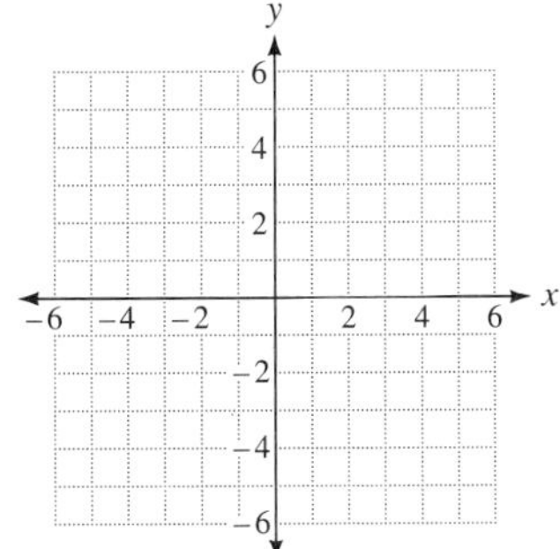

6. $4x - 3y \leq 12$

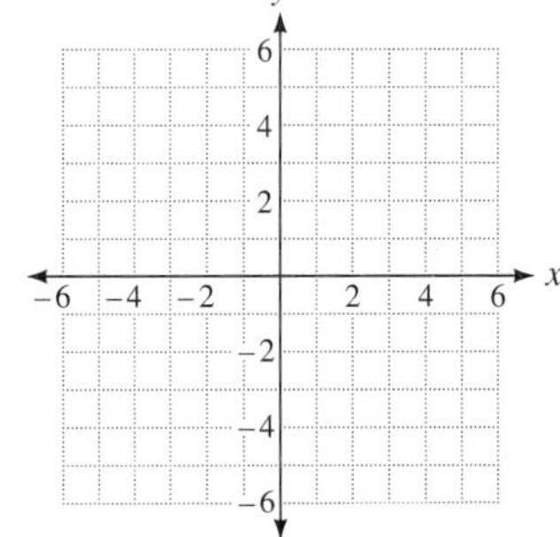

7. $3x - 9 \geq 0$

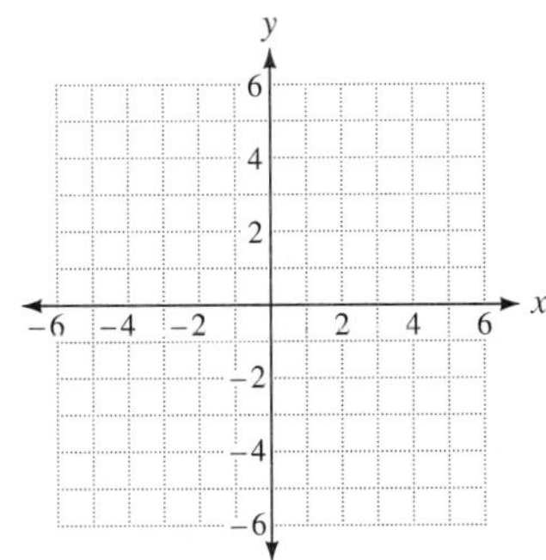

8. $3x - 9 < 0$

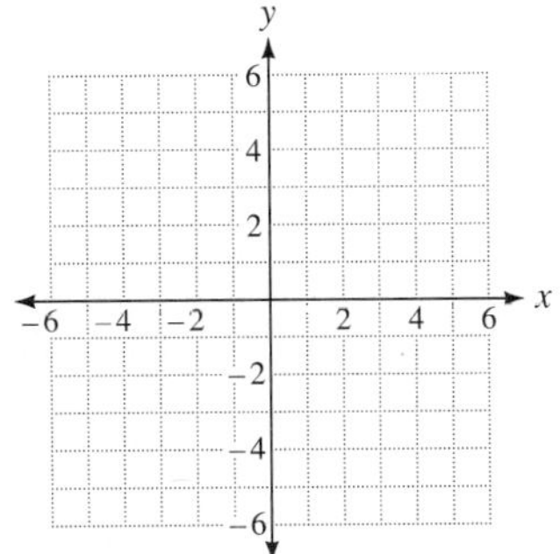

9. $2y - 8 < 0$

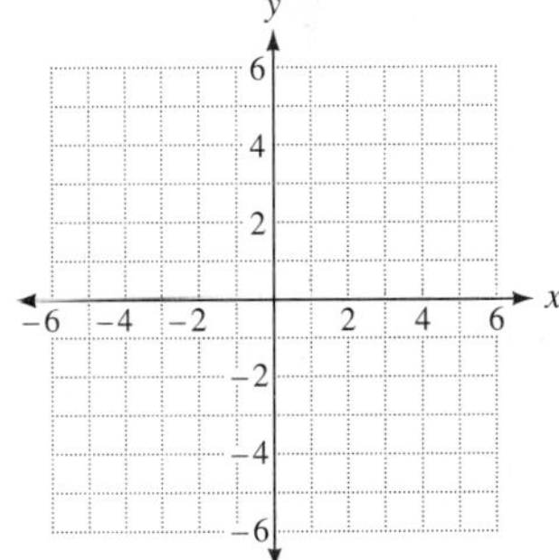

10. $2y - 8 \geq 0$

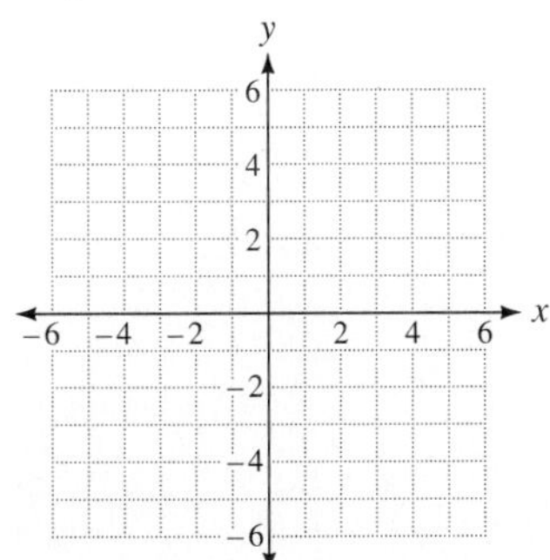

11. $4y + x \leq 0$

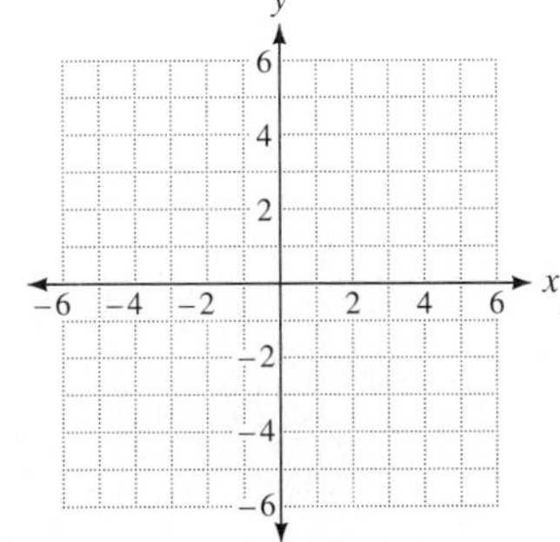

12. $4y + x > 0$

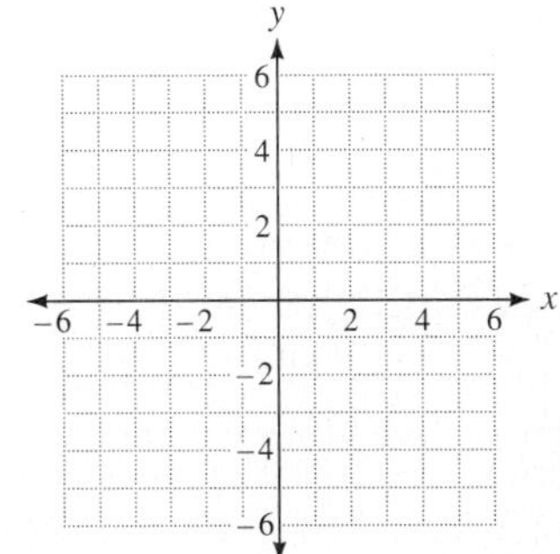

13. $3x + 2 \le 5x$

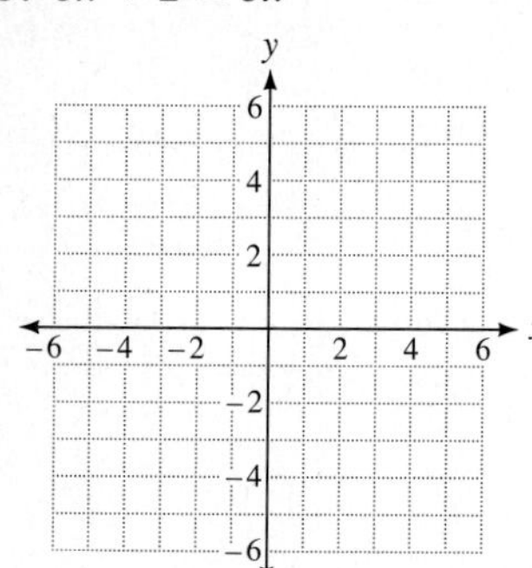

14. $3(x + y) < 2(x - y)$

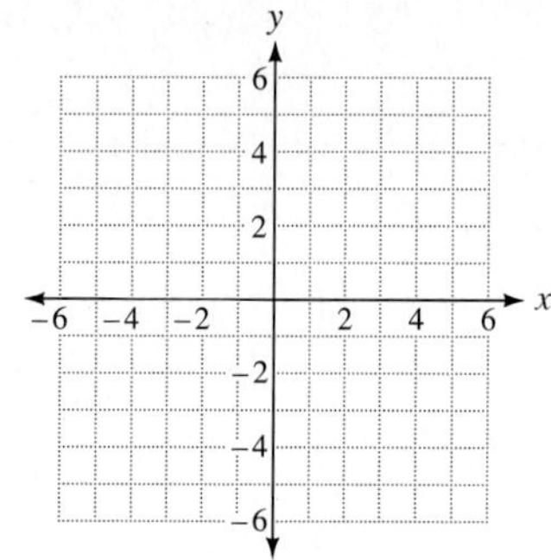

15. $5x + 6 \ge -3y - 9$

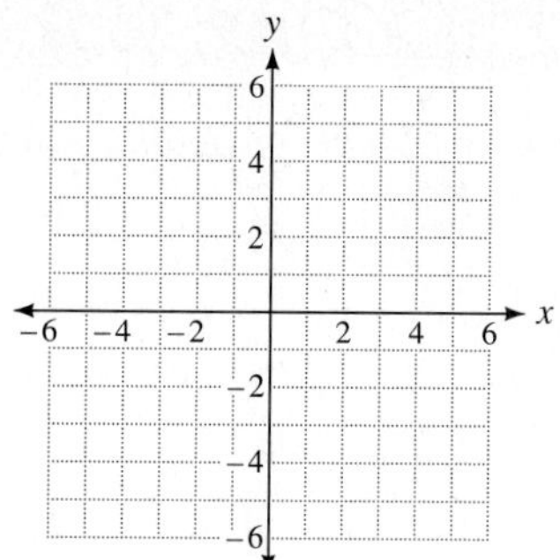

Graph each system of inequalities.

16. $2x - 5y \ge 10$
$\frac{1}{2}x + y > -1$

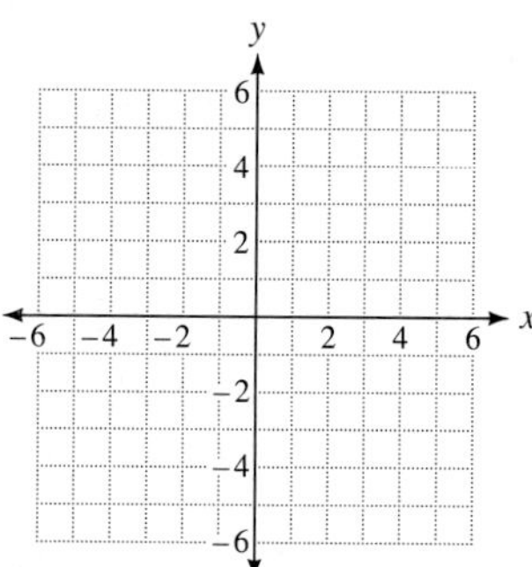

17. $2x - 5y < 10$
$\frac{1}{2}x + y \ge -1$

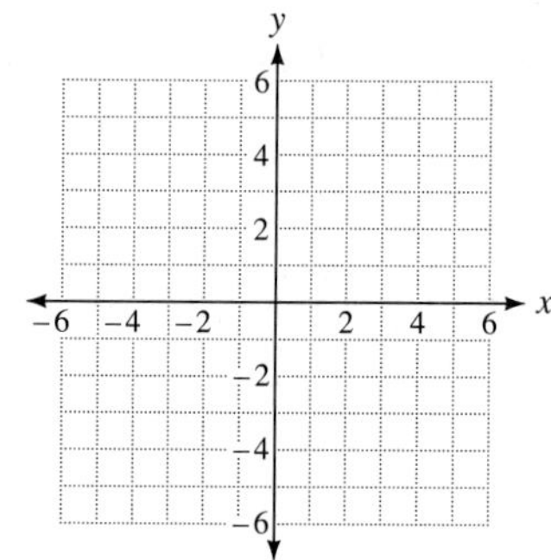

18. $2x - 5y \le 10$
$\frac{1}{2}x + y < -1$

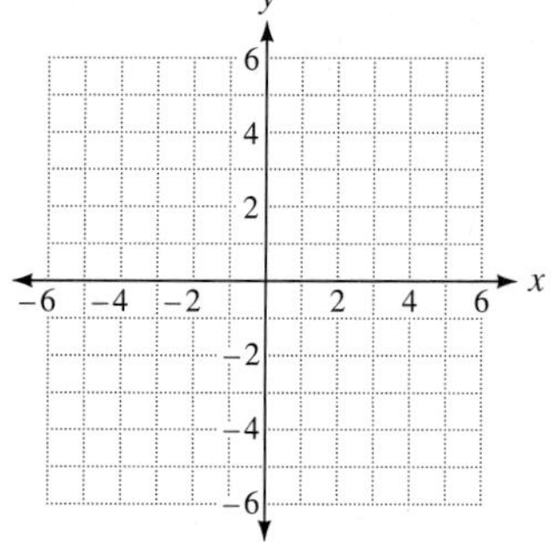

19. $x + y \ge -3$
$3x + 3y < 6$

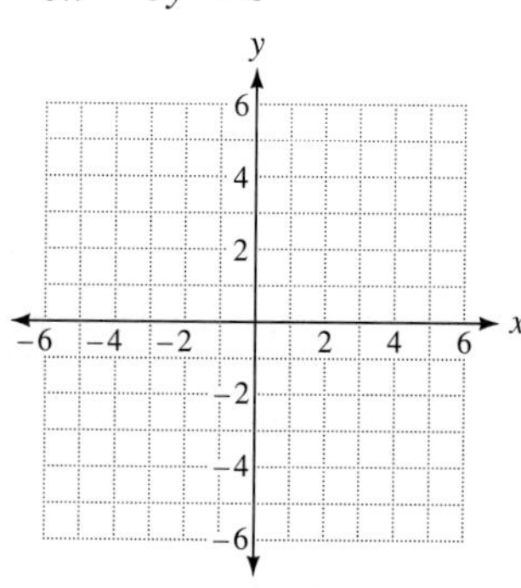

20. $2x + 4 > 0$
$y - 1 \le 0$

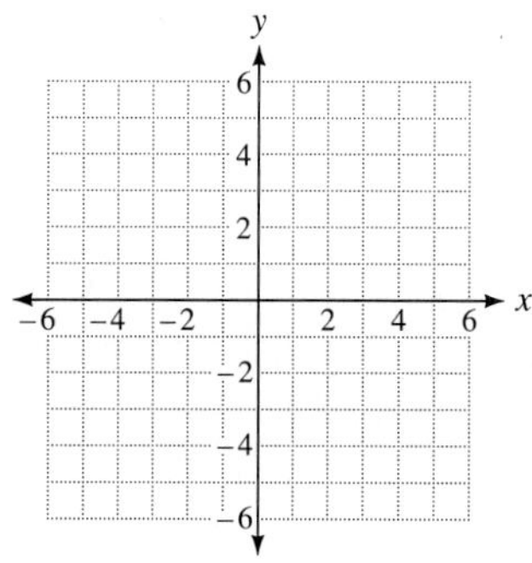

21. $4y - x \ge 4$
$y - 2 < 0$

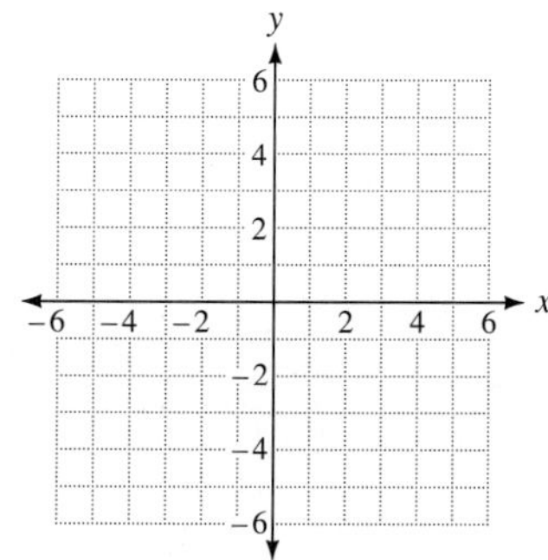

22. $2x + y \le 0$
$x - 2y \ge 0$

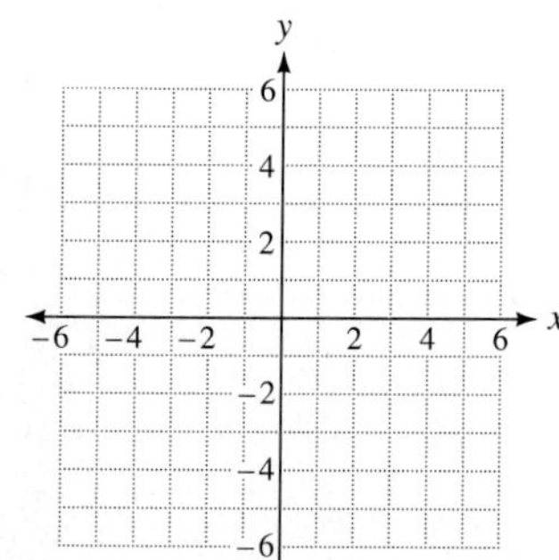

23. $x \le 0$
$y \ge 0$

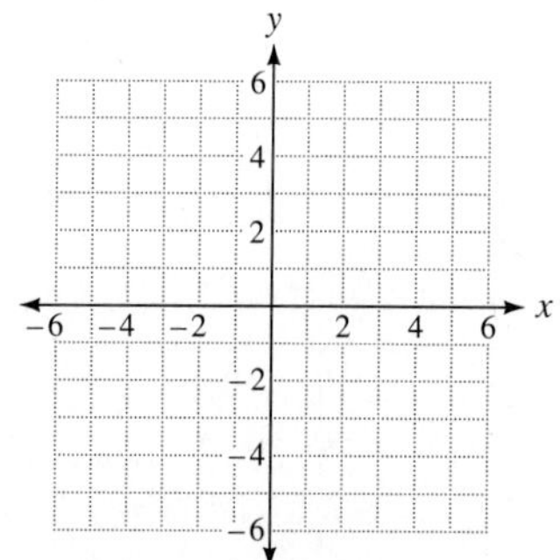

24. $x \le 0$
$y \ge 0$
$y - x \le 3$

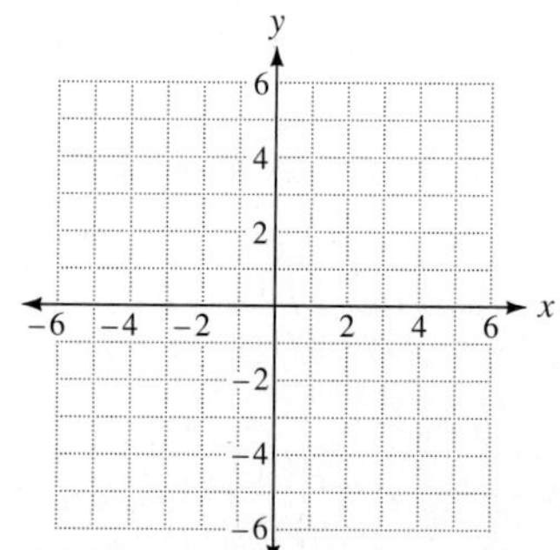

FOR REVIEW

Evaluate the following determinants.

25. $\begin{vmatrix} -4 & 5 \\ 2 & -3 \end{vmatrix}$

26. $\begin{vmatrix} 3 & 0 & 0 \\ 0 & -5 & 0 \\ 0 & 0 & 2 \end{vmatrix}$

Solve using Cramer's rule.

27. $2x - 9y = 15$
$4x + 7y = 5$

28. $2x + 2y - 3z = 1$
$5x + 5y + 4z = -9$
$3x \qquad + 2z = 7$

4.6 EXERCISES C

1. Evaluate the determinant and graph the resulting inequality.

$\begin{vmatrix} x & y \\ 5 & 1 \end{vmatrix} < 5$

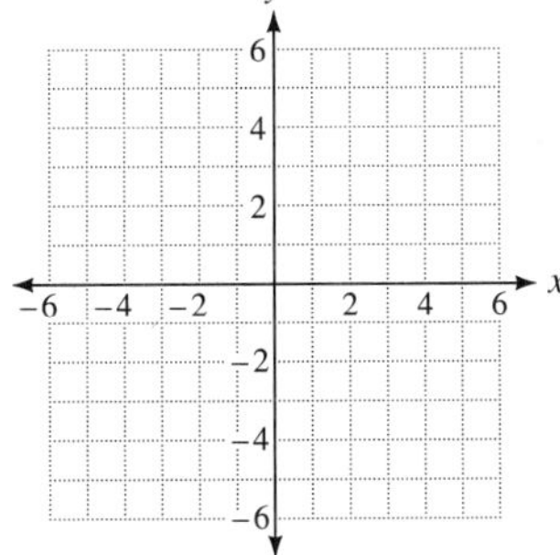

2. Graph the relation $\{(x, y) | x \geq 0,\ y \geq 0, \text{ and } 3x + 2y \leq 6\}$.

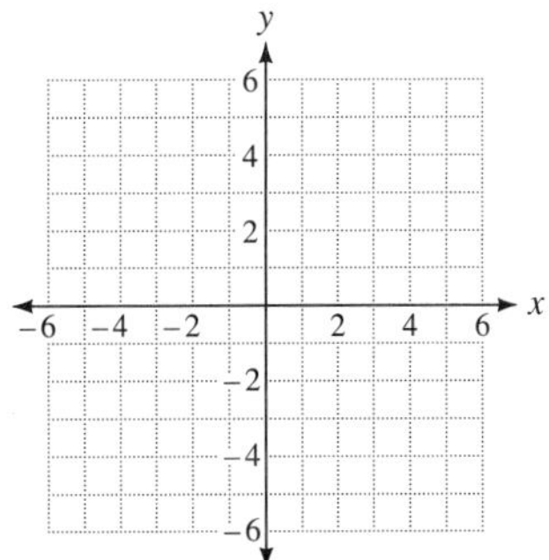

Graph each system of inequalities.

3. $x - y > 1$
$2x - 2y \leq -4$

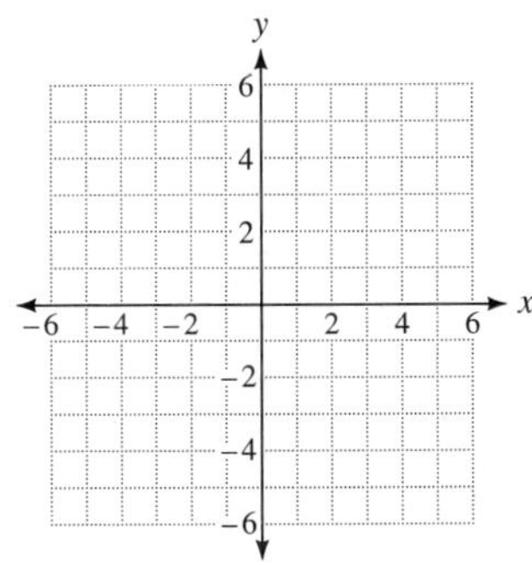

4. $0 \leq x \leq 3$
$y \geq 0$
$2y - x \leq 4$

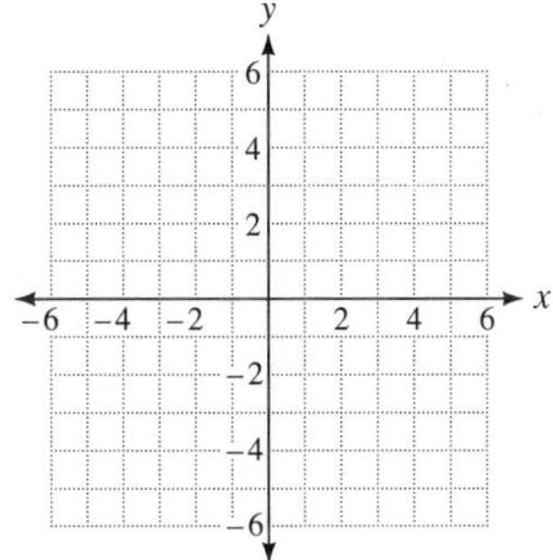

5. $x - 2 > 0$
$2 - x > 0$

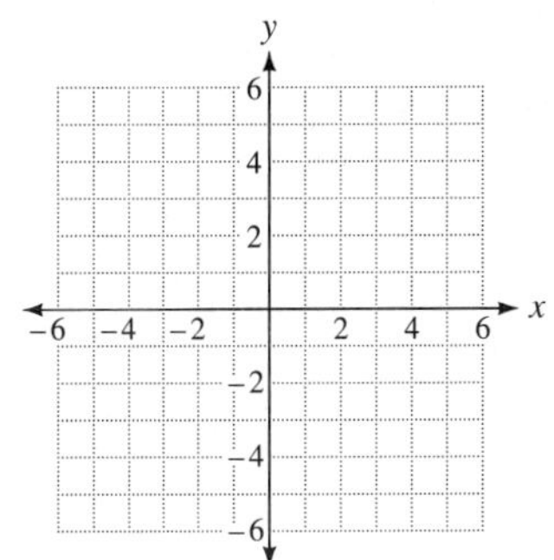

6. $y + 3 > 0$
$y + 4 \leq 0$

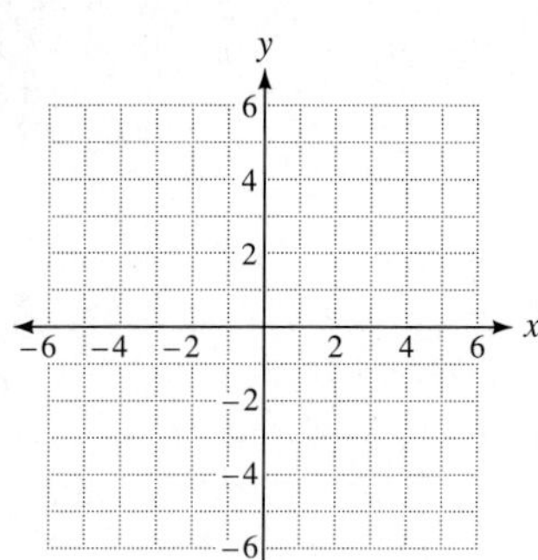

7. $x \leq 0$
$y \geq 0$
$2x - 5y \geq -10$
$x - y \geq -3$

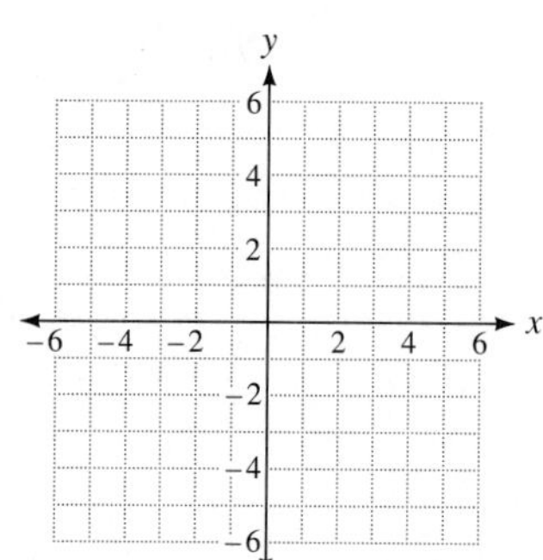

8. $x \leq 0$
$y \geq 0$
$2x - y \geq -6$
$x + 2y \leq 2$

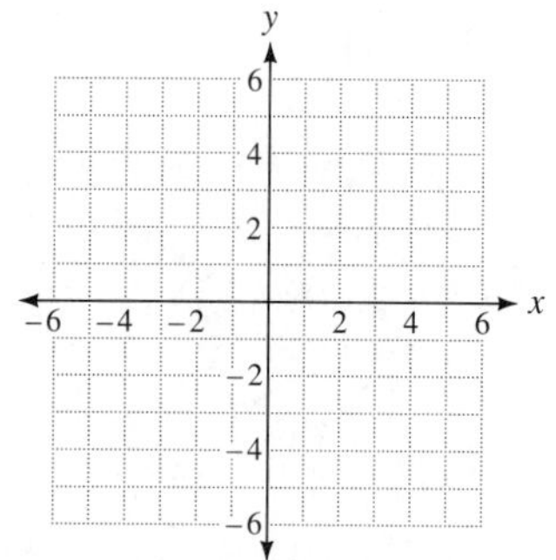

9. Recreation Furniture manufactures two types of picnic tables, one that seats four and another that seats six. The table below shows the production information. The number made weekly of each type of table should not exceed the workhours available in each manufacturing stage. Form the system of inequalities described by this information and graph the system.

	Workhours for four-seat table	Workhours for six-seat table	Maximum workhours per week
Construction	6	8	120
Finishing	1	3	30

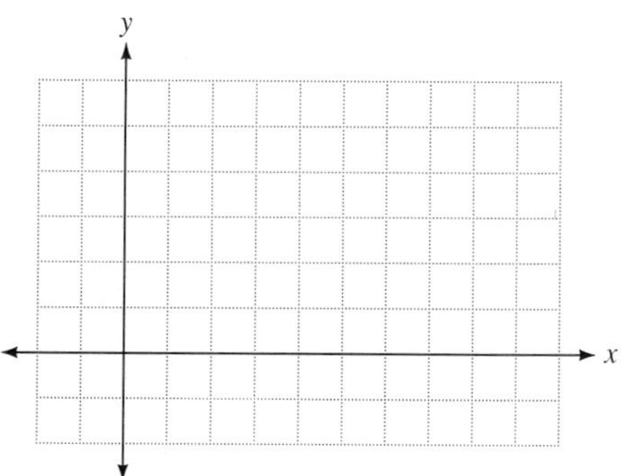

10. Repeat Exercise 9 using the table below.

	Workhours for four-seat table	Workhours for six-seat table	Maximum workhours per week
Construction	3	6	96
Finishing	1	4	36

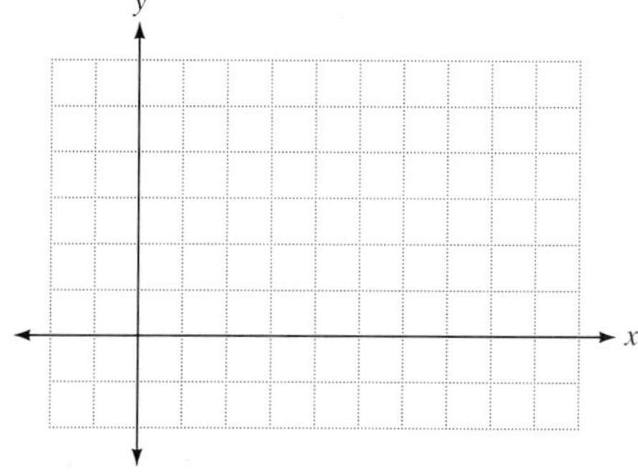

CHAPTER 4 REVIEW

KEY WORDS

4.1 A **system of equations** is a pair of linear equations in two variables.

A **solution** to a system of two linear equations in two variables is an ordered pair that solves each equation.

A system of equations is **dependent** if it has infinitely many solutions (the lines coincide).

A system of equations is **inconsistent** if it has no solution (the lines are parallel).

A system of equations is **independent** and **consistent** if it has exactly one solution (the lines intersect).

4.3 Two angles are **complementary** if their sum is 90°.

Two angles are **supplementary** if their sum is 180°.

4.4 An equation of the form $ax + by + cz = d$, where a, b, c, and d are real numbers, is a **linear equation in three variables.**

A **solution** to a system of three linear equations in three variables is an ordered triple of numbers that solves each equation in the system.

4.5 A **determinant** is a square array of numbers written within vertical bars whose value is a real number.

4.6 A **linear inequality in two variables** is an inequality that can be written in one of the forms $ax + by > c$, $ax + by < c$, $ax + by \geq c$, or $ax + by \leq c$.

A **system of inequalities** is two or more linear inequalities in two variables.

KEY CONCEPTS

4.1

1. A system of equations can be solved by the graphing method, but this method has limitations since it depends on the accuracy of the graphs.
2. By writing the equations in a system in slope-intercept form, it is easy to determine the relationship of the graphs and hence the number of solutions to the system.

4.2

1. To solve a system of two equations, use either the substitution method or the elimination method.
2. If an identity (such as $0 = 0$) results when solving a system of two equations, the system has infinitely many solutions.
3. If a contradiction (such as $5 = 0$) results when solving a system of two equations, the system has no solution.

4.3 To solve an applied problem that results in a system, make sure you define the variables accurately when you translate the words to equations. Sometimes a summary table can be useful.

4.4 When solving a system of three equations in three variables, if a contradiction results at any step, the system is inconsistent and has no solution. If an identity results after using only two equations, the system has either no solution *or* infinitely many solutions (is dependent); the third equation must be used to determine which case occurs.

4.5

1. Second-order determinant:

$$\begin{vmatrix} a_1 & b_1 \\ a_2 & b_2 \end{vmatrix} = a_1b_2 - a_2b_1$$

2. Third-order determinant:

$$\begin{vmatrix} a_1 & b_1 & c_1 \\ a_2 & b_2 & c_2 \\ a_3 & b_3 & c_3 \end{vmatrix} = a_1\begin{vmatrix} b_2 & c_2 \\ b_3 & c_3 \end{vmatrix} - a_2\begin{vmatrix} b_1 & c_1 \\ b_3 & c_3 \end{vmatrix} + a_3\begin{vmatrix} b_1 & c_1 \\ b_2 & c_2 \end{vmatrix}$$

3. Cramer's rule:

$$x = \frac{A_x}{A}, \quad y = \frac{A_y}{A}, \quad z = \frac{A_z}{A}$$

4.6

1. To graph a linear inequality in two variables, graph the equation using a solid line if the inequality is either $\leq$ or $\geq$ and a dashed line if it is either $<$ or $>$. Use a test point to determine which region should be shaded.
2. To graph a system of linear inequalities, graph each inequality in the system first. The graph of the system consists of all points in the region common to the graphs of the individual inequalities.

REVIEW EXERCISES

Part I

4.1 *Tell whether (1, −2) is a solution to the system.*

1. $2x + 3y + 4 = 0$
$-3x + y - 5 = 0$

2. $x - 2y = 5$
$4x + 2y = 0$

Without solving, give the number of solutions to the given system.

3. $2x - y = 7$
$x - 2y = 7$

4. $3x + 4y - 12 = 0$
$-6x - 8y + 12 = 0$

5. $2x + 2y = 2$
$5x + 5y = 5$

Solve using the graphing method.

6. $x + y = 3$
$2x - y = 3$

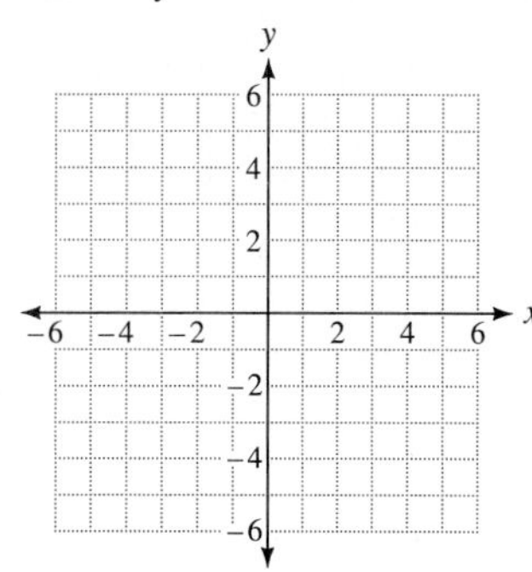

4.2 *Solve using either the substitution method or the elimination method, whichever seems appropriate.*

7. $3x - y = 2$
$-6x + 2y = 2$

8. $5x + 3y = -1$
$2x - 4y = -16$

9. $x + 7y = -1$
$-2x - 14y = 2$

4.3 *Solve using a system of equations.*

10. The sum of two numbers is 13 and their difference is 3. Find the numbers.

11. Janet McShane bought 2 shirts of the same value and 5 pairs of pants of the same value for \$94. Later she returned and bought 3 shirts and 1 pair of pants for \$50. How much does one shirt cost and how much does one pair of pants cost?

12. Chuck had \$5000 divided into two investments. The part invested in bonds at 8% together with the other part invested in certificates at 7% earned \$370 one year. How much was invested in each category?

13. An airplane flies 900 miles in 2 hours with the wind and 1050 miles in 3 hours against the wind. Find the speed of the plane in still air and the speed of the wind.

4.4 *Solve.*

14. $x + 2y = 3$
$y - 2z = -5$
$2x + z = 5$

15. $4x - 6y + 10z = -8$
$-2x + 3y - 5z = -4$
$7x - 6y + 3z = -5$

16. Determine whether $(-1, 2, 3)$ is a solution to the system.

$$\begin{aligned} x + y + z &= 4 \\ 3x - y + 2z &= 1 \\ -2x - y + z &= 3 \end{aligned}$$

Solve using a system of equations.

17. The average of a student's three scores is 80. The sum of the first and second scores is 150, and the third score is 5 more than the second. Find all three scores.

18. Joey has a collection of 75 coins consisting of pennies, nickels, and dimes with a value of \$4.78. If there are 5 fewer dimes than pennies and nickels combined, how many of each coin are there?

19. Find three numbers such that the sum of the first and second is twice the third, the first plus the third is 2, and the second minus the third is -4.

20. The largest angle of a triangle is twice the smallest, and the remaining angle is 20° less than the largest. Find the measure of each angle.

4.5 *Evaluate the determinants.*

21. $\begin{vmatrix} 3 & 7 \\ 12 & -2 \end{vmatrix}$

22. $\begin{vmatrix} 2 & -5 & 4 \\ 3 & -6 & -1 \\ -4 & 10 & -8 \end{vmatrix}$

Solve using Cramer's rule.

23. $\begin{aligned} 2x + 3y &= 6 \\ -x - 4y &= -3 \end{aligned}$

24. $\begin{aligned} x + 5y + 4z &= 7 \\ -2x - 8y + 7z &= -14 \\ 4y + 5z &= 0 \end{aligned}$

4.6 *Graph in a rectangular coordinate system.*

25. $x - 2y \le -2$

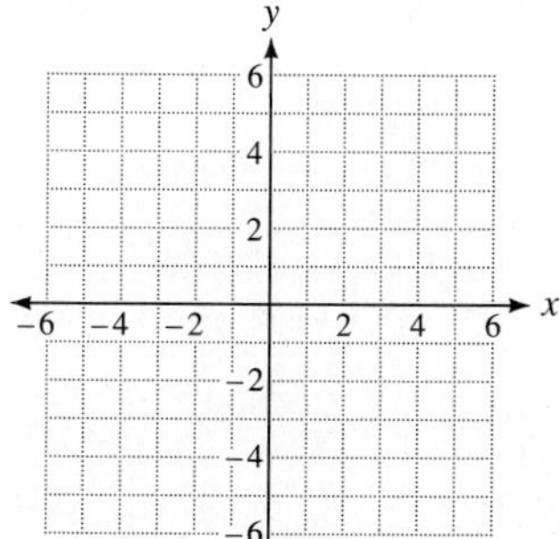

26. $2y + 4 > 0$

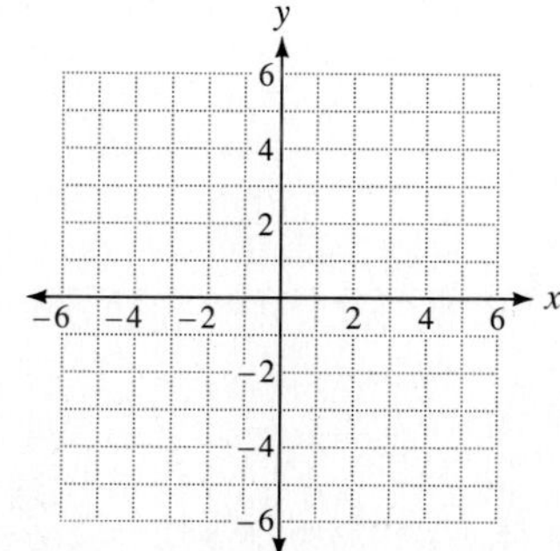

27. $3 - 3x \ge 0$

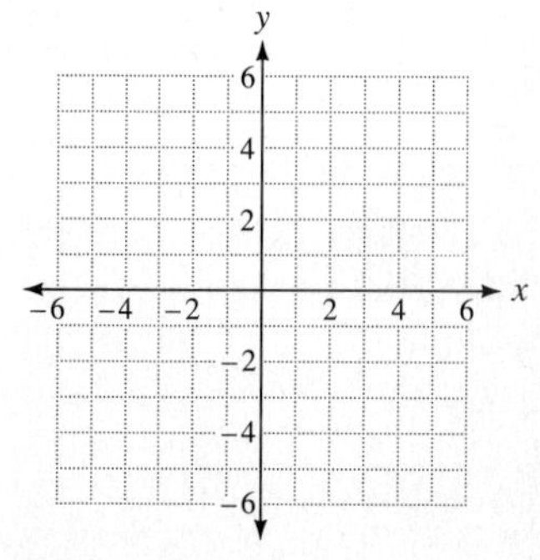

28. $3x + 2y < 6$

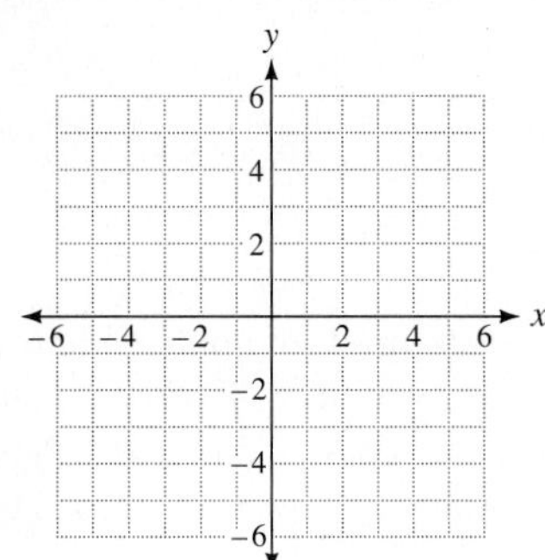

29. $3x + 2y < 6$
$2y + 4 > 0$

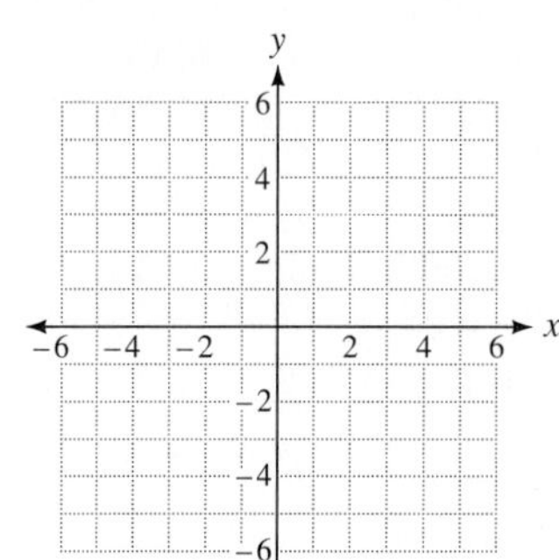

30. $y \le x + 1$
$y \ge x - 1$

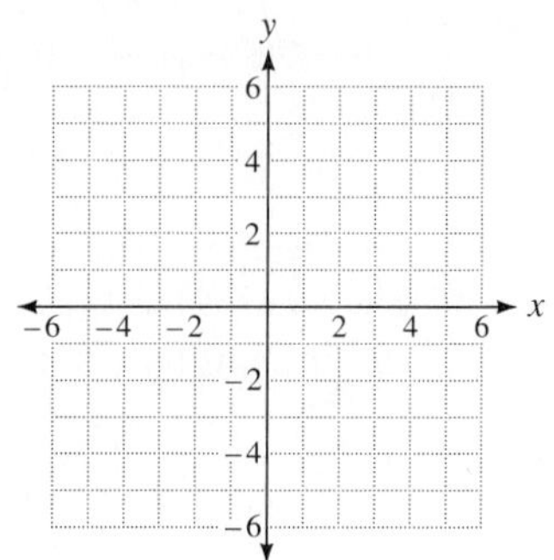

Part II

Solve.

31. $3x + y = -2$
$3x - y = -10$

32. $4x - 2y = 7$
$-2x + y = -1$

33. Dr. Gordon Johnson has one solution that is 15% salt and another that is 20% salt. How many gallons of each should he mix together to have 50 gallons of a solution that is 18% salt?

34. Gail has a collection of 80 coins made up of nickels and dimes. If the value of the collection is \$6.75, how many of each are there?

35. $x + y + z = 4$
$2x - y + z = -3$
$-x + 2y - 3z = 1$

36. $x - 3y + 2z = 6$
$4x - 2y + 3z = 14$
$2x + 4y - z = 2$

37. The sum of the ages of Mark, Larry, and Jane is 53. Jane is 5 years younger than Larry, and in two years, Mark will be the same age as Larry is now. How old is each?

38. $3x - 2y = 5$
$-3x + 2y = -5$

Graph in a rectangular coordinate system.

39. $2x + 3y \le -12$

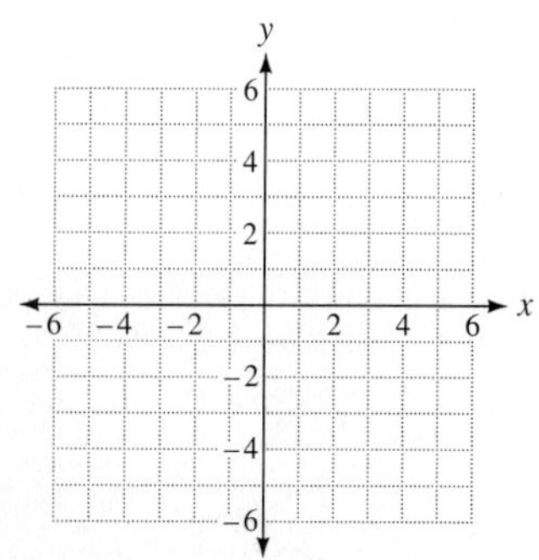

40. $x - y < 5$
$2x + y \ge 4$

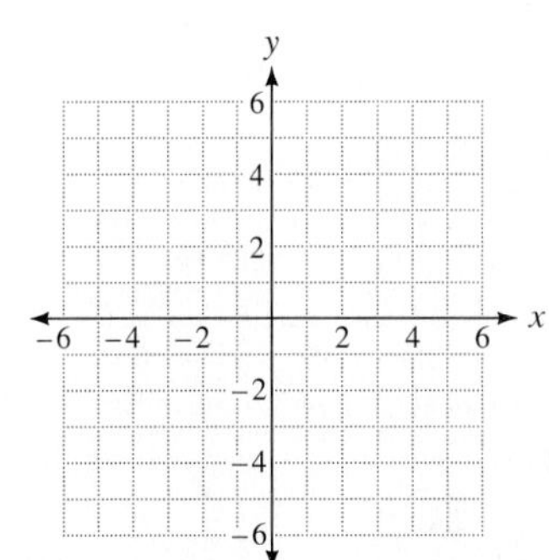

41. $x \ge 0$
$y \ge 0$
$x + y \le 4$

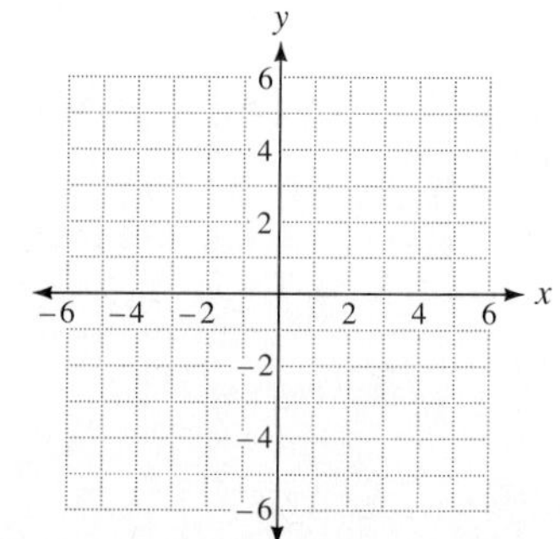

42. Without solving, tell the number of solutions to:

$$\begin{aligned} 2x + y &= 5 \\ 2x - y &= -5. \end{aligned}$$

43. Is $(3, -4)$ a solution to the following system?

$$\begin{aligned} 5x + y &= 11 \\ -3x + 2y &= -1 \end{aligned}$$

44. Solve using Cramer's rule.

$$\begin{aligned} 2x - 3y &= -5 \\ 6x + y &= 35 \end{aligned}$$

45. Evaluate the determinant.

$$\begin{vmatrix} 3 & -1 & 2 \\ 4 & 0 & 3 \\ -2 & -2 & 1 \end{vmatrix}$$

ANSWERS: **1.** no **2.** yes **3.** exactly one (lines intersect) **4.** no solution (lines are parallel) **5.** infinitely many (lines coincide) **6.** (2, 1) **7.** no solution **8.** $(-2, 3)$ **9.** infinitely many solutions of the form $\left(x, -\frac{1}{7}x - \frac{1}{7}\right)$ for x any real number **10.** 8, 5 **11.** shirt is \$12, pants are \$14 **12.** \$2000 in bonds, \$3000 in certificates **13.** plane: 400 mph; wind: 50 mph **14.** (1, 1, 3) **15.** no solution **16.** yes **17.** 65, 85, 90 **18.** 18 pennies, 22 nickels, 35 dimes **19.** 3, -5, -1 **20.** 40°, 60°, 80° **21.** -90 **22.** 0 **23.** (3, 0) **24.** (7, 0, 0)

25.

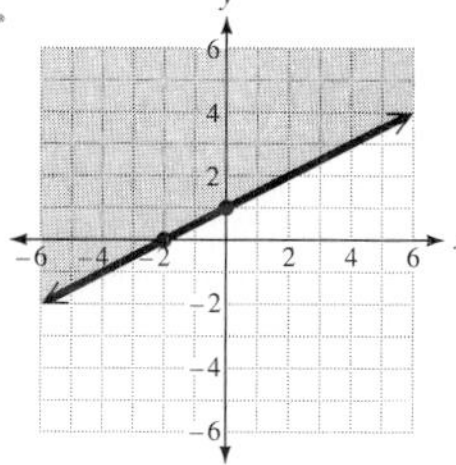

26.

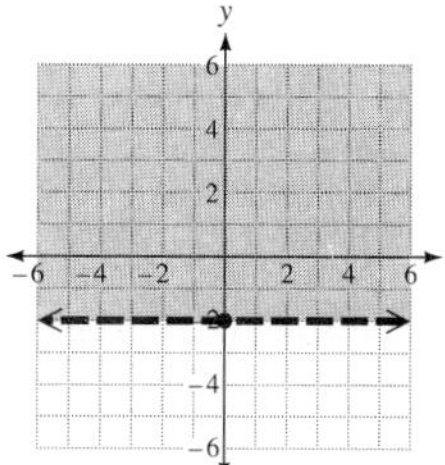

27.

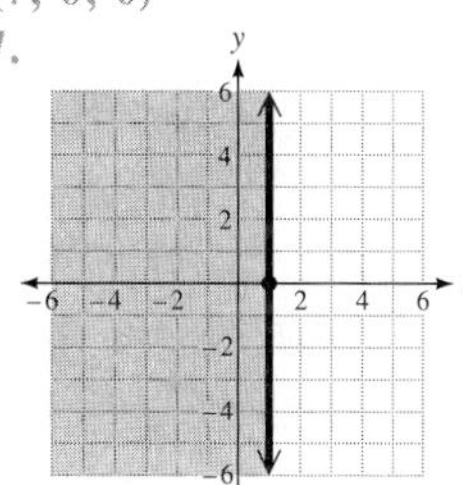

28.

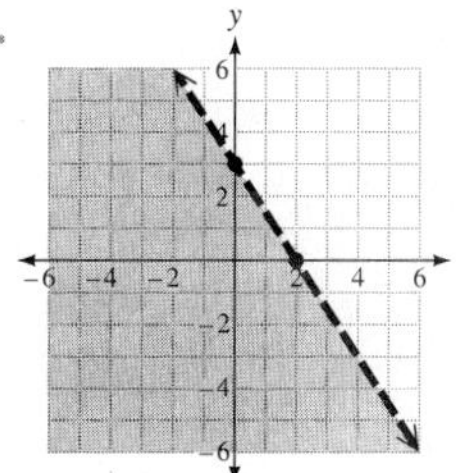

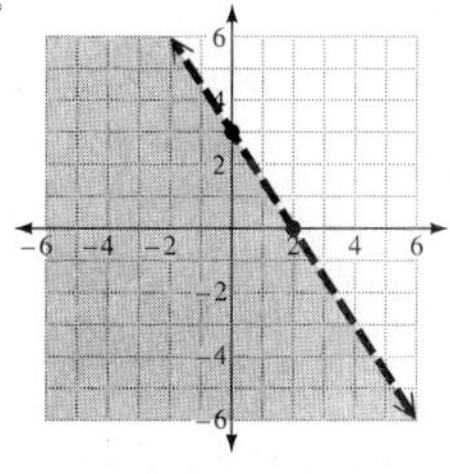

29.

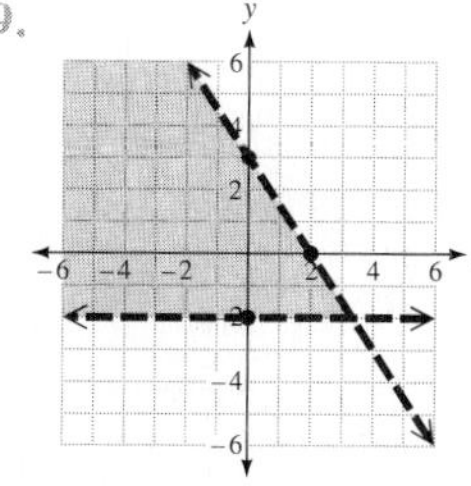

30.

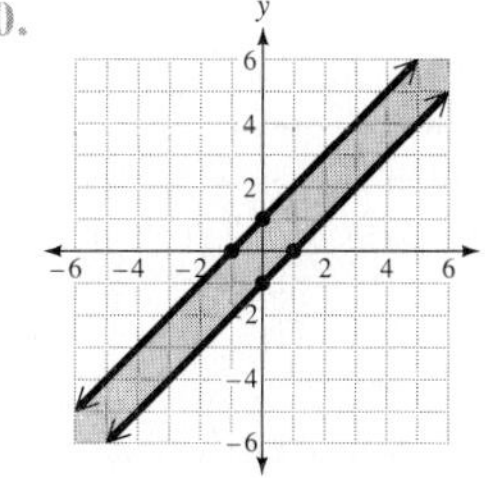

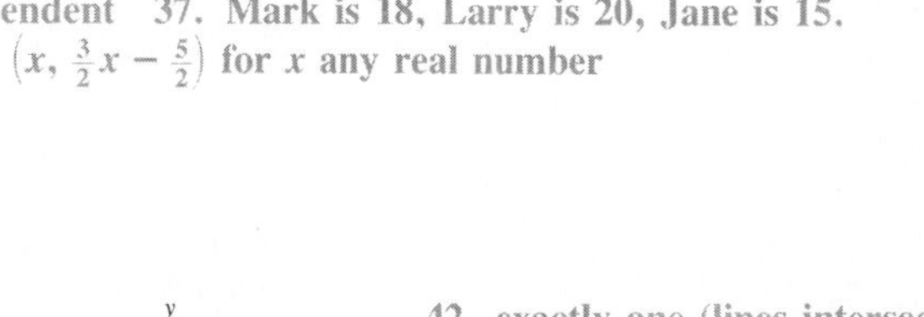

31. $(-2, 4)$ **32.** no solution **33.** 20 gal of 15% solution, 30 gal of 20% solution **34.** 55 dimes, 25 nickels **35.** $(-1, 3, 2)$ **36.** infinitely many solutions; the system is dependent **37.** Mark is 18, Larry is 20, Jane is 15. **38.** $\left(x, \frac{3}{2}x - \frac{5}{2}\right)$ for x any real number

39.

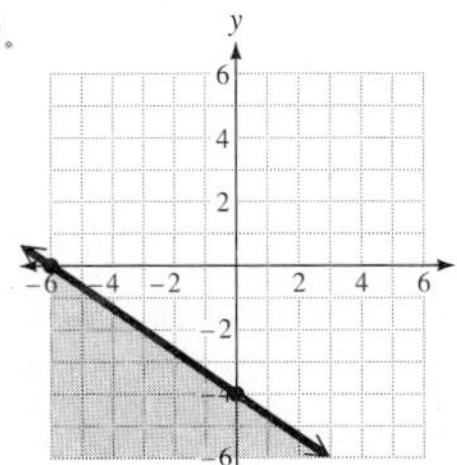

40.

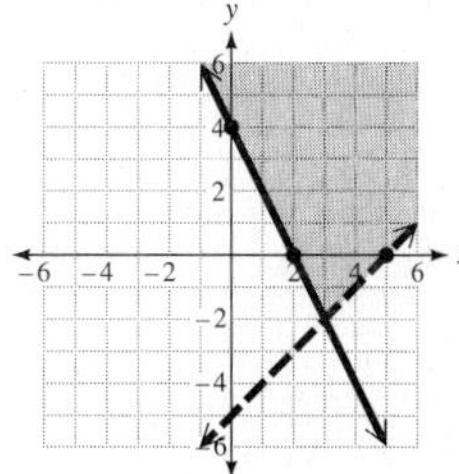

41.

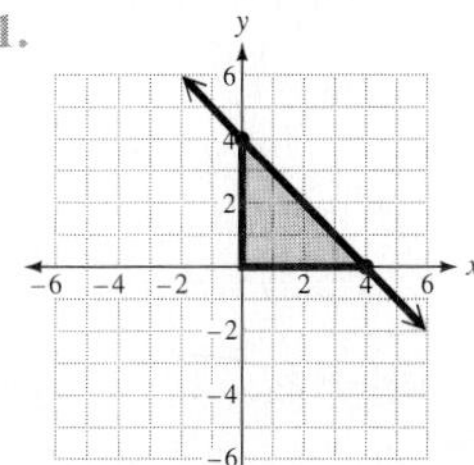

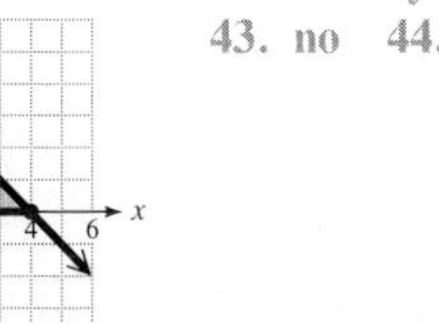

42. exactly one (lines intersect) **43.** no **44.** (5, 5) **45.** 12

CHAPTER 4 TEST

1. Tell whether $(-2, 4)$ is a solution of the system. 1. ______

$$\begin{aligned} 2x + 3y &= 8 \\ 3x - y &= 10 \end{aligned}$$

2. Without solving, give the number of solutions. 2. ______

$$\begin{aligned} -4x + 6y &= 10 \\ 2x - 3y &= -5 \end{aligned}$$

Solve.

3. $$\begin{aligned} 2x + 5y &= 5 \\ -4x - 8y &= -12 \end{aligned}$$ 3. ______

4. $$\begin{aligned} 2x - y &= 3 \\ -2x + y &= -3 \end{aligned}$$ 4. ______

5. The sum of two numbers is 17. Twice the first minus the second is 1. Find the numbers. 5. ______

6. A mixture of nuts sells for \$2.50 per pound. If the mix is composed of two kinds of nuts, one worth \$2.00 per pound and the other worth \$2.80 per pound, how many pounds of each are in 48 pounds of the mix? 6. ______

7. $$\begin{aligned} x + y - z &= 4 \\ x - y - z &= 2 \\ 3x - y - 3z &= -4 \end{aligned}$$ 7. ______

8. $$\begin{aligned} x + y - z &= 3 \\ 2x - y + 3z &= -4 \\ 3x + 2y - 2z &= 7 \end{aligned}$$ 8. ______

9. Bill has \$10,000 divided into three separate investments. Part is invested in stocks earning 8%, part is in a savings account earning 7%, and the rest is invested in a business. If the business does well, it will earn 10% and Bill's total earnings on the three investments will amount to \$810. If the business does poorly, he will lose 4% and his total earnings will amount to \$390. Set up the system of equations but do not solve. 9. ______

CHAPTER 4 TEST

CONTINUED

Evaluate.

10. $\begin{vmatrix} 6 & -2 \\ 5 & 1 \end{vmatrix}$

10. ______________

11. $\begin{vmatrix} 2 & 1 & 0 \\ -1 & 3 & 2 \\ 1 & 4 & 0 \end{vmatrix}$

11. ______________

Solve for ***x only*** *using Cramer's rule.*

12. $3x - y = 5$
$2x + 3y = -4$

12. ______________

13. $x + y + 3z = 5$
$y - z = 2$
$x + z = 3$

13. ______________

Graph the system.

14. $x - 3y < 6$
$2x + y \geq 4$

14.

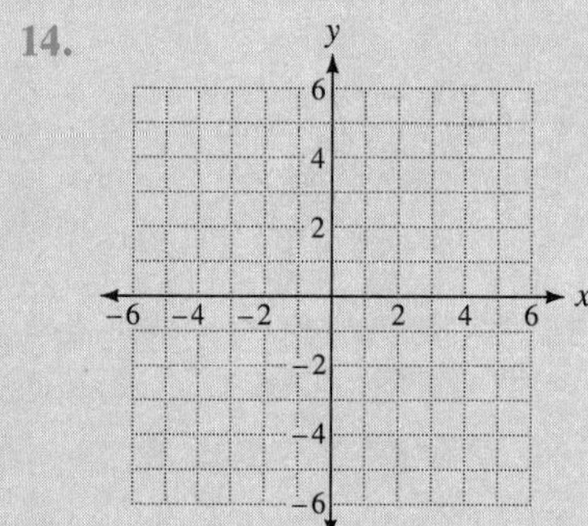

15. $x + y < 5$
$2x - 4 \geq 0$

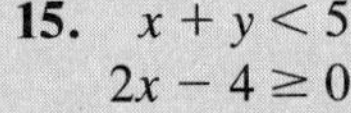

15.

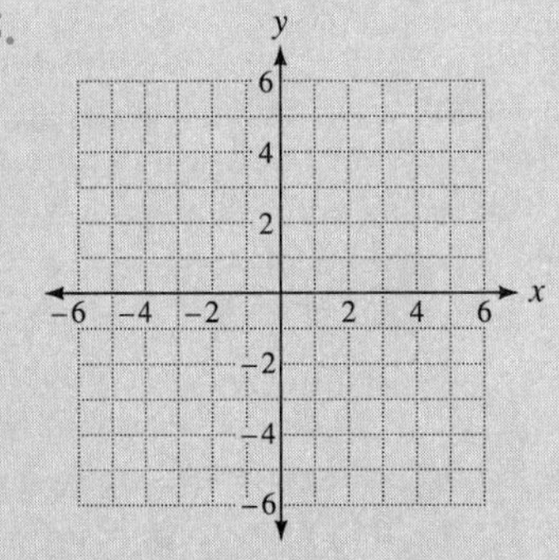

CHAPTER 5

Polynomials and Factoring

5.1 ADDITION AND SUBTRACTION OF POLYNOMIALS

STUDENT GUIDEPOSTS

1 Polynomials	4 Adding Polynomials
2 Collecting Like Terms	5 Subtracting Polynomials
3 Descending and Ascending Order	6 Evaluating Polynomials

1 POLYNOMIALS

In Chapter 1 we said that an algebraic expression involves sums, differences, products, or quotients of numbers and variables raised to various powers. Remember that within an algebraic expression, a term is a product of a number, called the numerical coefficient, and one or more variables raised to a power. A **polynomial** is an algebraic expression having as terms products of numbers and variables with whole-number exponents.

These Are Polynomials

$$-11x^2 + x^3 - x + 3, \quad x + y, \quad 3x^2y, \quad 2, \quad -8a^5b^4 + 6ab - 2, \quad \frac{3}{7}xy - \frac{4}{5}$$

The terms in the first polynomial are $-11x^2$, x^3, $-x$, and 3. The coefficients of these terms are -11, 1, -1, and 3, respectively.

These Are Not Polynomials

$$2\sqrt{x} - 1, \quad \frac{x + y}{x - y}, \quad 2^a, \quad 5|b| + 3$$

These examples illustrate algebraic expressions that are *not* polynomials. In a polynomial, a variable cannot appear within a radical sign, in the denominator, as an exponent, or within an absolute value.

A **monomial** is a polynomial with one term, while a **binomial** has two terms and a **trinomial** has three terms. A polynomial with more than three terms has no special name; we simply call it a polynomial.

A polynomial in one variable has the same variable in each of its terms, while a polynomial in several variables has two or more variables. We define the **degree of a term** to be the power on the variable, if there is only one variable in the term, and the sum of the powers on the variables for terms with more than one variable. The **degree of a polynomial** is equal to the degree of the term with the highest degree.

A polynomial that is a real number with no variables, such as 5, is said to have degree zero since 5 can be written $5x^0$ (remember $x^0 = 1$). However, the real number zero has no degree. We classify several polynomials by the number of terms and give the degree of each one in the following table.

Polynomial	*Type*	*Degree*	*Reason for degree*
$3x^4 + 2x^2 - 1$	Trinomial	4	Highest degree term is $3x^4$.
$5x$	Monomial	1	Highest degree term is $5x^1$.
-6	Monomial	0	$-6 = -6x^0$
0	Monomial	No degree	Zero has no degree.
$-2a^2b^3 + 8ab^5$	Binomial	6	Highest degree term is $8a^1b^5$ and $1 + 5 = 6$.
$2x^2 + xy + y^2 - 3$	Polynomial	2	Highest degree terms are $2x^2$, xy, and y^2.
$-3xyz + 2xy - 5z$	Trinomial	3	Highest degree term is $-3x^1y^1z^1$ and $1 + 1 + 1 = 3$.

As we saw in Chapter 1, like terms in a polynomial contain the same variables raised to the same powers. In the following table one term is like the original and the other one is unlike the original.

Term	*Like term*	*Unlike term*
$3x^4$	$-9x^4$	$3x^5$
5	-8	$-8x$
$7x^3y$	$-5x^3y$	$-5xy^3$
$-6a^2b^2c$	$15a^2b^2c$	$15a^2b^2c^2$
x^2b^3	$12x^2b^3$	$12x^2y^3$

2 COLLECTING LIKE TERMS

We can collect the like terms in a polynomial using the distributive law, as we saw in Section 1.5.

EXAMPLE 1 COLLECTING LIKE TERMS

Collect like terms in each of the polynomials.

(a) $6y^2 - 2y^3 + 3y - 4y^2 + 1 - 3y - 7$

$= 6y^2 - 4y^2 - 2y^3 + 3y - 3y + 1 - 7$ Commute so like terms are adjacent

$= (6 - 4)y^2 - 2y^3 + (3 - 3)y + (1 - 7)$ Use distributive law

$= 2y^2 - 2y^3 + 0 \cdot y - 6$

$= 2y^2 - 2y^3 - 6$

PRACTICE EXERCISE 1

Collect like terms.

(a) $2a^3 + 3a^4 - a^3 + a - a^4 + 2a$

(b) $7a^2b^3 + 3a^3b^2 - a^2b^2 - 4a^2b^3 + 9a^3b^2$

(b) $9x^4y^2 + x^4y^2 - 4xy^3 + 3 + 6xy^3 - 5$
$= 9x^4y^2 + x^4y^2 - 4xy^3 + 6xy^3 + 3 - 5$ Commute
$= (9 + 1)x^4y^2 + (-4 + 6)xy^3 + (3 - 5)$ Distributive law
$= 10x^4y^2 + 2xy^3 - 2$

Answers: (a) $a^3 + 2a^4 + 3a$
(b) $3a^2b^3 + 12a^3b^2 - a^2b^2$

3 DESCENDING AND ASCENDING ORDER

To add polynomials, it is helpful to arrange terms in order. Arrange the polynomial

$$3x^2 - 5x^3 + 4x^6 - 2 + 3x$$

with the highest degree term first, followed by the next highest, and so forth.

$4x^6 - 5x^3 + 3x^2 + 3x - 2$ Exponents get smaller from left to right

The polynomial is now in **descending order.** Note that any constant term will be last in this order since $-2 = -2x^0$. To write the polynomial in **ascending order** put the smallest power first, followed by the next higher power, and so forth.

$-2 + 3x + 3x^2 - 5x^3 + 4x^6$ Exponents get larger from left to right

For polynomials in several variables we need to choose the variable to be ordered. For example, for descending powers of x we write

$3x^3y + 7x^2y^3 - 5xy^2 + 12$ Descending powers of x

and for descending powers of y

$7x^2y^3 - 5xy^2 + 3x^3y + 12.$ Descending powers of y

The descending order of one of the variables is often used for adding or subtracting polynomials.

EXAMPLE 2 DESCENDING ORDER

Collect like terms and write in descending powers of a.

(a) $-3a^3 + a^7 - 7a - 4a^7 + a^3 + 1$
$= a^7 - 4a^7 - 3a^3 + a^3 - 7a + 1$
$= (1 - 4)a^7 + (-3 + 1)a^3 - 7a + 1$
$= -3a^7 - 2a^3 - 7a + 1$

(b) $2ab - 7a^2b^3 - b^2 + 4ab - 8a^3 + 2b^2 + 2a^2b^3$
$= -8a^3 - 7a^2b^3 + 2a^2b^3 + 2ab + 4ab - b^2 + 2b^2$
$= -8a^3 + (-7 + 2)a^2b^3 + (2 + 4)ab + (-1 + 2)b^2$
$= -8a^3 - 5a^2b^3 + 6ab + b^2$

PRACTICE EXERCISE 2

Collect like terms and write in descending powers of x.

(a) $4x - 3x^4 + 5x - 3x^5 + x^4 + 9x^5$

(b) $3x^2y^2 - 4x^3y + 7xy^3 - x^3y - 2x^2y^2$

Answers: (a) $6x^5 - 2x^4 + 9x$
(b) $-5x^3y + x^2y^2 + 7xy^3$

4 ADDING POLYNOMIALS

To add polynomials, we write the problem with descending powers of a variable, remove parentheses, and collect like terms.

EXAMPLE 3 Adding Polynomials

Add the polynomials.

(a) $-3x + x^4 - 5x^3 + 7$ and $3 + 3x - 5x^4 + 8x^2$

$(x^4 - 5x^3 - 3x + 7) + (-5x^4 + 8x^2 + 3x + 3)$ Arrange in descending order

$= x^4 - 5x^3 - 3x + 7 - 5x^4 + 8x^2 + 3x + 3$ Remove parentheses

$= (1 - 5)x^4 - 5x^3 + 8x^2 + (-3 + 3)x + (7 + 3)$ Collect like terms

$= -4x^4 - 5x^3 + 8x^2 + 0 \cdot x + 10$

$= -4x^4 - 5x^3 + 8x^2 + 10$

Although we usually add polynomials horizontally, we may also add vertically writing like terms in columns and leaving spaces for terms that are missing.

$$\begin{array}{rrrrr} x^4 & -\,5x^3 & & -\,3x & +\ \ 7 \\ -5x^4 & & +\,8x^2 & +\,3x & +\ \ 3 \\ \hline -4x^4 & -\,5x^3 & +\,8x^2 & +\,0x & +\,10 \end{array} = -4x^4 - 5x^3 + 8x^2 + 10$$

Leave spaces where terms are missing

(b) $x^4y - 5x^3y^2 - 3xy^3 + 7y$ and $-5x^4y + 8x^2y + 4xy^3 + 3$

$(x^4y - 5x^3y^2 - 3xy^3 + 7y) + (-5x^4y + 8x^2y + 4xy^3 + 3)$

$= x^4y - 5x^3y^2 - 3xy^3 + 7y - 5x^4y + 8x^2y + 4xy^3 + 3$ Remove parentheses

$= (1 - 5)x^4y - 5x^3y^2 + 8x^2y + (-3 + 4)xy^3 + 7y + 3$ Collect like terms

$= -4x^4y - 5x^3y^2 + 8x^2y + xy^3 + 7y + 3$

Practice Exercise 3

Add the polynomials.

(a) $9y - y^5 - 7y^2 + 2$ and $10y^2 + y^5 - 7y + 5$

(b) $x^2y^2 - 9xy + 8x + 3$ and $-4y + 2xy - 5 + 6x^2y^2$

Answers: (a) $3y^2 + 2y + 7$ (b) $7x^2y^2 - 7xy + 8x - 4y - 2$

With practice some steps shown in Example 3 can be omitted. We can add more than two polynomials the same way.

EXAMPLE 4 Adding Three Polynomials

Add $-3x^3y + 4x^2y^2 - 8x + y$, $5x^2y^2 - 7x + y + 5$, and $9x^3y + 4x - 3y$ using columns.

$$\begin{array}{rrrrr} -3x^3y & +\,4x^2y^2 & -\,8x & +\ \ y & \\ & 5x^2y^2 & -\,7x & +\ \ y & +\,5 \\ 9x^3y & & +\,4x & -\,3y & \\ \hline 6x^3y & +\,9x^2y^2 & -\,11x & -\ \ y & +\,5 \end{array}$$

Practice Exercise 4

Add $8ab - 2a^2b + 5ab^2$, $9a^2b - 4ab - 8ab^2$, and $-3ab^2 - 2ab + a^2b$.

Answer: $8a^2b + 2ab - 6ab^2$

5 SUBTRACTING POLYNOMIALS

Subtracting polynomials is the same as adding except that signs require more thought. We have to be careful to write the problem correctly and to change all signs on the subtrahend (polynomial being subtracted). For example, if we want to subtract $3x^2 - 5x + 7$ from $9x^2 - 4x - 8$, we write

$$(9x^2 - 4x - 8) - (3x^2 - 5x + 7).$$

To subtract, we change all the signs in $3x^2 - 5x + 7$ and then add.

$-(3x^2 - 5x + 7) = -3x^2 + 5x - 7$ Remove parentheses and change all three signs

Note that $3x^2$ becomes $-3x^2$, $-5x$ becomes $+5x$, and $+7$ becomes -7.

$(9x^2 - 4x - 8) - (3x^2 - 5x + 7)$

$= 9x^2 - 4x - 8 - 3x^2 + 5x - 7$ Remove parentheses and change all signs in the subtrahend

$= 9x^2 - 3x^2 - 4x + 5x - 8 - 7$ Collect like terms

$= (9 - 3)x^2 + (-4 + 5)x + (-8 - 7)$ Distributive law

$= 6x^2 + x - 15$

To do the same problem using columns, we add $-(3x^2 - 5x + 7) = -3x^2 + 5x - 7$ to $9x^2 - 4x - 8$. That is, we change all signs on the subtrahend and add.

$$\begin{array}{r} 9x^2 - 4x - 8 \\ -3x^2 + 5x - 7 \\ \hline 6x^2 + x - 15 \end{array}$$

Arrange like terms in vertical columns
Change all signs in the subtrahend and add

EXAMPLE 5 SUBTRACTING POLYNOMIALS

(a) Subtract $7x - 3x^2$ from $4 - 2x - 4x^2$.

$(-4x^2 - 2x + 4) - (-3x^2 + 7x)$

$= -4x^2 - 2x + 4 + 3x^2 - 7x$ Change signs

$= -4x^2 + 3x^2 - 2x - 7x + 4$ Collect like terms

$= (-4 + 3)x^2 + (-2 - 7)x + 4$ Distributive law

$= -x^2 - 9x + 4$

(b) Subtract $-3a^2b^3 + 9ab^2 - 5$ from $4a^2b^3 + 7ab^2 - 3b^2 + 8$.

$(4a^2b^3 + 7ab^2 - 3b^2 + 8) - (-3a^2b^3 + 9ab^2 - 5)$

$= 4a^2b^3 + 7ab^2 - 3b^2 + 8 + 3a^2b^3 - 9ab^2 + 5$ Change signs

$= (4 + 3)a^2b^3 + (7 - 9)ab^2 - 3b^2 + (8 + 5)$ Collect like terms

$= 7a^2b^3 - 2ab^2 - 3b^2 + 13$

To subtract $-3a^2b^3 + 9ab^2 - 5$ from $4a^2b^3 + 7ab^2 - 3b^2 + 8$ using columns, change the signs in $-3a^2b^3 + 9ab^2 - 5$ and add.

$$\begin{array}{r} 4a^2b^3 + 7ab^2 - 3b^2 + 8 \\ +3a^2b^3 - 9ab^2 \qquad\quad + 5 \\ \hline 7a^2b^3 - 2ab^2 - 3b^2 + 13 \end{array}$$

Change signs
Add

PRACTICE EXERCISE 5

(a) Subtract $-4x^2 + 5x^3$ from $2x + x^3 - 9x^2$.

(b) Subtract $6x^2y - 9xy^2 + 3xy$ from $5x^2y + 7xy^2 - 2xy$.

Answers: (a) $-5x^2 - 4x^3 + 2x$ (b) $-x^2y + 16xy^2 - 5xy$

Addition and subtraction of polynomials can be combined, as in the next example.

EXAMPLE 6 ADDING AND SUBTRACTING POLYNOMIALS

Perform the indicated operations.

(a) $(5a^3 + a^2 - 6) + (7a^2 - 3) - (-4a^3 + 7a - 5)$

$= 5a^3 + a^2 - 6 + 7a^2 - 3 + 4a^3 - 7a + 5$

$= 9a^3 + 8a^2 - 7a - 4$ Some steps deleted

By columns we proceed as follows.

$$\begin{array}{r} 5a^3 + a^2 \qquad\quad - 6 \\ 7a^2 \qquad\quad - 3 \\ +4a^3 \qquad\quad - 7a + 5 \\ \hline 9a^3 + 8a^2 - 7a - 4 \end{array}$$

Change signs

PRACTICE EXERCISE 6

Perform the indicated operations.

(a) $(4b^4 - 3b^2 + 1) + (-3b^4 + 2) - (b^4 - 9b^2 - 5)$

(b) $(9a^2b^2 - 2ab) + (-4ab + 8) - (-a^2b^2 + ab)$

(b) $(5x^3y^3 - 3x^2y) + (-7x^3y^3 + 4xy) - (9x^2y + 8xy)$
$= 5x^3y^3 - 3x^2y - 7x^3y^3 + 4xy - 9x^2y - 8xy$
$= (5 - 7)x^3y^3 + (-3 - 9)x^2y + (4 - 8)xy$
$= -2x^3y^3 - 12x^2y - 4xy$

Answers: (a) $6b^2 + 8$ (b) $10a^2b^2 - 7ab + 8$

6 EVALUATING POLYNOMIALS

In applied problems it is often necessary to evaluate polynomials for given values of the variable. Example 7 shows how this evaluation can be done.

EXAMPLE 7 Evaluating Polynomials

Evaluate the polynomials for the given values of the variables.

(a) $2x^2 - 7x + 1$ for $x = -3$

$2x^2 - 7x + 1 = 2(-3)^2 - 7(-3) + 1$ Substitute -3 for x using parentheses
$= 2(9) + 21 + 1$
$= 18 + 21 + 1 = 40$

(b) $5a^2b^3 - 2ab + 5$ for $a = 2$ and $b = -1$

$5a^2b^3 - 2ab + 5$
$= 5(2)^2(-1)^3 - 2(2)(-1) + 5$ Substitute $a = 2$ and $b = -1$ using parentheses
$= 5(4)(-1) + 4 + 5$
$= -20 + 4 + 5 = -11$

Practice Exercise 7

Evaluate the polynomials.

(a) $-3b^3 + 2b^2 - 5$ for $b = 2$

(b) $7x^3y - 4x^2y^2 - 6$ for $x = -1$ and $y = -3$

Answers: (a) -21 (b) -21

The following application to business involves addition, subtraction, and evaluation of polynomials.

EXAMPLE 8 Business Problem

In a small business the cost of manufacturing is $M(u) = 2u^2 - 3u + 5$ and the cost of the wholesale operation is $W(u) = u^2 + u + 6$ where u is the number of units involved.

(a) Find the polynomial which represents the total cost.

The total cost $C(u)$ is $M(u) + W(u)$.

$C(u) = M(u) + W(u) = (2u^2 - 3u + 5) + (u^2 + u + 6)$
$= 2u^2 + u^2 - 3u + u + 5 + 6$ Collect like terms
$= 3u^2 - 2u + 11$ Total cost equation

(b) Find the total cost when 10 units are produced and sold.

$C(u) = 3u^2 - 2u + 11$ Total cost equation from (a)
$C(10) = 3(10)^2 - 2(10) + 11$ Substitute $u = 10$
$= 3(100) - 2(10) + 11$
$= 300 - 20 + 11 = 291$

The cost for 10 units is \$291.

Practice Exercise 8

In a manufacturing operation,

$M(u) = 3u^2 - 3u + 2$ and
$W(u) = u^2 - u + 4$.

(a) Find $C(u) = M(u) + W(u)$.

(b) Find $C(u)$ when $u = 20$.

(c) If the total income from selling u units is $I(u) = 4u^2 - 3u$, find the profit $P(u)$.

The profit $P(u)$ is income minus cost.

$$P(u) = I(u) - C(u) = (4u^2 - 3u) - (3u^2 - 2u + 11)$$
$$= 4u^2 - 3u - 3u^2 + 2u - 11 \quad \text{Change signs}$$
$$= 4u^2 - 3u^2 - 3u + 2u - 11$$
$$= u^2 - u - 11 \quad \text{Profit equation}$$

(d) Find the profit when 15 units are produced and sold.

$$P(u) = u^2 - u - 11 \quad \text{Profit equation from (c)}$$
$$P(15) = (15)^2 - 15 - 11$$
$$= 225 - 15 - 11 = 199$$

The profit is \$199.

(c) If $I(u) = 5u^2 - 4u$, find $P(u) = I(u) - C(u)$.

(d) Find $P(u)$ when $u = 20$.

Answers: (a) $4u^2 - 4u + 6$ (b) \$1526 (c) $u^2 - 6$ (d) \$394

5.1 EXERCISES A

Tell whether the following are monomials, binomials, or trinomials and give the degree of each.

1. $-4x^2 - 3$ **2.** 10 **3.** $2x^4 - 3x^3$ **4.** $18a^5b^4$

5. $2xy + 3$ **6.** 0 **7.** $5xy^2 - 8x^2y - 17$ **8.** $9a^2b^4 - \frac{1}{3}$

9. $-14a^5 + b^4 - 7ab$ **10.** $-x - y$ **11.** $a^5b^6c^9 - 3a^2b^{20} - 3$ **12.** $27x^2y^2z^2$

Collect like terms and write in descending order.

13. $3y - 6 + 8y + 5$

14. $3x^2 - 2x + x^2 - 5x + 1$

15. $-6x^3 + 7x - 13x^5 - 7x + x^3$

16. $5y^8 - 6y^4 + 4y^3 - 6 + 7y^8 + 5y^4 + 1$

17. $-7 - 3a - 2a^2 + a^4 + 4a^2 + 3$

18. $5y^3 - 6y^7 - 12y^3 - 8y - 3y^7 + 1$

19. $0.5a^2 - 0.2 - 1.3a^2 - 0.7$

20 $\frac{1}{2}x^3 - \frac{1}{3}x - \frac{1}{4}x^3 - \frac{1}{9}x + 1$

Collect like terms and write in descending powers of the indicated variable.

21. $3xy - 2y - 5xy;\ x$

22. $7x^2y^3 - 2xy + 5xy - 5x^2y^3;\ x$

23. $5a^2b^2 - 2ab + 3 - 7a^2b^2 + 3ab - 1;\ a$

24. $8a^3b^2 - 2a^2b^3 + 3ab - 4a^2b^3 - ab;\ b$

25. $3x^4y^8 - 2xy^9 - 4x^4y^8 + 5xy^9 - 4x^{10}$; y

26. $-2abc^2 - 5a + bc + 3abc^2$; c

Add the polynomials.

27. $5x - 2$ and $-7x - 8$

28. $3xy + 4$ and $-7xy + 5$

29. $3y^2 - 5y + 2$ and $-7y^2 + y - 1$

30. $8a^2b - 3ab + 2$ and $9a^2b + 7ab - 5$

31. $-3x^4 - 7x^2 + 2$ and $x^3 + 5x - 5$

32. $10a^3b^3 - 7a^2b - 4ab^2$ and $-9a^3b^3 - 7a^2b + 5ab$

Subtract the polynomials.

33. $3y - 8$ from $2y + 5$

34. $-2xy + 3y$ from $9xy - 7y$

35. $-7a^2 + 2a - 5$ from $4a^2 - 5a - 3$

36. $7a^2b + ab - 5$ from $-10a^2b + ab + 8$

37. $-5y^4 - 6y^2 + 2y - 8$ from $y^4 + y^3 - y^2 + 1$

38. $-8a^3b^3 + 5a^2b + 10ab^2$ from $6a^3b^3 + 3a^2b - 4ab$

Perform the indicated operations and simplify.

39. $(3a^3 - 9a^4 + a^5 - 2) + (-6a + 7a^2 - a^3 + 2a^5)$

40. $(1.3a^2 - 5.2a + 3.7) + (2.3a^2 + 4.8a - 2.8)$

41. $\left(\frac{2}{3}z^2 - \frac{1}{5}z + 2\right) + \left(\frac{1}{2}z^2 + z - \frac{3}{4}\right)$

42. $(4y - 5y^3) + (-7 + 2y - y^2) - (7y^3 + 2 - 3y)$

43. $(7.3x^2y^2 - 8.7xy + 14.2) + (-6.7x^2y^2 + 2.3xy + 5.9)$

44 $(6x^3y^3 + 5x^2y^2) + (11x^2y^2 - 4xy) - (2x^3y^3 - 2x^2y^2 + 7xy)$

45. $3x - [4y - 7(5x - 2y) - (x - y) - 3(-4y - 2x)]$

46 $5a^2 - 4b(3a - [2b - b(a - 1) - (ab + b)])$

Add.

47.
$$\begin{array}{r} 5x^4 - 6x^3 - 2x^2 + 5 \\ -3x^4 + 8x^2 + 3x + 2 \\ \underline{-7x^4 + 8x^3 - 5x^2 + 9} \end{array}$$

48.
$$\begin{array}{r} 3.7a^3 - 8.6a + 9.3 \\ -5.8a^3 + 3.7a + 1.8 \\ \underline{8.5a^3 + 9.2a - 2.2} \end{array}$$

Evaluate the polynomial for the given values of the variables.

49. $10x^2 - 20x + 5;\ x = 2$

50. $5a^2b^2 - 2ab + 4;\ a = -1,\ b = 3$

51 $3uvw - 4u^2v;\ u = 5,\ v = -2,\ w = 3$

52. An index in a wholesale operation is given by the polynomial $x^2 + 8y^2 - 3xy + 2$. Find the index when $x = 5$ and $y = 2$.

In a business the manufacturing costs are given by the equation $M(u) = 3u^2 - 4u + 4$, and the cost of the wholesale operation is $W(u) = u^2 - u + 3$ where u is the number of units produced. See Example 8.

53 Find the polynomial $C(u)$ that represents the total cost of manufacturing and wholesale and use it in Exercises 54–55.

54. Find the total cost when 20 units are produced and sold.

55. Find the total cost when 5 units are produced and sold.

56 If the total income from the operation is $I(u) = 5u^2 - 2u$, find the polynomial that gives the profit $P(u)$ and use it to answer Exercises 57–58.

57. What is the profit when 20 units are produced and sold?

58. What is the profit when 5 units are produced and sold?

FOR REVIEW

Exercises 59–64 review material from Chapter 1 to help you prepare for the next section. Perform the indicated multiplication.

59. a^2a^4

60. $(a^4b^3)(ab^5)$

61. $-3(2x - y)$

62. $(x^3)^4$

63. $(x^2y^2)^3$

64. $12(x^2 + y^2)$

ANSWERS: 1. binomial; 2 2. monomial; 0 3. binomial; 4 4. monomial; 9 5. binomial; 2 6. monomial; no degree 7. trinomial; 3 8. binomial; 6 9. trinomial; 5 10. binomial; 1 11. trinomial; 22 12. monomial; 6 13. $11y - 1$ 14. $4x^2 - 7x + 1$ 15. $-13x^5 - 5x^3$ 16. $12y^8 - y^4 + 4y^3 - 5$ 17. $a^4 + 2a^2 - 3a - 4$ 18. $-9y^7 - 7y^3 - 8y + 1$ 19. $-0.8a^2 - 0.9$ 20. $\frac{1}{4}x^3 - \frac{4}{9}x + 1$ 21. $-2xy - 2y$ 22. $2x^2y^3 + 3xy$ 23. $-2a^2b^2 + ab + 2$ 24. $-6a^2b^3 + 8a^3b^2 + 2ab$ 25. $3xy^9 - x^4y^8 - 4x^{10}$ 26. $abc^2 + bc - 5a$ 27. $-2x - 10$ 28. $-4xy + 9$ 29. $-4y^2 - 4y + 1$ 30. $17a^2b + 4ab - 3$ 31. $-3x^4 + x^3 - 7x^2 + 5x - 3$ 32. $a^3b^3 - 14a^2b - 4ab^2 + 5ab$ 33. $-y + 13$ 34. $11xy - 10y$ 35. $11a^2 - 7a + 2$ 36. $-17a^2b + 13$ 37. $6y^4 + y^3 + 5y^2 - 2y + 9$ 38. $14a^3b^3 - 2a^2b - 10ab^2 - 4ab$ 39. $3a^5 - 9a^4 + 2a^3 + 7a^2 - 6a - 2$ 40. $3.6a^2 - 0.4a + 0.9$ 41. $\frac{7}{6}z^2 + \frac{4}{5}z + \frac{5}{4}$ 42. $-12y^3 - y^2 + 9y - 9$ 43. $0.6x^2y^2 - 6.4xy + 20.1$ 44. $4x^3y^3 + 18x^2y^2 - 11xy$ 45. $33x - 31y$ 46. $5a^2 - 12ab + 8b^2 - 8ab^2$ 47. $-5x^4 + 2x^3 + x^2 + 3x - 2$ 48. $6.4a^3 + 4.3a + 8.9$ 49. 5 50. 55 51. 110 52. 29 53. $4u^2 - 5u + 7$ 54. \$1507 55. \$82 56. $u^2 + 3u - 7$ 57. \$453 58. \$33 59. a^6 60. a^5b^8 61. $-6x + 3y$ 62. x^{12} 63. x^6y^6 64. $12x^2 + 12y^2$

5.1 EXERCISES B

Tell whether the following are monomials, binomials, or trinomials and give the degree of each.

1. $8x^2 - 2x + 1$ **2.** $3x - 1$ **3.** $3a^2b^2 - 2$ **4.** $6x^3y^7$

5. 42 **6.** $-abc + 2b + 3c$ **7.** 0 **8.** $\frac{2}{3}x^2b^3$

9. $x + 2y$ **10.** $a^2b^3c^4 + 2bc - 5a^{10}$ **11.** $5y^3 + x^2y^2$ **12.** $x^2 + y^2 + z^2$

Collect like terms and write in descending order.

13. $2y - 5 - 7y - 8$ **14.** $2x^3 - x^2 + 5x^3 - 2x + 3x^2$

15. $10x - 5x^3 + 3x + 8x^3 - x$ **16.** $-2y^3 + 4y^8 - 3y^3 + 7 - 1 + 2y^8$

17. $2a - 5 + 8a^2 - a^4 + 4 - 2a + 4a^4$ **18.** $-10y + 8y + 9y^{10} - y + 4y^{10}$

19. $0.5a^3 - 1.2a + 0.7a^3 + 0.6a$ **20.** $\frac{2}{3}x^2 - \frac{1}{3} + \frac{1}{9}x - \frac{1}{3}x^2 + \frac{2}{3}x - 1$

Collect like terms and write in descending powers of the indicated variable.

21. $2x^2y - xy + 3x^2y$; x **22.** $8x^2y^2 - 3xy^2 - 5x^2y^2 + 6xy^2$; x

23. $a^3b^2 - 5ab + 3 - 2a^3b^2 - 7ab + 2$; a **24.** $3a + 5a^2b - 3b^3 + ab^2 - 3a^2b + 2a$; b

25. $-xy^5 + 2x^9y^9 - 3y^{10} + xy^5 - 5x^9y^9 + 8y^{10}$; y **26.** $7a^2 - 3abc + 4a^2 + 9abc + 8c^2$; c

Add the polynomials.

27. $3y - 5$ and $7y + 8$ **28.** $-8xy - 4$ and $xy - 1$

29. $5a^3 - 3a + 7$ and $7a^3 - 10a - 8$ **30.** $-6x^3y + x^2y^2 - 5xy$ and $8x^3y - 10x^2y^2 + 6$

31. $x^2 - 2x + 1$ and $x^4 + x^3 - x^2 + 5$ **32.** $a^3b^2 - a^2b^3$ and $a^3b^2 - 5a^2b^3 + 7$

Subtract the polynomials.

33. $-8x - 7$ from $9x + 2$ **34.** $6xy - 3$ from $-2xy + 7$

35. $9x^3 - 7x^2 - 2x$ from $-7x^2 + 7x - 8$ **36.** $12x^3y - 7x^2y^2 + 14xy$ from $7x^3y + 6x^2y^2 - 5$

37. $x^4 - 3x^2 + 2x + 1$ from $3x^4 - 5x^3 + 7$

38. $-6x^3y^3 + x^2y^2 - xy$ from $x^3y^3 - x^2y^2 - xy$

Perform the indicated operations and simplify.

39. $(3a^3 - 9a^4 + a^5 - 2) - (-6a + 7a^2 - a^3 + 2a^5)$

40. $\left(\frac{3}{4}x^2 - \frac{1}{3}x - \frac{4}{5}\right) - \left(-\frac{1}{2}x^2 + \frac{4}{9}x - \frac{3}{25}\right)$

41. $(-5.8c^2 + 6.2c - 4.1) - (2.3c^2 + 4.8c - 1.3)$

42. $(2 - 2a^4 + a^3) - (a^3 - 3a^4 + 7a + 5) - (14a + a^3 - 3a^2)$

43. $\left(\frac{2}{9}a^3b^2 + \frac{3}{4}ab - \frac{2}{5}b\right) - \left(\frac{1}{3}a^3b^2 - \frac{1}{8}ab + \frac{3}{10}b\right)$

44. $(-8a^3b - 5a^2b^2) - (-3a^2b^2 - 10a) - (-7a^3b - 9a^2b^2 + 4a)$

45. $7a - [2b - 5(9a - 5b - 3) - 4 - 7(a - b)]$

46. $7x^2 - 2x\{3y - [-6y - x(y + 3) - y(x - 2)]\}$

Add.

47.
$$\begin{array}{rrrr} 180y^3 & -\ 24y^2 & -\ 85y & \\ -35y^3 & +\ 128y^2 & +\ 99y & -\ 240 \\ & 67y^2 & -\ 108y & +\ 560 \\ \hline \end{array}$$

48.
$$\begin{array}{rrrr} \frac{2}{7}x^3 & -\ 5x^2 & +\ \frac{3}{5}x & +\ 2 \\ \frac{3}{14}x^3 & -\ \frac{5}{6}x^2 & -\ \frac{7}{10}x & +\ \frac{3}{2} \\ \hline \end{array}$$

Evaluate the polynomial for the given values of the variables.

49. $3x^2 - 4x - 5;\ x = 3$

50. $2ab^2 - 3a^2b + ab;\ a = 2,\ b = -1$

51. $-6uvw + 7u - 5w;\quad u = 2,\ v = -3,\ w = 7$

52. A manufacturing index is given by the polynomial $2x^2 - 5xy + y^2 - 5$. Find the index when $x = 3$ and $y = 4$.

In a sales operation the wholesale cost is given by $W(u) = u^2 - 2u + 4$, *and the retail cost is* $R(u) = u^2 - u + 8$ *where* u *is the number of units sold.*

53. Find the polynomial $C(u)$ which represents the total cost of both operations and use it in Exercises 54–55.

54. Find the total cost when 2 units are sold.

55. Find the total cost when 30 units are sold.

56. If the total income from the operation is $I(u) = 3u^2 - 2u$, find the polynomial which gives the profit $P(u)$ and use it to answer Exercises 57–58.

57. What is the profit when 2 units are sold?

58. What is the profit when 30 units are sold?

FOR REVIEW

Exercises 59–64 review material from Chapter 1 to help you prepare for the next section.

59. x^3x^7 **60.** $(u^2v^5)(u^3v^4)$ **61.** $-4(a-3c)$

62. $(y^4)^2$ **63.** $(x^5y)^4$ **64.** $8(xy+z^2)$

5.1 EXERCISES C

Evaluate the polynomial for the given values of the variables.

1. $xy-3x+2y^2;\ x=\frac{1}{2},\ y=-\frac{1}{3}$

2. $0.0005a^3+0.001a^2+0.01a-0.007;\ a=0.3$

3. $x^2y-x^3+xy^3-5;\ x=2a,\ y=-a$

4. $\frac{4}{3}w^3-\frac{4}{5}w^2+w-\frac{1}{6};\ w=\frac{1}{2}$
[Answer: $\frac{3}{10}$]

Perform the indicated operations.

5. $(-10x^4y^3+3x^3y^4-7x^2z+5z)+(10x^4y^3+9x^3y^4-6z)-(12x^3y^4-z)$

6. $\frac{1}{4}(16a^3b^2-8ab^3+12ab)-\frac{1}{5}(25a^3b^2+5ab^3-10ab)+\frac{2}{3}(9ab-6ab^3)$
[Answer: $-a^3b^2-7ab^3+11ab$]

5.2 MULTIPLICATION OF POLYNOMIALS

STUDENT GUIDEPOSTS

1. Multiplying by Monomials
2. Multiplying Binomials
3. The FOIL Method
4. Special Products

1 MULTIPLYING BY MONOMIALS

To multiply two monomials, we use the commutative and associative laws with the law of exponents that we learned in Chapter 1,

$$a^ma^n=a^{m+n}.$$

EXAMPLE 1 MULTIPLYING MONOMIALS

(a) Multiply $6ab^2$ by $-a^3$.

$$(6ab^2)(-a^3)=(6)(-1)(aa^3)b^2 \qquad -a^3=(-1)a^3$$
$$=-6a^{1+3}b^2 \qquad a=a^1$$
$$=-6a^4b^2$$

PRACTICE EXERCISE 1

(a) Multiply $-x^4y^6$ by $9x^6y$.

(b) Multiply $2a^3b^4$ by $-8ab^5$.

(b) Multiply $-3x^2y^3$ by $5x^5y^2$.

$(-3x^2y^3)(5x^2y^2) = (-3)(5)(x^2x^5)(y^3y^2)$ Use the commutative law to change the order of factors

$= -15x^{2+5}y^{3+2}$ Use $a^ma^n = a^{m+n}$

$= -15x^7y^5$

Answers: (a) $-9x^{10}y^7$ (b) $-16a^4b^9$

The product of a monomial and a binomial or any other polynomial uses the distributive law as well as the methods for multiplying monomials.

EXAMPLE 2 Multiplying a Binomial by a Monomial

Multiply.

$-5x^2y(4x^2y^2 - 6y)$

$= (-5x^2y)(4x^2y^2) - (-5x^2y)(6y)$ Distributive property

$= (-5)(4)(x^2x^2)(yy^2) - (-5)(6)x^2(yy)$

$= -20x^{2+2}y^{1+2} - (-30)x^2y^{1+1}$ $a^ma^n = a^{m+n}$

$= -20x^4y^3 + 30x^2y^2$

Practice Exercise 2

Multiply $-3ab^3(2a^2b^2 - 7ab^4)$.

Answer: $-6a^3b^5 + 21a^2b^7$

2 MULTIPLYING BINOMIALS

We can multiply two binomials the way we multiply a monomial times a binomial if we think of one binomial as a single term and use the distributive law twice.

$(x + 2y)(5x - y) = (x + 2y)(5x) + (x + 2y)(-y)$ Distributive law

$= (x)(5x) + (2y)(5x) + (x)(-y) + (2y)(-y)$ Distributive law again, twice

$= 5x^2 + 10xy - xy - 2y^2$

$= 5x^2 + 9xy - 2y^2$

EXAMPLE 3 Multiplying Binomials

Multiply.

(a) $(2x - 1)(x + 3) = (2x - 1)(x) + (2x - 1)(3)$ Distributive law

$= (2x)(x) - (1)(x) + (2x)(3) - (1)(3)$ Distributive law

$= 2x^2 - x + 6x - 3$

$= 2x^2 + 5x - 3$ Collect like terms

(b) $(3a^2 + 2b)(2a^3 - 5b)$

$= (3a^2 + 2b)(2a^3) + (3a^2 + 2b)(-5b)$

$= (3a^2)(2a^3) + (2b)(2a^3) + (3a^2)(-5b) + (2b)(-5b)$

$= 6a^5 + 4a^3b - 15a^2b - 10b^2$

Practice Exercise 3

Multiply.

(a) $(4x + 3)(x - 2)$

(b) $(4u^3 - 3v)(2u^2 + 7v)$

Answers: (a) $4x^2 - 5x - 6$ (b) $8u^5 + 28u^3v - 6u^2v - 21v^2$

3 THE FOIL METHOD

Notice that when we multiplied $(x + 2y)(5x - y)$ above, each term in the first binomial was multiplied by each term in the second and then like terms were collected. A shorter way to find this product is to use the **FOIL method.** The letters FOIL stand for first terms Ⓕ, outside terms Ⓞ, inside terms Ⓘ, and last terms Ⓛ. If we multiply each of these pairs of terms, we obtain the same product as we did using the distributive law twice.

$$(x + 2y)(5x - y) = \overset{F}{(x)(5x)} + \overset{O}{(x)(-y)} + \overset{I}{(2y)(5x)} + \overset{L}{(2y)(-y)}$$
$$= 5x^2 - xy + 10xy - 2y^2$$
$$= 5x^2 + 9xy - 2y^2$$

Practice with the FOIL method will let us find products of binomials mentally. It is a time-saving technique.

EXAMPLE 4 Using the FOIL Method

Use the FOIL method to multiply the binomials.

$$(5a + 2b)(3a - 7b) = \overset{F}{(5a)(3a)} + \overset{O}{(5a)(-7b)} + \overset{I}{(2b)(3a)} + \overset{L}{(2b)(-7b)}$$
$$= 15a^2 - 35ab + 6ab - 14b^2$$
$$= 15a^2 - 29ab - 14b^2$$

Practice Exercise 4

Use the FOIL method to multiply.

$(9x - y)(6x - 5y)$

Answer: $54x^2 - 51xy + 5y^2$

Another method of multiplying involves writing one polynomial above the other and arranging like terms of the product in vertical columns. The column method is especially useful with trinomials or larger polynomials.

EXAMPLE 5 Multiplying Using Columns

Multiply using columns.

(a)

$$\begin{array}{rrrrl}
3x^2 & - 5xy & + 2y^2 & & \\
7x & - 8y & & & \\
\hline
21x^3 & - 35x^2y & + 14xy^2 & & 7x \text{ times the top polynomial} \\
 & - 24x^2y & + 40xy^2 & - 16y^3 & -8y \text{ times the top polynomial} \\
\hline
21x^3 & - 59x^2y & + 54xy^2 & - 16y^3 &
\end{array}$$

(b)

$$\begin{array}{rrrrrl}
6x^2 & - 7xy & + 5y^2 & & & \\
4x^2 & + 3xy & - 2y^2 & & & \\
\hline
24x^4 & - 28x^3y & + 20x^2y^2 & & & 4x^2 \text{ times the top polynomial} \\
 & 18x^3y & - 21x^2y^2 & + 15xy^3 & & 3xy \text{ times the top polynomial} \\
 & & - 12x^2y^2 & + 14xy^3 & - 10y^4 & -2y^2 \text{ times the top polynomial} \\
\hline
24x^4 & - 10x^3y & - 13x^2y^2 & + 29xy^3 & - 10y^4 &
\end{array}$$

Practice Exercise 5

Multiply.

(a)
$$\begin{array}{rrr}
2x^2 & + 3xy & - 4y^2 \\
6x & - y & \\
\hline
\end{array}$$

(b)
$$\begin{array}{rrr}
2x^2 & - 5xy & - 6y^2 \\
9x^2 & - xy & + 4y^2 \\
\hline
\end{array}$$

Answers: (a) $12x^3 + 16x^2y - 27xy^2 + 4y^3$ (b) $18x^4 - 47x^3y - 41x^2y^2 - 14xy^3 - 24y^4$

4 SPECIAL PRODUCTS

Products of certain binomials occur frequently and deserve special consideration. Look at the following three products.

(1) $(2x + 3y)(2x - 3y) = (2x)^2 - 6xy + 6xy - (3y)^2 = 4x^2 - 9y^2$

(2) $(2x + 3y)(2x + 3y) = (2x)^2 + 6xy + 6xy + (3y)^2 = 4x^2 + 12xy + 9y^2$

(3) $(2x - 3y)(2x - 3y) = (2x)^2 - 6xy - 6xy + (3y)^2 = 4x^2 - 12xy + 9y^2$

In (1) we have an example of the general formula

$$(a + b)(a - b) = a^2 - ab + ab - b^2 \quad \text{Using FOIL method}$$
$$(a + b)(a - b) = a^2 - b^2. \quad \text{Difference of squares}$$

We call this the difference of squares since the middle terms add to zero and the difference of squares is left. We can also write general formulas for both (2) and (3) above.

$$(a + b)^2 = a^2 + ab + ab + b^2 \quad \text{Using FOIL method}$$
$$(a + b)^2 = a^2 + 2ab + b^2 \quad \text{Perfect square binomial}$$
$$(\text{sum})^2 = (\text{first term})^2 + 2(\text{product}) + (\text{last term})^2$$

$$(a - b)^2 = a^2 - ab - ab + b^2 \quad \text{Using FOIL method}$$
$$(a - b)^2 = a^2 - 2ab + b^2 \quad \text{Perfect square binomial}$$
$$(\text{difference})^2 = (\text{first term})^2 + 2(\text{product}) + (\text{last term})^2$$

The products (2) and (3) fall into the pattern: the square of the first term, plus or minus twice the product of the first and last terms, plus the square of the last. These three formulas are special products that will be helpful when we factor polynomials in Section 5.4.

Special Product Formulas

The product of a sum and difference of two terms is the square of the first minus the square of the second.

1. $(a + b)(a - b) = a^2 - b^2$

The square of a binomial is the square of the first term plus twice the product of the terms plus the square of the second term.

2. $(a + b)(a + b) = (a + b)^2 = a^2 + 2ab + b^2$
3. $(a - b)(a - b) = (a - b)^2 = a^2 - 2ab + b^2$

CAUTION

Pay special attention to the difference between the first and third formulas: $(a - b)^2$ is not $a^2 - b^2$. Also note that $(a + b)^2 \neq a^2 + b^2$.

Now we can find certain products using the FOIL method or the special product formulas.

FOIL *Method*

$$(x + 6)(x - 6) = x^2 - 6x + 6x - 36$$
(F, O, I, L)
$$= x^2 - 36$$

Using $(a + b)(a - b) = a^2 - b^2$

$$(x + 6)(x - 6) = x^2 - 6^2 \qquad a = x,\ b = 6$$
$$= x^2 - 36$$

FOIL *Method*

$$(x + 6)(x + 6) = x^2 + 6x + 6x + 36$$
$$= x^2 + 12x + 36$$

Using $(a + b)^2 = a^2 + 2ab + b^2$

$$(x + 6)^2 = x^2 + 2(x)(6) + 6^2 \qquad a = x,\ b = 6$$
$$= x^2 + 12x + 36$$

FOIL *Method*

$$(x - 6)(x - 6) = x^2 - 6x - 6x + 36$$
$$= x^2 - 12x + 36$$

Using $(a - b)^2 = a^2 - 2ab + b^2$

$$(x - 6)^2 = x^2 - 2(x)(6) + 6^2 \qquad a = x,\ b = 6$$
$$= x^2 - 12x + 36$$

In the following examples we use the rules of exponents $(ab)^n = a^nb^n$ and $(a^m)^n = a^{mn}$.

EXAMPLE 6 Using Special Product Formulas

Find each product using the special product formulas.

(a) $(3x - 2y)(3x + 2y) = (3x)^2 - (2y)^2$ Use $(a - b)(a + b) = a^2 - b^2$, with $a = 3x$ and $b = 2y$

$= 3^2x^2 - 2^2y^2$ $(ab)^n = a^nb^n$

$= 9x^2 - 4y^2$

(b) $(3x + 2y)^2$

$= (3x)^2 + 2(3x)(2y) + (2y)^2$ Use $(a + b)^2 = a^2 + 2ab + b^2$, with $a = 3x$ and $b = 2y$

$= 9x^2 + 12xy + 4y^2$ $(ab)^n = a^nb^n$

(c) $(3x - 2y)^2$

$= (3x)^2 - 2(3x)(2y) + (2y)^2$ Use $(a - b)^2 = a^2 - 2ab + b^2$, with $a = 3x$ and $b = 2y$

$= 9x^2 - 12xy + 4y^2$ $(ab)^n = a^nb^n$

Practice Exercise 6

Find each product using the special product formulas.

(a) $(7a - 4b)(7a + 4b)$

(b) $(7a + 4b)^2$

(c) $(7a - 4b)^2$

Answers: (a) $49a^2 - 16b^2$ (b) $49a^2 + 56ab + 16b^2$ (c) $49a^2 - 56ab + 16b^2$

EXAMPLE 7 Using Special Product Formulas

Find each product using the special product formulas.

(a) $(5a^3 - 4b^2)(5a^3 + 4b^2) = (5a^3)^2 - (4b^2)^2$

Use $(a - b)(a + b) = a^2 - b^2$, with $a = 5a^3$ and $b = 4b^2$

$= 5^2(a^3)^2 - 4^2(b^2)^2$

$= 25a^6 - 16b^4$ $(a^m)^n = a^{mn}$

(b) $(5a^3 + 4b^2)^2 = (5a^3)^2 + 2(5a^3)(4b^2) + (4b^2)^2$

Use $(a + b)^2 = a^2 + 2ab + b^2$ with $a = 5a^3$ and $b = 4b^2$

$= 5^2(a^3)^2 + 40a^3b^2 + 4^2(b^2)^2$ $(ab)^n = a^nb^n$

$= 25a^6 + 40a^3b^2 + 16b^4$ $(a^m)^n = a^{mn}$

Practice Exercise 7

Find each product using the special product formulas.

(a) $(9x^4 - y^3)(9x^4 + y^3)$

(b) $(9x^4 - y^3)^2$

Answers: (a) $81x^8 - y^6$ (b) $81x^8 - 18x^4y^3 + y^6$

You should try to multiply polynomials mentally. Don't start multiplying, however, until you have examined the polynomials to determine the easiest method to use. At times, as in the next example, a special product formula might be applied more than once.

EXAMPLE 8 MULTIPLYING THREE POLYNOMIALS

Multiply $(x + 2y)(x - 2y)(x^2 + 4y^2)$.

Notice that $(x + 2y)(x - 2y)$ can be found using the special product $(a + b)(a - b)$. This product, $x^2 - 4y^2$, can be multiplied by $x^2 + 4y^2$ using the same formula.

$$\underbrace{(x + 2y)(x - 2y)}(x^2 + 4y^2)$$

$$= (x^2 - 4y^2)(x^2 + 4y^2)$$

$$= x^4 - 16y^4$$

PRACTICE EXERCISE 8

Multiply.

$(9u^2 + v^2)(3u - v)(3u + v)$

Answer: $81u^4 - v^4$

5.2 EXERCISES A

Multiply the polynomials.

1. $(3x)(7y)$ **2.** $(-5x^2)(-6xy)$ **3.** $(8a^2b)(-9ab^2)$

4. $-4x^3y(2xy - 5)$ **5.** $8a^2b^2(3ab - 2a + 5b)$ **6.** $-7a^3b(3a^2b^2 + 5a^2 - 6b^2)$

Multiply using the FOIL *method.*

7. $(x + y)(x + 2y)$ **8.** $(x - y)(x - 2y)$ **9.** $(x - y)(x + 2y)$

10. $(x + y)(x - 2y)$ **11.** $(2a + b)(a + 3b)$ **12.** $(2a - b)(a - 3b)$

13. $(2a + b)(a - 3b)$ **14.** $(2a - b)(a + 3b)$ **15.** $(5x - 2y)(x + 3y)$

16. $(10a + 3b)(7a - 4b)$ **17.** $(4x - 9y)(3x + 5y)$ **18.** $(4x + 9y)(3x - 5y)$

Multiply using columns.

19. $\begin{array}{l} 7a^2 - 3ab + 4b^2 \\ \underline{3a - 2b \qquad\quad} \end{array}$ **20.** $\begin{array}{l} 2x^2 + 4xy - 5y^2 \\ \underline{9x + 2y \qquad\quad} \end{array}$ **21.** $\begin{array}{l} 5x^2y^2 - 3xy - 2 \\ \underline{8xy - 7 \qquad\quad} \end{array}$

22. $\begin{array}{l} 2a - ab + 3b \\ \underline{4a + 2ab - 3b} \end{array}$ **23.** $\begin{array}{l} 7a - 2b + 5 \\ \underline{3a - b + 2} \end{array}$ **24.** $\begin{array}{l} a^2 - 3ab + 2b^2 \\ \underline{2a^2 + 4ab - 3b^2} \end{array}$

Multiply the polynomials.

25. $3(x - 5y)(2x + y)$ **26.** $a^2b(a + 2b)(a - 2b)$ **27.** $(x + 3y)^3$

28. $(a - 5b)(2a + b)(2a - 3b)$

29. $(x + y)(x^2 - xy + y^2)$

30. $(a - b)[3a^2 - (a + b)(a - 2b)]$

In an environmental study the final result $F(t)$ is given by $F(t) = N(t)S(t)$. Find $F(t)$ for the given $N(t)$ and $S(t)$.

31. $N(t) = 2t,\ S(t) = t^3 - t^2$

32. $N(t) = 2t - 5,\ S(t) = 3t + 2$

33. $N(t) = 2t^2 - 3t + 1,\ S(t) = 3t - 1$

34. $N(t) = t^2 - 2t - 5,\ S(t) = 2t^2 + t - 6$

35. Find $F(2)$ for $F(t)$ in Exercise 31.

36. Find $F(-2)$ for $F(t)$ in Exercise 32.

37. Find $F(3)$ for $F(t)$ in Exercise 33.

38. Find $F(-1)$ for $F(t)$ in Exercise 34.

Use the special product formulas to find the products.

39. $(x + 3)(x - 3)$

40. $(a - 4)^2$

41. $(b + 8)^2$

42. $(3x + 5)(3x - 5)$

43. $(3x + 5y)(3x - 5y)$

44. $(5a + 7b)^2$

45. $(8a - 5b)(8a - 5b)$

46. $(5x + 9y)^2$

47. $\left(\frac{1}{2}x - y\right)\left(\frac{1}{2}x + y\right)$

48. $(0.5x + 0.1y)^2$

49. $(x^2 + 1)(x^2 - 1)$

50. $(x^2 - 1)^2$

51. $(2a^2 + b^2)(2a^2 - b^2)$

52. $(3a^2 + b^2)^2$

53. $(3a^2 - 2b^4)(3a^2 + 2b^4)$

54. $(a^3 + 5b^2)^2$

55. $(x - y)^2 + (x + y)^2$

56. $[a - (b + c)][a + (b + c)]$

57. $[(2x - y)(2x + y)]^2$

58. $[2x^2 - (x - 2y)][2x^2 + (x - 2y)]$

59. $(u - 3v)(u + 3v)(u^2 + 9v^2)$

60. $(4x + 3y)(4x - 3y)(16x^2 + 9y^2)$

If Roy Troutman deposits A dollars in a savings account at i interest rate, the amount in his account at the end of 2 years is given by the equation $S = A(1 + i)^2$.

61. Obtain an equivalent equation by performing all the indicated multiplications.

62 Check your work in Exercise 61 by using both formulas to find S when $A = \$1000$ and $i = 0.12$.

FOR REVIEW

Tell whether the following are monomials, binomials, or trinomials and give the degree of each.

63. $2x^4 + 3y^3$ **64.** $a^2b^2 - c^2 + ac$ **65.** -6 **66.** $5x^2y^2z^2 - 2x^7y - 5z^4$

Perform the indicated operations.

67. $(x^2 + y^2 - 3) - (7x^2 - 4y^2 + 5) - (x^2 + 5y^2 - 3)$

68. $(6x^2y^2 - 2xy + 5) + (3x^2y^2 - 10) - (-10x^2y^2 + 5xy - 2)$

Exercises 69–72 review material from Section 1.5 to help you prepare for the next section. Use the distributive law to factor each expression.

69. $2x + 8y$ **70.** $uv + 3v$ **71.** $3x - 6y + 15z$ **72.** $aw - 2w + zw$

ANSWERS: 1. $21xy$ 2. $30x^3y$ 3. $-72a^3b^3$ 4. $-8x^4y^2 + 20x^3y$ 5. $24a^3b^3 - 16a^3b^2 + 40a^2b^3$ 6. $-21a^5b^3 - 35a^5b + 42a^3b^3$ 7. $x^2 + 3xy + 2y^2$ 8. $x^2 - 3xy + 2y^2$ 9. $x^2 + xy - 2y^2$ 10. $x^2 - xy - 2y^2$ 11. $2a^2 + 7ab + 3b^2$ 12. $2a^2 - 7ab + 3b^2$ 13. $2a^2 - 5ab - 3b^2$ 14. $2a^2 + 5ab - 3b^2$ 15. $5x^2 + 13xy - 6y^2$ 16. $70a^2 - 19ab - 12b^2$ 17. $12x^2 - 7xy - 45y^2$ 18. $12x^2 + 7xy - 45y^2$ 19. $21a^3 - 23a^2b + 18ab^2 - 8b^3$ 20. $18x^3 + 40x^2y - 37xy^2 - 10y^3$ 21. $40x^3y^3 - 59x^2y^2 + 5xy + 14$ 22. $8a^2 - 2a^2b^2 + 6ab + 9ab^2 - 9b^2$ 23. $21a^2 - 13ab + 29a + 2b^2 - 9b + 10$ 24. $2a^4 - 2a^3b - 11a^2b^2 + 17ab^3 - 6b^4$ 25. $6x^2 - 27xy - 15y^2$ 26. $a^4b - 4a^2b^3$ 27. $x^3 + 9x^2y + 27xy^2 + 27y^3$ 28. $4a^3 - 24a^2b + 17ab^2 + 15b^3$ 29. $x^3 + y^3$ 30. $2a^3 - a^2b + ab^2 - 2b^3$ 31. $F(t) = 2t^4 - 2t^3$ 32. $F(t) = 6t^2 - 11t - 10$ 33. $F(t) = 6t^3 - 11t^2 + 6t - 1$ 34. $F(t) = 2t^4 - 3t^3 - 18t^2 + 7t + 30$ 35. 16 36. 36 37. 80 38. 10 39. $x^2 - 9$ 40. $a^2 - 8a + 16$ 41. $b^2 + 16b + 64$ 42. $9x^2 - 25$ 43. $9x^2 - 25y^2$ 44. $25a^2 + 70ab + 49b^2$ 45. $64a^2 - 80ab + 25b^2$ 46. $25x^2 + 90xy + 81y^2$ 47. $\frac{1}{4}x^2 - y^2$ 48. $0.25x^2 + 0.1xy + 0.01y^2$ 49. $x^4 - 1$ 50. $x^4 - 2x^2 + 1$ 51. $4a^4 - b^4$ 52. $9a^4 + 6a^2b^2 + b^4$ 53. $9a^4 - 4b^8$ 54. $a^6 + 10a^3b^2 + 25b^4$ 55. $2x^2 + 2y^2$ 56. $a^2 - b^2 - 2bc - c^2$ 57. $16x^4 - 8x^2y^2 + y^4$ 58. $4x^4 - x^2 + 4xy - 4y^2$ 59. $u^4 - 81v^4$ 60. $256x^4 - 81y^4$ 61. $S = A + 2Ai + Ai^2$ 62. \$1254.40 63. binomial; 4 64. trinomial; 4 65. monomial; 0 66. trinomial; 8 67. $-7x^2 - 5$ 68. $19x^2y^2 - 7xy - 3$ 69. $2(x + 4y)$ 70. $v(u + 3)$ 71. $3(x - 2y + 5z)$ 72. $w(a - 2 + z)$

5.2 EXERCISES B

Multiply the polynomials.

1. $(5x)(8y)$ **2.** $(-8x^2)(-2xy^2)$ **3.** $(5ab^2)(-7a^2b)$

4. $-9xy^3(3xy - 2)$ **5.** $4a^3b^2(ab + 5a - 7b)$ **6.** $-10ab^3(5a^2b^2 + 2a^2 - 9b^2)$

Multiply using the FOIL *method.*

7. $(x + 3y)(x + y)$

8. $(x - 3y)(x - y)$

9. $(x - 3y)(x + y)$

10. $(x + 3y)(x - y)$

11. $(3a + b)(a + 2b)$

12. $(3a - b)(a - 2b)$

13. $(3a + b)(a - 2b)$

14. $(3a - b)(a + 2b)$

15. $(2x - 3y)(4x + y)$

16. $(12a + 5b)(3a - 7b)$

17. $(3x - 7y)(5x + 2y)$

18. $(3x + 7y)(5x - 2y)$

Multiply using columns.

19. $\begin{array}{l} 4a^2 - 5ab + 6b^2 \\ \underline{2a \quad - 3b \qquad\quad} \end{array}$

20. $\begin{array}{l} 3x^2 + 6xy - 2y^2 \\ \underline{7x \quad + 3y \qquad\quad} \end{array}$

21. $\begin{array}{l} 3x^2y^2 - 4xy - 5 \\ \underline{6xy \quad - 5 \qquad\quad} \end{array}$

22. $\begin{array}{l} 3a - 2ab + 5b \\ \underline{2a + 3ab - 4b} \end{array}$

23. $\begin{array}{l} 5a - 3b + 2 \\ \underline{2a - 2b + 1} \end{array}$

24. $\begin{array}{l} 3a^2 + \ ab - 5b^2 \\ \underline{\ a^2 - 3ab + 2b^2} \end{array}$

Multiply the polynomials.

25. $5(3x + 7y)(x - y)$

26. $ab^3(a - 3b)(a + 3b)$

27. $(2x + y)^3$

28. $(2a - 7b)(a + b)(5a - b)$

29. $(x - y)(x^2 + xy + y^2)$

30. $(2a - b)[a^2 - (2a - b)(a - b)]$

In an ecological study the final result $F(t)$ is given by $F(t) = E(t)H(t)$. Find $F(t)$ for the given $E(t)$ and $H(t)$.

31. $E(t) = t^2 - t,\ H(t) = 3t^2$

32. $E(t) = 3t + 4,\ H(t) = 2t - 3$

33. $E(t) = 3t^2 + 2t - 1,\ H(t) = 2t + 1$

34. $E(t) = -t^2 + 3t - 2,\ H(t) = t^2 + 3t - 4$

35. Find $F(3)$ for $F(t)$ in Exercise 31.

36. Find $F(-3)$ for $F(t)$ in Exercise 32.

37. Find $F(-1)$ for $F(t)$ in Exercise 33.

38. Find $F(2)$ for $F(t)$ in Exercise 34.

Use the special product formulas to find the products.

39. $(x + 7)(x - 7)$

40. $(a - 10)^2$

41. $(b + 9)^2$

42. $(2x + 7)(2x - 7)$

43. $(2x + 7y)(2x - 7y)$

44. $(9a + b)^2$

45. $(4a - 9b)^2$

46. $(10x + 3y)^2$

47. $\left(\frac{1}{3}x - y\right)\left(\frac{1}{3}x + y\right)$

48. $(0.2x + 0.5y)^2$

49. $(x^2 + 3)(x^2 - 3)$

50. $(x^2 - 3)^2$

51. $(a^2 - 2b^2)(a^2 + 2b^2)$

52. $(a^2 - 3b^2)^2$

53. $(5a^4 - 2b^2)(5a^4 + 2b^2)$

54. $(3a^2 + b^3)^2$

55. $(x + y)^2 - (x - y)^2$

56. $[a - (b - c)][a + (b - c)]$

57. $[(3x + 2y)(3x - 2y)]^2$

58. $[2x^2 + (x + 2y)][2x^2 - (x + 2y)]$

59. $(x + 5y)(x - 5y)(x^2 + 25y^2)$

60. $(2u + 3w)(2u - 3w)(4u^2 + 9w^2)$

If Susie Wong deposits A dollars in a savings account at i interest rate, the amount in her account at the end of 2 years is given by the equation $S = A(1 + i)^2$.

61. Obtain an equivalent equation by performing all the indicated multiplications.

62. Check your work in Exercise 61 by using both formulas to find S when $A = \$2000$ and $i = 0.14$.

FOR REVIEW

Tell whether the following are monomials, binomials, or trinomials and give the degree of each.

63. $7x^4 + 2x + 1$

64. $a^3b^2 - 5a^5b^3$

65. 0

66. 25

Perform the indicated operations.

67. $(3a^2 + 2b^2 - 5) - (-4a^2 + 3b^2 + 7) - (2a^2 - 8)$

68. $(3a^3b^3 + 5a^2b^2 - 4) - (-5a^3b^3 + 10ab + 7) - (6a^2b^2 - 3ab + 4)$

Exercises 69–72 review material from Section 1.5, to help you prepare for the next section. Use the distributive law to factor each expression.

69. $3x + 6y$

70. $ax + 7x$

71. $2x + 6y - 12z$

72. $xa - 6a + ya$

5.2 EXERCISES C

Multiply. Assume that m and n represent positive integers.

1. $[(x^n)^{2n}]^2$
[Answer: x^{4n^2}]

2. $(a^n)^{n^2}(a^{2n})^3$

3. $x^2y^n(x^ny^3 - xy^n)$

4. $(x^n + y^m)(x^n - 2y^m)$

5. $a^2(a^{m+n})^{m-n}$

6. $(a^2)^{mn}(a^{m-n})^{m-n}$
[Answer: $a^{m^2+n^2}$]

7. $(2x^n + 1)^2$
[Answer: $4x^{2n} + 4x^n + 1$]

8. $(2x^n - 1)^2$

9. $(2x^n - 1)(2x^n + 1)$

10. $(x^{2n} + y^{2n})(x^n + y^n)(x^n - y^n)$
[Answer: $x^{4n} - y^{4n}$]

11. $(x^n - y^n)^2(x^n + y^n)^2$

12. $(x^{n+1} + y^{n+1})(x^{n+1} - y^{n+1})$
[Answer: $x^{2n+2} - y^{2n+2}$]

13. $(a^n + b^n)^2$

14. $(2a^n - 3b^n)(a^n + b^n)$

5.3 COMMON FACTORS AND GROUPING

STUDENT GUIDEPOSTS

1 Greatest Common Factor (GCF)
2 Factoring by Grouping

1 GREATEST COMMON FACTOR (GCF)

We begin our study of factoring with polynomials having a common factor in each term. Remember that a factor of a term is a number or power of a variable that is a divisor of the term. For example,

2, x, x^2 are factors of $2x^2$

2, 3, 6, x, y are factors of $6xy$.

From the above we see that $2x^2$ and $6xy$ have *common factors* of 2 and x, while

$2x$ is the *greatest common factor*. The **greatest common factor (GCF)** of a polynomial is the common factor of each term containing the largest numerical factor and the highest power of each variable. We factor an expression like $2x^2 - 6xy$ by factoring out the GCF.

$$2x^2 - 6xy = 2x(x - 3y)$$

Notice that factoring uses the distributive law in reverse of the order used to multiply polynomials. Thus, factoring can be checked by multiplying. We shall see later that the process of factoring will be useful in such operations as simplifying fractions and solving equations.

To factor a polynomial such as $3x^2y + 6xy^2$, we could first factor each term into prime factors,

$$3x^2y + 6xy^2 = 3 \cdot x \cdot x \cdot y + 2 \cdot 3 \cdot x \cdot y \cdot y,$$

and then note that $3xy$ is the greatest common factor. To complete the factoring, we write

$$3x^2y + 6xy^2 = 3xy(x + 2y).$$

The step of factoring into primes may be omitted once we gain experience at recognizing common factors.

The expression $3x^2y + 6xy^2$ could also have been factored as $3(x^2y + 2xy^2)$ or $3x(xy + 2y^2)$, but the remaining binomial in each case still has a common factor in each term. When the instruction is given to factor an expression we will always factor completely by removing the greatest common factor.

EXAMPLE 1 FINDING THE GREATEST COMMON FACTOR

Find the GCF of each polynomial.

(a) $12x^2y^3 + 20xy^2 = 2 \cdot 2 \cdot 3 \cdot x \cdot x \cdot y^2 \cdot y + 2 \cdot 2 \cdot 5 \cdot x \cdot y^2$

The GCF is $2 \cdot 2 \cdot x \cdot y^2 = 4xy^2$.

(b) $27a^3b^3 - 18a^4b^2 = 3 \cdot 3 \cdot 3 \cdot a^3 \cdot b^2 \cdot b - 2 \cdot 3 \cdot 3 \cdot a^3 \cdot a \cdot b^2$

The GCF is $3 \cdot 3 \cdot a^3 \cdot b^2 = 9a^3b^2$.

PRACTICE EXERCISE 1

Find the GCF.

(a) $18x^4y + 24x^3y^2$

(b) $35a^5b^3 - 42a^2b^5$

Answers: (a) $6x^3y$ (b) $7a^2b^3$

EXAMPLE 2 FACTORING BY REMOVING THE GCF

Factor each polynomial.

(a) $5x^2 + 10xy = 5 \cdot x \cdot x + 2 \cdot 5 \cdot x \cdot y$ GCF is $5x$

$= 5x(x + 2y)$ Distributive law

(b) $3x^2y + 6xy^2 = 3 \cdot x \cdot x \cdot y + 2 \cdot 3 \cdot x \cdot y \cdot y$ GCF is $3xy$

$= 3xy(x + 2y)$ Distributive law

(c) $12x^2y^3 + 20xy^2$

$= 2 \cdot 2 \cdot 3 \cdot x \cdot x \cdot y^2 \cdot y + 2 \cdot 2 \cdot 5 \cdot x \cdot y^2$ GCF is $4xy^2$

$= 4xy^2(3xy + 5)$ Distributive law

With practice we can shorten our work.

(d) $27a^3b^3 - 18a^4b^2 = (9a^3b^2) \cdot 3b - (9a^3b^2) \cdot 2a$ GCF is $9a^3b^2$

$= 9a^3b^2(3b - 2a)$ Distributive law

(e) $2a^2b - ab = (ab) \cdot 2a - (ab) \cdot 1$ GCF is ab

$= ab(2a - 1)$ The 1 helps us remember what remains after ab is factored out

PRACTICE EXERCISE 2

Factor.

(a) $7x^3 + 14x^2y$

(b) $4x^3y + 12x^2y^2$

(c) $32x^4y^4 - 12x^2y^2$

(d) $6a^5b^3 - 24a^4b^5$

(e) $5x^2y^2 - 5xy$

(f) $7x^2 - y^2 = 7x^2 - y^2$ GCF is 1; cannot be factored

(g) $4a^2b^3 - 6a^2b^2 + 8ab^2$
$= (2ab^2) \cdot 2ab - (2ab^2) \cdot 3a + (2ab^2) \cdot 4$ GCF is $2ab^2$
$= 2ab^2(2ab - 3a + 4)$

(h) $-6x^3y^3 - 12x^2y^2 - 18xy$
$= -(6xy) \cdot x^2y^2 - (6xy) \cdot 2xy - (6xy) \cdot 3$ GCF is $6xy$
$= 6xy(-x^2y^2 - 2xy - 3)$
$= -6xy(x^2y^2 + 2xy + 3)$ -1 is factored out

(f) $3a^2b^2 + c^2$

(g) $2a^4b^2 - 10a^3b^3 - 6a^2b^4$

(h) $-5xy^2 - 15x^2y + 10x^2y^2$

Answers: (a) $7x^2(x + 2y)$
(b) $4x^2y(x + 3y)$
(c) $4x^2y^2(8x^2y^2 - 3)$
(d) $6a^4b^3(a - 4b^2)$
(e) $5xy(xy - 1)$
(f) cannot be factored
(g) $2a^2b^2(a^2 - 5ab - 3b^2)$
(h) $-5xy(y + 3x - 2xy)$

In Example 2(h) two answers are correct. Most of the time we factor out negative numbers, if necessary, to leave a plus sign in the first term of the remaining polynomial.

In some cases polynomials may contain symbols of grouping, or we may introduce grouping symbols to aid in factoring. For example, the polynomial

$$5x(x + y) + 7(x + y)$$

has the common factor $x + y$ in each of its terms. If we let the variable u equal $x + y$, the expression becomes $5xu + 7u$ and can be factored as follows.

$5xu + 7u = 5x \cdot u + 7 \cdot u$ GCF is u
$= (5x + 7)u$
$= (5x + 7)(x + y)$ Since $u = x + y$

We may omit the substitution of the variable u when we see that a grouped expression like $(x + y)$ can be considered as a single factor.

EXAMPLE 3 COMMON BINOMIAL FACTORS

Factor each polynomial.

(a) $2a(a + 2b) + 3b(a + 2b) = 2au + 3bu$ Let $u = a + 2b$
$= (2a + 3b)u$ The GCF is u
$= (2a + 3b)(a + 2b)$ Since $u = a + 2b$

(b) $7x(2x - 5y) - y(2x - 5y) = (7x - y)(2x - 5y)$ GCF is $2x - 5y$

(c) $5a(b - 1) - (b - 1) = 5a(b - 1) - (1)(b - 1)$ GCF is $b - 1$
$= (5a - 1)(b - 1)$ Distributive law

PRACTICE EXERCISE 3

Factor.

(a) $5x(2x - 7y) + 4y(2x - 7y)$

(b) $9a(a - 3b) - b(a - 3b)$

(c) $6x(y - 2) - (y - 2)$

Answers: (a) $(5x + 4y)(2x - 7y)$
(b) $(9a - b)(a - 3b)$
(c) $(6x - 1)(y - 2)$

2 FACTORING BY GROUPING

The ability to factor out a common binomial factor, as shown in Example 3, allows us to **factor by grouping.** With this process we group terms, factor the individual groups, and then factor the complete expression.

EXAMPLE 4 Factoring By Grouping

Factor by grouping.

(a) $3x^2 + 6xy + 5x + 10y$

$= (3x^2 + 6xy) + (5x + 10y)$

$= 3x(x + 2y) + 5(x + 2y)$ Factor $3x$ from first parentheses and 5 from second using the distributive law

$= (3x + 5)(x + 2y)$ Since $x + 2y$ is the GCF

(b) $6ab^2 - 3ab - 14b + 7$

$= (6ab^2 - 3ab) - (14b - 7)$ Note the -7, not 7

$= 3ab(2b - 1) - 7(2b - 1)$ $3ab$ from first parentheses and 7 from second

$= (3ab - 7)(2b - 1)$ Since $2b - 1$ is the GCF

Practice Exercise 4

Factor by grouping.

(a) $7x^3 + 14x^2y + 3xy + 6y^2$

(b) $12a^2b - 4ab - 15a + 5$

Answers: (a) $(7x^2 + 3y)(x + 2y)$ (b) $(4ab - 5)(3a - 1)$

Sometimes an expression can be grouped in more than one way. Consider the expression in Example 4(b).

$6ab^2 - 3ab - 14b + 7$

$= 6ab^2 - 14b - 3ab + 7$ Commutative law

$= 2b(3ab - 7) - 1(3ab - 7)$ Watch the sign here

$= (2b - 1)(3ab - 7)$

Notice that we obtain the same pair of factors as before; they are simply written in a different order.

Generally, we first try to find any common factors in all the terms of an expression. Once we have done this, if the expression has four terms, we try to factor by grouping, as illustrated in Example 5.

EXAMPLE 5 Factoring the GCF and by Grouping

Factor.

$4x^3y + 2x^2y^2 - 2x^2y - xy^2$

$= xy[4x^2 + 2xy - 2x - y]$ Remove greatest common factor

$= xy[(4x^2 + 2xy) - (2x + y)]$ $2x + y$, not $2x - y$

$= xy[2x(2x + y) - (1)(2x + y)]$ $2x$ from first parentheses and 1 from second

$= xy[(2x - 1)(2x + y)]$ Since $2x + y$ is the GCF

$= xy(2x - 1)(2x + y)$

Practice Exercise 5

Factor.

$6x^4y + 3x^3y^2 - 14x^3 - 7x^2y$

Answer: $x^2(3xy - 7)(2x + y)$

Factoring can simplify calculations in some applications, as in Example 6.

EXAMPLE 6 Physics Problem

If an object is propelled upward with initial velocity of 96 ft/sec from an initial height of 240 ft, ignoring air resistance, the height h of the object in feet after t seconds is given by $h = -16t^2 + 96t + 240$. Factor the

Practice Exercise 6

Refer to Example 6 to find the height of an object after 3 seconds.

greatest common factor from the right side of this equation and use the factored form to find the height of an object 5 seconds after it is propelled upward.

Factoring, we obtain

$$\begin{aligned} h &= -16t^2 + 96t + 240 \\ &= -16(t^2 - 6t - 15). \end{aligned}$$

Substitute 5 for t to obtain the height after 5 seconds.

$$\begin{aligned} h &= -16(5^2 - 6(5) - 15) \\ &= -16(25 - 30 - 15) \\ &= -16(-20) = 320 \end{aligned}$$

Thus, the object is 320 ft above the ground after 5 seconds.

Answer: 384 ft

5.3 EXERCISES A

Factor. First try to remove the GCF from each polynomial. If the expression has four terms, try to factor by grouping.

1. $5x + 10y$

2. $12a - 3b$

3. $14x + 7$

4. $5a^2b - 10ab^2$

5. $6x^2y^2 + 12xy$

6. $2ab^2 - 2ab$

7. $3a^2b^4 - 5$

8. $24x^8y^4 + 9x^3y^2$

9. $7a^2b - 14ab^2 + 28ab$

10. $-3x^5y^5 - 6x^4y^4 - 9x^6y^6$

11. $42a^5b^4 - 66a^4b^5 - 36a^4b^4$

12. $15a^3b^2 - 45a^2b^3 - 60a^2b$

13 $64xy^5z^2 - 128x^2y^3z^3 - 144xy^2z^4$

14 $3a(a + b) - 2b(a + b)$

15. $7x(5x - y) + 2y(5x - y)$

16. $2a^2(a + 4b) - 9b(a + 4b)$

17 $5x^2 + 2xy + 10x + 4y$

18. $a^2b - ab^2 + 5a - 5b$

19. $9x^2y - 18xy^2 - xy + 2y^2$

20 $2a^2b + 6ab - 3a - 9$

21. $x^3y^2 - y^3 + x^3y - y^2$

22. $a^3b^2c^3 - a^3c^2 - b^2c^4 + c^3$

23. $5x^2y - 5xy + 20x - 20$

24 $7x^3y^3 + 21x^2y^2 - 10x^3y^2 - 30x^2y$

An economist finds that certain economic conditions are related to the equation $E = 25c^2e^3 - 10c^3e^2$.

 25 Factor the expression for E.

26. Check your factoring in Exercise 25 by evaluating both expressions for $c = 2$ and $e = 3$.

If an object is propelled upward from a tower with initial velocity of v ft/sec from an initial height of s ft, ignoring air resistance, the height h of the object in feet after t seconds is given by $h = -16t^2 + vt + s$. Use the given values of v and s in Exercises 27–28, and express h as a product of factors. Use the factored form to find the height of an object that is propelled upward after **(a)** 3 *seconds and* **(b)** 10 *seconds.*

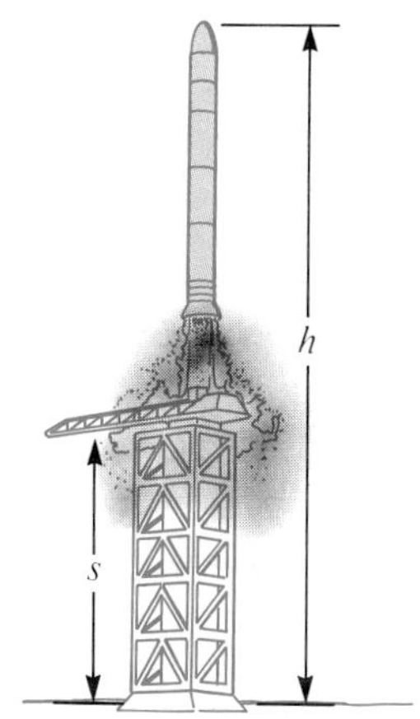

27. $v = 144$ ft/sec and $s = 160$ ft

28. $v = 112$ ft/sec and $s = 1920$ ft

FOR REVIEW

Multiply.

29. $6x^2y(5x^2 - 4xy + 10y^2)$

30. $(2x - y)(4x - 3y)$

31. $(5a - 3b)(10a + 7b)$

32. $(7x + 3y)(8x^2 - 4xy + 5y^2)$

33. $(3a + 2b)(3a - 2b)$

34. $(-4x + 3y)^2$

ANSWERS: 1. $5(x + 2y)$ 2. $3(4a - b)$ 3. $7(2x + 1)$ 4. $5ab(a - 2b)$ 5. $6xy(xy + 2)$ 6. $2ab(b - 1)$ 7. cannot be factored 8. $3x^3y^2(8x^5y^2 + 3)$ 9. $7ab(a - 2b + 4)$ 10. $-3x^4y^4(xy + 2 + 3x^2y^2)$ 11. $6a^4b^4(7a - 11b - 6)$ 12. $15a^2b(ab - 3b^2 - 4)$ 13. $16xy^2z^2(4y^3 - 8xyz - 9z^2)$ 14. $(3a - 2b)(a + b)$ 15. $(7x + 2y)(5x - y)$ 16. $(2a^2 - 9b)(a + 4b)$ 17. $(x + 2)(5x + 2y)$ 18. $(ab + 5)(a - b)$ 19. $y(9x - 1)(x - 2y)$ 20. $(2ab - 3)(a + 3)$ 21. $y(y + 1)(x^3 - y)$ 22. $c^2(a^3 - c)(b^2c - 1)$ 23. $5(xy + 4)(x - 1)$ 24. $x^2y(7y - 10)(xy + 3)$ 25. $E = 5c^2e^2(5e - 2c)$ 26. 1980 27. (a) 448 ft (b) 0 ft (the object is on the ground) 28. (a) 2112 ft (b) 1440 ft 29. $30x^4y - 24x^3y^2 + 60x^2y^3$ 30. $8x^2 - 10xy + 3y^2$ 31. $50a^2 + 5ab - 21b^2$ 32. $56x^3 - 4x^2y + 23xy^2 + 15y^3$ 33. $9a^2 - 4b^2$ 34. $16x^2 - 24xy + 9y^2$

5.3 EXERCISES B

Factor. First try to remove the GCF from each polynomial. If the expression has four terms, try to factor by grouping.

1. $6x - 9y$

2. $15a + 10b$

3. $18y - 9$

4. $14a^2b^2 + 21ab$

5. $10x^2y - 5xy^2$

6. $16a^2b^2 - 6ab$

7. $7a^3b^3 + 6$

8. $20x^9y^2 - 15x^6y^6$

9. $-6a^5b^5 - 8a^4b^4 - 4a^3b^2$

10. $11xy^2 - 22x^2y - 44xy$

11. $96a^4b^4 + 72a^9b^6 + 108a^3b^6$

12. $1000a^3b^3 + 10ab - 100a^2b^2$

13. $35x^9y^9z^9 - 49x^7y^7z^7 - 77x^5y^5z^5$

14. $5ab(2a - b) + 4b(2a - b)$

15. $9x(3x + 2y) - 5y(3x + 2y)$

16. $5a^2(3a - 4b) + 12b^2(3a - 4b)$

17. $7x^2 - 14xy + 4x - 8y$

18. $a^3 + a^2 + ab^2 + b^2$

19. $5x^2y^2 + 10x^2 - 7y^2 - 14$

20. $3a^2b^2 - 9ab^2 - 5a + 15$

21. $x^3y^3 + y^4 - x^3y^2 - y^3$

22. $a^2b^2c^2 - a^2b^2c - b^2c^3 + b^2c^2$

23. $6x^2y^2 + 6xy - 24xy^2 - 24y$

24. $6x^4y^4 - 6x^3y^2 + 8xy^2 - 8$

An environmentalist finds that $E = 20r^4s^2 - 35r^2s$.

25. Factor the expression for E.

26. Check your factoring in Exercise 25 by evaluating both expressions for $r = 3$ and $s = -1$.

If an object is propelled upward from a tower with initial velocity of v ft/sec from an initial height of s ft, ignoring air resistance, the height h of the object in feet after t seconds is given by $h = -16t^2 + vt + s$. *Use the given values of v and s in Exercises 27–28, and express h as a product of factors. Use the factored form to find the height of an object that is propelled upward after* ***(a)*** *3 seconds and* ***(b)*** *10 seconds.*

27. $v = 80$ ft/sec and $s = 1600$ ft

28. $v = 160$ ft/sec and $s = 2800$ ft

FOR REVIEW

Multiply.

29. $-8xy^2(4x^2y^2 + 3xy - 5)$

30. $(3x + y)(5x - 7y)$

31. $(4a - 5b)(9a - 2b)$

32. $(2x - 3y)\left(\frac{1}{3}x^2 + \frac{1}{2}xy + \frac{2}{3}y^2\right)$

33. $(5a + 2b)^2$

34. $(2x^2 - y)(2x^2 + y)$

5.3 EXERCISES C

Factor. Assume that n represents a positive integer.

1. $x^{n+2}y^{n+2} - x^ny^n$

2. $x^{3n}y^{n+1} + 5x^{2n}y^n$
[*Hint:* $x^{2n}y^n$ is a common factor of both terms.]

3. $x^ny + y + 2x^n + 2$
[Answer: $(x^n + 1)(y + 2)$]

4. $x^ny^n - y^n + 2x^n - 2$

5.4 FACTORING TRINOMIALS

STUDENT GUIDEPOSTS

1. Factoring $x^2 + bx + c$
2. Factoring $ax^2 + bx + c$, $a \neq 1$
3. Factoring Trinomials by Grouping
4. Factoring Using Substitution

1 FACTORING $x^2 + bx + c$

We begin our study of factoring trinomials of the form $ax^2 + bx + c$ by considering the case where a is 1. Suppose we use the FOIL method to obtain a trinomial product.

$$\overset{\text{F, O, I, L}}{(x + 5)(x - 7)} = \underset{\text{(F)}}{x \cdot x} + \underset{\text{(O)}}{x \cdot (-7)} + \underset{\text{(I)}}{5 \cdot x} + \underset{\text{(L)}}{5 \cdot (-7)}$$
$$= x^2 - 7x + 5x - 35$$
$$= \underset{\text{(F)}}{x^2} \quad \underset{\text{(O)+(I)}}{- 2x} \quad \underset{\text{(L)}}{- 35}$$

The product of the first terms, (F), gives x^2, the product of the last terms, (L), gives -35, and the sum of (O) and (I) gives the middle term, $-2x$. To factor $x^2 - 2x - 35$, we reverse the multiplication steps above and fill in the blanks in

$$x^2 - 2x - 35 = (x + \_\_)(x + \_\_).$$

The numbers in the blanks must multiply to give -35 and must add to give -2.

$$x^2 - 2x - 35 = (x + 5)(x - 7)$$

(sum: $5 + (-7) = -2$; product: $5 \cdot (-7) = -35$)

To Factor $x^2 + bx + c$

1. Write $x^2 + bx + c = (x + \_\_)(x + \_\_)$.
2. List all pairs of integers whose product is c.
3. Fill in the blanks with the pair from the list whose sum is b.
4. Check your work by multiplying the factors to see if their product equals the original trinomial.

EXAMPLE 1 FACTORING TRINOMIALS OF THE FORM $x^2 + bx + c$

Factor each trinomial.

(a) $x^2 + 7x + 10$ $\quad (b = 7$ and $c = 10)$

$x^2 + 7x + 10 = (x + \_\_)(x + \_\_)$

Factors of c = 10	*Sum of factors*
10, 1	$10 + 1 = 11$
2, 5	$2 + 5 = 7$
−10, −1	$-10 + (-1) = -11$
−2, −5	$-2 + (-5) = -7$

PRACTICE EXERCISE 1

Factor.

(a) $x^2 + 7x + 12$

Since the product of 2 and 5 is 10, which equals c, and the sum of 2 and 5 is 7, which equals b, we write

$$x^2 + 7x + 10 = (x + 2)(x + 5).$$

Note that since $c = 10 > 0$, the factors in each pair must have the same sign. Also, since $b = 7 > 0$, negative factors will not work. As a final check of our work, we use FOIL to multiply.

$$\begin{aligned}(x + 2)(x + 5) &= x^2 + 5x + 2x + 10 \\ &= x^2 + 7x + 10\end{aligned}$$

(b) $x^2 - 7x + 10 \quad (b = -7 \text{ and } c = 10)$

$x^2 - 7x + 10 = (x + \_\_)(x + \_\_)$

Factors of $c = 10$	*Sum of factors*
10, 1	$10 + 1 = 11$
2, 5	$2 + 5 = 7$
−10, −1	$-10 + (-1) = -11$
−2, −5	$-2 + (-5) = -7$

$x^2 - 7x + 10 = (x - 2)(x - 5)$ Check using FOIL

Since $b = -7$, positive factors will not work.

(c) $x^2 + 3x - 10 \quad (b = 3 \text{ and } c = -10)$

$x^2 + 3x - 10 = (x + \_\_)(x + \_\_)$

Factors of $c = -10$	*Sum of factors*
10, −1	$10 + (-1) = 9$
−10, 1	$-10 + 1 = -9$
2, −5	$2 + (-5) = -3$
−2, 5	$-2 + 5 = 3$

$x^2 + 3x - 10 = (x - 2)(x + 5)$ Check using FOIL

Since $c = -10 < 0$, the factors of c must have opposite signs.

(d) $x^2 - 3x - 10 \quad (b = -3 \text{ and } c = -10)$

$x^2 - 3x - 10 = (x + \_\_)(x + \_\_)$

Factors of $c = -10$	*Sum of factors*
10, −1	$10 + (-1) = 9$
−10, 1	$-10 + 1 = -9$
2, −5	$2 + (-5) = -3$
−2, 5	$-2 + 5 = 3$

$x^2 - 3x - 10 = (x + 2)(x - 5)$ Check using FOIL

(b) $x^2 - 7x + 12$

(c) $x^2 + x - 12$

(d) $x^2 - x - 12$

Answers: (a) $(x + 3)(x + 4)$ (b) $(x - 3)(x - 4)$ (c) $(x - 3)(x + 4)$ (d) $(x + 3)(x - 4)$

Our results in Example 1 would have been similar had the trinomials been in two variables. Compare Example 1 with the following.

$$\begin{aligned}x^2 + 7xy + 10y^2 &= (x + 2y)(x + 5y) \\ x^2 - 7xy + 10y^2 &= (x - 2y)(x - 5y) \\ x^2 + 3xy - 10y^2 &= (x - 2y)(x + 5y) \\ x^2 - 3xy - 10y^2 &= (x + 2y)(x - 5y)\end{aligned}$$

Thus, factoring with two or more variables can be as easy as with one variable. The only difference is that we are filling in the blanks in

$$x^2 + bxy + cy^2 = (x + \_\_y)(x + \_\_y).$$

EXAMPLE 2 Factoring Trinomials in Two Variables

Factor $x^2 - 8xy + 12y^2$.

$$x^2 - 8xy + 12y^2 = (x + \_\_y)(x + \_\_y)$$

Factors of $c = 12$	*Sum of factors*
12, 1	$12 + 1 = 13$
6, 2	$6 + 2 = 8$
4, 3	$4 + 3 = 7$
−12, −1	$-12 + (-1) = -13$
−6, −2	$-6 + (-2) = -8$
−4, −3	$-4 + (-3) = -7$

$$x^2 - 8xy + 12y^2 = (x - 6y)(x - 2y)$$

To check we multiply.

$$(x - 6y)(x - 2y) = x^2 - 2xy - 6xy + 12y^2 = x^2 - 8xy + 12y^2$$

Practice Exercise 2

Factor $x^2 - 11xy + 28y^2$.

Answer: $(x - 4y)(x - 7y)$

When factoring trinomials, remember to remove any common factors first, as in the next example.

EXAMPLE 3 Removing the GCF First

Factor $2a^2 - 10ab - 28b^2$.

First factor out the common factor 2.

$$2a^2 - 10ab - 28b^2 = 2(a^2 - 5ab - 14b^2)$$

Now factor the trinomial.

Factors of $c = -14$	*Sum of factors*
14, −1	$14 + (-1) = 13$
−14, 1	$-14 + 1 = -13$
7, −2	$7 + (-2) = 5$
−7, 2	$-7 + 2 = -5$

Thus we know $a^2 - 5ab - 14b^2 = (a - 7b)(a + 2b)$ and $2a^2 - 10ab - 28b^2 = 2(a - 7b)(a + 2b)$.

Practice Exercise 3

Factor $3x^2 + 12x - 36$.

Answer: $3(x - 2)(x + 6)$

CAUTION

When factoring a trinomial with a common factor, don't forget to include that factor in the final result. In Example 3 the common factor 2 is included with the factors $(a - 7b)$ and $(a + 2b)$.

2 FACTORING $ax^2 + bx + c$, $a \neq 1$

When factoring $ax^2 + bx + c$ with $a \neq 1$, we use the following trial and error method.

To Factor $ax^2 + bx + c$

1. Write $ax^2 + bx + c = (\_x + \_)(\_x + \_)$.
2. List all pairs of integers whose product is a and use these in the first blanks in each binomial factor.
3. List all pairs of integers whose product is c and use these in the second blanks.
4. Use trial and error to determine which pairs give the correct value for b.
5. Check your work by multiplying the factors.

EXAMPLE 4 Factoring $ax^2 + bx + c$ when $a \neq 1$

Factor $3x^2 + 17x + 10$. $(a = 3, b = 17, c = 10)$

Factors of 10

$$3x^2 + 17x + 10 = (\_x + \_)(\_x + \_)$$

Factors of 3

Since all terms of the trinomial are positive, list only the positive factors of a and c. List the factors of c in both orders as a reminder to try all possibilities.

Factors of a	*Factors of c*
3, 1	10, 1
	1, 10
	5, 2
	2, 5

$3x^2 + 17x + 10$

$= (\_x + \_)(\_x + \_)$

$= (3x + \_)(x + \_)$ The only factors of 3 are 3 and 1

$\stackrel{?}{=} (3x + 10)(x + 1)$ Does not work since $3x + 10x = 13x$

$\stackrel{?}{=} (3x + 1)(x + 10)$ Does not work since $30x + x = 31x$

$\stackrel{?}{=} (3x + 5)(x + 2)$ Does not work since $6x + 5x = 11x$

$\stackrel{?}{=} (3x + 2)(x + 5)$ This works since $15x + 2x = 17x$

$$3x^2 + 17x + 10 = (3x + 2)(x + 5)$$

To check we multiply.

$$(3x + 2)(x + 5) = 3x^2 + 15x + 2x + 10 = 3x^2 + 17x + 10$$

PRACTICE EXERCISE 4

Factor $3x^2 + 13x + 14$.

Answer: $(3x + 7)(x + 2)$

When factoring $ax^2 + bxy + cy^2$ with $a \neq 1$ we fill in the blanks in

$$ax^2 + bxy + cy^2 = (\_x + \_y)(\_x + \_y).$$

EXAMPLE 5 FACTORING $ax^2 + bxy + cy^2$ when $a \neq 1$

Factor $8x^2 - 13xy + 5y^2$. $(a = 8,\ b = -13,\ c = 5)$

Factors of 5

$$8x^2 - 13xy + 5y^2 = (\_\_x + \_\_y)(\_\_x + \_\_y)$$

Factors of 8

Since $c = 5 > 0$ and $b = -13 < 0$, the factors of c must both be negative.

Factors of a	*Factors of c*
8, 1	−1, −5
1, 8	
4, 2	
2, 4	

$8x^2 - 13xy + 5y^2$

$= (\_\_x + \_\_y)(\_\_x + \_\_y)$

$= (\_\_x - y)(\_\_x - 5y)$ The only factors of c are -1 and -5

$\stackrel{?}{=} (8x - y)(x - 5y)$ Does not work since $-40xy - xy = -41xy$

$\stackrel{?}{=} (x - y)(8x - 5y)$ This works because $-5xy - 8xy = -13xy$

$$8x^2 - 13xy + 5y^2 = (x - y)(8x - 5y)$$

To check we multiply.

$$(x - y)(8x - 5y) = 8x^2 - 5xy - 8xy + 5y^2$$
$$= 8x^2 - 13xy + 5y^2$$

PRACTICE EXERCISE 5

Factor $6x^2 - 31xy + 35y^2$.

Answer: $(3x - 5y)(2x - 7y)$

In general when factoring we insist that all factors be polynomials with integer coefficients. This means that there are many polynomials that cannot be factored. For example, $x^2 + x + 1$, $x^2 + 2xy + 2y^2$, $3u^2 - 4uv + 8v^2$, and $5x^2 - 2xy - y^2$ cannot be factored into binomials with integer coefficients. To show that $x^2 + 2xy + 2y^2$ cannot be factored, we can list the factors of $c = 2$ and see that they do not add to give $b = 2$.

Factors of $c = 2$	*Sum of factors*
1, 2	$1 + 2 = 3$

Since 1 and 2 are the only positive factors of $c = 2$, and $1 + 2 = 3 \neq 2$, the trinomial cannot be factored.

Always remove common factors first, as in the next example.

EXAMPLE 6 REMOVING THE GCF FIRST

Factor $-6x^2 - 21xy + 45y^2$.

We first factor out the common factor and make $a > 0$.

$$-6x^2 - 21xy + 45y^2 = -3(2x^2 + 7xy - 15y^2)$$

Next, we list the factors of a and c for the trinomial $2x^2 + 7xy - 15y^2$ $(a = 2,\ b = 7,\ c = -15)$.

PRACTICE EXERCISE 6

Factor $-10x^2 - 95xy + 50y^2$.

Factors of a	*Factors of c*		
2, 1	15, −1	and	−15, 1
	1, −15	and	−1, 15
	5, −3	and	−5, 3
	3, −5	and	−3, 5

$-6x^2 - 21xy + 45y^2 = -3(2x^2 + 7xy - 15y^2)$ First factor out the common factor

$= -3(2x + \_\_y)(x + \_\_y)$

$\stackrel{?}{=} -3(2x + 5y)(x - 3y)$ Does not work

$\stackrel{?}{=} -3(2x - 5y)(x + 3y)$ Does not work

$\stackrel{?}{=} -3(2x + 3y)(x - 5y)$ Does not work

$\stackrel{?}{=} -3(2x - 3y)(x + 5y)$ Works

$$-6x^2 - 21xy + 45y^2 = -3(2x - 3y)(x + 5y)$$

Check this by multiplying.

Answer: $-5(2x - y)(x + 10y)$

3 FACTORING TRINOMIALS BY GROUPING

Another method for factoring trinomials involves factoring by grouping. Consider the following.

$2x^2 + 11xy + 12y^2 = 2x^2 + 8xy + 3xy + 12y^2$ $\quad 11xy = 8xy + 3xy$

$= (2x^2 + 8xy) + (3xy + 12y^2)$ Group terms

$= 2x(x + 4y) + 3y(x + 4y)$ Factor the groups

$= (2x + 3y)(x + 4y)$ Factor out the common factor $x + 4y$

We have factored $2x^2 + 11xy + 12y^2$ by grouping the appropriate terms. But how do we decide to write $11xy = 8xy + 3xy$? Notice that 8 and 3 are factors of $2 \cdot 12 = 24$. That is, 8 and 3 are factors of the product ac in $ax^2 + bxy + cy^2$.

To Factor $ax^2 + bxy + cy^2$ by Grouping

1. Find the product ac.
2. List the factors of ac until a pair is found which add to give b.
3. Write bxy as a sum using these factors as coefficients of xy.
4. Factor the result by grouping.

EXAMPLE 7 Factoring a Trinomial by Grouping

Factor $3x^2 - 8xy + 5y^2$ by grouping.

Factors of $ac = 3 \cdot 5 = 15$	*Sum of factors*
15, 1	$15 + 1 = 16$
−15, −1	$-15 + (-1) = -16$
5, 3	$5 + 3 = 8$
−5, −3	$-5 + (-3) = -8$

Practice Exercise 7

Factor $2x^2 - 11xy + 9y^2$ by grouping.

$$3x^2 - 8xy + 5y^2 = 3x^2 - 5xy - 3xy + 5y^2$$

$= x(3x - 5y) - y(3x - 5y)$ Factor out x and $-y$

$= (x - y)(3x - 5y)$ Common factor is $3x - 5y$

Check this by multiplying.

Answer: $(2x - 9y)(x - y)$

EXAMPLE 8 FACTORING A TRINOMIAL BY GROUPING

Factor $7u^2 - 33uv - 10v^2$ by grouping.

We try those factors of $ac = 7(-10) = -70$ which seem most likely to add to $b = -33$.

Factors of ac $= -70$	*Sum of factors*
$35, -2$	$35 + (-2) = 33$
$-35, 2$	$-35 + 2 = -33$

$7u^2 - 33uv - 10v^2 = 7u^2 - 35uv + 2uv - 10v^2$ $7u^2 + 2uv - 35uv - 10v^2$ also works

$= 7u(u - 5v) + 2v(u - 5v)$ Factor groups

$= (7u + 2v)(u - 5v)$ Common factor is $u - 5v$

Check this by multiplying.

PRACTICE EXERCISE 8

Factor $5x^2 + 4xy - 33y^2$ by grouping.

Answer: $(5x - 11y)(x + 3y)$

4 FACTORING USING SUBSTITUTION

The factors of some polynomials can be more easily found if we make a simplifying substitution. For example, $2(x - y)^2 - 3(x - y) + 1$ can be factored by letting $u = x - y$. Thus,

$2(x - y)^2 - 3(x - y) + 1$

$= 2u^2 - 3u + 1$ Let $u = x - y$

$= (u - 1)(2u - 1)$

$= [(x - y) - 1][2(x - y) - 1]$ Return to the original variables

$= (x - y - 1)(2x - 2y - 1).$

EXAMPLE 9 FACTORING A TRINOMIAL BY SUBSTITUTION

Factor $x^4 + 5x^2 + 6$.

First we note that $x^4 = (x^2)^2$ and then let $u = x^2$.

$x^4 + 5x^2 + 6 = (x^2)^2 + 5x^2 + 6$

$= u^2 + 5u + 6$ $u = x^2$

$= (u + 2)(u + 3)$

$= (x^2 + 2)(x^2 + 3)$ Return to the original variable

PRACTICE EXERCISE 9

Factor $x^4 - 5x^2 + 6$.

Answer: $(x^2 - 2)(x^2 - 3)$

5.4 EXERCISES A

Factor completely.

1. $x^2 + 6x + 5$

2. $x^2 - 6x + 5$

3. $x^2 + 4x - 5$

4. $x^2 - 4x - 5$

5. $y^2 - 4y - 21$

6. $y^2 - 10y + 21$

7. $y^2 + 4y - 21$

8. $y^2 + 10y + 21$

9. $w^2 + 4w - 45$

10. $w^2 - 2w - 63$

11. $x^2 - x - 56$

12. $x^2 + 15x + 56$

13. $x^2 + 2x + 4$

14. $y^2 + 12y + 36$

15. $u^2 + 21u + 110$

16. $u^2 - u - 110$

17. $x^2 + 6xy + 5y^2$

18. $x^2 - 6xy + 5y^2$

19. $x^2 + 4xy - 5y^2$

20. $x^2 - 4xy - 5y^2$

21. $u^2 - 2uv - 35v^2$

22. $2w^2 + 19w + 24$

23. $2w^2 - 13w - 24$

24 $-4w^2 - 34w - 70$

25. $7w^2 - 14w + 7$

26. $2x^2 + 13xy - 24y^2$

27. $2x^2 - 19xy + 24y^2$

28. $2x^2 + 4xy + y^2$

29. $2u^2v^2 - 7uv - 30$

30. $5x^2 + 17xy - 12y^2$

31 $6x^2 + 23xy + 20y^2$

32. $-45u^2 + 150uv - 125v^2$

33. $6x^2 - 19xy - 20y^2$

34. $4x^2 - 32xy + 63y^2$

35. $2u^2 - 7uv + 6v^2$

36. $21u^2 - 23uv - 20v^2$

Factor by substitution.

37. $(x + 2)^2 - 3(x + 2) + 2$

38. $(x + y)^2 + 6(x + y) + 8$

39. $y^4 + y^2 - 6$

40. $u^6 - 7u^3 + 12$

A laboratory assistant finds that her work in evaluation of lab equations is made easy if she factors the equations. Factor each of the equations and evaluate for $u = 5$ *and* $v = -2$.

41. $K = u^2 + 6uv + 8v^2$

42. $L = u^2 - 2uv - 8v^2$

43. $M = 3u^2 - 5uv - 2v^2$

44. $N = 10u^2 - 23uv + 12v^2$

45. When a surface-to-air missile is fired with a velocity of 656 ft/sec from a submarine at a depth of −640 ft, the height h in feet of the missile is given by $h = -16t^2 + 656t - 640$. When h is negative, the missile is still under water; when h is zero, the missile is at water level; and when h is positive, the missile is above the surface of the water. Factor the right side of the equation giving h and use the factored form to determine the location of the missile **(a)** 0.5 seconds, **(b)** 1 second, and **(c)** 20 seconds after it has been fired.

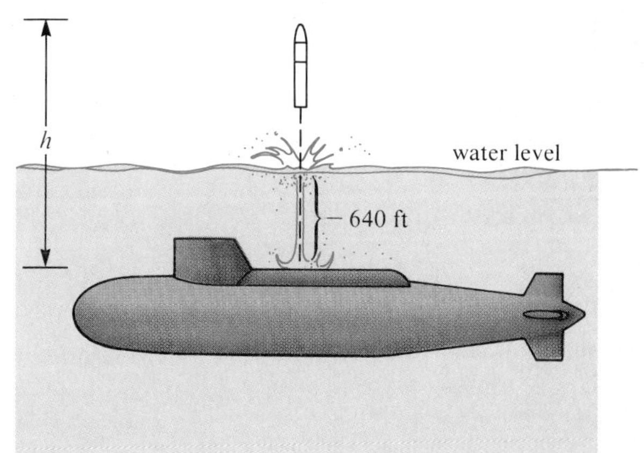

FOR REVIEW

Factor completely.

46. $20x^4y^2 + 28x^2y^3$

47. $25x^3y^3 - 35x^2y^3 + 45xy^3$

48. $2u^2 - 3uv - 4uv + 6v^2$

49. $21u^2 + 12uv - 35uv - 20v^2$

50. $4x^2 - 14xy + 10xy - 35y^2$

51. $12x^2 - 20xy - 9xy + 15y^2$

Exercises 52–55 review material from Section 5.2 to help you prepare for topics in the next section. Find each product.

52. $(y + 5)(y - 5)$ **53.** $(2x - 7y)(2x + 7y)$ **54.** $(3a + 4)^2$ **55.** $(5x - 2y)^2$

ANSWERS: 1. $(x + 1)(x + 5)$ 2. $(x - 1)(x - 5)$ 3. $(x - 1)(x + 5)$ 4. $(x + 1)(x - 5)$ 5. $(y + 3)(y - 7)$ 6. $(y - 3)(y - 7)$ 7. $(y - 3)(y + 7)$ 8. $(y + 3)(y + 7)$ 9. $(w - 5)(w + 9)$ 10. $(w + 7)(w - 9)$ 11. $(x + 7)(x - 8)$ 12. $(x + 7)(x + 8)$ 13. cannot be factored 14. $(y + 6)(y + 6)$ 15. $(u + 10)(u + 11)$ 16. $(u + 10)(u - 11)$ 17. $(x + y)(x + 5y)$ 18. $(x - y)(x - 5y)$ 19. $(x - y)(x + 5y)$ 20. $(x + y)(x - 5y)$ 21. $(u + 5v)(u - 7v)$ 22. $(w + 8)(2w + 3)$ 23. $(w - 8)(2w + 3)$ 24. $-2(w + 5)(2w + 7)$ 25. $7(w - 1)(w - 1)$ 26. $(x + 8y)(2x - 3y)$ 27. $(x - 8y)(2x - 3y)$ 28. cannot be factored 29. $(uv - 6)(2uv + 5)$ 30. $(x + 4y)(5x - 3y)$ 31. $(2x + 5y)(3x + 4y)$ 32. $-5(3u - 5v)(3u - 5v)$ 33. $(x - 4y)(6x + 5y)$ 34. $(2x - 7y)(2x - 9y)$ 35. $(u - 2v)(2u - 3v)$ 36. $(3u - 5v)(7u + 4v)$ 37. $x(x + 1)$ 38. $(x + y + 2)(x + y + 4)$ 39. $(y^2 - 2)(y^2 + 3)$ 40. $(u^3 - 3)(u^3 - 4)$ 41. $(u + 2v)(u + 4v)$; -3 42. $(u + 2v)(u - 4v)$; 13 43. $(3u + v)(u - 2v)$; 117 44. $(5u - 4v)(2u - 3v)$; 528 45. (a) -316 ft (b) 0 ft (c) 6080 ft 46. $4x^2y^2(5x^2 + 7y)$ 47. $5xy^3(5x^2 - 7x + 9)$ 48. $(u - 2v)(2u - 3v)$ 49. $(3u - 5v)(7u + 4v)$ 50. $(2x + 5y)(2x - 7y)$ 51. $(3x - 5y)(4x - 3y)$ 52. $y^2 - 25$ 53. $4x^2 - 49y^2$ 54. $9a^2 + 24a + 16$ 55. $25x^2 - 20xy + 4y^2$

5.4 EXERCISES B

Factor completely.

1. $x^2 + 8x + 7$ **2.** $x^2 - 8x + 7$ **3.** $x^2 + 6x - 7$

4. $x^2 - 6x - 7$ **5.** $y^2 + 7y - 18$ **6.** $y^2 - 11y + 18$

7. $y^2 - 7y - 18$ **8.** $y^2 + 11y + 18$ **9.** $w^2 + 5w - 50$

10. $w^2 + 2w - 63$ **11.** $x^2 + x - 56$ **12.** $x^2 - 15x + 56$

13. $x^2 - 2x + 4$ **14.** $y^2 + 14y + 48$ **15.** $u^2 - 18u + 72$

16. $u^2 - 6u - 72$ **17.** $x^2 + 8xy + 7y^2$ **18.** $x^2 - 8xy + 7y^2$

19. $x^2 + 6xy - 7y^2$ **20.** $x^2 - 6xy - 7y^2$ **21.** $u^2 + 2uv - 35v^2$

22. $2w^2 + 19w + 45$ **23.** $2w^2 - 19w + 45$ **24.** $-6w^2 + 51w - 105$

25. $5w^2 - 20w + 20$ **26.** $2x^2 - xy - 45y^2$ **27.** $2x^2 + xy - 45y^2$

28. $2x^2 - 4xy + y^2$ **29.** $2u^2v^2 + 7uv - 30$ **30.** $5x^2 - 23xy - 10y^2$

31. $6x^2 - 7xy - 20y^2$ **32.** $-36u^2 - 120uv - 100v^2$ **33.** $6x^2 - 11xy - 21y^2$

34. $4x^2 - 4xy - 63y^2$ **35.** $15u^2 - 14uv - 8v^2$ **36.** $12u^2 - 52uv + 35v^2$

Factor by substitution.

37. $(x - 5)^2 + 4(x - 5) + 3$ **38.** $(x + y)^2 - 5(x + y) - 6$

39. $y^4 - 7y^2 + 10$ **40.** $u^6 - u^3 - 30$

An assistant in an investment office finds that his work in evaluation of stock market equations is made easy by factoring the equations. Factor each of the equations and evaluate for $m = -3$ and $n = 2$.

41. $P = m^2 - 8mn + 15n^2$

42. $Q = m^2 + 2mn - 15n^2$

43. $R = 3m^2 + mn - 2n^2$

44. $S = 6m^2 - 19mn + 10n^2$

45. When a surface-to-air missile is fired with a velocity of 576 ft/sec from a submarine at a depth of -560 ft, the height h in feet of the missile is given by $h = -16t^2 + 576t - 560$. When h is negative, the missile is still under water; when h is zero, the missile is at water level; and when h is positive, the missile is above the surface of the water. Factor the right side of the equation giving h and use the factored form to determine the location of the missile **(a)** 0.5 seconds, **(b)** 1 second, and **(c)** 20 seconds after it has been fired.

FOR REVIEW

Factor completely.

46. $21x^2y^3 - 14xy^2$

47. $42x^4y^4 - 14x^4y^3 - 98x^4y^2$

48. $15u^2 + 6uv - 20uv - 8v^2$

49. $12u^2 - 42uv - 10uv + 35v^2$

50. $6x^2 + 20xy - 27xy - 90y^2$

51. $10x^2 - 16xy - 35xy + 56y^2$

Exercises 52–55 review material from Section 5.2 to help you prepare for topics in the next section. Find each product.

52. $(x - 3)(x + 3)$

53. $(4u + 5v)(4u - 5v)$

54. $(2x + 3)^2$

55. $(5u - 3v)^2$

5.4 EXERCISES C

Factor. Assume that n represents a positive integer.

1. $24x^2 - 65xy - 50y^2$

2. $198a^2 - 393ab + 189b^2$
[*Hint:* A number has 3 as a factor if 3 is a factor of the sum of its digits.

3. $x^{2n} + 5x^n + 6$
[Answer: $(x^n + 2)(x^n + 3)$]

4. $(x + 1)^{2n} + 3(x + 1)^n + 2$
[*Hint:* Substitute u for $x + 1$.]

5. $y^{2n+2} - y^{n+1} - 2$

6. $3x^{6n} + 2x^{5n} - x^{4n}$

5.5 SPECIAL FORMULAS AND SUMMARY OF FACTORING TECHNIQUES

STUDENT GUIDEPOSTS

1. Difference of Squares
2. Perfect Squares
3. Sum and Difference of Cubes
4. Factoring by Using Substitution

1 DIFFERENCE OF SQUARES

In Section 5.2 we saw that

$$(2x + 3y)(2x - 3y) = (2x)^2 - (3y)^2 = 4x^2 - 9y^2.$$

By reversing this we can factor $4x^2 - 9y^2$ as $(2x + 3y)(2x - 3y)$. This is an example of the general formula

$$a^2 - b^2 = (a + b)(a - b) \qquad \text{Difference of squares}$$

which was proved by multiplying $(a + b)(a - b)$. Notice that in this formula we are factoring a binomial which is the difference of two perfect squares.

EXAMPLE 1 FACTORING THE DIFFERENCE OF SQUARES

Factor using $a^2 - b^2 = (a + b)(a - b)$.

(a) $x^2 - 49 = x^2 - 7^2$ $x = a$ and $7 = b$ in $a^2 - b^2 = (a + b)(a - b)$
$= (x + 7)(x - 7)$

(b) $16u^2 - 25v^2 = 4^2u^2 - 5^2v^2$ $4^2u^2 = (4u)^2$
$= (4u)^2 - (5v)^2$ $4u = a$ and $5v = b$ in $a^2 - b^2 = (a + b)(a - b)$
$= (4u + 5v)(4u - 5v)$

(c) $18x^6 - 8y^4 = 2(9x^6 - 4y^4)$ Factor out common factor
$= 2[3^2(x^3)^2 - 2^2(y^2)^2]$
$= 2[(3x^3)^2 - (2y^2)^2]$ $3x^3 = a$ and $2y^2 = b$
$= 2(3x^3 + 2y^2)(3x^3 - 2y^2)$

PRACTICE EXERCISE 1

Factor using

$a^2 - b^2 = (a + b)(a - b)$.

(a) $y^2 - 81$

(b) $64u^2 - 49v^2$

(c) $75u^4 - 12v^6$

Answers: (a) $(y + 9)(y - 9)$
(b) $(8u + 7v)(8u - 7v)$
(c) $3(5u^2 + 2v^3)(5u^2 - 2v^3)$

CAUTION

Although $a^2 - b^2$ can be factored, $a^2 + b^2$ cannot. While $x^2 - 49$ was factored in Example 1, $x^2 + 49$ cannot be factored.

2 PERFECT SQUARES

We also considered in Section 5.2 the two results

$$(2x + 3y)^2 = (2x)^2 + 2(6xy) + (3y)^2 = 4x^2 + 12xy + 9y^2$$

and

$$(2x - 3y)^2 = (2x)^2 - 2(6xy) + (3y)^2 = 4x^2 - 12xy + 9y^2.$$

Again the process of multiplication can be reversed to factor the two expressions on the right. The general formulas to use are

$$a^2 + 2ab + b^2 = (a + b)^2 \qquad \text{Perfect square trinomial}$$

and

$$a^2 - 2ab + b^2 = (a - b)^2. \qquad \text{Perfect square trinomial}$$

Notice that these formulas apply only when the trinomial has two terms that are perfect squares and another that is plus or minus twice the product of the square roots of the squared terms. For example, in

$$x^2 + 8x + 16 = x^2 + 2(4x) + 4^2,$$

x^2 and 4^2 are perfect squares and $8x = 2 \cdot 4x = 2ab$ where $x = a$ and $4 = b$. Thus,

$$x^2 + 8x + 16 = (x + 4)^2.$$

EXAMPLE 2 FACTORING PERFECT SQUARE TRINOMIALS

Factor using the perfect square formulas.

(a) $u^2 + 10uv + 25v^2$

$= u^2 + 2(5uv) + 5^2v^2$ First and last terms are perfect squares

$= u^2 + 2u(5v) + (5v)^2$ $u = a,\ 5v = b$

$= (u + 5v)^2$

(b) $9x^2 + 42xy + 49y^2 = 3^2x^2 + 2(3)(7xy) + 7^2y^2$

$= (3x)^2 + 2(3x)(7y) + (7y)^2$ $3x = a,\ 7y = b$

$= (3x + 7y)^2$

(c) $u^2 - 10uv + 25v^2 = u^2 - 2(5uv) + 5^2v^2$

$= u^2 - 2u(5v) + (5v)^2$ $u = a,\ 5v = b$

$= (u - 5v)^2$

(d) $36x^4 - 84x^2y + 49y^2 = 6^2(x^2)^2 - 2(6)(7x^2y) + 7^2y^2$

$= (6x^2)^2 - 2(6x^2)(7y) + (7y)^2$ $6x^2 = a,\ 7y = b$

$= (6x^2 - 7y)^2$

PRACTICE EXERCISE 2

Factor using the perfect square formulas.

(a) $x^2 + 22x + 121$

(b) $16x^2 + 72xy + 81y^2$

(c) $x^2 - 22x + 121$

(d) $9x^4y^4 - 30x^2y^2 + 25$

Answers: (a) $(x + 11)^2$ (b) $(4x + 9y)^2$ (c) $(x - 11)^2$ (d) $(3x^2y^2 - 5)^2$

3 SUM AND DIFFERENCE OF CUBES

The next two formulas are used to factor binomials whose terms are perfect cubes.

$$a^3 + b^3 = (a + b)(a^2 - ab + b^2) \quad \text{Sum of cubes}$$
$$a^3 - b^3 = (a - b)(a^2 + ab + b^2) \quad \text{Difference of cubes}$$

To prove these we multiply.

$$\begin{array}{r} a^2 - ab + b^2 \\ \underline{a + b \qquad\qquad} \\ a^3 - a^2b + ab^2 \qquad \\ \underline{a^2b - ab^2 + b^3} \\ a^3 + 0a^2b + 0ab^2 + b^3 \\ = a^3 + b^3 \end{array} \qquad \begin{array}{r} a^2 + ab + b^2 \\ \underline{a - b \qquad\qquad} \\ a^3 + a^2b + ab^2 \qquad \\ \underline{-a^2b - ab^2 - b^3} \\ a^3 + 0a^2b + 0ab^2 - b^3 \\ = a^3 - b^3 \end{array}$$

Notice that $a^2 - ab + b^2$ and $a^2 + ab + b^2$ cannot be factored further.

EXAMPLE 3 FACTORING THE SUM OF CUBES

Factor using the sum of cubes formula.

(a) $x^3 + 8 = x^3 + 2^3$ $x = a$ and $2 = b$ in $a^3 + b^3 = (a + b)(a^2 - ab + b^2)$

$= (x + 2)(x^2 - 2x + 4)$

(b) $u^3 + 27v^3 = u^3 + 3^3v^3$

$= u^3 + (3v)^3$

$= (u + 3v)[u^2 - u(3v) + (3v)^2]$

$= (u + 3v)(u^2 - 3uv + 9v^2)$

PRACTICE EXERCISE 3

Factor using the sum of cubes formula.

(a) $x^3 + 125$

(b) $64u^3 + v^3$

(c) $27x^9 + 8y^3$

(c) $8x^3 + 125y^6$

$= 2^3x^3 + 5^3(y^2)^3$

$= (2x)^3 + (5y^2)^3$ $2x = a$, $5y^2 = b$

$= (2x + 5y^2)[(2x)^2 - (2x)(5y^2) + (5y^2)^2]$

$= (2x + 5y^2)(4x^2 - 10xy^2 + 25y^4)$

Answers: (a) $(x + 5)(x^2 - 5x + 25)$
(b) $(4u + v)(16u^2 - 4uv + v^2)$
(c) $(3x^3 + 2y)(9x^6 - 6x^3y + 4y^2)$

CAUTION

To factor a sum or difference of cubes, you must identify a and b and substitute these values into the appropriate formula. This means that *both formulas must be memorized.*

EXAMPLE 4 Factoring the Difference of Cubes

Factor using the difference of cubes formula.

(a) $y^3 - 64 = y^3 - 4^3$ $y = a$ and $4 = b$ in $a^3 - b^3 = (a - b)(a^2 + ab + b^2)$

$= (y - 4)(y^2 + 4y + 16)$

(b) $u^3 - 27v^3 = u^3 - (3v)^3$ $u = a$, $3v = b$

$= (u - 3v)[u^2 + u(3v) + (3v)^2]$

$= (u - 3v)(u^2 + 3uv + 9v^2)$

(c) $125x^3y^3 - 64 = 5^3x^3y^3 - 4^3$

$= (5xy)^3 - 4^3$ $5xy = a$, $4 = b$

$= (5xy - 4)[(5xy)^2 + (5xy)(4) + (4)^2]$

$= (5xy - 4)(25x^2y^2 + 20xy + 16)$

PRACTICE EXERCISE 4

Factor using the difference of cubes formula.

(a) $27 - x^3$

(b) $125u^3 - v^3$

(c) $8u^3 - 27v^3w^3$

Answers: (a) $(3 - x)(9 + 3x + x^2)$
(b) $(5u - v)(25u^2 + 5uv + v^2)$
(c) $(2u - 3vw)(4u^2 + 6uvw + 9v^2w^2)$

CAUTION

Do not write $a^2 - 2ab + b^2$ for $a^2 - ab + b^2$ or $a^2 + 2ab + b^2$ for $a^2 + ab + b^2$ when factoring the sum or difference of cubes.

In some factoring problems we need to use more than one of the formulas, as in the next example.

EXAMPLE 5 Using More Than One Formula

Factor.

(a) $x^6 - 1$

$= (x^3)^2 - 1^2$ Factor difference of squares first with $x^3 = a$ and $1 = b$

$= (x^3 + 1)(x^3 - 1)$

$= [(x + 1)(x^2 - x + 1)][(x - 1)(x^2 + x + 1)]$ Sum and difference of cubes

$= (x + 1)(x - 1)(x^2 - x + 1)(x^2 + x + 1)$

PRACTICE EXERCISE 5

Factor.

(a) $y^6 - 64$

(b) $u^2 - (v^2 - 2v + 1)$

(c) $u^4 - 16v^4$

(b) $x^2 + 2x + 1 - y^2$
$= (x^2 + 2x + 1) - y^2$
$= (x + 1)^2 - y^2$ $\quad x + 1 = a$ and $y = b$ in $a^2 - b^2 = (a + b)(a - b)$
$= [(x + 1) + y][(x + 1) - y]$
$= (x + 1 + y)(x + 1 - y)$

(c) $x^4 - 1 = (x^2)^2 - 1^2$ $\quad x^2 = a$ and $1 = b$ in $a^2 - b^2 = (a - b)(a + b)$
$= (x^2 - 1)(x^2 + 1)$
$= (x - 1)(x + 1)(x^2 + 1)$ $\quad x^2 + 1$ cannot be factored

Answers: (a) $(y + 2)(y - 2)(y^2 - 2y + 4)(y^2 + 2y + 4)$ (b) $(u + v - 1)(u - v + 1)$ (c) $(u + 2v)(u - 2v)(u^2 + 4v^2)$

4 FACTORING BY USING SUBSTITUTION

The substitution method used in Section 5.4 could have been used in Example 5. Example 6 illustrates this method.

EXAMPLE 6 Factoring by Using Substitution

Factor $25(x + 2y)^2 - 10(x + 2y) + 1$ by letting $u = x + 2y$.

$25(x + 2y)^2 - 10(x + 2y) + 1$
$= 25u^2 - 10u + 1$ $\quad$ Let $u = x + 2y$
$= (5u)^2 - 2(5u)(1) + 1$ $\quad 5u = a$ and $1 = b$
$= (5u - 1)^2$
$= [5(x + 2y) - 1]^2$ $\quad u = x + 2y$
$= (5x + 10y - 1)^2$

Practice Exercise 6

Factor $(5x + y)^2 - 8(5x + y) + 16$.

Answer: $(5x + y - 4)^2$

Here we summarize the special formulas used in factoring.

Special Formulas	
1. $a^2 - b^2 = (a + b)(a - b)$	Difference of squares
2. $a^2 + 2ab + b^2 = (a + b)^2$	Perfect square trinomial
3. $a^2 - 2ab + b^2 = (a - b)^2$	Perfect square trinomial
4. $a^3 + b^3 = (a + b)(a^2 - ab + b^2)$	Sum of cubes
5. $a^3 - b^3 = (a - b)(a^2 + ab + b^2)$	Difference of cubes

Some applications involve expressions that can be factored. This is illustrated in the next example.

EXAMPLE 7 Geometry Problem

A metal machine part in the shape of a cube with edge E has a rectangular solid removed from the center as shown in Figure 5.1. The volume of the remaining metal is given by $V = E^3 - Ee^2$. Find the factored form of this equation and use both formulas to compute the volume when $E =$ 12 cm and $e = 3$ cm.

Practice Exercise 7

The current I in amperes varies according to time t in seconds in an electrical circuit. If $I = 2t^2 - 20t +$ 50, find the factored form of this equation and use both formulas to compute the current when $t = 5$ sec.

Factoring, we have $V = E^3 - Ee^2$

$= E(E^2 - e^2)$

$= E(E + e)(E - e).$

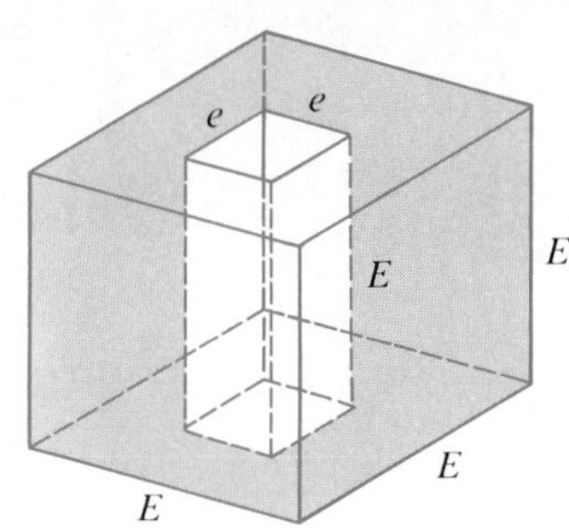

Figure 5.1

Substitute 12 for E and 3 for e in $V = E^3 - Ee^2$ first.

$$V = (12)^3 - (12)(3)^2$$
$$= 1728 - 108 = 1620$$

Now substitute in the factored formula.

$$V = 12(12 + 3)(12 - 3)$$
$$= 12(15)(9) = 1620$$

In both cases we obtain 1620 cm^3 for the volume of the machine part.

Answer: $I = 2(t - 5)^2$; 0 amperes

The exercises for this section start with applications of special formulas and conclude with a variety of factoring problems. As preparation we summarize the factoring techniques that we have learned in this chapter.

Factoring Techniques

1. Factor out any common factor. It is also helpful when factoring $ax^2 + bxy + cy^2$ to factor out (-1) if a is negative.
2. To factor a binomial, use

   $a^2 - b^2 = (a + b)(a - b)$ Difference of squares

   $a^3 + b^3 = (a + b)(a^2 - ab + b^2)$ Sum of cubes

   $a^3 - b^3 = (a - b)(a^2 + ab + b^2).$ Difference of cubes

3. To factor a trinomial, use the techniques of Section 5.4, but also consider perfect squares using the rules

   $a^2 + 2ab + b^2 = (a + b)^2$ Perfect square trinomial

   and $a^2 - 2ab + b^2 = (a - b)^2.$ Perfect square trinomial

   These may apply when the first and last terms of a trinomial are perfect squares.
4. To factor a polynomial with four terms, try to factor by grouping.

5.5 EXERCISES A

Factor using special formulas.

1. $x^2 - 9$

2. $y^2 - 6y + 9$

3. $u^3 - 8$

4. $y^2 - 64$

5. $u^2 - 12uv + 36v^2$

6. $7x^2 - 7y^2$

7. $8x^3 + 27y^3$

8. $4u^2 + 12uv + 9v^2$

9. $20x^2 - 45y^2$

10. $u^2 + 2uv + 2v^2$

11. $40x^3 + 5y^9$

12. $25u^2 - 9v^6$

13. $125u^3 + 27v^3$

14 $(x - y)^2 - 9$

15. $(x - 5)^2 - 4(x - 5) + 4$

Factor.

16. $x^2 - 14xy + 48y^2$

17. $x^2 - 2xy - 48y^2$

18. $6x^2 - 17xy + 10y^2$

19. $35x^2 - 2xy - y^2$

20. $-5x - 5y$

21. $27x^2y - 9x^2y^2 + 81xy^2$

22. $-10u^2 + 40v^2$

23 $7u^3 + 56v^3$

24. $2x^2 - 4xy + 2y^2$

25. $u^2 + 18uv + 80v^2$

26. $u^2 + 22uv + 121v^2$

27. $3x^2 - 36xy + 105y^2$

28. $4x^2y^2 - 4xy + 1$

29. $21u^2 - 4uv - v^2$

30. $250x^3 + 54y^3$

31. $3u^2 - 2uv - 16v^2$

32. $36u^2 + 60uv + 25v^2$

33. $x^2 + 5xy + 5y^2$

34. $-5x^2 + 70xy - 240y^2$

35. $u^6 - v^6$

36. $(3x + y)^2 - 9$

37 $(x + y)^2 + 2(x + y) + 1$

38. $27u^3v^3 - 8$

39 $3uv + u - 3v^2 - v$

40. $30x^2 + 31xy - 21y^2$

41. $x^3y^3 + z^3$

42. $9x^6 - 42x^3y + 49y^2$

43. $-16x^4 + 64y^2$

44 $-18u^2 - 69uv - 60v^2$

45. $5x^2 + 15xy + 15y^2$

46. $(2x + y)^2 - 6(2x + y) + 8$

47. $x^4 - x^2 - 6$

48 $(u + v)^2 + 3(u + v)(x + y) + 2(x + y)^2$

The area of a circular garden with a circular fountain in the center is given by the equation $A = \pi R^2 - \pi r^2$ *where R is the radius of the garden and r is the radius of the fountain.*

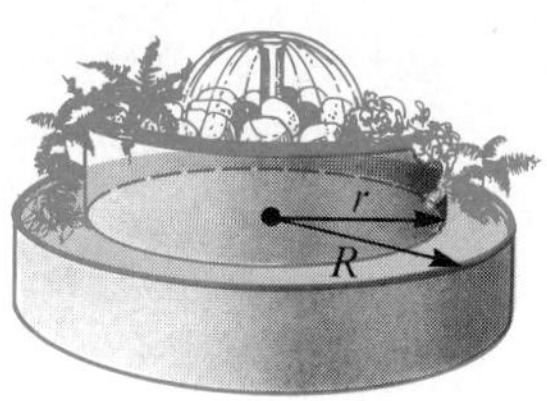

49. Find the factored form of the equation.

50. Use both formulas to find the area when $R = 10$ ft and $r = 4$ ft. (Use 3.14 for π.)

51. In an electrical circuit, the current I in amperes varies according to time t in seconds. If $I = 3t^2 - 36t + 108$, find the factored form of this equation and use both formulas to compute the current when $t = 6$ sec.

FOR REVIEW

The following exercises review material from Section 5.1 along with factoring to help you prepare for the next section. Evaluate each polynomial when $x = 2$. *Then factor the polynomial and evaluate the factored form when* $x = 2$. *Which form is easier to evaluate?*

52. $x^2 + x - 6$

53. $5x^2 - 20$

54. $3x^3 - 24$

ANSWERS: 1. $(x + 3)(x - 3)$ 2. $(y - 3)^2$ 3. $(u - 2)(u^2 + 2u + 4)$ 4. $(y + 8)(y - 8)$ 5. $(u - 6v)^2$ 6. $7(x + y)(x - y)$ 7. $(2x + 3y)(4x^2 - 6xy + 9y^2)$ 8. $(2u + 3v)^2$ 9. $5(2x + 3y)(2x - 3y)$ 10. cannot be factored 11. $5(2x + y^3)(4x^2 - 2xy^3 + y^6)$ 12. $(5u + 3v^3)(5u - 3v^3)$ 13. $(5u + 3v)(25u^2 - 15uv + 9v^2)$ 14. $(x - y + 3)(x - y - 3)$ 15. $(x - 7)^2$ 16. $(x - 6y)(x - 8y)$ 17. $(x + 6y)(x - 8y)$ 18. $(6x - 5y)(x - 2y)$ 19. $(7x + y)(5x - y)$ 20. $-5(x + y)$ 21. $9xy(3x - xy + 9y)$ 22. $-10(u + 2v)(u - 2v)$ 23. $7(u + 2v)(u^2 - 2uv + 4v^2)$ 24. $2(x - y)^2$ 25. $(u + 8v)(u + 10v)$ 26. $(u + 11v)^2$ 27. $3(x - 5y)(x - 7y)$ 28. $(2xy - 1)^2$ 29. $(3u - v)(7u + v)$ 30. $2(5x + 3y)(25x^2 - 15xy + 9y^2)$ 31. $(u + 2v)(3u - 8v)$ 32. $(6u + 5v)^2$ 33. cannot be factored 34. $-5(x - 6y)(x - 8y)$ 35. $(u + v)(u - v)(u^2 - uv + v^2)(u^2 + uv + v^2)$ 36. $(3x + y + 3)(3x + y - 3)$ 37. $(x + y + 1)^2$ 38. $(3uv - 2)(9u^2v^2 + 6uv + 4)$ 39. $(u - v)(3v + 1)$ 40. $(15x - 7y)(2x + 3y)$ 41. $(xy + z)(x^2y^2 - xyz + z^2)$ 42. $(3x^3 - 7y)^2$ 43. $-16(x^2 + 2y)(x^2 - 2y)$ 44. $-3(2u + 5v)(3u + 4v)$ 45. $5(x^2 + 3xy + 3y^2)$ 46. $(2x + y - 2)(2x + y - 4)$ 47. $(x^2 + 2)(x^2 - 3)$ 48. $(u + v + x + y)(u + v + 2x + 2y)$ 49. $A = \pi(R + r)(R - r)$ 50. 84π ft$^2 \approx 264$ ft^2 51. $I = 3(t - 6)^2$; 0 amperes 52. $(x - 2)(x + 3)$; 0 53. $5(x + 2)(x - 2)$; 0 54. $3(x - 2)(x^2 + 2x + 4)$; 0

5.5 EXERCISES B

Factor using special formulas.

1. $x^2 + 6x + 9$

2. $y^3 + 8$

3. $u^2 - 14u + 49$

4. $y^3 + 64$

5. $u^2 + 12uv + 36v^2$

6. $7x^3 + 7y^3$

7. $8x^3 - 27y^3$

8. $4u^2 - 12uv + 9v^2$

9. $x^2 + y^2$

10. $9u^2 + 6uv + v^2$

11. $49x^4 + 14x^2y^2 + y^4$

12. $9u^6 - 30u^3v^2 + 25v^4$

13. $5u^3 - 5000v^3$

14. $(x + y)^2 - 25$

15. $(x - y)^2 + 6(x - y) + 9$

Factor.

16. $x^2 + 13xy + 42y^2$

17. $x^2 - xy - 42y^2$

18. $5x^2 - 12xy - 32y^2$

19. $14x^2 + 25xy - 25y^2$

20. $4x^2y^2 - 2xy$

21. $21x^3y - 35x^2y^2 + 14x^2y$

22. $-108u^2 + 12v^2$

23. $40u^3 - 135v^3$

24. $8x^2 + 32xy + 32y^2$

25. $3u^2 - 8uv + 4v^2$

26. $144u^2 - 24uv + v^2$

27. $40x^2 - 28xy + 4y^2$

28. $9x^2y^2 - 6xy + 1$

29. $10u^2 + 99uv - 10v^2$

30. $24x^3 - 81y^3$

31. $30u^2 + 13uv - 3v^2$

32. $49u^2 - 42uv + 9v^2$

33. $8x^2 + 8xy + 3y^2$

34. $-28x^2 + 140xy - 175y^2$

35. $u^8 - v^8$

36. $9(x + 5y)^2 - 4$

37. $4(2x + y)^2 - 4(2x + y) + 1$

38. $u^3v^3 + x^3y^3$

39. $5u^2 - 3uv - 9v^2 + 15uv$

40. $36x^2 - 9xy - 10y^2$

41. $x^3y^3z^3 - 8$

42. $25x^4 + 30x^2y^2 + 9y^4$

43. $-288x^2 + 8y^4$

44. $25u^2 - 130uv + 25v^2$

45. $12x^2 - 12xy + 24y^2$

46. $(x^2 + y^2)^2 + 4(x^2 + y^2) + 4$

47. $x^4 - 25$

48. $(u^2v^2 + 1)^2 + 4(u^2v^2 + 1) + 3$

The volume of a hollow cylindrical column which is 10 m *high is given by the equation* $V = 10\pi R^2 - 10\pi r^2$ *where* R *is the outer radius and* r *is the inner radius.*

49. Find the factored form of the equation.

50. Use both formulas to find the volume when $R = 4$ m and $r = 2$ m. (Use 3.14 for π.)

51. In an electrical circuit, the current I in amperes varies according to time t in seconds. If $I = 4t^2 - 80t + 400$, find the factored form of this equation and use both formulas to compute the current when $t = 10$ sec.

FOR REVIEW

The following exercises review material from Section 5.1 along with factoring to help you prepare for the next section. Evaluate each polynomial when $x = -3$. *Then factor the polynomial and evaluate the factored form when* $x = -3$. *Which form is easier to evaluate?*

52. $x^2 - 2x - 15$

53. $7x^2 - 63$

54. $2x^3 + 54$

5.5 EXERCISES C

Factor. Assume that n represents a positive integer.

1. $x^{4n} - y^{4n}$

2. $x^{3n} - y^{3n}$

3. $x^{3n} + y^{3n}$

4. $x^{2n} + 2x^n y^n + y^{2n}$

5. $x^{2n} - 2x^n y^n + y^{2n}$

6. $x^n y^n + y^n + x^n + 1$
[Answer: $(x^n + 1)(y^n + 1)$]

7. $8x^{3n} - 27y^{3n}$
[Answer: $(2x^n - 3y^n)(4x^{2n} + 6x^n y^n + 9y^{2n})$]

8. $8x^{3n} + 27y^{3n}$

5.6 APPLICATIONS USING FACTORING

STUDENT GUIDEPOSTS

1. The Zero-Product Rule
2. Quadratic Equations
3. Applied Problems

1 THE ZERO-PRODUCT RULE

Factoring can be used to solve equations and applied problems. The polynomial in an equation such as

$$x^2 + 2x - 15 = 0$$

can be factored to give

$$(x - 3)(x + 5) = 0.$$

This equation can then be solved by using a property of the real number system called the **zero-product rule,** which states that if a product of two or more factors is zero, then at least one of the factors must be zero.

Zero-Product Rule

If a and b are numbers or expressions and

if $ab = 0$, then $a = 0$ or $b = 0$.

Using the zero-product rule, the solutions of an equation like

$$(x - 3)(x + 5) = 0$$

are found by setting each factor equal to zero and solving for x.

EXAMPLE 1 USING THE ZERO-PRODUCT RULE

Solve $(x - 3)(x + 5) = 0$.

By the zero-product rule,

$$x - 3 = 0 \quad \text{or} \quad x + 5 = 0$$
$$x = 3 \qquad\qquad x = -5.$$

The solutions are 3 and -5. Check these by substituting into the original equation.

PRACTICE EXERCISE 1

Solve $(x + 4)(x - 10) = 0$.

Answer: -4, 10

The zero-product rule can be extended to products of three or more factors, as shown in the next example.

EXAMPLE 2 Extending the Zero-Product Rule

Solve $x(x - 3)(2x + 3) = 0$.

$$x = 0 \quad \text{or} \quad x - 3 = 0 \quad \text{or} \quad 2x + 3 = 0 \qquad \text{Zero-product rule}$$
$$x = 0 \qquad x = 3 \qquad 2x = -3$$
$$x = -\frac{3}{2}$$

The solutions are 0, 3, and $-\frac{3}{2}$. Check by substituting in the original equation.

Practice Exercise 2

Solve $(x - 2)(x + 9)(2x - 5) = 0$.

Answer: 2, -9, $\frac{5}{2}$

CAUTION

Note that the zero-product rule applies only to *zero-products*. It cannot be used to solve equations like $ab = 6$, $a + b = 0$, or $a - b = 0$. For example,

$$(x - 2) - (2x - 1) = 0$$

is solved the same way we solved linear equations in Chapter 2. We first clear parentheses and then combine like terms. Also, if $(x - 1)(x + 5) = 6$, this does not mean that each factor is 6.

2 QUADRATIC EQUATIONS

In Example 1, if we had started with

$$x^2 + 2x - 15 = 0,$$

we could have factored to get

$$(x - 3)(x + 5) = 0$$

and then continued as in the example. We have, in effect, solved a *quadratic equation*. A **quadratic equation** or **second-degree equation** is an equation that can be written in the form

$$ax^2 + bx + c = 0,$$

where x is the variable and a, b, and c are constants with $a \neq 0$. We will solve equations of this type using factoring in this section but will wait for a complete discussion of quadratic equations until Chapter 8.

EXAMPLE 3 Solving a Quadratic Equation

Solve the quadratic equation by factoring.

$$x^2 - x - 12 = 0$$
$$(x - 4)(x + 3) = 0 \qquad \text{Factor}$$
$$x - 4 = 0 \quad \text{or} \quad x + 3 = 0 \qquad \text{Zero-product rule}$$
$$x = 4 \qquad x = -3$$

Practice Exercise 3

Solve $x^2 - 2x - 63 = 0$.

Check: $(4)^2 - 4 - 12 \stackrel{?}{=} 0$ $\quad (-3)^2 - (-3) - 12 \stackrel{?}{=} 0$

$16 - 4 - 12 \stackrel{?}{=} 0 \quad 9 + 3 - 12 \stackrel{?}{=} 0$

$16 - 16 = 0 \quad 12 - 12 = 0$

The solutions are 4 and -3.

Answer: 9, -7

When we solve applied problems, some equations may involve fractions or may require collecting terms. The following examples illustrate these techniques.

EXAMPLE 4 Solving Quadratic Equations

Solve.

(a) $5x^2 - 2x + \frac{1}{5} = 0$

$25x^2 - 10x + 1 = 0$ Multiply by 5 to clear fractions

$(5x - 1)^2 = 0$

$5x - 1 = 0$ or $5x - 1 = 0$ Zero-product rule

$5x = 1 \quad 5x = 1$

$x = \frac{1}{5} \quad x = \frac{1}{5}$

There is only one solution, $\frac{1}{5}$. Check by substituting.

(b) $2x^2 - 2x = 12$

$2x^2 - 2x - 12 = 0$ Collect terms on left side

$2(x^2 - x - 6) = 0$ Factor out 2

$x^2 - x - 6 = 0$ Divide both sides by 2

$(x - 3)(x + 2) = 0$ Factor

$x - 3 = 0$ or $x + 2 = 0$ Zero-product rule

$x = 3 \quad x = -2$

The solutions are 3 and -2. Check by substituting.

(c) $5x^2 = 3x$

$5x^2 - 3x = 0$ Collect terms on left side

$x(5x - 3) = 0$ Factor out x

$x = 0$ or $5x - 3 = 0$ Zero-product rule

$5x = 3$

$x = \frac{3}{5}$

The solutions are 0 and $\frac{3}{5}$. Check by substituting.

PRACTICE EXERCISE 4

Solve.

(a) $\frac{4}{3}x^2 + 4x + 3 = 0$

(b) $3x^2 - 18x = 48$

(c) $9x^2 = -6x$

Answers: (a) $-\frac{3}{2}$ (b) 8, -2 (c) 0, $-\frac{2}{3}$

CAUTION

When solving an equation such as $5x^2 = 3x$ in Example 4(c), students sometimes make the mistake of dividing both sides by x. This gives $5x = 3$, leading to $x = \frac{3}{5}$, but the solution $x = 0$ is lost. Never divide both sides of an equation by an expression containing the variable.

The steps for solving an equation using the zero-product rule are summarized below.

To Solve an Equation Using the Zero-Product Rule

1. If necessary, rewrite the equation making one side 0 by collecting like terms and clearing fractions.
2. Factor the nonzero side.
3. Use the zero-product rule to set each factor equal to 0.
4. Solve each equation obtained in step 3.
5. Check all solutions in the original equation.

3 APPLIED PROBLEMS

Factoring is used to solve many applied problems. In the following examples, write out detailed descriptions of the variables and draw and label any geometric figures.

EXAMPLE 5 CONSECUTIVE ODD INTEGER PROBLEM

One-third the product of two consecutive positive odd integers is 85. Find the integers.

Let x = the first positive odd integer,

$x + 2$ = the next consecutive odd integer,

$\frac{1}{3}x(x+2) = \frac{x(x+2)}{3}$ = one-third their product.

The equation we must solve is

$$\frac{x(x+2)}{3} = 85.$$

Multiply by 3 to clear fractions.

$$x(x+2) = 255$$
$$x^2 + 2x = 255$$
$$x^2 + 2x - 255 = 0$$
$$(x-15)(x+17) = 0$$
$$x - 15 = 0 \quad \text{or} \quad x + 17 = 0$$
$$x = 15 \qquad\qquad x = -17$$

Since x must be a positive integer, we rule out -17. Thus, 15 is the first integer, and the next odd integer is 17 ($x + 2 = 17$). Since 15 and 17 are consecutive positive odd integers and $\frac{1}{3}(15)(17) = 85$, these solutions check.

PRACTICE EXERCISE 5

Five times the product of two consecutive positive even integers is 840. Find the integers.

Let x = the first even integer,

$x + 2$ = the second even integer.

Answer: 12, 14

EXAMPLE 6 GEOMETRY PROBLEM

Find the length and width of a rectangle if the length is 5 cm more than the width and the area is 84 cm^2. See the sketch in Figure 5.2.

Let w = width of the rectangle,

$w + 5$ = length of the rectangle.

PRACTICE EXERCISE 6

Find the length and width of a rectangle if the width is 8 ft less than the length and the area is 105 ft^2.

$$w(w + 5) = 84$$
$$w^2 + 5w = 84$$
$$w^2 + 5w - 84 = 0$$
$$(w - 7)(w + 12) = 0$$
$$w - 7 = 0 \quad \text{or} \quad w + 12 = 0$$
$$w = 7 \qquad\qquad w = -12$$

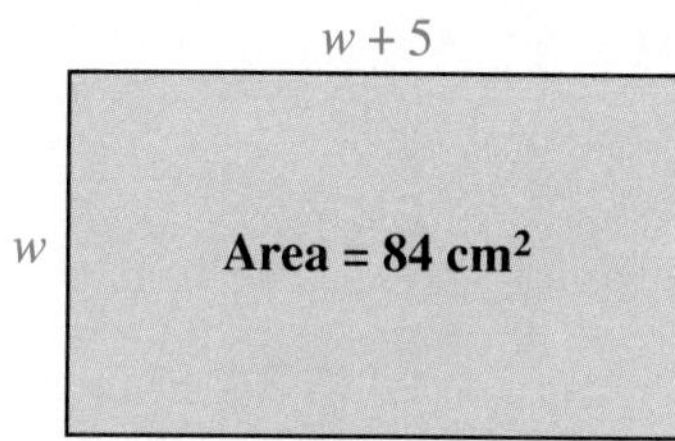

Figure 5.2

Since -12 cannot be a width of a rectangle, the width is 7 cm and the length is $w + 5 = 7 + 5 = 12$ cm.

Answer: 15 ft, 7 ft

EXAMPLE 7 Business Problem

A wholesale warehouse uses the equation $P(n) = 2n^2 - 7n - 15$ to calculate the profit P on the number n of refrigerators sold.

(a) Find the profit when 20 refrigerators are sold.
This is an evaluation problem. We calculate $P(20)$.

$$P(n) = 2n^2 - 7n - 15 \qquad \text{Profit equation}$$
$$P(20) = 2(20)^2 - 7(20) - 15 \qquad n = 20$$
$$= 2(400) - 140 - 15 = 645$$

The profit is \$645 when 20 are sold.

(b) Find the number sold when the profit was \$57.
Since $P(n) = \$57$ we need to solve the following equation.

$$P(n) = 2n^2 - 7n - 15 = 57 \qquad \text{Profit was \$57}$$
$$2n^2 - 7n - 72 = 0 \qquad \text{Collect like terms}$$
$$(2n + 9)(n - 8) = 0 \qquad \text{Factor}$$
$$2n + 9 = 0 \quad \text{or} \quad n - 8 = 0 \qquad \text{Zero-product rule}$$
$$n = -\frac{9}{2} \quad \text{or} \quad n = 8$$

Since n is a number of refrigerators, we rule out $-\frac{9}{2}$. Thus, 8 refrigerators were sold.

Practice Exercise 7

A retail store uses the equation $P(n) = 2n^2 - 5n - 50$ to calculate the profit P on the number n of dresses sold.

(a) Find the profit when 15 dresses are sold.

(b) Find the number sold when the profit was \$100.

Answers: (a) \$325 (b) 10

5.6 EXERCISES A

Solve.

1. $(x - 3)(x + 6) = 0$

2. $(2x - 1)(x + 7) = 0$

3. $(5y - 1)(2y - 3) = 0$

4. $(x + 2.3)(x - 5.7) = 0$

5. $\left(y - \frac{3}{2}\right)\left(y + \frac{7}{2}\right) = 0$

6. $5(x + 3)(x - 9) = 0$

7. $y(y - 1)(y - 3) = 0$

8. $(x + 1)(x - 2)(x + 3) = 0$

9. $(y - 5)(y - 8)(2y + 1) = 0$

10. $(x + 3) - (2x - 5) = 0$

11. $(3y + 1)(y - 5)^2 = 0$

12. $x^2 - 5x + 6 = 0$

13. $x^2 - x - 30 = 0$

14. $y^2 + 2y - 63 = 0$

15. $\frac{1}{3}x^2 + \frac{13}{3}x = -10$

16 $3u^2 = 5u$

17. $v^2 - 10v = -25$

18. $x^2 + \frac{3}{2}x = \frac{9}{2}$

19. $3x^2 + 2x = 1$

20. $5u^2 = 17u - 6$

21. $\frac{1}{2}y^2 - 3y + 4 = 0$

22. $0.1v^2 - 0.2v + 0.1 = 0$

23. $x(x + 5) = -6$

24. $u(u - 2) = 3u - 6$

25. $3y^2 = 30y - 48$

26 $y^3 - 2y^2 + y = 0$

27. $2x^2 - 5 = x^2 + 4$

Solve.

28. The square of a number is 5 more than four times the number. Find the number.

Let x = the number,
x^2 = the square of the number,
$4x$ = four times the number.
$x^2 = 5 +$ ________

29. The product of Samantha's present age and her age 7 years ago is 170. How old is Sam?

Let x = Sam's present age.
$x - 7 =$

30. The product of two consecutive positive even integers is 440. Find the integers.

31. One more than a number times 1 less than the number is 35. Find the number.

32. Jose is 12 years older than Mike. The product of their ages is 220. How old is each?

33. The hypotenuse of a right triangle is 5 cm, and one leg is 1 cm longer than the other. Find the length of each leg.

Let x = length of one leg.

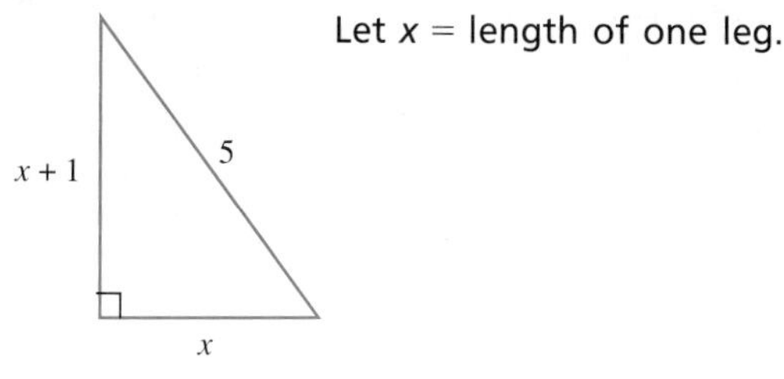

34. The number of square inches in the area of a square is 12 more than the number of inches in its perimeter. Find the length of each side.

35. Wanda's age 4 years ago times her age in 6 years is 144. How old is Wanda?

36. A rectangular garden is 6 yd longer than it is wide. What are the dimensions of the garden if its area is 520 yd^2?

Let w = width of garden.

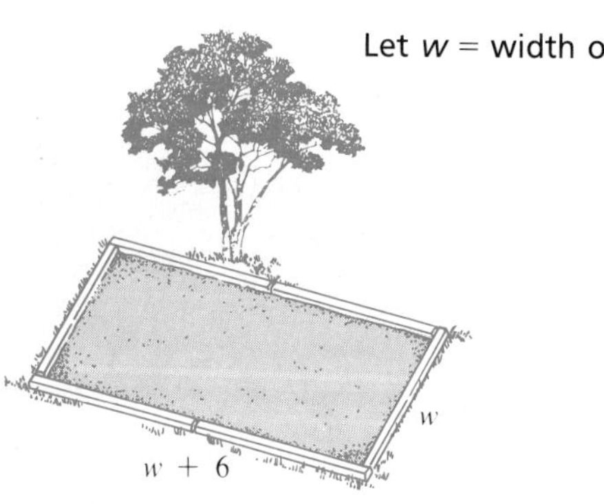

37. The sum of the squares of two positive consecutive even integers is 100. Find the integers.

38. If twice Alana's age is squared, the result is the same as the square of her age 3 years from now. How old is Alana?

39. The area of a parallelogram is 55 ft^2. If the base is 1 ft greater than twice the altitude, find the base and the altitude.

40 A box is 12 cm high. The length is 7 cm more than the width. If the volume is 1728 cm^3, find the length and width.

Let w = width of box.

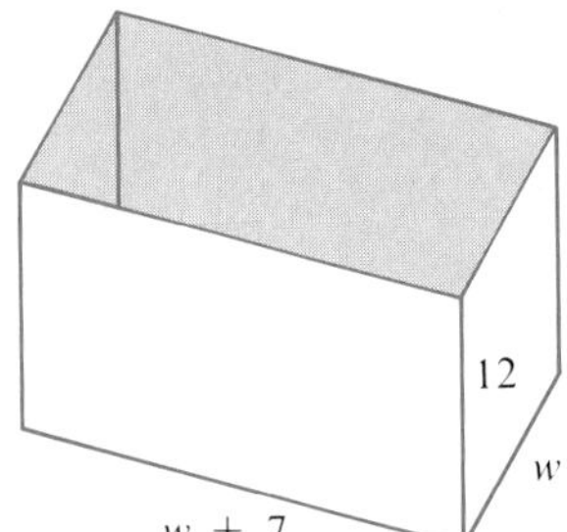

41 The area of a triangle is 42 cm^2. If the base is 8 cm less than the height, find the base and height.

The current I in amperes in an electrical circuit varies according to time t in seconds by the equation $I = 5t^2 - 50t + 125$.

42. What is the initial amperage (when $t = 0$)?

43. What is the amperage in 2 seconds (when $t = 2$)?

44. In how many seconds will the amperage be 0 amperes?

45. In how many seconds will the amperage be 80 amperes?

The profit P of a retail operation is given by the equation $P(n) = 3n^2 - 2n - 10$ *where n is the number of suits sold.*

46. Find the profit when $n = 9$.

47. Find the profit when 20 suits are sold.

48. How many suits were sold on a day when the profit was $-\$10$ (there was a \$10 loss)?

49 How many suits were sold on a day when the profit was \$30?

FOR REVIEW

Factor.

50. $9x^2 - 64y^2$

51. $121x^2 - 44xy + 4y^2$

52. $54u^3 - 16v^3$

53. $125u^3 + v^6$

Exercises 54–56 review material from Section 1.6 to help you prepare for the next chapter. Simplify and write without negative exponents.

54. $\frac{x^2}{x^5}$ **55.** $\frac{3x^3y}{6xy^4}$ **56.** $\left(\frac{4^0xy^2}{2x^2y}\right)^{-2}$

ANSWERS: **1.** 3, −6 **2.** $\frac{1}{2}$, −7 **3.** $\frac{1}{5}$, $\frac{3}{2}$ **4.** −2.3, 5.7 **5.** $\frac{3}{2}$, $-\frac{7}{2}$ **6.** −3, 9 **7.** 0, 1, 3 **8.** −1, 2, −3 **9.** 5, 8, $-\frac{1}{2}$ **10.** 8 **11.** $-\frac{1}{3}$, 5 **12.** 2, 3 **13.** −5, 6 **14.** 7, −9 **15.** −3, −10 **16.** 0, $\frac{5}{3}$ **17.** 5 **18.** $\frac{3}{2}$, −3 **19.** $\frac{1}{3}$, −1 **20.** $\frac{2}{5}$, 3 **21.** 2, 4 **22.** 1 **23.** −2, −3 **24.** 2, 3 **25.** 2, 8 **26.** 0, 1 **27.** 3, −3 **28.** 5 or −1 **29.** 17 yr **30.** 20, 22 **31.** 6 or −6 **32.** Jose: 22 yr; Mike: 10 yr **33.** 3 cm, 4 cm **34.** 6 in **35.** 12 yr **36.** 26 yd, 20 yd **37.** 6, 8 **38.** 3 yr **39.** 11 ft, 5 ft **40.** 16 cm, 9 cm **41.** 6 cm, 14 cm **42.** 125 amperes **43.** 45 amperes **44.** 5 seconds **45.** 1 second and 9 seconds **46.** $215 **47.** $1150 **48.** 0 **49.** 4 **50.** $(3x + 8y)(3x - 8y)$ **51.** $(11x - 2y)^2$ **52.** $2(3u - 2v)(9u^2 + 6uv + 4v^2)$ **53.** $(5u + v^2)(25u^2 - 5uv^2 + v^4)$ **54.** $\frac{1}{x^3}$ **55.** $\frac{x^2}{2y^3}$ **56.** $\frac{4x^2}{y^2}$

5.6 EXERCISES B

Solve.

1. $(x - 5)(x + 9) = 0$

2. $(3x - 2)(x + 5) = 0$

3. $(4y - 3)(y - 8) = 0$

4. $(x + 4.9)(x - 8.8) = 0$

5. $\left(y - \frac{8}{3}\right)\left(y + \frac{2}{3}\right) = 0$

6. $8(x - 2)(x + 10) = 0$

7. $y(y + 3)(y - 2) = 0$

8. $(x + 3)(x - 2)(x + 5) = 0$

9. $(y + 6)(y + 9)(2y + 3) = 0$

10. $(x + 4) - (3x - 4) = 0$

11. $(2y - 3)^2(y + 6)^2 = 0$

12. $x^2 - 4x + 3 = 0$

13. $x^2 - 3y - 18 = 0$

14. $y^2 + 4y - 32 = 0$

15. $\frac{1}{4}x^2 + x = \frac{21}{4}$

16. $7u = 5u^2$

17. $v^2 = 16v - 64$

18. $x^2 - \frac{3}{2}x = \frac{5}{2}$

19. $3x^2 = 10x - 8$

20. $5u^2 + 6 = -17u$

21. $\frac{1}{3}y^2 + \frac{4}{3}y + 1 = 0$

22. $0.1v^2 - 0.6v + 0.9 = 0$

23. $x(x - 6) = -8$

24. $u(u + 5) = -2u - 10$

25. $5y^2 = 40y - 75$

26. $y^3 + 4y^2 + 4y = 0$

27. $3x^2 - 8 = 2x^2 + 8$

Solve.

28. If the square of Waldo's age is decreased by 300, the result is twenty times his age. How old is Waldo?

29. The product of 1 less than a number and 7 more than the number is 105. Find the number.

30. Find two consecutive negative odd integers whose product is 143.

31. The product of Ed's age in two years and his age two years ago is 320. What is Ed's present age?

32. The length of a rectangle is 8 ft more than the width. Find the dimensions if its area is 384 ft^2.

33. Adding 4 to the square of Zeke's age is the same as subtracting 3 from eight times his age. How old is Zeke?

34. The length of a rectangle is 10 m more than the width. Find the length and width if the area is 264 m^2.

35. Five times the product of two positive consecutive integers is 550. Find the integers.

36. The square of a number is 48 more than 8 times the number. Find the number.

37. The area of a triangle is 114 ft^2 and the height is 7 ft more than the base. Find the height and base.

38. If the sides of a square are lengthened by 3 cm, the area will be 196 cm^2. Find the length of a side.

39. What number has the property that the number times itself is the same as the number plus itself?

40. A ladder 13 ft long leans against a wall. If the base of the ladder is 5 ft from the wall, how far would the lower end have to be pulled out so that the upper end is pulled down the same amount?

41. The sum of the squares of two positive consecutive integers is 113. Find the integers.

The current I in amperes in an electrical circuit varies according to time t in seconds by $I = 3t^2 - 48t + 192$.

42. What is the initial amperage (when $t = 0$)?

43. What is the amperage in 5 seconds (when $t = 5$)?

44. In how many seconds will the amperage be 0 amperes?

45. In how many seconds will the amperage be 108 amperes?

The profit P of a wholesale operation is given by the equation $P(n) = 2n^2 - 3n - 12$, *where n is the number of appliances sold.*

46. Find the profit when $n = 7$.

47. Find the profit when 15 appliances are sold.

48. How many appliances were sold on a day when the profit was $-\$13$ (a \$13 loss)?

49. How many appliances were sold on a day when the profit was \$123?

FOR REVIEW

Factor.

50. $125x^2 - 20y^2$

51. $25x^2 + 30xy + 9y^2$

52. $27u^3 - v^6$

53. $64u^3 + 125v^3$

Exercises 54–56 review material from Section 1.6 to help you prepare for the next chapter. Simplify and write without negative exponents.

54. $\dfrac{a^4}{a^{11}}$

55. $\dfrac{2u^4v^2}{8u^2v^5}$

56. $\left(\dfrac{3^\circ x^2y}{3xy^3}\right)^{-2}$

5.6 EXERCISES C

Solve for x. Assume that a and b are constant real numbers.

1. $x(x - 1)(x + 3)(x + 7) = 0$

2. $(x^2 - x - 2)(x^3 - x^2 - 20x) = 0$
[Answer: 2, −1, 0, 5, −4]

3. $(x - a)(x + 3b) = 0$

4. $2x^2 + ax - 2bx - ab = 0$

5. $(x - 3)^2 - a^2 = 0$

6. $(x - 3)(x + 2) - a^2 = (2x - 1)(x + 2) - 2a^2$
[Answer: $a - 2$, $-a - 2$]

CHAPTER REVIEW

KEY WORDS

5.1 A **polynomial** is an algebraic expression with terms that are products of numbers and variables raised to whole-number exponents.

A **monomial** is a polynomial with one term.

A **binomial** is a polynomial with two terms.

A **trinomial** is a polynomial with three terms.

The **degree of a term** in a polynomial is the exponent on the variable if only one variable is present. When more variables are present in a term, the **degree of the term** is the sum of the exponents on the variables.

The **degree of a polynomial** is equal to the degree of the term with the highest degree.

Terms in a polynomial that have the same variables raised to the same powers are called **like terms.**

When like terms in a polynomial are combined using the distributive law we are **collecting like terms.**

A polynomial is written in **descending (ascending) order** when the terms are arranged with the highest (lowest) degree term first, followed by the next highest (lowest), and so forth.

5.3 The **greatest common factor (GCF)** of a polynomial is the common factor of each term that contains the largest numerical factor and the highest power of each variable.

5.6 A **quadratic equation (second-degree equation)** is an equation that can be written in the form $ax^2 + bx + c = 0$ with $a \neq 0$.

KEY CONCEPTS

5.1 **1.** The degree of a nonzero monomial such as 5 is 0. The zero monomial 0 has no degree.

2. Only like terms can be collected. For example, $3x^2 - 2x$ cannot be simplified since $3x^2$ and $-2x$ are not like terms.

3. Change *all* signs when removing parentheses preceded by a minus sign in a subtraction problem. For example, $3x^3 - (2x^2 - 5x + 6) = 3x^3 - 2x^2 + 5x - 6$.

5.2 **1.** Use the FOIL method to multiply binomials or the column method to multiply polynomials with several terms.

2. Special product formulas:

(a) $(a + b)(a - b) = a^2 - b^2$

(b) $(a + b)(a + b) = (a + b)^2$
$= a^2 + 2ab + b^2$ (*not* $a^2 + b^2$)

(c) $(a - b)(a - b) = (a - b)^2$
$= a^2 - 2ab + b^2$ (*not* $a^2 - b^2$)

5.3 **1.** In any factoring problem always remove the greatest common factor (GCF) first.

2. Polynomials with four terms can often be factored by grouping.

5.4 **1.** To factor a trinomial of the form $x^2 + bx + c$, form all pairs of integers whose product is c. Use the pair whose sum is b in $x^2 + bx + c = (x + \_\_)(x + \_\_)$.

2. To factor $ax^2 + bx + c$, $a \neq 1$, list all pairs of integers whose product is a and list all pairs whose product is c. Use trial and error to determine which pairs give the correct value for b.

3. You should never give a wrong answer to a factoring problem because you can always check your work by multiplying the factors.

5.5 **1.** Difference of squares:

$$a^2 - b^2 = (a + b)(a - b)$$

2. Perfect square trinomials:

$$a^2 + 2ab + b^2 = (a + b)^2$$

and

$$a^2 - 2ab + b^2 = (a - b)^2$$

3. Sum of cubes:

$$a^3 + b^3 = (a + b)(a^2 - ab + b^2)$$

4. Difference of cubes:

$$a^3 - b^3 = (a - b)(a^2 + ab + b^2)$$

5.6 The zero-product rule states that if $ab = 0$, then $a = 0$ or $b = 0$. This rule is used to solve equations that can be factored.

REVIEW EXERCISES

Part I

5.1 *For each polynomial, (a) tell whether it is a monomial, binomial, or trinomial, (b) give the term of degree 2 and its coefficient, and (c) give the degree of the polynomial.*

1. $x^2 + 2x + 5$

2. $-4ab$

3. $6x^4y^5z^2 - 7x^4 + 9y^2$

Collect like terms and write in descending powers of x.

4. $5x^2y - 3xy - 4x^2y + 5$

5. $6x^2y^2 - 8x^3y + 4xy - 6x^3y - x^2y^2$

Evaluate the polynomial for the given values of the variables.

6. $-2x^2 - 5x + 10$; $x = -2$

7. $3a^2b^3 - 7ab^2 + 5b$; $a = -1$, $b = 2$

8. Add $4y^2 - 16y + y^3 - 2$ and $-3y + 7y^3 + 2 - y^2$.

9. Subtract $-10x^3y^2 + 6x^2y^2 - 5xy$ from $5x^3y^2 - 4x^2y^2 + 12$.

Perform the indicated operations.

10. $(7a^3b^2 - 3a^2b) + (6a^3b^2 + 5ab^2) - (-2a^3b^2 + 7ab^2)$

11. $(a^3b^3 - 5a + 2b) + (-2a^3b^3 + 4a - 8b) - (5a^3b^3 - 6b)$

Add.

12.
$$\begin{array}{r} 36x^2y^2 - 6xy + 5 \\ \underline{22x^2y^2 + 18xy - 17} \end{array}$$

13.
$$\begin{array}{r} 105x^3y^2 + 71x^2y^3 - 8xy + 5 \\ \underline{-82x^3y^2 - 92x^2y^3 - 9xy + 18} \end{array}$$

5.2 *Multiply.*

14. $(3u^2v)(-8uv^2)$

15. $-18uv^3(2u - 5v)$

16. $(x - 2y)(2x - y)$

17. $(u + 8)^2$

18. $(u - 8)^2$

19. $(3x - 5y)(7x + 4y)$

20. $(4u + 11v)(2u + 3v)$

21. $(2u - 5v)(2u + 5v)$

22. $(7u - v)^2$

23. $(3x^2 + 2y)^2$

24. $(2x^2 - 5y^2)(2x^2 + 5y^2)$

25. $(3u - v)(u - 5v)(u + 5v)$

26. $[3x^2 - (x - y)][3x^2 + (x - y)]$

27. $(2x + 3y)^3$

28. $\begin{array}{r} 6x^2y^2 - 7xy + 5 \\ 4xy + 3 \\ \hline \end{array}$

5.3–5.5 *Factor.*

29. $5u^2v^2 - 10uv$

30. $-11u^3v^2 - 77u^2v^2 - 88uv^2$

31. $15x^2 - 10xy + 3xy - 2y^2$

32. $x^2 - 4x - 21$

33. $3y^2 + 23y + 14$

34. $u^2 + 10uv + 21v^2$

35. $-5x^2 - 15xy + 50y^2$

36. $6x^2 - 54y^2$

37. $8u^3 + v^3$

38. $8u^3 - v^3$

39. $4x^2 - 12xy + 9y^2$

40. $9x^2 + 30xy + 25y^2$

41. $2x^2 + xy - 8x - 4y$

42. $(x - y)^2 - 4$

43. $(x^2 + y^2)^2 - 3(x^2 + y^2) + 2$

44. $(x + y)^2 - (u - v)^2$

45. $x^4 - 5x^2 + 4$

46. $(3x + 5)^2 - 7(3x + 5) + 12$

5.6 *Solve.*

47. $(x + 2)(x - 10) = 0$

48. $y\left(y - \frac{1}{2}\right)(y + 7) = 0$

49. $3x^2 - 2x = 5$

50. $3u(u + 1) = u^2 - 2u - 2$

Solve.

51. Robert is 6 years older than Wong. The product of their ages is 160. How old is each?

52. The area of a triangle is 20 m^2 and the base is 3 m more than the height. Find the base and height.

The profit P of a wholesale operation is given by the equation $P(n) = 2n^2 - 4n - 10$ where n is the number of widgets sold.

53. Find the profit when 20 widgets are sold.

54. How many widgets were sold on a day when the profit was $150?

Part II

Factor.

55. $16u^4 - v^4$

56. $6y^2 + y - 40$

57. $6x^2 - 14xy - 21xy + 49y^2$

58. $9x^2 + 6x + 1$

59. $27a^3 - w^3$

60. $4u^2 - 20u + 25$

Perform the indicated operations.

61. $(10 - 12y^2 + y^3 + 8y) - (6y + 5 - 7y^3 + 8y^2)$

62. $(x - 2y)(x + 2y)$

63. $(6a + 7b)(a + 2b)$

64. $(3u - 5)^2$

65. $(4x + 3y)^2$

66. $(5a^2b^2 - 5ab + 14) + (6a^2b^2 + 7ab - 5)$

In a business the manufacturing cost is $M(u) = 2u^2 - 3u + 5$, the wholesale cost is $W(u) = u^2 - 2u + 4$, and the income is $I(u) = 4u^2 - 2u$ where u is the number of units manufactured and sold.

67. Find the total cost $C(u) = M(u) + W(u)$.

68. Find the total cost when u is 20.

69. Find the profit $P(u) = I(u) - C(u)$.

70. Find the number of units manufactured and sold when the profit was $9.

71. Collect like terms and write in descending powers of x. $6x^4y^3 - 7x^6y - 3xy + 7x^6y$

72. Multiply.
$$\begin{array}{l} 5x^2 + 11xy - 6y^2 \\ \underline{3x - 7y \qquad\qquad} \end{array}$$

The area of seating in a circular auditorium with a circular stage in the center is $A = \pi R^2 - \pi r^2$ where R is the radius of the auditorium and r is the radius of the stage.

73. Find the factored form of this equation.

74. Use the factored form to find the area of seating when $R = 20$ m and $r = 4$ m. (Use 3.14 for π.)

Solve.

75. $(2x - 1)(5x - 3) = 0$

76. $3x^2 - 22x - 16 = 0$

77. $x^2 = \frac{7}{3}x$

78. $(3x - 1)(2x + 1)^2 = 0$

79. The product of 1 less than a number and 7 more than the number is 153. Find the number.

80. The current I in amperes in an electrical circuit varies according to time t in seconds by the equation $I = 6t^2 - 48t + 96$. **(a)** What is the initial amperage in the circuit? **(b)** In how many seconds will the amperage be 0 amperes?

ANSWERS: **1.** (a) trinomial (b) x^2, 1 (c) 2 **2.** (a) monomial (b) $-4ab$, -4 (c) 2 **3.** (a) trinomial (b) $9y^2$, 9 (c) 11 **4.** $x^2y - 3xy + 5$ **5.** $-14x^3y + 5x^2y^2 + 4xy$ **6.** 12 **7.** 62 **8.** $8y^3 + 3y^2 - 19y$ **9.** $15x^3y^2 - 10x^2y^2 + 5xy + 12$ **10.** $15a^3b^2 - 3a^2b - 2ab^2$ **11.** $-6a^3b^3 - a$ **12.** $58x^2y^2 + 12xy - 12$ **13.** $23x^3y^2 - 21x^2y^3 - 17xy + 23$ **14.** $-24u^3v^3$ **15.** $-36u^2v^3 + 90uv^4$ **16.** $2x^2 - 5xy + 2y^2$ **17.** $u^2 + 16u + 64$ **18.** $u^2 - 16u + 64$ **19.** $21x^2 - 23xy - 20y^2$ **20.** $8u^2 + 34uv + 33v^2$ **21.** $4u^2 - 25v^2$ **22.** $49u^2 - 14uv + v^2$ **23.** $9x^4 + 12x^2y + 4y^2$ **24.** $4x^4 - 25y^4$ **25.** $3u^3 - u^2v - 75uv^2 + 25v^3$ **26.** $9x^4 - x^2 + 2xy - y^2$ **27.** $8x^3 + 36x^2y + 54xy^2 + 27y^3$ **28.** $24x^3y^3 - 10x^2y^2 - xy + 15$ **29.** $5uv(uv - 2)$ **30.** $-11uv^2(u^2 + 7u + 8)$ **31.** $(5x + y)(3x - 2y)$ **32.** $(x - 7)(x + 3)$ **33.** $(3y + 2)(y + 7)$ **34.** $(u + 3v)(u + 7v)$ **35.** $-5(x + 5y)(x - 2y)$ **36.** $6(x - 3y)(x + 3y)$ **37.** $(2u + v)(4u^2 - 2uv + v^2)$ **38.** $(2u - v)(4u^2 + 2uv + v^2)$ **39.** $(2x - 3y)^2$ **40.** $(3x + 5y)^2$ **41.** $(2x + y)(x - 4)$ **42.** $(x - y - 2)(x - y + 2)$ **43.** $(x^2 + y^2 - 1)(x^2 + y^2 - 2)$ **44.** $(x + y + u - v)(x + y - u + v)$ **45.** $(x + 1)(x - 1)(x + 2)(x - 2)$ **46.** $(3x + 2)(3x + 1)$ **47.** -2, 10 **48.** 0, $\frac{1}{2}$, -7 **49.** $\frac{5}{3}$, -1 **50.** $-\frac{1}{2}$, -2 **51.** Robert is 16, Wong is 10. **52.** 8 m, 5 m **53.** \$710 **54.** 10 **55.** $(2u - v)(2u + v)(4u^2 + v^2)$ **56.** $(2y - 5)(3y + 8)$ **57.** $(2x - 7y)(3x - 7y)$ **58.** $(3x + 1)^2$ **59.** $(3a - w)(9a^2 + 3aw + w^2)$ **60.** $(2u - 5)^2$ **61.** $8y^3 - 20y^2 + 2y + 5$ **62.** $x^2 - 4y^2$ **63.** $6a^2 + 19ab + 14b^2$ **64.** $9u^2 - 30u + 25$ **65.** $16x^2 + 24xy + 9y^2$ **66.** $11a^2b^2 + 2ab + 9$ **67.** $C(u) = 3u^2 - 5u + 9$ **68.** \$1109 **69.** $P(u) = u^2 + 3u - 9$ **70.** 3 **71.** $6x^4y^3 - 3xy$ **72.** $15x^3 - 2x^2y - 95xy^2 + 42y^3$ **73.** $A = \pi(R + r)(R - r)$ **74.** approximately 1206 m^2 **75.** $\frac{1}{2}$, $\frac{3}{5}$ **76.** $-\frac{2}{3}$, 8 **77.** 0, $\frac{7}{3}$ **78.** $-\frac{1}{2}$, $\frac{1}{3}$ **79.** 10, -16 **80.** (a) 96 amperes (b) 4 seconds

CHAPTER 5 TEST

Given the polynomial $2x^4 - 5x^2y + y^2$:

1. Give the term which has degree 3. 1. ______

2. What is the degree of the polynomial? 2. ______

3. Is the polynomial a monomial, binomial, or trinomial? 3. ______

4. Evaluate the polynomial when $x = 2$ and $y = -1$. 4. ______

Perform the indicated operations and simplify.

5. $(3x^3 - 2x^2 + x - 7) + (2x^2 - x + 3)$ 5. ______

6. $(3a^3b^2 - 2ab + b^4) - (4a^3b^2 - 2ab - 3b^4)$ 6. ______

7. $(2x^2y^3 - xy) + (3x^2y^3 + 7) - (7 - xy)$ 7. ______

8. Add. $\begin{array}{r} 2ab - 3a^2b^2 + 8 \\ \underline{6ab + 3a^2b^2 - 7} \end{array}$ 8. ______

Multiply.

9. $-4u^2v(3u - 2v)$ 9. ______

10. $(a - 3b)(a + 3b)$ 10. ______

11. $(z + 5)^2$ 11. ______

12. $(z - 5)^2$ 12. ______

13. $(3x - 2y)(4x + 5y)$ 13. ______

14. $(a^2 + b^2)(a - b)(a + b)$ 14. ______

Factor.

15. $-11x^2y^3 + 66xy^2 - 44x^2y^2$ 15. ______

16. $6u^2 - uv - v^2$ 16. ______

17. $2x^2 - 8$ 17. ______

18. $a^3 + 8b^3$ 18. ______

19. $x^2 - 6xy + 9y^2$ 19. ______

20. $bx - 2b - xy + 2y$ 20. ______

21. $(x + y)^2 - 5(x + y) + 6$ 21. ______

Solve.

22. $(x + 1)(x - 7) = 0$ 22. ______

23. $2y^2 + 9y = 5$ 23. ______

24. Mack is 3 years older than Betty and the product of their ages is 180. How old is each? 24. ______

25. A retailer's profit is given by the equation $P(n) = 2n^2 - 59n - 5$ where n is the number of items sold. How many items were sold on Tuesday when the profit was \$25? 25. ______

CHAPTER 6

Rational Expressions

6.1 BASIC CONCEPTS OF ALGEBRAIC FRACTIONS

STUDENT GUIDEPOSTS

1. Rational Expressions
2. Fundamental Principle of Fractions
3. Reducing to Lowest Terms

In Chapter 1 we saw that a rational number or fraction is a quotient of integers. In this chapter we define an **algebraic fraction** as a fraction that contains a variable in the numerator, the denominator, or both, such as

$$\frac{2x}{x-1}, \quad \frac{\sqrt{y}+1}{2y+1}, \quad \frac{(a+b)^2}{2}, \quad \frac{7}{(x-y)(y+1)}.$$

1 RATIONAL EXPRESSIONS

An algebraic fraction that is the ratio of two polynomials is called a **rational expression.** Of the algebraic fractions listed above, only

$$\frac{\sqrt{y}+1}{2y+1}$$

is *not* a rational expression since $\sqrt{y}+1$ is not a polynomial. Most of the algebraic fractions that we will consider are rational expressions.

Since division by zero is undefined, with rational expressions we must be careful to exclude replacements for the variable that make the denominator equal to zero.

To Find the Values for Which a Rational Expression Is Not Defined

1. Set the denominator equal to zero.
2. Solve the resulting equation; any solution must be excluded as a possible value of the variable.

EXAMPLE 1 FINDING VALUES TO EXCLUDE

Find the values for the variable that must be excluded.

(a) $\dfrac{3x}{x-5}$

Setting $x - 5 = 0$ and solving, we obtain $x = 5$. Then 5 must be excluded since if x is 5, the denominator is zero.

PRACTICE EXERCISE 1

Find the values for the variable that must be excluded.

(a) $\dfrac{w}{2w+1}$

(b) $\dfrac{3y}{y^2 - y - 6}$

$$y^2 - y - 6 = 0$$
$$(y - 3)(y + 2) = 0 \qquad \text{Factor}$$
$$y - 3 = 0 \quad \text{or} \quad y + 2 = 0 \qquad \text{Zero-product rule}$$
$$y = 3 \qquad\qquad y = -2$$

The values that must be excluded are 3 and -2.

(c) $\dfrac{x^2 - y^2}{5}$

Since the denominator can never be zero, there are no values that have to be excluded. The fraction is defined for all values of x and y.

(b) $\dfrac{a + 3}{27}$

(c) $\dfrac{z - 8}{z^2 - 1}$

Answers: (a) $-\frac{1}{2}$ (b) none (c) 1 and -1

2 FUNDAMENTAL PRINCIPLE OF FRACTIONS

In Chapter 1, we saw that multiplying or dividing the numerator and denominator of a fraction by the same nonzero number gives us equivalent fractions.

$$\frac{4}{10} = \frac{4 \cdot 5}{10 \cdot 5} = \frac{20}{50} \qquad \tfrac{4}{10} \text{ is equivalent to } \tfrac{20}{50}$$
$$\frac{4}{10} = \frac{4 \div 2}{10 \div 2} = \frac{2}{5} \qquad \tfrac{4}{10} \text{ is equivalent to } \tfrac{2}{5}$$

We can generalize this idea to include algebraic fractions.

Fundamental Principle of Fractions

If the numerator and denominator of an algebraic fraction are multiplied or divided by the same nonzero expression or number, the resulting algebraic fraction is **equivalent** to the original. That is,

$$\frac{a}{b} = \frac{ac}{bc} \quad \text{and} \quad \frac{ac}{bc} = \frac{a}{b} \quad (b \neq 0 \text{ and } c \neq 0).$$

EXAMPLE 2 DETERMINING EQUIVALENT FRACTIONS

Are the fractions equivalent?

(a) $\dfrac{x}{x - y}$ and $\dfrac{x\,(x + y)}{(x - y)\,(x + y)}$

These are equivalent (if $x + y \neq 0$). Note that $\frac{x(x + y)}{(x - y)(x + y)}$ has been built up from $\frac{x}{x - y}$ by multiplying numerator and denominator by $(x + y)$, while $\frac{x}{x - y}$ has been reduced from $\frac{x(x + y)}{(x - y)(x + y)}$ by dividing numerator and denominator by $(x + y)$.

(b) $\dfrac{3x}{x + 2}$ and $\dfrac{9}{x - 3}$

These are *not* equivalent since there is no expression that both the numerator and denominator of one fraction can be multiplied by to give the other.

PRACTICE EXERCISE 2

Are the fractions equivalent?

(a) $\dfrac{2a}{a + 1}$ and $\dfrac{2a^2}{a^2 + a}$

(b) $\dfrac{2w}{2w - 5}$ and $\dfrac{w}{w - 5}$.

Answers: (a) yes (b) no

3 REDUCING TO LOWEST TERMS

A fraction is **reduced to lowest terms** when 1 (or -1) is the only number or expression that divides both numerator and denominator. For example, we might reduce $\frac{-12}{42}$ to lowest terms as follows.

$$\frac{-12}{42} = \frac{(-1)\cdot\cancel{2}\cdot 2\cdot\cancel{3}}{\cancel{2}\cdot\cancel{3}\cdot 7} = \frac{-2}{7}$$ Factor completely; divide out common factors

When we factor completely, we factor into **primes,** numbers or expressions that can be divided only by themselves and 1. Sometimes we can reduce a fraction without factoring all the way to primes. This can happen when we recognize larger common factors in the numerator and denominator. For example, we might reduce $\frac{-12}{42}$ as follows.

$$\frac{-12}{42} = \frac{(-1)\cdot\cancel{6}\cdot 2}{\cancel{6}\cdot 7} = \frac{-2}{7}$$

Since we recognize 6 as a factor common to the numerator and denominator, it was not necessary to factor 6 into primes.

CAUTION

Divide out *factors* only, not terms. Only expressions *multiplied by* (not added to) every other expression in the numerator or denominator can be divided out. For example,

$$\frac{2\cdot 4}{2} = \frac{\cancel{2}\cdot 4}{\cancel{2}} \quad \text{but} \quad \frac{2+4}{2} \neq \frac{\cancel{2}+4}{\cancel{2}}.$$

The term "cancel" is sometimes used for "divide out." However, to avoid errors like the one shown in the caution statement above, it is probably better to use "divide out" since division is the inverse operation to multiplication.

To Reduce an Algebraic Fraction to Lowest Terms

1. Factor numerator and denominator completely.
2. Divide out all common factors.
3. Multiply the remaining factors.

EXAMPLE 3 Reducing Fractions to Lowest Terms

Reduce to lowest terms.

(a) $$\frac{20x^2y}{50x^3y^4} = \frac{\cancel{2}\cdot 2\cdot\cancel{5}\cdot\cancel{x}\cdot\cancel{x}\cdot\cancel{y}}{\cancel{2}\cdot\cancel{5}\cdot 5\cdot\cancel{x}\cdot\cancel{x}\cdot x\cdot\cancel{y}\cdot y\cdot y\cdot y} = \frac{2}{5x\cdot y\cdot y\cdot y}$$
$$= \frac{2}{5xy^3}$$

Compare this with the result obtained by applying the rules of exponents to x and y.

$$\frac{20x^2y}{50x^3y^4} = \frac{20}{50}x^{2-3}y^{1-4} = \frac{2\cdot\cancel{10}}{5\cdot\cancel{10}}x^{-1}y^{-3} = \frac{2}{5}\,\frac{1}{x}\,\frac{1}{y^3} = \frac{2}{5xy^3}$$

PRACTICE EXERCISE 3

Reduce to lowest terms.

(a) $\frac{15ab^2}{25a^3b^3}$

Here we reduced $\frac{20}{50}$ by dividing out the common factor, 10. With practice, our work could be shortened as follows:

$$\frac{20x^2y}{50x^3y^4} = \frac{2 \cdot \cancel{10} \cdot \cancel{x^2} \cdot \cancel{y}}{5 \cdot \cancel{10} \cdot \cancel{x^2} \cdot x \cdot \cancel{y} \cdot y^3} = \frac{2}{5xy^3}$$ Divide out common factor $10x^2y$

(b) $\dfrac{6x^3 + 2x}{2x} = \dfrac{\cancel{2x}(3x^2 + 1)}{\cancel{2x}} = \dfrac{3x^2 + 1}{1} = 3x^2 + 1$

Never divide out terms, as in $\dfrac{6x^3 + \cancel{2x}}{\cancel{2x}}$. This is wrong

(c) $\dfrac{x + xy}{y} = \dfrac{x(1 + y)}{y}$ Cannot be reduced; it is already in lowest terms

(d) $\dfrac{x^2 + x - 2}{x^2 - 1} = \dfrac{\cancel{(x - 1)}(x + 2)}{\cancel{(x - 1)}(x + 1)}$

$= \dfrac{x + 2}{x + 1}$

(e) $\dfrac{2a^3 + a^2b - ab^2}{a^2 + ab} = \dfrac{a(2a^2 + ab - b^2)}{a(a + b)}$

$= \dfrac{\cancel{a(a + b)}(2a - b)}{\cancel{a(a + b)}}$

$= 2a - b$

(b) $\dfrac{3w^2 + 9w}{3w}$

(c) $\dfrac{2y}{3y + 6}$

(d) $\dfrac{a^2 - 4}{a^2 + a - 6}$

(e) $\dfrac{x^3 - x^2y - 2xy^2}{x^3 - 3x^2y + 2xy^2}$

Answers: (a) $\frac{3}{5a^2b}$ (b) $w + 3$ (c) cannot be reduced (d) $\frac{a+2}{a+3}$ (e) $\frac{x+y}{x-y}$

CAUTION

When dividing out factors, remember that the quotient is 1 (not zero). For example,

$$\frac{x}{3x^2} = \frac{\cancel{x}}{3 \cdot \cancel{x} \cdot x} = \frac{1}{3 \cdot 1 \cdot x} = \frac{1}{3x},$$

not $\dfrac{0}{3x}$. We can write small 1's by factors that are divided out, as shown below, to remind us of this.

$$\frac{x}{3x^2} = \frac{\overset{1}{\cancel{x}}}{3\underset{1}{\cancel{x}}x} = \frac{1}{3x}$$

Suppose we wish to reduce $\frac{y-x}{x-y}$ to lowest terms. Notice that $y - x$ and $x - y$ are negatives of each other.

$\dfrac{y - x}{x - y} = \dfrac{(-1)(-y + x)}{(x - y)}$ Factor -1 from each term in the numerator

$= \dfrac{(-1)\cancel{(x - y)}}{\cancel{(x - y)}}$ $-y + x = x - y$ by the commutative law

$= -1$

In general we have shown that

$$\frac{a - b}{b - a} = -1, \quad \text{when } a \neq b.$$

EXAMPLE 4 Reducing a Fraction with Negative Factors

Reduce $\frac{y^2 - xy}{x^2 - xy}$ to lowest terms.

$$\frac{y^2 - xy}{x^2 - xy} = \frac{y(y - x)}{x(x - y)} = \frac{y}{x} \cdot \frac{y - x}{x - y} = \frac{y}{x} \cdot (-1) = -\frac{y}{x}$$

Practice Exercise 4

Reduce $\frac{m^2 - n^2}{mn - m^2}$ to lowest terms.

Answer: $\frac{m + n}{-m}$

Finding values to exclude in a rational expression in an applied problem can be slightly different. This is shown in the next example.

EXAMPLE 5 Interest Problem

The formula

$$P = \frac{A}{1 + 0.08t}$$

gives the relationship between a principal P that must be deposited in an account earning 8% simple interest to accumulate a specified amount A at the end of t years.

(a) Find the value of P if A is to be \$1160 after 2 years.
Substitute 1160 for A and 2 for t.

$$P = \frac{1160}{1 + 0.08(2)} = \frac{1160}{1 + 0.16} = \frac{1160}{1.16} = 1000$$

Thus, the principal must be \$1000.

(b) What values of t must be excluded in the formula for P?

Usually we only want to exclude values for the variable that make a denominator zero. For $1 + 0.08t$ to equal 0, t would be $\frac{-1}{0.08} = -12.5$. However, since a negative number for time in years would not make sense in this problem, every negative value for t must be excluded.

Practice Exercise 5

The base of a triangle with area A and height h is given by the formula $b = \frac{2A}{h}$.

(a) Find b if $A = 18.9$ cm^2 and $h =$ 5.4 cm.

(b) What values of h must be excluded in the formula?

Answers: (a) 7.0 cm (b) all values of $h \leq 0$

6.1 EXERCISES A

Find the values for the variable that must be excluded.

1. $\frac{x}{3x + 1}$ **2.** $\frac{y^2 + 3}{y + 3}$ **3.** $\frac{3x + 1}{x}$ **4.** $\frac{a + 1}{(a - 1)(a + 2)}$

5. $\frac{5}{y^2 - 2y + 1}$ **6.** $\frac{a + 1}{2a^3 + 7a^2 + 6a}$ **7.** $\frac{4x + 5}{x^2 - x}$ **8.** $\frac{5 - w}{5 + w^2}$

Are the given fractions equivalent? Explain. (Assume values that make the denominator zero are excluded.)

9. $\frac{2}{3}, \frac{6}{9}$ **10.** $\frac{4}{x}, \frac{x}{4}$ **11.** $\frac{5}{a}, \frac{-5}{-a}$

12. $\frac{-7}{x}, \frac{7}{-x}$

13. $\frac{x+1}{y}, \frac{2(x+1)}{2y}$

14. $\frac{a+b}{3}, \frac{5a+5b}{15}$

15. $\frac{2}{x^3}, \frac{2+y}{x^3+y}$

16. $\frac{a+b}{a^2-b^2}, \frac{1}{a-b}$

17. $\frac{a^2-b^2+c^2}{b-c}, \frac{-a^2+b^2-c^2}{c-b}$

Reduce to lowest terms. (Assume values that make the denominator zero are excluded.)

18. $\frac{45}{105}$

19. $\frac{91}{39}$

20. $\frac{9x^7y^4}{18x^3y^5}$

21. $\frac{7a^2b^3z^5}{21ab^5z^2}$

22. $\frac{2x}{2x^2+2x}$

23. $\frac{y+4}{y^2+4}$

24. $\frac{a^2-b^2}{a^2-2ab+b^2}$

25. $\frac{x^3-y^3}{x-y}$

26. $\frac{a-z}{z-a}$

27. $\frac{x-2y}{2y-x}$

28. $\frac{x+2y}{2y+x}$

29. $\frac{x-2y}{2y+x}$

30. $\frac{a^4-b^4}{a^2+b^2}$

31. $\frac{xz+xw+yz+yw}{z^2+zw}$

32. $\frac{15x+7x^2-2x^3}{x^2-8x+15}$

33. $\frac{a^2-2a}{2-a}$

34. $\frac{9a^2-15a+25}{27a^3+125}$

35. $\frac{3z^2-2z-1}{z^3-z^2+2z-2}$

36. $\frac{8u^3-27}{2u-3}$

37. $\frac{(x+2)(x^2-3x+2)}{(x-2)(x^2+x-2)}$

38. $\frac{y^2-x^2+2x-1}{y^2-x^2+y+x}$

Dr. Gordon Johnson's assistant, Burford, worked Exercises 39–40 as indicated. Why was Burford fired?

39. $\frac{2+4x}{4x} = \frac{2+\cancel{4x}}{\cancel{4x}} = 2$

40. $\frac{x+2}{x^2-4} = \frac{\cancel{x+2}}{\cancel{(x+2)}(x-2)} = \frac{0}{x-2} = 0$

41. Reduce $\frac{5-xy}{xy-5}$ to lowest terms.

42. Reduce $\frac{5+xy}{xy+5}$ to lowest terms.

43. Reduce $\frac{x^2 - xy - 2y^2}{x + y}$ to lowest terms. Substitute 2 for x and -1 for y in both the original and the reduced fraction to show that the two are equivalent.

In Exercises 44–45, use a calculator to find the value of P, to the nearest cent, if $P = \frac{A}{1 + 0.07t}$ *and A and t have the given values.*

44 $A = \$5000$ and $t = 6$ yr

45. $A = \$10{,}000$ and $t = 12$ yr

46. You have probably observed the change in the pitch of a train whistle as the train moves quickly past you. The normal frequency (number of vibrations per second) of the whistle appears to increase as the train approaches and decrease as the train moves away. Physicists call this the **Doppler effect,** named for Austrian physicist C. J. Doppler (1803–53). The formula that describes the Doppler effect is

$$F = \frac{1100f}{1100 - V},$$

where F is the observed frequency of a sound, f is the actual frequency of a sound, and V is the velocity of the train (a positive number if the train is approaching and a negative number if it is going away). Suppose a train whistle produces a sound with a frequency of 612.0 vibrations per second. What is the observed frequency **(a)** as the train is approaching at 80 mph? **(b)** as the train is going away at 80 mph?

FOR REVIEW

Exercises 47–49 review multiplication and division of fractions to help you prepare for the next section.

47. $\frac{3}{8} \cdot \frac{4}{27}$

48. $\frac{4}{5} \div \frac{8}{35}$

49. $\frac{9}{5} \div 3$

50. Factor 13,650 into a product of prime integers.

ANSWERS: 1. $-\frac{1}{3}$ 2. -3 3. 0 4. 1, -2 5. 1 6. 0, -2, $-\frac{3}{2}$ 7. 0, 1 8. none 9. yes 10. no 11. yes 12. yes 13. yes 14. yes 15. no 16. yes 17. yes 18. $\frac{3}{7}$ 19. $\frac{7}{3}$ 20. $\frac{x^4}{2y}$ 21. $\frac{az^3}{3b^2}$ 22. $\frac{1}{x+1}$ 23. $\frac{y+4}{y^2+4}$ 24. $\frac{a+b}{a-b}$ 25. $x^2 + xy + y^2$ 26. -1 27. -1 28. 1 29. $\frac{x-2y}{2y+x}$ 30. $a^2 - b^2$ 31. $\frac{x+y}{z}$ 32. $\frac{x(2x+3)}{3-x}$ 33. $-a$ 34. $\frac{1}{3a+5}$ 35. $\frac{3z+1}{z^2+2}$ 36. $4u^2 + 6u + 9$ 37. 1 38. $\frac{y+x-1}{y+x}$ 39. Terms have been divided out, *not* factors. 40. The numerator is 1, *not* 0. 41. -1 42. 1 43. $x - 2y$; both values are 4 44. \$3521.13 45. \$5434.78 46. (a) 660.0 vibrations/sec (b) 570.5 vibrations/sec 47. $\frac{1}{18}$ 48. $\frac{7}{2}$ 49. $\frac{3}{5}$ 50. $2 \cdot 3 \cdot 5 \cdot 5 \cdot 7 \cdot 13$

6.1 EXERCISES B

Find the values for the variable that must be excluded.

1. $\frac{y}{2y - 1}$

2. $\frac{a^2 + 7}{a - 2}$

3. $\frac{4z - 3}{z}$

4. $\frac{x + 5}{(x + 1)(x - 3)}$

5. $\frac{16}{a^2 - 4a + 4}$

6. $\frac{x - 1}{x^3 + x^2 - 6x}$

7. $\frac{3a + 2}{a^2 + a}$

8. $\frac{3c + 2}{3c^2 + 2}$

Are the given fractions equivalent? Explain. (Assume values that make the denominator zero are excluded.)

9. $\frac{3}{4}, \frac{15}{20}$

10. $\frac{5}{y}, \frac{y}{5}$

11. $\frac{-3}{x^2}, \frac{3}{-x^2}$

12. $\frac{-4}{-y^2}, \frac{4}{y^2}$

13. $\frac{a+3}{b}, \frac{4(a+3)}{4b}$

14. $\frac{x-y}{7}, \frac{3x-3y}{21}$

15. $\frac{5+x}{y^3+x}, \frac{5}{y^3}$

16. $\frac{x+2}{x^2-4}, \frac{1}{x-2}$

17. $\frac{x^2-5+y}{x-2y}, \frac{5-x^2-y}{2y-x}$

Reduce to lowest terms. (Assume values that make the denominator zero are excluded.)

18. $\frac{39}{52}$

19. $\frac{60}{135}$

20. $\frac{15a^3b^8}{10ab^5}$

21. $\frac{16x^3y^7u^5}{12xy^8u^{10}}$

22. $\frac{5u}{5u^2+5u}$

23. $\frac{a+2}{a^2+4}$

24. $\frac{x^2-y^2}{x^2+2xy+y^2}$

25. $\frac{u^3+v^3}{u+v}$

26. $\frac{2a-3w}{3w-2a}$

27. $\frac{2x-y}{y-2x}$

28. $\frac{2x+y}{y+2x}$

29. $\frac{2x-y}{y+2x}$

30. $\frac{x^2-y^2}{x^4-y^4}$

31. $\frac{ab-ac+db-dc}{ca+cd}$

32. $\frac{u^2-10u}{10-u}$

33. $\frac{2a^2-a-3}{2a^3-3a^2+2a-3}$

34. $\frac{8u^3-27}{2u-3}$

35. $\frac{(x+2)(x^2-3x+2)}{(x-2)(x^2+x-2)}$

36. $\frac{9a^2-15a+25}{27a^3+125}$

37. $\frac{a^2-2ab+b^2-4}{a^2-b^2-2a-2b}$

38. $\frac{u^3+1+2u^2+2u}{u^3-1}$

Dr. Gordon Johnson's assistant Burford, worked Exercises 39–40 as indicated. Why was Burford fired?

39. $\frac{x+3z}{3z} = \frac{x+\cancel{3z}}{\cancel{3z}} = x$

40. $\frac{y-1}{y^2-1} = \frac{\cancel{y-1}}{(y+1)\cancel{(y-1)}} = \frac{0}{y+1} = 0$

41. Reduce $\frac{2-3a}{3a-2}$ to lowest terms.

42. Reduce $\frac{2+3a}{3a+2}$ to lowest terms.

43. Substitute -1 for x and 3 for y in $x-y$ and $\frac{x^2-3xy+2y^2}{x-2y}$. Why are both results equal to -4?

In Exercises 44–45, use a calculator to find the value of P, to the nearest cent, if $P = \frac{A}{1+0.07t}$ *and A and t have the given values.*

44. $A = \$2500$ and $t = 5$ yr

45. $A = \$10{,}000$ and $t = 7$ yr

46. By the Doppler effect, the frequency of a sound such as a train whistle appears to increase as the train approaches and decrease as the train moves away. The formula that describes the Doppler effect is

$$F = \frac{1100f}{1100-V},$$

where F is the observed frequency of a sound, f is the actual frequency of a sound, and V is the velocity of the train (a positive number if the train is approaching and a negative number if it is going away). Suppose a car horn produces a sound with a frequency of 414.0 vibrations per second. What is the observed frequency **(a)** as the car is approaching at 65 mph? **(b)** as the car is going away at 65 mph?

FOR REVIEW

Exercises 47–49 review multiplication and division of fractions to help you prepare for the next section.

47. $\frac{6}{7} \cdot 28$

48. $\frac{3}{4} \div \frac{5}{4}$

49. $12 \div \frac{4}{5}$

50. Factor 48,510 into a product of prime integers.

6.1 EXERCISES C

Find the values for the variable that must be excluded.

1. $\dfrac{ab}{ab + b - 5a - 5}$
[*Hint:* Factor the denominator by grouping terms.]

2. $\dfrac{x}{(x^2 - 4x + 4) - 25}$
[Answer: 7, −3]

Reduce to lowest terms. (Assume values that make the denominator zero are excluded.)

3. $\dfrac{y^2 + 6y + 5}{(y - 1)^2 - 4}$

4. $\dfrac{a^2 + ab + b^2}{a^4 - b^3a + a^3b - b^4}$

6.2 MULTIPLICATION AND DIVISION OF FRACTIONS

STUDENT GUIDEPOSTS

1 Multiplying Fractions

2 Dividing Fractions

1 MULTIPLYING FRACTIONS

Remember that the product of numerical fractions is equal to the product of all numerators divided by the product of all denominators. That is,

$$\frac{a}{b} \cdot \frac{c}{d} = \frac{ac}{bd}, \quad b \neq 0 \text{ and } d \neq 0$$

The same is true for algebraic fractions. In both cases, the resulting fraction should be reduced to lowest terms, which is easy to do if all common factors in the numerator and denominator are divided out before multiplying. For example,

$$\frac{6}{35} \cdot \frac{7}{3} = \frac{6 \cdot 7}{35 \cdot 3} = \frac{2 \cdot \cancel{3} \cdot \cancel{7}}{5 \cdot \cancel{7} \cdot \cancel{3}} = \frac{2}{5}$$

Had we multiplied first to obtain $\frac{42}{105}$, we would have had to factor (undo this multiplication) to reduce the product.

To Multiply Two or More Algebraic Fractions

1. Factor all numerators and denominators and place the indicated product of all numerator factors over the product of all denominator factors.
2. Divide out common factors to reduce to lowest terms.
3. Multiply the remaining numerator factors, then multiply the remaining denominator factors.

EXAMPLE 1 MULTIPLYING FRACTIONS

Multiply and simplify.

(a) $$\frac{3a}{8} \cdot \frac{2}{9a^2} = \frac{\cancel{3} \cdot \cancel{a} \cdot \cancel{2}}{\cancel{2} \cdot 2 \cdot 2 \cdot \cancel{3} \cdot 3 \cdot \cancel{a} \cdot a} = \frac{1}{2 \cdot 2 \cdot 3 \cdot a} = \frac{1}{12a}$$

In a problem like this, we can take several shortcuts without first factoring to primes. The work might appear as

$$\frac{\overset{1 \cdot 1 \cdot 1}{\cancel{3} \cdot \cancel{a} \cdot \cancel{2}}}{\underset{4 \cdot 3 \cdot a}{\cancel{8} \cdot \cancel{9}\, \cancel{a^2}}} = \frac{1}{12a},$$

in which 3 was divided into 9, 2 into 8, and a into a^2.

(b) $$\frac{x^2 - 5x + 6}{x^2 + x - 6} \cdot \frac{x + 3}{x - 2} = \frac{\cancel{(x-2)}(x-3)\cancel{(x+3)}}{\cancel{(x+3)}\cancel{(x-2)}(x-2)} = \frac{x-3}{x-2}$$

(c) $$\frac{a^2 - 4b^2}{49a^3 - 4ab^2} \cdot \frac{7a^2 - 2ab}{a + 2b} = \frac{(a-2b)\cancel{(a+2b)} \cdot \cancel{a}\cancel{(7a-2b)}}{\cancel{a}\cancel{(7a-2b)}(7a+2b)\cancel{(a+2b)}}$$
$$= \frac{a - 2b}{7a + 2b}$$

(d) $$\frac{y^2 - 2y - 3}{y^2 - y - 2} \cdot \frac{y^2 + 3y - 10}{(5 + y)(1 - 2y)} \cdot \frac{2y^2 - 3y + 1}{y^2 - 4y + 3}$$
$$= \frac{\cancel{(y-3)}\cancel{(y+1)}\cancel{(y-2)}\cancel{(y+5)}(2y-1)\cancel{(y-1)}}{\cancel{(y+1)}\cancel{(y-2)}\cancel{(5+y)}(1-2y)\cancel{(y-1)}\cancel{(y-3)}}$$ Cancel common factors
$$= \frac{2y - 1}{1 - 2y} = -1$$ $\frac{a-b}{b-a} = -1$

PRACTICE EXERCISE 1

Multiply and simplify.

(a) $\frac{5x^3}{7} \cdot \frac{14}{25x}$

(b) $\frac{a^2 + 3a - 4}{a^2 - 16} \cdot \frac{a - 4}{a + 1}$

(c) $\frac{u^2 - 3uv}{3u^2 - 8uv - 3v^2} \cdot \frac{3u^2 + 10uv + 3v^2}{u^2 + 3uv}$

(d) $\frac{a^2 - 1}{a^2 + a} \cdot \frac{a - 2}{a^2 - 3a + 2} \cdot \frac{a^2 + 5a}{2a + 1}$

Answers: (a) $\frac{2x^2}{5}$ (b) $\frac{a-1}{a+1}$ (c) 1 (d) $\frac{a+5}{2a+1}$

CAUTION

There will be a lot of extra work if we multiply numerators and denominators without factoring first. For example, in Exercise 1(b), if we multiply first we obtain

$$\frac{x^2 - 5x + 6}{x^2 + x - 6} \cdot \frac{x + 3}{x - 2} = \frac{x^3 - 2x^2 - 9x + 18}{x^3 - x^2 - 8x + 12}.$$

To reduce the product to lowest terms, we would now have to factor the two third-degree polynomials in the numerator and denominator. *Never multiply without first factoring and dividing out common factors.*

2 DIVIDING FRACTIONS

When one numerical fraction is divided by another, for example,

$$\frac{2}{3} \div \frac{2}{15} \quad \text{or} \quad \frac{\frac{2}{3}}{\frac{2}{15}},$$

the fraction $\frac{2}{15}$ is called the **divisor.** Recall that to calculate the quotient, we multiply $\frac{2}{3}$ by the reciprocal of $\frac{2}{15}$, $\frac{15}{2}$, the fraction formed by interchanging the numerator and the denominator.

$$\frac{2}{3} \div \frac{2}{15} = \frac{2}{3} \cdot \frac{15}{2} = \frac{2}{3} \cdot \frac{3 \cdot 5}{2} = \frac{\cancel{2} \cdot \cancel{3} \cdot 5}{\cancel{3} \cdot \cancel{2}} = 5$$

Thus, if $\frac{a}{b}$ and $\frac{c}{d}$ are two fractions with $\frac{c}{d} \neq 0$, then

$$\frac{a}{b} \div \frac{c}{d} = \frac{a}{b} \cdot \frac{d}{c}.$$

The following table shows several fractions and their reciprocals.

Fraction	$\frac{3}{y}$	$\frac{x+y}{(x-y)^2}$	$5 = \frac{5}{1}$	$y = \frac{y}{1}$	$\frac{1}{2}$	$a + b$
Reciprocal	$\frac{y}{3}$	$\frac{(x-y)^2}{x+y}$	$\frac{1}{5}$	$\frac{1}{y}$	$2 = \frac{2}{1}$	$\frac{1}{a+b}$ $\left(\text{not } \frac{1}{a} + \frac{1}{b}\right)$

Notice that in all cases the product of a fraction and its reciprocal is 1. Division of algebraic fractions is similar to division of numerical fractions.

To Divide One Algebraic Fraction by Another

1. Find the reciprocal of the divisor.
2. Multiply following the same procedure as for multiplying fractions.

EXAMPLE 2 Dividing Fractions

Divide and simplify.

(a) $\frac{9x^5}{20y^7} \div \frac{27x^3}{10y^4} = \frac{9x^5}{20y^7} \cdot \frac{10y^4}{27x^3}$ The reciprocal of $\frac{27x^3}{10y^4}$ is $\frac{10y^4}{27x^3}$

$$= \frac{\cancel{9x^3}x^2 \cdot \cancel{10y^4}}{2 \cdot \cancel{10y^4}y^3 \cdot 3 \cdot \cancel{9x^3}}$$

$$= \frac{x^2}{6y^3}$$

(b) $\frac{a}{a+2} \div \frac{a}{5} = \frac{a}{a+2} \cdot \frac{5}{a}$ The reciprocal of $\frac{a}{5}$ is $\frac{5}{a}$

$$= \frac{\cancel{a} \cdot 5}{(a+2) \cdot \cancel{a}} = \frac{5}{a+2}$$

(c) $\frac{x+y}{x} \div x = \frac{x+y}{x} \cdot \frac{1}{x}$ The reciprocal of x is $\frac{1}{x}$

$$= \frac{x+y}{x^2}$$

(d) $\frac{x^2-6x+9}{x^2-4} \div \frac{x-3}{x+2} = \frac{x^2-6x+9}{x^2-4} \cdot \frac{x+2}{x-3}$

$$= \frac{\cancel{(x-3)}(x-3)\cancel{(x+2)}}{(x-2)\cancel{(x+2)}\cancel{(x-3)}} = \frac{x-3}{x-2}$$

PRACTICE EXERCISE 2

Divide and simplify.

(a) $\frac{5a^3}{16b^5} \div \frac{25a^2}{12b^3}$

(b) $\frac{w}{6} \div \frac{w}{6+w}$

(c) $\frac{a}{a-b} \div \frac{a}{b}$

(d) $\frac{m^2-3mn+2n^2}{m^2-n^2} \div \frac{m+2n}{m+n}$

(e) $\dfrac{a^2-b^2}{a^2+a-2}\cdot\dfrac{a^2+2a}{a^2+2ab+b^2}\div\dfrac{a^3-b^3}{a^2-a+ba-b}$

$$=\frac{(a+b)(a-b)}{(a-1)(a+2)}\cdot\frac{a(a+2)}{(a+b)(a+b)}\cdot\frac{a(a-1)+b(a-1)}{(a-b)(a^2+ab+b^2)}$$

$$=\frac{(a+b)(a-b)a(a+2)(a+b)(a-1)}{(a-1)(a+2)(a+b)(a+b)(a-b)(a^2+ab+b^2)}$$

$$=\frac{a}{a^2+ab+b^2}$$

(e) $\dfrac{x+y}{x-y}\cdot\dfrac{x^2-xy}{x^2-xy+y^2}\div\dfrac{x^2+xy}{x^3+y^3}$

Answers: (a) $\frac{3a}{20b^2}$ (b) $\frac{6+w}{6}$ (c) $\frac{b}{a-b}$ (d) $\frac{m-2n}{m+2n}$ (e) $x+y$

6.2 EXERCISES A

Multiply and simplify.

1. $\dfrac{2ab}{5}\cdot\dfrac{25}{4a^3}$

2. $\dfrac{-4a^3b^2}{3ab^3}\cdot\dfrac{9a^5b^7}{8a^3b^3}$

3. $\dfrac{2c}{7}\cdot\dfrac{7b+14}{4(b+2)}$

4. $\dfrac{3x^2}{x+2}\cdot\dfrac{x^2-4}{3x}$

5. $\dfrac{4x+8}{9x+9}\cdot\dfrac{x^2-1}{2x+4}$

6. $\dfrac{y}{y^2-yz-12z^2}\cdot\dfrac{y^2+3yz}{y-4z}$

7. $\dfrac{a^2+ab}{a-5b}\cdot\dfrac{5b-a}{b^2-a^2}$

8. $\dfrac{9a^2-6a+1}{28+7a}\cdot\dfrac{a^2-16}{3a^2-13a+4}$

9. $\dfrac{a^2-6a-16}{2a^2-128}\cdot\dfrac{a^2+16a+64}{3a^2+30a+48}$

10. $\dfrac{4-a^2}{c^2+2cd+d^2}\cdot\dfrac{c^2-d^2}{2-a-a^2}\cdot\dfrac{c+d-ca-da}{2a-a^2}$

Find the reciprocal of each expression.

11. $\dfrac{4}{5}$

12. 8

13. $\dfrac{x}{3}$

14. $2z$

15. $\dfrac{u}{v+u}$

16. $x+3$

17. $4x+3$

18. 0.27

Divide and simplify.

19. $\dfrac{a}{a+3}\div\dfrac{a}{5}$

20. $\dfrac{3(a+b)}{7}\div\dfrac{9(a+b)}{28}$

21. $\dfrac{(u^2v^3)^2}{3uv}\div\dfrac{(uv^2)^3}{12u^3v}$

22. $\dfrac{x^2 - x - 2}{y^2} \div \dfrac{x^2 - 1}{y^2}$

23. $\dfrac{4}{a^2 - 9} \div \dfrac{4a^2 - 4a - 24}{a^2 - 6a + 9}$

24. $\dfrac{a^3 - 27}{a^2 - 3a} \div \dfrac{a^2 + 3a + 9}{a}$

25. $\dfrac{2a^2 - 2a}{2a + 4} \div \dfrac{a^2 - 1}{a^2 + 3a + 2}$

26. $\dfrac{a^2 - 4a + 3}{a^2 - 2a - 15} \div \dfrac{2a^2 - 6a + 4}{a^2 + 3a - 10}$

Perform the indicated operations.

27. $\dfrac{u^2 - v^2}{u^2 - uv + v^2} \cdot \dfrac{2u^2 - 3uv - 2v^2}{u^2 + 2uv + v^2} \div \dfrac{2u^2 - uv - v^2}{u^3 + v^3}$

28. $\dfrac{3s^2 - 5st - 2t^2}{t^2 + st - 2s^2} \cdot \dfrac{t^2 + 4st - 5s^2}{12s^2 + 7st + t^2} \div \dfrac{5s^2 - 9st - 2t^2}{8s^2 - 2st - t^2}$

FOR REVIEW

29. Burford worked the following problem. What is wrong with it?

$$\frac{2 + (x + y)}{3(x + y)} = \frac{2 + \cancel{(x + y)}}{3\cancel{(x + y)}} = \frac{2}{3}$$

Find the values of the variable that must be excluded.

30. $\dfrac{x + 7}{x^2 + 5x + 6}$

31. $\dfrac{3}{x^2}$

Are the given fractions equivalent?

32. $\dfrac{x + y}{x^2 - y^2}, \dfrac{1}{x - y}$

33. $\dfrac{x^2}{4}, \dfrac{x^2 + 5}{9}$

Reduce to lowest terms.

34. $\dfrac{x^2 + 3x}{x^2 + 5x + 6}$

35. $\dfrac{a^3 - ab^2}{a^2 - ab}$

Exercises 36–41 review adding and subtracting fractions to help you prepare for the next section. Perform the indicated operations.

36. $\dfrac{7}{6} + \dfrac{1}{6}$

37. $\dfrac{7}{10} - \dfrac{3}{10}$

38. $\dfrac{2}{15} + \dfrac{2}{21}$

39. $\frac{3}{35} - \frac{4}{63}$ **40.** $8 + \frac{6}{7}$ **41.** $\frac{41}{3} - 3$

ANSWERS: 1. $\frac{5b}{2a^2}$ 2. $\frac{-3a^4b^3}{2}$ 3. $\frac{c}{2}$ 4. $x(x-2)$ 5. $\frac{2(x-1)}{9}$ 6. $\frac{y^2}{(y-4z)^2}$ 7. $\frac{-a}{b-a}$ or $\frac{a}{a-b}$ 8. $\frac{3a-1}{7}$ 9. $\frac{1}{6}$ 10. $\frac{c-d}{a}$ 11. $\frac{5}{4}$ 12. $\frac{1}{8}$ 13. $\frac{3}{x}$ 14. $\frac{1}{2z}$ 15. $\frac{v+u}{u}$ 16. $\frac{1}{x+3}$ 17. $\frac{1}{4x+3}$ 18. $\frac{100}{27}$ 19. $\frac{5}{a+3}$ 20. $\frac{4}{3}$ 21. $4u^3$ 22. $\frac{x-2}{x-1}$ 23. $\frac{1}{(a+3)(a+2)}$ 24. 1 25. a 26. $\frac{(a-3)(a+5)}{2(a-5)(a+3)}$ 27. $u-2v$ 28. $\frac{2s-t}{2s+t}$ 29. Terms have been divided out, *not* factors. 30. $-2, -3$ 31. 0 32. yes 33. no 34. $\frac{x}{x+2}$ 35. $a+b$ 36. $\frac{4}{3}$ 37. $\frac{2}{5}$ 38. $\frac{8}{35}$ 39. $\frac{1}{45}$ 40. $\frac{62}{7}$ 41. $\frac{32}{3}$

6.2 EXERCISES B

Multiply and simplify.

1. $\frac{5x^3y^2}{x^2y^5} \cdot \frac{x^5y^7}{10x^3y}$

2. $\frac{-2x^2y}{7x} \cdot \frac{14xy^3}{6x^2y^5}$

3. $\frac{u^2v^2}{a-b} \cdot \frac{2b-2a}{4uv}$

4. $\frac{ab}{a-3b} \cdot \frac{(a-3b)^2}{a^2b-3ab^2}$

5. $\frac{a^3+a^2b}{5a} \cdot \frac{25}{3a+3b}$

6. $\frac{x^3-y^3}{2xy} \cdot \frac{4x^2y^2}{x^2+xy+y^2}$

7. $\frac{x^2-4}{x^2-4x+4} \cdot \frac{x^2-9x+14}{x^3+2x^2}$

8. $\frac{2x^2-5xy-3y^2}{4x^2-y^2} \cdot \frac{8x^2+10xy+3y^2}{4x^2-9xy-9y^2}$

9. $\frac{x^2+4x-5}{2x^2+16x} \cdot \frac{x^3-64x}{x^2-9x+8}$

10. $\frac{x^2+3xy+2y^2}{x^2-x} \cdot \frac{x^3-x^2y+xy^2}{x^2+xy-2y^2} \cdot \frac{x^2-x-xy+y}{x^3+y^3}$

Find the reciprocal of each expression.

11. $\frac{1}{7}$

12. $\frac{7}{y}$

13. y

14. $\frac{1}{y+z}$

15. $\frac{a-b}{b}$

16. $\frac{3x-5}{2y+1}$

17. $2\frac{2}{3}$

18. 0

Divide and simplify.

19. $\frac{3x}{x-1} \div \frac{6x}{x-1}$

20. $\frac{20a^2b}{18ab^3} \div \frac{5a^4b}{9a^2b^2}$

21. $\frac{(xy^2)^4}{5x^3y^5} \div \frac{x^7y^7}{(5xy^2)^3}$

22. $\frac{a^2-ab}{a+b} \div \frac{a-b}{a^2+2ab+b^2}$

23. $\frac{uv-uw+xv-xw}{v-w} \div \frac{v^2-2vw+w^2}{xv-xw}$

24. $\frac{x-y}{4x+4y} \div \frac{x^2-2xy+y^2}{x^2-y^2}$

25. $\frac{z^2-z-6}{z^2-81} \div \frac{z^2-9z+18}{4z+36}$

26. $\frac{x^2-y^2}{x^3-y^3} \div \frac{x^2+2xy+y^2}{x^2+xy+y^2}$

Perform the indicated operations.

27. $\frac{y^2+xy-y-x}{x^4-x^3+x^2-x} \cdot \frac{2x^4+2x^3+2x^2+2x}{x^3-x+x^2y-y} \div \frac{2y-2}{x^2-2x+1}$

28. $\dfrac{ax - ay + bx - by}{ax + ay - bx - by} \cdot \dfrac{ax + ay - 3bx - 3by}{ax - ay - 3bx - 3by} \div \dfrac{a + b}{b - a}$

FOR REVIEW

29. Burford worked the following problem. What is wrong with it?

$$\frac{a-5}{a^2-4a-5} = \frac{a-5}{(a-5)(a+1)} = \frac{\cancel{a-5}}{\cancel{(a-5)}(a+1)} = \frac{0}{a+1} = 0$$

Find the values of the variable that must be excluded.

30. $\dfrac{x - 8}{x^2 - 4x + 3}$

31. $\dfrac{3 + y}{y^2}$

Are the given fractions equivalent?

32. $\dfrac{x - y}{x^2 - y^2}, \dfrac{1}{x + y}$

33. $\dfrac{y^3}{2}, \dfrac{y^3 + 3}{5}$

Reduce to lowest terms.

34. $\dfrac{x^2 - 5x}{x^2 - 8x + 15}$

35. $\dfrac{a^2b - a^3}{a^3 - ab^2}$

Exercises 36–41 review adding and subtracting fractions to help you prepare for the next section. Perform the indicated operations.

36. $\dfrac{3}{10} + \dfrac{9}{10}$

37. $\dfrac{5}{6} - \dfrac{1}{6}$

38. $\dfrac{5}{6} + \dfrac{1}{10}$

39. $\dfrac{4}{15} - \dfrac{2}{21}$

40. $2 + \dfrac{3}{5}$

41. $\dfrac{32}{5} - 4$

6.2 EXERCISES C

Perform the indicated operations.

1. $\dfrac{x^3 + x^2 - 20x}{(x - 1)^2 - 9} \cdot \dfrac{x^3 + 8}{ax + 5a + x + 5} \div \dfrac{x^3 - 2x^2 + 4x}{a^2 - 6a - 7}$

2. $\dfrac{ax - bx + a - b}{b^2 - 3ab - 10a^2} \div \left[\dfrac{b^2 - a^2}{2b^2 + 7ab + 6a^2} \div \dfrac{b^2 - 4ab - 5a^2}{x^3 + 1}\right]$ $\left[\text{Answer: } -\dfrac{2b + 3a}{x^2 - x + 1}\right]$

6.3 ADDITION AND SUBTRACTION OF FRACTIONS

STUDENT GUIDEPOSTS

1. Adding and Subtracting Like Fractions
2. Least Common Denominator (LCD)
3. Rewriting Fractions
4. Adding and Subtracting Unlike Fractions

1 ADDING AND SUBTRACTING LIKE FRACTIONS

As with multiplication and division, addition and subtraction of algebraic fractions parallel the corresponding operations on numerical fractions. Fractions that have the same denominator are called **like fractions.** We add or subtract like fractions such as $\frac{a}{b}$ and $\frac{c}{b}$ as follows.

$$\frac{a}{b}+\frac{c}{b}=\frac{a+c}{b} \quad \text{and} \quad \frac{a}{b}-\frac{c}{b}=\frac{a-c}{b}$$

For example,

$$\frac{3}{5}+\frac{1}{5}=\frac{3+1}{5}=\frac{4}{5} \quad \text{and} \quad \frac{3}{5}-\frac{1}{5}=\frac{3-1}{5}=\frac{2}{5}.$$

The resulting sum or difference should always be reduced to lowest terms.

EXAMPLE 1 ADDING AND SUBTRACTING LIKE FRACTIONS

Perform the indicated operation.

(a) $\frac{4}{9}+\frac{2}{9}=\frac{4+2}{9}=\frac{6}{9}=\frac{\cancel{3}\cdot 2}{\cancel{3}\cdot 3}=\frac{2}{3}$ Reduce to lowest terms

(b) $\frac{a+b}{5a}+\frac{a^2-b}{5a}=\frac{(a+b)+(a^2-b)}{5a}$

$$=\frac{a+a^2}{5a}=\frac{\cancel{a}(1+a)}{5\cancel{a}}=\frac{1+a}{5}$$

(c) $\frac{2x+y}{x-1}-\frac{x-y}{x-1}=\frac{(2x+y)-(x-y)}{x-1}$ Use parentheses and watch signs

$$=\frac{2x+y-x+y}{x-1}=\frac{x+2y}{x-1}$$

PRACTICE EXERCISE 1

Perform the indicated operations.

(a) $\frac{1}{12}+\frac{7}{12}$

(b) $\frac{x-y}{xy}+\frac{y^2-x}{xy}$

(c) $\frac{b+3a}{a+2}-\frac{b-a}{a+2}$

Answers: (a) $\frac{2}{3}$ (b) $\frac{y-1}{x}$ (c) $\frac{4a}{a+2}$

CAUTION

Parentheses are extremely important, especially in a subtraction problem. In the above example, $-(x-y)$ becomes $-x+y$ when the parentheses are cleared. If you do not enclose the numerator in parentheses, it is easy to make a sign error.

Often, two fractions have denominators that are negatives of each other. When this occurs, multiply both the numerator and denominator of *one* of the fractions by -1. Since we are multiplying the fraction by the multiplicative identity, 1, in the form $\frac{-1}{-1}$, the result is equivalent to the original and has the same denominator as the other fraction.

EXAMPLE 2 FRACTIONS WHOSE DENOMINATORS ARE NEGATIVES

Perform the indicated operation.

(a) $$\frac{x}{3} + \frac{x-y}{-3} = \frac{x}{3} + \frac{(-1)(x-y)}{(-1)(-3)}$$
$$= \frac{x}{3} + \frac{-x+y}{3} = \frac{x+(-x+y)}{3} = \frac{y}{3}$$

(b) $$\frac{4y}{y-2} - \frac{y+4}{2-y} = \frac{4y}{y-2} - \frac{(-1)(y+4)}{(-1)(2-y)}$$
$$= \frac{4y}{y-2} - \frac{-y-4}{-2+y}$$
$$= \frac{4y}{y-2} - \frac{(-y-4)}{y-2}$$
$$= \frac{4y-(-y-4)}{y-2} = \frac{4y+y+4}{y-2}$$ Watch signs
$$= \frac{5y+4}{y-2}$$

PRACTICE EXERCISE 2

Perform the indicated operation.

(a) $\frac{2-a}{-5} + \frac{2}{5}$

(b) $\frac{3u}{u-v} - \frac{v-2u}{v-u}$

Answers: (a) $\frac{a}{5}$ (b) $\frac{u+v}{u-v}$

2 LEAST COMMON DENOMINATOR (LCD)

To add or subtract fractions with different denominators, called **unlike fractions,** we need to convert them to equivalent fractions with a common denominator. For example, to add $\frac{4}{15}$ and $\frac{5}{6}$, one common denominator is $15 \cdot 6 = 90$.

$$\frac{4}{15} + \frac{5}{6} = \frac{4 \cdot 6}{15 \cdot 6} + \frac{5 \cdot 15}{6 \cdot 15} = \frac{24}{90} + \frac{75}{90} = \frac{99}{90} = \frac{\cancel{9} \cdot 11}{\cancel{9} \cdot 10} = \frac{11}{10}$$

If we use the **least common denominator,** it takes less effort to reduce the final sum or difference to lowest terms. For example, we could have used 30 as the least common denominator above.

$$\frac{4}{15} + \frac{5}{6} = \frac{4 \cdot 2}{15 \cdot 2} + \frac{5 \cdot 5}{6 \cdot 5} = \frac{8}{30} + \frac{25}{30} = \frac{33}{30} = \frac{\cancel{3} \cdot 11}{\cancel{3} \cdot 10} = \frac{11}{10}$$

The method for finding the least common denominator (LCD) of two or more fractions, including algebraic fractions, is summarized here.

To Find the LCD of Two or More Fractions

1. Factor all denominators and reduce all fractions to lowest terms.
2. When there are no common factors in any two denominators, the LCD is the product of *all* denominators.
3. When there are common factors in two or more denominators, each factor appears in the LCD as many times as it appears in the denominator where it is found the greatest number of times.

EXAMPLE 3 Finding the Least Common Denominator

Find the LCD of the fractions.

(a) $\dfrac{7}{150}$ and $\dfrac{2}{315}$

Factor the denominators.

$$150 = 2 \cdot 3 \cdot 5 \cdot 5 \quad \text{and} \quad 315 = 3 \cdot 3 \cdot 5 \cdot 7$$

The LCD must consist of one 2, two 3s, two 5s, and one 7. Thus, the LCD is $2 \cdot 3 \cdot 3 \cdot 5 \cdot 5 \cdot 7 = 3150$.

(b) $\dfrac{3}{x}$ and $\dfrac{5}{x + y}$

Since x and $x + y$ are already completely factored, and since there are no common factors in the two denominators, the LCD is $x(x + y)$.

(c) $\dfrac{3a + 1}{a^2 - b^2}$ and $\dfrac{7b}{2a - 2b}$

Factor the denominators.

$$a^2 - b^2 = (a - b)(a + b) \quad \text{and} \quad 2a - 2b = 2(a - b)$$

The LCD must consist of one $(a - b)$, one $(a + b)$, and one 2. Thus, the LCD is $2(a - b)(a + b)$.

Practice Exercise 3

Find the LCD of the fractions.

(a) $\dfrac{2}{105}$ and $\dfrac{7}{60}$

(b) $\dfrac{4}{a + 1}$ and $\dfrac{3}{a}$

(c) $\dfrac{2x}{x^2 - 4}$ and $\dfrac{x + 2}{x^2 - 2x}$

Answers: (a) 420 (b) $a(a + 1)$ (c) $x(x + 2)(x - 2)$

Be sure the fractions are reduced to lowest terms before finding the LCD.

EXAMPLE 4 Reducing Fractions before Finding the LCD

Find the LCD of $\dfrac{a + 1}{a^2 + 2a + 1}$ and $\dfrac{5 - a}{a^2 - 4a - 5}$.

Factor the denominators.

$$a^2 + 2a + 1 = (a + 1)(a + 1) \quad \text{and}$$
$$a^2 - 4a - 5 = (a + 1)(a - 5)$$

We might almost conclude that the LCD is $(a + 1)^2(a - 5)$. However, we can reduce the given fractions to lowest terms as follows.

$$\frac{a + 1}{a^2 + 2a + 1} = \frac{\cancel{(a + 1)}}{(a + 1)\cancel{(a + 1)}} = \frac{1}{a + 1}$$

$$\frac{5 - a}{a^2 - 4a - 5} = \frac{5 - a}{(a - 5)(a + 1)} = \frac{(-1)\cancel{(a - 5)}}{\cancel{(a - 5)}(a + 1)} = \frac{-1}{a + 1}$$

$5 - a$ and $a - 5$ are negatives

Thus, to find the LCD of the given fractions, all we need to do is find the LCD of $\frac{1}{a + 1}$ and $\frac{-1}{a + 1}$, which is simply $a + 1$.

Practice Exercise 4

Find the LCD of $\dfrac{2y - x}{x^2 - xy - 2y^2}$ and $\dfrac{x - y}{x^2 - y^2}$.

Answer: $x + y$

3 REWRITING FRACTIONS

Before we add and subtract unlike fractions, we reverse the process of reducing a fraction to lowest terms and consider rewriting a fraction as an equivalent fraction with a specified denominator.

EXAMPLE 5 Rewriting Fractions

Write each fraction as an equivalent fraction with the specified denominator.

(a) $\frac{3y}{2x}$ with denominator $10x^2y$

Since $10x^2y = 2 \cdot 5 \cdot xxy = (2x)(5xy)$, we multiply numerator and denominator by $5xy$.

$$\frac{3y}{2x} = \frac{3y\,(5xy)}{2x\,(5xy)} = \frac{15xy^2}{10x^2y}$$

(b) $\frac{4a+b}{2a-b}$ with denominator $4a^2 - b^2$

Since $4a^2 - b^2 = (2a-b)(2a+b)$, we multiply numerator and denominator by $2a + b$.

$$\frac{4a+b}{2a-b} = \frac{(4a+b)\,(2a+b)}{(2a-b)\,(2a+b)} = \frac{8a^2+6ab+b^2}{4a^2-b^2}$$

Practice Exercise 5

Write each fraction as an equivalent fraction with the specified denominator.

(a) $\frac{4a}{x^2}$ with denominator $6x^3y$

(b) $\frac{x+y}{x-y}$ with denominator $x^2 - y^2$

Answers: (a) $\frac{24axy}{6x^3y}$ (b) $\frac{x^2+2xy+y^2}{x^2-y^2}$

4 ADDING AND SUBTRACTING UNLIKE FRACTIONS

To add or subtract unlike fractions, we first find their LCD and then convert each fraction to an equivalent fraction with the LCD as the denominator. The technique is illustrated by adding two numerical fractions.

$$\frac{2}{21} + \frac{3}{35} = \frac{2}{3 \cdot 7} + \frac{3}{5 \cdot 7}$$ 21 = 3 · 7 and 35 = 5 · 7; the LCD is 3 · 5 · 7 = 105

$$= \frac{2 \cdot 5}{3 \cdot 7 \cdot 5} + \frac{3 \cdot 3}{5 \cdot 7 \cdot 3}$$ Multiply numerator and denominator of $\frac{2}{21}$ by 5, of $\frac{3}{35}$ by 3

$$= \frac{2 \cdot 5 + 3 \cdot 3}{3 \cdot 5 \cdot 7}$$ Denominators are now the same; add numerators and place the sum over the LCD

$$= \frac{10 + 9}{3 \cdot 5 \cdot 7}$$ Leave denominator factored and simplify the numerator

$$= \frac{19}{105}$$ Since 19 has no factor of 3, 5, or 7, the resulting fraction is in lowest terms

The same procedure applies to algebraic fractions.

To Add or Subtract Algebraic Fractions

1. Express denominators in completely factored form and reduce all fractions.
2. Find the LCD of the fractions.
3. Multiply the numerator and denominator of each fraction by all factors present in the LCD but missing in the denominator of the particular fraction, so the fractions have the same denominator.
4. Indicate the sum or difference of all numerators, using parentheses if needed, and place the result over the LCD.
5. Simplify the numerator and reduce the resulting fraction.

EXAMPLE 6 Adding Unlike Fractions

Add.

$$\frac{5}{6a}+\frac{a+1}{15a^2}$$

$$=\frac{5}{2\cdot 3a}+\frac{a+1}{3\cdot 5\cdot a\cdot a}$$ Factor denominators

$$=\frac{5\cdot(5\cdot a)}{2\cdot 3\cdot a\cdot(5\cdot a)}+\frac{(a+1)(2)}{3\cdot 5\cdot a\cdot a\cdot(2)}$$ Supply missing factors: LCD is $2\cdot 3\cdot 5\cdot a\cdot a$

$$=\frac{5\cdot 5\cdot a+(a+1)\cdot 2}{2\cdot 3\cdot a\cdot 5\cdot a}$$ Add numerators over the LCD

$$=\frac{25a+2a+2}{2\cdot 3\cdot 5\cdot a\cdot a}$$

$$=\frac{27a+2}{30a^2}$$ No common factors; the sum is in lowest terms

Practice Exercise 6

Add.

$$\frac{y^2+1}{25y^3}+\frac{7}{10y}$$

Answer: $\frac{37y^2+2}{50y^3}$

EXAMPLE 7 Subtracting Unlike Fractions

Subtract.

(a) $\frac{3}{a+b}-\frac{2}{a-b}$ Denominators already factored; LCD = $(a+b)(a-b)$

$$=\frac{3(a-b)}{(a+b)(a-b)}-\frac{2(a+b)}{(a+b)(a-b)}$$ Supply missing factors

$$=\frac{3(a-b)-2(a+b)}{(a+b)(a-b)}$$ Subtract numerators over LCD

$$=\frac{3a-3b-2a-2b}{(a+b)(a-b)}$$ Watch signs

$$=\frac{a-5b}{(a+b)(a-b)}$$ Since no common factors exist, this is in lowest terms

(b) $\frac{2xy}{x^2-y^2}-\frac{y}{x+y}$

$$=\frac{2xy}{(x-y)(x+y)}-\frac{y}{x+y}$$ LCD = $(x-y)(x+y)$

$$=\frac{2xy}{(x-y)(x+y)}-\frac{y(x-y)}{(x+y)(x-y)}$$

$$=\frac{2xy-y(x-y)}{(x-y)(x+y)}$$

$$=\frac{2xy-yx+y^2}{(x-y)(x+y)}$$

$$=\frac{xy+y^2}{(x-y)(x+y)}$$ This one can be simplified

$$=\frac{y\cancel{(x+y)}}{(x-y)\cancel{(x+y)}}=\frac{y}{x-y}$$

Practice Exercise 7

Subtract.

(a) $\frac{5}{x+1}-\frac{8}{x-1}$

(b) $\frac{3a}{a-b}-\frac{6ab}{a^2-b^2}$

Answers: (a) $\frac{-3x-13}{(x+1)(x-1)}$ (b) $\frac{3a}{a+b}$

EXAMPLE 8 Reducing Fractions before Subtracting

Subtract.

$$\frac{y^2-1}{y^3-1}-\frac{y}{2y^2+2y+2}$$

$$=\frac{\cancel{(y-1)}(y+1)}{\cancel{(y-1)}(y^2+y+1)}-\frac{y}{2(y^2+y+1)}$$ Factor and reduce first fraction

$$=\frac{y+1}{y^2+y+1}-\frac{y}{2(y^2+y+1)}$$ LCD = $2(y^2+y+1)$

$$=\frac{2(y+1)}{2(y^2+y+1)}-\frac{y}{2(y^2+y+1)}$$ Supply missing factor

$$=\frac{2(y+1)-y}{2(y^2+y+1)}$$ Subtract numerators over LCD

$$=\frac{2y+2-y}{2(y^2+y+1)}=\frac{y+2}{2(y^2+y+1)}$$ Already reduced to lowest terms

PRACTICE EXERCISE 8

Add.

$$\frac{1-x}{x}+\frac{x^2-y^2}{x^2-2xy+y^2}$$

Answer: $\frac{2xy+x-y}{x(x-y)}$

Adding or subtracting three or more fractions uses the same technique. This is shown in the next example.

EXAMPLE 9 Adding and Subtracting Three Fractions

Perform the indicated operations.

$$\frac{2}{x^2-4}+\frac{1}{x+2}-\frac{x}{x^2-x-2}$$

$$=\frac{2}{(x+2)(x-2)}+\frac{1}{x+2}-\frac{x}{(x+1)(x-2)}$$ LCD = $(x+2)(x-2)(x+1)$

$$=\frac{2(x+1)}{(x+2)(x-2)(x+1)}+\frac{(1)(x-2)(x+1)}{(x+2)(x-2)(x+1)}-\frac{x(x+2)}{(x+1)(x-2)(x+2)}$$ Supply missing factors

$$=\frac{2(x+1)+(1)(x-2)(x+1)-x(x+2)}{(x+2)(x-2)(x+1)}$$ Add and subtract numerators over LCD

$$=\frac{2x+2+x^2-x-2-x^2-2x}{(x+2)(x-2)(x+1)}$$ Multiply and remove parentheses

$$=\frac{-x}{(x+2)(x-2)(x+1)}$$ Collect like terms

PRACTICE EXERCISE 9

Perform the indicated operations.

$$\frac{2y}{y^2+3y-4}-\frac{1}{y+3}-\frac{y}{y^2+2y-3}$$

Answer: $\frac{4-y}{(y-1)(y+4)(y+3)}$

6.3 EXERCISES A

Perform the indicated operation.

1. $\frac{3}{x} + \frac{8}{x}$
2. $\frac{3y + 2}{y - 4} + \frac{-2 - 3y}{y - 4}$
3. $\frac{a}{2} + \frac{2a - 1}{-2}$
4. $\frac{2a}{a - b} + \frac{3a}{b - a}$
5. $\frac{x}{8x - 6} - \frac{5}{2(3 - 4x)}$
6. $\frac{z}{1 - z^2} - \frac{1 - z}{(z - 1)(z + 1)}$

Find the LCD of each pair of fractions.

7. $\frac{1}{39}$ and $\frac{5}{33}$
8. $\frac{2}{y}$ and $\frac{3}{5y}$
9. $\frac{a + b}{a}$ and $\frac{3}{4a^5}$
10. $\frac{6}{5b}$ and $\frac{3}{5b + 5}$
11. $\frac{a}{3a + 3b}$ and $\frac{b}{a^2 - b^2}$
12. $\frac{3}{b^2 + 5b + 6}$ and $\frac{b}{b^2 - 5b - 6}$
13. $\frac{2}{x^2 - y^2}$ and $\frac{3}{x^2 - 2xy + y^2}$
14. $\frac{2}{a - b}$ and $\frac{a}{b - a}$
15. $\frac{z + 2}{z^2 + 4z + 4}$ and $\frac{z - 1}{z^2 + z - 2}$
16. $\frac{a + 3}{a^3 - 27}$ and $\frac{a}{a^2 + 3a + 9}$
17. $\frac{x + 3}{5x^4 - 15x^3 - 50x^2}$ and $\frac{x - 7}{10x^3 - 100x^2 + 250x}$

Write each fraction as an equivalent fraction with the specified denominator.

18. $\frac{8}{3y}$; $27y^3z$
19. $\frac{a}{a - 1}$; $4a - 4$
20. $\frac{y}{y + 4}$; $y^2 - 16$

Perform the indicated operation.

21. $\frac{3}{5x} + \frac{5 - 6x}{10x^2}$
22. $\frac{3}{a + 4} - \frac{2}{a - 4}$
23. $\frac{2}{y + 3} - \frac{7}{y - 3}$
24. $\frac{4}{3z} + \frac{1 - 12z}{9z^2}$
25. $\frac{2}{x - 2} + \frac{5}{x}$
26. $\frac{7}{a^2 - 25} + \frac{2}{a - 5}$

27. $\dfrac{2x}{9-x^2} - \dfrac{1}{3-x}$

28. $\dfrac{5}{3y-3} + \dfrac{3}{1-y}$

29. $\dfrac{xy}{3x-3y} + \dfrac{xy}{5y-5x}$

30. $\dfrac{3}{2a^2-2a} + \dfrac{5}{2-2a}$

31. $\dfrac{y^2}{y-1} - 2$

32. $\dfrac{2x+1}{x+2} + x$

33 $a - b - \dfrac{-ab^2}{a^2+ab}$

34. $\dfrac{y}{y^2-y-20} + \dfrac{2}{y+4}$

35 $\dfrac{2a-1}{a^2-4a+4} + \dfrac{2a+3}{4-a^2}$

36. $\dfrac{-2z}{8z^3+27} + \dfrac{1}{4z^2-6z+9}$

37. $\dfrac{2}{x^2-x-6} - \dfrac{1}{x^2-2x-3}$

38. $\dfrac{2}{a^2+ab-2b^2} + \dfrac{2}{a^2+3ab+2b^2}$

39. $\dfrac{x+2}{x^2+5x+6} - \dfrac{x+1}{x^2+4x+3}$

40. $\dfrac{4-3a}{3a^2+6a+12} + \dfrac{a^2-4}{a^3-8}$

41. $\dfrac{u}{u^2-v^2} + \dfrac{v}{u^2-2uv+v^2} - \dfrac{1}{u+v}$

42. $\dfrac{-x}{x^2+x-2} - \dfrac{1}{x^2-2x+1} + \dfrac{1}{x+2}$

43. $\dfrac{a}{a^2-5a+6} - \dfrac{2}{a^2-4a+3} - \dfrac{3}{a^2-3a+2}$

44 $\dfrac{x+y}{x^2+2xy+y^2} - \dfrac{x}{x^2-xy-2y^2} + \dfrac{y}{x^2-4xy-5y^2}$

The early Greeks were aware of the formula

$$M = 180 - \frac{360}{n}$$

which gives the degree measure M of each interior angle of a regular polygon with n sides. Express M as a single fraction and use the result in Exercises 45–46.

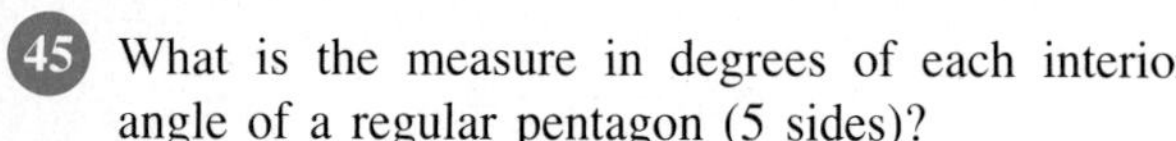

45 What is the measure in degrees of each interior angle of a regular pentagon (5 sides)?

46 What is the measure in degrees of each interior angle of a regular octagon (8 sides)?

FOR REVIEW

Perform the indicated operation.

47. $\frac{a^3 - b^3}{a^2 - b^2} \cdot \frac{a^2 + 2ab + b^2}{a^2 + ab + b^2}$

48. $\frac{x^2 - 2xy + y^2}{xy - y^2} \div \frac{x^3 - xy^2}{x^2 + 2xy + y^2}$

The following exercises review material covered in Chapter 2 to help you prepare for the next section. Solve each equation.

49. $2 - (2a - 1) = 3(a - 3)$

50. $\frac{x - 1}{3} = \frac{3x - 2}{12}$

ANSWERS: 1. $\frac{11}{x}$ 2. 0 3. $\frac{1-a}{2}$ 4. $\frac{-a}{a-b}$ or $\frac{a}{b-a}$ 5. $\frac{x+5}{8x-6}$ 6. $\frac{1}{1-z^2}$ 7. 429 8. $5y$ 9. $4a^5$ 10. $5b(b+1)$ 11. $3(a-b)(a+b)$ 12. $(b+2)(b+3)(b-6)(b+1)$ 13. $(x+y)(x-y)^2$ 14. $a-b$ 15. $z+2$ 16. $(a-3)(a^2+3a+9)$ 17. $10x^2(x-5)^2(x+2)$ 18. $\frac{72y^2z}{27y^3z}$ 19. $\frac{4a}{4a-4}$ 20. $\frac{y^2-4y}{y^2-16}$ 21. $\frac{1}{2x^2}$ 22. $\frac{a-20}{(a+4)(a-4)}$ 23. $\frac{-5y-27}{(y+3)(y-3)}$ 24. $\frac{1}{9z^2}$ 25. $\frac{7x-10}{x(x-2)}$ 26. $\frac{2a+17}{(a-5)(a+5)}$ 27. $\frac{-1}{x+3}$ 28. $\frac{-4}{3(y-1)}$ 29. $\frac{2xy}{15(x-y)}$ 30. $\frac{3-5a}{2a(a-1)}$ 31. $\frac{y^2-2y+2}{y-1}$ 32. $\frac{x^2+4x+1}{x+2}$ 33. $\frac{a^2}{a+b}$ 34. $\frac{3y-10}{(y-5)(y+4)}$ 35. $\frac{4(a+1)}{(a-2)^2(a+2)}$ 36. $\frac{3}{8z^3+27}$ 37. $\frac{x}{(x-3)(x+2)(x+1)}$ 38. $\frac{4a}{(a-b)(a+b)(a+2b)}$ 39. 0 40. $\frac{10}{3(a^2+2a+4)}$ 41. $\frac{2uv}{(u-v)^2(u+v)}$ 42. $\frac{-2x-1}{(x-1)^2(x+2)}$ 43. $\frac{a^2-6a+13}{(a-1)(a-2)(a-3)}$ 44. $\frac{y(8y-x)}{(x+y)(x-2y)(x-5y)}$ 45. 108° 46. 135° 47. $a+b$ 48. $\frac{x+y}{xy}$ 49. $\frac{12}{5}$ 50. 2

6.3 EXERCISES B

Perform the indicated operation.

1. $\frac{4}{2y} - \frac{5}{2y}$

2. $\frac{3b}{b+1} - \frac{2b-1}{b+1}$

3. $\frac{3b}{4} + \frac{4b-3}{-4}$

4. $\frac{4x}{3-x} - \frac{x+1}{x-3}$

5. $\frac{2x}{10-5x} - \frac{3}{5(x-2)}$

6. $\frac{z-1}{(1-z)(1+z)} - \frac{1+z}{z^2-1}$

Find the LCD of each pair of fractions.

7. $\frac{5}{14}$ and $\frac{7}{38}$

8. $\frac{1}{4}$ and $\frac{3}{x}$

9. $\frac{3}{a}$ and $\frac{2}{9a}$

10. $\frac{3}{2x}$ and $\frac{1}{2x+2}$

11. $\frac{u}{2u-2v}$ and $\frac{3}{u^2-v^2}$

12. $\frac{x}{x^2+4x+3}$ and $\frac{5}{x^2+6x-7}$

13. $\frac{6}{a^2-b^2}$ and $\frac{7}{a^2+2ab+b^2}$

14. $\frac{3}{5a-b}$ and $\frac{b}{b-5a}$

15. $\frac{a+3}{a^2+6a+9}$ and $\frac{a-4}{a^2-a-12}$

16. $\frac{x-2}{x^3+8}$ and $\frac{2x}{x^2-2x+4}$

17. $\frac{3y+2}{3y^5-6y^4+3y^3}$ and $\frac{y+1}{6y^4+24y^3-30y^2}$

Write each fraction as an equivalent fraction with the specified denominator.

18. $\frac{5}{4a}$; $12a^2b$

19. $\frac{a}{a-b}$; a^2-b^2

20. $\frac{2xy}{3x-y}$; $3x^2+2xy-y^2$

Perform the indicated operation.

21. $\frac{7}{2x}+\frac{2-14x}{4x^2}$

22. $\frac{5}{a-3}-\frac{4}{a+3}$

23. $\frac{3}{y+5}-\frac{6}{y-5}$

24. $\frac{6}{7z}+\frac{3-18z}{21z^2}$

25. $\frac{5}{x+4}+\frac{6}{x}$

26. $\frac{3}{a^2-16}+\frac{2}{a-4}$

27. $\frac{6y}{4-y^2}-\frac{1}{2+y}$

28. $\frac{8}{2x-2}+\frac{5}{1-x}$

29. $\frac{a^2b^2}{4a-4b}+\frac{a^2b^2}{7b-7a}$

30. $\frac{4}{3x^2-3x}+\frac{5}{3-3x}$

31. $\frac{x^2}{x-2}-3$

32. $\frac{3y-2}{y-5}+y$

33. $x+y-\frac{-xy^2}{x^2-xy}$

34. $\frac{2x}{x^2-3x-4}+\frac{3}{x+1}$

35. $\frac{5u-3}{u^2-10u+25}+\frac{u-8}{25-u^2}$

36. $\frac{-6x}{27x^3-8}+\frac{2}{9x^2+6x+4}$

37. $\frac{-3}{a^2+4a-5}-\frac{2}{a^2+6a+5}$

38. $\dfrac{3}{u^2 + 5uv + 6v^2} + \dfrac{-1}{u^2 + 3uv + 2v^2}$

39. $\dfrac{a + 3}{a^2 + 10a + 21} - \dfrac{a - 2}{a^2 + 5a - 14}$

40. $\dfrac{2x - 5}{4x^2 - 12x + 36} + \dfrac{x^2 - 9}{x^3 + 27}$

41. $\dfrac{b}{a^2 - b^2} - \dfrac{1}{a - b} + \dfrac{a}{a^2 + 2ab + b^2}$

42. $\dfrac{5}{x^2 - 5x + 6} - \dfrac{x}{x^2 - 4x + 4} + \dfrac{1}{x - 2}$

43. $\dfrac{u}{u^2 - 2u - 3} - \dfrac{u}{u^2 + 3u + 2} - \dfrac{5}{u^2 - u - 6}$

44. $\dfrac{x - y}{x^2 - 2xy + y^2} - \dfrac{y}{x^2 + xy - 2y^2} + \dfrac{x}{x^2 + 2xy - 3y^2}$

The early Greeks were aware of the formula

$$M = 180 - \frac{360}{n}$$

which gives the degree measure M of each interior angle of a regular polygon with n sides. Express M as a single fraction and use the result in Exercises 45–46.

45. What is the measure in degrees of each interior angle of a regular hexagon (6 sides)?

46. What is the measure in degrees of each interior angle of a regular decagon (10 sides)?

FOR REVIEW

Perform the indicated operations.

47. $\dfrac{2x^2 + 9x - 5}{2x^2 - 3x + 1} \div \dfrac{x + 5}{x - 1}$

48. $\dfrac{3a + 6b}{a^2 + ab - 2b^2} \cdot \dfrac{a^3 - b^3}{a^2 + ab + b^2}$

The following exercises review material covered in Chapter 2 to help you prepare for the next section. Solve each equation.

49. $7 - 4x = -5$

50. $\dfrac{x + 2}{4} + \dfrac{x + 7}{8} = 1$

6.3 EXERCISES C

Perform the indicated operation.

1. $\dfrac{-5}{5 - x} - \dfrac{x}{x - 5}$

2. $\dfrac{a}{(a - b)(a - 2b)} + \dfrac{b}{(b - a)(2b - a)}$

Find the LCD of each group of fractions.

3. $\dfrac{1}{x^2+x}$, $\dfrac{2x}{x^2-x-2}$, and $\dfrac{3}{x^2-4x+4}$

4. $\dfrac{a-b}{a^2-b^2}$, $\dfrac{x+1}{ax+bx+a+b}$, and $\dfrac{a-3b}{a^2-2ab-3b^2}$
[Answer: $a+b$]

5. $\dfrac{2b}{a^2-b^2}-\dfrac{b}{a^2+ab}-\dfrac{a^2+2ab+b^2}{a^3-ab^2}$
$\left[\text{Answer: } \dfrac{1}{b-a}\right]$

6. $\dfrac{-x}{x^3-y^3}-\dfrac{1}{y^2-x^2}+\dfrac{1}{x^2+xy+y^2}$

6.4 SOLVING FRACTIONAL EQUATIONS

STUDENT GUIDEPOSTS

1. Fractional Equations
2. The Cross-Product Equation
3. Literal Fractional Equations

1 FRACTIONAL EQUATIONS

An equation that contains one or more algebraic fractions or rational expressions is called a **fractional equation.** To **clear the fractions,** we multiply both sides of the equation by the LCD of all fractions and then solve the equation using the techniques of Chapter 2. Every solution to this new equation must be checked in the *original* equation; if the solution makes one of the denominators equal to zero, it must be discarded. If all solutions are discarded, the equation has no solution.

EXAMPLE 1 Solving a Fractional Equation

Solve $\dfrac{5}{x}-\dfrac{7}{6}=\dfrac{3}{2x}$.

$6x\left[\dfrac{5}{x}-\dfrac{7}{6}\right]=6x\left[\dfrac{3}{2x}\right]$ Multiply both sides by the LCD, $6x$

$6\cancel{x}\cdot\dfrac{5}{\cancel{x}}-\cancel{6}x\cdot\dfrac{7}{\cancel{6}}=6\cancel{x}\cdot\dfrac{3}{2\cancel{x}}$ Distribute

$30-7x=9$ Clear fractions

$-7x=-21$ Subtract 30 from both sides

$x=3$ Divide both sides by -7

Check: $\dfrac{5}{3}-\dfrac{7}{6}\stackrel{?}{=}\dfrac{3}{2\cdot 3}$

$\dfrac{2\cdot 5}{2\cdot 3}-\dfrac{7}{6}\stackrel{?}{=}\dfrac{3}{6}$

$\dfrac{10}{6}-\dfrac{7}{6}\stackrel{?}{=}\dfrac{3}{6}$

$\dfrac{3}{6}=\dfrac{3}{6}$

The solution is 3.

PRACTICE EXERCISE 1

Solve $\dfrac{1}{x}-\dfrac{1}{3}=\dfrac{1}{3x}$.

Answer: 2

To Solve a Fractional Equation

1. Find the LCD of all fractions in the equation.
2. Multiply both sides of the equation by the LCD to clear fractions. Make sure that *all* terms are multiplied.
3. Solve this equation.
4. Check all possible solutions in the original equation. THIS IS IMPORTANT!

EXAMPLE 2 SOLVING A FRACTIONAL EQUATION

Solve $\frac{2}{y} = \frac{3}{4-y}$.

$$y(4-y)\left[\frac{2}{y}\right] = y(4-y)\left[\frac{3}{4-y}\right]$$ Multiply both sides by the LCD, $y(4-y)$ to clear fractions

$$2(4-y) = 3y$$

$$8-2y = 3y$$

$$8 = 5y$$ Add $2y$ to both sides

$$\frac{8}{5} = y$$ Divide both sides by 5

Check: $\frac{2}{\frac{8}{5}} \stackrel{?}{=} \frac{3}{4-\frac{8}{5}}$

$$\frac{10}{8} \stackrel{?}{=} \frac{3}{\frac{12}{5}}$$

$$\frac{5}{4} \stackrel{?}{=} \frac{15}{12}$$

$$\frac{5}{4} \stackrel{?}{=} \frac{5}{4}$$

The solution is $\frac{8}{5}$.

PRACTICE EXERCISE 2

Solve $\frac{3}{z} = \frac{6}{z-4}$.

Answer: -4

2 THE CROSS-PRODUCT EQUATION

Equations that consist of two equated fractions, such as the one in Example 2, can be solved using a shortcut. We know that $\frac{a}{b} = \frac{c}{d}$ if and only if the cross-products ad and bc are equal. As a result, to solve

$$\frac{2}{y} = \frac{3}{4-y}$$

we can set equal the cross-products $2(4-y)$ and $3y$, forming the equation

$$2(4-y) = 3y.$$

Notice that we immediately arrive at the second step in Example 2 above. However, this technique works only for these special equations of two fractions. Do not try to use it, for example, to solve an equation like the one in Example 1.

CAUTION

Check all possible solutions. Multiplying both sides of an equation by a variable can produce "solutions" which are values for the variable that make a denominator zero in the original equation. This is shown in the next example.

EXAMPLE 3 Solving a Fractional Equation with No Solution

Solve $\dfrac{a^2+9}{a^2-9} - \dfrac{3}{a+3} = \dfrac{-a}{3-a}$.

We first observe that $a^2 - 9 = (a-3)(a+3)$. If we multiply the numerator and denominator of $\frac{-a}{3-a}$ by -1, we obtain the equivalent fraction $\frac{a}{a-3}$. Then the LCD is $(a-3)(a+3)$, and we multiply both sides by it.

$$(a-3)(a+3)\left[\frac{a^2+9}{(a-3)(a+3)} - \frac{3}{(a+3)}\right] = (a-3)(a+3)\left[\frac{a}{(a-3)}\right]$$

$$(a-3)(a+3)\cdot\frac{a^2+9}{(a-3)(a+3)} - (a-3)(a+3)\cdot\frac{3}{(a+3)}$$

$$= (a-3)(a+3)\cdot\frac{a}{(a-3)}$$

$$a^2+9-(a-3)\cdot 3 = (a+3)\cdot a$$

$$a^2+9-(3a-9) = a^2+3a$$

$$a^2+9-3a+9 = a^2+3a$$

$$-3a+18 = 3a$$

$$18 = 6a$$

$$3 = a$$

Check: $\dfrac{3^2+9}{3^2-9} - \dfrac{3}{3+3} \stackrel{?}{=} \dfrac{-3}{3-3}$

$$\frac{18}{0} - \frac{3}{6} = \frac{-3}{0}$$

Since division by 0 is undefined, 3 must be discarded as a possible value for a. There is no solution.

PRACTICE EXERCISE 3

Solve $\dfrac{15}{w^2+5w} + \dfrac{w+4}{w+5} = \dfrac{w+3}{w}$.

Answer: no solution (0 does not check)

Some fractional equations reduce to a quadratic equation, as in the example on the following page.

EXAMPLE 4 Solving a Quadratic Fractional Equation

Solve $\frac{y}{3} - \frac{6}{y} = 1.$

$$3y\left[\frac{y}{3} - \frac{6}{y}\right] = 3y\,[1] \qquad \text{Multiply by LCD, } 3y$$

$$3y \cdot \frac{y}{3} - 3y \cdot \frac{6}{y} = 3y \qquad \text{Distribute}$$

$$y^2 - 18 = 3y \qquad \text{This time we have a quadratic equation}$$

$$y^2 - 3y - 18 = 0$$

$$(y + 3)(y - 6) = 0 \qquad \text{Factor}$$

$$y + 3 = 0 \quad \text{or} \quad y - 6 = 0$$

$$y = -3 \qquad\qquad y = 6$$

Check: $\frac{-3}{3} - \frac{6}{-3} \stackrel{?}{=} 1$ $\qquad \frac{6}{3} - \frac{6}{6} \stackrel{?}{=} 1$

$-1 + 2 \stackrel{?}{=} 1$ $\qquad 2 - 1 \stackrel{?}{=} 1$

$1 = 1$ $\qquad 1 = 1$

-3 checks. $\qquad$ 6 checks.

The solutions are -3 and 6.

Practice Exercise 4

Solve $\frac{x}{4} + \frac{3}{x} = 2.$

Answer: 2 and 6

CAUTION

A common mistake is to confuse addition or subtraction of algebraic fractions with solving fractional equations. Compare the two problems:

Solve: $\frac{2}{y} + \frac{y}{y+1} = 1$

$$y(y+1)\left[\frac{2}{y} + \frac{y}{y+1}\right] = y(y+1)(1)$$

$$y(y+1)\frac{2}{y} + y(y+1)\frac{y}{y+1} = y^2 + y$$

$$2y + 2 + y^2 = y^2 + y$$

$$2y + 2 = y$$

$$y = -2$$

Add: $\frac{2}{y} + \frac{y}{y+1} + 1$

$$\frac{2(y+1)}{y(y+1)} + \frac{y \cdot y}{y(y+1)} + \frac{y(y+1)}{y(y+1)}$$

$$\frac{2(y+1) + y^2 + y(y+1)}{y(y+1)}$$

$$\frac{2y + 2 + y^2 + y^2 + y}{y(y+1)}$$

$$\frac{2y^2 + 3y + 2}{y(y+1)}$$

The LCD $y(y + 1)$ must be found in both cases, but it is used in different ways. In the equation, we multiply both sides by $y(y + 1)$ to clear the fractions and eventually find the numerical solution -2. In the addition problem, each term is written as an equivalent fraction with denominator $y(y + 1)$. We then find the sum of these fractions, obtaining an algebraic expression.

EXAMPLE 5 NUMBER PROBLEM

If the numerator of a fraction exceeds the denominator by 6, and the value of the fraction is $\frac{5}{3}$, find the fraction.

Let x = the denominator of the fraction,
$x + 6$ = the numerator of the fraction. (Why?)

The fraction is $\frac{x+6}{x}$, so the equation we must solve is

$$\frac{x+6}{x} = \frac{5}{3}. \quad \text{The LCD is } 3x$$

$$3x \cdot \frac{x+6}{x} = 3x \cdot \frac{5}{3}$$

$$3(x+6) = x \cdot 5 \quad \text{Use parentheses}$$

$$3x + 18 = 5x$$

$$18 = 2x$$

$$9 = x$$

$$x + 6 = 15$$

The fraction is $\frac{15}{9}$. Since the numerator 15 is 6 more than the denominator 9, and $\frac{15}{9}$ reduces to $\frac{5}{3}$, $\frac{15}{9}$ checks.

PRACTICE EXERCISE 5

If the denominator of a fraction exceeds the numerator by 9 and the value of the fraction is $\frac{4}{7}$, find the fraction.

Let x = numerator,
$x + 9$ = denominator.

Answer: $\frac{12}{21}$

3 LITERAL FRACTIONAL EQUATIONS

Many literal equations are fractional equations that can be solved using the method of this section.

EXAMPLE 6 LITERAL FRACTIONAL EQUATION

Solve $\frac{1}{a} + \frac{1}{b} = c$ for a.

$$\frac{1}{a} + \frac{1}{b} = c$$

$$ab\left(\frac{1}{a} + \frac{1}{b}\right) = abc \quad \text{Multiply both sides by the LCD, } ab$$

$$ab \cdot \frac{1}{a} + ab \cdot \frac{1}{b} = abc \quad \text{Distributive law}$$

$$b + a = abc$$

$$b = abc - a \quad \text{Subtract } a \text{ to get all terms with the desired variable on the same side}$$

$$b = a(bc - 1) \quad \text{Factor out } a \text{ using the distributive law}$$

$$\frac{b}{bc - 1} = a \quad \text{Divide by } bc - 1\text{, the coefficient of } a$$

PRACTICE EXERCISE 6

Solve $\frac{1}{w} - \frac{1}{u} = \frac{1}{v}$ for w.

Answer: $w = \frac{uv}{u+v}$

EXAMPLE 7 Electricity Problem

The resistance R of an electrical circuit is given by $\frac{1}{R} = \frac{1}{R_1} + \frac{1}{R_2}$ where R_1 and R_2 are resistors in parallel, as shown in Figure 6.1. Solve this equation for R_1.

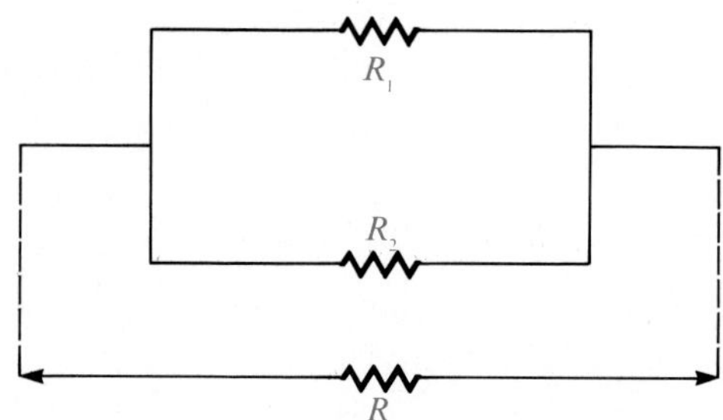

Figure 6.1

$$\frac{1}{R} = \frac{1}{R_1} + \frac{1}{R_2}$$

$$RR_1R_2\left[\frac{1}{R}\right] = RR_1R_2\left[\frac{1}{R_1} + \frac{1}{R_2}\right] \quad \text{Multiply both sides by the LCD, } RR_1R_2$$

$$R_1R_2 = RR_2 + RR_1$$

$$R_1R_2 - RR_1 = RR_2 \quad \text{Subtract to get both terms with } R_1 \text{ on the left}$$

$$R_1[R_2 - R] = RR_2 \quad \text{Factor out } R_1$$

$$R_1 = \frac{RR_2}{R_2 - R} \quad \text{Divide by } R_2 - R\text{, the coefficient of } R_1$$

Practice Exercise 7

Solve $\frac{1}{R} = \frac{1}{R_1} + \frac{1}{R_2}$ for R.

Answer: $R = \frac{R_1R_2}{R_1 + R_2}$

6.4 EXERCISES A

Solve.

1. $\frac{2}{x} - \frac{1}{x} = -6$

2. $\frac{7}{y} = \frac{5}{y-2}$

3. $\frac{1}{y-1} + \frac{2}{y+1} = 0$

4. $\frac{a-2}{a+3} = \frac{a+3}{a+2}$

5. $\frac{1}{a+2} + \frac{1}{a-2} = \frac{1}{a^2-4}$

6. $\frac{11}{c^2-25} - \frac{2}{c-5} = \frac{1}{c+5}$

7. $\frac{x}{x+4} = \frac{4}{x-4} + \frac{x^2+16}{x^2-16}$

8. $\frac{1}{z^2-2z+1} + \frac{1}{z-1} = 2$

9. $\frac{1}{a} = \frac{-6}{a^2+5}$

10. $\dfrac{2y}{y-1} - 2 = \dfrac{5}{2y}$

11. $\dfrac{x^2}{x+12} = 1$

12. $5 + \dfrac{x}{x-3} = \dfrac{-6}{x-3}$

13. $\dfrac{25}{y-5} - \dfrac{25}{y} = \dfrac{1}{4}$

14. $\dfrac{-1}{a^2-3a} = \dfrac{1}{a} + \dfrac{a}{a-3}$

In Exercises 15–18, either solve the given equation or perform the indicated operation.

15. $\dfrac{3x+5}{6} = \dfrac{4+3x}{5}$

16. $\dfrac{3x+5}{6} + \dfrac{4+3x}{5}$

17. $\dfrac{2}{y+1} - \dfrac{3}{y} = 0$

18. $\dfrac{2}{y+1} - \dfrac{3}{y}$

19. Find two numbers whose sum is 45 and whose quotient is $\frac{2}{3}$.

20. Harriet is three-fourths as old as Lorraine, and the difference of their ages is 10 years. How old is each?

21. The reciprocal of 5 less than a number is twice the reciprocal of the number. Find the number.

22. The numerator of a fraction is 4 less than the denominator. If both the numerator and denominator are increased by 2, the value of the fraction is $\frac{2}{3}$. Find the fraction.

23. A student made 75, 80, and 82 on the first three of four tests in a course. What must she make on the fourth test for an average score of 83?

24. If five times the reciprocal of a number is equal to the reciprocal of 4 more than the number, find the number.

25. The sum of the reciprocals of two consecutive integers is nine times the reciprocal of their product. Find the integers.

26. If Brenda is 5 years older than Jane, and the quotient of their ages is $\frac{5}{4}$, how old is each?

Solve each equation for the indicated variable.

27. $\dfrac{1}{a} + \dfrac{1}{b} = c$ for b

28. $\dfrac{1}{a} + \dfrac{1}{b} = \dfrac{1}{c}$ for c

29. $\dfrac{a^{-1}}{c} = \dfrac{d^{-1}}{b}$ for a

[*Hint:* Substitute $\frac{1}{a}$ for a^{-1} and $\frac{1}{d}$ for d^{-1}.]

30 $a^{-1} = b + c^{-1}$ for c

31. Solve the simple interest formula

$$P = \frac{A}{1 + rt} \text{ for } r.$$

32. Solve Doppler's formula

$$F = \frac{1100f}{1100 - V} \text{ for } V.$$

FOR REVIEW

Perform the indicated operations.

33. $\dfrac{2y - 1}{y^2 - 4y + 4} + \dfrac{2y + 1}{4 - y^2}$

34. $\dfrac{12}{z^3 + 8} + \dfrac{z}{z^2 - 2z + 4} - \dfrac{1}{z + 2}$

ANSWERS: 1. $-\frac{1}{6}$ 2. 7 3. $\frac{1}{3}$ 4. $-\frac{13}{6}$ 5. $\frac{1}{2}$ 6. 2 7. no solution 8. $\frac{1}{2}$, 2 9. −5, −1 10. 5 11. 4, −3 12. $\frac{3}{2}$ 13. 25, −20 14. −2, 1 15. $\frac{1}{3}$ 16. $\frac{33x + 49}{30}$ 17. −3 18. $\frac{-y - 3}{y(y + 1)}$ 19. 18, 27 20. Harriet is 30, Lorraine is 40. 21. 10 22. $\frac{6}{10}$ 23. 95 24. −5 25. 4, 5 26. Brenda is 25, Jane is 20. 27. $b = \frac{a}{ac - 1}$ 28. $c = \frac{ab}{a + b}$ 29. $a = \frac{bd}{c}$ 30. $c = \frac{a}{1 - ab}$ 31. $r = \frac{A - P}{Pt}$ 32. $V = \frac{1100(F - f)}{F}$ 33. $\frac{6y}{(y - 2)^2(y + 2)}$ 34. $\frac{4}{z^2 - 2z + 4}$

6.4 EXERCISES B

Solve.

1. $\dfrac{3}{a + 1} = \dfrac{5}{a}$

2. $\dfrac{3}{z} + \dfrac{1}{z} = -4$

3. $\dfrac{x + 1}{x + 2} = \dfrac{x + 2}{x - 1}$

4. $\dfrac{1}{z - 5} + \dfrac{7}{z + 3} = 0$

5. $\dfrac{6}{x + 3} + \dfrac{2}{x - 3} = \dfrac{20}{x^2 - 9}$

6. $\dfrac{12}{z^2 - 9} + \dfrac{3}{z + 3} = \dfrac{2}{z - 3}$

7. $\dfrac{9 + y}{16 + 2y} = \dfrac{7}{16 + 2y}$

8. $\dfrac{8}{x^2 + 8x + 16} + \dfrac{2}{x + 4} = 3$

9. $\dfrac{10}{y - 1} = \dfrac{16}{y - 1} - 3y$

10. $\dfrac{u^2}{35 - 2u} = 1$

11. $\dfrac{a + 2}{a - 1} + \dfrac{3}{a} = 1$

12. $7 + \dfrac{5}{z - 2} = \dfrac{z}{z - 2}$

13. $\dfrac{3x}{2x + 3} = \dfrac{8x^2 + 12}{4x^2 - 9} - \dfrac{x}{2x - 3}$

14. $\dfrac{-10}{y^2 + 3y - 10} = \dfrac{y}{y - 2} + \dfrac{1}{y + 5}$

In Exercises 15–18, either solve the given equation or perform the indicated operation.

15. $\frac{5x-2}{3} = \frac{2+7x}{4}$

16. $\frac{5x-2}{3} - \frac{2+7x}{4}$

17. $\frac{5}{y+2} + \frac{2}{y} = 0$

18. $\frac{5}{y+2} + \frac{2}{y}$

19. If the numerator of a fraction exceeds the denominator by 6, and the value of the fraction is $\frac{5}{3}$, find the fraction.

20. Sandi is four-fifths as old as Richard, and the difference in their ages is 7 years. How old is each?

21. Twice the reciprocal of a number, increased by the number itself, is -3. Find the number.

22. The denominator of a fraction is 1 more than the numerator. If both the numerator and denominator are decreased by 3, the value of the fraction is $\frac{4}{5}$. Find the fraction.

23. If the average weight of the four linemen for the Rams is 243 lb, and the weights of three of them are 220 lb, 240 lb, and 260 lb, how much does the fourth one weigh?

24. When 5 is added to the reciprocal of a number, the result is the same as 16 times the reciprocal of the number. Find the number.

25. The reciprocal of the smaller of two consecutive even integers, less the reciprocal of the larger integer, is equal to the reciprocal of the larger integer. Find the integers.

26. Walter is 8 years younger than Lupe and the quotient of their ages is $\frac{21}{17}$. How old is each?

Solve each equation for the indicated variable.

27. $\frac{1}{a} + \frac{1}{b} = \frac{1}{c}$ for a

28. $\frac{ab}{c} = \frac{wx}{d}$ for d

29. $(ab)^{-1} = a^{-1} + b^{-1}$ for b

30. $b^{-1} + c^{-1} = 2$ for b

31. Solve the temperature conversion formula
$$\frac{C}{F-32} = \frac{5}{9} \text{ for } F.$$

32. Solve the simple interest formula
$$t = \frac{A-P}{Pr} \text{ for } P.$$

FOR REVIEW

Perform the indicated operations.

33. $\frac{3a+2}{a^2-6a+9} + \frac{3a-2}{9-a^2}$

34. $\frac{3}{z^3+1} + \frac{z}{z^2-z+1} - \frac{1}{z+1}$

6.4 EXERCISES C

Solve.

1. $\frac{1}{x^2-x-2} = \frac{3}{x^2-2x-3} + \frac{1}{x^2-5x+6}$

2. $\frac{x+1}{x+2} - \frac{x+3}{x+5} = \frac{x+2}{x+5} - \frac{x+1}{x+3}$
[Answer: $1, -\frac{5}{2}$]

3. Solve the focal length formula

$$\frac{1}{f} = \frac{1}{d_1} + \frac{1}{d_2} \text{ for } d_1.$$

4. Solve the gas law

$$\frac{PV}{T} = \frac{pv}{t} \text{ for } P.$$

In Exercises 5–6, either solve the equation or perform the indicated operation.

5. $\dfrac{3}{x^2 + x - 2} + \dfrac{2}{x^2 - 1} = \dfrac{-1}{x^2 + 3x + 2}$

6. $\dfrac{3}{x^2 + x - 2} + \dfrac{2}{x^2 - 1} - \dfrac{1}{x^2 + 3x + 2}$

6.5 RATIO, PROPORTION, AND VARIATION

STUDENT GUIDEPOSTS

1. Ratios and Proportions
2. Solving Proportions
3. Direct Variation
4. Inverse Variation
5. Joint Variation
6. Combined Variation

1 RATIOS AND PROPORTIONS

Numerous applied problems can be simplified using the concepts of ratio, proportion, and variation. The **ratio** of one number a to another number b is the quotient $\frac{a}{b}$, sometimes expressed by the notation $a:b$. Ratios occur in applications such as percent, rate of speed, and cost. Consider the following examples.

Applications	*Ratio involved*
6% sales tax	$\dfrac{\$6}{\$100}$ = \$6 per \$100 tax
240 miles in 4 hours	$\dfrac{240 \text{ mi}}{4 \text{ hr}}$ = 60 mph
\$80 for 4 shirts	$\dfrac{\$80}{4 \text{ shirts}}$ = \$20 per shirt

Ratios are used to define *proportions*.

Proportions

An equation stating that two ratios are equal is called a **proportion.** That is, if $\frac{a}{b}$ and $\frac{c}{d}$ are two ratios, the equation

$$\frac{a}{b} = \frac{c}{d}$$

is a proportion. It is read "*a* is to *b* as *c* is to *d*."

2 SOLVING PROPORTIONS

In the proportion

$$\frac{3}{5} = \frac{18}{30},$$

the numbers 3 and 30 are the **extremes** of the proportion, and 5 and 18 are the **means.** Consider the proportion

$$\frac{a}{b} = \frac{c}{d}$$

with extremes a and d and means b and c. If we multiply both sides by the LCD, bd, we obtain

$$bd\frac{a}{b} = bd\frac{c}{d}$$
$$ad = bc,$$

the cross-product equation. We have just verified the following property of proportions.

Means-Extremes Property of Proportions

In any proportion, the product of the extremes equals the product of the means.

The means-extremes property allows us to solve for an unknown term in a proportion, called **solving the proportion.** Suppose we solve the proportion

$$\frac{x}{3} = \frac{8}{27}.$$

$27x = 3 \cdot 8$ Product of extremes = product of means

$27x = 24$

$x = \frac{24}{27} = \frac{8}{9}$

Ratio and proportion are useful in many applied problems. Suppose that it takes 3 hours to grade 24 tests. How many hours would it take to grade 56 tests? Letting x be the number of hours to grade 56 tests, the proportion

time to grade 24 tests → $\frac{3}{24} = \frac{x}{56}$ ← time to grade 56 tests

describes the problem. That is, the ratios of the number of hours to the number of tests are equal. Using the means-extremes property,

$$3 \cdot 56 = 24x$$
$$7 = x.$$

It would take 7 hours to grade 56 tests.

EXAMPLE 1 Using Proportions

If a boat uses 14 gallons of gas to go 102 miles, how many gallons would be needed to go 510 miles?

Let x = the number of gallons needed to go 510 miles.

Practice Exercise 1

If a scout troop can pick up 7 bags of trash in 2 hr, how many bags can they pick up in 8 hr?

The equation to solve is

$$\frac{14}{102} = \frac{x}{510}$$

miles on 14 gallons → 102 510 ← miles on x gallons

$$102x = 510 \cdot 14 \quad \text{Cross-product equation}$$

$$x = \frac{510 \cdot 14}{102}$$

$$x = \frac{\cancel{102} \cdot 5 \cdot 14}{\cancel{102}} = 70$$

It would take 70 gallons to go 510 miles.

Answer: 28 bags

Two triangles are **similar** if the angles of one are equal respectively to the angles of the other. To show similarity, all that is necessary is to show that two angles of one triangle are equal to two angles of the other. (Why?) Similar triangles have the property that corresponding sides (sides opposite the equal angles) are proportional. If $\angle A = \angle A'$, $\angle B = \angle B'$, and $\angle C = \angle C'$, as in Figure 6.2 ($\angle$ is the symbol for *angle*), then $\triangle ABC$ is similar to $\triangle A'B'C'$ ($\triangle$ is the symbol for *triangle*). The fact that the sides are proportional is expressed by the following equations.

$$\frac{a}{a'} = \frac{b}{b'} = \frac{c}{c'}$$

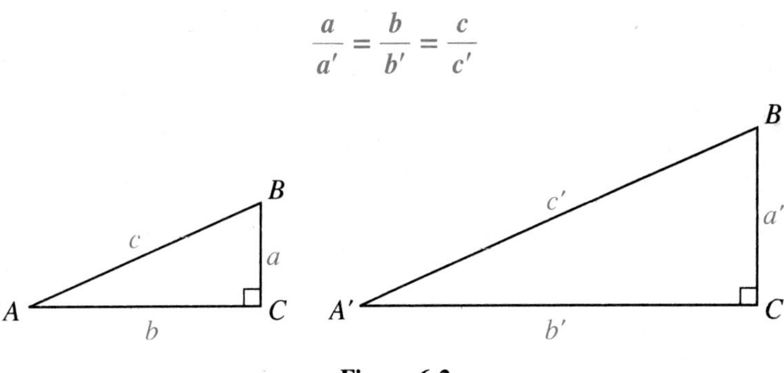

Figure 6.2

EXAMPLE 2 Similar Triangle Application

If a man who is 6 ft tall stands in the sunlight and casts a shadow 14 ft long, while a flagpole casts a shadow 35 ft long, how tall is the pole?

It is best to make a sketch, as in Figure 6.3.

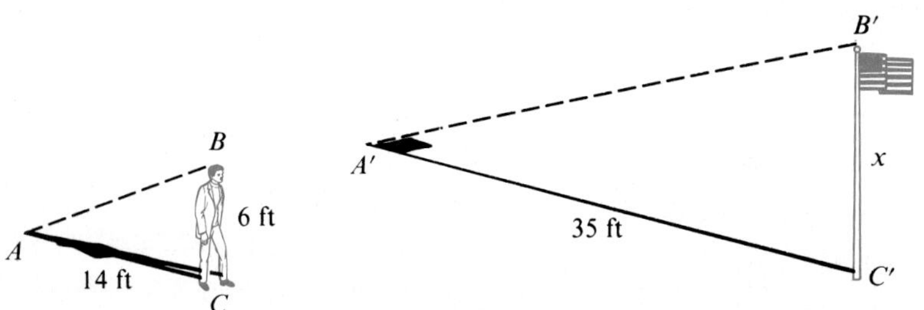

Figure 6.3

We assume that $\angle ACB = \angle A'C'B' = 90°$ (a right angle). Since both measurements are taken at the same time, the sun's rays make the same angle with level ground, that is, $\angle BAC = \angle B'A'C'$. Thus, $\triangle ABC$ is similar to $\triangle A'B'C$.

Practice Exercise 2

A tree casts a shadow 4 ft long at the same time a building, which is 85 ft tall, casts a shadow 17 ft in length. How tall is the tree?

$$\frac{AC}{A'C'} = \frac{BC}{B'C'}$$

$$\frac{14}{35} = \frac{6}{x}$$

$$14x = 6 \cdot 35$$

$$x = \frac{6 \cdot 35}{14} = 15$$

The pole is 15 ft tall.

Answer: 20 ft

EXAMPLE 3 NUMBER PROBLEM

What number must be added to each of 20, 8, 32, and 14 to give four numbers which are in proportion in that order?

Let $x =$ the number to be added.

We need to solve

$$\frac{20 + x}{8 + x} = \frac{32 + x}{14 + x}.$$

$$(20 + x)(14 + x) = (8 + x)(32 + x)$$

$$280 + 34x + x^2 = 256 + 40x + x^2$$

$$280 + 34x = 256 + 40x$$

$$24 = 6x$$

$$4 = x$$

The desired number is 4. Check.

PRACTICE EXERCISE 3

What number must be subtracted from each of 8, 12, 20, and 40 to give four numbers which are in proportion in that order?

Let $x =$ number to be subtracted.

Answer: 5

3 DIRECT VARIATION

Three types of relationships identified with the term **variation** provide another way of looking at proportion problems. For example, the distance a car travels **varies** or changes in a given period of time as the average speed varies. Similarly, the strength of a wooden beam varies as we vary the length, width, and thickness of the beam, and the amount of illumination from a light source varies as we vary the distance from the source. The study of these and other types of variation is important to the scientist since many laws in engineering, chemistry, physics, and biology are stated using variation.

The first type of relationship, *direct variation,* can be written as a ratio equal to a constant.

Direct Variation

If two variables x and y have a constant ratio c,

$$\frac{y}{x} = c \quad \text{or} \quad y = cx,$$

we say that **y varies directly as x, y is proportional to x,** or simply **y varies as x.** The number c is called the **constant of variation** or **constant of proportionality.**

Perhaps the simplest example of direct variation is found in the formula for the circumference of a circle. The circumference is equal to π times the diameter.

$$C = \pi d$$

Thus, π is the constant of variation and, in fact, π is defined as the ratio of the circumference of any circle to its diameter. We are often interested in finding the constant of variation and the equation of variation that describe a given situation.

EXAMPLE 4 Using Direct Variation

Find the equation of variation if y varies directly as x, and $y = 16$ when $x = 4$.

The first step is to write the equation described by the phrase *y varies directly as x*.

$$y = cx$$

Next we substitute the given values for y and x.

$$16 = c(4)$$
$$4 = c$$

The equation of variation is $y = 4x$.

Practice Exercise 4

Find the equation of variation if y varies directly as x, and $y = 25$ when $x = 2$.

Answer: $y = 12.5x$

EXAMPLE 5 Electricity Problem

According to Ohm's law, the voltage V in an electrical circuit varies directly as the current I in the circuit. If the voltage is 20 volts when the current is 3 amps, find the variation equation and use it to determine the voltage when the current is 27 amps.

The equation described is $V = cI$. Substitute $V = 20$ and $I = 3$.

$$20 = c(3)$$
$$\frac{20}{3} = c$$

The equation of variation is $V = \frac{20}{3}I$. When $I = 27$, $V = \frac{20}{3} \cdot 27 = 20 \cdot 9 = 180$ volts.

Practice Exercise 5

If the voltage in a circuit is 110 volts when the current is 9 amps, use Ohm's law to determine the voltage when the current is 18 amps.

Answer: 220 volts

4 INVERSE VARIATION

The second kind of variation, *inverse variation,* can be written as a product equal to a constant.

Inverse Variation

If two variables x and y have a constant product c,

$$xy = c \quad \text{or} \quad y = \frac{c}{x},$$

we say that **y varies inversely as x** or **y is inversely proportional to x.** Again **c** is the **constant of variation.**

If $y = \frac{c}{x}$, then if y becomes large, x must get small, and if y becomes small, x must get large. Perhaps the simplest example of inverse variation is the relationship between the time t and the rate r it takes for a vehicle to travel a fixed distance d. If the rate is increased, the time decreases, and if the rate is decreased, the time increases. This variation is given by

$$d = rt, \quad t = \frac{d}{r}, \quad r = \frac{d}{t},$$

where, in all three equations, we assume that d is constant (the constant of variation).

As with direct variation, we must be able to calculate the constant of variation and obtain the variation equation in an inverse variation problem.

EXAMPLE 6 Physics Problem

The volume V of a gas kept at a constant temperature varies inversely as the pressure P. If the volume is 150 ft^3 when the pressure is 18 lb/ft^2, find the variation equation and use it to determine the volume when the pressure is 12 lb/ft^2.

The equation described is

$$V = \frac{c}{P}.$$

Substitute $V = 150$ and $P = 18$.

$$150 = \frac{c}{18}$$

$$2700 = c$$

The equation of variation is $V = \frac{2700}{P}$, and when $P = 12$, $V = \frac{2700}{12} =$ 225 ft^3.

Practice Exercise 6

If the volume of a gas is 200 ft^3 when the pressure is 25 lb/ft^2, use the gas equation to find the pressure when the volume is 500 ft^3.

Answer: 10 lb/ft^2

5 JOINT VARIATION

Often, a quantity will vary directly as the product of two quantities. When this occurs we have a *joint variation* problem.

Joint Variation

If three variables x, y, and z are so related that

$$z = cxy$$

where c is a constant, we say that z varies directly as x and directly as y or that z **varies jointly as x and y.**

EXAMPLE 7 Using Joint Variation

Find the equation of variation if z varies jointly as x and y, and $z = 36$ when $x = 3$ and $y = 2$.

The first step is to write the equation described by the phrase *z varies jointly as x and y*.

$$z = cxy$$

Practice Exercise 7

Find the equation of variation if u varies jointly as v and w, and $u = 45$ when $v = 3$ and $w = 5$.

Next we substitute the given values for z, x, and y.

$$36 = c \cdot 3 \cdot 2$$
$$6 = c$$

The equation of variation is $z = 6xy$.

Answer: $u = 3vw$

6 COMBINED VARIATION

Any of the preceding types of variation can be combined to form more complex variation equations. These **combined variations** are best understood by considering the following:

Statement	*Translates to*
y varies directly as x and inversely as z (There is only one constant of variation, c, and only one equation.)	$y = \frac{cx}{z}$
a varies directly as the square of b and inversely as the cube of t	$a = \frac{cb^2}{t^3}$
u varies jointly as v and the square of w and inversely as the square root of x	$u = \frac{cvw^2}{\sqrt{x}}$

To Solve a Variation Problem

1. Translate the variation word description to an equation involving the constant of variation.
2. Using given values for the variables, substitute into the variation equation and determine the constant of variation.
3. If needed, using the variation equation with the calculated constant of variation, find the missing value of one of the variables.

6.5 EXERCISES A

Solve.

1. If a boat uses 20 gal of gas to go 75 mi, how many miles can the boat travel on 36 gal of gas?

2 In an election the winning candidate won by a 5 to 3 ratio. If he received 820 votes, how many votes did the loser receive?

3. If 150 ft of wire weighs 45 lb, what will 240 ft of the same wire weigh?

4. If on a map $\frac{3}{4}$ in represents 20 mi, how many miles will be represented by 9 in?

5. If 3 lb of steak cost $8.25, how much will 5 lb of the steak cost?

6. What number must be added to each of 25, 35, 10, and 15 to give four numbers that are in proportion in that order?

7 A rope 65 ft long is cut into two pieces which have the ratio of 8:5. How long is each piece?

8. A boy 5 ft tall casts a shadow 12 ft long at the same time that a tower casts a shadow 252 ft long. How tall is the tower?

Write the equation of variation for each of the following.

9. u varies directly as t.

10. x varies directly as y and inversely as z.

11. u varies jointly as x, y, and z.

12. a varies directly as b and inversely as the product of f and g.

13 a varies directly as u and inversely as the product of v and w, and $a = 4$ when $u = 3$, $v = 3$, and $w = 4$.

14. z varies directly as the cube of x and inversely as the square root of y, and $z = 4$ when $x = 2$ and $y = 9$.

15. The current I in an electrical circuit having constant voltage varies inversely as the resistance R of the circuit.

16. The frequency f of a guitar string of a given length varies directly as the square root of the tension t of the string.

Solve.

17 The gravitational attraction A between two masses varies inversely as the square of the distance d between them. If the force of attraction is 64 lb when the masses are 9 ft apart, find the attraction when the masses are 24 ft apart.

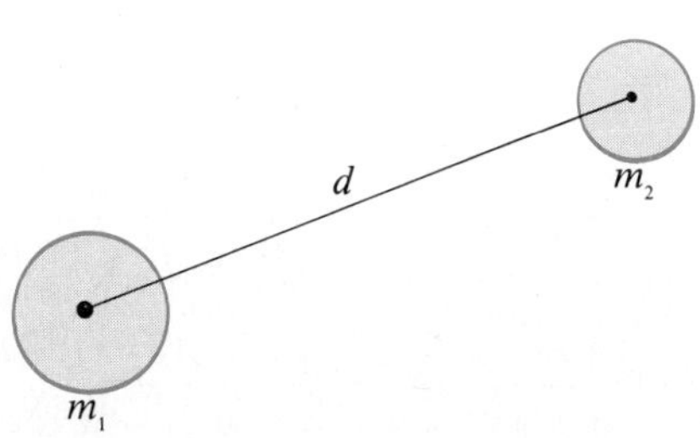

18. The period of vibration P of a pendulum varies directly as the square root of its length l. If the period of vibration of a pendulum of length 64 cm is $\frac{1}{3}$ sec, find the length of a pendulum with a period of $\frac{5}{24}$ sec.

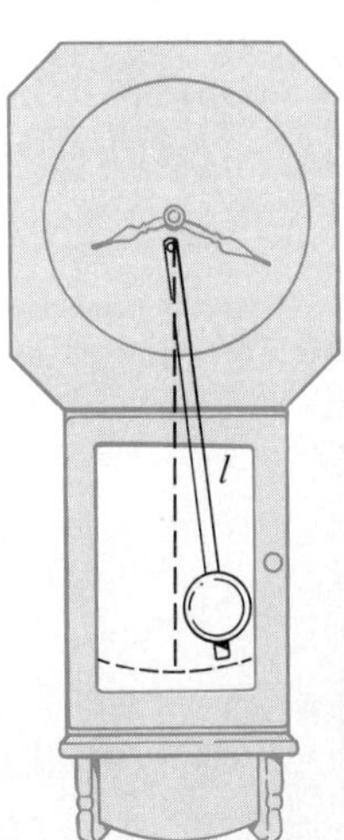

19. The intensity I of light varies inversely as the square of the distance d from the source. If the intensity of illumination on a screen 8 ft from a light is 5 foot-candles, find the intensity on a screen 16 ft from the light.

20 The weight w of a body above the surface of the earth varies inversely as the square of its distance d from the center of the earth. If an object weighs 150 lb on the surface of the earth, what would the object weigh 1000 mi above the surface of the earth? [*Hint:* Use 4000 mi for the radius of the earth.]

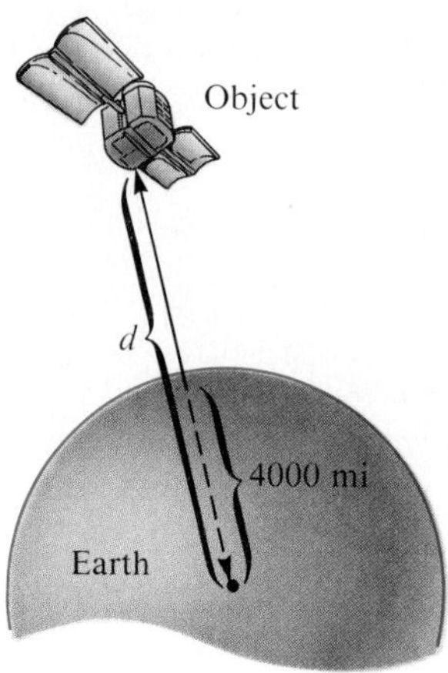

FOR REVIEW

Solve.

21. $\dfrac{1}{x^2-1} = \dfrac{1}{x-1}$

22. $\dfrac{4z+2}{z^2-z} + \dfrac{5}{z-1} = \dfrac{1}{z}$

Solve for the indicated variable.

23. $A = \dfrac{ra}{r+a}$ for r

24. $I = \dfrac{nm}{A+np}$ for p

Exercises 25–26 will help you prepare for the next section.

25. If Mike can do a job in 8 hours, how much of the job will he be able to do in **(a)** 1 hr, **(b)** 3 hr, **(c)** 8 hr?

26. Let x be the speed of a boat in miles per hour in still water. What would represent the actual speed of a boat traveling downstream if the speed of the current is 4 mph? What would represent the actual speed of the boat traveling upstream?

ANSWERS: 1. 135 mi 2. 492 votes 3. 72 lb 4. 240 mi 5. $13.75 6. 5 7. 40 ft; 25 ft 8. 105 ft 9. $u = ct$ 10. $x = \frac{cy}{z}$ 11. $u = cxyz$ 12. $a = \frac{cb}{fg}$ 13. $a = \frac{16u}{vw}$ 14. $z = \frac{3x^3}{2\sqrt{y}}$ 15. $I = \frac{c}{R}$ 16. $f = c\sqrt{t}$ 17. 9 lb 18. 25 cm 19. $\frac{5}{4}$ foot-candles 20. 96 lb 21. 0 22. $-\frac{3}{8}$ 23. $r = \frac{aA}{a-A}$ 24. $p = \frac{nm - AI}{In}$ 25. (a) $\frac{1}{8}$ of the job (b) $\frac{3}{8}$ of the job (c) $\frac{8}{8}$ of the job, that is, the whole job 26. $x + 4$ mph; $x - 4$ mph

6.5 EXERCISES B

Solve.

1. If it takes 2 hours to type 10 pages, how many hours would it take to type 45 pages?

2. In a recent election, Representative Wettaw won by a 7 to 4 ratio. If his opponent got 6100 votes, how many votes did he get?

3. In a sample of 92 tires, 3 were defective. How many defective tires would you expect in a sample of 644 tires?

4. On a trail map, $\frac{1}{2}$ in represents 15 mi. How many miles are represented by $2\frac{1}{2}$ in?

5. If 4 lb of candy cost \$5.40, how much will 7 lb of the candy cost?

6. What number when subtracted from 12, 22, 3, and 4 will make the resulting numbers, in that order, proportional?

7. A chemist wishes to make a mixture of concentrated hydrochloric acid and distilled water so that the ratio of acid to water is 7:3. If he starts with 21 L of acid, how much water should be mixed with the acid?

8. In two similar triangles, the sides of one are 6 in, 7 in, and 12 in. If the shortest side in the other is 18 in, find the other two sides.

Write the equation of variation for each of the following.

9. m varies jointly as n and p.

10. a varies jointly as b and the square root of d.

11. u varies directly with v and inversely with w.

12. y varies jointly as w and the square of x and inversely as the square root of z.

13. y varies jointly as x and z and inversely as the square of w, and $y = 18$ when $x = 2$, $z = 6$, and $w = 2$.

14. a varies jointly as s and the square of t and inversely as the cube of u, and $a = 1$ when $s = 2$, $t = 3$, and $u = 4$.

15. The volume V of a rectangular box with a fixed height varies jointly as its length l and width w.

16. The volume V of a cone varies jointly as its height h and the square of the radius r of its base.

Solve.

17. *The horsepower P* needed to move a ship varies directly as the cube of the speed s. If the horsepower needed for a speed of 15 mph is 23,625, find the horsepower needed for a speed of 20 mph.

18. The volume V of a given mass of gas varies directly as the temperature T and inversely as the pressure P. If $V = 360$ in^3 when $T = 450°$ and $P = 25$ lb/in^2, find the volume when $T = 350°$ and $P = 20$ lb/in^2.

19. The time T it takes to make an enlargement of a photo negative varies directly as the area A of the enlargement. If it takes 32 sec to make an 8-by-10 enlargement, find the time it takes to make a 5-by-7 enlargement.

20. The force of the wind blowing against the side of a building varies jointly as the surface area of the side and the square of the velocity of the wind. If a 4 mph wind exerts a force of 100 lb against the side of a building with surface area 100 ft^2, how much force will a wind of 20 mph exert against a surface with area 400 ft^2.

FOR REVIEW

Solve.

21. $\dfrac{a^2}{12 - a} = 1$

22. $\dfrac{14}{y^2 + 8y - 33} = \dfrac{y}{y - 3} - \dfrac{1}{y + 11}$

Solve for the indicated variable.

23. $\dfrac{A}{a} = \dfrac{B + b}{b}$ for b

24. $I = \dfrac{nm}{A + nP}$ for n

Exercises 25–26 will help you prepare for the next section.

25. If Stephanie can do a job in x hours, how much of the job will she be able to do in **(a)** 1 hr, **(b)** 4 hr, **(c)** x hr?

26. An airplane can fly 450 mph in calm air. What would represent the ground speed of the airplane if it is flying into a headwind with velocity x mph? What would represent the ground speed flying with the wind?

6.5 EXERCISES C

Solve.

1. To estimate the number of antelope in a game preserve, a ranger catches 40 antelope, tags their ears, and returns them to the preserve. Later on, he catches 25 antelope and finds 4 of them tagged. Use a proportion to estimate the number of antelope in the preserve.

2. The uniformly distributed safe load L that a horizontal beam can carry varies jointly as its width w and the square of its depth d, and inversely as its length l. If a 10-ft beam with width 2 ft and depth $\frac{1}{2}$ ft can carry a weight of 500 lb, how much can a 10-ft beam support if its width is 2 ft and its depth is 1 ft?
[Answer: 2000 lb]

3. When the means of a proportion are equal, as in $\frac{a}{b} = \frac{b}{c}$, b is called the **mean proportional** between a and c. The altitude h drawn to the hypotenuse c of a right triangle is the mean proportional between the segments of the hypotenuse (x and y). Find h if $x = 9$ inches and $y = 16$ inches. See the figure.

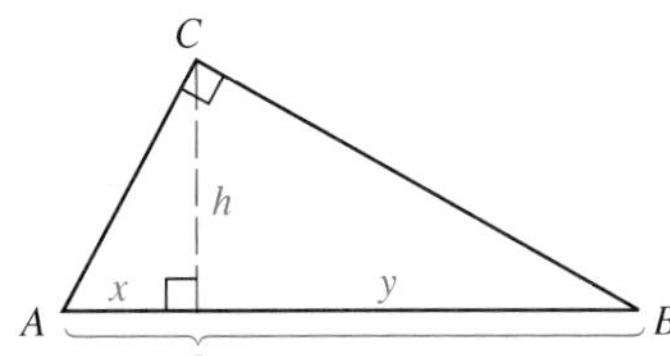

6.6 MORE APPLICATIONS USING RATIONAL EXPRESSIONS

STUDENT GUIDEPOSTS

1 Motion Problems
2 Work Problems

1 MOTION PROBLEMS

Some motion problems that use the formula

$$d = rt \quad \text{(distance = rate} \cdot \text{time)}$$

translate into fractional equations. When solving motion problems that involve two distances, two times, and two rates, it is helpful if we precisely describe each of these six quantities. Study the following examples carefully.

EXAMPLE 1 Motion Problem

The speed of a stream is 4 mph. A boat travels 48 mi upstream in the same time it takes to travel 72 mi downstream. What is the speed of the boat in still water?

Let x = speed of the boat in still water.

When the boat goes downstream, its speed relative to the stream bank is its speed in still water *increased* by the speed of the stream. Thus,

$x + 4$ = speed of the boat when traveling downstream.

When the boat travels upstream, its speed relative to the bank is its speed in still water *decreased* by the speed of the stream. Thus,

$x - 4$ = speed of the boat when traveling upstream,
72 = distance traveled downstream,
48 = distance traveled upstream.

Again we calculate the two times using $t = \frac{d}{r}$.

$$\frac{72}{x+4} = \text{time of travel downstream}$$

$$\frac{48}{x-4} = \text{time of travel upstream}$$

Since the two times are equal, we need to solve the equation

$$\frac{72}{x+4} = \frac{48}{x-4}.$$

$$72(x-4) = 48(x+4) \quad \text{Cross-product equation}$$

$$72x - 288 = 48x + 192$$

$$24x - 288 = 192 \quad \text{Subtract } 48x \text{ from both sides}$$

$$24x = 480 \quad \text{Add 288 to both sides}$$

$$x = 20$$

The speed of the boat in still water is 20 mph.

Practice Exercise 1

The speed of a plane in still air is 250 mph. If the plane can travel 846 mi with the wind in the same time that it can travel 654 mi against the wind, what is the wind speed?

Let x = wind speed,
$250 + x$ = rate with wind,
$250 - x$ = rate against wind.

Answer: 32 mph

CAUTION

In Example 1, a common mistake is to conclude that the speed upstream is $4 - x$ rather than $x - 4$. Which number, x or 4, must be larger if the boat actually makes progress upstream? Do you see why $x - 4$ is the correct rate?

EXAMPLE 2 Motion Problem

On a business trip, Terry McGinnis traveled 300 miles from Chicago. She averaged 65 mph for most of the trip but was slowed to 40 mph during a violent storm. If the total time of the trip was 5 hours, how many miles did she drive at 40 mph?

Let x = the number of miles driven at 40 mph,
$300 - x$ = the number of miles driven at 65 mph.

Practice Exercise 2

Kim Doane rode a bicycle 960 miles on a vacation. She averaged 50 miles per day most of the trip but was slowed to 35 miles per day after she pulled a muscle. If the total time of the trip was 21 days, how many days did she average 35 miles per day?

We can calculate the two different times driven at each rate using $t = \frac{d}{r}$. Then

$$\frac{x}{40} = \text{number of hours driven at 40 mph,}$$

$$\frac{300 - x}{65} = \text{number of hours driven at 65 mph.}$$

Since the total time of travel was 5 hours, we need to solve the following equation.

$$\frac{x}{40} + \frac{300 - x}{65} = 5$$

$$(5)(8)(13)\left[\frac{x}{40} + \frac{300 - x}{65}\right] = (5)(8)(13)(5) \quad \text{The LCD is (5)(8)(13)}$$

$$13x + 8(300 - x) = 2600$$

$$13x + 2400 - 8x = 2600$$

$$5x = 200$$

$$x = 40$$

Thus, Terry drove 40 miles of the trip at the reduced rate of 40 mph.

Answer: 6 days

2 WORK PROBLEMS

Consider the following problem: Jim can do a job in 3 hr and Dave can do the same job in 7 hr. How long would it take them to do the job if they worked together? This type of problem is usually called a *work problem* and three principles must be kept in mind.

1. The time it takes to do a job when the individuals work together must be less than the time it takes for the fastest worker to complete the job alone. Thus the time together is *not* the average of the two times (which would be 5 hr in this case). Since Jim can do the job alone in 3 hr, with help the time must clearly be less than 3 hr.
2. If a job can be done in t hr, in 1 hr $\frac{1}{t}$ of the job will be completed. For example, since Jim can do the job in 3 hr, in 1 hr he would do $\frac{1}{3}$ of the job. Similarly, in 1 hr Dave would do $\frac{1}{7}$ of the job since he can do it all in 7 hr.
3. The work done by Jim in 1 hr added to the work done by Dave in 1 hr equals the amount of work done together in 1 hr. Thus, if t is the time it takes to complete the job working together,

   (amount Jim does in 1 hr) + (amount Dave does in 1 hr)
   = (amount done together in 1 hr)

   translates to

$$\frac{1}{3} + \frac{1}{7} = \frac{1}{t}.$$

We see that this word problem translates to a fractional equation. Multiply both sides by the LCD, $21t$.

$$21t\left[\frac{1}{3} + \frac{1}{7}\right] = 21t\left(\frac{1}{t}\right)$$

$$21t\left(\frac{1}{3}\right) + 21t\left(\frac{1}{7}\right) = 21$$

$$7t + 3t = 21$$

$$10t = 21$$

$$t = \frac{21}{10}$$

It would take Jim and Dave $\frac{21}{10}$ hr (2 hr 6 min) to complete the job working together. Does this seem reasonable in view of (1) above? Check.

A variation of this type of problem is to look for the time it takes one individual when given the time for the other and the time working together.

EXAMPLE 3 WORK PROBLEM

G & L Painters have the painting contract for all the new homes in a subdivision. When Gary and Lew work together, it takes 5 days to paint one house. If Gary paints one house by himself, it takes 7 days to complete the job. How long would it take Lew to paint one house if he worked alone?

We have the following information.

5 = the number of days to paint one house together,

$\frac{1}{5}$ = the amount painted together in 1 day,

7 = the number of days for Gary to paint one house,

$\frac{1}{7}$ = the amount painted by Gary in 1 day.

Let x = the number of days for Lew to paint one house,

$\frac{1}{x}$ = the amount painted by Lew in 1 day.

(amount by Gary) + (amount by Lew) = (amount together)

$$\frac{1}{7} + \frac{1}{x} = \frac{1}{5}$$

msp,509Multiply both sides of this equation by the LCD, $35x$.

$$35x\left(\frac{1}{7} + \frac{1}{x}\right) = 35x\left(\frac{1}{5}\right)$$

$$5x + 35 = 7x$$

$$35 = 2x$$

$$\frac{35}{2} = x$$

Thus, it would take Lew $17\frac{1}{2}$ days to paint one house by himself. Check.

PRACTICE EXERCISE 3

Bob and Lennie can rebuild an engine in 3 days when they work together. If Bob can do the same job alone in 4 days, how long would it take Lennie to do it if he works alone?

Let t = time for Lennie to do the job working alone

$\frac{1}{4} + \frac{1}{t} =$ ________.

Answer: 12 days

CAUTION

When solving work problems (as with any word problem), be neat and complete. Do not take shortcuts, especially when writing down the pertinent information. Writing detailed descriptions of the variables can eliminate errors. Pattern your work after the examples.

The work problem on the next page results in a quadratic equation.

EXAMPLE 4 Work Problem

It takes pipe A 9 days longer to fill a reservoir than pipe B. If the two pipes are turned on together, they can fill the reservoir in 20 days. How long would it take each pipe to fill the reservoir alone?

Let t = the number of days required for B to fill the reservoir,
$t + 9$ = the number of days required for A to fill the reservoir,
20 = the number of days required to fill the reservoir together,
$\frac{1}{t}$ = the amount filled by B in 1 day,
$\frac{1}{t+9}$ = the amount filled by A in 1 day,
$\frac{1}{20}$ = the amount filled by A and B working together in 1 day.

We must solve $\frac{1}{t} + \frac{1}{t+9} = \frac{1}{20}$. The LCD $= 20t(t+9)$

$$20t(t+9)\frac{1}{t} + 20t(t+9)\frac{1}{t+9} = 20t(t+9)\frac{1}{20}$$
$$20(t+9) + 20t = t(t+9)$$
$$20t + 180 + 20t = t^2 + 9t$$
$$40t + 180 = t^2 + 9t$$
$$0 = t^2 - 31t - 180$$

This time we obtain a quadratic equation to solve.

$$(t+5)(t-36) = 0$$
$$t + 5 = 0 \quad \text{or} \quad t - 36 = 0$$
$$t = -5 \qquad\qquad t = 36$$
$$t + 9 = 45$$

Since -5 is meaningless for a time in a work problem, we have $t = 36$ as the only solution. Therefore it takes pipe B 36 days to fill the reservoir alone, and pipe A takes 45 days to fill it. Check.

Practice Exercise 4

When Matt and Wendy work together it takes 6 days to wash the windows in a building. If it takes Matt 9 days longer than Wendy to wash them by himself, how long would it take each to complete the job working alone?

Answer: Matt: 18 days; Wendy: 9 days

6.6 EXERCISES A

Solve.

1. A freight train travels 5 mph slower than a passenger train. If the freight train travels 260 mi in the same time that the passenger train travels 280 mi, find the speed of each.

Let x = speed of the passenger train,
$x - 5$ = speed of the freight train.
$260 =$
$280 =$
$\frac{280}{x} =$
$\frac{260}{x-5} =$

2. The speed of a stream is 5 mph. If a boat travels 50 mi downstream in the same time that it takes to travel 25 mi upstream, what is the speed of the boat in still water?

Let x = speed of the boat in still water,

$x - 5$ = speed of the boat going upstream.

$x + 5 =$

$50 =$

$25 =$

$\frac{50}{x+5} =$

$\frac{25}{x-5} =$

3. A plane flies 480 mi with the wind and 330 mi against the wind in the same length of time. If the speed of the wind is 25 mph, what is the speed of the plane in still air?

Let x = speed of the plane in still air.

$x + 25 =$

$x - 25 =$

$480 =$

$330 =$

4. The speed of a boat in still water is 24 mph. If the boat travels 54 mi upstream in the same time that it takes to travel 90 mi downstream, what is the speed of the stream?

5. A man walks a distance of 12 mi at a rate 8 mph slower than the rate he rides a bicycle for a distance of 24 mi. If the total time of the trip is 5 hr, how fast does he walk? How fast does he ride? [*Hint:* $\frac{12}{x} + \frac{24}{x+8} = 5$]

6 A woman flies from Phoenix to Denver (a distance of 800 mi) at a rate 40 mph faster than on the return trip. If the total time of the trip is 9 hr, what was her rate going to Denver, and what was her rate returning to Phoenix?

7. The time of a plane trip of 450 mi, with the wind is three-fifths the time of the return trip against the wind. If the plane travels 120 mph in still air, what is the wind speed?

8 Dennis and Alberta rent a boat at the Riverside Campground at 9:00 A.M. They are told that due to the current of the river, the boat will travel 6 mph upstream and 12 mph downstream. If they must return the boat at 1:00 P.M. the same day, how far upstream can they travel, and what time must they turn back?

9. If Clyde can do a job in 8 days and Irv can do the same job in 3 days, how long would it take them to do the job together?

8 = the number of days for Clyde to do the job

$\frac{1}{8}$ = the amount Clyde does in 1 day

$3 =$

$\frac{1}{3} =$

Let t = the number of days to do the job together,

$\frac{1}{t}$ = the amount done in 1 day working together.

10. If pipe A can fill a tank in 20 hr and pipe B can fill the same tank in 15 hr, how long would it take to fill the tank if both pipes fill together?

$20 =$

$\frac{1}{20} =$

$15 =$

$\frac{1}{15} =$

Let $t =$ the number of hours to fill tank together.

$\frac{1}{t} =$

11. On the day that the Wisconsin cheese shipment arrives at Perko's Delicatessen, it takes Perko 3 hr to slice and display the cheese. When Perko and his assistant Graydon work together, it takes only 2 hr to process the same shipment. How long would it take Graydon to slice and display the cheese if he worked alone?

Let $t =$ time it takes Graydon to cut the cheese.

12. When Barb does a job by herself, it takes her 7 hr. If Barb and Wilma work together it takes 6 hr to do the same job. How long does it take Wilma to do the job if she works alone?

Let $t =$ time it takes Wilma to do the job.

13. When each works alone, Burford can mow a lawn in 3 hr less time than Ernie. When they work together, it takes 2 hr. How long does it take each to do the job by himself?

14. It takes Sybil 6 hr longer to paint a room than it takes Joanne. Working together, they can paint the room in 4 hr. How long would it take each working alone to paint the room?

15 A tank can be filled by an inlet pipe in 4 hr. It can be drained by an outlet pipe in 10 hr. If the inlet and outlet pipes are opened simultaneously, how long will it take to fill the empty tank?

16. When Arnie was a small child, he ''helped'' his mother plant onions. She could plant the entire garden in 3 hr. However, Arnie, without his mother realizing it, followed behind her, picking up the onion sets at a rate that would have completely unplanted the garden in 5 hr. Under these circumstances, how long would it have taken to plant the onions?

FOR REVIEW

Solve.

17. If a car used 16 gal of gas for a trip of 264 mi, how much gas will be used for a trip of 1100 mi?

18. A girl casts a shadow 5.5 ft long at the same time that a tree which is 35 ft tall casts a shadow 55 ft long. How tall is the girl?

Write the equation of variation for each of the following.

19. w varies directly as a and inversely as the square root of b.

20. u varies jointly as w and the cube of z and inversely as the square root of x.

Solve.

21. The gravitational attraction A between two masses varies inversely as the square of the distance d between them. If the force of attraction is 20 lb when the masses are 5 ft apart, find the attraction when the masses are 10 ft apart.

22. The resistance R of a wire at a constant temperature varies directly as the length L and inversely as the square of the diameter D. A section of wire having a diameter of 0.01 in and a length of 1 ft has a resistance of 8.2 ohms. Find the approximate resistance of a wire 1 mi in length with diameter 0.05 in.

Exercises 23 and 24 will help you prepare for the next section. Simplify each expression and write as a single fraction.

23. $\dfrac{\frac{1}{2}+1}{\frac{1}{2}-1}$

24. $\dfrac{\frac{3}{4}+\frac{1}{8}}{\frac{1}{2}-\frac{1}{4}}$

ANSWERS: 1. freight train: 65 mph; passenger train: 70 mph 2. 15 mph 3. 135 mph 4. 6 mph 5. walk: 4 mph; ride: 12 mph 6. 200 mph going; 160 mph returning 7. 30 mph 8. 16 miles; 11:40 A.M. 9. $\frac{24}{11}$ days 10. $\frac{60}{7}$ hr 11. 6 hr 12. 42 hr 13. Burford: 3 hr; Ernie: 6 hr 14. Sybil: 12 hr; Joanne: 6 hr 15. $\frac{20}{3}$ hr 16. $\frac{15}{2}$ hr 17. $66\frac{2}{3}$ gal 18. 3.5 ft 19. $w = \frac{ca}{\sqrt{b}}$ 20. $u = \frac{cwz^3}{\sqrt{x}}$ 21. 5 lb 22. 1732 ohms 23. -3 24. $\frac{7}{2}$

6.6 EXERCISES B

Solve.

1. Sam and Carolyn agree to meet in Las Vegas for the weekend. Sam travels 240 mi while Carolyn travels 220 mi. If Sam's rate of travel is 5 mph more than Carolyn's, and they travel the same length of time, at what speed does each travel?

2. John drives 12 mph faster than his mother. If John travels 310 mi in the same time that his mother travels 250 mi, find the speed of each.

3. A plane flies 1160 km with the wind and 840 km against the wind in the same length of time. If the speed of the plane in still air is 250 km/hr, what is the wind speed?

4. Merrilee can fly her plane 100 mph in still air. When the wind is blowing, she flies 345 mi with the wind in the same time that she flies 255 mi against the wind. What is the wind speed?

5. The Bells drive their car a distance of 150 mi at a rate of 30 mph faster than the rate they drive their boat a distance of 50 mi. If the total time of the trip is $5\frac{1}{2}$ hr, how fast do they drive each vehicle?

6. Craig Slack traveled from Newark, New Jersey to a convention in another city 300 mi away. Going, he traveled at a rate of 10 mph more than the rate returning. If the total time of the trip was 11 hr, what was his rate going and his rate returning?

7. A boat takes $1\frac{1}{2}$ times longer to go 72 mi upstream than to return. If the boat cruises at 30 mph in still water, what is the rate of the current?

8. Dick's boat can travel 35 mph in still water. If he leaves at 8:00 A.M., how far downstream can Dick travel, in a river with current of 5 mph, if he must be back to where he starts in 7 hr? At what time must he turn back upstream?

9. If Wylie Smith, Sports Information Director at a leading state university, can prepare a game program in 3 days and his student assistant can do the same job in 8 days, how long will it take them to prepare the program if they work together?

10. Nancy, a file clerk, can file a stack of folders in 75 min. Her assistant can file the same stack in 125 min. How long will it take them to do the job if they start together?

11. Ken can make a machine part in 5 hr. If he and his assistant work together it requires 3 hr to make the part. How long would it take the assistant by herself to make the part?

12. Pipe A can fill a tank in 8 hr. When both pipe A and B are turned on, the tank is filled in 2 hr. How long would be required for pipe B to fill the tank when it is turned on alone?

13. A new machine can produce 1000 units in 6 hr less time than the old machine. When they are both used, the 1000 units are produced in 4 hr. How long does it take each to produce 1000 units working alone?

14. It takes Howard twice as long to stock the freezer in a market as it does Mary. Together they can do the job in 3 hr. How long would it take each working alone?

15. A pond can be filled by an inlet pipe in 10 days and drained by an outlet pipe in 20 days. If the inlet and outlet pipes are opened simultaneously, how long will it take to fill the empty pond?

16. A barrel can be filled by an inlet pipe in 12 min, while it can be drained by an outlet pipe in 16 min. Through an error, both pipes are opened simultaneously. How long will it take to fill the empty barrel?

FOR REVIEW

Solve.

17. A man 6 ft tall casts a shadow 20 ft long. What is the length of the shadow cast at the same time by a building which is 51 ft tall?

18. In an election, the winning candidate won by a 5:3 margin. If he received 1250 votes, how many did the loser receive?

Write the equation of variation for each of the following.

19. y varies directly with x and inversely with the square of z.

20. m varies jointly as the square root of a and the square of b, and inversely as the cube of t.

Solve.

21. The period of vibration of a pendulum varies directly as the square root of its length l. If the period of a pendulum of length 25 cm is $\frac{1}{4}$ sec, find the period of a pendulum of length 100 cm.

22. The time T it takes an elevator to lift a weight w varies jointly as the weight and the distance d to be lifted and inversely as the power p of the motor. If it takes 24 sec for a 3-horsepower motor to lift 500 lb 60 ft, what power is needed to lift 700 lb 100 ft in 35 sec?

Exercises 23 and 24 will help you prepare for the next section. Simplify each expression and write as a single fraction.

23. $\dfrac{\frac{1}{4} - 1}{\frac{1}{4} + 1}$

24. $\dfrac{\frac{2}{5} + \frac{3}{10}}{\frac{1}{6} + \frac{5}{12}}$

6.6 EXERCISES C

Solve.

1. One of the events in the University Days Games involves swimming, running, and riding a bike over a fixed distance. The winner of the event, Mike Ratliff, averaged 2 mph swimming, 10 mph running, and 28 mph riding. The total distance was broken into three parts so that he swam $\frac{1}{20}$ of the total, ran $\frac{1}{4}$ of the total, and rode the remaining portion. If the winning time was 7.5 hr, what was the total distance covered? [Answer: 100 mi]

2. A large reservoir can be filled by one inlet pipe in 5 days. It takes 8 days to fill it using a second pipe. The reservoir can be drained by an outlet pipe in 10 days. If all three pipes are opened simultaneously, how long will it take to fill the empty reservoir?

3. The Walters traveled to their mountain cabin located 330 miles away. For most of the trip they traveled 65 mph on an interstate highway, but for the final portion they drove on a winding mountain road averaging only 15 mph. If the total time of the trip was 7 hours, how many hours did they drive on the interstate?

4. Cynthia Allen traveled 340 miles west of Fargo. For most of the trip she averaged 60 mph, but for one period of time she was slowed to 20 mph due to a major accident. If the total time of travel was 7 hours, how many miles did she drive at the reduced speed?

6.7 SIMPLIFYING COMPLEX FRACTIONS

A fractional expression that contains at least one other fraction within it is called a **complex fraction.** The following are examples of complex fractions.

$$\frac{\frac{1}{5}}{\frac{2}{3}}, \quad \frac{a}{\frac{3}{4}}, \quad \frac{\frac{x}{y}}{7}, \quad \frac{2+\frac{1}{y}}{3}, \quad \frac{2+\frac{a}{b}}{2-\frac{a}{b}}$$

A complex fraction is **simplified** when all of its component fractions have been eliminated and a simple fraction obtained. There are two basic methods for simplifying a complex fraction. The first involves writing both the numerator and the denominator as single fractions and then dividing.

To Simplify a Complex Fraction (Method 1)

1. Change the numerator and denominator to single fractions.
2. Divide the two fractions.
3. Reduce to lowest terms.

EXAMPLE 1 Simplifying a Complex Fraction

Simplify $\dfrac{\frac{x}{y}+2}{2-\frac{x}{y}}$.

$$\frac{x}{y}+2=\frac{x}{y}+\frac{2}{1}=\frac{x}{y}+\frac{2\cdot y}{1\cdot y}=\frac{x+2y}{y}$$ Add the numerator fractions

$$2-\frac{x}{y}=\frac{2}{1}-\frac{x}{y}=\frac{2\cdot y}{1\cdot y}-\frac{x}{y}=\frac{2y-x}{y}$$ Subtract the denominator fractions

Practice Exercise 1

Simplify $\dfrac{5+\frac{a}{b}}{\frac{a}{b}-5}$.

Thus,

$$\frac{\frac{x}{y}+2}{2-\frac{x}{y}} = \frac{\frac{x+2y}{y}}{\frac{2y-x}{y}} = \frac{x+2y}{y} \div \frac{2y-x}{y} = \frac{x+2y}{y} \cdot \frac{y}{2y-x}$$

$$= \frac{(x+2y)\not{y}}{\not{y}(2y-x)} = \frac{x+2y}{2y-x}.$$

Answer: $\frac{5b+a}{a-5b}$

The second method involves clearing all fractions within the expression by multiplying numerator and denominator by the LCD of all the internal fractions.

To Simplify a Complex Fraction (Method 2)

1. Find the LCD of all fractions within the complex fraction.
2. Multiply numerator and denominator of the complex fraction by the LCD to obtain an equivalent fraction.
3. Reduce to lowest terms.

EXAMPLE 2 SIMPLIFYING A COMPLEX FRACTION

Simplify $\dfrac{1+\dfrac{y}{x-y}}{\dfrac{y}{x+y}-1}$.

$$\frac{\left(1+\frac{y}{x-y}\right)(x-y)(x+y)}{\left(\frac{y}{x+y}-1\right)(x-y)(x+y)}$$ Multiply by LCD, $(x-y)(x+y)$, and distribute

$$= \frac{(1)(x-y)(x+y) + \frac{y}{\cancel{x-y}}\cancel{(x-y)}(x+y)}{\frac{y}{\cancel{(x+y)}}(x-y)\cancel{(x+y)} - (1)(x-y)(x+y)}$$

$$= \frac{(x^2-y^2)+y(x+y)}{y(x-y)-(x^2-y^2)}$$ Use parentheses

$$= \frac{x^2-y^2+yx+y^2}{yx-y^2-x^2+y^2}$$ Watch signs

$$= \frac{x^2+yx}{yx-x^2} = \frac{\cancel{x}(x+y)}{\cancel{x}(y-x)}$$

$$= \frac{x+y}{y-x}$$

PRACTICE EXERCISE 2

Simplify $\dfrac{\dfrac{a}{a-b}-1}{1-\dfrac{a}{a+b}}$.

Answer: $\frac{a+b}{a-b}$

When we evaluate an algebraic expression, the result is often a complex fraction that must be simplified.

EXAMPLE 3 Evaluation Leading to a Complex Fraction

Evaluate $\dfrac{a-2}{a+3}$ when $a = \dfrac{x-y}{y}$.

Substitute $\frac{x-y}{y}$ for a.

$$\frac{\frac{x-y}{y}-2}{\frac{x-y}{y}+3} = \frac{\left(\frac{x-y}{y}-2\right)\cdot y}{\left(\frac{x-y}{y}+3\right)\cdot y} \quad \text{Multiply by the LCD, } y$$

$$= \frac{\frac{x-y}{\not y}\cdot \not y - 2\cdot y}{\frac{x-y}{\not y}\cdot \not y + 3\cdot y} \quad \text{Distribute}$$

$$= \frac{x-y-2y}{x-y+3y} = \frac{x-3y}{x+2y}$$

Practice Exercise 3

Evaluate $\dfrac{u-1}{u+1}$ when $u = \dfrac{a+b}{a}$.

Answer: $\frac{b}{2a+b}$

We often have complex fractions when using negative exponents. After applying the definition to remove the negative exponent, we simplify the result using Method 1 in the next example.

EXAMPLE 4 Physics Problem

The focal length f of a simple convex lens is given by

$$f = (d_1^{-1} + d_2^{-1})^{-1}$$

where d_1 is the distance from an object to the lens and d_2 is the distance from the lens to the image of the object. Remove negative exponents, simplify the complex fraction, and use the result to find f when $d_1 = 20$ inches and $d_2 = 12$ inches.

$$f = (d_1^{-1} + d_2^{-1})^{-1} = \frac{1}{d_1^{-1} + d_2^{-1}} \qquad u^{-1} = \frac{1}{u}$$

$$= \frac{1}{\frac{1}{d_1} + \frac{1}{d_2}}$$

$$= \frac{1}{\frac{d_2}{d_1 d_2} + \frac{d_1}{d_1 d_2}} \qquad \text{The LCD of } \frac{1}{d_1} \text{ and } \frac{1}{d_2} \text{ is } d_1 d_2.$$

$$= \frac{1}{\frac{d_2 + d_1}{d_1 d_2}} = 1 \div \frac{d_2 + d_1}{d_1 d_2}$$

$$= 1 \cdot \frac{d_1 d_2}{d_2 + d_1} \qquad \text{Multiply by the reciprocal of the divisor}$$

$$= \frac{d_1 d_2}{d_2 + d_1}$$

Practice Exercise 4

Simplify $(x^{-1} - y^{-1})^{-1}$. Evaluate when $x = -2$ and $y = -6$.

To find f, substitute 20 for d_1 and 12 for d_2.

$$f = \frac{(20)(12)}{12 + 20} = \frac{240}{32} = 7.5$$

The focal length is 7.5 inches.

Answer: $\frac{xy}{y - x}$; -3

6.7 EXERCISES A

Simplify.

1. $\dfrac{\frac{x}{2}}{\frac{3}{x + 1}}$

2. $\dfrac{\frac{a + 1}{3}}{\frac{2a - 1}{6}}$

3. $\dfrac{\frac{a}{b}}{\frac{b^2}{a + 1}}$

4. $\dfrac{\frac{y + 3}{5}}{\frac{(y + 3)^2}{10}}$

5. $\dfrac{\frac{1}{a} + 1}{\frac{1}{a} - 1}$

6. $\dfrac{1 - \frac{2}{3y}}{y - \frac{4}{9y}}$

7. $\dfrac{\frac{2}{a} - \frac{2}{b}}{\frac{1}{a^2} - \frac{1}{b^2}}$

8. $\dfrac{1 - \frac{x}{y}}{y - \frac{x^2}{y}}$

9. $\dfrac{\frac{1}{a} - \frac{1}{b}}{\frac{a}{b} - \frac{b}{a}}$

10. $\dfrac{x - \frac{4}{x}}{1 + \frac{2}{x}}$

11. $\dfrac{a - 3 + \frac{2}{a}}{a - 4 + \frac{3}{a}}$

12. $\dfrac{\frac{1}{x + y} - \frac{1}{x - y}}{\frac{-2}{x - y}}$

Evaluate and simplify.

13. $\dfrac{a + 1}{a - 1}$ for $a = \dfrac{x + y}{y}$

14. $\dfrac{u - 1}{u + 1}$ for $u = \dfrac{a}{b}$

15. $\dfrac{x^2 + y^2}{xy}$ for $x = \dfrac{1}{u}$, $y = \dfrac{1}{v}$

Simplify.

16. $1 - \dfrac{1}{1 - \dfrac{1}{a}}$

17. $a - \dfrac{a}{1 - \dfrac{a}{1 - a}}$

18. $\dfrac{\dfrac{1}{xy} + \dfrac{1}{yz} + \dfrac{1}{xz}}{\dfrac{x + y + z}{xyz}}$

19. $x^{-1} + 1$

20. $u^{-1} + v^{-1}$

21. $a^{-2} + b^{-2}$

22. $(x^{-1} + 1)^{-1}$

23. $\dfrac{2x^{-1} + 2y^{-1}}{(xy)^{-1}}$

24. $\dfrac{u^{-1} + v^{-1}}{u^{-1} - v^{-1}}$

Use the formula $f = \dfrac{d_1 d_2}{d_2 + d_1}$ *to find the focal length f, to the nearest tenth of a centimeter, for the values of* d_1 *and* d_2 *given in Exercises 25–26.*

25. $d_1 = 8.0$ cm and $d_2 = 14.0$ cm

26. $d_1 = 16.0$ cm and $d_2 = 9.0$ cm

27. Calculating an average rate of speed uses a complex fraction. Suppose Marla hikes to the bottom of the Grand Canyon at a rate of 4 mph and returns to the top at a rate of 2 mph. What is her average rate for the hike? At first glance we might assume the average rate is 3 mph, the average of 4 and 2. However, her rate must be closer to 2 mph than to 4 mph since she walks for a longer period of time at this slower rate. Let D represent the distance hiked one direction. From the distance formula distance = (rate)(time), we have

$$\text{time} = \frac{\text{distance}}{\text{rate}}.$$

$$\text{time hiking down} = \frac{\text{distance down}}{\text{rate down}} = \frac{D}{4}$$

$$\text{time hiking up} = \frac{\text{distance up}}{\text{rate up}} = \frac{D}{2}$$

$$\text{average rate} = \frac{\text{total distance}}{\text{total time}} = \frac{2D}{\dfrac{D}{4} + \dfrac{D}{2}}$$

Simplify the complex fraction and find Marla's average rate on her hike. Give answer correct to the nearest tenth.

28. Repeat Exercise 27 assuming that Marla hiked down the canyon at a rate of 5 mph and returned to the top at a rate of 1.5 mph. Give answer correct to the nearest tenth.

FOR REVIEW

Solve.

29. The speed of a boat in still water is 30 km/hr. If the boat travels 72 km upstream in the same time that it takes to travel 108 km downstream, what is the speed of the stream?

30. Winston can prepare a window display in 4 hr and Wanda can do the same job in 3 hr. How long will it take to prepare the display if they work together?

Perform the indicated operations and simplify.

31. $\dfrac{x}{x-5}+\dfrac{2x+1}{5-x}$

32. $\dfrac{12}{z^3+8}+\dfrac{z}{z^2-2z+4}-\dfrac{1}{z+2}$

Exercises 33–36 review topics from Sections 1.6 and 5.1 to help you prepare for the next section. Perform the indicated operations.

33. $\dfrac{15x^3y^2}{5x^2y}$

34. $\dfrac{-22a^3b^4}{11ab^5}$

35. Subtract.
$$\begin{array}{r} 3x^2+2x \\ \underline{3x^2+3x} \end{array}$$

36. Subtract.
$$\begin{array}{r} 2y^3-5x^2 \\ \underline{2y^3-4x^2} \end{array}$$

ANSWERS: 1. $\frac{x(x+1)}{6}$ 2. $\frac{2(a+1)}{2a-1}$ 3. $\frac{a(a+1)}{b^3}$ 4. $\frac{2}{y+3}$ 5. $\frac{1+a}{1-a}$ 6. $\frac{3}{3y+2}$ 7. $\frac{2ab}{b+a}$ 8. $\frac{1}{y+x}$ 9. $\frac{-1}{a+b}$ 10. $x-2$ 11. $\frac{a-2}{a-3}$ 12. $\frac{y}{x+y}$ 13. $\frac{x+2y}{x}$ 14. $\frac{a-b}{a+b}$ 15. $\frac{v^2+u^2}{uv}$ 16. $\frac{-1}{a-1}$ 17. $\frac{-a^2}{1-2a}$ 18. 1 19. $\frac{1+x}{x}$ 20. $\frac{u+v}{uv}$ 21. $\frac{a^2+b^2}{a^2b^2}$ 22. $\frac{x}{1+x}$ 23. $2(x+y)$ 24. $\frac{v+u}{v-u}$ 25. 5.1 cm 26. 5.8 cm 27. 2.7 mph 28. 2.3 mph 29. 6 km/hr 30. $\frac{12}{7}$ hr 31. $\frac{x+1}{5-x}$ 32. $\frac{4}{z^2-2z+4}$ 33. $3xy$ 34. $-\frac{2a^2}{b}$ 35. $-x$ 36. $-x^2$

6.7 EXERCISES B

Simplify.

1. $\dfrac{\dfrac{3}{y}}{\dfrac{5}{y-1}}$

2. $\dfrac{\dfrac{b+3}{6}}{\dfrac{4b-5}{3}}$

3. $\dfrac{\dfrac{x}{2y}}{\dfrac{xy}{4}}$

4. $\dfrac{\dfrac{(a-1)^2}{3}}{\dfrac{a-1}{9}}$

5. $\dfrac{1+\dfrac{1}{x}}{1-\dfrac{1}{x^2}}$

6. $\dfrac{\dfrac{2}{z}+3}{\dfrac{4}{z^2}-9}$

7. $\dfrac{1-\dfrac{a^2}{b^2}}{1-\dfrac{a}{b}}$

8. $\dfrac{\dfrac{2}{x}-\dfrac{5}{y}}{\dfrac{4}{x^2}-\dfrac{25}{y^2}}$

9. $\dfrac{\dfrac{2}{a-2}+1}{\dfrac{2}{a+2}-1}$

10. $\dfrac{1-\dfrac{2}{x}-\dfrac{3}{x^2}}{1+\dfrac{1}{x}}$

11. $\dfrac{1-\dfrac{7}{x}+\dfrac{10}{x^2}}{1+\dfrac{1}{x}-\dfrac{6}{x^2}}$

12. $\dfrac{\dfrac{3}{2u-v}-\dfrac{3}{2u+v}}{\dfrac{-6v}{2u+v}}$

Evaluate and simplify.

13. $\dfrac{1-x}{x}$ for $x=\dfrac{a}{b}$

14. $\dfrac{x+1}{x-1}$ for $x=\dfrac{u}{u-v}$

15. $\dfrac{a+b}{a^2+b^2}$ for $a=\dfrac{1}{x}$, $b=\dfrac{1}{y}$

Simplify.

16. $1-\dfrac{1}{1-\dfrac{1}{x+1}}$

17. $\dfrac{\dfrac{x+1}{x-1}-\dfrac{x-1}{x+1}}{\dfrac{x+1}{x-1}+\dfrac{x-1}{x+1}}$

18. $\dfrac{\dfrac{a}{bc}+\dfrac{b}{ac}+\dfrac{c}{ab}}{\dfrac{1}{b^2c^2}+\dfrac{1}{a^2c^2}+\dfrac{1}{a^2b^2}}$

19. $a^{-1}-1$

20. $x^{-1}-y^{-1}$

21. $u^{-2}-v^{-2}$

22. $(a^{-1}-1)^{-1}$

23. $\dfrac{2a^{-1}-2b^{-1}}{(ab)^{-1}}$

24. $\dfrac{(x^{-1}-y^{-1})^{-1}}{xy}$

Use the formula $f=\dfrac{d_1d_2}{d_2+d_1}$ *to find the focal length f, to the nearest tenth of a centimeter, for the values of* d_1 *and* d_2 *given in Exercises 25–26.*

25. $d_1=5.0$ cm and $d_2=11.0$ cm

26. $d_1=18.0$ cm and $d_2=7.0$ cm

27. Finding an average rate of speed results in a complex fraction. Suppose a hiker walks a distance D at a rate of 4 mph and returns distance D at a rate of 3 mph. The average for the trip is

$$\frac{2D}{\dfrac{D}{4}+\dfrac{D}{3}}$$

Simplify the complex fraction to find the average rate walked by the hiker. Give answer correct to the nearest tenth.

28. Repeat Exercise 27 assuming the hiker walked one way at a rate of 5.2 mph and returned at a rate of 3.6 mph. Give answer correct to the nearest tenth.

FOR REVIEW

Solve.

29. A plane flies 1650 km with the wind and 1230 km against the wind in the same length of time. If the speed of the wind is 70 km/hr, what is the speed of the plane in still air?

30. When Simon paints a room by himself, it takes 5 hr. If Simon and Jane work together it takes 2 hr to paint the room. How long does it take Jane to do the job if she works alone?

Perform the indicated operations and simplify.

31. $\dfrac{1}{a^2-4a+4}-\dfrac{3-a}{a^2-4a+4}$

32. $\dfrac{2y-1}{y^2-4y+4}+\dfrac{2y+1}{4-y^2}$

Exercises 33–36 review topics from Sections 1.6 and 5.1 to help you prepare for the next section. Perform the indicated operations.

33. $\dfrac{12a^4b^6}{3a^5b^3}$

34. $\dfrac{-30x^3y^7}{10x^4y^2}$

35. Subtract.
$$\begin{array}{r} 2y^2 + 3y \\ \underline{2y^2 + 7y} \end{array}$$

36. Subtract.
$$\begin{array}{r} 4x^3 - 6x^2 \\ \underline{4x^3 - 3x^2} \end{array}$$

6.7 EXERCISES C

Simplify.

1. $$\dfrac{\dfrac{1}{\dfrac{1}{x} - 1} + x}{\dfrac{1}{\dfrac{1}{x} + 1} + x}$$

2. $$\left[\dfrac{1 + \dfrac{a-1}{a+1}}{1 - \dfrac{a-1}{a+1}}\right]^{-2}$$

3. $\left[\dfrac{(x^{-1} - y^{-1})^{-1}}{(x^{-1} + y^{-1})^{-1}}\right]^{-1}$ $\left[\text{Answer: } \dfrac{y-x}{y+x}\right]$

4. $\left[1 - \dfrac{1 - \dfrac{1}{a+1}}{1 + \dfrac{1}{a-1}}\right]^{-1}$ $\left[\text{Answer: } \dfrac{a+1}{2}\right]$

6.8 DIVISION OF POLYNOMIALS

STUDENT GUIDEPOSTS

1. Dividing by a Monomial
2. Dividing by a Binomial
3. Dividing by a Polynomial

1 DIVIDING BY A MONOMIAL

We learned how to add, subtract, and multiply polynomials in Chapter 5. Since the quotient of two polynomials is actually a rational expression, division was saved until now. We begin with dividing a polynomial by a monomial (a polynomial with only one term).

To Divide a Polynomial by a Monomial

Divide each term of the polynomial by the monomial and simplify the resulting sum or difference of fractions.

EXAMPLE 1 DIVIDING BY A MONOMIAL

Divide.

(a) $\dfrac{5x^3 + 10x^2 - 20x}{5x} = \dfrac{5x^3}{5x} + \dfrac{10x^2}{5x} - \dfrac{20x}{5x} = x^2 + 2x - 4$

Notice that the first step is the reverse of adding fractions with the same denominators.

PRACTICE EXERCISE 1

Divide.

(a) $\dfrac{12y^4 - 24y^3 + 6y}{6y}$

(b) $\dfrac{3a^3b^4 + 6a^2b^3 - 9ab^5 + 3ab}{3a^2b^2}$

$$= \frac{3a^3b^4}{3a^2b^2} + \frac{6a^2b^3}{3a^2b^2} - \frac{9ab^5}{3a^2b^2} + \frac{3ab}{3a^2b^2}$$

$$= ab^2 + 2b - \frac{3b^3}{a} + \frac{1}{ab}$$ Remember the rules of exponents

(b) $\dfrac{4u^4v^5 + 16u^3v^3 - 8u^2v^4 + 12u^5v^5}{4u^4v^4}$

Answers: (a) $2y^3 - 4y^2 + 1$
(b) $v + \dfrac{4}{uv} - \dfrac{2}{u^2} + 3uv$

When dividing by a monomial, practice will allow you to skip the middle step and proceed directly to the final answer.

2 DIVIDING BY A BINOMIAL

Dividing a polynomial by a binomial (a polynomial with two terms) is more complex than dividing by a monomial. The method is shown in the examples that follow.

To Divide a Polynomial by a Binomial

1. Arrange the terms of both in descending order.
2. Divide the first term of the polynomial (the dividend) by the first term of the binomial (the divisor) to obtain the first term of the quotient.
3. Multiply the first (new) term of the quotient by the binomial and subtract the result from the dividend. Bring down the next term to obtain a new polynomial which becomes the new dividend.
4. Divide the first term of the new dividend polynomial by the first term of the binomial and continue this process until the variable in the first term of the remainder dividend is raised to a lower power than the variable in the first term of the divisor.

EXAMPLE 2 DIVIDING BY A BINOMIAL

Divide $x^2 - 15 + 2x$ by $x - 3$.

1. Arrange terms in descending order. $x - 3\overline{)x^2 + 2x - 15}$
2. Divide the first term of the polynomial by the first term of the binomial.

equals

$$\begin{array}{r} x \\ x - 3\overline{)x^2 + 2x - 15} \end{array}$$ $x^2 \div x = x$

divided by

3. Multiply the first term of the quotient by the binomial and subtract the results from the dividend. Bring down the next term to obtain a new dividend.

$$\begin{array}{r} x \\ x - 3\overline{)x^2 + 2x - 15} \\ \underline{x^2 - 3x} \\ 5x - 15 \end{array}$$

$x(x - 3) = x^2 - 3x$

Subtract $x^2 - 3x$ from $x^2 + 2x$ by changing the signs on $x^2 - 3x$ and adding; then bring down -15

PRACTICE EXERCISE 2

Divide $a^2 - 5a - 14$ by $a + 2$.

4. Divide the new dividend polynomial by the binomial using Steps 2 and 3.

$$\begin{array}{rrl}
 & x + 5 & \text{Divide } 5x - 15 \text{ by } x - 3 \\
x - 3 & \overline{)x^2 + 2x - 15} & \\
 & \underline{x^2 - 3x} & \\
 & 5x - 15 & \\
 & \underline{5x - 15} & \text{Multiply } x - 3 \text{ by } 5 \\
 & 0 & \text{No variable in the new dividend; the process terminates}
\end{array}$$

The quotient is $x + 5$. To check, we multiply the quotient $x + 5$ by the divisor $x - 3$ to obtain the dividend $x^2 + 2x - 15$.

Answer: $a - 7$

EXAMPLE 3 Dividing by a Binomial

Divide $21y + 18y^2 + 40 + 20y^3$ by $7 + 5y$.

First, arrange terms in descending order of y.

$$\begin{array}{rrl}
 & 4y^2 - 2y + 7 & \leftarrow 4y^2\text{: The quotient of } 20y^3 \text{ and } 5y;\ -2y\text{: The quotient of } -10y^2 \text{ and } 5y;\ 7\text{: The quotient of } 35y \text{ and } 5y \\
5y + 7 & \overline{)20y^3 + 18y^2 + 21y + 40} & \\
 & \underline{20y^3 + 28y^2} & \leftarrow \text{The quotient } 4y^2 \text{ times the divisor } 5y + 7 \\
 & -10y^2 + 21y & \text{Subtract } 20y^3 + 28y^2 \text{ from } 20y^3 + 18y^2 \text{ and bring down } 21y \\
 & \underline{-10y^2 - 14y} & \leftarrow \text{The quotient } -2y \text{ times } 5y + 7 \\
 & 35y + 40 & \text{Subtract } -10y^2 - 14y \text{ from } -10y^2 + 21y \text{ and bring down } 40 \\
 & \underline{35y + 49} & \leftarrow \text{The quotient 7 times } 5y + 7 \\
 & -9 & \text{Subtract; no variable in new dividend}
\end{array}$$

The answer can be expressed as either $4y^2 - 2y + 7$ with remainder -9, or $4y^2 - 2y + 7 - \frac{9}{5y+7}$. To check this, multiply the quotient $4y^2 - 2y + 7$ by the divisor $5y + 7$ and add the remainder -9. The result is the dividend $20y^3 + 18y^2 + 21y + 40$.

Practice Exercise 3

Divide $20x^2 - 6x + 16x^3 - 10$ by $3 + 4x$.

Answer: $4x^2 + 2x - 3 - \frac{1}{4x+3}$

When terms are missing, either leave space for them or write them with zero coefficients. This is illustrated in the next example.

EXAMPLE 4 Dividing by a Binomial

Divide $a^3 + 8$ by $a + 2$.

Either leave space for the missing terms or write them with zero coefficients.

$$\begin{array}{rr}
 & a^2 - 2a + 4 \\
a + 2 & \overline{)a^3 \qquad\qquad + 8} \\
 & \underline{a^3 + 2a^2} \\
 & -2a^2 \\
 & \underline{-2a^2 - 4a} \\
 & 4a + 8 \\
 & \underline{4a + 8} \\
 & 0
\end{array}
\qquad\qquad
\begin{array}{rr}
 & a^2 - 2a + 4 \\
a + 2 & \overline{)a^3 + 0a^2 + 0a + 8} \\
 & \underline{a^3 + 2a^2} \\
 & -2a^2 + 0a \\
 & \underline{-2a^2 - 4a} \\
 & 4a + 8 \\
 & \underline{4a + 8} \\
 & 0
\end{array}$$

You should recognize that $a + 2$ and $a^2 - 2a + 4$ are the factors of the sum of cubes $a^3 + 8$.

Practice Exercise 4

Divide $y^3 - 64$ by $y - 4$.

Answer: $y^2 + 4y + 16$

To divide polynomials with two variables, we arrange the terms of each in order of descending powers of one variable and divide as before.

EXAMPLE 5 DIVIDING POLYNOMIALS IN TWO VARIABLES

$(2x^2 + 5xy - 3y^2) \div (2x - y)$

Here the terms are arranged in descending order relative to the variable x.

$$\begin{array}{r}
x + 3y \\
2x - y\,\overline{)\,2x^2 + 5xy - 3y^2} \\
\underline{2x^2 - xy} \\
6xy - 3y^2 \\
\underline{6xy - 3y^2} \\
0
\end{array}$$

We could also arrange in descending order relative to y and then divide.

$$\begin{array}{r}
3y + x \\
-y + 2x\,\overline{)\,-3y^2 + 5xy + 2x^2} \\
\underline{-3y^2 + 6xy} \\
-xy + 2x^2 \\
\underline{-xy + 2x^2} \\
0
\end{array}$$

In either case, the quotient is $x + 3y$, and $2x^2 + 5xy - 3y^2 = (2x - y)(x + 3y)$.

PRACTICE EXERCISE 5

Divide $6a^2 + 5ab + b^2$ by $3a + b$.

Answer: $2a + b$

3 DIVIDING BY A POLYNOMIAL

The procedure for dividing a polynomial by a binomial extends to division by other polynomials. This is illustrated in the next example.

EXAMPLE 6 DIVIDING BY A POLYNOMIAL

Divide $z^4 - 3z^3 + 2z - 5$ by $z^2 - z + 1$.

$$\begin{array}{r}
z^2 - 2z - 3 \\
z^2 - z + 1\,\overline{)\,z^4 - 3z^3 \qquad + 2z - 5} \\
\underline{z^4 - z^3 + z^2} \\
-2z^3 - z^2 + 2z \\
\underline{-2z^3 + 2z^2 - 2z} \\
-3z^2 + 4z - 5 \\
\underline{-3z^2 + 3z - 3} \\
z - 2
\end{array}$$

Thus, the quotient is $z^2 - 2z - 3$ with remainder of $z - 2$. Alternatively, we could give the answer as

$$z^2 - 2z - 3 + \frac{z - 2}{z^2 - z + 1}.$$

To check this, multiply and add $z - 2$.

PRACTICE EXERCISE 6

Divide $x^4 + 3x^3 + x^2 - 2x - 5$ by $x^2 + x - 2$.

Answer: $x^2 + 2x + 1 + \frac{x - 3}{x^2 + x - 2}$

6.8 EXERCISES A

Divide.

1. $\dfrac{28a^5 + 44a^4 - 8a^3}{-4a}$

2. $\dfrac{5x^5 - 30x^4 + 25x^3 - 5}{5x^2}$

3. $(-27a^5 + 9a^3 - 81a^2) \div (-3a)$

4. $\dfrac{-72y^3 + 24y^2 + 88y}{-8y}$

5. $\dfrac{15a^4b^3 - 3a^2b^2 + 9a^3b}{-3a^2b^2}$

6. $\dfrac{5x^2y^3z - 10x^3y^2z^5 + 20xyz^4}{-5xy^2z^3}$

7. $a - 2\overline{)3a^2 - 8a + 4}$

8. $(4a - 12 + a^2) \div (a + 6)$

9. $(8y^3 - 1) \div (2y - 1)$

10. $2x - 3\overline{)4x^4 \qquad - 7x^2 + x + 3}$

11. $\dfrac{4y^4 - 7y^2 + y + 1}{2y - 1}$

12. $x + 2\overline{)x^3 + 2x^2 - x - 2}$

13. $\dfrac{9a^3 + 3a^2 + 7a + 1}{3a + 2}$

14. $(5xy + x^2 + 6y^2) \div (x + 2y)$

15. $(a^3 + b^3) \div (a + b)$

16. $\dfrac{x^4 + 3x^3 + 2x^2 - x + 5}{x^2 - x + 1}$

17. $\dfrac{z^4 + z^3 + 4}{z^3 - z + 1}$

18. $\dfrac{2y^4 + 4y^3 - y + 2}{2y^3 + 1}$

19. $3y^2 - 1\overline{)9y^4 + 12y^2 - 5}$

20. $2x^2 + 3\overline{)6x^4 + 2x^3 + 5x^2 - x + 1}$

21. Find a value for p so that the polynomial $y^3 - y^2 + 2y + p$ is divisible by $y - 1$ with remainder zero.

22. Find a value for m so that when the polynomial $2x^3 + x^2 - 4x + m$ is divided by $2x + 1$, the remainder is 5.

23. The area of a rectangle is $6m^2 + 13m + 6$. Find the width if the length is $3m + 2$.

24. If a boat travels a distance of $a^3 + 2a^2 - 40a + 25$ mi in $a - 5$ hr, find its rate of speed.

FOR REVIEW

Simplify.

25. $\dfrac{y - \dfrac{9}{y}}{1 + \dfrac{3}{y}}$

26. $\dfrac{\dfrac{3}{a} + \dfrac{7}{b}}{\dfrac{9b}{a} - \dfrac{49a}{b}}$

27. Evaluate $\dfrac{1 - a}{a - 1}$ for $a = \dfrac{x + y}{x - y}$.

28. Simplify $\dfrac{(u^{-1} + 2v^{-1})^{-1}}{2uv}$.

ANSWERS: 1. $-7a^4 - 11a^3 + 2a^2$ 2. $x^3 - 6x^2 + 5x - \frac{1}{x^2}$ 3. $9a^4 - 3a^2 + 27a$ 4. $9y^2 - 3y - 11$ 5. $-5a^2b + 1 - \frac{3a}{b}$ 6. $\frac{-xy}{z^2} + 2x^2z^2 - \frac{4z}{y}$ 7. $3a - 2$ 8. $a - 2$ 9. $4y^2 + 2y + 1$ 10. $2x^3 + 3x^2 + x + 2 + \frac{9}{2x - 3}$ 11. $2y^3 + y^2 - 3y - 1$ 12. $x^2 - 1$ 13. $3a^2 - a + 3 - \frac{5}{3a + 2}$ 14. $x + 3y$ 15. $a^2 - ab + b^2$ 16. $x^2 + 4x + 5$ 17. $z + 1 + \frac{z^2 + 3}{z^3 - z + 1}$ 18. $y + 2 - \frac{2y}{2y^3 + 1}$ 19. $3y^2 + 5$ 20. $3x^2 + x - 2 + \frac{-4x + 7}{2x^2 + 3}$ 21. $p = -2$ 22. $m = 3$ 23. $2m + 3$ 24. $a^2 + 7a - 5$ 25. $y - 3$ 26. $\frac{1}{3b - 7a}$ 27. -1 28. $\frac{1}{2(v + 2u)}$

6.8 EXERCISES B

Divide.

1. $\dfrac{45y^3 - 15y^2 + 20y}{-5y}$

2. $\dfrac{30a^6 - 36a^5 + 24a^4 - 6}{6a^3}$

3. $\dfrac{13z^5 + 39z^4 - 169}{-13z^2}$

4. $(-16x^3 + 28x^2 - 8x) \div (-4x)$

5. $\dfrac{35x^5y^3 - 49x^4y^5 + 28x^2y}{-7x^3y^2}$

6. $\dfrac{49a^3b^2c - 21a^3bc^3 - 28a^2b}{-7a^2bc^2}$

7. $(x^3 - 27) \div (x - 3)$

8. $x - 1\overline{)x^3 \qquad + x - 3}$

9. $c + 3\overline{)4c^2 + 7c - 15}$

10. $2x - 5\overline{)12x^3 + 4x^2 - 41x + 140}$

11. $\dfrac{6c^4 - 18c^2 - 8c - 8}{2c - 4}$

12. $3a + 2\overline{)6a^3 + 13a^2 + 24a + 12}$

13. $\dfrac{16x^3 + 8x^2 + x - 3}{4x + 3}$

14. $3a + b\overline{)3a^3 + a^2b - 6a - 2b}$

15. $(x^3 - y^3) \div (x - y)$

16. $\dfrac{y^4 + 5y^3 + y^2 - 13y + 6}{y^2 + 2y - 3}$

17. $\dfrac{a^4 + 3a^2 + 7}{a^3 + a^2 - 2}$

18. $\dfrac{6x^4 + 3x^3 - 1}{3x^3 - 1}$

19. $4z^2 + 1\overline{)12z^4 + 7z^2 + 1}$

20. $3y^2 - 2\overline{)9y^4 + 6y^3 + 12y^2 + 2y + 7}$

21. Find a value for m so that the polynomial $x^3 + x^2 + x + m$ is divisible by $x + 1$ with remainder zero.

22. Find a value for p so that when the polynomial $3x^3 + x^2 + 4x + p$ is divided by $3x - 2$, the remainder is -3.

23. The area of a rectangle is $8u^2 - 10u - 3$. Find the length if the width is $2u - 3$.

24. Linda walks a distance of $2x^3 + 3x^2 - 5x - 3$ km in $2x + 1$ hr. Find her walking rate.

FOR REVIEW

Simplify.

25. $\dfrac{x + \dfrac{5}{x}}{\dfrac{6}{x} - 1}$

26. $\dfrac{\dfrac{5a}{b} - \dfrac{2b}{a}}{\dfrac{8}{a} + \dfrac{11}{b}}$

27. Evaluate $\dfrac{u^2}{u + 1}$ for $u = \dfrac{x}{y}$.

28. Simplify $\dfrac{x^{-1} + y^{-1}}{x^{-1} - y^{-1}}$.

6.8 EXERCISES C

Divide.

1. $a^2 + ab + b^2\overline{)a^5 + 2a^4b + 2a^3b^2 + 2a^2b^3 + ab^4 + b^5}$

2. Determine the value of a so that $x^4 + 2x^2 + a$ is divisible by $x - 1$ with remainder of -1. [Answer: $a = -4$]

3. Evaluate $x^3 - 2x^2 + x - 3$ when $x = 2$. Divide $x^3 - 2x^2 + x - 3$ by $x - 2$ and compare the value of the remainder with the value obtained first.

4. Evaluate $x^4 - 2x^2 + 7$ when $x = -1$. Divide $x^4 - 2x^2 + 7$ by $x + 1$ and compare the value of the remainder with the value obtained first. [Answer: Both are 6.]

6.9 SYNTHETIC DIVISION

Ordinary long division of a polynomial by a binomial of the form $x - a$, presented in the previous section, can be shortened using **synthetic division.** Suppose that we divide $2x^3 - 3x^2 + 3x - 4$ by $x - 2$.

$$
\begin{array}{r|l}
 & 2x^2 + x + 5 \\
x - 2 & 2x^3 - 3x^2 + 3x - 4 \\
 & \underline{2x^3 - 4x^2} \\
 & \quad x^2 + 3x \\
 & \quad \underline{x^2 - 2x} \\
 & \qquad 5x - 4 \\
 & \qquad \underline{5x - 10} \\
 & \qquad\qquad 6
\end{array}
$$

The quotient is $2x^2 + x + 5$

The remainder is 6

We could shorten our work by writing only the coefficients since the position of each coefficient tells us what power of x is associated with it.

$$
\begin{array}{r|l}
 & 2 + 1 + 5 \\
1 - 2 & 2 - 3 + 3 - 4 \\
 & \underline{2 - 4} \\
 & \quad 1 + 3 \\
 & \quad \underline{1 - 2} \\
 & \qquad 5 - 4 \\
 & \qquad \underline{5 - 10} \\
 & \qquad\qquad 6
\end{array}
$$

$2 + 1 + 5$ corresponds to $2x^2 + x + 5$

The first term in each line above (in color) is unnecessary since it is exactly the same as the number directly above it. Hence, we abbreviate further by eliminating these numbers.

$$
\begin{array}{r|l}
 & 2 + 1 + 5 \\
1 - 2 & 2 - 3 + 3 - 4 \\
 & \underline{\quad - 4} \\
 & \quad 1 + 3 \\
 & \quad \underline{\quad - 2} \\
 & \qquad 5 - 4 \\
 & \qquad \underline{\quad - 10} \\
 & \qquad\qquad 6
\end{array}
$$

The 1 in the divisor $1 - 2$ is unnecessary since we are no longer writing the first number in other lines which it was used to obtain. Also, the process of "bringing down" the next term merely wastes space since all subtractions could be performed in one line, as shown in color below.

$$
\begin{array}{r|l}
 & 2 + 1 + 5 \\
-2 & 2 - 3 + 3 - 4 \\
 & \underline{\quad - 4 - 2 - 10} \\
 & \quad 1 + 5 + 6
\end{array}
$$

The initial 2 in the quotient $2 + 1 + 5$ is always the same as the initial term in the dividend $2 - 3 + 3 - 4$, and the remaining terms in the quotient $(1 + 5)$ are duplicated in front of the remainder (which is 6) in the bottom line $1 + 5 + 6$. Thus, we can abbreviate further by writing the quotient only in the bottom line by first "bringing down" the 2.

$$
\begin{array}{r|l}
-2 & 2 - 3 + 3 - 4 \\
 & \underline{\quad - 4 - 2 - 10} \\
 & 2 + 1 + 5 + 6
\end{array}
$$

Finally, since the division lines are unnecessary, we omit them. Also, since most of us are able to add more quickly than we subtract, by changing the sign of the divisor (from -2 to $+2$) we can obtain the coefficients in the quotient by addition instead of subtraction.

$$\begin{array}{r|rrrr} +2 & 2 & -\,3 & +\,3 & -\,4 \\ & & +\,4 & +\,2 & +\,10 \\ \hline & 2 & +\,1 & +\,5 & +\,6 \end{array}$$

The first 2 is brought down below the line and multiplied by the divisor $+2$, and the product, 4, is placed above the line. Add -3 to 4, obtaining 1, multiply 1 by the divisor $+2$, and place the product, 2, above the line. Add 3 to 2, obtaining 5, multiply 5 by the divisor $+2$, and place the product, 10, above the line. The final addition of -4 and 10 gives 6, the last entry below the line, which is the remainder, while $2 + 1 + 5$ corresponds to the quotient $2x^2 + x + 5$.

We now summarize the steps to follow for synthetic division.

Synthetic Division

To divide a polynomial by $x - a$:

Step 1: Write the coefficients of the polynomial in descending order with a 0 for any missing term.

Step 2: Write $\underline{a|}$ to the left of the coefficients.

Step 3: Bring down the first coefficient, multiply it by a, place the product under the second coefficient, and add it to the second coefficient.

Step 4: Multiply the sum found in Step 3 by a, place the product under the next coefficient, and add.

Step 5: Continue this process until finished.

Step 6: The quotient has coefficients found in the last row with the first term of degree one less than the degree of the polynomial. The last number in the row is the remainder.

Suppose we illustrate these steps by dividing $x^3 + 5x^2 + 1$ by $x + 3$.

$$\begin{array}{rrrr} 1 & 5 & 0 & 1 \end{array}$$

Step 1: Since the x^2-term is missing, write 0.

$$\begin{array}{r|rrrr} -3 & 1 & 5 & 0 & 1 \end{array}$$

Step 2: Since $x - a = x + 3 = x - (-3)$, $a = -3$.

$$\begin{array}{r|rrrr} -3 & 1 & 5 & 0 & 1 \\ & & -3 & & \\ \hline & 1 & 2 & & \end{array}$$

Step 3: Bring down 1, multiply 1 by -3, write the product -3 under 5 and add to obtain 2.

$$\begin{array}{r|rrrr} -3 & 1 & 5 & 0 & 1 \\ & & -3 & -6 & \\ \hline & 1 & 2 & -6 & \end{array}$$

Step 4: Multiply 2 by -3, write the product -6 under 0 and add to obtain -6.

$$\begin{array}{r|rrrr} -3 & 1 & 5 & 0 & 1 \\ & & -3 & -6 & 18 \\ \hline & 1 & 2 & -6 & 19 \end{array}$$

Step 5: Continue the process. The final number 19 is the remainder.

$x^2 + 2x - 6$

Step 6: Write the quotient and remainder.

The quotient is $x^2 + 2x - 6$ with a remainder of 19.

Of course all of these steps are performed in one writing so your work will appear as in Step 5.

CAUTION

Remember that in synthetic division the dividing number, a, comes from $x - a$, and a may be positive or negative. For example, to divide by $x - 3$, $a = 3$ and is positive; and to divide by $x + 5 = x - (-5)$, $a = -5$ and is negative. Simply change the sign of the number in the dividing binomial and place it in the position next to the coefficients of the polynomial.

EXAMPLE 1 Using Synthetic Division

Use synthetic division to find $(x^4 - 2x^2 + x + 3) \div (x - 1)$.

Be sure to supply the zero in the position of the missing x^3 term. The polynomial is $x^4 + 0x^3 - 2x^2 + x + 3$, and $a = 1$.

$$\begin{array}{r|l} 1 & 1 + 0 - 2 + 1 + 3 \\ & \quad\;\; 1 + 1 - 1 + 0 \\ \hline & 1 + 1 - 1 + 0 + 3 \end{array}$$

The quotient is $x^3 + x^2 - x$ and the remainder is 3.

Practice Exercise 1

Use synthetic division to find $(y^4 + y^3 - 10y - 4) \div (y - 2)$.

Answer: $y^3 + 3y^2 + 6y + 2$

CAUTION

When dividing a polynomial by a binomial using synthetic division, arrange the polynomial in descending powers of the variable and insert a zero for each of the terms with zero coefficients.

EXAMPLE 2 Using Synthetic Division

Use synthetic division to find $(3x^5 - 2x^3 + 7 - x + x^2) \div (x + 2)$.

The divisor is $x + 2$ so we have $x + 2 = x - (-2)$. Be sure to supply the coefficient zero and rearrange the terms in descending order. The polynomial is $3x^5 + 0x^4 - 2x^3 + x^2 - x + 7$.

$$\begin{array}{r|l} -2 & 3 + 0 - 2 + 1 - 1 + 7 \\ & \quad\; - 6 + 12 - 20 + 38 - 74 \\ \hline & 3 - 6 + 10 - 19 + 37 - 67 \end{array}$$

The quotient is $3x^4 - 6x^3 + 10x^2 - 19x + 37$, and the remainder is -67.

With practice, the process of synthetic division becomes quite mechanical, and can be completed in a very short time.

Practice Exercise 2

Use synthetic division to find $(5a^4 + 2a^2 + 4a^5 + 4a + 5) \div (a + 1)$.

Answer: The quotient is $4a^4 + a^3 - a^2 + 3a + 1$ with remainder 4.

Remember that synthetic division works only when the divisor binomial is of the form $x - a$. For example, we could not use synthetic division to find $(3x^4 - x^3 + x^2 + 4x - 7) \div (x^2 - 1)$ since the divisor has an x^2-term.

6.9 EXERCISES A

Use synthetic division to find the quotient and remainder in each of the following.

1. $(x^3 - x^2 + 2x - 3) \div (x - 1)$

2. $(2x^3 - 3x + 5) \div (x - 2)$

3. $(x^3 - x^2 - 5x + 2) \div (x + 2)$

4. $(x^3 + 5x^2 + 4x - 6) \div (x + 3)$

5. $(x^4 - 2x^3 + x^2 - x + 5) \div (x + 2)$

6. $(x^4 + 2x^3 - 4x^2 + 2x - 2) \div (x - 1)$

7 $(x^3 - x + x^4 - 3) \div (x + 1)$

8. $(2x - 2x^3 - x^2 + x^4 - 5) \div (x - 2)$

9. $(2x^5 - 3x^4 - 6x^3 + x + 8) \div (x - 3)$

10. $(x^5 - 2x^4 - 10x^3 + 9x^2 - 5x + 4) \div (x - 4)$

11. $(6x^3 - x + 4) \div \left(x - \frac{1}{2}\right)$

12. $(8x^3 + 2x + 3) \div \left(x + \frac{1}{2}\right)$

13 $(3x^5 + x^2 - 5) \div (x + 1)$

14. $(2x^4 - 3x^2 + x + 7) \div (x^2 - 1)$

FOR REVIEW

Divide.

15. $\dfrac{42ab^2c - 60a^2bc + 6ab}{-6a^2b^2}$

16. $2x + 3\overline{)4x^3 + 6x^2 + 10x + 8}$

17. $(x^4 - 1) \div (x - 1)$

18. $\dfrac{x^4 + 3x^2 + 6}{x^2 - x + 3}$

Exercises 19–24 review material covered in Section 1.2 to help you prepare for the next section. Find each absolute value.

19. $|0|$

20. $|7|$

21. $|-7|$

22. $|x|$ if $x > 0$

23. $|x|$ if $x < 0$

24. $|x|$ if $x = 0$

ANSWERS: 1. $x^2 + 2; -1$ 2. $2x^2 + 4x + 5; 15$ 3. $x^2 - 3x + 1; 0$ 4. $x^2 + 2x - 2; 0$ 5. $x^3 - 4x^2 + 9x - 19; 43$ 6. $x^3 + 3x^2 - x + 1; -1$ 7. $x^3 - 1; -2$ 8. $x^3 - x; -5$ 9. $2x^4 + 3x^3 + 3x^2 + 9x + 28; 92$ 10. $x^4 + 2x^3 - 2x^2 + x - 1; 0$ 11. $6x^2 + 3x + \frac{1}{2}; \frac{17}{4}$ 12. $8x^2 - 4x + 4; 1$ 13. $3x^4 - 3x^3 + 3x^2 - 2x + 2; -7$ 14. cannot be done using synthetic division 15. $-\frac{7c}{a} + \frac{10c}{b} - \frac{1}{ab}$ 16. $2x^2 + 5 - \frac{7}{2x+3}$ 17. $x^3 + x^2 + x + 1$ 18. $x^2 + x + 1 + \frac{-2x+3}{x^2 - x + 3}$ 19. 0 20. 7 21. 7 22. x 23. $-x$ 24. 0

6.9 EXERCISES B

Use synthetic division to find the quotient and remainder in each of the following.

1. $(x^3 - 2x^2 + x - 5) \div (x - 1)$

2. $(2x^3 + 5x - 8) \div (x - 2)$

3. $(x^3 + 2x^2 - 3x - 6) \div (x + 2)$

4. $(x^3 + 8x^2 + 10x - 15) \div (x + 3)$

5. $(x^4 - 3x^3 + 2x^2 + 2x - 3) \div (x + 2)$

6. $(x^4 + 4x^3 - 2x^2 - 5x + 7) \div (x - 1)$

7. $(x^2 - 8x + x^4 - 2) \div (x + 1)$

8. $(9 - 3x^2 + x^3 - x^4 + 2x) \div (x - 2)$

9. $(x^5 - 6x^4 + 4x^3 + x^2 - x + 3) \div (x - 3)$

10. $(2x^5 - 7x^4 - 15x^2 - x + 5) \div (x - 4)$

11. $(4x^3 - 2x^2 + 8x - 2) \div \left(x - \frac{1}{2}\right)$

12. $(8x^3 + 4x - 5) \div \left(x + \frac{1}{2}\right)$

13. $(6x^6 + x^5 - 2) \div (x + 1)$

14. $(3x^5 - 5x^4 + x^2 - 8) \div (x^2 + 1)$

FOR REVIEW

Divide.

15. $\dfrac{35x^2y^2z - 49x^2z^2 + 14yz^2}{7xy^2}$

16. $3x + 2\overline{)6x^3 - 5x^2 - 8}$

17. $(x^5 - 1) \div (x - 1)$

18. $\dfrac{x^4 + 5x^3 + x^2 + 1}{x^2 + x - 5}$

Exercises 19–24 review material from Section 1.2 to help prepare for the next section. Find each absolute value.

19. $|5 - 5|$

20. $|10|$

21. $|-10|$

22. $|y|$ if $y = 0$

23. $|y|$ if $y < 0$

24. $|y|$ if $y > 0$

6.9 EXERCISES C

Use synthetic division to find the quotient and remainder in each of the following.

1. $(5.6x^3 + 6.2x^2 - 7.2x + 1.35) \div (x - 0.5)$ [Answer: $5.6x^2 + 9x - 2.7; 0$]

2. $(1.05x^4 - 8.04672x^2 + 80.604x - 80.0236) \div (x + 5.08)$

3. Use synthetic division to find the remainder when $4x^4 - 3x^2 + 2x - 1$ is divided by $x + 2$. Evaluate the polynomial when $x = -2$ and compare this result with the remainder. [Answer: Both are 47.]

4. Use synthetic division to show that $x - 1$ is a factor of $x^4 + x^3 - x^2 - 2x + 1$.

CHAPTER 6 REVIEW

KEY WORDS

6.1 An **algebraic fraction** is a fraction that contains a variable in the numerator, the denominator, or both.

A **rational expression** is an algebraic fraction that is the ratio of two polynomials.

Two algebraic fractions are **equivalent** if one can be obtained from the other by multiplying or dividing both numerator and denominator by the same nonzero expression.

A fraction is **reduced to lowest terms** when 1 (or -1) is the only number or expression that divides both numerator and denominator.

A factor is **prime** when its only divisors are itself and 1.

6.2 When one expression is divided by a second, the second expression is called the **divisor.**

6.3 **Like fractions** have the same denominator. **Unlike fractions** have different denominators.

The **least common denominator (LCD)** of two fractions is the smallest expression that has both denominators as factors.

6.4 A **fractional equation** is an equation that contains one or more algebraic fractions or rational expressions.

6.5 The **ratio** of one number a to another number b is the quotient $\frac{a}{b}$.

An equation stating that two ratios are equal is called a **proportion.**

Two triangles are **similar** if the angles of one are equal respectively to the angles of the other.

6.7 A fractional expression that contains at least one other fraction within it is called a **complex fraction.**

KEY CONCEPTS

6.1 **1.** Fundamental principle of fractions: If the numerator and denominator of an algebraic fraction are multiplied or divided by the same nonzero expression or number, the resulting fraction is equivalent to the original.

2. When reducing fractions, divide out factors only, *never divide out terms.*

6.2 **1.** When multiplying or dividing fractions, factor numerators and denominators and divide out common factors. Do *not* find the LCD.

2. The reciprocal of $a + b$ is $\frac{1}{a+b}$ and *not* $\frac{1}{a} + \frac{1}{b}$.

6.3 **1.** When subtracting fractions, use parentheses to avoid sign errors.

2. Before finding the LCD of several fractions, reduce all fractions to lowest terms.

3. Do not divide out terms when adding or subtracting fractions. Only factors can be divided out.

4. Form the cross-product equation only when two fractions are equated.

6.4 **1.** To solve a fractional equation, multiply both sides by the LCD of all fractions. Check all answers in the original equation and exclude any that result in division by zero.

2. Do not confuse simplifying an algebraic expression with solving a fractional equation.

6.5 **1.** Means-extremes property: In a proportion, the product of the extremes is equal to the product of the means.

2. The means-extremes property is used to solve a proportion by forming the cross-product equation.

3. y varies directly as x: $y = cx$

4. y varies inversely as x: $y = \frac{c}{x}$

5. z varies jointly as x and y: $z = cxy$

6.7 To simplify a complex fraction, multiply all terms by the LCD of all fractions involved.

6.8 When dividing a polynomial by a binomial, remember to subtract at each intermediate step. *Do not add.*

6.9 When using synthetic division, remember to add at each step. *Do not subtract.*

REVIEW EXERCISES

Part I

6.1 **1.** Find the values for the variable that must be excluded.

$$\frac{x+2}{x(x-5)}$$

2. Are the given fractions equivalent?

$$\frac{2}{x-y}, \frac{-2}{y-x}$$

Reduce to lowest terms. (Assume values that make the denominator zero are excluded.)

3. $\dfrac{6x^3y^3z^2}{2xy^5z^2}$

4. $\dfrac{8u^3-1}{1-2u}$

6.2 *Multiply.*

5. $\dfrac{x^2+2xy+y^2}{x-2} \cdot \dfrac{x^2-4}{(x+y)^2}$

6. $\dfrac{a^2+2ab+b^2}{a^2-b^2} \cdot \dfrac{b-a}{4b+4a}$

Divide.

7. $\dfrac{x^2-x-6}{x^2-36} \div \dfrac{2x-6}{x^2-4x-12}$

8. $\dfrac{a^2-ab-2b^2}{a^2-3ab+2b^2} \div \dfrac{2a+b}{a-b}$

Perform the indicated operations.

9. $\dfrac{2u^2+uv-v^2}{u^2-uv} \cdot \dfrac{u^2-v^2}{4u^2-v^2} \div \dfrac{u^2+2uv+v^2}{v^2+2uv}$

6.3 *Perform the indicated operations.*

10. $\dfrac{3a}{a+2} - \dfrac{2a}{a+2}$

11. $\dfrac{4x+2}{x-6} + \dfrac{3x-5}{6-x}$

Find the LCD of each pair of fractions.

12. $\dfrac{6}{5x^2}, \dfrac{4}{3x}$

13. $\dfrac{2y}{y^2-5y+6}, \dfrac{y+3}{y^2-6y+9}$

Perform the indicated operations.

14. $\dfrac{b}{a^2+ab+b^2} + \dfrac{a^2-ab}{a^3-b^3}$

15. $\dfrac{7x}{x^2-2x+1} - \dfrac{3}{x-1}$

16. $\dfrac{a^2 - 3ab}{a^2 - b^2} - \dfrac{-2ab - b^2}{2a^2 - ab - b^2}$

17. $\dfrac{-8}{y^2 - y - 6} + \dfrac{y}{y^2 - 6y + 9} - \dfrac{1}{y + 2}$

6.4 *Solve.*

18. $1 - \dfrac{a + 2}{a - 1} = \dfrac{3}{a}$

19. $3y = \dfrac{16}{y - 1} - \dfrac{10}{y - 1}$

20. The reciprocal of 8 more than a number is three times the reciprocal of the number. Find the number.

21. Three times the reciprocal of a number is the same as the number increased by 2. Find the number.

Solve each equation for the indicated variable.

22. $\dfrac{1}{a + b} = c$ for a

23. $x^{-1} + \dfrac{1}{y} = \dfrac{1}{z}$ for y

6.5 *Solve.*

24. If 150 ft of wire weighs 45 lb, what will 270 ft of the same wire weigh?

25. If a car travels 700 mi on 40 gal of gas, how many miles can it travel on 56 gal of gas?

26. The volume V of a given mass of gas varies directly as the temperature T and inversely as the pressure P. If $V = 400$ in^3 when $T = 350°$ and $P = 25$ lb/in^2, find the pressure when $T = 420°$ and $V = 600$ in^3.

27. The intensity I of light varies inversely as the square of the distance d from the source. If the intensity of illumination on a screen 15 ft from a light is 3.5 foot-candles, find the intensity on a screen 5 ft from the light.

6.6 *Solve.*

28. A plane takes $2\frac{1}{2}$ times as long to fly between two cities against the wind as with the wind. If the cities are 500 mi apart and the speed of the plane in still air is 280 mph, find the speed of the wind.

29. The speed of a river is 2 mph. If a boat can travel 104 mi upstream in the same time it can travel 120 mi downstream, what is the speed of the boat in still water?

30. If pipe A can fill a tank in 5 hr and pipe B requires 2 hr to fill the same tank, how long would it take to fill the tank if the pipes are turned on together?

31. If Lizette can do a job in 12 hr less time than Bo, and together they can do the job in 8 hr, how long would it take each to do it alone?

6.7 *Simplify.*

32. $\dfrac{\dfrac{x}{y}-\dfrac{y}{x}}{\dfrac{1}{y}-\dfrac{1}{x}}$

33. $\dfrac{(a^{-1}+2b^{-1})^{-1}}{\dfrac{ab}{2}}$

6.8 *Divide.*

34. $\dfrac{75z^5-50z^4+100z^2}{-25z^3}$

35. $\dfrac{64a^3b^3c^3-8a^2bc^4+24a^2b^2}{8a^2b^2c^2}$

36. $x-4\overline{)5x^3-18x^2-14x+24}$

37. $2x+1\overline{)4x^3-4x^2-x+5}$

6.9 *Use synthetic division to find the quotient and remainder in each of the following.*

38. $(x^4-3x^3+2x^2-1)\div(x+1)$

39. $(x^5-5x^3-5x+2)\div(x-2)$

Part II

40. If $\frac{3}{4}$ in on a map represents 20 mi, how many miles will be represented by 12 in?

41. Are the given fractions equivalent?

$$\frac{x}{a}, \frac{x+7}{a+7}$$

42. Simplify.

$$\frac{x-1-\dfrac{6}{x}}{1+\dfrac{2}{x}}$$

43. Find the LCD of the pair of fractions.

$$\frac{x+3}{x^3-8}, \frac{3x}{x^2+2x+4}$$

44. Suppose y varies jointly as x and the square of z and inversely as the square root of w. Find the equation of variation if $y = 10$ when $x = 3$, $z = 2$, and $w = 36$.

45. Find the values for the variable that must be excluded.

$$\frac{a+8}{a^3+a^2-2a}$$

46. Reduce to lowest terms.

$$\frac{b-w}{w-b}$$

47. Add.

$$\frac{6}{x+y}+\frac{3x+y}{x^2-y^2}$$

48. Perform the indicated operations and simplify.

$$\frac{x^3 - y^3}{x^2 - 3x + xy - 3y} \cdot \frac{x^2 + 2xy - 3x - 6y}{x^3 + x^2y + xy^2} \div \frac{x^2 + xy - 2y^2}{x^3 - xy^2}$$

49. It takes 5 hours to fill the Bonnett's swimming pool and 10 hours to drain it. If the pool is empty, the filler pipe is turned on but the drain is accidentally left open, how many hours will it take to fill the pool?

50. Larry is 6 years older than Amy and the ratio of their ages is $\frac{7}{6}$. How old is each?

51. Solve. $\dfrac{1}{a} + \dfrac{a}{a+1} = \dfrac{a^2}{a^2 + a}$

52. Solve the gravitational force formula

$$F = G\frac{m_1 m_2}{d^2} \quad \text{for } G.$$

53. Perform the indicated operations and simplify.

$$\frac{x}{x^2 + x - 6} + \frac{2}{2 + x - x^2} + \frac{3}{x^2 + 4x + 3}$$

54. Divide. $x - 3\overline{)x^3 - 2x^2 + 5x - 8}$

55. The wind speed is 50 mph. An airplane can travel a distance of 1750 miles with the wind in the same time that it can travel 1250 miles against the wind. What is the speed of the airplane in calm air?

56. The stopping distance d of a car varies directly as the square of the speed v it is traveling. If a car traveling 50 mph needs 150 ft to stop, what is the stopping distance of a car traveling 100 mph?

57. Multiply. $\dfrac{x^2 + x - 2}{x^2 + 6x + 5} \cdot \dfrac{x^2 - 2x - 3}{x^2 - x - 6}$

58. Use synthetic division to find the quotient and remainder.

$$(x^4 - 3x^2 + 2x - 5) \div (x + 2)$$

59. Write $\dfrac{3x}{x-2}$ as an equivalent fraction with denominator $3x^2 - 12$.

60. Divide. $\dfrac{42x^3y^2 + 36x^4y - 72xy^5}{-6x^3y^4}$

ANSWERS: 1. 0, 5 2. yes 3. $\frac{3x^2}{y^2}$ 4. $-4u^2 - 2u - 1$ 5. $x + 2$ 6. $-\frac{1}{4}$ 7. $\frac{(x+2)^2}{2(x+6)}$ 8. $\frac{a+b}{2a+b}$ 9. $\frac{v}{u}$ 10. $\frac{a}{a+2}$ 11. $\frac{x+7}{x-6}$ 12. $15x^2$ 13. $(y-2)(y-3)^2$ 14. $\frac{a+b}{a^2+ab+b^2}$ 15. $\frac{4x+3}{(x-1)^2}$ 16. $\frac{a-b}{a+b}$ 17. $\frac{15}{(y-3)^2(y+2)}$ 18. $\frac{1}{2}$ 19. 2, -1 20. -12 21. 1 or -3 22. $a = \frac{1-bc}{c}$ 23. $y = \frac{xz}{x-z}$ 24. 81 lb 25. 980 mi 26. 20 lb/in^2 27. 31.5 foot-candles 28. 120 mph 29. 28 mph 30. $\frac{10}{7}$ hr 31. Lizette: 12 hr; Bo: 24 hr 32. $x + y$ 33. $\frac{2}{b+2a}$ 34. $-3z^2 + 2z - \frac{4}{z}$ 35. $8abc - \frac{c^2}{b} + \frac{3}{c^2}$ 36. $5x^2 + 2x - 6$ 37. $2x^2 - 3x + 1 + \frac{4}{2x+1}$ 38. $x^3 - 4x^2 + 6x - 6$; 5 39. $x^4 + 2x^3 - x^2 - 2x - 9$; -16 40. 320 mi 41. no 42. $x - 3$ 43. $x^3 - 8$ 44. $y = \frac{5xz^2}{\sqrt{w}}$ 45. 0, 1, -2 46. -1 47. $\frac{9x-5y}{(x-y)(x+y)}$ 48. $x - y$ 49. 10 hr 50. Larry is 42, Amy is 36. 51. no solution 52. $G = \frac{d^2F}{m_1m_2}$ 53. $\frac{x^2+2x-12}{(x-2)(x+3)(x+1)}$ 54. $x^2 + x + 8 + \frac{16}{x-3}$ 55. 300 mph 56. 600 ft 57. $\frac{x-1}{x+5}$ 58. $x^3 - 2x^2 + x$; -5 59. $\frac{9x(x+2)}{3x^2-12}$ 60. $-\frac{7}{y^2} - \frac{6x}{y^3} + \frac{12y}{x^2}$

CHAPTER 6 TEST

1. Find the values for the variable that must be excluded in $\frac{3x}{x^2 - 1}$. 1. ______

2. Reduce $\frac{x^3 - y^3}{x - y}$ to lowest terms. 2. ______

Perform the indicated operations.

3. $\frac{a^2 - a - 2}{a^2 + 3a} \cdot \frac{a^2 - 3a}{a^2 - 5a + 6}$ 3. ______

4. $\frac{u^2 - v^2}{u^3 + v^3} \div \frac{u - v}{u^2 - uv + v^2}$ 4. ______

5. $\frac{x^3 - y^3}{x^2 - xy + 5x - 5y} \cdot \frac{x^2 + 3x - 10}{x^2 + 7x} \div \frac{x^3 + x^2y + xy^2}{x^2 + 5x - 14}$ 5. ______

6. $\frac{3}{x + 1} + \frac{4 - 2x}{x^2 - 1}$ 6. ______

7. $\frac{3a}{a^2 - 2ab + b^2} - \frac{3}{a - b}$ 7. ______

8. $\frac{a}{a^2 + 5a + 6} + \frac{2}{4 - a^2} - \frac{1}{a + 3}$ 8. ______

Solve.

9. $a + \frac{1}{a + 2} = \frac{4}{a + 2}$ 9. ______

10. Jeff is 2 years older than Sue and the ratio of their ages is $\frac{6}{5}$. How old is each? 10. ______

11. Solve $\dfrac{2}{x + y} = w$ for y. 11. ____________

12. Suppose y varies jointly as x and the square root of z, and inversely as w. Find the equation of variation if $y = 5$ when $x = 3$, $z = 25$, and $w = 6$. 12. ____________

13. An older machine can do a job in 104 minutes; it takes 40 minutes when a newer machine works simultaneously. How long would it take the newer machine working alone to do the job? 13. ____________

14. The speed of a stream is 3 mph. If a canoe can travel 27 mi upstream in the same time it can travel 45 mi downstream, what is the speed of the canoe in still water? 14. ____________

15. Simplify. $\dfrac{x + 2 - \dfrac{15}{x}}{1 + \dfrac{5}{x}}$ 15. ____________

Divide.

16. $\dfrac{24a^5 - 16a^3 - 4a^2}{-4a^2}$ 16. ____________

17. $2a + 1\overline{)8a^3 + 2a^2 + 5a + 3}$ 17. ____________

18. Use synthetic division to find the quotient and remainder.
$(x^4 + 2x^3 - 2x^2 - 4x - 3) \div (x + 1)$ 18. ____________

CHAPTER 7

Radicals and Exponents

7.1 ROOTS AND RADICALS

STUDENT GUIDEPOSTS

1. Square Roots
2. Cube Roots
3. Higher Order Roots
4. Approximating Square Roots

1 SQUARE ROOTS

We begin this section by reviewing several concepts presented in Chapter 1. Remember that if a and x are real numbers such that

$$x = a \cdot a = a^2,$$

then a is a square root of x. If a is an integer, then x is a perfect square. Every *positive* real number has two square roots since if a is a square root of x, then

$$x = a^2 \quad \textbf{and} \quad x = (-a)^2.$$

Thus, both a and $-a$ are square roots of x. We called the positive square root the principal square root and indicated it by using a radical, $\sqrt{\ }$. For example, the principal square root of 9 is $\sqrt{9}$ or 3. To indicate the negative square root of 9 using a radical, we write $-\sqrt{9} = -3$. For both square roots of 9 we write $\pm\sqrt{9}$ or ± 3, where ± 3 means both 3 and -3.

Even though every positive real number has two square roots, 0 has only one, since $-0 = 0$. Negative real numbers have no real square roots. For example, $\sqrt{-25}$ is not a real number since no real number squared is -25. In general, since $a^2 \geq 0$ for each real number a, *no negative number can have a real square root*.

Notice that

$$\sqrt{7^2} = \sqrt{49} = 7 \quad \text{and} \quad \sqrt{(-7)^2} = \sqrt{49} = 7.$$

Thus, if $x = 7$, a positive number, then $\sqrt{x^2} = x$, but if $x = -7$, a negative number, then $\sqrt{x^2} = -x = -(-7) = 7$. Remember the definition of absolute value from Chapter 1.

$$|x| = \begin{cases} x \text{ if } x \geq 0 \\ -x \text{ if } x < 0 \end{cases}$$

Since $\sqrt{x^2} = x$ or $\sqrt{x^2} = -x$, with absolute value we can define square roots in a single expression.

Square Roots

Let x be any real number. Then

$$\sqrt{x^2} = |x|.$$

EXAMPLE 1 SIMPLIFYING SQUARE ROOTS

Simplify the square roots.

(a) $\sqrt{36} = \sqrt{6^2} = |6| = 6$ $\quad 6^2 = 36$

(b) $\sqrt{(-6)^2} = |-6| = 6$

(c) $\sqrt{-36}$ is not a real number.

(d) $\sqrt{0^2} = |0| = 0$ $\quad 0^2 = 0$

(e) $\sqrt{(2x)^2} = |2x| = |2||x| = 2|x|$
This follows by a property of absolute value: $|ab| = |a||b|$

(f) $\sqrt{(a+1)^2} = |a+1|$

(g) $\sqrt{y^2 - 4y + 4} = \sqrt{(y-2)^2} = |y-2|$

(h) $-\sqrt{u^2} = -|u|$

PRACTICE EXERCISE 1

Simplify the square roots.

(a) $\sqrt{100}$

(b) $\sqrt{(-10)^2}$

(c) $\sqrt{-100}$

(d) $-\sqrt{0^2}$

(e) $\sqrt{(3y)^2}$

(f) $\sqrt{(x-3)^2}$

(g) $\sqrt{a^2 + 10a + 25}$

(h) $-\sqrt{y^2}$

Answers: (a) 10 (b) 10 (c) not a real number (d) 0 (e) $3|y|$ (f) $|x-3|$ (g) $|a+5|$ (h) $-|y|$

2 CUBE ROOTS

We can also define a cube root of a number. If x and a are real numbers such that

$$x = a^3,$$

then a is a **cube root** of x. If a is an integer, we call x a **perfect cube.** Unlike square roots, there is only one cube root of a number, and negative numbers have cube roots. We denote the cube root by $\sqrt[3]{}$. Thus,

$$\sqrt[3]{8} = \sqrt[3]{2^3} = 2 \quad \text{and} \quad \sqrt[3]{-8} = \sqrt[3]{(-2)^3} = -2.$$

Cube Roots

Let x be any real number. Then

$$\sqrt[3]{x^3} = x.$$

Notice that when finding cube roots, absolute value is not needed since a real number has only one cube root. The cube root of a negative number is negative, while the cube root of a positive number is positive.

EXAMPLE 2 SIMPLIFYING CUBE ROOTS

Simplify the cube roots.

(a) $\sqrt[3]{27} = \sqrt[3]{3^3} = 3$ $\quad 3^3 = 27$

(b) $\sqrt[3]{-27} = \sqrt[3]{(-3)^3} = -3$ $\quad (-3)^3 = -27$

(c) $\sqrt[3]{125} = \sqrt[3]{5^3} = 5$ $\quad 5^3 = 125$

PRACTICE EXERCISE 2

Simplify the cube roots.

(a) $\sqrt[3]{8}$

(b) $\sqrt[3]{-8}$

(c) $\sqrt[3]{216}$

(d) $\sqrt[3]{-64} = \sqrt[3]{(-4)^3} = -4$ $\quad (-4)^3 = -64$

(e) $\sqrt[3]{1} = \sqrt[3]{1^3} = 1$ $\quad 1^3 = 1$

(f) $\sqrt[3]{0} = \sqrt[3]{0^3} = 0$ $\quad 0^3 = 0$

(g) $\sqrt[3]{(4a)^3} = 4a$

(d) $\sqrt[3]{-216}$
(e) $-\sqrt[3]{1}$
(f) $-\sqrt[3]{0}$
(g) $-\sqrt[3]{(4a)^3}$

Answers: (a) 2 (b) -2 (c) 6 (d) -6 (e) -1 (f) 0 (g) $-4a$

3 HIGHER ORDER ROOTS

We may also define fourth roots, fifth roots, and so on. In general, if x and a are real numbers such that

$$x = a^k, \qquad k \text{ is a natural number greater than 1}$$

then a is a ***k*th root** of x. If a is an integer, x is a **perfect *k*th power.** The symbol $\sqrt[k]{x}$ indicates the **principal *k*th root** of x. k is the **index** on the radical. The index must be specified on all radical expressions except square roots, where the index is understood to be 2.

Just as with square roots, when k is an even number, only nonnegative numbers can have kth roots. Likewise, when k is even, principal kth roots are defined using absolute value. For example,

$$\sqrt[4]{2^4} = |2| = 2 \quad \text{and} \quad \sqrt[4]{(-2)^4} = |-2| = 2.$$

However, we saw that negative numbers can have cube roots, and cube roots do not involve absolute value. The same is true for all kth roots where k is an odd number.

Higher Order Roots

Let x be any real number, with k a natural number.

1. **If k is even, $\sqrt[k]{x^k} = |x|$.**
2. **If k is odd, $\sqrt[k]{x^k} = x$.**

EXAMPLE 3 SIMPLIFYING HIGHER ORDER ROOTS

Simplify.

(a) $\sqrt[4]{81} = \sqrt[4]{3^4} = |3| = 3$ $\quad 3^4 = 81$

(b) $\sqrt[4]{-81}$ is not a real number. Notice that no real number can be multiplied by itself four times and have a negative product like -81.

(c) $\sqrt[4]{(-3)^4} = |-3| = 3$

(d) $\sqrt[5]{243} = \sqrt[5]{3^5} = 3$ $\quad 3^5 = 243$

(e) $\sqrt[5]{-243} = \sqrt[5]{(-3)^5} = -3$ $\quad (-3)^5 = -243$

(f) $\sqrt[6]{-64}$ is not a real number.

(g) $-\sqrt[6]{64} = -\sqrt[6]{2^6} = -|2| = -2$ $\quad 2^6 = 64$

(h) $\sqrt[5]{(2x)^5} = 2x$

PRACTICE EXERCISE 3

Simplify.

(a) $\sqrt[4]{625}$

(b) $\sqrt[4]{-625}$

(c) $\sqrt[4]{(-5)^4}$

(d) $\sqrt[5]{5^5}$

(e) $\sqrt[5]{(-5)^5}$

(f) $\sqrt[6]{-1}$

(g) $-\sqrt[6]{1}$

(h) $\sqrt[5]{(3u)^5}$

(i) $\sqrt[4]{(3y)^4} = |3y| = |3||y| = 3|y|$

(j) $\sqrt[6]{(a-3)^6} = |a-3|$

(k) $\sqrt[4]{0} = \sqrt[4]{0^4} = |0| = 0 \qquad 0^4 = 0$

(i) $\sqrt[4]{(5x)^4}$

(j) $\sqrt[5]{(a-3)^5}$

(k) $\sqrt[5]{0^5}$

Answers: (a) 5 (b) not a real number (c) 5 (d) 5 (e) -5 (f) not a real number (g) -1 (h) $3u$ (i) $5|x|$ (j) $a-3$ (k) 0

When expressions under radicals are nonnegative, taking even roots is much simpler since absolute value need not be used. As a result, *unless we specifically state otherwise,* we will make the following assumption from now on.

Taking Roots from This Point Forward

All variables and algebraic expressions under even-indexed radicals represent nonnegative numbers so that for any index k,

$$\sqrt[k]{x^k} = x.$$

EXAMPLE 4 Simplifying Roots

Simplify. (Variables and expressions under even-indexed radicals are nonnegative.)

(a) $\sqrt{(2x)^2} = 2x$
We do not need to write $2|x|$ as in Example 1(e) since here $|x| = x$.

(b) $\sqrt{(a+1)^2} = a+1$
Now $|a+1| = a+1$ so absolute value is not necessary.

(c) $\sqrt[4]{(3y)^4} = 3y$

(d) $\sqrt[6]{(a-3)^6} = a-3$

(e) $\sqrt{y^2-4y+4} = \sqrt{(y-2)^2} = y-2$

Practice Exercise 4

Simplify.

(a) $\sqrt{(3y)^2}$

(b) $\sqrt{(x-3)^2}$

(c) $\sqrt[4]{(5x)^4}$

(d) $\sqrt[8]{(u+5)^8}$

(e) $\sqrt{a^2+10a+25}$

Answers: (a) $3y$ (b) $x-3$ (c) $5x$ (d) $u+5$ (e) $a+5$

4 APPROXIMATING SQUARE ROOTS

Square roots of perfect squares such as $\sqrt{9} = \sqrt{3^2} = 3$ and $\sqrt{0.25} = \sqrt{(0.5)^2} = 0.5$ are rational numbers. In many applications we have to approximate square roots of numbers such as $\sqrt{2}$ and $\sqrt{0.3}$ that are not perfect squares. The best way to do this is with a calculator that has a square root key $\boxed{\sqrt{\ }}$. For example, to approximate $\sqrt{2}$, we use the following steps on a calculator*:

$$2\boxed{\sqrt{\ }} \longrightarrow \boxed{1.4142136}$$

The display shows 1.4142136 which is a rational number approximation of the irrational number $\sqrt{2}$. In some situations we might use 1.4 for $\sqrt{2}$, correct to the nearest tenth, or 1.41 correct to the nearest hundredth.

Scientific notation and significant digits, introduced in Section 1.6, are sometimes used for approximating roots.

*Calculator steps are given using algebraic logic. If you have a calculator that uses Reverse Polish Notation (RPN), consult your operator's manual.

EXAMPLE 5 Approximating Roots Using a Calculator

(a) Approximate $\frac{-2 + \sqrt{53}}{4}$, correct to the nearest hundredth.
The steps to use on a calculator are:

2 [±] [+] 53 [√] [=] [÷] 4 [=] ⟶ 1.3200275

The display shows 1.3200275, which rounded to the nearest hundredth is 1.32. Note that [±] is the change sign key which made 2 into −2 in the first step.

(b) Approximate $\sqrt{243{,}000{,}000}$, correct to three significant digits.
Since very large or very small numbers do not fit on a calculator display, such numbers must first be expressed in scientific notation. To evaluate $\sqrt{243{,}000{,}000}$ on a scientific calculator we use these steps:

2.43 [EE] 8 [√] ⟶ 1.5588 04

The display gives the answer in scientific notation, 1.5588×10^4. This can be written as 15,600, correct to three significant digits.

Practice Exercise 5

(a) Approximate $\frac{4 - \sqrt{61}}{3}$, correct to the nearest hundredth.

(b) Approximate $\sqrt{0.0000000195}$, correct to three significant digits.

Answers: (a) −1.27
(b) $1.3964 \times 10^{-4} \approx 0.000140$

Many applied problems involve radicals. Exact answers that are irrational numbers must often be approximated to give us a better idea of their value, as in the next example.

EXAMPLE 6 Physics Problem

The distance in miles that can be seen on the surface of the ocean can be approximated by

$$d = 1.4\sqrt{h},$$

where h is the height of the viewer in feet above the surface of the water. From a blimp at an altitude of 3000 ft, is it possible to see a ship at a distance of 90 mi?

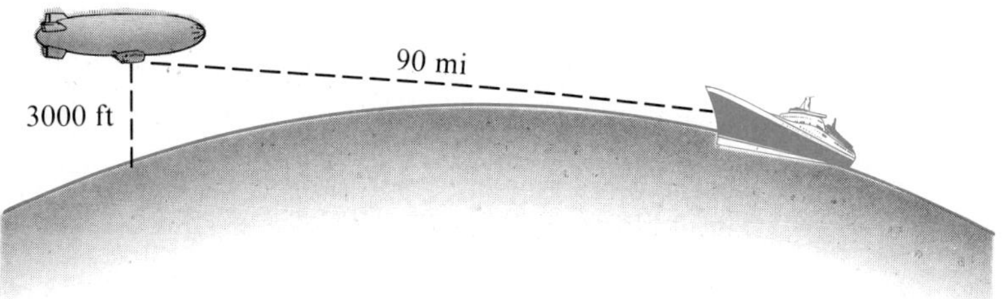

Figure 7.1

Substitute 3000 for h and evaluate d.

$$\begin{aligned} d &= 1.4\sqrt{h} \\ &= 1.4\sqrt{3000} \\ &\approx 76.7 \end{aligned}$$

Since the viewing distance is about 77 miles and since the ship is 90 miles away, the ship is below the horizon and cannot be seen from the blimp. The exact viewing distance is $1.4\sqrt{3000}$ miles, but to most of us, this does not have the same meaning as ''about 77 miles.''

Practice Exercise 6

What is the approximate viewing distance of an airplane flying at an altitude of 33,000 ft?

Answer: 254 mi

7.1 EXERCISES A

Simplify each radical, writing it as an integer or a rational number.

1. $\sqrt{36}$
2. $\sqrt{-16}$
3. $\sqrt[3]{27}$
4. $\sqrt[3]{-64}$
5. $-\sqrt[3]{8}$
6. $-\sqrt{4}$
7. $(\sqrt{25})^3$
8. $(\sqrt[3]{27})^2$
9. $\sqrt{\frac{1}{4}}$
10. $\sqrt[3]{\frac{1}{8}}$
11. $\sqrt[4]{\frac{1}{16}}$
12. $\sqrt[5]{-32}$
13. $\sqrt[6]{64}$
14. $\sqrt[7]{128}$
15. $\sqrt[7]{-128}$
16. $\sqrt[5]{(-6)^5}$

Simplify each radical in Exercises 17–32. For these exercises do not assume that expressions under even-indexed radicals necessarily represent nonnegative numbers. That is, use absolute value when necessary.

17. $\sqrt{x^2}$
18. $\sqrt[3]{x^3}$
19. $\sqrt{-x^2}$
20. $-\sqrt{x^2}$
21. $-\sqrt[3]{x^3}$
22. $\sqrt[3]{-x^3}$
23. $\sqrt[4]{y^4}$
24. $\sqrt[8]{y^8}$
25. $\sqrt[5]{w^5}$
26. $\sqrt[5]{(6xy)^5}$
27. $\sqrt[4]{(-2x)^4}$
28. $\sqrt[5]{(-2x)^5}$
29. $-\sqrt[4]{(-2x)^4}$
30. $-\sqrt[5]{(-2x)^5}$
31. $\sqrt{(x+7)^2}$
32. $\sqrt[3]{(x+7)^3}$

Simplify each radical in Exercises 33–52. For these exercises, return to the assumption that expressions under even-indexed radicals represent nonnegative numbers. That is, absolute value will not be necessary.

33. $\sqrt{u^2}$
34. $\sqrt[3]{u^3}$
35. $\sqrt{(2a)^2}$
36. $\sqrt[3]{(2a)^3}$
37. $\sqrt{49a^2}$
38. $\sqrt{(xy)^2}$
39. $\sqrt{(x-9)^2}$
40. $\sqrt[3]{(x-9)^3}$
41. $\sqrt[3]{(-u)^3}$
42. $\sqrt[4]{(7y)^4}$
43. $\sqrt[3]{-8x^3}$
44. $\sqrt[4]{-x^4}$
45. $\sqrt[7]{x^7}$
46. $\sqrt[5]{(-2x)^5}$
47. $\sqrt[8]{(6xy)^8}$
48. $\sqrt[6]{(-2)^6a^6}$
49. $\sqrt{x^2-14x+49}$
50. $\sqrt[3]{(-1)^3x^3}$
51. $\sqrt[4]{(-1)^4x^4}$
52. $\sqrt[5]{(-7)^5a^5w^5}$

Use a calculator to approximate each number correct to the nearest hundredth.

53. $\sqrt{291}$
54. $3+\sqrt{5}$
55. $\dfrac{-4+\sqrt{60}}{3}$
56. $\dfrac{3\pm\sqrt{122}}{5}$

Use a calculator and scientific notation to approximate each number. Give answers correct to three significant digits.

57. $\sqrt{693{,}000{,}000}$ **58.** $\sqrt{0.0000000355}$ **59** $\dfrac{\sqrt{1{,}250{,}000{,}000}}{0.00623}$

In a retail outlet the cost c is related to the number of items sold n by the equation $c = \sqrt[3]{n}$.

60. Find c when $n = 27$. **61.** Find c when $n = 5^3$. **62.** Find n when $c = 10$.

In a scientific study a quantity p is related to another variable w by the equation $p = \sqrt[4]{w}$.

63. Find p when $w = 81$. **64.** Find p when $w = -81$. **65.** Find w when $p = 5$.

The distance in miles that can be seen on the surface of the ocean can be approximated by $d = 1.4\sqrt{h}$, *where h is the height of the viewer in feet above the surface of the water. Use this formula in Exercises 66–67.*

66 From a balloon at an altitude of 1800 ft, is it possible to view a ship at a distance of 50 mi?

67. From a small airplane at an altitude of 4200 ft, is it possible to view an island at a distance of 100 mi?

The time T in seconds required for a pendulum of length L feet to make one swing is given by $T = 2\pi\sqrt{\frac{L}{32}}$. *In Exercises 68–69, find the time, to the nearest tenth of a second, that is needed for a pendulum of the given length to make one swing. Use 3.14 for* π.

68. 10 ft **69.** 32.5 ft

FOR REVIEW

The following exercises review material from Chapter 1 and will help you prepare for the next section. Evaluate each pair of radical expressions and compare the results.

70. $\sqrt{4 \cdot 9}$, $\sqrt{4}\,\sqrt{9}$ **71.** $\dfrac{\sqrt{144}}{\sqrt{36}}$, $\sqrt{\dfrac{144}{36}}$ **72.** $\dfrac{\sqrt{81}\,\sqrt{49}}{\sqrt{9}}$, $\sqrt{\dfrac{(81)(49)}{9}}$

ANSWERS: 1. 6 2. not a real number 3. 3 4. −4 5. −2 6. −2 7. 125 8. 9 9. $\frac{1}{2}$ 10. $\frac{1}{2}$ 11. $\frac{1}{2}$ 12. −2 13. 2 14. 2 15. −2 16. −6 17. $|x|$ 18. x 19. not a real number unless $x = 0$, then $\sqrt{-0^2} = \sqrt{0} = 0$ 20. $-|x|$ 21. $-x$ 22. $-x$ 23. $|y|$ 24. $|y|$ 25. w 26. $6xy$ 27. $2|x|$ 28. $-2x$ 29. $-2|x|$ 30. $2x$ 31. $|x+7|$ 32. $x+7$ 33. u 34. u 35. $2a$ 36. $2a$ 37. $7a$ 38. xy 39. $x-9$ 40. $x-9$ 41. $-u$ 42. $7y$ 43. $-2x$ 44. not a real number unless $x = 0$, then $\sqrt[4]{-0^4} = \sqrt[4]{0} = 0$ 45. x 46. $-2x$ 47. $6xy$ 48. $2a$ 49. $x-7$ 50. $-x$ 51. x 52. $-7aw$ 53. 17.06 54. 5.24 55. 1.25 56. 2.81 and −1.61 57. 26,300 58. 0.000188 59. 5,680,000 60. 3 61. 5 62. 1000 63. 3 64. not a real number 65. 625 66. yes, the viewing distance is about 59.4 mi 67. no, the viewing distance is about 90.7 mi 68. 3.5 sec 69. 6.3 sec 70. 6, 6; equal 71. 2, 2; equal 72. 21, 21; equal

7.1 EXERCISES B

Simplify each radical, writing it as an integer or a rational number.

1. $\sqrt{16}$ **2.** $\sqrt{-25}$ **3.** $\sqrt[3]{8}$ **4.** $\sqrt[3]{-27}$

5. $-\sqrt[3]{64}$ **6.** $-\sqrt{9}$ **7.** $(\sqrt{36})^3$ **8.** $(\sqrt[3]{8})^2$

9. $\sqrt{\frac{1}{9}}$ **10.** $\sqrt[3]{\frac{1}{27}}$ **11.** $\sqrt[4]{\frac{1}{81}}$ **12.** $\sqrt[5]{-243}$

13. $-\sqrt[6]{64}$ **14.** $\sqrt[7]{-128}$ **15.** $\sqrt[7]{(-3)^7}$ **16.** $\sqrt[6]{(-3)^6}$

Simplify each radical in Exercises 17–32. For these exercises do not assume that expressions under even-indexed radicals necessarily represent nonnegative numbers. That is, use absolute value when necessary.

17. $\sqrt{y^2}$ **18.** $\sqrt[3]{y^3}$ **19.** $\sqrt{-y^2}$ **20.** $-\sqrt{y^2}$

21. $-\sqrt[3]{a^3}$ **22.** $\sqrt[3]{-a^3}$ **23.** $\sqrt[6]{y^6}$ **24.** $\sqrt[8]{z^8}$

25. $\sqrt[7]{w^7}$ **26.** $\sqrt[3]{(2ab)^3}$ **27.** $\sqrt[6]{(-5w)^6}$ **28.** $\sqrt[3]{(-5w)^3}$

29. $-\sqrt{(-3x)^2}$ **30.** $-\sqrt[3]{(-3x)^3}$ **31.** $\sqrt{(y-4)^2}$ **32.** $\sqrt[5]{(y-4)^5}$

Simplify each radical in Exercises 33–52. For these exercises, return to the assumption that expressions under even-indexed radicals represent nonnegative numbers. That is, absolute value will not be necessary.

33. $\sqrt{a^2}$ **34.** $\sqrt[5]{u^5}$ **35.** $\sqrt{(5x)^2}$ **36.** $\sqrt[3]{(5x)^3}$

37. $\sqrt{81y^2}$ **38.** $\sqrt{(ab)^2}$ **39.** $\sqrt{(y+10)^2}$ **40.** $\sqrt[3]{(y+10)^3}$

41. $\sqrt[3]{(-x)^3}$ **42.** $\sqrt[6]{(8w)^6}$ **43.** $\sqrt[3]{-27y^3}$ **44.** $\sqrt[6]{-x^6}$

45. $\sqrt[9]{u^9}$ **46.** $\sqrt[3]{(-4x)^3}$ **47.** $\sqrt[8]{(2wy)^8}$ **48.** $\sqrt[4]{(-2)^4a^4}$

49. $\sqrt{x^2+12x+36}$ **50.** $\sqrt[5]{(-1)^5w^5}$ **51.** $\sqrt[6]{(-1)^6y^6}$ **52.** $\sqrt[3]{(-3)^3x^3y^3}$

Use a calculator to approximate each number correct to the nearest hundredth.

53. $\sqrt{483}$ **54.** $2-\sqrt{3}$ **55.** $\frac{-3+\sqrt{20}}{4}$ **56.** $\frac{2 \pm \sqrt{141}}{7}$

Use a calculator and scientific notation to approximate each number. Give answers correct to three significant digits.

57. $\sqrt{392{,}000{,}000}$ **58.** $\sqrt{0.00000000515}$ **59.** $\frac{\sqrt{2{,}340{,}000{,}000}}{0.000442}$

In a retail outlet the cost c is related to the number of items sold n by the equation $c = \sqrt[3]{n}$.

60. Find c when $h = 8$. **61.** Find c when $n = 7^3$. **62.** Find n when $c = 3$.

In a scientific study a quantity p is related to another variable w by the equation $p = \sqrt[4]{w}$.

63. Find p when $w = 16$. **64.** Find p when $w = -16$. **65.** Find w when $p = 3$.

The distance in miles that can be seen on the surface of the ocean can be approximated by $d = 1.4\sqrt{h}$, *where h is the height of the viewer in feet above the surface of the water. Use this formula in Exercises 66–67.*

66. From a blimp at an altitude of 2300 ft, is it possible to view a ship at a distance of 75 mi?

67. From a jet airplane at an altitude of 25,000 ft, is it possible to view the shoreline of a country at a distance of 200 mi?

The time T in seconds required for a pendulum of length L feet to make one swing is given by $T = 2\pi\sqrt{\frac{L}{32}}$. *In Exercises 68–69, find the time, to the nearest tenth of a second, that is required for a pendulum of the given length to make one swing. Use* 3.14 *for* π.

68. 24 ft

69. 7.5 ft

FOR REVIEW

The following exercises review material from Chapter 1 and will help you prepare for the next section. Evaluate each pair of radical expressions and compare the results.

70. $\sqrt{36 \cdot 16}$, $\sqrt{36}\,\sqrt{16}$

71. $\frac{\sqrt{25}}{\sqrt{100}}$, $\sqrt{\frac{25}{100}}$

72. $\frac{\sqrt{64}}{\sqrt{4}\,\sqrt{16}}$, $\sqrt{\frac{64}{(4)(16)}}$

7.1 EXERCISES C

For Exercises 1–10 assume that x is a negative number.

1. Is $-x$ positive or negative?

2. True or false: $|x| = -x$?

Simplify.

3. $\sqrt{x^2}$
[Answer: $-x$]

4. $\sqrt[3]{x^3}$

5. $\sqrt{x^2} - \sqrt[3]{x^3}$

6. $\sqrt{x^2} + \sqrt[3]{x^3}$
[Answer: 0]

7. $\sqrt[4]{x^4}$
[Answer: $-x$]

8. $\sqrt[5]{x^5}$

9. $\sqrt{(-x)^2}$

10. $\sqrt[3]{(-x)^3}$

7.2 SIMPLIFYING RADICALS

STUDENT GUIDEPOSTS

1. Simplifying Products
2. Simplifying Quotients
3. Rationalizing a Denominator
4. Rationalizing a Numerator
5. Simplest Form

1 SIMPLIFYING PRODUCTS

In Section 7.1 we used the rule $\sqrt[k]{x^k} = x$ to simplify radicals such as

$$\sqrt{25} = 5, \quad \sqrt[3]{27x^3} = 3x, \quad \text{and} \quad \sqrt[6]{y^6} = y.$$

To simplify radicals involving radicands that are not perfect powers we can use the following rule.

Simplifying Rule for Products

If a and b are any real numbers or expressions, where a and b are nonnegative if k is even, then

$$\sqrt[k]{ab} = \sqrt[k]{a}\,\sqrt[k]{b}.$$

The kth root of a product is equal to the product of the kth roots.

This rule can be used to simplify radicals when one or more of the factors of the radicand is a perfect kth power. For example,

$$\begin{aligned} \sqrt{50} &= \sqrt{25 \cdot 2} && \text{25 is a perfect square} \\ &= \sqrt{25}\,\sqrt{2} && \sqrt{ab} = \sqrt{a}\,\sqrt{b} \\ &= 5\sqrt{2}. && \text{Simplified form} \end{aligned}$$

In the case of square roots, we look for perfect squares, isolate each one using $\sqrt{ab} = \sqrt{a}\,\sqrt{b}$, and simplify using $\sqrt{x^2} = x$. If you have trouble recognizing perfect square factors, factor the radicand into primes. In the above example, we could have written $50 = 2 \cdot 5 \cdot 5 = 2 \cdot 5^2 = 2 \cdot 25$.

EXAMPLE 1 Simplifying Square Roots

Simplify the square roots.

(a) $\sqrt{75} = \sqrt{25 \cdot 3}$ 25 is a perfect square
$= \sqrt{25}\,\sqrt{3}$ $\sqrt{ab} = \sqrt{a}\,\sqrt{b}$
$= 5\sqrt{3}$ $\sqrt{25} = \sqrt{5^2} = 5$

(b) $\sqrt{7^3} = \sqrt{7^2 \cdot 7}$ $7^3 = 7^{2+1} = 7^2 \cdot 7$
$= \sqrt{7^2}\,\sqrt{7}$ $\sqrt{ab} = \sqrt{a}\,\sqrt{b}$
$= 7\sqrt{7}$

(c) $\sqrt{35} = \sqrt{5 \cdot 7}$ Cannot be simplified since neither 5 nor 7 is a perfect square
$= \sqrt{35}$

(d) $5\sqrt{x^6} = 5\sqrt{(x^3)^2}$ $x^6 = x^{3 \cdot 2} = (x^3)^2$
$= 5x^3$

(e) $5\sqrt{x^7} = 5\sqrt{x^6 \cdot x}$ $x^7 = x^{6+1} = x^6 \cdot x$
$= 5\sqrt{x^6}\,\sqrt{x}$ $\sqrt{ab} = \sqrt{a}\,\sqrt{b}$
$= 5\sqrt{(x^3)^2}\,\sqrt{x}$
$= 5x^3\sqrt{x}$

(f) $\sqrt{45a^3} = \sqrt{9 \cdot 5 \cdot a^2 \cdot a}$ 9 and a^2 are perfect squares
$= \sqrt{9 \cdot a^2 \cdot 5a}$ $5a$ will be left under the radical
$= \sqrt{9}\,\sqrt{a^2}\,\sqrt{5a}$ Rule extended to $\sqrt{abc} = \sqrt{a}\,\sqrt{b}\,\sqrt{c}$
$= 3a\sqrt{5a}$

PRACTICE EXERCISE 1

Simplify the square roots.

(a) $\sqrt{88}$

(b) $\sqrt{11^3}$

(c) $\sqrt{42}$

(d) $4\sqrt{y^8}$

(e) $4\sqrt{y^9}$

(f) $\sqrt{108u^3}$

Answers: (a) $2\sqrt{22}$ (b) $11\sqrt{11}$ (c) cannot be simplified (d) $4y^4$ (e) $4y^4\sqrt{y}$ (f) $6u\sqrt{3u}$

To simplify cube roots we look for perfect cubes under the radical, use $\sqrt[3]{ab} = \sqrt[3]{a}\,\sqrt[3]{b}$, and then $\sqrt[3]{x^3} = x$. As with square roots, prime factorization may be used to find perfect cubes.

EXAMPLE 2 Simplifying Cube Roots

Simplify the cube roots.

(a) $\sqrt[3]{54} = \sqrt[3]{27 \cdot 2}$ 27 is a perfect cube
$= \sqrt[3]{27} \cdot \sqrt[3]{2}$ $\sqrt[3]{ab} = \sqrt[3]{a}\,\sqrt[3]{b}$
$= 3\sqrt[3]{2}$ $\sqrt[3]{27} = \sqrt[3]{3^3} = 3$

(b) $7\sqrt[3]{x^5} = 7\sqrt[3]{x^3 \cdot x^2}$ $x^5 = x^{3+2} = x^3 \cdot x^2$
$= 7\sqrt[3]{x^3}\,\sqrt[3]{x^2}$ $\sqrt[3]{ab} = \sqrt[3]{a}\,\sqrt[3]{b}$
$= 7x\sqrt[3]{x^2}$ Simplest form

PRACTICE EXERCISE 2

Simplify the cube roots.

(a) $\sqrt[3]{88}$

(b) $6\sqrt[3]{y^8}$

(c) $\sqrt[3]{8x^7y^{11}} = \sqrt[3]{8 \cdot x^6 \cdot x \cdot y^9 \cdot y^2}$ 8, x^6, and y^9 are perfect cubes

$= \sqrt[3]{8 \cdot x^6 \cdot y^9 \cdot xy^2}$ xy^2 will be left under the radical

$= \sqrt[3]{8}\sqrt[3]{x^6}\sqrt[3]{y^9}\sqrt[3]{xy^2}$ $\sqrt[3]{abcd} = \sqrt[3]{a}\sqrt[3]{b}\sqrt[3]{c}\sqrt[3]{d}$

$= 2x^2y^3\sqrt[3]{xy^2}$ $x^6 = x^{2\cdot 3} = (x^2)^3$ and $y^9 = y^{3\cdot 3} = (y^3)^3$

(c) $\sqrt[3]{27a^5b^{13}}$

Answers: (a) $2\sqrt[3]{11}$ (b) $6y^2\sqrt[3]{y^2}$ (c) $3ab^4\sqrt[3]{a^2b}$

In general, to simplify a radical with index k, we look for perfect kth powers, use $\sqrt[k]{ab} = \sqrt[k]{a}\sqrt[k]{b}$, and then $\sqrt[k]{x^k} = x$.

EXAMPLE 3 Simplifying Higher Order Roots

Simplify.

(a) $\sqrt[4]{48} = \sqrt[4]{16 \cdot 3}$ 16 is a perfect fourth power

$= \sqrt[4]{16}\sqrt[4]{3}$ $\sqrt[4]{ab} = \sqrt[4]{a}\sqrt[4]{b}$

$= \sqrt[4]{2^4}\sqrt[4]{3}$ $16 = 2^4$

$= 2\sqrt[4]{3}$ Don't forget the index; $\sqrt{3} \neq \sqrt[4]{3}$

(b) $\sqrt[5]{y^7} = \sqrt[5]{y^5 \cdot y^2}$ $y^7 = y^{5+2} = y^5 \cdot y^2$

$= \sqrt[5]{y^5}\sqrt[5]{y^2}$ $\sqrt[5]{ab} = \sqrt[5]{a}\sqrt[5]{b}$

$= y\sqrt[5]{y^2}$

(c) $y\sqrt[6]{64x^3y^8} = y\sqrt[6]{2^6x^3y^6y^2}$ $y^8 = y^{6+2} = y^6y^2$

$= y\sqrt[6]{2^6y^6x^3y^2}$ x^3y^2 will be left under the radical

$= y\sqrt[6]{2^6}\sqrt[6]{y^6}\sqrt[6]{x^3y^2}$ $\sqrt[6]{abc} = \sqrt[6]{a}\sqrt[6]{b}\sqrt[6]{c}$

$= 2y^2\sqrt[6]{x^3y^2}$

Practice Exercise 3

Simplify.

(a) $\sqrt[5]{96}$

(b) $\sqrt[4]{u^{11}}$

(c) $xy\sqrt[6]{12x^7y^{14}}$

Answers: (a) $2\sqrt[5]{3}$ (b) $u^2\sqrt[4]{u^3}$ (c) $x^2y^3\sqrt[6]{12xy^2}$

2 SIMPLIFYING QUOTIENTS

When a radicand involves a quotient, the following simplifying rule may be applied. To use it we look for perfect powers in both the numerator and denominator.

Simplifying Rule for Quotients

If a and b are any real numbers ($b \neq 0$) or expressions, where a is nonnegative and b is positive if k is even, then

$$\sqrt[k]{\frac{a}{b}} = \frac{\sqrt[k]{a}}{\sqrt[k]{b}}.$$

The kth root of a quotient is equal to the quotient of the kth roots.

EXAMPLE 4 Simplifying Roots of Quotients

Simplify.

(a) $\sqrt{\frac{49}{25}} = \frac{\sqrt{49}}{\sqrt{25}}$ $\sqrt{\frac{a}{b}} = \frac{\sqrt{a}}{\sqrt{b}}$

$= \frac{7}{5}$

Practice Exercise 4

Simplify.

(a) $\sqrt{\frac{121}{64}}$

(b) $\sqrt[3]{\dfrac{54}{125}} = \dfrac{\sqrt[3]{54}}{\sqrt[3]{125}}$ $\sqrt[3]{\dfrac{a}{b}} = \dfrac{\sqrt[3]{a}}{\sqrt[3]{b}}$

$= \dfrac{\sqrt[3]{27 \cdot 2}}{\sqrt[3]{125}}$ 27 and 125 are perfect cubes

$= \dfrac{\sqrt[3]{3^3}\,\sqrt[3]{2}}{\sqrt[3]{5^3}}$ $\sqrt[3]{ab} = \sqrt[3]{a}\,\sqrt[3]{b}$

$= \dfrac{3\sqrt[3]{2}}{5}$

(c) $\sqrt[4]{\dfrac{48x^4}{3y^8}} = \sqrt[4]{\dfrac{16 \cdot \cancel{3} \cdot x^4}{\cancel{3} \cdot y^8}}$ Simplify under radical first

$= \dfrac{\sqrt[4]{2^4x^4}}{\sqrt[4]{(y^2)^4}}$ $\sqrt[4]{\dfrac{a}{b}} = \dfrac{\sqrt[4]{a}}{\sqrt[4]{b}}$, $y^8 = (y^2)^4$

$= \dfrac{\sqrt[4]{2^4}\,\sqrt[4]{x^4}}{\sqrt[4]{(y^2)^4}}$ $\sqrt[4]{ab} = \sqrt[4]{a}\,\sqrt[4]{b}$

$= \dfrac{2x}{y^2}$

(b) $\sqrt[3]{\dfrac{56}{125}}$

(c) $\sqrt[4]{\dfrac{162x^5}{2y^{12}}}$

Answers: (a) $\frac{11}{8}$ (b) $\frac{2\sqrt[3]{7}}{5}$ (c) $\dfrac{3x\sqrt[4]{x}}{y^3}$

3 RATIONALIZING A DENOMINATOR

In some problems the denominator is not a perfect power. For example, $\frac{6x}{\sqrt{5}}$ has a square root in the denominator (the denominator is not a rational number). The process of removing radicals from the denominator is called **rationalizing the denominator.** A radical expression is not considered completely simplified until all radicals are removed from the denominator.

To Rationalize a Denominator

1. Simplify the fraction.
2. Multiply both numerator and denominator of the fraction by a radical that makes the radicand in the denominator a perfect power, equal to the index on the radical.

To rationalize the fraction $\frac{6x}{\sqrt{5}}$, multiply both numerator and denominator by $\sqrt{5}$.

$$\frac{6x}{\sqrt{5}} = \frac{6x\sqrt{5}}{\sqrt{5}\,\sqrt{5}}$$

$$= \frac{6x\sqrt{5}}{5} \qquad \sqrt{a}\,\sqrt{a} = \sqrt{a^2} = a$$

To see why this works and to find the necessary multipliers for radicals with a higher index, we use $\sqrt[k]{ab} = \sqrt[k]{a}\,\sqrt[k]{b}$ in reverse, that is,

$$\sqrt[k]{a}\,\sqrt[k]{b} = \sqrt[k]{ab},$$

to construct the following table.

Radical in denominator	*Multiply and divide by*	*Reason*
$\sqrt{7}$	$\sqrt{7}$	$\sqrt{7}\sqrt{7} = \sqrt{7^2} = 7$
$\sqrt[3]{7}$	$\sqrt[3]{7^2}$	$\sqrt[3]{7}\sqrt[3]{7^2} = \sqrt[3]{7 \cdot 7^2} = \sqrt[3]{7^3} = 7$
$\sqrt[4]{x}$	$\sqrt[4]{x^3}$	$\sqrt[4]{x}\sqrt[4]{x^3} = \sqrt[4]{x \cdot x^3} = \sqrt[4]{x^4} = x$
$\sqrt[4]{x^3}$	$\sqrt[4]{x}$	$\sqrt[4]{x^3}\sqrt[4]{x} = \sqrt[4]{x^3 \cdot x} = \sqrt[4]{x^4} = x$
$\sqrt[5]{y^2}$	$\sqrt[5]{y^3}$	$\sqrt[5]{y^2}\sqrt[5]{y^3} = \sqrt[5]{y^2y^3} = \sqrt[5]{y^5} = y$

EXAMPLE 5 Rationalizing Denominators

Rationalize the denominator.

(a) $\sqrt{\dfrac{36x^2}{y}} = \dfrac{\sqrt{36x^2}}{\sqrt{y}}$ $\qquad \sqrt{\dfrac{a}{b}} = \dfrac{\sqrt{a}}{\sqrt{b}}$

$= \dfrac{\sqrt{36}\sqrt{x^2}}{\sqrt{y}}$

$= \dfrac{6x}{\sqrt{y}}$

$= \dfrac{6x\sqrt{y}}{\sqrt{y}\sqrt{y}}$ $\qquad$ Multiply numerator and denominator by $\sqrt{y}$

$= \dfrac{6x\sqrt{y}}{y}$ $\qquad \sqrt{y}\sqrt{y} = y$

(b) $\dfrac{\sqrt[3]{x}}{\sqrt[3]{5}} = \dfrac{\sqrt[3]{x}\sqrt[3]{5^2}}{\sqrt[3]{5}\sqrt[3]{5^2}}$ $\qquad$ Multiply numerator and denominator by $\sqrt[3]{5^2}$

$= \dfrac{\sqrt[3]{x \cdot 5^2}}{\sqrt[3]{5^3}}$ $\qquad \sqrt[3]{a}\sqrt[3]{b} = \sqrt[3]{ab}$

$= \dfrac{\sqrt[3]{25x}}{5}$

(c) $\dfrac{\sqrt[5]{2y}}{\sqrt[5]{4x^2y^3}} = \dfrac{\sqrt[5]{2y}}{\sqrt[5]{2^2x^2y^3}}$

$= \dfrac{\sqrt[5]{2y}\sqrt[5]{2^3x^3y^2}}{\sqrt[5]{2^2x^2y^3}\sqrt[5]{2^3x^3y^2}}$

$= \dfrac{\sqrt[5]{2y \cdot 2^3x^3y^2}}{\sqrt[5]{2^2x^2y^3 \cdot 2^3x^3y^2}}$

$= \dfrac{\sqrt[5]{2^4x^3y^3}}{\sqrt[5]{2^5x^5y^5}}$

$= \dfrac{\sqrt[5]{16x^3y^3}}{2xy}$

Practice Exercise 5

Rationalize the denominator.

(a) $\sqrt{\dfrac{49u^2}{v}}$

(b) $\dfrac{\sqrt[3]{xy}}{\sqrt[3]{25}}$

(c) $\dfrac{\sqrt[4]{7x^2}}{\sqrt[4]{2x^3y}}$

Answers: (a) $\frac{7u\sqrt{v}}{v}$ (b) $\frac{\sqrt[3]{5xy}}{5}$ (c) $\frac{\sqrt[4]{56x^3y^3}}{2xy}$

We could have rationalized the denominator in Example 5(a) in another way. If we multiply the numerator and denominator of the radicand by y, we will obtain the same result.

$$\sqrt{\frac{36x^2}{y}} = \sqrt{\frac{36x^2 \cdot y}{y \cdot y}} = \frac{\sqrt{36}\sqrt{x^2}\sqrt{y}}{\sqrt{y^2}} = \frac{6x\sqrt{y}}{y}$$

4 RATIONALIZING A NUMERATOR

Sometimes in more advanced courses it is necessary to change the form of a radical expression by **rationalizing the numerator.** The procedure is similar to rationalizing denominators.

EXAMPLE 6 Rationalizing Numerators

Rationalize the numerator.

(a) $\dfrac{\sqrt{6}}{\sqrt{10}} = \dfrac{\sqrt{6}\,\sqrt{6}}{\sqrt{10}\,\sqrt{6}}$ Multiply numerator and denominator by $\sqrt{6}$

$= \dfrac{6}{\sqrt{10 \cdot 6}}$ $\sqrt{a}\,\sqrt{b} = \sqrt{ab}$

$= \dfrac{6}{\sqrt{2 \cdot 5 \cdot 2 \cdot 3}}$ Factor and look for perfect squares

$= \dfrac{6}{\sqrt{2^2 \cdot 15}} = \dfrac{6}{\sqrt{2^2}\,\sqrt{15}}$

$= \dfrac{6}{2\sqrt{15}} = \dfrac{3}{\sqrt{15}}$

(b) $\dfrac{\sqrt[4]{x^3}}{\sqrt[4]{y}} = \dfrac{\sqrt[4]{x^3}\,\sqrt[4]{x}}{\sqrt[4]{y}\,\sqrt[4]{x}}$ Multiply numerator and denominator by $\sqrt[4]{x}$

$= \dfrac{\sqrt[4]{x^3 \cdot x}}{\sqrt[4]{yx}}$ $\sqrt[4]{a}\,\sqrt[4]{b} = \sqrt[4]{ab}$

$= \dfrac{\sqrt[4]{x^4}}{\sqrt[4]{yx}} = \dfrac{x}{\sqrt[4]{yx}}$

Practice Exercise 6

Rationalize the numerator.

(a) $\dfrac{\sqrt{x}}{\sqrt{y}}$

(b) $\dfrac{\sqrt[3]{y^2}}{\sqrt[3]{3}}$

Answers: (a) $\frac{x}{\sqrt{xy}}$ (b) $\frac{y}{\sqrt[3]{3y}}$

5 SIMPLEST FORM

The techniques in this section can be used to change a radical expression into *simplest form*.

Simplest Form of a Radical Expression

A radical expression is written in **simplest form** provided:

1. The radicand contains no factor raised to a power greater than or equal to the index.
2. There are no radicals in a denominator.
3. There are no fractions in a radicand.

The radicals $\sqrt{a^3}$ and $\sqrt[3]{24}$ are not in simplest form since they violate property 1.

$$\sqrt{a^3} = \sqrt{a^2 a} = a\sqrt{a} \quad \text{and} \quad \sqrt[3]{24} = \sqrt[3]{8 \cdot 3} = 2\sqrt[3]{3}$$

The radicals $\frac{1}{\sqrt{2}}$ and $\frac{\sqrt[3]{3}}{\sqrt[3]{x}}$ are not in simplest form since they violate property 2.

$$\frac{1}{\sqrt{2}} = \frac{1 \cdot \sqrt{2}}{\sqrt{2}\,\sqrt{2}} = \frac{\sqrt{2}}{2} \quad \text{and} \quad \frac{\sqrt[3]{3}}{\sqrt[3]{x}} = \frac{\sqrt[3]{3}\,\sqrt[3]{x^2}}{\sqrt[3]{x}\,\sqrt[3]{x^2}} = \frac{\sqrt[3]{3x^2}}{x}$$

Finally, the radicals $\sqrt{\frac{2}{3}}$ and $\sqrt[3]{\frac{x^2}{y}}$ are not in simplest form since they violate property 3.

$$\sqrt{\frac{2}{3}} = \frac{\sqrt{2}}{\sqrt{3}} = \frac{\sqrt{2}\sqrt{3}}{\sqrt{3}\sqrt{3}} = \frac{\sqrt{6}}{3} \quad \text{and} \quad \sqrt[3]{\frac{x^2}{y}} = \frac{\sqrt[3]{x^2}\sqrt[3]{y^2}}{\sqrt[3]{y}\sqrt[3]{y^2}} = \frac{\sqrt[3]{x^2y^2}}{y}$$

We now look at an application that requires simplifying radicals.

EXAMPLE 7 Physics Problem

An object dropped from a height of h ft takes t seconds to reach the ground where $t = \sqrt{\frac{h}{16}}$. How long would it take a rock to hit the water if dropped from a bridge 100 ft above the surface of the water?

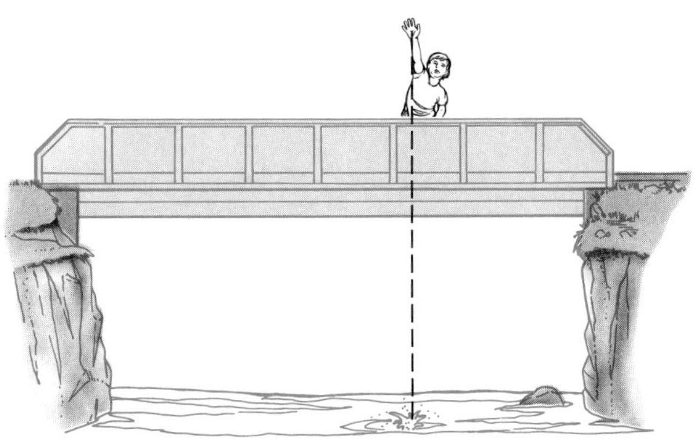

Figure 7.2

Substitute 100 for h and simplify.

$$t = \sqrt{\frac{h}{16}}$$
$$= \sqrt{\frac{100}{16}}$$
$$= \frac{\sqrt{100}}{\sqrt{16}} = \frac{\sqrt{10^2}}{\sqrt{4^2}} = \frac{10}{4} = \frac{5}{2}$$

It would take $\frac{5}{2}$ or 2.5 seconds for the rock to hit the water.

Practice Exercise 7

How long would it take a ball to reach the ground if it is dropped from a window in a building from a height of 320 ft? Give answer to the nearest tenth of a second.

Answer: 4.5 sec

7.2 EXERCISES A

Simplify each of the radicals. Assume all variables and expressions under even-indexed radicals are nonnegative.

1. $\sqrt{125}$

2. $6\sqrt{98}$

3. $\sqrt{75y^3}$

4. $3\sqrt{x^8}$

5. $7\sqrt{50x^3y^5}$

6. $\sqrt[3]{24}$

7. $\sqrt[3]{27x^4}$

8. $5\sqrt[3]{y^8}$

9. $\sqrt[3]{16x^4y^6}$ **10.** $\sqrt[3]{-125x^9y^5}$ **11.** $\sqrt[4]{32}$ **12.** $\sqrt[4]{81x^7}$

13. $4\sqrt[3]{81x^3}$ **14.** $\sqrt[4]{-16x^4}$ **15.** $\sqrt{20x^7y^9}$ **16.** $\sqrt{15xy}$

17. $\sqrt[4]{625x^8y^{12}}$ **18.** $\sqrt[5]{-243x^6y^7}$ **19.** $\sqrt[6]{a^5b^7c^{12}}$ **20** $\sqrt[3]{-27x^4y^5z^8}$

21. $\sqrt{\dfrac{45}{16}}$ **22.** $\sqrt{\dfrac{125}{y^2}}$ **23.** $\sqrt[3]{\dfrac{125}{8}}$ **24.** $\sqrt[3]{\dfrac{56x^5}{7y^3}}$

25. $\sqrt[3]{\dfrac{-27x^4}{8}}$ **26.** $\sqrt[3]{\dfrac{-27x^4}{8x}}$ **27.** $\sqrt[4]{\dfrac{16}{81}}$ **28.** $\sqrt[5]{\dfrac{32x}{y^5}}$

29 $\dfrac{3x}{y}\sqrt[5]{\dfrac{x^5y^6}{243}}$ **30.** $\sqrt[3]{\dfrac{10x^6y^7}{-27z^9}}$ **31.** $\sqrt{\dfrac{x^2+2x+1}{x^2-2x+1}}$ **32.** $\sqrt[3]{\dfrac{-16x^8y^{12}}{z^{15}}}$

If the given radical is a denominator, determine the radical which must be multiplied by numerator and denominator to rationalize the denominator.

33. $\sqrt{3x}$ **34** $\sqrt[3]{5x^2}$ **35.** $\sqrt[5]{x^3}$ **36.** $\sqrt[4]{x^3y}$

Simplify. Rationalize all denominators.

37. $\dfrac{3}{\sqrt{2}}$ **38.** $\dfrac{8}{\sqrt{x}}$ **39.** $\sqrt{\dfrac{25}{y}}$ **40.** $\dfrac{3}{\sqrt[3]{7}}$

41. $\dfrac{x^2}{\sqrt[3]{y}}$ **42.** $\dfrac{x^2}{\sqrt[3]{y^2}}$ **43** $\dfrac{7}{\sqrt[4]{y}}$ **44.** $\dfrac{7}{\sqrt[4]{y^3}}$

45. $\sqrt[3]{\dfrac{81x^3}{y^2}}$ **46.** $\sqrt[5]{\dfrac{-32x^7}{y^3}}$ **47.** $\dfrac{\sqrt[3]{3z^4}}{\sqrt[3]{2x^2y}}$ **48** $\dfrac{\sqrt[3]{x^2z^4}}{\sqrt[3]{2y^2}}$

Rationalize the numerator.

49. $\dfrac{\sqrt{5}}{\sqrt{2x}}$ **50.** $\dfrac{2\sqrt{x}}{\sqrt{6y}}$ **51.** $\sqrt{\dfrac{2}{5}}$ **52.** $\dfrac{\sqrt[3]{3}}{\sqrt[3]{x}}$

53. $\dfrac{\sqrt[4]{x^3}}{\sqrt[4]{7}}$ **54.** $\sqrt[3]{\dfrac{2}{3}}$ **55.** $\dfrac{\sqrt[3]{4}}{\sqrt[3]{4a}}$ **56.** $\dfrac{\sqrt{xy}}{6}$

In a manufacturing process, the three quantities p, r, and s are related by the equation $p = \sqrt{\frac{r}{s}}$.

57. Find p when $r = 45$ and $s = 36$.

58. Find p when $r = 1$ and $s = 3$.

59. Find r when $p = 5$ and $s = 1$.

60. Find s when $p = 5$ and $r = 1$.

If an object is dropped from a cliff of height h in feet, the time t in seconds it takes the object to reach the ground is given by $t = \sqrt{\frac{h}{16}}$. *Use this information in Exercises 61–62.*

61. Find the time it would take for a rock to reach the water if it is dropped from a cliff 169 ft above the water level.

62. Find the time it would take for a rock to reach the water if it is dropped from a cliff 200 ft above the water level. Give answer correct to the nearest tenth of a second.

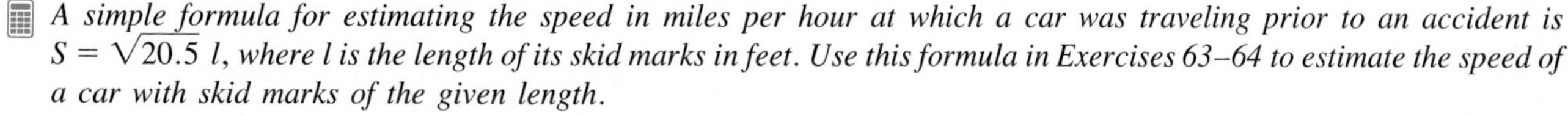

A simple formula for estimating the speed in miles per hour at which a car was traveling prior to an accident is $S = \sqrt{20.5\,l}$, *where l is the length of its skid marks in feet. Use this formula in Exercises 63–64 to estimate the speed of a car with skid marks of the given length.*

63. 50 ft

64. 85 ft

FOR REVIEW

Simplify the radicals in Exercises 65–70. Use absolute value bars when necessary. (That is, do not assume variables are nonnegative for these six exercises.)

65. $\sqrt{36y^2}$ **66.** $\sqrt[3]{64x^3y^3}$ **67.** $\sqrt[5]{-32(x+y)^5}$

68. $\sqrt{a^6}$ **69.** $\sqrt[5]{32a^{20}b^{10}}$ **70.** $\sqrt[4]{(2xy^2)^4}$

ANSWERS: 1. $5\sqrt{5}$ 2. $42\sqrt{2}$ 3. $5y\sqrt{3y}$ 4. $3x^4$ 5. $35xy^2\sqrt{2xy}$ 6. $2\sqrt[3]{3}$ 7. $3x\sqrt[3]{x}$ 8. $5y^2\sqrt[3]{y^2}$ 9. $2xy^2\sqrt[3]{2x}$ 10. $-5x^3y\sqrt[3]{y^2}$ 11. $2\sqrt[4]{2}$ 12. $3x\sqrt[4]{x^3}$ 13. $12x\sqrt[3]{3}$ 14. not a real number 15. $2x^3y^4\sqrt{5xy}$ 16. $\sqrt{15xy}$ 17. $5x^2y^3$ 18. $-3xy\sqrt[5]{xy^2}$ 19. $bc^2\sqrt[6]{a^5b}$ 20. $-3xyz^2\sqrt[3]{xy^2z^2}$ 21. $\frac{3\sqrt{5}}{4}$ 22. $\frac{5\sqrt{5}}{y}$ 23. $\frac{5}{2}$ 24. $\frac{2x\sqrt[3]{x^2}}{y}$ 25. $\frac{-3x\sqrt[3]{x}}{2}$ 26. $\frac{-3x}{2}$ 27. $\frac{2}{3}$ 28. $\frac{2\sqrt[5]{x}}{y}$ 29. $x^2\sqrt[5]{y}$ 30. $-\frac{x^2y^2\sqrt[3]{10y}}{3z^3}$ 31. $\frac{x+1}{x-1}$ 32. $\frac{-2x^2y^4\sqrt[3]{2x^2}}{z^5}$ 33. $\sqrt{3x}$ 34. $\sqrt[3]{5^2x}$ 35. $\sqrt[5]{x^2}$ 36. $\sqrt[4]{xy^3}$ 37. $\frac{3\sqrt{2}}{2}$ 38. $\frac{8\sqrt{x}}{x}$ 39. $\frac{5\sqrt{y}}{y}$ 40. $\frac{3\sqrt[3]{49}}{7}$ 41. $\frac{x^2\sqrt[3]{y^2}}{y}$ 42. $\frac{x^2\sqrt[3]{y}}{y}$ 43. $\frac{7\sqrt[4]{y^3}}{y}$ 44. $\frac{7\sqrt[4]{y}}{y}$ 45. $\frac{3x\sqrt[3]{3y}}{y}$ 46. $\frac{-2x\sqrt[5]{x^2y^2}}{y}$ 47. $\frac{z\sqrt[3]{12xy^2z}}{2xy}$ 48. $\frac{z\sqrt[3]{4x^2yz}}{2y}$ 49. $\frac{5}{\sqrt{10x}}$ 50. $\frac{2x}{\sqrt{6xy}}$ 51. $\frac{2}{\sqrt{10}}$ 52. $\frac{3}{\sqrt[3]{9x}}$ 53. $\frac{x}{\sqrt[4]{7x}}$ 54. $\frac{2}{\sqrt[3]{12}}$ 55. $\frac{1}{\sqrt[3]{a}}$ 56. $\frac{xy}{6\sqrt{xy}}$ 57. $\frac{\sqrt{5}}{2}$ 58. $\frac{\sqrt{3}}{3}$ 59. 25 60. $\frac{1}{25}$ 61. $\frac{13}{4}$ sec 62. 3.5 sec 63. 32 mph 64. 42 mph 65. $6|y|$ 66. $4xy$ 67. $-2(x+y)$ 68. $|a^3|$ 69. $2a^4b^2$ 70. $2|x|y^2$

7.2 EXERCISES B

Simplify each of the radicals. Assume all variables under even indexed radicals are nonnegative.

1. $\sqrt{175}$ **2.** $10\sqrt{147}$ **3.** $\sqrt{18x^3}$ **4.** $8\sqrt{x^{10}}$

5. $9\sqrt{75x^3y^7}$ **6.** $\sqrt[3]{40}$ **7.** $\sqrt[3]{8x^4}$ **8.** $3\sqrt[3]{z^{10}}$

9. $\sqrt[3]{81x^5y^3}$ **10.** $\sqrt[3]{-216x^5y^9}$ **11.** $\sqrt[4]{243}$ **12.** $\sqrt[4]{16x^5}$

13. $5\sqrt[3]{16x^6}$ **14.** $\sqrt[4]{-81x^4}$ **15.** $\sqrt{45x^3y^7}$ **16.** $\sqrt{6xyz}$

17. $\sqrt[4]{81x^{12}y^8}$ **18.** $\sqrt[5]{-32x^8y^6}$ **19.** $\sqrt[6]{a^{10}b^5c^{12}}$ **20.** $\sqrt[3]{-8x^5y^8z^4}$

21. $\sqrt{\frac{20}{49}}$ **22.** $\sqrt{\frac{50}{y^2}}$ **23.** $\sqrt[3]{\frac{64}{125}}$ **24.** $\sqrt[3]{\frac{54x^4}{2y^3}}$

25. $\sqrt[3]{\frac{-8x^5}{27}}$ **26.** $\sqrt[3]{\frac{-8x^5}{27x^2}}$ **27.** $\sqrt[4]{\frac{625}{16}}$ **28.** $\sqrt[5]{\frac{243x^2}{y^5}}$

29. $\frac{2x}{y}\sqrt[5]{\frac{x^6y^5}{32}}$ **30.** $\sqrt[3]{\frac{15x^7y^6}{-8z^{12}}}$ **31.** $\sqrt{\frac{x^2-6x+9}{x^2+6x+9}}$ **32.** $\sqrt[3]{\frac{-81x^9y^{11}}{z^9}}$

If the given radical is a denominator, determine the radical which must be multiplied by numerator and denominator to rationalize the denominator.

33. $\sqrt{5x}$ **34.** $\sqrt[3]{3x^2}$ **35.** $\sqrt[5]{x^2}$ **36.** $\sqrt[4]{xy^3}$

Simplify. Rationalize all denominators.

37. $\frac{5}{\sqrt{3}}$ **38.** $\frac{7}{\sqrt{x}}$ **39.** $\sqrt{\frac{16}{y}}$ **40.** $\frac{7}{\sqrt[3]{5}}$

41. $\frac{2x^3}{\sqrt[3]{y}}$ **42.** $\frac{2x^3}{\sqrt[3]{y^2}}$ **43.** $\frac{8}{\sqrt[4]{y}}$ **44.** $\frac{8}{\sqrt[4]{y^3}}$

45. $\sqrt[3]{\dfrac{54x^3}{y}}$ **46.** $\sqrt[5]{\dfrac{-243x^6}{y^2}}$ **47.** $\dfrac{\sqrt[3]{4x^4}}{\sqrt[3]{3yz^2}}$ **48.** $\dfrac{\sqrt[3]{x^4z^2}}{\sqrt[3]{3y^2}}$

Rationalize the numerator.

49. $\dfrac{\sqrt{3}}{\sqrt{5y}}$ **50.** $\dfrac{4\sqrt{y}}{\sqrt{2a}}$ **51.** $\sqrt{\dfrac{3}{7}}$ **52.** $\dfrac{\sqrt[3]{2}}{\sqrt[3]{x}}$

53. $\dfrac{\sqrt[4]{y}}{\sqrt[4]{5}}$ **54.** $\sqrt[3]{\dfrac{3}{5}}$ **55.** $\dfrac{\sqrt[3]{9}}{\sqrt[3]{18a}}$ **56.** $\dfrac{\sqrt{ab}}{5}$

In an ecological study, the three quantities u, v, and w are related by the equation $u = \sqrt[3]{vw}$.

57. Find u when $v = 4$ and $w = 10$.

58. Find u when $v = 2$ and $w = \dfrac{1}{5}$.

59. Find v when $u = 4$ and $w = 1$.

60. Find w when $u = 4$ and $v = 2$.

If an object is dropped from a cliff of height h in feet, the time t in seconds it takes the object to reach the ground is given by $t = \sqrt{\dfrac{h}{16}}$. *Use this information in Exercises 61–62.*

61. Find the time it would take for a rock to reach the water if it is dropped from a cliff 225 ft above the water level.

62. Find the time it would take for a rock to reach the water if it is dropped from a cliff 300 ft above the water level. Give answer correct to the nearest tenth of a second.

A simple formula for estimating the speed in miles per hour at which a car was traveling prior to an accident is $S = \sqrt{20.5\, l}$, *where l is the length of its skid marks in feet. Use this formula in Exercises 63–64 to estimate the speed of a car with skid marks of the given length.*

63. 45 ft

64. 100 ft

FOR REVIEW

Simplify the radicals in Exercises 65–70. Use absolute value bars when necessary. (That is, do not assume variables are nonnegative for these six exercises.)

65. $\sqrt{49a^2}$ **66.** $\sqrt[3]{125x^3y^3}$ **67.** $\sqrt[5]{-243(x-y)^5}$

68. $\sqrt[4]{a^4}$ **69.** $\sqrt[3]{27x^9y^6}$ **70.** $\sqrt[6]{(2uv^2)^6}$

7.2 EXERCISES C

Simplify. Rationalize all denominators.

1. $\sqrt{800x^5y^3}$ **2.** $(\sqrt[3]{a^3b^6})^2$ **3.** $\sqrt{x^2 + 4x^3}$

4. $\sqrt[5]{\dfrac{2a}{81a^2}}$ $\left[\text{Answer: } \dfrac{\sqrt[5]{6a^4}}{3a}\right]$

5. $\dfrac{1}{\sqrt[3]{(x+y)^2}}$

6. $\sqrt[4]{4x\sqrt{16x^6}}$ [Answer: $2x$]

7. Give an example to show that $\sqrt{a+b} \neq \sqrt{a} + \sqrt{b}$.

8. Give an example to show that $\sqrt{a^2 + b^2} \neq |a| + |b|$.

7.3 MULTIPLICATION AND DIVISION OF RADICALS

STUDENT GUIDEPOSTS

1 Multiplying Radicals
2 Dividing Radicals

1 MULTIPLYING RADICALS

In the previous section we used the product rule $\sqrt[k]{ab} = \sqrt[k]{a}\sqrt[k]{b}$ to simplify radicals. The emphasis was on writing a radical as a product of radicals. Using the symmetric law of equality, we can change the order and emphasize writing a product of radicals as a single radical.

$$\sqrt[k]{a}\sqrt[k]{b} = \sqrt[k]{ab}$$

After applying this product rule to combine radicals, we can sometimes simplify the resulting radical by removing perfect powers.

EXAMPLE 1 Multiplying Radicals

Multiply and simplify.

(a) $\sqrt{5}\,\sqrt{45} = \sqrt{5 \cdot 45}$ $\quad \sqrt{a}\,\sqrt{b} = \sqrt{ab}$

$= \sqrt{5 \cdot 5 \cdot 9}$ $\quad$ Factor to find perfect squares

$= \sqrt{25}\,\sqrt{9}$ $\quad \sqrt{ab} = \sqrt{a}\,\sqrt{b}$

$= 5 \cdot 3 = 15$

(b) $\sqrt{2x}\,\sqrt{8x} = \sqrt{2x \cdot 8x}$ $\quad \sqrt{a}\,\sqrt{b} = \sqrt{ab}$

$= \sqrt{16x^2}$ $\quad$ 16 and x^2 are perfect squares

$= \sqrt{16}\,\sqrt{x^2}$ $\quad \sqrt{ab} = \sqrt{a}\,\sqrt{b}$

$= 4x$

(c) $\sqrt{2xy}\,\sqrt{6x^3y^3} = \sqrt{2xy \cdot 6x^3y^3}$ $\quad \sqrt{a}\,\sqrt{b} = \sqrt{ab}$

$= \sqrt{2^2 \cdot x^4 \cdot y^4 \cdot 3}$ $\quad x \cdot x^3 = x^{1+3} = x^4$ and $y \cdot y^3 = y^4$

$= \sqrt{2^2}\,\sqrt{(x^2)^2}\,\sqrt{(y^2)^2}\,\sqrt{3}$ $\quad x^4 = (x^2)^2$ and $y^4 = (y^2)^2$

$= 2x^2y^2\sqrt{3}$

(d) $5\sqrt[3]{3a^2}\,\sqrt[3]{9a} = 5\sqrt[3]{3a^2 \cdot 9a}$ $\quad \sqrt[3]{a}\,\sqrt[3]{b} = \sqrt[3]{ab}$

$= 5\sqrt[3]{27a^3}$ $\quad$ 27 and a^3 are perfect cubes

$= 5\sqrt[3]{27}\,\sqrt[3]{a^3}$ $\quad \sqrt[3]{ab} = \sqrt[3]{a}\,\sqrt[3]{b}$

$= 15a$

(e) $\sqrt[4]{8x^2}\,\sqrt[4]{4x^3} = \sqrt[4]{2^3x^2 \cdot 2^2x^3}$ $\quad \sqrt[4]{a}\,\sqrt[4]{b} = \sqrt[4]{ab}$

$= \sqrt[4]{2^5x^5}$

$= \sqrt[4]{2^4x^4 \cdot 2x}$ $\quad$ $2x$ will be left under the radical

$= \sqrt[4]{2^4}\,\sqrt[4]{x^4}\,\sqrt[4]{2x}$ $\quad \sqrt[4]{abc} = \sqrt[4]{a}\,\sqrt[4]{b}\,\sqrt[4]{c}$

$= 2x\sqrt[4]{2x}$

PRACTICE EXERCISE 1

Multiply and simplify.

(a) $\sqrt{7}\,\sqrt{175}$

(b) $\sqrt{3y}\,\sqrt{27y^3}$

(c) $\sqrt{12x^2y}\,\sqrt{9x^2y^3}$

(d) $4\sqrt[3]{4a^4}\,\sqrt[3]{16a^2}$

(e) $\sqrt[4]{9y^6}\,\sqrt[4]{27y^3}$

Answers: (a) 35 (b) $9y^2$ (c) $6x^2y^2\sqrt{3}$ (d) $16a^2$ (e) $3y^2\sqrt[4]{3y}$

2 DIVIDING RADICALS

When an expression involves the quotient of two radicals, it may be simplified in a similar manner.

$$\frac{\sqrt[k]{a}}{\sqrt[k]{b}} = \sqrt[k]{\frac{a}{b}}$$

Here we are writing a quotient of radicals as a single radical. It can sometimes be simplified by dividing out common factors and removing perfect powers.

EXAMPLE 2 Dividing Radicals

Divide and simplify.

(a) $\frac{\sqrt{27a^3}}{\sqrt{3a}} = \sqrt{\frac{27a^3}{3a}}$ $\qquad \frac{\sqrt{a}}{\sqrt{b}} = \sqrt{\frac{a}{b}}$

$= \sqrt{\frac{9 \cdot \cancel{3} \cdot a^2 \cdot \cancel{a}}{\cancel{3}\cancel{a}}}$ $\qquad$ Divide out common factors

$= \sqrt{9a^2}$ $\qquad$ 9 and a^2 are perfect squares

$= \sqrt{9}\sqrt{a^2} = 3a$

(b) $\frac{\sqrt{5x^3y}}{\sqrt{9xy^3}} = \sqrt{\frac{5x^3y}{9xy^3}}$ $\qquad \frac{\sqrt{a}}{\sqrt{b}} = \sqrt{\frac{a}{b}}$

$= \sqrt{\frac{5x^2 \cdot \cancel{x} \cdot \cancel{y}}{9\cancel{x} \cdot y^2 \cdot \cancel{y}}}$ $\qquad$ x and y are common factors

$= \sqrt{\frac{5x^2}{9y^2}}$

$= \frac{\sqrt{x^2}\sqrt{5}}{\sqrt{9}\sqrt{y^2}}$ $\qquad \sqrt{\frac{a}{b}} = \frac{\sqrt{a}}{\sqrt{b}}$ and $\sqrt{ab} = \sqrt{a}\sqrt{b}$

$= \frac{x\sqrt{5}}{3y}$

(c) $\frac{\sqrt[3]{16x^7y}}{\sqrt[3]{2xy^4}} = \sqrt[3]{\frac{16x^7y}{2xy^4}}$ $\qquad \frac{\sqrt[3]{a}}{\sqrt[3]{b}} = \sqrt[3]{\frac{a}{b}}$

$= \sqrt[3]{\frac{8 \cdot \cancel{2} \cdot x^6 \cdot \cancel{x} \cdot \cancel{y}}{\cancel{2} \cdot \cancel{x} \cdot y^3 \cdot \cancel{y}}}$ $\qquad$ Divide out common factors

$= \sqrt[3]{\frac{8x^6}{y^3}}$

$= \frac{\sqrt[3]{8}\sqrt[3]{(x^2)^3}}{\sqrt[3]{y^3}}$ $\qquad \sqrt[3]{\frac{a}{b}} = \frac{\sqrt[3]{a}}{\sqrt[3]{b}}$ and $\sqrt[3]{ab} = \sqrt[3]{a}\sqrt[3]{b}$

$= \frac{2x^2}{y}$

(d) $\frac{\sqrt{3x^4}}{\sqrt{15y}} = \sqrt{\frac{3x^4}{15y}}$ $\qquad$ Quotient rule

$= \sqrt{\frac{x^4}{5y}}$ $\qquad$ Divide out the common factor 3

$= \frac{\sqrt{x^4}}{\sqrt{5y}}$

$= \frac{x^2}{\sqrt{5y}}$

$= \frac{x^2\sqrt{5y}}{\sqrt{5y}\sqrt{5y}}$ $\qquad$ Rationalize the denominator

$= \frac{x^2\sqrt{5y}}{5y}$

Practice Exercise 2

Divide and simplify.

(a) $\frac{\sqrt{125b^5}}{\sqrt{5b^3}}$

(b) $\frac{\sqrt{8x^4y^3}}{\sqrt{9x^2y^5}}$

(c) $\frac{\sqrt[3]{81x^9y^3}}{\sqrt[3]{3x^3y^9}}$

(d) $\frac{\sqrt{7x^3}}{\sqrt{14y}}$

Answers: (a) $5b$ (b) $\frac{2x\sqrt{2}}{3y}$ (c) $\frac{3x^2}{y^2}$ (d) $\frac{x\sqrt{2xy}}{2y}$

CAUTION

The multiplication and division rules apply only when the radicals have the same index. For example,

$$\sqrt{2}\,\sqrt[3]{5} \quad \text{and} \quad \frac{\sqrt{3}}{\sqrt[4]{2}}$$

cannot be simplified using the above rules.

EXAMPLE 3 Simplifying Products and Quotients

Perform the indicated operation and simplify.

(a)
$$\begin{aligned}\sqrt{4a^2y}\,\sqrt[3]{8a^3y^2} &= \sqrt{2^2a^2y}\,\sqrt[3]{2^3a^3y^2}\\ &= 2a\sqrt{y}\cdot 2a\sqrt[3]{y^2}\\ &= 4a^2\sqrt{y}\,\sqrt[3]{y^2}\end{aligned}$$

Do not try to use the product rule on $\sqrt{y}\,\sqrt[3]{y^2}$ since the indexes are different.

(b)
$$\begin{aligned}\frac{\sqrt[4]{16x^4y}}{\sqrt[3]{8x^3y}} &= \frac{\sqrt[4]{2^4x^4y}}{\sqrt[3]{2^3x^3y}}\\ &= \frac{2x\sqrt[4]{y}}{2x\sqrt[3]{y}} = \frac{\sqrt[4]{y}}{\sqrt[3]{y}}\end{aligned}$$

We cannot use the quotient rule at this point since the indexes are different. We can, however, rationalize the denominator.

$$\frac{\sqrt[4]{y}\,\sqrt[3]{y^2}}{\sqrt[3]{y}\,\sqrt[3]{y^2}} = \frac{\sqrt[4]{y}\,\sqrt[3]{y^2}}{\sqrt[3]{y^3}} = \frac{\sqrt[4]{y}\,\sqrt[3]{y^2}}{y}.$$

PRACTICE EXERCISE 3

Perform the indicated operation and simplify.

(a) $\sqrt{9xw^2}\,\sqrt[4]{16a^4x}$

(b) $\dfrac{\sqrt[3]{27a^9}}{\sqrt{9a}}$

Answers: (a) $6aw\sqrt{x}\,\sqrt[4]{x}$
(b) $a^2\sqrt{a}$

Many types of geometry problems, such as the one in the following example, involve radicals.

EXAMPLE 4 Geometry Problem

The radius r of a circle inscribed in a triangle with sides a, b, and c (see Figure 7.3) is given by

$$r = \frac{\sqrt{s-a}\,\sqrt{s-b}\,\sqrt{s-c}}{\sqrt{s}}$$

where $s = \frac{1}{2}(a+b+c)$. Express r as a single radical and use the result to find the radius of a circle inscribed in a triangle with sides 12.0 cm, 14.0 cm, and 20.0 cm.

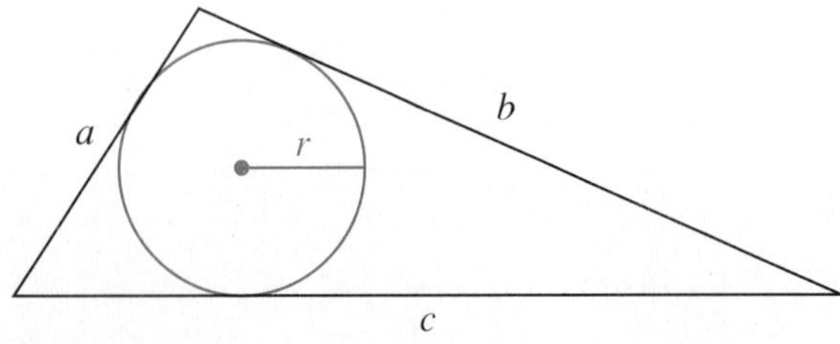

Figure 7.3

PRACTICE EXERCISE 4

Find the radius of a circle inscribed in a triangle with sides 13.1 ft, 17.3 ft, and 18.4 ft.

Use the product rule to write r as

$$r = \frac{\sqrt{(s-a)(s-b)(s-c)}}{\sqrt{s}}.$$

Now use the quotient rule.

$$r = \sqrt{\frac{(s-a)(s-b)(s-c)}{s}}$$

Next we find the value of s using 12.0 for a, 14.0 for b, and 20.0 for c.

$$s = \frac{1}{2}(a+b+c)$$

$$= \frac{1}{2}(12.0 + 14.0 + 20.0) = \frac{1}{2}(46) = 23$$

Then
$$r = \sqrt{\frac{(23-12)(23-14)(23-20)}{23}}$$

$$= \sqrt{\frac{(11)(9)(3)}{23}} = \sqrt{\frac{297}{23}} \approx 3.5935.$$

The radius of the circle is approximately 3.6 cm.

Answer: 4.4 ft

7.3 EXERCISES A

Multiply and simplify.

1. $\sqrt{18}\sqrt{98}$
2. $2\sqrt{44}\sqrt{33}$
3. $\sqrt{7x}\sqrt{21x}$
4. $\sqrt{5x}\sqrt{5x}$
5. $\sqrt{5x^2y}\sqrt{35xy^2}$
6. $\sqrt{3a^3b^3}\sqrt{12a^3b^2}$
7. $(\sqrt{3xy})^2$
8. $\sqrt{5xy}\sqrt{15x^2y}\sqrt{3xy^2}$
9. $\sqrt{7a^2b}\sqrt{14ab^2}\sqrt{ab^2}$
10. $7\sqrt[3]{6}\sqrt[3]{9}$
11. $\sqrt[3]{2}\sqrt[3]{32}$
12. $\sqrt[4]{4}\sqrt[4]{4}$
13. $\sqrt[3]{3x}\sqrt[3]{9x^2}$
14. $\sqrt[4]{9x^2}\sqrt[4]{9x^2}$
15. $\sqrt[5]{16x^2}\sqrt[5]{4x^3}$
16. $\sqrt[3]{5x^2y^4}\sqrt[3]{50x^3y^2}$
17. $\sqrt[3]{6ab}\sqrt[3]{18a^2b^3}$
18. $\sqrt[4]{9x^2y^5}\sqrt[4]{45x^3y^2}$
19. $\sqrt{(x+y)^3}\sqrt{(x+y)}$
20. $\sqrt{16x^2y^2}\sqrt[3]{16x^3y^3}$
21. $\sqrt[3]{27x^6y^3}\sqrt[4]{48x^8y^5}$

Divide and simplify.

22. $\frac{\sqrt{2}}{\sqrt{50}}$
23. $\frac{10\sqrt{15}}{\sqrt{75}}$
24. $\frac{\sqrt{50x^3}}{\sqrt{2x}}$

25. $\dfrac{\sqrt{75x^2}}{\sqrt{3x}}$

26. $\dfrac{3\sqrt{5x^2y^3}}{\sqrt{4x^4y}}$

27. $\dfrac{\sqrt{98x^3y^5}}{\sqrt{18x^2y^7}}$

28. $\dfrac{\sqrt{25(x+y)^3}}{\sqrt{x+y}}$

29. $\dfrac{\sqrt[3]{81}}{\sqrt[3]{3}}$

30. $\dfrac{\sqrt[3]{6x^2y^3}}{\sqrt[3]{48x^5y^2}}$

31. $\dfrac{\sqrt[4]{48}}{\sqrt[4]{243}}$

32. $\dfrac{\sqrt[4]{a^7b^9}}{\sqrt[4]{16a^3b^2}}$

33. $\dfrac{\sqrt[5]{96x^8}}{\sqrt[5]{3x^2}}$

34. $\dfrac{\sqrt{3x^2}}{\sqrt{3y}}$

35. $\dfrac{\sqrt[3]{27x^3}}{\sqrt[3]{2y}}$

36. $\dfrac{\sqrt[4]{16y^5}}{\sqrt[4]{x^3y}}$

37. $\dfrac{\sqrt[3]{(2x-y)^4(x-2y)^2}}{\sqrt[3]{(2x-y)(x-2y)^5}}$

38. $\dfrac{\sqrt[3]{9x^4y^3}}{\sqrt[3]{8x^6y^9}}$

39. $\dfrac{\sqrt[4]{16x^{12}y^9}}{\sqrt[3]{27x^{12}y^9}}$

During a scientific experiment, Dr. Gordon Johnson found that the three quantities p, q, and r are related by the equation $r = \dfrac{\sqrt{p}}{\sqrt{q}}$.

40. Find r when $p = 98$ and $q = 50$.

41. Find p when $r = 2$ and $q = 9$.

42. Find r when $p = 1$ and $q = 3$.

43. Find r when $p = 25q$.

Use the formula for the radius of a circle inscribed in a triangle obtained in Example 4 to find r for the values of a, b, and c given in Exercises 44–45.

44. $a = 8.0$ ft, $b = 30.0$ ft, $c = 26.0$ ft

45. $a = 11.2$ cm, $b = 16.4$ cm, $c = 21.8$ cm

46. Three circles with radii 3.0 cm, 4.0 cm and 5.0 cm are tangent externally as shown in the figure at right. If their centers are joined to form a triangle, what is the radius of the inscribed circle?

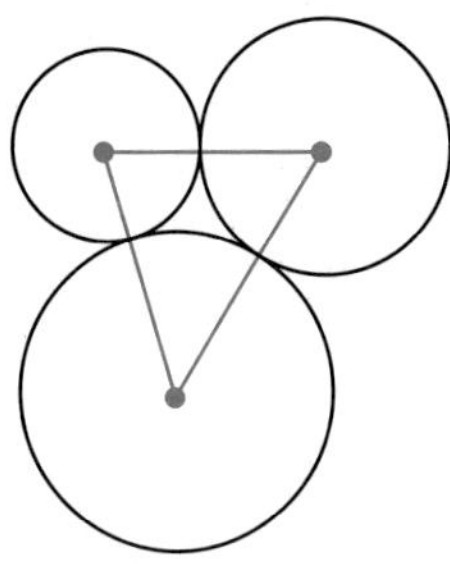

FOR REVIEW

Simplify.

47. $\sqrt{45x^2y^3}$

48. $\sqrt[3]{54a^4b^4}$

49. $\sqrt[4]{16x^9y^4z^3}$

50. $\sqrt{\dfrac{8x^2y^3}{2y}}$ **51.** $\sqrt[3]{\dfrac{27x^2y^2}{x^5y}}$ **52.** $\sqrt{\dfrac{25x^2}{y}}$

Exercises 53–60 review material covered in Sections 1.5 and 5.2 to help you prepare for the next section. Use the distributive law to collect like terms.

53. $7x + 3x$ **54.** $3y - 5y$ **55.** $2x - 5x + 8x$ **56.** $6x - 3y + 2x + y$

Multiply the polynomials.

57. $3x(5 + 2x)$ **58.** $(x - 2y)^2$ **59.** $(x + 5y)(x - 5y)$ **60.** $(3x - 2y)(2x + 5y)$

ANSWERS: 1. 42 2. $44\sqrt{3}$ 3. $7x\sqrt{3}$ 4. $5x$ 5. $5xy\sqrt{7xy}$ 6. $6a^3b^2\sqrt{b}$ 7. $3xy$ 8. $15x^2y^2$ 9. $7a^2b^2\sqrt{2b}$ 10. $21\sqrt[3]{2}$ 11. 4 12. 2 13. $3x$ 14. $3x$ 15. $2x\sqrt[5]{2}$ 16. $5xy^2\sqrt[3]{2x^2}$ 17. $3ab\sqrt[3]{4b}$ 18. $3xy\sqrt[4]{5xy^3}$ 19. $(x + y)^2$ 20. $8x^2y^2\sqrt[3]{2}$ 21. $6x^4y^2\sqrt[4]{3y}$ 22. $\frac{1}{5}$ 23. $2\sqrt{5}$ 24. $5x$ 25. $5\sqrt{x}$ 26. $\frac{3y\sqrt{5}}{2x}$ 27. $\frac{7\sqrt{x}}{3y}$ 28. $5(x + y)$ 29. 3 30. $\frac{\sqrt[3]{y}}{2x}$ 31. $\frac{2}{3}$ 32. $\frac{ab\sqrt[4]{b^3}}{2}$ 33. $2x\sqrt[5]{x}$ 34. $\frac{x\sqrt{y}}{y}$ 35. $\frac{3x\sqrt[3]{4y^2}}{2y}$ 36. $\frac{2y\sqrt[4]{x}}{x}$ 37. $\frac{2x - y}{x - 2y}$ 38. $\frac{3\sqrt{y}}{2y^2}$ 39. $\frac{2\sqrt[4]{y}}{3xy}$ 40. $\frac{7}{5}$ 41. 36 42. $\frac{\sqrt{3}}{3}$ 43. 5 44. 3.0 ft 45. 3.6 cm 46. 2.2 cm 47. $3xy\sqrt{5y}$ 48. $3ab\sqrt[3]{2ab}$ 49. $2x^2y\sqrt[4]{xz^3}$ 50. $2xy$ 51. $\frac{3\sqrt[3]{y}}{x}$ 52. $\frac{5x\sqrt{y}}{y}$ 53. $10x$ 54. $-2y$ 55. $5x$ 56. $8x - 2y$ 57. $15x + 6x^2$ 58. $x^2 - 4xy + 4y^2$ 59. $x^2 - 25y^2$ 60. $6x^2 + 11xy - 10y^2$

7.3 EXERCISES B

Multiply and simplify.

1. $\sqrt{50}\,\sqrt{18}$ **2.** $5\sqrt{63}\,\sqrt{21}$ **3.** $\sqrt{3x}\,\sqrt{21x}$

4. $\sqrt{8x}\,\sqrt{8x}$ **5.** $\sqrt{3xy^2}\,\sqrt{15x^2y}$ **6.** $\sqrt{5a^2b^3}\,\sqrt{45a^3b^3}$

7. $(\sqrt{5xy})^2$ **8.** $\sqrt{4xy}\,\sqrt{2xy^2}\,\sqrt{8x^2y}$ **9.** $\sqrt{3ab^2}\,\sqrt{12a^2b}\,\sqrt{3a^2b}$

10. $8\sqrt[3]{15}\,\sqrt[3]{75}$ **11.** $\sqrt[3]{9}\,\sqrt[3]{81}$ **12.** $\sqrt[4]{25}\,\sqrt[4]{25}$

13. $\sqrt[3]{5x}\,\sqrt[3]{25x^2}$ **14.** $\sqrt[4]{4x^2}\,\sqrt[4]{4x^2}$ **15.** $\sqrt[5]{8x^2}\,\sqrt[5]{8x^3}$

16. $\sqrt[3]{3x^4y^2}\,\sqrt[3]{54x^2y^3}$ **17.** $\sqrt[3]{15ab}\,\sqrt[3]{36a^2b^3}$ **18.** $\sqrt[4]{8x^5y^2}\,\sqrt[4]{10x^2y^3}$

19. $\sqrt{3(x - y)}\,\sqrt{3(x - y)^5}$ **20.** $\sqrt[3]{27x^6y^9}\,\sqrt{27x^3y^3}$ **21.** $\sqrt[3]{24x^5y^7}\,\sqrt[4]{16x^4y^{12}}$

Divide and simplify.

22. $\dfrac{\sqrt{3}}{\sqrt{75}}$ **23.** $\dfrac{7\sqrt{12}}{\sqrt{147}}$ **24.** $\dfrac{\sqrt{75x^3}}{\sqrt{3x}}$

25. $\dfrac{\sqrt{50x^2}}{\sqrt{2x}}$ **26.** $\dfrac{6\sqrt{7x^3y^2}}{\sqrt{9xy^4}}$ **27.** $\dfrac{\sqrt{50x^5y^3}}{\sqrt{98x^7y^2}}$

28. $\dfrac{\sqrt{36(x - y)^2}}{\sqrt{x - y}}$ **29.** $\dfrac{\sqrt[3]{250}}{\sqrt[3]{2}}$ **30.** $\dfrac{\sqrt[3]{15x^3y^2}}{\sqrt[3]{120x^2y^5}}$

31. $\dfrac{\sqrt[4]{32}}{\sqrt[4]{162}}$

32. $\dfrac{\sqrt[4]{81a^9b^7}}{\sqrt[4]{16a^2b^3}}$

33. $\dfrac{\sqrt[5]{3x^{10}}}{\sqrt[5]{96x^4}}$

34. $\dfrac{\sqrt{5y^2}}{\sqrt{5x}}$

35. $\dfrac{\sqrt[3]{27y^3}}{\sqrt[3]{2x^2}}$

36. $\dfrac{\sqrt[4]{81x^5}}{\sqrt[4]{xy^3}}$

37. $\dfrac{\sqrt[3]{(3x+y)(x-3y)^4}}{\sqrt[3]{(3x+y)^4(x-3y)}}$

38. $\dfrac{\sqrt[3]{27x^9y^4}}{\sqrt{121x^4y^6}}$

39. $\dfrac{\sqrt[3]{16x^4y^5}}{\sqrt[4]{16x^8y^{12}}}$

In a wholesale operation the three quantities c, n, and k are related by the equation $c = \sqrt{n}\,\sqrt{k}$.

40. Find c when $n = 8$ and $k = 18$.

41. Find k when $c = 6$ and $n = 4$.

42. Find c when $n = 1$ and $k = \dfrac{1}{5}$.

43. Find c when $n = \dfrac{25}{k}$.

Use the formula for the radius of a circle inscribed in a triangle obtained in Example 4 to find r for the values of a, b, and c given in Exercises 44–45.

44. $a = 7.0$ ft, $b = 25.0$ ft, $c = 20.0$ ft

45. $a = 14.3$ cm, $b = 17.1$ cm, $c = 19.3$ cm

46. Ben Whitney owns a triangular pasture with sides 75.0 yd, 100.0 yd, and 115.0 yd. He wishes to build a circular exercise track for his horses so that the circuit is as large as possible. What will be the radius of the track? [*Hint:* The largest circle in a triangle is the inscribed circle.]

FOR REVIEW

Simplify.

47. $\sqrt{20x^3y^2}$

48. $\sqrt[3]{40a^5b^5}$

49. $\sqrt[4]{81x^4y^3z^9}$

50. $\sqrt{\dfrac{45x^3y^2}{5x}}$

51. $\sqrt[3]{\dfrac{125x^4y^4}{x^3y^7}}$

52. $\sqrt{\dfrac{36y^2}{x}}$

Exercises 53–60 review material covered in Sections 1.5 and 5.2 to help you prepare for the next section. Use the distributive law to collect like terms.

53. $6y + 11y$

54. $5x - 13x$

55. $8y - 3y + 12y$

56. $4x + 7y + x - 10y$

Multiply the polynomials.

57. $2y(3 - 4y)$

58. $(y + 3x)^2$

59. $(y + 4x)(y - 4x)$

60. $(2x - 3y)(4x + 5y)$

7.3 EXERCISES C

Perform the indicated operations and simplify.

1. $\sqrt{70a^3b^4c}\,\sqrt{42ab^2c^7}$

2. $\dfrac{\sqrt[3]{256xyz^2}}{\sqrt[3]{4xy^2z^4}}$

3. $\sqrt{6x^{-3}}\,\sqrt{2x}$

4. $\dfrac{\sqrt{3x^{-1}y^3}}{\sqrt{15xy^{-1}}}$

[Answer: $\dfrac{y^2\sqrt{5}}{5x}$]

5. $\sqrt[3]{5x^{-2}y^5}\,\sqrt[3]{50x^{-1}y^{-1}}$

6. $\dfrac{\sqrt{x^2y^3}\,\sqrt{36xy^2}}{\sqrt{x^{-1}y^{-2}}\,\sqrt{25x^{-2}y^3}}$

[Answer: $\dfrac{6x^3y^2}{5}$]

7.4 MORE OPERATIONS ON RADICALS

STUDENT GUIDEPOSTS

1. Like and Unlike Radicals
2. Adding and Subtracting Radicals
3. Summary of Simplifying Techniques
4. Multiplying Binomial Radicals
5. Rationalizing a Binomial Denominator
6. Rationalizing a Binomial Numerator

1 LIKE AND UNLIKE RADICALS

There are no rules for addition and subtraction similar to the rules for multiplication and division. In general,

$$\sqrt{a+b} \neq \sqrt{a} + \sqrt{b},$$

since, for example,

$$5 = \sqrt{25} = \sqrt{9+16} \neq \sqrt{9} + \sqrt{16} = 3 + 4 = 7.$$

Also, in general,

$$\sqrt{a-b} \neq \sqrt{a} - \sqrt{b},$$

since, for example,

$$4 = \sqrt{16} = \sqrt{25-9} \neq \sqrt{25} - \sqrt{9} = 5 - 3 = 2.$$

We may, however, use the distributive laws and the rules for multiplication and division to simplify some expressions involving addition and subtraction. Before we can add or subtract radicals, they must be *like radicals*. Two radicals that have the same index and the same radicand are **like radicals.** If the indexes or the radicands differ, they are **unlike radicals.** The following tables give examples.

Like radicals	*Index*	*Radicand*
$3\sqrt{7},\ -\sqrt{7},\ 8\sqrt{7}$	2	7
$6\sqrt[3]{x},\ 9\sqrt[3]{x},\ -3\sqrt[3]{x}$	3	x
$-3\sqrt[4]{2xy^3},\ 5\sqrt[4]{2xy^3}$	4	$2xy^3$

Unlike radicals	*Reason*
$5\sqrt{11},\ -3\sqrt[3]{11}$	Indexes are different
$4\sqrt{5},\ 3\sqrt{5x}$	Radicands are different
$2\sqrt[3]{xy^2},\ 2\sqrt[5]{x^2y}$	Indexes and radicands are different

2 ADDING AND SUBTRACTING RADICALS

We add or subtract like radicals using the distributive law just as we collect like terms of polynomials. Recall,

$$6xy - 2xy + 3xy = (6 - 2 + 3)xy = 7xy.$$

The process is the same for like radicals.

$$6\sqrt{xy} - 2\sqrt{xy} + 3\sqrt{xy} = (6 - 2 + 3)\sqrt{xy} = 7\sqrt{xy}$$

EXAMPLE 1 Adding and Subtracting Like Radicals

Add or subtract and simplify.

(a) $2\sqrt{3} + 7\sqrt{3} = (2 + 7)\sqrt{3} = 9\sqrt{3}$ Distributive law

(b) $\sqrt{x} - 3\sqrt{x} = (1 - 3)\sqrt{x} = -2\sqrt{x}$ Distributive law

(c) $5\sqrt{2x} + 3\sqrt{x}$ cannot be simplified since the radicands are different.

(d) $5\sqrt[3]{y} + 3\sqrt[4]{y}$ cannot be simplified since the indexes are different.

(e) $6\sqrt[4]{xy} - 3\sqrt[4]{xy} + 2\sqrt[3]{xy} = (6 - 3)\sqrt[4]{xy} + 2\sqrt[3]{xy}$ Distributive law

$= 3\sqrt[4]{xy} + 2\sqrt[3]{xy}$ These two terms cannot be combined

Practice Exercise 1

Add or subtract and simplify.

(a) $7\sqrt{5} + 10\sqrt{5}$

(b) $2\sqrt{y} - 6\sqrt{y}$

(c) $-3\sqrt[3]{5y} + 8\sqrt[3]{6y}$

(d) $9\sqrt[4]{11x} - 5\sqrt[3]{11x}$

(e) $6\sqrt[3]{4xy} - 2\sqrt[4]{4xy} + 7\sqrt[3]{4xy}$

Answers: (a) $17\sqrt{5}$ (b) $-4\sqrt{y}$ (c) cannot be simplified (d) cannot be simplified (e) $13\sqrt[3]{4xy} - 2\sqrt[4]{4xy}$

Some unlike radicals can be changed into like radicals by simplifying one or both. For example $\sqrt{12}$ and $\sqrt{75}$ are not like radicals, but if we simplify

$$\sqrt{12} = \sqrt{4 \cdot 3} = 2\sqrt{3} \quad \text{and} \quad \sqrt{75} = \sqrt{25 \cdot 3} = 5\sqrt{3},$$

we obtain like radicals. Thus

$$\sqrt{12} + \sqrt{75} = 2\sqrt{3} + 5\sqrt{3} = 7\sqrt{3}.$$

To Add or Subtract Radicals

1. Simplify each radical as much as possible.
2. Use the distributive law to collect any like radicals.

EXAMPLE 2 Adding and Subtracting Radicals

Add or subtract.

(a) $6\sqrt{125} + 3\sqrt{20} = 6\sqrt{25 \cdot 5} + 3\sqrt{4 \cdot 5}$ Factor to obtain perfect squares

$= 6\sqrt{25}\,\sqrt{5} + 3\sqrt{4}\,\sqrt{5}$ $\sqrt{a \cdot b} = \sqrt{a} \cdot \sqrt{b}$

$= 6 \cdot 5 \cdot \sqrt{5} + 3 \cdot 2 \cdot \sqrt{5}$ $6\sqrt{25 \cdot 5} = 6 \cdot 5\sqrt{5}$ *not* $(6 + 5)\sqrt{5}$

$= 30\sqrt{5} + 6\sqrt{5}$

$= (30 + 6)\sqrt{5}$ Distributive law

$= 36\sqrt{5}$

(b) $-7\sqrt[3]{54} + 3\sqrt[3]{16} = -7\sqrt[3]{27 \cdot 2} + 3\sqrt[3]{8 \cdot 2}$ 27 and 8 are perfect cubes

$= -7\sqrt[3]{27} \cdot \sqrt[3]{2} + 3\sqrt[3]{8}\,\sqrt[3]{2}$ $\sqrt[3]{ab} = \sqrt[3]{a}\,\sqrt[3]{b}$

$= -7 \cdot 3 \cdot \sqrt[3]{2} + 3 \cdot 2 \cdot \sqrt[3]{2}$

$= -21\sqrt[3]{2} + 6\sqrt[3]{2}$

$= (-21 + 6)\sqrt[3]{2}$

$= -15\sqrt[3]{2}$

Practice Exercise 2

Add or subtract.

(a) $4\sqrt{75} + 8\sqrt{12}$

(b) $9\sqrt[3]{40} - 3\sqrt[3]{135}$

(c) $8\sqrt{2x^3} - 2x\sqrt{18x} - 5\sqrt{8x^3}$
$= 8\sqrt{x^2 \cdot 2x} - 2x\sqrt{9 \cdot 2x} - 5\sqrt{4x^2 \cdot 2x}$
$= 8\sqrt{x^2}\,\sqrt{2x} - 2x\sqrt{9}\,\sqrt{2x} - 5\sqrt{4}\,\sqrt{x^2}\,\sqrt{2x}$
$= 8x\sqrt{2x} - 2x \cdot 3\sqrt{2x} - 5 \cdot 2 \cdot x\sqrt{2x}$
$= 8x\sqrt{2x} - 6x\sqrt{2x} - 10x\sqrt{2x}$
$= (8x - 6x - 10x)\sqrt{2x}$
$= \ -8x\sqrt{2x}$

(d) $4\sqrt[3]{8y^5} - 3\sqrt[3]{27y^5} = 4\sqrt[3]{8 \cdot y^3 \cdot y^2} - 3\sqrt[3]{27 \cdot y^3 \cdot y^2}$
$= 4\sqrt[3]{8}\,\sqrt[3]{y^3}\,\sqrt[3]{y^2} - 3\sqrt[3]{27}\,\sqrt[3]{y^3}\,\sqrt[3]{y^2}$
$= 4 \cdot 2 \cdot y\sqrt[3]{y^2} - 3 \cdot 3 \cdot y\sqrt[3]{y^2}$
$= 8y\sqrt[3]{y^2} - 9y\sqrt[3]{y^2}$
$= \ -y\sqrt[3]{y^2}$

(c) $-3\sqrt{5y^5} - 2y^2\sqrt{5y} + 8y\sqrt{5y^3}$

(d) $7\sqrt[3]{27x^7} - 5x\sqrt[3]{27x^4}$

Answers: (a) $36\sqrt{3}$ (b) $9\sqrt[3]{5}$ (c) $3y^2\sqrt{5y}$ (d) $6x^2\sqrt[3]{x}$

Sometimes a radical expression can be simplified by first rationalizing a denominator and then adding or subtracting.

EXAMPLE 3 Rationalizing before Adding or Subtracting

Rationalize the denominator and then add or subtract.

(a) $3\sqrt{2x} + \dfrac{5x}{\sqrt{2x}} = 3\sqrt{2x} + \dfrac{5x\sqrt{2x}}{\sqrt{2x}\,\sqrt{2x}}$ Rationalize the denominator

$= 3\sqrt{2x} + \dfrac{5x\sqrt{2x}}{2x}$

$= 3\sqrt{2x} + \dfrac{5\sqrt{2x}}{2}$ Divide out the common factors

$= \left(3 + \dfrac{5}{2}\right)\sqrt{2x}$

$= \dfrac{11}{2}\sqrt{2x} = \dfrac{11\sqrt{2x}}{2}$

(b) $\dfrac{3}{\sqrt[3]{y}} - \dfrac{7\sqrt[3]{y^2}}{y} = \dfrac{3\sqrt[3]{y^2}}{\sqrt[3]{y}\,\sqrt[3]{y^2}} - \dfrac{7\sqrt[3]{y^2}}{y}$ Rationalize the denominator

$= \dfrac{3\sqrt[3]{y^2}}{y} - \dfrac{7\sqrt[3]{y^2}}{y}$

$= \left(\dfrac{3}{y} - \dfrac{7}{y}\right)\sqrt[3]{y^2}$

$= -\dfrac{4}{y}\sqrt[3]{y^2} = \dfrac{-4\sqrt[3]{y^2}}{y}$

Practice Exercise 3

Rationalize the denominator and then add or subtract.

(a) $\dfrac{7y}{\sqrt{5y}} + 6\sqrt{5y}$

(b) $\dfrac{9\sqrt[3]{2y}}{2y} - \dfrac{13}{\sqrt[3]{4y^2}}$

Answers: (a) $\frac{37\sqrt{5y}}{5}$ (b) $-\frac{2\sqrt[3]{2y}}{y}$

3 SUMMARY OF SIMPLIFYING TECHNIQUES

On the next page we present a summary of all the techniques used to simplify radical expressions.

To Simplify a Radical Expression

1. Combine all like radicals by adding and subtracting.
2. When indicated, use the rules for multiplication and division,

$$\sqrt[k]{a}\sqrt[k]{b} = \sqrt[k]{ab} \quad \text{and} \quad \frac{\sqrt[k]{a}}{\sqrt[k]{b}} = \sqrt[k]{\frac{a}{b}}.$$

3. Remove all perfect powers from under the radicals.
4. Rationalize all denominators.

These techniques, as well as the distributive laws, are used in the following multiplication problems.

EXAMPLE 4 MULTIPLYING RADICAL EXPRESSIONS

Multiply.

(a) $\sqrt{7}(\sqrt{35} + \sqrt{7}) = \sqrt{7}\sqrt{35} + \sqrt{7}\sqrt{7}$ Distributive law

$= \sqrt{7 \cdot 35} + \sqrt{7 \cdot 7}$ $\sqrt{a}\sqrt{b} = \sqrt{ab}$

$= \sqrt{7 \cdot 7 \cdot 5} + \sqrt{7 \cdot 7}$ Factor

$= \sqrt{7^2 \cdot 5} + \sqrt{7^2}$ 7^2 is a perfect square

$= \sqrt{7^2}\sqrt{5} + \sqrt{7^2}$

$= 7\sqrt{5} + 7$

(b) $\sqrt{3}(\sqrt{15} - \sqrt{60}) = \sqrt{3}\sqrt{15} - \sqrt{3}\sqrt{60}$ Distributive law

$= \sqrt{3 \cdot 15} - \sqrt{3 \cdot 60}$ $\sqrt{a}\sqrt{b} = \sqrt{ab}$

$= \sqrt{3^2 \cdot 5} - \sqrt{3^2 \cdot 2^2 \cdot 5}$ 3^2 and 2^2 are perfect squares

$= \sqrt{3^2}\sqrt{5} - \sqrt{3^2}\sqrt{2^2}\sqrt{5}$ $\sqrt{ab} = \sqrt{a}\sqrt{b}$

$= 3\sqrt{5} - 3 \cdot 2\sqrt{5}$ Remove perfect squares

$= 3\sqrt{5} - 6\sqrt{5} = -3\sqrt{5}$

(c) $\sqrt[3]{x}(2\sqrt[3]{8x^2} - \sqrt[3]{27x^2})$

$= 2 \cdot \sqrt[3]{x}\sqrt[3]{8x^2} - \sqrt[3]{x}\sqrt[3]{27x^2}$ Distributive law

$= 2\sqrt[3]{x \cdot 8x^2} - \sqrt[3]{x \cdot 27x^2}$ $\sqrt[3]{a}\sqrt[3]{b} = \sqrt[3]{ab}$

$= 2\sqrt[3]{8x^3} - \sqrt[3]{27x^3}$

$= 2\sqrt[3]{8}\sqrt[3]{x^3} - \sqrt[3]{27}\sqrt[3]{x^3}$ $\sqrt[3]{ab} = \sqrt[3]{a}\sqrt[3]{b}$

$= 2 \cdot 2 \cdot x - 3 \cdot x$ Remove perfect cubes

$= 4x - 3x = x$

PRACTICE EXERCISE 4

Multiply.

(a) $\sqrt{11}(\sqrt{22} + \sqrt{11})$

(b) $\sqrt{5}(\sqrt{35} - \sqrt{140})$

(c) $\sqrt[3]{2y}(4\sqrt[3]{32y^2} - \sqrt[3]{108y^2})$

Answers: (a) $11\sqrt{2} + 11$ (b) $-5\sqrt{7}$ (c) $10y$

4 MULTIPLYING BINOMIAL RADICALS

Many radical expressions can be multiplied just as we multiplied binomials in Chapter 5. The FOIL method can be used.

EXAMPLE 5 MULTIPLYING BINOMIAL RADICALS

Multiply.

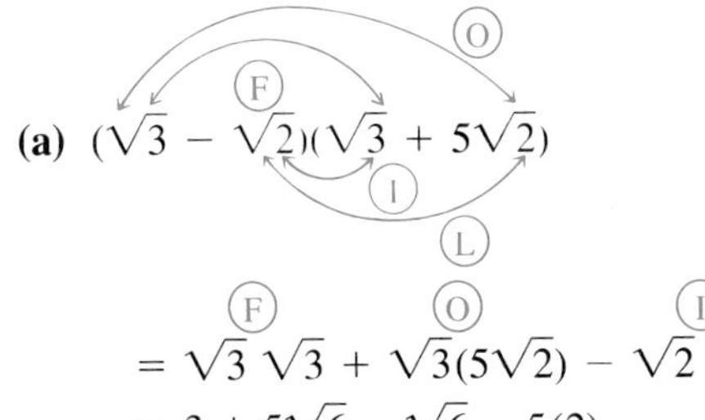

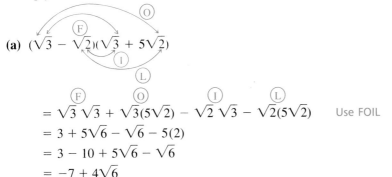

(a) $(\sqrt{3} - \sqrt{2})(\sqrt{3} + 5\sqrt{2})$

$$= \overset{F}{\sqrt{3}\,\sqrt{3}} + \overset{O}{\sqrt{3}(5\sqrt{2})} - \overset{I}{\sqrt{2}\,\sqrt{3}} - \overset{L}{\sqrt{2}(5\sqrt{2})} \quad \text{Use FOIL}$$
$$= 3 + 5\sqrt{6} - \sqrt{6} - 5(2)$$
$$= 3 - 10 + 5\sqrt{6} - \sqrt{6}$$
$$= -7 + 4\sqrt{6}$$

(b) $(\sqrt{x} - 5)^2 = (\sqrt{x} - 5)(\sqrt{x} - 5)$
$$= \sqrt{x}\,\sqrt{x} - 5\sqrt{x} - 5\sqrt{x} + 5 \cdot 5 \quad \text{Use FOIL}$$
$$= \sqrt{x^2} - 10\sqrt{x} + 25$$
$$= x - 10\sqrt{x} + 25$$

(c) $(\sqrt{5} + \sqrt{7})(\sqrt{5} - \sqrt{7})$
$$= \sqrt{5}\,\sqrt{5} - \sqrt{5}\,\sqrt{7} + \sqrt{5}\,\sqrt{7} - \sqrt{7}\,\sqrt{7} \quad \text{Use FOIL}$$
$$= \sqrt{5^2} - \sqrt{35} + \sqrt{35} - \sqrt{7^2}$$
$$= 5 - 7 = -2$$

PRACTICE EXERCISE 5

Multiply.

(a) $(\sqrt{5} + \sqrt{11})(\sqrt{5} - 3\sqrt{11})$

(b) $(\sqrt{2y} + 7)^2$

(c) $(\sqrt{13} + \sqrt{3})(\sqrt{13} - \sqrt{3})$

Answers: (a) $-28 - 2\sqrt{55}$ (b) $2y + 14\sqrt{2y} + 49$ (c) 10

5 RATIONALIZING A BINOMIAL DENOMINATOR

Notice that in Example 5(c) the product was a rational number. This occurred because the expressions multiplied fit the formula, $(a + b)(a - b) = a^2 - b^2$, which always gives a rational number.

$$(\sqrt{5} + \sqrt{7})(\sqrt{5} - \sqrt{7}) = (\sqrt{5})^2 - (\sqrt{7})^2 \quad \sqrt{5} = a \text{ and } \sqrt{7} = b$$
$$= 5 - 7 = -2$$

Square root expressions such as $\sqrt{5} + \sqrt{7}$ and $\sqrt{5} - \sqrt{7}$ are called **conjugates** of each other. Thus $\sqrt{5} - \sqrt{7}$ is the conjugate of $\sqrt{5} + \sqrt{7}$, and $\sqrt{5} + \sqrt{7}$ is the conjugate of $\sqrt{5} - \sqrt{7}$. Since the product of conjugates is a rational number (-2 is our example), we can use this concept to rationalize binomial denominators (denominators with two terms). First consider several examples.

Binomial	*Conjugate*	*Product*
$\sqrt{7} - \sqrt{5}$	$\sqrt{7} + \sqrt{5}$	$(\sqrt{7} - \sqrt{5})(\sqrt{7} + \sqrt{5}) = (\sqrt{7})^2 - (\sqrt{5})^2 = 7 - 5 = 2$
$2\sqrt{3} + \sqrt{2}$	$2\sqrt{3} - \sqrt{2}$	$(2\sqrt{3} + \sqrt{2})(2\sqrt{3} - \sqrt{2}) = (2\sqrt{3})^2 - (\sqrt{2})^2 = 4 \cdot 3 - 2 = 10$
$\sqrt{x} - \sqrt{y}$	$\sqrt{x} + \sqrt{y}$	$(\sqrt{x} - \sqrt{y})(\sqrt{x} + \sqrt{y}) = (\sqrt{x})^2 - (\sqrt{y})^2 = x - y$
$3\sqrt{x} + 2$	$3\sqrt{x} - 2$	$(3\sqrt{x} + 2)(3\sqrt{x} - 2) = (3\sqrt{x})^2 - (2)^2 = 9x - 4$

To Rationalize a Binomial Denominator

1. Simplify the fraction as much as possible.
2. Multiply both numerator and denominator of the fraction by the conjugate of the denominator and simplify.

EXAMPLE 6 RATIONALIZING A BINOMIAL DENOMINATOR

Rationalize the denominator.

(a) $\frac{2}{\sqrt{7}-\sqrt{5}} = \frac{2(\sqrt{7}+\sqrt{5})}{(\sqrt{7}-\sqrt{5})(\sqrt{7}+\sqrt{5})}$ Multiply numerator and denominator by $\sqrt{7}+\sqrt{5}$

$= \frac{2(\sqrt{7}+\sqrt{5})}{(\sqrt{7})^2-(\sqrt{5})^2}$ $(a-b)(a+b)=a^2-b^2$

$= \frac{2(\sqrt{7}+\sqrt{5})}{7-5}$

$= \frac{\cancel{2}(\sqrt{7}+\sqrt{5})}{\cancel{2}} = \sqrt{7}+\sqrt{5}$

(b) $\frac{\sqrt{3}}{5+2\sqrt{3}} = \frac{\sqrt{3}(5-2\sqrt{3})}{(5+2\sqrt{3})(5-2\sqrt{3})}$ Multiply numerator and denominator by $5-2\sqrt{3}$

$= \frac{\sqrt{3}(5-2\sqrt{3})}{(5)^2-(2\sqrt{3})^2}$ $(a+b)(a-b)=a^2-b^2$

$= \frac{\sqrt{3}(5-2\sqrt{3})}{25-4\cdot 3}$ $(2\sqrt{3})^2 = 2^2(\sqrt{3})^2 = 4\cdot 3$

$= \frac{5\sqrt{3}-2\sqrt{3}\sqrt{3}}{25-12}$

$= \frac{5\sqrt{3}-6}{13}$ $2\sqrt{3}\sqrt{3} = 2(\sqrt{3})^2 = 2\cdot 3 = 6$

(c) $\frac{x-9}{\sqrt{x}-3} = \frac{(x-9)(\sqrt{x}+3)}{(\sqrt{x}-3)(\sqrt{x}+3)}$ Multiply numerator and denominator by $\sqrt{x}+3$

$= \frac{(x-9)(\sqrt{x}+3)}{(\sqrt{x})^2-(3)^2}$ $(a-b)(a+b)=a^2-b^2$

$= \frac{\cancel{(x-9)}(\sqrt{x}+3)}{\cancel{(x-9)}}$ Divide out common factors

$= \sqrt{x}+3$

PRACTICE EXERCISE 6

Rationalize the denominator.

(a) $\frac{5}{\sqrt{13}+\sqrt{3}}$

(b) $\frac{\sqrt{7}}{3\sqrt{7}-2}$

(c) $\frac{x-25}{\sqrt{x}+5}$

Answers: (a) $\frac{\sqrt{13}-\sqrt{3}}{2}$ (b) $\frac{21+2\sqrt{7}}{59}$ (c) $\sqrt{x}-5$

CAUTION

The technique for rationalizing a binomial denominator works only when the radicals involved are square roots. For example, it will not rationalize the denominator in

$$\frac{1}{\sqrt[3]{x}+\sqrt[3]{y}}$$

if we multiply numerator and denominator by $\sqrt[3]{x}-\sqrt[3]{y}$ or if we try $\sqrt[3]{x^2}-\sqrt[3]{y^2}$.

6 RATIONALIZING A BINOMIAL NUMERATOR

A binomial numerator can also be rationalized using a similar technique. This time, however, we multiply the numerator and denominator by the conjugate of the *numerator*.

$\frac{\sqrt{2}+\sqrt{5}}{3} =$

$\frac{(\sqrt{2}+\sqrt{5})(\sqrt{2}-\sqrt{5})}{3(\sqrt{2}-\sqrt{5})}$ Multiply numerator and denominator by $\sqrt{2}-\sqrt{5}$, the conjugate of the numerator

$= \frac{(\sqrt{2})^2-(\sqrt{5})^2}{3(\sqrt{2}-\sqrt{5})}$ Use $(a+b)(a-b) = a^2 - b^2$

$= \frac{2-5}{3(\sqrt{2}-\sqrt{5})}$ Simplify

$= \frac{-\cancel{3}}{\cancel{3}(\sqrt{2}-\sqrt{5})}$ Divide out common factor 3 to reduce the fraction

$= \frac{-1}{\sqrt{2}-\sqrt{5}}$

CAUTION

Notice that we did not use the distributive law to write $3(\sqrt{2}-\sqrt{5})$ in the form $3\sqrt{2}-3\sqrt{5}$ in the denominators above. It is best to leave expressions in factored form so we can reduce the fraction (if possible) more easily.

Some equations we will study in the next chapter have solutions similar in form to

$$\frac{-2+2\sqrt{5}}{6}.$$

Such expressions can often be reduced to lowest terms by factoring the numerator and dividing out all common factors. For example,

$$\frac{-2+2\sqrt{5}}{6} = \frac{\cancel{2}(-1+\sqrt{5})}{\cancel{2}\cdot 3} = \frac{-1+\sqrt{5}}{3}.$$

7.4 EXERCISES A

Add or subtract.

1. $\sqrt{5}+3\sqrt{5}$

2. $4\sqrt{7}-9\sqrt{7}$

3. $3\sqrt{5x}-4\sqrt{5x}$

4. $6\sqrt[3]{5}-4\sqrt[4]{5}$

5. $6\sqrt[3]{x^2y}-7\sqrt[3]{x^2y}$

6. $3\sqrt[4]{3y^3}+8\sqrt[4]{3x^3}$

7. $\frac{1}{2}\sqrt{32}-\frac{1}{3}\sqrt{162}$

8. $2\sqrt[3]{2}+6\sqrt[3]{16}$

9 $9\sqrt[3]{250}-4\sqrt[3]{128}$

10. $3\sqrt[4]{48}+\sqrt[4]{243}$

11. $\sqrt[4]{625}-\sqrt[3]{125}$

12. $4\sqrt{98}-3\sqrt{72}+2\sqrt{32}$

13. $-6\sqrt{16y^3} + 2\sqrt{9y^3}$

14. $6\sqrt{48y^4} + 4\sqrt{12y^4}$

15. $y\sqrt{x^3} + x\sqrt{xy^2}$

16. $7\sqrt{5xy^2} - 3\sqrt{5xy^2}$

17. $6\sqrt[3]{8x} - 5\sqrt[3]{27x}$

18. $5\sqrt[3]{3x^4} + 7x\sqrt[3]{3x}$

19. $-7\sqrt[3]{40a^5} + 10a\sqrt[3]{135a^2}$

20. $-\sqrt[4]{x^5y^7} + 2x\sqrt[4]{xy^7}$

21. $3\sqrt[5]{a^{10}b^{12}} - ab\sqrt[5]{a^5b^7}$

22. $-8\sqrt{75a^2b^5} + 3b\sqrt{147a^2b^3}$

23. $2\sqrt[3]{27x^4y^6} - 4y\sqrt[3]{125x^4y^3}$

24. $8\sqrt[3]{54} - 4\sqrt[3]{16} - 5\sqrt[3]{250}$

Rationalize the denominator and then add or subtract.

25. $3\sqrt{5} - \dfrac{1}{\sqrt{5}}$

26. $\dfrac{6}{\sqrt{7}} - 2\sqrt{7}$

27. $3\sqrt{12} - \dfrac{5}{\sqrt{3}}$

28. $6\sqrt{5x} - \dfrac{x}{\sqrt{5x}}$

29. $\dfrac{8\sqrt{xy}}{x} + \dfrac{7y}{\sqrt{xy}}$

30. $\dfrac{\sqrt[3]{x^2y^2}}{xy} + \dfrac{1}{\sqrt[3]{xy}}$

An engineer finds that P, s, and t are related by the equation $P = 2\sqrt{s} + \frac{6}{\sqrt{t}}$.

31. Find P when $s = 27$ and $t = 3$.

32. Find s when $P = 3$ and $t = 36$.

Multiply and simplify.

33. $\sqrt{3}(\sqrt{5} + \sqrt{3})$

34. $\sqrt{5}(\sqrt{15} - \sqrt{35})$

35. $\sqrt{3}(2 + \sqrt{6})$

36. $\sqrt{3}\left(\sqrt{3} + \dfrac{2}{\sqrt{3}}\right)$

37. $\sqrt{8}(\sqrt{2} + 3\sqrt{12})$

38. $\sqrt{xy}\left(\dfrac{1}{\sqrt{x}} + \dfrac{1}{\sqrt{y}}\right)$

39. $\sqrt{3x}(\sqrt{3} - \sqrt{x})$

40. $\sqrt[4]{x^3}(\sqrt[4]{xy^3} + \sqrt[4]{xy^4})$

41. $\sqrt[3]{5}(\sqrt[3]{25} - \sqrt[3]{200})$

42. $(\sqrt{5} + \sqrt{2})(\sqrt{5} + \sqrt{2})$

43. $(3\sqrt{3} + \sqrt{7})(3\sqrt{3} - \sqrt{7})$

44 $(2\sqrt{5} - \sqrt{2})(3\sqrt{5} + 4\sqrt{2})$

45. $(\sqrt{12} - \sqrt{5})^2$

46. $(\sqrt{x} + 1)(\sqrt{x} - 1)$

47. $(\sqrt{x} + 1)^2$

48. $(\sqrt{y} - 3)^2$

49. $(\sqrt{y} + 5)(\sqrt{y} - 3)$

50. $(2\sqrt{x} + \sqrt{y})(2\sqrt{x} - \sqrt{y})$

Rationalize the denominators.

51. $\dfrac{1}{2 + \sqrt{3}}$

52. $\dfrac{4}{1 - \sqrt{5}}$

53. $\dfrac{\sqrt{7}}{\sqrt{2} - \sqrt{3}}$

54 $\dfrac{\sqrt{2}}{2\sqrt{2} + \sqrt{3}}$

55. $\dfrac{\sqrt{8} - \sqrt{3}}{\sqrt{2} + \sqrt{3}}$

56. $\dfrac{\sqrt{x} + 1}{\sqrt{x} - 1}$

57. $\dfrac{\sqrt{x} + \sqrt{y}}{\sqrt{x} - \sqrt{y}}$

58 $\dfrac{4x - 25}{2\sqrt{x} - 5}$

Rationalize the numerators.

59. $\dfrac{\sqrt{3} + 2}{4}$

60. $\dfrac{2 - \sqrt{2}}{2}$

61. $\dfrac{\sqrt{5} + \sqrt{3}}{\sqrt{3}}$

62. $\dfrac{\sqrt{2} - 1}{\sqrt{2} + 1}$

63. $\dfrac{\sqrt{y} - 1}{\sqrt{y} + 1}$

64. $\dfrac{\sqrt{y} - \sqrt{x}}{\sqrt{y} + \sqrt{x}}$

65. $\dfrac{\sqrt{x} + 2}{x - 4}$

66 $\dfrac{\sqrt{x + h} - \sqrt{x}}{h}$

Reduce to lowest terms.

67. $\dfrac{3 + 6\sqrt{2}}{3}$

68 $\dfrac{5 + \sqrt{125}}{10}$

69. $\dfrac{4 - \sqrt{8}}{2}$

Dr. Gordon Johnson's assistant, Burford, obtained the false results in Exercises 70–72. How would you correct the right side of each equation to keep Burford from being fired?

70. $5\sqrt{2y} = \sqrt{10y}$

71. $(\sqrt{y} + 2)^2 = y + 4$

72. $(4\sqrt{3})^2 = 12$

FOR REVIEW

Simplify.

73. $\sqrt[3]{5xy^3}\sqrt[3]{20x^2y}\sqrt[3]{5x^2y}$

74. $\dfrac{\sqrt{98x^7y^3}}{\sqrt{50x^5y^2}}$

75. $\dfrac{\sqrt[3]{250x^4}}{\sqrt[3]{2y}}$

The following exercises review material from Section 1.6 and will help you prepare for the next section. Use the rules of exponents to simplify.

76. $(x^2y)^3$

77. $\left(\dfrac{x^3y^5}{xy^2}\right)^4$

78. $\left(\dfrac{x^{-4}y}{x^{-1}y^3}\right)^{-2}$

ANSWERS: 1. $4\sqrt{5}$ 2. $-5\sqrt{7}$ 3. $-\sqrt{5x}$ 4. cannot be simplified 5. $-\sqrt[3]{x^2y}$ 6. cannot be simplified 7. $-\sqrt{2}$ 8. $14\sqrt[3]{2}$ 9. $29\sqrt[3]{2}$ 10. $9\sqrt[4]{3}$ 11. 0 12. $18\sqrt{2}$ 13. $-18y\sqrt{y}$ 14. $32y^2\sqrt{3}$ 15. $2xy\sqrt{x}$ 16. $4y\sqrt{5x}$ 17. $-3\sqrt[3]{x}$ 18. $12x\sqrt[3]{3x}$ 19. $16a\sqrt[3]{5a^2}$ 20. $xy\sqrt[4]{xy^3}$ 21. $2a^2b^2\sqrt[5]{b^2}$ 22. $-19ab^2\sqrt{3b}$ 23. $-14xy^2\sqrt[3]{x}$ 24. $-9\sqrt[3]{2}$ 25. $\frac{14\sqrt{5}}{5}$ 26. $\frac{-8\sqrt{7}}{7}$ 27. $\frac{13\sqrt{3}}{3}$ 28. $\frac{29\sqrt{5x}}{5}$ 29. $\frac{15\sqrt{xy}}{x}$ 30. $\frac{2\sqrt[3]{x^2y^2}}{xy}$ 31. $8\sqrt{3}$ 32. 1 33. $\sqrt{15}+3$ 34. $5\sqrt{3}-5\sqrt{7}$ 35. $2\sqrt{3}+3\sqrt{2}$ 36. 5 37. $4+12\sqrt{6}$ 38. $\sqrt{y}+\sqrt{x}$ 39. $3\sqrt{x}-x\sqrt{3}$ 40. $x\sqrt[4]{y^3}+xy$ 41. -5 42. $7+2\sqrt{10}$ 43. 20 44. $22+5\sqrt{10}$ 45. $17-4\sqrt{15}$ 46. $x-1$ 47. $x+2\sqrt{x}+1$ 48. $y-3$ 49. $y+2\sqrt{y}-15$ 50. $4x-y$ 51. $2-\sqrt{3}$ 52. $-1-\sqrt{5}$ 53. $-\sqrt{14}-\sqrt{21}$ 54. $\frac{4-\sqrt{6}}{5}$ 55. $3\sqrt{6}-7$ 56. $\frac{x+2\sqrt{x}+1}{x-1}$ 57. $\frac{x+2\sqrt{xy}+y}{x-y}$ 58. $2\sqrt{x}+5$ 59. $\frac{-1}{4(\sqrt{3}-2)}$ 60. $\frac{1}{2+\sqrt{2}}$ 61. $\frac{2}{\sqrt{15}-3}$ 62. $\frac{1}{(\sqrt{2}+1)^2}$ 63. $\frac{y-1}{(\sqrt{y}+1)^2}$ 64. $\frac{y-x}{(\sqrt{y}+\sqrt{x})^2}$ 65. $\frac{1}{\sqrt{x}-2}$ 66. $\frac{1}{\sqrt{x+h}+\sqrt{x}}$ 67. $1+2\sqrt{2}$ 68. $\frac{1+\sqrt{5}}{2}$ 69. $2-\sqrt{2}$ 70. $\sqrt{50y}$ 71. $y+4\sqrt{y}+4$ 72. 48 73. $5xy\sqrt[3]{4x^2y^2}$ 74. $\frac{7x\sqrt{y}}{5}$ 75. $\frac{5x\sqrt[3]{xy^2}}{y}$ 76. x^6y^3 77. x^8y^{12} 78. x^6y^4

7.4 EXERCISES B

Add or subtract.

1. $\sqrt{3}+8\sqrt{3}$

2. $8\sqrt{5}-11\sqrt{5}$

3. $4\sqrt{3x}-6\sqrt{3x}$

4. $7\sqrt[3]{11}-3\sqrt[4]{11}$

5. $5\sqrt[3]{xy^2}-9\sqrt[3]{xy^2}$

6. $3\sqrt[4]{7x}+9\sqrt[4]{7y}$

7. $\frac{2}{3}\sqrt{45}-\frac{3}{4}\sqrt{80}$

8. $8\sqrt[3]{3}+2\sqrt[3]{24}$

9. $7\sqrt[3]{128}-\sqrt[3]{54}$

10. $2\sqrt[4]{162}+\sqrt[4]{32}$

11. $\sqrt[3]{64}-\sqrt[3]{216}$

12. $3\sqrt{147}+2\sqrt{108}-2\sqrt{48}$

13. $-2\sqrt{25y^3}+5\sqrt{49y^3}$

14. $3\sqrt{8y^6}+2\sqrt{18y^6}$

15. $5x\sqrt{y^3}-y\sqrt{x^2y}$

16. $6\sqrt{3x^2y}-2\sqrt{3x^2y}$

17. $8\sqrt[3]{125x}-10\sqrt[3]{64x}$

18. $3x\sqrt[3]{2x}+5\sqrt[3]{2x^4}$

19. $-5\sqrt[3]{24a^5}+3a\sqrt[3]{81a^2}$

20. $-2\sqrt[4]{x^7y^5}+3y\sqrt[4]{x^7y}$

21. $2ab\sqrt[5]{2a^7b^5}-\sqrt[5]{2a^{12}b^{10}}$

22. $-4\sqrt{125a^5b^2}+5a\sqrt{20a^3b^2}$

23. $5\sqrt[3]{8x^6y^4}-3x\sqrt[3]{27x^3y^4}$

24. $2\sqrt[3]{81}+3\sqrt[3]{24}-4\sqrt[3]{375}$

Rationalize the denominator and then add or subtract.

25. $2\sqrt{7}-\dfrac{3}{\sqrt{7}}$

26. $\dfrac{8}{\sqrt{5}}-3\sqrt{5}$

27. $2\sqrt{27}-\dfrac{8}{\sqrt{3}}$

28. $2\sqrt{7x}-\dfrac{3x}{\sqrt{7x}}$

29. $\dfrac{5\sqrt{xy}}{y}+\dfrac{3x}{\sqrt{xy}}$

30. $\dfrac{\sqrt[3]{xy}}{xy}+\dfrac{1}{\sqrt[3]{x^2y^2}}$

An economist finds that C, n, and r are related by the equation $C = 3\sqrt{n} - \frac{10}{\sqrt{r}}$.

31. Find C when $n = 18$ and $r = 2$.

32. Find n when $C = 2$ and $r = 100$.

Multiply and simplify.

33. $\sqrt{5}(\sqrt{3} + \sqrt{5})$

34. $\sqrt{3}(\sqrt{21} - \sqrt{15})$

35. $\sqrt{5}(2 + \sqrt{10})$

36. $\sqrt{2}\left(\sqrt{2} - \dfrac{3}{\sqrt{2}}\right)$

37. $\sqrt{27}(\sqrt{3} + 2\sqrt{12})$

38. $\sqrt{xy}\left(\dfrac{1}{\sqrt{y}} - \dfrac{1}{\sqrt{x}}\right)$

39. $\sqrt{5x}(\sqrt{5} + \sqrt{x})$

40. $\sqrt[4]{y^3}(\sqrt[4]{x^3y} - \sqrt[4]{x^4y})$

41. $\sqrt[3]{3}(\sqrt[3]{9} + \sqrt[3]{72})$

42. $(\sqrt{7} - \sqrt{3})(\sqrt{7} - \sqrt{3})$

43. $(2\sqrt{5} + \sqrt{3})(2\sqrt{5} - \sqrt{3})$

44. $(\sqrt{3} - 4\sqrt{7})(2\sqrt{3} + \sqrt{7})$

45. $(\sqrt{18} - \sqrt{2})^2$

46. $(\sqrt{x} - 2)(\sqrt{x} + 2)$

47. $(\sqrt{x} + 2)^2$

48. $(\sqrt{y} - 5)^2$

49. $(\sqrt{x} + 1)(\sqrt{x} - 3)$

50. $(\sqrt{x} - 3\sqrt{y})(\sqrt{x} + 3\sqrt{y})$

Rationalize the denominators.

51. $\dfrac{3}{1 - \sqrt{2}}$

52. $\dfrac{2}{2 + \sqrt{3}}$

53. $\dfrac{\sqrt{2}}{\sqrt{3} + \sqrt{7}}$

54. $\dfrac{\sqrt{3}}{3\sqrt{3} + \sqrt{2}}$

55. $\dfrac{\sqrt{8} - 3}{3\sqrt{2} - 2}$

56. $\dfrac{\sqrt{y} - 1}{\sqrt{y} + 1}$

57. $\dfrac{\sqrt{x} - \sqrt{y}}{\sqrt{x} + \sqrt{y}}$

58. $\dfrac{4x - 9}{2\sqrt{x} + 3}$

Rationalize the numerators.

59. $\dfrac{\sqrt{3} - 2}{4}$

60. $\dfrac{2 + \sqrt{2}}{2}$

61. $\dfrac{\sqrt{2} - \sqrt{5}}{\sqrt{3}}$

62. $\dfrac{\sqrt{3} + 1}{\sqrt{3} - 1}$

63. $\dfrac{\sqrt{x} + 1}{\sqrt{x} - 1}$

64. $\dfrac{\sqrt{a} + \sqrt{b}}{a - b}$

65. $\dfrac{\sqrt{y} - 2}{y - 4}$

66. $\dfrac{\sqrt{3(x + h)} - \sqrt{3x}}{h}$

Reduce to lowest terms.

67. $\dfrac{8 - \sqrt{48}}{6}$

68. $\dfrac{7 - 14\sqrt{3}}{7}$

69. $\dfrac{3 - \sqrt{18}}{3}$

Dr. Gordon Johnson's assistant, Burford, obtained the false results in Exercises 70–72. How would you correct the right side of each equation to keep Burford from being fired?

70. $3(2\sqrt{xy}) = 6\sqrt{3xy}$

71. $(\sqrt{x} - 5)^2 = x - 25$

72. $(2 + \sqrt{x - 1})^2 = 4 + (x - 1)$

FOR REVIEW

Simplify.

73. $\sqrt[3]{7x^2y}\,\sqrt[3]{28xy^2}\,\sqrt[3]{7xy^2}$

74. $\dfrac{\sqrt{32x^8y^5}}{\sqrt{162x^7y^3}}$

75. $\dfrac{\sqrt[3]{54y^4}}{\sqrt[3]{2x}}$

The following exercises review material from Section 1.6 and will help you prepare for the next section. Use the rules of exponents to simplify.

76. $(xy^3)^4$

77. $\left(\dfrac{x^6y^7}{x^5y}\right)^3$

78. $\left(\dfrac{x^{-1}y^{-6}}{x^3y^{-2}}\right)^{-3}$

7.4 EXERCISES C

Add or subtract.

1. $3a\sqrt{(a-b)^3} + 2ab\sqrt{a-b} + b\sqrt{(a-b)^3}$
[Answer: $(3a^2 - b^2)\sqrt{a-b}$]

2. $\sqrt[5]{64x^7y^5} - x\sqrt[5]{486x^2y^5}$

Rationalize the denominator and add or subtract.

3. $\sqrt{20} + \sqrt{\dfrac{1}{5}} - \dfrac{\sqrt{5}}{5}$

4. $\dfrac{\sqrt{x} + \frac{1}{\sqrt{x}}}{\frac{1}{\sqrt{x}}}$ [*Hint:* Multiply numerator and denominator by $\sqrt{x}$.]

5. $\sqrt{5x^2 + 10x + 5} + \sqrt{20x^2 + 40x + 20}$

6. $5x\sqrt[3]{16xy^3} + 3\sqrt[3]{54x^4y^3} - xy\sqrt[3]{2x}$

Multiply and simplify.

7. $(\sqrt[4]{a} - \sqrt[4]{b^3})(\sqrt[4]{a^3} + \sqrt[4]{b})$

8. $(\sqrt{3+\sqrt{5}} - \sqrt{3-\sqrt{5}})^2$
[Answer: 2]

Rationalize the denominators.

9. $\dfrac{5}{5 - \sqrt{x-5}}$

10. $\dfrac{x - \sqrt{y}}{x - y - \sqrt{y}}$

11. Multiply.
$(\sqrt[3]{a} + \sqrt[3]{b})(\sqrt[3]{a^2} - \sqrt[3]{ab} + \sqrt[3]{b^2})$
[Answer: $a + b$]

12. Rationalize the denominator.
$\dfrac{1}{\sqrt[3]{a} + \sqrt[3]{b}}$
[*Hint*: Consider Exercise 11.]

7.5 RATIONAL EXPONENTS

STUDENT GUIDEPOSTS

1. Defining $a^{1/n}$
2. Defining $a^{m/n}$
3. Simplifying Radicals Using Exponents
4. Rationalizing Denominators
5. Evaluating Expressions with Rational Exponents

In Chapter 1 we discussed the exponential expression a^n, where *base a* is a real number and *exponent n* is an integer. The properties of integer exponents that were established then are reviewed below.

Properties of Exponents

For real numbers a and b and integers m and n,

1. $a^n = \underbrace{a \cdot a \cdot a \ldots a}_{n \text{ factors}} \quad (n \geq 1)$
2. $a^m a^n = a^{m+n}$ (Product rule)
3. $a^m \div a^n = \dfrac{a^m}{a^n} = a^{m-n} \quad (a \neq 0)$ (Quotient rule)
4. $(a^m)^n = a^{mn}$ (Power rule)
5. $a^0 = 1$ for $a \neq 0$
6. $a^{-n} = \dfrac{1}{a^n}$ and $\dfrac{1}{a^{-n}} = a^n$ if $a \neq 0$
7. $(ab)^n = a^n b^n$
8. $\left(\dfrac{a}{b}\right)^n = \dfrac{a^n}{b^n} \quad (b \neq 0)$.

1 DEFINING $a^{1/n}$

We now look at exponential expressions in which the exponent is any rational number (not just an integer). We begin by considering exponents of the form $1/n$, where n is a natural number. For our definition of $a^{1/n}$ to be useful, the properties of exponents should still apply when one or more of the exponents is any rational number. For example, consider $a^{1/2}$. If the power rule $(a^m)^n = a^{mn}$ is to apply, then

$$(a^{1/2})^2 = a^{(1/2)2} = a^1 = a.$$

Similarly, $$(a^{1/3})^3 = a^{(1/3)3} = a^1 = a.$$

As a result, $a^{1/2}$ appears to be a square root of a while $a^{1/3}$ is a cube root of a. In general,

$$a^{1/n} \text{ must be an } n\text{th root of } a$$

since $$(a^{1/n})^n = a^{(1/n)n} = a^{n/n} = a^1 = a.$$

Definition of $a^{1/n}$

Let a be a real number, with n a natural number. Then

$$a^{1/n} = \sqrt[n]{a}.$$

Since $\sqrt[n]{a}$ is not a real number when $a < 0$ and n is even, the definition of $a^{1/n}$ does not apply when a is negative and n is even.

EXAMPLE 1 WRITING $a^{1/n}$ AS A RADICAL

Express each as a radical and simplify.

(a) $3^{1/2} = \sqrt{3}$

(b) $5^{1/3} = \sqrt[3]{5}$

(c) $9^{1/2} = \sqrt{9} = 3 \qquad 9 = 3^2$

(d) $125^{1/3} = \sqrt[3]{125} = 5 \qquad 125 = 5^3$

(e) $16^{1/4} = \sqrt[4]{16} = 2 \qquad 16 = 2^4$

(f) $(243)^{1/5} = \sqrt[5]{243} = 3 \qquad 243 = 3^5$

PRACTICE EXERCISE 1

Express each as a radical and simplify.

(a) $11^{1/2}$

(b) $7^{1/3}$

(c) $25^{1/2}$

(d) $64^{1/3}$

(e) $81^{1/4}$

(f) $32^{1/5}$

(g) $x^{1/5} = \sqrt[5]{x}$

(h) $(xy)^{1/3} = \sqrt[3]{xy}$

(g) $y^{1/3}$

(h) $(2ab)^{1/4}$

Answers: (a) $\sqrt{11}$ (b) $\sqrt[3]{7}$ (c) 5 (d) 4 (e) 3 (f) 2 (g) $\sqrt[3]{y}$ (h) $\sqrt[4]{2ab}$

2 DEFINING $a^{m/n}$

We can now extend the definition of exponents to include rational numbers of the form $\frac{m}{n}$. Suppose we consider the radical $\sqrt[3]{a^2}$. This expression can be rewritten using exponents.

$$\sqrt[3]{a^2} = (a^2)^{1/3} = a^{2 \cdot 1/3} = a^{2/3} \qquad (a^m)^n = a^{m \cdot n}$$

Also, $$(\sqrt[3]{a})^2 = (a^{1/3})^2 = a^{1/3 \cdot 2} = a^{2/3}.$$

Thus, $$\sqrt[3]{a^2} = a^{2/3} = (\sqrt[3]{a})^2.$$

This illustrates the following definition.

Definition of $a^{m/n}$

Let a be a real number, with m and n natural numbers. Then

$$a^{m/n} = \sqrt[n]{a^m} = (\sqrt[n]{a})^m$$

provided that $a^{1/n}$ exists. (That is, if a is not negative when n is even.) Also,

$$a^{-m/n} = \frac{1}{a^{m/n}}$$

provided that $a \neq 0$ and $a^{m/n}$ exists.

All of the properties of integer exponents also hold for rational number exponents whenever the individual exponential expressions exist.

EXAMPLE 2 Expressions with Rational-Number Exponents

Simplify and write with only positive exponents.

(a) $2^{1/3} \cdot 2^{1/4} = 2^{1/3+1/4} = 2^{7/12} \qquad a^m a^n = a^{m+n}$

(b) $\frac{3^{2/3}}{3^{5/3}} = 3^{2/3-5/3} = 3^{-3/3} = 3^{-1} = \frac{1}{3} \qquad \frac{a^m}{a^n} = a^{m-n}$

(c) $(5^{1/3})^{3/4} = 5^{(1/3)(3/4)} = 5^{1/4} \qquad (a^m)^n = a^{mn}$

(d) $81^{-3/4} = \frac{1}{81^{3/4}} \qquad a^{-m/n} = \frac{1}{a^{m/n}}$

$= \frac{1}{(\sqrt[4]{81})^3} = \frac{1}{3^3} = \frac{1}{27} \qquad$ Also using $a^{3/4} = (\sqrt[4]{a})^3$

(e) $(3 \cdot 5)^{2/3} = 3^{2/3} \cdot 5^{2/3} \qquad (ab)^n = a^n b^n$

(f) $\left(\frac{8}{27}\right)^{2/3} = \frac{8^{2/3}}{27^{2/3}} \qquad \left(\frac{a}{b}\right)^n = \frac{a^n}{b^n}$

$= \frac{(\sqrt[3]{8})^2}{(\sqrt[3]{27})^2} = \frac{2^2}{3^2} = \frac{4}{9} \qquad$ Also using $a^{2/3} = (\sqrt[3]{a})^2$

PRACTICE EXERCISE 2

Simplify and write with only positive exponents.

(a) $3^{1/5} \cdot 3^{1/3}$

(b) $\frac{2^{1/4}}{2^{5/4}}$

(c) $(7^{2/3})^{3/4}$

(d) $27^{-4/3}$

(e) $(4 \cdot 7)^{3/2}$

(f) $\left(\frac{81}{16}\right)^{3/4}$

Answers: (a) $3^{8/15}$ (b) $\frac{1}{2}$ (c) $7^{1/2}$ (d) $\frac{1}{81}$ (e) $8 \cdot 7^{3/2}$ (f) $\frac{27}{8}$

Notice that in Example 2(f) we could have written

$$\left(\frac{8}{27}\right)^{2/3} = \frac{8^{2/3}}{27^{2/3}} = \frac{\sqrt[3]{8^2}}{\sqrt[3]{27^2}} = \frac{\sqrt[3]{64}}{\sqrt[3]{729}} = \frac{4}{9}.$$

However, to square first and then find the cube root of a large number is more difficult. Thus, in general we take the root first and then raise to the power using

$$a^{m/n} = (\sqrt[n]{a})^m.$$

EXAMPLE 3 Simplifying Exponential Expressions

Simplify. Assume all variables are positive and write with only positive exponents.

(a) $(8x^3)^{2/3} = (\sqrt[3]{8x^3})^2 = (2x)^2 = 4x^2$

(b) $x^{2/3}x^{1/3} = x^{2/3+1/3} = x^1 = x$

(c) $x^{2/3}x^{-1/3} = x^{2/3-1/3} = x^{1/3}$

(d) $(y^2)^{-1/3} = y^{2\cdot(-1/3)} = y^{-2/3} = \dfrac{1}{y^{2/3}}$

(e) $\dfrac{y^{1/2}}{y^{1/4}} = y^{1/2-1/4} = y^{1/4}$

(f) $(9y^2)^{3/2} = (\sqrt[2]{9y^2})^3 = (3y)^3 = 27y^3$

(g) $(x^6)^{1/8} = x^{6\cdot 1/8} = x^{6/8} = x^{3/4}$

Practice Exercise 3

Simplify and write with only positive exponents.

(a) $(27x^6)^{1/3}$

(b) $y^{4/3}y^{2/3}$

(c) $y^{4/3}y^{-1/3}$

(d) $(x^3)^{-2/3}$

(e) $\dfrac{y^{3/5}}{y^{1/5}}$

(f) $(32y^{10})^{2/5}$

(g) $(u^{12})^{3/4}$

Answers: (a) $3x^2$ (b) y^2 (c) y (d) $\frac{1}{x^2}$ (e) $y^{2/5}$ (f) $4y^4$ (g) u^9

3 SIMPLIFYING RADICALS USING EXPONENTS

We sometimes use rational exponents to simplify radical expressions.

EXAMPLE 4 Simplifying Radicals Using Exponents

Simplify.

(a)
$$\begin{aligned}\sqrt{25x^8y^6} &= (25x^8y^6)^{1/2} && \sqrt{a} = a^{1/2}\\ &= (5^2)^{1/2}(x^8)^{1/2}(y^6)^{1/2} && (ab)^n = a^nb^n\\ &= 5^{2\cdot 1/2}x^{8\cdot 1/2}y^{6\cdot 1/2} && (a^m)^n = a^{mn}\\ &= 5x^4y^3\end{aligned}$$

(b)
$$\begin{aligned}\sqrt[3]{\frac{8x^{10}}{27y^9}} &= \left(\frac{2^3x^{10}}{3^3y^9}\right)^{1/3} && \sqrt[3]{a} = a^{1/3}\\ &= \frac{(2^3)^{1/3}(x^{10})^{1/3}}{(3^3)^{1/3}(y^9)^{1/3}} && (ab)^n = a^nb^n \text{ and } \left(\frac{a}{b}\right)^n = \frac{a^n}{b^n}\\ &= \frac{2x^{10/3}}{3y^{9/3}} && (a^m)^n = a^{mn}\\ &= \frac{2x^{9/3}x^{1/3}}{3y^3} && a^{10/3} = a^{9/3+1/3} = a^{9/3}\cdot a^{1/3}\\ &= \frac{2x^3\sqrt[3]{x}}{3y^3}\end{aligned}$$

Practice Exercise 4

Simplify.

(a) $\sqrt{36x^{10}y^{12}}$

(b) $\sqrt[3]{\dfrac{81y^8}{125x^{15}}}$

Answers: (a) $6x^5y^6$ (b) $\frac{3y^2\sqrt[3]{3y^2}}{5x^5}$

4 RATIONALIZING DENOMINATORS

To rationalize denominators using rational exponents, multiply numerator and denominator by an expression that makes the power in the denominator a whole number.

EXAMPLE 5 Rationalizing Denominators

Rationalize the denominator.

(a) $\dfrac{1}{\sqrt[3]{7^2}} = \dfrac{1}{7^{2/3}} = \dfrac{1 \cdot 7^{1/3}}{7^{2/3} \cdot 7^{1/3}}$

$= \dfrac{7^{1/3}}{7^{2/3+1/3}} = \dfrac{7^{1/3}}{7^1} = \dfrac{7^{1/3}}{7} = \dfrac{\sqrt[3]{7}}{7}$

(b) $\dfrac{3}{\sqrt[4]{x}} = \dfrac{3}{x^{1/4}} = \dfrac{3 \cdot x^{3/4}}{x^{1/4} \cdot x^{3/4}} = \dfrac{3x^{3/4}}{x^{1/4+3/4}} = \dfrac{3x^{3/4}}{x} = \dfrac{3\sqrt[4]{x^3}}{x}$

PRACTICE EXERCISE 5

Rationalize the denominator.

(a) $\dfrac{1}{\sqrt[4]{11^3}}$

(b) $\dfrac{\sqrt[3]{y}}{\sqrt[3]{x}}$

Answers: (a) $\dfrac{\sqrt[4]{11}}{11}$ (b) $\dfrac{\sqrt[3]{x^2y}}{x}$

5 EVALUATING EXPRESSIONS WITH RATIONAL EXPONENTS

Some expressions involving rational number exponents are themselves rational numbers and can be simplified, while others represent irrational numbers and cannot be simplified, only approximated. For example,

$$4^{1/2} = 2, \quad 8^{2/3} = 4, \quad \text{and} \quad \left(\frac{8}{27}\right)^{2/3} = \frac{4}{9}$$

all represent rational numbers, whereas

$$5^{1/4}, \quad 2^{13/12}, \quad \text{and} \quad \left(\frac{8}{27}\right)^{3/2}$$

represent irrational numbers and cannot be simplified further. To approximate numbers such as these, we can use a calculator with a $\boxed{y^x}$ key. For example, if we use the following steps on a calculator,

$$5 \ \boxed{y^x} \ 4 \ \boxed{1/x} \ \boxed{=} \rightarrow \boxed{1.49535},$$

The display shows 1.49535 as an approximation of $5^{1/4}$ or $\sqrt[4]{5}$. Similarly, if we use the steps

$$2 \ \boxed{y^x} \ \boxed{(} \ 13 \ \boxed{\div} \ 12 \ \boxed{)} \ \boxed{=} \rightarrow \boxed{2.11893},$$

the display shows 2.11893 as an approximation of $2^{13/12}$. Finally, $\left(\frac{8}{27}\right)^{3/2}$ can be approximated as follows:

$$8 \div 27 \ \boxed{=} \ \boxed{y^x} \ \boxed{(} \ 3 \ \boxed{\div} \ 2 \ \boxed{)} \ \boxed{=} \rightarrow \boxed{0.161283}.$$

To evaluate expressions with rational number exponents, a calculator with a $\boxed{y^x}$ key is necessary in many applications.

EXAMPLE 6 Investment Problem

The annual rate of return r on an investment of P dollars that is worth A dollars after t years is given by $r = \left(\frac{A}{P}\right)^{1/t} - 1$. Find the approximate rate of return on a painting which was purchased for \$1500 and sold 8 years later for \$9400.

PRACTICE EXERCISE 6

Find the rate of return on a necklace that was purchased for \$2110 and sold 6 years later for \$3950.

We use the given formula with $P = \$1500$, $A = \$9400$, and $t = 8$ years.

$$r = \left(\frac{9400}{1500}\right)^{1/8} - 1$$

Use the following calculator steps.

9400 [÷] 1500 [=] [y^x] 8 [1/x] [=] [−] 1 [=] → [0.257852]

The return is approximately 25.8%.

Answer: 11.0%

7.5 EXERCISES A

Simplify by writing each expression as an integer or a rational number, reduced to lowest terms.

1. $36^{1/2}$ **2.** $(-16)^{1/2}$ **3.** $-144^{1/2}$ **4.** $27^{1/3}$

5. $(-64)^{1/3}$ **6.** $-8^{1/3}$ **7.** $(-4)^{1/2}$ **8.** $25^{3/2}$

9. $16^{-1/2}$ **10.** $\left(\frac{9}{16}\right)^{1/2}$ **11.** $\left(\frac{1}{16}\right)^{-1/4}$ **12.** $\frac{1}{25^{-1/2}}$

13. $(2^3)^{4/3}$ **14.** $4^{2/3}4^{1/3}$ **15.** $(4^{-6})^{-1/12}$ **16.** $\frac{81^{3/4}}{81^{1/4}}$

In Exercises 17–28 simplify by writing the expression so that each variable occurs only once and all exponents are positive. All variables represent positive real numbers.

17. $x^{2/3}x^{1/3}$ **18.** $\frac{w^{3/4}}{w^{1/4}}$ **19.** $(x^{1/2})^3$

20. $x^{-1/3}x^{4/3}$ **21.** $(w^3y^6)^{1/3}$ **22.** $(a^{1/3}b)^{1/2}$

23 $\left(\frac{a^3}{b^{-6}}\right)^{1/3}$ **24.** $(8a^3b^{-6})^{1/3}$ **25.** $(3a^{1/2}b^{3/2})^2$

26. $\left(\frac{x^{-1}y^{-1/2}z}{x^{-2/3}y^{1/3}z^{2/3}}\right)^6$ **27** $\left(\frac{96a^3b^{-2}}{3a^{-2}b^8}\right)^{1/5}$ **28.** $\left(\frac{x^{-2/3}}{y^{-1/4}}\right)^{12}(x^{-1/3}y)^{-3}$

Write all radicals in exponential notation, simplify, and give answers in radical notation.

29. $\sqrt[3]{125y^6}$ **30.** $\sqrt[4]{a^8b^{12}}$ **31.** $\sqrt{8x^4y^8}$

32. $\sqrt[3]{8x^4}$

33. $\sqrt[3]{\frac{16x^9}{y^{12}}}$

34 $\sqrt[5]{\frac{a^{20}}{625b^{15}}}$

Use rational exponents to rationalize the denominator. Give answers in radical form.

35. $\frac{1}{\sqrt{5}}$

36 $\frac{1}{\sqrt[3]{5^2}}$

37. $\frac{3}{\sqrt[4]{3}}$

38. $\frac{2}{\sqrt[3]{x}}$

39. $\frac{6}{\sqrt[4]{y^3}}$

40. $\frac{2}{\sqrt[5]{ab}}$

A chemist learns that C and r are related by the equation $C = 3r^{2/3}$.

41. Find C when $r = 8$.

42. Find C when $r = 0$.

43. Find C when $r = 125$.

44. Find C when $r = -8$.

Use a calculator with a y^x key to approximate each number correct to the nearest tenth.

45. $7^{1/3}$

46. $3^{3/4}$

47. $5^{-2/3}$

48. $(1.08)^{7/3}$

49. $(1.06)^{-5/4}$

50. $\left(\frac{3}{5}\right)^{5/3}$

51. The length of an edge of a cube with volume V is given by the formula $e = V^{1/3}$. A storage bin to be constructed in the shape of a cube must hold 50 ft^3 of grain. To the nearest tenth of a foot, what must be the length of an edge of the bin?

52. The radius r of a sphere with volume V is given by $r = \sqrt[3]{\frac{3V}{4\pi}}$. Find the radius (to the nearest tenth of an inch) of a sphere with volume 65.5 in^3. Use 3.14 for π.

53. The annual rate of return r on an investment of P dollars that is worth A dollars after t years is given by $r = \left(\frac{A}{P}\right)^{1/t} - 1$. Find the approximate annual rate of return on a set of rare books purchased for \$1000 and sold 5 years later for \$2400.

Burford worked Exercises 54 and 55. What is wrong with each of them?

54. $-2 = (-2)^1 = (-2)^{2/2} = [(-2)^2]^{1/2} = 4^{1/2} = 2$

55. $3 = 3^1 = 3^{3/3} = (3^3)^{1/3} = 27^{1/3} = \frac{1}{3}(27) = 9$

FOR REVIEW

Add or subtract.

56. $-3\sqrt{147} - 5\sqrt{48}$

57. $2\sqrt{9x^3} - 3\sqrt{4x^3}$

58. $\sqrt[3]{8x^4y^6} + y\sqrt[3]{27x^4y^3}$

59. $\sqrt{3x} - \dfrac{x}{\sqrt{3x}}$

Simplify. Rationalize denominators.

60. $\sqrt{5}(\sqrt{125} - \sqrt{45})$

61. $(\sqrt{x} - \sqrt{2})(\sqrt{x} + 2\sqrt{2})$

62. $\dfrac{\sqrt{3}}{\sqrt{3} - \sqrt{2}}$

63. $\dfrac{x - 25}{\sqrt{x} + 5}$

Exercises 64 and 65 will help you prepare for the next section. Find each power.

64. $(1 + \sqrt{y - 2})^2$

65. $(1 - 2\sqrt{x + 1})^2$

ANSWERS: 1. 6 2. not a real number 3. -12 4. 3 5. -4 6. -2 7. not a real number 8. 125 9. $\frac{1}{4}$ 10. $\frac{3}{4}$ 11. 2 12. 5 13. 16 14. 4 15. 2 16. 9 17. x 18. $w^{1/2}$ 19. $x^{3/2}$ 20. x 21. wy^2 22. $a^{1/6}b^{1/2}$ 23. ab^2 24. $\frac{2a}{b^2}$ 25. $9ab^3$ 26. $\frac{z^2}{x^2y^5}$ 27. $\frac{2a}{b^2}$ 28. $\frac{1}{x^7}$ 29. $5y^2$ 30. a^2b^3 31. $2x^2y^4\sqrt{2}$ 32. $2x\sqrt[3]{x}$ 33. $\frac{2x^3\sqrt[3]{2}}{y^4}$ 34. $\frac{a^4\sqrt[5]{5}}{5b^3}$ 35. $\frac{\sqrt{5}}{5}$ 36. $\frac{\sqrt[3]{5}}{5}$ 37. $\sqrt[4]{27}$ 38. $\frac{2\sqrt[3]{x^2}}{x}$ 39. $\frac{6\sqrt[4]{y}}{y}$ 40. $\frac{2\sqrt[5]{a^4b^4}}{ab}$ 41. 12 42. 0 43. 75 44. 12 45. 1.9 46. 2.3 47. 0.3 48. 1.2 49. 0.9 50. 0.4 51. 3.7 ft 52. 2.5 in 53. 19.1% 54. -2 cannot be written as $[(-2)^2]^{1/2}$ which is $4^{1/2} = 2$. 55. $27^{1/3}$ is $\sqrt[3]{27}$ *not* $\frac{1}{3}(27) = 9$ 56. $-41\sqrt{3}$ 57. 0 58. $5xy^2\sqrt[3]{x}$ 59. $\frac{2\sqrt{3x}}{3}$ 60. 10 61. $x + \sqrt{2x} - 4$ 62. $3 + \sqrt{6}$ 63. $\sqrt{x} - 5$ 64. $-1 + 2\sqrt{y - 2} + y$ 65. $5 - 4\sqrt{x + 1} + 4x$

7.5 EXERCISES B

Simplify by writing each expression as an integer or a rational number, reduced to lowest terms.

1. $64^{1/2}$

2. $(-49)^{1/2}$

3. $-9^{1/2}$

4. $64^{1/3}$

5. $(-27)^{1/3}$

6. $-27^{1/3}$

7. $(-9)^{1/2}$

8. $64^{2/3}$

9. $27^{-1/3}$

10. $\left(\dfrac{27}{8}\right)^{2/3}$

11. $\left(\dfrac{1}{81}\right)^{-1/4}$

12. $\dfrac{1}{9^{-1/2}}$

13. $(3^2)^{5/2}$

14. $7^{2/5}7^{3/5}$

15. $(3^{-8})^{-1/4}$

16. $\dfrac{27^{1/3}}{27^{2/3}}$

Simplify by writing the expression so that each variable occurs only once and all exponents are positive. All variables represent positive real numbers.

17. $z^{1/4}z^{3/4}$

18. $\dfrac{y^{5/6}}{y^{1/6}}$

19. $(y^5)^{3/5}$

20. $z^{-1/5}z^{2/5}$

21. $(a^4b^8)^{1/4}$

22. $(u^{1/4}v)^{1/3}$

23. $\left(\dfrac{w^4}{z^{-8}}\right)^{1/4}$

24. $(4u^2v^{-6})^{1/2}$

25. $(5x^{1/3}y^{2/3})^3$

26. $\left(\dfrac{a^{-2}b^{-1/4}c^3}{a^{-1/2}b^{3/4}c^0}\right)^4$

27. $\left(\dfrac{b^{-1/12}}{8b^{1/3}b^{-1/4}}\right)^2$

28. $\left(\dfrac{a^{1/2}}{b^{-3/4}}\right)^8(a^{-1/5}b)^5$

Write all radicals in exponential notation, simplify, and give answers in radical notation.

29. $\sqrt[3]{64x^9}$

30. $\sqrt[5]{a^{10}b^{15}}$

31. $\sqrt{12x^{10}y^6}$

32. $\sqrt[3]{27x^5}$

33. $\sqrt[3]{\dfrac{81y^{12}}{x^{15}}}$

34. $\sqrt[5]{\dfrac{a^{25}}{81b^{15}}}$

Use rational exponents to rationalize each denominator. Give answers in radical form.

35. $\dfrac{1}{\sqrt{7}}$

36. $\dfrac{1}{\sqrt[3]{7^2}}$

37. $\dfrac{5}{\sqrt[4]{5}}$

38. $\dfrac{-6}{\sqrt[3]{x^2}}$

39. $\dfrac{9}{\sqrt[5]{y^4}}$

40. $\dfrac{3}{\sqrt[4]{ab}}$

A recording company finds that R and s are related by the equation $R = 5s^{3/4}$.

41. Find R when $s = 16$.

42. Find R when $s = 0$.

43. Find R when $s = 81$.

44. Find R when $s = -81$.

Use a calculator with a $\boxed{y^x}$ key to approximate each number correct to the nearest tenth.

45. $12^{1/5}$

46. $7^{2/3}$

47. $9^{-3/4}$

48. $(1.09)^{7/3}$

49. $(1.07)^{-9/4}$

50. $\left(\dfrac{6}{7}\right)^{7/6}$

51. The length of an edge of a cube with volume V is given by the formula $e = V^{1/3}$. What is the length of an edge, to the nearest hundredth of an inch, of a cubical container that holds 1 quart of water? Use the fact that 1 qt = 57.75 in^3.

52. The radius r of a sphere with volume V is given by $r = \sqrt[3]{\frac{3V}{4\pi}}$. Find the radius, to the nearest tenth of a foot, of a sphere with volume 126.4 ft^3.

53. The annual rate of return r on an investment of P dollars that is worth A dollars after t years is given by $r = \left(\frac{A}{P}\right)^{1/t} - 1$. Find the approximate annual rate of return on an antique car purchased for \$6000 and sold 4 years later for \$8500.

Burford worked Exercises 54 and 55. What is wrong with each of them?

54. $-4 = (-4)^1 = (-4)^{2/2} = [(-4)^2]^{1/2} = 16^{1/2} = 4$

55. $2 = 2^1 = 2^{3/3} = (2^3)^{1/3} = 8^{1/3} = \dfrac{1}{3}(8) = \dfrac{8}{3}$

FOR REVIEW

Add or subtract.

56. $-4\sqrt{80} - 3\sqrt{45}$

57. $-7\sqrt{16x^3} + 5\sqrt{49x^3}$

58. $x\sqrt[3]{x^5y^9} - y\sqrt[3]{8x^8y^6}$

59. $\sqrt{5x} + \dfrac{x}{\sqrt{5x}}$

Simplify. Rationalize denominators.

60. $\sqrt{3}(\sqrt{12} + \sqrt{147})$

61. $(2\sqrt{y} + \sqrt{3})(\sqrt{y} - \sqrt{3})$

62. $\dfrac{\sqrt{5}}{\sqrt{5} - 2}$

63. $\dfrac{x - 36}{\sqrt{x} - 6}$

Exercises 64 and 65 will help you prepare for the next section. Find each power.

64. $(2 - \sqrt{x + 1})^2$

65. $(1 + 2\sqrt{x - 3})^2$

7.5 EXERCISES C

Simplify and give answers in radical form.

1. Is $(a + b)^{1/2}$ the same as $a^{1/2} + b^{1/2}$?

2. Is $(a + b)^{1/2}$ the same as $\frac{1}{2}(a + b)$?

3. For what values of x does $(x^2)^{1/2} = x$?

4. For what values of x does $(x^2)^{1/2} = -x$?

Multiply and simplify expressing all answers with positive exponents. All variables represent positive numbers.

5. $a^{1/4}(a^{3/4} + 2a^{1/4})$

6. $2b^{1/3}(b^{-1/3} - b^{1/3})$

7. $(a^{1/2} + b^{1/2})(a^{1/2} - b^{1/2})$

8. $(a^{1/2} - b^{1/2})^2$

Simplify each expression. Assume that m is a positive integer and that x and y are positive real numbers.

9. $x^m x^{m/3}$ [Answer: $x^{4m/3}$]

10. $(x^3)^{m/3}(y^2)^{m/2}$

11. $\dfrac{x^{m/2}}{x^{m/4}}$

12. $\left(\dfrac{x^3}{y}\right)^m\left(\dfrac{y^m}{x^3}\right)^m$

13. $\dfrac{x^{m/3}y^{m/2}}{(xy)^m}$

14. $\left(\dfrac{x^{2m}y^{4m}}{x^m}\right)^{1/m}$

15. Factor $a^{1/2}$ from $a^{3/2} - a^{1/2}$
[Answer: $a^{1/2}(a - 1)$]

16. Factor $x^{1/2} + x^{1/4} - 2$ as if it were a trinomial.
[*Hint:* Let $u = x^{1/4}$.] [Answer: $(x^{1/4} - 1)(x^{1/4} + 2)$]

7.6 SOLVING RADICAL EQUATIONS

STUDENT GUIDEPOSTS

1. Radical Equations
2. Solving Radical Equations
3. Literal Equations

1 RADICAL EQUATIONS

An equation containing radicals with variables in the radicand is called a **radical equation.** To solve a radical equation we change it to an equation containing no radicals by using the rule on the following page.

Power Rule

If a and b are algebraic expressions such that $a = b$, then $a^n = b^n$ for any natural number n.

When this rule is applied to an equation involving a variable, the resulting equation may have more solutions than the original. For example, the equation

$$x = 5$$

has only one solution, 5, but if we square both sides,

$$x^2 = 25$$

has two solutions, 5 and -5. Thus, since $a = b$ and $a^n = b^n$ may not be equivalent, we need to check any possible solutions in the original equation after using the power rule. Solutions that do not check in the original equation are called **extraneous roots.**

2 SOLVING RADICAL EQUATIONS

We now present a method for solving radical equations.

To Solve a Radical Equation

1. If only one radical is present, isolate this radical on one side of the equation, simplify, and proceed to step 3.
2. If two radicals are present, isolate one of the radicals on one side of the equation.
3. Raise both sides to a power equal to the index on the radical.
4. Solve the resulting equation. If a radical remains in the equation, isolate it and raise to the power again.
5. Check all possible solutions in the *original* equation since the equations involved may not be equivalent. If all roots are extraneous, the equation has no solution.

CAUTION

It is necessary to isolate the radicals, as stated in step 1 above, since the entire quantity on each side of the equation must be raised to the power. *You cannot square term by term.*

$(\sqrt{4})^2 + (\sqrt{9})^2 = 4 + 9 = 13 \neq (\sqrt{4} + \sqrt{9})^2 = (2 + 3)^2 = 5^2 = 25$

EXAMPLE 1 Solving a Radical Equation

Solve.

$$3\sqrt{2y - 5} - \sqrt{y + 23} = 0$$

$$3\sqrt{2y - 5} = \sqrt{y + 23} \quad \text{Isolate the radicals}$$

$$(3\sqrt{2y - 5})^2 = (\sqrt{y + 23})^2 \quad \text{Use the power rule}$$

$$9(2y - 5) = y + 23 \quad \text{Do not forget to square 3}$$

$$18y - 45 = y + 23$$

$$17y = 68$$

$$y = 4$$

Practice Exercise 1

Solve.

$$2\sqrt{4y - 3} - \sqrt{2y + 9} = 0$$

Check: $3\sqrt{2(4) - 5} - \sqrt{4 + 23} \stackrel{?}{=} 0$

$3\sqrt{3} - \sqrt{27} \stackrel{?}{=} 0$

$3\sqrt{3} - 3\sqrt{3} \stackrel{?}{=} 0$, so the solution is 4.

Answer: $\frac{3}{2}$

To stress the need for checking solutions we consider an equation with no solution.

EXAMPLE 2 Solving a Radical Equation

Solve.

$\sqrt{x^2 + 5} - x + 5 = 0$

$\sqrt{x^2 + 5} = x - 5$ Isolate the radical

$(\sqrt{x^2 + 5})^2 = (x - 5)^2$ Use the power rule

$x^2 + 5 = x^2 - 10x + 25$ $(x - 5)^2$ is *not* $x^2 - 25$; do not forget the middle term $-10x$

$5 = -10x + 25$ Subtract x^2 from both sides

$-20 = -10x$

$2 = x$

Check: $\sqrt{2^2 + 5} - 2 + 5 \stackrel{?}{=} 0$

$\sqrt{9} - 2 + 5 \stackrel{?}{=} 0$

$3 - 2 + 5 \stackrel{?}{=} 0$

$6 \neq 0$, so the equation has no solution.

Practice Exercise 2

Solve.

$\sqrt{x^2 - 9} + x - 1 = 0$

Answer: no solution

EXAMPLE 3 Squaring Twice to Solve an Equation

Solve.

$\sqrt{y + 3} - \sqrt{y - 2} = 1$

$\sqrt{y + 3} = 1 + \sqrt{y - 2}$ Isolate $\sqrt{y + 3}$

$(\sqrt{y + 3})^2 = (1 + \sqrt{y - 2})^2$ Use the power rule

$y + 3 = 1 + 2\sqrt{y - 2} + y - 2$ Don't forget the middle term, $2\sqrt{y - 2}$

$y + 3 = 2\sqrt{y - 2} + y - 1$

$3 = 2\sqrt{y - 2} - 1$ A radical remains

$4 = 2\sqrt{y - 2}$ To isolate $\sqrt{y - 2}$, add 1 to both sides

$2 = \sqrt{y - 2}$ Simplify; divide both sides by 2

$2^2 = (y - 2)^2$ Use power rule again

$4 = y - 2$

$6 = y$

The solution is 6. Check this in the original equation.

Practice Exercise 3

Solve.

$\sqrt{y + 6} - \sqrt{y - 1} = 1$

Answer: 10

Some radical equations result in equations that have to be solved by the factoring technique studied in Chapter 5.

EXAMPLE 4 Solving a Radical Equation that Is Quadratic

Solve.

$\sqrt{x+15} - x + 5 = 0$

$$\sqrt{x+15} = x - 5 \quad \text{Isolate the radical}$$
$$(\sqrt{x+15})^2 = (x-5)^2 \quad \text{Square both sides}$$
$$x + 15 = x^2 - 10x + 25 \quad (a-b)^2 = a^2 - 2ab + b^2$$
$$0 = x^2 - 11x + 10 \quad \text{Subtract } x \text{ and } 15$$
$$0 = (x-1)(x-10) \quad \text{Factor}$$
$$x - 1 = 0 \quad \text{or} \quad x - 10 = 0 \quad \text{Zero-product rule}$$
$$x = 1 \qquad\qquad x = 10$$

Check:
$$\sqrt{1+15} - 1 + 5 \stackrel{?}{=} 0 \qquad \sqrt{10+15} - 10 + 5 \stackrel{?}{=} 0$$
$$\sqrt{16} - 1 + 5 \stackrel{?}{=} 0 \qquad \sqrt{25} - 10 + 5 \stackrel{?}{=} 0$$
$$4 - 1 + 5 \stackrel{?}{=} 0 \qquad 5 - 10 + 5 \stackrel{?}{=} 0$$
$$8 \neq 0 \qquad 0 = 0$$

1 does not check. 10 does check.

The only solution is 10.

PRACTICE EXERCISE 4

Solve.

$\sqrt{x+14} - x + 6 = 0$

Answer: 11

EXAMPLE 5 A Radical Equation Involving Cube Roots

Solve.

$\sqrt[3]{z+2} - \sqrt[3]{3z+8} = 0$

$$\sqrt[3]{z+2} = \sqrt[3]{3z+8} \quad \text{Isolate radicals}$$
$$(\sqrt[3]{z+2})^3 = (\sqrt[3]{3z+8})^3 \quad \text{Cube both sides}$$
$$z + 2 = 3z + 8$$
$$-2z = 6$$
$$z = -3$$

The solution is -3. Check this.

PRACTICE EXERCISE 5

Solve.

$\sqrt[3]{2x-5} - \sqrt[3]{5x+1} = 0$

Answer: -2

If the radicals in an equation are expressed using rational exponents, the solution technique is the same.

EXAMPLE 6 Solving Equations with Rational Exponents

Solve.

(a) $x^{1/2} = 5$

$$(x^{1/2})^2 = 5^2 \quad \text{Square both sides}$$
$$x = 25 \quad \text{This does check}$$

(b) $(x-2)^{1/3} = 3$

$$[(x-2)^{1/3}]^3 = 3^3 \quad \text{Cube both sides}$$
$$x - 2 = 27 \quad (a^{1/3})^3 = a$$
$$x = 29 \quad \text{This does check}$$

PRACTICE EXERCISE 6

Solve.

(a) $x^{1/4} = 3$

(b) $(2x-1)^{1/3} = 2$

Answers: (a) 81 (b) $\frac{9}{2}$

3 LITERAL EQUATIONS

In Chapter 2 we solved various literal equations that were linear in the variables. In this section we consider literal equations that involve powers and radicals. We shall assume that all radical expressions are defined.

EXAMPLE 7 Solving Literal Equations

Solve each literal equation.

(a) $r = \sqrt{\frac{A}{\pi}}$ for A

Similar numerical equation

$r = \sqrt{\frac{A}{\pi}}$		$5 = \sqrt{\frac{A}{3}}$
$r^2 = \frac{A}{\pi}$	Square both sides	$5^2 = \frac{A}{3}$
$\pi r^2 = \frac{A}{\pi} \cdot \pi$	Multiply by π	$3 \cdot 5^2 = \frac{A}{3} \cdot 3$
$\pi r^2 = A$		$75 = A$

(b) $a = \sqrt{b - c}$ for c

Similar numerical equation

$a = \sqrt{b - c}$	$3 = \sqrt{5 - c}$
$a^2 = (\sqrt{b - c})^2$	$(3)^2 = (\sqrt{5 - c})^2$
$a^2 = b - c$	$9 = 5 - c$
$a^2 - b = b - b - c$	$9 - 5 = 5 - 5 - c$
$a^2 - b = -c$	$9 - 5 = -c$
$(-1)(a^2 - b) = (-1)(-c)$	$(-1)(4) = (-1)(-c)$
$b - a^2 = c$	$-4 = c$

PRACTICE EXERCISE 7

Solve each literal equation.

(a) $u = \sqrt{\frac{5}{v}}$ for v

(b) $2m = \sqrt{n - k}$ for n

Answers: (a) $v = \frac{5}{u^2}$ (b) $n = 4m^2 + k$

We conclude this section with an example of an applied problem that involves a radical equation.

EXAMPLE 8 Physics Problem

The time T in seconds it takes for a pendulum to make one swing is related to the length L in feet of the pendulum by

$$T = 2\pi\sqrt{\frac{L}{32}}.$$

Find the length of a pendulum (correct to the nearest hundredth of a foot) if it makes one swing in 4 seconds.

PRACTICE EXERCISE 8

Find the length of a pendulum (correct to the nearest tenth of a foot) if it makes one swing in 1.5 seconds.

Substitute 4 for T and solve.

$$4 = 2\pi\sqrt{\frac{L}{32}} \qquad T = 4$$

$$\frac{4}{2\pi} = \sqrt{\frac{L}{32}} \qquad \text{Divide by } 2\pi$$

$$\frac{2}{\pi} = \sqrt{\frac{L}{32}} \qquad \text{Reduce fraction}$$

$$\left(\frac{2}{\pi}\right)^2 = \frac{L}{32} \qquad \text{Square both sides}$$

$$32\left(\frac{4}{\pi^2}\right) = L \qquad \text{Multiply by 32}$$

We use the following calculator steps to find L.

32 [×] 4 [÷] [π] [x^2] [=] → 12.969112

The pendulum is approximately 12.97 ft in length.

Answer: 1.8 ft

7.6 EXERCISES A

Solve.

1. $\sqrt{z-3} + 3 = 6$
2. $4\sqrt{2y-1} - 2 = 0$
3. $8\sqrt{3x-1} + 7 = 0$
4. $\sqrt{5x-1} - 2\sqrt{x+1} = 0$
5. $\sqrt{x^2-5} + x - 5 = 0$ [*Hint:* $(5-x)^2 = 25 - 10x + x^2$, *not* $25 - x^2$.]
6. $\sqrt{a^2+9} + a + 1 = 0$
7. $\sqrt{y+6} + \sqrt{y+11} = 5$ [*Hint*: $(5-\sqrt{y-11})^2 = 25 - 10\sqrt{y-11} + y - 11$, *not* $25 - (y-11)$.]
8. $\sqrt{x+1} + \sqrt{x+7} = -1$
9. $\sqrt{w+2} + \sqrt{w+6} = 4$
10. $\dfrac{8}{\sqrt{a}} = 2$
11. $\dfrac{15}{\sqrt{x}} = \dfrac{20}{\sqrt{x+7}}$
12. $\sqrt{x-1} = x - 1$
13. $\sqrt{y-2} - y + 2 = 0$
14. $\sqrt{2x+1} - \sqrt{x-3} = 2$
15. $\sqrt{3x-2} - \sqrt{2x+5} = -1$

16. $\sqrt[3]{a+1} = \sqrt[3]{2a+7}$

17. $\sqrt[3]{x^2+x} = \sqrt[3]{2+x^2}$

18 $\sqrt[4]{w+2} = \sqrt[4]{5w}$

19. $\sqrt[3]{3a+2} + 4 = 6$

20. $\sqrt{x^2+x-3} = x$

21. $\sqrt{a^2+a+1} = a$

22. $(x+1)^{1/2} = 2$

23. $(x-5)^{1/3} = -3$

24. $(y+7)^{1/4} = -2$

Solve. Assume all variables are suitably chosen so that all radicals are meaningful.

25. $z = y\sqrt{x}$ for x

26. $x + \sqrt{y} = z$ for y

27. $a = b\sqrt{1+x}$ for x

28 $a = \sqrt{1 + \frac{b}{x}}$ for x

29. $\sqrt{ab+d} = c$ for b

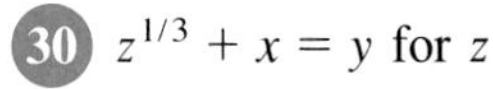

30 $z^{1/3} + x = y$ for z

Solve.

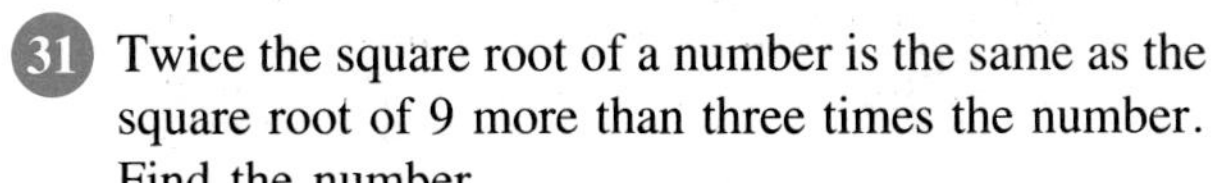

31 Twice the square root of a number is the same as the square root of 9 more than three times the number. Find the number.

32. If the square root of 1 less than a number is subtracted from the square root of 15 more than the number, the result is 2. Find the number.

A manufacturer finds that the cost per day of producing an airplane part is given by $c = 100\sqrt[3]{n} + 1200$ *where c is the cost and n is the number of parts produced.*

33. What is the cost when no parts (overhead) are produced?

34 Find the cost when 8 parts are produced.

35. Find the cost when 125 parts are produced.

36. How many parts were made on a day when the cost was $2800?

In an environmental study it is found that A and n are related by the equation $A = \sqrt{n+16} - \sqrt{n}$.

37. Find A when $n = 0$.

38. Find A when $n = 20$.

39. Find n when $A = 2$.

40. Find n when $A = 0$.

The head trainer at NAU, Mike Nesbitt, has found that he can estimate the number of boxes of tape he needs for a season of competition from the formula $t = 50\sqrt[3]{n - 11}$, where n is the number of members on the team.

41. Find the number of boxes needed when the team has 12 members.

42. If at the end of the season Mike had used a total of 200 boxes, how many team members were there?

43 The time T in seconds it takes a pendulum of length L feet to make one swing is given by $T = 2\pi\sqrt{\frac{L}{32}}$. How long is a pendulum (to the nearest hundredth of a foot) if it makes one swing in 1.5 seconds? Use 3.14 for π.

44. The distance d in miles that can be seen on the surface of the ocean is given by $d = 1.4\sqrt{h}$ where h is the height in feet above the surface of the water. How high (to the nearest foot) would the crow's nest on a ship need to be in order to view another ship at a distance of 18.5 miles?

45. The speed at which a car was traveling at the time of an accident can be estimated from the length of the skid mark l as follows: A test car (the car involved if possible) is driven under the same conditions at a test speed S, and skidded to a stop. If L is the length of the test skid, $s = S\sqrt{\frac{l}{L}}$. At the scene of an accident, a policeman determines that $S = 35$ mph, $l = 170$ ft, and $L = 51$ ft. What was the approximate speed of the car involved in the accident?

FOR REVIEW

Simplify by writing the expression so that each variable occurs only once and all exponents are positive.

46. $x^{-1/5}x^{6/5}$

47. $\dfrac{y^{3/4}}{y^{1/2}}$

48. $(27u^3v^{-9})^{1/3}$

Write in exponential notation, simplify, and give answers in radical notation. Rationalize denominators.

49. $\sqrt[3]{64x^3y^{12}}$

50. $\sqrt[4]{\dfrac{16u^5}{v^8}}$

51. $\dfrac{5}{\sqrt[5]{ab}}$

52. Rationalize the denominator.

$$\frac{\sqrt{2}}{\sqrt{2} + \sqrt{11}}$$

53. Rationalize the numerator.

$$\frac{\sqrt{2 + h} - \sqrt{2}}{h}$$

The following exercises will help you prepare for the material in the next section.

54. If $c = \sqrt{a^2 + b^2}$, find c to the nearest tenth if $a = 2.1$ and $b = 4.5$.

55. If $a = \sqrt{c^2 - b^2}$, find a to the nearest tenth if $c = 8.3$ and $b = 6.1$.

Evaluate. Leave answers in radical form but simplify.

56. $\sqrt{(2 - 5)^2 + (-1 - 6)^2}$

57. $\sqrt{(-1 - (-3))^2 + (4 - (-2))^2}$

ANSWERS: 1. 12 2. $\frac{5}{8}$ 3. no solution 4. 5 5. 3 6. no solution 7. -2 8. no solution 9. $\frac{1}{4}$ 10. 16 11. 9 12. 2; 1 13. 2; 3 14. 4; 12 15. 2 16. -6 17. 2 18. $\frac{1}{2}$ 19. 2 20. 3 21. no solution 22. 3 23. -22 24. no solution 25. $x = \frac{z^2}{y^2}$ 26. $y = (z - x)^2$ 27. $x = \frac{a^2}{b^2} - 1$ 28. $x = \frac{b}{a^2 - 1}$ 29. $b = \frac{c^2 - d}{a}$ 30. $z = (y - x)^3$ 31. 9 32. 10 33. \$1200 34. \$1400 35. \$1700 36. 4096 37. 4 38. $6 - 2\sqrt{5}$ 39. 9 40. no solution 41. 50 42. 75 43. 1.83 ft 44. 175 ft 45. 64 mph 46. x 47. $y^{1/4}$ 48. $\frac{3u}{v^3}$ 49. $4xy^4$ 50. $\frac{2u\sqrt[4]{u}}{v^2}$ 51. $\frac{5\sqrt[5]{a^4b^4}}{ab}$ 52. $\frac{\sqrt{22} - 2}{9}$ 53. $\frac{1}{\sqrt{2 + h} + \sqrt{2}}$ 54. 5.0 55. 5.6 56. $\sqrt{58}$ 57. $2\sqrt{10}$

7.6 EXERCISES B

Solve.

1. $\sqrt{4x - 3} - 2 = 3$

2. $5\sqrt{a - 3} + 10 = 0$

3. $3\sqrt{x - 4} - 9 = 0$

4. $\sqrt{2z + 9} = \sqrt{4 + 2z}$

5. $\sqrt{y^2 + 2} - y - 2 = 0$

6. $\sqrt{v^2 - 13} - v + 1 = 0$

7. $\sqrt{z + 20} - \sqrt{z + 4} = 2$

8. $\sqrt{a - 1} + \sqrt{a + 8} = -2$

9. $\sqrt{y + 10} - \sqrt{y - 6} = 2$

10. $\dfrac{9}{\sqrt{x}} = 3$

11. $\dfrac{1}{\sqrt{y}} = \dfrac{2}{\sqrt{y + 12}}$

12. $\sqrt{x + 1} = x + 1$

13. $\sqrt{y - 2} - y + 4 = 0$

14. $\sqrt{2x + 3} - \sqrt{x + 5} = -1$

15. $\sqrt{3x + 1} - \sqrt{2x - 1} = 1$

16. $\sqrt[3]{z + 2} = \sqrt[3]{2z - 23}$

17. $\sqrt[3]{1 - y^2} = \sqrt[3]{y + 1 - y^2}$

18. $\sqrt[4]{3x} = \sqrt[4]{4 - x}$

19. $3\sqrt[3]{3z - 1} - 2 = 4$

20. $\sqrt{y^2 - y + 2} = y$

21. $\sqrt{z^2 + 2z + 2} = z$

22. $(x - 1)^{1/2} = 2$

23. $(x + 6)^{1/3} = -3$

24. $(y + 5)^{1/4} = -3$

Solve. Assume all variables are suitably chosen so that all radicals are meaningful.

25. $z = x\sqrt{y}$ for y

26. $\sqrt{x} + y = z$ for x

27. $c = d\sqrt{2 - x}$ for x

28. $a = \sqrt{1 + \dfrac{b}{x}}$ for b

29. $\sqrt{ab + d} = c$ for d

30. $a^{1/4} - b = c$ for a

31. Three times the square root of a number is the same as the square root of 32 more than the number. Find the number.

32. If the square root of 3 less than a number is added to the square root of 8 less than the number, the result is 5. Find the number.

In a retail outlet the cost per day, c, is related to the number of small appliances sold, n, by $c = 10\sqrt[3]{n} + 200$.

33. What is the cost when no appliances (overhead) are sold?

34. Find the cost when 8 appliances are sold.

35. Find the cost when 64 appliances are sold.

36. How many appliances were sold on a day when the cost was $230?

A scientist finds that R and s are related by the equation $R = \sqrt{s + 25} + \sqrt{s + 9}$.

37. Find R when $s = 0$.

38. Find R when $s = 16$.

39. Find s when $R = 4$.

40. Find s when $R = 0$.

A manufacturer of puzzles has found that his daily expenses for production can be estimated by $e = 100\sqrt[3]{n + 8}$, *where e represents total expenses and n is the number of puzzles produced.*

41. Find the total expenses when no puzzles are produced.

42. How many puzzles were made on a day when the expenses were $300?

43. The time T in seconds it takes for a pendulum of length L feet to make one swing is given by $T = 2\pi\sqrt{\frac{L}{32}}$. How long is a pendulum (to the nearest hundredth of a foot) if it makes one swing in 7.2 seconds? Use 3.14 for π.

44. The distance in miles d that can be seen on the surface of the ocean is given by $d = 1.4\sqrt{h}$ where h is the height in feet above the surface of the water. How far would a periscope on a submarine have to be raised in order to view a destroyer located 2.9 miles away? (Give answer to the nearest tenth of a foot.)

45. The speed at which a car was traveling at the time of an accident can be estimated from the length of the skid mark l as follows: A test car (the car involved if possible) is driven under the same conditions at a test speed S, and skidded to a stop. If L is the length of the test skid, $s = S\sqrt{\frac{l}{L}}$. Estimate the speed of an accident vehicle when $S = 30$ mph, $l = 150$ ft, and $L = 24$ ft.

FOR REVIEW

Simplify by writing the expression so that each variable occurs only once and all exponents are positive.

46. $x^{-2/3}x^{5/3}$

47. $\frac{y^{5/4}}{y^{3/4}}$

48. $(81u^4v^{-12})^{1/4}$

Write in exponential notation, simplify, and give answers in radical notation. Rationalize denominators.

49. $\sqrt[3]{125x^6y^9}$

50. $\sqrt[4]{\frac{625u^9}{v^{16}}}$

51. $\frac{3}{\sqrt[5]{a^2b^2}}$

52. Rationalize the denominator.
$$\frac{\sqrt{2} + 1}{\sqrt{2} - 1}$$

53. Rationalize the numerator.
$$\frac{\sqrt{7 + h} - \sqrt{7}}{h}$$

The following exercises will help you prepare for the material in the next section.

54. If $c = \sqrt{a^2 + b^2}$, find c to the nearest tenth if $a = 3.6$ and $b = 7.2$.

55. If $b = \sqrt{c^2 - a^2}$, find b to the nearest tenth if $c = 10.4$ and $a = 7.3$.

Evaluate. Leave answers in radical form but simplify.

56. $\sqrt{(3-6)^2 + (-1-3)^2}$

57. $\sqrt{(-2-(-4))^2 + (7-(-1))^2}$

7.6 EXERCISES C

Solve.

1. $\sqrt{\sqrt{x} + 5} = \sqrt{x} - 1$

[Answer: 16]

2. $\sqrt{x+2} - \dfrac{6}{\sqrt{x+7}} = 0$

3. $\dfrac{x+1}{\sqrt{x^2 + 2x + 1}} = \dfrac{1}{3}$

[Answer: no solution]

4. $\sqrt{x+5} + \sqrt{x-3} = \sqrt{4x}$

5. $\sqrt{\sqrt{x} + \sqrt{x+1}} + 5 = 0$

6. Solve $c^{1/2} = \sqrt{b + \sqrt{a}}$ for a.
[Answer: $a = b^2 - 2bc + c^2$]

Assume that $c = 5\sqrt[3]{n} + 3.6$ for Exercises 7 and 8.

7. Find c when n is 2.8.

8. Find n when c is 7.4.
[Answer: approximately 0.44]

7.7 THE PYTHAGOREAN THEOREM AND THE DISTANCE AND MIDPOINT FORMULAS

STUDENT GUIDEPOSTS

1. The Pythagorean Theorem
2. Applications of the Pythagorean Theorem
3. The Distance Formula
4. The Midpoint Formula

1 THE PYTHAGOREAN THEOREM

Many applications involve a property of right triangles called *the Pythagorean theorem*. Remember that the longest side of a right triangle, the side opposite the right angle, is the **hypotenuse** of the triangle and the other two sides are called **legs.** (See Figure 7.4 on the following page.)

The Pythagorean Theorem

In a right triangle with hypotenuse of length c and legs of lengths a and b,

$$c^2 = a^2 + b^2.$$

We can remember the Pythagorean theorem in words as:

In a right triangle, the square of the hypotenuse is equal to the sum of the squares of the legs.

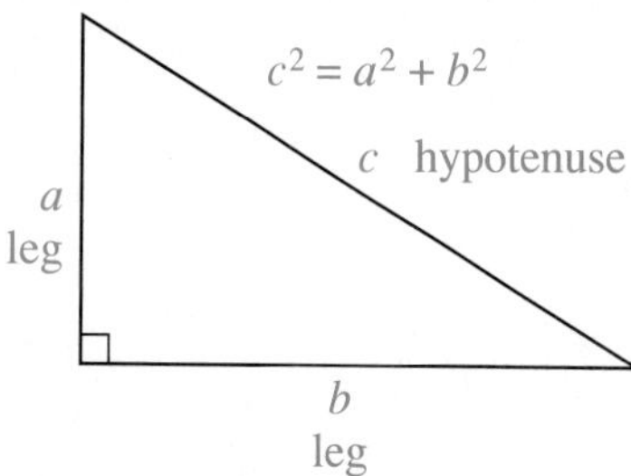

Figure 7.4 The Pythagorean Theorem

We can solve $c^2 = a^2 + b^2$ for c by taking the principle square root of both sides.

$$c = \sqrt{a^2 + b^2}$$

To solve the equation for a, first subtract b^2 from both sides.

$$c^2 = a^2 + b^2$$
$$c^2 - b^2 = a^2 + b^2 - b^2 \quad \text{Subtract } b^2$$
$$c^2 - b^2 = a^2$$
$$\sqrt{c^2 - b^2} = a \quad \text{Take the principal square root of both sides}$$

Similarly, $\sqrt{c^2 - a^2} = b$.

We now use these formulas to solve problems involving right triangles.

EXAMPLE 1 Using the Pythagorean Theorem

Find the missing side in each right triangle.

(a) Find the hypotenuse c when $a = 3$ and $b = 4$. See Figure 7.5.

$$c^2 = a^2 + b^2$$
$$c = \sqrt{a^2 + b^2}$$
$$= \sqrt{3^2 + 4^2}$$
$$= \sqrt{9 + 16}$$
$$= \sqrt{25} = 5$$

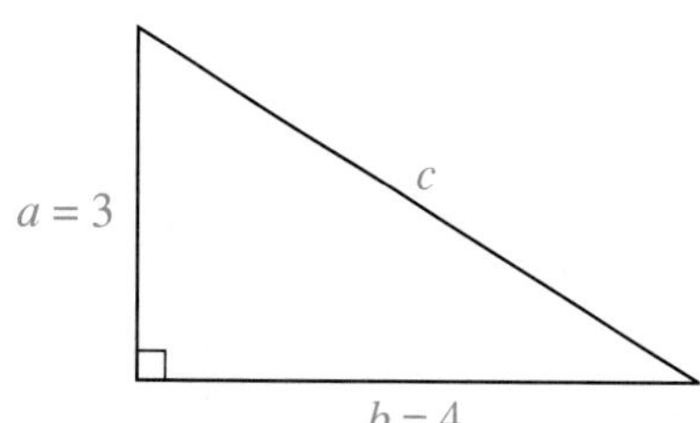

Figure 7.5

(b) Find the leg b when $c = 10$ and $a = 4$. See Figure 7.6.

$$c^2 = a^2 + b^2$$
$$b^2 = c^2 - a^2$$
$$b = \sqrt{c^2 - a^2}$$
$$= \sqrt{10^2 - 4^2}$$
$$= \sqrt{100 - 16}$$
$$= \sqrt{84}$$
$$= \sqrt{4 \cdot 21}$$
$$= 2\sqrt{21}$$

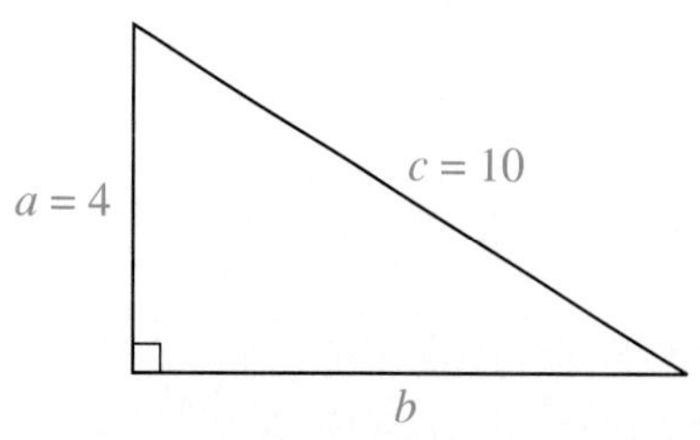

Figure 7.6

Practice Exercise 1

Find the missing side in each right triangle.

(a) Find the hypotenuse c when $a = 1$ and $b = 3$.

(b) Find the leg a when $c = 5$ and $b = 2$.

Answers: (a) $\sqrt{10}$ (b) $\sqrt{21}$

CAUTION

Mistakes can be made when calculating square roots in problems involving the Pythagorean theorem. In Example 1(a), $\sqrt{9 + 16} = \sqrt{25} = 5$ *not* $\sqrt{9} + \sqrt{16} = 3 + 4 = 7$. *Do not take roots term by term in a radicand.*

In Example 1(b) we left the value of b in exact irrational form, $2\sqrt{21}$. In many problems we will approximate such values using a calculator. For example, $2\sqrt{21} \approx 9.2$, and 9.2 gives us a better idea of the length of a side than $2\sqrt{21}$.

2 APPLICATIONS OF THE PYTHAGOREAN THEOREM

In most applied problems involving the Pythagorean theorem, a sketch of the appropriate triangle is helpful.

EXAMPLE 2 Navigation Problem

An airplane flew 20 mi west and then 30 mi south from an airport. To the nearest mile, how many miles was it from the starting point? See Figure 7.7.

$$
\begin{aligned}
d^2 &= (20)^2 + (30)^2 \\
d &= \sqrt{(20)^2 + (30)^2} \\
&= \sqrt{400 + 900} \\
&= \sqrt{1300} \\
&= \sqrt{100 \cdot 13} \\
&= 10\sqrt{13} \\
&\approx 36.1
\end{aligned}
$$

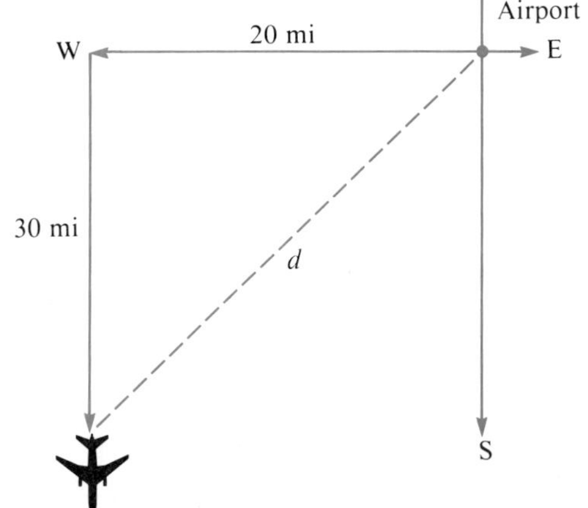

The airplane was about 36 mi from the airport.

Figure 7.7

Practice Exercise 2

A small boat left an island and sailed 25 miles north and then 10 miles east. To the nearest mile, how far was the boat from the island?

Answer: 27 mi

EXAMPLE 3 Geometry Problem

A baseball diamond is a square 90 feet in length on each side. How far is it directly across the diamond from first base to third base? See Figure 7.8.

$$
\begin{aligned}
x^2 &= 90^2 + 90^2 \\
x &= \sqrt{90^2 + 90^2} \\
&= \sqrt{2(90)^2} \\
&= \sqrt{(90)^2}\sqrt{2} \\
&= 90\sqrt{2} \\
&\approx 127.3
\end{aligned}
$$

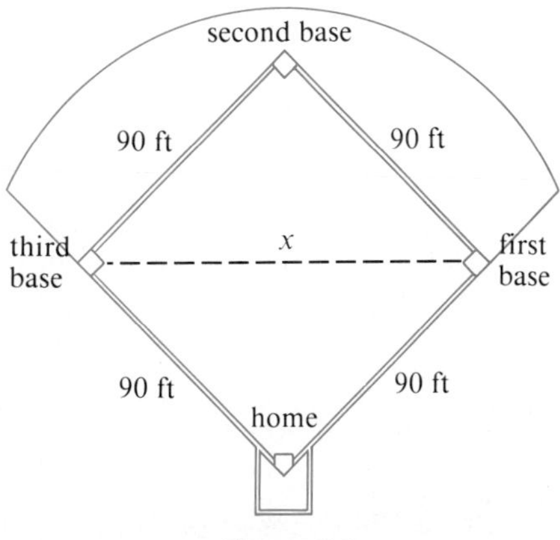

Figure 7.8

It's about 127.3 ft from first base to third base.

Practice Exercise 3

A rectangular lot is to be divided by a fence across the diagonal. Approximately how long will the fence need to be if the lot is 40 m wide and 60 m long?

Answer: approximately 72 m

3 THE DISTANCE FORMULA

Finding the distance between two points in a rectangular coordinate system is a direct application of the Pythagorean theorem. Suppose $P(x_1, y_1)$ and $Q(x_2, y_2)$ are two points as shown in Figure 7.9. Then the numbers $x_2 - x_1$ and $y_2 - y_1$ are the lengths of the legs of the right triangle with vertices P, Q, and R. Applying the Pythagorean theorem, we obtain

$$d^2 = (x_2 - x_1)^2 + (y_2 - y_1)^2,$$

and since d is positive,

$$d = \sqrt{(x_2 - x_1)^2 + (y_2 - y_1)^2}$$

gives the length of the hypotenuse of the triangle—that is, the distance between the points (x_1, y_1) and (x_2, y_2). Since $(x_2 - x_1)$ and $(y_2 - y_1)$ are both squared, $(x_1 - x_2)^2 = (x_2 - x_1)^2$ and $(y_1 - y_2)^2 = (y_2 - y_1)^2$, and it is immaterial which point we label (x_1, y_1) and which point we label (x_2, y_2). Although our particular points are in quadrants I and II, the same results hold regardless of the location of the two points.

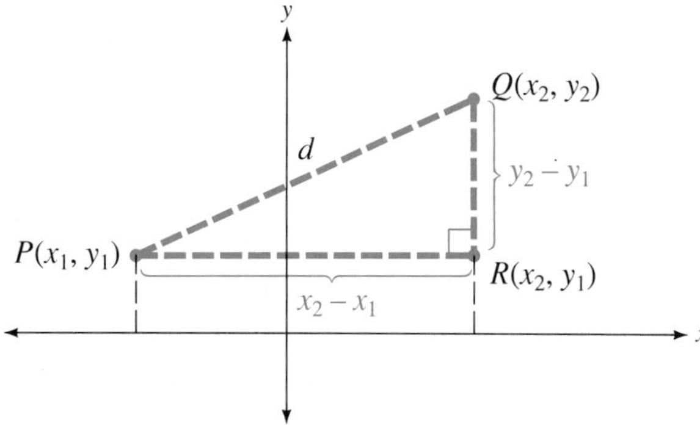

Figure 7.9 Distance between Two Points

These results are summarized in the following definition.

Distance Formula

The distance between two points with coordinates (x_1, y_1) and (x_2, y_2) is given by

$$d = \sqrt{(x_2 - x_1)^2 + (y_2 - y_1)^2}.$$

EXAMPLE 4 USING THE DISTANCE FORMULA

Find the distance between $(-5, 1)$ and $(3, 7)$. Substitute into the distance formula using $(x_1, y_1) = (-5, 1)$ and $(x_2, y_2) = (3, 7)$.

$$\begin{aligned} d &= \sqrt{[3 - (-5)]^2 + (7 - 1)^2} \\ &= \sqrt{8^2 + 6^2} \\ &= \sqrt{64 + 36} = \sqrt{100} = 10 \end{aligned}$$

PRACTICE EXERCISE 4

Find the distance between (2, 5) and (6, 8).

Answer: 5

4 THE MIDPOINT FORMULA

Another formula, the *midpoint formula,* can be established using similar triangles. It shows that the coordinates of a point midway between two given points are found by averaging the corresponding x-coordinates and y-coordinates.

Midpoint Formula

The coordinates $(\bar{x}, \bar{y})$ of the midpoint of a line segment joining (x_1, y_1) and (x_2, y_2) are given by the **midpoint formula**

$$(\bar{x}, \bar{y}) = \left(\frac{x_1 + x_2}{2}, \frac{y_1 + y_2}{2}\right).$$

That is, the x-coordinate of the midpoint is the average of the x-coordinates of the points, and the y-coordinate of the midpoint is the average of the y-coordinates of the points.

CAUTION

In the distance formula we subtract x-coordinates and y-coordinates while in the midpoint formula we add x-coordinates and y-coordinates. Do not confuse these operations when working with these two formulas.

EXAMPLE 5 Using the Midpoint Formula

Find the midpoint between the points $(-2, 1)$ and $(6, -3)$.

Substitute into the midpoint formula.

$$(\bar{x}, \bar{y}) = \left(\frac{x_1 + x_2}{2}, \frac{y_1 + y_2}{2}\right) = \left(\frac{-2 + 6}{2}, \frac{1 + (-3)}{2}\right)$$

$$= \left(\frac{4}{2}, \frac{-2}{2}\right) = (2, -1)$$

The midpoint is $(2, -1)$.

Practice Exercise 5

Find the midpoint between the points $(-3, -7)$ and $(5, 4)$.

Answer: $\left(1, -\frac{3}{2}\right)$

The Pythagorean theorem, the distance formula, and the midpoint formula are frequently used tools in algebra, and should be committed to memory.

7.7 EXERCISES A

Find the unknown leg or hypotenuse in each right triangle with the given sides. Leave answer in radical form, where appropriate.

1. $b = 12$, $c = 13$

2. $a = 7$, $c = 10$

3. $a = \sqrt{2}$, $c = 3$

4 $b = 5$, $c = 5\sqrt{3}$

Find the missing side in each right triangle, correct to the nearest tenth of a unit.

5. $a = 1.7$ cm, $b = 3.4$ cm

6. $a = 15.2$ ft, $c = 21.3$ ft

7. $b = 6.5$ in, $c = 17.3$ in

8. $a = 12.4$ m, $c = 10.6$ m

Solve.

9 A field in the shape of a square has a diagonal 2 mi in length. What is the length of a side?

10. A machinist has a rectangular part that measures 15 cm by 20 cm. What is the length of a diagonal?

11. The base of a ladder which is 16 ft long is placed 4 ft from a wall of a building. How high up the wall is the top of the ladder?

12. An airplane left an airport and flew 120 km north and then 200 km west. How far was the plane from the airport?

13. The diagonal of a square measures 16 cm. What is the length of each equal side, to the nearest tenth of a centimeter?

14 The triangle below is an isosceles right triangle. Find a formula for the hypotenuse h in terms of a.

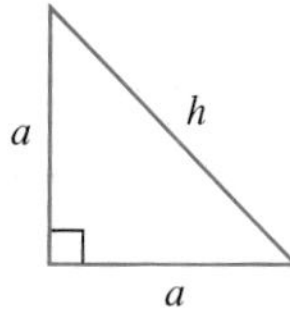

15. Bridges are built with expansion joints to prevent buckling during the heat of summer. Assume, however, that a bridge 1 mile long (5280 ft) has no such joint and expands 2 ft in length during July and that it bulges up in the middle. How high is the bulge? The answer may be surprising!

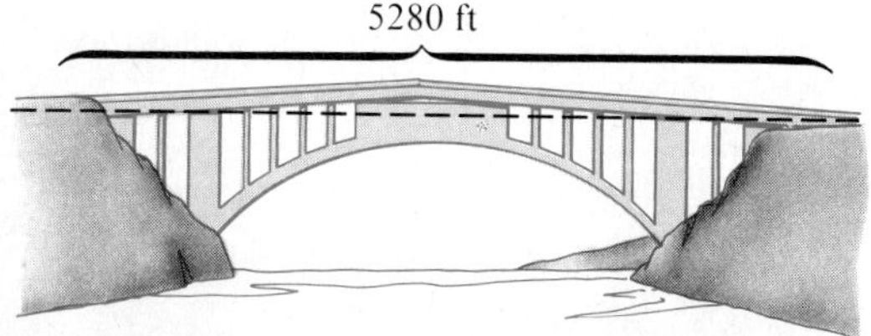

16. Find the area of an isosceles triangle with equal sides of 20 ft and third side 30 ft. [*Hint:* Find the height of the triangle first.]

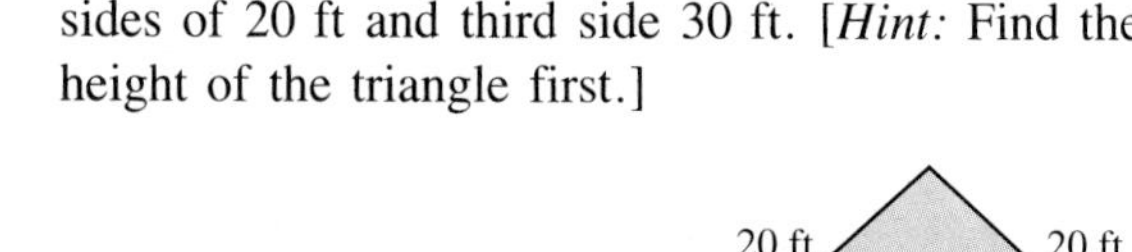

Find the distance between the given points.

17. (3, 6) and (−3, −2)

18. (3, 2) and (−2, −10)

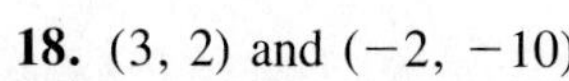

19. $(-2, 1)$ and $(5, -3)$

20. $(2, 2)$ and $(-2, -2)$

The converse of the Pythagorean theorem is also true. That is, if $c^2 = a^2 + b^2$, then a triangle with sides a, b, and c is a right triangle. Find the lengths of the sides of the triangle ABC with the given vertices. Is the triangle a right triangle?

21. $A(-2, 1)$; $B(2, 5)$; $C(1, -2)$

22. $A(1, 2)$; $B(1, -3)$; $C(-2, 1)$

Find the midpoint of the segment between each pair of points.

23. $(2, 3)$ and $(8, 1)$

24. $(-4, 7)$ and $(5, -3)$

25. $(-8, -3)$ and $(-8, 11)$

26. $\left(-5, \frac{1}{2}\right)$ and $\left(3, -\frac{3}{2}\right)$

27 Find the values of a and b if the midpoint of the segment joining $(a, 3)$ and $(-2, b + 1)$ is $(2, 1)$.

28. Find the value of a if the distance between the points $(a, 1)$ and $(-1, 4)$ is 5.

FOR REVIEW

Solve.

29. $\sqrt{y^2 + 9} - y - 1 = 0$

30. $\sqrt{2y + 3} - \sqrt{y - 2} = 2$

The following exercises review material from Chapter 5 and this chapter to help you prepare for the next section. Simplify each radical in Exercises 31–33.

31. $\sqrt{16}$

32. $\sqrt{8}$

33. $\sqrt{108}$

Perform the indicated operations in Exercises 34 and 35.

34. $(2x - y) - (5x - 7y)$

35. $(5 + x\sqrt{2})(5 - x\sqrt{2})$

36. Rationalize the denominator. $\dfrac{7}{3 - \sqrt{2}}$

ANSWERS: 1. 5 2. $\sqrt{51}$ 3. $\sqrt{7}$ 4. $5\sqrt{2}$ 5. 3.8 cm 6. 14.9 ft 7. 16.0 in 8. There is no triangle since c must be longer than a. 9. $\sqrt{2}$ mi $\approx$ 1.4 mi 10. 25 cm 11. $4\sqrt{15}$ ft $\approx$ 15.5 ft 12. $40\sqrt{34}$ km $\approx$ 233 km 13. $8\sqrt{2}$ cm $\approx$ 11.3 cm 14. $h = a\sqrt{2}$ 15. about 72.7 ft 16. about 198.4 ft^2 17. 10 18. 13 19. $\sqrt{65}$ 20. $4\sqrt{2}$ 21. yes 22. no 23. $(5, 2)$ 24. $\left(\frac{1}{2}, 2\right)$ 25. $(-8, 4)$ 26. $\left(-1, -\frac{1}{2}\right)$ 27. $a = 6$; $b = -2$ 28. $a = 3$ or $a = -5$ 29. 4 30. 3; 11 31. 4 32. $2\sqrt{2}$ 33. $6\sqrt{3}$ 34. $-3x + 6y$ 35. $25 - 2x^2$ 36. $3 + \sqrt{2}$

7.7 EXERCISES B

Find the unknown leg or hypotenuse in each right triangle with the given sides. Leave answer in radical form, where appropriate.

1. $a = 6$, $c = 10$ **2.** $b = 2$, $c = 5$ **3.** $b = \sqrt{3}$, $c = 3$ **4.** $a = 7$, $c = 7\sqrt{3}$

Find the missing side in each right triangle, correct to the nearest tenth of a unit.

5. $a = 5.2$ cm, $b = 6.9$ cm

6. $a = 24.3$ ft, $c = 36.7$ ft

7. $b = 2.4$ in, $c = 23.5$ in

8. $b = 16.5$ m, $c = 11.5$ m

Solve.

9. A machine part in the shape of a square has a diagonal 6 cm in length. What is the length of a side?

10. A field in the shape of a rectangle is 300 yd long and 200 yd wide. What is the length of a diagonal?

11. How long must a ladder be to reach 8 m up the side of a building when the base of the ladder is 2 m from the building?

12. An airplane is 350 mi from the airport. To get to the present position, it flew 200 mi due south and then flew due west. How far west has it gone?

13. A guy wire reaches from the top of a 25-ft pole to the ground 12 ft from the pole. How long is the wire?

14. In a right triangle, one leg is x and the other is $x + 1$. Find a formula for the hypotenuse h in terms of x.

15. The softball diamond in a city park is a square 70 ft on a side. How far is it from home plate to second base?

16. Find the area of an equilateral triangle with sides of length 10 cm.

Find the distance between the given points.

17. $(3, -1)$ and $(-9, 4)$ **18.** $(-2, 5)$ and $(2, 2)$ **19.** $(-3, 2)$ and $(4, 1)$ **20.** $(0, -3)$ and $(7, 2)$

The converse of the Pythagorean theorem is also true. That is, if $c^2 = a^2 + b^2$, then a triangle with sides a, b, and c is a right triangle. Find the lengths of the sides of the triangle ABC with the given vertices. Is the triangle a right triangle?

21. $A(-2, -1)$; $B(1, -2)$; $C(3, 3)$

22. $A(-3, 2)$; $B(4, 1)$; $C(1, -2)$

Find the midpoint of the segment between each pair of points.

23. $(5, 2)$ and $(3, 12)$ **24.** $(-6, 5)$ and $(8, -3)$ **25.** $(6, -2)$ and $(-5, -2)$ **26.** $\left(3, -\frac{1}{2}\right)$ and $\left(-3, \frac{7}{2}\right)$

27. Find the values of a and b if the midpoint of the segment joining $(a - 1, 5)$ and $(-6, b)$ is $\left(-\frac{3}{2}, 2\right)$.

28. Find the value of a if the distance between the points $(4, -2)$ and $(a, -6)$ is 5.

FOR REVIEW

Solve.

29. $\sqrt{y^2 + 7} - y - 1 = 0$

30. $\sqrt{3x + 1} - \sqrt{2x - 1} = 1$

The following exercises will help you prepare for the next section. Simplify each radical in Exercises 31–33.

31. $\sqrt{81}$ **32.** $\sqrt{12}$ **33.** $\sqrt{300}$

Perform the indicated operations in Exercises 34 and 35.

34. $(3x - 2y) - (4x - 10y)$

35. $(2 - x\sqrt{3})(2 + x\sqrt{3})$

36. Rationalize the denominator. $\dfrac{2}{\sqrt{11} - 3}$

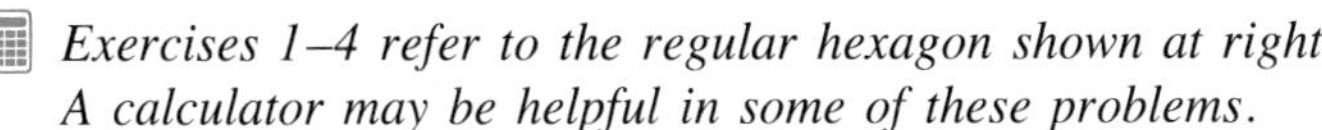

7.7 EXERCISES C

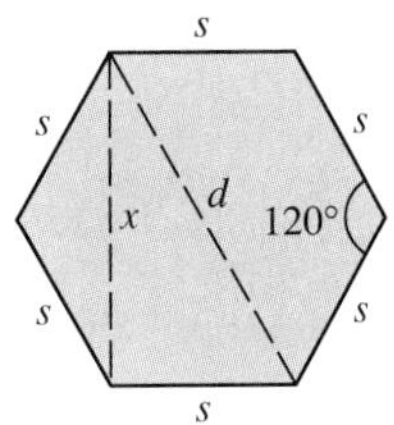

Exercises 1–4 refer to the regular hexagon shown at right. A calculator may be helpful in some of these problems.

1. Find the distance between parallel sides, x, in terms of the equal sides s. [*Hint:* The side opposite the 30° acute angle in a right triangle is half the hypotenuse.]

2. Find the length of diagonal d in terms of the equal sides s.
[Answer: $d = 2s$]

3. What is the value of x when $s = 5$ cm? What is the value of d?

4. What is the area of the hexagon when $s = 5$ cm?

5. Find the equation of the perpendicular bisector of the line segment joining the two points $(-3, 2)$ and $(5, -4)$. [Answer: $4x - 3y - 7 = 0$]

6. A square has area 144 in^2. What is the area of the square formed by joining the midpoints of the sides of this square?

7.8 COMPLEX NUMBERS

STUDENT GUIDEPOSTS

1. Imaginary Numbers
2. Complex Numbers
3. Equality of Complex Numbers
4. Operations on Complex Numbers

1 IMAGINARY NUMBERS

Until now we have discussed only square roots of *nonnegative* numbers. The square root of a negative number, such as -4, is not a real number since no real number a, when squared, can be -4. That is, if a is any real number, $a^2 \geq 0$ so that $a^2 \neq -4$. We now develop a new set of numbers in which square roots of negative numbers are defined. These numbers, imaginary numbers, play an important role in mathematics with many applications in physics and engineering.

Heron of Alexandria (circa 150 B.C. to 250 A.D.) first noticed that the square root of a negative number posed an interesting problem. René Descartes (1596–1650) first used the terms *real* and *imaginary numbers*. Don't let the term *imaginary* mislead you, however, since imaginary numbers are no more "imaginary" than real numbers.

By agreeing to follow rules for radicals similar to those we followed with real numbers, only one new numeral is needed to develop the imaginary numbers. We use i as the numeral corresponding to $\sqrt{-1}$.

$$\sqrt{-1} = i \quad \text{so that} \quad i^2 = -1.$$

Remember that the product rule for radicals,

$$\sqrt{a}\sqrt{b} = \sqrt{ab}$$

applies only when $a \geq 0$ and $b \geq 0$. However, if we agree to extend this rule to the case when $a \geq 0$ and $b = -1$, every imaginary number can be written as the product of a real number and the number i. For example,

$$\sqrt{-4} = \sqrt{(4)(-1)} = \sqrt{4}\sqrt{-1} = 2i.$$

EXAMPLE 1 SIMPLIFYING IMAGINARY NUMBERS

Simplify.

(a) $\sqrt{-9} = \sqrt{9(-1)} = \sqrt{9}\sqrt{-1} = 3i$ $\quad \sqrt{-1} = i$

(b) $\sqrt{-2}\sqrt{-5} = \sqrt{2(-1)}\sqrt{5(-1)} = \sqrt{2} \cdot i \cdot \sqrt{5} \cdot i$
$= \sqrt{2}\sqrt{5}i^2 = \sqrt{10}(-1) = -\sqrt{10}$ $\quad i^2 = -1$

PRACTICE EXERCISE 1

Simplify.

(a) $\sqrt{-144}$

(b) $\sqrt{-6}\sqrt{-11}$

Answers: (a) $12i$ (b) $-\sqrt{66}$

CAUTION

Do not write $\sqrt{-4}\sqrt{-9} = \sqrt{(-4)(-9)} = \sqrt{36} = 6$ since $\sqrt{-4}\sqrt{-9} = (2i)(3i) = 6i^2 = -6$. The rule $\sqrt{a}\sqrt{b} = \sqrt{ab}$ does not apply when a and b are both negative. We can use this rule when $a \geq 0$ and $b = -1$ to express square roots of negative numbers in terms of the imaginary number i.

EXAMPLE 2 MORE ON IMAGINARY NUMBERS

Simplify.

(a) $\dfrac{\sqrt{-30}}{\sqrt{-6}} = \dfrac{\sqrt{30(-1)}}{\sqrt{6(-1)}} = \dfrac{\sqrt{30}\cancel{i}}{\sqrt{6}\cancel{i}} = \sqrt{\dfrac{30}{6}} = \sqrt{5}$

(b) $\sqrt{-16} + \sqrt{-49} = \sqrt{16(-1)} + \sqrt{49(-1)}$
$= \sqrt{16}i + \sqrt{49}i = 4i + 7i = 11i$

PRACTICE EXERCISE 2

Simplify.

(a) $\dfrac{\sqrt{-45}}{\sqrt{-5}}$

(b) $\sqrt{-4} + \sqrt{-121}$

Answers: (a) 3 (b) $13i$

Powers of i cycle through the same four values: i, -1, $-i$, and 1. This is shown in the following table that lists the first sixteen powers of i.

Powers of the Imaginary Number i

$i^1 = i$	$i^5 = i$	$i^9 = i$	$i^{13} = i$
$i^2 = -1$	$i^6 = -1$	$i^{10} = -1$	$i^{14} = -1$
$i^3 = -i$	$i^7 = -i$	$i^{11} = -i$	$i^{15} = -i$
$i^4 = 1$	$i^8 = 1$	$i^{12} = 1$	$i^{16} = 1$

We can use this cyclical pattern to find any higher power of i by writing the expression using factors of $i^4 = 1$.

EXAMPLE 3 FINDING POWERS OF i

Find each power of i.

(a) $i^{21} = i^{20}i = (i^4)^5(i)$ $\quad$ Express 21 as a multiple of 4(20)
$= (1)^5 i = (1)i = i$ $\quad$ plus 1 (a number less than 4)

PRACTICE EXERCISE 3

Find each power of i.

(a) i^{32}

(b) i^{53}

(b) $i^{40} = (i^4)^{10} = (1)^{10} = 1$

(c) $i^{122} = i^{120}i^2 = (i^4)^{30}i^2 = (1)^{30}i^2 = i^2 = -1$

(d) $i^{363} = i^{360}i^3 = (i^4)^{90}i^3 = (1)^{90}i^3 = i^3 = -i$

(c) i^{146}

(d) i^{411}

Answers: (a) 1 (b) i (c) -1 (d) $-i$

2 COMPLEX NUMBERS

Leonhard Euler (1707–83) is credited as being the first to use i for $\sqrt{-1}$. But Carl Friedrich Gauss (1777–1855) first saw that the real numbers and the imaginary numbers could be combined to form a powerful number system, the *complex numbers*. A number of the form

$$a + bi \qquad a \text{ and } b \text{ real numbers}$$

is called a **complex number.** The number a is the **real part** of $a + bi$, and b is the **imaginary part.** If $b = 0$, then the complex number $a + bi = a + 0i = a$ is simply the real number a. If $a = 0$, it is the imaginary number bi. Thus, the complex numbers include both the real numbers and the imaginary numbers.

Figure 7.10 summarizes the relationships among the sets of numbers we have studied. Some examples of each type of number are listed.

Complex Numbers ($a + bi$): $2 + 3i, \sqrt{2} - i, 3 + 0i, 0 - i\sqrt{7}, 12 + 57i$

- Real Numbers (a): $3, -\frac{1}{2}, -\pi, \sqrt{5}, \frac{5}{2}, e$
  - Rational Numbers: $\frac{5}{2}, -\frac{1}{2}, \frac{101}{7}, 3, -17, 0$
  - Irrational Numbers: $\sqrt{5}, -\pi, \pi, e, -\sqrt{13}, \sqrt{2}$
- Imaginary Numbers (bi): $2i, -17i, -i\sqrt{7}, \pi i, -\frac{2}{3}i, 4i$

Figure 7.10 Sets of Numbers

3 EQUALITY OF COMPLEX NUMBERS

Before we can do arithmetic on the set of complex numbers, we need to decide when two complex numbers are *equal*.

Equality of Complex Numbers

Two complex numbers $a + bi$ and $c + di$ are **equal,**

$$a + bi = c + di,$$

if and only if their real parts are equal ($a = c$) and their imaginary parts are equal ($b = d$).

EXAMPLE 4 Equality of Complex Numbers

(a) If $a + bi = 3 + 2i$, then $a = 3$ and $b = 2$.

Practice Exercise 4

(a) If $x + yi = 5 - 2i$, find x and y.

(b) If $c + di = 5$, then $c = 5$ and $d = 0$ since $5 = 5 + 0i$.

(c) If $u + vi = -\sqrt{3}i$, then $u = 0$ and $v = -\sqrt{3}$ since $-\sqrt{3}i = 0 - \sqrt{3}i$.

(d) If $2x + yi = 3x - 2 + 7i$, then

$$2x = 3x - 2 \quad \text{and} \quad y = 7.$$
$$-x = -2$$
$$x = 2$$

(b) If $a + bi = -8$, find a and b.

(c) If $w + ui = i\sqrt{5}$, find w and u.

(d) If $4a - bi = a + 3 - 9i$, find a and b.

Answers: (a) $x = 5, y = -2$ (b) $a = -8, b = 0$ (c) $w = 0, u = \sqrt{5}$ (d) $a = 1, b = 9$

CAUTION

When a radical is the coefficient of i, as in $-\sqrt{3}\,i$ in Example 4(c), notice that i is not under the radical. Since it is a common mistake to put i under the radical, we often write $-\sqrt{3}\,i$ as $-i\sqrt{3}$.

4 OPERATIONS ON COMPLEX NUMBERS

The basic operations on complex numbers are easy to learn since the number i acts in the same way as any other literal number in calculations.

Addition and Subtraction of Complex Numbers

Let $a + bi$ and $c + di$ be two complex numbers. Their **sum** is

$$(a + bi) + (c + di) = (a + c) + (b + d)i,$$

and their **difference** is

$$(a + bi) - (c + di) = (a - c) + (b - d)i.$$

Notice that for addition and subtraction, we add (or subtract) the real parts and then add (or subtract) the imaginary parts.

EXAMPLE 5 Adding and Subtracting Complex Numbers

(a) Add.

$(3 + 2i) + (-4 - i) = 3 - 4 + 2i - i$ Use commutative rule

$= (3 - 4) + (2 - 1)i$ Add real parts and imaginary parts

$= -1 + i$

(b) Subtract.

$(2 - i\sqrt{3}) - (4 + 3i\sqrt{3}) = 2 - 4 - i\sqrt{3} - 3i\sqrt{3}$

$= (2 - 4) - i(\sqrt{3} + 3\sqrt{3})$

$= -2 - 4i\sqrt{3}$

Practice Exercise 5

(a) Add.
$(5 - 3i) + (-2 - i)$

(b) Subtract.
$(8 + i\sqrt{5}) - (-3 - 2i\sqrt{5})$

Answers: (a) $3 - 4i$ (b) $11 + 3i\sqrt{5}$

Multiplication of Complex Numbers

Let $a + bi$ and $c + di$ be two complex numbers. Their **product** is

$$(a + bi)(c + di) = (ac - bd) + (ad + bc)i.$$

Although the definition of multiplication of complex numbers may appear strange at first, it seems more reasonable when we multiply as if the numbers are binomials using the FOIL method.

$$\begin{aligned}(a + bi)(c + di) &= \overset{F}{ac} + \overset{O}{adi} + \overset{I}{bci} + \overset{L}{bdi^2}\\ &= ac + bdi^2 + adi + bci\\ &= ac + bd(-1) + adi + bci \qquad i^2 = -1\\ &= (ac - bd) + (ad + bc)i\end{aligned}$$

In fact, rather than memorizing the definition, it is probably better to use the FOIL method in particular cases, keeping in mind that $i^2 = -1$.

EXAMPLE 6 MULTIPLYING COMPLEX NUMBERS

Multiply.

(a) $(4 + 2i)(3 - 5i)$

$$\begin{aligned}(4 + 2i)(3 - 5i) &= \overset{F}{4 \cdot 3} - \overset{O}{4 \cdot 5i} + \overset{I}{2 \cdot 3i} - \overset{L}{2 \cdot 5 \cdot i^2}\\ &= 12 - 20i + 6i - 10\,i^2\\ &= 12 - 14i - 10\,(-1)\\ &= 12 + 10 - 14i\\ &= 22 - 14i\end{aligned}$$

(b) $(5 + i\sqrt{2})(5 - i\sqrt{2}) = 5 \cdot 5 - 5i\sqrt{2} + 5i\sqrt{2} - i^2\sqrt{2}\,\sqrt{2}$

$$\begin{aligned}&= 25 - i^2 \cdot 2\\ &= 25 - (-1) \cdot 2\\ &= 25 + 2 = 27\end{aligned}$$

PRACTICE EXERCISE 6

Multiply.

(a) $(2 - i)(5 + 7i)$

(b) $(8 + i\sqrt{5})(8 - i\sqrt{5})$

Answers: (a) $17 + 9i$ (b) 69

In Example 6(b), a special product was found. The numbers $5 + i\sqrt{2}$ and $5 - i\sqrt{2}$ are *conjugates,* and whenever two conjugates are multiplied, the result is a real number. The **conjugate** of the complex number $a + bi$ is $a - bi$, and the **conjugate** of $a - bi$ is $a + bi$. Thus, $a + bi$ and $a - bi$ are **conjugates** of each other. (Compare this definition to the one for conjugates of irrational numbers in Section 7.4.) One application of conjugates is in division of complex numbers.

Division of Complex Numbers

Let $a + bi$ and $c + di$ be two complex numbers. The **quotient** of $a + bi$ and $c + di$ can be simplified by multiplying numerator and denominator by the conjugate of $c + di$, $c - di$. That is,

$$\frac{(a + bi)}{(c + di)} = \frac{(a + bi)(c - di)}{(c + di)(c - di)} = \frac{(ac + bd) + (bc - ad)i}{c^2 + d^2}$$

As with multiplying complex numbers, when dividing complex numbers it is better to use the method of multiplying numerator and denominator by the conjugate of the denominator rather than memorizing the definition.

EXAMPLE 7 Dividing Complex Numbers

Divide $(3 + 7i)$ by $(2 - 5i)$.

$$\frac{3+7i}{2-5i} = \frac{(3+7i)(2+5i)}{(2-5i)(2+5i)}$$

$$= \frac{6 + 15i + 14i + 35i^2}{4 + 10i - 10i - 25i^2}$$

$$= \frac{6 + 29i - 35}{4 + 25} \qquad i^2 = -1$$

$$= \frac{-29 + 29i}{29}$$

$$= \frac{-29}{29} + \frac{29i}{29} = -1 + i$$

Practice Exercise 7

Divide $(2 - 3i)$ by $(1 + i)$.

Answer: $\frac{-1-5i}{2}$

EXAMPLE 8 Finding the Reciprocal of a Complex Number

Find the reciprocal of $3 + 2i$ and express it in the form $a + bi$.

The reciprocal of $3 + 2i$ is $\frac{1}{3+2i}$. We need to divide the complex number $1 = 1 + 0i$ by the complex number $3 + 2i$. Thus, finding a reciprocal is simply a division problem.

$$\frac{1}{3+2i} = \frac{1(3-2i)}{(3+2i)(3-2i)} = \frac{3-2i}{9 - 6i + 6i - 4i^2}$$

$$= \frac{3-2i}{9+4} = \frac{3-2i}{13} = \frac{3}{13} - \frac{2}{13}i$$

Practice Exercise 8

Find the reciprocal of $4 - 3i$ and express it in the form $a + bi$.

Answer: $\frac{4}{25} + \frac{3}{25}i$

Many applications of complex numbers are found in the study of electricity.

EXAMPLE 9 Electricity Problem

Ohm's law of electricity is $E = IR$ where E is electromotive force in volts, I is current in amperes, and R is resistance in ohms. Find the current if $E = 1 + 4i$ volts and $R = 1 + i$ ohms.

Since $E = IR$, solving for I gives

$$I = \frac{E}{R}.$$

Substitute $1 + 4i$ for E and $1 + i$ for R and divide.

$$I = \frac{1+4i}{1+i} = \frac{(1+4i)(1-i)}{(1+i)(1-i)} = \frac{1 - i + 4i - 4i^2}{1 - i^2}$$

$$= \frac{1 + 3i + 4}{1 + 1} \qquad i^2 = -1$$

$$= \frac{5+3i}{2} = \frac{5}{2} + \frac{3}{2}i$$

Thus, the current is $\frac{5}{2} + \frac{3}{2}i$ amperes.

Practice Exercise 9

Use Ohm's law to find I when $E = 1 + 2i$ volts and $R = 3 + i$ ohms.

Answer: $\frac{1}{2} + \frac{1}{2}i$ amperes

7.8 EXERCISES A

Express in terms of the imaginary number i.

1. $\sqrt{-9}$ **2.** $\sqrt{-36}$ **3.** $-\sqrt{-36}$

4. $\sqrt{-8}$ **5.** $-\sqrt{-3}$ **6.** $\sqrt{-\pi^2}$

Simplify and express each answer as a real number or in terms of i.

7. $\sqrt{-4} + \sqrt{-9}$ **8.** $\sqrt{-4} - \sqrt{-9}$ **9.** $\sqrt{-7} + 2\sqrt{-7}$ **10** $\sqrt{-4}\,\sqrt{-9}$

11. $\sqrt{-3}\,\sqrt{-3}$ **12.** $\sqrt{-7}\,\sqrt{-3}$ **13** $\dfrac{\sqrt{-4}}{\sqrt{-9}}$ **14.** $\dfrac{\sqrt{-7}}{\sqrt{-3}}$

Using the definition of equality of complex numbers, determine x and y.

15. $x + yi = 2 - 3i$ **16.** $7x + 3yi = 21$ **17** $(4x + 1) + 7yi = 5 - i$

Perform the indicated operations.

18. $(2 + 3i) + (5 + 8i)$ **19.** $(2 + 3i) - (5 + 8i)$ **20.** $3 + (5 - 2i)$

21. $-6i + (8 + 2i)$ **22.** $(-3 - 2i) - (6 - 7i)$ **23.** $(-5 + 4i) + (5 - 4i)$

24. $7(2 - 3i)$ **25.** $7i(2 - 3i)$ **26.** $(1 - 2i)(3 + 4i)$

27. $(8 + 4i)(-7 + 3i)$ **28.** $(2 + 3i)(2 - 3i)$ **29.** $(2 + 3i)^2$

30. i^{85} **31.** $i^{15} + i^{33}$ **32.** $i^{40} - i^{22}$

Give the conjugate of each of the following complex numbers.

33. $4 + 6i$ **34.** $-4 - 6i$ **35.** 8 **36.** $6i$ **37.** $-7i$

Find the following quotients.

38. $\dfrac{3 + 5i}{1 + i}$ **39.** $\dfrac{7}{4 - 3i}$ **40.** $\dfrac{3i}{5 - 4i}$ **41** $\dfrac{8 - 7i}{5 + 4i}$

Find the reciprocal of each number and express it in the form $a + bi$.

42. $1 - i$ **43.** $-4 + 8i$ **44.** $-2i$ **45.** 3

Determine whether each statement in Exercises 46–50 is true or false.

46. Every rational number is a complex number.

47. Every complex number is a real number.

48. Every imaginary number is a real number.

49. Every complex number is a rational number.

50. The product of a complex number and its conjugate is always a real number.

In Exercises 51–52, use Ohm's law $E = IR$, given in Example 9, to find the missing variable.

51. $E = 3 + i$, $I = 2 - i$; find R. **52.** $I = 2 + i$, $R = 3 - i$; find E.

FOR REVIEW

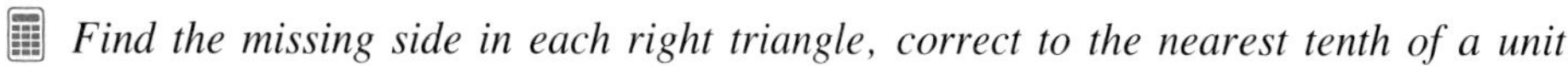

Find the missing side in each right triangle, correct to the nearest tenth of a unit.

53. $a = 19.3$ m, $b = 14.5$ m **54.** $b = 12.9$ ft, $c = 34.4$ ft

55. Find the distance between the points $(4, -7)$ and $(-3, -1)$.

56. Find the midpoint of the segment between the points $(-4, 3)$ and $(12, -6)$.

57. Find the length of x in the square shown at right. Give answer correct to the nearest tenth of a centimeter.

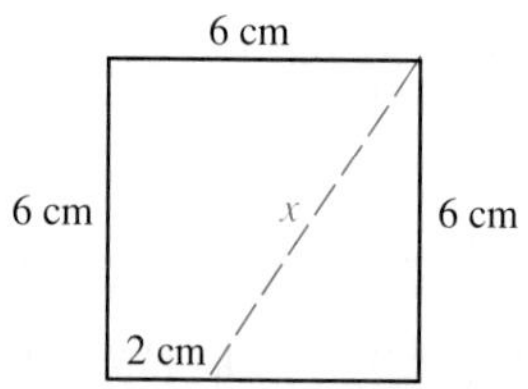

Exercises 58–60 review material from Chapter 5 to help you prepare for the next chapter. Factor each polynomial.

58. $x^2 + 2x - 15$ **59.** $x^2 - 7x$ **60.** $4x^2 + 4x + 1$

ANSWERS: 1. $3i$ 2. $6i$ 3. $-6i$ 4. $2i\sqrt{2}$ 5. $-i\sqrt{3}$ 6. πi 7. $5i$ 8. $-i$ 9. $3i\sqrt{7}$ 10. -6 11. -3 12. $-\sqrt{21}$ 13. $\frac{2}{3}$ 14. $\frac{\sqrt{7}}{\sqrt{3}}$ 15. $x = 2$; $y = -3$ 16. $x = 3$; $y = 0$ 17. $x = 1$; $y = -\frac{1}{7}$ 18. $7 + 11i$ 19. $-3 - 5i$ 20. $8 - 2i$ 21. $8 - 4i$ 22. $-9 + 5i$ 23. 0 24. $14 - 21i$ 25. $21 + 14i$ 26. $11 - 2i$ 27. $-68 - 4i$ 28. 13

29. $-5 + 12i$ **30.** i **31.** 0 **32.** 2 **33.** $4 - 6i$ **34.** $-4 + 6i$ **35.** 8 **36.** $-6i$ **37.** $7i$ **38.** $4 + i$ **39.** $\frac{28}{25} + \frac{21}{25}i$ **40.** $-\frac{12}{41} + \frac{15}{41}i$ **41.** $\frac{12}{41} - \frac{67}{41}i$ **42.** $\frac{1}{2} + \frac{1}{2}i$ **43.** $-\frac{1}{20} - \frac{1}{10}i$ **44.** $\frac{1}{2}i$ **45.** $\frac{1}{3}$ **46.** true **47.** false **48.** false **49.** false **50.** true **51.** $1 + i$ ohms **52.** $7 + i$ volts **53.** 24.1 m **54.** 31.9 ft **55.** $\sqrt{85}$ **56.** $\left(4, -\frac{3}{2}\right)$ **57.** 7.2 cm **58.** $(x - 3)(x + 5)$ **59.** $x(x - 7)$ **60.** $(2x + 1)^2$

7.8 EXERCISES B

Express in terms of the imaginary number i.

1. $\sqrt{-16}$ **2.** $\sqrt{-121}$ **3.** $-\sqrt{-121}$ **4.** $\sqrt{-12}$ **5.** $-\sqrt{-5}$ **6.** $-\sqrt{-\pi^2}$

Simplify and express each answer as a real number or in terms of i.

7. $\sqrt{-16} + \sqrt{-25}$ **8.** $\sqrt{-16} - \sqrt{-25}$ **9.** $\sqrt{-7} - 2\sqrt{-7}$ **10.** $\sqrt{-25}\,\sqrt{-16}$

11. $\sqrt{-7}\,\sqrt{-7}$ **12.** $\sqrt{-5}\,\sqrt{-11}$ **13.** $\dfrac{\sqrt{-25}}{\sqrt{-16}}$ **14.** $\dfrac{\sqrt{-5}}{\sqrt{-11}}$

Using the definition of equality of complex numbers, determine x and y.

15. $x + yi = 5 - 15i$ **16.** $2x - 8yi = 10$ **17.** $(2x - 1) + 8yi = 3 - 16i$

Perform the indicated operation.

18. $(2 - 3i) + (3 + 7i)$ **19.** $(2 - 3i) - (3 + 7i)$ **20.** $-7 + (2 - 3i)$

21. $-5i + (3 + i)$ **22.** $(-2 - 11i) - (5 - i)$ **23.** $(-8 + 2i) + (8 - 2i)$

24. $5(3 - 5i)$ **25.** $11i(1 - 5i)$ **26.** $(2 - 3i)(1 + 4i)$

27. $(1 + i)(1 - i)$ **28.** $(2 - 3i)^2$ **29.** $(1 + i\sqrt{2})^2$

30. i^{82} **31.** $i^{12} + i^{49}$ **32.** $i^{19} - i^{25}$

Give the conjugate of each of the following complex numbers.

33. $7 + 3i$ **34.** $-7 - 3i$ **35.** -12 **36.** $11i$ **37.** $-15i$

Find the following quotients.

38. $\dfrac{4 + 3i}{2 + i}$ **39.** $\dfrac{6}{5 - 3i}$ **40.** $\dfrac{2i}{4 - 3i}$ **41.** $\dfrac{6 - 5i}{1 + 3i}$

Find the reciprocal of each number and express it in the form $a + bi$.

42. $2 - i$ **43.** $-4 - 8i$ **44.** $5i$ **45.** 5

Determine whether each statement in Exercises 46–50 is true or false.

46. Every real number is a complex number.

47. Every imaginary number is a complex number.

48. Every irrational number is a complex number.

49. The sum of a complex number and its conjugate is always a real number.

50. The difference between a complex number and its conjugate is always a real number.

In Exercises 51–52, use Ohm's law $E = IR$, given in Example 9, to find the missing variable.

51. $E = 3 + i$, $I = 2 + i$; find R. **52.** $I = 2 + i$, $R = 4 - i$; find E.

FOR REVIEW

Find the missing side in each right triangle, correct to the nearest tenth of a unit.

53. $a = 27.3$ cm, $b = 8.6$ cm

54. $a = 17.1$ yd, $c = 42.3$ yd

55. Find the distance between the points $(-2, 5)$ and $(3, -5)$.

56. Find the midpoint of the segment between the points $(-9, -3)$ and $(5, 2)$

57. Which triangle at right, ABC or PQR has the greater area?

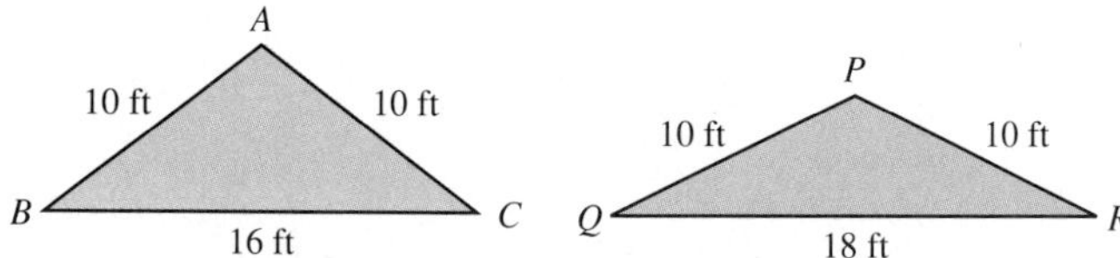

Exercises 58–60 review material from Chapter 5 to help you prepare for the next chapter. Factor each polynomial.

58. $3x^2 + x - 2$

59. $4x^2 - 4x + 1$

60. $16x^2 - 49$

7.8 EXERCISES C

Perform the indicated operations and express the result in the form $a + bi$.

1. $1 + \dfrac{1}{1 + \dfrac{1}{i}}$ $\left[\text{Answer: } \dfrac{3}{2} + \dfrac{1}{2}i\right]$

2. $(2 - i)^3(2 + i)^3$

3. $\dfrac{2}{1 - i} + \dfrac{3}{2 + i}$

CHAPTER 7 REVIEW

KEY WORDS

7.1 A number a is a **cube root** of x if $a^3 = x$. In this case, x is called a **perfect cube.**

A number a is called a **kth root** of x if $a^k = x$. In this case x is a **perfect kth power.**

7.2 The process of removing radicals from the denominator (or numerator) of a fraction is called **rationalizing the denominator (or numerator).**

7.4 **Like radicals** have the same index and the same radicand.

Unlike radicals have either different indexes or different radicands, or both.

Square root expressions such as $\sqrt{5} + \sqrt{3}$ and $\sqrt{5} - \sqrt{3}$ are called **conjugates** of each other.

7.6 A **radical equation** contains radicals with variables in the radicand.

A solution to a radical equation that has been introduced by raising both sides to a power and that does not check in the original equation is called an **extraneous root.**

7.8 **Imaginary numbers** come from taking square roots of negative real numbers.

A number of the form $a + bi$, where a and b are real and $i = \sqrt{-1}$ is called a **complex number.**

The **real part** of $a + bi$ is a, and the **imaginary part** is b.

Two complex numbers $a + bi$ and $c + di$ are **equal** if and only if $a = c$ and $b = d$.

The complex numbers $a + bi$ and $a - bi$ are **conjugates** of each other.

KEY CONCEPTS

7.1 The **principal kth root** of x is denoted by $\sqrt[k]{x}$. If $x \geq 0$, then $\sqrt[k]{x} \geq 0$.

If $x < 0$, then (a) $\sqrt[k]{x} < 0$ when k is odd. (b) $\sqrt[k]{x}$ is not a real number when k is even.

7.2
1. If $\sqrt[k]{a}$ and $\sqrt[k]{b}$ are defined, then $\sqrt[k]{a}\sqrt[k]{b} = \sqrt[k]{ab}$.
2. If $\sqrt[k]{a}$ and $\sqrt[k]{b}$ are defined and $b \neq 0$, then $\dfrac{\sqrt[k]{a}}{\sqrt[k]{b}} = \sqrt[k]{\dfrac{a}{b}}$.

7.4
1. $\sqrt[k]{a+b}$ *is not* $\sqrt[k]{a} + \sqrt[k]{b}$.
2. $\sqrt[k]{a-b}$ *is not* $\sqrt[k]{a} - \sqrt[k]{b}$.
3. To rationalize a binomial denominator, multiply both the numerator and denominator by the conjugate of the denominator.

7.5
1. If a is a real number and n is a natural number, then
$$a^{1/n} = \sqrt[n]{a}.$$
Also, if m is a natural number and $a^{1/n}$ exists, then
$$a^{m/n} = \sqrt[n]{a^m} = (\sqrt[n]{a})^m.$$
2. All rules of integer exponents apply to rational exponents.

7.6
1. To solve a radical equation, isolate a radical on one side of the equation and raise both sides to a power equal to the index on the radical. Check all solutions in the original equation.
2. When solving a literal equation, it is sometimes helpful to solve a similar numerical equation first.

7.7
1. The Pythagorean theorem: In a right triangle with hypotenuse of length c and legs of length a and b,
$$c^2 = a^2 + b^2.$$
2. The distance formula: The distance between two points (x_1, y_1) and (x_2, y_2) is
$$d = \sqrt{(x_2 - x_1)^2 + (y_2 - y_1)^2}.$$
3. The midpoint formula: The coordinates of the midpoint of the segment joining (x_1, y_1) and (x_2, y_2) are
$$(\bar{x}, \bar{y}) = \left(\frac{x_1 + x_2}{2}, \frac{y_1 + y_2}{2}\right).$$
4. Remember that in the distance formula the coordinates are subtracted, and in the midpoint formula, they are added.

7.8
1. $i = \sqrt{-1}$; that is $i^2 = -1$.
2. We can add and subtract complex numbers just like we added and subtracted binomials.
3. Use the FOIL method to multiply complex numbers. Remember that $i^2 = -1$.
4. To divide one complex number by another, multiply numerator and denominator by the conjugate of the denominator.

REVIEW EXERCISES

Part I

7.1 *Simplify each radical in Exercises 1–8. For these exercises do not assume that variables and algebraic expressions under even-indexed radicals necessarily represent nonnegative numbers.*

1. $\sqrt[3]{w^3}$
2. $\sqrt{y^2}$
3. $-\sqrt[3]{(2x)^3}$
4. $\sqrt[4]{(3w)^4}$
5. $\sqrt{(-5x)^2}$
6. $-\sqrt[5]{(-4y)^5}$
7. $\sqrt{(x+10)^2}$
8. $-\sqrt[3]{(y-1)^3}$

For the remainder of these exercises, return to the assumption that all variables and algebraic expressions under even-indexed radicals are positive. Simplify each radical.

9. $\sqrt{(-3)^2}$
10. $\sqrt{-16}$
11. $\sqrt{81a^2}$
12. $\sqrt{9x^2y^2}$
13. $\sqrt[3]{a^3b^3}$
14. $\sqrt[3]{-8x^3}$
15. $\sqrt[4]{16u^4v^4}$
16. $\sqrt[6]{(-3)^6u^6}$

Use a calculator to approximate each number correct to the nearest hundredth.

17. $\sqrt{453}$
18. $2 - \sqrt{7}$
19. $\dfrac{-3 + \sqrt{29}}{2}$

20. The distance in miles that can be seen on the surface of the ocean can be approximated by $d = 1.4\sqrt{h}$ where h is the height of the viewer in feet above the surface of the water. To the nearest mile, what is the viewing distance from a helicopter at an altitude of 3500 ft?

7.2 *Simplify each radical. Assume all variables under even-indexed radicals are positive.*

21. $\sqrt{50}$

22. $2\sqrt{x^4y^6}$

23. $4\sqrt{20x^3y^5}$

24. $\sqrt[3]{40}$

25. $\sqrt[4]{16x^8y^{12}}$

26. $\sqrt[5]{32x^6}$

27. $\sqrt{\dfrac{45}{y^2}}$

28. $\sqrt[3]{\dfrac{-8x^3y^4}{125}}$

29. $\sqrt[4]{\dfrac{625x^5}{y^8}}$

Rationalize the denominators.

30. $\dfrac{5}{\sqrt{3}}$

31. $\dfrac{y}{\sqrt[3]{x}}$

32. $\dfrac{\sqrt{x^3}}{\sqrt{2y}}$

Rationalize the numerators.

33. $\dfrac{\sqrt{2}}{5}$

34. $\dfrac{\sqrt[3]{x^2}}{3}$

35. $\dfrac{\sqrt{2}}{\sqrt{6}}$

36. The formula $S = \sqrt{20.5\,l}$, where l is the length of the skid marks, can be used to approximate the speed of a vehicle prior to an accident. What was the approximate speed of a car that left 75 ft of skid marks before a collision?

7.3 *Multiply and simplify.*

37. $\sqrt{54}\,\sqrt{50}$

38. $\sqrt[3]{6ab}\,\sqrt[3]{18a^2b^3}$

39. $\sqrt[4]{27x^3y}\,\sqrt[4]{9x^2y^3}$

Divide and simplify.

40. $\dfrac{\sqrt{3}}{\sqrt{75}}$

41. $\dfrac{\sqrt{48ab^7}}{\sqrt{3a^5}}$

42. $\dfrac{\sqrt[3]{27a^3}}{\sqrt[3]{2a^2b^2}}$

7.4 *Add or subtract. Rationalize all denominators.*

43. $2\sqrt{5x} + 9\sqrt{5x}$

44. $5\sqrt{147} - 4\sqrt{48}$

45. $-6\sqrt[3]{-54} + 8\sqrt[3]{250}$

46. $5a\sqrt{63a} - 4\sqrt{175a^3}$

47. $\sqrt[3]{27x^5y^9} - 3y\sqrt[3]{125x^5y^6}$

48. $4\sqrt{3x} + \dfrac{x}{\sqrt{3x}}$

Multiply and simplify.

49. $\sqrt{5}(\sqrt{15} - \sqrt{3})$

50. $\sqrt{5}\left(\sqrt{5} - \frac{3}{\sqrt{5}}\right)$

51. $\sqrt[3]{xy}(\sqrt[3]{x^2} + \sqrt[3]{y^2})$

52. $(\sqrt{3} - 2)^2$

53. $(\sqrt{x} - 3)(\sqrt{x} + 2)$

54. $(\sqrt{x} - \sqrt{y})(\sqrt{x} - 2\sqrt{y})$

Rationalize the denominators.

55. $\dfrac{2}{\sqrt{5} - \sqrt{3}}$

56. $\dfrac{\sqrt{3}}{\sqrt{3} - 1}$

57. $\dfrac{x - 9}{\sqrt{x} - 3}$

7.5 *Simplify, writing the expression so that each variable occurs only once and all exponents are positive.*

58. $x^{-2/3}x^{5/3}$

59. $\left(\dfrac{a^{-6}}{b^6}\right)^{1/3}$

60. $(y^{-3/4})^{-8}$

61. $\left(\dfrac{x^{2/3}y^{-4/3}}{z^{-5/3}}\right)^3$

62. $\left(\dfrac{25x^5y^{-3}}{x^{-5}y^{-9}}\right)^{1/2}$

63. $\left(\dfrac{54a^{-6}b^2}{9a^{-3}b^8}\right)^{-2}$

Write in exponential notation, simplify, and give answers in radical notation. Rationalize all denominators.

64. $\sqrt{45x^4y^6}$

65. $\sqrt[3]{\dfrac{40x^4}{y^9}}$

66. $\dfrac{2x}{\sqrt[3]{3y}}$

Use a calculator with a $\boxed{y^x}$ *key to approximate each number correct to the nearest tenth.*

67. $13^{1/4}$

68. $6^{2/5}$

69. $(1.08)^{-11/4}$

70. $\left(\dfrac{7}{8}\right)^{8/7}$

71. The length of an edge of a cube with volume V is given by $e = \sqrt[3]{V}$. To the nearest tenth of an inch, what is the length of an edge of a cube that holds 400 in^3?

7.6 *Solve.*

72. $\sqrt{x + 5} - \sqrt{x - 4} = 1$

73. $\sqrt{y^2 + 4} + y + 2 = 0$

74. $(y - 5)^{1/4} = 2$

75. $\sqrt{ab + c} = 5$ for a

76. $z^{1/3}x^{1/3} = y$ for z

77. $x = z\sqrt{1 - y}$ for y

78. The formula $T = 2\pi\sqrt{\frac{L}{32}}$ gives the time of swing for a pendulum of length L. If $T = 5$ sec, find L correct to the nearest tenth of a foot. Use 3.14 for π.

79. Twice the square root of one more than a number is the same as three times the square root of 4 less than the number. Find the number.

7.7 **80.** In a right triangle with leg $b = 50$ and hypotenuse $c = 80$, find a.

81. What is the length of a diagonal in a rectangle with sides 4.5 cm and 7.8 cm? Give answer to the nearest tenth of a centimeter.

82. Find the distance between the points $(-7, 2)$ and $(4, -5)$.

83. Find the midpoint of the segment joining $(-2, -5)$ and $(6, -1)$.

7.8 *Simplify.*

84. $\sqrt{-16} - 4\sqrt{-9}$

85. $\sqrt{-16} \cdot \sqrt{-9}$

86. $\dfrac{\sqrt{-16}}{\sqrt{-9}}$

Perform the indicated operations.

87. $(7 - 4i) + (3 + 6i)$

88. $(7 - 4i) - (3 + 6i)$

89. $2i - (3 - 6i)$

90. $-6i(5 - 2i)$

91. $(4 - 3i)(4 + 3i)$

92. $(4 - 3i)^2$

93. i^{25}

94. $\dfrac{1}{2 + i}$

95. $\dfrac{3 - 2i}{4 + 5i}$

Part II

Perform the indicated operations. Simplify each radical. Assume all variables under even-indexed radicals are positive. Rationalize all denominators.

96. $\sqrt{75}$

97. $\sqrt[3]{54}$

98. $\dfrac{\sqrt{8xy^5}}{\sqrt{2x^2y}}$

99. $3a\sqrt{45a} + 2\sqrt{20a^3}$

100. $\sqrt{3}(\sqrt{3} - \sqrt{6})$

101. $(\sqrt{2} + 1)^2$

102. $\sqrt[3]{25x^2y}\,\sqrt[3]{10xy^4}$

103. $\dfrac{\sqrt{7}}{\sqrt{7} + 1}$

104. $\sqrt[4]{\dfrac{32a^6}{a^2b^5}}$

Use a calculator to approximate each number correct to the nearest tenth.

105. $40^{1/3}$

106. $\sqrt[7]{65}$

107. $\dfrac{-2 - \sqrt{17}}{3}$

Perform the indicated operations.

108. $(2 - 3i) - (7 - 4i)$

109. $\dfrac{1 - 4i}{2 + i}$

110. $(3 - 5i)(2 - i)$

111. Find the midpoint of the segment joining $(4, -3)$ and $(-2, 6)$.

112. If x is any real number, positive, negative, or zero, simplify $\sqrt{(-6x)^2}$.

Solve.

113. $\sqrt{2x - 3} - \sqrt{x - 2} = 1$

114. $T = 2\pi\sqrt{\dfrac{L}{32}}$ for L.

115. What is the length of a diagonal, to the nearest tenth of an inch, in a square with sides 2.5 inches?

116. Find the distance between the points $(3, -3)$ and $(5, 1)$.

117. Rationalize the numerator.
$\dfrac{\sqrt{x} + 4}{x - 16}$

118. A researcher finds that M and d are related by the formula
$M = \sqrt{d + 2} - \sqrt{d - 3}$. Find d when $M = 1$.

Simplify, writing the expression so that each variable occurs only once and all exponents are positive. Write answers in radical notation. Rationalize denominators.

119. $\dfrac{(x^{-2/3})^2}{x^{1/3}}$

120. $\left(\dfrac{36a^4b^3}{a^2b^4}\right)^{1/2}$

ANSWERS: 1. w 2. $|y|$ 3. $-2x$ 4. $3|w|$ 5. $5|x|$ 6. $4y$ 7. $|x + 10|$ 8. $1 - y$ 9. 3 10. not a real number 11. $9a$ 12. $3xy$ 13. ab 14. $-2x$ 15. $2uv$ 16. $3u$ 17. 21.28 18. −0.65 19. 1.19 20. 83 mi 21. $5\sqrt{2}$ 22. $2x^2y^3$ 23. $8xy^2\sqrt{5xy}$ 24. $2\sqrt[3]{5}$ 25. $2x^2y^3$ 26. $2x\sqrt[5]{x}$ 27. $\frac{3\sqrt{5}}{y}$ 28. $\frac{-2xy\sqrt[3]{y}}{5}$ 29. $\frac{5x\sqrt[4]{x}}{y^2}$ 30. $\frac{5\sqrt{3}}{3}$ 31. $\frac{y\sqrt[3]{x^2}}{x}$ 32. $\frac{x\sqrt{2xy}}{2y}$ 33. $\frac{2}{5\sqrt{2}}$ 34. $\frac{x}{3\sqrt[3]{x}}$ 35. $\frac{1}{\sqrt{3}}$ 36. 39 mph 37. $30\sqrt{3}$ 38. $3ab\sqrt[3]{4b}$ 39. $3xy\sqrt[4]{3x}$ 40. $\frac{1}{5}$ 41. $\frac{4b^3\sqrt{b}}{a^2}$ 42. $\frac{3\sqrt[3]{4ab}}{2b}$ 43. $11\sqrt{5x}$ 44. $19\sqrt{3}$ 45. $58\sqrt[3]{2}$ 46. $-5a\sqrt{7a}$ 47. $-12xy^3\sqrt[3]{x^2}$ 48. $\frac{13\sqrt{3x}}{3}$ 49. $5\sqrt{3} - \sqrt{15}$ 50. 2 51. $x\sqrt[3]{y} + y\sqrt[3]{x}$ 52. $7 - 4\sqrt{3}$ 53. $x - \sqrt{x} - 6$ 54. $x - 3\sqrt{xy} + 2y$ 55. $\sqrt{5} + \sqrt{3}$ 56. $\frac{3 + \sqrt{3}}{2}$ 57. $\sqrt{x} + 3$ 58. x 59. $\frac{1}{a^2b^2}$ 60. y^6 61. $\frac{x^2z^5}{y^4}$ 62. $5x^5y^3$ 63. $\frac{a^6b^{12}}{36}$ 64. $3x^2y^3\sqrt{5}$ 65. $\frac{2x\sqrt[3]{5x}}{y^3}$ 66. $\frac{2x\sqrt[3]{9y^2}}{3y}$ 67. 1.9 68. 2.0 69. 0.8 70. 0.9 71. 7.4 in 72. 20 73. no solution 74. 21 75. $a = \frac{25 - c}{b}$ 76. $z = \frac{y^3}{x}$ 77. $y = \frac{z^2 - x^2}{z^2}$ 78. 20.3 ft 79. 8 80. $10\sqrt{39}$ 81. 9.0 cm 82. $\sqrt{170}$ 83. $(2, -3)$ 84. $-8i$ 85. −12 86. $\frac{4}{3}$ 87. $10 + 2i$ 88. $4 - 10i$ 89. $-3 + 8i$ 90. $-12 - 30i$ 91. 25 92. $7 - 24i$ 93. i 94. $\frac{2 - i}{5}$ 95. $\frac{2 - 23i}{41}$ 96. $5\sqrt{3}$ 97. $3\sqrt[3]{2}$ 98. $\frac{2y^2\sqrt{x}}{x}$ 99. $13a\sqrt{5a}$ 100. $3 - 3\sqrt{2}$ 101. $3 + 2\sqrt{2}$ 102. $5xy\sqrt[3]{2y^2}$ 103. $\frac{7 - \sqrt{7}}{6}$ 104. $\frac{2a\sqrt[4]{2b^3}}{b^2}$ 105. 3.4 106. 1.8 107. −2.0 108. $-5 + i$ 109. $\frac{-2 - 9i}{5}$ 110. $1 - 13i$ 111. $\left(1, \frac{3}{2}\right)$ 112. $6|x|$ 113. 2; 6 114. $L = \frac{8T^2}{\pi^2}$ 115. 3.5 in 116. $2\sqrt{5}$ 117. $\frac{1}{\sqrt{x} - 4}$ 118. 7 119. $\frac{\sqrt[3]{x}}{x^2}$ 120. $\frac{6a\sqrt{b}}{b}$

CHAPTER 7 TEST

Assume that algebraic expressions under even-indexed radicals are positive. Evaluate.

1. $\sqrt{(-7)^2}$ 1. ______

2. $-\sqrt{7^2}$ 2. ______

3. $\sqrt{16a^2b^2}$ 3. ______

4. $\sqrt[3]{8x^3}$ 4. ______

5. $\sqrt[5]{(-2)^5y^5}$ 5. ______

Simplify.

6. $\sqrt{75}$ 6. ______

7. $\sqrt[3]{54}$ 7. ______

8. $\sqrt[4]{16y^5}$ 8. ______

9. $\sqrt[4]{\dfrac{81x^7}{y^8}}$ 9. ______

Multiply and simplify.

10. $\sqrt{5}(\sqrt{10} - \sqrt{2})$ 10. ______

11. $(\sqrt{3} + 1)^2$ 11. ______

12. $\sqrt[4]{27a^2b^5}\ \sqrt[4]{3a^2b^2}$ 12. ______

13. $(\sqrt{x} - \sqrt{a})(\sqrt{x} + 3\sqrt{a})$ 13. ______

Add or subtract.

14. $\sqrt{18} + 2\sqrt{2} - 4\sqrt{8}$ 14. ______

15. $\sqrt[3]{8x^4y^5} - 2y\sqrt[3]{x^4y^2}$ 15. ______

16. Express $27^{4/3}$ as a radical and simplify. 16. ______

17. Evaluate $\dfrac{-9 + \sqrt{20}}{3}$ using a calculator, correct to the nearest tenth. 17. ______

18. Simplify and rationalize the denominator. $\dfrac{\sqrt{50xy^5}}{\sqrt{2x^3}}$ 18. ________

19. Rationalize the denominator. $\dfrac{3}{\sqrt{2}-\sqrt{5}}$ 19. ________

20. Rationalize the numerator. $\dfrac{\sqrt{6}+3}{-3}$ 20. ________

Solve.

21. $\sqrt{3x+1} - \sqrt{4x-3} = -1$ 21. ________

22. $(y-4)^{1/3} = -2$ 22. ________

23. Given the two points $(2, -1)$ and $(-4, 1)$,
(a) find the distance between them,
(b) find the midpoint of the line segment joining them.

23.(a) ________
23.(b) ________

24. A hiker walked 3 mi south, then 4 mi west. How far was he from the starting point? 24. ________

25. Solve $a^{1/2}b^{1/2} = c$ for a. 25. ________

Perform the indicated operations.

26. $(-2+5i) - (-5+4i)$ 26. ________

27. $(7-2i)(-2+4i)$ 27. ________

28. $\dfrac{2+i}{5-i}$ 28. ________

29. At Buster's Restaurant, the number n of seats needed in the waiting area can be approximated by $n = 1.5\sqrt{a}$, where a is the number of expected arrivals for dinner. About how many seats are needed when 120 people will dine in an evening? 29. ________

30. If a is any real number, positive, negative, or zero, simplify $\sqrt{(-5a)^2}$. 30. ________

CHAPTER 8

Quadratic Equations and Inequalities

8.1 FACTORING AND TAKING ROOTS

STUDENT GUIDEPOSTS

1 Zero-Product Rule	4 Number of Solutions
2 Quadratic Equations in General Form	5 Equations of the Form $ax^2 + bx = 0$
3 Method of Factoring	6 Method of Taking Roots

1 ZERO-PRODUCT RULE

In Section 5.6 we solved quadratic equations such as $2x^2 - x - 15 = 0$ by factoring and using the zero-product rule, repeated below.

Zero-Product Rule

If a and b are numbers or expressions and

$$\textbf{if } ab = 0, \quad \textbf{then} \quad a = 0 \textbf{ or } b = 0.$$

The zero-product rule tells us that the solutions of an equation, such as $(x - 3)(2x + 5) = 0$, can be found by setting each factor equal to zero and solving both equations for x. This is reviewed in Example 1.

EXAMPLE 1 USING THE ZERO-PRODUCT RULE

Solve $(x - 3)(2x + 5) = 0$.

$$(x - 3)(2x + 5) = 0$$

$$x - 3 = 0 \quad \text{or} \quad 2x + 5 = 0 \qquad \text{Zero-product rule}$$

$$x = 3 \qquad\qquad 2x = -5$$

$$x = -\frac{5}{2}$$

The solutions are 3 and $-\frac{5}{2}$. These can be checked easily by substituting the values into the original equation.

PRACTICE EXERCISE 1

Solve $(y + 1)(5y - 4) = 0$.

Answer: $-1, \frac{4}{5}$

Remember that the zero-product rule can be extended to three or more factors. For example, to solve $x(x-3)(2x+3)=0$,

$$x=0 \quad \text{or} \quad x-3=0 \quad \text{or} \quad 2x+3=0 \qquad \text{Zero-product rule}$$
$$x=3 \qquad 2x=-3$$
$$x=-\frac{3}{2}$$

The solutions are 0, 3, and $-\frac{3}{2}$.

2 QUADRATIC EQUATIONS IN GENERAL FORM

The equation in Example 1 is equivalent to the quadratic equation

$$2x^2-x-15=0.$$

Before solving more equations of this type, we consider the form of a quadratic equation. Any quadratic equation (second-degree equation) can be written in the **general form**

$$ax^2+bx+c=0, \quad a \neq 0$$

where x is the variable and a, b, and c are constants. Since a quadratic equation must have an x^2 term, $a \neq 0$. If we also require $a>0$, it is easier to factor the trinomial. In a given equation, if the coefficient of x^2 is negative, we can always multiply both sides by -1 to make $a>0$.

EXAMPLE 2 WRITING EQUATIONS IN GENERAL FORM

(a) Write $3x^2=-5x+4$ in general form.
Adding $5x$ and -4 to both sides of the equation, we obtain

$$3x^2+5x-4=0.$$

This is in general form with $a=3$, $b=5$, and $c=-4$ (not 4).

(b) Write $2x^2+9=7x^2-4x$ in general form.
Adding $-7x^2$ and $4x$ to both sides, we have

$$-5x^2+4x+9=0.$$

If we multiply both sides of the equation by (-1), we obtain the general form

$$5x^2-4x-9=0,$$

where $a=5$, $b=-4$, and $c=-9$.

PRACTICE EXERCISE 2

Write each equation in general form and identify a, b, and c.

(a) $2x^2=5+x$

(b) $4y+y^2=1-2y+y^2$

Answers: (a) $2x^2-x-5=0$; $a=2$, $b=-1$, $c=-5$ (b) not a quadratic equation

3 METHOD OF FACTORING

The method for solving some quadratic equations by factoring and using the zero-product rule is summarized below.

To Solve a Quadratic Equation by Factoring

1. Write the equation in general form by clearing any fractions, dividing by any constant factors, and making the coefficient of x^2 positive by multiplying the equation by -1, if necessary.
2. Factor the left side into a product of two factors.
3. Use the zero-product rule to solve the resulting equation.
4. Check your answers in the original equation.

EXAMPLE 3 Solving a Quadratic Equation by Factoring

Solve $\frac{2}{3}x^2 - \frac{1}{3}x = 1$.

$\frac{2}{3}x^2 - \frac{1}{3}x - 1 = 0$ Subtract 1 from both sides

$2x^2 - x - 3 = 0$ Multiply by 3 to clear fractions and obtain general form

$(2x - 3)(x + 1) = 0$ Factor

$2x - 3 = 0$ or $x + 1 = 0$ Zero-product rule

$2x = 3$ $\quad x = -1$

$x = \frac{3}{2}$

Check: $x = \frac{3}{2}$ $\qquad x = -1$

$\frac{2}{3}\left(\frac{3}{2}\right)^2 - \frac{1}{3}\left(\frac{3}{2}\right) \stackrel{?}{=} 1 \qquad \frac{2}{3}(-1)^2 - \frac{1}{3}(-1) \stackrel{?}{=} 1$

$\frac{3}{2} - \frac{1}{2} \stackrel{?}{=} 1 \qquad \frac{2}{3} + \frac{1}{3} \stackrel{?}{=} 1$

$1 = 1 \qquad 1 = 1$

The solutions are $\frac{3}{2}$ and -1.

Practice Exercise 3

Solve $\frac{3}{2}y^2 + \frac{5}{2}y = 1$.

Answer: $\frac{1}{3}, -2$

4 NUMBER OF SOLUTIONS

Every quadratic equation has two solutions. Sometimes a solution occurs twice as in the next example. In more advanced courses, we often refer to such a solution as a **root of multiplicity two,** or a **double root.**

EXAMPLE 4 Solving an Equation with a Double Root

Solve $x^2 = 6x - 9$.

$x^2 - 6x + 9 = 0$ Write in general form

$(x - 3)^2 = 0$ Factor

$x - 3 = 0$ or $x - 3 = 0$ Zero-product rule

$x = 3$ $\quad x = 3$

Both solutions are 3, that is, 3 is a root of multiplicity two. We will usually say simply that the solution is 3. Check 3 in the original equation.

Practice Exercise 4

Solve $y^2 + 14y = -49$.

Answer: -7

More complicated-looking equations can often be transformed into the general form and solved as in Example 5 on the next page.

EXAMPLE 5 Solving a More Complex Quadratic Equation

Solve $3(y^2 - 6) = y(y + 7) - 3$.

$$3y^2 - 18 = y^2 + 7y - 3 \quad \text{First clear parentheses}$$
$$2y^2 - 18 = 7y - 3$$
$$2y^2 - 7y - 18 = -3$$
$$2y^2 - 7y - 15 = 0 \quad \text{Write in general form}$$
$$(2y + 3)(y - 5) = 0 \quad \text{Factor}$$
$$2y + 3 = 0 \quad \text{or} \quad y - 5 = 0 \quad \text{Zero-product rule}$$
$$2y = -3 \qquad y = 5$$
$$y = -\frac{3}{2}.$$

The solutions are $-\frac{3}{2}$ and 5. Check in the original equation.

Practice Exercise 5

Solve $2(z^2 - 7) = 3(z + 2)$.

Answer: 4, $-\frac{5}{2}$

5 EQUATIONS OF THE FORM $ax^2 + bx = 0$

If a quadratic equation in general form has no constant term, then $c = 0$, and the equation has the form $ax^2 + bx = 0$. For example,

$$4x^2 - 7x = 0$$

is a quadratic equation in general form for which $a = 4$, $b = -7$, and $c = 0$. Factoring out an x, we obtain

$$x(4x - 7) = 0,$$

so that $x = 0$ and $x = \frac{7}{4}$. In general, *any quadratic equation for which* $c = 0$ *will always have* 0 *as one of the solutions.*

EXAMPLE 6 Solving an Equation in which $c = 0$

Solve $2x^2 = 9x$.

$$2x^2 - 9x = 0 \quad \text{Note that } c = 0$$
$$x(2x - 9) = 0 \quad \text{Factor}$$
$$x = 0 \quad \text{or} \quad 2x - 9 = 0 \quad \text{Zero-product rule}$$
$$2x = 9$$
$$x = \frac{9}{2}$$

The solutions are 0 and $\frac{9}{2}$. Check in the original equation.

Practice Exercise 6

Solve $5y^2 = -3y$.

Answer: 0, $-\frac{3}{5}$

CAUTION

Is $2x^2 = 9x$ equivalent to the equation $2x = 9$, which is obtained by dividing both sides by x? The answer is no since $2x = 9$ has only one solution, $x = \frac{9}{2}$, while $2x^2 = 9x$ has two solutions, $\frac{9}{2}$ *and* 0. Never divide both sides of an equation by the variable or by an expression containing the variable.

6 METHOD OF TAKING ROOTS

If a quadratic equation in general form has no x term, then $b = 0$, and the equation takes the form

$$ax^2 + c = 0.$$

Recall that

$$\sqrt{x^2} = |x| = \begin{cases} x \text{ if } x \geq 0 \\ -x \text{ if } x < 0. \end{cases}$$

Thus, taking the square root of both sides of an equation such as

$$x^2 = 9$$

we obtain

$$\sqrt{x^2} = \sqrt{9}$$
$$|x| = 3.$$

But since $|x| = 3$ is equivalent to $x = 3$ or $-x = 3$, then

$$x = 3 \quad \text{or} \quad x = -3.$$

Thus, the equation $x^2 = 9$ has solutions 3 and -3. To solve an equation such as $x^2 - 9 = 0$, our work would appear as follows:

$$x^2 - 9 = 0$$
$$x^2 = 9 \qquad \text{Solve the equation for } x^2$$
$$x = \pm\sqrt{9} \qquad \text{Take the square root of both sides}$$
$$x = \pm 3. \qquad x = +3 \text{ (or 3) } \textit{and } x = -3$$

The method for solving a quadratic equation when $b = 0$, shown above, is called the **method of taking roots.** We summarize this method below.

To Solve the Quadratic Equation $ax^2 + c = 0$ by Taking Roots

1. Subtract c from both sides, then divide both sides by a to obtain

$$x^2 = -\frac{c}{a}.$$

2. Take the square root of both sides to obtain

$$x = \pm\sqrt{-\frac{c}{a}}.$$

The equation $x^2 = 9$, solved above by taking roots, could be written as $x^2 - 9 = 0$ and solved by factoring, since $x^2 - 9 = (x - 3)(x + 3) = 0$ also gives 3 and -3 as the solutions. However, not all equations of this type can be factored using integers, as shown in Example 7 (b) below.

EXAMPLE 7 SOLVING QUADRATIC EQUATIONS BY TAKING ROOTS

(a) Solve $3x^2 - 48 = 0$.

$$3x^2 = 48 \qquad \text{Add 48 to both sides}$$
$$x^2 = 16 \qquad \text{Divide by 3}$$
$$x = \pm\sqrt{16} = \pm 4 \qquad \text{Take square root of both sides}$$

The solutions are 4 and -4. Check.

PRACTICE EXERCISE 7

(a) Solve $2y^2 - 50 = 0$.

(b) Solve $6y^2 - 72 = 0$.

$6y^2 = 72$ Add 72

$y^2 = 12$ Divide by 6

$y = \pm\sqrt{12} = \pm 2\sqrt{3}$ Take square root of both sides

The solutions are $2\sqrt{3}$ and $-2\sqrt{3}$. Check.

(b) Solve $5z^2 - 40 = 0$.

Answers: (a) 5, −5 (b) $\pm 2\sqrt{2}$

Sometimes solutions to a quadratic equation are not real numbers but are complex or imaginary numbers. One reason we introduced complex numbers in Chapter 7 was to provide a number system in which *all* quadratic equations have solutions.

EXAMPLE 8 Quadratic Equations with Complex Solutions

(a) Solve $3x^2 + 12 = 0$.

$3x^2 = -12$ Subtract 12 from both sides

$x^2 = -4$ Divide both sides by 3

$x = \pm\sqrt{-4}$ Take square root of both sides

$x = \pm\sqrt{4(-1)} = \pm\sqrt{4}\sqrt{-1} = \pm 2i$

Check: $x = 2i$

$3(2i)^2 + 12 \stackrel{?}{=} 0$

$3(-4) + 12 \stackrel{?}{=} 0$

$-12 + 12 = 0$

$0 = 0$

$x = -2i$

$3(-2i)^2 + 12 \stackrel{?}{=} 0$

$3(-4) + 12 \stackrel{?}{=} 0$

$-12 + 12 \stackrel{?}{=} 0$

$0 = 0$

The solutions are $\pm 2i$, that is, $2i$ and $-2i$.

(b) Solve $(y - 2)^2 + 3 = 0$.

Although this equation is not exactly in the form for taking roots, we can still use that method and solve for $y - 2$ first.

$(y - 2)^2 = -3$ Subtract 3 from both sides

$y - 2 = \pm\sqrt{-3}$ Take square root of both sides

$y - 2 = \pm i\sqrt{3}$ $\sqrt{-3} = \sqrt{(-1)(3)} = \sqrt{-1}\sqrt{3} = i\sqrt{3}$

$y = 2 \pm i\sqrt{3}$ Add 2 to both sides

The solutions are $2 + i\sqrt{3}$ and $2 - i\sqrt{3}$, which do check in the original equation.

Practice Exercise 8

(a) Solve $9x^2 + 4 = 0$.

(b) Solve $(y + 1)^2 + 16 = 0$

Answers: (a) $\pm\frac{2}{3}i$ (b) $-1 \pm 4i$

Many applied problems involve quadratic equations.

EXAMPLE 9 Interest Problem

The amount of money A which will result if a principal P is invested at r percent interest compounded annually for two years is given by $A = P(1 + r)^2$. If \$2000 grows to \$2599.20 in 2 years using this formula, what is the rate of interest?

The formula above is a quadratic equation that can be solved by tak-

Practice Exercise 9

Use $A = P(1 + r)^2$ to find the rate of interest used for \$5000 to grow to \$5832 in 2 years.

ing roots. In this case we first solve for $1 + r$ and then for r.

$A = P(1 + r)^2$	Formula
$2599.20 = 2000(1 + r)^2$	$A = \$2599.20$ and $P = \$2000$
$1.2996 = (1 + r)^2$	Divide by 2000
$\pm\sqrt{1.2996} = 1 + r$	Take the square root
$1.14 = 1 + r$	Use only the positive square root obtained with a calculator
$0.14 = r$	

The interest rate is 14%. Had we used the $1 + r = -1.14$, we would have $r = -2.14$, which has no meaning in this problem.

Answer: 8%

8.1 EXERCISES A

Solve using the zero-product rule.

1. $(x - 1)(x + 2) = 0$
2. $(4y - 1)(4y + 1) = 0$
3. $x(x + 8) = 0$
4. $\frac{1}{3}y\left(y - \frac{2}{3}\right) = 0$
5. $(8.2x - 24.6)(7.2x - 14.4) = 0$
6. $(3x + 2)(5x - 1) = 0$
7. $(y - 1) - (2y + 3) = 0$
8. $(x - 1)(x + 1)(x - 2) = 0$
9. $3z(z - 1)(z + 7) = 0$

Simplify each equation, express it in general form, and identify the constants a, b, and c.

10. $2x^2 + x = 5$
11. $z^2 = 3z$
12. $2(x + x^2) = 3x - 1$
13. $z^2 + 1 = 0$
14. $3z = 3(z^2 + 1)$
15. $(x - 2)(x + 2) = 5$

Solve the following quadratic equations. All of these equations have real number solutions.

16. $y^2 - y - 2 = 0$
17. $2x^2 - 2x - 12 = 0$
18. $5z^2 - 15z = 0$
19. $4y^2 - 20y + 25 = 0$
20. $z(8 - 2z) = 6$
21. $5x^2 - 4x = 3x^2 + 9x + 7$

22. $(3y - 1)(2y + 1) = 3(2y + 1)$

23. $9(x^2 + 1) = 24x - 7$

24. $5 + z^2 - 2z = z(z - 2)$

25. $5y(y + 1) = 5y^2 + 5y$

26. $x^2 = 16$

27. $3z^2 - 10 = -10$

28. $x^2 = \frac{1}{4}$

29. $(x - 2)^2 = 25$

30. $4y^2 = 72$

31. $2z^2 = 5 + z^2$

32. $y^2 + 2y = 2y + 4$

33. $x^3 + 3x^2 + 2x = 0$

Solve the following quadratic equations. All of these equations have complex number solutions.

34. $x^2 + 9 = 0$

35. $4y^2 + 1 = 0$

36. $25z^2 + 75 = 0$

37. $(x + 1)^2 = -144$

38. $(2x - 1)^2 + 4 = 0$

39. $(z - 8)^2 + 7 = 0$

Solve the following quadratic equations. Solutions to these are either real or complex numbers.

40. $6x = -12x^2$

41. $-4y^2 + 64 = 0$

42. $4z^2 + 64 = 0$

43. $(x - 4)^2 + 5 = 0$

44. $(x - 4)^2 - 5 = 0$

45. $6z^3 - z^2 - 12z = 0$

Solve.

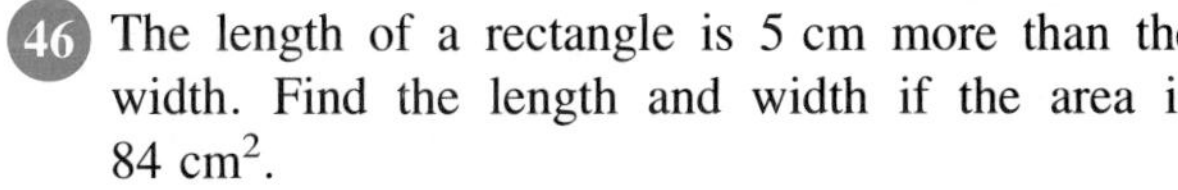

46. The length of a rectangle is 5 cm more than the width. Find the length and width if the area is 84 cm^2.

Let w = the width
w + 5 = the length

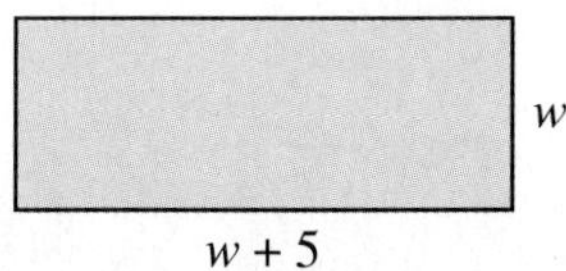

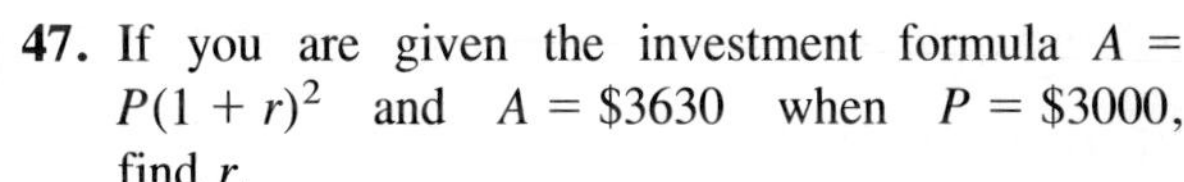

47. If you are given the investment formula $A = P(1 + r)^2$ and $A = \$3630$ when $P = \$3000$, find r.

48. The pressure P in pounds per square foot of a wind blowing V miles per hour can be approximated by $P = \frac{3V^2}{1000}$. Suppose the pressure of the wind against the side of a building is 12.25 lb/ft^2. What is the approximate velocity of the wind?

49 During the winter, the demand equation for heating oil is $D = \frac{500}{P}$ while the supply equation is $S = p - 95$, where p is the price in cents per gallon, D is the number of gallons demanded (in thousands), and S is the number of gallons supplied (also in thousands). Find the price per gallon when the supply equals the demand.

Determine what is wrong with each of the "solutions" in Exercises 50–51.

50.
$$\begin{aligned} x^2 + 5x &= 0 \\ x(x + 5) &= 0 \\ \frac{x(x + 5)}{x} &= \frac{0}{x} \\ x + 5 &= 0 \\ x &= -5 \end{aligned}$$

51.
$$\begin{aligned} x^2 - x &= 5 \\ x(x - 1) &= 5 \\ x = 5 \quad \text{or} \quad x - 1 &= 5 \\ x &= 6 \end{aligned}$$

FOR REVIEW

The following exercises review material that we covered in Chapters 5 and 7. They will help you prepare for the next section. Find the products in Exercises 52–53.

52. $(x + 7)^2$

53. $\left(x - \frac{1}{2}\right)^2$

Factor the trinomials in Exercises 54–55.

54. $x^2 - 6x + 9$

55. $x^2 + x + \frac{1}{4}$

Simplify the square roots in Exercises 56–58.

56. $\sqrt{12}$

57. $\sqrt{\frac{81}{16}}$

58. $\sqrt{\frac{10}{9}}$

Evaluate each expression in Exercises 59–64 if $a = 3$, $b = -2$, and $c = -5$.

59. $-b$

60. $4ac$

61. $b^2 - 4ac$

62. $\sqrt{b^2 - 4ac}$

63. $-b + \sqrt{b^2 - 4ac}$

64. $-b - \sqrt{b^2 - 4ac}$

ANSWERS: 1. $1, -2$ 2. $\frac{1}{4}, -\frac{1}{4}$ 3. $0, -8$ 4. $0, \frac{2}{3}$ 5. $3, 2$ 6. $\frac{1}{5}, -\frac{2}{3}$ 7. -4 8. $1, -1, 2$ 9. $0, 1, -7$ 10. $2x^2 + x - 5 = 0$; $a = 2$, $b = 1$, $c = -5$ 11. $z^2 - 3z = 0$; $a = 1$, $b = -3$, $c = 0$ 12. $2x^2 - x + 1 = 0$; $a = 2$, $b = -1$, $c = 1$ 13. $z^2 + 1 = 0$; $a = 1$, $b = 0$, $c = 1$ 14. $z^2 - z + 1 = 0$; $a = 1$, $b = -1$, $c = 1$ 15. $x^2 - 9$; $a = 1$, $b = 0$, $c = -9$ 16. $2, -1$ 17. $3, -2$ 18. $0, 3$ 19. $\frac{5}{2}$ 20. $1, 3$ 21. $7, -\frac{1}{2}$ 22. $\frac{4}{3}, -\frac{1}{2}$ 23. $\frac{4}{3}$ 24. no solution 25. Any number is a solution. 26. ± 4 27. 0 28. $\pm\frac{1}{2}$ 29. $-3, 7$ 30. $\pm 3\sqrt{2}$ 31. $\pm\sqrt{5}$ 32. ± 2 33. $0, -1, -2$ 34. $\pm 3i$ 35. $\pm\frac{1}{2}i$ 36. $\pm i\sqrt{3}$ 37. $-1 \pm 12i$ 38. $\frac{1}{2} \pm i$ 39. $8 \pm i\sqrt{7}$ 40. $0, -\frac{1}{2}$ 41. ± 4 42. $\pm 4i$ 43. $4 \pm i\sqrt{5}$ 44. $4 \pm \sqrt{5}$ 45. $0, -\frac{4}{3}, \frac{3}{2}$ 46. 12 cm, 7 cm 47. 10% 48. 64 mph 49. 100¢/gal 50. The solution $x = 0$ has been lost by incorrectly dividing both sides by x. 51. Do not use the zero-product rule on a product equal to 5. 52. $x^2 + 14x + 49$ 53. $x^2 - x + \frac{1}{4}$ 54. $(x - 3)^2$ 55. $\left(x + \frac{1}{2}\right)^2$ 56. $2\sqrt{3}$ 57. $\frac{9}{4}$ 58. $\frac{\sqrt{10}}{3}$ 59. 2 60. -60 61. 64 62. 8 63. 10 64. -6

8.1 EXERCISES B

Solve using the zero-product rule.

1. $(y - 2)(y + 5) = 0$
2. $(6z + 5)(6z - 5) = 0$
3. $(y - 7)y = 0$
4. $\dfrac{3}{4}z(z + 7) = 0$
5. $(1.1y - 3.3)(2.2y + 6.6) = 0$
6. $(4x - 3)(7x + 1) = 0$
7. $(z + 3) + (z - 3) = 0$
8. $(x - 2)(x + 5)(x - 3) = 0$
9. $7z(z - 2)(z + 10) = 0$

Simplify each equation, express it in general form, and identify the constants a, b, and c.

10. $y - y^2 = 5$
11. $x^2 = -7x - 12$
12. $2y^2 + 18y = 4$
13. $x^2 = 5$
14. $5y^2 = 8(y^2 - y)$
15. $(x + 3)(x - 3) = 4x$

Solve the following quadratic equations. All of these equations have real-number solutions.

16. $x^2 - 5x + 6 = 0$
17. $2z^2 - 8z - 24 = 0$
18. $4y^2 - 8y = 0$
19. $4x^2 + 12x + 9 = 0$
20. $\dfrac{1}{2}z^2 - 3z + 4 = 0$
21. $4y^2 + 4 = 7y^2 + 11y + 14$
22. $(x + 1)(x - 5) = -5x + 7$
23. $2z(z + 1) = 1 + 3z$
24. $3y^2 - 6y + 2 = 3y(y - 2)$
25. $2x^2 + 6x = 2x(x + 3)$
26. $(2z - 3)(z + 1) = 4(2z - 3)$
27. $y^2 - 81 = 0$
28. $3z^2 - 27 = 0$
29. $(y + 3)^2 = 81$
30. $2x^2 - 150 = 0$
31. $7y^2 - \dfrac{1}{7} = 0$
32. $3 - x^2 = x^2 - 3$
33. $3z^2 - 8z = 147 - 8z$

Solve the following quadratic equations. All of these equations have complex-number solutions.

34. $x^2 + 49 = 0$
35. $9y^2 + 1 = 0$
36. $10z^2 + 50 = 0$
37. $(x + 2)^2 = -100$
38. $(2x - 1)^2 + 1 = 0$
39. $(z + 6)^2 + 6 = 0$

Solve the following quadratic equations. Solutions to these equations are either real or complex numbers.

40. $y^2 = 6 + y$
41. $-3z^2 + 6 = 0$
42. $3z^2 + 6 = 0$
43. $(x + 5)^2 + 8 = 0$
44. $(x + 5)^2 - 8 = 0$
45. $y^3 + 6y^2 + 8y = 0$

Solve.

46. The area of a triangle is 30 in^2. Find the base and height if the base is 4 in more than the height.

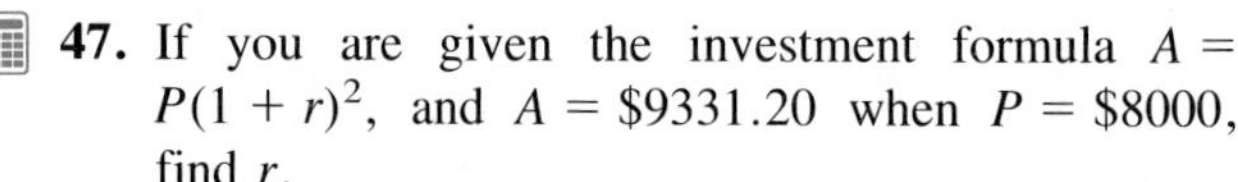

47. If you are given the investment formula $A = P(1 + r)^2$, and $A = \$9331.20$ when $P = \$8000$, find r.

48. The pressure P in pounds per square foot of a wind blowing V miles per hour can be approximated by $P = \frac{3V^2}{1000}$. Suppose the pressure of the wind against the side of a bridge is 1.88 lb/ft^2. What is the approximate velocity of the wind?

49. The demand equation for gasoline is $D = \frac{1100}{P}$ and the supply equation is $S = p - 100$, where p is the price in cents per gallon, D is the number of gallons demanded (in ten thousands), and S is the number of gallons supplied (also in ten thousands). Find the price per gallon when the supply equals the demand.

Tell what is wrong with each of the "solutions" in Exercises 50–51.

50.
$$x^2 + 6x = 0$$
$$x(x + 6) = 0$$
$$\frac{x(x + 6)}{x} = \frac{0}{x}$$
$$x + 6 = 0$$
$$x = -6$$

51.
$$x^2 + x = 5$$
$$x(x + 1) = 5$$
$$x = 5 \quad \text{or} \quad x + 1 = 5$$
$$x = 4$$

FOR REVIEW

The following exercises review material that we covered in Chapters 5 and 7. They will help you prepare for the next section. Find the products in Exercises 52–53.

52. $(x - 10)^2$

53. $\left(x + \frac{1}{2}\right)^2$

Factor the trinomials in Exercises 54–55.

54. $x^2 + 6x + 9$

55. $x^2 - x + \frac{1}{4}$

Simplify the square roots in Exercises 56–58.

56. $\sqrt{20}$

57. $\sqrt{\frac{121}{25}}$

58. $\sqrt{\frac{3}{16}}$

Evaluate each expression in Exercises 59–64 if $a = 2$, $b = -1$, and $c = -4$.

59. $-b$

60. $4ac$

61. $b^2 - 4ac$

62. $\sqrt{b^2 - 4ac}$

63. $-b + \sqrt{b^2 - 4ac}$

64. $-b - \sqrt{b^2 - 4ac}$

8.1 EXERCISES C

Solve for x. Assume that a and b are positive constants.

1. $x^2 - 9a^2 = 0$

2. $x^2 - (b + a)^2 = 0$
[Answer: $\pm(b + a)$]

3. $x^2 - bx - 2b^2 = 0$

4. $(20x^2 + 23x - 7)(21x^2 + 5x - 6) = 0$
$\left[\text{Answer: } -\frac{7}{5}, \frac{1}{4}, -\frac{2}{3}, \frac{3}{7}\right]$

5. $(x - 1)(x + 2) + (x - 3)(x + 2) = 0$

The distance h in feet that a free-falling object falls without being given an initial velocity is given by the formula $h = 16t^2$. *Use this formula in Exercises 6 and 7.*

6. How far will an object fall in 2.75 sec?

7. How long would it take an object to fall from a bridge to the bottom of a gorge if the bridge is 650 ft above the level of the river? Give answer to the nearest tenth of a second.

8.2 COMPLETING THE SQUARE AND THE QUADRATIC FORMULA

STUDENT GUIDEPOSTS

1. Completing the Square
2. The Quadratic Formula
3. Choosing the Method to Use

1 COMPLETING THE SQUARE

Factoring and taking roots are the easiest and best ways to solve many quadratic equations, but some quadratic equations cannot be solved by these techniques. For such equations, we usually use the *quadratic formula*, which is derived by a process called *completing the square*.

Suppose we want to solve an equation of the form

$$(x + 5)^2 = 4.$$

By the method of taking roots from Section 8.1, we take the square root of both sides obtaining

$$x + 5 = \pm\sqrt{4}$$
$$x + 5 = \pm 2$$
$$x = -5 \pm 2. \quad \text{Subtract 5 from both sides}$$

Thus, $x = -5 + 2 = -3$ or $x = -5 - 2 = -7$.

If a general quadratic equation

$$ax^2 + bx + c = 0$$

is changed into an equation of the form

$$(x - A)^2 = B,$$

where A and B are real numbers, the solutions are

$$x = A \pm \sqrt{B}. \quad \text{One is } A + \sqrt{B}\text{, the other is } A - \sqrt{B}$$

To see how this transformation can be made, we begin by considering the square of the binomial $x + d$.

$$(x + d)^2 = x^2 + 2dx + d^2$$

Notice that in the expression $x^2 + 2dx + d^2$, the constant term, d^2, is the square of one half the coefficient of x, $2d$. Had we been given

$$x^2 + 2dx,$$

we could **complete the square** by adding the square of one half the coefficient of x,

$$\left[\frac{1}{2}(2d)\right]^2 = d^2.$$

Several examples are given in the following table.

To complete square on:	*Add (one-half the coefficient of x)²:*	*To obtain the perfect square:*
$x^2 + 6x$	$9\left[9 = \left(\frac{1}{2} \cdot 6\right)^2\right]$	$x^2 + 6x + 9 = (x + 3)^2$
$x^2 - 10x$	$25\left[25 = \left(\frac{1}{2} \cdot (-10)\right)^2\right]$	$x^2 - 10x + 25 = (x - 5)^2$
$x^2 - x$	$\frac{1}{4}\left[\frac{1}{4} = \left(\frac{1}{2} \cdot (-1)\right)^2\right]$	$x^2 - x + \frac{1}{4} = \left(x - \frac{1}{2}\right)^2$

We now use the process of completing the square to solve an equation.

EXAMPLE 1 Solving an Equation by Completing the Square

Solve $x^2 - 6x + 8 = 0$ by completing the square.

First we isolate the constant term on the right side of the equation, leaving space as indicated.

$$x^2 - 6x \qquad = -8$$

Then we complete the square on the left side by adding $\left[\frac{1}{2} \cdot (-6)\right]^2 = (-3)^2 = 9$. But if 9 is added on the left side, to keep both sides equal, 9 must be added to the right side.

$$x^2 - 6x + 9 = -8 + 9$$

$$(x - 3)^2 = 1 \qquad \text{Check by squaring the left side}$$

To complete the solution, we take the square root of both sides of this equation.

$x - 3 = \pm\sqrt{1}$ $\qquad$ $\pm\sqrt{1}$ represents two numbers, $\sqrt{1}$ and $-\sqrt{1}$

$x - 3 = \pm 1$ $\qquad$ $\sqrt{1} = 1$

$x = 3 \pm 1$ $\qquad$ Solve for x by adding 3 to both sides

$x = 3 + 1$ or $x = 3 - 1$ $\qquad$ $x = 3 \pm 1$ means $x = 3 + 1$ or $x = 3 - 1$

$x = 4 \qquad x = 2$

Thus, the solutions are 2 and 4. We can also solve this equation by the factoring method: $(x - 2)(x - 4) = 0$, which gives us the same solutions of 2 and 4. Notice that the factoring method is much easier and would be preferred in this case.

Practice Exercise 1

Solve $y^2 - 4y - 5 = 0$ by completing the square.

First write the equation in the form

$$y^2 - 4y \qquad = 5.$$

Then add $\left[\frac{1}{2}(-4)\right]^2 = [-2]^2 = 4$ to both sides.

Answer: 5, −1

If the coefficient of x^2 in a quadratic equation is not 1, we *cannot* complete the square by taking one-half the coefficient of x and squaring it. For example, $(2x + 3)^2 = 4x^2 + 12x + 9$ but $\left(\frac{1}{2} \cdot 12\right)^2 = 6^2 \neq 9$. In such a case, we need to divide both sides of the equation by a number that will change the coefficient of x^2 to 1, as in the following example.

EXAMPLE 2 Solving an Equation by Completing the Square

Solve $2x^2 + 2x - 3 = 0$ by completing the square.

$2x^2 + 2x = 3$ Isolate the constant

$x^2 + x = \frac{3}{2}$ Divide by 2 to make the coefficient of x^2 equal to 1

$x^2 + x + \frac{1}{4} = \frac{3}{2} + \frac{1}{4}$ $\left(\frac{1}{2} \cdot 1\right)^2 = \frac{1}{4}$

$\left(x + \frac{1}{2}\right)^2 = \frac{7}{4}$ $\frac{3}{2} + \frac{1}{4} = \frac{6}{4} + \frac{1}{4} = \frac{7}{4}$

$x + \frac{1}{2} = \pm\sqrt{\frac{7}{4}} = \pm\frac{\sqrt{7}}{\sqrt{4}}$ $\sqrt{\frac{a}{b}} = \frac{\sqrt{a}}{\sqrt{b}}$

$= \frac{\pm\sqrt{7}}{2}$

$x = -\frac{1}{2} \pm \frac{\sqrt{7}}{2} = \frac{-1 \pm \sqrt{7}}{2}$

The solutions are $\frac{-1 + \sqrt{7}}{2}$ and $\frac{-1 - \sqrt{7}}{2}$.

Practice Exercise 2

Solve $3x^2 - x - 5 = 0$ by completing the square.

First isolate the constant.

$$3x^2 - x = 5$$

Then divide each term by 3.

$$x^2 - \frac{1}{3}x = \frac{5}{3}$$

Answer: $\frac{1 \pm \sqrt{61}}{6}$

To Solve a Quadratic Equation by Completing the Square

1. Simplify the equation as much as possible and write it with both variable terms on the left side and the constant term on the right side.
2. If the coefficient of x^2 is not 1, divide each term by that coefficient.
3. Add the square of one-half the coefficient of x to both sides.
4. The left side is now the perfect square of a binomial.
5. Use the method of taking roots to complete the solution.

2 THE QUADRATIC FORMULA

Rather than continuing to solve particular quadratic equations by completing the square, we can solve a general quadratic equation using this method, and in the process, derive the *quadratic formula*.

$ax^2 + bx + c = 0$ $a \neq 0$

$ax^2 + bx = -c$ Isolate the constant

$\frac{\not{a}x^2}{\not{a}} + \frac{b}{a}x = -\frac{c}{a}$ Divide by a to make the coefficient of x^2 equal to 1

$x^2 + \frac{b}{a}x + \left(\frac{b}{2a}\right)^2 = -\frac{c}{a} + \left(\frac{b}{2a}\right)^2$ $\left(\frac{1}{2} \cdot \frac{b}{a}\right)^2 = \left(\frac{b}{2a}\right)^2$

$\left(x + \frac{b}{2a}\right)^2 = -\frac{c}{a} + \frac{b^2}{4a^2}$

$= -\frac{4ac}{4a^2} + \frac{b^2}{4a^2}$ $4a^2$ is LCD

$= \frac{b^2 - 4ac}{4a^2}$ Subtract

$$x + \frac{b}{2a} = \pm\sqrt{\frac{b^2 - 4ac}{4a^2}} = \frac{\pm\sqrt{b^2 - 4ac}}{\sqrt{4a^2}} \qquad \sqrt{\frac{a}{b}} = \frac{\sqrt{a}}{\sqrt{b}}$$

$$= \frac{\pm\sqrt{b^2 - 4ac}}{2a}$$

$$x = -\frac{b}{2a} \pm \frac{\sqrt{b^2 - 4ac}}{2a} \qquad \text{Subtract } \frac{b}{2a}$$

$$x = \frac{-b \pm \sqrt{b^2 - 4ac}}{2a} \qquad \text{2a is the denominator of the entire expression}$$

This last formula is called the **quadratic formula** and it must be memorized. To solve a quadratic equation, identify the constants a, b, and c and substitute them into the quadratic formula.

To Solve a Quadratic Equation Using the Quadratic Formula

1. Write the equation in general form, $ax^2 + bx + c = 0$. Be sure that all common factors have been divided out and all fractions have been removed.
2. Identify the constants a, b, and c.
3. Substitute the values of a, b, and c into the quadratic formula,
$$x = \frac{-b \pm \sqrt{b^2 - 4ac}}{2a}.$$
4. Simplify the numerical expression to obtain the solutions.

EXAMPLE 3 Solving an Equation by the Quadratic Formula

Solve $x^2 - 6x + 8 = 0$ using the quadratic formula.

The equation is already in general form with $a = 1$, $b = -6$ (not 6), and $c = 8$.

$$x = \frac{-b \pm \sqrt{b^2 - 4ac}}{2a}$$

$$= \frac{-(-6) \pm \sqrt{(-6)^2 - 4(1)(8)}}{2(1)} \qquad \text{Multiply 4(1)(8), then subtract the product from } (-6)^2$$

$$= \frac{6 \pm \sqrt{36 - 32}}{2}$$

$$= \frac{6 \pm \sqrt{4}}{2} = \frac{6 \pm 2}{2}$$

$$x = \frac{6 + 2}{2} = \frac{8}{2} = 4 \quad \text{or} \quad x = \frac{6 - 2}{2} = \frac{4}{2} = 2$$

The solutions are 4 and 2.

Practice Exercise 3

Solve $x^2 + 2x - 15 = 0$ using the quadratic formula.

Answer: 3, −5

It would have been easier to solve the equation in Example 3 by factoring, since $x^2 - 6x + 8 = (x - 2)(x - 4)$. By using the zero-product rule we obtain solutions 2 and 4, the same as were obtained using the quadratic formula.

EXAMPLE 4 Solving an Equation with Irrational Solutions

Solve $6x^2 - 10 = -8x$ using the quadratic formula.

First write the equation in general form and simplify.

$$6x^2 + 8x - 10 = 0 \quad \text{Add } 8x \text{ to both sides}$$
$$2(3x^2 + 4x - 5) = 0 \quad \text{Factor out 2}$$
$$3x^2 + 4x - 5 = 0 \quad \text{Divide by 2 to simplify}$$

Thus $a = 3$, $b = 4$, and $c = -5$. $c = -5$, not 5

$$x = \frac{-b \pm \sqrt{b^2 - 4ac}}{2a} = \frac{-4 \pm \sqrt{(4)^2 - 4(3)(-5)}}{2(3)}$$
$$= \frac{-4 \pm \sqrt{16 + 60}}{6} \quad \text{Watch the signs}$$
$$= \frac{-4 \pm \sqrt{76}}{6} = \frac{-4 \pm \sqrt{4 \cdot 19}}{6} \quad \text{4 is a perfect square}$$
$$= \frac{-4 \pm \sqrt{4}\sqrt{19}}{6} \quad \sqrt{ab} = \sqrt{a}\sqrt{b}$$
$$= \frac{-4 \pm 2\sqrt{19}}{6} \quad \sqrt{4} = 2$$
$$= \frac{\cancel{2}(-2 \pm \sqrt{19})}{\cancel{2} \cdot 3} \quad \text{Factor and divide out 2}$$
$$= \frac{-2 \pm \sqrt{19}}{3} \quad \text{This is simplified}$$

The solutions are $\dfrac{-2 + \sqrt{19}}{3}$ and $\dfrac{-2 - \sqrt{19}}{3}$.

Practice Exercise 4

Solve $5x^2 + 2x - 1 = 0$ using the quadratic formula.

Answer: $\frac{-1 \pm \sqrt{6}}{5}$

3 CHOOSING THE METHOD TO USE

In general, to solve a quadratic equation, the methods of factoring and taking roots are the easiest and quickest to apply. If these two techniques are not appropriate, it is better to go directly to the quadratic formula rather than to use the method of completing the square.

We now solve a quadratic equation with complex solutions using the quadratic formula.

EXAMPLE 5 A Quadratic Equation with Complex Solutions

Solve $x^2 - 2x + 3 = 0$.

Use the quadratic formula with $a = 1$, $b = -2$, and $c = 3$.

$$x = \frac{-(-2) \pm \sqrt{(-2)^2 - 4(1)(3)}}{2(1)}$$
$$= \frac{2 \pm \sqrt{4 - 12}}{2} = \frac{2 \pm \sqrt{-8}}{2} = \frac{2 \pm 2i\sqrt{2}}{2}$$
$$= \frac{\cancel{2}(1 \pm i\sqrt{2})}{\cancel{2}} = 1 \pm i\sqrt{2}$$

Practice Exercise 5

Solve $x^2 - 6x + 10 = 0$.

Answer: $3 \pm i$

Many applied problems can be described by quadratic equations and solved using the quadratic formula. We often approximate solutions using a calculator to make the values more meaningful.

EXAMPLE 6 Geometry Problem

If the hypotenuse of a right triangle is 7 inches long and one leg is 3 inches longer than the other, find the measure of each leg.

The triangle is shown in Figure 8.1. Recall that the Pythagorean theorem states that the sum of the squares of the legs of a right triangle is equal to the square of the hypotenuse.

Let x = the measure of one leg,
$x + 3$ = the measure of the other leg.

Use the Pythagorean theorem.

$$x^2 + (x + 3)^2 = 7^2$$
$$x^2 + x^2 + 6x + 9 = 49$$
$$2x^2 + 6x + 9 = 49$$
$$2x^2 + 6x - 40 = 0$$
$$x^2 + 3x - 20 = 0$$

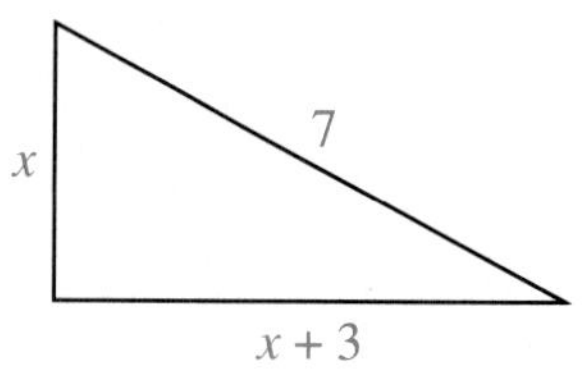

Figure 8.1

Since we cannot factor, we use the quadratic formula.

$$x = \frac{-b \pm \sqrt{b^2 - 4ac}}{2a}$$

$$x = \frac{-3 \pm \sqrt{(3)^2 - 4(1)(-20)}}{2(1)} \qquad a = 1,\ b = 3,\ c = -20$$

$$= \frac{-3 \pm \sqrt{9 + 80}}{2} = \frac{-3 \pm \sqrt{89}}{2}$$

We obtain $\frac{-3+\sqrt{89}}{2}$ and $\frac{-3-\sqrt{89}}{2}$ as solutions to the quadratic equation. But since $\frac{-3-\sqrt{89}}{2}$ is negative, it is discarded so that

$$x = \frac{-3 + \sqrt{89}}{2} \quad \text{and} \quad x + 3 = \frac{\sqrt{89} - 3}{2} + 3$$

$$= \frac{\sqrt{89} - 3}{2} \qquad\qquad = \frac{\sqrt{89} - 3}{2} + \frac{6}{2} = \frac{\sqrt{89} + 3}{2}.$$

Although the legs are exactly $\frac{\sqrt{89}-3}{2}$ inches and $\frac{\sqrt{89}+3}{2}$ inches, in a practical situation, we would probably give approximations to these lengths. Since $\sqrt{89} \approx 9.4$, the legs are approximately $\frac{9.4-3}{2} = 3.2$ inches and $\frac{9.4+3}{2} = 6.2$ inches.

Practice Exercise 6

The hypotenuse of a right triangle is 11 km long and one leg is 4 km shorter than the other. Find the length of each leg to the nearest tenth of a kilometer.

Answer: 9.5 km, 5.5 km

EXAMPLE 7 Physics Problem

If an object is propelled upward from an initial height of s feet with initial velocity of v ft/sec, ignoring air resistance, the height h of the object t seconds after it is released is given by

$$h = -16t^2 + vt + s.$$

A child throws a rock upward from the edge of a building 30 ft high with initial velocity 32 ft/sec. At what time, to the nearest tenth of a second,

Practice Exercise 7

A rocket is fired from an underground silo with initial velocity 450 ft/sec. If the silo is 50 ft below ground level, at what time, to the nearest tenth of a second, will the rocket be half a mile high? [*Hint:* $s = -50$ and 1 mile = 5280 ft.]

will the rock be 20 ft above ground level? See Figure 8.2.

Substitute 30 for s, 32 for v, and 20 for h into $h = -16t^2 + vt + s$.

$$20 = -16t^2 + 32t + 30$$
$$0 = -16t^2 + 32t + 10$$
$$0 = 8t^2 - 16t - 5$$

Use the quadratic formula with $a = 8$, $b = -16$, and $c = -5$.

$$t = \frac{-b \pm \sqrt{b^2 - 4ac}}{2a}$$
$$= \frac{-(-16) \pm \sqrt{(-16)^2 - 4(8)(-5)}}{2(8)}$$
$$= \frac{16 \pm \sqrt{256 + 160}}{16} = \frac{16 \pm \sqrt{416}}{16}$$
$$= \frac{16 \pm 4\sqrt{26}}{16} = \frac{4 \pm \sqrt{26}}{4}$$

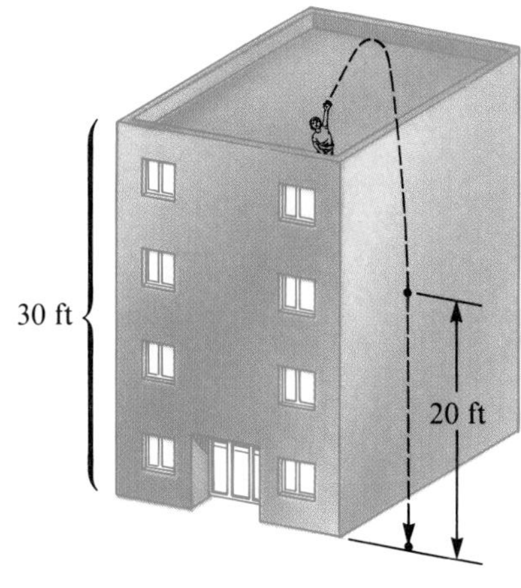

Figure 8.2

If we approximate $\sqrt{26}$ with a calculator, $\sqrt{26} \approx 5.1$. Then

$$t = \frac{4 + 5.1}{4} \approx 2.3 \quad \text{or} \quad t = \frac{4 - 5.1}{4} \approx -0.3.$$

Since t represents time, we discard the solution -0.3. Thus, the rock will be 20 ft from the ground about 2.3 seconds after it is tossed.

Answer: 8.6 sec (on the way up) and 19.5 sec (on the way down)

8.2 EXERCISES A

Solve.

1. $(z - 5)^2 = 16$

2. $(y + 8)^2 - 3 = 0$

3. $(2x + 3)^2 = 9$

4. $(5z + 1)^2 - 81 = 0$

5. $(2y - 3)^2 + 11 = 0$

6. $(3x - 7)^2 = 8$

7 $(3z - 1)^2 = \frac{4}{9}$

8. $(3x + 2)^2 + 6 = 0$

What must be added to complete the square in each expression?

9. $y^2 - 6y$

10. $x^2 + x$

11 $z^2 + \frac{8}{5}z$

Solve by completing the square.

12. $x^2 + 6x + 5 = 0$

13. $z^2 - 2z - 8 = 0$

14. $2u^2 + 3u - 1 = 0$

15. $x^2 + 2x + 3 = 0$

Solve by using the quadratic formula.

16. $v^2 - 8v + 15 = 0$

17. $y^2 - 25 = 0$

18. $3u^2 + 10u = -3u - 4$

19. $4x(x + 3) = 9x$

20. $5u^2 - u = 1$

21. $6x^2 - 9x - 12 = 0$

22. $8u^2 = 4(u^2 + 4)$

23. $\frac{1}{3}x^2 - \frac{4}{9}x - \frac{2}{3} = 0$

24. $(2u - 1)(u - 2) = 2(u + 4) + 3$

Solve. (First try factoring or taking roots, then use the quadratic formula if necessary.)

25. $x^2 + 6x + 8 = 0$

26. $3x^2 + 4x = 5$

27. $z^2 - 6 = 1 - 6z$

28. $3u^2 - 3u + 2 = 1 + 2u$

29. $x^2 + 2x + 2 = 0$

30. $10w^2 = 12w + 6$

31. $2x^2 - 3x + 4 = 0$

32. $(2z - 1)(z - 2) - 11 = 2(z + 4) - 8$

33. $4w^2 + 9w + 3 = 0$

Solve.

34. The hypotenuse of a right triangle is 11 m, and one leg is 2 m longer than the other. Rounded to the nearest tenth, find the length of each leg.

35. The sum of the squares of a number and 2 more than the number is 6. Find the number.

36. Ivan Danielson, the water commissioner in southwestern Colorado, uses a special gauge to measure stream velocity. When this gauge, an L-shaped tube, is placed in a stream in such a way that the water flows into one end, the height h (in feet) of the water in the tube is related to the velocity v (in miles per hour) of the stream by the formula $v^2 = 64h$. What is the approximate velocity of West Deer Creek if Mr. Danielson has taken a reading of $h = 0.4$ ft?

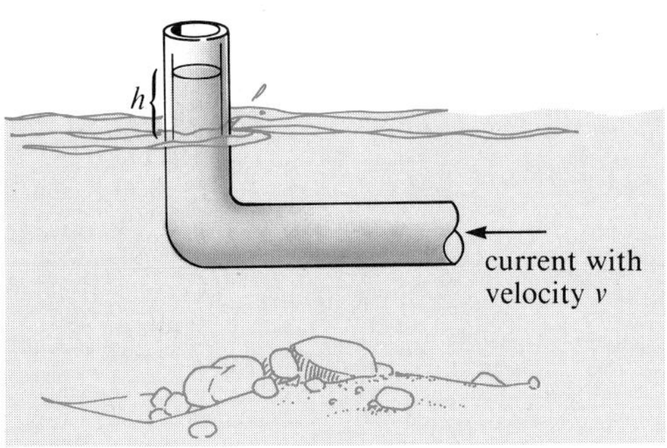

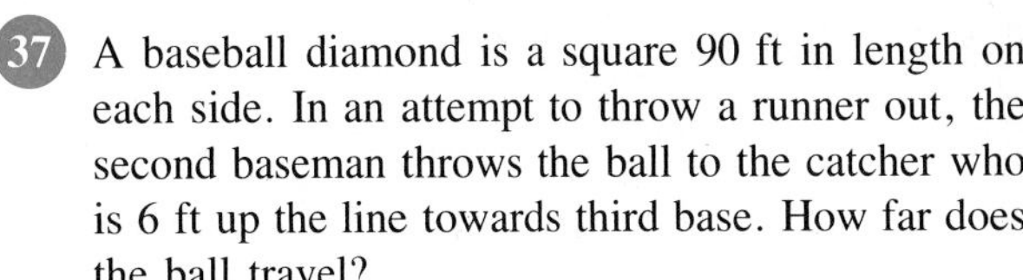

37 A baseball diamond is a square 90 ft in length on each side. In an attempt to throw a runner out, the second baseman throws the ball to the catcher who is 6 ft up the line towards third base. How far does the ball travel?

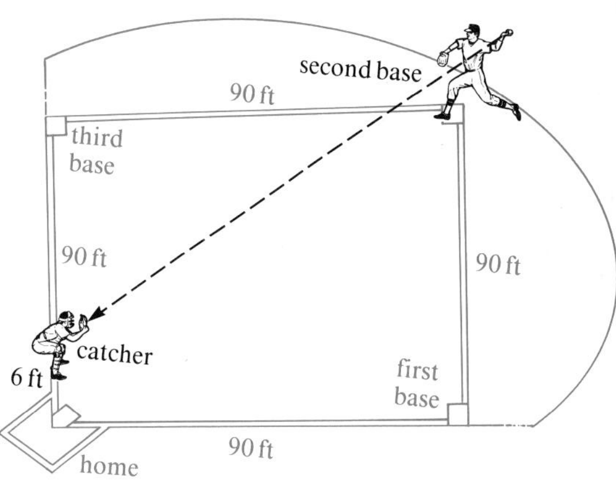

38. If an object is shot upward that has initial velocity v_0 ft/sec, in t seconds its height above ground level is given by $h = v_0t - 16t^2$. If a man fires a gun with muzzle velocity (initial velocity) of 2500 ft/sec at a helicopter 4936 ft directly above him, how long will it take for the bullet to hit the helicopter?

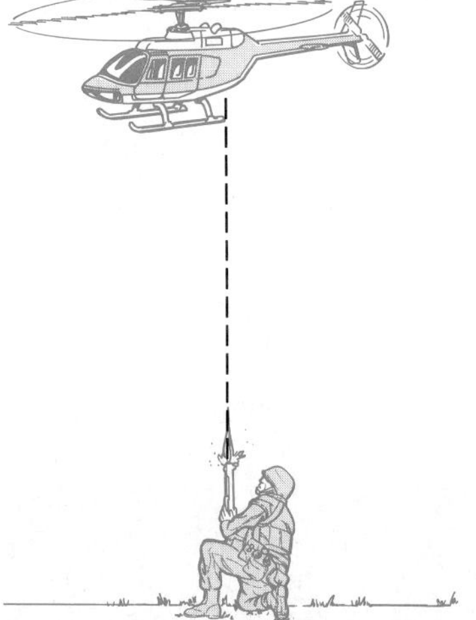

39. The square of a number is 5 more than four times the number. Find the number.

FOR REVIEW

Solve each equation in Exercises 40–42.

40. $(x - 5)(x + 12) = 0$

41. $z(z - 5)(z + 8) = 0$

42. $(y - 3) - (2y + 7) = 0$

The solutions to the quadratic equation $ax^2 + bx + c = 0$ *are*

$$x = \frac{-b \pm \sqrt{b^2 - 4ac}}{2a}.$$

Use this information in Exercises 43–46 to help you prepare for the next section.

43. Under what conditions will the solutions reduce to one real-number solution, $-\frac{b}{2a}$?

44. Under what conditions will the solutions be two different real numbers?

45. Under what conditions will the solutions be two different complex numbers?

46. Under what conditions will the solutions be two different imaginary numbers?

ANSWERS: 1. 9, 1 2. $-8 \pm \sqrt{3}$ 3. -3, 0 4. -2, $\frac{8}{5}$ 5. $\frac{3 \pm i\sqrt{11}}{2}$ 6. $\frac{7 \pm 2\sqrt{2}}{3}$ 7. $\frac{5}{9}$, $\frac{1}{9}$ 8. $\frac{-2 \pm i\sqrt{6}}{3}$ 9. 9 10. $\frac{1}{4}$ 11. $\frac{16}{25}$ 12. -1, -5 13. 4, -2 14. $\frac{-3 \pm \sqrt{17}}{4}$ 15. $-1 \pm i\sqrt{2}$ 16. 3, 5 17. 5, -5 18. $-\frac{1}{3}$, -4 19. 0, $-\frac{3}{4}$ 20. $\frac{1 \pm \sqrt{21}}{10}$ 21. $\frac{3 \pm \sqrt{41}}{4}$ 22. ± 2 23. $\frac{2 \pm \sqrt{22}}{3}$ 24. $\frac{9}{2}$, -1 25. -2, -4 26. $\frac{-2 \pm \sqrt{19}}{3}$ 27. 1, -7 28. $\frac{5 \pm \sqrt{13}}{6}$ 29. $-1 \pm i$ 30. $\frac{3 \pm 2\sqrt{6}}{5}$ 31. $\frac{3 \pm i\sqrt{23}}{4}$ 32. $\frac{9}{2}$, -1 33. $\frac{-9 \pm \sqrt{33}}{8}$ 34. 6.7 m, 8.7 m 35. $-1 \pm \sqrt{2}$ 36. 5 mph 37. 123.1 ft 38. 2 sec 39. 5 or -1 40. 5, -12 41. 0, 5, -8 42. -10 43. when $b^2 - 4ac = 0$ 44. when $b^2 - 4ac > 0$ 45. when $b^2 - 4ac < 0$ 46. when $b = 0$ and $-4ac < 0$

8.2 EXERCISES B

Solve.

1. $(x - 7)^2 = 49$

2. $(z + 2)^2 - 5 = 0$

3. $(3y - 1)^2 = 4$

4. $(3x + 5)^2 - 7 = 0$

5. $(4z + 7)^2 + 8 = 0$

6. $(2y + 1)^2 = \frac{1}{4}$

7. $(2x - 5)^2 = 1$

8. $(3y + 1)^2 + 10 = 0$

What must be added to complete the square in each expression?

9. $x^2 - 10x$

10. $z^2 + 3z$

11. $y^2 - \frac{1}{3}y$

Solve by completing the square.

12. $y^2 - 10y + 21 = 0$

13. $2x^2 = 7 - 3x$

14. $3v^2 + 12v = 135$

15. $n^2 + 5n = -12$

Solve by using the quadratic formula.

16. $x^2 + 5x - 14 = 0$

17. $z^2 - 16 = 0$

18. $x^2 = 4x + 21$

19. $4v^2 = 20v - 25$

20. $y(y + 1) + 2y = 4$

21. $4v^2 + 1 = 2v(v + 2)$

22. $(3u + 1)(u - 2) = 3(u + 1)$

23. $\frac{1}{2}y^2 - \frac{5}{4}y + \frac{1}{4} = 0$

24. $10v^2 = 5(v^2 + 25)$

Solve. (First try factoring or taking roots, then use the quadratic formula if necessary.)

25. $x^2 - 6x + 8 = 0$

26. $2x^2 + 2x = 3$

27. $y^2 - 9 = 3y$

28. $5x^2 + 3x = 2$

29. $y^2 + 4y + 5 = 0$

30. $3x(x - 2) = 12 - 6x$

31. $\frac{2}{3}v^2 - \frac{1}{3}v - 1 = 0$

32. $-3z^2 + 2z - 4 = 0$

33. $(y + 1)^2 = 5 + 3y$

Solve.

34. The hypotenuse of a right triangle is 9 in, and one leg is 3 in longer than the other. Rounded to the nearest tenth, find the length of each leg.

35. The number of square yards in the area of a square is 3.75 less than the number of yards in its perimeter. Find the length of each side.

36. Find the length of a side of an equilateral triangle if the height is 12 m.

37. Two boats leave an island at the same time, one traveling due north and the other traveling due west. If both are traveling at a rate of 30 mph, approximately how far apart will the two be in 2 hr?

38. A child with a slingshot fires a rock upward from the edge of a building 20 ft high with an initial velocity of 70 ft/sec. Ignoring air resistance, when will the rock be 30 ft high? Give answer correct to the nearest tenth of a second. Use the formula in Example 7.

39. If the square of Waldo's age is decreased by 300, the result is twenty times his age. How old is Waldo?

FOR REVIEW

Solve each equation in Exercises 40–42.

40. $(x + 3)(2x - 7) = 0$

41. $y(y + 2)(y - 10) = 0$

42. $(z - 3) + (z - 5) = 0$

The solutions to the quadratic equation with $a = 1$, $x^2 + bx + c = 0$, are

$$x = \frac{-b \pm \sqrt{b^2 - 4c}}{2}.$$

Use this information in Exercises 43–46 to help you prepare for the next section.

43. Under what conditions will the solutions reduce to one real-number solution, $-\frac{b}{2}$?

44. Under what conditions will the solutions be two different real numbers?

45. Under what conditions will the solutions be two different complex numbers?

46. Under what conditions will the solutions be two different imaginary numbers?

8.2 EXERCISES C

Solve by completing the square. Assume that b and c are constant real numbers.

1. $4x^2 + 7x - 2 = 0$

2. $x^2 + bx + c = 0$

Solve by using the quadratic formula.

3. $x^2 + \sqrt{2}x - 3 = 0$

4. $\sqrt{3}x^2 + 3x - \frac{1}{2} = 0$

$\left[\text{Answer: } \dfrac{-3\sqrt{3} \pm \sqrt{27 + 6\sqrt{3}}}{6}\right]$

8.3 USING THE DISCRIMINANT

STUDENT GUIDEPOSTS

1 The Discriminant

2 Finding an Equation Given the Solutions

In the following example we use the quadratic formula to solve three different quadratic equations with three different types of solutions. (The equations in the first two parts would normally be solved by the method of factoring, but we are using the formula to illustrate a point.)

EXAMPLE 1 Solving Equations Using the Quadratic Formula

Solve each equation using the quadratic formula.

(a) $x^2 - 2x - 3 = 0$

$$x = \frac{-b \pm \sqrt{b^2 - 4ac}}{2a}$$

$$= \frac{-(-2) \pm \sqrt{(-2)^2 - 4(1)(-3)}}{2(1)} \qquad a = 1, b = -2, c = -3$$

$$= \frac{2 \pm \sqrt{4 + 12}}{2} = \frac{2 \pm \sqrt{16}}{2}$$

$$= \frac{2 \pm 4}{2} = 1 \pm 2$$

In this case we obtain exactly two real solutions, $x = 3$ and $x = -1$.

(b) $x^2 - 2x + 1 = 0$

$$x = \frac{-(-2) \pm \sqrt{(-2)^2 - 4(1)(1)}}{2(1)} \qquad a = 1, b = -2, c = 1$$

$$= \frac{2 \pm \sqrt{4 - 4}}{2} = \frac{2 \pm \sqrt{0}}{2}$$

$$= \frac{2}{2} = 1$$

In this case we obtain exactly one real solution, $x = 1$, a double root.

PRACTICE EXERCISE 1

Solve each equation using the quadratic formula.

(a) $x^2 + 7x - 8 = 0$

(b) $x^2 - 6x + 9 = 0$

(c) $x^2 - 2x + 2 = 0$

$$x = \frac{-(-2) \pm \sqrt{(-2)^2 - 4(1)(2)}}{2(1)} \qquad a = 1,\ b = -2,\ c = 2$$

$$= \frac{2 \pm \sqrt{4 - 8}}{2}$$

$$= \frac{2 \pm \sqrt{-4}}{2}$$

$$= \frac{2 \pm 2i}{2}$$

$$= \frac{2(1 \pm i)}{2} = 1 \pm i$$

In this case we obtain two complex solutions which are conjugates, $x = 1 + i$ and $x = 1 - i$.

(c) $x^2 - 2x + 5 = 0$

Answers: (a) 1, −8 (b) 3 (c) $1 \pm 2i$

1 THE DISCRIMINANT

If we examine the solutions given by the quadratic formula in each part of Example 1,

$$x = \frac{-b \pm \sqrt{b^2 - 4ac}}{2a},$$

we see that the kind of solution is determined by the number under the radical,

$$b^2 - 4ac.$$

This number is called the **discriminant** of the quadratic equation $ax^2 + bx + c = 0$.

If $b^2 - 4ac > 0$, then there are two real solutions,

$$x_1 = \frac{-b + \sqrt{b^2 - 4ac}}{2a} \quad \text{and} \quad x_2 = \frac{-b - \sqrt{b^2 - 4ac}}{2a}$$

In Example 1(a), $x^2 - 2x - 3 = 0$ has two real solutions and the discriminant is 16, which is greater than zero. In (b), the discriminant of $x^2 - 2x + 1 = 0$ is zero, and there is exactly one real solution (a double root). In (c), the discriminant of $x^2 - 2x + 2 = 0$ is -4 which is less than zero, and there are two complex solutions that are conjugates of each other. These results are generalized in the following.

Types of Solutions to a Quadratic Equation

The quadratic equation $ax^2 + bx + c = 0$ has one of the following.

1. Two real-number solutions if $b^2 - 4ac > 0$.
2. Exactly one real-number solution if $b^2 - 4ac = 0$ (sometimes considered to be two real but equal solutions, that is, a double root).
3. Two nonreal complex-number solutions if $b^2 - 4ac < 0$. (The solutions are conjugates of each other.)

EXAMPLE 2 Finding Types of Solutions to an Equation

Tell the nature of the solutions to the following quadratic equations by calculating the discriminants.

(a) $x^2 - 6x + 7 = 0$ $\quad a = 1, b = -6, c = 7$

$$\begin{aligned} b^2 - 4ac &= (-6)^2 - 4(1)(7) \\ &= 36 - 28 \\ &= 8 > 0 \end{aligned}$$

There are two real-number solutions.

(b) $2x^2 + 9 = 0$ $\quad a = 2, b = 0, c = 9$

$$\begin{aligned} b^2 - 4ac &= (0)^2 - 4(2)(9) \\ &= 0 - 72 \\ &= -72 < 0 \end{aligned}$$

There are two nonreal complex-number solutions.

(c) $-x^2 + 10x - 25 = 0$ $\quad a = -1, b = 10, c = -25$

$$\begin{aligned} b^2 - 4ac &= (10)^2 - 4(-1)(-25) \\ &= 100 - 100 = 0 \end{aligned}$$

There is only one real-number solution, a double root.

Practice Exercise 2

Use the discriminant to tell the nature of the solutions.

(a) $4x^2 - 20x + 25 = 0$

(b) $5x^2 + 45 = 0$

(c) $7x^2 - 20x - 3 = 0$

Answers: (a) one real solution (b) two complex solutions (c) two real solutions

2 FINDING AN EQUATION GIVEN THE SOLUTIONS

We now turn to the problem of finding a quadratic equation when we are given its solutions. To see how this is done, solve the following equation by factoring.

$$\begin{aligned} x^2 - 2x - 3 &= 0 \\ (x + 1)(x - 3) &= 0 \\ x + 1 = 0 \quad &\text{or} \quad x - 3 = 0 \\ x = -1 \quad & \quad\quad\; x = 3 \end{aligned}$$

We need only reverse the steps in the problem above to construct the quadratic equation from the solutions. Suppose that we are given two solutions, -1 and 3.

$$\begin{aligned} x = -1 \quad &\text{or} \quad x = 3 \\ x - (-1) = 0 \quad &\text{or} \quad x - 3 = 0 \\ x + 1 = 0 \quad &\text{or} \quad x - 3 = 0 \\ (x + 1)(x - 3) &= 0 \\ x^2 - 2x - 3 &= 0 \end{aligned}$$

Finding a Quadratic Equation with Given Solutions

If x_1 and x_2 are two real numbers or two conjugate complex numbers, the equation

$$(x - x_1)(x - x_2) = 0$$

is a quadratic equation with real coefficients having solutions x_1 and x_2.

EXAMPLE 3 FINDING EQUATIONS GIVEN THE SOLUTIONS.

Find the quadratic equation that has the given solutions.

(a) $x_1 = -2$ and $x_2 = 5$

$$(x - x_1)(x - x_2) = 0$$

$$(x - (-2))(x - 5) = 0 \quad \text{Substitute the given solutions}$$

$$(x + 2)(x - 5) = 0$$

$$x^2 - 5x + 2x - 10 = 0 \quad \text{Multiply the binomial factors}$$

$$x^2 - 3x - 10 = 0$$

(b) $x_1 = 1 + 3i$ and $x_2 = 1 - 3i$

Notice that x_1 and x_2 are conjugates of each other.

$$(x - x_1)(x - x_2) = 0$$

$$[x - (1 + 3i)][x - (1 - 3i)] = 0 \quad \text{Substitute the solutions}$$

$$x^2 - (1 - 3i)x - (1 + 3i)x + (1 + 3i)(1 - 3i) = 0 \quad \text{Multiply the binomial factors}$$

$$x^2 - 2x + 10 = 0 \quad \text{Simplify}$$

PRACTICE EXERCISE 3

Find the quadratic equation that has the given solutions.

(a) $x_1 = 3$ and $x_2 = -8$

(b) $x_1 = 2 - i$ and $x_2 = 2 + i$

Answers: (a) $x^2 + 5x - 24 = 0$ (b) $x^2 - 4x + 5 = 0$

8.3 EXERCISES A

Use the discriminant to find the nature of the solutions.

1. $x^2 - 3x + 2 = 0$

2. $x^2 + 4x + 5 = 0$

3. $3x^2 - 2x + 5 = 0$

4. $x^2 + 6x + 9 = 0$

5. $-2x^2 - 3x + 1 = 0$

6. $5x^2 + 7 = 0$

7. $-3x^2 + 2x - 4 = 0$

8. $4x^2 - 4x + 1 = 0$

9. $7x^2 - 3 = 0$

Find the quadratic equation that has the following solutions.

10. 2, 3

11. -2, 3

12. $\sqrt{3}$, $-\sqrt{3}$

13. $2i$, $-2i$

14 $2 + i, 2 - i$ **15.** $0, 4$ **16.** $\frac{2}{3}, -1$ **17** $2 + 3i, 2 - 3i$

FOR REVIEW

What must be added to complete the square in each of the following?

18. $x^2 + 12x$ **19.** $z^2 + \frac{3}{4}z$ **20.** $y^2 + \frac{2}{3}y$

Solve.

21. $3(x^2 - 1) = 4x$ **22.** $z^2 - 5z + 2 = 2 - z$ **23.** $3(y^2 + 2y) - 8 = 0$

24. The pressure P in lb/ft^2 of a wind blowing V mph can be approximated by the equation

$$P = \frac{3V^2}{1000}.$$

What is the approximate velocity of a wind which generates a pressure of 8.4 lb/ft^2 against the side of Jean Benning's mobile home?

25. Solve the equation $3x^2 - 2x + 7 = 0$. It has complex solutions.

The following exercises review material from Chapters 6 and 7 to help you prepare for the next section. Solve each equation in Exercises 26 and 27.

26. $\frac{1}{x-2} + \frac{x}{x+2} = \frac{x^2}{x^2-4}$ **27.** $\sqrt{x^2+1} - x = 1$

28. Anthony Burgess can paint a shed in 6 hours and his father can paint the same shed in 3 hours. How long would it take them to paint the shed if they worked together?

ANSWERS: 1. two real solutions 2. two complex solutions 3. two complex solutions 4. one real solution 5. two real solutions 6. two complex solutions 7. two complex solutions 8. one real solution 9. two real solutions 10. $x^2 - 5x + 6 = 0$ 11. $x^2 - x - 6 = 0$ 12. $x^2 - 3 = 0$ 13. $x^2 + 4 = 0$ 14. $x^2 - 4x + 5 = 0$ 15. $x^2 - 4x = 0$ 16. $3x^2 + x - 2 = 0$ 17. $x^2 - 4x + 13 = 0$ 18. 36 19. $\frac{9}{64}$ 20. $\frac{1}{9}$ 21. $\frac{2 \pm \sqrt{13}}{3}$ 22. 0, 4 23. $\frac{-3 \pm \sqrt{33}}{3}$ 24. 52.9 mph 25. $\frac{1 \pm 2i\sqrt{5}}{3}$ 26. no solution 27. 0 28. 2 hr

8.3 EXERCISES B

Use the discriminant to determine the nature of the solutions.

1. $x^2 - 6x + 8 = 0$
2. $x^2 + 6x + 10 = 0$
3. $x^2 - 6x + 9 = 0$
4. $2x^2 - 3x + 7 = 0$
5. $-3x^2 - 4x + 2 = 0$
6. $4x^2 + 9 = 0$
7. $-4x^2 + x - 3 = 0$
8. $25x^2 + 10x + 1 = 0$
9. $4x^2 - 17 = 0$

Find the quadratic equation that has the following solutions.

10. 3, 5
11. −3, 5
12. $5i, -5i$
13. $\sqrt{5}, -\sqrt{5}$
14. $5 - i, 5 + i$
15. $\frac{1}{4}, -1$
16. −8, 0
17. $3 - 4i, 3 + 4i$

FOR REVIEW

What must be added to complete the square in each of the following?

18. $z^2 + 3z$
19. $y^2 + \frac{1}{2}y$
20. $x^2 - x$

Solve.

21. $2(y^2 - 2y) = 3$
22. $9x^2 + 12x + 4 = 0$
23. $4(z^2 - 1) = 7z$

24. If an object is thrown upward with initial velocity v_0 ft/sec, in t sec its height above ground level is given by $h = v_0t - 16t^2$. If the initial velocity is 128 ft/sec, how long will it take for an object thrown upward to reach a height of 240 ft?

25. Solve the equation

$$4x^2 + 81 = 0.$$

It has complex solutions.

The following exercises review material from Chapters 6 and 7 to help you prepare for the next section. Solve each equation in Exercises 26 and 27.

26. $\frac{1}{x+1} + \frac{2}{x-4} = \frac{1}{2}$

27. $\sqrt{x^2 + 5} - 1 = x$

28. John Babbitt can repair his Blazer in 4 hours, and when his sister Jane helps him, the job takes 3 hours. How long would it take Jane to do the job by herself?

8.3 EXERCISES C

1. Find the discriminant and use it to tell the nature of the solutions. $x^2 + \sqrt{2}x + 3 = 0$

2. Find the quadratic equation that has the following solutions. $\sqrt{2}, 3\sqrt{2}$
[Answer: $x^2 - 4\sqrt{2}x + 6 = 0$]

3. Verify that the product of the solutions of $ax^2 + bx + c = 0$ is $\frac{c}{a}$.

4. Verify that the sum of the solutions of $ax^2 + bx + c = 0$ is $-\frac{b}{a}$.

5. Find a quadratic equation that has $-\frac{7}{2}$ as the sum of its solutions and $\frac{13}{2}$ as the product of its solutions.

6. One solution of $mx^2 - 14x - 5 = 0$ is 5. Find m and then find the other solution.
$\left[\text{Answer:} \quad m = 3; -\frac{1}{3}\right]$

The discriminant can also be used for factoring a trinomial with integer coefficients. It can be shown that $ax^2 + bx + c$, with a, b, and c integers, is factorable (with integer coefficient factors) whenever the discriminant satisfies $b^2 - 4ac \geq 0$ and $b^2 - 4ac$ is a perfect square. Use this in Exercises 7–12 to tell whether the given polynomial is factorable. If it is, find the factors.

7. $3x^2 - 26x - 77$ [Answer: $b^2 - 4ac = (40)^2$; $(3x + 7)(x - 11)$]

8. $4x^2 - 28x + 21$

9. $49x^2 + 84x + 36$

10. $2x^2 - 3$ [Answer: $b^2 - 4ac = 24$; cannot factor]

11. $11x^2 - 99$

12. $63x^2 + 19x - 182$

8.4 EQUATIONS REDUCIBLE TO QUADRATIC EQUATIONS

STUDENT GUIDEPOSTS

1. Equations Quadratic in Form
2. The u-Substitution Method
3. Solving Fractional Equations
4. Solving Radical Equations
5. Literal Equations

1 EQUATIONS QUADRATIC IN FORM

Some equations that are not quadratic can be solved using the techniques of Sections 8.1 and 8.2 by making an appropriate substitution. For example,

$$x^4 - 13x^2 + 36 = 0$$

is not a quadratic equation (it contains x^4), but if we substitute u for x^2, we obtain the equation

$$u^2 - 13u + 36 = 0,$$

which is quadratic in u. Equations like this are **quadratic in form.**

EXAMPLE 1 Solving an Equation Quadratic in Form

Solve.

$$x^4 - 13x^2 + 36 = 0$$

$$(x^2)^2 - 13x^2 + 36 = 0 \qquad x^4 = (x^2)^2$$

$$u^2 - 13u + 36 = 0 \qquad \text{Let } u = x^2\text{, then } u^2 = (x^2)^2 = x^4$$

$$(u - 4)(u - 9) = 0 \qquad \text{Use the zero-product rule}$$

$$u - 4 = 0 \quad \text{or} \quad u - 9 = 0$$

$$u = 4 \qquad u = 9$$

Since $u = x^2$,

$$x^2 = 4 \qquad x^2 = 9 \qquad \text{Solve for the original variable}$$

$$x = \pm 2 \qquad x = \pm 3.$$

Practice Exercise 1

Solve $x^6 - 9x^3 + 8 = 0$.
Let $u = x^3$ and substitute.

Check: $x^4 - 13x^2 + 36 = 0$

$$(2)^4 - 13(2)^2 + 36 \stackrel{?}{=} 0 \qquad (3)^4 - 13(3)^2 + 36 \stackrel{?}{=} 0$$
$$16 - 52 + 36 \stackrel{?}{=} 0 \qquad 81 - 117 + 36 = 0$$
$$0 = 0$$

3 checks and -3 will check also.

2 checks and -2 will check also.
The solutions are 2, -2, 3, and -3.

Answer: 1, 2

CAUTION

Notice in Example 1 that the solutions to the original equation are not 4 and 9, which are the values of the variable u and not of x. Always remember to find the solutions for the original variable by back-substituting.

2 THE *u*-SUBSTITUTION METHOD

The technique for solving an equation quadratic in form, illustrated in Example 1, is sometimes called the ***u*-substitution method.**

EXAMPLE 2 Using the *u*-Substitution Method

Solve.

$$y^{2/3} - 5y^{1/3} - 6 = 0$$
$$(y^{1/3})^2 - 5y^{1/3} - 6 = 0 \qquad \text{Let } u = y^{1/3}\text{; then } u^2 = y^{2/3}$$
$$u^2 - 5u - 6 = 0$$
$$(u + 1)(u - 6) = 0$$
$$u + 1 = 0 \quad \text{or} \quad u - 6 = 0$$
$$u = -1 \qquad u = 6$$

Since $u = y^{1/3}$, $y^{1/3} = -1$ $\qquad y^{1/3} = 6$ $\qquad$ Back-substitute

$$(y^{1/3})^3 = (-1)^3 \qquad (y^{1/3})^3 = 6^3 \qquad \text{Cube both sides}$$
$$y = -1 \qquad y = 216.$$

Check:

$$(-1)^{2/3} - 5(-1)^{1/3} - 6 \stackrel{?}{=} 0 \qquad (216)^{2/3} - 5(216)^{1/3} - 6 \stackrel{?}{=} 0$$
$$(-1^{1/3})^2 - 5(-1) - 6 \stackrel{?}{=} 0 \qquad (216^{1/3})^2 - 5(6) - 6 \stackrel{?}{=} 0$$
$$(-1)^2 + 5 - 6 \stackrel{?}{=} 0 \qquad 6^2 - 30 - 6 \stackrel{?}{=} 0$$
$$1 + 5 - 6 = 0 \qquad 36 - 30 - 6 = 0$$

The solutions are -1 and 216.

Practice Exercise 2

Solve $(a + 5) - 5(a + 5)^{1/2} + 4 = 0$.
Let $u = (a + 5)^{1/2}$.
Then $u^2 = (a + 5)$

Answer: 11, -4

3 SOLVING FRACTIONAL EQUATIONS

In Chapters 6 and 7 when a fractional or radical equation resulted in a quadratic equation, the solution was obtained by using the method of factoring. Now that we have studied other solution techniques, we can solve a greater variety of problems. First, we review the solution method for fractional equations.

To Solve a Fractional Equation

1. Find the LCD of all fractions in the equation.
2. Multiply each term on both sides of the equation by the LCD to clear fractions.
3. Solve the resulting equation which is free of fractions.
4. Check to make sure that the solutions are not values that make a denominator zero in one of the fractions in the original equation.

EXAMPLE 3 SOLVING FRACTIONAL EQUATIONS

(a) Solve. $\frac{3}{x+1} + \frac{5}{x-1} = 1$

$$(x+1)(x-1)\left[\frac{3}{x+1} + \frac{5}{x-1}\right] = 1(x+1)(x-1) \quad \text{Multiply by LCD}$$

$$\frac{3(x+1)(x-1)}{x+1} + \frac{5(x+1)(x-1)}{x-1} = x^2 - 1 \quad \text{Distribute and divide out common factors}$$

$$3(x-1) + 5(x+1) = x^2 - 1$$

$$3x - 3 + 5x + 5 = x^2 - 1$$

$$x^2 - 8x - 3 = 0 \quad \text{Write in general form}$$

$$x = \frac{-(-8) \pm \sqrt{(-8)^2 - 4(1)(-3)}}{2(1)} \quad \text{Use the quadratic formula}$$

$$= \frac{8 \pm \sqrt{64 + 12}}{2}$$

$$= \frac{8 \pm \sqrt{76}}{2} = \frac{8 \pm 2\sqrt{19}}{2} = 4 \pm \sqrt{19}$$

The solutions are $4 + \sqrt{19}$ and $4 - \sqrt{19}$. Since the only values that make a denominator zero in the original equation are 1 and -1, these two solutions will check.

(b) $1 + \frac{3}{x-5} = \frac{1}{x}$

$$x(x-5)\left[1 + \frac{3}{x-5}\right] = x(x-5)\left[\frac{1}{x}\right] \quad \text{Multiply by LCD}$$

$$x(x-5) + 3x = x - 5 \quad \text{Distribute}$$

$$x^2 - 5x + 3x = x - 5$$

$$x^2 - 3x + 5 = 0 \quad \text{Write in general form}$$

$$x = \frac{-(-3) \pm \sqrt{(-3)^2 - 4(1)(5)}}{2(1)} \quad \text{Use the quadratic formula}$$

$$= \frac{3 \pm \sqrt{9 - 20}}{2} = \frac{3 \pm \sqrt{-11}}{2} = \frac{3 \pm i\sqrt{11}}{2}$$

This time we obtain complex solutions $\frac{3 + i\sqrt{11}}{2}$ and $\frac{3 - i\sqrt{11}}{2}$. Since 0 and 5 are the only numbers that make a denominator zero in the original equation, these two solutions will check.

PRACTICE EXERCISE 3

(a) Solve $\frac{1}{y+2} - \frac{3}{y-2} = 4$

(b) $\frac{2}{x} - \frac{1}{x-3} = \frac{4x}{x-3}$

Answers: (a) $\frac{-1 \pm \sqrt{33}}{4}$ (b) $\frac{1 \pm i\sqrt{95}}{8}$

Some applied problems, when translated to a mathematical model, result in quadratic equations. Since most quadratic equations have two distinct solutions, and since it is possible that only one of these is an appropriate solution to the given problem, care must be taken to *check all possible solutions in the words of the original problem*. Remember when solving applied problems, write complete and detailed definitions of the variables.

Many work problems similar to those we studied in Section 6.6 translate into quadratic equations, as in the next example.

EXAMPLE 4 WORK PROBLEM

When Laura and Will work together, they can complete a lab test in 2 hr less time than Laura takes to do it alone. If Will can do the experiment in 6 hr working alone, find the time Laura needs if she works alone and the time needed when they work together. Report the answers to the nearest tenth of an hour.

Let x = time Laura needs working alone,

$x - 2$ = time working together,

6 = time Will needs working alone,

$\frac{1}{x}$ = amount done by Laura in one hour,

$\frac{1}{x-2}$ = amount done working together in one hour,

$\frac{1}{6}$ = amount done by Will in one hour.

(amount by Laura) + (amount by Will) = (amount together)

$$\frac{1}{x} + \frac{1}{6} = \frac{1}{x-2}$$

$$6x(x-2)\left[\frac{1}{x} + \frac{1}{6}\right] = \frac{6x(x-2)}{x-2} \quad \text{The LCD is } 6x(x-2)$$

$$\frac{6x(x-2)}{x} + \frac{6x(x-2)}{6} = \frac{6x(x-2)}{x-2} \quad \text{Distribute and divide out common factors}$$

$$6(x-2) + x(x-2) = 6x$$

$$6x - 12 + x^2 - 2x = 6x$$

$$x^2 - 2x - 12 = 0 \quad \text{General form}$$

$$x = \frac{-(-2) \pm \sqrt{(-2)^2 - 4(1)(-12)}}{2} \quad \text{Quadratic formula}$$

$$= \frac{2 \pm \sqrt{4 + 48}}{2}$$

$$= \frac{2 \pm \sqrt{52}}{2}$$

$$= \frac{2 \pm 2\sqrt{13}}{2}$$

$$= 1 \pm \sqrt{13}$$

Since $1 - \sqrt{13}$ is negative it must be discarded as a value for time. The solution is $1 + \sqrt{13}$ or approximately 4.6 hr for Laura. Thus,

$$x = 4.6 \text{ hr} \quad \text{Laura's time}$$

$$x - 2 = 2.6 \text{ hr.} \quad \text{Time together}$$

PRACTICE EXERCISE 4

Julie and her father can complete a task in 4 hr working together. If it takes Julie 3 hr longer than her father to do the same task working alone, to the nearest tenth of an hour, how long does it take each alone?

Answer: Julie: 9.8 hr; her father: 6.8 hr

Some motion problems, involving the formula $d = rt$, translate into fractional equations that reduce to quadratic equations.

EXAMPLE 5 Motion Problem

A trucker transported a heavy load of steel girders from Los Angeles to Albuquerque, a distance of 825 miles, and returned to Los Angeles empty. His rate of speed going was 11 mph slower than his rate returning. If the total time of travel was 27.5 hours, what was his rate of speed with the full load?

Let x = the rate returning to Los Angeles,
$x - 11$ = the rate traveling with a full load.

We use the distance formula $d = rt$, solved for $t = \frac{d}{r}$, to express the two times. Since the distance going and the distance returning are both 825, we have

$$\frac{825}{x} = \text{time spent returning to Los Angeles,}$$

$$\frac{825}{x-11} = \text{time spent traveling with a full load.}$$

Since the total time of the trip was 27.5 hours, we have the following equation.

$$\frac{825}{x} + \frac{825}{x-11} = 27.5 \qquad \text{The LCD is } x(x-11)$$

$$x(x-11)\left[\frac{825}{x} + \frac{825}{x-11}\right] = 27.5\,x(x-11) \qquad \text{Multiply both sides by the LCD}$$

$$x(x-11)\frac{825}{x} + x(x-11)\frac{825}{x-11} = 27.5\,x(x-11) \qquad \text{Use distributive law}$$

$$825(x-11) + 825x = 27.5x^2 - 302.5x$$

$$825x - 9075 + 825x = 27.5x^2 - 302.5x$$

$$1650x - 9075 = 27.5x^2 - 302.5x$$

$$16{,}500x - 90{,}750 = 275x^2 - 3025x \qquad \text{Clear decimals}$$

$$0 = 275x^2 - 19{,}525x + 90{,}750$$

$$0 = 275(x^2 - 71x + 330) \qquad \text{Factor out 275}$$

$$0 = x^2 - 71x + 330 \qquad \text{Divide out 275}$$

$$0 = (x-5)(x-66)$$

$$x - 5 = 0 \quad \text{or} \quad x - 66 = 0$$

$$x = 5 \qquad\qquad x = 66$$

We can discard 5 as a solution since $x - 11 = 5 - 11 = -6$, which cannot be a rate. Thus, the rate returning to Los Angeles was 66 mph, making the rate traveling with a full load 55 mph ($x - 11 = 66 - 11 = 55$).

Practice Exercise 5

A boat sailed downstream 270 mi and then returned to the dock. The rate of travel downstream was 3 mph faster than the rate upstream. If the total time of the trip was 19 hr, what was the rate downstream?

Answer: 30 mph

4 SOLVING RADICAL EQUATIONS

We now solve equations involving radicals that reduce to quadratic equations. Before we restate the method for solving these equations, recall the power rule, that is needed in their solution.

If $a = b$, then $a^n = b^n$.

To Solve a Radical Equation

1. If only one radical is present, isolate this radical on one side of the equation, simplify, and proceed to step 3.
2. If two radicals are present, isolate one of the radicals on one side of the equation.
3. Raise both sides to a power equal to the index on the radical.
4. Solve the resulting equation. If a radical remains in the equation, isolate it and raise to the power again.
5. Check all possible solutions in the *original* equation since the equations involved may not be equivalent. If all extraneous roots are discarded, the equation has no solution.

EXAMPLE 6 Solving a Radical Equation

Solve $\sqrt{3x+1} - \sqrt{x+9} = 2$.

$$\sqrt{3x+1} = 2 + \sqrt{x+9}$$ Isolate one radical

$$(\sqrt{3x+1})^2 = (2+\sqrt{x+9})^2$$ Square both sides

$$3x+1 = 4 + 4\sqrt{x+9} + x + 9$$ Use $(a+b)^2 = a^2 + 2ab + b^2$: $4\sqrt{x+9}$ is $2ab$

$$2x - 12 = 4\sqrt{x+9}$$ Isolate the radical

$$x - 6 = 2\sqrt{x+9}$$ Factor out 2 and divide by 2

$$(x-6)^2 = (2\sqrt{x+9})^2$$ Square both sides

$$x^2 - 12x + 36 = 4(x+9)$$ Square 2 on the right

$$x^2 - 12x + 36 = 4x + 36$$

$$x^2 - 16x = 0$$

$$x(x-16) = 0$$

$$x = 0 \quad \text{or} \quad x - 16 = 0$$

$$x = 16$$

Check: $\sqrt{3(0)+1} - \sqrt{0+9} \stackrel{?}{=} 2$ $\qquad$ $\sqrt{3(16)+1} - \sqrt{16+9} \stackrel{?}{=} 2$

$\sqrt{1} - \sqrt{9} \stackrel{?}{=} 2$ $\qquad$ $\sqrt{49} - \sqrt{25} \stackrel{?}{=} 2$

$1 - 3 \neq 2$ $\qquad$ $7 - 5 = 2$

The only solution is 16.

Practice Exercise 6

Solve $\sqrt{2y+5} - \sqrt{y+6} = 1$.

Answer: 10

CAUTION

Is $(2+\sqrt{x+9})^2$ equal to $4 + (x+9)$? The answer is no. Remember that $(a+b)^2 = a^2 + 2ab + b^2$, so that

$$(2+\sqrt{x+9})^2 = 2^2 + 2(2)\sqrt{x+9} + (\sqrt{x+9})^2 \quad a = 2 \text{ and } b = \sqrt{x+9}$$
$$= 4 + 4\sqrt{x+9} + x + 9.$$

Also, remember that you *must check* possible solutions in the *original* equation.

Sometimes we can recognize that a radical equation has no solution before going through all the steps used to obtain "solutions" that will not check. For example, an equation such as

$$\sqrt{4x-3}+\sqrt{2x+2}=-1$$

can have no solution since the left side must be nonnegative but the right side is -1. Remember that $\sqrt{4x-3}$ and $\sqrt{2x+2}$ represent the principal (nonnegative) square roots of numbers, so their sum is also nonnegative and cannot equal -1. If we did not see this at first, solving would give possible solutions of 1 and 7, which do not check.

5 LITERAL EQUATIONS

Many literal equations require the methods of this chapter for solution.

EXAMPLE 7 Solving a Literal Equation

Solve $V=\pi r^2h$ (volume of a cylinder) for r.

We can use the method of taking roots on this formula.

$$V=\pi r^2 h \quad \text{Volume of a cylinder}$$

$$\frac{V}{\pi h}=r^2 \quad \text{Divide by } \pi h$$

$$\pm\sqrt{\frac{V}{\pi h}}=r \quad \text{Square root of both sides}$$

$$\sqrt{\frac{V}{\pi h}}=r \quad \text{Use positive radical since } r \text{ is the radius of the cylinder}$$

Practice Exercise 7

Solve $d=\frac{1}{2}gt^2$ for t, where t is positive.

Answer: $t=\sqrt{\frac{2d}{g}}$

8.4 EXERCISES A

Solve, using the u-substitution method.

1. $x^4-5x^2+4=0$
2. $(z+3)^2-6(z+3)+8=0$
3. $y-5\sqrt{y}+4=0$
4. $x-6-3\sqrt{x-6}+2=0$
5. $2(y-5)^2+5(y-5)=3$
6. $x^4-16=0$
7. $z^4+20=9z^2$
8. $y-11\sqrt{y}+28=0$
9. $x^{-2}-x^{-1}-12=0$
[*Hint:* Let $u=x^{-1}$.]

10. $(z-7)^4 - 13(z-7)^2 + 42 = 0$

11. $y^6 - 26y^3 - 27 = 0$

12. $x^4 - x^2 - 12 = 0$

Solve.

13. $\frac{3y}{2} = \frac{y^2+1}{y}$

14. $\frac{1}{x} = \frac{-5}{x^2+6}$

15. $\frac{2v}{v+2} - 1 = \frac{3}{v}$

16. $\frac{x^2}{2x+4} = 1$

17. $\frac{2x-1}{x+1} = \frac{x-3}{x+2}$

18. $\frac{2}{x+3} + \frac{3}{x-3} = 1$

19. $\frac{2}{x+3} - \frac{1}{x-2} = 1$

20. $\frac{-2}{v^2-2v} = \frac{1}{v} + \frac{3v}{v-2}$

21. $\frac{1}{2} + \frac{1}{u+5} = \frac{1}{u-1}$

22. $\sqrt{x^2+3} - 2 = 0$

23. $\sqrt{x} = 2 - x$

24. $z\sqrt{2} = \sqrt{6-4z}$

25. $\sqrt{x+4} - x + 8 = 0$

26. $\sqrt{5z-1} + z - 3 = 0$

27. $\sqrt{y^2-2y+10} - \sqrt{3y+4} = 0$

28. $\sqrt{x^2-3x} + \sqrt{3x+7} = 0$

29. $\sqrt{2z+5} - \sqrt{z+2} = 1$

30. $\sqrt{3y+3} + \sqrt{y-1} = 4$

31. $(x-5)^{1/2} = x^{1/2} + 1$

32. $\sqrt[3]{z^2} = 2$

33. If the reciprocal of the square of a number is the same as the reciprocal of 5 more than the number, find the number. Use your calculator to give answers correct to three decimal places.

34. A number divided by 6 less than the number is equal to -2 divided by the number. Find the number. Give answers as decimals correct to three decimal places.

35. Maria can do a job in 3 hr less time than Wanda. If they work together they can do the job in 4 hr. How long will it take each working alone to do the job? Report the answer to the nearest tenth of an hour.

36. If Les and Chip work together they can paint a room in 2 hr less time than Les can do the job working alone. If it takes Chip 8 hr to paint the room working alone, how long does it take Les by himself and how long does it take them working together? Report the answer to the nearest tenth of an hour.

37. When Alene and Zelda work together, they can build a room on the house in 3 days less time than Alene can do the job by herself. If it takes Zelda 10 days to build the room working alone, how long does it take Alene by herself and how long does it take them working together? Report the answer to the nearest tenth of a day.

38. The amount of money A which will result if a principal P is invested at r percent interest compounded annually for two years is given by $A = P(1 + r)^2$. If \$1000 grows to \$1254.40 in 2 years using this formula, what is the rate of interest?

The distance d in miles to the horizon from an object h miles above the surface of the earth can be approximated by $d = \sqrt{h(h + 8000)}$. *Use this formula in Exercises 39–40.*

39. What is the approximate distance to the horizon from an aircraft cruising at an altitude of 6.5 miles?

40. What is the approximate altitude of an airplane if the horizon is 130 miles away?

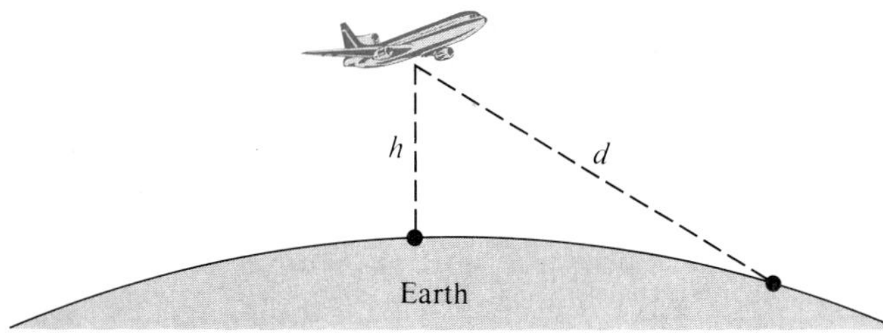

41. On a vacation, Shirley Cicero traveled 900 mi to her destination and returned home. Going, her average rate was 10 mph faster than the rate returning. If the total time of travel going and returning was 33 hr, what was the rate going?

42. An airplane traveled a distance of 2000 miles and returned to the airport. If the rate going was 25 mph faster than the rate returning, and if the total time of travel was 9.5 hr, to the nearest mile per hour, how fast did the airplane travel on the return trip?

If a car is traveling at a rate of v mph, the shortest distance in feet, d, needed to bring the car to a full stop under ideal conditions (including the reaction time of the driver) can be approximated by the formula $d = 0.05v^2 + 1.05v$.

43. If a child chases a ball into the street 60 ft in front of a car traveling 25 mph, can the driver avoid a tragedy?

44. If a car is traveling 50 mph, what is the shortest distance needed to bring the car to a full stop?

Solve for the indicated variable.

45. $A = \pi r^2$ for r $(r > 0)$

46. $\sqrt{c^2 - a^2} = b$ for c $(c > 0)$

47. $\dfrac{u^2}{1 + v^2} = 2$ for v

48 $\dfrac{1}{x} + \dfrac{n}{x + 1} = 1$ for x

FOR REVIEW

Use the discriminant to tell the nature of the solutions to the following quadratic equations (two real, one real, or two complex).

49. $x^2 - 3x - 4 = 0$

50. $9x^2 + 6x + 1 = 0$

51. $2x^2 - 2x + 5 = 0$

Find the quadratic equation that has the following solutions.

52. $3, -11$

53. $\dfrac{1}{3}, -\dfrac{1}{2}$

54. $7i, -7i$

The following exercises review material from Chapter 2 to help you prepare for the next section. Solve each compound inequality.

55. $3x - 2 < 1 - (1 - x)$ and $4x + 1 > 2x - (x + 1)$

56. $4x + 3 > 5x - (4 - x)$ or $6(x - 5) > 2x + 6$

ANSWERS: 1. $\pm 1, \pm 2$ 2. $1, -1$ 3. $1, 16$ 4. $7, 10$ 5. $2, \frac{11}{2}$ 6. $\pm 2, \pm 2i$ 7. $\pm 2, \pm\sqrt{5}$ 8. $16, 49$ 9. $\frac{1}{4}, -\frac{1}{3}$ 10. $7 \pm \sqrt{6}, 7 \pm \sqrt{7}$ 11. $3, -1$ 12. $\pm i\sqrt{3}, \pm 2$ 13. $\pm\sqrt{2}$ 14. $-2, -3$ 15. $6, -1$ 16. $1 \pm \sqrt{5}$ 17. $\frac{-5 \pm \sqrt{21}}{2}$ 18. $\frac{5 \pm \sqrt{73}}{2}$ 19. $\pm i$ 20. $-\frac{1}{3}$ 21. $-2 \pm \sqrt{21}$ 22. $1, -1$ 23. 1 24. 1 25. 12 26. 1 27. 2, 3 28. no solution 29. $2, -2$ 30. 2 31. no solution 32. $\pm 2\sqrt{2}$ 33. $2.791, -1.791$ 34. $2.606, -4.606$ 35. Maria: 6.8 hr; Wanda: 9.8 hr 36. Les: 5.1 hr; together: 3.1 hr 37. Alene: 7.2 days; together: 4.2 days 38. 12% 39. 228 mi 40. 2.1 mi 41. 60 mph 42. 409 mph 43. Yes; he can stop in 57.5 ft. 44. 177.5 ft 45. $r = \sqrt{\frac{A}{\pi}}$ 46. $c = \sqrt{a^2 + b^2}$ 47. $v = \pm\sqrt{\frac{u^2 - 2}{2}}$ 48. $x = \frac{n \pm \sqrt{n^2 + 4}}{2}$ 49. two real solutions 50. one real solution 51. two complex solutions 52. $x^2 + 8x - 33 = 0$ 53. $6x^2 + x - 1 = 0$ 54. $x^2 + 49 = 0$

8.4 EXERCISES B

Solve using the u-substitution method.

1. $y^4 - 29y^2 + 100 = 0$

2. $(x - 1)^2 - 2(x - 1) - 15 = 0$

3. $z + 3\sqrt{z} - 10 = 0$

4. $y + 1 - 8\sqrt{y + 1} + 7 = 0$

5. $3(z + 4)^2 - 7(z + 4) = 6$

6. $y^4 - 81 = 0$

7. $x^4 + 24 = 10x^2$

8. $z^{2/3} + 2z^{1/3} = 15$

9. $y^{-2} + 3y^{-1} - 40 = 0$

10. $(x - 1)^4 - 8(x - 1)^2 + 15 = 0$

11. $z^6 - 3z^3 - 10 = 0$

12. $y^4 + 6y^2 + 5 = 0$

Solve.

13. $\dfrac{2y^2 + 3}{y} = \dfrac{5y}{2}$

14. $\dfrac{3}{x^2 - 4} = \dfrac{1}{x}$

15. $\dfrac{3u}{u - 2} - 2 = \dfrac{1}{u}$

16. $\dfrac{1}{x^2} = \dfrac{1}{3x - 1}$

17. $\dfrac{2x + 1}{x - 1} = \dfrac{x + 2}{x - 3}$

18. $\dfrac{3}{x - 4} + \dfrac{2}{x + 4} = 1$

19. $\dfrac{2}{x + 2} - \dfrac{1}{x - 1} = 1$

20. $\dfrac{1}{3} - \dfrac{1}{v + 2} = \dfrac{1}{v + 3}$

21. $\dfrac{3}{x^2 + 3x} + \dfrac{4}{x} = \dfrac{2x}{x + 3}$

22. $\sqrt{y^2 - 7} - 3 = 0$

23. $6 - y = \sqrt{y}$

24. $x\sqrt{3} = \sqrt{9x + 30}$

25. $\sqrt{2y + 11} - y + 2 = 0$

26. $\sqrt{3x + 1} - 2x + 6 = 0$

27. $\sqrt{z^2 + 6z} + \sqrt{2z + 21} = 0$

28. $\sqrt{10y - 3} - \sqrt{3y^2 - 7y + 7} = 0$

29. $\sqrt{3x + 1} - \sqrt{x - 4} = 3$

30. $\sqrt{5z + 4} - \sqrt{3z - 2} = 2$

31. $(y - 3)^{1/2} = y^{1/2} + 1$

32. $\sqrt[3]{y^2 + 1} = 3$

33. The reciprocal of 1 more than the square of a number is equal to the reciprocal of 3 times the number. Find the number. Give answers correct to three decimal places.

34. The ratio of 2 more than a number and 3 less than the number is equal to 4 times the number. Find the number. Give answers correct to three decimal places.

35. José needs 4 days more to build a garage than his father needs. If working together they can do the job in 6 days, how long does it take each working alone? Report the answer to the nearest tenth of a day.

36. If Heidi and Heather work together they can prepare a display in 3 hr less time than Heidi can do the job working alone. If Heather can prepare the display in 5 hr, how long would it take Heidi to do the job working alone and how long does it take them working together? Report the answer to the nearest tenth of an hour.

37. Art can complete a lab test in 2 hr less time than Wes. If they work together they can do the test in 3 hr. How long will it take each working alone to do the test? Report the answer to the nearest tenth of an hour.

38. The amount of money A which will result if a principal P is invested at r percent interest compounded annually for two years is given by $A = P(1 + r)^2$. What is the rate of interest if \$5000 grows to \$6612.50 in 2 years?

The distance d in miles to the horizon from an object h miles above the surface of the earth can be approximated by $d = \sqrt{h(h + 8000)}$. *Use this formula in Exercises 39–40.*

39. If an airplane is flying at an altitude of 5 miles, approximately how far is it from the airplane to the horizon?

40. What is the approximate altitude of an aircraft if the horizon is 155 miles away?

41. A man out for exercise walks a distance of 12 miles at a rate of 8 mph slower than the rate he rides a bicycle for a distance of 24 miles. If the total time of the trip is 5 hours, how fast does he walk? How fast does he ride?

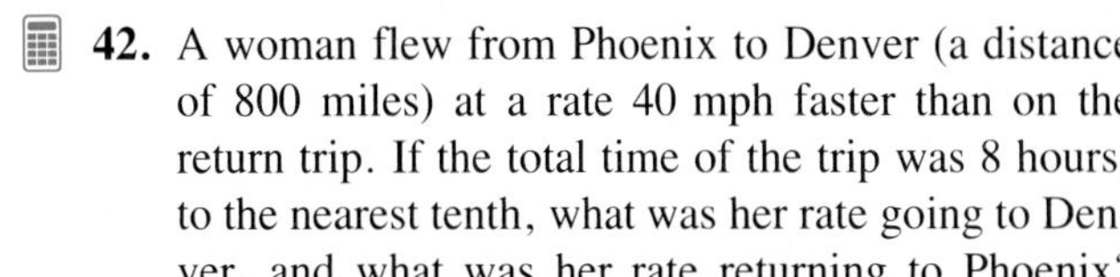

42. A woman flew from Phoenix to Denver (a distance of 800 miles) at a rate 40 mph faster than on the return trip. If the total time of the trip was 8 hours, to the nearest tenth, what was her rate going to Denver, and what was her rate returning to Phoenix?

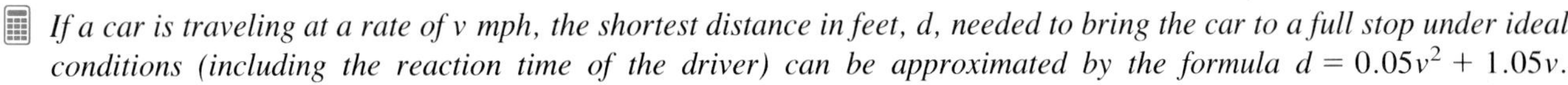

If a car is traveling at a rate of v mph, the shortest distance in feet, d, needed to bring the car to a full stop under ideal conditions (including the reaction time of the driver) can be approximated by the formula $d = 0.05v^2 + 1.05v$.

43. If a dog runs into the street 40 ft in front of a car traveling 18 mph, can the driver avoid hitting the dog?

44. If a car is traveling 60 mph, what is the shortest distance needed to bring the car to a full stop?

Solve for the indicated variable.

45. $A = 4\pi r^2$ for r $(r > 0)$

46. $\sqrt{c^2 - b^2} = a$ for b $(b > 0)$

47. $\dfrac{4v}{u^2 + v^2} = 3$ for u

48. $\dfrac{n}{x} - \dfrac{n}{x-1} = 1$ for x

FOR REVIEW

Use the discriminant to tell the nature of the solutions to the following quadratic equations (two real, one real, or two complex).

49. $x^2 - 5x + 6 = 0$

50. $36x^2 + 12x + 1 = 0$

51. $3x^2 + 2x + 3 = 0$

Find the quadratic equation which has the following solutions.

52. $5, -7$

53. $\dfrac{2}{3}, -\dfrac{1}{5}$

54. $-9i, 9i$

The following exercises review material from Chapter 2 to help you prepare for the next section. Solve each compound inequality.

55. $4(1 - 2x) > 3 - (x - 1)$ and $(x - 4) - (3x + 2) < 5$

56. $6x - (3 - x) < -(2x - 6)$ or $x - (3x + 1) > -3x + 5$

8.4 EXERCISES C

Solve.

1. $\left(\dfrac{x^2-3}{x}\right)^2 - 5\left(\dfrac{x^2-3}{x}\right) + 6 = 0$

$\left[\text{Answer: } -1, 3, \dfrac{3 \pm \sqrt{21}}{2}\right]$

2. $\sqrt[3]{x+1} + \sqrt[6]{x+1} - 2 = 0$

3. $\dfrac{x+3}{x-5} + \dfrac{x}{x-3} = 1$

4. $\dfrac{1}{x+\sqrt{2}} + \dfrac{1}{x-\sqrt{2}} = \dfrac{x^2}{x^2-2}$

[*Hint:* $x^2 - 2 = (x - \sqrt{2})(x + \sqrt{2})$]

5. $\sqrt{2x+3} + \sqrt{7-x} = \sqrt{8x+1}$
[Answer: 3]

6. $(x^6 - 2)^{1/3} = x$

7. $\dfrac{1}{x^2 - x - 2} = \dfrac{3}{x^2 - 2x - 3} + \dfrac{1}{x^2 - 5x + 6}$
[Answer: $\frac{2}{3}$]

8. $\dfrac{1}{\sqrt{x+1}} + \dfrac{3}{2} = 2$

9. If a train is traveling at v mph, the shortest distance in feet, d, needed to bring the train to a full stop under ideal conditions can be approximated by $d = 0.05v^2 + 1.2v$. A car, stalled on the railroad tracks, is seen by the engineer at a distance of 450 ft. If the crew applies the brakes immediately and stops the train within inches of the car, what was the velocity of the train prior to braking? Give answer correct to the nearest tenth.

8.5 SOLVING QUADRATIC AND RATIONAL INEQUALITIES

STUDENT GUIDEPOSTS

1. Solving Quadratic Inequalities
2. Solving Rational Inequalities

1 SOLVING QUADRATIC INEQUALITIES

Inequalities such as

$$x^2 + x - 6 < 0 \quad \text{and} \quad 2x^2 - 3x + 7 \ge 0$$

are called **quadratic inequalities.** In both instances, if the inequality symbol were replaced by an equal sign, a quadratic equation would result. There are many techniques for solving these inequalities, but the method we present is perhaps the simplest. We begin by replacing the inequality symbol with an equal sign, then solve and graph the solution as in Figure 8.3.

$$x^2 + x - 6 = 0$$
$$(x - 2)(x + 3) = 0 \qquad \text{Factor}$$
$$x - 2 = 0 \quad \text{or} \quad x + 3 = 0 \qquad \text{Zero-product rule}$$
$$x = 2 \qquad\qquad x = -3$$

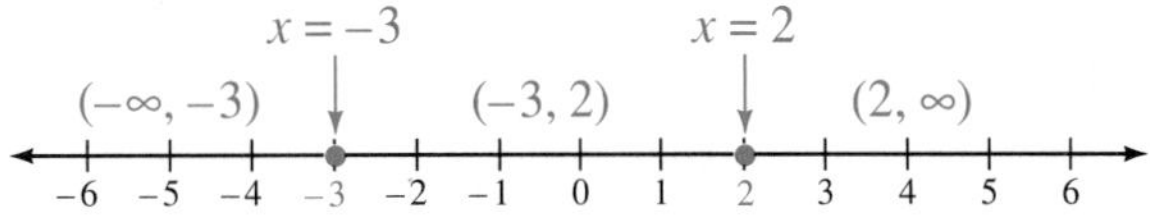

Figure 8.3

The solutions to the equation separate the number line into three intervals:

$$(-\infty, -3), \quad (-3, 2), \quad \text{and} \quad (2, \infty).$$

We know that -3 and 2 are *not* solutions to $x^2 + x - 6 < 0$ since they *are* solutions to the equation $x^2 + x - 6 = 0$. Thus, solutions to the inequality must be found in one or more of the three intervals.

It can be shown that all numbers in a given interval produce the same results when substituted into the inequality; that is, either they all solve the inequality or else none of them solves the inequality. To get a feeling for this, place the value of the quadratic expression directly above the number on the number line used to obtain it, as in Figure 8.4. For example, if $x = 1$, $x^2 + x - 6 = (1)^2 + (1) - 6 = -4$. We place -4 directly above 1 on the number line. The other values, 14, 6, 0, -4, -6, -6, -4, 0, 6, and 14 are obtained the same way.

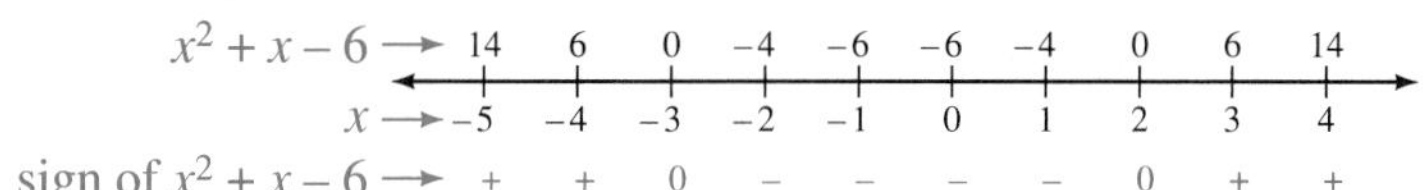

Figure 8.4

Notice that:

$$\text{for } x \text{ in } (-\infty, -3), \quad x^2 + x - 6 > 0 \text{ (is +)},$$
$$\textbf{for } x \textbf{ in } (-3, 2), \quad x^2 + x - 6 < 0 \text{ (is −)},$$
$$\text{for } x \text{ in } (2, \infty), \quad x^2 + x - 6 > 0 \text{ (is +)}.$$

Thus, the solution to $x^2 + x - 6 < 0$ is

$$(-3, 2) \quad \text{or} \quad -3 < x < 2.$$

On the other hand, the inequality $x^2 + x - 6 > 0$ has solution

$$(-\infty, -3) \text{ or } (2, \infty), \quad \textit{or} \quad x < -3 \text{ or } 2 < x.$$

Similarly, the solution to $x^2 + x - 6 \le 0$ is

$$[-3, 2] \quad \text{or} \quad -3 \le x \le 2.$$

Notice that equality holds this time since -3 and 2 solve the equation $x^2 + x - 6 = 0$, hence the inequality $x^2 + x - 6 \le 0$. Also, the solution to $x^2 + x - 6 \ge 0$ is

$$(-\infty, -3] \text{ or } [2, \infty), \quad \textit{or} \quad x \le -3 \text{ or } 2 \le x.$$

In general, to solve a quadratic inequality, we solve the corresponding equation, the solutions to which are the **critical points** of the inequality. The critical points separate the number line into intervals. Since all numbers in an interval either solve or do not solve the given inequality, it is necessary to select only one **test point** in each interval. We illustrate this procedure in the next examples.

EXAMPLE 1 A Quadratic Inequality (Two Critical Points)

Solve $2x^2 + 5x - 3 \ge 0$.

First we find the critical points by solving $2x^2 + 5x - 3 = 0$.

$$(2x - 1)(x + 3) = 0 \qquad \text{Factor}$$
$$2x - 1 = 0 \quad \text{or} \quad x + 3 = 0 \qquad \text{Zero-product rule}$$
$$2x = 1 \qquad x = -3$$
$$x = \frac{1}{2}$$

The critical points separate the number line into the three intervals shown in Figure 8.5.

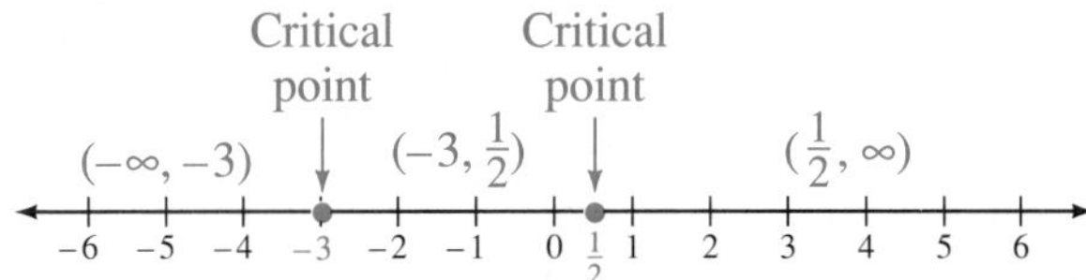

Figure 8.5

The numbers $\frac{1}{2}$ and -3 are solutions to $2x^2 + 5x - 3 \ge 0$ since they *are* solutions to $2x^2 + 5x - 3 = 0$. We find other solutions by choosing test points, such as -4, 0, and 1, from each interval.

Practice Exercise 1

(a) Solve $x^2 - x - 6 \le 0$.

First solve $x^2 - x - 6 = 0$ to obtain the critical points. Use -3, 0, and 4 as test points.

Let $x = -4$: $2x^2 + 5x - 3 = 2(-4)^2 + 5(-4) - 3 = 32 - 20 - 3 = 9 > 0$.

Let $x = 0$: $2x^2 + 5x - 3 = 2(0)^2 + 5(0) - 3 = 0 + 0 - 3 = -3 < 0$.

Let $x = 1$: $2x^2 + 5x - 3 = 2(1)^2 + 5(1) - 3 = 2 + 5 - 3 = 4 > 0$.

Then, for any x in $(-\infty, -3)$, $2x^2 + 5x - 3 > 0$. Similarly, if x is in $\left(-3, \frac{1}{2}\right)$, $2x^2 + 5x - 3 < 0$, and if x is in $\left(\frac{1}{2}, \infty\right)$, $2x^2 + 5x - 3 > 0$. We summarize this by writing the appropriate inequality above each interval on a number line, as in Figure 8.6.

Critical point | Critical point

$2x^2 + 5x - 3 > 0$ | $2x^2 + 5x - 3 < 0$ | $2x^2 + 5x - 3 > 0$

−6 −5 −4 −3 −2 −1 0 $\frac{1}{2}$ 1 2 3 4

Figure 8.6

The solution to the inequality $2x^2 + 5x - 3 \geq 0$ is

$$(-\infty, -3] \text{ or } \left[\frac{1}{2}, \infty\right), \quad \text{or} \quad x \leq -3 \text{ or } \frac{1}{2} \leq x.$$

The graph of the solution is shown in Figure 8.7. (Remember from Chapter 2 that the square brackets mean that the critical points are part of the solution. Parentheses are used when they are not part of the solution.)

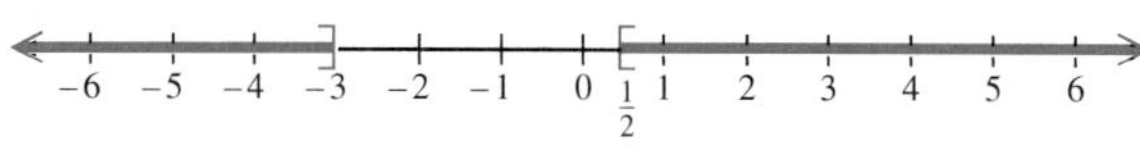

Figure 8.7 $(-\infty, -3]$ or $\left[\frac{1}{2}, \infty\right)$

In view of the results above, if we had been asked to solve $2x^2 + 5x - 3 \leq 0$, the solution would have been

$$\left[-3, \frac{1}{2}\right] \quad \text{or} \quad -3 \leq x \leq \frac{1}{2},$$

as shown in Figure 8.8.

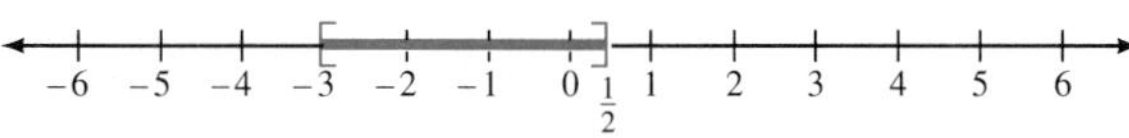

Figure 8.8 $\left[-3, \frac{1}{2}\right]$

(b) Using your calculations in the first part, what is the solution to $x^2 - x - 6 \geq 0$?

Answers: (a) $[-2, 3]$ or $-2 \leq x \leq 3$ (b) $(-\infty, -2]$ or $[3, \infty)$, *or* $x \leq -2$ or $x \geq 3$.

EXAMPLE 2 A QUADRATIC INEQUALITY (ONE CRITICAL POINT)

Solve $x^2 - 4x + 4 > 0$.

We first solve $x^2 - 4x + 4 = 0$.

$$(x - 2)(x - 2) = 0$$

$$x - 2 = 0 \quad \text{or} \quad x - 2 = 0$$

$$x = 2 \qquad\qquad x = 2$$

With only one critical point, the number line is divided into only two intervals,

$$(-\infty, 2) \quad \text{and} \quad (2, \infty).$$

PRACTICE EXERCISE 2

(a) Solve $x^2 + 6x + 9 < 0$.

First solve $x^2 + 6x + 9 = 0$ and show there is only one critical point. Use -4 and 0 as test points.

Then evaluate at test points 0 and 3.

Let $x = 0$: $0^2 - 4(0) + 4 = 4 > 0$.
Let $x = 3$: $3^2 - 4(3) + 4 = 9 - 12 + 4 = 1 > 0$.

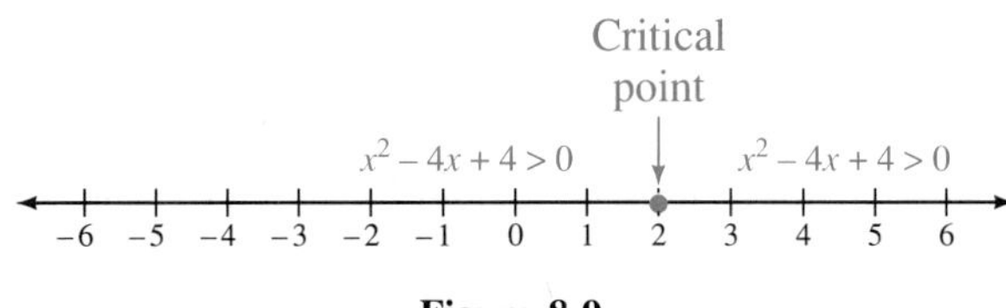

Figure 8.9

The results are shown in Figure 8.9. The solution to $x^2 - 4x + 4 > 0$ is

$$(-\infty, 2) \text{ or } (2, \infty) \quad \textit{or} \quad x < 2 \text{ or } x > 2,$$

and it is graphed in Figure 8.10.

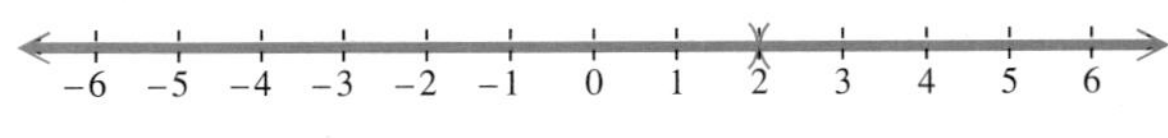

Figure 8.10 $(-\infty, 2)$ or $(2, \infty)$

In view of the results above, we can see that the solution to the inequality

$$x^2 - 4x + 4 \geq 0$$

is the entire collection of real numbers, or $(-\infty, \infty)$. Also, the solution to

$$x^2 - 4x + 4 \leq 0$$

is $x = 2$, and the inequality

$$x^2 - 4x + 4 < 0$$

has no solution.

(b) Solve $x^2 + 6x + 9 > 0$.

(c) Solve $x^2 + 6x + 9 \leq 0$.

(d) Solve $x^2 + 6x + 9 \geq 0$.

Answers: (a) no solution (b) $(-\infty, -3)$ or $(-3, \infty)$, *or* $x < -3$ *or* $x > -3$ (c) $x = -3$ (d) Every real number is a solution, or $(-\infty, \infty)$.

EXAMPLE 3 A Quadratic Inequality (No Critical Points)

Solve $x^2 + x + 2 \geq 0$.

We first solve $x^2 + x + 2 = 0$.

$$x = \frac{-1 \pm \sqrt{(1)^2 - 4(1)(2)}}{2(1)} \qquad \text{This time we cannot factor}$$

$$= \frac{-1 \pm \sqrt{-7}}{2} = \frac{-1 \pm i\sqrt{7}}{2}$$

Since only real (not complex) numbers are identified with points on a number line, there are no points on the line associated with the solutions $(-1 \pm i\sqrt{7})/2$. As a result, the solutions in this case do not divide the number line into separate parts, so $x^2 + x + 2$ has the same sign regardless of the choice of value for the variable x. Since

$$0^2 + 0 + 2 = 2 > 0,$$

every real number is a solution to the given inequality, as graphed in Figure 8.11 on the next page. In interval notation this is $(-\infty, \infty)$.

Practice Exercise 3

(a) Solve $x^2 + 3x + 5 \leq 0$.

Show that the solutions to $x^2 + 3x + 5 = 0$ are not real. Use 0 as a test point.

(b) Solve $x^2 + 3x + 5 < 0$.

Figure 8.11 $(-\infty, \infty)$

In view of the above, the solution to

$$x^2 + x + 2 > 0$$

is also the entire set of real numbers, or $(-\infty, \infty)$, and the inequalities

$$x^2 + x + 2 < 0 \quad \text{and} \quad x^2 + x + 2 \leq 0$$

have no solution.

(c) Solve $x^2 + 3x + 5 \geq 0$.

(d) Solve $x^2 + 3x + 5 > 0$.

Answers: (a) no solution (b) no solution (c) Every real number is a solution, or $(-\infty, \infty)$. (d) Every real number is a solution, or $(-\infty, \infty)$.

We now summarize the method for solving a quadratic inequality.

Method for Solving a Quadratic Inequality

1. Replace the inequality symbol with an equal sign and solve the resulting quadratic equation, obtaining the critical points.
2. If there are two real critical points, they separate the number line into three distinct intervals. Evaluate the quadratic expression at an arbitrary test point chosen from each interval and, using these values, specify the solution.
3. If there is only one real critical point, it separates the number line into two distinct intervals. Evaluate at an arbitrary test point in each interval and specify the solution.
4. If there are no real critical points (only complex solutions to the equation), either every number is a solution or else there is no solution, as determined by trying a single test point.

Sometimes the critical points are irrational numbers and we use rational approximations of them to determine possible test points.

EXAMPLE 4 Quadratic Inequality (Irrational Critical Points)

Solve $(x + 3)(x - 1) \leq -2$.
Simplify $x^2 + 2x - 3 \leq -2$.

$$x^2 + 2x - 1 \leq 0$$

We need to solve the following equation.

$x^2 + 2x - 1 = 0$ Writing in general form

$$x = \frac{-2 \pm \sqrt{2^2 - 4(1)(-1)}}{2(1)}$$

$$= \frac{-2 \pm \sqrt{8}}{2} = \frac{-2 \pm 2\sqrt{2}}{2}$$

$$= \frac{\cancel{2}(-1 \pm \sqrt{2})}{\cancel{2}} = -1 \pm \sqrt{2}$$

Since $\sqrt{2} \approx 1.4$, $-1 \pm \sqrt{2} \approx -1 \pm 1.4$. Then the critical points are, approximately, 0.4 and -2.4. Knowing this, we choose test points -3, 0, and 1.

Practice Exercise 4

Solve $(x - 7)(x + 1) \geq -13$.

Let $x = -3$: $(-3)^2 + 2(-3) - 1 = 9 - 6 - 1 = 2 > 0$

Let $x = 0$: $0^2 + 2(0) - 1 = -1 < 0$

Let $x = 1$: $1^2 + 2(1) - 1 = 2 > 0$

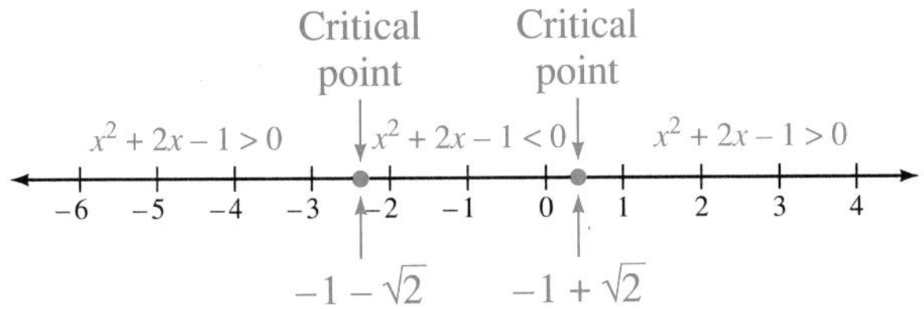

Figure 8.12

The solution, $[-1 - \sqrt{2}, -1 + \sqrt{2}]$ or $-1 - \sqrt{2} \le x \le -1 + \sqrt{2}$, is shown in Figure 8.13.

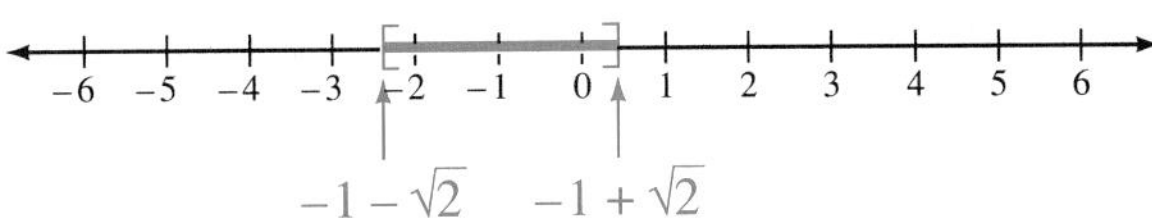

Figure 8.13 $[-1 - \sqrt{2}, -1 + \sqrt{2}]$

Answer: $(-\infty, 3 - \sqrt{3}]$ or $[3 + \sqrt{3}, \infty)$, *or* $x \le 3 - \sqrt{3}$ or $x \ge 3 + \sqrt{3}$

2 SOLVING RATIONAL INEQUALITIES

Inequalities such as $\frac{x-1}{x+2} \ge 0$, $\frac{3y-7}{y} \le 0$, $\frac{z^2 - z - 2}{z+3} < 0$ are called **rational inequalities.** By adjusting the definition of a critical point to include any value of the variable that makes the numerator zero *or* the denominator zero, rational inequalities can be solved using a method similar to that for quadratic inequalities. The critical points of a rational inequality divide the number line into various intervals, and a test point is chosen from each. The solution to the inequality is determined by the sign of the rational expression at each test point.

EXAMPLE 5 Solving a Rational Inequality

Solve $\frac{x-1}{x+2} \ge 0$ and graph the solution.

To find the critical points, set both the numerator and the denominator equal to zero and solve each equation. Here the critical points are 1 and −2. These separate the number line into three intervals:

$$(-\infty, -2), \quad (-2, 1), \quad (1, \infty).$$

In this case, 1 *is* a solution but −2 *is not*. In general, critical points that result from setting the denominator equal to zero will never be solutions to the inequality since division by zero is not defined.

We use test points −3, 0, and 2.

Let $x = -3$: $\frac{-3-1}{-3+2} = \frac{-4}{-1} = 4 > 0$

Let $x = 0$: $\frac{0-1}{0+2} = -\frac{1}{2} < 0$

Let $x = 2$: $\frac{2-1}{2+2} = \frac{1}{4} > 0$

These results are shown in Figure 8.14 on the next page.

PRACTICE EXERCISE 5

Solve.

$$\frac{x-1}{x+2} < 0$$

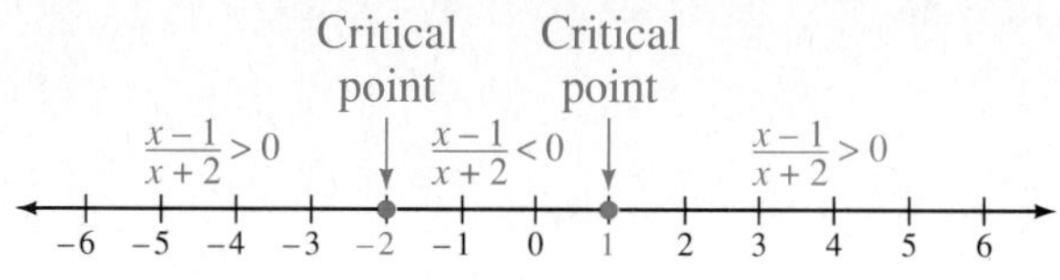

Figure 8.14

Thus, the solution to the given inequality is

$$(-\infty, -2) \text{ or } [1, \infty), \quad or \quad x < -2 \text{ or } 1 \le x.$$

Notice that 1 is included in the solution but -2 is not. The graph of the solution is given in Figure 8.15.

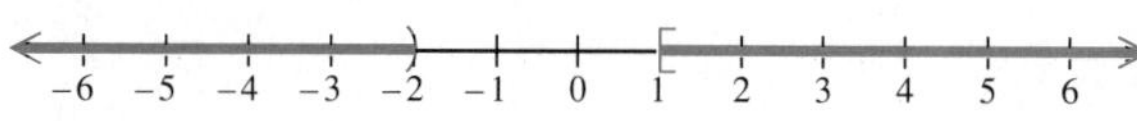

Figure 8.15

Answer: $(-2, 1)$ or $-2 < x < 1$

CAUTION

To solve rational inequalities such as the one in Example 5, can we multiply both sides by the denominator to clear the fraction? The answer is no. For example,

$$\frac{x-1}{x+2} \ge 0$$

$$(x+2)\frac{x-1}{x+2} \ge 0(x+2) \qquad \text{THIS IS WRONG}$$

$$x - 1 \ge 0$$

$$x \ge 1.$$

Remember that inequalities change signs when multiplied by a negative expression, but the inequality symbol remains the same when the expression is positive. Since $x + 2$ can be either positive or negative, we cannot simply multiply both sides by it.

If the right side of a rational inequality is not zero, we must change it to an equivalent inequality with a zero on the right.

EXAMPLE 6 Solving a Rational Inequality

Solve and graph $\frac{2x+1}{x-1} > 3$.

To solve an inequality of this form, we need to change the inequality to an equivalent inequality having a fractional expression greater than zero. In this case, subtract 3 from both sides and simplify on the left to obtain

$$\frac{4-x}{x-1} > 0.$$

The critical points are 4 and 1, so we can use 0, 2, and 5 as test points.

Practice Exercise 6

Solve.

$$\frac{x-3}{x+1} \le 2$$

Let $x = 0$: $\dfrac{4-0}{0-1} = \dfrac{4}{-1} = -4 < 0$

Let $x = 2$: $\dfrac{4-2}{2-1} = \dfrac{2}{1} = 2 > 0$

Let $x = 5$: $\dfrac{4-5}{5-1} = \dfrac{-1}{4} = -\dfrac{1}{4} < 0$

We summarize these results in Figure 8.16.

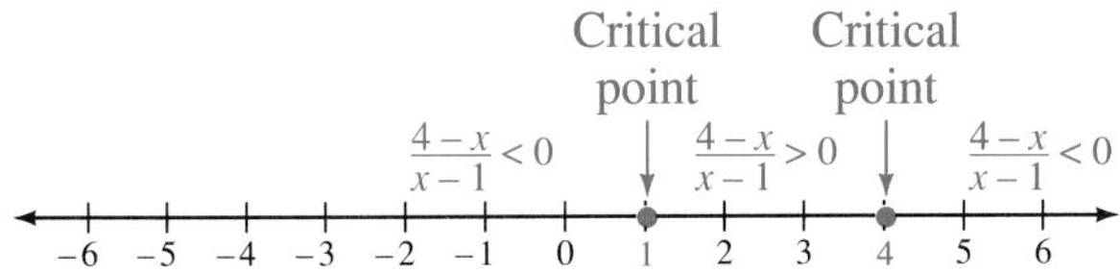

Figure 8.16

Thus, the solution is

$$(1, 4) \quad \text{or} \quad 1 < x < 4,$$

which is graphed in Figure 8.17.

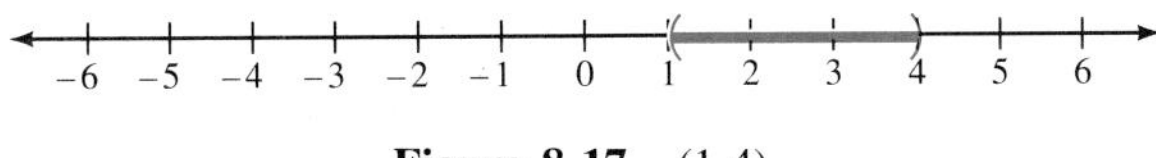

Figure 8.17 (1,4)

Answer: $(-\infty, -5]$ or $(-1, \infty)$, *or* $x \le -5$ or $x > -1$

EXAMPLE 7 Solving a More Complex Rational Inequality

Solve $\dfrac{z^2 - z - 2}{z + 3} \le 0$ and graph the solution.

Solve $z^2 - z - 2 = 0$ and $z + 3 = 0$ to obtain critical points -1, 2, and -3, and notice that -1 and 2 are solutions to the given inequality while -3 is not. This time the number line is separated into four intervals:

$$(-\infty, -3), \quad (-3, -1), \quad (-1, 2), \quad (2, \infty).$$

Evaluate at test points -4, -2, 0, and 3.

Let $z = -4$: $\dfrac{(-4)^2 - (-4) - 2}{(-4) + 3} = -18 < 0$

Let $z = -2$: $\dfrac{(-2)^2 - (-2) - 2}{(-2) + 3} = 4 > 0$

Let $z = 0$: $\dfrac{(0)^2 - (0) - 2}{(0) + 3} = -\dfrac{2}{3} < 0$

Let $z = 3$: $\dfrac{(3)^2 - (3) - 3}{(3) + 3} = \dfrac{2}{3} > 0$

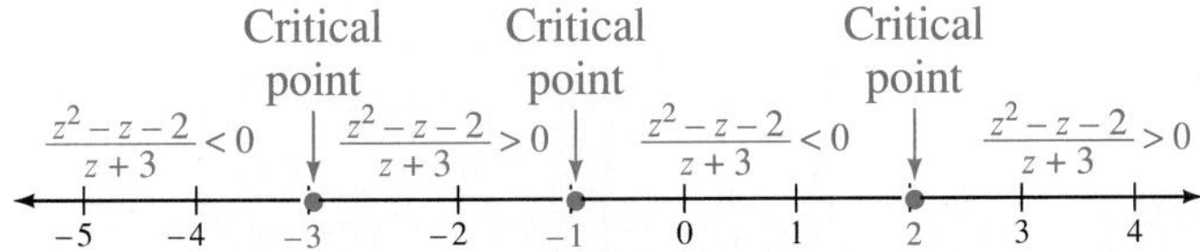

Figure 8.18

PRACTICE EXERCISE 7

Solve.

$$\frac{x^2 - x - 12}{x - 2} \ge 0$$

The results, shown in Figure 8.18, indicate that the solution to $\dfrac{z^2 - z - 2}{z + 3} \le 0$ is

$$(-\infty, -3) \text{ or } [-1, 2] \quad \textit{or} \quad z < -3 \text{ or } -1 \le z \le 2,$$

which is graphed in Figure 8.19.

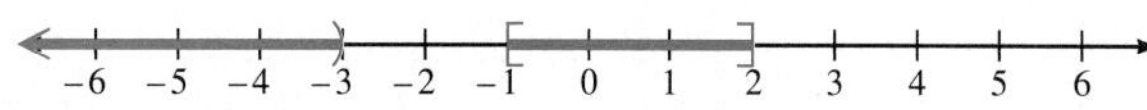

Figure 8.19 $(-\infty, -3)$ or $[-1, 2]$

Answer: $[-3, 2)$ or $[4, \infty)$, *or* $-3 \le x < 2$ or $x \ge 4$

Looking at Figure 8.18, we can also see that the solution to $\dfrac{z^2 - z - 2}{z + 3} > 0$ is

$$(-3, -1) \text{ or } (2, \infty) \quad \textit{or} \quad -3 < z < -1 \text{ or } 2 < z.$$

Notice that this time the numerator critical points, -1 and 2, are not included in the solution.

Other types of inequalities, such as

$$(x - 1)(x - 2)(x - 5) \ge 0 \quad \text{and} \quad \frac{y^2 - 4}{y^2 + 9} < 0$$

can be solved using exactly the same critical-test point technique.

8.5 EXERCISES A

Solve each inequality and give the answer using inequalities and interval notation. Graph the solution.

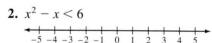

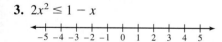

1. $x^2 - 5x + 6 > 0$

−5 −4 −3 −2 −1 0 1 2 3 4 5

2. $x^2 - x < 6$

−5 −4 −3 −2 −1 0 1 2 3 4 5

3. $2x^2 \le 1 - x$

−5 −4 −3 −2 −1 0 1 2 3 4 5

4 $2x^2 + x \ge 15$

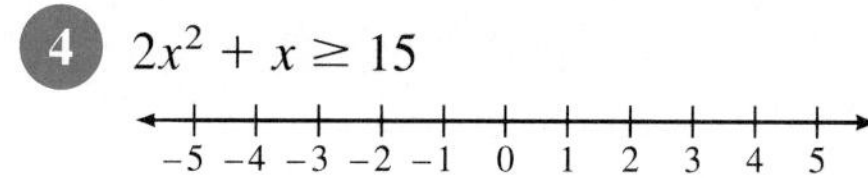

5. $-x^2 + 2x + 2 < 0$

−5 −4 −3 −2 −1 0 1 2 3 4 5

6. $x^2 + 2x + 2 > 0$

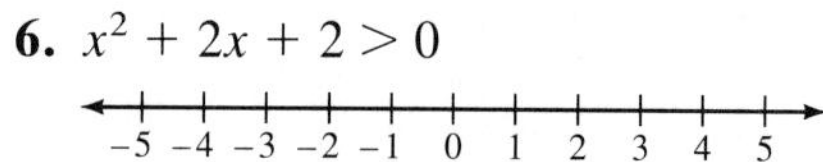

7. $4x^2 - 12x + 9 \le 0$

−5 −4 −3 −2 −1 0 1 2 3 4 5

8. $4x^2 - 12x + 9 < 0$

−5 −4 −3 −2 −1 0 1 2 3 4 5

9. $3x^2 - 2x \leq 2$

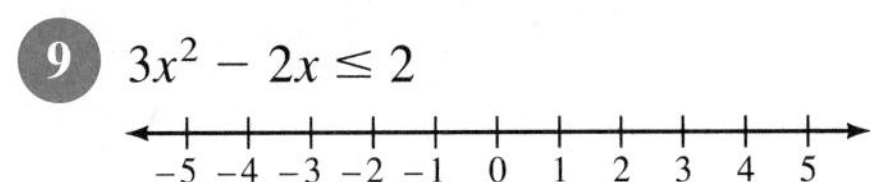

10. $(x - 2)(x + 2) < 3x$

11. $x(x - 5) > 0$

12. $x^2 - 9 \leq 0$

13. $x^2 + 9 \leq 0$

14. $x(x - 3) < x + 5$

15. $\dfrac{x - 3}{x + 1} < 0$

16. $\dfrac{x - 2}{x + 3} \geq 0$

17. $\dfrac{2x + 1}{x - 1} < 3$

18. $\dfrac{2x - 1}{x + 3} \geq 1$

19. $\dfrac{3x - 2}{x - 3} > 2$

20. $\dfrac{3}{x - 1} \leq 1$

21. $\dfrac{x}{x + 1} < 2$

22. $\dfrac{(x - 3)(x + 2)}{x - 1} > 0$

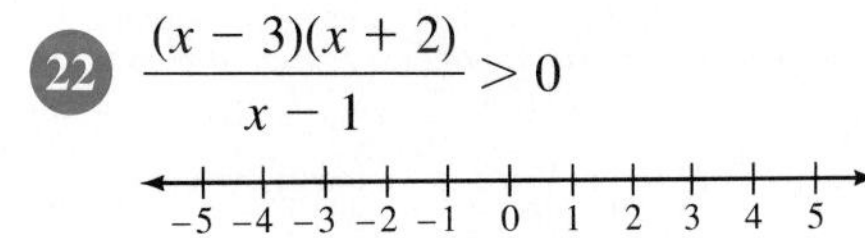

23. $\dfrac{x^2 - 25}{x} \leq 0$

24. $(x - 1)(x + 4)(x - 5) \geq 0$

25. Burford worked this problem. What is wrong with his "solution"?

$$\frac{x-3}{x+5} \leq 0$$

$$(x+5)\frac{x-3}{x+5} \leq (x+5)(0)$$

$$x-3 \leq 0$$

$$x \leq 3$$

FOR REVIEW

Solve each equation or applied problem in Exercises 26–32.

26. $6x^{-2} + x^{-1} = 2$

27. $(y-8)^2 + 5(y-8) - 6 = 0$

28. $\frac{1}{x^2-1} = \frac{1}{x-1}$

29. $\frac{14}{y^2+8y-33} = \frac{y}{y-3} - \frac{1}{y+11}$

30. $\sqrt{7y+23} - \sqrt{3y+7} = 2$

31. $\sqrt{x+8} - x + 4 = 0$

32. When each works alone, Ernie can mow a lawn in 3 hours less time than Burford. When they work together, it takes 2 hours. How long does it take each to do the job by himself?

The following exercises, reviewing material from Section 8.2, will help you prepare for the next chapter. In Exercises 33 and 34, what must be added to complete the square?

33. $x^2 + 4x$

34. $x^2 - x$

Solve the quadratic equation in Exercises 35 and 36 by completing the square.

35. $x^2 + 4x - 21 = 0$

36. $x^2 - 2x - 2 = 0$

37. What is the smallest possible value y can assume if x is a real number and $y = (x-1)^2 + 4$?

38. What is the largest possible value y can assume if x is a real number and $y = -(x+1)^2 + 2$?

ANSWERS:

1. $x < 2$ or $3 < x$; $(-\infty, 2)$ or $(3, \infty)$

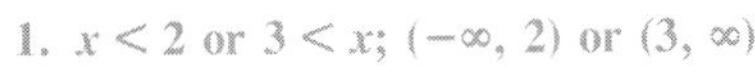

2. $-2 < x < 3$; $(-2, 3)$

3. $-1 \le x \le \frac{1}{2}$; $\left[-1, \frac{1}{2}\right]$

4. $x \le -3$ or $x \ge \frac{5}{2}$; $(-\infty, -3]$ or $\left[\frac{5}{2}, \infty\right)$

5. $x > 1 + \sqrt{3}$ or $x < 1 - \sqrt{3}$; $(-\infty, 1 - \sqrt{3})$ or $(1 + \sqrt{3}, \infty)$

6. every real number is a solution; $(-\infty, \infty)$

7. $x = \frac{3}{2}$; $\left[\frac{3}{2}, \frac{3}{2}\right]$

8. no solution

9. $\frac{1-\sqrt{7}}{3} \le x \le \frac{1+\sqrt{7}}{3}$; $\left[\frac{1-\sqrt{7}}{3}, \frac{1+\sqrt{7}}{3}\right]$

10. $-1 < x < 4$; $(-1, 4)$

11. $x < 0$ or $5 < x$; $(-\infty, 0)$ or $(5, \infty)$

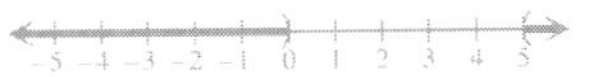

12. $-3 \le x \le 3$; $[-3, 3]$

13. no solution

14. $-1 < x < 5$; $(-1, 5)$

15. $-1 < x < 3$; $(-1, 3)$

16. $x < -3$ or $2 \le x$; $(-\infty, -3)$ or $[2, \infty)$

17. $x < 1$ or $x > 4$; $(-\infty, 1)$ or $(4, \infty)$

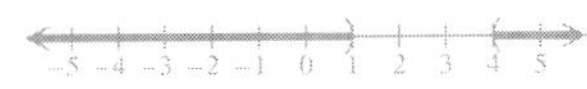

18. $x < -3$ or $4 \le x$; $(-\infty, -3)$ or $[4, \infty)$

19. $x < -4$ or $3 < x$; $(-\infty, -4)$ or $(3, \infty)$

20. $x < 1$ or $4 \le x$; $(-\infty, 1)$ or $[4, \infty)$

21. $x < -2$ or $-1 < x$; $(-\infty, -2)$ or $(-1, \infty)$

22. $-2 < x < 1$ or $x > 3$; $(-2, 1)$ or $(3, \infty)$

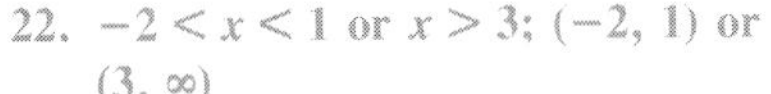

23. $x \le -5$ or $0 < x \le 5$; $(-\infty, -5]$ or $(0, 5]$

24. $-4 \le x \le 1$ or $5 \le x$; $[-4, 1]$ or $[5, \infty)$

25. It is incorrect to multiply both sides by $x + 5$ since $x + 5$ can be positive or negative. 26. 2, $-\frac{3}{2}$ 27. 2, 9 28. 0 29. 1 30. -1, -2 31. 8 32. Ernie: 3 hr; Burford: 6 hr 33. 4 since $x^2 + 4x + 4 = (x + 2)^2$ 34. $\frac{1}{4}$ since $x^2 - x + \frac{1}{4} = \left(x - \frac{1}{2}\right)^2$ 35. 3, -7 36. $1 \pm \sqrt{3}$ 37. 4 when $x = 1$ 38. 2 when $x = -1$

8.5 EXERCISES B

Solve each inequality and give the answer using inequalities and interval notation. Graph the solution.

1. $x^2 - 5x + 6 \le 0$

2. $x^2 - x \ge 6$

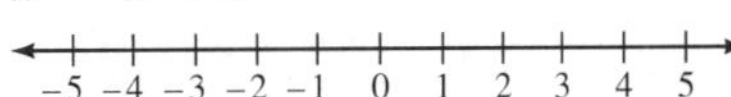

3. $2x^2 > 1 - x$

4. $2x^2 + x < 15$

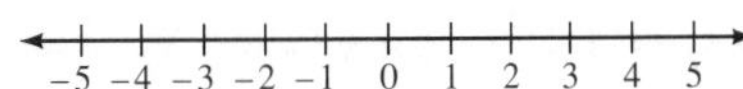

5. $-x^2 + 2x + 2 \ge 0$

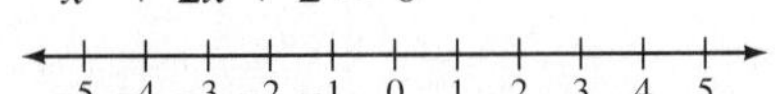

6. $x^2 + 2x + 2 \le 0$

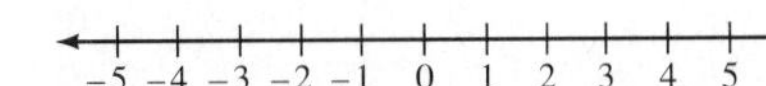

7. $4x^2 - 12x + 9 \geq 0$

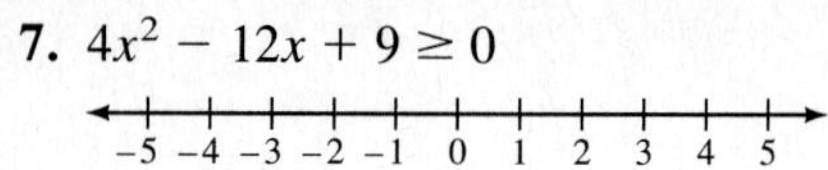

8. $4x^2 - 12x + 9 > 0$

9. $3x^2 - 2x > 2$

10. $(x - 2)(x + 2) \geq 3x$

11. $x(x - 5) \leq 0$

12. $x^2 - 9 > 0$

13. $x^2 + 9 > 0$

14. $x(x - 3) \geq x + 5$

15. $\dfrac{x - 3}{x + 1} \geq 0$

16. $\dfrac{x - 2}{x + 3} < 0$

17. $\dfrac{2x + 1}{x - 1} \geq 3$

18. $\dfrac{2x - 1}{x + 3} < 1$

19. $\dfrac{3x - 2}{x - 3} \leq 2$

20. $\dfrac{3}{x - 1} > 1$

21. $\dfrac{x}{x + 1} \geq 2$

22. $\dfrac{(x - 3)(x + 2)}{x - 1} \leq 0$

23. $\dfrac{x^2 - 25}{x} > 0$

24. $(x - 1)(x + 4)(x - 5) < 0$

25. Burford worked this problem. What is wrong with his "solution"?

$$\frac{x^2 - 4}{x} > 0$$

$$x\left(\frac{x^2 - 4}{x}\right) > x(0)$$

$$x^2 - 4 > 0$$

$$x^2 > 4$$

$$x > 2 \text{ or } x > -2$$

FOR REVIEW

Solve each equation or applied problem in Exercises 26–32.

26. $x^4 - 5x^2 = -6$

27. $(y + 1)^2 - 2(y + 1) - 35 = 0$

28. $\dfrac{x}{3} - \dfrac{6}{x} = 1$

29. $\dfrac{1}{x + 2} - \dfrac{2}{x + 1} = 5$

30. $\sqrt{2x - 1} - \sqrt{x - 1} = 1$

31. $x - 3 - 4\sqrt{x - 3} + 3 = 0$

32. The distance d in miles from an object h miles above the surface of the earth to the horizon can be approximated by $d = \sqrt{h(h + 8000)}$. If an airplane is flying at an altitude of 4.2 mi, approximately how far is it from the airplane to the horizon?

The following exercises, reviewing material from Section 8.2, will help you prepare for the next chapter. In Exercises 33 and 34, what must be added to complete the square?

33. $x^2 + 8x$

34. $x^2 + x$

Solve the quadratic equation in Exercises 35 and 36 by completing the square.

35. $x^2 - 2x - 15 = 0$

36. $x^2 - x - 5 = 0$

37. What is the smallest possible value y can assume if x is a real number and $y = (x + 2)^2 + 5$?

38. What is the largest possible value y can assume if x is a real number and $y = -(x + 2)^2 + 3$?

8.5 EXERCISES C

Solve each inequality and give the answer using inequalities and interval notation. Graph the solution.

1. $\dfrac{x^2 - 4}{x^2 + 9} < 0$

−5 −4 −3 −2 −1 0 1 2 3 4 5

2. $\dfrac{x^2 - 3}{2x} < 1$

−5 −4 −3 −2 −1 0 1 2 3 4 5

3. $\dfrac{2}{x - 2} \geq \dfrac{1}{x + 1}$

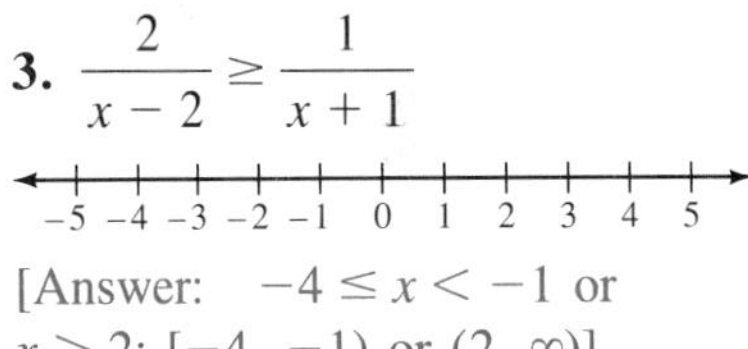

[Answer: $-4 \leq x < -1$ or $x > 2$; $[-4, -1)$ or $(2, \infty)$]

CHAPTER 8 REVIEW

KEY WORDS

8.1 The **general form** of a quadratic equation is $ax^2 + bx + c = 0$, $a > 0$.

When a solution to a quadratic equation occurs twice, it is a **root of multiplicity two** or a **double root.**

8.3 The **discriminant** of the quadratic equation $ax^2 + bx + c = 0$ is the number $b^2 - 4ac$.

8.5 An inequality of the form $ax^2 + bx + c < 0$ or $ax^2 + bx + c \geq 0$ is a **quadratic inequality.**

The **critical points** are solutions to the quadratic equation that is formed by changing the inequality symbol to an equal sign.

A **test point** is a number in one of the intervals formed by critical points that is used to determine the sign of the quadratic expression in that interval.

KEY CONCEPTS

8.1

1. Zero-product rule: If $ab = 0$, then $a = 0$ or $b = 0$. This rule only applies when one side of the equation is zero, and the other side is a product.
2. Do not use the zero-product rule on a zero-sum or zero-difference equation. For example, $(2x + 1) - (x + 5) = 0$ is *not* a zero-product equation. To solve it, clear parentheses and combine like terms. *Do not* set $2x + 1$ and $x + 5$ equal to zero!
3. The method of taking roots applies only when $b = 0$; that is, to a quadratic equation of the form $ax^2 + c = 0$.
4. Every quadratic equation in which $c = 0$, that is, of the form $ax^2 + bx = 0$, has 0 as one solution.

8.2

1. To complete the square on $x^2 + dx$, add $\left(\frac{1}{2}d\right)^2$.
2. The method of completing the square by adding the square of half the coefficient of the x term applies only when the coefficient of x^2 is $+1$.
3. Quadratic formula: The solutions of $ax^2 + bx + c = 0$ are given by
$$x = \frac{-b \pm \sqrt{b^2 - 4ac}}{2a}.$$
4. When solving a quadratic equation, always try factoring or taking roots (when applicable) first. If these methods fail, go directly to the quadratic formula.

8.3

1. The discriminant of a quadratic equation is used to determine the nature of the solutions.
If $b^2 - 4ac > 0$, the quadratic equation has two real solutions.
If $b^2 - 4ac = 0$, the quadratic equation has exactly one real solution, a double root.
If $b^2 - 4ac < 0$, the quadratic equation has two complex solutions.
2. If x_1 and x_2 are two real numbers or two conjugate complex numbers, then
$$(x - x_1)(x - x_2) = 0$$
is the quadratic equation with real coefficients having solutions x_1 and x_2.

8.4

1. When solving an equation quadratic in form using a u-substitution, be sure to give solutions for the original variable and not for the substituted variable u.
2. Solving radical equations depends on the power rule:
$$\text{If } a = b \text{ then } a^n = b^n.$$
3. When solving an equation involving square roots, *do not* square term by term; rather, square both sides and remember the middle term if one side is a binomial.
4. Be sure to check your answers in the original equation.

8.5

1. To solve $ax^2 + bx + c > 0$ or $ax^2 + bx + c < 0$, first solve $ax^2 + bx + c = 0$, obtaining the critical points. Then use test points, finding the value of $ax^2 + bx + c$ in the intervals formed by the critical points, to solve the inequality.
2. Rational inequalities are solved in the same way as quadratic inequalities using the critical-test point technique.

REVIEW EXERCISES

Part I

8.1, 8.2 *Solve.*

1. $y^2 - 2y - 48 = 0$
2. $3x(x + 5) = -5x - 25$
3. $5x^2 - 125 = 0$
4. $y^2 + 100 = 20y$
5. $y^2 - \frac{15}{2}y + 14 = 0$
6. $(x - 5)(3x + 7) = -30$

7. $x^2 - 4x = -7$

8. $x^2 + 12 = 0$

9. $3x^2 = 15x$

10. The amount of money A that will result if a principal P is invested at r percent interest compounded annually for 2 years is given by $A = P(1 + r)^2$. If \$1000 grows to \$1188.10 in 2 years using this formula, what is the rate of interest?

11. The hypotenuse of a right triangle is 16 meters, and one leg is 3 meters longer than the other. Rounded to the nearest tenth, find the length of each leg.

12. If an object is thrown downward with initial velocity of v_0 ft/sec, in t seconds its distance traveled is given by $h = v_0t + 16t^2$. If the initial velocity is 64 ft/sec, how long will it take for an object thrown from a balloon 960 feet high to reach the ground?

13. A radio tower was bent over one third of the distance from the bottom during a heavy windstorm. If the top of the tower now rests on the ground 50 feet from the base of the tower, to the nearest foot, how tall was the tower?

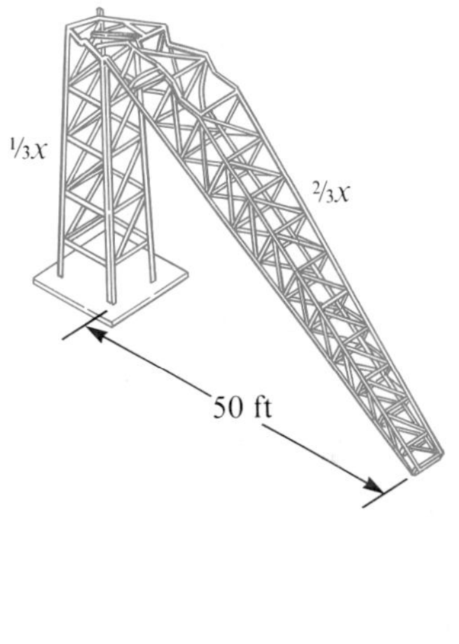

8.3 *Use the discriminant to tell the nature of the solutions.*

14. $9x^2 + 6x + 1 = 0$

15. $2x^2 - 3x = 7$

16. $2x^2 = 3x - 7$

Give the quadratic equation that has the following solutions.

17. 4, 1

18. $1 + i,\ 1 - i$

19. $-1,\ \frac{3}{2}$

8.4 *Solve.*

20. $x^4 - 11x^2 + 28 = 0$

21. $y + 4 - 5\sqrt{y+4} + 6 = 0$

22. $(y+2)^2 - (y+2) = 6$

23. $\dfrac{3}{x+2} + \dfrac{2}{x-2} = 2$

24. $\dfrac{1}{3} - \dfrac{2}{y-4} = \dfrac{1}{y+1}$

25. $\dfrac{u+2}{u-3} - \dfrac{u}{u+4} = -1$

26. $\sqrt{3x^2 - 5} - 4 = 0$

27. $\sqrt{4x+1} - x + 5 = 0$

28. $\sqrt{2y^2 - 5y + 4} - \sqrt{y^2 + 6y - 20} = 0$

29. Fred can complete a lab test in 2 hr less time than Walt. If they work together they can do the test in 3 hr. How long will it take each working alone to complete the test? Report the answer to the nearest tenth of an hour?

30. On a vacation, the Hagoods drive their car a distance of 420 miles at a rate 40 mph faster than the rate they drive their boat a distance of 50 miles. If the total time of the trip is $9\frac{1}{2}$ hours, how fast do they drive each vehicle?

In a physics experiment the equation $P = \sqrt{3y+1} - \sqrt{y-1}$ is used.

31. Find P when y is 9. Give both the exact answer and an approximate answer correct to the nearest hundredth.

32. Find y when P is 2.

Solve for the indicated variable.

33. $E = mc^2$ for c

34. $\dfrac{n}{x+1} + \dfrac{n}{x-1} = -1$ for x

8.5 *Solve. Give answer using inequalities and interval notation. Graph the solution.*

35. $9x^2 - 6x + 1 > 0$

−5 −4 −3 −2 −1 0 1 2 3 4 5

36. $3x^2 + 4x - 4 \le 0$

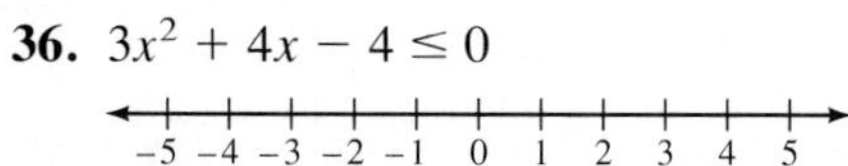

37. $\dfrac{x-5}{x+3} \le 0$

38. $\dfrac{(x+1)(x-4)}{x+2} > 0$

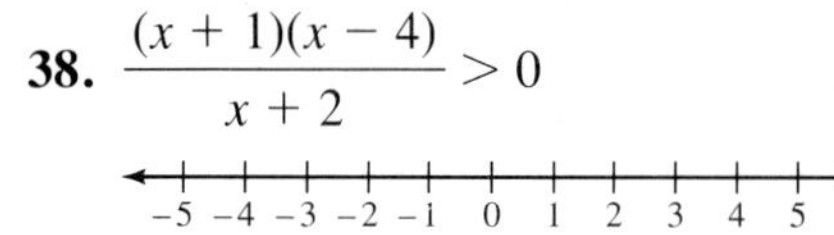

Part II

Solve.

39. $v^4 - 6v^2 + 8 = 0$

40. $y^2 = 4y - 2$

41. $7x^2 + 21 = 0$

42. $\sqrt{6z+4} - \sqrt{2z+5} = 3$

43. $\dfrac{x}{x-1} - \dfrac{1}{x+1} = \dfrac{2}{x^2-1}$

44. $(z+1)^2 - 5(z+1) + 6 = 0$

45. $x^2 + 2x + 2 = 0$

46. $x^2 - 7x = 0$

47. Twice the product of two positive consecutive odd integers is 390. Find the integers.

48. Ten less than the square of Ezra's age is the same as 8 more than three times his age. How old is Ezra?

49. Five times the number of square centimeters in the area of a square is the same as the number of centimeters in the perimeter increased by 3. Find the length of a side correct to the nearest hundredth of a centimeter.

50. A polygon having n sides has d diagonals where $d = \frac{n^2 - 3n}{2}$. If a polygon has 14 diagonals, how many sides does it have?

51. It takes Sybil 6 hours longer to paint a room than it takes Joanna. Working together, they can paint the room in 4 hours. How long would it take each working alone to paint the room?

52. Jeff walked a distance of 15 miles at a rate of 4 mph slower than the rate he jogged 10 miles. If the total time of the trip was 6 hours, to the nearest tenth, how fast did he jog?

53. Use the discriminant to tell the nature of the solutions to $2x^2 - 5x + 7 = 0$.

54. Solve by completing the square.

$$3x^2 - 4x + 1 = 0$$

55. Give the quadratic equation that has -2 and $\frac{1}{2}$ as solutions.

56. Solve $W = gt^2 + 5$ for t.

Solve. Give answer using inequalities and interval notation. Graph the solution.

57. $x^2 - 7x + 10 \leq 0$

−5 −4 −3 −2 −1 0 1 2 3 4 5

58. $\dfrac{x + 8}{x + 4} \leq 3$

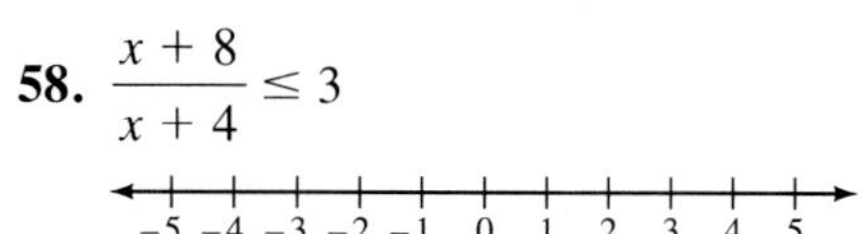

ANSWERS: 1. 8; −6 2. $-\frac{5}{3}$; −5 3. 5; −5 4. 10 5. $\frac{7}{2}$; 4 6. $\frac{4 \pm \sqrt{31}}{3}$ 7. $2 \pm i\sqrt{3}$ 8. $\pm 2i\sqrt{3}$ 9. 0; 5 10. 9% 11. 9.7 m; 12.7 m 12. 6 sec 13. 87 ft 14. one real 15. two real 16. two complex 17. $x^2 - 5x + 4 = 0$ 18. $x^2 - 2x + 2 = 0$ 19. $2x^2 - x - 3 = 0$ 20. ± 2; $\pm\sqrt{7}$ 21. 0; 5 22. 1; −4 23. $\frac{5 \pm \sqrt{73}}{4}$ 24. $6 \pm \sqrt{34}$ 25. $-5 \pm \sqrt{29}$ 26. $\pm\sqrt{7}$ 27. 12 28. 3; 8 29. Fred: 5.2 hr; Walt: 7.2 hr 30. car: 60 mph; boat: 20 mph 31. $2\sqrt{7} - 2\sqrt{2}$; 2.46 32. 1; 5 33. $c = \pm\sqrt{\frac{E}{m}}$ 34. $-n \pm \sqrt{n^2 + 1}$ 35. $x < \frac{1}{3}$ or $x > \frac{1}{3}$; $\left(-\infty, \frac{1}{3}\right)$ or $\left(\frac{1}{3}, \infty\right)$ 36. $-2 \leq x \leq \frac{2}{3}$; $\left[-2, \frac{2}{3}\right]$ 37. $-3 < x \leq 5$; $(-3, 5]$ 38. $-2 < x < -1$ or $x > 4$; $(-2, -1)$ or $(4, \infty)$ 39. ± 2; $\pm\sqrt{2}$ 40. $2 \pm \sqrt{2}$ 41. $\pm i\sqrt{3}$ 42. 10 43. no solution 44. 1; 2 45. $-1 \pm i$ 46. 0; 7 47. 13; 15 48. 6 yr 49. 1.27 cm 50. 7 sides 51. Sybil: 12 hr; Joanna: 6 hr 52. 7.2 mph 53. two complex 54. $\frac{1}{3}$; 1 55. $2x^2 + 3x - 2 = 0$ 56. $t = \pm\sqrt{\frac{W - 5}{g}}$ 57. $2 \leq x \leq 5$; $[2, 5]$ 58. $x < -4$ or $x \geq -2$; $(-\infty, -4)$ or $[-2, \infty)$

CHAPTER 8 TEST

Solve.

1. $2x^2 = 5 - 9x$ 1. ______

2. $y^2 + 2y - 7 = 0$ 2. ______

3. $x^2 + 2x + 3 = 0$ 3. ______

4. $(y + 3)^2 - 4(y + 3) - 12 = 0$ 4. ______

5. $\dfrac{x^2}{4 - 3x} = 1$ 5. ______

6. The base of a triangle is 2 cm more than the height. Find the base and height if the area is 24 cm^2. 6. ______

7. Three times the reciprocal of the square of a number is the same as twice the reciprocal of the number. Find the number. 7. ______

8. In an experiment the equation $P = \sqrt{x + 1} - \sqrt{2x}$ is used. Find x when P is -1. 8. ______

9. Solve $a + b = 9c^2$ for c. 9. ______

10. It takes a 5-inch pipe 9 days longer to fill a reservoir than a 7-inch pipe. If the two pipes are turned on together, they can fill the reservoir in 20 days. How long would it take the 7-inch pipe to fill the reservoir alone? 10. ______

CHAPTER 8 TEST

CONTINUED

11. The amount of money A that will result if a principal P is invested at r percent interest compounded annually for two years is given by $A = P(1 + r)^2$. If \$10,000 grows to \$11,556.25 in 2 years, what is the rate of interest?

11. ____________________

12. When an object is launched upward with initial velocity of 96 ft/sec from an initial height of 112 ft, ignoring air resistance, the height h of the object after t seconds is given by $h = -16t^2 + 96t + 112$. Use the discriminant of the appropriate quadratic equation to tell the number of times that the object is at a height of 256 ft. At what time(s) is the object actually at that height?

12. ____________________

13. Use the discriminant to find the nature of the solutions of $4x^2 - 5x + 8 = 0$.

13. ____________________

14. Give the quadratic equation that has solutions $\frac{3}{5}$, -2.

14. ____________________

15. Solve $-x^2 + 4x + 5 < 0$. Graph the solution.

15. −5 −4 −3 −2 −1 0 1 2 3 4 5

16. Solve $\dfrac{y - 1}{y - 4} \leq 0$. Graph the solution.

16. −5 −4 −3 −2 −1 0 1 2 3 4 5

CHAPTER 9

More on Functions

9.1 QUADRATIC FUNCTIONS

STUDENT GUIDEPOSTS

1 Quadratic Functions	3 Graphing Quadratic Functions
2 Parabolas	4 Maximum and Minimum Problems

When a baseball player hits a pop fly, the path taken by the ball is a curve like the one shown in Figure 9.1. In this section we study *quadratic functions* whose graphs are curves similar to this.

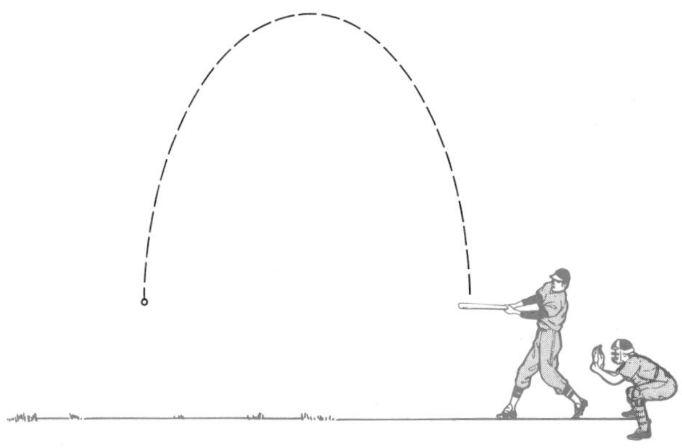

Figure 9.1

1 QUADRATIC FUNCTIONS

In Chapter 3 we discussed linear functions defined by the equation $f(x) = mx + b$. We now consider another class of functions that have many applications in mathematics. Any function that can be described by an equation of the form

$$f(x) = ax^2 + bx + c \qquad (a, b, c \text{ constants}, a \neq 0)$$

is a **quadratic function.** If $a = 0$, we would lose the squared term and have a linear function. Like linear functions, quadratic functions have the set of real numbers for their domain.

2 PARABOLAS

The graph of a quadratic function is a U-shaped curve called a **parabola.** Consider the two special quadratic functions $f(x) = x^2$ $(a = 1, b = 0, c = 0)$ and $g(x) = -x^2$ $(a = -1, b = 0\ c = 0)$. Remember that $f(x)$ and $g(x)$ are names for y. The graphs of these two functions, obtained by making a table of values, are given in Figures 9.2 and 9.3, respectively. Notice that for $f(x) = x^2$, $a = 1 > 0$

and the parabola opens upward; for $g(x) = -x^2$, $a = -1 < 0$ and the parabola opens downward.

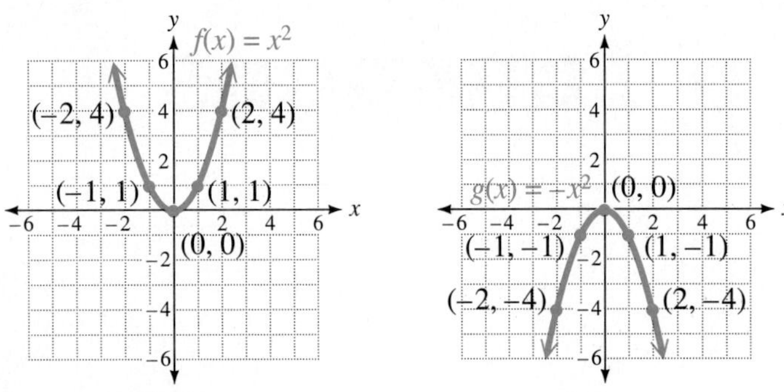

Figure 9.2 **Figure 9.3**

When a parabola opens upward, it has a low point or minimum and when it opens downward, it has a high point or maximum. In both cases, this point is called the **vertex** of the parabola. In our two example functions the vertex of each is at the origin (0, 0).

If we memorize the graphs of $f(x) = x^2$ and $g(x) = -x^2$, we can use them to find the graphs of any quadratic function of the form

$$f(x) = x^2 + c \quad \text{or} \quad f(x) = -x^2 + c.$$

Study the graphs in Figure 9.4. Notice that the graph of $f(x) = x^2 + 2$ can be obtained by "sliding" the graph of $f(x) = x^2$ up 2 units. The graph of $f(x) = x^2 - 1$ is obtained by sliding the graph of $f(x) = x^2$ down 1 unit since $c = -1$. Similarly, we can graph $f(x) = -x^2 + 3$ by sliding the graph of $g(x) = -x^2$ up 3 (+3) units, and graph $f(x) = -x^2 - 2$ by sliding the graph of $g(x) = -x^2$ down 2 (−2) units. By using this information, we can quickly sketch the graphs of a multitude of quadratic functions by simply moving or sliding a memorized graph.

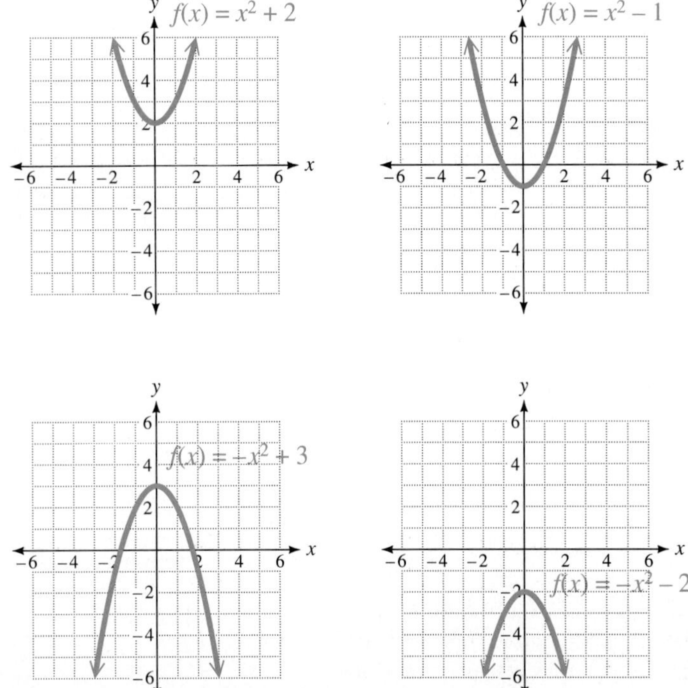

Figure 9.4 Quadratic Functions

We now consider graphing general quadratic functions. Suppose we begin by making a table of values and plot the graph of $f(x) = x^2 - 2x - 3$ in Figure 9.5. Notice that the graph crosses the x-axis when $f(x)$ (or y) is 0 and x is -1 and 3. The points $(-1, 0)$ and $(3, 0)$ are the ***x*-intercepts.** The parabola opens up (notice that $a = 1 > 0$) and has a low point at the vertex $(1, -4)$. Also, there are pairs of points at the same distance on either side of the line $x = 1$, the **line of symmetry** of the parabola.

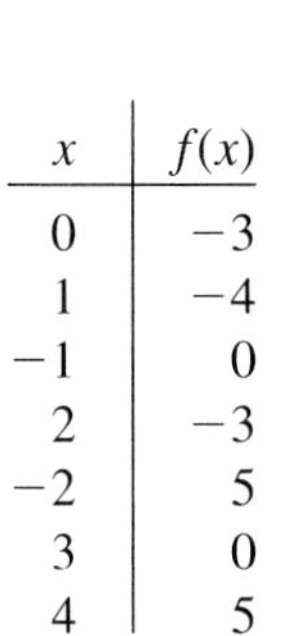

x	$f(x)$
0	-3
1	-4
-1	0
2	-3
-2	5
3	0
4	5

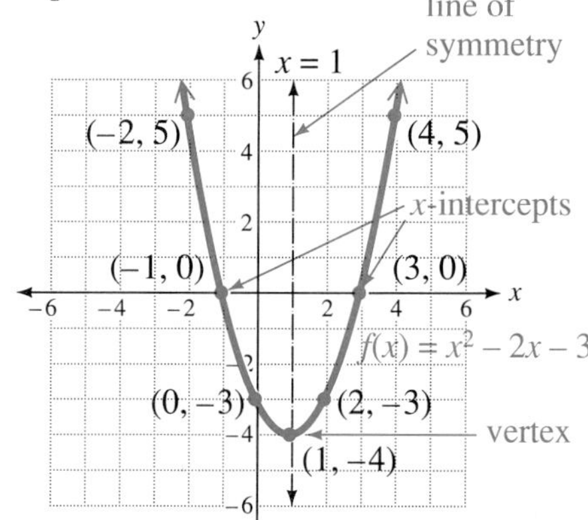

Figure 9.5

3 GRAPHING QUADRATIC FUNCTIONS

Knowing that the graph of a general quadratic function $f(x) = ax^2 + bx + c$ is a parabola, we can minimize the number of points to plot by determining whether the graph opens up or down and finding the x-intercepts, the vertex, and the line of symmetry. In view of the preceding discussion we can conclude, as shown in Figure 9.6, that the graph of a parabola $f(x) = ax^2 + bx + c$

opens **up** when the coefficient of x^2, a, is **positive**, or

opens **down** when the coefficient of x^2, a, is **negative**.

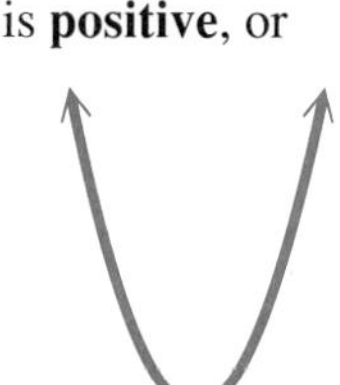

$a > 0$

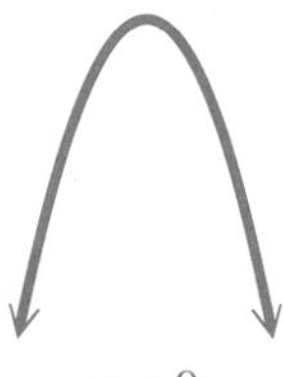

$a < 0$

Figure 9.6

To find the x-intercepts, we set $f(x)$ equal to zero and solve the resulting quadratic equation. In our example, $f(x) = x^2 - 2x - 3$, so we solve

$$x^2 - 2x - 3 = 0 \quad f(x) = 0$$
$$(x + 1)(x - 3) = 0$$
$$x + 1 = 0 \quad \text{or} \quad x - 3 = 0$$
$$x = -1 \quad \text{or} \quad x = 3.$$

The intercepts are $(-1, 0)$ and $(3, 0)$.

One way to find the vertex is to complete the square in $f(x) = x^2 - 2x - 3$.

$$f(x) = x^2 - 2x \qquad -3 \quad \text{Rewrite leaving space as indicated}$$

To complete the square on $x^2 - 2x$, we add the square of half the coefficient of x, $\left[\frac{1}{2}(-2)\right]^2 = \left[-1\right]^2 = 1$. But to maintain equality with $f(x)$, we must also subtract this value as shown below.

$f(x) = x^2 - 2x + 1 - 3 - 1$ — Add 1 and subtract 1, the square of half the coefficient of x

$= (x^2 - 2x + 1) - 4$ — Group the terms in the trinomial

$= (x - 1)^2 - 4$ — Factor the trinomial

Since the graph opens up, the smallest value that $f(x)$ can assume occurs when $(x - 1)^2$ is as small as possible. This is when $(x - 1)^2$ is 0, or when $x = 1$. When $x = 1$,

$$f(1) = (1 - 1)^2 - 4 = 0 - 4 = -4.$$

Thus, the vertex has coordinates $(1, -4)$, which agrees with our findings in Figure 9.5. Also, the line of symmetry has equation

$$x = x\text{-coordinate of the vertex}$$

which in our example is $x = 1$.

Rather than complete the square to find the vertex each time we have a quadratic function, we can complete the square on a general function $f(x) = ax^2 + bx + c$, obtaining the x-coordinate of the vertex as $-\frac{b}{2a}$. In our example, since $b = -2$ and $a = 1$,

$$-\frac{b}{2a} = -\frac{(-2)}{2(1)} = \frac{2}{2} = 1.$$ Substitute -2 for b and 1 for a

To find the y-coordinate, we substitute the value found for the x-coordinate into the formula for $f(x)$. Again for our particular example,

$$f(1) = (1)^2 - 2(1) - 3 = 1 - 2 - 3 = -4.$$ Substitute 1 for x

Thus, the vertex is $\left(-\frac{b}{2a}, f\left(-\frac{b}{2a}\right)\right) = (1, -4)$. The line of symmetry is always the vertical line

$$x = -\frac{b}{2a}.$$ Which is $x = 1$ when $b = -2$ and $a = 1$

One other point that is easy to find is the y-intercept $(0, c)$. In our example $c = -3$, so the y-intercept is $(0, -3)$.

We summarize these remarks in the following.

To Graph the Quadratic Function $f(x) = ax^2 + bx + c$

1. Find whether the graph opens up (if $a > 0$) or down (if $a < 0$).
2. Find the vertex and line of symmetry using the fact that the x-coordinate of the vertex is $-\frac{b}{2a}$ and the line of symmetry is $x = -\frac{b}{2a}$. The y-coordinate of the vertex is found by evaluating $f\left(-\frac{b}{2a}\right)$.
3. If the vertex is located above the x-axis and the graph opens down, or if the vertex is located below the x-axis and the graph opens up, find the x-intercepts by solving the equation

$$ax^2 + bx + c = 0.$$

4. Plot the vertex and the x-intercepts (or two other points on either side of the line of symmetry if there are no x-intercepts) and sketch the parabola.

EXAMPLE 1 Graphing a Quadratic Function

Use the x-intercepts, the vertex, and the line of symmetry to graph $g(x) = -2x^2 + 12x - 10$.

Since the coefficient of x^2 is negative ($a = -2$), the parabola opens down. With $a = -2$ and $b = 12$,

$$-\frac{b}{2a} = -\frac{12}{2(-2)} = -\frac{12}{-4} = 3.$$

Thus, the x-coordinate of the vertex is 3, and the y-coordinate is

$$\begin{aligned} g(3) &= -2(3)^2 + 12(3) - 10 \\ &= -2(9) + 36 - 10 \\ &= -18 + 36 - 10 = 8 \end{aligned}$$

With the vertex at (3, 8) and the graph opening down, we know that it will also cross the x-axis. Thus we look for the x-intercepts.

$$\begin{aligned} -2x^2 + 12x - 10 &= 0 && g(x) = 0 \text{ gives intercepts} \\ -2(x^2 - 6x + 5) &= 0 && \text{Factor out } -2\text{, change all signs} \\ x^2 - 6x + 5 &= 0 && \text{Divide by } -2 \\ (x-1)(x-5) &= 0 && \text{Factor} \end{aligned}$$

$$x - 1 = 0 \quad \text{or} \quad x - 5 = 0$$
$$x = 1 \qquad\qquad x = 5$$

The x-intercepts are (1, 0) and (5, 0). Finally, the line of symmetry, $x = -\frac{b}{2a}$, is $x = 3$, and the graph is given in Figure 9.7.

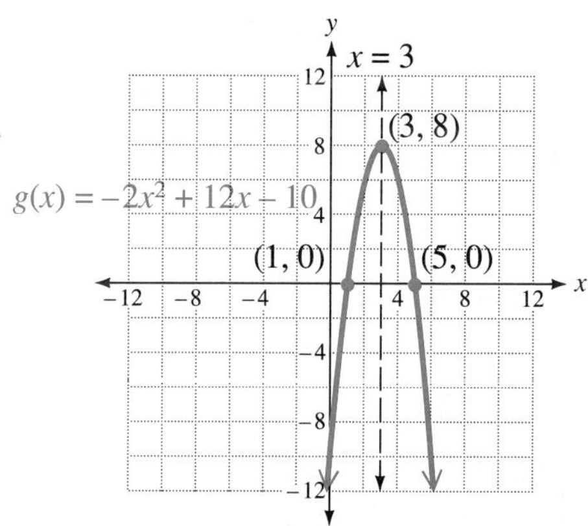

Figure 9.7

PRACTICE EXERCISE 1

Use the x-intercepts, the vertex, and the line of symmetry to graph $f(x) = x^2 - 2x - 8$. Since $a = 1 > 0$, the graph opens ________.

The x-coordinate of the vertex is given by $-\frac{b}{2a} =$ ________.

The y-coordinate of the vertex is given by $f\left(-\frac{b}{2a}\right) =$ ________. To find the x-intercepts, set $f(x) = 0$ and solve.

$$x^2 - 2x - 8 = 0$$

Finally, the line of symmetry is

$$x = -\frac{b}{2a} = \text{\_\_\_\_\_\_\_\_}.$$

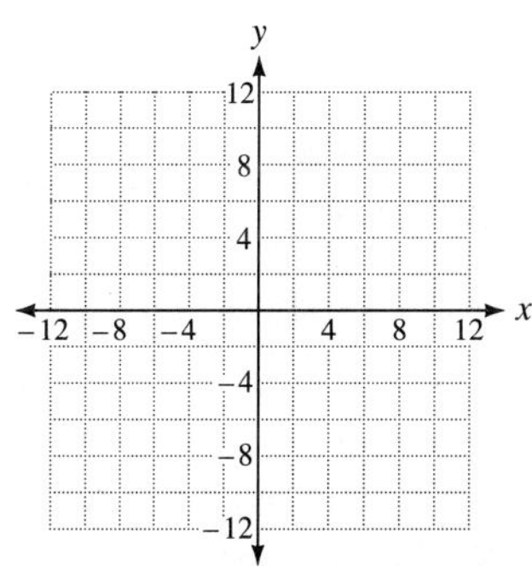

Answer: The graph is a parabola that opens up, has vertex $(1, -9)$, has x-intercepts $(-2, 0)$ and $(4, 0)$, and line of symmetry $x = 1$.

EXAMPLE 2 Graphing a Quadratic Function

(a) Graph $f(x) = x^2 + 5x + 4$.

Since $a = 1 > 0$, the parabola opens up. With $a = 1$ and $b = 5$, $-\frac{b}{2a} = -\frac{5}{2(1)} = -\frac{5}{2}$, so that the line of symmetry is $x = -\frac{5}{2}$. Also,

$$f\left(-\frac{5}{2}\right) = \left(-\frac{5}{2}\right)^2 + 5\left(-\frac{5}{2}\right) + 4$$
$$= \frac{25}{4} - \frac{25}{2} + 4 = -\frac{9}{4},$$

PRACTICE EXERCISE 2

(a) Graph $g(x) = -x^2 - 5x - 4$.

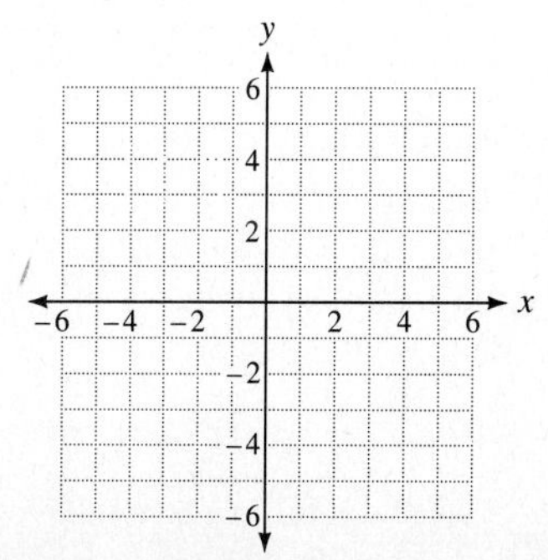

so that the vertex is $\left(-\frac{5}{2}, -\frac{9}{4}\right)$. With the vertex located below the x-axis, and the parabola opening up, we look for the x-intercepts.

$$x^2 + 5x + 4 = 0$$
$$(x + 4)(x + 1) = 0$$
$$x + 4 = 0 \quad \text{or} \quad x + 1 = 0$$
$$x = -4 \qquad\qquad x = -1$$

Thus, the x-intercepts are $(-4, 0)$ and $(-1, 0)$, and the graph is given in Figure 9.8.

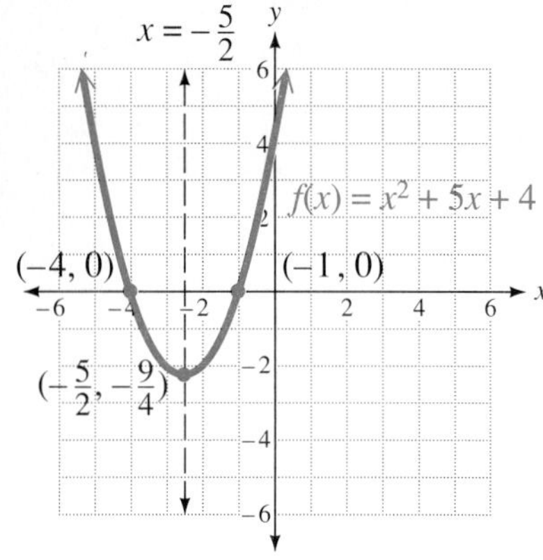

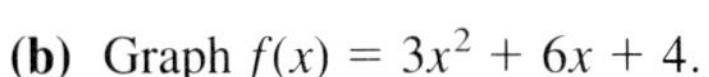
Figure 9.8

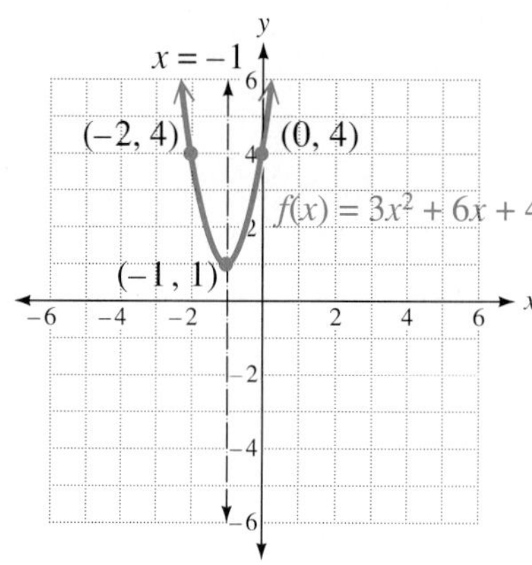

Figure 9.9

(b) Graph $f(x) = 3x^2 + 6x + 4$.
Since $a = 3 > 0$, the parabola opens up. With $a = 3$ and $b = 6$, $-\frac{b}{2a} = -\frac{6}{2(3)} = -1$. Thus, the line of symmetry is $x = -1$ and the vertex is $(-1, f(-1)) = (-1, 1)$. With the vertex above the x-axis and the parabola opening up, there will be no x-intercepts. If we were to solve $3x^2 + 6x + 4 = 0$, the solutions we obtain would be

$$\frac{-3 \pm i\sqrt{3}}{3},$$

which are not real numbers. (Remember that the coordinate axes are only used to plot real numbers.) Since there are no x-intercepts, we determine two additional points on either side of the line of symmetry $x = -1$. For example, when $x = 0$, $f(0) = 4$, and when $x = -2$, $f(-2) = 4$. This gives us the points $(0, 4)$ and $(-2, 4)$. The graph is given in Figure 9.9.

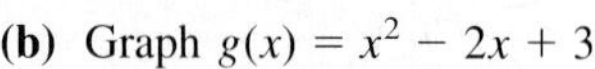
(b) Graph $g(x) = x^2 - 2x + 3$.

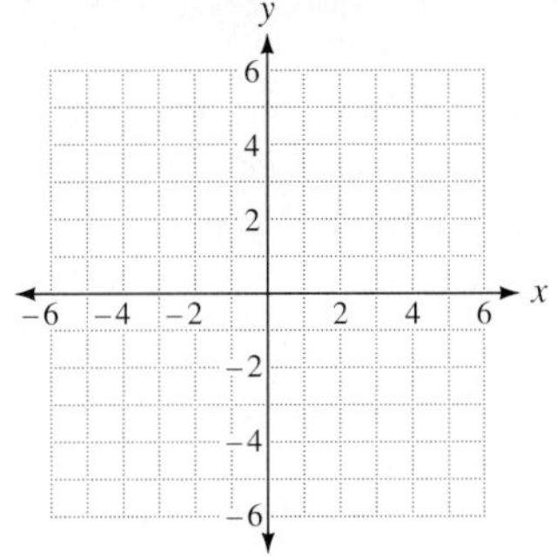

Answers: (a) The graph is a parabola that opens down, has vertex $\left(-\frac{5}{2}, \frac{9}{4}\right)$, has x-intercepts $(-4, 0)$ and $(-1, 0)$, and line of symmetry $x = -\frac{5}{2}$. (b) The graph is a parabola that opens up, has vertex $(1, 2)$, has no x-intercepts, has line of symmetry $x = 1$, and passes through $(0, 3)$ and $(2, 3)$.

4 MAXIMUM AND MINIMUM PROBLEMS

Remember that the vertex is a point at which a quadratic function $f(x) = ax^2 + bx + c$ assumes a largest or maximum value when $a < 0$ and a smallest or minimum value when $a > 0$. This fact can be used to solve a variety of applied problems, called **maximum** or **minimum problems,** that can be translated into quadratic functions for which the maximum or minimum value is desired. If the parabola opens up it has a minimum at the vertex; if it opens down it has a maximum at the vertex.

EXAMPLE 3 Maximum Problem

A man has 36 yd of fencing material. What are the dimensions of the largest rectangular pen that he can enclose? What is the maximum area?

Let l = the length of the desired pen,
w = the width of the desired pen.

Make a sketch as in Figure 9.10. We know that

$$2l + 2w = 36. \quad \text{The perimeter} = 36 \text{ yd}$$

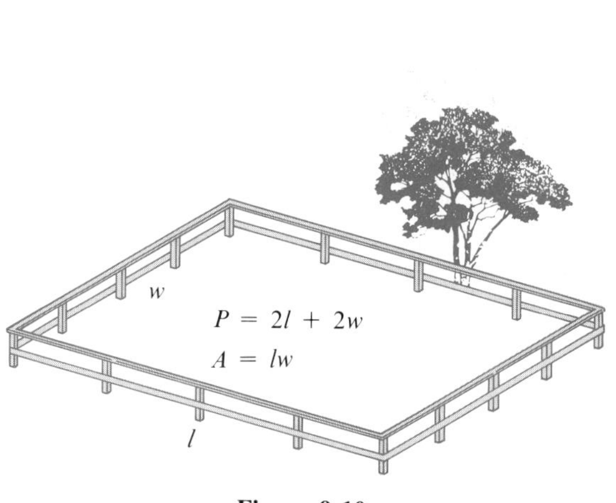

Figure 9.10

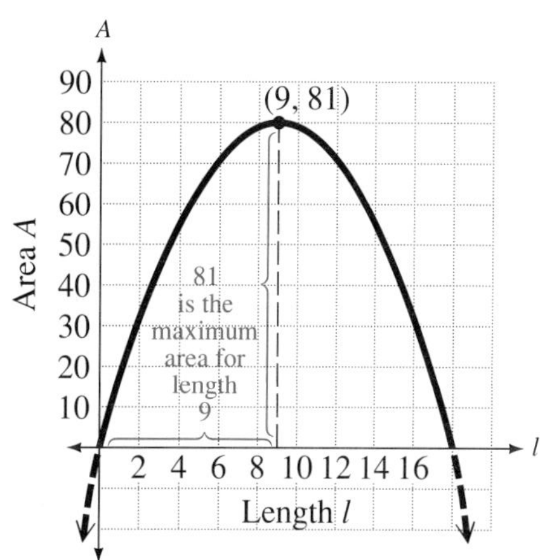

Figure 9.11

The quantity we wish to maximize is the area, A, of the pen.

$$A = lw$$

By solving $2l + 2w = 36$ for w,

$$w = 18 - l,$$

and substituting, we obtain

$$A = lw = l(18 - l) = -l^2 + 18l.$$

Thus, quadratic function $A = -l^2 + 18l$ gives the area of the pen as a function of the length l. We can graph this function as in Figure 9.11, where the area A is shown along the y-axis and the length l along the x-axis. The maximum point of the graph (the vertex) will give the maximum area (the y-coordinate) determined by a particular length (the x-coordinate). With $a = -1$, and $b = 18$, $-\frac{b}{2a} = -\frac{18}{2(-1)} = 9$. Thus, the y-coordinate of the vertex is

$$-(9)^2 + 18(9) = -81 + 162 = 81. \quad \text{Substitute 9 for } x$$

Since $a = -1 < 0$, the parabola $A = -l^2 + 18l$ opens down. Then 81 is the maximum value of A, occurring when $l = 9$. Since, when $l = 9$, $w = 18 - l = 9$, the dimensions that give the maximum area of 81 yd^2 are 9 yd by 9 yd.

Practice Exercise 3

What is the minimum product of two numbers whose difference is 4? What are the numbers?

Let x = one number,
y = second number.

We know that $x - y = 4$ or that $x = y + 4$. We must minimize the product $P = xy = (y + 4)y$.

Answer: The minimum product is -4 and the numbers are 2 and -2.

9.1 EXERCISES A

In Exercises 1–4, sketch the graph of each quadratic function by sliding the graph of $f(x) = x^2$ or $g(x) = -x^2$ up or down.

1. $h(x) = x^2 + 1$

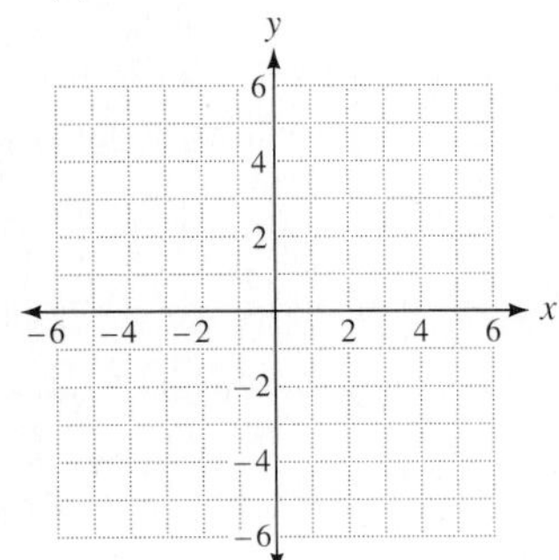

2. $h(x) = x^2 - 2$

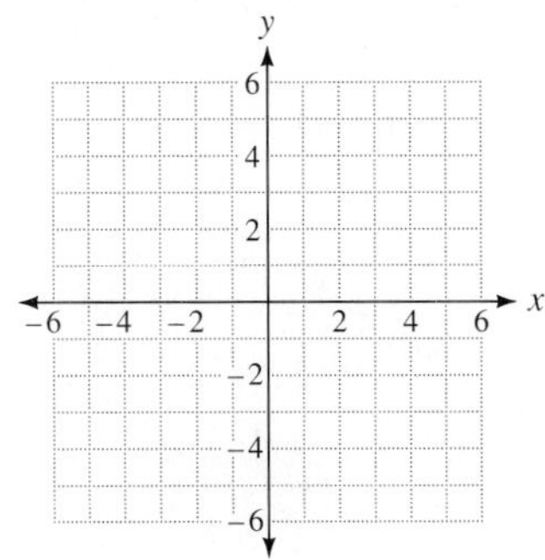

3. $h(x) = -x^2 + 1$

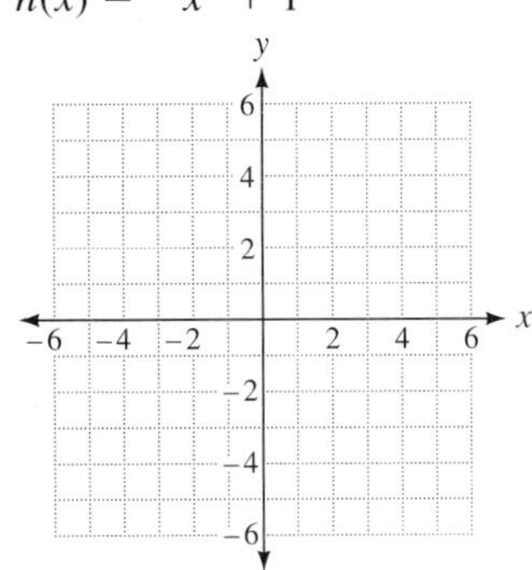

4. $h(x) = -x^2 - 3$

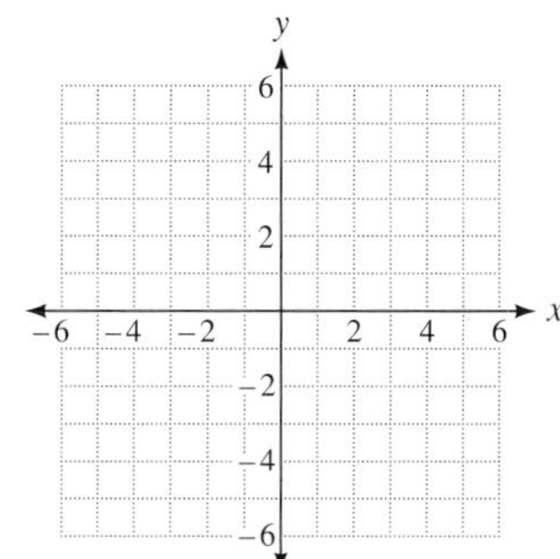

Without graphing, give the x-intercepts (if they exist) and the vertex, and tell whether the graph of each quadratic function opens up or down.

5. $f(x) = x^2 - 4x + 3$

6. $g(x) = x^2 + 4x + 3$

7. $f(x) = x^2 + 2x + 2$

8. $g(x) = x^2 + 8x$

9. $f(x) = -x^2 + 5x - 6$

10. $g(x) = 3x^2 - 6x + 3$

11. $f(x) = -2x^2 + 4x - 3$

12. $g(x) = 3x^2 + 5x + 1$

Find the x-intercepts (if they exist), the vertex, and graph of the function.

13. $f(x) = x^2 + 2x - 3$

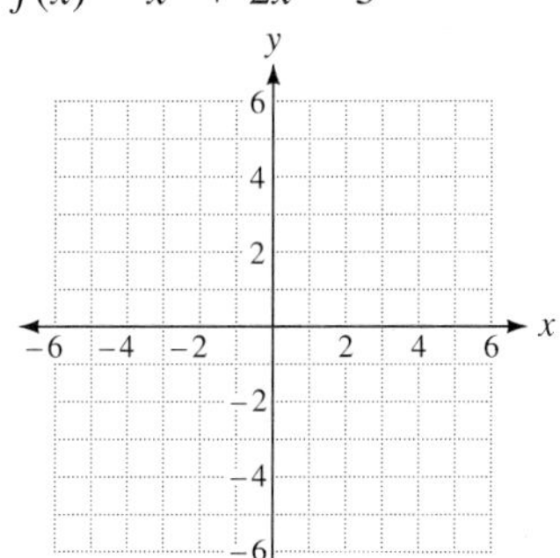

14. $g(x) = -x^2 - 2x + 3$

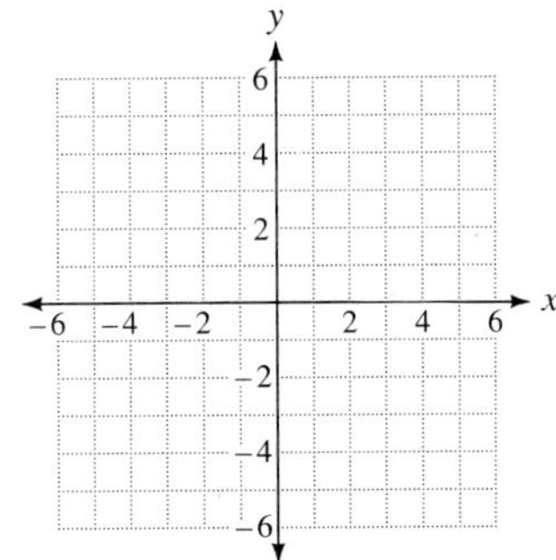

15. $f(x) = x^2 - 4x$

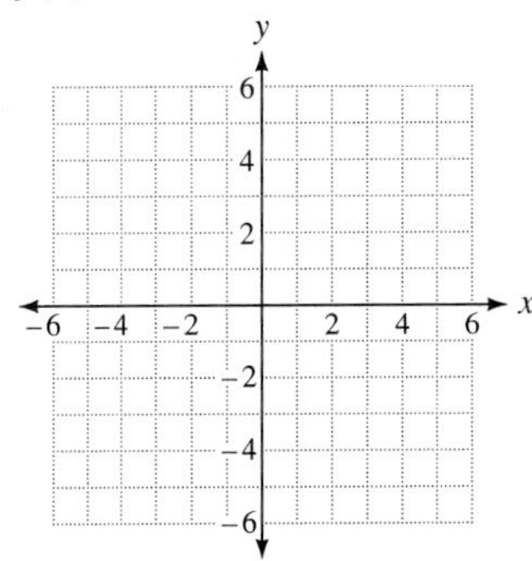

16. $g(x) = -x^2 + 4x$

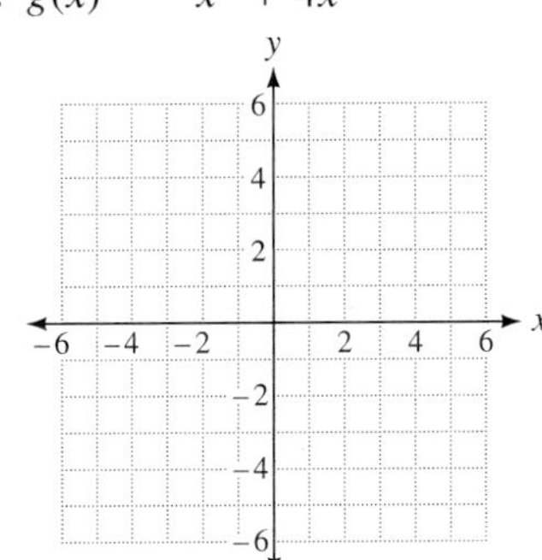

17. $f(x) = -4x^2 - 8x - 4$

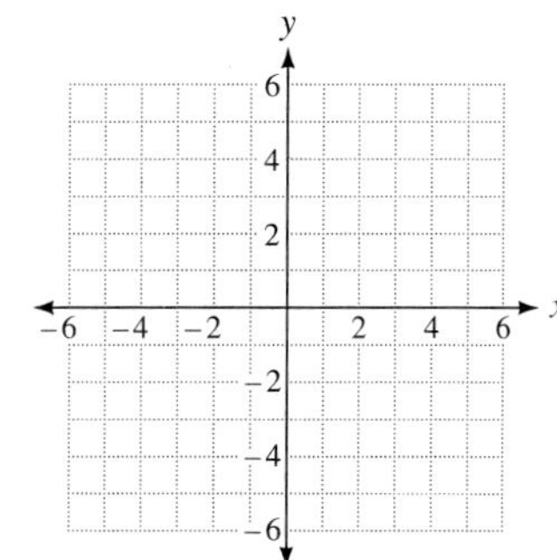

18. $f(x) = 2x^2 - 8x + 11$

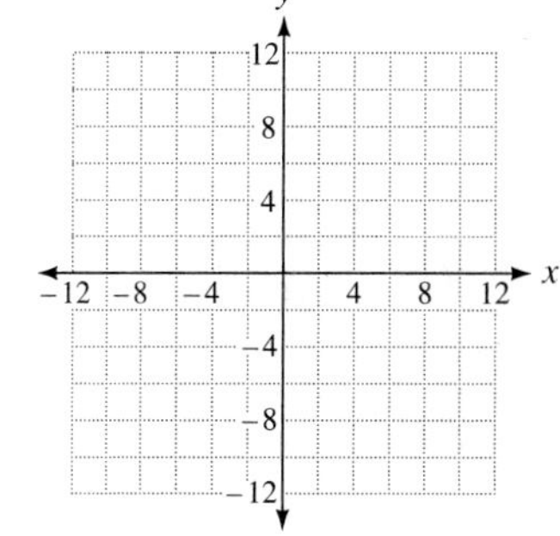

Solve.

19. A man has 40 m of fencing material. What are the dimensions of the largest rectangular pen which he can enclose? What is the maximum area?

20. What is the minimum product of two numbers whose difference is 8? What are the numbers?

21 Over a period of years, by studying the daily sales at Perko's Delicatessen, Perko has discovered that his profit can be approximated by $P = -x^2 + 80x - 1500$, where x represents the number of sandwiches sold. How many sandwiches should he sell daily to maximize his profit? What is the profit under these ideal conditions?

22. A ball is thrown upward with initial velocity of 32 ft/sec. If its height after x sec is given by $h = -16x^2 + 32x$, find the maximum height that it will attain. At what time will the ball hit the ground?

FOR REVIEW

The following exercises review material from Chapter 3 to help you prepare for the next section. Specify the domain of each function defined by the equation given in Exercises 23–25.

23. $y = 2x + 8$ **24.** $y = x^2 + 5$ **25.** $y = \dfrac{1}{x + 9}$

For each function in Exercises 26–27, evaluate: ***(a)*** $g(0)$ ***(b)*** $g(-1)$ ***(c)*** $g(2)$ ***(d)*** $g(x + h)$ ***(e)*** $g(x + h) - g(x)$.

26. $g(x) = 5x - 1$ **27.** $g(x) = x^2 - 1$

ANSWERS: 1.

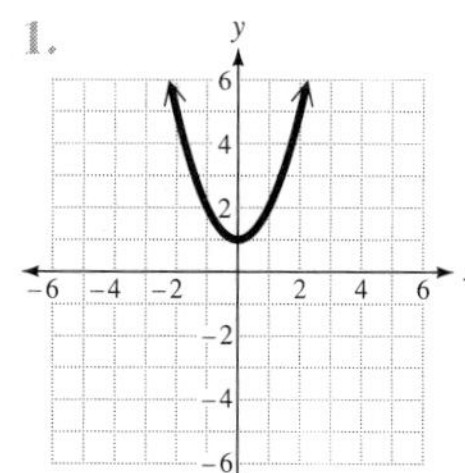

2.

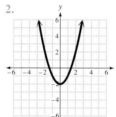

3.

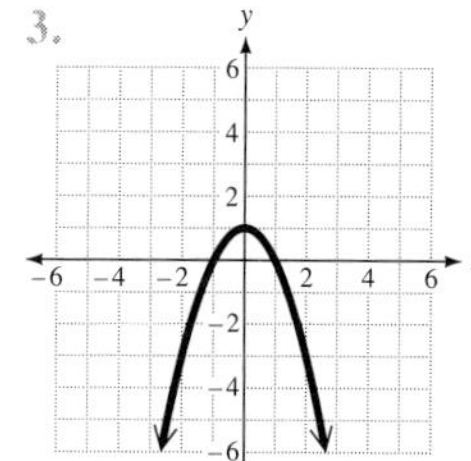

4.

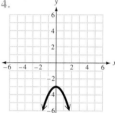

5. (1, 0) and (3, 0); (2, −1); up 6. (−1, 0) and (−3, 0); (−2, −1); up 7. no x-intercepts; (−1, 1); up 8. (0, 0) and (−8, 0); (−4, −16); up 9. (2, 0) and (3, 0); $\left(\frac{5}{2}, \frac{1}{4}\right)$; down 10. (1, 0) is the only x-intercept; (1, 0); up 11. no x-intercepts; (1, −1); down 12. $\left(\frac{-5+\sqrt{13}}{6}, 0\right)$ and $\left(\frac{-5-\sqrt{13}}{6}, 0\right)$; $\left(-\frac{5}{6}, -\frac{13}{12}\right)$; up

13.

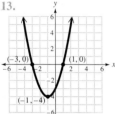

14.

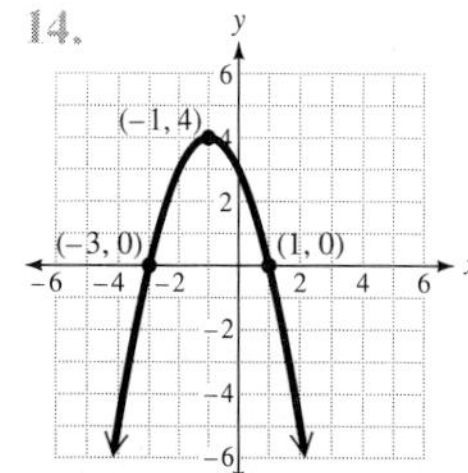

15.

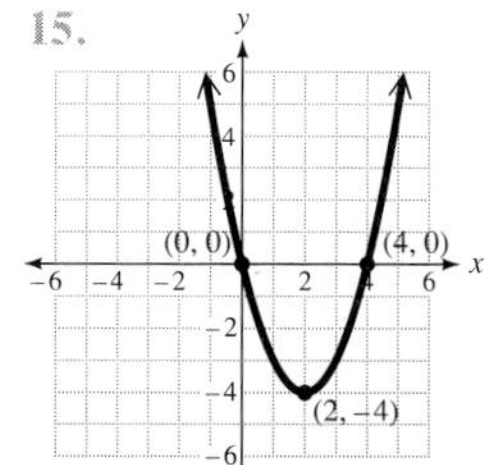

16.

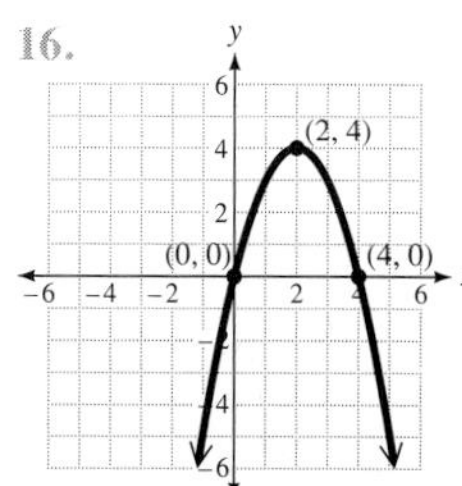

17.

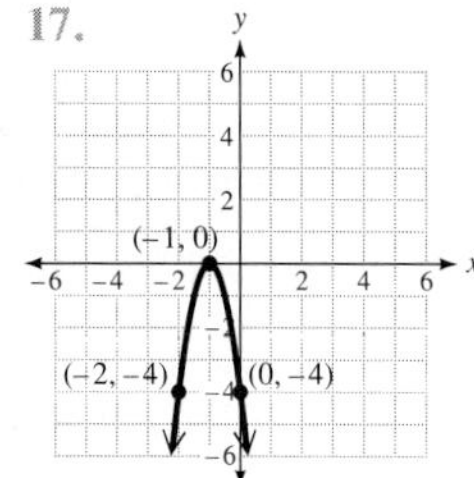

18.

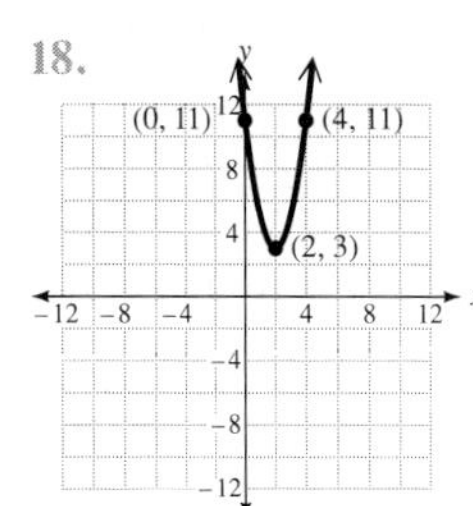

19. 10 m by 10 m; 100 m^2 20. −16; 4 and −4 21. 40 sandwiches; $100 22. 16 ft; 2 sec after it is thrown upward 23. all real numbers 24. all real numbers 25. all real numbers except −9 26. (a) −1 (b) −6 (c) 9 (d) $5x + 5h - 1$ (e) $5h$ 27. (a) −1 (b) 0 (c) 3 (d) $x^2 + 2xh + h^2 - 1$ (e) $2xh + h^2$

9.1 EXERCISES B

In Exercises 1–4, sketch the graph of each quadratic function by sliding the graph of $f(x) = x^2$ or $g(x) = -x^2$ up or down.

1. $h(x) = x^2 + 3$

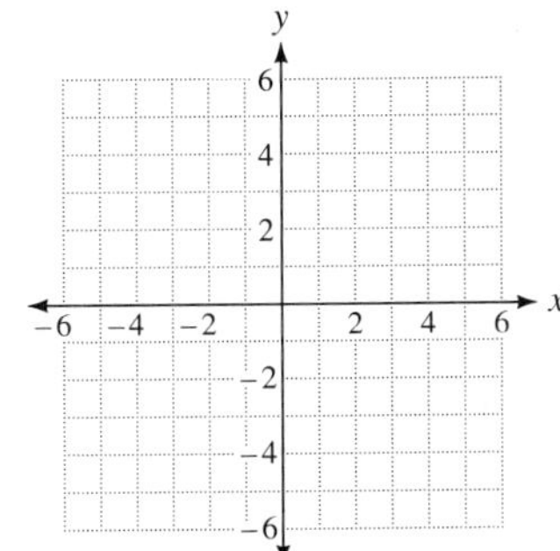

2. $h(x) = x^2 - 3$

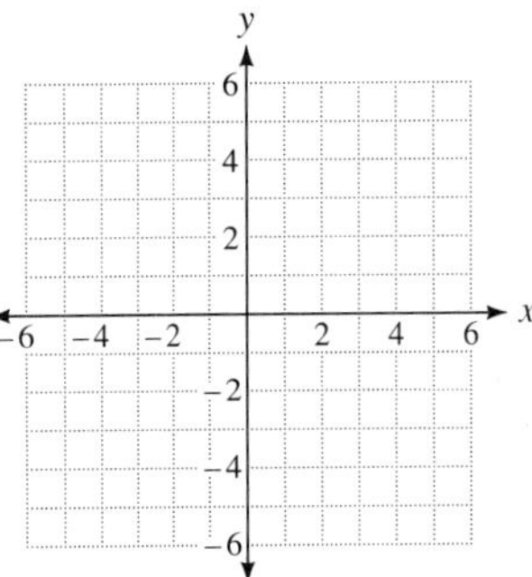

3. $h(x) = -x^2 + 2$

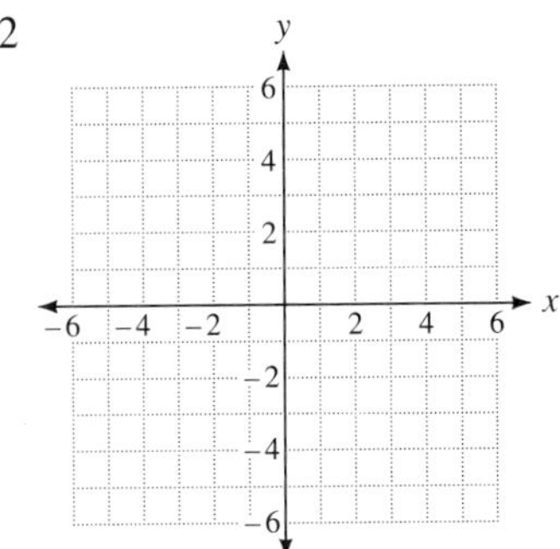

4. $h(x) = -x^2 - 1$

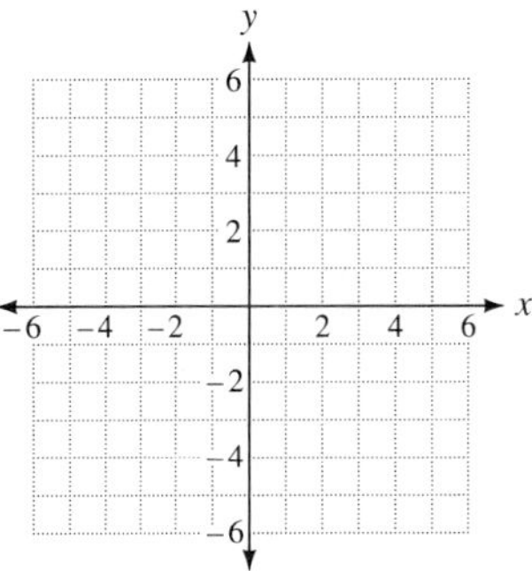

Without graphing, give the x-intercepts (if they exist) and the vertex, and tell whether the graph of each quadratic function opens up or down.

5. $f(x) = -x^2 + 4x - 3$

6. $g(x) = -x^2 - 4x - 3$

7. $h(x) = -x^2 - 2x - 2$

8. $f(x) = -x^2 - 8x$

9. $g(x) = x^2 - 5x + 6$

10. $h(x) = -3x^2 + 6x - 3$

11. $f(x) = 2x^2 - 4x + 3$

12. $g(x) = -3x^2 - 5x - 1$

Find the x-intercepts (if they exist), the vertex, and graph of the function.

13. $f(x) = x^2 - 2x - 3$

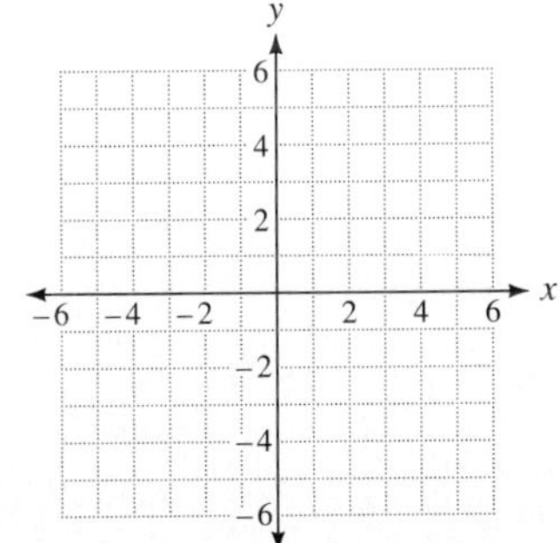

14. $g(x) = -x^2 + 2x + 3$

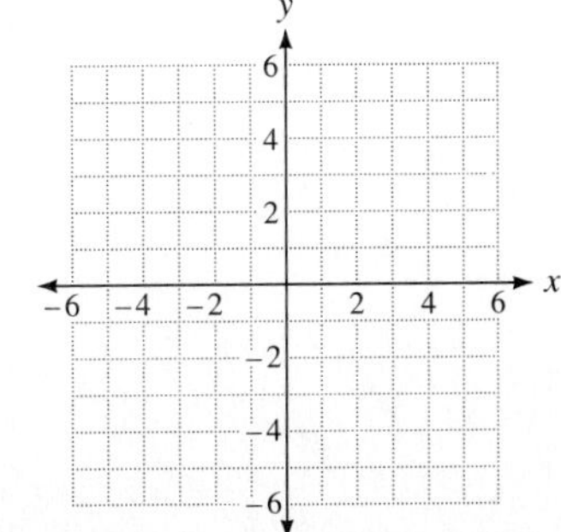

15. $f(x) = x^2 + 4x$

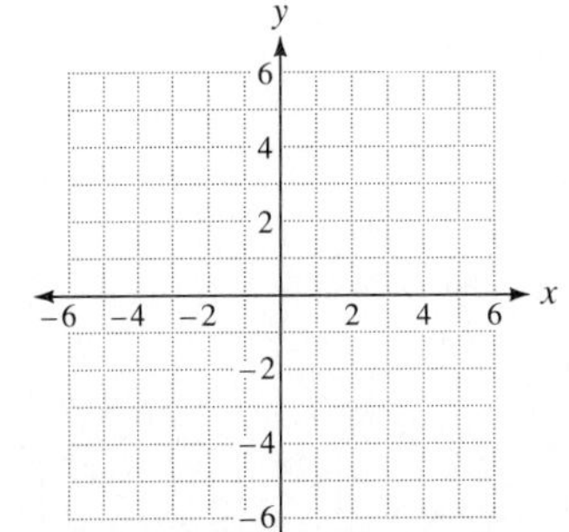

16. $g(x) = -x^2 - 4x$

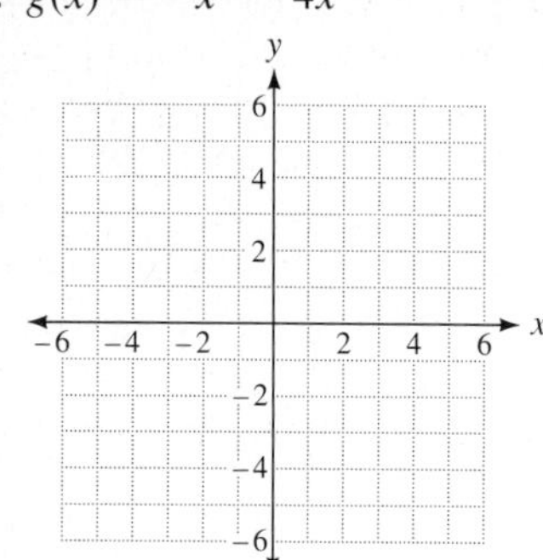

17. $f(x) = -4x^2 + 8x - 4$

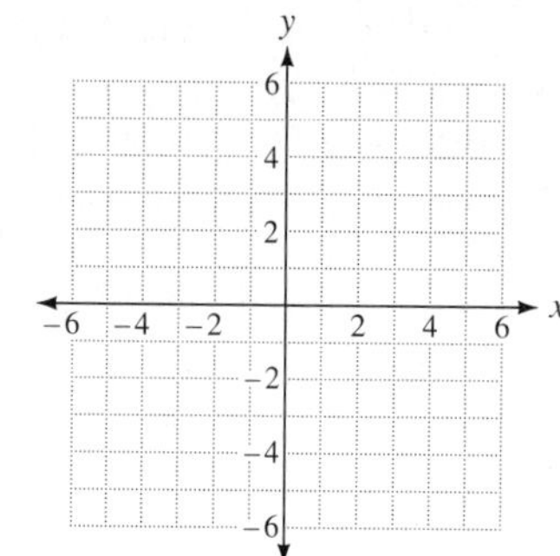

18. $g(x) = 4x^2 - 8x + 4$

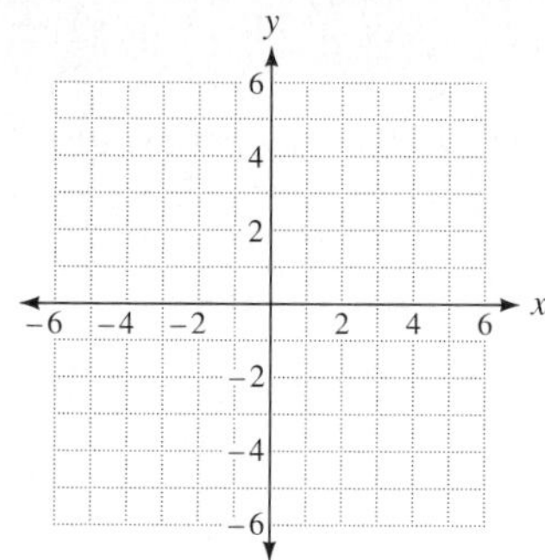

Solve.

19. A woman has 60 yd of fencing material. What are the dimensions of the largest rectangular pen which she can enclose? What is the maximum area?

20. What is the maximum product of two numbers whose sum is 50? What are the numbers?

21. Larsen's Sharpening Shop specializes in sharpening saws. Terry Larsen has determined that the cost of operating the shop is given by $c = x^2 - 40x + 430$, where x is the number of saws sharpened daily. What is the number of saws Terry must sharpen each day in order to minimize the operational cost? What is the cost under these ideal conditions?

22. An arrow is shot upward with an initial velocity of 128 ft/sec. If its height after x sec is given by $h = -16x^2 + 128x$, find the maximum height that it will attain. At what time will the arrow hit the ground?

FOR REVIEW

The following exercises review material from Chapter 3 to help you prepare for the next section. Specify the domain of each function defined by the equation given in Exercises 23–25.

23. $y = -3x + 12$

24. $y = -3x^2 + x$

25. $y = \dfrac{2}{x - 5}$

For each function in Exercises 26–27, evaluate: **(a)** $g(0)$ **(b)** $g(-1)$ **(c)** $g(2)$ **(d)** $g(x + h)$ **(e)** $g(x + h) - g(x)$.

26. $g(x) = 2x + 3$

27. $g(x) = -x^2 + 2$

9.1 EXERCISES C

Find the vertex, the x-intercepts, and graph of the function.

1. $f(x) = 2x^2 - 8x + 5$

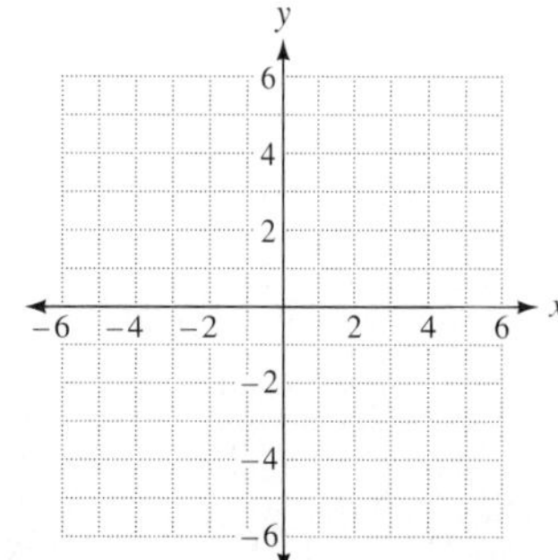

2. $f(x) = -2x^2 + 5x + 7$

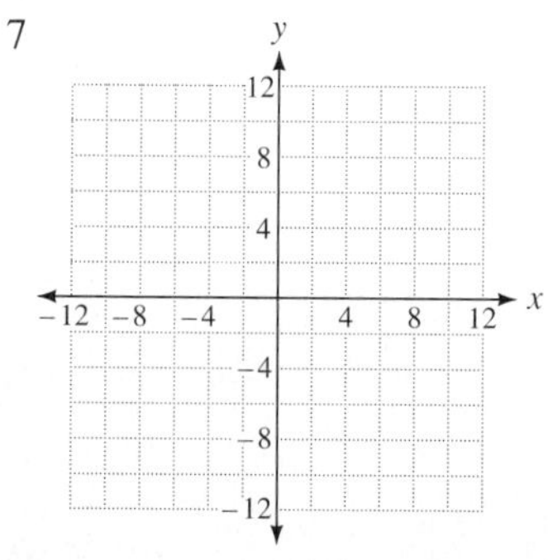

3. Suppose that (1, 0), (−2, −3), and (4, 39) are three ordered pairs in the quadratic function with equation $y = f(x) = ax^2 + bx + c$. Find the values of a, b, and c by solving a system of three linear equations. Use these values to find the value of y when x is 2. [Answer: $a = 2$, $b = 3$, $c = -5$; $y = 9$ when $x = 2$]

4. When a department store opens, the number of people in the store increases from zero to a maximum number then down to zero again at closing time. Suppose the number N of people in a store is given as a function of time t by $N = -15t^2 + 180t$, where $t = 0$ corresponds to the time the store opens at 10:00 A.M. At what time will the store have the maximum number of customers? What is the maximum number of customers? At what times does the store close?

5. Chuck Little plans to build a rectangular pen adjacent to a stream. What is the maximum area he can enclose with 100 ft of fencing if no fence is required next to the stream?

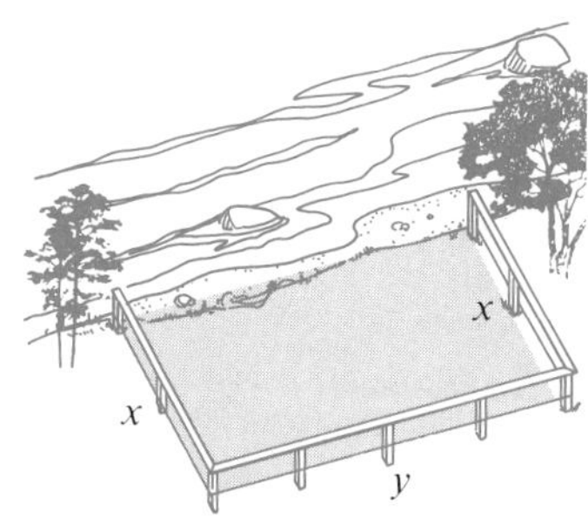

9.2 PROPERTIES OF FUNCTIONS AND SPECIAL FUNCTIONS

STUDENT GUIDEPOSTS

1 The Domain of a Function	4 Polynomial Functions
2 Algebra of Functions	5 Square Root Functions
3 Composition of Functions	6 Absolute Value Functions

1 THE DOMAIN OF A FUNCTION

Remember from Section 3.5 that the domain of a function given by an equation $y = f(x)$ is the largest set of real numbers for which the expression in x is defined. For example,

$$f(x) = 2x + 1 \quad \text{and} \quad g(x) = \frac{2}{x - 3}$$

are functions with domains the set of all real numbers and the set of all real numbers except 3, respectively. Since division by zero is undefined, 3 must be excluded from the domain of g. When a radical is used in the definition of a function, such as

$$h(x) = \sqrt{x + 2},$$

the domain of the function includes only those real numbers for which the radicand is nonnegative. Thus, $x + 2$ must be nonnegative so that $x + 2 \geq 0$ or $x \geq -2$. The domain of h is then all real numbers $x \geq -2$.

EXAMPLE 1 Finding Domains

Give the domain of each function.

(a) $f(x) = x^2 + 2x - 3$
Since $x^2 + 2x - 3$ is defined for every real number x, the domain of f is the set of all real numbers.

(b) $g(x) = \sqrt{x - 5}$
Since $x - 5$ must be nonnegative,

$$x - 5 \geq 0$$
$$x \geq 5.$$

The domain of g is all real numbers $x \geq 5$.

(c) $h(x) = \dfrac{1}{\sqrt{2 - x}}$
Since $2 - x$ must be nonnegative,

$$2 - x \geq 0.$$
$$-x \geq -2$$
$$x \leq 2 \quad \text{Reverse the inequality}$$

However in this case, x cannot be 2 since the denominator $\sqrt{2 - x}$ would be zero. Thus, the domain of h is all real numbers $x < 2$.

Practice Exercise 1

Give the domain of each function.

(a) $f(x) = -x^2 + 8$

(b) $g(x) = \sqrt{x + 7}$

(c) $h(x) = \dfrac{5}{\sqrt{x - 8}}$

Answers: (a) all real numbers (b) all real numbers $x \geq -7$ (c) all real numbers $x > 8$

2 ALGEBRA OF FUNCTIONS

When two functions are given, new functions can be formed by algebraically combining them as defined below.

Algebra of Functions

Let f and g be two functions.

$(f + g)(x) = f(x) + g(x)$	Sum of f and g
$(f - g)(x) = f(x) - g(x)$	Difference of f and g
$(fg)(x) = f(x) \cdot g(x)$	Product of f and g
$\left(\dfrac{f}{g}\right)(x) = \dfrac{f(x)}{g(x)}, \quad g(x) \neq 0$	Quotient of f and g

EXAMPLE 2 Finding Algebraic Combinations of Functions

Let $f(x) = 2x + 1$ and $g(x) = x - 5$.

(a) Find $(f + g)(x) = f(x) + g(x) = (2x + 1) + (x - 5)$ $f(x) = 2x + 1$ and $g(x) = x - 5$

$$= 3x - 4$$

(b) Find $(f + g)(-3)$. Since $(f + g)(x) = 3x - 4$ from (a), substitute -3 for x.

$$(f + g)(-3) = 3(-3) - 4$$
$$= -9 - 4 = -13$$

Practice Exercise 2

Let $f(x) = x + 5$ and $g(x) = x^2 - 1$.

(a) Find $(f + g)(x)$.

(b) Find $(f + g)(-2)$.

(c) Find $(f-g)(-2)$. Since $(f-g)(x)=f(x)-g(x)$ $=(2x+1)-(x-5)=x+6$,

$$(f-g)(-2)=(-2)+6=4.$$

(d) Find $(fg)(0)$. Since $(fg)(x)=f(x)\cdot g(x)$ $=(2x+1)(x-5)=2x^2-9x-5$,

$$(fg)(0)=2(0)^2-9(0)-5$$
$$=0-0-5=-5.$$

(e) Find $\left(\frac{f}{g}\right)(5)$. Since $\left(\frac{f}{g}\right)(x)=\frac{f(x)}{g(x)}=\frac{2x+1}{x-5}$, we must substitute 5 for x to find $\left(\frac{f}{g}\right)(5)$.

$$\left(\frac{f}{g}\right)(5)=\frac{2(5)+1}{(5)-5}=\frac{11}{0}, \text{ which is not defined.}$$

Notice that the quotient function is not defined at any value of x that makes the denominator 0.

(c) Find $(f-g)(-3)$.

(d) Find $(fg)(1)$.

(e) Find $\left(\frac{f}{g}\right)(-5)$.

Answers: (a) $(f+g)(x)=x^2+x+4$ (b) $(f+g)(-2)=6$ (c) $(f-g)(-3)=-6$ (d) $(fg)(1)=0$ (e) $\left(\frac{f}{g}\right)(-5)=0$

3 COMPOSITION OF FUNCTIONS

Suppose that $f(x)=3x+2$ and $g(x)=1-x$. Then

$$g(3)=1-(3)=-2$$
$$\text{and}\quad f(-2)=3(-2)+2=-6+2=-4.$$

If these results are combined, we have actually found

$$f[g(3)]=f[-2] \quad \text{Since } g(3)=-2$$
$$=-4. \quad \text{Since } f(-2)=-4$$

Combining two functions in this way is defined below.

Composition of Functions

Let f and g be two functions. The expression $f[g(x)]$ is the **composition of f with g**, and $g[f(x)]$ is the **composition of g with f**.

EXAMPLE 3 COMPOSITION OF FUNCTIONS

Let $f(x)=5x+2$ and $g(x)=x^2-1$.

(a) Find $f[g(2)]$.
We first find $g(2)$.

$$g(2)=(2)^2-1=4-1=3$$

Now we find $f[g(2)]$.

$$f[g(2)]=f[3] \quad g(2)=3$$
$$=5(3)+2$$
$$=15+2=17$$

(b) Find $g[f(0)]$.
First find $f(0)$.

$$f(0)=5(0)+2=2$$

PRACTICE EXERCISE 3

Let $f(x)=2x^2+3$ and $g(x)=x-4$.

(a) Find $f[g(4)]$.

(b) Find $g[f(0)]$.

Now we find $g[f(0)]$.

$$g[f(0)] = g[2] \qquad f(0) = 2$$
$$= (2)^2 - 1$$
$$= 4 - 1 = 3$$

Answers: (a) 3 (b) −1

4 POLYNOMIAL FUNCTIONS

Linear and quadratic functions studied in Sections 3.5 and 9.1 are actually special cases of *polynomial functions*. A function of the form

$$f(x) = a_n x^n + a_{n-1}x^{n-1} + \cdots + a_1 x + a_0$$

where $a_n, a_{n-1}, \ldots, a_1, a_0$ are real numbers, with $a_n \neq 0$, is a **polynomial function** of degree n. Notice that if $n = 1$,

$$f(x) = a_1 x + a_0$$

is a linear function, and if $n = 2$

$$f(x) = a_2 x^2 + a_1 x + a_0$$

is a quadratic function. The domain of every polynomial function is the entire set of real numbers.

Although graphing polynomial functions is a topic usually considered in more advanced courses, we can obtain a rough graph by plotting points.

EXAMPLE 4 GRAPHING A POLYNOMIAL FUNCTION

Graph the function $f(x) = x^3 - 2x^2 - 5x + 6$.

Construct a table of values, plot the corresponding points, and join them with a smooth curve. The graph, which is typical of many third-degree polynomials, is given in Figure 9.12.

x	$f(x)$
−2	0
−1	8
0	6
1	0
2	−4
3	0
4	18

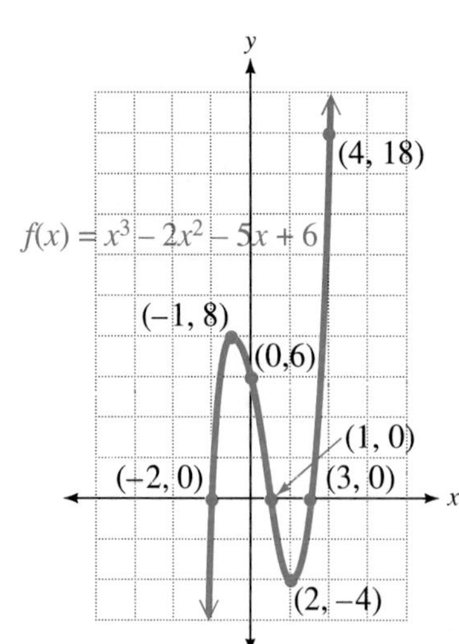

Figure 9.12 Third-Degree Polynomial Function

PRACTICE EXERCISE 4

Graph $f(x) = -x^4 + 5x^2 - 4$.
Complete the table below.

x	$f(x)$
−3	
−2	
−1	
0	
1	
2	
3	

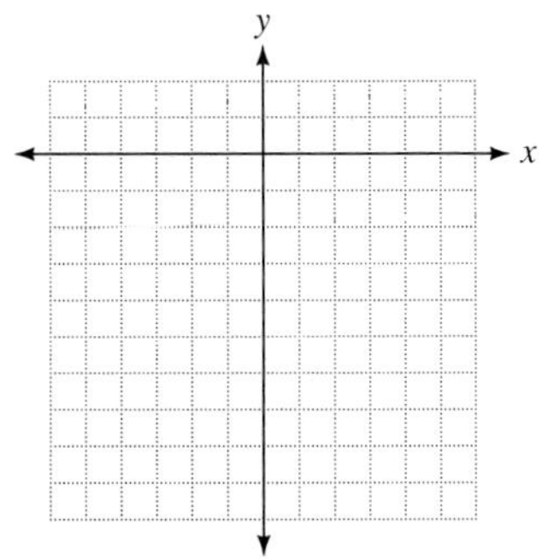

Answer: The graph passes through the points (−3, −40), (−2, 0), (−1, 0), (0, −4), (1, 0), (2, 0), and (3, −40).

5 SQUARE ROOT FUNCTIONS

Another special function is the square root function. Remember that the radical denotes the principal (nonnegative) square root of a number. For example, $\sqrt{4}$ represents 2, not -2 nor ± 2. The function

$$f(x) = \sqrt{x}, \quad \textbf{for } x \geq 0$$

is called the **square root function.** Notice that the domain of the square root function is the set of all nonnegative real numbers. The graph of this function, given in Figure 9.13, is actually half of a parabola opening to the right with vertex (0, 0). We will consider relations like this in more detail in Chapter 11.

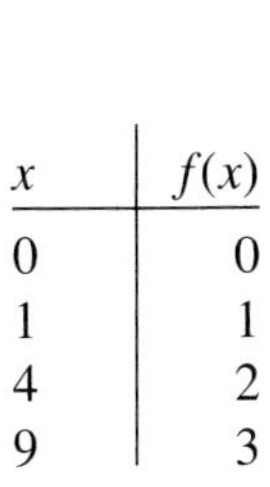

x	$f(x)$
0	0
1	1
4	2
9	3

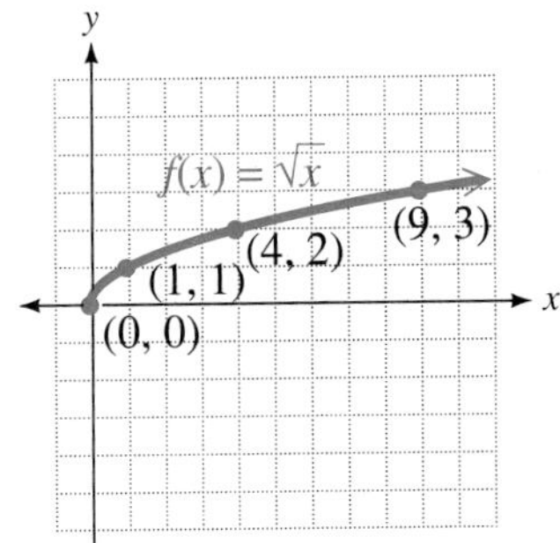

Figure 9.13 Square Root Function

EXAMPLE 5 GRAPHING A SQUARE ROOT FUNCTION

Graph the function $f(x) = \sqrt{x + 3}$.

Since $x + 3$ must be nonnegative, the domain of this function is all real numbers $x \geq -3$. Its graph, given in Figure 9.14, has the same shape as the square root function, but shifted to the left 3 units.

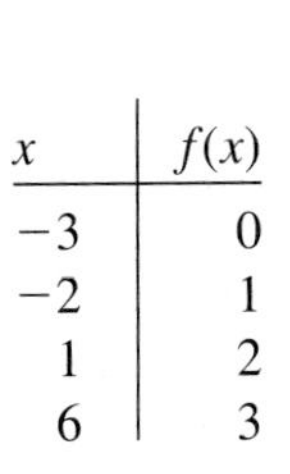

x	$f(x)$
-3	0
-2	1
1	2
6	3

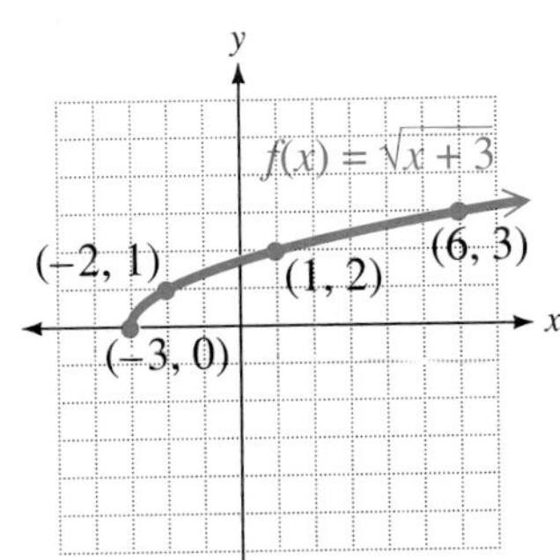

Figure 9.14

PRACTICE EXERCISE 5

Graph $f(x) = \sqrt{x - 1}$.

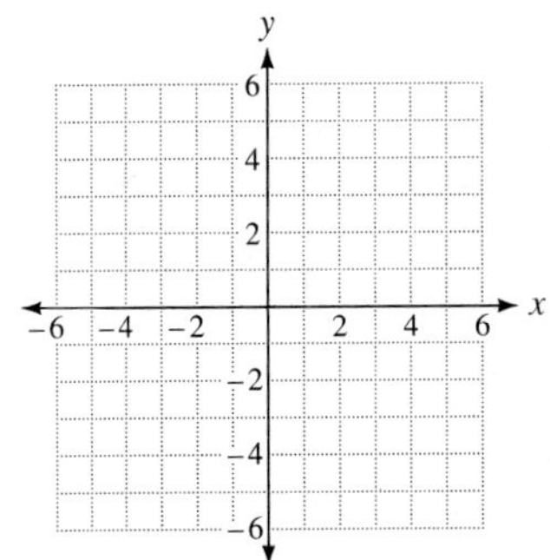

Answer: The graph has the same shape as the square root function, but shifted right 1 unit.

6 ABSOLUTE VALUE FUNCTIONS

Recall the definition of absolute value, given in Chapter 1.

$$|x| = \begin{cases} x \text{ if } x \geq 0 \\ -x \text{ if } x < 0 \end{cases}$$

We use this to define the **absolute value function**

$$f(x) = |x|.$$

In Section 3.1 we graphed the equation $y = |x|$ which gives us the graph of the absolute value function, a V-shaped curve shown in Figure 9.15.

x	$f(x)$
0	0
1	1
−1	1
2	2
−2	2
3	3
−3	3

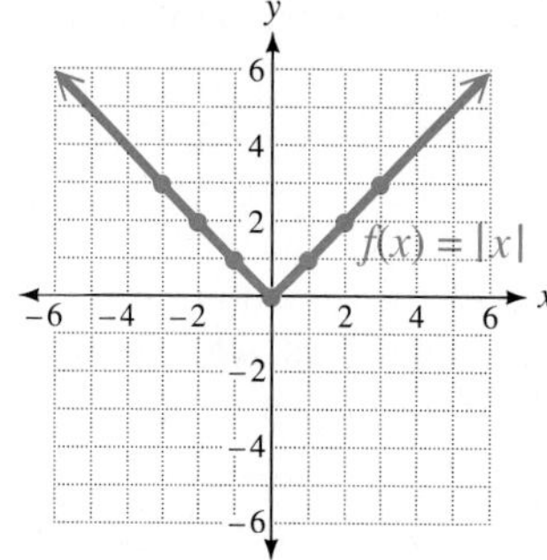

Figure 9.15 Absolute Value Function

Variations of the absolute value function have graphs similar in shape to that of $f(x) = |x|$, as shown in the next example.

EXAMPLE 6 GRAPHING AN ABSOLUTE VALUE FUNCTION

Graph the function $f(x) = |x + 1|$.

Construct a table of values like the one below. The graph of $f(x) = |x + 1|$ is given in Figure 9.16. Notice that the graph could be obtained by shifting the graph of the absolute value function to the left 1 unit.

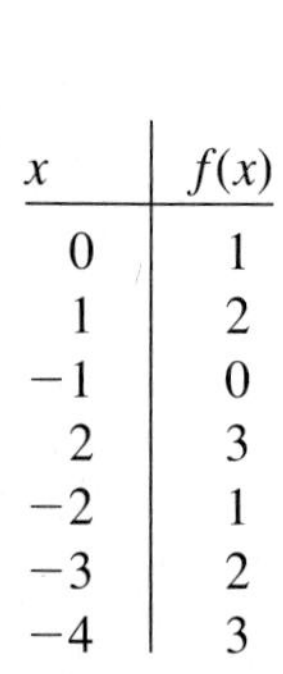

x	$f(x)$
0	1
1	2
−1	0
2	3
−2	1
−3	2
−4	3

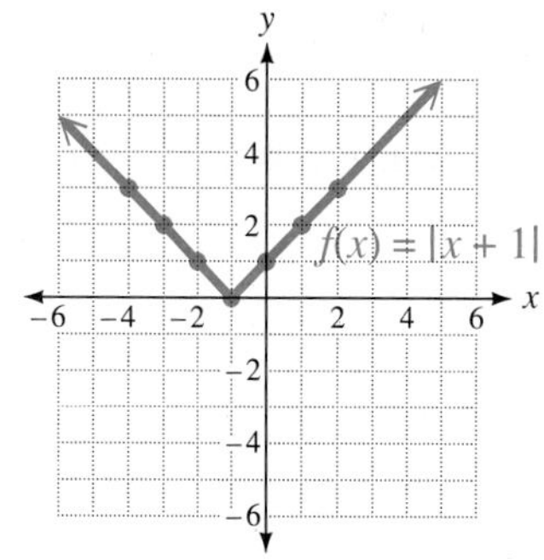

Figure 9.16

PRACTICE EXERCISE 6

Graph $f(x) = |x - 3|$.

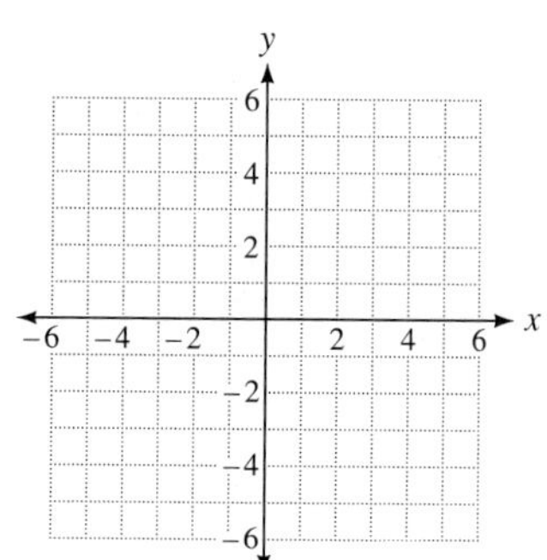

Answer: The graph is similar to $f(x) = |x|$ but shifted right 3 units.

9.2 EXERCISES A

Find the domain of the function defined by each equation.

1. $y = \sqrt{4 - x}$

2. $y = \dfrac{5}{x + 5}$

3. $y = \dfrac{x}{x - 1}$

4. $y = \dfrac{2}{\sqrt{x + 2}}$

5. $y = \sqrt{x^2 + 10}$

6. $f(x) = \sqrt{x^2 - 1}$

7. $f(x) = \dfrac{1}{\sqrt{x^2 - 1}}$

8 $f(x) = \sqrt{x^2 - 8x + 12}$

Assume that $f(x) = 4 + 3x$ and $g(x) = x^2 - 3$. Find each function value.

9. $(f + g)(-1)$ **10.** $(f + g)(-4)$ **11.** $(f - g)(2)$ **12.** $(f - g)(3)$

13. $(fg)(-2)$ **14.** $(fg)(-1)$ **15.** $\left(\frac{f}{g}\right)(-3)$ **16.** $\left(\frac{f}{g}\right)(-4)$

17 $f[g(-1)]$ **18.** $f[g(-4)]$ **19.** $g[f(2)]$ **20.** $g[f(-2)]$

Make a table of values and sketch the graph of each function.

21. $f(x) = x^3 - 4x$

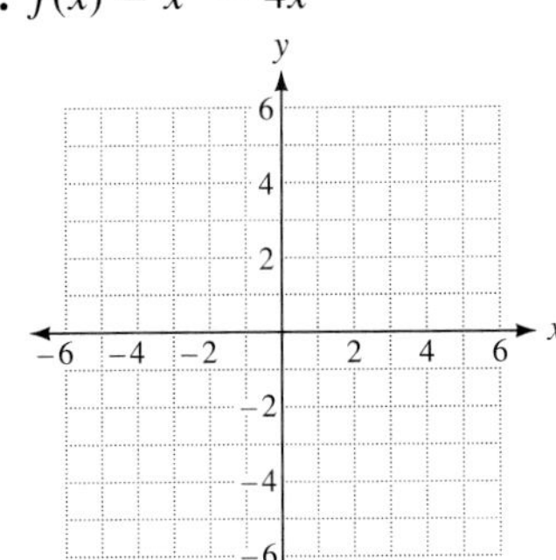

22. $f(x) = -x^3 + x^2 + 9x - 9$

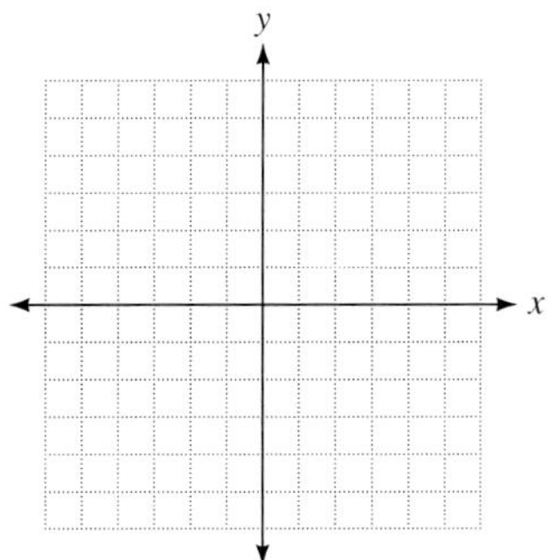

23. $f(x) = x^4 - 20x^2 + 64$

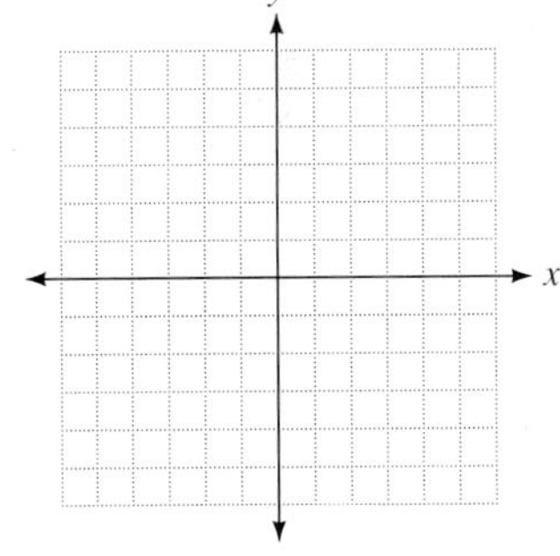

24. $f(x) = \sqrt{x + 2}$

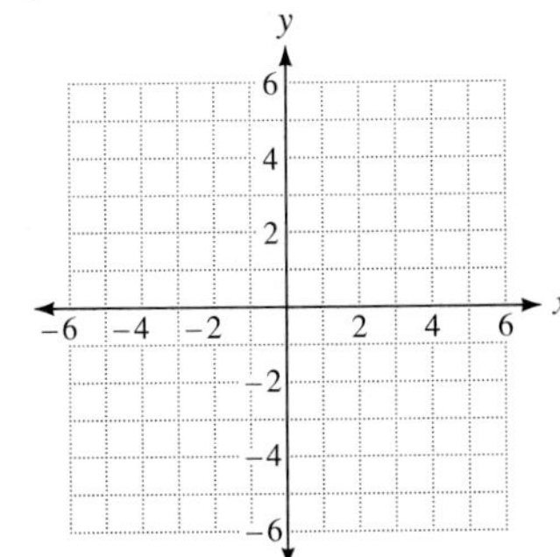

25. $f(x) = \sqrt{x - 3}$

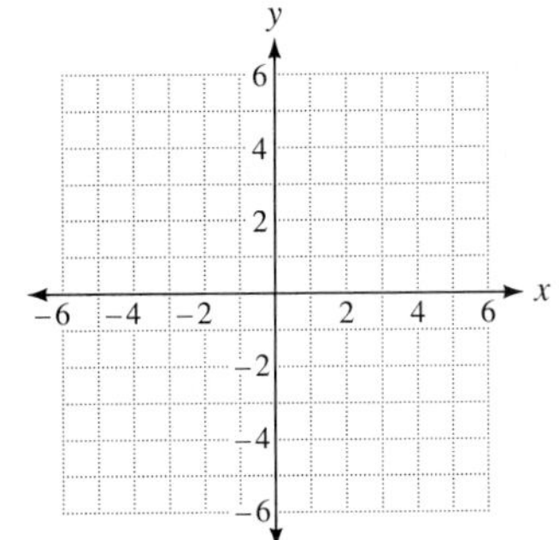

26. $f(x) = -\sqrt{x}$

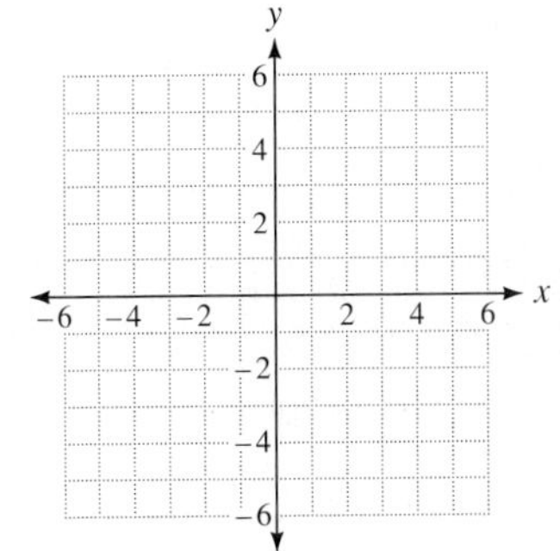

27. $f(x) = |x - 1|$

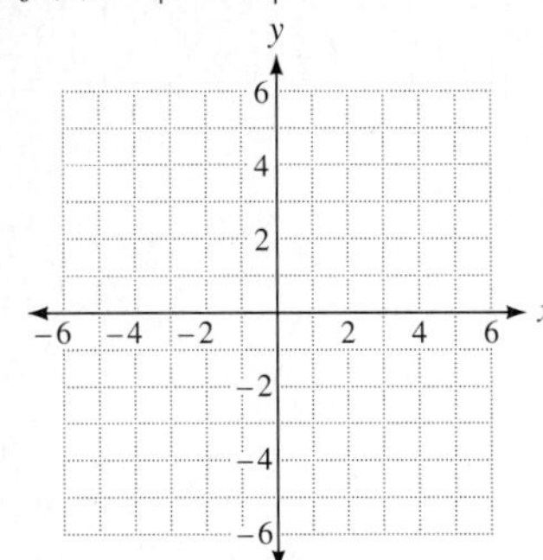

28. $f(x) = |x + 2|$

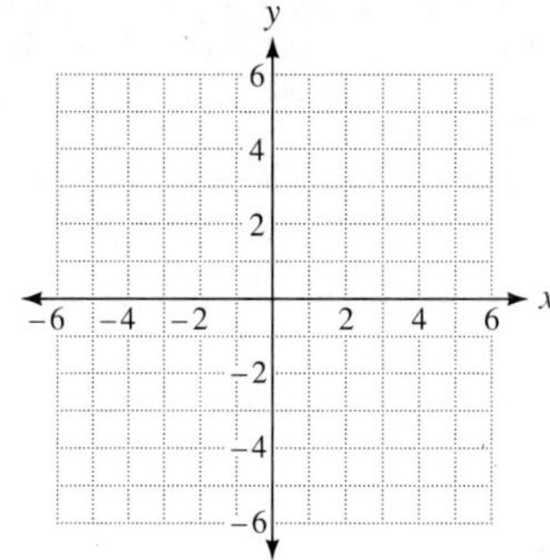

29. $f(x) = |x| + 1$

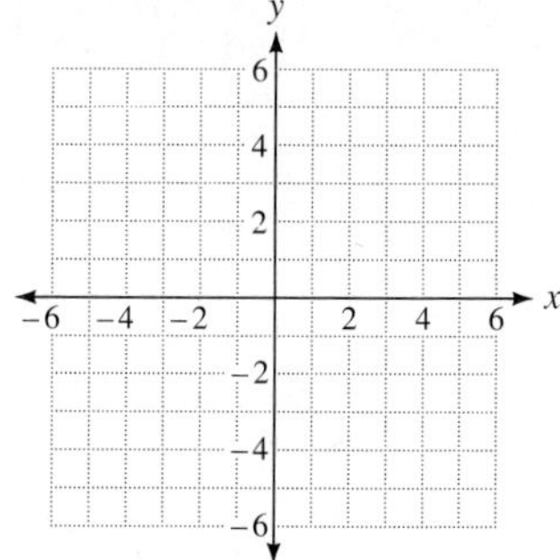

30. $f(x) = |x| - 1$

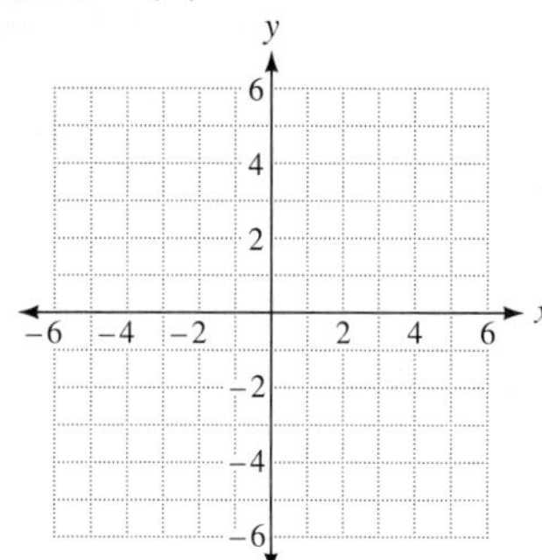

31. $f(x) = -|x|$

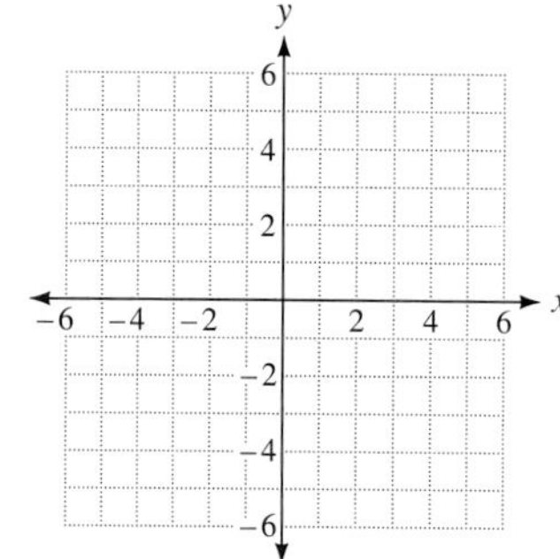

32. $f(x) = -|x + 2|$

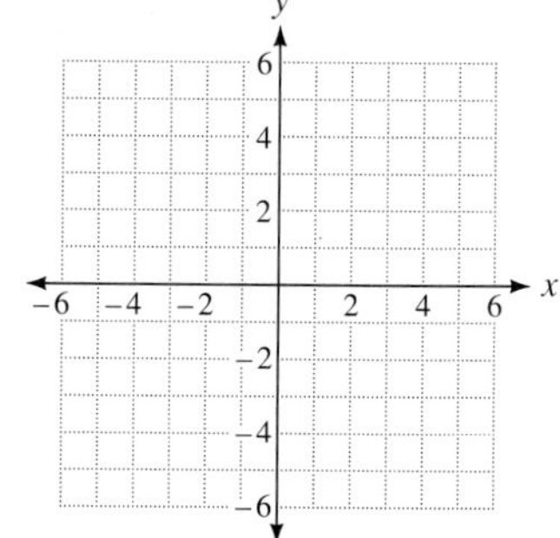

FOR REVIEW

Without graphing, tell the x-intercepts, the vertex, the line of symmetry, and whether the graph of each quadratic function opens up or down in Exercises 33–34.

33. $f(x) = x^2 - 2x - 15$

34. $g(x) = -x^2 + 2x + 15$

Find the vertex, the x-intercepts, and the graph of each function in Exercises 35–36.

35. $f(x) = x^2 - 2x - 3$

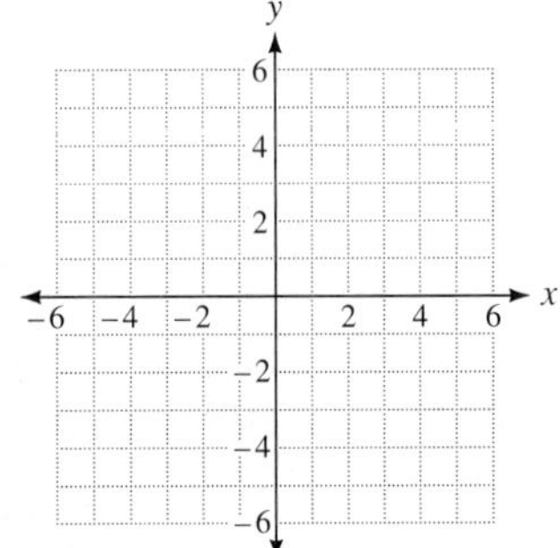

36. $g(x) = -x^2 - 2x + 3$

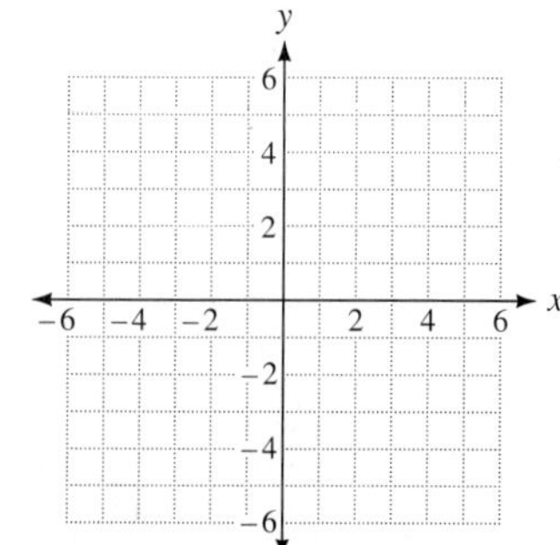

37. The length and width of a rectangle have a sum of 100 feet. What width will produce a maximum area?

The following exercise reviews material from Chapter 3 to help you prepare for the next section.

38. Consider the relation $r = \{(1, 3), (2, 1), (3, 1), (4, 2)\}$.

(a) Graph the relation r.

(b) Is the relation r a function? Explain.

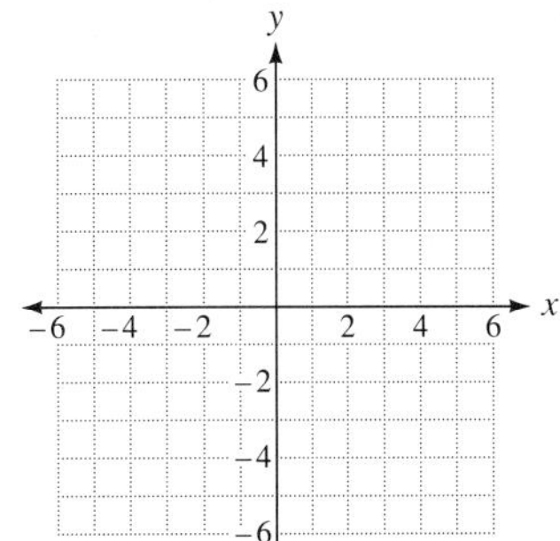

(c) Give the relation s formed by interchanging the x and y coordinates of each ordered pair in r.

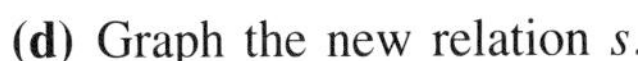

(d) Graph the new relation s.

(e) Is the new relation s a function? Explain.

(f) Could you eliminate one ordered pair from r so that the relation s formed by interchanging the coordinates is a function?

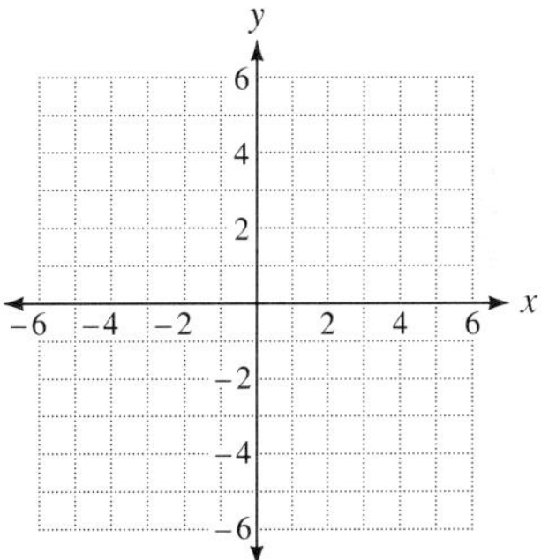

ANSWERS: 1. all real numbers $x \le 4$ 2. all real numbers except -5 3. all real numbers except 1 4. all real numbers $x > -2$ 5. all real numbers 6. all real numbers $x \le -1$ or $x \ge 1$ 7. all real numbers $x < -1$ or $x > 1$ 8. all real numbers $x \le 2$ or $x \ge 6$ 9. -1 10. 5 11. 9 12. 7 13. -2 14. -2 15. $-\frac{5}{6}$ 16. $-\frac{8}{13}$ 17. -2 18. 43 19. 97 20. 1

21.
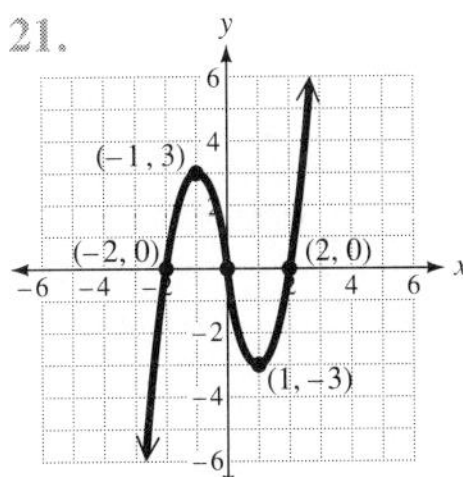

22.
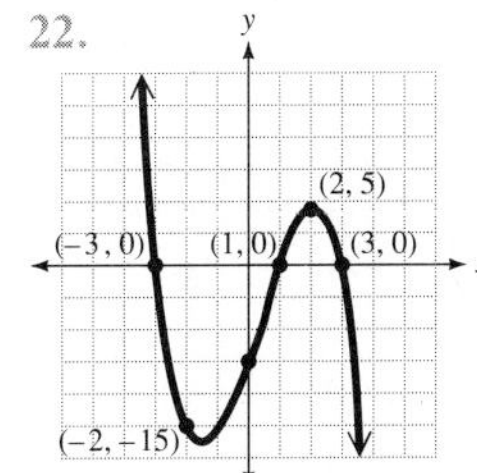

23.
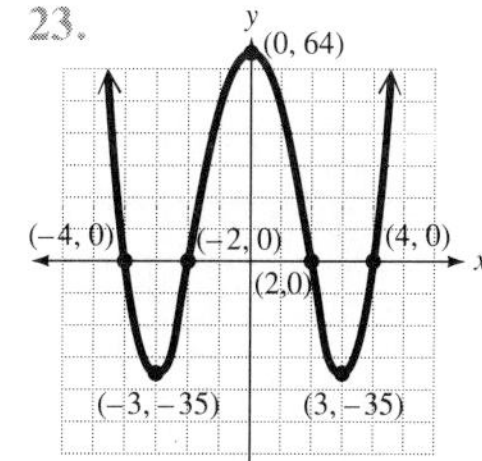

24.
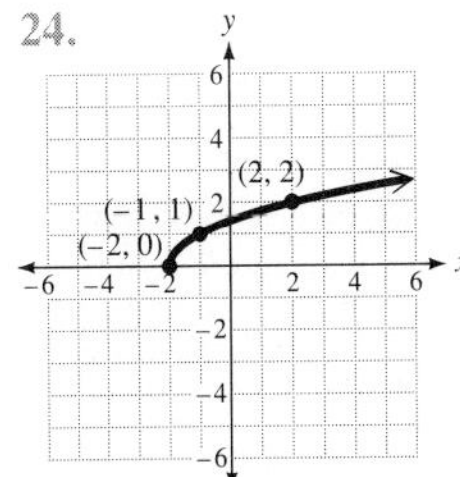

25.
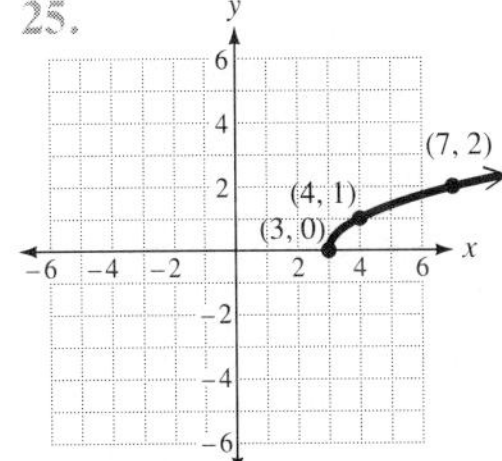

26.
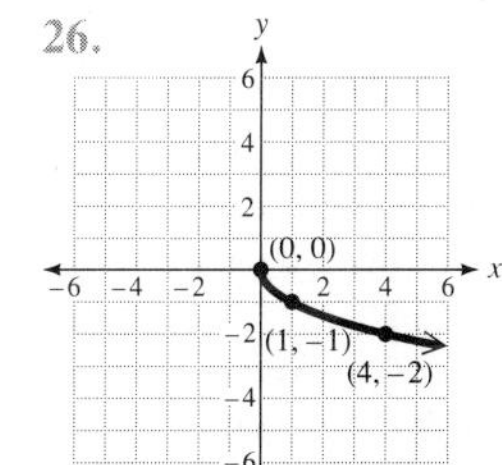

27.
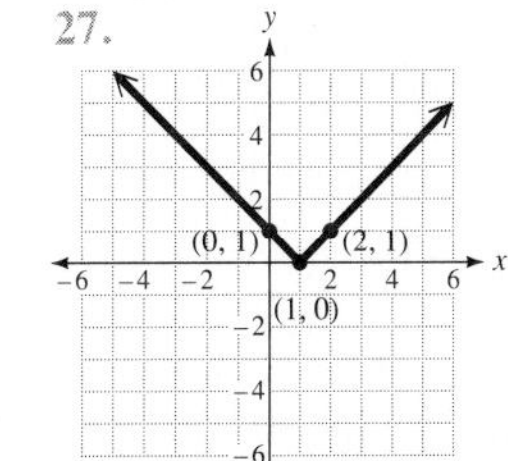

28.
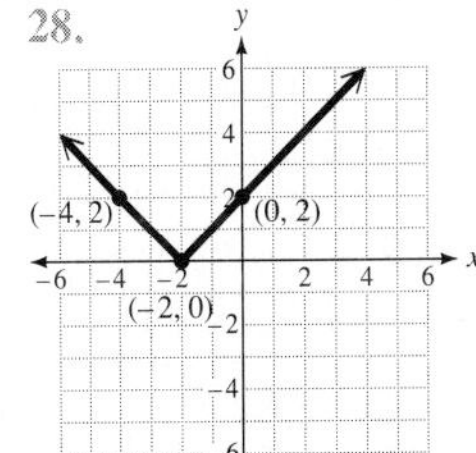

29.
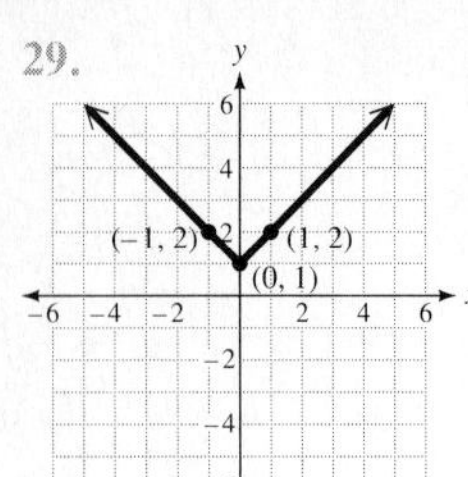

30.
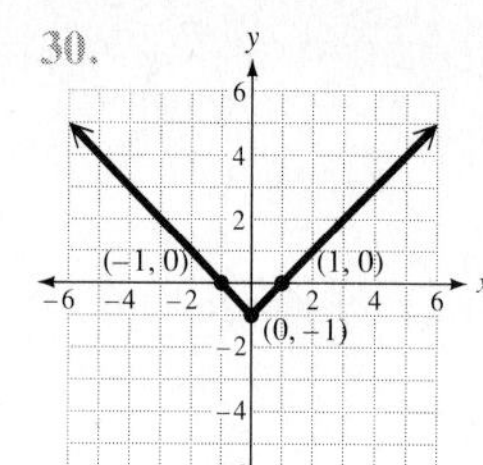

31.
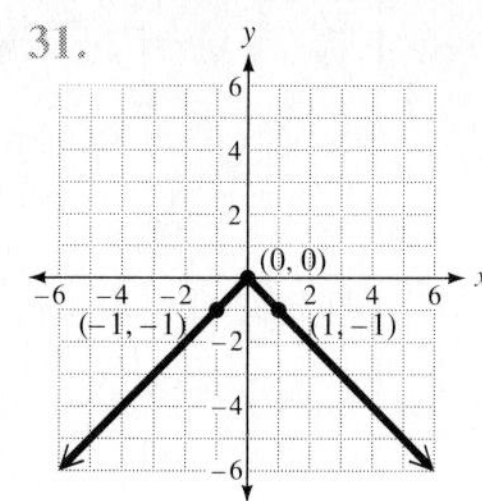

32.
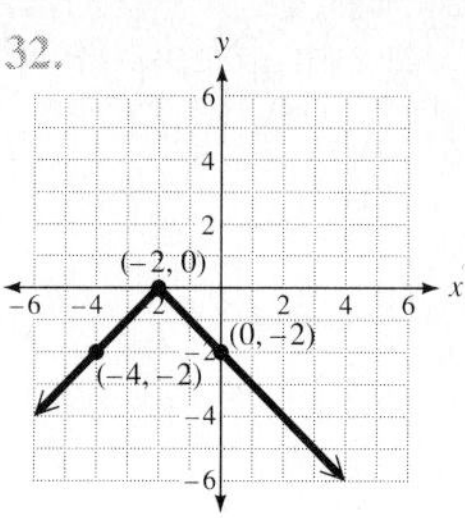

33. (5, 0) and (−3, 0); (1, −16); $x = 1$; up 34. (5, 0) and (−3, 0); (1, 16); $x = 1$; down

35.
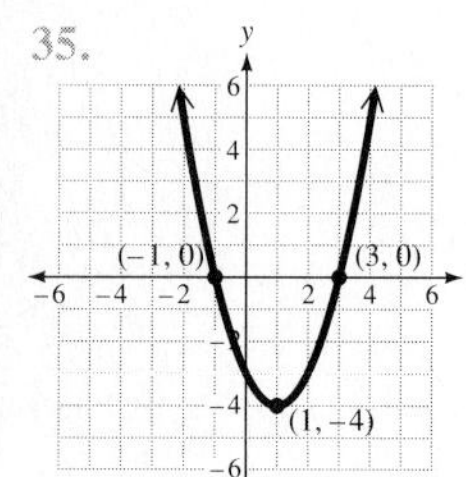

36.
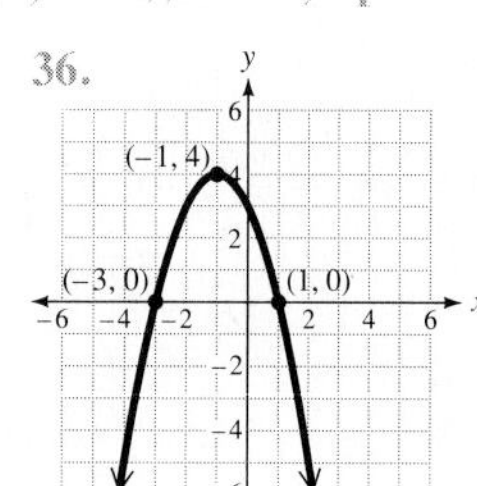

37. 50 ft 38. (a)
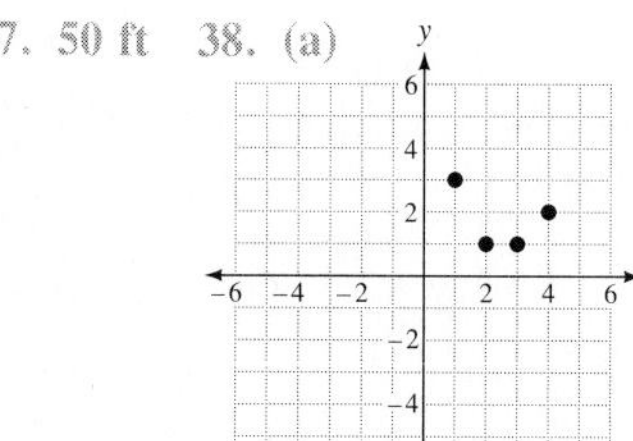

(b) yes (c) $s = \{(3, 1), (1, 2), (1, 3), (2, 4)\}$

(d)
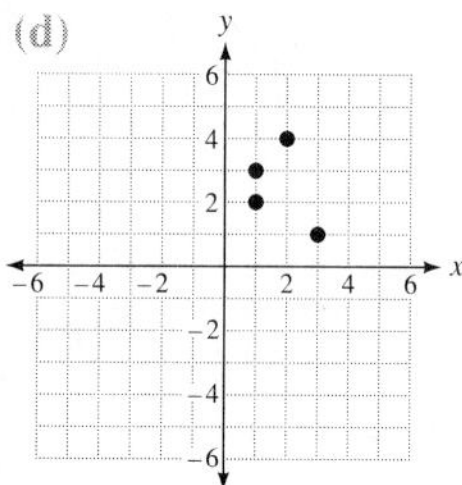

(e) no (f) Yes; if either (2, 1) or (3, 1) were omitted from r then s would be a function.

9.2 EXERCISES B

Find the domain of the function defined by each equation.

1. $y = \sqrt{x - 2}$
2. $y = \dfrac{x}{x + 5}$
3. $y = \dfrac{2}{x - 3}$
4. $y = \dfrac{7}{\sqrt{x - 3}}$
5. $y = \dfrac{1}{\sqrt{x^2 + 8}}$
6. $g(x) = \sqrt{x^2 - 4}$
7. $g(x) = \dfrac{1}{\sqrt{x^2 - 4}}$
8. $g(x) = \sqrt{x^2 - 5x - 14}$

Assume that $f(x) = x^2 + 2$ and $g(x) = 1 - 3x$. Find each function value.

9. $(f + g)(-1)$
10. $(f + g)(-4)$
11. $(f - g)(2)$
12. $(f - g)(3)$
13. $(fg)(-2)$
14. $(fg)(-1)$
15. $\left(\dfrac{f}{g}\right)(-3)$
16. $\left(\dfrac{f}{g}\right)(-4)$
17. $f[g(-1)]$
18. $f[g(-4)]$
19. $g[f(2)]$
20. $g[f(-2)]$

Make a table of values and sketch the graph of each function.

21. $f(x) = x^3 - 9x$

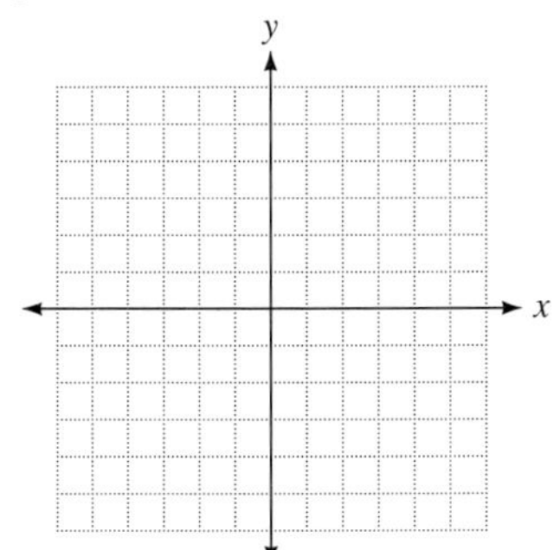

22. $f(x) = -x^3 + 4x^2 + 4x - 16$

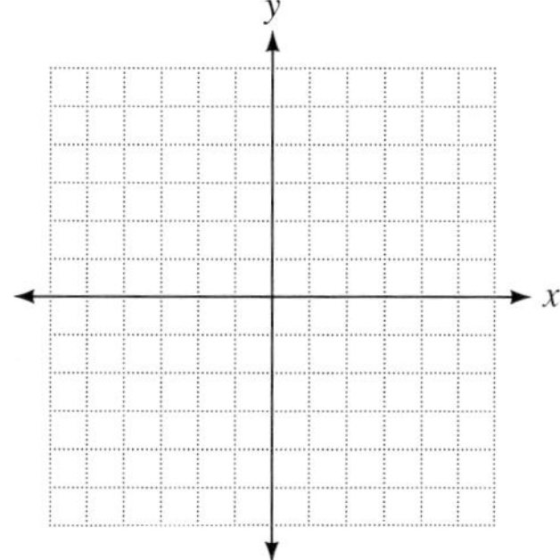

23. $f(x) = x^4 - 10x^2 + 9$

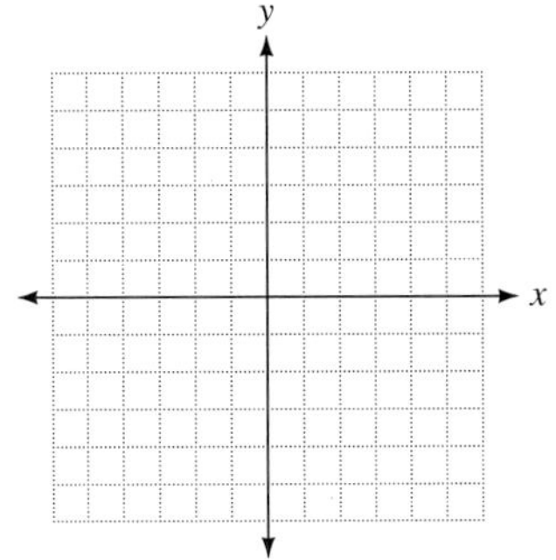

24. $f(x) = \sqrt{x + 4}$

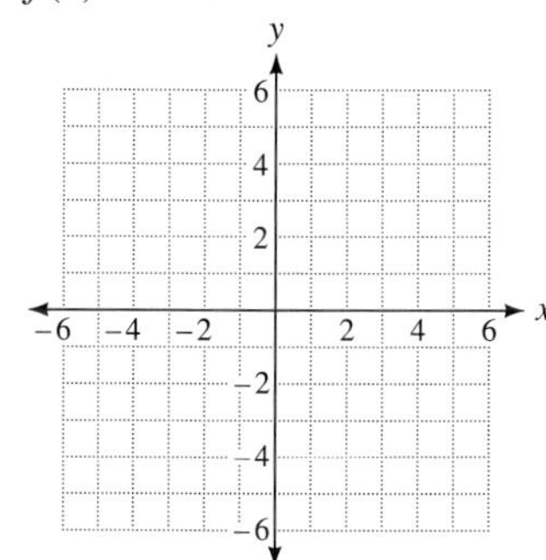

25. $f(x) = \sqrt{x - 2}$

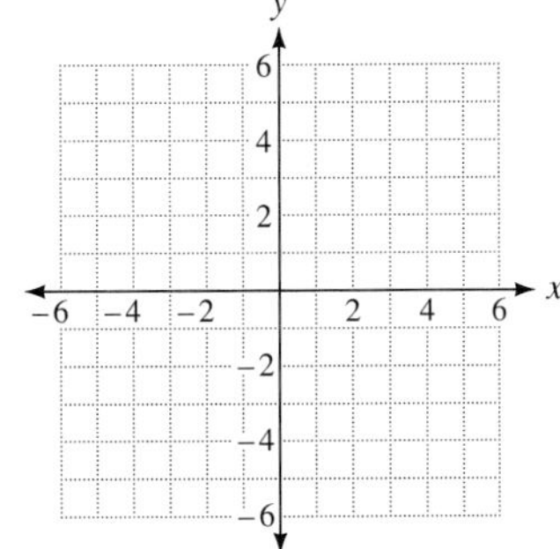

26. $f(x) = -\sqrt{x + 4}$

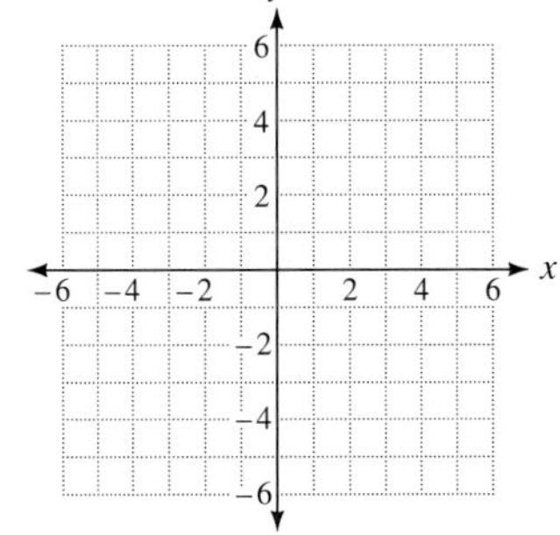

27. $f(x) = |x - 2|$

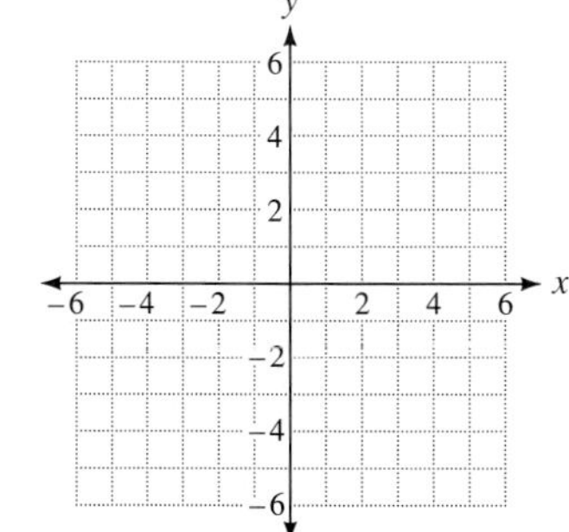

28. $f(x) = |x + 3|$

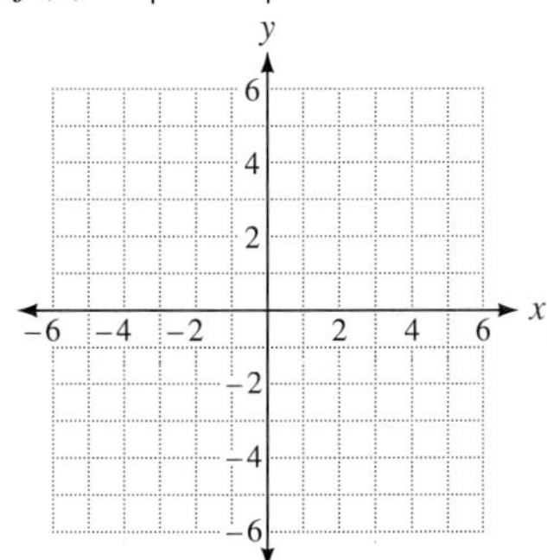

29. $f(x) = |x| + 2$

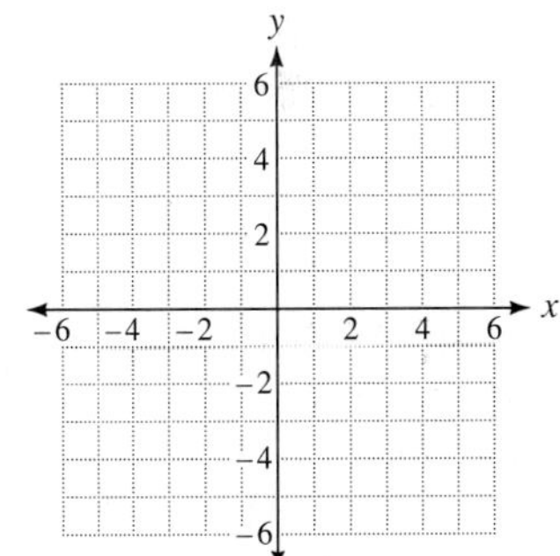

30. $f(x) = |x| - 2$

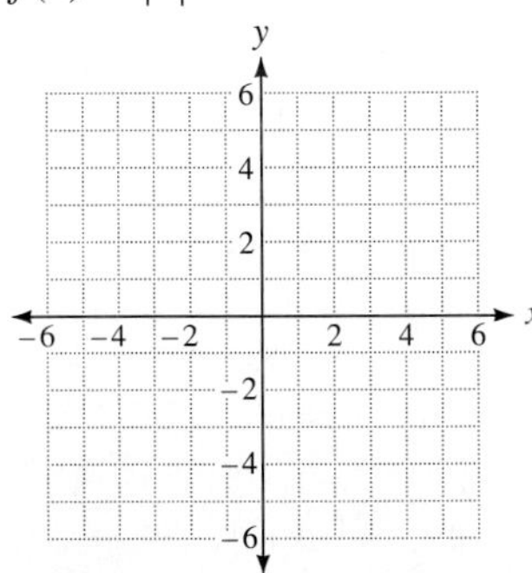

31. $f(x) = 2|x|$

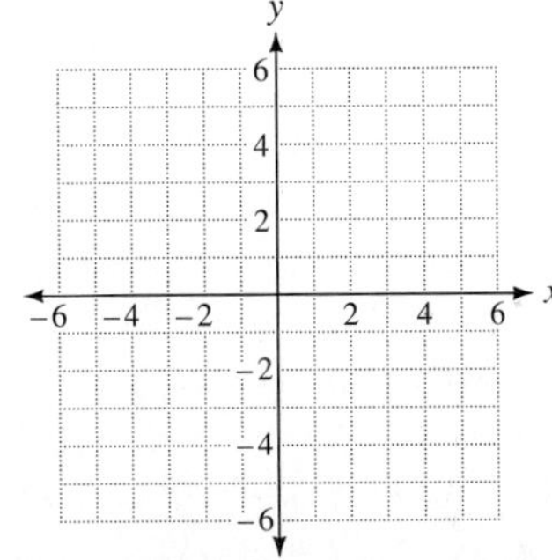

32. $f(x) = -|x - 1|$

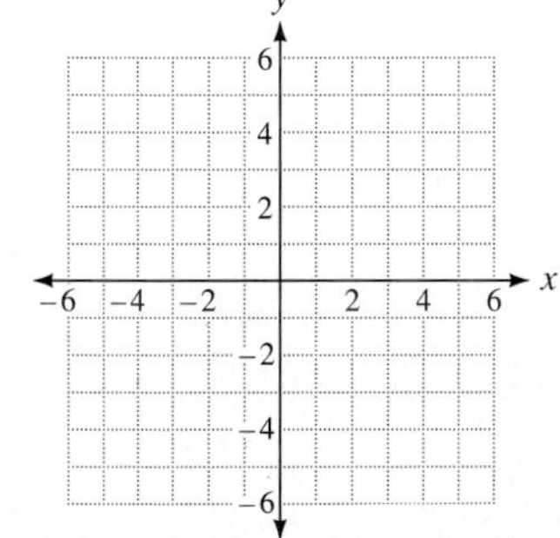

FOR REVIEW

Without graphing, tell the x-intercepts, the vertex, the line of symmetry, and whether the graph of each quadratic function opens up or down in Exercises 33–34.

33. $g(x) = 4x^2 - 25$

34. $f(x) = -3x^2 + 2x - 1$

Find the vertex, the x-intercepts, and the graph of each function in Exercises 35–36.

35. $f(x) = x^2 - x + 5$

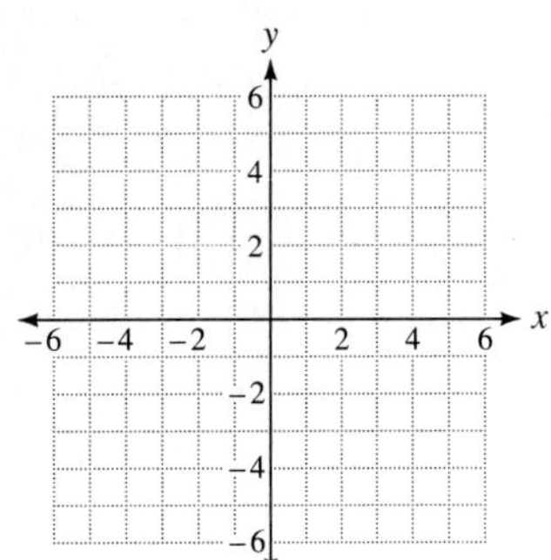

36. $g(x) = -x^2 + 2x + 2$

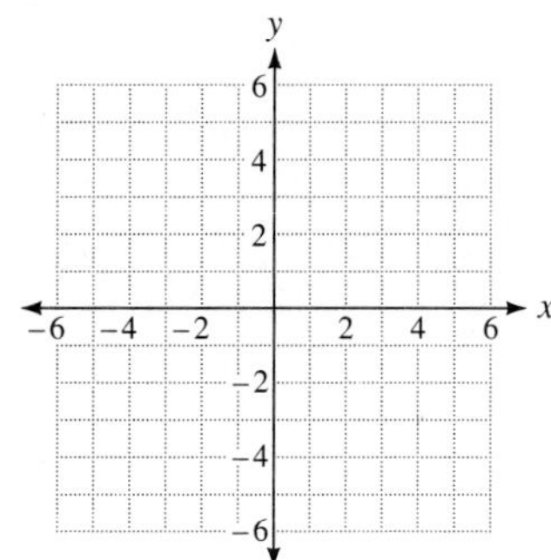

37. What is the minimum product of two numbers whose difference is 10? What are the numbers?

The following exercise reviews material from Chapter 3 to help you prepare for the next section.

38. Consider the relation $r = \{(1, 2), (2, 2), (3, 4), (4, 3)\}$.

(a) Graph the relation r.

(b) Is the relation r a function? Explain.

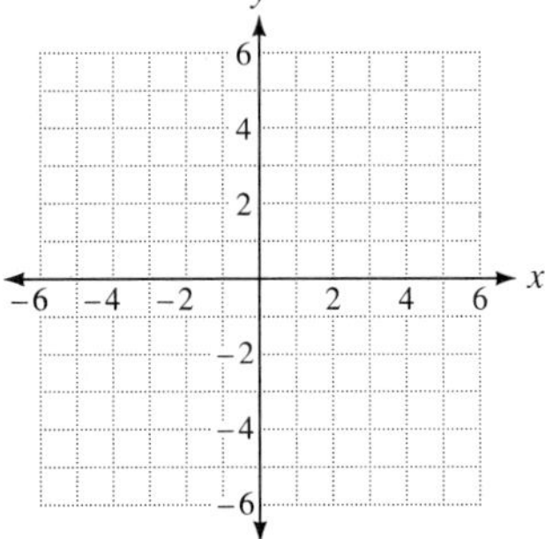

(c) Determine the relation s formed by interchanging the x and y coordinates of each ordered pair in r.

(d) Graph the new relation s.

(e) Is the new relation s a function? Explain.

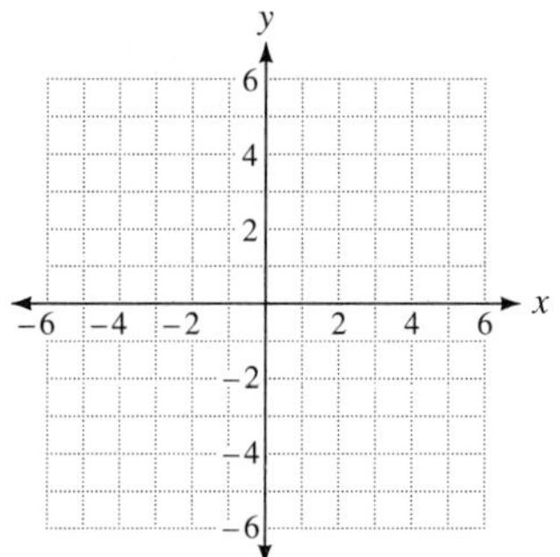

(f) Could you eliminate one ordered pair from r so that the relation s formed by interchanging the coordinates is a function?

9.2 EXERCISES C

1. Assume that $f(x) = 5x + 4$ and $g(x) = x^2 - x + 1$. Find **(a)** $f[g(x)]$, **(b)** $g[f(x)]$, and **(c)** $f[g(x + h)]$.

For each function in Exercises 2–4, evaluate **(a)** $f(x + h)$ **(b)** $f(x + h) - f(x)$ **(c)** $\frac{f(x + h) - f(x)}{h}$. *(Assume that $h \neq 0$.) Expressions like these are needed in calculus.*

2. $f(x) = x + 1$ **3.** $f(x) = x^2 - 1$ **4.** $f(x) = 5$

5. A tray is to be made from a rectangular piece of aluminum measuring 12 inches by 12 inches by cutting equal-sized squares from each corner and bending up the sides. Assume that the measure of the side of each square is x. Express the volume V of the tray as a function of x. Give the domain of this function. Graph the function on its domain and use the graph to estimate the size of the square to be cut from each corner to maximize the volume. What is the maximum volume?

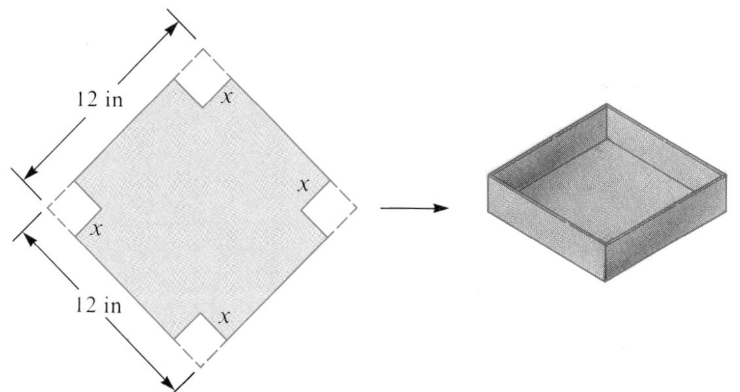

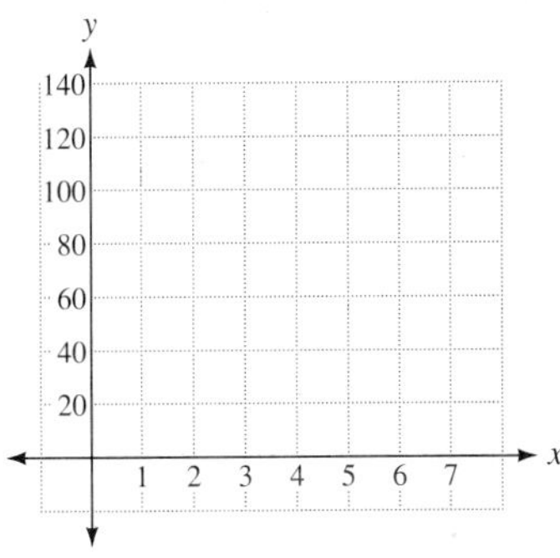

9.3 INVERSE FUNCTIONS

STUDENT GUIDEPOSTS

1. Inverse Relations
2. One-to-One Functions
3. Horizontal Line Test
4. Graphical Interpretation of Inverses
5. Finding Inverse Functions

1 INVERSE RELATIONS

Suppose we interchange the x and y coordinates of each ordered pair in a relation to obtain a new relation. The resulting relation and the original are called *inverses* of each other. For example, consider the relation

$$\{(1, 2), (2, 3), (3, 2)\}.$$

Then the inverse relation is

$$\{(2, 1), (3, 2), (2, 3)\}.$$

Using an arrow diagram, the given relation is

1 ⟶ 2
2 ⟶ 3
3 ↗ (to 2)

while the inverse relation is

2 ⟶ 1
3 ⟶ 2
2 ↘ 3.

The inverse could have been found by simply reversing the direction of each arrow.

Inverse of a Relation

If R is a relation, the **inverse** of R, denoted by R^{-1}, is given by

$$R^{-1} = \{(y, x) | (x, y) \in R\}.$$

The domain of R^{-1} is equal to the range of R, and the range of R^{-1} is equal to the domain of R.

CAUTION

The symbol for an inverse relation, R^{-1}, is read "R inverse." This symbol should not be confused with $\frac{1}{R} = R^{-1}$. We can tell them apart only by context.

EXAMPLE 1 FINDING INVERSE RELATIONS

Consider the relation $R = \{(0, 2), (1, 1), (2, 3), (3, 2)\}$. Find R^{-1}. Is R a function? Is R^{-1} a function?

To find R^{-1}, we interchange the coordinates of each pair in R. Thus,

$$R^{-1} = \{(2, 0), (1, 1), (3, 2), (2, 3)\}.$$

R is a function since no two ordered pairs have the same x-coordinate and different y-coordinates. The same is not true for R^{-1} since (2, 0) and (2, 3) are in R^{-1} with the same x-coordinate, 2, but different y-coordinates, 0 and 3.

PRACTICE EXERCISE 1

Consider the relation $S = \{(0, 1), (0, 4), (2, 3), (3, 4)\}$. Find S^{-1}. Is S a function? Is S^{-1} a function?

Answer: $S^{-1} = \{(1, 0), (4, 0), (3, 2), (4, 3)\}$; S is not a function. S^{-1} is not a function.

2 ONE-TO-ONE FUNCTIONS

It is clear from Example 1 that even though a relation is a function, its inverse need not be a function. For a relation to be a function, each element in the domain must correspond to exactly one element in the range. In order for the inverse of a function to be a function (not just a relation), the reverse must be true. Functions with this reverse property are called *one-to-one*.

One-to-One Functions

A function is a **one-to-one function** is each element in the range of the function corresponds to exactly one element in the domain.

EXAMPLE 2 FINDING ONE-TO-ONE FUNCTIONS

Tell whether each function is one-to-one.

(a) $f = \{(1, 2), (2, -1), (3, 0), (4, 3)\}$

The elements in the range of f are 2, -1, 0, and 3. Since 2 corresponds only to 1, -1 corresponds only to 2, 0 corresponds only to 3, 3 corresponds only to 4, f is one-to-one.

PRACTICE EXERCISE 2

Tell whether each function is one-to-one.

(a) $h = \{(1, 5), (2, 5), (3, 4)\}$

(b) $g = \{(1, 2), (2, 3), (3, 2)\}$
Since 2 in the range of g corresponds to both 1 *and* 3 in the domain, g is not one-to-one.

(b) $f = \{(2, 3), (3, 4), (4, 5)\}$
Answers: (a) h is not one-to-one. (b) f is one-to-one.

3 HORIZONTAL LINE TEST

It is easy to tell if a function is one-to-one by looking at its graph. Remember that for a relation to be a function, its graph must pass the vertical line test. Like this, for a function to be one-to-one, its graph must pass the **horizontal line test.**

Horizontal Line Test

If a horizontal line can be passed through the graph of a function in such a way that the line intersects two or more points on the graph, the function *is not* one-to-one.

The function $g = \{(1, 2), (2, 3), (3, 2)\}$ in Example 2(b) has its graph given in Figure 9.17(a). The horizontal line l through (1, 2) and (3, 2) shows that the function is not one-to-one.

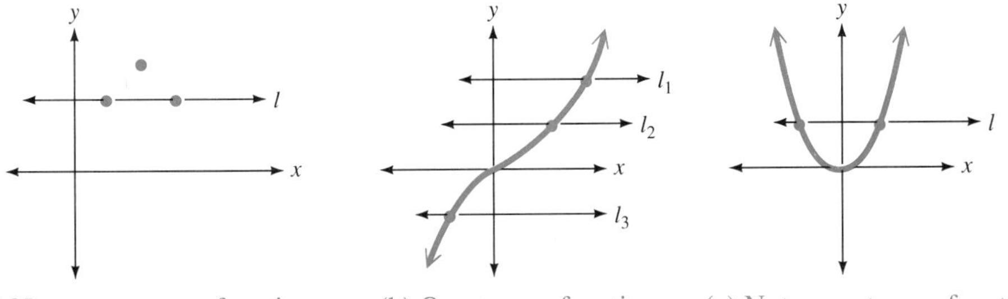

(a) Not a one-to-one function (b) One-to-one function (c) Not a one-to-one function
Figure 9.17 Horizontal Line Test

The function graphed in Figure 9.17(b) is one-to-one since no matter what horizontal line is passed through the graph, it will intersect the curve in no more than one point. On the other hand, the function graphed in Figure 9.17(c) is not one-to-one since the line l intersects the graph in two points.

4 GRAPHICAL INTERPRETATION OF INVERSES

One way to tell if two functions f and g are inverses is to look at their graphs. The function g is the inverse of function f provided the graph of g is symmetric to the graph of f with respect to the line $y = x$. Two points that are symmetric with respect to the line $y = x$ have coordinates of the form (a, b) and (b, a), as in Figure 9.18.

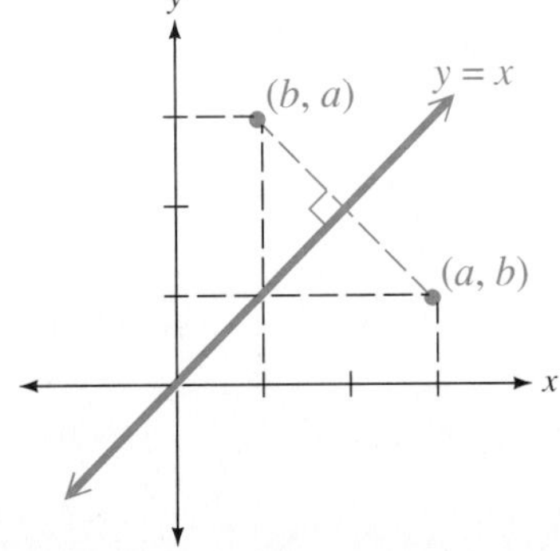

Figure 9.18 Symmetry with Respect to $y = x$

EXAMPLE 3 Graphing Inverse Functions

Graph $f(x) = x^2$ $(x \geq 0)$ and $g(x) = \sqrt{x}$, and consider symmetry with respect to $y = x$.

A careful check of the graphs of these functions in Figure 9.19 shows symmetry with respect to $y = x$. Thus, f and g are inverses.

x	$f(x)$
0	0
$\frac{1}{2}$	$\frac{1}{4}$
1	1
2	4

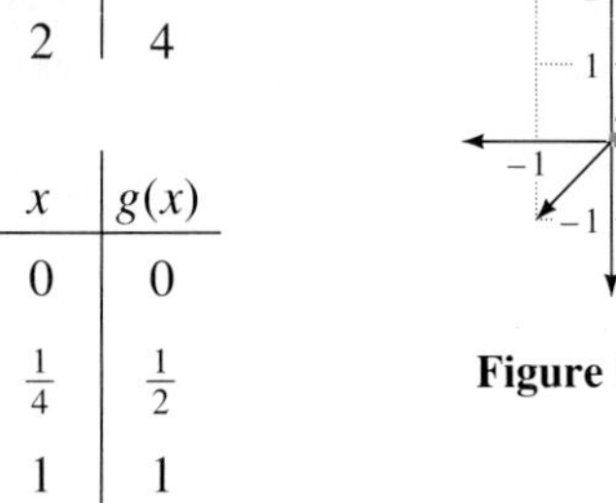

x	$g(x)$
0	0
$\frac{1}{4}$	$\frac{1}{2}$
1	1
4	2

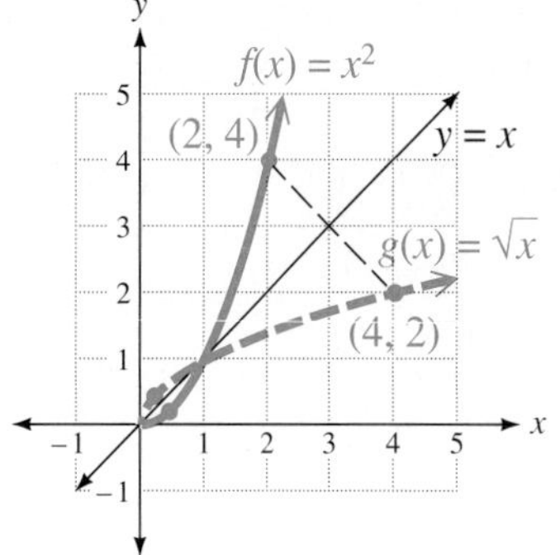

Figure 9.19 Inverse Functions

The reason for restricting the domain of $f(x) = x^2$ to $x \geq 0$ is to force the function to be one-to-one, so that its inverse exists as a function. (The graph of $f(x) = x^2$ for all real numbers x, given in Figure 9.17(c), shows that without this restriction the function is not one-to-one.)

Practice Exercise 3

Consider the graphs of $f(x) = 3x$ and $g(x) = \frac{1}{3}x$ shown below. Are f and g inverse functions?

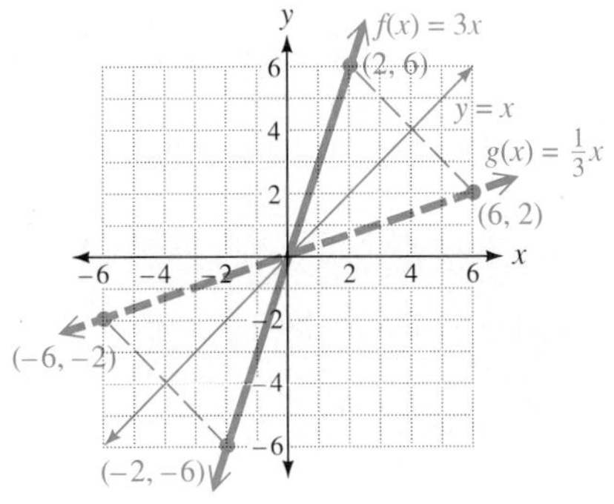

Answer: yes

5 FINDING INVERSE FUNCTIONS

Notice that in Example 3, the points plotted for the inverse function pairs involve an interchange of the x and y coordinates. For example, (2, 4) is on $f(x) = x^2$ while (4, 2) is on $g(x) = \sqrt{x}$. This observation leads to a method for finding the equation of the inverse of a given function.

Consider the function $f(x) = 3x$, and write it as $y = 3x$. If we interchange x and y and solve for y, we obtain $y = \frac{1}{3}x$.

$$y = 3x$$

$$x = 3y \qquad \text{Interchange } x \text{ and } y$$

$$\frac{1}{3}x = y \qquad \text{Solve for } y$$

Also, if we write $y = x^2$, assume that $x \geq 0$ (to make the function one-to-one) as in Example 3, and follow the same procedure, we obtain $y = \sqrt{x}$.

$$y = x^2$$

$$x = y^2 \qquad \text{Interchange } x \text{ and } y$$

$$\sqrt{x} = y \qquad \text{Take square root of both sides; remember } x \text{ and } y \text{ are nonnegative}$$

These examples illustrate the method for finding the equation of the inverse of a given one-to-one function.

Finding the Inverse of a Function

To find the inverse of a one-to-one function f defined by $y = f(x)$, interchange x and y, and then solve this new equation for y. The inverse of f is denoted by $y = f^{-1}(x)$.

EXAMPLE 4 Finding the Inverse of a Function

Find the inverse of each function.

(a) $f(x) = 3x - 5$.

$y = 3x - 5$ Replace $f(x)$ with y

$x = 3y - 5$ Interchange x and y

$x - 5 = 3y$ Solve for y

$\frac{x-5}{3} = y$ The defining relation for the inverse of f

Thus, $$f^{-1}(x) = \frac{1}{3}x - \frac{5}{3}.$$

(b) $g(x) = x^3 + 2$.

$y = x^3 + 2$ $y = g(x)$

$x = y^3 + 2$ Interchange x and y

$x - 2 = y^3$ Solve for y

$\sqrt[3]{x-2} = y$ Take cube root of both sides

Thus, $g^{-1}(x) = \sqrt[3]{x-2}$.

Practice Exercise 4

Find the inverse of each function.

(a) $f(x) = \frac{1}{2}x + \frac{3}{2}$

(b) $g(x) = 4 - 8x^3$

Answers: (a) $f^{-1}(x) = 2x - 3$ (b) $g^{-1}(x) = \frac{1}{2}\sqrt[3]{4-x}$

9.3 EXERCISES A

Find the inverse of each relation. Is the original relation a function? Is the inverse relation also a function?

1. $\{(0, 1), (1, 2)\}$

2. $\{(-1, 1), (0, 1), (1, 1)\}$

3. $\{(-3, 1), (-2, 2), (-1, 0), (1, 5), (2, 4)\}$

4. $\{(0, 1), (0, 2), (1, 5), (2, 7)\}$

5. $\{(1, 5), (1, -5), (2, 5), (2, -5)\}$

6. $\{(2, -1), (3, -1), (4, -1), (5, -1), (6, -1)\}$

Tell whether each graph is the graph of a one-to-one function.

7. 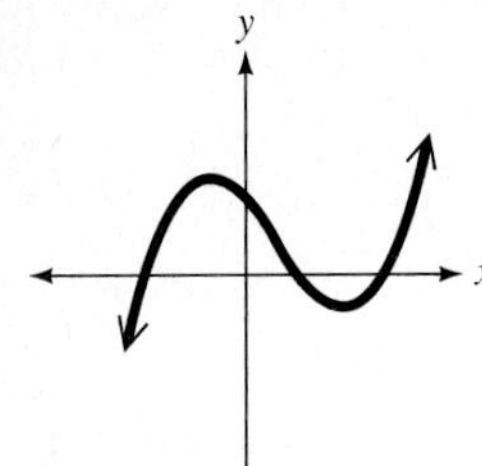

8. 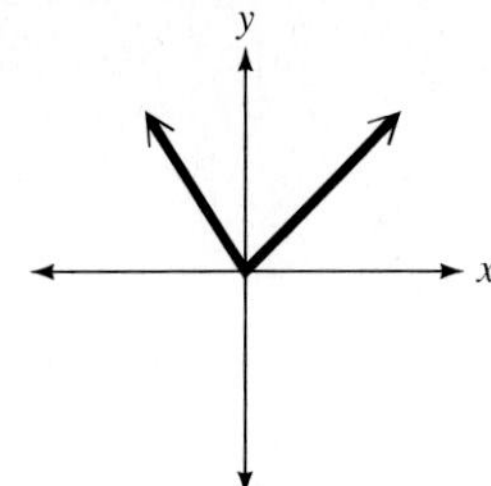

9. 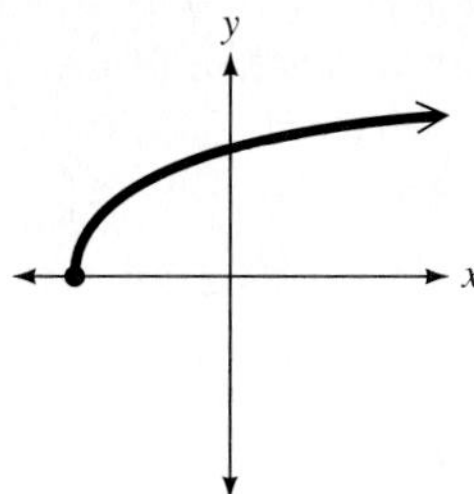

10. 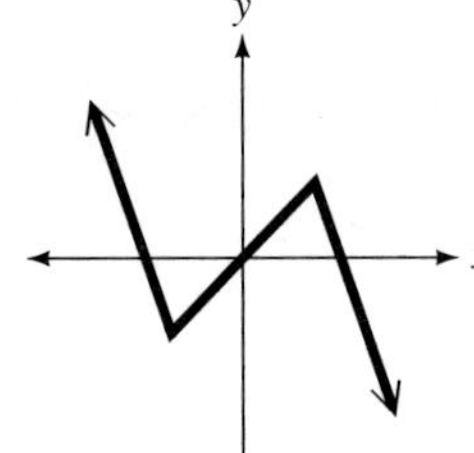

11. 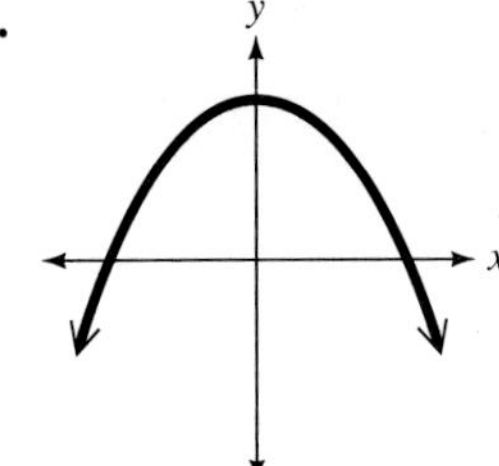

12. 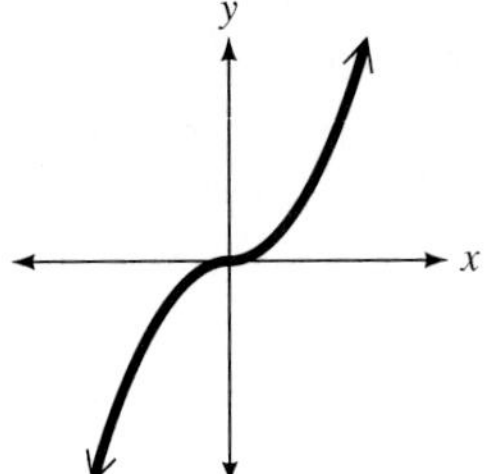

Graph each pair of functions in the same coordinate system. Are the functions inverses of each other?

13. $f(x) = 2x$ and $g(x) = -2x$

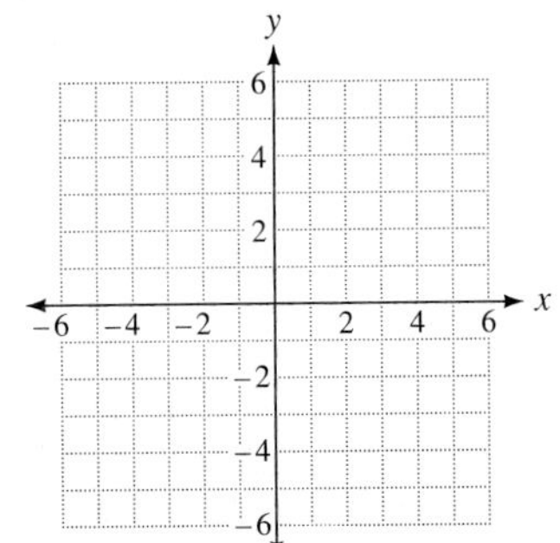

14. $f(x) = \sqrt{x - 2}$ and $g(x) = x^2 + 2,\ x \geq 0$

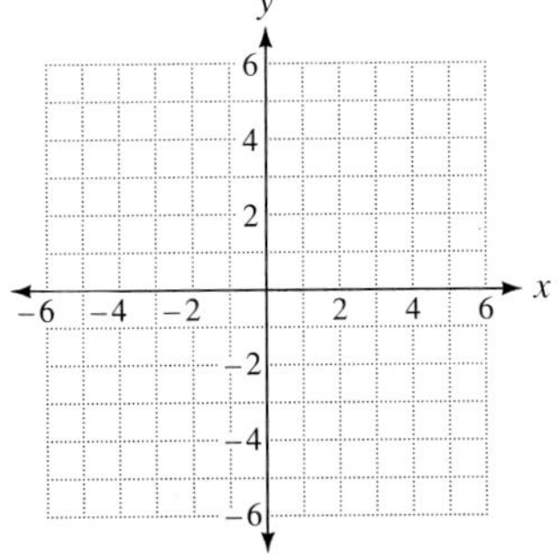

Without graphing, find the inverse of each function.

15. $f(x) = 2x + 3$

16 $f(x) = 2x^3$

17. $f(x) = \frac{1}{3}x + 2$

18. $f(x) = -x$

19. $f(x) = 2$

20 $g(x) = x^2 + 1,\ x \geq 0$

In Exercises 21–22, the graph of a function is given. Sketch the graph of its inverse on the same grid.

21.

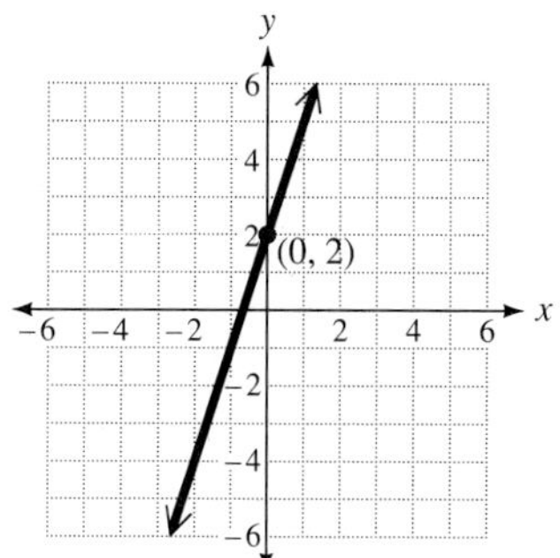

22.

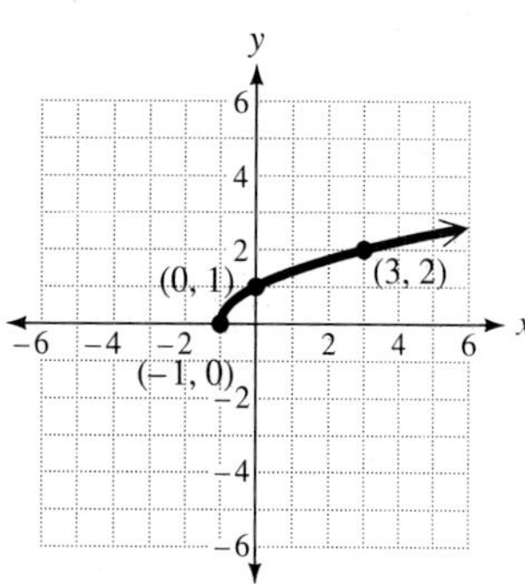

23. Consider the function $f(x) = 2x + 1$. Find the equation for the inverse of f. Use these two functions to find the following.

(a) $f(3)$ **(b)** $f^{-1}(3)$ **(c)** $f[f^{-1}(3)]$ **(d)** $f^{-1}[f(3)]$

24. Consider the function $g(x) = 8x^3$. Find the equation for the inverse of g. Use these two functions to find the following.

(a) $g(-2)$ **(b)** $g^{-1}(-2)$ **(c)** $g[g^{-1}(-2)]$ **(d)** $g^{-1}[g(-2)]$

FOR REVIEW

Sketch the graph of each function.

25. $f(x) = |x - 3|$

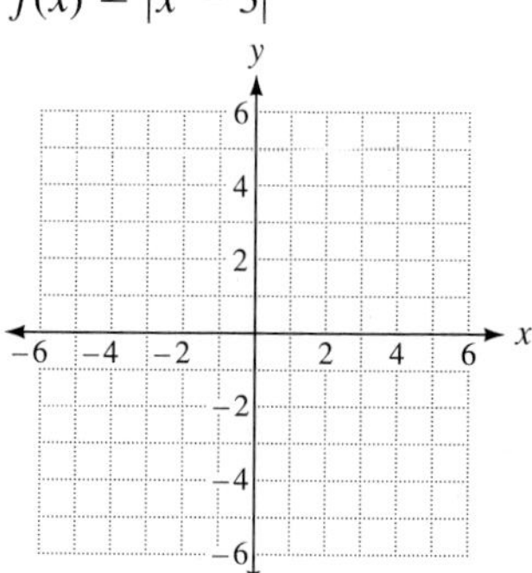

26. $f(x) = |x| - 3$

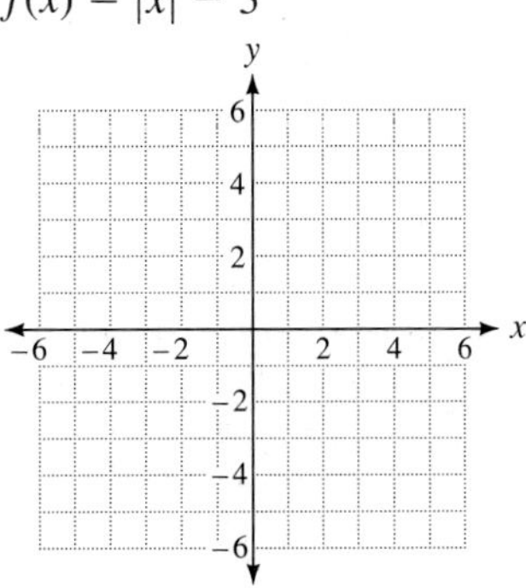

27. $f(x) = \sqrt{x + 4}$

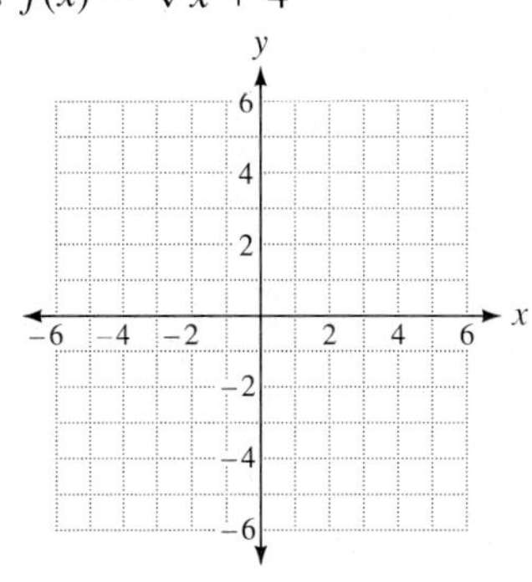

Exercises 28–30 will help you prepare for the material in Chapter 10. Give the value of x in each equation.

28. $x^2 = 25$ **29.** $2^x = 4$ **30.** $3^x = \frac{1}{9}$

ANSWERS: 1. {(1, 0), (2, 1)}; yes; yes 2. {(1, −1), (1, 0), (1, 1)}; yes; no 3. {(1, −3), (2, −2), (0, −1), (5, 1), (4, 2)}; yes; yes 4. {(1, 0), (2, 0), (5, 1), (7, 2)}; no; yes 5. {(5, 1), (−5, 1), (5, 2), (−5, 2)}; no; no 6. {(−1, 2), (−1, 3), (−1, 4), (−1, 5), (−1, 6)}; yes; no 7. not one-to-one 8. not one-to-one 9. is one-to-one 10. not one-to-one 11. not one-to-one 12. is one-to-one

13. no;

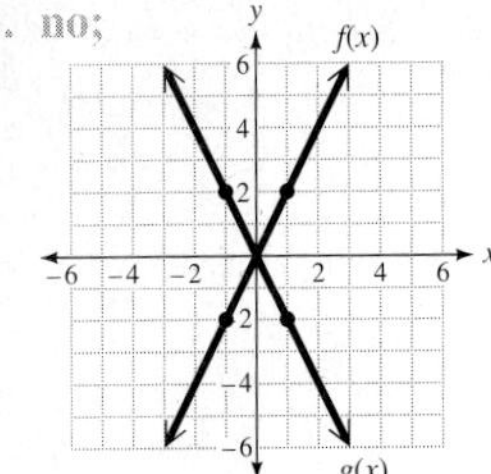

14. yes;

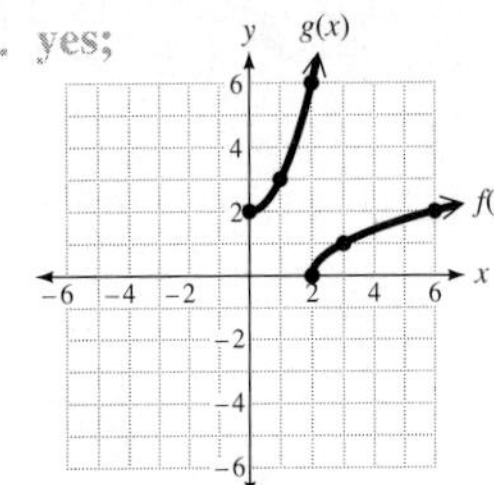

15. $f^{-1}(x) = -\frac{1}{2}x - \frac{3}{2}$ 16. $f^{-1}(x) = \sqrt[3]{\frac{x}{2}}$ 17. $f^{-1}(x) = 3x - 6$ 18. $f^{-1}(x) = -x$ 19. $f^{-1}(x)$ does not exist since $f(x) = 2$ is not one-to-one 20. $g^{-1}(x) = \sqrt{x - 1}$

21.

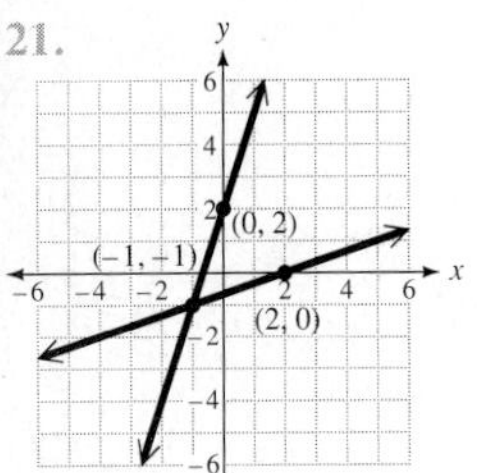

22.

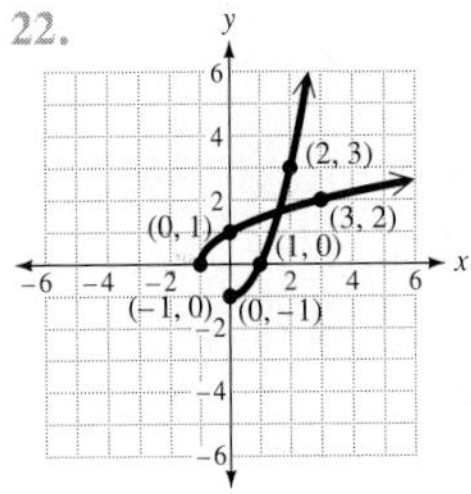

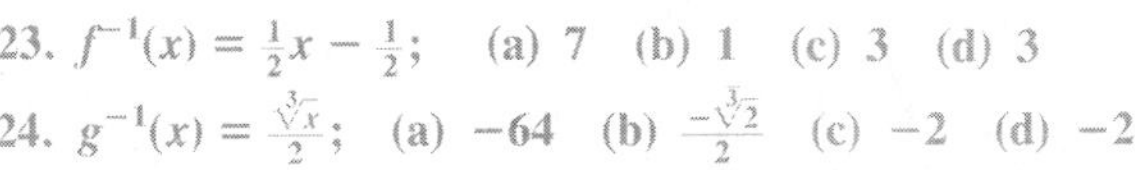

23. $f^{-1}(x) = \frac{1}{2}x - \frac{1}{2}$; (a) 7 (b) 1 (c) 3 (d) 3

24. $g^{-1}(x) = \frac{\sqrt[3]{x}}{2}$; (a) -64 (b) $\frac{-\sqrt[3]{2}}{2}$ (c) -2 (d) -2

25.

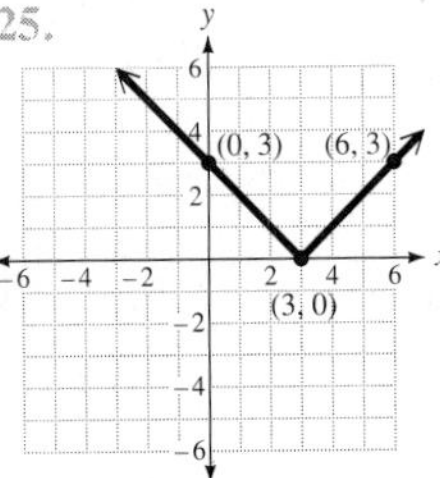

26.

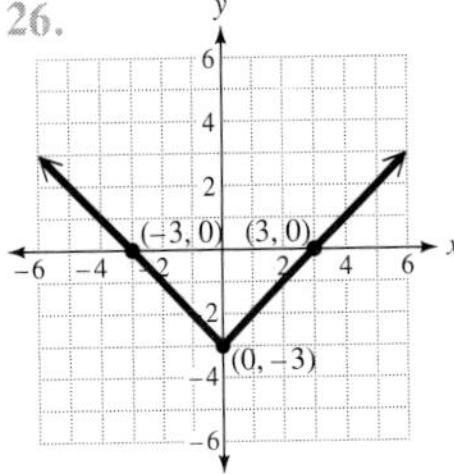

27.

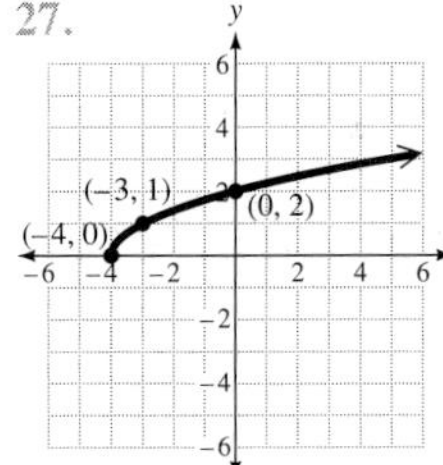

28. 5 or -5 29. 2 30. -2

9.3 EXERCISES B

Find the inverse of each relation. Is the original relation a function? Is the inverse relation also a function?

1. $\{(2, 1), (5, 4)\}$

2. $\{(5, 0), (10, 0), (20, 0)\}$

3. $\{(3, -1), (2, 5), (1, 7), (0, 8), (-1, 4)\}$

4. $\{(-1, 0), (-1, 1), (-1, 2), (-1, 3)\}$

5. $\{(-3, 0), (-3, 2), (0, -2), (2, -2)\}$

6. $\{(-2, -2), (-1, -1), (0, 0), (1, 1), (2, 2)\}$

Tell whether each graph is the graph of a one-to-one function.

7.

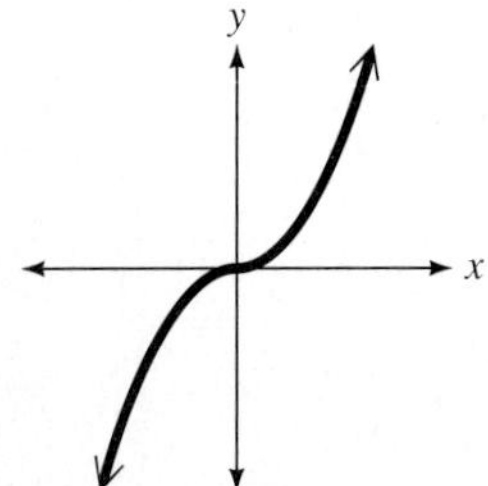

8.

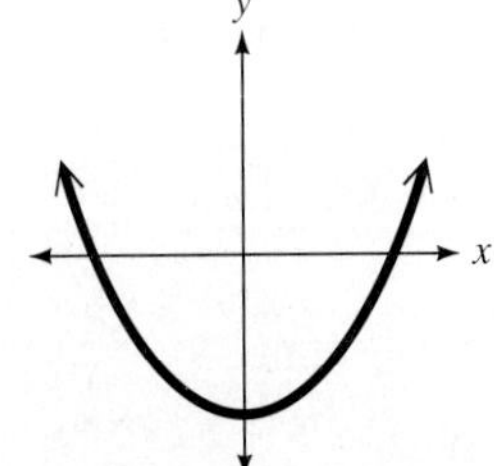

9.

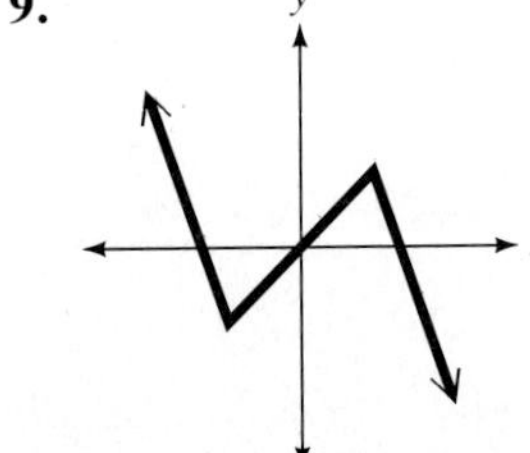

10.

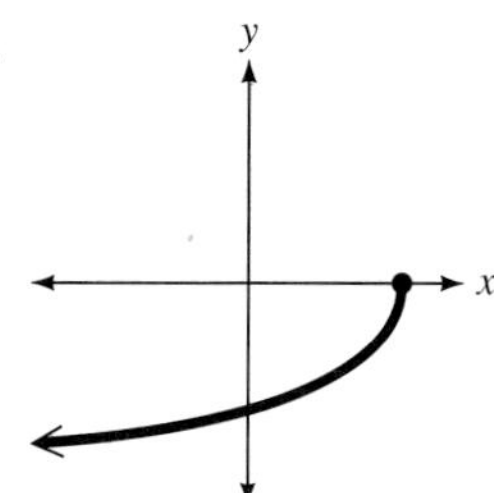

11.

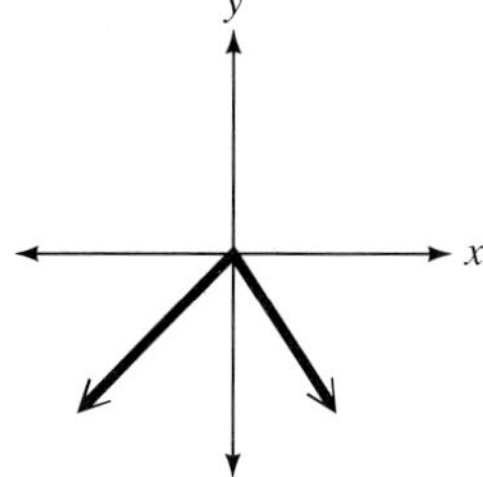

12.

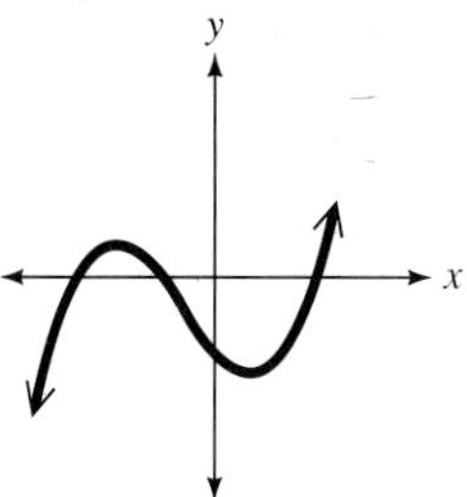

Graph each pair of functions in the same coordinate system. Are the functions inverses of each other?

13. $f(x) = -3x$ and $g(x) = 3x$

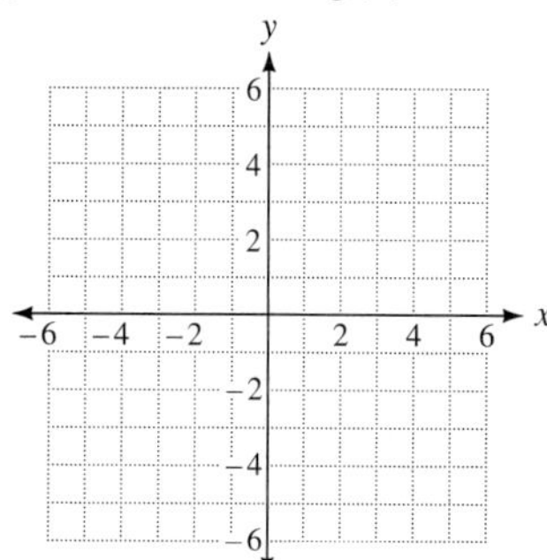

14. $f(x) = \sqrt{x + 4}$ and $g(x) = x^2 - 4,\ x \geq 0$

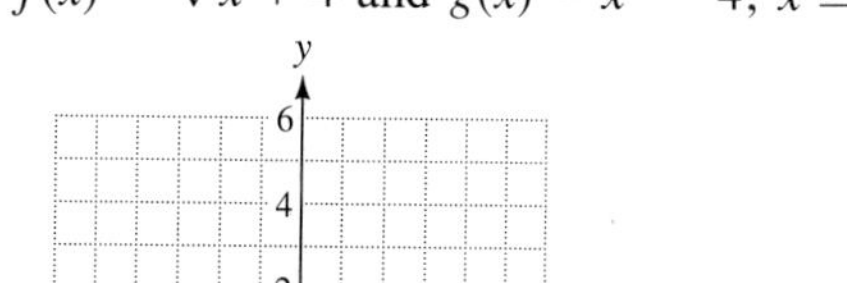

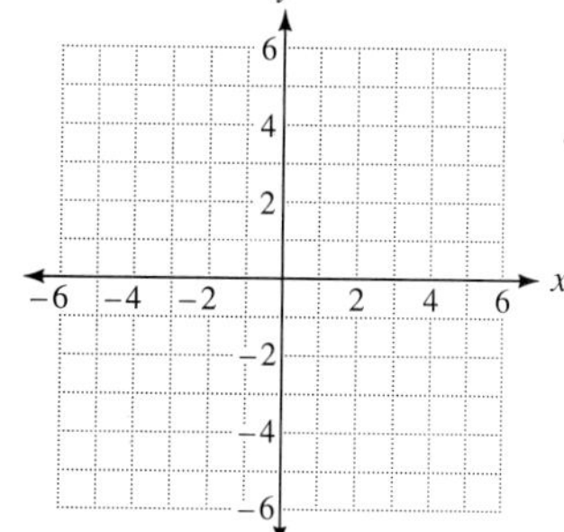

Without graphing, find the inverse of each function.

15. $g(x) = \dfrac{1}{2}x - 1$

16. $g(x) = -x^3$

17. $g(x) = 5x - 6$

18. $g(x) = x$

19. $f(x) = -3$

20. $f(x) = -x^2 + 1,\ x \geq 0$

In Exercises 21–22, the graph of a function is given. Sketch the graph of its inverse on the same grid.

21.

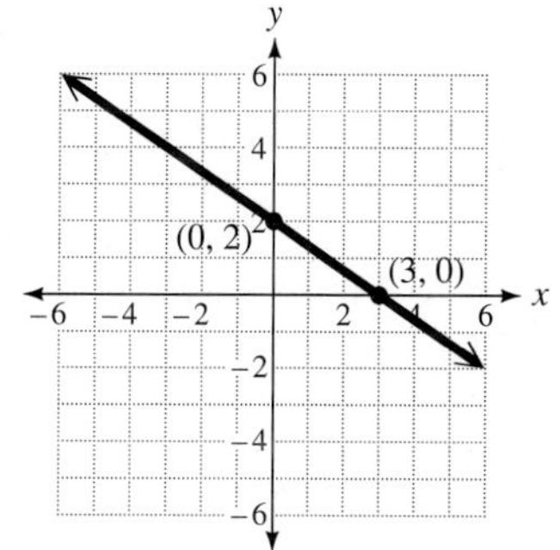

22.

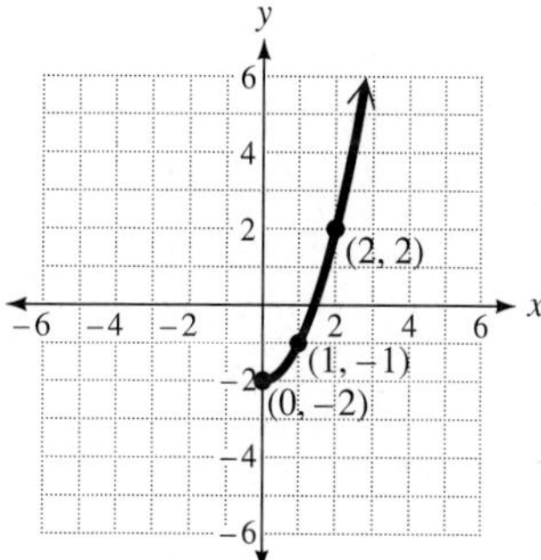

23. Consider the function $g(x) = \frac{1}{3}x - 1$. Find the equation for the inverse of g. Use these two functions to find the following.

(a) $g(-2)$ **(b)** $g^{-1}(-2)$ **(c)** $g[g^{-1}(-2)]$ **(d)** $g^{-1}[g(-2)]$

24. Consider the function $f(x) = -8x^3$. Find the equation for the inverse of f. Use these two functions to find the following.

(a) $f(-1)$ **(b)** $f^{-1}(-1)$ **(c)** $f[f^{-1}(-1)]$ **(d)** $f^{-1}[f(-1)]$

FOR REVIEW

Sketch the graph of each function.

25. $f(x) = |x - 4|$

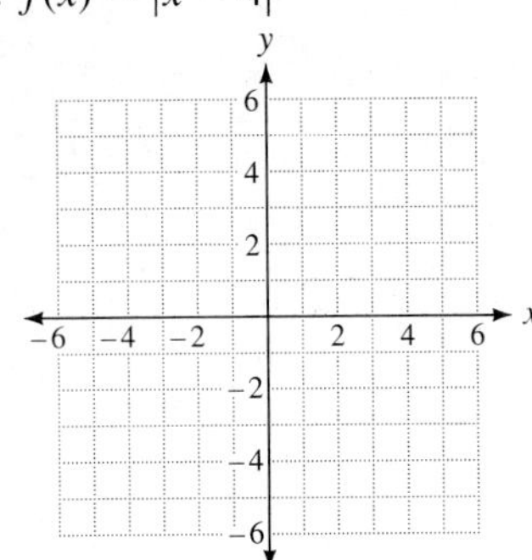

26. $f(x) = x^3 - 16x$

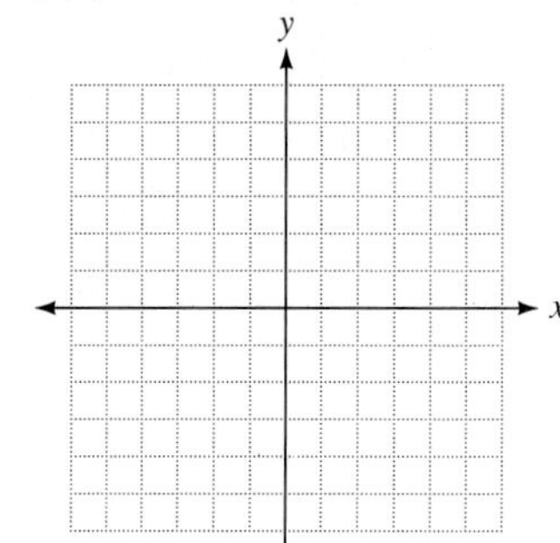

27. $f(x) = -\sqrt{x + 4}$

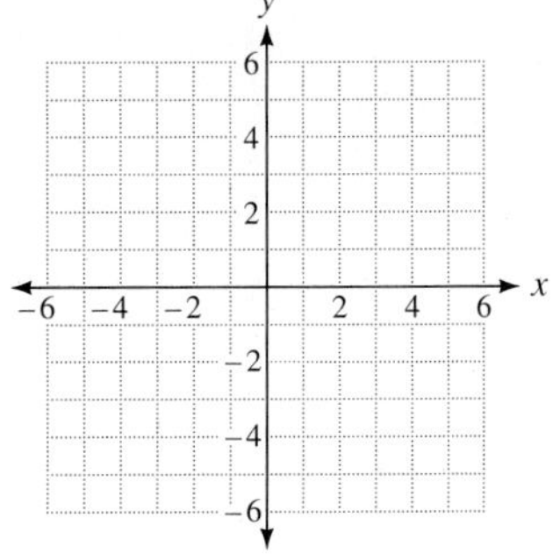

Exercises 28–30 will help you prepare for the material in Chapter 10. Find the value of x in each equation.

28. $x^2 = 9$ **29.** $3^x = 27$ **30.** $2^x = \frac{1}{8}$

9.3 EXERCISES C

A function f is called ***even*** *if $f(x) = f(-x)$ for every x in the domain of f, and* ***odd*** *if $f(x) = -f(-x)$ for every x in the domain of f. Tell whether each function is even, odd, or neither.*

1. $f(x) = x$ **2.** $f(x) = x^2$ **3.** $f(x) = x^3$

4. $f(x) = 2x^3 - x$ **5.** $f(x) = x^2 - x$ **6.** $f(x) = 5$

7. The function $f(x) = [x]$, where $[x]$ is the largest integer n such that $n \le x$, is called the **greatest integer function.** For example, $\left[\frac{1}{2}\right] = 0$, $[1] = 1$, $[1.5] = 1$, $[2] = 2$, etc. Sketch the graph of the greatest integer function. Do you see why this function is sometimes called the *stairstep function?*

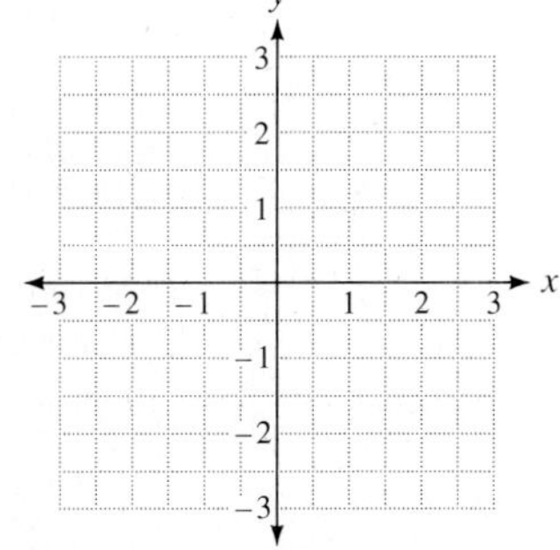

CHAPTER 9 REVIEW

KEY WORDS

9.1 A function defined by the equation $f(x) = ax^2 + bx + c$, $a \neq 0$, is a **quadratic function.**

The graph of a quadratic function is a **parabola.**

The high point or low point on a parabola is its **vertex.**

The line through the vertex of a parabola about which the curve is symmetric is called the **line of symmetry** of the parabola.

9.2 If f and g are two functions, the expression $f[g(x)]$ is the **composition of** f **with** g, and $g[f(x)]$ is the **composition of** g **with** f.

A **polynomial function** of degree n is a function defined by $f(x) = a_nx^n + a_{n-1}x^{n-1} + \ldots + a_1x + a_0$, $a_n \neq 0$.

The **square root function** is defined by $f(x) = \sqrt{x}$, $x \geq 0$.

The **absolute value function** is defined by $f(x) = |x|$.

9.3 The **inverse** of a relation R is $R^{-1} = \{(y, x) | (x, y) \in R\}$.

A function is **one-to-one** if each element in the range corresponds to exactly one element in the domain.

KEY CONCEPTS

9.1 Consider the quadratic function $f(x) = ax^2 + bx + c$, $a \neq 0$, whose graph is a parabola.

1. If $a > 0$ the parabola opens up; if $a < 0$ the parabola opens down.
2. The x-coordinate of the vertex is $-\frac{b}{2a}$, and the y-coordinate is $f\left(-\frac{b}{2a}\right)$.
3. The line of symmetry is the vertical line $x = -\frac{b}{2a}$.
4. If the parabola opens up and the vertex is above the x-axis, there are no x-intercepts. Likewise, if the parabola opens down and the vertex is below the x-axis there are no x-intercepts.
5. When there are x-intercepts, they can be found by solving the quadratic equation $ax^2 + bx + c = 0$.
6. The y-intercept is easy to find and has coordinates $(0, c)$.
7. Some problems that require finding a maximum or minimum value of a quadratic expression can be solved by finding the vertex of the parabola.

9.2

1. When a function is defined by an equation, the domain is the largest set of real numbers for which the expression in x is defined. This means that values of x that make a denominator zero or a radicand negative must be excluded.
2. If f and g are two functions,
   (a) $(f + g)(x) = f(x) + g(x)$
   (b) $(f - g)(x) = f(x) - g(x)$
   (c) $(fg)(x) = f(x) \cdot g(x)$
   (d) $\left(\frac{f}{g}\right)(x) = \frac{f(x)}{g(x)}$, $g(x) \neq 0$.
3. The graph of every square root function is half of a parabola.
4. The graph of every absolute value function is a V-shaped curve. Do not confuse the graph of a quadratic function (a U-shaped parabola) with the graph of an absolute value function.

9.3

1. To find the inverse of a relation given as a set of ordered pairs, form the set of ordered pairs obtained by interchanging the x- and y-coordinates.
2. The horizontal line test is used to tell whether a function is one-to-one in the same way the vertical line test is used to tell whether a relation is a function.
3. If two functions are inverses, their graphs are symmetric with respect to the line $y = x$.
4. To find the inverse of a function f defined by the equation $y = f(x)$, interchange x and y and solve the resulting equation for y to obtain the equation of the inverse, $y = f^{-1}(x)$.

REVIEW EXERCISES

Part I

9.1 *Sketch the graph of each function by sliding the graph of $f(x) = x^2$ or $g(x) = -x^2$ up or down.*

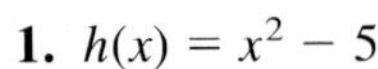

1. $h(x) = x^2 - 5$

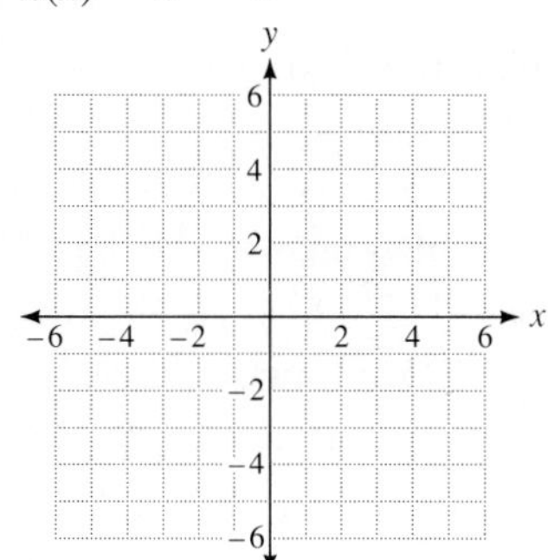

2. $h(x) = -x^2 + 4$

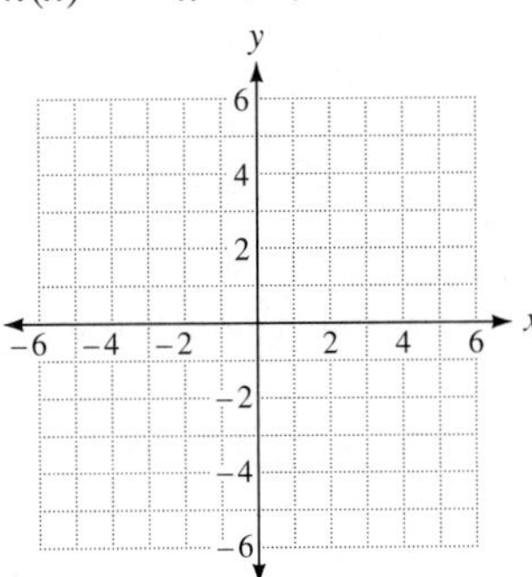

Without graphing, give the x-intercepts (if they exist) and the vertex, and tell whether the graph of each quadratic function opens up or down.

3. $f(x) = x^2 + 7x - 30$

4. $f(x) = -x^2 + 2x - 5$

Find the x-intercepts (if they exist), the vertex, and graph of the function.

5. $f(x) = x^2 - 2x - 3$

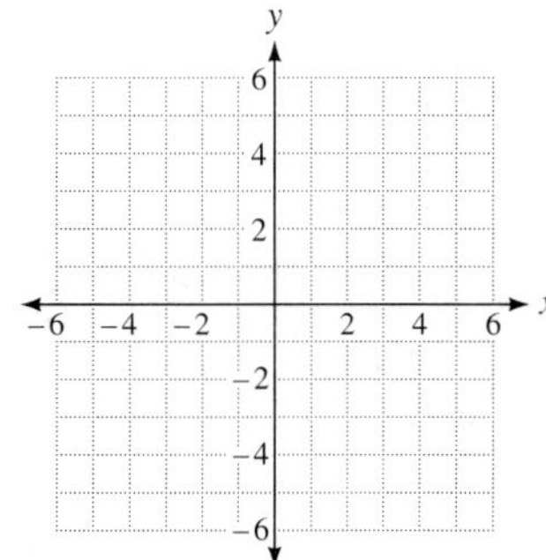

6. $g(x) = -x^2 + 4x - 5$

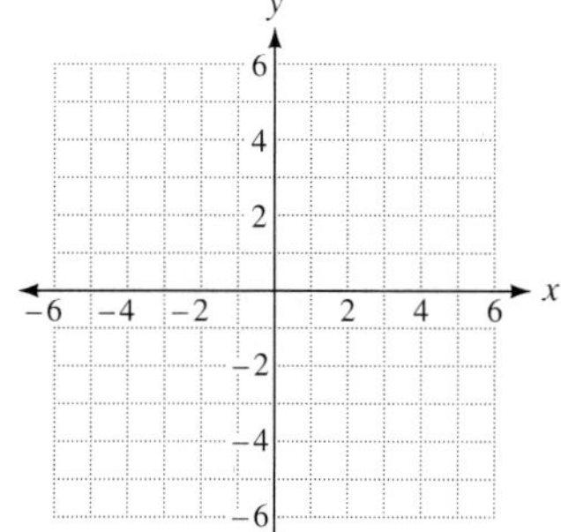

7. Jack Pritchard is building a rectangular room with a fixed perimeter of 72 ft. What dimensions would give the maximum area? What is the maximum area?

9.2 *Find the domain of the function defined by each equation.*

8. $y = \dfrac{3}{x - 8}$

9. $y = \sqrt{2x - 1}$

10. $f(x) = \dfrac{7}{\sqrt{2x - 1}}$

Assume that $f(x) = 2 - x$ and $g(x) = x^2 - 2$. Find each function value.

11. $(f + g)(3)$

12. $(f - g)(-2)$

13. $(fg)(1)$

14. $\left(\frac{f}{g}\right)(-1)$

15. $f[g(2)]$

16. $g[f(-3)]$

Make a table of values and sketch the graph of each function.

17. $f(x) = x^3 + 1$

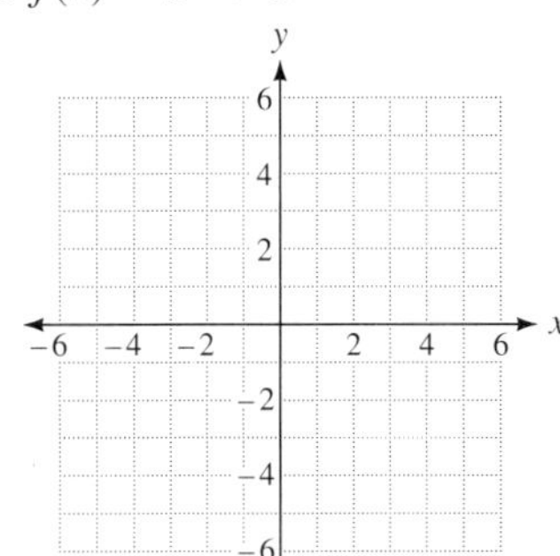

18. $f(x) = -\sqrt{x + 6}$

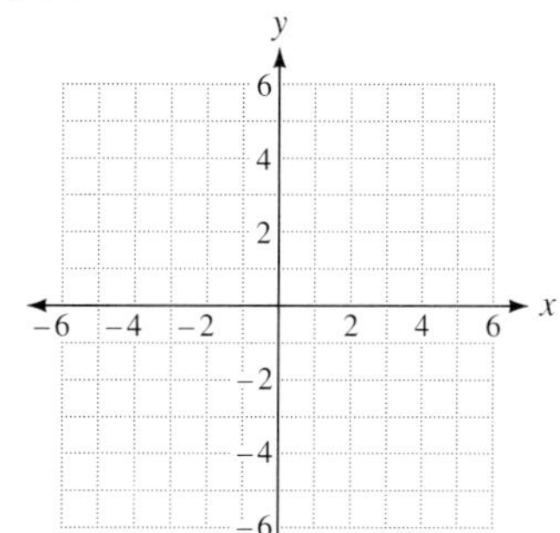

19. $f(x) = |x - 5|$

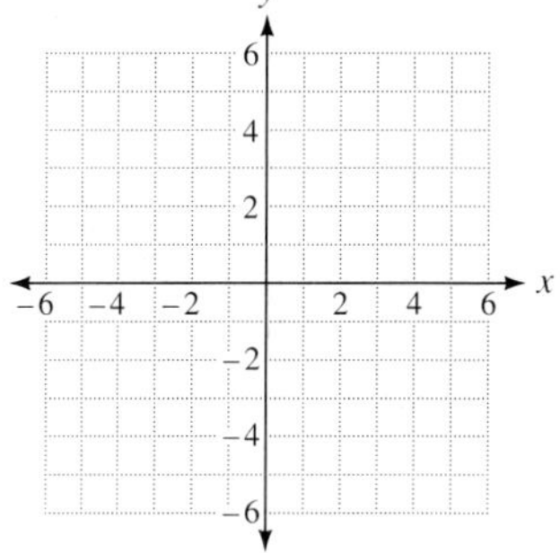

9.3 **20.** Find the inverse of the relation $\{(-2, 5), (0, 3), (1, 3)\}$. Is the original relation a function? Is the inverse a function?

21. Is this the graph of a one-to-one function?

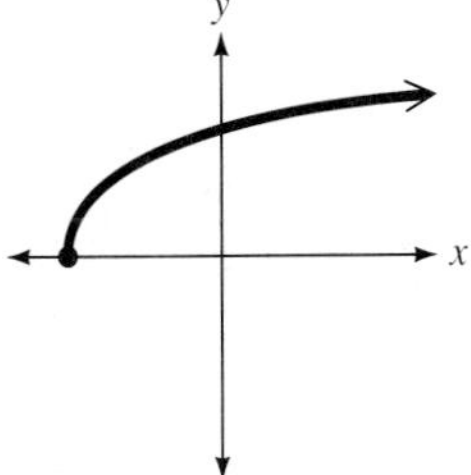

22. Graph the functions $f(x) = \sqrt{x - 3}$ and $g(x) = x^2 + 3,\ x \geq 0$. Are these inverses of each other?

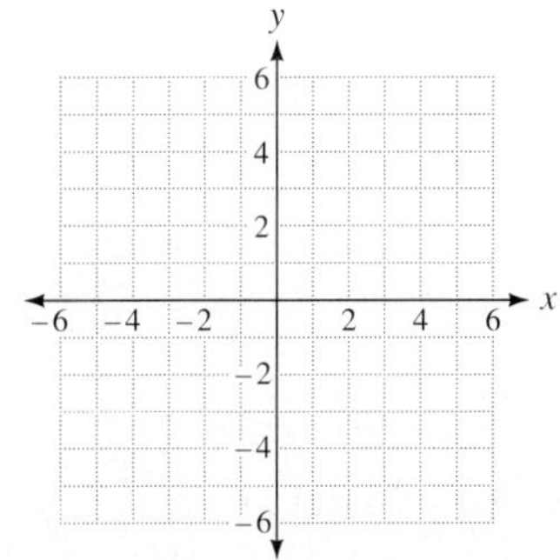

23. Consider the function $f(x) = 4x - 1$. Find the equation for the inverse of f. Use these to find the following.

(a) $f(-2)$
(b) $f^{-1}(-2)$
(c) $f[f^{-1}(-2)]$
(d) $f^{-1}[f(-2)]$

Part II

Graph each function and tell whether it is one-to-one.

24. $f(x) = 2x^2 - 5$

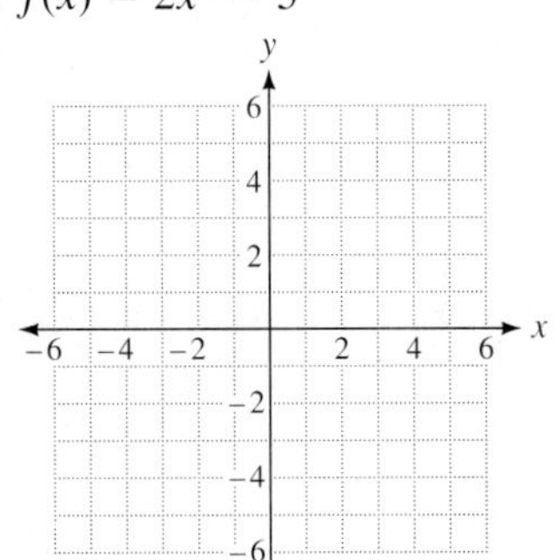

25. $f(x) = |x| - 6$

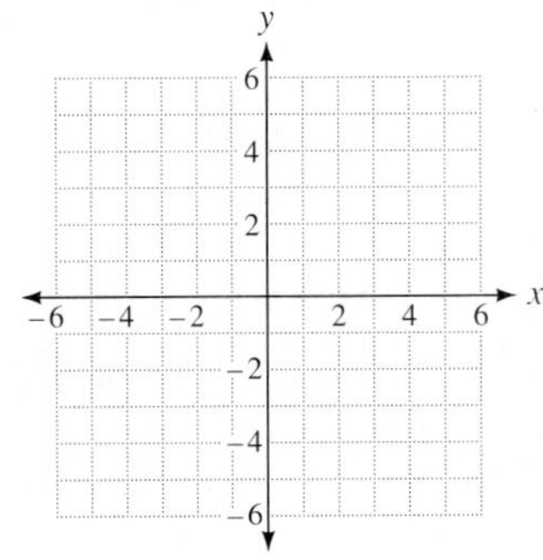

26. $f(x) = \sqrt{x + 5}$

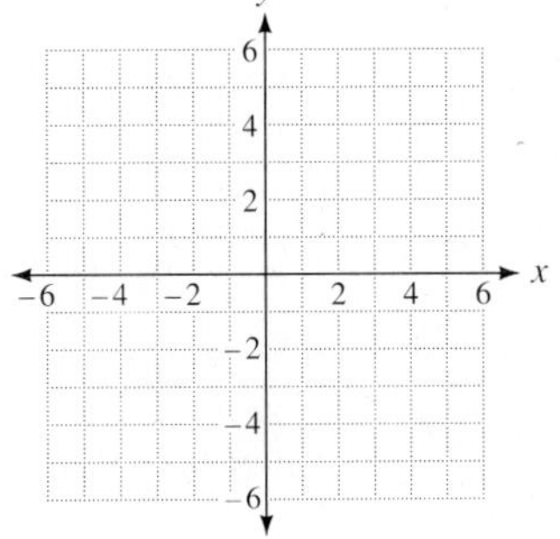

27. What is the domain of the function $f(x) = \frac{6}{\sqrt{6 - x}}$?

28. What is the vertex of the graph of $g(x) = -x^2 + 2x + 35$? What are the x-intercepts? Does the graph open up or down?

If $f(x) = 1 + 2x$ *and* $g(x) = \frac{1}{3}x - \frac{2}{3}$, *find each of the following.*

29. $f^{-1}(x)$

30. $g^{-1}(x)$

31. $f^{-1}(-1)$

32. $g^{-1}(-1)$

33. $f[g(-1)]$

34. $f^{-1}[f(-2)]$

35. $g[g^{-1}(7)]$

36. $(f + g)(3)$

37. $\left(\frac{f}{g}\right)(0)$

38. $(fg)(-1)$

39. $(f - g)(-2)$

40. $f^{-1}[g^{-1}(-1)]$

41. What is the minimum product of two numbers whose difference is 30? What are the two numbers?

ANSWERS: 1.

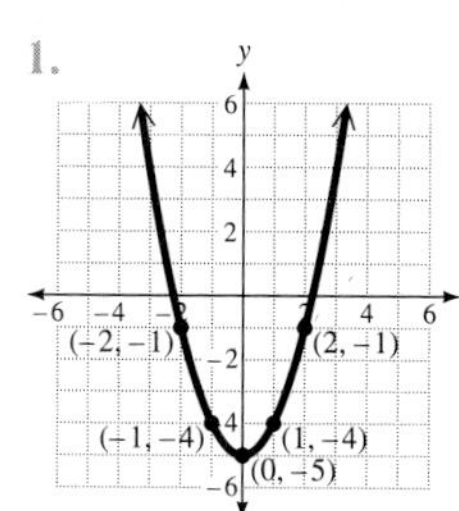

2.

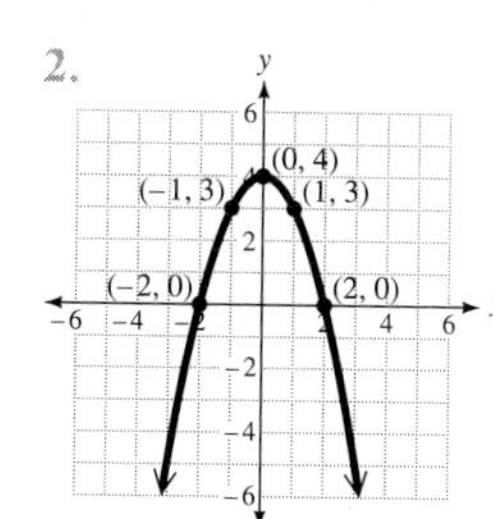

3. (3, 0) and (−10, 0); $\left(-\frac{7}{2}, -\frac{169}{4}\right)$; up
4. no x-intercepts; (1, −4); down

5.

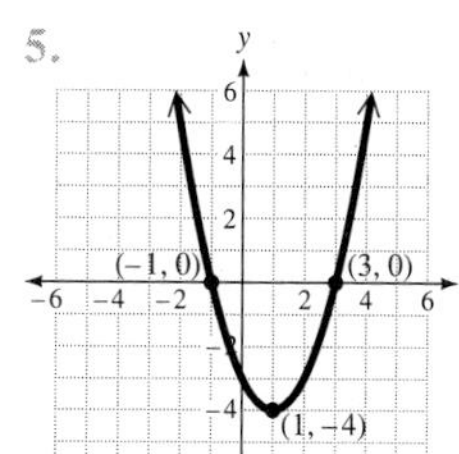

6.

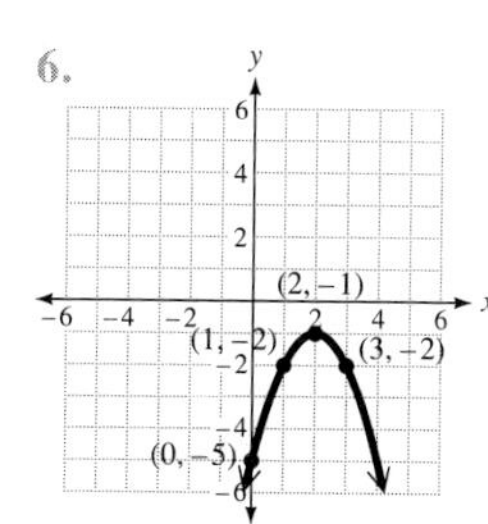

no x-intercepts 7. 18 ft by 18 ft; 324 ft^2 8. all real numbers except 8 9. all real numbers $x \geq \frac{1}{2}$ 10. all real numbers $x > \frac{1}{2}$ 11. 6 12. 2 13. −1 14. −3 15. 0 16. 23

17.

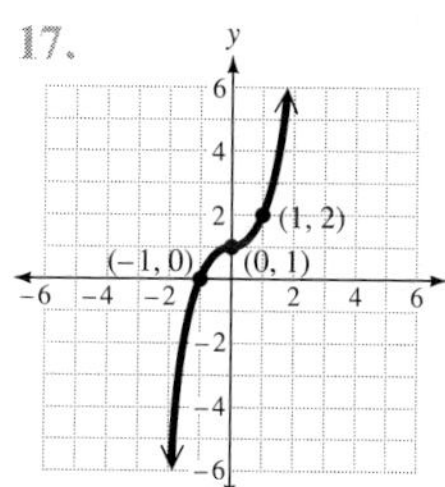

18.

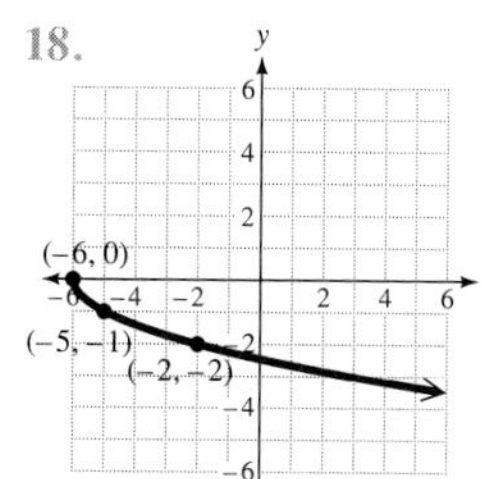

19.

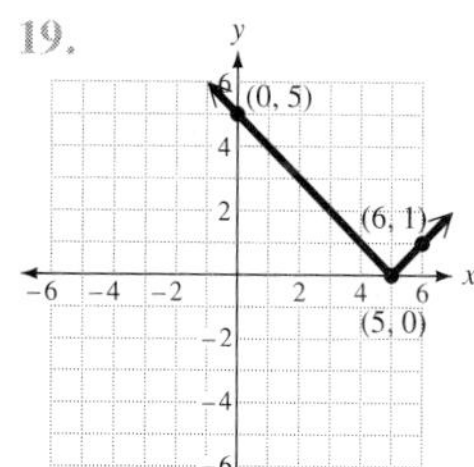

20. {(5, −2), (3, 0), (3, 1)}; yes; no 21. yes

22. yes;

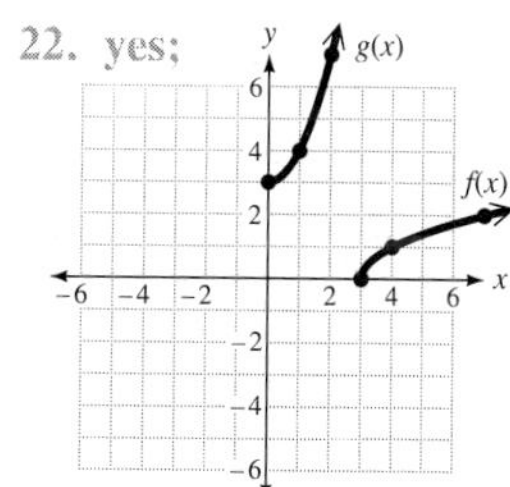

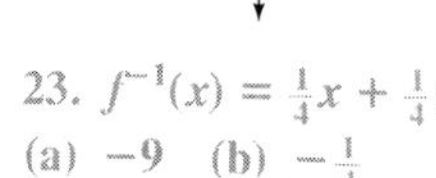

23. $f^{-1}(x) = \frac{1}{4}x + \frac{1}{4}$;
(a) −9 (b) $-\frac{1}{4}$
(c) −2 (d) −2

24.

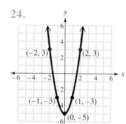

not one-to-one

25.

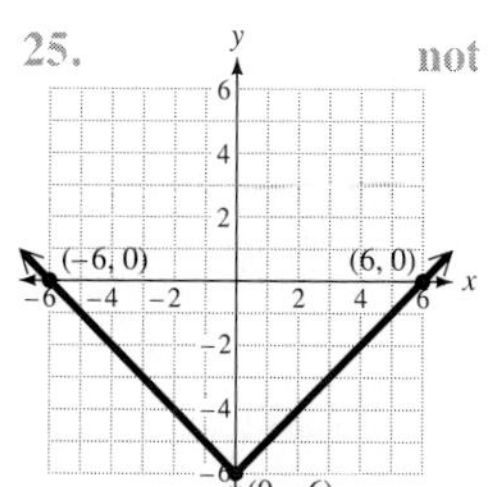

not one-to-one

26.

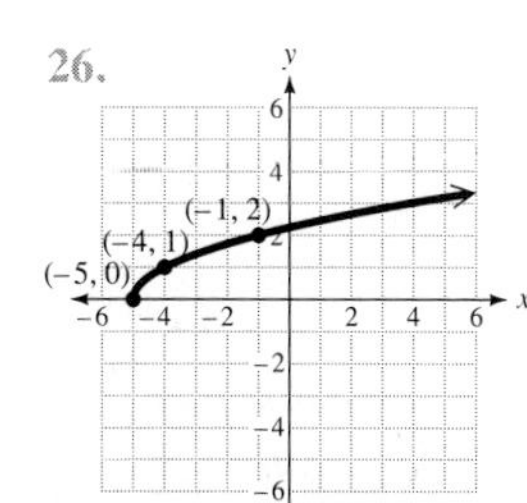

is one-to-one 27. all real numbers $x < 6$ 28. (1, 36); (−5, 0) and (7, 0); down 29. $f^{-1}(x) = \frac{1}{2}x - \frac{1}{2}$ 30. $g^{-1}(x) = 3x + 2$ 31. −1 32. −1 33. −1 34. −2 35. 7 36. $\frac{22}{3}$ 37. $-\frac{3}{2}$ 38. 1 39. $-\frac{5}{3}$ 40. −1 41. −225; 15 and −15

CHAPTER 9 TEST

1. How would you obtain the graph of $h(x) = x^2 + 3$ from the graph of $f(x) = x^2$?

1. 

2. Find the x-intercepts, the vertex, and graph of the function $f(x) = x^2 - 4x + 3$.

2. 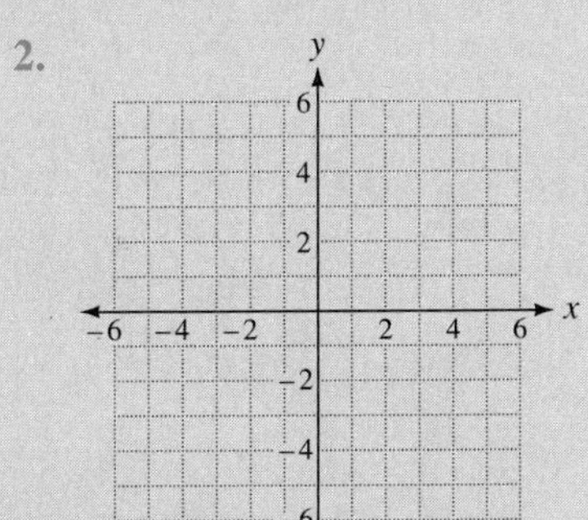

3. What is the maximum product of two numbers whose sum is 24?

3. ______

4. Find the domain of the function $f(x) = \dfrac{1}{\sqrt{x + 10}}$.

4. ______

Assume that $f(x) = x - 7$ and $g(x) = -x^2 + 3$. Find each function value in problems 5–10.

5. $(f + g)(-3)$

5. ______

6. $\left(\dfrac{f}{g}\right)(2)$

6. ______

7. $(f - g)(-1)$

7. ______

8. $(fg)(-2)$

8. ______

9. $f[g(-3)]$

9. ______

10. $g[f(5)]$

10. ______

11. Is this the graph of a one-to-one function?

11. ______

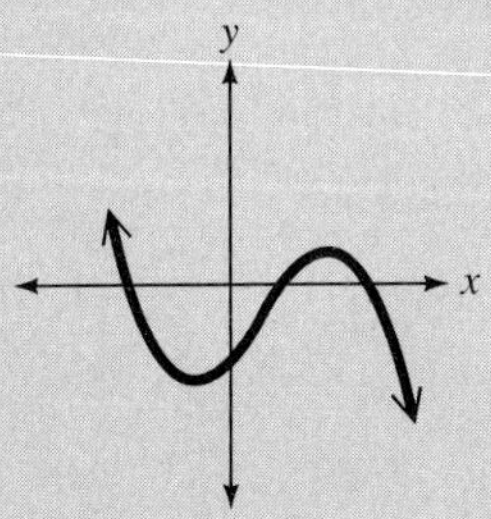

CHAPTER 9 TEST

CONTINUED

12. Give the equation for the inverse of the function $f(x) = \frac{1}{4}x - \frac{3}{4}$ and use it to find **(a)** $f^{-1}(2)$, and **(b)** $f[f^{-1}(2)]$.

12. ____________
12. (a) ____________
12. (b) ____________

Graph the following functions.

13. $f(x) = -\sqrt{x + 2}$

13.

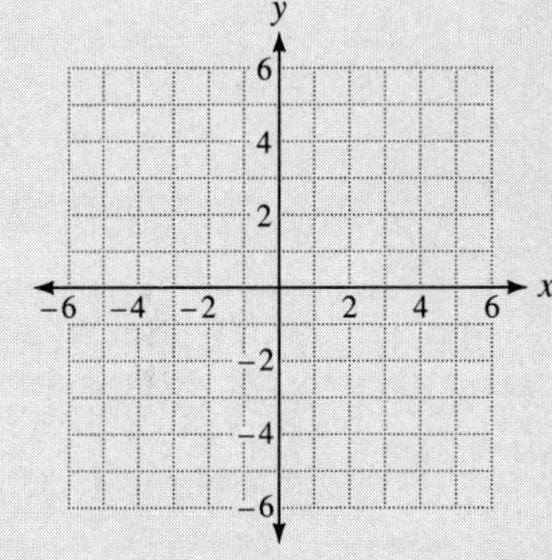

14. $f(x) = |x| + 2$

14.

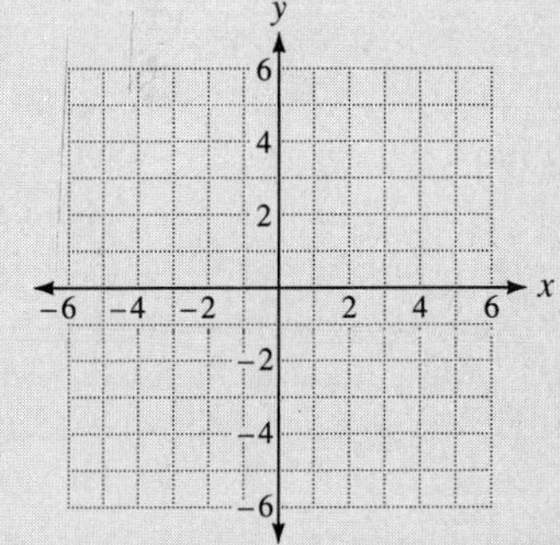

15. $f(x) = |x + 2|$

15.

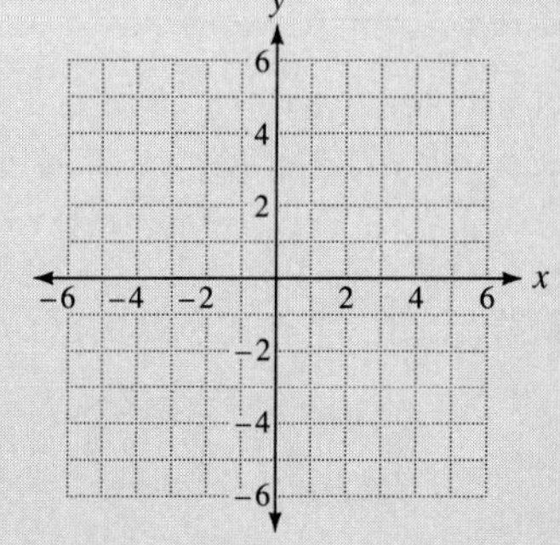

CHAPTER 10

Exponential and Logarithmic Functions

10.1 BASIC CONCEPTS

STUDENT GUIDEPOSTS

1. Logarithms
2. Exponential and Logarithmic Forms
3. Exponential and Logarithmic Properties
4. Calculating Exponentials

Historically, logarithms were developed to help carry out complicated numerical calculations. With the advent of computers and calculators, however, computing with logarithms is no longer a necessary skill. Nevertheless, logarithmic and exponential equations and functions remain important, with numerous applications in mathematics.

Calculators are invaluable tools for working with exponentials and logarithms; hence, our presentation is geared toward their use. If you prefer to use a table of logarithms, the appendix at the back of the text provides detailed instructions. In all calculator computations, we have stored the intermediate results with no rounding until the final step. There can be small differences in answers you get from a logarithmic table and from various calculators. But answers should be close enough to verify the accuracy of your work.

1 LOGARITHMS

In previous chapters we solved equations such as

$$3x + 2 = 7 \quad \text{and} \quad 2x^2 - 5x - 3 = 0.$$

In these equations, the variable does not appear in an exponent. We now consider equations such as

$$3^x = 9$$

in which the variable does occur in an exponent. The equation-solving rules we have already learned are of little help for these equations. At present, we must discover the solution by inspection. In this case, the problem is not too difficult, and we would probably recognize that when 3 is raised to the second power, the result is 9. Thus, the solution to the equation $3^x = 9$ is

$$x = 2.$$

Alternatively, we might give the solution in words.

x is the exponent on 3 which gives the number 9

The word *logarithm* can be used instead of *exponent*. The solution might be written

x is the logarithm on 3 which gives the number 9.

Since 3 is called the *base* in the expression 3^x, the solution can also be written

x is the logarithm to the base 3 of 9.

This final statement is usually symbolized by

$$x = \log_3 9.$$

2 EXPONENTIAL AND LOGARITHMIC FORMS

We showed above that

$$3^x = 9 \quad \text{and} \quad x = \log_3 9$$

are two forms of the same equation, with $3^x = 9$ called the **exponential form** and $x = \log_3 9$ called the **logarithmic form.** That is, they are equivalent equations, with the logarithmic form being "solved" for the variable x.

Definition of Logarithm

Let a, x, and y be real numbers, $a > 0$, $a \neq 1$, that satisfy the exponential equation

$$a^x = y.$$

Then x is the **logarithm to the base a of y,** and the equivalent logarithmic equation is

$$x = \log_a y.$$

Notice in the definition, that we restrict a to values greater than 0 since if $a < 0$, then a^x would not be defined for numbers such as $x = \frac{1}{2}$. Also, if $a = 1$ we have $a^x = a^y$ for all x and y. Remember also that the word *logarithm* means *exponent*.

EXAMPLE 1 CONVERTING FORMS

Give an equivalent equation for each of the following.

Equation	*Equivalent equation*	
$2^x = 32$	$x = \log_2 32$	$x = 5$ in this case
$7^x = c$	$x = \log_7 c$	
$x = \log_3 8$	$3^x = 8$	
$x = \log_5 125$	$5^x = 125$	$x = 3$ in this case

PRACTICE EXERCISE 1

Give an equivalent equation for each of the following.

(a) $8^x = 64$

(b) $b^x = 5$

(c) $x = \log_4 37$

(d) $x = \log_9 81$

Answers: (a) $x = \log_8 64$ (b) $x = \log_b 5$ (c) $4^x = 37$ (d) $9^x = 81$

To convert an equation from exponential to logarithmic form, or vice versa, it may help to remember that in both cases the base is written below the level of the logarithm (exponent), as indicated in the diagram on the next page.

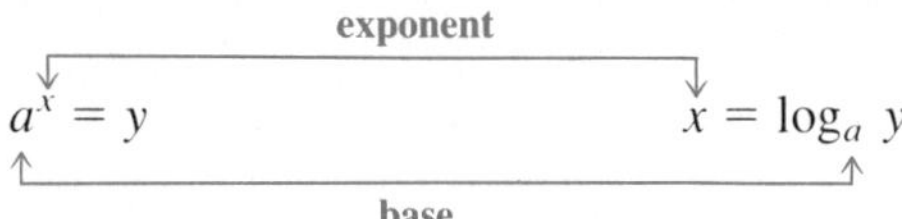

Since solutions to simple exponential equations can often be found by direct inspection, solutions to some logarithmic equations are more easily found by converting to exponential form.

EXAMPLE 2 FINDING THE VALUE OF A LOGARITHM

Give the numerical value of x in $x = \log_4 16$.

We first convert $x = \log_4 16$ to exponential form: $4^x = 16$. At this point, it might be clear that x must be 2 since $4^2 = 16$. By writing 16 as a power of 4, $16 = 4^2$, we have

$$4^x = 4^2.$$

Since for any a ($a \neq 0$ and $a \neq 1$), $a^x = a^y$ implies that $x = y$, we conclude that $x = 2$.

PRACTICE EXERCISE 2

Give the numerical value of x in $x = \log_3 81$.

Answer: 4

EXAMPLE 3 SOLVING A LOGARITHMIC EQUATION

Find the numerical value of x in $\log_x \frac{1}{8} = -3$.

Convert to exponential form.

$$x^{-3} = \frac{1}{8}$$

$$x^{-3} = 2^{-3} \qquad \frac{1}{8} = \frac{1}{2^3} = 2^{-3}$$

It is now clear that x is 2 since the exponents in both sides of the equation are -3, forcing the bases in both sides to be equal. That is, in general, if $a^x = b^x$, then $a = b$ ($x \neq 0$).

PRACTICE EXERCISE 3

Find the numerical value of x in $\log_x \frac{1}{16} = -2$.

Answer: 4

3 EXPONENTIAL AND LOGARITHMIC PROPERTIES

Notice in Examples 2 and 3, we worked toward writing an equation in which each side was an exponential form, and then equated either exponents or bases. These properties of exponentials, summarized below, will become even more apparent when we discuss exponential functions in Section 10.2.

Properties of Exponentials

Let x and y be real numbers, a and b positive real numbers with $a \neq 1$ and $b \neq 1$.

1. $a^x = a^y$ if and only if $x = y$.

2. $a^x = b^x$ if and only if $a = b$.

We now develop two important properties of logarithms. Since

$$a^0 = 1 \quad \text{(for } a \text{ any number except 0)},$$

rewriting this equation in logarithmic form, we have

$$\log_a 1 = 0.$$

Likewise, since $a^1 = a$ (for a any number),

converting to logarithmic form, we have

$$\log_a a = 1.$$

These two results are summarized next.

Properties of Logarithms

If a is any base ($a > 0$, $a \neq 1$),

1. $\log_a a = 1$ $(a^1 = a)$
2. $\log_a 1 = 0.$ $(a^0 = 1)$

For example, $\log_5 5 = 1$ (since $5^1 = 5$), $\log_5 1 = 0$ (since $5^0 = 1$), and $\log_{10} 10 = 1$ (since $10^1 = 10$).

4 CALCULATING EXPONENTIALS

A calculator with a [y^x] key can be helpful for calculating exponentials. For example, if we wanted to approximate 2^π, we would use this sequence of keys:

2 [y^x] [π] [=] ⟶ [8.82498]

The display would show 8.82498.

EXAMPLE 4 CALCULATING EXPONENTIALS

Use a calculator to approximate each number correct to four decimal places.

(a) $2^{\sqrt{2}}$

2 [y^x] 2 [$\sqrt{\ }$] [=] ⟶ [2.66514]

Thus, $2^{\sqrt{2}} \approx 2.6651$.

(b) $(1.04)^{60}$

1.04 [y^x] 60 [=] ⟶ [10.5196]

Thus, $(1.04)^{60} \approx 10.5196$.

PRACTICE EXERCISE 4

Calculate each number correct to three decimal places.

(a) $\pi^{\sqrt{3}}$

(b) $(1.085)^{40}$

Answers: (a) 7.263 (b) 26.133

Many applied problems, such as the one in the next example, involve exponentials.

EXAMPLE 5 BIOLOGY PROBLEM

The number of bacteria N in a culture can be approximated by $N = (1000)2^t$ where t is time in hours. How many bacteria are present in 12.5 hours?

Substitute 12.5 for t and find the value of N using a calculator.

$$N = (1000)2^{12.5} = 5{,}792{,}620$$

1000 [×] 2 [y^x] 12.5 [=] ⟶ [5792620]

There are about 5,792,620 bacteria after 12.5 hours.

PRACTICE EXERCISE 5

Use the formula $N = (1000)2^t$ to find the number of bacteria present after 20 hours.

Answer: 1.0486×10^9

10.1 EXERCISES A

Convert each equation to logarithmic form.

1. $2^3 = 8$

2. $5^x = 100$

3. $m = n^p$

4. $8^0 = 1$

5. $7^{1/2} = c$

6. $10^{-2} = 0.01$

Convert each equation to exponential form.

7. $\log_a 3 = x$

8. $\log_{10} 100 = 2$

9. $\log_8 4 = \frac{2}{3}$

10. $\log_a 1 = 0$

11. $u = \log_v 7$

12. $\log_b \left(\frac{1}{b}\right) = -1$

Give the numerical value of x in each of the following exponential equations.

13. $2^x = 8$

14. $3^x = 81$

15. $5^x = \frac{1}{5}$

16. $10^x = 0.0001$

17. $x^{-1} = \frac{1}{4}$

18. $x^{1/2} = 3$

Find the numerical value of x by first converting to the equivalent exponential equation.

19. $x = \log_2 16$

20. $x = \log_8 2$

21. $\log_x \frac{1}{4} = -1$

22. $\log_x 11 = 1$

23. $\log_4 x = \frac{1}{2}$

24. $\log_2 x = 5$

25. If we know that $5^x = 5^{3.7}$, then what is x?

26. If we know that $7^{2.3} = y^{2.3}$, then what is y?

Evaluate each logarithm.

27. $\log_4 4$

28. $\log_7 7$

29. $\log_3 1$

30. $\log_8 1$

**Use a calculator to approximate each number, correct to three decimal places.*

31. $5^{\sqrt{2}}$

32. 5^{π}

33. $(1.08)^{50}$

34. Change $y = \log_3 (-9)$ to exponential form. Is there a value of y that makes this true?

35. In a scientific study, it has been shown that the variables w and v satisfy the equation $w = \log_{100} v$. **(a)** What is the value of v when w is $\frac{1}{2}$? **(b)** What is the value of w when v is 10,000?

*Since a calculator may be helpful in working many of the problems throughout this chapter, individual problems will not be designated as calculator exercises.

36. An electrical engineer uses the formula $g = 10\log_{10}\frac{a}{b}$ to calculate the gain on an amplifier. What is the value of g when a is 200 and b is 20?

37. A banker uses the formula $A = P(1.06)^n$ to calculate interest compounded annually. What is the value of A when P is \$1000 and n is 20 years?

38 The number of bacteria N in a culture is given in terms of time t, in hours, by $N = (2000)2^t$. How many bacteria are present initially when $t = 0$? How many are present in 7 hours?

39. The atmospheric pressure P on an object, in pounds per square inch, can be approximated by $P = 14.7(2.7)^{-0.2x}$ where x is the height of the object in miles above sea level. What is the pressure on an object 2 miles high?

FOR REVIEW

Exercises 40–42 review material from Chapter 3 to help you prepare for the next section. Graph each function.

40. $y = 2x$

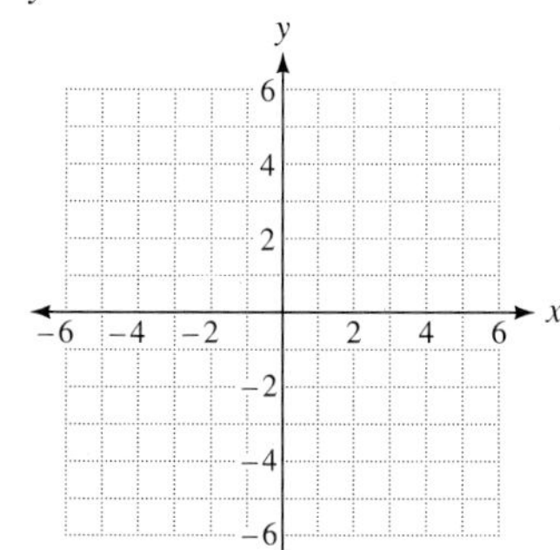

41. $y = -2x$

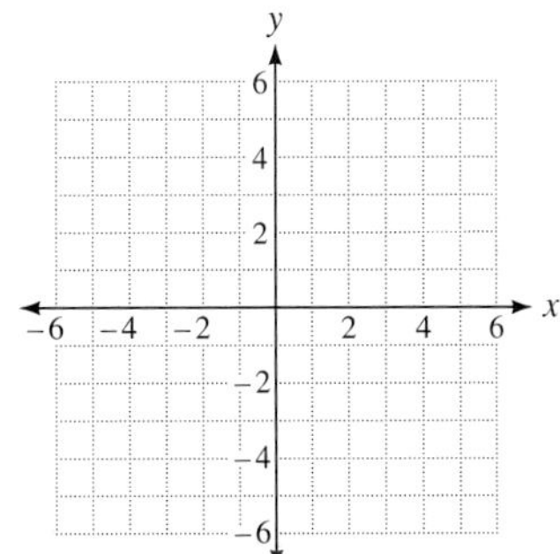

42. $y = x^2 + 1$

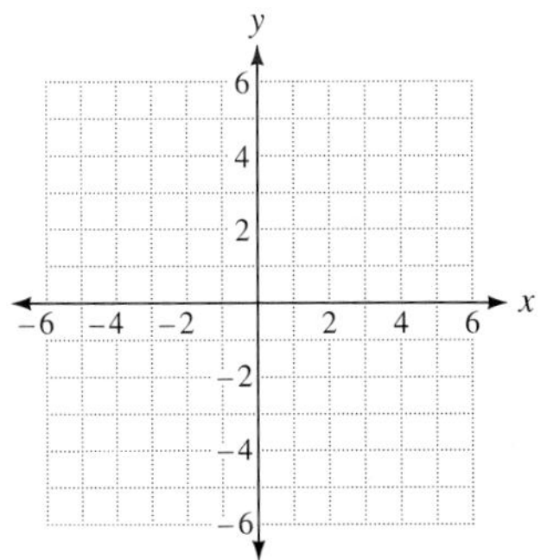

ANSWERS: 1. $\log_2 8 = 3$ 2. $\log_5 100 = x$ 3. $\log_n m = p$ 4. $\log_8 1 = 0$ 5. $\log_7 c = \frac{1}{2}$ 6. $\log_{10}(0.01) = -2$ 7. $a^x = 3$ 8. $10^2 = 100$ 9. $8^{2/3} = 4$ 10. $a^0 = 1$ 11. $v^u = 7$ 12. $b^{-1} = \frac{1}{b}$ 13. $x = 3$ 14. $x = 4$ 15. $x = -1$ 16. $x = -4$ 17. $x = 4$ 18. $x = 9$ 19. $x = 4$ 20. $x = \frac{1}{3}$ 21. $x = 4$ 22. $x = 11$ 23. $x = 2$ 24. $x = 32$ 25. 3.7 26. 7 27. 1 28. 1 29. 0 30. 0 31. 9.739 32. 156.993 33. 46.902 34. $3^y = -9$; no 35. (a) 10 (b) 2 36. 10 37. \$3207.14 38. 2000; 256,000 39. 9.88 lb/in^2

40.

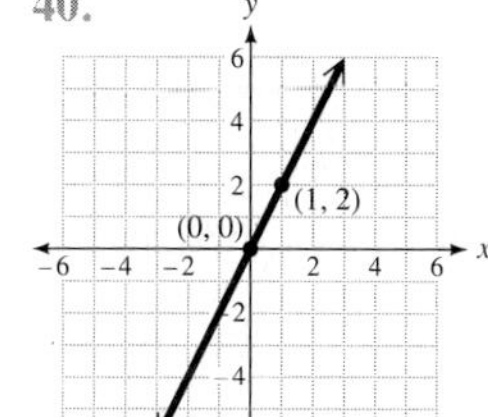

41.

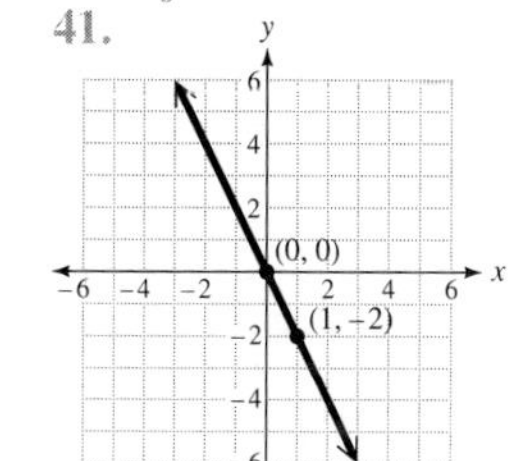

42.

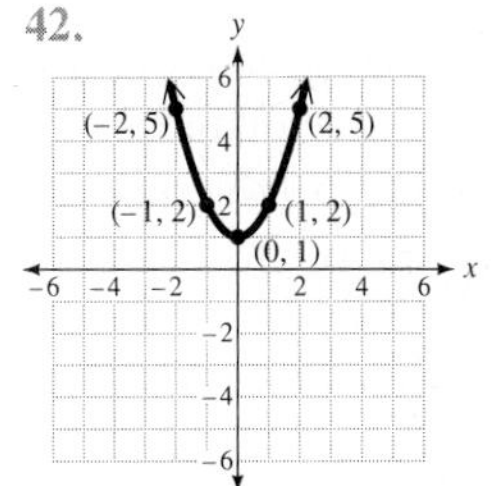

10.1 EXERCISES B

Convert each equation to logarithmic form.

1. $2^4 = 16$

2. $3^x = 200$

3. $u = v^w$

4. $9^0 = 1$

5. $11^{1/2} = a$

6. $10^{-3} = 0.001$

Convert each equation to exponential form.

7. $\log_a 5 = x$

8. $\log_{10} 1000 = 3$

9. $\log_{27} 9 = \dfrac{2}{3}$

10. $\log_b 1 = 0$

11. $w = \log_v 11$

12. $\log_a \left(\frac{1}{a}\right) = -1$

Give the numerical value of x in each of the following exponential equations.

13. $2^x = 4$

14. $3^x = 27$

15. $7^x = \frac{1}{7}$

16. $10^x = 100$

17. $x^5 = 243$

18. $x^{1/3} = 2$

Find the numerical value of x by first converting to the equivalent exponential equation.

19. $x = \log_2 8$

20. $x = \log_{16} 2$

21. $\log_x \frac{1}{3} = -1$

22. $\log_x 15 = 1$

23. $\log_9 x = \frac{1}{2}$

24. $\log_2 x = 6$

25. If we know that $9^x = 9^{4.1}$, then what is x?

26. If we know that $12^{-3.9} = y^{-3.9}$, then what is y?

Evaluate each logarithm.

27. $\log_8 8$

28. $\log_\pi \pi$

29. $\log_7 1$

30. $\log_{\sqrt{2}} 1$

Use a calculator to approximate each number, correct to three decimal places.

31. $4^{\sqrt{3}}$

32. $\pi^{\sqrt{2}}$

33. $(1.20)^{25}$

34. Change $y = \log_4 (-4)$ to exponential form. Is there a value of y that makes this true?

35. An economist has found that the variables c and p satisfy the equation $c = \log_{25} p$. **(a)** What is the value of p when c is $\frac{1}{2}$? **(b)** What is the value of c when p is $\frac{1}{25}$?

36. An electrical engineer uses the formula $g = 10 \log_{10} \frac{a}{b}$ to calculate the gain on an amplifier. What is the value of g when a is 300 and b is 30?

37. A banker uses the formula $A = P(1.06)^n$ to calculate interest compounded annually. What is the value of A when P is \$1000 and n is 30 years?

38. The number of bacteria N in a culture is given in terms of time t, in hours, by $N = (4000)2^t$. How many bacteria are present initially when $t = 0$? How many are present in 7 hours?

39. The atmospheric pressure P on an object, in pounds per square inch, can be approximated by $P = 14.7(2.7)^{-0.2x}$ where x is the height of the object in miles above sea level. What is the pressure on an object 5 miles high?

FOR REVIEW

Exercises 40–42 review material from Chapter 3 to help you prepare for the next section. Graph each function.

40. $y = 3x$

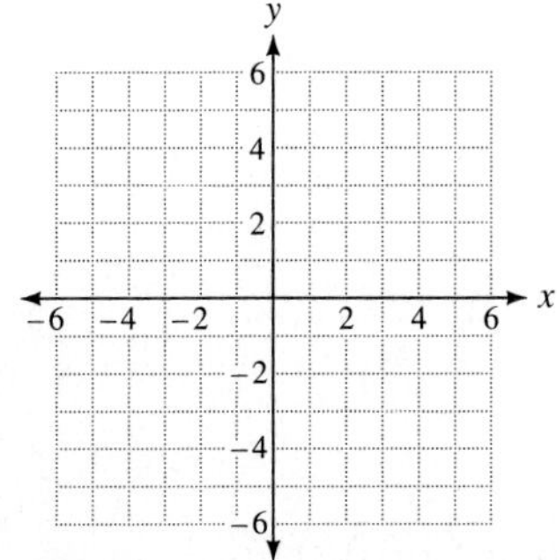

41. $y = -3x$

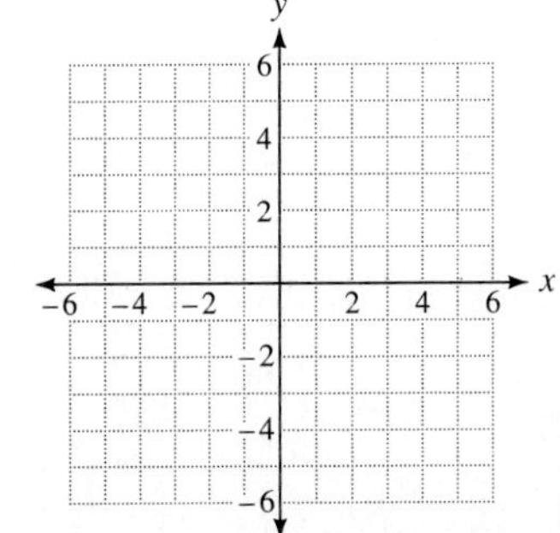

42. $y = x^2 - 1$

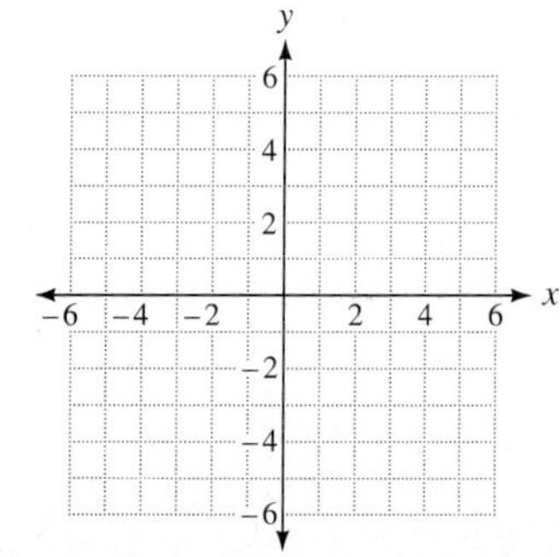

10.1 EXERCISES C

Give the numerical value of x.

1. $\log_3 (x^2 + 8x) = 2$
2. $\log_{\sqrt{2}} x = -2$
3. $\log_3 \sqrt[3]{9} = x$
4. $\log_x \sqrt{7} = \frac{1}{2}$
5. $\log_x \sqrt[3]{5} = \frac{1}{3}$
6. $|\log_5 x^2| = 2$

[Answer: $\pm 5, \pm \frac{1}{5}$]

10.2 LOGARITHMIC AND EXPONENTIAL FUNCTIONS

STUDENT GUIDEPOSTS

1. Exponential Functions
2. Logarithmic Functions

1 EXPONENTIAL FUNCTIONS

In Chapters 3 and 9 we considered several functions, including constant, linear, and quadratic functions, and their graphs. Two other important kinds of functions are exponential and logarithmic functions. A function defined by the equation

$$y = a^x, \quad a > 0,\ a \neq 1$$

is an **exponential function,** and the number a is called the **base.** It is not difficult to calculate the value of y when x is a rational number p/q since $a^{p/q}$ is defined to be the qth root of a raised to the pth power. Although we have not defined a^x for irrational x such as $x = \sqrt{2}$ or $x = \pi$, we shall assume that $a^{\sqrt{2}}$ can be approximated as closely as we wish by using rational approximations of $\sqrt{2}$. For example, since $\sqrt{2} \approx 1.414$,

$$a^{\sqrt{2}} \approx a^{1.414} = a^{1414/1000},$$

which does have meaning for us. The base a to other irrational powers can be approximated in a similar way, or by using a calculator with a $\boxed{y^x}$ key.

x	y
0	1
-1	$\frac{1}{2}$
1	2
-2	$\frac{1}{4}$
2	4
-3	$\frac{1}{8}$
3	8
-4	$\frac{1}{16}$

We now graph $y = 2^x$ and show that our assumption about irrational powers is reasonable. We select values of x and calculate the corresponding values of y to obtain the table of values in Figure 10.1. We can use 1.41 as an approximation for $\sqrt{2}$, and from the graph, we see that 2.7 is an approximation for $2^{\sqrt{2}}$. Similarly, using 3.14 as an approximation for π, we can estimate 2^π to be about 8.8. A calculator gives 2.66514 for $2^{\sqrt{2}}$ and 8.82498 for 2^π. These values are of course more accurate than the values we are able to read from the graph.

To compare exponential functions, we graph $y = 2^x$, $y = \left(\frac{1}{2}\right)^x$, $y = 3^x$, $y = \left(\frac{1}{3}\right)^x$, $y = 10^x$, and $y = \left(\frac{1}{10}\right)^x$ in the same coordinate system. First we construct a table of values. The scale for our graphs in Figure 10.2 does not allow us to plot all of the points listed in the table, but the values are included there for purposes of comparison.

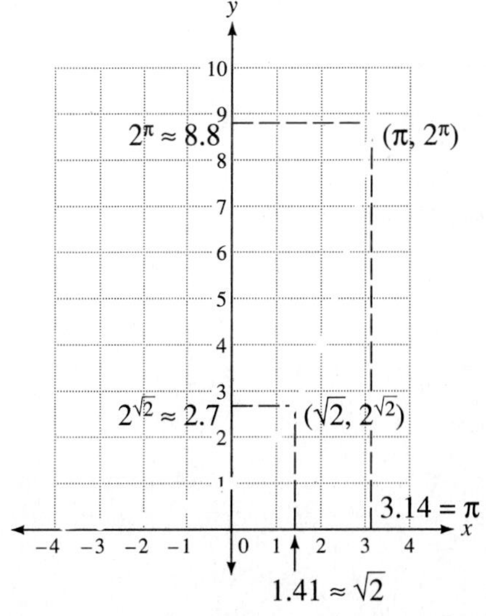

Figure 10.1

x	$y = 2^x$	$y = \left(\frac{1}{2}\right)^x$	$y = 3^x$	$y = \left(\frac{1}{3}\right)^x$	$y = 10^x$	$y = \left(\frac{1}{10}\right)^x$
0	1	1	1	1	1	1
−1	$\frac{1}{2}$	2	$\frac{1}{3}$	3	$\frac{1}{10}$	10
1	2	$\frac{1}{2}$	3	$\frac{1}{3}$	10	$\frac{1}{10}$
−2	$\frac{1}{4}$	4	$\frac{1}{9}$	9	$\frac{1}{100}$	100
2	4	$\frac{1}{4}$	9	$\frac{1}{9}$	100	$\frac{1}{100}$
−3	$\frac{1}{8}$	8	$\frac{1}{27}$	27	$\frac{1}{1000}$	1000
3	8	$\frac{1}{8}$	27	$\frac{1}{27}$	1000	$\frac{1}{1000}$
−4	$\frac{1}{16}$	16	$\frac{1}{81}$	81	$\frac{1}{10,000}$	10,000
4	16	$\frac{1}{16}$	81	$\frac{1}{81}$	10,000	$\frac{1}{10,000}$

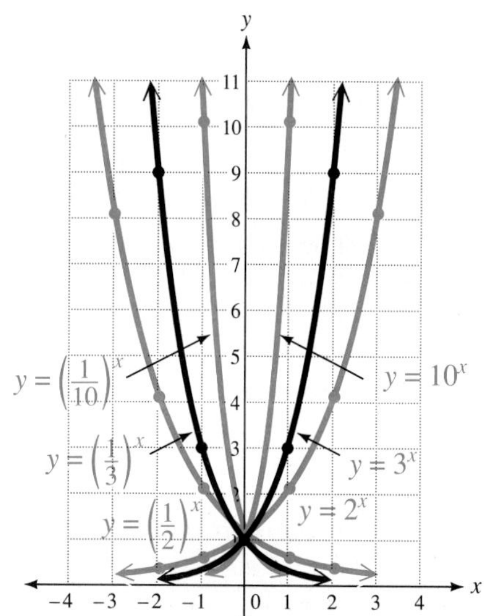

Figure 10.2 Exponential Functions

❷ LOGARITHMIC FUNCTIONS

Several observations can be made about the functions above. In all cases the graph passes through the point (0, 1), since $a^0 = 1$ for any value of a. For every base a greater than 1, the graph rises from left to right, and toward the left it gets closer and closer to (but never crosses) the x-axis. For every base a less than 1, the graph decreases from left to right, and toward the right it gets closer and closer to (but never crosses) the x-axis. Since each exponential function

$$y = a^x \qquad a > 0 \text{ and } a \neq 1$$

is either increasing or decreasing, it is one-to-one and has an inverse. To find the equation of that inverse, we interchange x and y,

$$x = a^y \qquad x \text{ and } y \text{ have been interchanged}$$

and solve for y. To solve this equation for y we simply convert it to logarithmic form.

$$y = \log_a x.$$

A **logarithmic function** is a function defined by an equation of the form

$$y = \log_a x.$$

To graph a logarithmic function we can graph its equivalent exponential form, keeping in mind that the roles of x and y have been interchanged. Suppose we graph $y = \log_2 x$ by plotting $x = 2^y$.

$x = 2^y$	y
1	0
$\frac{1}{2}$	−1
2	1
$\frac{1}{4}$	−2
4	2
$\frac{1}{8}$	−3
8	3

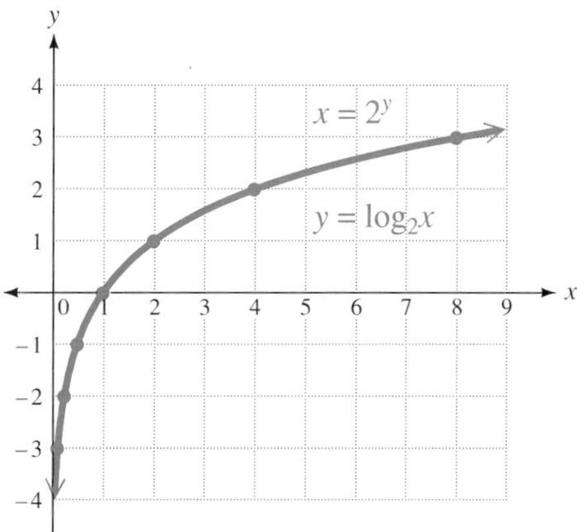

Figure 10.3

To construct the table of values in Figure 10.3, we find x for a given value of y, then plot the points, and join them with a smooth curve.

If we restrict ourselves to $a > 1$, the graph of each exponential function $y = a^x$ has the same basic shape, shown in Figure 10.4, and passes through the point (0, 1). The graph of every logarithmic function $y = \log_a x$ has the same basic shape, shown in Figure 10.5, and passes through (1, 0). With this in mind, we can quickly graph an exponential or logarithmic function by plotting a few points.

We can also observe, by plotting the graphs of

$$y = \log_a x \quad \text{and} \quad y = a^x$$

in the same coordinate system, as in Figure 10.6, that each graph is the reflection of the other across the line $y = x$. This illustrates the fact that these functions are inverses of each other. If the graph of either is known, the graph of the other can easily be found by reflection in the line $y = x$.

There are many applications of logarithmic and exponential functions. The chemistry problem on the next page is one example.

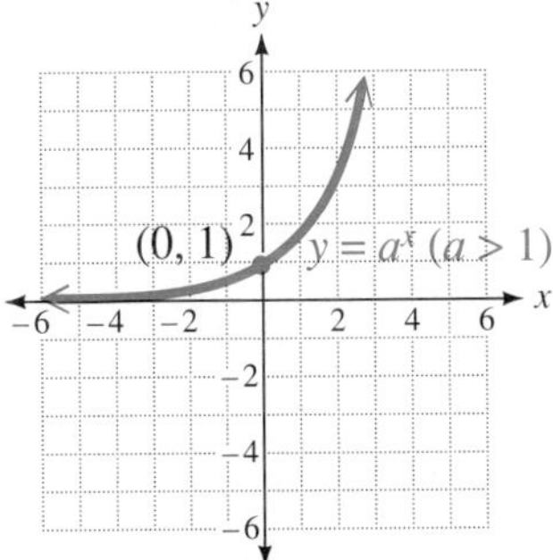

Figure 10.4
General Exponential Function

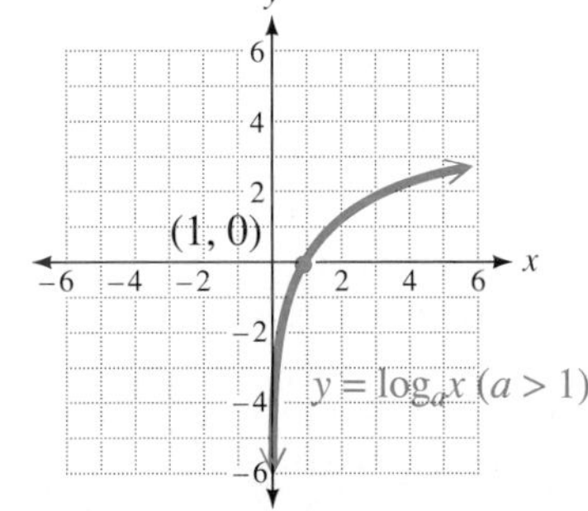

Figure 10.5
General Logarithmic Function

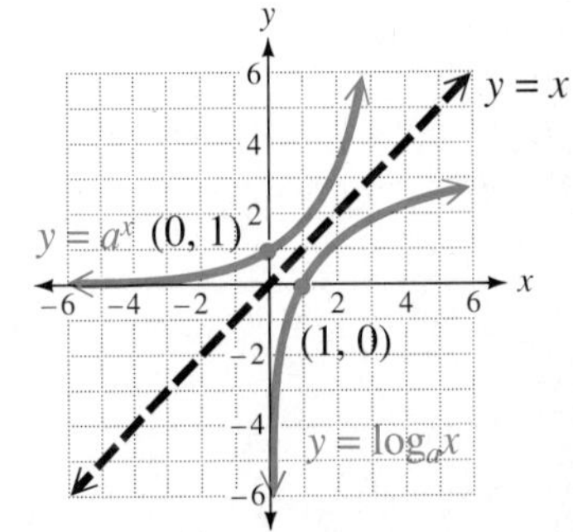

Figure 10.6
Inverse Functions

EXAMPLE 1 Chemistry Problem

A radioactive isotope has a half-life of 1200 years. This means that if 1000 grams of the isotope are allowed to decay, the amount A that remains after t years is given by the exponential function $A = (1000)2^{-t/1200}$. Find the amount remaining at the end of 2400 years and sketch the graph of the function for $t \geq 0$.

To find the amount remaining after 2400 years, substitute 2400 for t.

$$A = (1000)2^{-2400/1200}$$
$$= (1000)2^{-2}$$
$$= (1000)\frac{1}{2^2}$$
$$= (1000)\frac{1}{4}$$
$$= 250$$

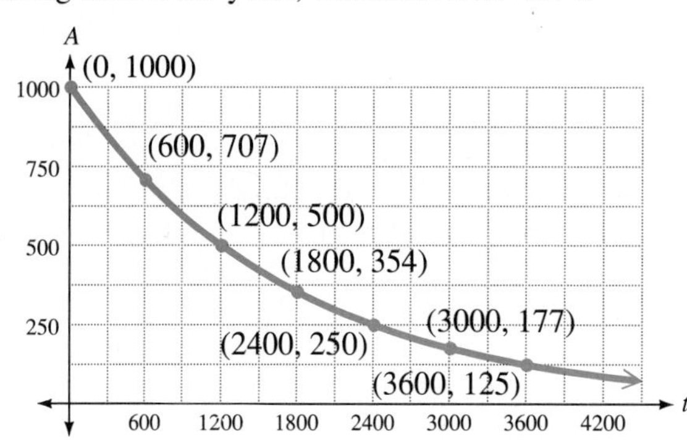

Figure 10.7 $A = (1000)2^{-t/1200}$

Thus 250 grams remain after 2400 years. The graph of the function for $t \geq 0$ is given in Figure 10.7.

Practice Exercise 1

Refer to Example 1 and find the amount of radioactive isotope that would remain after 5000 years.

Answer: About 55.7 grams

10.2 EXERCISES A

1. Given an exponential function $y = a^x$, a is called the ________________.
2. The graph of the logarithmic function $y = \log_a x$ can be found by plotting the points that satisfy the exponential equation ________________.
3. The graph of $y = \log_3 x$ can be found by reflecting the graph of $y = 3^x$ across the line with equation ________________.
4. The graph of $y = 5^x$ and $y =$ ________________ are reflections of each other across the line $y = x$.
5. Complete the following table and sketch the graphs of the functions $y = 2^x$, $y = \left(\frac{1}{3}\right)^x$, $y = 3^x$, $y = 4^x$ in the given coordinate system.

x	$y = 2^x$	$y = \left(\frac{1}{3}\right)^x$	$y = 3^x$	$y = 4^x$
0				
−1				
1				
−2				
2				
−3				
3				

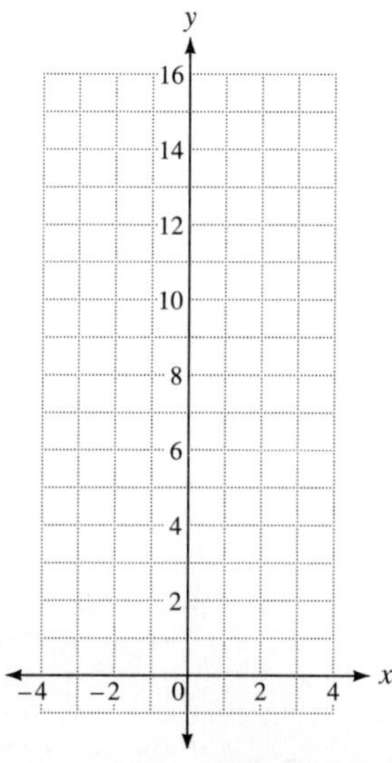

6. Sketch the graph of $y = \log_3 x$ by plotting $x = 3^y$.

x	y
	0
	1
	-1
	2
	-2

7. Having plotted $y = 3^x$ in Exercise 5, could you have used this information to sketch the graph required in Exercise 6? Explain.

8. The half-life of an antibiotic in the bloodstream is about 8 hours. This means that when 64 milligrams of the drug are absorbed in your bloodstream, the number D of milligrams remaining after t hours is given by the function $D = (64)2^{-t/8}$. Graph this function for $t \geq 0$ and find the number of milligrams of the antibiotic remaining in the bloodstream after **(a)** 0 hours, **(b)** 8 hours, **(c)** 16 hours, and **(d)** 24 hours.

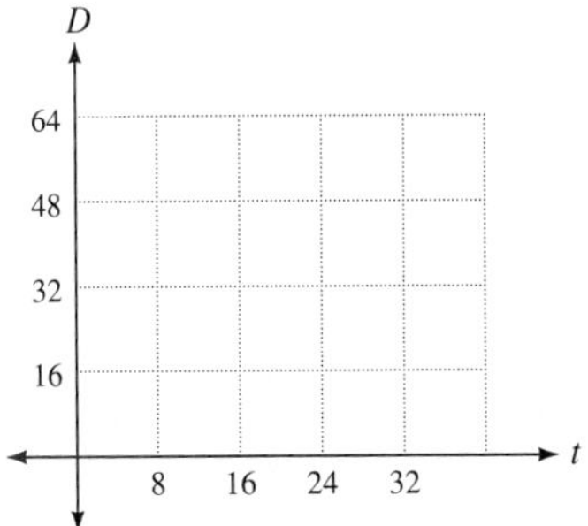

9. A bacteria culture containing 100 cells doubles in number each hour. The number N of bacteria present after t hours is given by the exponential function $N = (100)2^t$. Graph this function for $t \geq 0$ and find the number of bacteria present after **(a)** 0 hours, **(b)** 2 hours, **(c)** 4 hours, and **(d)** 6 hours.

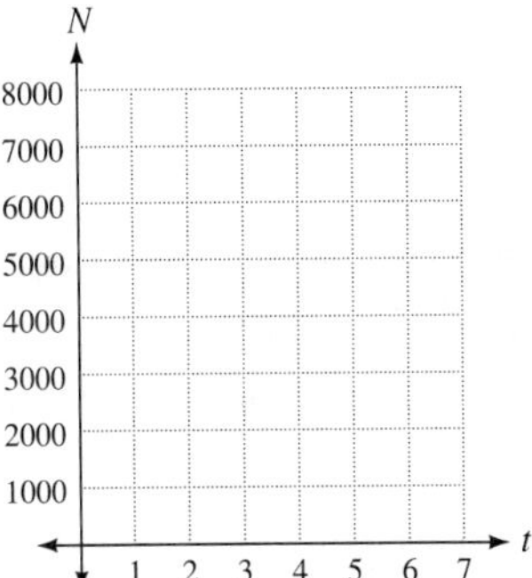

FOR REVIEW

Exercises 10–13 review properties of exponents from Section 1.6 to help you prepare for the next section. Simplify each expression.

10. a^3a^5 **11.** $\dfrac{a^8}{a^2}$ **12.** $(a^5)^3$ **13.** $\dfrac{a^2a^{-4}}{a}$

ANSWERS:

1. base
2. $x = a^y$
3. $y = x$
4. $\log_5 x$

5.

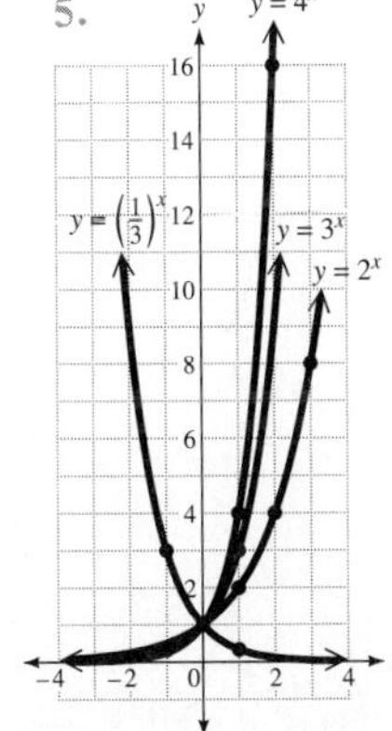

6.

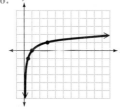

7. yes; by reflecting the graph across the line $y = x$

8.

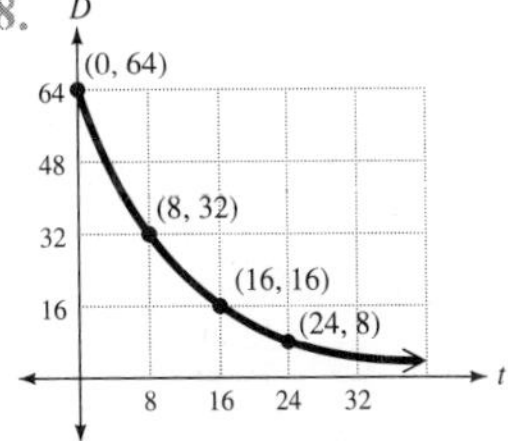

(a) 64 mg (b) 32 mg (c) 16 mg (d) 8 mg

9.

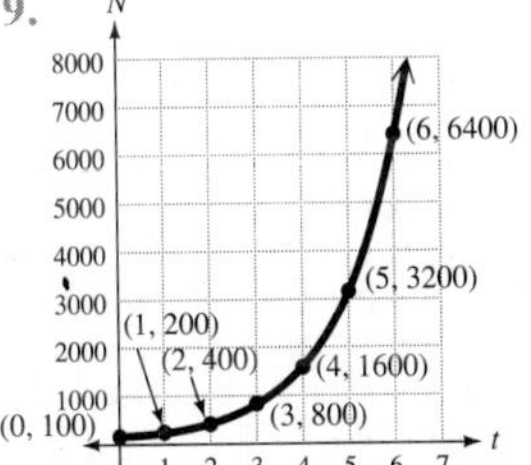

(a) 100 (b) 400 (c) 1600 (d) 6400

10. a^8 11. a^6 12. a^{15} 13. $\frac{1}{a^3}$

10.2 EXERCISES B

1. A function described by the equation $y = a^x$, $a > 0$, is called a(n) ________________ function.

2. The graph of the logarithmic function, ________________, can be found by plotting the points that satisfy the exponential equation $x = a^y$.

3. The graph of $y = \log_5 x$ can be found by reflecting the graph of ________________ across the line with equation $y = x$.

4. The graph of ________________ and $y = \log_{12} x$ are reflections of each other across the line $y = x$.

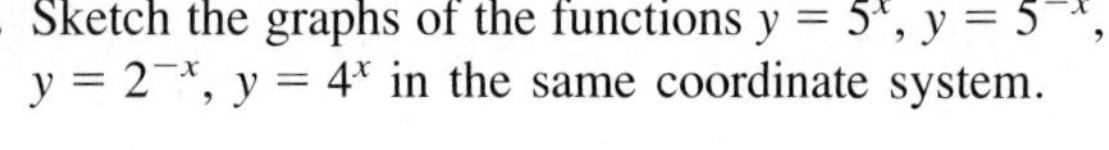

5. Sketch the graphs of the functions $y = 5^x$, $y = 5^{-x}$, $y = 2^{-x}$, $y = 4^x$ in the same coordinate system.

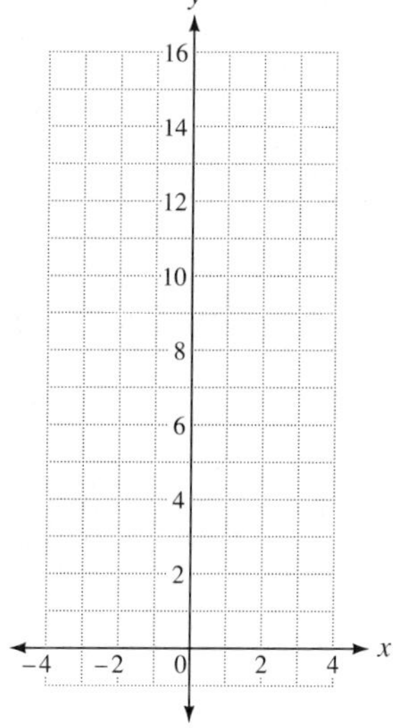

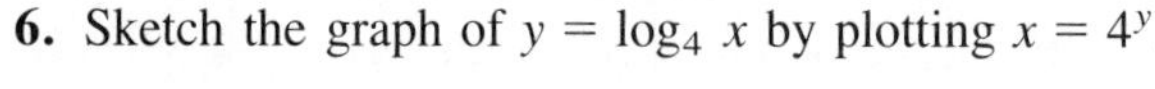

6. Sketch the graph of $y = \log_4 x$ by plotting $x = 4^y$.

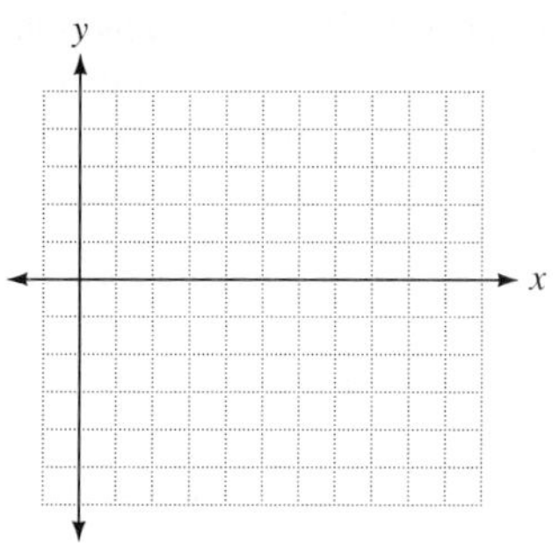

7. Having plotted $y = 4^x$ in Exercise 5, could you have used this information to sketch the graph required in Exercise 6? Explain.

8. The half-life of radioactive strontium-90, used for fuel in nuclear reactors, is approximately 28 years. This means that if 500 grams of strontium-90 are allowed to decay, the number A of grams remaining after t years is given by the exponential equation $A = (500)2^{-t/28}$. Graph this function for $t \geq 0$ and find the number of grams of strontium-90 remaining after **(a)** 0 years, **(b)** 28 years, **(c)** 56 years, and **(d)** 140 years.

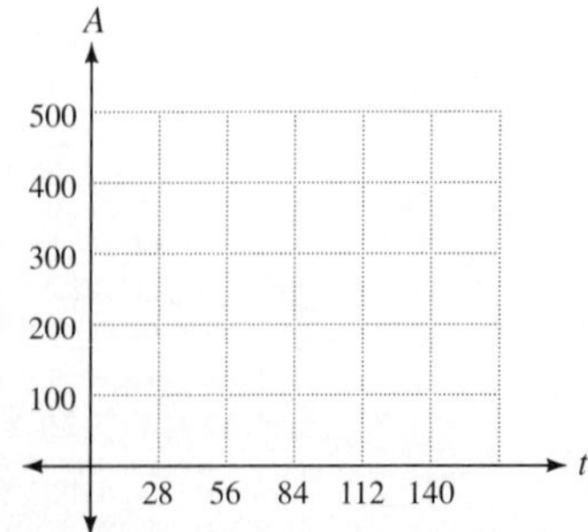

9. The value of a rare coin purchased for \$100 triples each year. This means that the value V after t years is given by the exponential function $V = (100)3^t$. Graph this function for $t \geq 0$ and determine the value of the coin after **(a)** 0 years, **(b)** 1 year, **(c)** 2 years, and **(d)** 5 years.

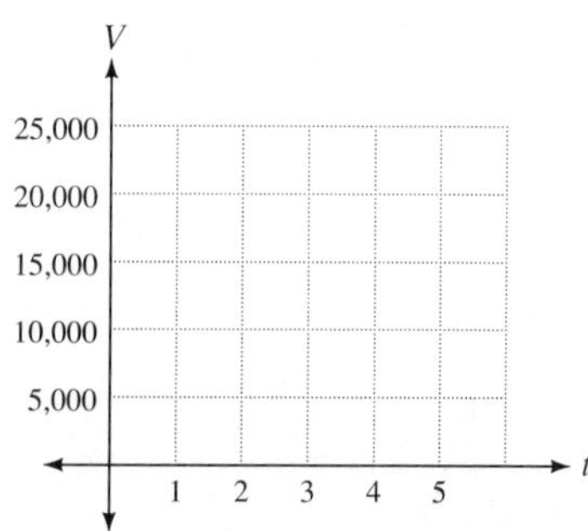

FOR REVIEW

Exercises 10–13 review properties of exponents from Section 1.6 to help you prepare for the next section. Simplify each expression.

10. a^4a^{10} **11.** $\frac{a^7}{a^3}$ **12.** $(a^6)^4$ **13.** $\frac{a^3a^{-2}}{a^{-1}}$

10.2 EXERCISES C

Graph the functions.

1. $y = 2^{x+1}$

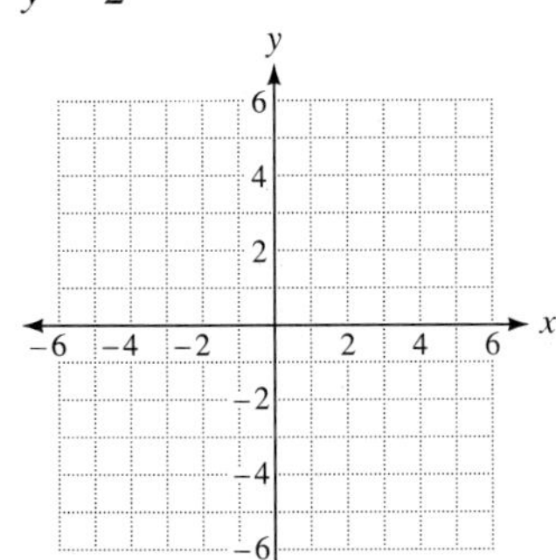

2. $y = \log_2 (x + 1)$

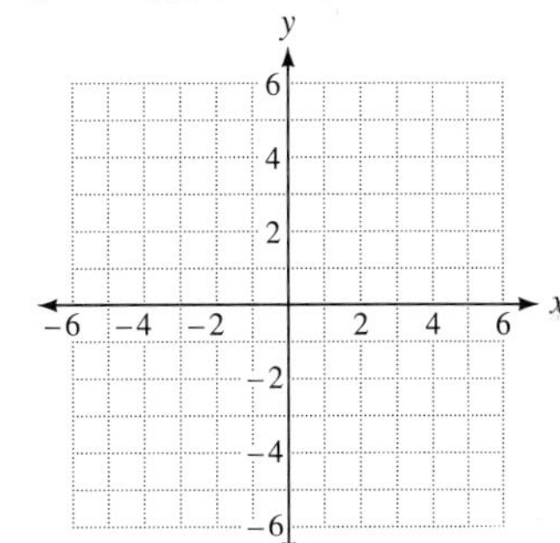

3. $y = 3^{2x}$

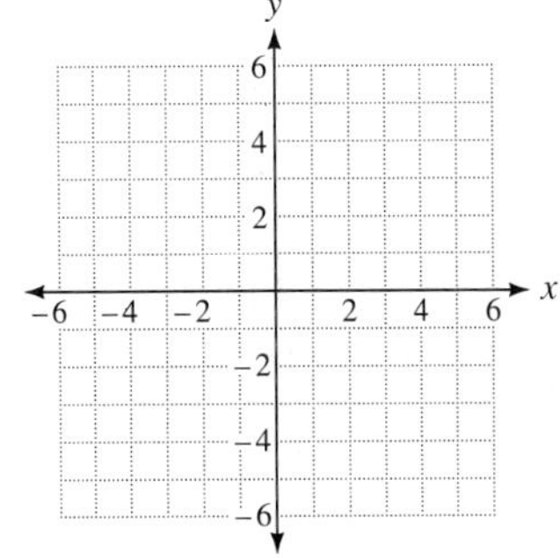

10.3 PROPERTIES OF LOGARITHMS

STUDENT GUIDEPOSTS

1. The Product Rule
2. The Quotient Rule
3. The Power Rule
4. Using a Combination of Rules

1 THE PRODUCT RULE

Since logarithms are exponents, the basic rules of exponents can be used to develop several useful properties of logarithms.

Suppose that a, x, and y are numbers, and a is a suitable base for a logarithm ($a > 0$, $a \neq 1$). The product rule for exponents is

$$a^x \cdot a^y = a^{x+y}.$$

Set $\quad u = a^x \quad \text{and} \quad v = a^y$

and write each in logarithmic form.

$$x = \log_a u \quad \text{and} \quad y = \log_a v$$

Then $\quad u \cdot v = a^x \cdot a^y = a^{x+y}.$

Write $u \cdot v = a^{x+y}$ in logarithmic form.

$$\log_a (u \cdot v) = x + y$$

Substitute $\log_a u$ for x and $\log_a v$ for y.

$$\log_a (u \cdot v) = \log_a u + \log_a v$$

This gives us the **product rule** for logarithms.

Product Rule

For any base a ($a > 0$, $a \neq 1$) and positive numbers u and v,

$$\log_a (u \cdot v) = \log_a u + \log_a v.$$

(The log of a product is the sum of the logs.)

EXAMPLE 1 USING THE PRODUCT RULE

Express $\log_2 (4 \cdot 8)$ as a sum of logarithms.

$$\log_2 (4 \cdot 8) = \log_2 4 + \log_2 8 \quad \text{Product rule}$$

In this case, we can verify the equation directly.

$$\log_2 (4 \cdot 8) = \log_2 32 = 5 \quad 2^5 = 32$$
$$\log_2 4 = 2 \quad 2^2 = 4$$
$$\log_2 8 = 3 \quad 2^3 = 8$$

Hence, $\log_2 (4 \cdot 8) = 5 = 2 + 3 = \log_2 4 + \log_2 8.$

PRACTICE EXERCISE 1

Express $\log_2 (16 \cdot 32)$ as a sum of logarithms.

Answer: $\log_2 16 + \log_2 32$

EXAMPLE 2 USING THE PRODUCT RULE IN REVERSE

Express $\log_a 3 + \log_a w$ as a single logarithm.

$$\log_a 3 + \log_a w = \log_a (3 \cdot w) \quad \text{Product rule in reverse}$$

PRACTICE EXERCISE 2

Express $\log_a 17 + \log_a m$ as a single logarithm.

Answer: $\log_a (17 \cdot m)$

CAUTION

Do not conclude that the sum of two logs is the log of a sum.

$$\log_a u + \log_a v \neq \log_a (u + v)$$

For example,

$$\log_a 6 = \log_a (2 \cdot 3) = \log_a 2 + \log_a 3 \neq \log_a (2 + 3) = \log_a 5.$$

Also note that $\log_a (u \cdot v) \neq \log_a u \cdot \log_a v$.

2 THE QUOTIENT RULE

If a is a base for a logarithm, and x and y are numbers, the quotient rule for exponents is

$$\frac{a^x}{a^y} = a^{x-y}.$$

As before, we set

$$u = a^x \quad \text{and} \quad v = a^y$$

and express these in logarithmic form.

$$x = \log_a u \quad \text{and} \quad y = \log_a v$$

Then
$$\frac{u}{v} = \frac{a^x}{a^y} = a^{x-y}.$$

Writing $\frac{u}{v} = a^{x-y}$ in logarithmic form, we obtain

$$\log_a \frac{u}{v} = x - y.$$

Substituting $\log_a u$ for x and $\log_a v$ for y,

$$\log_a \frac{u}{v} = \log_a u - \log_a v.$$

This is the **quotient rule** for logarithms.

Quotient Rule

For any base a ($a > 0$, $a \neq 1$) and positive numbers u and v,

$$\log_a \frac{u}{v} = \log_a u - \log_a v.$$

(The log of a quotient is the difference of the logs.)

EXAMPLE 3 Using the Quotient Rule

Express $\log_2 \frac{32}{8}$ as a difference of logarithms.

$$\log_2 \frac{32}{8} = \log_2 32 - \log_2 8 \qquad \text{Quotient rule}$$

In this case, we can verify the equation directly.

$$\log_2 \frac{32}{8} = \log_2 4 = 2 \qquad 2^2 = 4$$

$$\log_2 32 = 5 \qquad 2^5 = 32$$

$$\log_2 8 = 3 \qquad 2^3 = 8$$

Hence, $$\log_2 \frac{32}{8} = 2 = 5 - 3 = \log_2 32 - \log_2 8.$$

Practice Exercise 3

Express $\log_4 \frac{4}{64}$ as a difference of logarithms.

Answer: $\log_4 4 - \log_4 64$

EXAMPLE 4 Using the Quotient Rule in Reverse

Express $\log_a 3 - \log_a w$ as a single logarithm.

$$\log_a 3 - \log_a w = \log_a \frac{3}{w} \qquad \text{Quotient rule in reverse}$$

Practice Exercise 4

Express $\log_a 17 - \log_a m$ as a single logarithm.

Answer: $\log_a \frac{17}{m}$

CAUTION

Do not conclude that the difference of two logs is the log of the difference.

$$\log_a u - \log_a v \neq \log_a (u - v)$$

For example,

$$\log_a 3 = \log_a \frac{6}{2} = \log_a 6 - \log_a 2 \neq \log_a (6 - 2) = \log_a 4.$$

Note also that $\log_a \frac{u}{v} \neq \frac{\log_a u}{\log_a v}$.

3 THE POWER RULE

Assume that $u = a^x$

and raise both sides of this equation to the cth power.

$$u^c = (a^x)^c$$

By the power rule for exponents,

$$u^c = (a^x)^c = a^{x \cdot c}.$$

Converting $u^c = a^{x \cdot c}$ to logarithmic form, we obtain

$$\log_a u^c = x \cdot c.$$

But since $u = a^x$, $x = \log_a u$ so that

$$\log_a u^c = x \cdot c = (\log_a u) \cdot c = c \log_a u.$$

Thus, we have the **power rule** for logarithms.

Power Rule

For any base a ($a > 0$, $a \neq 1$), any positive number u and any number c,

$$\log_a u^c = c \log_a u.$$

(The log of a number to a power is the power times the log of the number.)

EXAMPLE 5 Using the Product Rule

(a) Express $\log_2 4^3$ as a product.

$$\log_2 4^3 = 3 \log_2 4 \quad \text{Power rule}$$

We can verify the above equation directly.

$$\log_2 4^3 = \log_2 64 = 6 \quad 2^6 = 64$$
$$3 \log_2 4 = 3 \cdot 2 = 6 \quad 2^2 = 4$$

(b) Express $\log_2 \sqrt[3]{64}$ as a product.

First we write $\sqrt[3]{64}$ as $64^{1/3}$. Then

$$\log_2 \sqrt[3]{64} = \log_2 64^{1/3} = \frac{1}{3} \log_2 64. \quad \text{Power rule}$$

The truth of this equation is clear since

$$\log_2 \sqrt[3]{64} = \log_2 4 = 2 \quad \sqrt[3]{64} = 4 \text{ and } 2^2 = 4$$

$$\text{and } \frac{1}{3} \log_2 64 = \frac{1}{3} \cdot 6 = 2. \quad 2^6 = 64$$

PRACTICE EXERCISE 5

(a) Express $\log_7 8^2$ as a product.

(b) Express $\log_7 27^{1/3}$ as a product.

Answers: (a) $2 \log_7 8$
(b) $\frac{1}{3} \log_7 27$

CAUTION

Notice the difference between $\log_a u^c$ and $(\log_a u)^c$. For example,

$$\log_a u^2 = 2 \log_a u \quad \text{but} \quad (\log_a u)^2 = (\log_a u)(\log_a u).$$

4 USING A COMBINATION OF RULES

Logarithms of expressions that involve more complicated products, quotients, or powers can be simplified using a combination of the rules.

EXAMPLE 6 Using a Combination of Rules

Express $\log_a \frac{x\sqrt{y}}{z^3}$ in terms of logarithms of x, y, and z.

$$\log_a \frac{x\sqrt{y}}{z^3} = \log_a \frac{x \cdot y^{1/2}}{z^3} \qquad \text{Convert radical to fractional exponent}$$
$$= \log_a (x \cdot y^{1/2}) - \log_a z^3 \qquad \text{Quotient rule}$$
$$= \log_a x + \log_a y^{1/2} - \log_a z^3 \qquad \text{Product rule}$$
$$= \log_a x + \frac{1}{2} \log_a y - 3 \log_a z \qquad \text{Power rule}$$

Practice Exercise 6

Express $\log_a \frac{u^3}{v^2\sqrt{w}}$ in terms of logarithms of u, v, and w.

Answer: $3 \log_a u - 2 \log_a v - \frac{1}{2} \log_a w$

Notice in Example 6 that the quotient and product rules are used first with the power rule last. When a combination of logarithms of this nature is simplified to a single logarithm using the rules in reverse, the power rule is used first followed by the quotient and product rules.

EXAMPLE 7 Using a Combination of Rules in Reverse

Express $\frac{1}{3} \log_a x - 5 \log_a y + \log_a z$ as a single logarithm.

$$\frac{1}{3} \log_a x - 5 \log_a y + \log_a z$$
$$= \log_a x^{1/3} - \log_a y^5 + \log_a z \qquad \text{Always use power rule first}$$
$$= \log_a \frac{x^{1/3}}{y^5} + \log_a z \qquad \text{Quotient rule}$$
$$= \log_a \left(\frac{x^{1/3}}{y^5} \cdot z\right) \qquad \text{Product rule}$$
$$= \log_a \frac{z\sqrt[3]{x}}{y^5} \qquad \text{Simplify}$$

Practice Exercise 7

Express $-4 \log_a u + \frac{1}{2} \log_a v - 8 \log_a w$ as a single logarithm.

Answer: $\log_a \frac{\sqrt{v}}{u^4w^8}$

EXAMPLE 8 Using the Rules to Evaluate Logarithms

Given that $\log_a 2 = 0.3010$ and $\log_a 3 = 0.4771$, find the following logarithms.

(a) $\log_a 6 = \log_a (2 \cdot 3) = \log_a 2 + \log_a 3$
$= 0.3010 + 0.4771$
$= 0.7781$

(b) $\log_a 9 = \log_a 3^2 = 2 \log_a 3 = 2(0.4771) = 0.9542$

(c) $\log_a \frac{3}{2} = \log_a 3 - \log_a 2 \doteq 0.4771 - 0.3010 = 0.1761$

Practice Exercise 8

Given that $\log_a 2 = 0.3010$ and $\log_a 5 = 0.6990$, find the following logarithms.

(a) $\log_a 10$

(b) $\log_a 50$

(c) $\log_a \frac{25}{4}$

Answers: (a) 1.0000 (b) 1.6990 (c) 0.7960

10.3 EXERCISES A

Express as a sum of logarithms.

1. $\log_3 (8 \cdot 80)$

2. $\log_5 (20 \cdot u)$

3. $\log_c (a \cdot b)$

Express as a difference of logarithms.

4. $\log_4 \frac{7}{30}$

5. $\log_2 \frac{12}{v}$

6. $\log_c \frac{a}{b}$

Express as a product.

7. $\log_2 x^5$

8. $\log_3 \sqrt{y}$

9. $\log_a b^{-3}$

Express as a single logarithm.

10. $\log_4 3 + \log_4 8$

11. $\log_b x + \log_b y$

12. $\log_2 5 - \log_2 9$

13. $\log_c x - \log_c y$

Express in terms of logarithms of x, y, and z.

14. $\log_a \frac{x^3 y}{z}$

15 $\log_a xy^2\sqrt{z}$

16. $\log_a \sqrt[3]{\frac{xy^2}{z}}$

Express as a single logarithm.

17 $\frac{1}{2} \log_a y - \log_a x + 2 \log_a z$

18. $3 \log_a x - \log_a y - \log_a z$

19. $\frac{1}{2}(\log_a x - \log_a y + 3 \log_a z)$

Given that $\log_a 3 = 0.4771$ *and* $\log_a 5 = 0.6990$, *find the following logarithms.*

20. $\log_a 15$

21. $\log_a \frac{5}{3}$

22. $\log_a 5^2$

23. $\log_a 75$

24. $\log_a \frac{25}{3}$

25 $\log_a 0.2$

26. $\log_a \sqrt[3]{5}$

27. $\log_a \sqrt{15}$

Tell whether the following are true or false.

28. $\frac{\log_a u}{\log_a v} = \log_a u - \log_a v$

29. $\log_a a = 1$

30. $\log_a 1 = 0$

31. $\log_a u^c = (\log_a c)(\log_a u)$

32. $\log_a (u + v) = \log_a u + \log_a v$

33. $\log_a (u \cdot v) = (\log_a u)(\log_a v)$

Solve.

34. Dr. Gordon Johnson, a well-known meteorologist, uses the formula $W = \log_2 (m + n)$. Calculate the value of W when m is 10 and n is 6.

35. When Dr. Johnson's assistant, Burford, was asked to find the value of W (Exercise 34) when m and n are both 8, he performed the following calculation:

$$W = \log_2 (m + n) = \log_2 (8 + 8)$$
$$= \log_2 8 + \log_2 8 = 3 + 3 = 6.$$

Why was Burford fired? What is the actual value of W?

36. If $Y = \log_2 (a - b)$, calculate the value of Y when $a = 8$ and $b = 4$, and show that $Y \neq \log_2 a - \log_2 b$.

37. If $V = \log_3 \frac{a}{b}$, calculate the value of V when $a = 27$ and $b = 3$, and show that $V \neq \frac{\log_3 a}{\log_3 b}$.

FOR REVIEW

38. Convert $x^a = y$ to logarithmic form.

39. Convert $\log_b m = p$ to exponential form.

Give the numerical value of x in the following.

40. $2^x = 128$

41. $x = \log_3 81$

42. $\log_x \dfrac{1}{8} = -3$

43. If you have the graph of $y = 5^x$, how could you use it to obtain the graph of $y = \log_5 x$?

44. The number n of deer in a newly opened game preserve is given by $n = 75 \log_4 (3t + 1)$, where t is time in years after the opening. Find the number of deer in the preserve when t is 5.

ANSWERS: 1. $\log_3 8 + \log_3 80$ 2. $\log_5 20 + \log_5 u$ 3. $\log_c a + \log_c b$ 4. $\log_4 7 - \log_4 30$ 5. $\log_2 12 - \log_2 v$ 6. $\log_c a - \log_c b$ 7. $5 \log_2 x$ 8. $\frac{1}{2} \log_3 y$ 9. $-3 \log_a b$ 10. $\log_4 (3 \cdot 8)$ 11. $\log_b xy$ 12. $\log_2 \frac{5}{9}$ 13. $\log_c \frac{x}{y}$ 14. $3 \log_a x + \log_a y - \log_a z$ 15. $\log_a x + 2 \log_a y + \frac{1}{2} \log_a z$ 16. $\frac{1}{3} (\log_a x + 2 \log_a y - \log_a z)$ 17. $\log_a \frac{z^2\sqrt{y}}{x}$ 18. $\log_a \frac{x^3}{yz}$ 19. $\log_a \sqrt{\frac{xz^3}{y}}$ 20. 1.1761 21. 0.2219 22. 1.3980 23. 1.8751 24. 0.9209 25. −0.6990 26. 0.2330 27. 0.5881 28. false 29. true (convert to exponential form) 30. true (convert to exponential form) 31. false 32. false 33. false 34. $W = \log_2 16 = 4$ 35. He was fired because he assumed that the logarithm of a sum is the sum of the logarithms, which is *not true*. The actual value of W when m and n are both 8 is 4. 36. $Y = 2 \neq 3 - 2 = 1$ 37. $V = 2 \neq \frac{3}{1} = 3$ 38. $a = \log_x y$ 39. $b^p = m$ 40. $x = 7$ 41. $x = 4$ 42. $x = 2$ 43. Reflect it across the line $y = x$ since the functions are inverses. 44. 150

10.3 EXERCISES B

Express as a sum of logarithms.

1. $\log_4 (5 \cdot 6)$

2. $\log_8 (a \cdot 9)$

3. $\log_b (u \cdot v)$

Express as a difference of logarithms.

4. $\log_5 \frac{8}{9}$

5. $\log_3 \frac{u}{7}$

6. $\log_b \frac{x}{y}$

Express as a product.

7. $\log_3 x^4$

8. $\log_4 \sqrt{z}$

9. $\log_a u^{-2}$

Express as a single logarithm.

10. $\log_3 4 + \log_3 9$

11. $\log_a w + \log_a y$

12. $\log_2 8 - \log_2 3$

13. $\log_a w - \log_a u$

Express in terms of logarithms of x, y, and z.

14. $\log_a \frac{x^2 y}{z^2}$

15. $\log_a xy^3\sqrt{z}$

16. $\log_a \sqrt[3]{\frac{x}{y^2 z}}$

Express as a single logarithm.

17. $\frac{1}{2} \log_a x - \log_a y + 3 \log_a z$

18. $2 \log_a z - \log_a y - \log_a x$

19. $\frac{1}{2} (\log_a y + \log_a x - 5 \log_a z)$

Given that $\log_a 3 = 0.4771$ and $\log_a 7 = 0.8451$, find the following logarithms.

20. $\log_a 21$

21. $\log_a \frac{7}{3}$

22. $\log_a 7^2$

23. $\log_a 147$

24. $\log_a \frac{49}{3}$

25. $\log_a \frac{15}{35}$

26. $\log_a \sqrt[3]{49}$

27. $\log_a \sqrt{63}$

Tell whether the following are true or false.

28. $\log_a 0 = 1$

29. $\log_a (x + y) = (\log_a x)(\log_a y)$

30. $\log_a 1 = a$

31. $x - y = \log_a x - \log_a y$

32. $\log_a x^c = (\log_a x)^c$

33. $c \log_a x = \frac{c}{\log_a x}$

Solve.

34. An environmental chemist might use the formula $H = \log_2 (a - p)$. Calculate the value of H when a is 10 and p is 6.

35. When Margo used the formula in Exercise 34 to find H when a is 8 and p is 4, she did the following calculation:

$$H = \log_2 (8 - 4) = \log_2 8 - \log_2 4$$
$$= 3 - 2 = 1.$$

What did she do wrong? What is the actual value of H?

36. If $U = \log_2 (a + b)$, calculate the value of U when $a = 8$ and $b = 8$, and show that $U \neq \log_2 a + \log_2 b$.

37. If $W = \log_3 ab$, calculate the value of W when $a = 1$ and $b = 9$, and show that $W \neq (\log_3 a)(\log_3 b)$.

FOR REVIEW

38. Convert $m^w = t$ to logarithmic form.

39. Convert $\log_p k = d$ to exponential form.

Find the numerical value of x in the following.

40. $2^x = 256$

41. $x = \log_5 625$

42. $\log_x \frac{1}{25} = -2$

43. What exponential equation would you graph to obtain the graph of $y = \log_3 x$?

44. The number n of items in stock in a distribution center is given by $n = 25 \log_3 (2t + 1)$, where t is time in days following the opening of the center. Find the number of items in stock when t is 4.

10.3 EXERCISES C

1. Express $\log_3 (a^2 - ab + b^2) + \log_3 (a + b)$ as a single logarithm and simplify.

2. Express $\log_a \sqrt[3]{\frac{x^2\sqrt{y}}{(x+y)^3}}$ as a sum or difference of logarithms.

3. Find the numerical value of x if $\frac{1}{3} \log_a 8 - \frac{5}{2} \log_a 4 - \log_a \frac{1}{16} = \log_a x$.
[Answer: 1]

4. Find the value of $\log_{10} [\log_4 (\log_2 16)]$.
[Answer: 0]

10.4 COMMON AND NATURAL LOGARITHMS

STUDENT GUIDEPOSTS

1. Common Logarithms
2. Base Conversion Formulas
3. Antilogarithms
4. Natural Logarithms
5. Natural Logarithmic Function

1 COMMON LOGARITHMS

Since $1^y = x$ can be true only if $x = 1$ ($1^y = 1$ for every y), $y = \log_1 x$ has very little importance. Thus, 1 is not used as a base for logarithms. However, every positive number except 1 can be used. Since our number system is based on 10, logarithms to the base 10 were commonly used for computation. As a result, base 10 logarithms are called **common logarithms.** For convenience, we omit the base number 10 in common logarithmic expressions. For example,

$$\log_{10} x \quad \text{can be written as} \quad \log x.$$

Thus, when you see a logarithm written without the base, the base is understood to be 10.

The fundamental property that makes common logarithms useful is the fact that every positive number can be expressed as a power of 10. Consider the list of integer powers of 10 on the following page and their logarithmic form.

Power of 10		*Logarithmic form*		
10^5	$= 100{,}000$	$\log_{10} 100{,}000$	$= \log 100{,}000$	$= 5$
10^4	$= 10{,}000$	$\log_{10} 10{,}000$	$= \log 10{,}000$	$= 4$
10^3	$= 1000$	$\log_{10} 1000$	$= \log 1000$	$= 3$
10^2	$= 100$	$\log_{10} 100$	$= \log 100$	$= 2$
10^1	$= 10$	$\log_{10} 10$	$= \log 10$	$= 1$
10^0	$= 1$	$\log_{10} 1$	$= \log 1$	$= 0$
10^{-1}	$= 0.1$	$\log_{10} 0.1$	$= \log 0.1$	$= -1$
10^{-2}	$= 0.01$	$\log_{10} 0.01$	$= \log 0.01$	$= -2$
10^{-3}	$= 0.001$	$\log_{10} 0.001$	$= \log 0.001$	$= -3$
10^{-4}	$= 0.0001$	$\log_{10} 0.0001$	$= \log 0.0001$	$= -4$
10^{-5}	$= 0.00001$	$\log_{10} 0.00001$	$= \log 0.00001$	$= -5$

Several observations can be made from the list. The logarithm of 1 is 0, the logarithm of a number greater than 1 is positive, and the logarithm of a number less than 1 is negative. On the assumption that numbers between two integer powers of 10 have logarithms between the logarithms of the integer powers of 10 (that is, between the two integers), the log of a number between 1 and 10 must be a number between 0 and 1. For example,

$$10^0 = 1 < 5 < 10 = 10^1$$

so that $$0 = \log 1 < \log 5 < \log 10 = 1.$$

Thus, the logarithm of 5 must be a positive decimal between 0 and 1. Similarly, the logarithm of a number between 10 and 100 must be a number between 1 and 2 and so forth. Prior to the availability of calculators, tables like that at the back of this book were used to approximate logarithms. The appendix of this text provides instructions and examples for these tables.

As an example, suppose we find the value of log 726. This means we are looking for the exponent (logarithms are exponents) on 10 that would give 726. That is, if $a = \log 726$, a is an exponent, and in fact, $10^a = 726$. With $10^2 = 100$, and $10^3 = 1000$, we would anticipate that since

$$10^2 < 10^a < 10^3$$
$$100 < 726 < 1000$$

a is a number between 2 and 3. We find this number in Example 1(a).

EXAMPLE 1 Using a Calculator to Find Common Logarithms

Use a calculator to find the following common logarithms.

(a) log 726

Enter 726 and press the key labeled [LOG], as in the sequence below.

726 [LOG] → [2.8609366]

The display shows log 726. Remember that this means 10 raised to the power 2.8609366 is approximately 726.

(b) log 0.00459

Use the following sequence.

0.00459 [LOG] → [−2.3381873]

Correct to four decimal places, $\log 0.00459 \approx -2.3382$.

Practice Exercise 1

Use a calculator to find the following common logarithms, correct to four decimal places.

(a) log 249

(b) log 0.000641

(c) log (−4562)

The key [+/−] changes the sign of 4562 in the steps below.

4562 [+/−] [LOG] → [ERROR]

In this case, the display shows *error* since logarithms of negative numbers are not defined.

(c) log (−0.0543)

Answers: (a) 2.3962 (b) −3.1931 (c) none

2 BASE CONVERSION FORMULAS

Now that we know how to find common logarithms using a calculator, we might wonder how to find logarithms to other bases. To answer this we will explore the relationship among logarithms to different bases. For example, consider $\log_a x$ and $\log_b x$. If we let

$$u = \log_b x,$$

then

$$b^u = x.$$

Take the logarithm to the base a of both sides.

$$\log_a b^u = \log_a x \quad \text{If } w = z \text{ then } \log_a w = \log_a z$$

$$u \log_a b = \log_a x \quad \text{Power rule}$$

$$(\log_b x)(\log_a b) = \log_a x \quad \text{Substitute } \log_b x \text{ for } u$$

$$\log_b x = \frac{\log_a x}{\log_a b} \quad \textit{Not } \log_a \frac{x}{b} = \log_a x - \log_a b$$

A special case of this formula results if $x = a$.

$$\log_b x = \frac{\log_a x}{\log_a b}$$

$$\log_b a = \frac{\log_a a}{\log_a b} \quad \text{Substitute } a \text{ for } x$$

$$\log_b a = \frac{1}{\log_a b} \quad \text{Log}_a\, a = 1$$

We have just established the following conversion formulas.

Base Conversion Formulas

If a and b are any two bases, and $x > 0$, then

$$\log_b x = \frac{\log_a x}{\log_a b} \quad \text{and} \quad \log_b a = \frac{1}{\log_a b}.$$

EXAMPLE 2 USING BASE CONVERSION FORMULAS

(a) Express $\log_{100} x$ in terms of $\log_{10} x$ and simplify.

$$\log_{100} x = \frac{\log_{10} x}{\log_{10} 100} = \frac{\log x}{\log 10^2} = \frac{\log x}{2} = \frac{1}{2} \log x$$

PRACTICE EXERCISE 2

(a) Express $\log_{16} 4$ in terms of $\log_2 4$ and simplify.

(b) Express $\log_5 9$ in terms of base 9 logarithms.

$$\log_5 9 = \frac{1}{\log_9 5}$$

(b) Express $\log_3 7$ in terms of base 7 logarithms.

Answers: (a) $\frac{1}{2}$ (b) $\frac{1}{\log_7 3}$

We can now find the logarithm of a number to any base using the first base conversion formula, common logarithms, and a calculator.

EXAMPLE 3 Finding Logarithms to Other Bases

Find $\log_7 428$ correct to four decimal places.

Using the first base conversion formula with $x = 428$, $b = 7$, and $a = 10$, we have the following.

$$\log_7 428 = \frac{\log 428}{\log 7} \approx 3.1138$$

The calculator steps are:

428 [LOG] [÷] 7 [LOG] [=] → [3.1137734].

Practice Exercise 3

Find $\log_5 743$, correct to four decimal places.

Answer: 4.1075

CAUTION

In the formula $\log_b x = \frac{\log_a x}{\log_a b}$, we divide the two logarithms $\log_a x$ and $\log_a b$; we don't subtract them. Do not confuse this with $\log_a \frac{x}{b}$, which is $\log_a x - \log_a b$. Notice in Example 3 that we divided log 428 by log 7 to obtain 3.1138; we did not subtract.

In Example 3 we found that $\log_7 428 \approx 3.1138$. Remembering that a logarithm is an exponent, this means that $7^{3.1138} \approx 428$. We can use the [y^x] key on a calculator to verify this result.

7 [y^x] 3.1138 [=] → [428.022]

Notice that due to round-off errors, the display is not exactly 428, but rather an approximation of it.

3 ANTILOGARITHMS

If we are given a number, we now know how to find its common logarithm. Suppose we reverse this procedure and find a number when its logarithm is given. The number that corresponds to a given logarithm is called its **antilogarithm (antilog).** In the equation

$$x = \log n$$

x is the logarithm of n and n is the antilogarithm of x. In exponential form, $n = 10^x$. To find an antilogarithm using a calculator, we use either the [10^x] key or the [INV] key followed by the [LOG] key. Both sequences are given below where we find the antilogarithm of 2.8609366.

2.8609366 [10^x] → [725.99997]

2.8609366 [INV] [LOG] → [725.99997]

In Example 1(a), we found log 726 to be 2.8609366. Notice that the antilogarithm of 2.8609366 found above does not give the exact value 726, but rather a close approximation to it due to round-off differences in the calculator.

EXAMPLE 4 Finding Antilogarithms

Use a calculator to find the following antilogs correct to three significant digits.

(a) $\log n = 0.6278$

We need to find n where $n = 10^{0.6278}$.

0.6278 [10^x] → 4.2442406

or 0.6278 [INV] [LOG] → 4.2442406

Thus, n is approximately 4.24.

(b) $\log n = -4.3205$

4.3205 [+/−] [10^x] → 0.0000478

or 4.3205 [+/−] [INV] [LOG] → 0.0000478

Thus, n is approximately 0.0000478.

Practice Exercise 4

Use a calculator to find each antilog, correct to three significant digits.

(a) $\log n = 1.6275$

(b) $\log n = -2.0058$

Answers: (a) 42.4 (b) 0.00987

4 NATURAL LOGARITHMS

Logarithm bases other than 10 occur frequently in science and mathematics. One of the most widely used base numbers is e. The number e is an irrational number, approximately equal to 2.71828, and originates in a natural way in calculus. For this reason, logarithms to the base e are called **natural logarithms.** The standard notation for the natural logarithm of a number x is

$$\ln x \text{ instead of } \log_e x.$$

The [ln x] key on a calculator is used to find natural logarithms the same way the [LOG] key is used to find common logarithms.

EXAMPLE 5 Finding Natural Logarithms

Use a calculator to find the following natural logarithms, correct to four decimal places.

(a) ln 329

The steps to follow are given below.

329 [ln x] → 5.7960578

Thus, $\ln x$ is approximately 5.7961, correct to four decimal places.

(b) ln 0.00541

0.00541 [ln x] → −5.2195062

Thus, $\ln 0.00541 \approx -5.2195$.

Practice Exercise 5

Find the following natural logarithms, correct to four decimal places.

(a) ln 1598

(b) ln 0.000641

Answers: (a) 7.3765 (b) −7.3525

To find antilogarithms of natural logarithms, we use either the $\boxed{e^x}$ key or the $\boxed{\text{INV}}$ and $\boxed{\ln x}$ keys, much as we did for common logarithms.

EXAMPLE 6 FINDING NATURAL ANTILOGARITHMS

Use a calculator to find the following antilogs correct to three significant digits.

(a) $\ln x = 3.2541$

We need to find x where $x = e^{3.2541}$

$$3.2541 \ \boxed{e^x} \rightarrow \boxed{25.896297}$$

$$\text{or} \quad 3.2541 \ \boxed{\text{INV}} \ \boxed{\ln x} \rightarrow \boxed{25.896297}$$

Thus, x is approximately 25.9.

(b) $\ln x = -1.4738$

$$1.4738 \ \boxed{+/-} \ \boxed{e^x} \rightarrow \boxed{0.2290534}$$

$$\text{or} \quad 1.4738 \ \boxed{+/-} \ \boxed{\text{INV}} \ \boxed{\ln x} \rightarrow \boxed{0.2290534}$$

Thus, x is approximately 0.229.

PRACTICE EXERCISE 6

Use a calculator to find each antilog, correct to three significant digits.

(a) $\ln x = 2.6117$

(b) $\ln x = -3.7542$

Answers: (a) 13.6 (b) 0.0234

5 NATURAL LOGARITHMIC FUNCTION

In Section 10.2 we graphed exponential functions $y = a^x$ and their inverse logarithmic functions $y = \log_a x$. The **natural logarithmic function** $y = \ln x$ and its inverse $y = e^x$ are important functions in mathematics. So important is $y = e^x$ that it is recognized above all other exponential functions and sometimes called *the* **exponential function.** Representative values of each function are given in the tables with the graphs shown in Figure 10.8.

x	$\ln x$
0.5	−0.7
1	0
2	0.7
3	1.1
4	1.4
5	1.6

x	e^x
−3	0.05
−2	0.1
−1	0.4
0	1
1	2.7
2	7.4

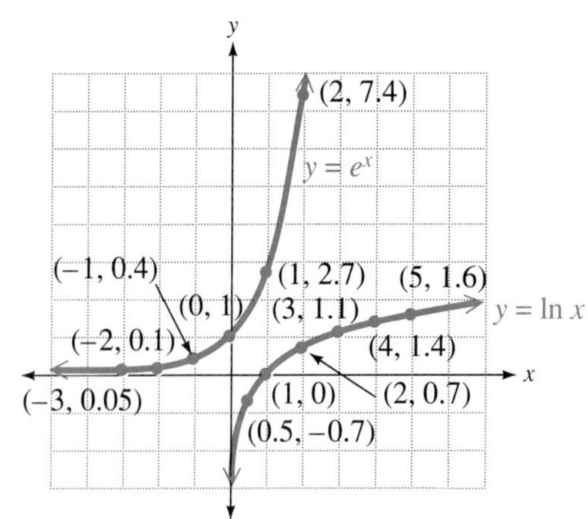

Figure 10.8 Natural Logarithmic and Exponential Functions

Variations of the exponential function appear in numerous applications. Two of these variations are graphed in Figure 10.9.

x	e^{-x}	$e^{-x/2}$
0	1.00	1.00
−1	2.72	1.65
1	0.37	0.61
−2	7.40	2.72
2	0.14	0.37
−3	20.1	4.49
3	0.05	0.22
−4	54.7	7.40
4	0.02	0.14

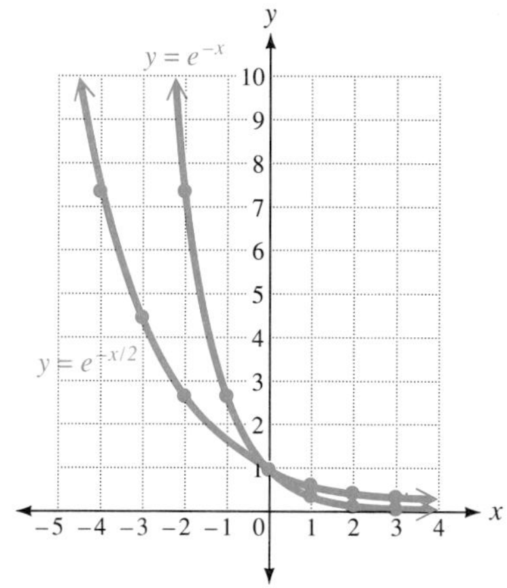

Figure 10.9 Variations of Exponential Function

10.4 EXERCISES A

Tell whether the statement is true *or* false. *If the statement is false, explain why.*

1. Common logarithms use 10 as the base number.
2. Instead of writing $\log_{10} x$ we usually write $\log x$.
3. In the equation $y = \log m$, y is called the antilogarithm of m.

Find each common logarithm, correct to four decimal places.

4. $\log 253$
5. $\log 0.159$
6. $\log 17.6$
7. $\log(-43)$
8. $\log 1{,}265{,}000$
9. $\log 0.00391$

Express the given logarithms in terms of the indicated base, and simplify.

10. $\log_{1000} x$; 10
11. $\log_{32} 8$; 2
12. $\log_7 12$; 12
13. $\log_{100} 10^6$; 10
14. $\log_{36} x$; 6
15. $\log_6 x$; 36

Evaluate each logarithm using common logarithms.

16. $\log_{100} 10$
17. $\log_7 12$
18. $\log_5 30$
19. $\log_7 1.32$
20. $\log_2 0.0352$
21. $\log_{4.5} 275$

Find each common antilogarithm, correct to three significant digits.

22. $\log n = 2.0359$ **23.** $\log n = 0.6298$ **24.** $\log n = -1.7645$ **25.** $\log n = 3.6415$

Find each natural logarithm, correct to four decimal places.

26. $\ln 627$ **27.** $\ln 0.654$ **28.** $\ln 18.5$ **29.** $\ln 23{,}400$

Find each natural antilogarithm, correct to three significant digits.

30. $\ln x = 1.4532$ **31.** $\ln x = 0.6428$ **32.** $\ln x = 4.6565$ **33.** $\ln x = -3.4115$

34 The demand equation for x units of a product relative to the price per unit p is given by $p = 200 - 0.4\, e^{0.004x}$. Find the price if the demand is 100 units.

35. The atmospheric pressure P, measured in pounds per square inch, at an altitude x, measured in miles above sea level, can be approximated by $P = 14.7\, e^{-0.2x}$. What is the atmospheric pressure on the outside of a jet airplane cruising at an altitude of 6 miles?

36. In chemistry, the **pH (hydrogen potential)** of a substance is defined by

$$\text{pH} = -\log\,[\text{H}^+],$$

where $[\text{H}^+]$ is the concentration of hydrogen ions in the solution. The pH scale varies from 0 to 14, with the pH of distilled water equal to 7. A substance is an acid if its pH is less than 7, and a base if its pH is greater than 7. Find the pH of a soft drink with $[\text{H}^+] = 3.45 \times 10^{-7}$.

37 Stars are categorized by brightness on a scale measured in magnitudes with the faintest star visible to the naked eye assigned a magnitude of 6. Telescopes are designed to view stars beyond a magnitude of 6. The limiting magnitude L of a telescope depends on the diameter d of its lens and is given by $L = 8.8 + 2.2 \ln d$. What is the limiting magnitude of a telescope with a lens 15 inches in diameter?

FOR REVIEW

Express as a single logarithm.

38. $\frac{1}{3}(\log_a x - 2\log_a y)$ **39.** $3\log_a x - \log_a y - 4\log_a z$

Express in terms of logarithms of x, y, and z.

40. $\log_a \frac{x^3\sqrt{z}}{y^5}$ **41.** $\log_a \sqrt{xyz}$

Exercises 42–51 review some of the properties of logarithms.

42. $\log 2 =$ ______ **43.** $\log 3 =$ ______ **44.** $\log 5 =$ ______ **45.** $\log 6 =$ ______

46. Does $\log (2 + 3) = \log 2 + \log 3$?

47. Does $\log (2 \cdot 3) = (\log 2)(\log 3)$?

48. Does $\log (2 \cdot 3) = \log 2 + \log 3$?

49. Does $\log \frac{6}{3} = \frac{\log 6}{\log 3}$?

50. Does $\log \frac{6}{3} = \log 6 - \log 3$?

51. Does $\log_2 5 = \log 5 - \log 2$?

The following exercises will help you prepare for the next section. Tell whether the statement in Exercises 52–54 is true *or* false.

52. If $x = y$ then $\log_a x = \log_a y$.

53. If $2^x = 2^y$ then $x = 2y$.

54. If $\log_a x = \log_a y$ then $x = y$.

Solve each equation.

55. $\frac{x + 10}{x + 1} = 10$

56. $x(x + 1) = 2$

ANSWERS: 1. true 2. true 3. false, y is the logarithm of m 4. 2.4031 5. −0.7986 6. 1.2455 7. Logarithms of negative numbers are not defined. 8. 6.1021 9. −2.4078 10. $\frac{1}{3} \log x$ 11. $\frac{3}{5}$ 12. $\frac{1}{\log_{12} 7}$ 13. 3 14. $\frac{1}{2} \log_6 x$ 15. $2 \log_{36} x$ 16. $\frac{1}{2}$ 17. 1.2770 18. 2.1133 19. 0.1427 20. −4.8283 21. 3.7344 22. 109 23. 4.26 24. 0.0172 25. 4380 26. 6.4409 27. −0.4246 28. 2.9178 29. 10.0605 30. 4.28 31. 1.90 32. 105 33. 0.0330 34. \$199.40 35. approximately 4.43 lb/in^2 36. approximately 6.46 37. approximately 14.8 38. $\log_a \sqrt[3]{\frac{x}{y^2}}$ 39. $\log_a \frac{x^3}{yz^4}$ 40. $3 \log_a x + \frac{1}{2} \log_a z - 5 \log_a y$ 41. $\frac{1}{2} (\log_a x + \log_a y + \log_a z)$ 42. 0.3010 43. 0.4771 44. 0.6990 45. 0.7782 46. no 47. no 48. yes 49. no 50. yes 51. no 52. true 53. false 54. true 55. 0 56. −2, 1

10.4 EXERCISES B

Tell whether the statement is true *or* false. *If the statement is false, explain why.*

1. Natural logarithms use 2 as the base number.

2. The natural logarithm of x is written $\ln x$.

3. If n is the common antilogarithm of m, then $n = 10^m$.

Find each common logarithm, correct to four decimal places.

4. $\log 478$ **5.** $\log 0.987$ **6.** $\log 38.4$

7. $\log (-67)$ **8.** $\log 3{,}415{,}000$ **9.** $\log 0.00628$

Express the given logarithms in terms of the indicated base, and simplify.

10. $\log_{100} x$; 10 **11.** $\log_{64} 8$; 2 **12.** $\log_5 8$; 8

13. $\log_{1000} 10^6$; 10 **14.** $\log_{81} x$; 3 **15.** $\log_3 x$; 81

Evaluate each logarithm using common logarithms.

16. $\log_{1000} 10$

17. $\log_{12} 7$

18. $\log_8 24$

19. $\log_4 62.4$

20. $\log_3 0.000915$

21. $\log_{1.6} 0.134$

Find each common antilogarithm, correct to three significant digits.

22. $\log n = 4.1378$

23. $\log n = 0.4134$

24. $\log n = -2.6643$

25. $\log n = 5.1167$

Find each natural logarithm, correct to four decimal places.

26. $\ln 431$

27. $\ln 0.238$

28. $\ln 27.6$

29. $\ln 0.00395$

Find each natural antilogarithm, correct to three significant digits.

30. $\ln x = 2.3597$

31. $\ln x = 0.7415$

32. $\ln x = 5.2138$

33. $\ln x = -2.0015$

34. The demand equation for x units of a product relative to the price per unit p is given by $p = 200 - 0.4\, e^{0.004x}$. Find the price if the demand is 150 units.

35. The atmospheric pressure P, measured in pounds per square inch, at an altitude x, measured in miles above sea level, can be approximated by $P = 14.7\, e^{-0.2x}$. What is the atmospheric pressure on the outside of an airplane cruising at an altitude of 4 miles?

36. In chemistry, the **pH (hydrogen potential)** of a substance is defined by

$$\text{pH} = -\log\, [\text{H}^+],$$

where $[\text{H}^+]$ is the concentration of hydrogen ions in the solution. The pH scale varies from 0 to 14, with the pH of distilled water equal to 7. A substance is an acid if its pH is less than 7, and a base if its pH is greater than 7. Find the pH of a spa with $[\text{H}^+] = 2.15 \times 10^{-8}$.

37. Stars are categorized by brightness on a scale measured in magnitudes with the faintest star visible to the naked eye assigned a magnitude of 6. Telescopes are designed to view stars beyond a magnitude of 6. The limiting magnitude L of a telescope depends on the diameter d of its lens and is given by $L = 8.8 + 2.2 \ln d$. What is the limiting magnitude of a telescope with a lens 10 inches in diameter?

Express as a single logarithm.

38. $\dfrac{1}{3}(\log_a x + 2 \log_a y)$

39. $3 \log_a x + \log_a y - 4 \log_a z$

Express in terms of logarithms of x, y, and z.

40. $\log_a \dfrac{xy^3}{\sqrt{z}}$

41. $\log_a \sqrt[3]{xyz}$

Exercises 42–51 review some of the properties of logarithms.

42. $\log 2 =$ ______

43. $\log 4 =$ ______

44. $\log 6 =$ ______

45. $\log 8 =$ ______

46. Does $\log(2 + 4) = \log 2 + \log 4$?

47. Does $\log(2.4) = (\log 2)(\log 4)$?

48. Does $\log(2 \cdot 4) = \log 2 + \log 4$?

49. Does $\log \dfrac{8}{4} = \dfrac{\log 8}{\log 4}$?

50. Does $\log \dfrac{8}{2} = \log 8 - \log 2$?

51. Does $\log_4 8 = \log 8 - \log 4$?

The following exercises will help you prepare for the next section. Tell whether the statement in Exercises 52–54 is true *or* false.

52. If $x \neq y$ then $\log_a x \neq \log_a y$.

53. If $2^x = 2^y$ then $x = y$.

54. If $\log_a x = \log_a y$ then $x = ay$.

Solve each equation.

55. $\dfrac{x+2}{x-2} = 4$

56. $x(x - 3) = 4$

10.4 EXERCISES C

Sketch the graph of each function.

1. $y = e^{x+1}$

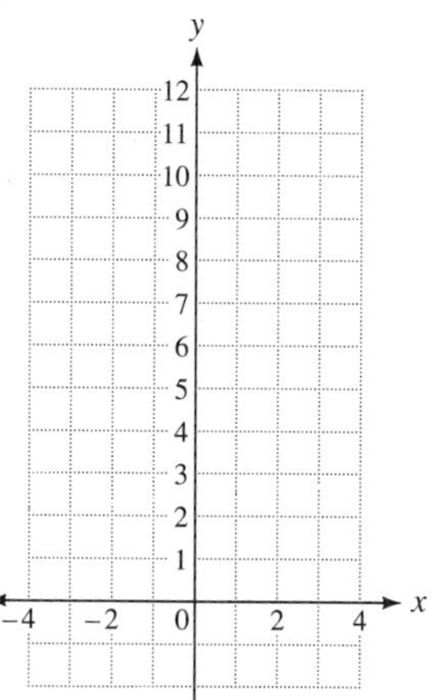

2. $y = e^{x-1}$

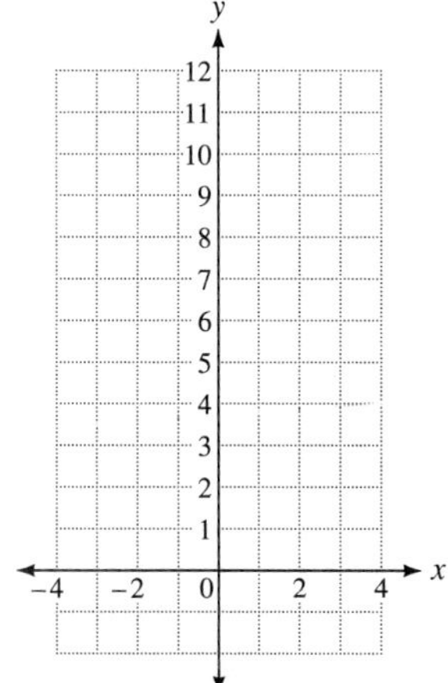

3. $y = e^{3x}$

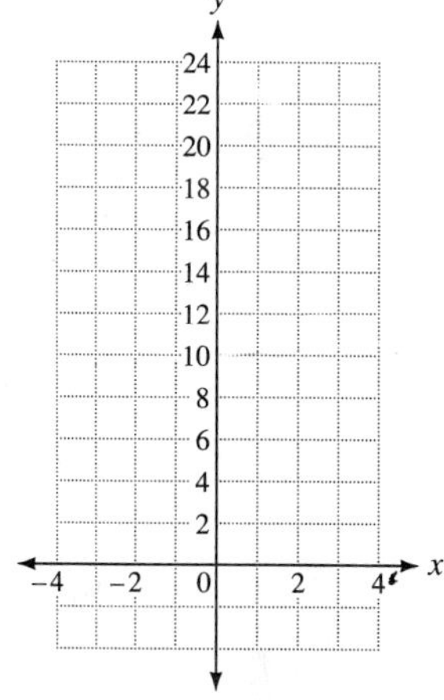

4. $y = e^{x^2}$

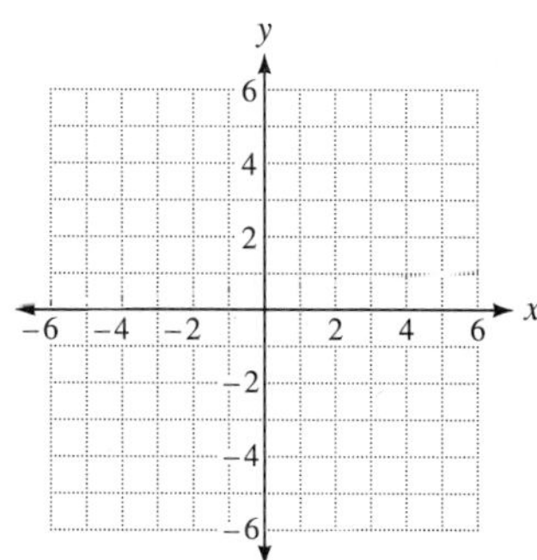

5. $y = \ln (x + 1)$

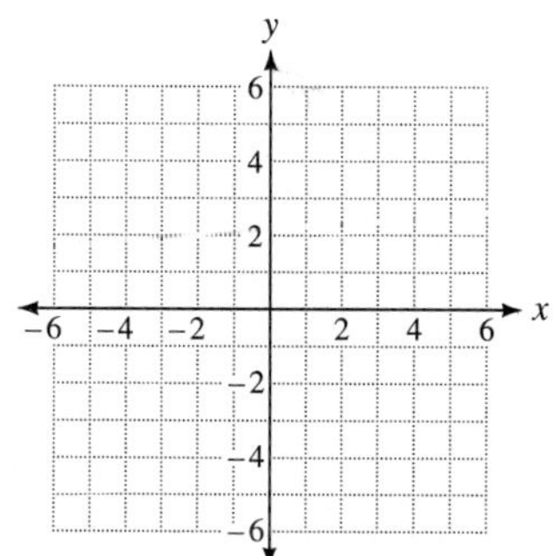

6. $y = \ln x^2$

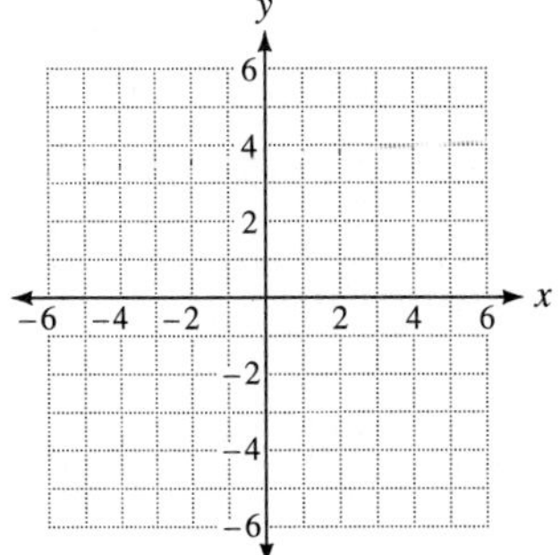

10.5 LOGARITHMIC AND EXPONENTIAL EQUATIONS

STUDENT GUIDEPOSTS

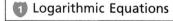

1. Logarithmic Equations
2. Exponential Equations

1 LOGARITHMIC EQUATIONS

Equations that contain logarithms of expressions involving the variable are called **logarithmic equations.** Solving these equations requires a thorough knowledge of the properties of logarithms.

EXAMPLE 1 SOLVING A LOGARITHMIC EQUATION

Solve $\log_5 x = \log_5 (3x - 8)$.
If the logarithms of two numbers using the same base are equal, then the numbers are equal. Thus,

$$x = 3x - 8. \quad \text{If } \log_a n = \log_a m, \text{ then } n = m$$
$$-2x = -8$$
$$x = 4$$

Check: $\log_5 4 \stackrel{?}{=} \log_5 (3(4) - 8)$
$\log_5 4 \stackrel{?}{=} \log_5 (12 - 8)$
$\log_5 4 = \log_5 4$

The solution is 4.

PRACTICE EXERCISE 1

Solve $\log_3 2x = \log_3 (x - 4)$.
Be sure to check!

Answer: no solution

CAUTION

Always check possible solutions in the original equation. Remember that any value that results in the logarithm of a negative number or zero must be excluded. This happened in Practice Exercise 1 above. Solving as in Example 1, we would obtain $x = -4$, but $\log 2(-4) = \log (-8)$ is not defined. Since this solution must be discarded, we concluded that the equation has no solution.

EXAMPLE 2 SOLVING A LOGARITHMIC EQUATION

Solve.

$$\log (x + 10) - \log (x + 1) = 1$$

$$\log \frac{x + 10}{x + 1} = 1 \quad \text{Use quotient rule on left side}$$

$$\frac{x + 10}{x + 1} = 10^1 = 10 \quad \text{Convert to exponential form}$$

$$x + 10 = 10(x + 1) \quad \text{Clear fraction}$$
$$x + 10 = 10x + 10$$
$$0 = 9x$$
$$0 = x$$

Check: $\log (0 + 10) - \log (0 + 1) \stackrel{?}{=} 1$
$\log 10 - \log 1 \stackrel{?}{=} 1$
$1 - 0 \stackrel{?}{=} 1$
$1 = 1$

The solution is 0.

PRACTICE EXERCISE 2

Solve.

$$\log (x + 8) - \log (x - 1) = 1$$

Answer: 2

To Solve a Logarithmic Equation

1. Obtain a single logarithmic expression using the same base on one side of the equation, or write each side as a logarithm using the same base.
2. Convert the result to an exponential equation or find the antilog of both sides and solve the result.
3. Check all possible solutions in the original equation. (Negative numbers and zero do not have logarithms.)

EXAMPLE 3 SOLVING A LOGARITHMIC EQUATION

Solve.

$\log x + \log (x + 1) = \log 2$

$\log x(x + 1) = \log 2$ Product rule

$x(x + 1) = 2$ If $\log n = \log m$ then $n = m$

$x^2 + x - 2 = 0$ Write the quadratic equation in general form

$(x + 2)(x - 1) = 0$ Factor

$x + 2 = 0$ or $x - 1 = 0$ Use the zero-product rule

$x = -2$ $x = 1$

When we try to check -2, we obtain

$$\log (-2) + \log (-2 + 1) \stackrel{?}{=} \log 2$$

which means that -2 is not a solution since $\log (-2)$ and $\log (-1)$ are not defined. The only solution is 1. (It does check.)

PRACTICE EXERCISE 3

Solve.

$\log (x - 2) + \log (x + 2) = \log 5$

Answer: 3

2 EXPONENTIAL EQUATIONS

Equations with variable exponents are called **exponential equations.** Such equations can often be solved by taking the logarithm (using an appropriate base) of both sides and then applying the power rule.

EXAMPLE 4 SOLVING AN EXPONENTIAL EQUATION

Solve the exponential equation for x.

$10^{2x+1} = 100$

$\log 10^{2x+1} = \log 100$ Take common log of both sides

$(2x + 1) \log 10 = \log 100$ Power rule

$(2x + 1)(1) = 2$ $\log 10 = 1$ and $\log 100 = 2$

$2x + 1 = 2$

$2x = 1$

$x = \frac{1}{2}$

The solution is $\frac{1}{2}$. Check in the original equation.

PRACTICE EXERCISE 4

Solve the exponential equation for x.

$5^{3x-2} = 25$

Answer: $\frac{4}{3}$

When both sides of an exponential equation can be expressed as a power having the same base, it is much faster to equate the corresponding exponents (set them equal to each other) and solve. Suppose we work Example 4 using this method.

$$10^{2x+1} = 100$$
$$10^{2x+1} = 10^2 \quad \text{Express as powers with same base}$$
$$2x + 1 = 2 \quad \text{Equate exponents, since } 10^n = 10^m \text{ means } n = m$$
$$2x = 1$$
$$x = \frac{1}{2}$$

In such cases we can eliminate a few steps.

To Solve an Exponential Equation

1. Try to express each side as a power using the same base, and equate the resulting exponents.
2. If Step 1 fails, take the logarithm of each side and use the power rule to eliminate the variable exponents.
3. Solve the resulting equation and check in the original.

EXAMPLE 5 Solving an Exponential Equation

Solve $3^{x+1} = \frac{1}{27}$.

Since $\frac{1}{27} = \frac{1}{3^3} = 3^{-3}$, we need to solve the equation below.

$$3^{x+1} = 3^{-3} \quad \text{Express as powers with same base}$$
$$x + 1 = -3 \quad \text{Equate exponents}$$
$$x = -4 \quad \text{This does check}$$

The solution is -4.

Practice Exercise 5

Solve $7^{2x-1} = \frac{1}{49}$.

Answer: $-\frac{1}{2}$

EXAMPLE 6 Solving an Exponential Equation

Solve $5^{2x} = 8$.

Since 8 is not easily expressible as a power with base 5, we take the common logarithm of both sides.

$$\log 5^{2x} = \log 8$$
$$2x \log 5 = \log 8 \quad \text{Power rule}$$
$$x = \frac{\log 8}{2 \log 5} \quad \text{Divide by } 2 \log 5$$

This is the exact form of the answer. If we need a decimal approximation, such as in an applied problem, we would use a calculator. The steps are given below.

8 [LOG] [÷] 5 [LOG] [÷] 2 [=] → [0.6460148]

Thus, $x \approx 0.646$, correct to three decimal places.

Practice Exercise 6

Solve $9^{3x} = 4$.

Answer: $\frac{\log 4}{3 \log 9} \approx 0.210$

Many applied problems use formulas that involve natural logarithmic or exponential equations.

EXAMPLE 7 PHYSICS PROBLEM

When a beam of light passes through water, as shown in Figure 10.10, the intensity of the light I is diminished by the number of feet x from the surface according to the formula $I = I_0 e^{-kx}$, where I_0 is the intensity of the light at the surface and k is the coefficient of extinction depending on the nature of the water. If the coefficient of extinction of Lake Mary is 0.06, at what depth (to the nearest tenth of a foot) will the intensity of a light be reduced to 50% of that at the surface?

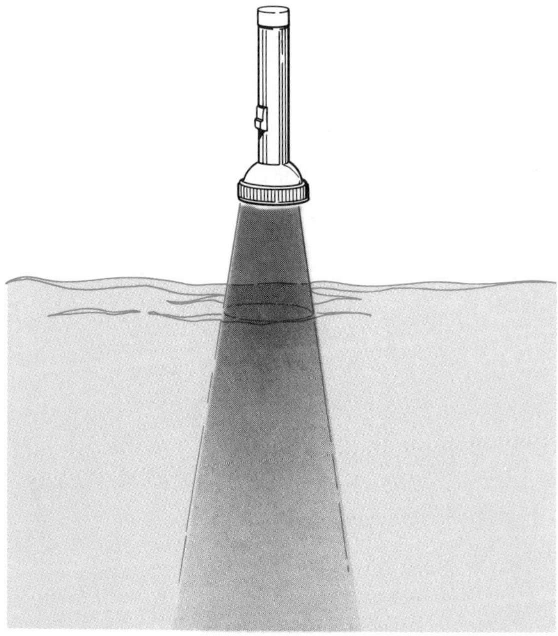

Figure 10.10

We are asked to find the depth at which $I = 0.5I_0$ (50% of the original intensity) when $k = 0.06$. Substitute these values in the formula.

$$0.5I_0 = I_0 e^{-0.06x}$$

$$0.5 = e^{-0.06x} \quad \text{Divide both sides by } I_0$$

$$\ln 0.5 = \ln e^{-0.06x} \quad \text{Take the natural log of both sides}$$

$$\ln 0.5 = -0.06x \quad \ln e^u = u$$

$$\frac{\ln 0.5}{-0.06} = x \quad \text{Divide both sides by } -0.06$$

Using a calculator we find $x \approx 11.552453$.

At a depth of about 11.6 ft, the light has been diminished by 50%.

PRACTICE EXERCISE 7

Refer to Example 7. At what depth will the intensity of light be 25% of the original intensity?

Answer: 23.1 ft

10.5 EXERCISES A

Solve the following for x. When appropriate, give answer using three significant digits.

1. $\log (3x + 1) = \log (2x + 5)$

2. $\log_3 (x + 6) = 2 \log_3 x$

3. $\log (9x + 1) - \log x = 1$

4. $\log(x-3)+\log x=1$

5 $\log_2(x+1)+\log_2(x-1)=3$

6. $\log(x^2+1)-\log(x-2)=1$

7. $\log_5 x+\log_5(x-1)=\log_5 12$

8. $\log x-\log(7x+6)=-1$

9. $\ln x+\ln(x+2)=\ln 3$

10. $10^x=1000$

11. $3^x=16$

12. $10^{3x+5}=100$

13. $2^{x+7}=8$

14. $5^{x^2-1}=125$

15. $2^{3x}=7$

16 $3^{x^2}9^x=\frac{1}{3}$

17. $6^{x+1}=27$

18 $2^{2x}3^{x+1}=5$

19. If the coefficient of extinction of Crystal Lake is 0.008, use the formula in Example 7 and find the depth (to the nearest tenth of a foot) at which light will be reduced by 50% from that at the surface.

20 Repeat Exercise 19 for an 80% reduction of the light intensity at the surface.

FOR REVIEW

Express each logarithm in terms of the indicated base and simplify.

21. $\log_8 128$; 2

22. $\log_6 4$; 4

Evaluate each logarithm.

23. $\log_2 18$

24. $\log 38.6$

25. $\ln 0.233$

ANSWERS: 1. 4 2. 3 3. 1 4. 5 5. 3 6. 3, 7 7. 4 8. 2 9. 1 10. 3 11. 2.52 12. −1 13. −4 14. ±2 15. 0.936 16. −1 17. 0.839 18. 0.206 19. 86.6 ft 20. 201.2 ft 21. $\frac{7}{3}$ 22. $\frac{1}{\log_4 6}$ 23. 4.1699 24. 1.5866 25. −1.4567

10.5 EXERCISES B

Solve the following for x. When appropriate, give answer using three significant digits.

1. $\log(x+5)=\log(1-5x)$

2. $2\log(x+1)=\log 4x$

3. $\log(3x+1)-\log(4-x)=1$

4. $\log(6x+2)+\log 4x=1$

5. $\log(x+1)-\log(x^2+19)=-1$

6. $\log_3(2x+1)+\log_3(x-1)=3$

7. $\log_8 (x + 2) + \log_8 (x - 3) = \log_8 6$

8. $\log (2x - 1) - \log (9x + 1) = -1$

9. $\ln (x + 1) + \ln (x - 1) = \ln 3$

10. $100^x = 10$

11. $8^x = 22$

12. $1000^{x+2} = 10$

13. $3^{2x+4} = 27$

14. $2^{2x^2+3} = 32$

15. $8^{3x}2^{x^2} = 2^{10}$

16. $5^{x+2} = 18$

17. $3^{2x} = 4^{x+1}$

18. $4^{x-5}3^{2x} = 10$

19. If the coefficient of extinction of Lake Pleasant is 0.09, use the formula in Example 7 and find the depth (to the nearest tenth of a foot) at which light will be reduced by 50% from that at the surface.

20. Repeat Exercise 19 for an 80% reduction of the light intensity at the surface.

FOR REVIEW

Express each logarithm in terms of the indicated base and simplify.

21. $\log_{32} 64$; 2

22. $\log_5 x$; 25

Evaluate each logarithm.

23. $\log_3 41$

24. $\log_8 0.0159$

25. $\log 0.0045$

26. $\ln 62.9$

10.5 EXERCISES C

Solve for x.

1. $(\log_5 x)^2 = 2 \log_5 x$

2. $7^{2x}3^x = 10$

3. $2^{x^2}3^{x^2} = 6^{5x-6}$
[Answer: 2, 3]

Solve for n.

4. $3170 = 1000(1.08)^n$

5. $397 = 1000(1.08)^{-n}$

6. $S = A(1 + i)^n$

7. $A = S(1 + i)^{-n}$

8. $y = 2.5e^{-3n}$

9. $y = 800e^{2n}$

10. $y = ce^{n/3}$
[Answer: $n = 3(\ln y - \ln c)$]

11. $y = ce^{-n/2}$

12. $y = 500(1 + x)^{n/2}$
$\left[\text{Answer: } n = \dfrac{2 \log y - 2 \log 500}{\log (1 + x)}\right]$

10.6 MORE APPLICATION PROBLEMS

Many applied problems in science and business use formulas that involve logarithmic or exponential expressions. The compound interest formula is one of the most common in this category. The sum S of an amount A invested for n compounding periods at a rate of interest i per period is given by the formula

$$S = A(1 + i)^n.$$

Solving this formula for A gives us the initial or present value necessary for accumulating with interest the future sum S.

$$A = S(1 + i)^{-n}$$

EXAMPLE 1 Compound Interest Problem

If \$1000 is invested at an annual rate of 8% compounded quarterly, what is the value of the account at the end of 10 years?

The formula we use is

$$S = A(1 + i)^n,$$

where $A = \$1000$, the quarterly interest rate i is $\frac{0.08}{4} = 0.02$, and the number of compounding periods n is $4(10) = 40$ (4 per year for 10 years).

$$S = 1000(1 + 0.02)^{40}$$
$$= 1000(1.02)^{40}$$

We can use a calculator with a $\boxed{y^x}$ key to evaluate S as follows:

1.02 $\boxed{y^x}$ 40 $\boxed{\times}$ 1000 $\boxed{=}$ → 2208.04

In 10 years the investment will be worth \$2208.04.

Practice Exercise 1

Repeat Example 1. However, this time assume compounding is semiannual (twice a year).

Answer: \$2191.12

EXAMPLE 2 Compound Interest Problem

The Barkers wish to have \$50,000 available for their newborn child's college education. How much should they deposit today in a savings account that pays 9% interest compounded semiannually, to accumulate the needed amount by the end of 18 years?

The formula to use is

$$A = S(1 + i)^{-n}$$

where $S = \$50{,}000$, the semiannual interest rate i is $\frac{0.09}{2} = 0.045$, and the number of compounding periods n is $2(18) = 36$ (2 per year for 18 years).

$$A = 50{,}000(1 + 0.045)^{-36}$$
$$= 50{,}000(1.045)^{-36}$$

1.045 $\boxed{y^x}$ 36 $\boxed{+/-}$ $\boxed{\times}$ 50,000 → 10251.41

The Barkers should deposit \$10,251.41 to have the necessary \$50,000 in 18 years.

Practice Exercise 2

Repeat Example 2. This time assume compounding is quarterly.

Answer: \$10,074.21

Many scientific applications can be described by an equation of the form

$$y = c \log_a \frac{x}{x_0}.$$

For the measurement of sound and earthquakes, the base a is 10, while for cell-generation problems, the base a is 2.

EXAMPLE 3 BIOLOGY PROBLEM

The equation that describes the number of cell fissions that occur in a given period of time is the number y given by

$$y = \log_2 \frac{x}{x_0},$$

where x_0 is the initial cell count and x is the final cell count. Find the number of cell fissions in a given time if the initial cell count is 25 and the final cell count is 630,000.

Since $x_0 = 25$ and $x = 630{,}000$, we evaluate the following expression.

$$\begin{aligned} y &= \log_2 \frac{630{,}000}{25} \\ &= \log_2 25{,}200 \qquad \text{Simplify fraction} \\ &= \frac{\log 25{,}200}{\log 2} \approx 14.6 \end{aligned}$$

25,200 [LOG] [÷] 2 [LOG] [=] → 14.621136

There are approximately 14.6 fissions during that period.

PRACTICE EXERCISE 3

Find the number of cell fissions if the initial cell count is 150 and the final cell count is 325,000.

Answer: approximately 11.1 fissions

EXAMPLE 4 PHYSICS PROBLEM

The loudness of sound in decibels is given by the formula

$$y = 10 \log \frac{x}{x_0}$$

where x_0 is the weakest sound detected by the observer and x is the intensity of a given sound. Find the value of y if x is 51,000 times x_0.

$$\begin{aligned} y &= 10 \log \frac{51{,}000x_0}{x_0} \qquad x = 51{,}000x_0 \\ &= 10 \log 51{,}000 \\ &\approx 47.1 \end{aligned}$$

The loudness of sound is about 47.1 decibels.

PRACTICE EXERCISE 4

What is the loudness of a sound 200,000 times more intense than the weakest sound detected by an observer?

Answer: 53.0 decibels

The growth of a population of organisms that is not limited by predators, food supply, or living space, can be approximated exponentially by

$$y = Ie^{kt},$$

where y is the number of organisms after time t starting with an initial population I. The value of k is the percent of growth per unit of time t.

EXAMPLE 5 POPULATION GROWTH PROBLEM

The population of Upstate, Maine, varies according to the equation

$$y = 320e^{0.12t}$$

where t is measured in years. Find the time it will take for the population to double.

PRACTICE EXERCISE 5

How long will it take for the population of Upstate, Maine to triple?

Letting $t = 0$, we get $y = 320e^0 = 320(1) = 320$. Thus, when $t = 0$, the population is 320. We need to find t when $y = 2(320) = 640$.

$$640 = 320e^{0.12t}$$
$$\frac{640}{320} = e^{0.12t}$$
$$2 = e^{0.12t}$$
$$\ln 2 = \ln e^{0.12t}$$
$$\ln 2 = 0.12t \quad \text{Power rule}$$
$$t = \frac{\ln 2}{0.12} \approx 5.78$$

The population will double in about 5.78 years.

Answer: about 9.2 yr

10.6 EXERCISES A

Solve.

1. If \$1250 is invested at an annual rate of 12% compounded quarterly, what will be the value of the account at the end of 10 years?

2. What amount must Cathy invest at 8% interest compounded semiannually, to accumulate \$12,000 at the end of 15 years?

3. Margot invested \$685 in an account 25 years ago. What is the value of the account if the annual interest is 8% compounded **(a)** semiannually? **(b)** quarterly?

4. What amount must Tim invest at 8% interest compounded quarterly, to accumulate \$4000 at the end of 12 years?

5. Use the equation $y = \log_2 \frac{x}{x_0}$ to find the number of cell fissions that occur during the time of an experiment if there are 125 cells initially and 9525 cells finally.

6. Use the equation $y = \log_2 \frac{x}{x_0}$ to find the final number of cells in an experiment if there were 420 cells initially and the number of fissions was 12.

7. Use the sound equation $y = 10 \log \frac{x}{x_0}$ to determine the loudness of a sound in decibels if the given sound has intensity 63,000 times the weakest sound detected.

8. Use the equation $y = 10 \log \frac{x}{x_0}$ to find $\frac{x}{x_0}$ if the loudness of the sound is 90 decibels.

9. The magnitude y on the Richter scale of the intensity x of an earthquake is given by $y = \log \frac{x}{x_0}$ where x_0 is the minimum intensity used for comparison. If an earthquake has intensity 50,100 times the minimum intensity, what is its Richter scale magnitude?

10. Use the equation in Exercise 9 to find the ratio $\frac{x}{x_0}$ if the magnitude of an earthquake is 7.0 on the Richter scale.

11. The population of Warm, Florida, is increasing according to the equation $y = 5000e^{0.02t}$, where t is in years. Find the population 12 years after the initial count.

12. How long will it take for the population in Exercise 11 to double?

13. Use the equation $y = 250e^{-0.5t}$ to find the half-life (the time for one-half of a substance to decay) of a radioactive substance where y is the amount of the substance remaining and t is in years.

14. Use the equation $y = 1200e^{-0.08t}$ to determine when one-fourth of a given radioactive substance will be gone.

15. Given that $\text{pH} = -\log [H_3O^+]$ and that $[H_3O^+] = 10^{-8.2}$, find the pH of the solution.

16. If the pH of an acid solution is 3.2, find $[H_3O^+]$ using $\text{pH} = -\log [H_3O^+]$.

FOR REVIEW

Solve for x.

17. $\log x + \log (x - 2) = \log 3$

18. $2^{3x+1} = 16$

The following exercises review material from Chapters 3, 8, and 9 to help you prepare for the next chapter. Graph each function.

19. $x + 3y - 6 = 0$

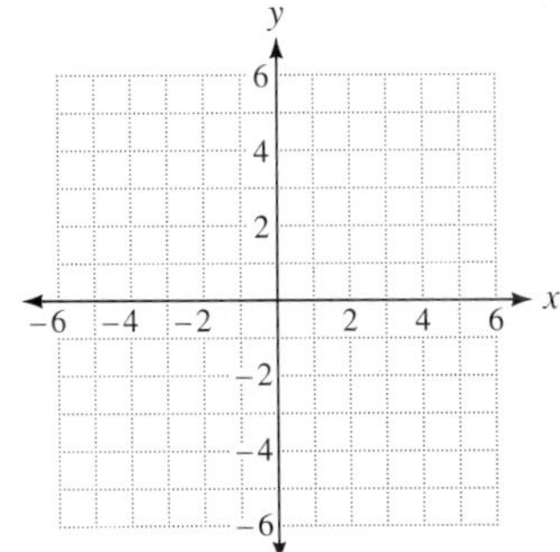

20. $y = x^2 - 4x - 5$

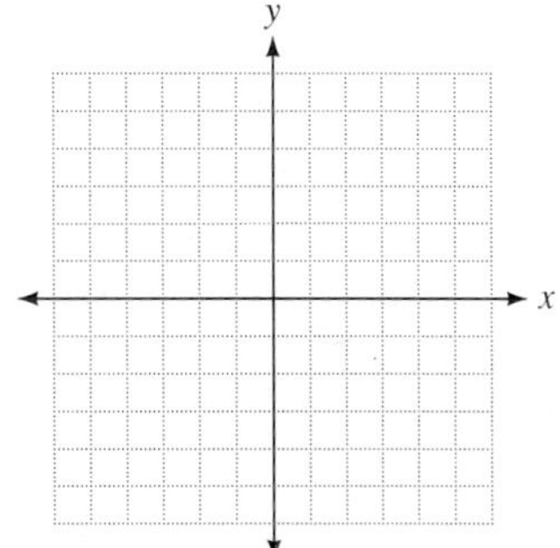

21. $y = \sqrt{x - 5}$

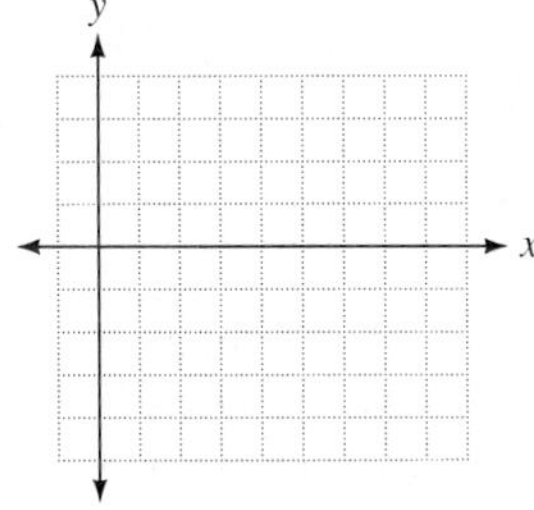

What must be added to complete the square in each expression?

22. $x^2 - 8x$

23. $y^2 + 10y$

24. $x^2 - x$

ANSWERS: 1. \$4077.55 2. \$3699.83 3. (a) \$4868.08 (b) \$4962.58 4. \$1546.15 5. 6.25 fissions 6. 1,720,320 7. 48.0 decibels 8. 1.0×10^9 9. 4.7 10. 10,000,000 11. 6356 12. 34.7 years 13. 1.39 years 14. 3.6 years 15. 8.2 16. $10^{-3.2}$ 17. 3 18. 1 19. 20.

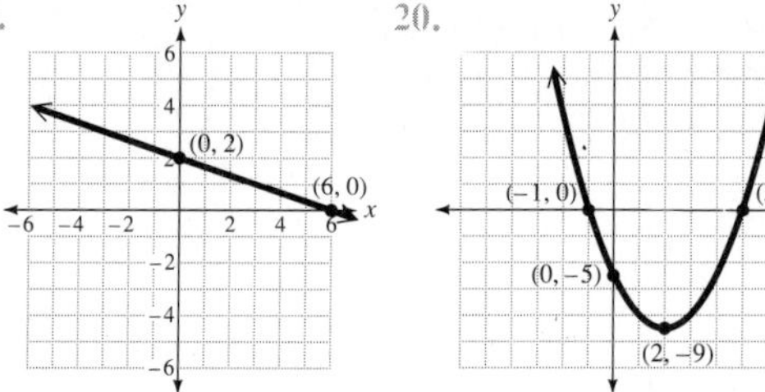

21.

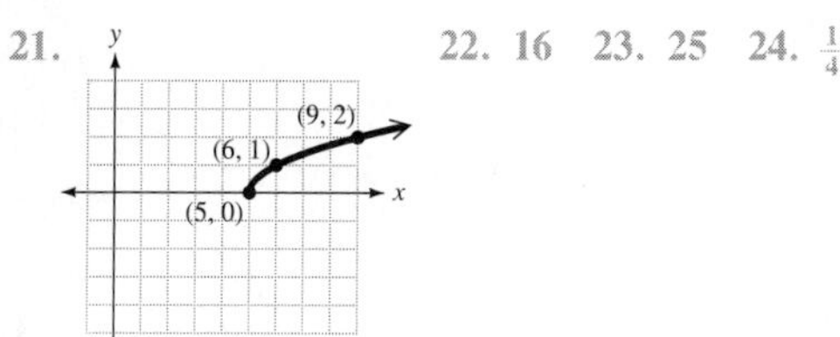

22. 16 23. 25 24. $\frac{1}{4}$

10.6 EXERCISES B

Solve.

1. Ms. Rutter invests \$5000 at 16% interest compounded semiannually. How much will she have in her account at the end of 9 years?

2. Larry wants to have \$9000 in his account in three years so he can buy a new car. How much must he invest now at 12% compounded quarterly to accumulate the \$9000?

3. Walter invested \$4360 in an account 15 years ago. What is the value of the account if interest is 12% compounded **(a)** quarterly? **(b)** monthly?

4. Susan must have \$750 in two years to pay a debt. How much must she invest today to have this amount in her account at the end of that time if interest is 12% compounded monthly?

5. Use the equation $y = \log_2 \frac{x}{x_0}$ to find the number of cell fissions that occur in a given period of time if there are 30 cells initially and 437,000 cells finally.

6. Use the equation $y = \log_2 \frac{x}{x_0}$ to find the number of cell fissions that occur if at the end of the experiment there are 320 times as many cells as there were at the beginning.

7. Use the sound equation $y = 10 \log\frac{x}{x_0}$ to determine the loudness of a sound in decibels if the given sound has intensity 210,000 times the weakest sound detected.

8. Use the equation $y = 10 \log \frac{x}{x_0}$ to find $\frac{x}{x_0}$ if the loudness of the sound is 36 decibels.

9. The magnitude y on the Richter scale of the intensity x of an earthquake is given by $y = \log \frac{x}{x_0}$ where x_0 is the minimum intensity used for comparison. If an earthquake has intensity 75,000 times the minimum intensity, what is its Richter scale magnitude?

10. Use the equation in Exercise 9 to find the ratio $\frac{x}{x_0}$ if the magnitude of an earthquake is 6.0 on the Richter scale.

11. The population of a state is decreasing according to the equation $y = 3{,}200{,}000e^{-0.03t}$. Find the population after 6 years.

12. How long will it take for the population in Exercise 11 to be one-half the original amount?

13. Use the equation $y = 1000e^{-0.4t}$ to find the half-life (the time for one-half of a substance to decay) of a radioactive substance where y is the amount of the substance remaining and t is in years.

14. Use the equation $y = 500e^{-0.05t}$ to determine when one-fourth of a given radioactive substance will be gone.

15. Given that $\text{pH} = -\log\,[H_3O^+]$ and that $[H_3O^+] = 10^{-7.3}$, find the pH of the solution.

16. If the pH of a solution is 4.5, find $[H_3O^+]$ using $\text{pH} = -\log\,[H_3O^+]$.

FOR REVIEW

Solve for x.

17. $\log_2 (x + 5) - \log_2 (x - 7) = 2$

18. $2^x3^x = 11$

The following exercises review material from Chapters 3, 8, and 9 to help you prepare for the next chapter. Graph each function.

19. $2x - 3y = 0$

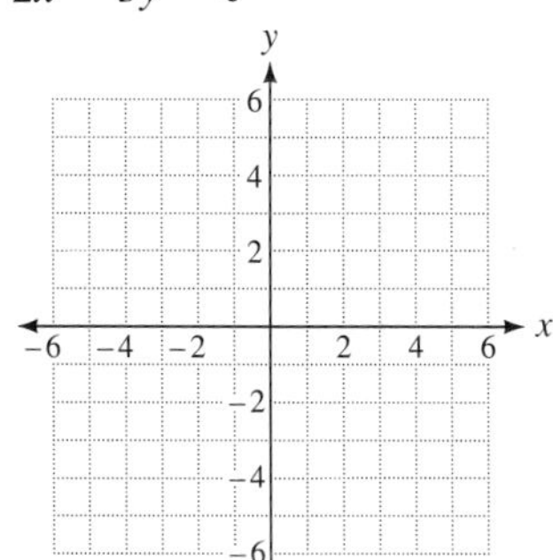

20. $y = -x^2 + 4x + 5$

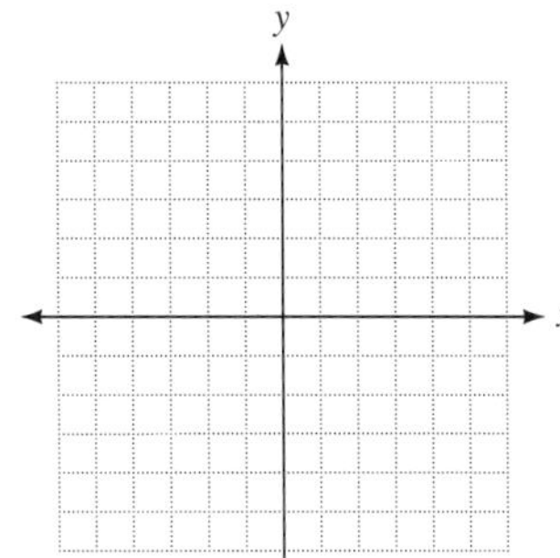

21. $y = -\sqrt{x - 5}$

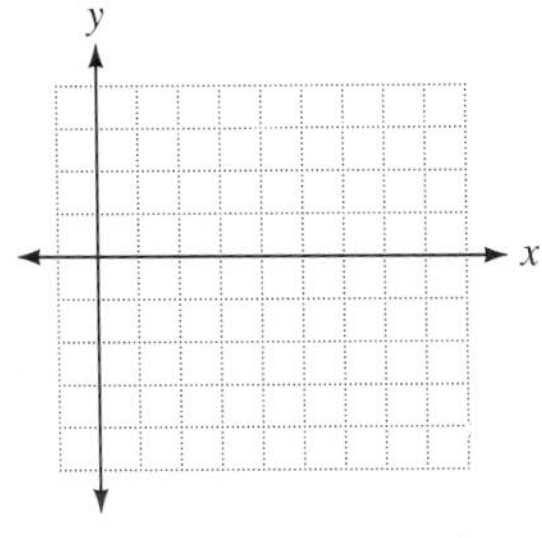

What must be added to complete the square in each expression?

22. $x^2 - 4x$

23. $y^2 + 16y$

24. $x^2 + x$

10.6 EXERCISES C

The formula used for population growth, $y = Ie^{kt}$, is called the Malthusian model after Thomas Malthus who, in the 18th century, made predictions about growing populations. Remember that k is the percent of growth per unit of time. Use this information in the following exercises.

1. The population of the United States was 237 million in 1985. If the annual growth rate is approximately 0.9%, what will be the population in the year 2000?

2. When the interstate freeway bypassed a small town in New Mexico, the population began to decline from 12,000 residents at an annual rate of 4.1%. How many years will it take for the population to reach 8000?

3. The population of Alpine, Oregon was 800 in 1970 and 1150 in 1990. Assuming the same growth rate find the value of k (correct to three decimal places) and use it to predict the population of Alpine in the year 2010.

4. The population of the southwestern United States is expected to double in 30 years. What will be the approximate annual growth rate of this region?

CHAPTER 10 REVIEW

KEY WORDS

10.1 A **logarithm** is an exponent. In fact, if $a^x = y$ where $a > 0$, $a \neq 1$, then x is the **logarithm to the base *a* of *y*.**

10.2 A function defined by the equation $y = a^x$, $a > 0$, $a \neq 1$, is an **exponential function** with **base *a*.**

A function defined by the equation $y = \log_a x$, $a > 0$, $a \neq 1$, is a **logarithmic function.**

10.4 Logarithms using 10 as a base are **common logarithms.**

A number that corresponds to a given logarithm is its **antilogarithm (antilog).**

Logarithms using e as a base are **natural logarithms.**

The function defined by the equation $y = \ln x$ is the **natural logarithmic function.**

10.5 An equation that contains logarithms of expressions involving the variable is a **logarithmic equation.**

An equation that contains variable exponents is an **exponential equation.**

KEY CONCEPTS

10.1 **1.** If a, x, and y are numbers, $a > 0$ $a \neq 1$, then the equation $y = a^x$ is equivalent to the equation $x = \log_a y$, with the first being the exponential form and the second the logarithmic form.

2. Suppose $a, b > 0$, $a \neq 1$, and $b \neq 1$. Then

$a^x = a^y$ if and only if $x = y$, and
$a^x = b^x$ if and only if $a = b$.

3. If a is any base ($a > 0$, $a \neq 1$), $\log_a a = 1$ and $\log_a 1 = 0$.

10.2 **1.** The logarithmic function $y = \log_a x$ and the exponential function $y = a^x$ are inverses. The graph of each is the reflection of the graph of the other across the line $y = x$.

2. To graph a logarithmic function $y = \log_a x$, it is easiest to plot a few points determined by the equivalent exponential equation $x = a^y$.

10.3 **1.** Product rule: $\log_a (uv) = \log_a u + \log_a v$ [which is not $\log_a (u + v)$]

2. Quotient rule: $\log_a \frac{u}{v} = \log_a u - \log_a v$ [which is not $\log_a (u - v)$]

3. Power rule: $\log_a u^c = c \log_a u$ [which is not $\log_a cu$]

10.4 **1.** Common logarithms are written as $\log x$ rather than $\log_{10} x$.

2. If a and b are any two bases, and $x > 0$, then

$$\log_b x = \frac{\log_a x}{\log_a b} \text{ and } \log_b a = \frac{1}{\log_a b}.$$

3. If $x = \log n$, x is the logarithm of n and n is the antilogarithm (antilog) of x. In exponential form, $n = 10^x$.

4. The number e is an irrational number that can be approximated by 2.71828.

5. Natural logarithms are written as $\ln x$ rather than $\log_e x$.

10.5 **1.** To solve a logarithmic equation use one of the following techniques.

(a) Obtain a single logarithmic expression using the same base on one side of the equation and convert the result to an exponential equation.

(b) Write each side as a logarithm using the same base and find the antilog of both sides.

2. To solve an exponential equation: try to express each side as a power using the same base and equate exponents; if this fails, take the log of both sides and use the power rule to eliminate the variable exponents.

3. When solving logarithmic and exponential equations, always check your answers in the original equation. Remember that the logarithm of a negative number or zero is not defined so that "solutions" that require finding such logarithms must be discarded.

REVIEW EXERCISES

Part I

10.1 **1.** Another term for the word *logarithm* is ________________.

2. The exponential form of $y = \log_a x$ is ________________.

3. The logarithmic form of $u = b^v$ is ________________.

4. Every positive number except ________________ can be used as a base for logarithms.

Convert to logarithmic form.

5. $2^x = 50$

6. $x^3 = 27$

7. $u^v = q$

Convert to exponential form.

8. $\log_3 a = b$

9. $\log_a a = 1$

10. $\log_c w = m$

Give the numerical value of x.

11. $x^4 = 81$

12. $x^{-1} = \frac{1}{7}$

13. $x^{1/2} = 4$

14. $x = \log_2 8$

15. $\log_x 125 = 3$

16. $\log_3 x = -2$

17. For any base a, $\log_a a =$ ____________.

18. For any base a, $\log_a 1 =$ ____________.

10.2 **19.** Sketch the graph of $y = 5^x$ and use it to sketch the graph of $y = \log_5 x$.

x	$y = 5^x$
0	
1	
−1	
2	
−2	

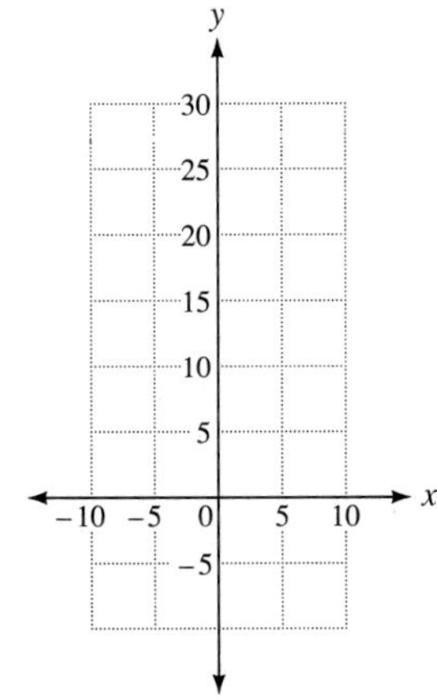

10.3 **20.** If a is any base and u and v are positive numbers, $\log_a (u \cdot v) =$ ________________.

21. If a is any base and u and v are positive numbers, $\log_a \left(\frac{u}{v}\right) =$ ________________.

22. If a is any base, u is a positive number, and c is any number, $\log_a u^c =$ ________________.

23. The rule given in Exercise 20 is called the ______________ rule.

24. The rule given in Exercise 21 is called the ______________ rule.

25. The rule given in Exercise 22 is called the ______________ rule.

26. Express in terms of logarithms of x, y, and z.
$\log_a \dfrac{x^3\sqrt{z}}{y^4}$

27. Express as a single logarithm.
$\dfrac{1}{3}\log_a x + 5\log_a y$

Decide whether the following are true or false. If false, change the right side to make it true.

28. $\log_a \dfrac{u}{v} = \dfrac{\log_a u}{\log_a v}$

29. $\log_a a = a$

30. $\log_a (u \cdot v) = (\log_a u)(\log_a v)$

31. $\log_a u^c = u \log_a c$

10.4 *Find each logarithm.*

32. $\log 4135$

33. $\log 0.000789$

34. $\log_3 6.29$

35. $\ln 125.6$

36. $\ln 0.0625$

37. $\log_8 0.00788$

Find each antilogarithm, correct to three significant digits.

38. $\log x = 2.2547$

39. $\log x = -1.6554$

40. $\ln x = 1.7983$

10.5 *Solve the following equations.*

41. $\log (x + 3) + \log x = 1$

42. $\log x - \log (3x + 7) = \log \dfrac{1}{10}$

43. $3^{5x-1} = 81$

44. $5^{2x} = 8$

10.6 *Solve each applied problem.*

45. If \$875 is invested at an annual rate of 4% compounded quarterly, what will be the value of the account at the end of 20 years?

46. What amount must Paul invest at 12% interest compounded semiannually, to accumulate \$2750 at the end of 10 years?

47. Use $y = \log_2 \frac{x}{x_0}$ to find the number of cell fissions that occur during an experiment if there were 540 cells initially and the final count was 8960 cells.

48. An earthquake measured 5.2 on the Richter scale. Use $y = \log \frac{x}{x_0}$ to determine the ratio $\frac{x}{x_0}$.

Part II

49. Express as a single logarithm.
$\frac{1}{2}(\log_a x - \log_a z - 3 \log_a y)$

50. Express in terms of logarithms of x, y, and z.
$\log_a \sqrt[5]{\frac{xy}{z^2}}$

Find each logarithm.

51. log 0.00378

52. ln 17.2

53. $\log_5 2.59$

Solve the following equations.

54. $\ln x + \ln (x + 4) = \ln 5$

55. $2^x 2^{x^2} = 4$

Solve.

56. If \$8000 is invested at an annual rate of 14% compounded semiannually, what is the value of the account at the end of 5 years?

57. Use the sound equation $y = 10 \log \frac{x}{x_0}$ to find the loudness of a sound in decibels if the given sound has intensity 85,000 times the weakest sound detected.

Find each antilogarithm, correct to three significant digits.

58. $\log x = 2.4137$

59. $\log x = -0.3742$

60. $\ln x = -2.4558$

61. Convert to logarithmic form.
$8^{-3} = x$

62. Convert to exponential form.
$\ln 6 = x$

Solve.

63. The population of a state is increasing according to the equation $y = 2{,}200{,}000e^{0.04t}$. Find the time it will take for the population to double (t is in years).

64. Annette wants to have \$20,000 in her account in four years to have a down payment for a piece of property. How much must she invest now at 10% compounded quarterly to accumulate the \$20,000?

Answer true *or* false *in the following. If false, explain why.*

65. The exponential form of $y = \log_a x$ is $a^x = y$.

66. The logarithmic form of $u = b^v$ is $v = \log_b u$.

67. If a is any base and u and v are positive numbers, then $\log_a (u + v) = (\log_a u)(\log_a v)$.

68. If a is any base and u and v are positive numbers, then $\log_a \frac{u}{v} = \log_a u - \log_a v$.

69. If a is any base and u is a positive number, then $\log_a \sqrt{u} = 2 \log_a u$.

70. Base 10 logarithms are called common logarithms.

71. In the equation $x = \log n$, n is called the antilogarithm of x.

72. The graph of $y = \log_a x$ can be obtained from the graph of $y = a^x$ by reflection in the line $y = x$.

73. If a and b are bases, then $\dfrac{1}{\log_b a} = \log_a b$.

74. If a and b are bases and $x > 0$, then $\dfrac{\log_a x}{\log_a b} = \log_x a$.

75. If $y = \ln x$ then $e^y = x$.

ANSWERS: 1. exponent 2. $x = a^y$ 3. $v = \log_b u$ 4. 1 5. $x = \log_2 50$ 6. $3 = \log_x 27$ 7. $v = \log_u q$ 8. $3^b = a$ 9. $a^1 = a$ 10. $c^m = w$ 11. 3 12. 7 13. 16 14. 3 15. 5 16. $\frac{1}{9}$ 17. 1 18. 0 19. (Reflect the graph of $y = 5^x$ across the line $y = x$ to obtain the graph of $y = \log_5 x$.)

x	$y = 5^x$
0	1
1	5
-1	$\frac{1}{5}$
2	25
-2	$\frac{1}{25}$

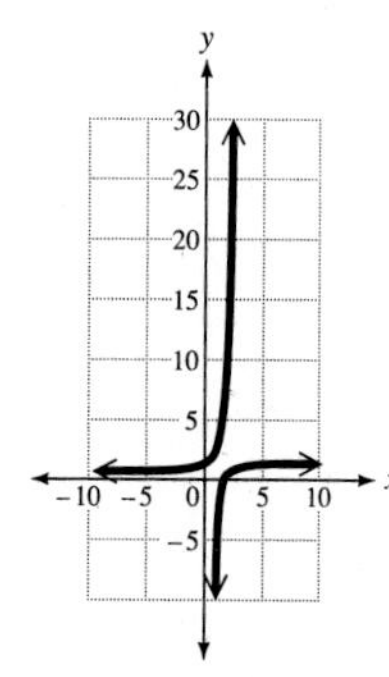

20. $\log_a u + \log_a v$ 21. $\log_a u - \log_a v$ 22. $c \log_a u$ 23. product 24. quotient 25. power 26. $3 \log_a x + \frac{1}{2} \log_a z - 4 \log_a y$ 27. $\log_a y^5 \sqrt[3]{x}$ 28. false; $\log_a u - \log_a v$ 29. false; 1 30. false; $\log_a u + \log_a v$ 31. false; $c \log_a u$ 32. 3.6165 33. -3.1029 34. 1.6739 35. 4.8331 36. -2.7726 37. -2.3292 38. 180 39. 0.0221 40. 6.04 41. 2 42. 1 43. 1 44. 0.646 45. \$1939.63 46. \$857.46 47. 4.1 fissions 48. $10^{5.2}$ 49. $\log_a \sqrt{\frac{x}{zy^3}}$ 50. $\frac{1}{5} (\log_a x + \log_a y - 2 \log_a z)$ 51. -2.4225 52. 2.8449 53. 0.5913 54. 1 55. 1, -2 56. \$15,737.20 57. 49.3 decibels 58. 259 59. 0.422 60. 0.0858 61. $-3 = \log_8 x$ 62. $e^x = 6$ 63. 17.3 years 64. \$6312.42 65. false; $x = a^y$ 66. true 67. false; $\log_a (u + v)$ cannot be simplified 68. true 69. false; $\log_a \sqrt{u} = \frac{1}{2} \log_a u$ 70. true 71. true 72. true 73. true 74. false; $\frac{\log_a x}{\log_a b} = \log_b x$ 75. true

CHAPTER 10 TEST

1. Convert $\log_b 5 = x$ to exponential form. 1. ______

2. Convert $u^v = 7$ to logarithmic form. 2. ______

Give the numerical value of x.

3. $x^{-3} = \dfrac{1}{64}$ 3. ______

4. $\log_5 \dfrac{1}{5} = x$ 4. ______

5. $\log_9 x = \dfrac{1}{2}$ 5. ______

6. $\log_x \sqrt{11} = \dfrac{1}{2}$ 6. ______

7. Express $\log_a \dfrac{x^2y^3}{z}$ as logarithms of x, y, and z. 7. ______

8. Express $2 \log_a z - \dfrac{1}{3} \log_a y$ as a single logarithm. 8. ______

Decide whether the following are true or false.

9. $\log_a (u + v) = \log_a u + \log_a v$ 9. ______

10. $\log_a u^{1/2} = \dfrac{\log_a u}{2}$ 10. ______

11. For any base a, $\log_a a =$ ____. 11. ______

12. Find log 0.612. 12. ______

13. Find ln 6.35. 13. ______

14. Find the antilogarithm, correct to three significant digits.
$\log x = 3.2923$

14. ______

15. Find the antilogarithm, correct to three significant digits.
$\ln x = -0.4351$

15. ______

Solve.

16. $2^{x^2} = 16$

16. ______

17. $\log x - \log (x + 1) = -1$

17. ______

18. Use the sound equation $y = 10 \log \frac{x}{x_0}$ to find the loudness of a sound in decibels if the given sound has intensity 120,000 times the weakest sound detected.

18. ______

19. The loan officer at the First Local Bank uses the formula $A = P(1.02)^n$. Find the value of A when P is \$5620 and n is 20.

19. ______

20. The population of a city is increasing according to $y = 120{,}000e^{0.15t}$. In how many years will the population be 10 times the original count?

20. ______

21. Graph $y = \log_2 x$.

21.

y
6
4
2
−6 −4 −2 2 4 6 x
−2
−4
−6

CHAPTER 11

Conic Sections

11.1 THE PARABOLA

STUDENT GUIDEPOSTS

1. Second-Degree Equations in Two Variables
2. Conic Sections
3. Parabolas Opening Up or Down
4. Parabolas Opening Left or Right
5. Standard Forms of a Parabola

The study of conics, including parabolas, circles, ellipses, and hyperbolas, dates back to the early Greeks who used these curves to solve various construction problems. However, applications of conics continue to grow today. The reflecting properties of parabolic surfaces are used in telescopes, signal lights, radar units, television antennae, and navigational systems. Planets and satellites travel in elliptical orbits, and atomic particles and comets travel in hyperbolic paths.

1 SECOND-DEGREE EQUATIONS IN TWO VARIABLES

In Chapter 3 we studied linear equations in two variables of the form $ax + by + c = 0$, with a and b not both zero, whose graphs are straight lines. Linear equations are sometimes called first-degree equations since the variables x and y appear only to the first power. An equation of the form

$$\boldsymbol{Ax^2 + Bxy + Cy^2 + Dx + Ey + F = 0},$$

where at least one of A, B, or C is not zero, is called a **second-degree equation in two variables** x and y.

2 CONIC SECTIONS

Many second degree equations have graphs that can be obtained by intersecting a cone and a plane. These curves are called **conic sections,** or **conics.** The four primary conic sections are parabolas, circles, ellipses, and hyperbolas. See Figure 11.1 on the following page.

3 PARABOLAS OPENING UP OR DOWN

Since we are already familiar with parabolas from our work in Chapter 9, it is appropriate to begin our study of the conic sections with the parabola. Geometrically, a **parabola** is the set of all points in a plane that are equidistant from a point (called the **focus**) and a straight line not containing the point. Although equations of parabolas can be derived from this definition, such a development is usually deferred to more advanced courses.

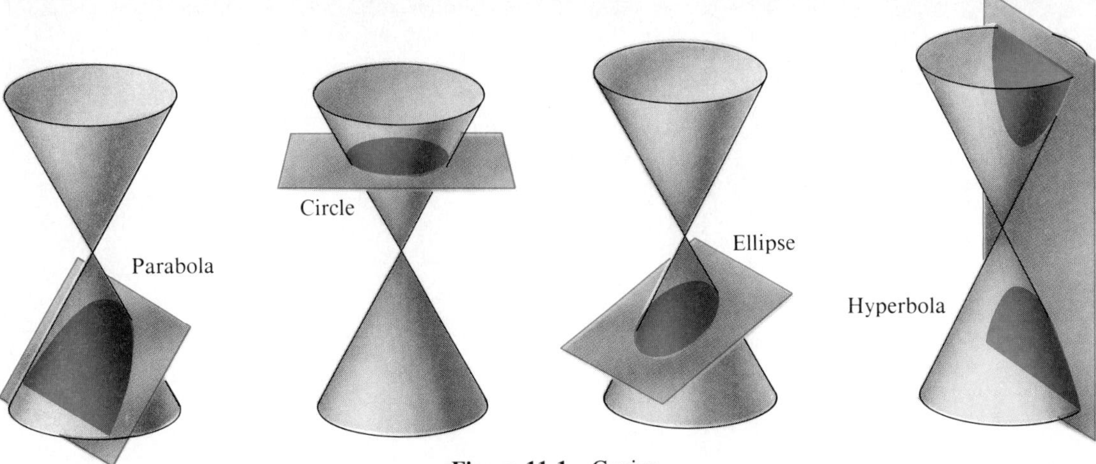

Figure 11.1 Conics

Recall from Section 9.1 that the graph of a quadratic function

$$y = ax^2 + bx + c \qquad a \neq 0$$

is a parabola that opens up if $a > 0$ and down if $a < 0$. Notice that such equations are special types of second-degree equations in the two variables x and y. If we complete the square in the variable x, the equation can be written in the form

$$y = a(x - h)^2 + k. \qquad a \neq 0$$

For example, consider the equation $y = 2x^2 - 4x - 1$. Suppose we complete the square on x. We first factor a 2 from the terms involving x and rewrite the equation leaving space as shown.

$$y = 2(x^2 - 2x \qquad) - 1$$

To complete the square on x inside the parentheses, we must add the square of half the coefficient of x, $\left[\frac{1}{2} \cdot (-2)\right]^2 = \left[-1\right]^2 = 1$. But if 1 is added inside the parentheses, since it is multiplied by 2, to maintain equality we must subtract 2 as shown.

$$y = 2(x^2 - 2x + 1) - 1 - 2 \qquad 2(+1) = 2 \text{ so subtract 2 to maintain equality}$$

Thus, we have $y = 2(x - 1)^2 - 3$. $\qquad x^2 - 2x + 1 = (x - 1)^2$

Since $a = 2 > 0$, the parabola opens up and has vertex at the lowest point on the curve where the value of y is as small as possible. Since $2(x - 1)^2 \geq 0$, the smallest value y can assume occurs when $2(x - 1)^2 = 0$, that is, when $x = 1$. If x is 1, then $y = 2(1 - 1)^2 - 3 = 0 - 3 = -3$. Thus, the vertex is at $(1, -3)$ and the line of symmetry is $x = 1$. Since the vertex is below the x-axis and the parabola opens up, the curve crosses the x-axis at x-intercepts found by solving

$$2x^2 - 4x - 1 = 0.$$

Using the quadratic formula, we can obtain the solutions

$$x = \frac{2 \pm \sqrt{6}}{2},$$

which are approximately 2.2 and −0.2. The graph is given in Figure 11.2.

4 PARABOLAS OPENING LEFT OR RIGHT

The graph of a second-degree equation of the form

$$x = ay^2 + by + c$$

is also a parabola, but this relation was not considered in Section 9.1 since it is

Figure 11.2 Parabola Opening Up

not a function. The graph of $y = ax^2 + bx + c$ will open up or down, and the graph of $x = ay^2 + by + c$ will open left or right.

EXAMPLE 1 Graphing a Parabola Opening Right

Graph $x = y^2 - 2y - 1$.

To make a table of values, we substitute values of y into the equation and simplify to obtain the corresponding values of x. The points in the table are plotted in Figure 11.3.

x	y
-1	0
-2	1
2	-1
-1	2
7	-2
2	3
7	4

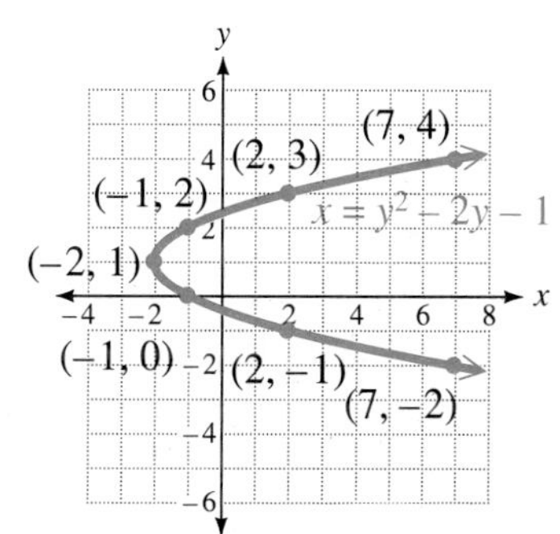

Figure 11.3 Parabola Opening Right

Practice Exercise 1

Graph $x = -y^2 + 2y + 1$.

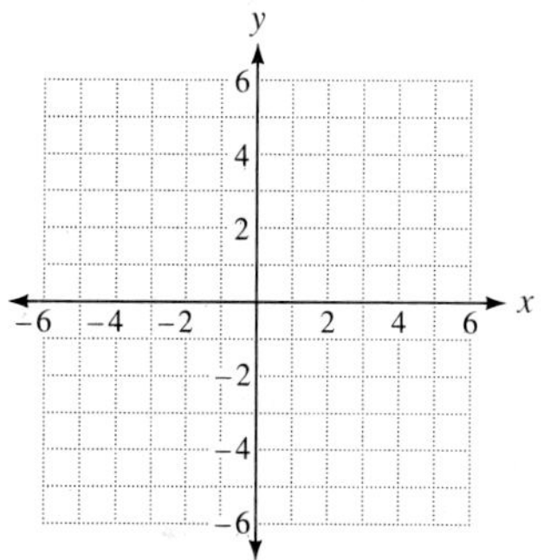

Answer: The graph is a parabola opening left with vertex (2, 1) and passing through (1, 2) and (1, 0).

5 STANDARD FORMS OF A PARABOLA

Notice that $x = y^2 - 2y - 1$ is not a function since for a given value of x, 2 for example, there are two values of y, 3 and -1. Thus, this relation does not pass the vertical line test. Rather than simply plotting points, we can identify the graph more directly by completing the square in y.

$$x = (y^2 - 2y \qquad) - 1 \quad \text{Rewrite leaving space}$$

Add the square of half the coefficient of y, $\left[\frac{1}{2} \cdot (-2)\right]^2 = [-1]^2 = 1$, inside the parentheses and balance this out by subtracting 1 outside the parentheses.

$$\begin{aligned} x &= (y^2 - 2y + 1) - 1 - 1 \\ &= (y - 1)^2 - 2 \end{aligned}$$

We have now written the equation in the form

$$x = a(y - k)^2 + h,$$

where $a = 1 > 0$ and $(h, k) = (-2, 1)$. In general, whenever $a > 0$, the parabola opens to the right and the vertex is at (h, k). This agrees with what we found in Example 1. Also, if $a < 0$, the parabola will open to the left. We summarize these remarks in the following.

Standard Forms of the Equation of a Parabola

The second-degree equation $y = ax^2 + bx + c$ can be written in the **standard form** $y = a(x - h)^2 + k$ by completing the square on x. The graph has one of the following forms.

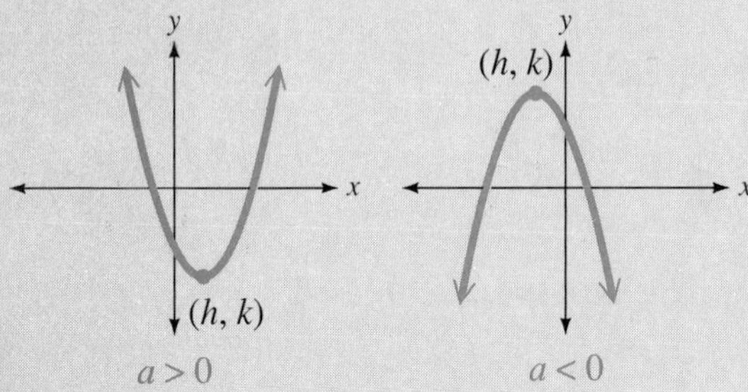

Figure 11.4

The second-degree equation $x = ay^2 + by + c$ can be written in the **standard form** $x = a(y - k)^2 + h$ by completing the square on y. The graph has one of the following forms.

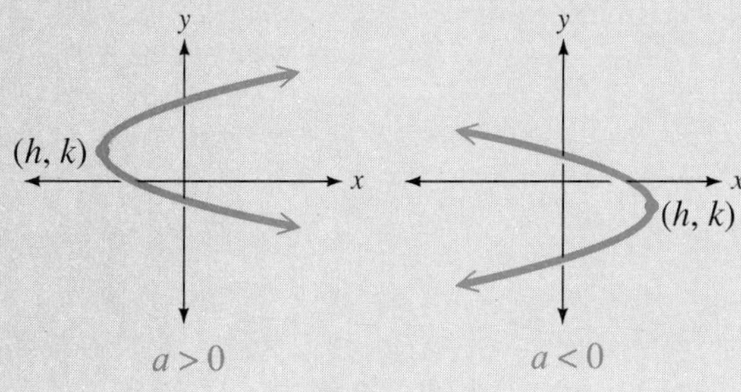

Figure 11.5

EXAMPLE 2 Using Standard Form to Graph a Parabola

Graph $x = -y^2 - 4y - 7$ by writing the equation in standard form.

We complete the square on y.

$$x = -(y^2 + 4y \quad\quad) - 7 \qquad \text{Factor out } -1 \text{ and leave space}$$
$$= -(y^2 + 4y + 4) - 7 + 4$$

The square of half the coefficient of y is $\left(\frac{1}{2} \cdot 4\right)^2 = (2)^2 = 4$. Since adding 4 inside the parentheses is the same as adding -4 to the right side (it is multiplied by $a = -1$), we add 4 outside the parentheses to maintain equality. Thus

$$x = -(y + 2)^2 - 3.$$

The vertex is $(-3, -2)$ since $y + 2 = y - k$ means that $k = -2$. Also, $a = -1 < 0$ so the parabola opens to the left. There are no y-intercepts (which would normally be found by solving $-y^2 - 4y - 7 = 0$). To find two additional points on the graph, let $x = -4$ and solve for y.

$$-4 = -y^2 - 4y - 7$$

Practice Exercise 2

Graph $y = -x^2 + 4x - 1$ by writing the equation in standard form.

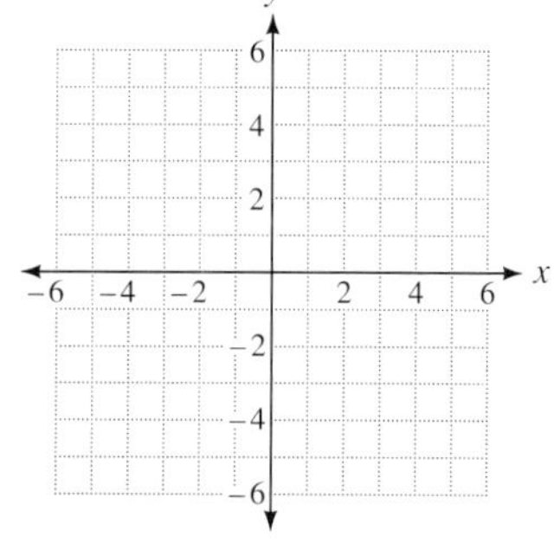

$$y^2 + 4y + 3 = 0$$
$$(y + 3)(y + 1) = 0$$
$$y + 3 = 0 \quad \text{or} \quad y + 1 = 0$$
$$y = -3 \qquad\qquad y = -1$$

Thus, $(-4, -3)$ and $(-4, -1)$ are two points on the parabola along with the x-intercept $(-7, 0)$. The graph is given in Figure 11.6.

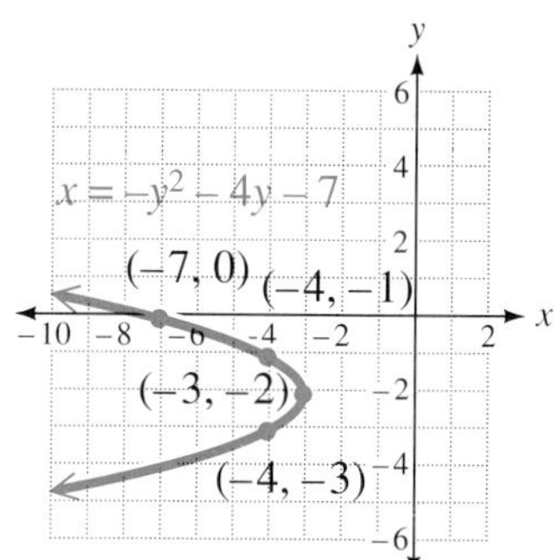

Figure 11.6

Answer: $y = -(x - 2)^2 + 3$, the parabola opens down with vertex $(2, 3)$ and passes through $(0, -1)$ and $(4, -1)$.

The next example illustrates one of the many applications of parabolas.

EXAMPLE 3 Architecture Problem

The entry to a shopping mall is a parabolic arch 20 ft high and 16 ft wide at the base. Find the equation for the parabola if the vertex is placed at the origin of a coordinate system. See Figure 11.7. What is the vertical clearance 4 ft from the center of the entryway?

Since the vertex is $(h, k) = (0, 0)$, and since the parabola opens down, the equation has the form

$$y = a(x - h)^2 + k$$
$$= a(x - 0)^2 + 0 = ax^2.$$

One point on the parabola is $(8, -20)$. Substitute 8 for x and -20 for y to find the value of a.

$$y = ax^2$$
$$-20 = a(8)^2$$
$$-\frac{20}{64} = a$$
$$-\frac{5}{16} = a$$

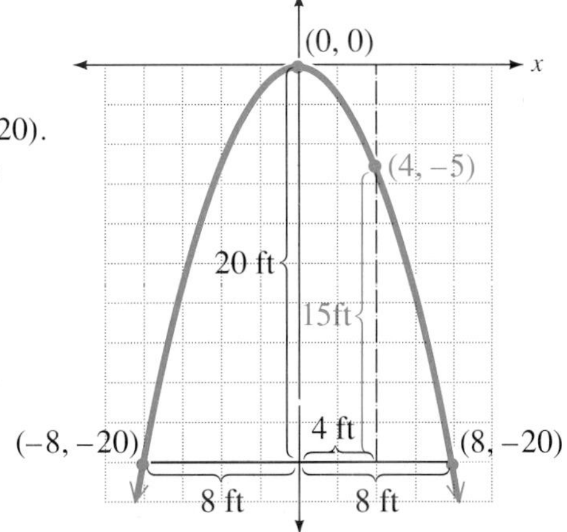

Figure 11.7

Thus, the equation of the parabola is $y = -\frac{5}{16}x^2$. To find the vertical clearance 4 ft from the center, we need to find y when x is 4.

$$y = -\frac{5}{16}x^2 = -\frac{5}{16}(4)^2 = -5$$

Hence, the vertical clearance 4 ft from the center of the arch is 5 ft lower than the total height 20 ft, or 15 ft.

Practice Exercise 3

Can a delivery truck with a load 9 ft wide and 13.5 ft tall drive through the arch in Example 3?

Answer: Yes; the vertical clearance 4.5 ft from the center of the arch is about 13.7 ft.

11.1 EXERCISES A

*For each parabola, **(a)** give the vertex, and **(b)** state whether the parabola opens up, down, left, or right.*

1. $y = 3(x - 1)^2 + 5$

2. $y = -2(x + 1)^2 - 3$

3. $x = -2(y - 3)^2 + 4$

4. $x = 5(y + 4)^2 - 7$

5. $y = x^2 - 5$

6. $x = -y^2 + 6$

7. $y = x^2 - 2x + 9$

8. $x = -y^2 + 6y - 15$

9. $x = -2y^2 - 4y - 7$

Graph each parabola by writing the equation in standard form.

10. $x = y^2$

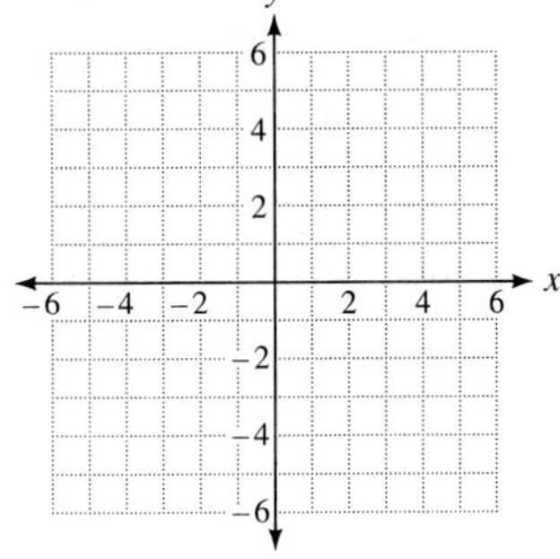

11. $y = -x^2$

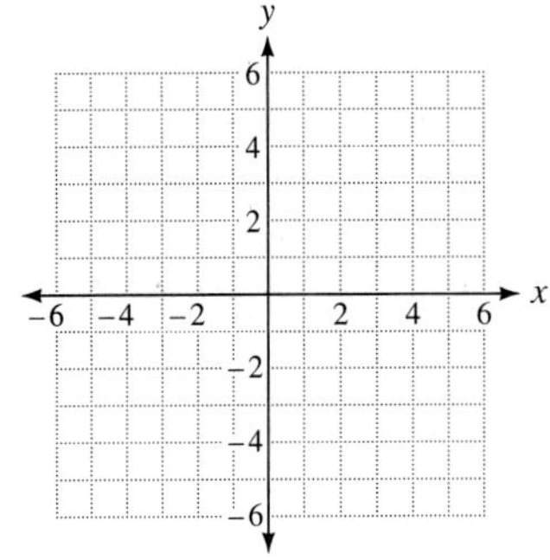

12. $x = -y^2$

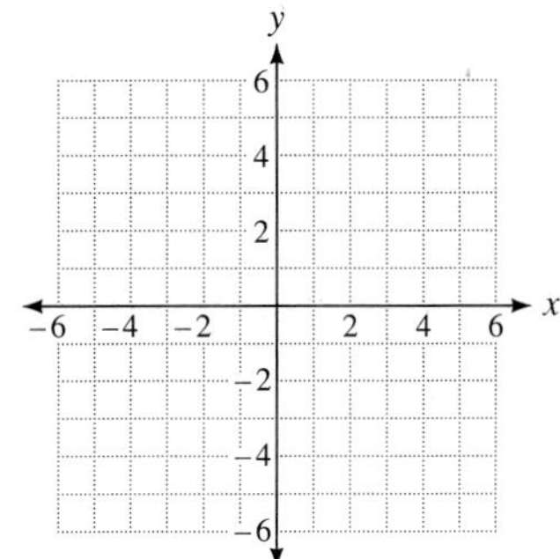

13. $y = x^2 - 2x + 3$

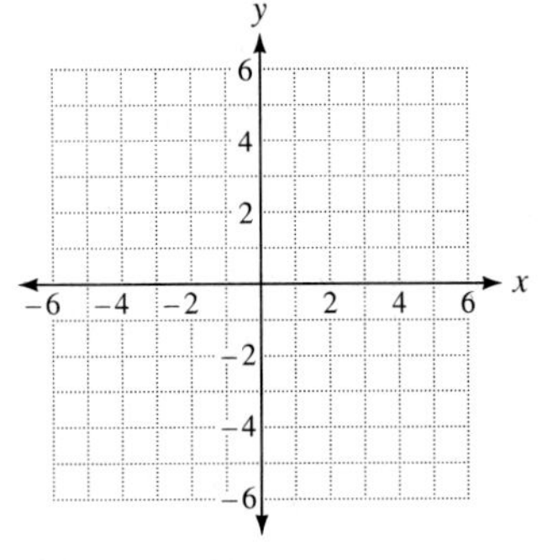

14. $x = y^2 - 2$

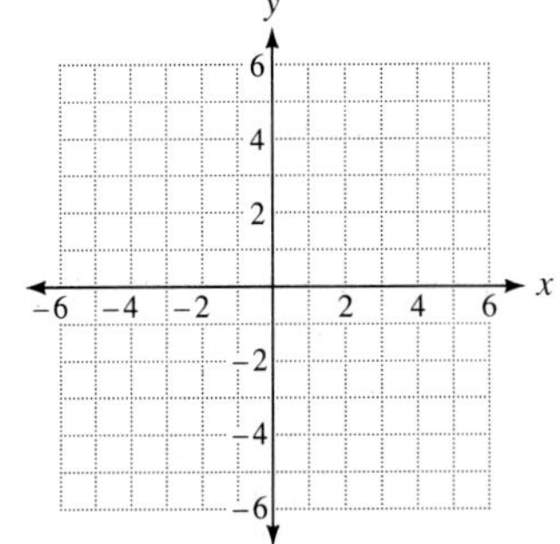

15. $x = y^2 - 2y - 3$

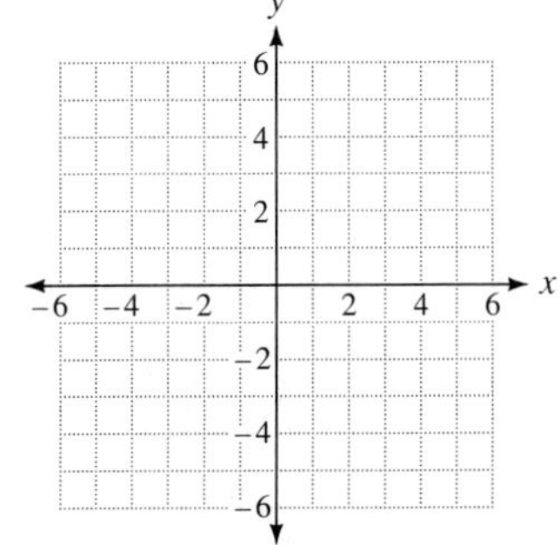

16. $4x = -y^2$

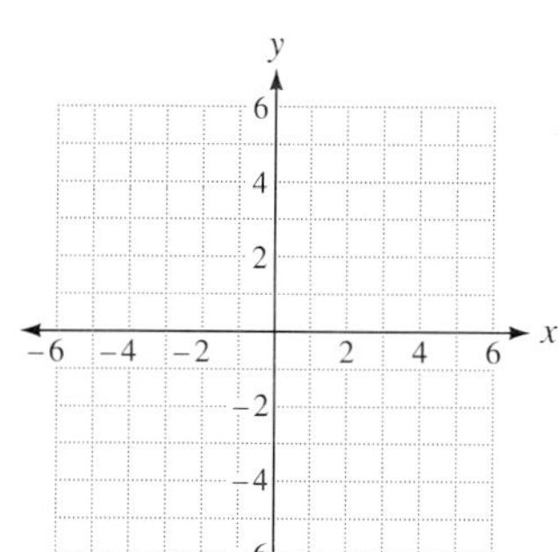

17. $2y = -x^2 + 2x + 1$

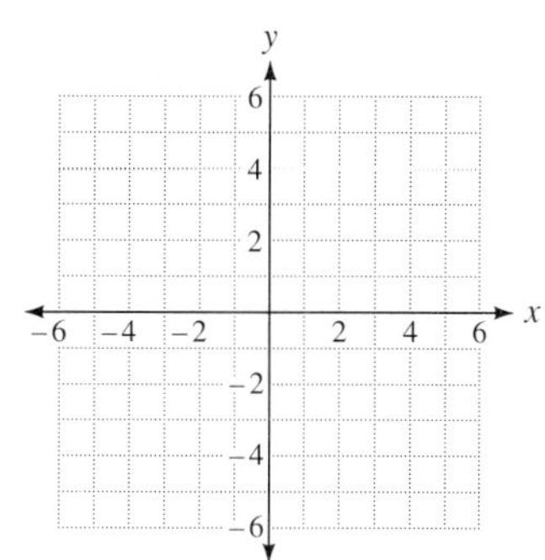

18. $8x = y^2 - 16$

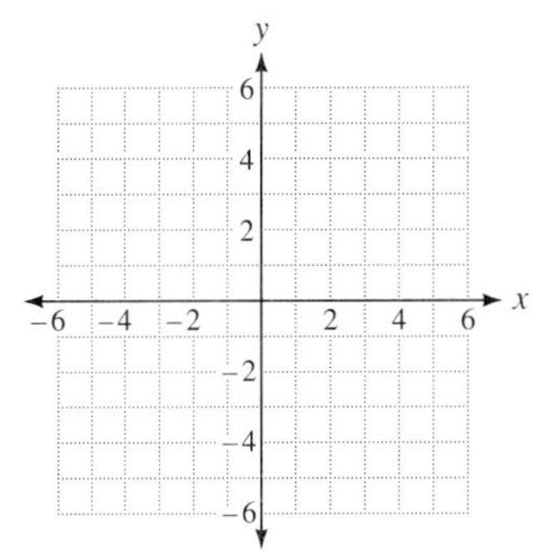

19. The cross-section of a ditch is a parabola. If the surface of the water in a full ditch is 10 ft wide and the ditch is 6 ft deep at the center, how deep is the water 2 ft from the edge of the ditch?

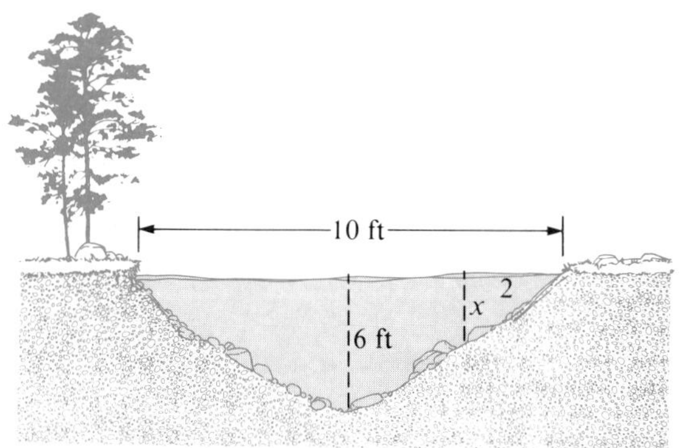

20. A footbridge over a creek is in the shape of a parabola. If the span of the bridge is 32 ft and the arch rises 4 ft above the bank of the creek, find the equation of the parabola assuming that its vertex is at the origin and it opens down.

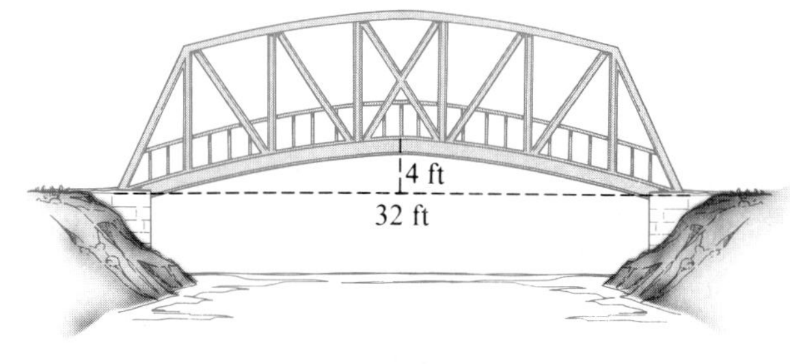

FOR REVIEW

Exercises 21–24 review topics covered in Chapter 10.

21. If $x = 439$, find **(a)** $\log x$, and **(b)** $\ln x$, correct to four decimal places.

22. If $x = -2.4351$, find **(a)** the common antilogarithm of x, and **(b)** the natural antilogarithm of x, correct to three significant digits.

23. Solve. $\log (x - 3) + \log x = 1$

24. Express $\log_a \frac{x^2\sqrt{y}}{z}$ using sums, differences, or multiples of logarithms of a single variable.

Exercises 25 and 26 review the distance formula from Section 7.7 to help you prepare for the next section.

25. Find the distance between the two points $(3, -7)$ and $(-1, 4)$.

26. The point (x, y) is 2 units from $(-1, 3)$. Find an equation involving x and y that describes this relationship.

ANSWERS: 1. (a) $(1, 5)$ (b) up 2. (a) $(-1, -3)$ (b) down 3. (a) $(4, 3)$ (b) left 4. (a) $(-7, -4)$ (b) right 5. (a) $(0, -5)$ (b) up 6. (a) $(6, 0)$ (b) left 7. (a) $(1, 8)$ (b) up 8. (a) $(-6, 3)$ (b) left 9. (a) $(-5, -1)$ (b) left

10. (1, 1) (4, 2) (1, −1) (4, −2)

11.

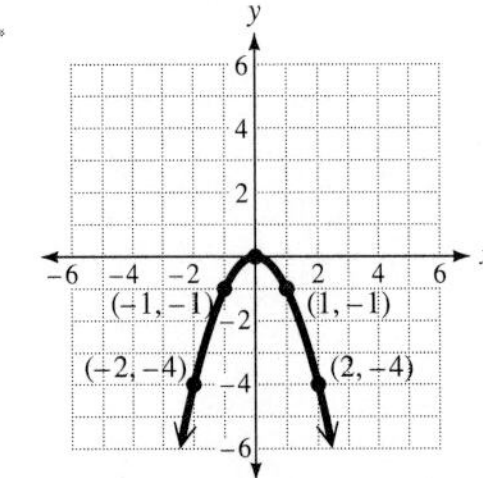

12.

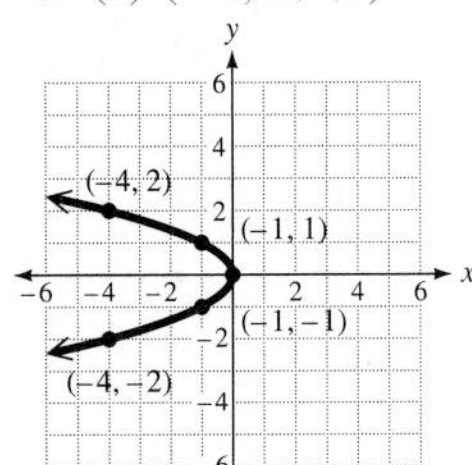

13.

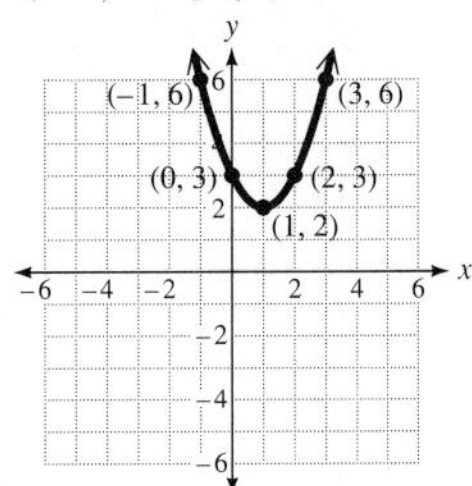

14. (−1, 1) (2, 2) (−2, 0) (−1, −1) (2, −2)

15.

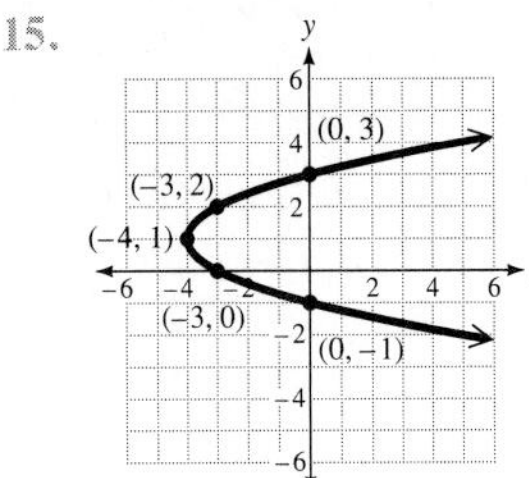

16.

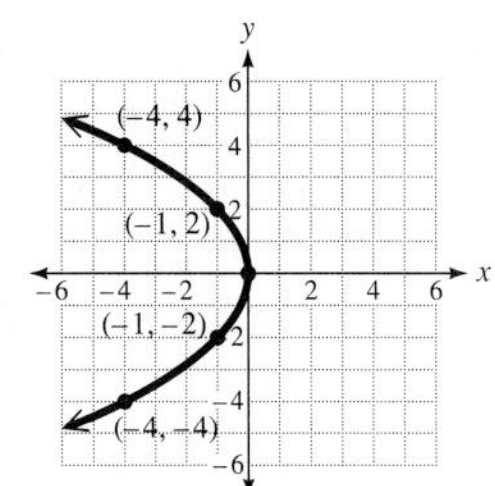

17.

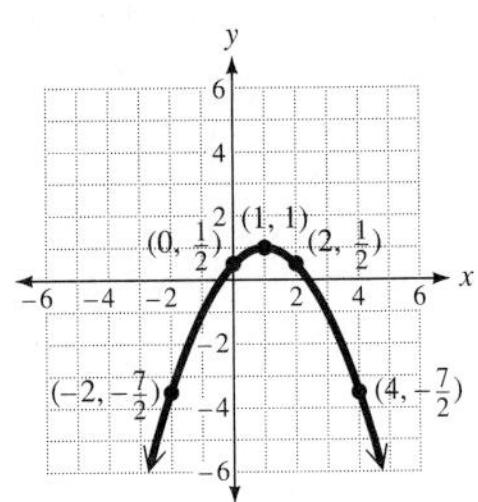

18. (9/8, 5) (0, 4) (−2, 0) (0, −4) (9/8, −5)

19. 3.84 ft 20. $y = -\frac{1}{64}x^2$ 21. (a) 2.6425 (b) 6.0845 22. (a) 0.00367 (b) 0.0876 23. 5 24. $2 \log_a x + \frac{1}{2} \log_a y - \log_a z$ 25. $\sqrt{137}$ 26. $(x + 1)^2 + (y - 3)^2 = 4$

11.1 EXERCISES B

*For each parabola, **(a)** give the vertex, and **(b)** state whether the parabola opens up, down, left, or right.*

1. $y = 2(x + 1)^2 - 5$
2. $y = -4(x - 2)^2 + 6$
3. $x = -5(y + 3)^2 - 5$
4. $x = 3(y - 4)^2 + 5$
5. $x = y^2 + 2$
6. $y = -x^2 - 3$
7. $x = y^2 + 10y + 24$
8. $y = 3x^2 - 12x + 20$
9. $2y = -x^2 - 2x - 7$

Graph each parabola by writing the equation in standard form.

10. $x = 2y^2$

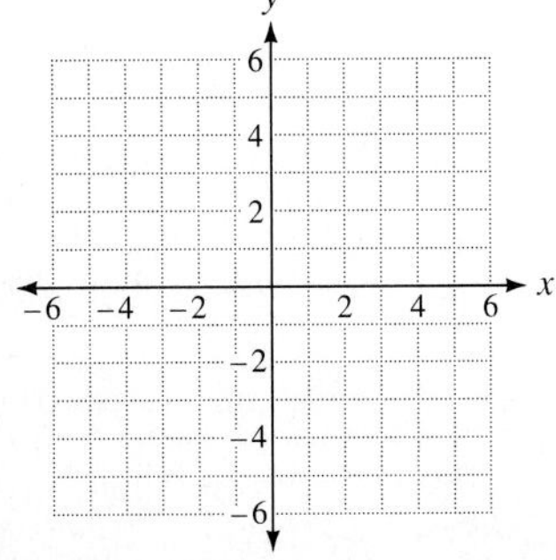

11. $y = -\dfrac{1}{2}x^2$

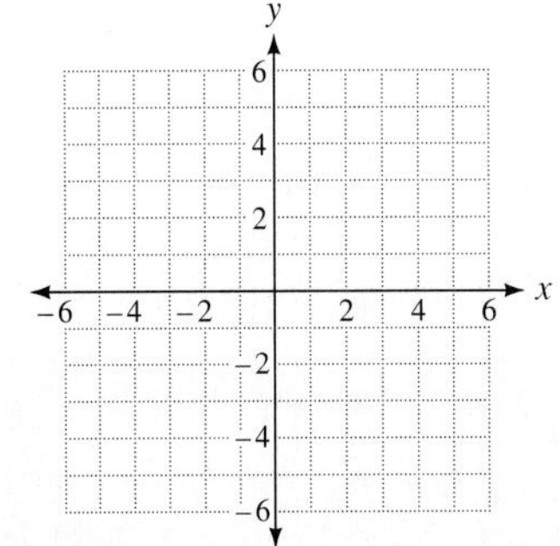

12. $x = -y^2 + 2$

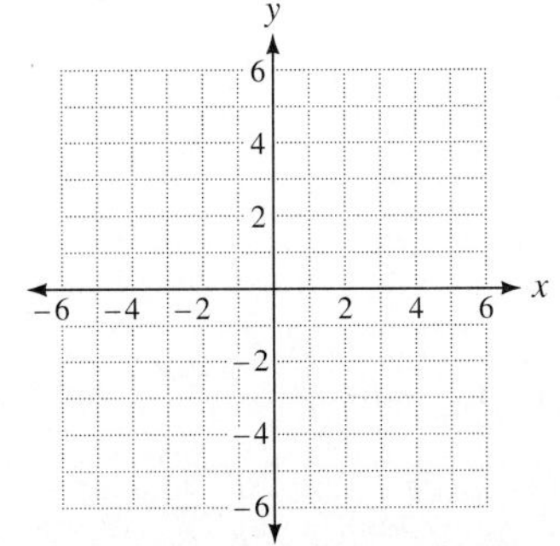

13. $y = x^2 - 4x + 3$

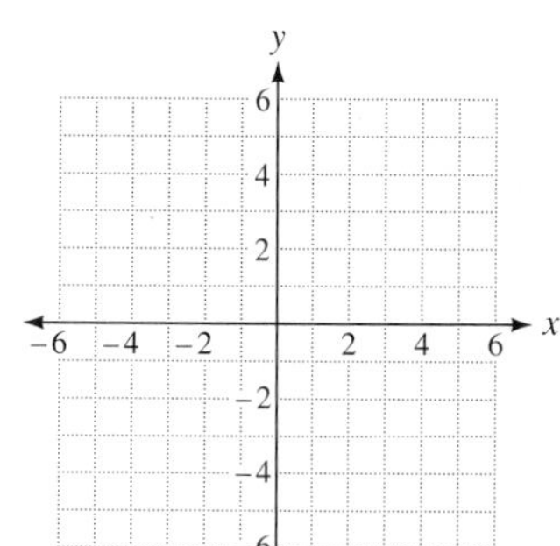

14. $x = y^2 + 1$

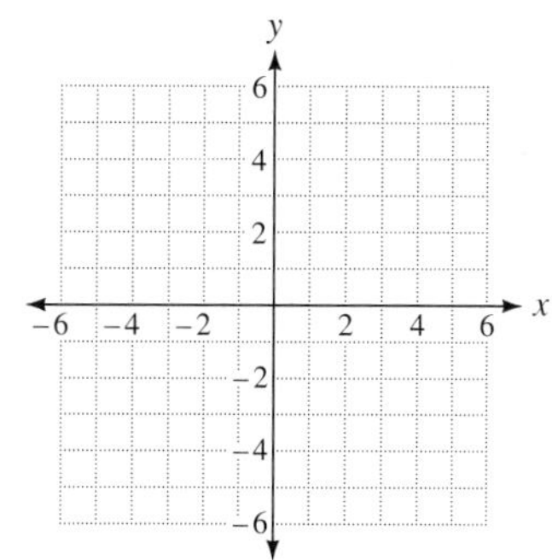

15. $x = -y^2 - 2y + 3$

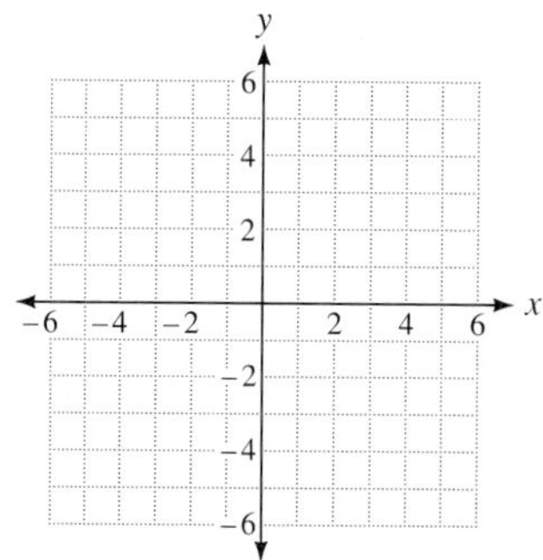

16. $2y = x^2$

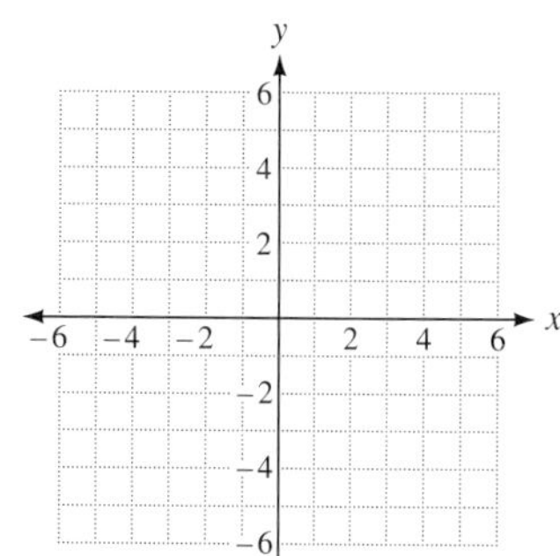

17. $4x = -y^2 - 2y + 3$

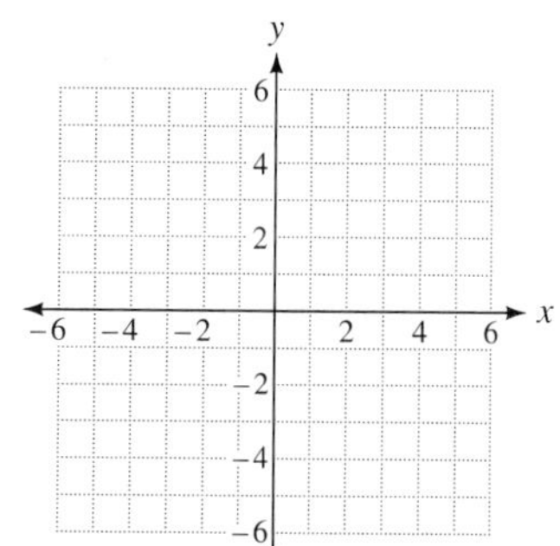

18. $8y = -x^2 - 6x + 7$

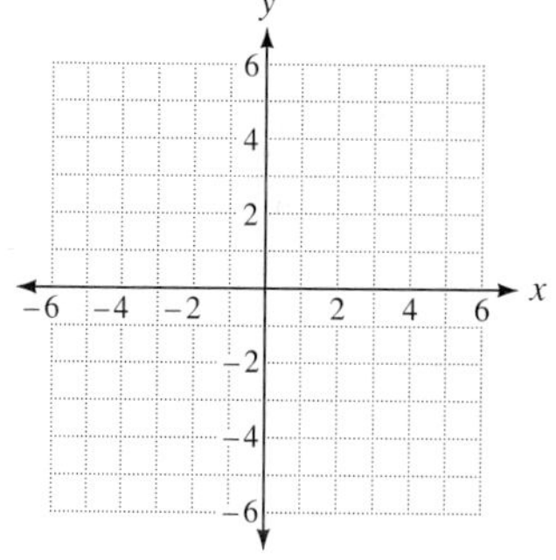

19. A tunnel is in the shape of a parabola. The maximum height is 30 ft and it is 24 ft wide at the base. Place the vertex at the origin and find the equation of the parabola. What is the vertical clearance 4 ft from the edge of the tunnel?

20. A television antenna is formed by rotating a parabola around its axis of symmetry. If the diameter of the antenna is 6 meters and it is $\frac{3}{4}$ meter deep, find the equation of the parabola by placing the vertex at the origin and assuming that it opens to the right.

FOR REVIEW

Exercises 21–24 review topics covered in Chapter 10.

21. If $x = 0.385$, find **(a)** $\log x$, and **(b)** $\ln x$, correct to four decimal places.

22. If $x = -1.5496$, find **(a)** the common antilogarithm of x, and **(b)** the natural antilogarithm of x, correct to three significant digits.

23. Solve. $2^{3x^2} = 16^x$

24. Express $\log_a \frac{x^3y^4}{\sqrt{z}}$ using sums, differences, or multiples of logarithms of a single variable.

Exercises 25 and 26 review the distance formula from Section 7.7 to help you prepare for the next section.

25. Find the distance between the two points $(-2, 3)$ and $(4, -1)$.

26. The point (x, y) is 3 units from $(2, -1)$. Find an equation involving x and y that describes this relationship.

11.1 EXERCISES C

Write each equation in standard form and identify the vertex. Does the parabola open up, down, left, or right?

1. $4y^2 + 4y + 12x + 9 = 0$

2. $36x^2 - 24x - 180y + 49 = 0$

11.2 THE CIRCLE

STUDENT GUIDEPOSTS

 Standard Form of a Circle

 General Form of a Circle

To study the next conic section, the circle, we begin with its definition and then use the distance formula developed in Section 7.7,

$$d = \sqrt{(x_2 - x_1)^2 + (y_2 - y_1)^2},$$

which gives the distance between two points with coordinates (x_1, y_1) and (x_2, y_2). From these we find the standard form of the equation of a circle.

A **circle** is the set of all points in a plane that are a fixed distance r from a given point (h, k) in that plane. The given point is the **center** of the circle and the fixed distance is the **radius** of the circle.

1 STANDARD FORM OF A CIRCLE

Let (x, y) be an arbitrary point on the circle with radius r centered at the point (h, k), as in Figure 11.8. Since every point (x, y) is r units from (h, k), by the distance formula we have

$$r = \sqrt{(x - h)^2 + (y - k)^2}.$$

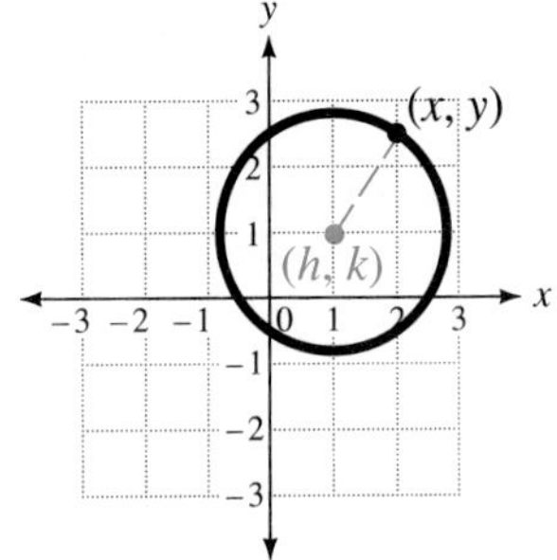

Figure 11.8 Circle

Squaring both sides yields the standard form of the equation of a circle.

Standard Form of the Equation of a Circle

The **standard form** of the equation of a circle with center (h, k) and radius r is

$$(x - h)^2 + (y - k)^2 = r^2.$$

Circles centered at the origin have a simplified standard form. Consider the equation

$$x^2 + y^2 = 4.$$

We can write this equation in the form

$$(x - 0)^2 + (y - 0)^2 = 2^2$$

to see that the center is indeed (0, 0) and the radius is 2. See the graph in Figure 11.9.

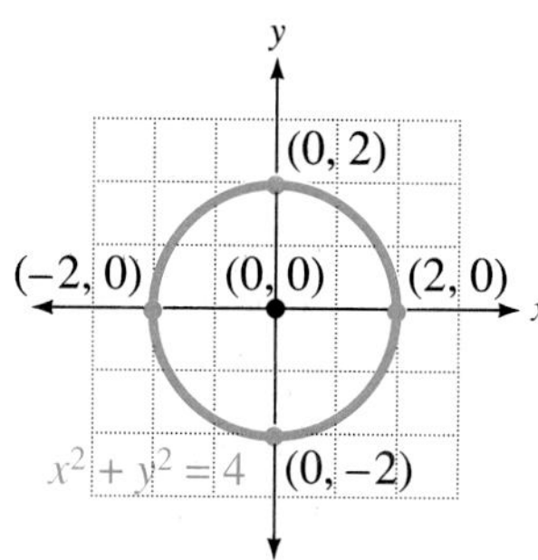

Figure 11.9

EXAMPLE 1 Graphing a Circle

Graph $(x - 1)^2 + (y + 2)^2 = 9$.

We rewrite the equation in the form $(x - h)^2 + (y - k)^2 = r^2$.

$$(x - 1)^2 + [y - (-2)]^2 = 3^2$$

Thus, the center is $(1, -2)$, and the radius is 3. See Figure 11.10.

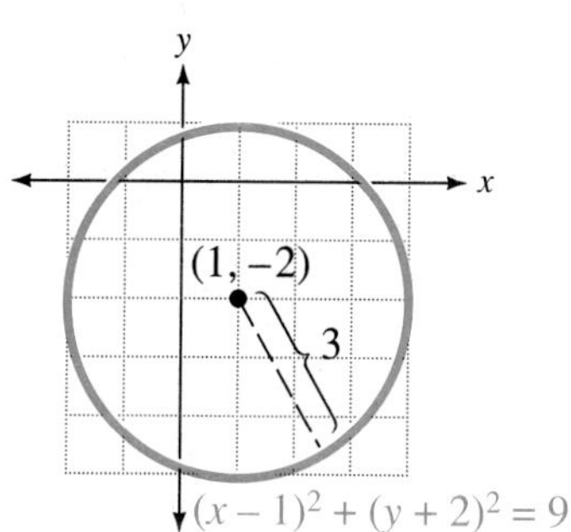

Figure 11.10

Practice Exercise 1

Graph $(x + 1)^2 + (y - 2)^2 = 4$.

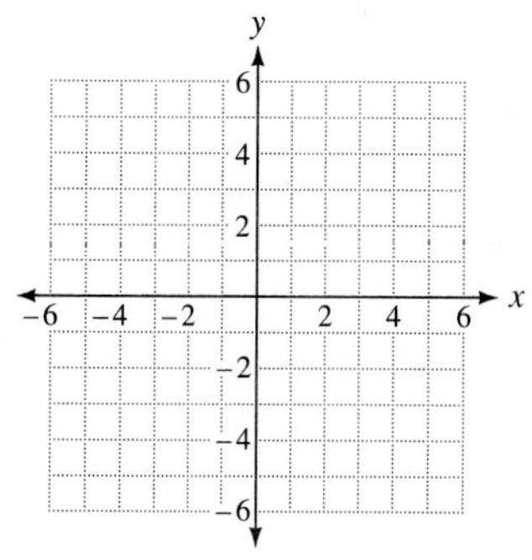

Answer: The graph is a circle centered at $(-1, 2)$ with radius 2.

If we are given the center and radius of a circle, we can find the standard form of its equation by substituting h, k, and r into $(x - h)^2 + (y - k)^2 = r^2$.

EXAMPLE 2 FINDING THE STANDARD FORM

Find the standard form of the equation of the circle centered at $(2, -1)$ with radius 3.

Substitute $h = 2$, $k = -1$, and $r = 3$ in the standard form.

$$(x - h)^2 + (y - k)^2 = r^2$$

$$(x - 2)^2 + (y - (-1))^2 = 3^2 \quad \text{Watch the sign on } k$$

$$(x - 2)^2 + (y + 1)^2 = 9$$

PRACTICE EXERCISE 2

Find the standard form of the equation of the circle centered at $(-4, 3)$ with radius $\sqrt{5}$.

Answer: $(x + 4)^2 + (y - 3)^2 = 5$

2 GENERAL FORM OF A CIRCLE

When the standard form of the equation of a circle is expanded, the resulting equation, called the **general form of the equation of a circle,** is actually a second-degree equation of the form

$$x^2 + y^2 + Dx + Ey + F = 0.$$

EXAMPLE 3 FINDING THE GENERAL FORM

Find the general form of the equation of the circle centered at $(2, -1)$ with radius 3.

From Example 2, the standard form of this equation is

$$(x - 2)^2 + (y + 1)^2 = 9.$$

Expanding, we have

$$x^2 - 4x + 4 + y^2 + 2y + 1 = 9.$$

The general form is

$$x^2 + y^2 - 4x + 2y - 4 = 0.$$

PRACTICE EXERCISE 3

Find the general form of the equation of the circle centered at $(-4, 3)$ with radius $\sqrt{5}$.

Answer: $x^2 + y^2 + 8x - 6y + 20 = 0$

Finding the general form of the equation of a circle from the standard form is simply a matter of squaring and collecting like terms. The reverse procedure, finding the standard form from the general form, requires completing the squares in both variables x and y. This is shown in the next example.

EXAMPLE 4 FINDING THE STANDARD FORM

Find the standard form of the equation of the circle with general form

$$x^2 + y^2 - 6x + 10y + 29 = 0.$$

Give the center and radius, and graph the equation.

Start by writing the equation with the variable terms on the left side and the constant term on the right, leaving space as shown.

$$x^2 - 6x \qquad + y^2 + 10y \qquad = -29$$

Next, to complete the square in x, add $\left[\frac{1}{2}(-6)\right]^2 = 9$, and to complete the square in y, add $\left[\frac{1}{2}(10)\right]^2 = 25$ to the left side. Then, to maintain equality add 9 and 25 to the right side.

PRACTICE EXERCISE 4

Find the standard form of the equation of the circle with general form $x^2 + y^2 + 4x - 6y + 9 = 0$.

What is the center of the circle?

What is the radius?

$$x^2 - 6x + 9 + y^2 + 10y + 25 = -29 + 9 + 25$$
$$(x^2 - 6x + 9) + (y^2 + 10y + 25) = 5$$
$$(x - 3)^2 + (y + 5)^2 = 5 \quad \text{Factor}$$

This is the standard form of the equation of the circle with center $(3, -5)$. Notice that since $r^2 = 5$, the radius is $\sqrt{5}$. See Figure 11.11.

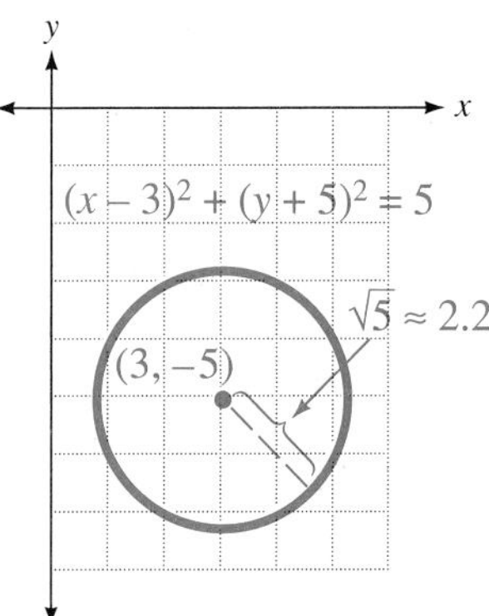

Figure 11.11

Graph the circle.

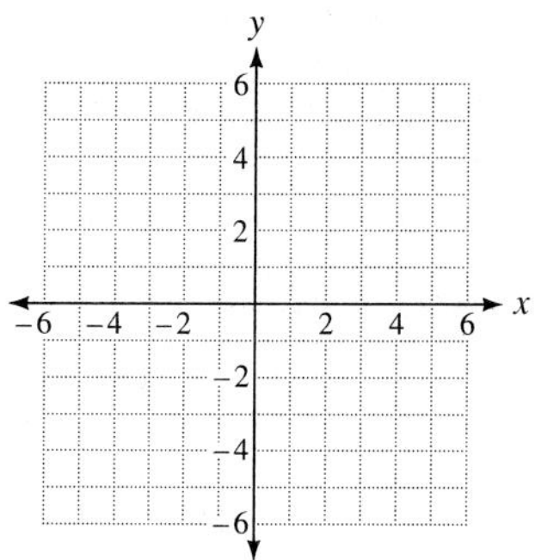

Answer: $(x + 2)^2 + (y - 3)^2 = 4$; $(-2, 3)$; $r = 2$

The circle is used in many applications. The following example illustrates its use in construction.

EXAMPLE 5 ENGINEERING PROBLEM

A railroad tunnel for a small-scale train is designed in the shape of a semicircle with diameter 6 meters. Give the equation for this semicircle and find the vertical clearance 1 meter from the centerline of the tunnel.

Suppose we set a coordinate system with the origin (0, 0) at the centerline of the tunnel as in Figure 11.12.

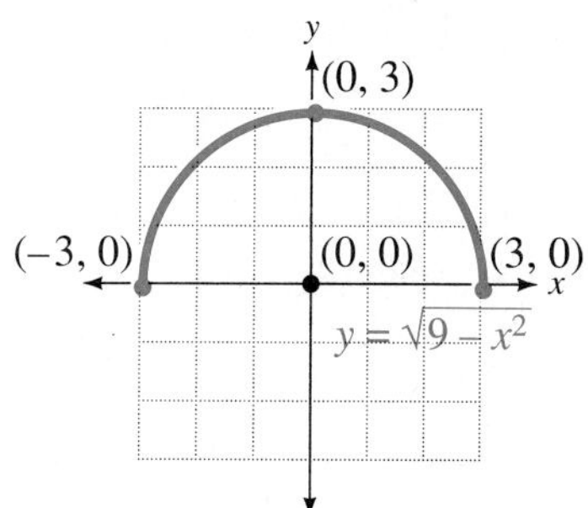

Figure 11.12 Railroad Tunnel

The equation of the circle, the top half of which is our semicircle is

$$(x - 0)^2 + (y - 0)^2 = 3^2 \quad r = 3 \text{ since the diameter is } 6$$
$$x^2 + y^2 = 9$$

If we solve this equation for y,

$$y^2 = 9 - x^2$$
$$y = \pm\sqrt{9 - x^2},$$

PRACTICE EXERCISE 5

Can a boxcar 3 meters wide and 3 meters high pass through the tunnel if the tracks are centered in the tunnel?

the top half of the circle is described with the positive value for y,

$$y = \sqrt{9 - x^2}.$$

To find the vertical clearance 1 meter from the center line, let $x = 1$ in this equation.

$$y = \sqrt{9 - (1)^2} = \sqrt{9 - 1} = \sqrt{8} \approx 2.83$$

The clearance is approximately 2.83 meters.

Answer: No; the vertical clearance 1.5 meters from the center is only about 2.6 meters.

The equation of the bottom half of the circle in Example 5 is $y = -\sqrt{9 - x^2}$. In Section 9.2 we graphed square root functions of the forms $y = \sqrt{x - 1}$ and $y = -\sqrt{x - 1}$ which were the upper and lower halves, respectively, of the parabola $y^2 = x - 1$ or $x = y^2 + 1$. Although circles and parabolas opening left and right are not functions, the upper and lower halves of each can be described as functions using variations of the square root function.

11.2 EXERCISES A

Find the standard and general forms of the equation of a circle centered at the given point and with the given radius.

1. $(-3, 2)$, $r = 1$

2. $(5, -3)$, $r = 2$

3. $(-1, -1)$, $r = 3$

4. $\left(-\frac{1}{2}, 0\right)$, $r = 1$

5. $(0, 0)$, $r = 1$

6. $(-2, -4)$, $r = 6$

Find the standard form of the equation of each circle and identify the center and the radius.

7. $x^2 + y^2 - 2x + 4y - 20 = 0$

8. $x^2 + y^2 + 6x = 0$

9 $4x^2 + 4y^2 + 40x - 4y + 37 = 0$

Graph each equation.

10. $x^2 + y^2 = 25$

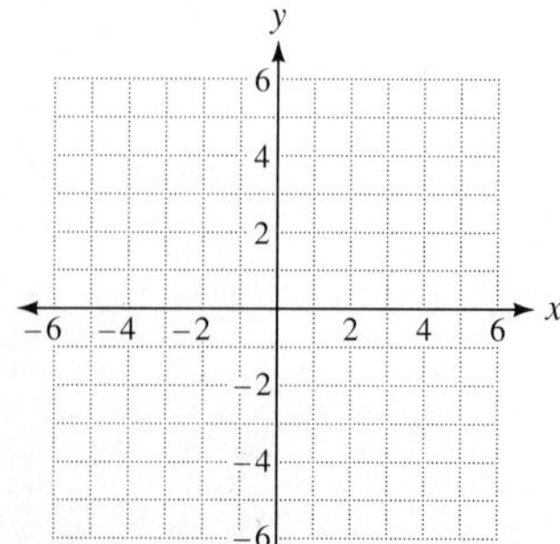

11. $(x - 2)^2 + (y + 3)^2 = 4$

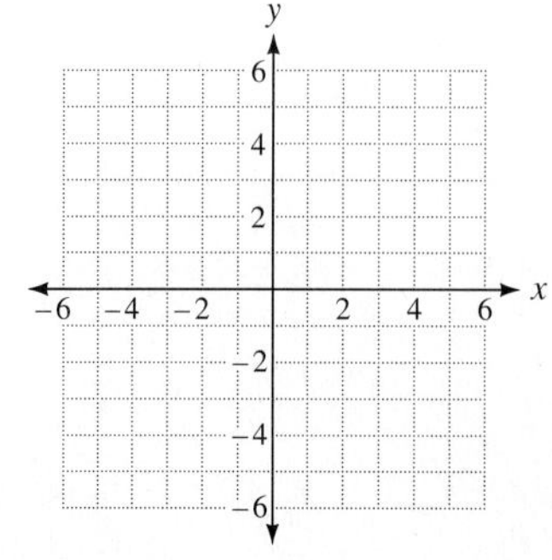

12. $(x - 1)^2 + (y - 2)^2 = 9$

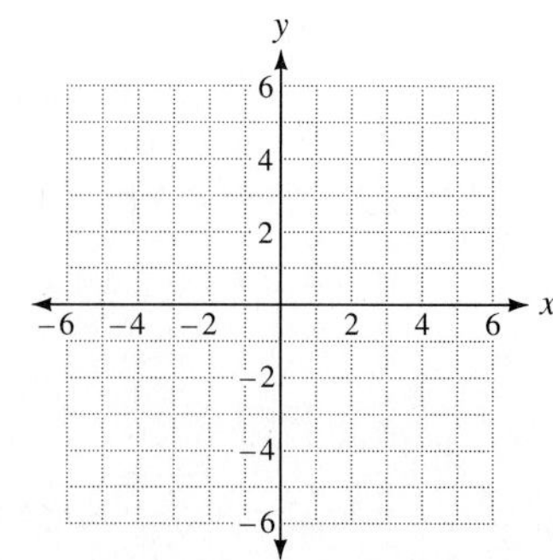

13. $x^2 + (y - 3)^2 = 4$

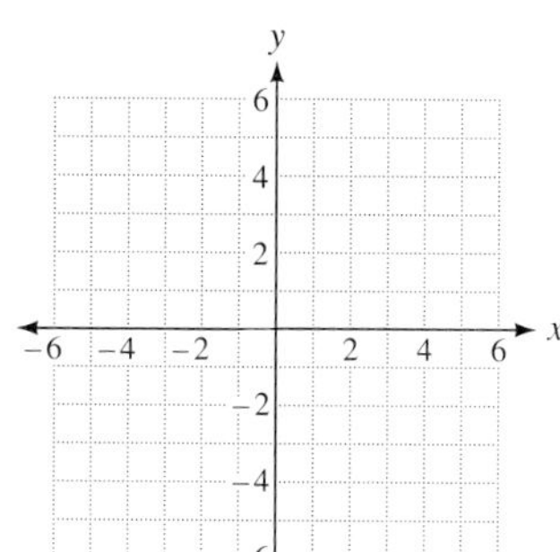

14. $y = \sqrt{25 - x^2}$

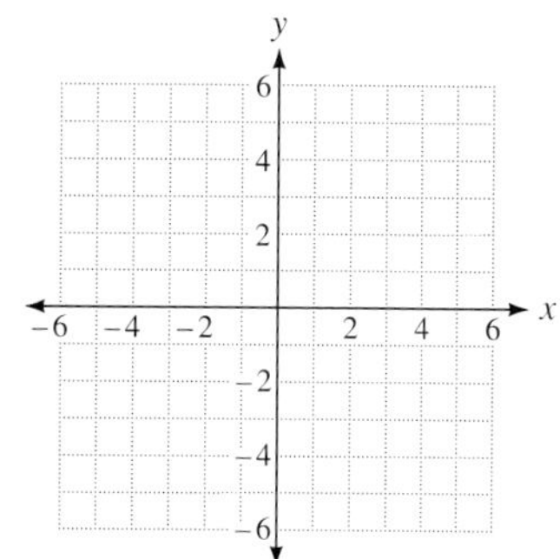

15. $y = -\sqrt{16 - x^2}$

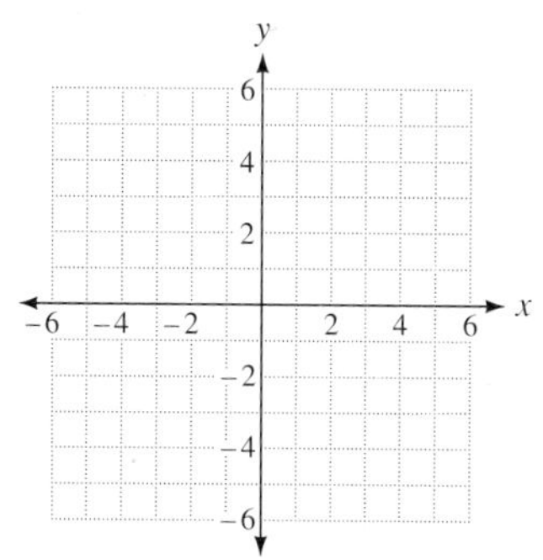

16. A highway tunnel in the shape of a semicircle has a diameter measuring 30 ft. Find the equation of this semicircle using a coordinate system with origin at the center line of the roadway. What is the vertical clearance 5 ft from the edge of the tunnel?

17. A canal with cross-section a semicircle is to have a depth of 3 ft at a distance of 4 ft from the center of the waterline in a full canal. Find the equation of the semicircle. What is the maximum depth of the water in the canal?

FOR REVIEW

Graph each equation.

18. $y^2 = 4x$

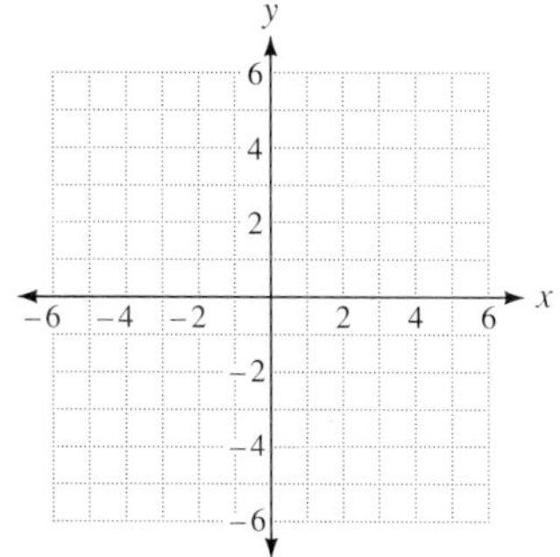

19. $x^2 - 4x + 8y + 4 = 0$

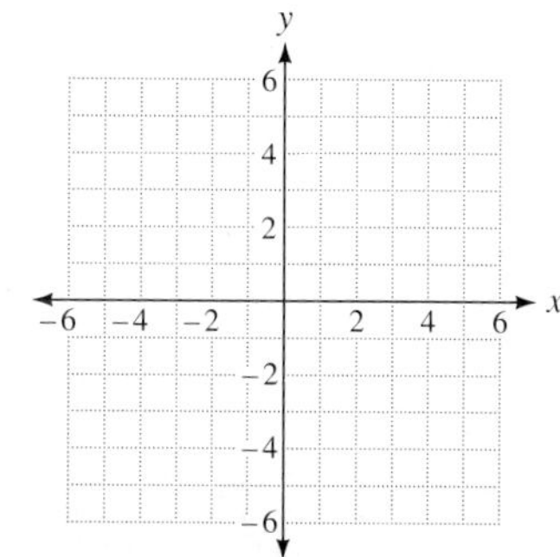

20. $y = -\sqrt{x - 3}$

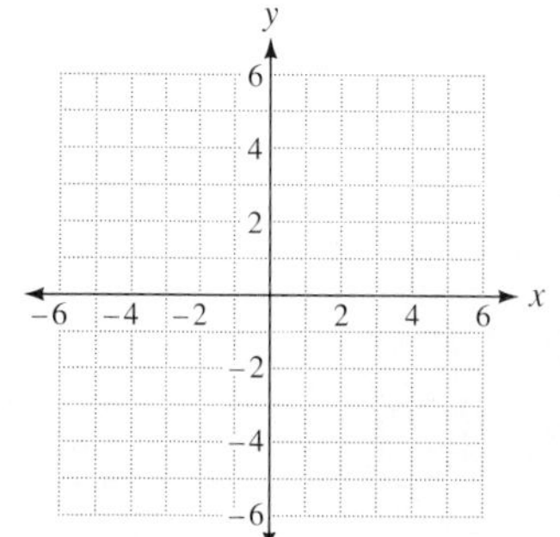

Exercises 21–23 will help you prepare for the next section.

21. For the equation $\frac{x^2}{4} + \frac{y^2}{9} = 1$, **(a)** find the value(s) of y when $x = 0$, and **(b)** find the value(s) of x when $y = 0$.

22. For the equation $\frac{x^2}{4} - \frac{y^2}{9} = 1$, **(a)** find the value(s) of y when $x = 0$, and **(b)** find the value(s) of x when $y = 0$.

23. Solve the equation $\frac{x^2}{4} - \frac{y^2}{9} = 0$ for y. Discuss the graphs of the two solutions.

ANSWERS: 1. $(x+3)^2 + (y-2)^2 = 1^2$; $x^2 + y^2 + 6x - 4y + 12 = 0$ 2. $(x-5)^2 + (y+3)^2 = 2^2$; $x^2 + y^2 - 10x + 6y + 30 = 0$ 3. $(x+1)^2 + (y+1)^2 = 3^2$; $x^2 + y^2 + 2x + 2y - 7 = 0$ 4. $\left(x + \frac{1}{2}\right)^2 + y^2 = 1^2$; $x^2 + y^2 + x - \frac{3}{4} = 0$ or $4x^2 + 4y^2 + 4x - 3 = 0$ 5. $x^2 + y^2 = 1^2$; $x^2 + y^2 - 1 = 0$ 6. $(x+2)^2 + (y+4)^2 = 6^2$; $x^2 + y^2 + 4x + 8y - 16 = 0$ 7. $(x-1)^2 + (y+2)^2 = 5^2$; $(1, -2)$; $r = 5$ 8. $(x+3)^2 + y^2 = 3^2$; $(-3, 0)$; $r = 3$ 9. $(x+5)^2 + \left(y - \frac{1}{2}\right)^2 = 4^2$; $\left(-5, \frac{1}{2}\right)$; $r = 4$

10.
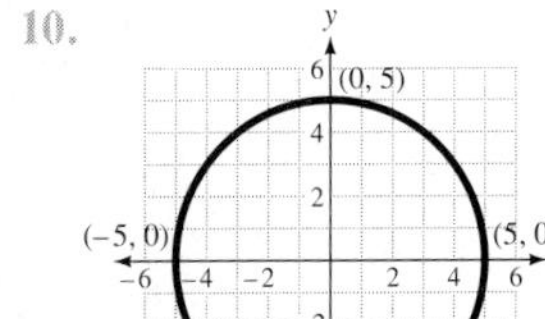

11.
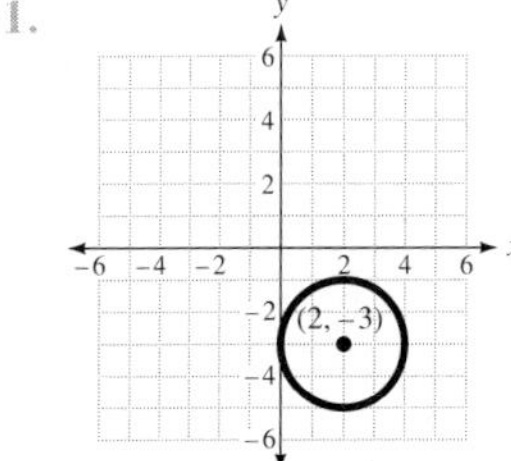

12.
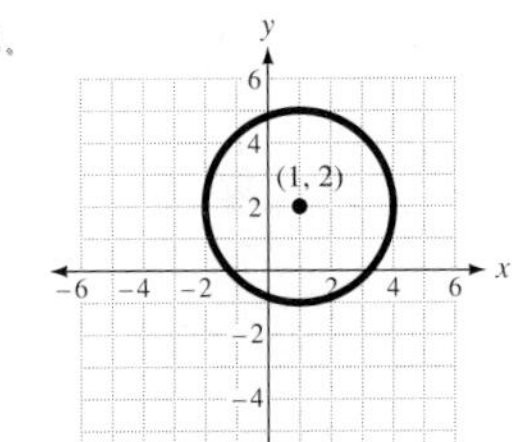

13.
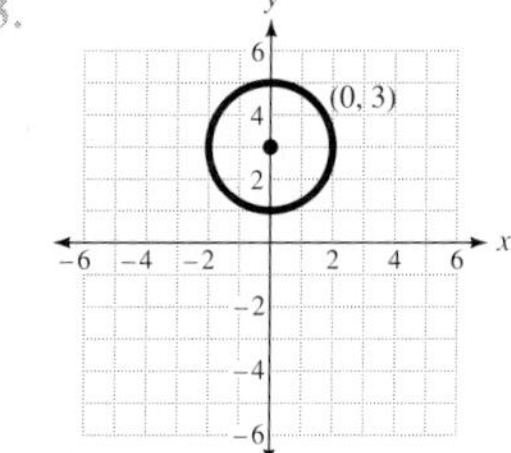

14.
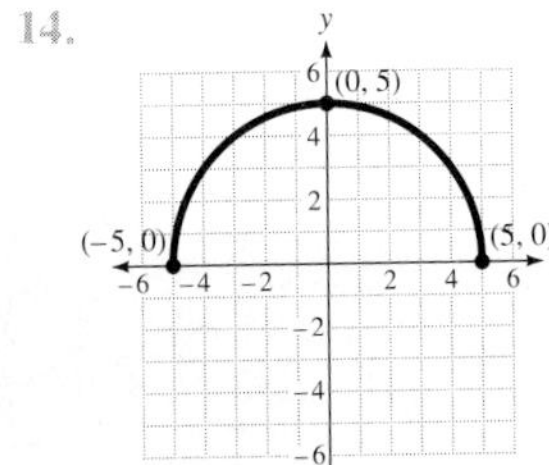

15.
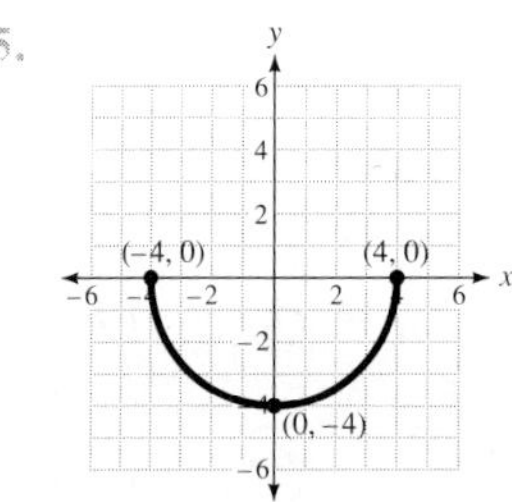

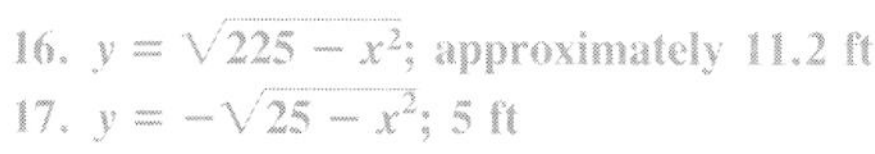

16. $y = \sqrt{225 - x^2}$; approximately 11.2 ft
17. $y = -\sqrt{25 - x^2}$; 5 ft

18.
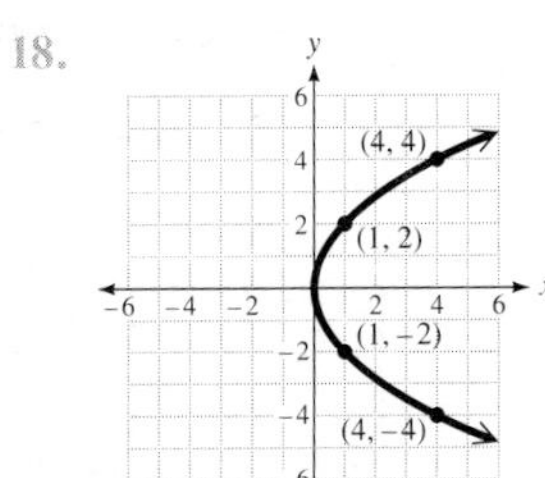

19.
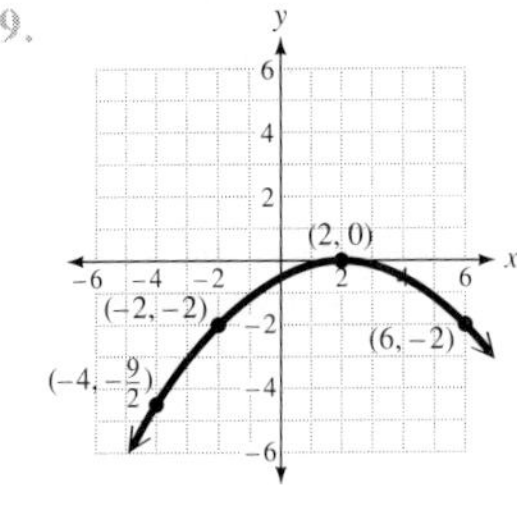

20.
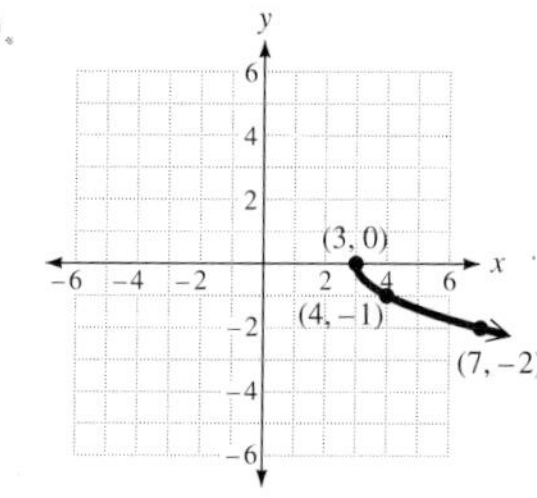

21. (a) ± 3 (b) ± 2 22. (a) The values are imaginary, $\pm 3i$ (b) ± 2 23. $y = \pm\frac{3}{2}x$; the graphs are intersecting lines passing through the origin $(0, 0)$ with slopes $\frac{3}{2}$ and $-\frac{3}{2}$.

11.2 EXERCISES B

Find the standard and general forms of the equation of a circle centered at the given point and with the given radius.

1. $(2, -1)$, $r = 3$

2. $(1, 4)$, $r = 5$

3. $(-3, -2)$, $r = 6$

4. $(0, 4)$, $r = 2$

5. $\left(2, \frac{1}{2}\right)$, $r = 5$

6. $(0, 0)$, $r = 7$

Find the standard form of the equation of each circle and identify the center and the radius.

7. $x^2 + y^2 - 10x + 4y + 20 = 0$

8. $x^2 + y^2 + 8y = 0$

9. $4x^2 + 4y^2 + 4x - 8y + 1 = 0$

Graph each equation.

10. $x^2 + y^2 = 9$

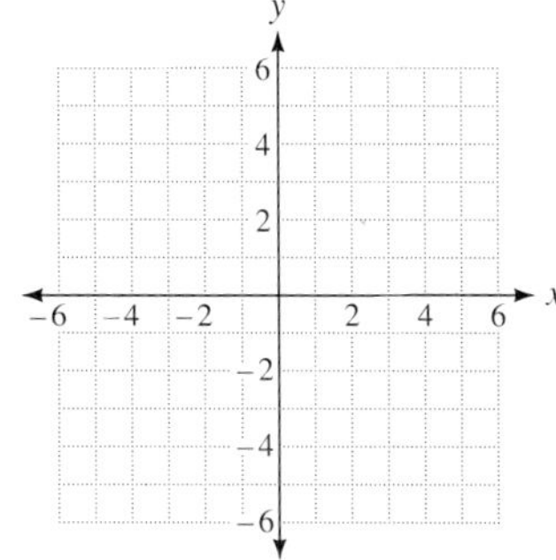

11. $(x - 1)^2 + (y + 1)^2 = 4$

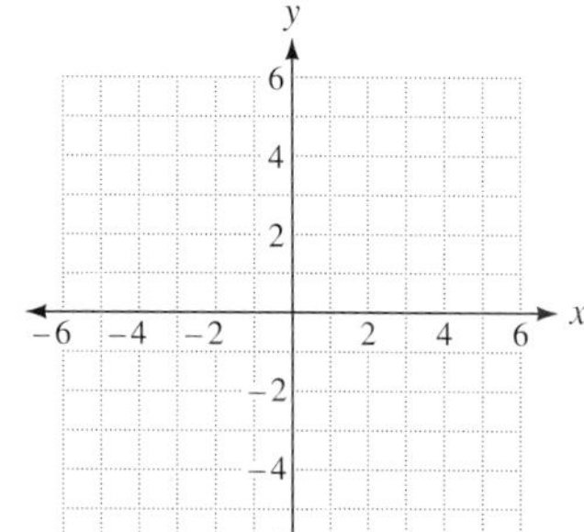

12. $(x + 2)^2 + (y + 2)^2 = 4$

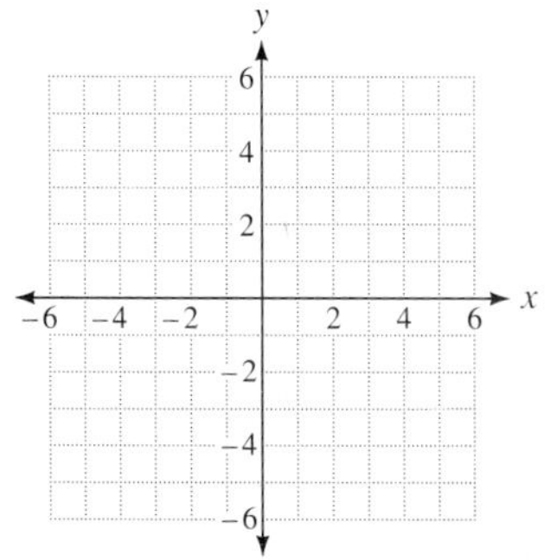

13. $(x - 3)^2 + y^2 = 9$

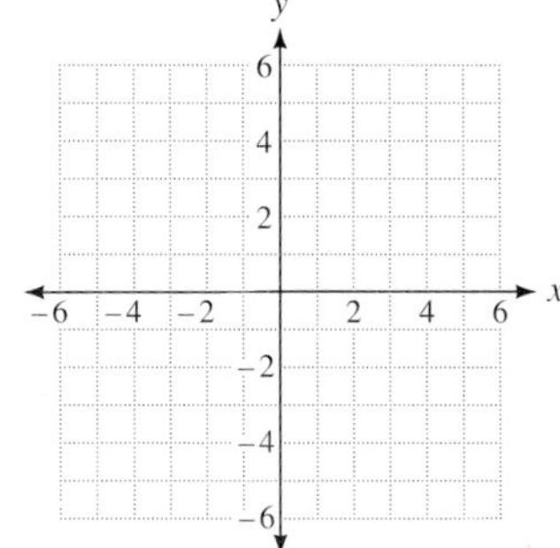

14. $y = -\sqrt{25 - x^2}$

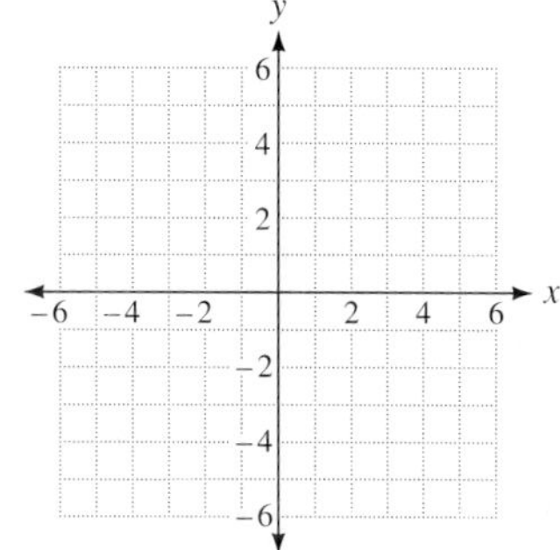

15. $y = \sqrt{16 - x^2}$

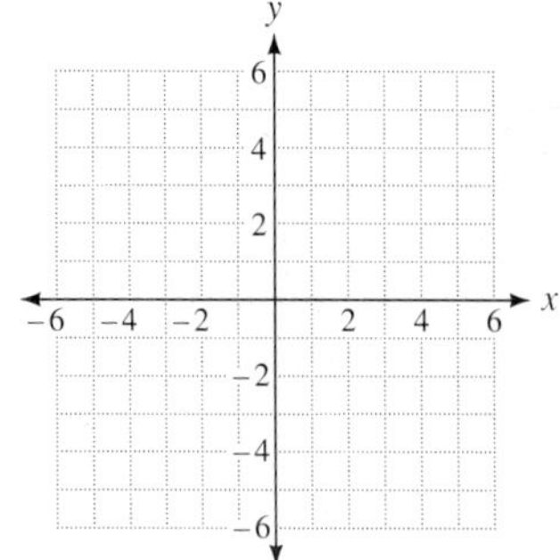

16. A canal with cross-section a semicircle is 8 ft deep at the center. Find the equation of this semicircle by placing a coordinate system with the origin at the center of the water level in a full canal. What is the depth of the canal 2 ft from the edge of the canal?

17. A railroad tunnel is in the shape of a semicircle. If the vertical clearance of the tunnel is 12 ft at a distance of 5 ft from the centerline, what is the equation of the semicircle? What is the maximum width of the tunnel at ground level?

FOR REVIEW

Graph each equation.

18. $x = 4y^2$

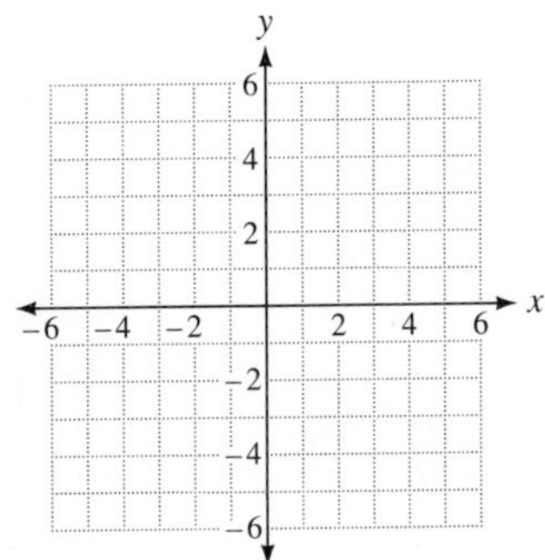

19. $y^2 - 4y + 8x + 4 = 0$

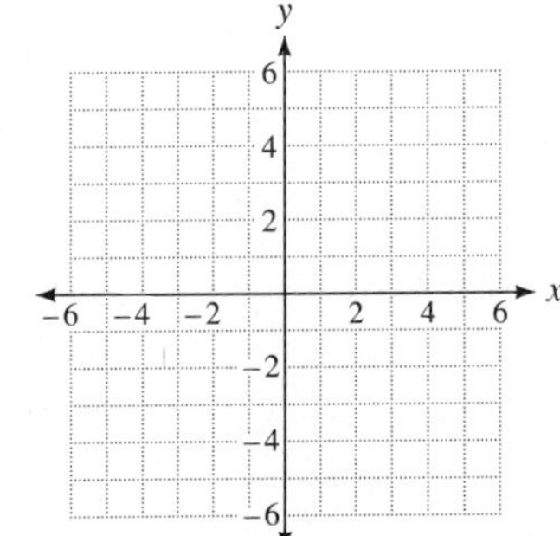

20. $y = \sqrt{x + 5}$

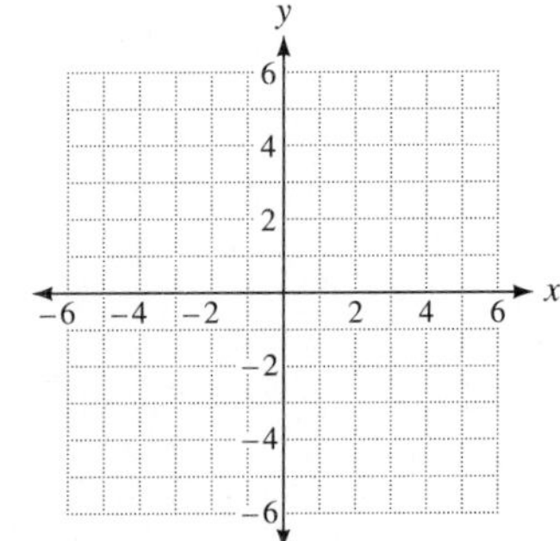

Exercises 21–23 will help you prepare for the next section.

21. For the equation $\frac{x^2}{16} + \frac{y^2}{4} = 1$, **(a)** find the value(s) of y when $x = 0$, and **(b)** find the value(s) of x when $y = 0$.

22. For the equation $\frac{x^2}{16} - \frac{y^2}{4} = 1$, **(a)** find the value(s) of y when $x = 0$, and **(b)** find the value(s) of x when $y = 0$.

23. Solve the equation $\frac{x^2}{16} - \frac{y^2}{4} = 0$ for y. Discuss the graphs of the two solutions.

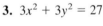

11.2 EXERCISES C

1. Find the equation of the circle whose diameter has endpoints $(-2, -3)$ and $(6, -3)$.
[Answer: $(x - 2)^2 + (y + 3)^2 = 16$]

2. The circle centered at $(2, 3)$ with radius 5 intersects the x-axis at what two points?

Graph each equation.

3. $3x^2 + 3y^2 = 27$

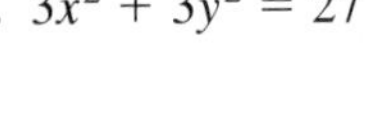

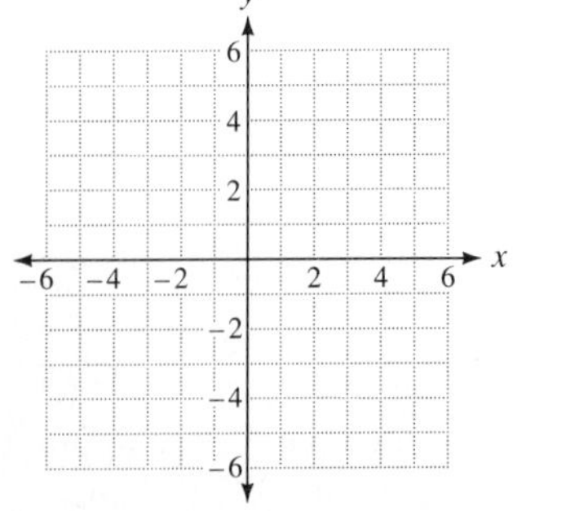

4. $x^2 + y^2 - 4x = 0$

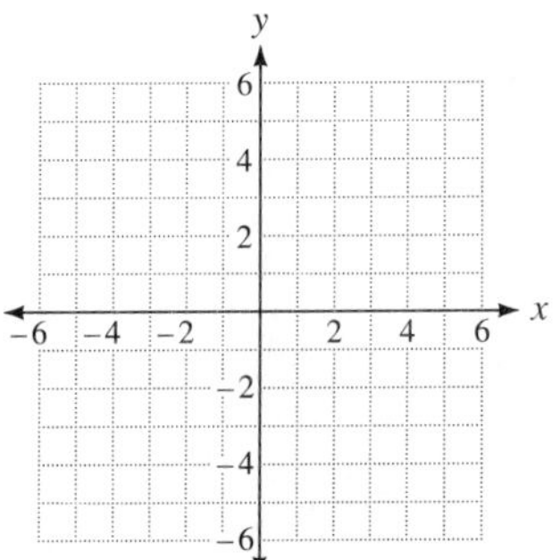

5. $x^2 + y^2 + 6y = 0$

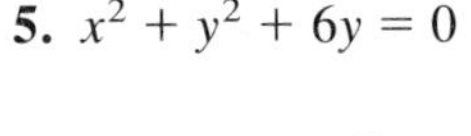

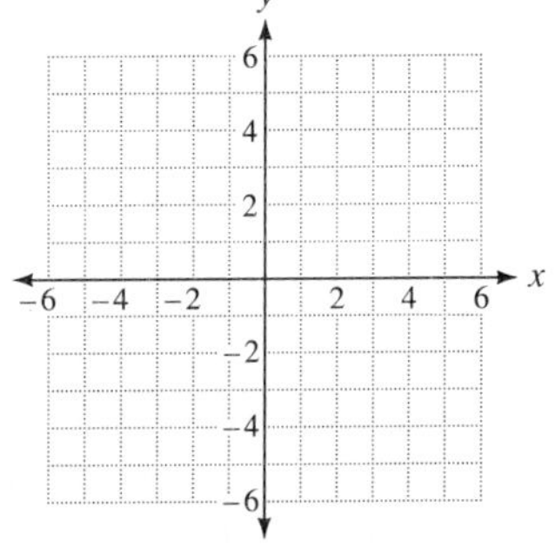

11.3 THE ELLIPSE AND THE HYPERBOLA

STUDENT GUIDEPOSTS

1. Standard Form of an Ellipse
2. Standard Forms of a Hyperbola
3. Asymptotes
4. Alternate Form of a Hyperbola
5. Summary of Conics

A circle is the set of all points in a plane whose distance from *one* fixed point (the center) is a constant. An **ellipse** is the set of all points in a plane the sum of whose distances from *two* fixed points (called **foci,** plural of **focus**) is a constant. If a string is attached to two pins and drawn taut using a pencil, the oval-shaped curve traced out by moving the pencil around the pins while keeping the string taut is an ellipse. See Figure 11.13.

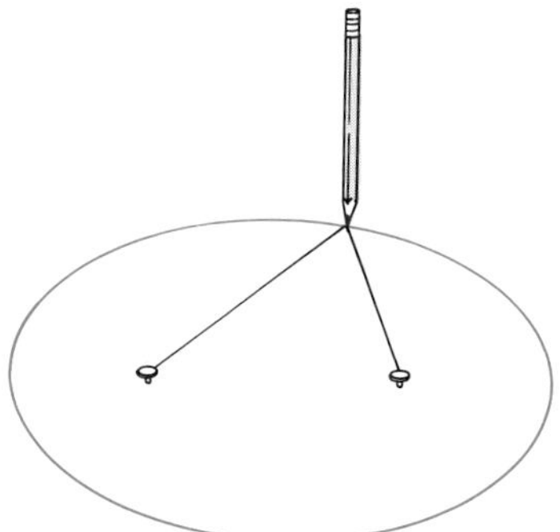

Figure 11.13 Construction of Ellipse

1 STANDARD FORM OF AN ELLIPSE

The orbit through which the earth travels around the sun and the orbits of communication satellites around the earth can all be approximated by ellipses. To simplify the discussion, we will consider only ellipses whose graphs are centered at the origin of a coordinate system.

Standard Form of the Equation of an Ellipse

The **standard form** of the equation of an ellipse centered at the origin with x-intercepts $(a, 0)$ and $(-a, 0)$ and y-intercepts $(0, b)$ and $(0, -b)$ is

$$\frac{x^2}{a^2} + \frac{y^2}{b^2} = 1. \quad a > 0, b > 0, a \neq b$$

If $a = b$ in $\frac{x^2}{a^2} + \frac{y^2}{b^2} = 1$, the equation can be written as $\frac{x^2}{a^2} + \frac{y^2}{a^2} = 1$ or as $x^2 + y^2 = a^2$. This is the equation of a circle centered at the origin, not an ellipse.

EXAMPLE 1 FINDING THE EQUATION OF AN ELLIPSE

Find the standard form of the equation of the ellipse centered at the origin with x-intercepts $(3, 0)$ and $(-3, 0)$ and y-intercepts $(0, 2)$ and $(0, -2)$.

Substitute $a = 3$ and $b = 2$ in the standard form.

PRACTICE EXERCISE 1

Find the standard form of the equation of the ellipse centered at the origin with intercepts $(6, 0)$, $(-6, 0)$, $(0, 1)$ and $(0, -1)$.

$$\frac{x^2}{3^2} + \frac{y^2}{2^2} = 1$$

$$\frac{x^2}{9} + \frac{y^2}{4} = 1$$

Answer: $\frac{x^2}{36} + \frac{y^2}{1} = 1$

Graphing an ellipse when its equation is given in standard form is simply a matter of plotting the four intercepts.

EXAMPLE 2 GRAPHING AN ELLIPSE

Graph the ellipse with equation $\frac{x^2}{25} + \frac{y^2}{4} = 1$.

We write the equation as $\frac{x^2}{5^2} + \frac{y^2}{2^2} = 1$.

Thus, the x-intercepts are (5, 0) and (−5, 0), and the y-intercepts (0, 2) and (0, −2). The graph is given in Figure 11.14.

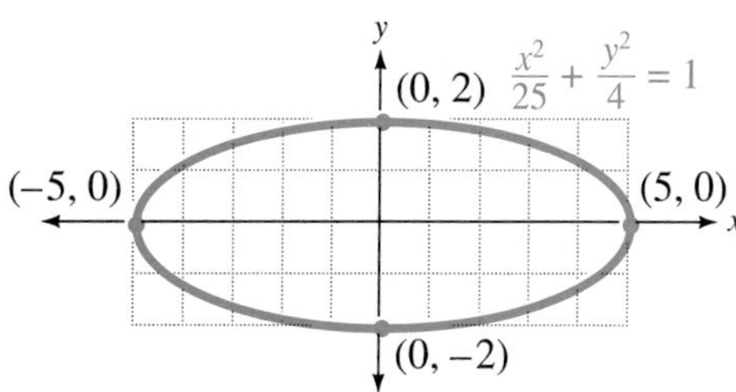

Figure 11.14

PRACTICE EXERCISE 2

Graph $\frac{x^2}{4} + \frac{y^2}{25} = 1$.

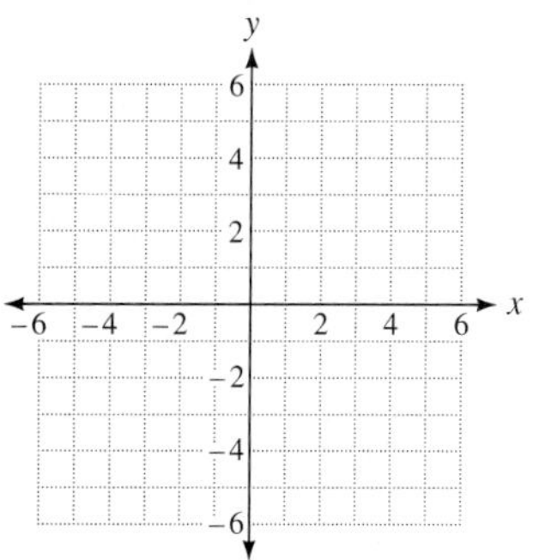

Answer: The graph is the ellipse with intercepts (2, 0), (−2, 0), (0, 5), and (0, −5).

A *hyperbola* is defined much like an ellipse. A **hyperbola** is the set of all points in a plane the *difference* of whose distances from two fixed points (called foci) is constant. Many modern buildings have cross-sectional structures in the shape of a hyperbola.

2 STANDARD FORMS OF A HYPERBOLA

As we did with ellipses, we will concentrate on hyperbolas centered at the origin of a coordinate system.

Standard Forms of the Equation of a Hyperbola

The **standard form** of the equation of a hyperbola centered at the origin with x-intercepts $(a, 0)$ and $(-a, 0)$ and opening left and right is

$$\frac{x^2}{a^2} - \frac{y^2}{b^2} = 1. \quad a > 0 \text{ and } b > 0$$

The **standard form** of the equation of a hyperbola centered at the origin with y-intercepts $(0, b)$ and $(0, -b)$ and opening up and down is

$$\frac{y^2}{b^2} - \frac{x^2}{a^2} = 1. \quad a > 0 \text{ and } b > 0$$

3 ASYMPTOTES

To plot an ellipse centered at the origin, the four intercepts are sufficient. For a hyperbola centered at the origin, however, there are only two intercepts, so additional information is needed. Associated with every hyperbola are two intersecting lines with the property that points on the hyperbola get closer to the lines as they get farther from the center. Any graph that approaches a line in this way is said to be **asymptotic** to the line, and the line is an **asymptote** of the graph.

Asymptotes of a Hyperbola

The asymptotes of either of the hyperbolas

$$\frac{x^2}{a^2} - \frac{y^2}{b^2} = 1 \quad \text{or} \quad \frac{y^2}{b^2} - \frac{x^2}{a^2} = 1$$

are the lines containing the diagonals of the rectangle with sides parallel to the axes, passing through the points $(a, 0)$, $(-a, 0)$, $(0, b)$, and $(0, -b)$.

Figure 11.15 shows the graphs of the two kinds of hyperbolas centered at the origin. To graph any hyperbola, plot the points $(a, 0)$, $(-a, 0)$, $(0, b)$, and $(0, -b)$, construct the rectangle through these points with sides parallel to the axes, and draw the diagonals of the rectangle, extending them to form the asymptotes. The hyperbola is the pair of curves that pass through the intercepts and approach the asymptotes. Notice that the hyperbola opens left and right when the coefficient of the x^2-term is positive, as in Figure 11.15(a), and opens up and down when the coefficient of the y^2-term is positive, as in Figure 11.15(b).

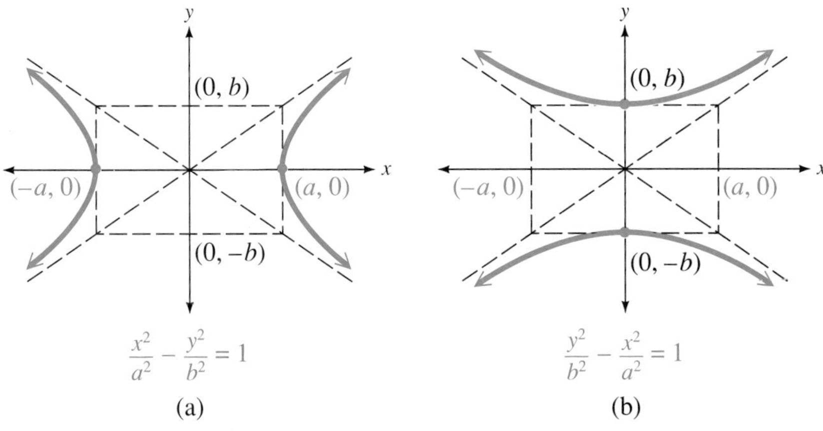

Figure 11.15 Asymptotes

EXAMPLE 3 GRAPHING A HYPERBOLA

Graph the hyperbola with equation $\dfrac{x^2}{4} - \dfrac{y^2}{9} = 1$.

This can be written $\dfrac{x^2}{2^2} - \dfrac{y^2}{3^2} = 1$.

Since $a = 2$ and $b = 3$, the rectangle passes through the points $(2, 0)$, $(-2, 0)$, $(0, 3)$, and $(0, -3)$. We construct the rectangle, extend the diagonals to get the asymptotes, plot the intercepts, and sketch the hyperbola (it opens right and left since the x^2 term is positive) as in Figure 11.16.

PRACTICE EXERCISE 3

Graph $\dfrac{y^2}{9} - \dfrac{x^2}{4} = 1$.

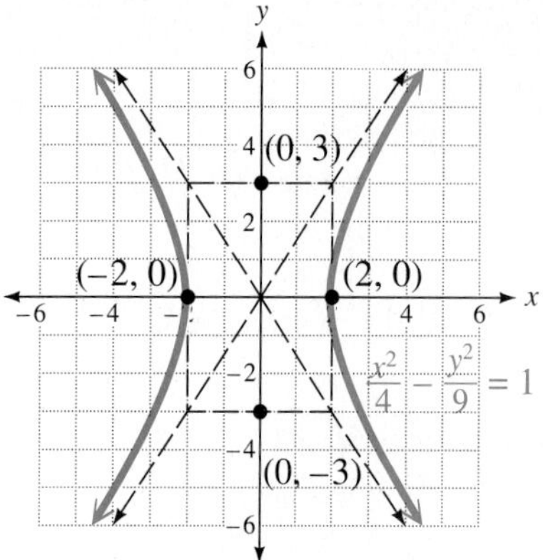

Figure 11.16

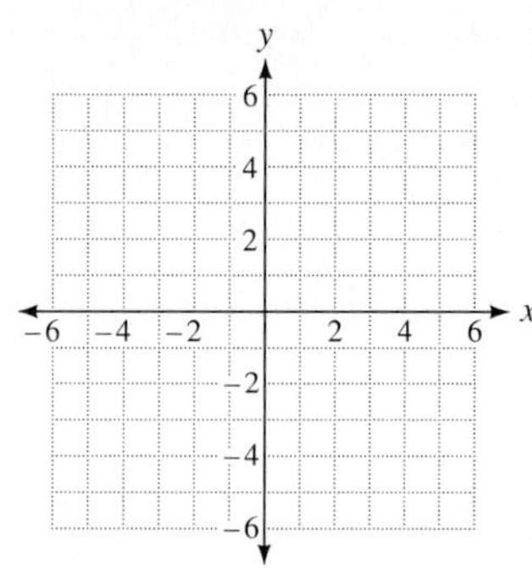

Answer: The hyperbola opens up and down with y-intercepts $(0, 3)$ and $(0, -3)$ and same asymptotes as in Figure 11.16.

4 ALTERNATE FORM OF A HYPERBOLA

The hyperbolas we have studied thus far open up and down or right and left with intercepts on the y-axis or x-axis, respectively. Other hyperbolas have the axes as asymptotes, and these are considered now.

Alternate Form of the Equation of a Hyperbola

Hyperbolas having the x-axis and y-axis as asymptotes have equations

$$xy = c, \quad c \text{ a constant.}$$

EXAMPLE 4 GRAPHING A HYPERBOLA WITH EQUATION $xy = c$

Graph $xy = 6$.

First we solve for y,

$$y = \frac{6}{x},$$

and then make a table of values. Notice that we cannot use 0 for x. The graph is given in Figure 11.17.

x	y
6	1
3	2
2	3
1	6
−1	−6
−2	−3
−3	−2
−6	−1

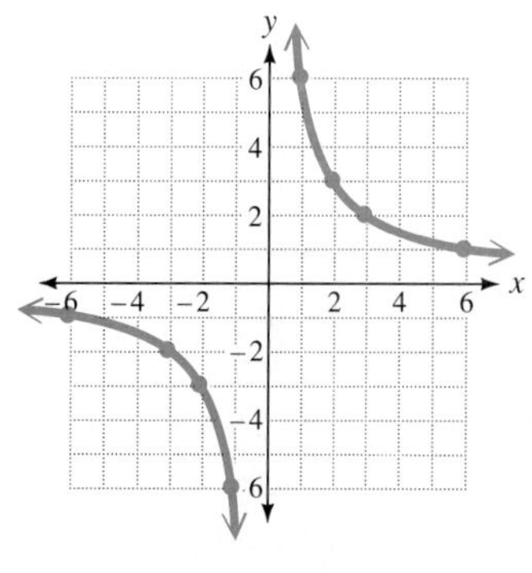

Figure 11.17

PRACTICE EXERCISE 4

Graph $xy = -6$.

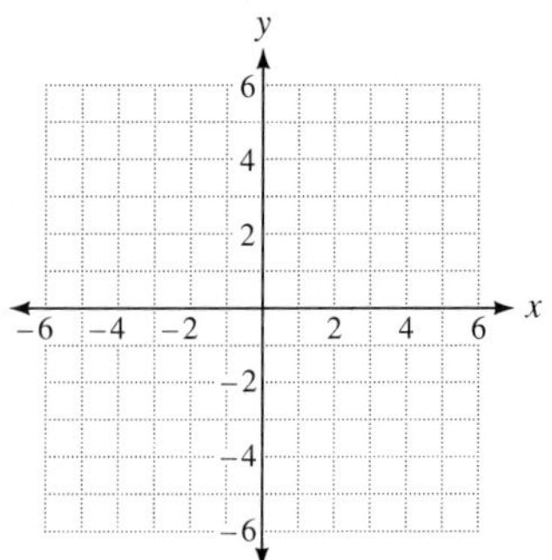

Answer: The graph is much like that in Figure 11.17 but in quadrants II and IV instead of I and III.

The two parts of the hyperbola in Example 4 are in quadrants I and III. This happens when $c > 0$ in $xy = c$. When $c < 0$, the two parts are in quadrants II and IV as in Practice Exercise 4. Hyperbolas of the form $xy = c$ have the same shape as hyperbolas with standard-form equations but they are rotated 45°.

5 SUMMARY OF CONICS

We conclude this section with a summary of the equations and graphs of the conic sections studied in this chapter.

Summary of the Conic Sections

Equation	*Description of graph*	*What to look for*
$y = a(x - h)^2 + k$	Parabola: opens up if $a > 0$, opens down if $a < 0$, vertex at (h, k)	x is squared but y is not
$x = a(y - k)^2 + h$	Parabola: opens right if $a > 0$, opens left if $a < 0$, vertex at (h, k)	y is squared but x is not
$(x - h)^2 + (y - k)^2 = r^2$	Circle: center at (h, k) with radius r	x^2 and y^2 have the same coefficient
$\frac{x^2}{a^2} + \frac{y^2}{b^2} = 1$	Ellipse: center at $(0, 0)$, x-intercepts $(\pm a, 0)$ and y-intercepts $(0, \pm b)$	x^2 and y^2 have different coefficients with the same sign
$\frac{x^2}{a^2} - \frac{y^2}{b^2} = 1$	Hyperbola: center at $(0, 0)$, x-intercepts $(\pm a, 0)$, opens left and right	x^2 has positive coefficient and y^2 has negative coefficient
$\frac{y^2}{b^2} - \frac{x^2}{a^2} = 1$	Hyperbola: center at $(0, 0)$, y-intercepts $(0, \pm b)$, opens up and down	y^2 has positive coefficient and x^2 has negative coefficient
$xy = c$	Hyperbola: in quadrants I and III if $c > 0$, and in II and IV if $c < 0$	xy-term

EXAMPLE 5 Identifying Conics

Identify the graph of each equation.

(a) $y^2 = x + 5$

Since y is squared and x is not, the graph is a parabola. In fact, since $x = y^2 - 5$, $a > 0$ and the parabola opens to the right.

(b) $2x^2 = 5 - 2y^2$

We can write the equation in the form $2x^2 + 2y^2 = 5$ or $x^2 + y^2 = \frac{5}{2}$. Since x and y are both squared and have the same coefficient, the graph is a circle centered at the origin with radius $\sqrt{\frac{5}{2}} = \frac{\sqrt{10}}{2}$.

Practice Exercise 5

Identify the graph of each equation.

(a) $9x^2 + 4y^2 = 36$

[*Hint:* Divide both sides by 36.]

(b) $\frac{x}{5} = \frac{1}{y}$

Answers: (a) ellipse
(b) hyperbola

11.3 EXERCISES A

Give the standard form of the equation of the ellipse centered at the origin with the given x-intercepts and y-intercepts.

1. (2, 0), (−2, 0) and (0, 3), (0, −3)

2. (9, 0), (−9, 0) and (0, 11), (0, −11)

3. (7, 0), (−7, 0) and (0, 1), (0, −1)

4. (10, 0), (−10, 0) and (0, 2), (0, −2)

Give the standard form of the equation of the hyperbola centered at the origin with the given a, b, and direction of opening.

5. $a = 2$, $b = 3$, up and down

6. $a = 9$, $b = 11$, left and right

7. $a = 7$, $b = 1$, left and right

8. $a = 10$, $b = 2$, up and down

Graph.

9. $\frac{x^2}{16} + \frac{y^2}{4} = 1$

10. 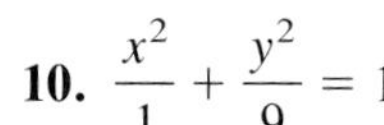

11. 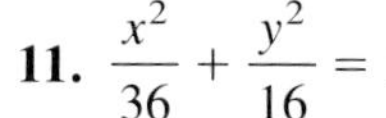

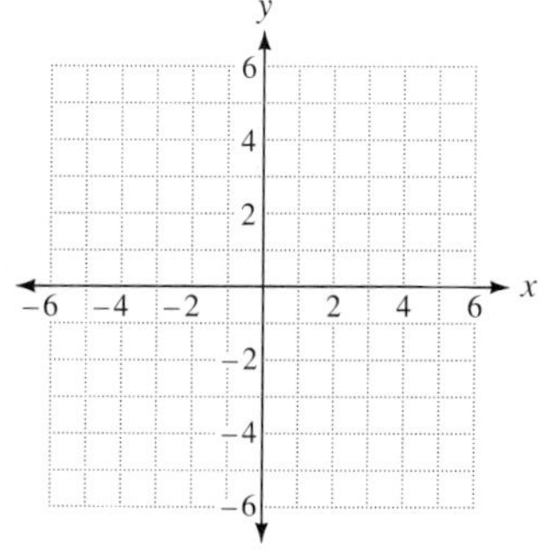

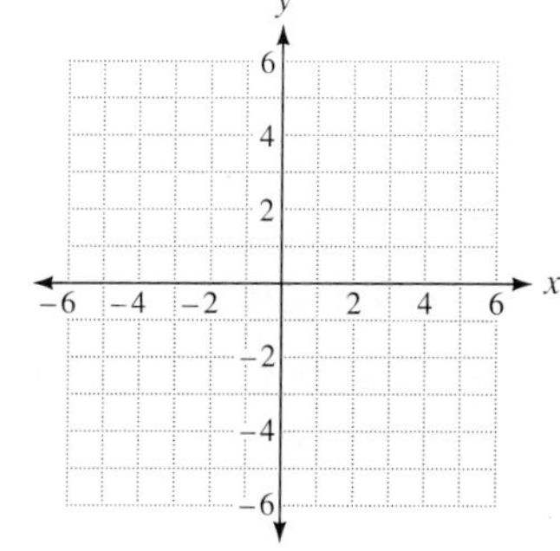

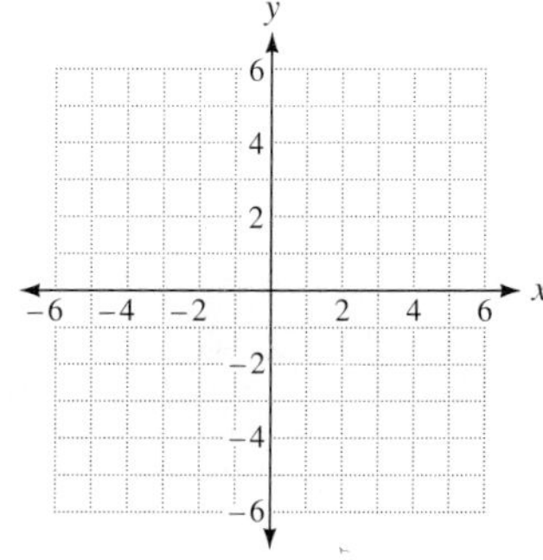

12. $25x^2 + 9y^2 = 225$

13. 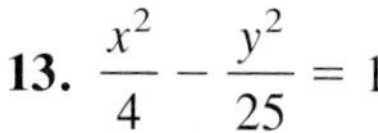

14. 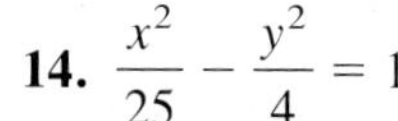

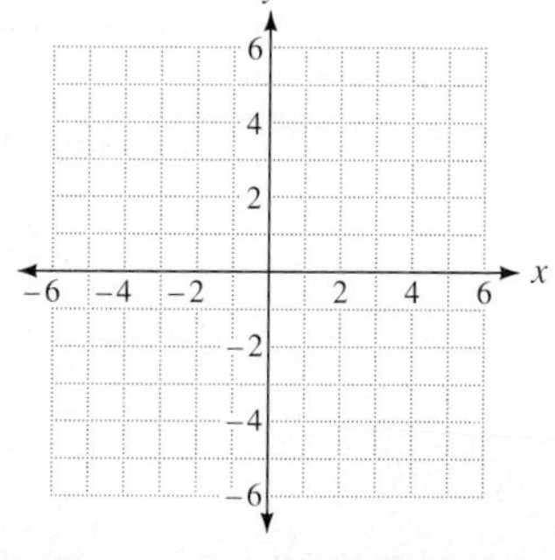

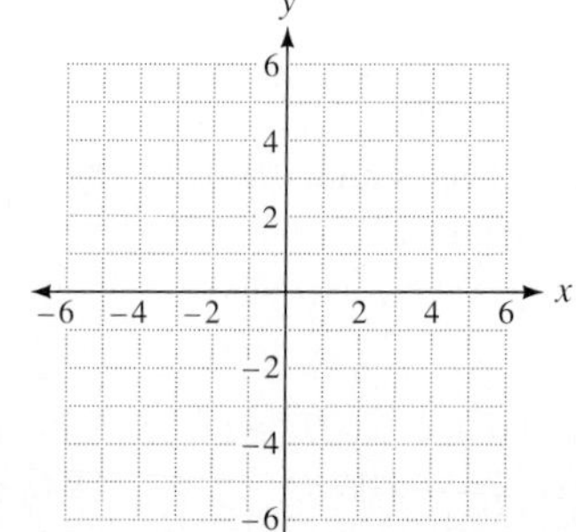

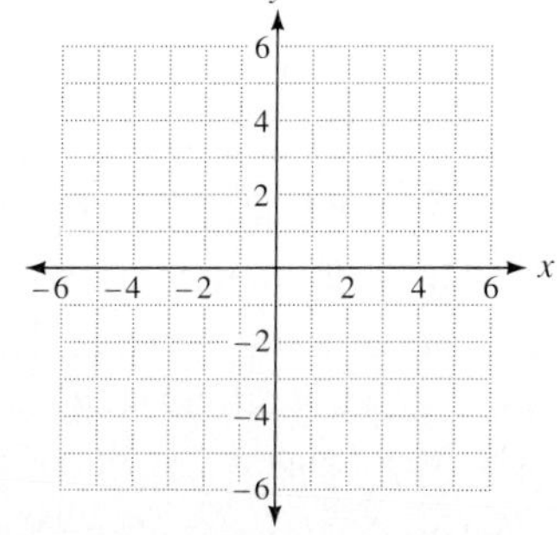

15. $\dfrac{y^2}{4} - \dfrac{x^2}{25} = 1$

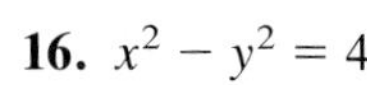

16. $x^2 - y^2 = 4$

17. $y^2 - 4x^2 = 16$

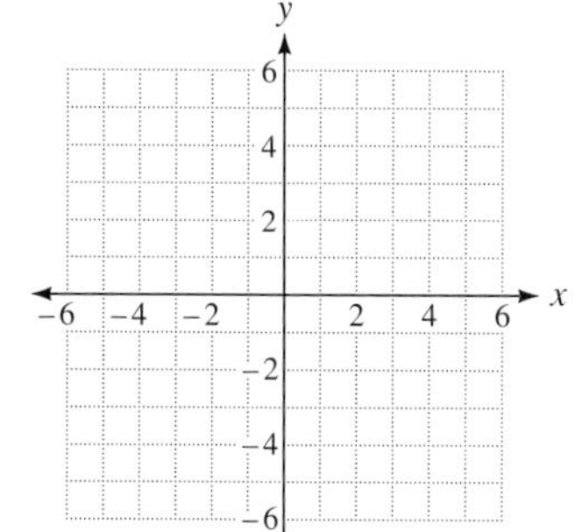

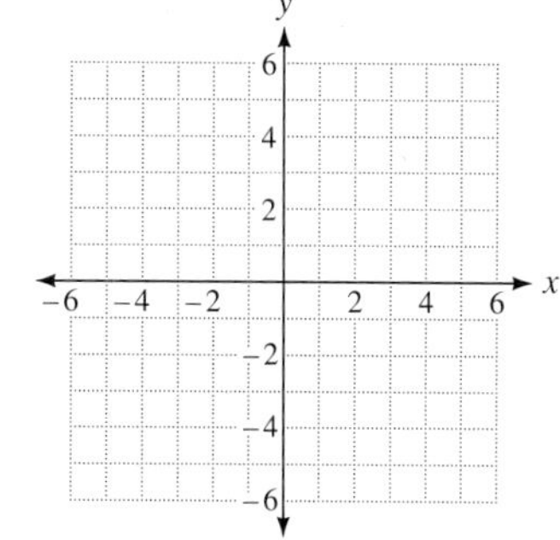

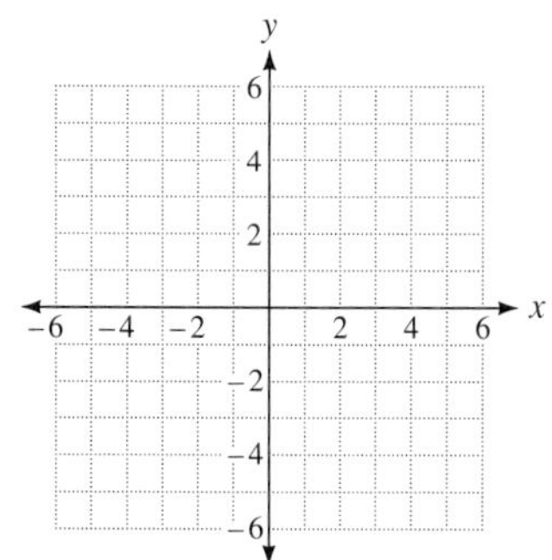

18. $xy = 4$

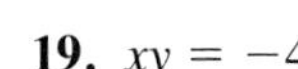

19. $xy = -4$

20. $xy = 12$

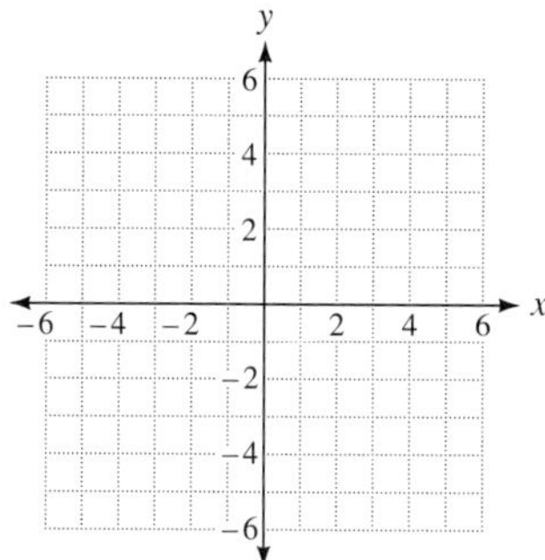

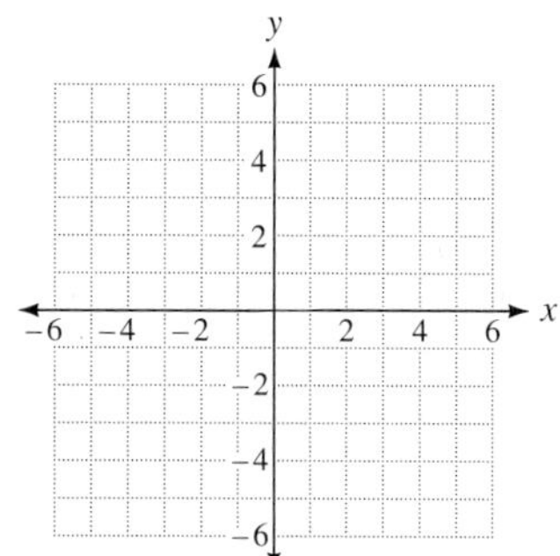

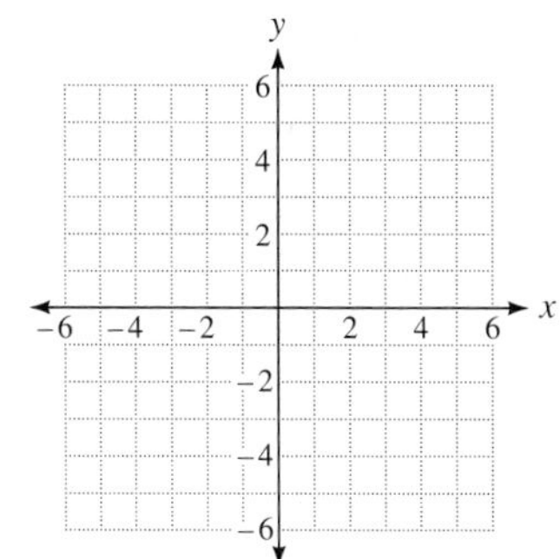

Identify each of the following as the equation of a parabola, circle, ellipse, or hyperbola.

21. $x^2 + y^2 = 25$

22. $x^2 = 2y$

23. $\dfrac{x^2}{4} - \dfrac{y^2}{49} = 1$

24. $(x - 1)^2 + (y + 3)^2 = 16$

25. $\dfrac{x^2}{36} + \dfrac{y^2}{16} = 1$

26. $y^2 - x^2 = 16$

27. $y = \dfrac{1}{x}$

28. $y^2 - 2x + y - 5 = 0$

29. $x^2 + y^2 - 2x + 2y - 2 = 0$

30. A point is sometimes called a *degenerate circle* or *degenerate ellipse* since a plane can intersect a cone in a single point. Make a sketch to show how this can happen.

31. What is the graph of $x^2 + y^2 = 0$. Compare this with Exercise 30.

FOR REVIEW

Find the standard and general forms of the equation of a circle centered at the given point and with the given radius.

32. $(2, 2)$, $r = 2$ **33.** $(0, 0)$, $r = 8$ **34.** $(-1, 5)$, $r = 7$

Graph.

35. $x = y^2 + 4y$ **36.** $(x - 1)^2 + (y + 1)^2 = 4$ **37.** $y = -\sqrt{36 - x^2}$

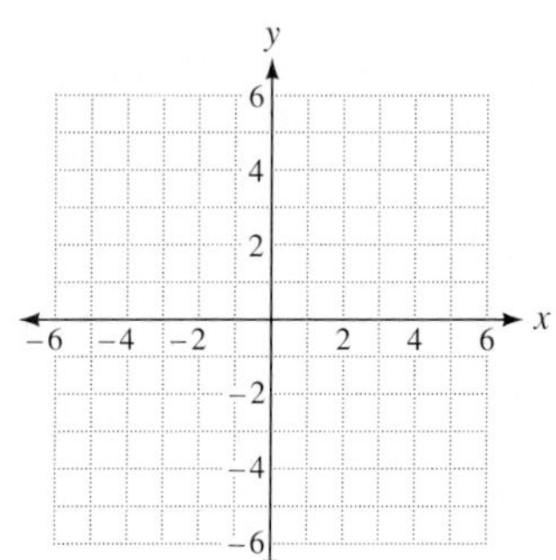

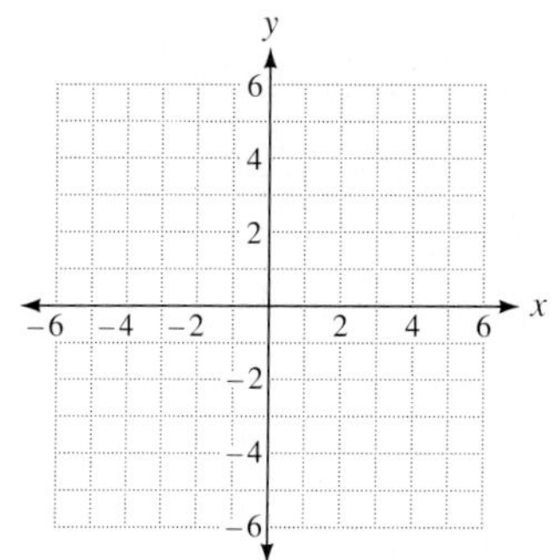

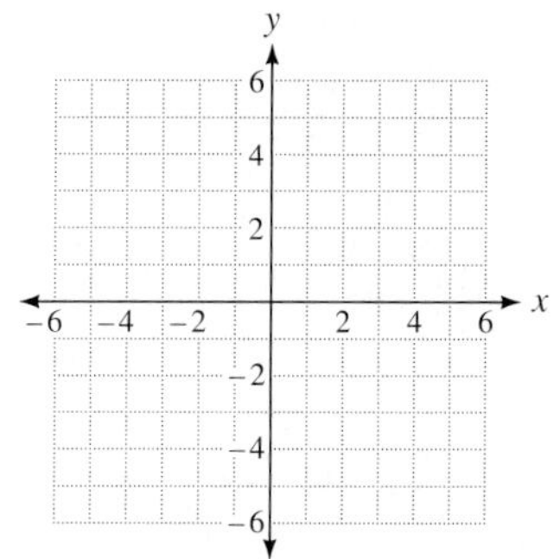

Exercises 38–40 review material from Chapter 8 to help you prepare for the next section. Solve each equation.

38. $x^2 - x - 12 = 0$ **39.** $9 + y^2 = 14$ **40.** $x^4 - 25x^2 + 144 = 0$

ANSWERS: 1. $\frac{x^2}{4} + \frac{y^2}{9} = 1$ 2. $\frac{x^2}{81} + \frac{y^2}{121} = 1$ 3. $\frac{x^2}{49} + \frac{y^2}{1} = 1$ 4. $\frac{x^2}{100} + \frac{y^2}{4} = 1$ 5. $\frac{y^2}{9} - \frac{x^2}{4} = 1$ 6. $\frac{x^2}{81} - \frac{y^2}{121} = 1$ 7. $\frac{x^2}{49} - \frac{y^2}{1} = 1$ 8. $\frac{y^2}{4} - \frac{x^2}{100} = 1$

9.

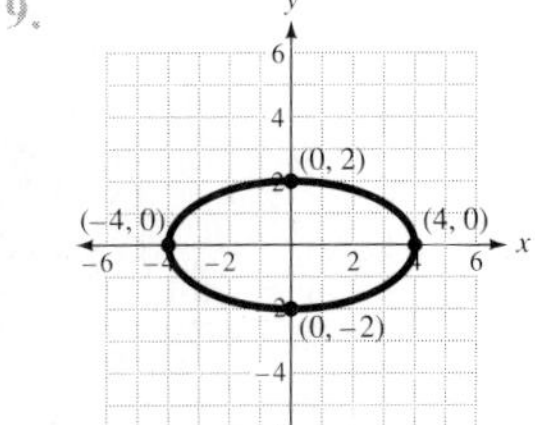

10.

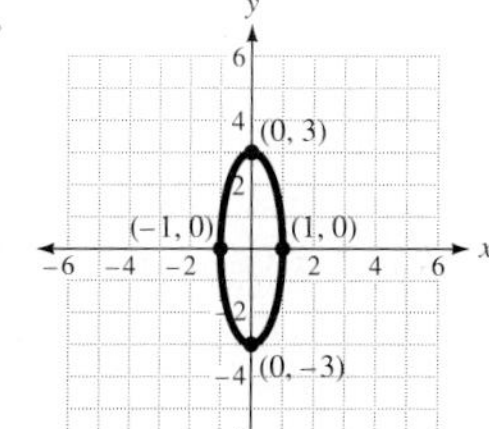

11.

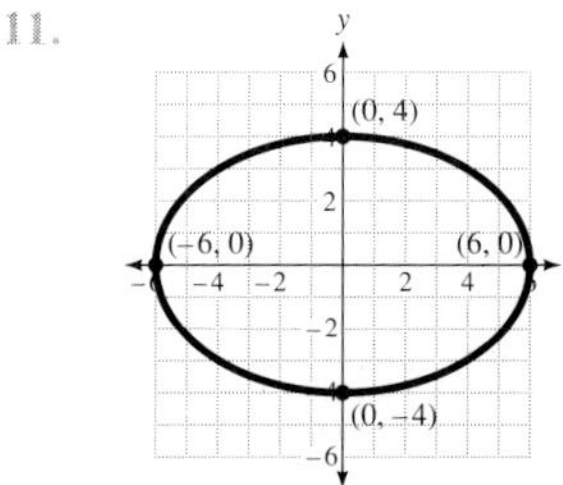

12.

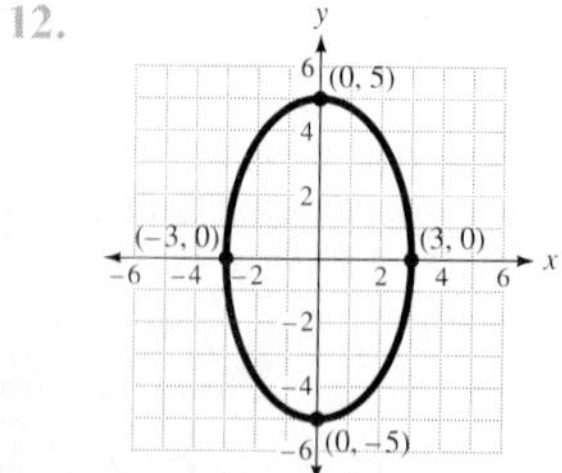

13.

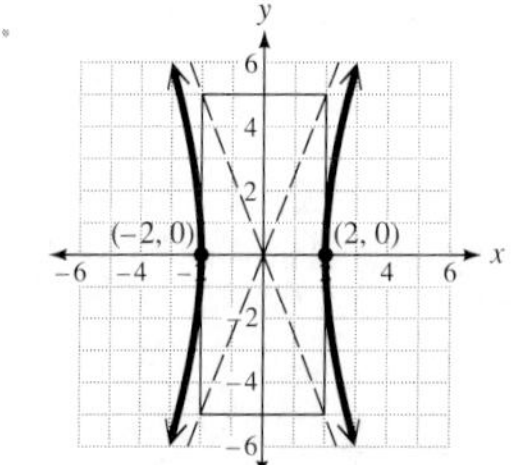

14.

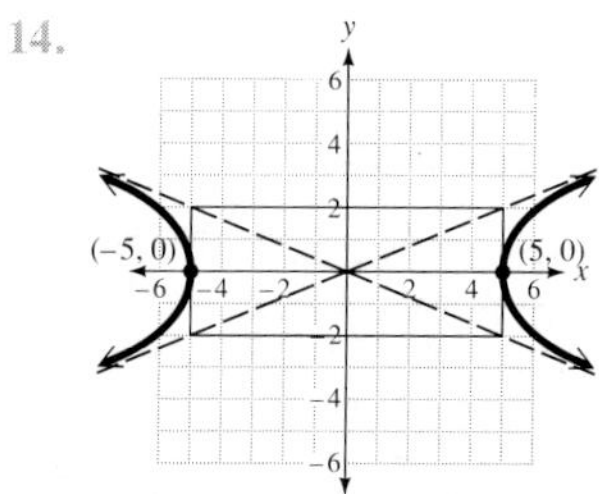

15.

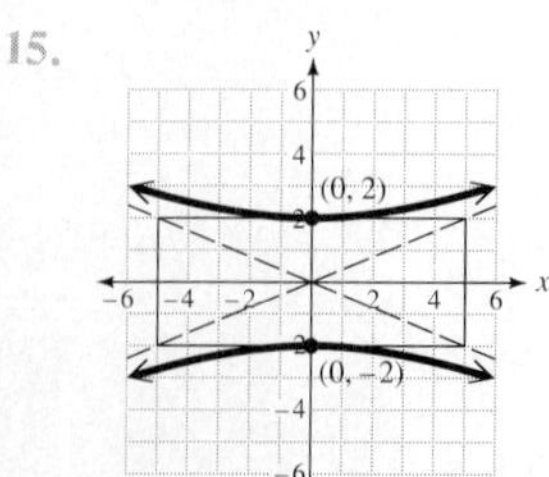

16.

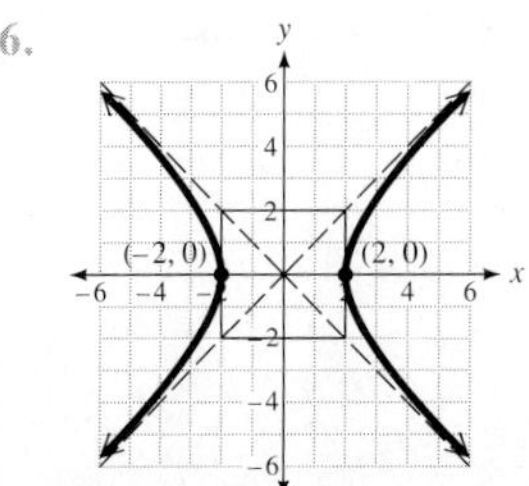

17.

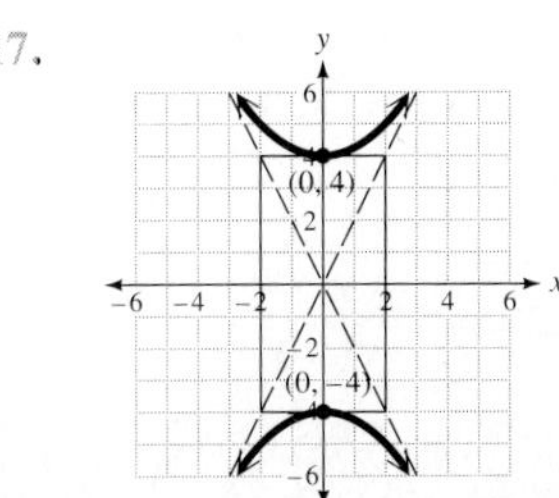

18.

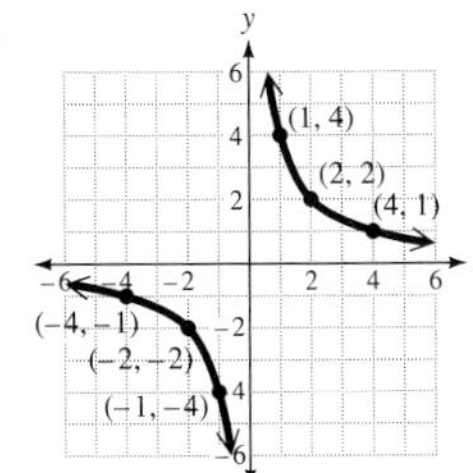

19.

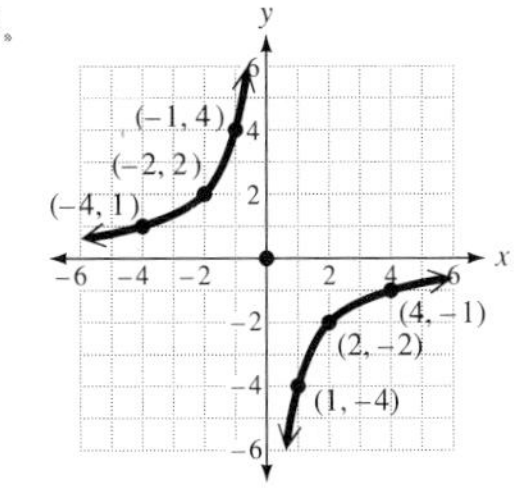

20.

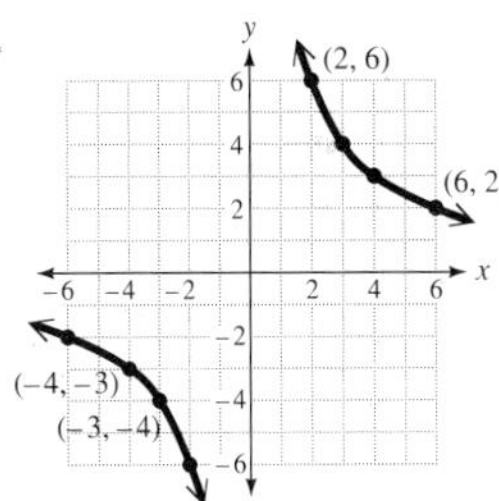

21. circle 22. parabola 23. hyperbola 24. circle 25. ellipse 26. hyperbola 27. hyperbola 28. parabola 29. circle 30. Have the plane pass through the vertex of the cone and only through this point. 31. The graph is the point (0, 0). Since the equation has the appearance of the equation of a circle, this could be called a point circle. 32. $(x-2)^2+(y-2)^2=2^2$; $x^2+y^2-4x-4y+4=0$ 33. $x^2+y^2=8^2$; $x^2+y^2-64=0$ 34. $(x+1)^2+(y-5)^2=7^2$; $x^2+y^2+2x-10y-23=0$

35.

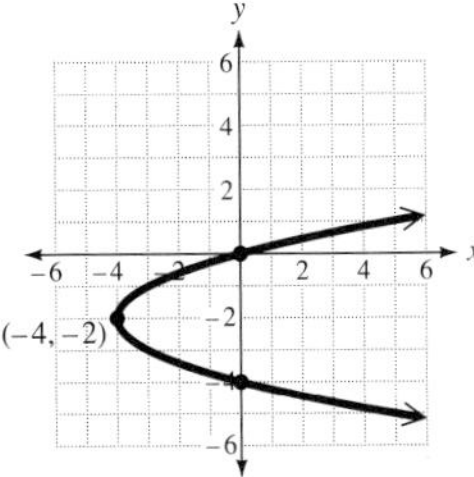

36.

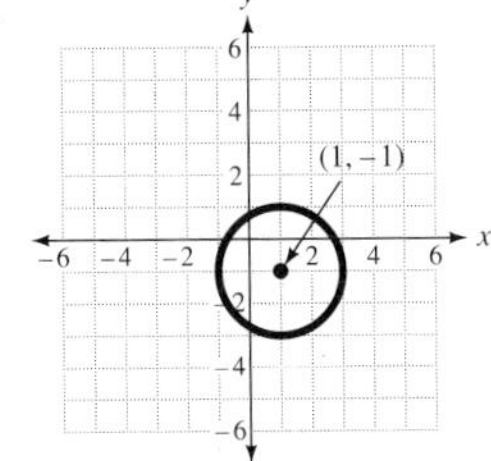

37.

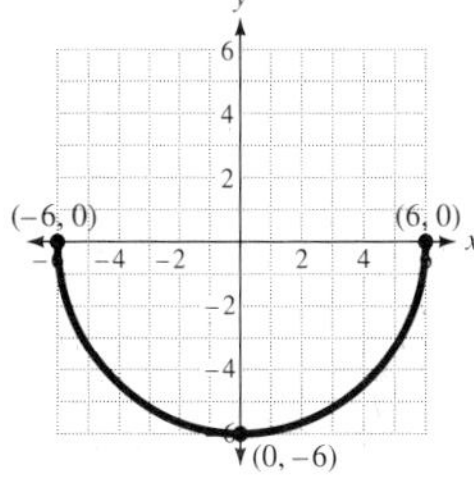

38. −3, 4 39. $\pm\sqrt{5}$ 40. ±3, ±4

11.3 EXERCISES B

Give the standard form of the equation of the ellipse centered at the origin with the given x-intercepts and y-intercepts.

1. (6, 0), (−6, 0) and (0, 4), (0, −4)

2. (12, 0), (−12, 0) and (0, 9), (0, −9)

3. (5, 0), (−5, 0) and (0, 8), (0, −8)

4. (1, 0), (−1, 0) and (0, 7), (0, −7)

Give the standard form of the equation of the hyperbola centered at the origin with the given a, b, and direction of opening.

5. $a=6$, $b=4$, up and down

6. $a=12$, $b=9$, up and down

7. $a=5$, $b=8$, left and right

8. $a=1$, $b=7$, left and right

Graph.

9. $\dfrac{x^2}{25}+\dfrac{y^2}{9}=1$

10. 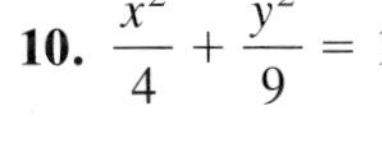 $\dfrac{x^2}{4}+\dfrac{y^2}{9}=1$

11. $\dfrac{x^2}{16}+\dfrac{y^2}{25}=1$

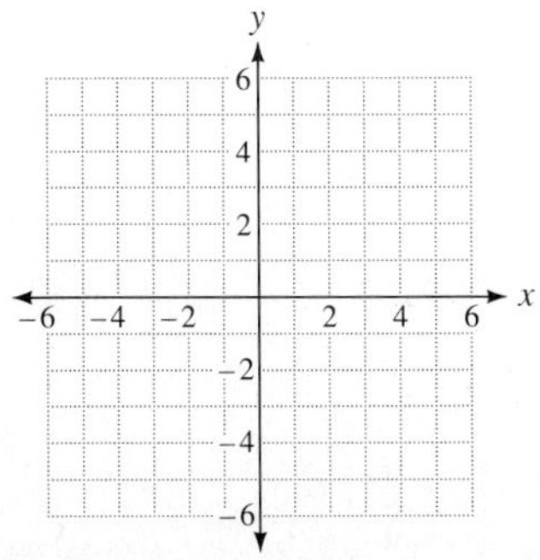

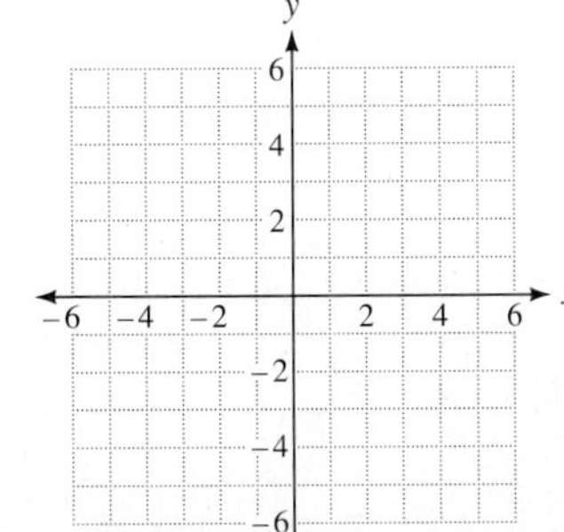

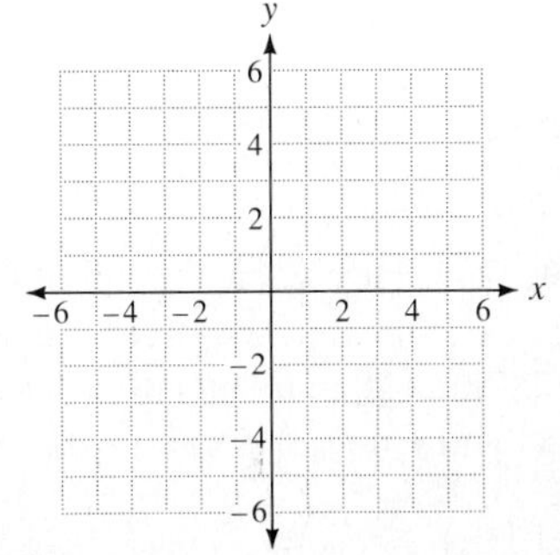

12. $x^2 + 25y^2 = 25$

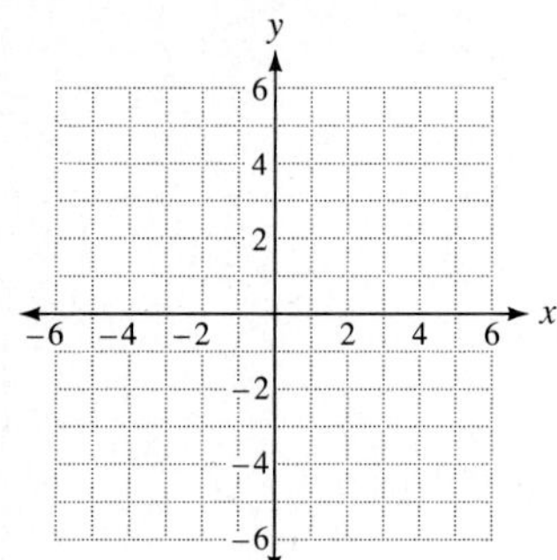

13. $\dfrac{x^2}{16} - \dfrac{y^2}{9} = 1$

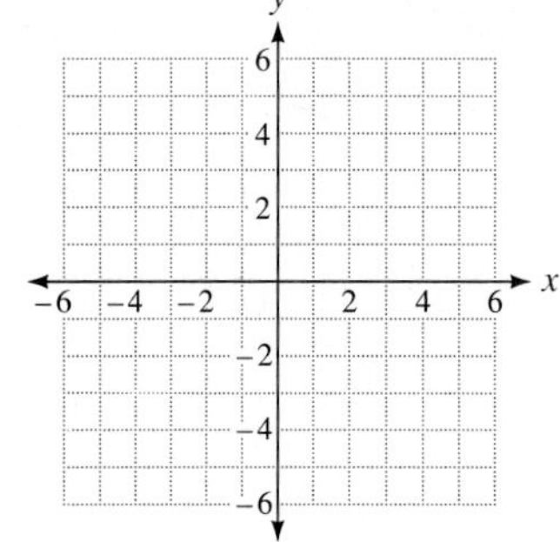

14. $\dfrac{y^2}{16} - \dfrac{x^2}{9} = 1$

15. $x^2 - y^2 = 1$

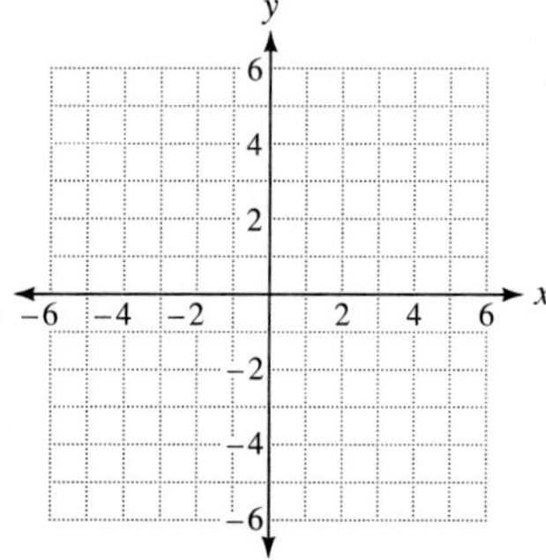

16. $y^2 - x^2 = 1$

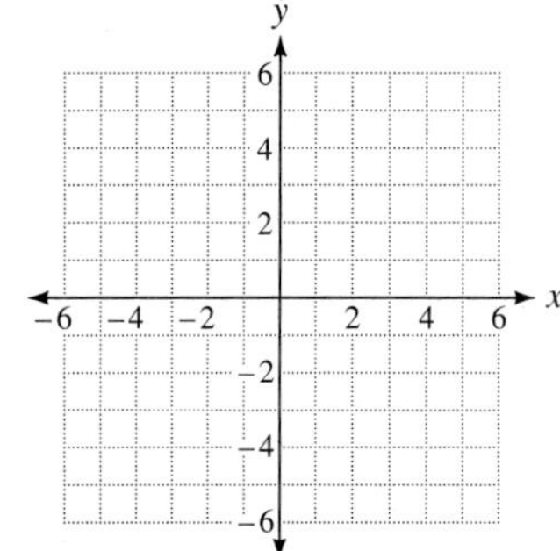

17. $9x^2 - 4y^2 = 36$

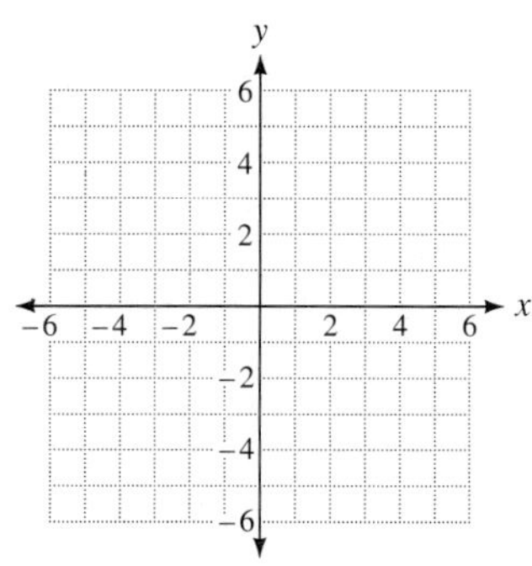

18. $xy = 8$

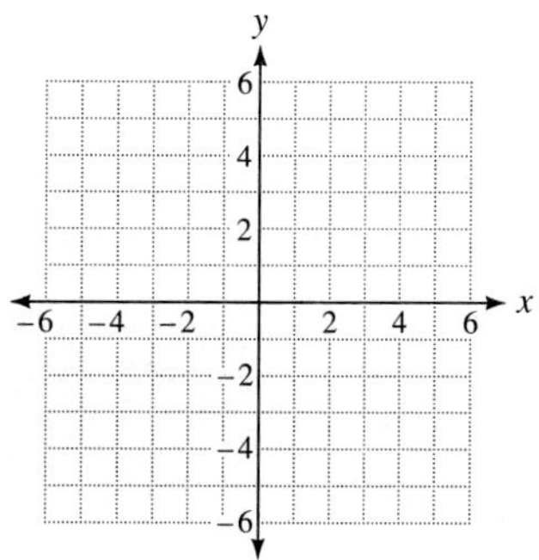

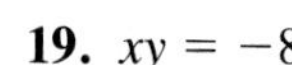

19. $xy = -8$

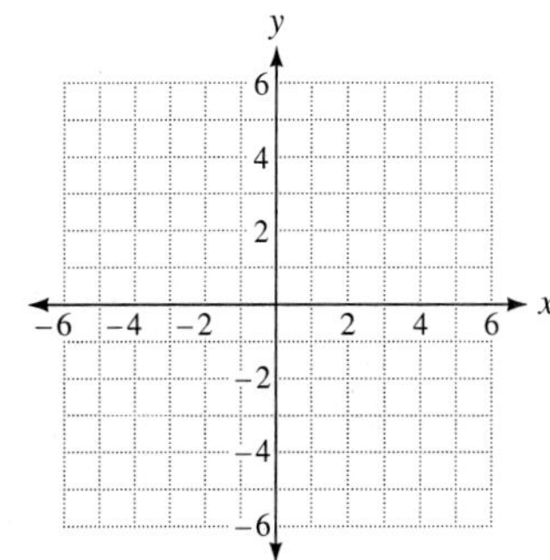

20. $xy = -12$

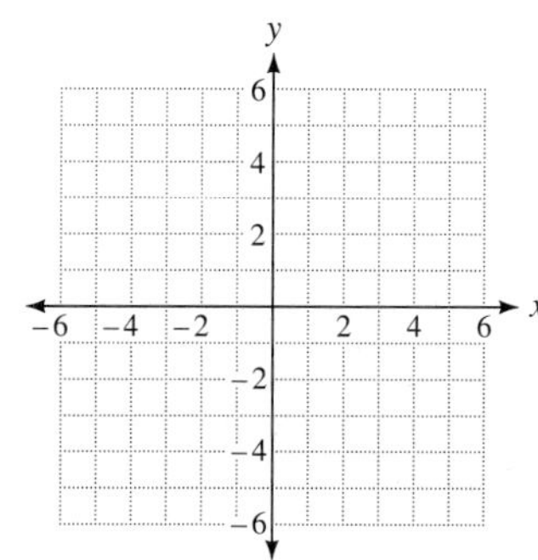

Identify each of the following as the equation of a parabola, circle, ellipse, or hyperbola.

21. $(x - 1)^2 + (y + 5)^2 = 49$

22. $x^2 + y^2 = 49$

23. $\dfrac{x^2}{49} - \dfrac{y^2}{4} = 1$

24. $\dfrac{x^2}{49} + \dfrac{y^2}{4} = 1$

25. $x = y^2 - y + 1$

26. $y^2 - x^2 = 36$

27. $x = \dfrac{7}{y}$

28. $8x^2 + 3y^2 = 24$

29. $x^2 + y^2 - 2x - 8 = 0$

30. Two intersecting straight lines are sometimes called a *degenerate hyperbola* since a plane can intersect a cone in two intersecting lines. Make a sketch to show how this can happen.

31. What is the graph of $x^2 - y^2 = 0$? Compare this with Exercise 30.

FOR REVIEW

Find the standard and general forms of the equation of a circle centered at the given point and with the given radius.

32. $(3, 4)$, $r = 1$

33. $(0, 0)$, $r = 11$

34. $(-2, 3)$, $r = \sqrt{5}$

Graph.

35. $x = -y^2 + 4$

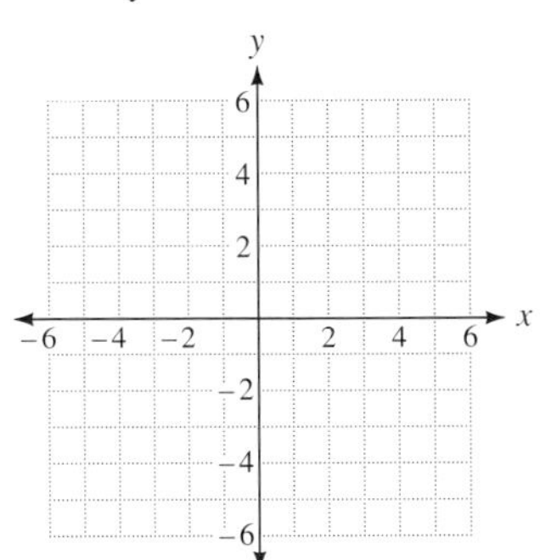

36. $x^2 + y^2 = 1$

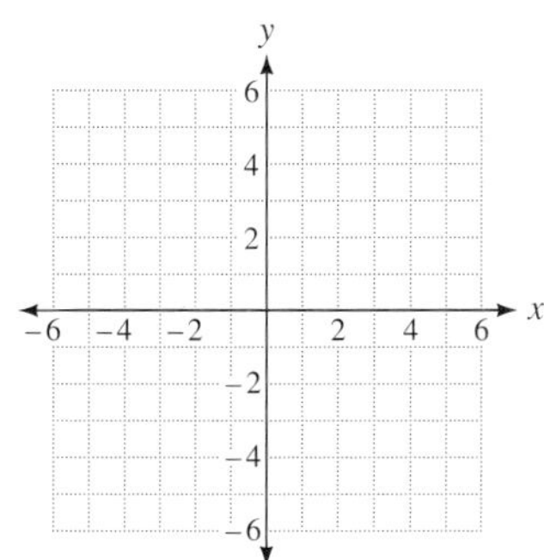

37. $y = -\sqrt{9 - x^2}$

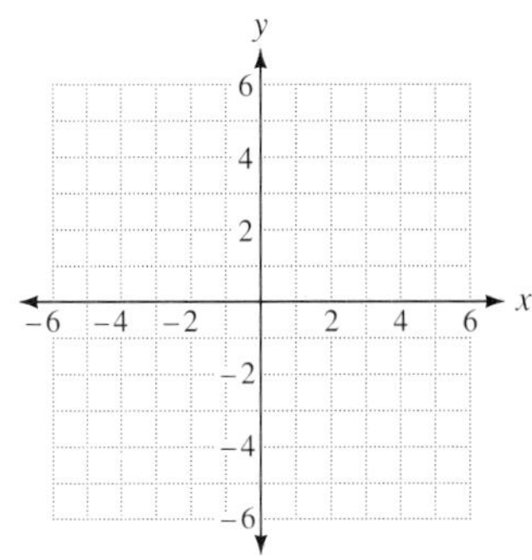

Exercises 38–40 review material from Chapter 8 to help you prepare for the next section. Solve each equation.

38. $2x^2 + x - 15 = 0$

39. $y^2 + 6 = 13$

40. $w^4 - 11w^2 + 18 = 0$

11.3 EXERCISES C

1. Give the intercepts of the ellipse with equation $4x^2 + 25y^2 = 100$.

2. Give the intercepts of the hyperbola with equation $9x^2 - 4y^2 = 72$.

3. Find the equation of the ellipse passing through $\left(2, \frac{2\sqrt{5}}{3}\right)$ with x-intercepts $(3, 0)$ and $(-3, 0)$.
$\left[\text{Answer: } \frac{x^2}{9} + \frac{y^2}{4} = 1\right]$

4. Find the equation of the hyperbola passing through $(2\sqrt{2}, 1)$ and $(-2\sqrt{2}, 1)$ with x-intercepts $(2, 0)$ and $(-2, 0)$.

5. A straight line is sometimes called a *degenerate parabola* since a plane can intersect a cone in a straight line. Make a sketch to show how this can happen.

11.4 NONLINEAR SYSTEMS OF EQUATIONS

STUDENT GUIDEPOSTS

1. Nonlinear Systems
2. Solving Nonlinear Systems
3. The Elimination Method

1 NONLINEAR SYSTEMS

When we studied systems of two linear (first-degree) equations in Chapter 4, the graphs of the equations gave useful information about the solutions. The same is true for **nonlinear systems** that involve second-degree equations. A solution to the nonlinear system

$$x^2 + y^2 = 25$$
$$x + y = 1$$

is an ordered pair of numbers that satisfies both equations. Graphically, a solution is a point of intersection of the two graphs. In the system above, the graph of the first equation is a circle, and the graph of the second is a line. In general, the graphs of a circle and a line can be related in one of three different ways, as shown in Figure 11.18.

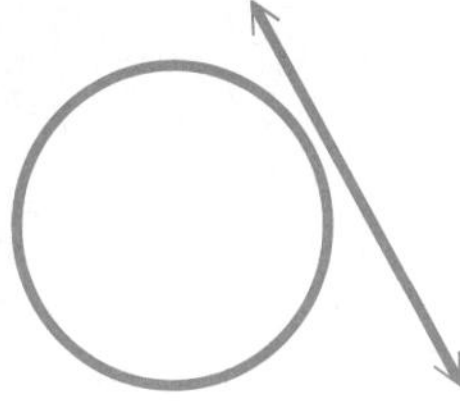

no points of intersection
no real solutions

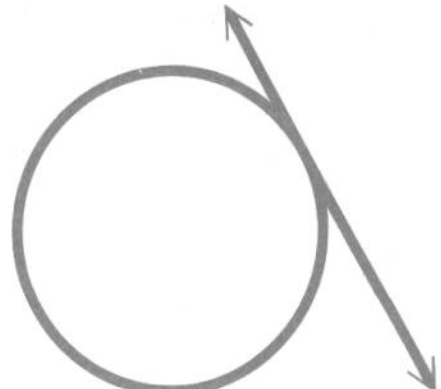

one point of intersection
one real solution

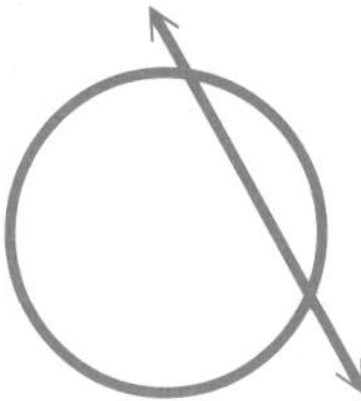

two points of intersection
two real solutions

Figure 11.18 Intersecting a Circle and a Line

If we graph both equations in the same coordinate system, as in Figure 11.19, the points of intersection appear to have coordinates $(-3, 4)$ and $(4, -3)$. By substitution, it can be shown that these pairs of numbers satisfy both equations, hence are solutions to the system.

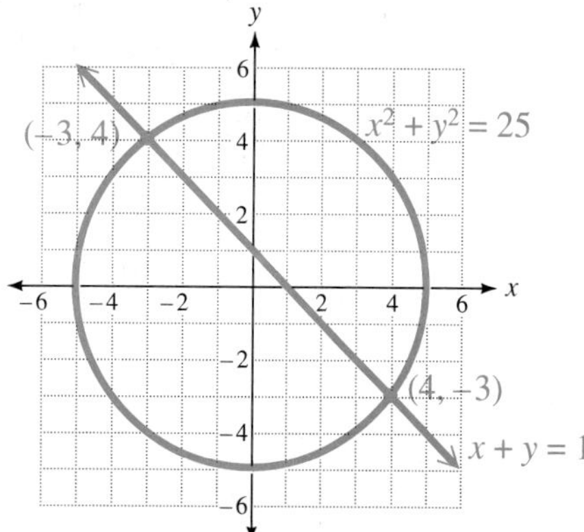

Figure 11.19

2 SOLVING NONLINEAR SYSTEMS

It is better to solve systems algebraically rather than graphically. When one equation is a first-degree equation, the following method is easy to apply.

To Solve a System of One First- and One Second-Degree Equation

1. Solve the first-degree equation for either variable.
2. Substitute in the second-degree equation and solve the resulting single-variable equation.
3. Substitute each of these values in the first-degree equation to find the corresponding values of the second variable.
4. Check your answer by substitution in both equations.

EXAMPLE 1 Solving a Nonlinear System

Solve $x^2 + y^2 = 25$
$x + y = 1.$

Solve $x + y = 1$ for y.

$$y = 1 - x$$
$$x^2 + (1 - x)^2 = 25 \quad \text{Substitute } 1 - x \text{ for } y$$
$$x^2 + 1 - 2x + x^2 = 25$$
$$2x^2 - 2x - 24 = 0$$
$$x^2 - x - 12 = 0 \quad \text{Divide by 2}$$
$$(x - 4)(x + 3) = 0$$
$$x - 4 = 0 \quad \text{or} \quad x + 3 = 0$$
$$x = 4 \qquad x = -3$$
$$y = 1 - x \qquad y = 1 - x$$
$$= 1 - 4 = -3 \qquad = 1 - (-3) = 4$$

Check: $(4, -3)$ $\qquad$ $(-3, 4)$

$$4^2 + (-3)^2 \stackrel{?}{=} 25 \quad 4 + (-3) \stackrel{?}{=} 1 \qquad (-3)^2 + 4^2 \stackrel{?}{=} 25 \quad (-3) + 4 \stackrel{?}{=} 1$$
$$16 + 9 \stackrel{?}{=} 25 \quad 1 = 1 \qquad 9 + 16 \stackrel{?}{=} 25 \quad 1 = 1$$
$$25 = 25 \qquad 25 = 25$$

The solutions are $(4, -3)$ and $(-3, 4)$, which were obtained previously by graphing.

PRACTICE EXERCISE 1

Solve $2x^2 + y^2 = 9$
$y - 2x = 3.$

Answer: $(0, 3)$, $(-2, -1)$

3 THE ELIMINATION METHOD

Many systems of second-degree equations can be solved using an **elimination method** similar to the one used for systems of linear equations in Chapter 4. Such systems have zero, one, two, three, or four solutions.

The Elimination Method

1. If the equations can be added or subtracted to eliminate a variable, add or subtract and proceed to (3). Otherwise,
2. Multiply one or both equations by a constant, then add or subtract them to eliminate a variable.
3. Solve the resulting single-variable equation. Substitute each value into one of the original equations and solve. The resulting pair(s) of numbers are solution(s) to the system.
4. Check your answer by substitution in both equations.

EXAMPLE 2 Using the Elimination Method

(a) Solve $x^2 + y^2 = 14$
$x^2 - y^2 = 4.$

Add to eliminate y.

$$2x^2 = 18$$
$$x^2 = 9$$
$$x = \pm 3$$

PRACTICE EXERCISE 2

(a) Solve $x^2 + 2y^2 = 8$
$x^2 - 2y^2 = 0.$

Substitute 3 for x in the first equation.

$$(3)^2 + y^2 = 14$$
$$9 + y^2 = 14$$
$$y^2 = 5$$
$$y = \pm\sqrt{5}$$

Two of the solutions are $(3, \sqrt{5})$ and $(3, -\sqrt{5})$. When we substitute -3 for x in the first equation we obtain $y = \pm\sqrt{5}$. Thus, there are four solutions: $(3, \sqrt{5})$, $(3, -\sqrt{5})$, $(-3, \sqrt{5})$, and $(-3, -\sqrt{5})$. Figure 11.20 shows the four points of intersection of the circle and the hyperbola.

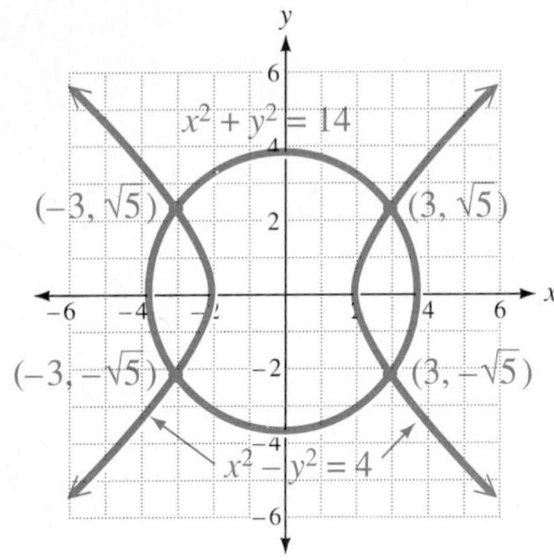

Figure 11.20

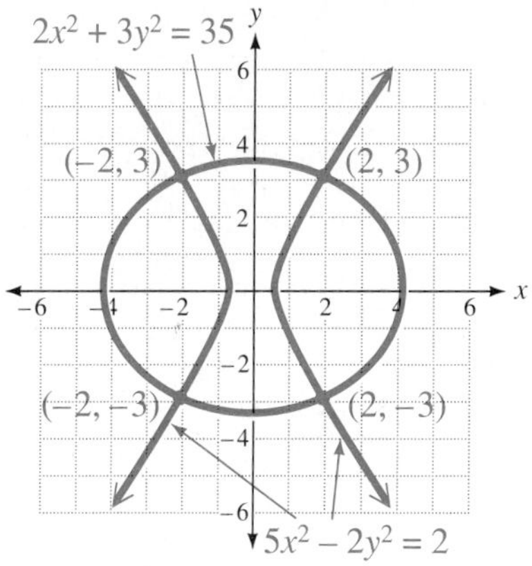

Figure 11.21

(b) Solve $5x^2 + y^2 = 10$
$2x^2 + 3y^2 = 17$.

(b) Solve $2x^2 + 3y^2 = 35$
$5x^2 - 2y^2 = 2$.

If we multiply the first equation by 2 and the second equation by 3, we obtain the following system.

$$4x^2 + 6y^2 = 70$$
$$15x^2 - 6y^2 = 6$$

Solving this system by the elimination method, we obtain four solutions: $(2, 3)$, $(2, -3)$, $(-2, 3)$, and $(-2, -3)$. Figure 11.21 shows the four points of intersection of the ellipse and the hyperbola.

Answers: (a) $(2, \sqrt{2})$, $(2, -\sqrt{2})$, $(-2, \sqrt{2})$, $(-2, -\sqrt{2})$ (b) $(1, \sqrt{5})$, $(1, -\sqrt{5})$, $(-1, \sqrt{5})$, $(-1, -\sqrt{5})$

When one of the equations in a system is of the form $xy = c$, it is best to solve this equation for one of the variables and substitute in the other equation.

EXAMPLE 3 Solving a System with One Equation $xy = c$

Solve $x^2 + y^2 = 25$
$xy = 12$.

$$x = \frac{12}{y} \quad \text{Solve } xy = 12 \text{ for } x$$

$$\left(\frac{12}{y}\right)^2 + y^2 = 25 \quad \text{Substitute in the first equation}$$

$$\frac{144}{y^2} + y^2 = 25$$

$$144 + y^4 = 25y^2 \quad \text{Multiply by } y^2$$

Practice Exercise 3

Solve $x^2 - y^2 = -5$
$xy = 6$.

$$\begin{aligned} y^4 - 25y^2 + 144 &= 0 \\ u^2 - 25u + 144 &= 0 \qquad \text{Let } u = y^2 \\ (u-9)(u-16) &= 0 \end{aligned}$$

$$\begin{aligned} u - 9 &= 0 \quad \text{or} \quad & u - 16 &= 0 \\ u &= 9 & u &= 16 \\ y^2 &= 9 & y^2 &= 16 \qquad y^2 = u \\ y &= \pm 3 & y &= \pm 4 \end{aligned}$$

When $y = 3$, $x = \frac{12}{3} = 4$; when $y = -3$, $x = \frac{12}{-3} = -4$;
when $y = 4$, $x = \frac{12}{4} = 3$; when $y = -4$, $x = \frac{12}{-4} = -3$.
Thus, the solutions are $(4, 3)$, $(-4, -3)$, $(3, 4)$, and $(-3, -4)$. Check.

Answer: $(2, 3)$, $(-2, -3)$

Remember that the equation $xy = 12$ is the graph of a hyperbola with asymptotes the x- and y-axes. In Example 3 we found that this hyperbola intersects the circle $x^2 + y^2 = 25$ in the four points $(4, 3)$, $(-4, -3)$, $(3, 4)$, and $(-3, -4)$.

In the next example we solve one of the many applied problems involving nonlinear systems of equations.

EXAMPLE 4 BUSINESS PROBLEM

Outdoor Industries manufactures and sells gas barbeque grills. If x represents the number of grills made and sold each week, the total weekly cost is given by

$$C = 50x + 1000,$$

and the total weekly revenue produced is given by

$$R = 100x - 0.2x^2.$$

Find the break-even value of x, the value of x for which the weekly cost and revenue are equal.

We need to find the value of x when $C = R$, that is, when

$$\begin{aligned} 50x + 1000 &= 100x - 0.2x^2. \\ 0.2x^2 - 50x + 1000 &= 0 \qquad \text{Collect like terms} \\ 2x^2 - 500x + 10{,}000 &= 0 \qquad \text{To clear the decimal, multiply by 10} \\ x^2 - 250x + 5000 &= 0 \qquad \text{Factor out 2 and divide both sides by 2.} \end{aligned}$$

Since the left side cannot be factored, we use the quadratic formula.

$$\begin{aligned} x &= \frac{-b \pm \sqrt{b^2 - 4ac}}{2a} = \frac{-(-250) \pm \sqrt{(-250)^2 - 4(1)(5000)}}{2(1)} \\ &= \frac{250 \pm \sqrt{62{,}500 - 20{,}000}}{2} = \frac{250 \pm \sqrt{42{,}500}}{2} \end{aligned}$$

Using a calculator, we find that $\sqrt{42{,}500}$ is approximately 206.2. Thus,

$$x \approx \frac{250 \pm 206.2}{2} = 228.1 \text{ and } 21.9.$$

Since x is the number of grills manufactured and sold weekly we use 228 and 22 for the approximate break-even points. That is, if either 228 or 22 grills are made and sold each week, the cost of production and the revenue realized will be about the same.

PRACTICE EXERCISE 4

The area of a rectangular pasture must be 30,000 yd^2, and the length must be three times the width. Find the dimensions of the pasture.

Answer: 300 yd by 100 yd

11.4 EXERCISES A

Solve each system.

1. $x^2 + y^2 = 10$
$x - y = 2$

2. $x + y = 3$
$x^2 - y^2 = 3$

3. $xy = 20$
$x + y = 9$

4. $5x^2 + xy - y^2 = -1$
$y - 2x = 1$

5. $xy - x^2 = 5$
$x - 2y = 3$

6. $x^2 + y^2 = 25$
$x^2 - y^2 = 25$

7. $x^2 + 3y^2 = 37$
$2x^2 - y^2 = 46$

8. $x^2 + y^2 = 5$
$xy = 2$

9. $3xy - y^2 = -13$
$2xy + y^2 = -2$
[*Hint:* Eliminate the xy term.]

10. $x^2 + y^2 = 4$
$xy = 10$

11. $x^2 + 2xy + y^2 = 9$
$x^2 - 2xy - y^2 = 9$

12. $x^2 + y^2 = 169$
$x + y = 17$

Write as a nonlinear system and solve.

13. The sum of the squares of two numbers is 100. The difference of the numbers is 2. Find the numbers.

14. The product of two numbers is 12 and the difference of their squares is 7. Find the numbers.

15. The area of a rectangular field must be 4050 m^2 and the length must be twice the width. Find the dimensions.

16. At what points do the ellipse $\frac{x^2}{1} + \frac{y^2}{4} = 1$ and the hyperbola $x^2 - y^2 = 1$ intersect?

17. Larson Manufacturing makes and sells beehives. If x represents the number of hives made and sold each week, the total weekly cost and revenue produced are $C = 10x + 80$ and $R = 25x - 0.2x^2$. Find the break-even value of x (when $C = R$).

FOR REVIEW

Graph.

18. $\frac{x^2}{9} + \frac{y^2}{25} = 1$

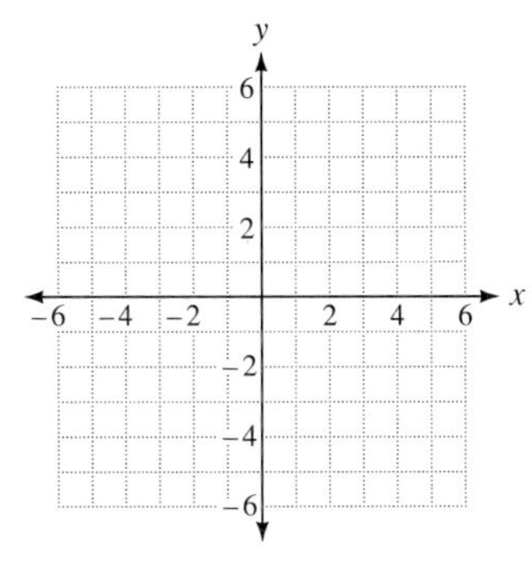

19. $\frac{y^2}{25} - \frac{x^2}{9} = 1$

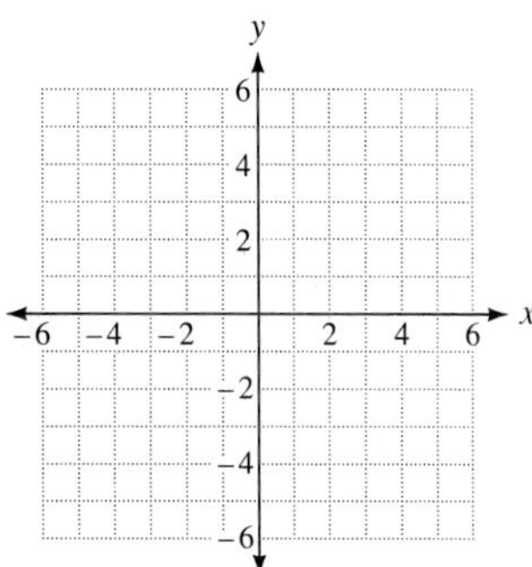

Exercises 20–23 review material from Section 3.5 and will help you prepare for Chapter 12. Find $f(1)$, $f(3)$, and $f(10)$ for each function.

20. $f(n) = 2n + 1$ **21.** $f(n) = n^2 - 1$ **22.** $f(n) = (-1)^n n^2$ **23.** $f(n) = \frac{n+1}{n}$

ANSWERS: 1. (3, 1), (−1, −3) 2. (2, 1) 3. (5, 4), (4, 5) 4. (0, 1), (1, 3) 5. no real solutions 6. (5, 0), (−5, 0) 7. (5, 2), (5, −2), (−5, 2), (−5, −2) 8. (2, 1), (−2, −1), (1, 2), (−1, −2) 9. $\left(-\frac{3}{2}, 2\right)$, $\left(\frac{3}{2}, -2\right)$ 10. no real solutions 11. (3, 0), (3, −6), (−3, 0), (−3, 6) 12. (5, 12), (12, 5) 13. (8, 6); (−6, −8) 14. (4, 3); (−4, −3) 15. 45 m by 90 m 16. (1, 0), (−1, 0) 17. approximately 6 and 69 18.

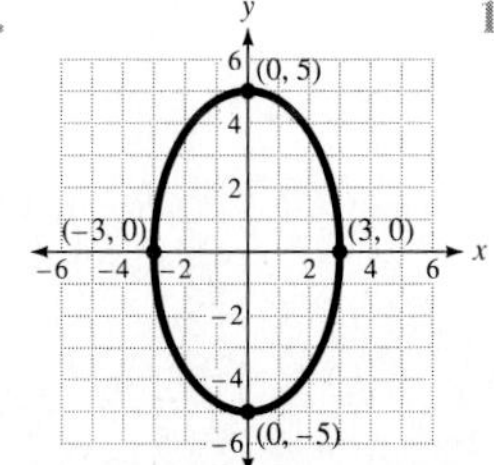

19.

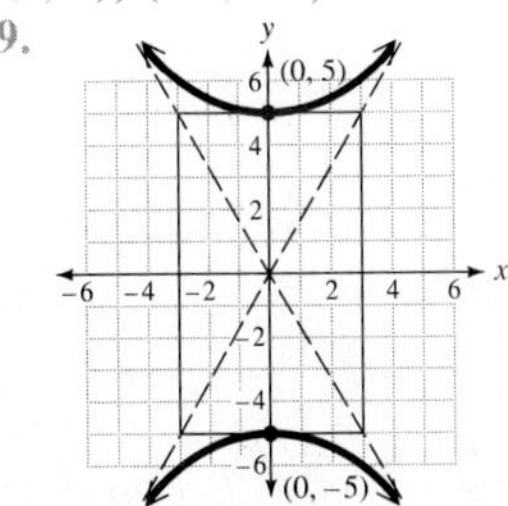

20. 3; 7; 21 21. 0; 8; 99 22. −1; −9; 100 23. 2; $\frac{4}{3}$; $\frac{11}{10}$

11.4 EXERCISES B

Solve each system.

1. $x^2 + y^2 = 8$
   $x + y = 4$

2. $x^2 - 2y^2 = 7$
   $2x + y = 7$

3. $x - y = -4$
   $x^2 - y^2 = -16$

4. $xy = 12$
   $2x - y = -2$

5. $x^2 - xy + y^2 = 1$
   $3x - y = 2$

6. $y^2 - xy = 3$
   $2x + y = 7$

7. $x^2 - 3y^2 = -8$
   $x^2 + 4y^2 = 13$

8. $y^2 - x^2 = 0$
   $x^2 + y^2 = 8$

9. $x^2 + 2y^2 = 1$
   $xy = -2$

10. $x^2 - 3xy = 18$
    $x^2 - 2xy = 15$

11. $2x^2 - 3xy + y^2 = 6$
    $x^2 + 3xy - y^2 = 69$

12. $x^2 + y^2 = 200$
    $x - 2y = -10$

Write as a nonlinear system and solve.

13. The sum of the squares of two numbers is 106. If the sum of the numbers is 4, find the numbers.

14. The product of two real numbers is -20 and the difference of their squares is 9. Find the numbers.

15. The area of a large room is 2800 ft^2 and the length is 30 ft more than the width. Find the dimensions.

16. At what points do the circle $x^2 + y^2 = 16$ and the parabola $y^2 = -6x$ intersect?

17. Tim Chandler makes and sells decorative patio tables. If x represents the number of tables made and sold each week, the total cost and revenue produced are $C = 30x + 200$ and $R = 50x - 0.2x^2$. Find the break-even value of x (when $C = R$).

FOR REVIEW

Graph.

18. $\dfrac{x^2}{36} + \dfrac{y^2}{9} = 1$

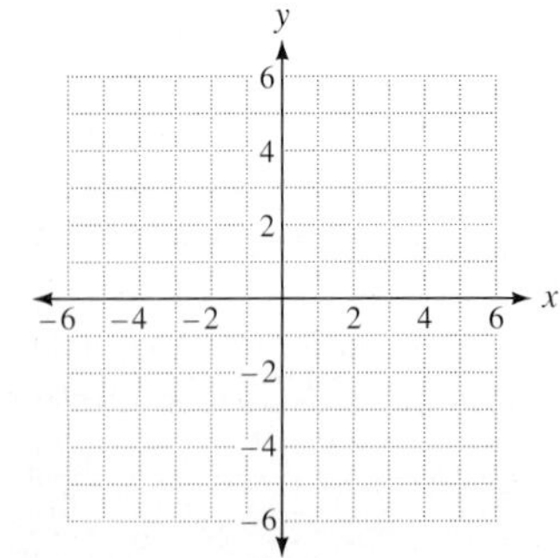

19. $\dfrac{x^2}{4} - \dfrac{y^2}{1} = 1$

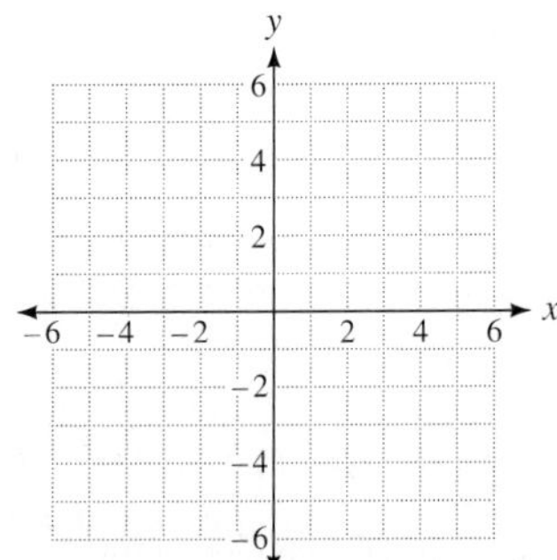

Exercises 20–23 review material from Section 3.5 and will help you prepare for Chapter 12. Find $f(1)$, $f(3)$, and $f(10)$ for each function.

20. $f(n) = 3n - 1$

21. $f(n) = n^2 + 1$

22. $f(n) = (-1)^n(n + 1)$

23. $f(n) = \dfrac{n - 1}{n}$

11.4 EXERCISES C

Write as a nonlinear system and solve.

1. The area of a rectangle is 21 cm^2, and its diagonal is $\sqrt{58}$ cm long. Find the dimensions of the rectangle. [Answer: 7 cm by 3 cm]

2. The perimeter of a rectangular field is 13.8 mi, and its area is 11 mi^2. Find the dimensions of the field.

Solve.

3. $x^2 + xy + y^2 = 9$
   $3x^2 - xy = 0$

4. $x^2 + xy + y^2 = 6$
   $x^2 - y^2 = 0$ [Answer: $(\sqrt{2}, \sqrt{2})$, $(-\sqrt{2}, -\sqrt{2})$, $(\sqrt{6}, -\sqrt{6})$, $(-\sqrt{6}, \sqrt{6})$]

CHAPTER 11 REVIEW

KEY WORDS

11.1 A **second-degree equation in two variables** x and y is an equation of the form $Ax^2 + Bxy + Cy^2 + Dx + Ey + F = 0$.

Conic sections (or **conics**) are curves that can be formed by intersecting a plane and a cone.

A **parabola** is the set of all points in a plane equidistant from a point (the **focus**) and a straight line not containing the point.

11.2 A **circle** is a set of all points in a plane that are a fixed distance r, called the **radius,** from a given point (h, k) in the plane, called the **center** of the circle.

11.3 An **ellipse** is the set of all points in a plane the sum of whose distances from two fixed points (called **foci**) is constant.

A **hyperbola** is the set of all points in a plane the difference of whose distances from two fixed points (called **foci**) is constant.

The lines approached by a hyperbola are called **asymptotes.**

11.4 A system of equations in which at least one equation is not linear is called a **nonlinear system.**

KEY CONCEPTS

11.1 Parabola:

1. $y = ax^2 + bx + c$ opens up or down;
   $y = a(x - h)^2 + k$ Standard form
   vertex (h, k), opens up if $a > 0$, opens down if $a < 0$.
2. $x = ay^2 + by + c$ opens left or right;
   $x = a(y - k)^2 + h$ Standard form
   vertex (h, k), opens right if $a > 0$, opens left if $a < 0$.

11.2

1. Circle with center (h, k) and radius r:
   $(x - h)^2 + (y - k)^2 = r^2$ Standard form
2. When the standard form is expanded, the general form of a circle is obtained.
   $x^2 + y^2 + Dx + Ey + F = 0$ General form

11.3

1. Ellipse with center (0, 0): $\dfrac{x^2}{a^2} + \dfrac{y^2}{b^2} = 1$
2. Hyperbola with center (0, 0):
   $\dfrac{x^2}{a^2} - \dfrac{y^2}{b^2} = 1$ opens left and right;
   $\dfrac{y^2}{b^2} - \dfrac{x^2}{a^2} = 1$ opens up and down.
3. Hyperbola with the axes as asymptotes:
   $xy = c$
   If $c > 0$, curves are in quadrants I and III.
   If $c < 0$, curves are in quadrants II and IV.

11.4

1. A system of one linear and one quadratic equation may have zero, one, or two real solutions. To solve, use the substitution method.
2. A system of two second-degree equations may have zero, one, two, three, or four real solutions. To solve, use the elimination method or the substitution method.

REVIEW EXERCISES

Part I

11.1 *Graph.*

1. $y = x^2 - 2x - 2$

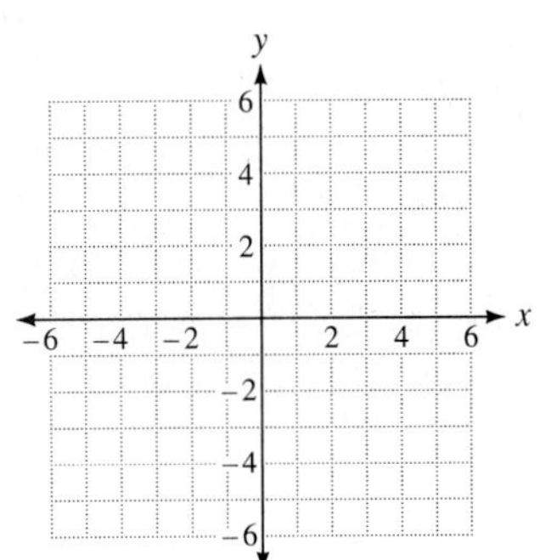

2. $x = -y^2 + 2y + 3$

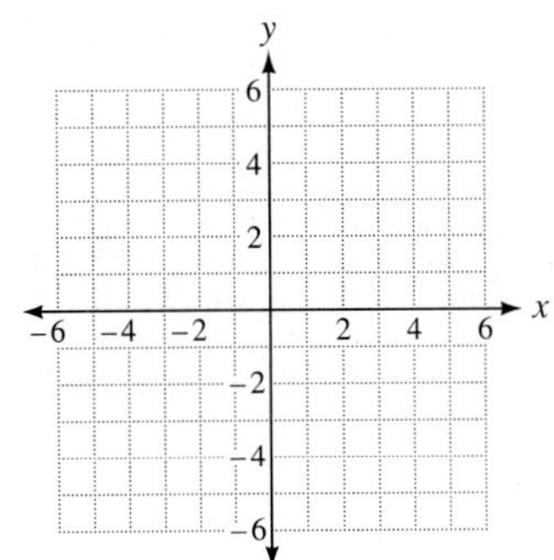

3. $8x = y^2 - 4y - 4$

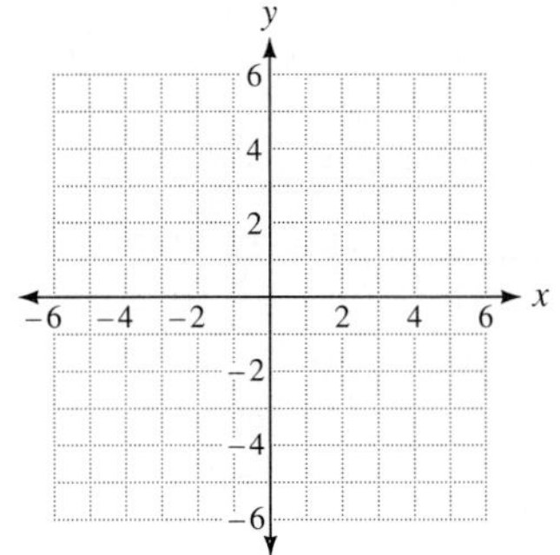

11.2 *Find the standard and general forms of the equation of a circle centered at the given point with the given radius.*

4. $(-5, 2)$, $r = 4$

5. $(2, 6)$, $r = 3$

6. $\left(\frac{1}{2}, \frac{1}{2}\right)$, $r = 2$

7. $(0, 5)$, $r = 1$

Find the standard form of the equation of each circle and give the center and radius.

8. $x^2 + y^2 - 4x + 16y + 19 = 0$

9. $4x^2 + 4y^2 + 4x - 24y + 33 = 0$

Graph.

10. $(x + 2)^2 + (y - 2)^2 = 16$

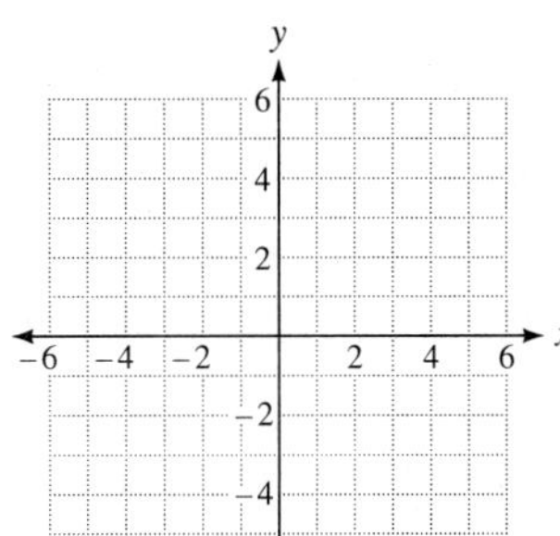

11. $y = \sqrt{9 - x^2}$

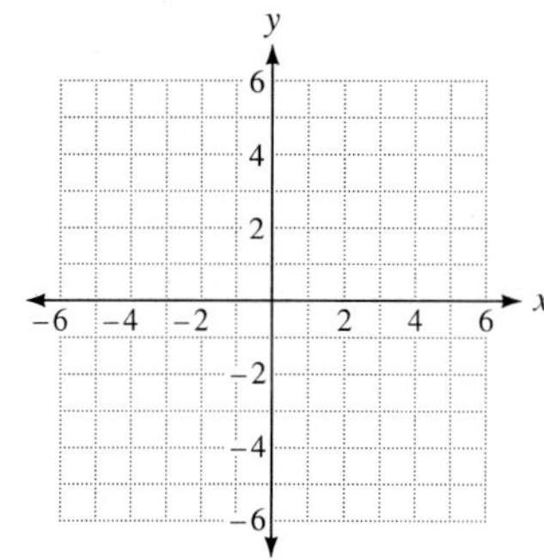

12. $y = -\sqrt{4 - x^2}$

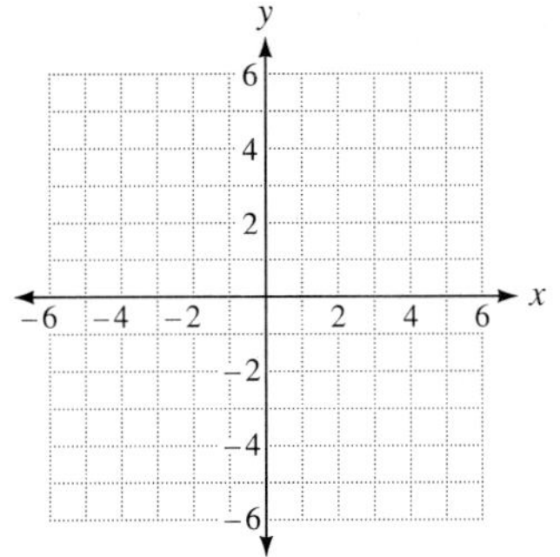

13. A tunnel in the shape of a semicircle must have a vertical clearance of 8 ft at a distance of 6 ft on either side of the center line. Find the equation of the semicircle by placing the origin at the centerline, and give the maximum height of the tunnel.

11.3 *Give the standard form of each equation.*

14. Ellipse: x-intercepts $(1, 0)$, $(-1, 0)$
y-intercepts $(0, 5)$, $(0, -5)$

15. Ellipse: x-intercepts $(4, 0)$, $(-4, 0)$
y-intercepts $(0, 3)$, $(0, -3)$

16. Hyperbola: $a = 5$, $b = 3$, opens up and down

17. Hyperbola: $a = 2$, $b = 7$, opens left and right

Graph.

18. $\dfrac{x^2}{9} + \dfrac{y^2}{4} = 1$

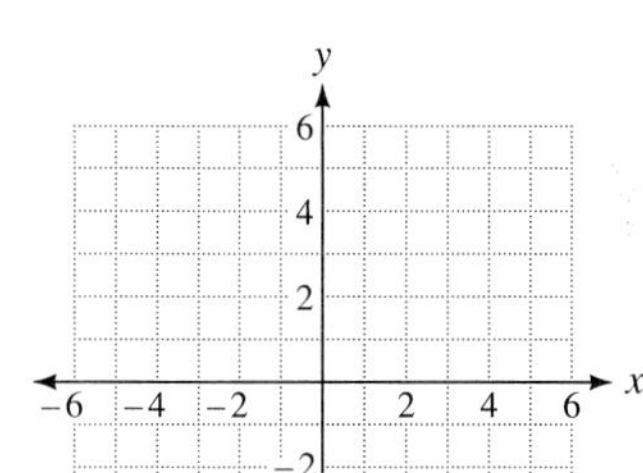

19. $\dfrac{x^2}{16} - \dfrac{y^2}{4} = 1$

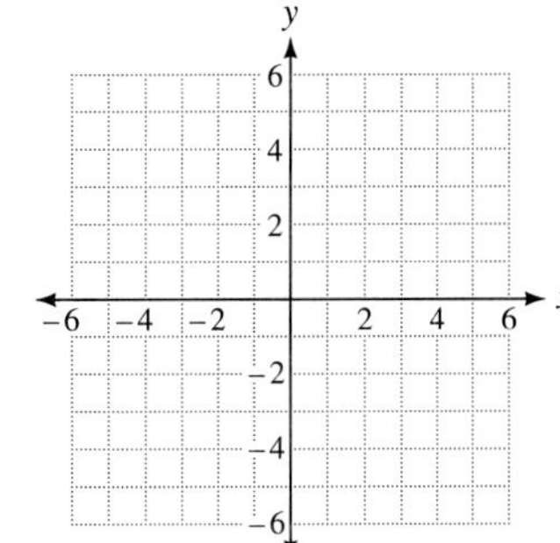

20. $\dfrac{y^2}{4} - \dfrac{x^2}{16} = 1$

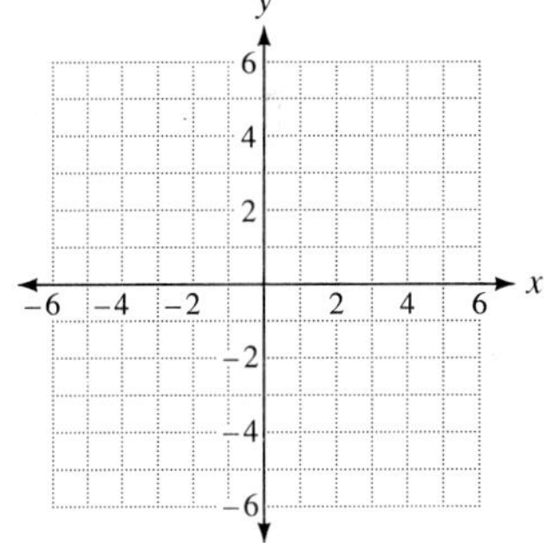

Identify each as the equation of a parabola, circle, ellipse, or hyperbola.

21. $x^2 + y^2 = 49$

22. $x^2 - 8y = 0$

23. $(x + 4)^2 + (y - 1)^2 = 20$

24. $\dfrac{x^2}{4} + \dfrac{y^2}{9} = 1$

25. $\dfrac{x^2}{4} - \dfrac{y^2}{9} = 1$

26. $x^2 = y - 2$

27. $y^2 + 2x - 4y = 0$

28. $y^2 - 9x^2 = 9$

29. $16x^2 + 25y^2 = 400$

11.4 *Solve the following systems.*

30. $2x^2 + y^2 = 3$
$x - y = 2$

31. $x^2 + y^2 = 20$
$xy = 8$

32. $3x^2 + y^2 = 3$
$x^2 - 2y^2 = 1$

Write as nonlinear systems and solve.

33. At what points do the parabola $y^2 = 3x$ and the hyperbola $\frac{x^2}{4} - \frac{y^2}{4} = 1$ intersect?

34. A triangular reflecting surface is in the shape of a right triangle with hypotenuse 5 ft. If one leg must be one foot longer than the other, what are the lengths of the legs?

Part II

Graph.

35. $4y = x^2 + 2x + 13$

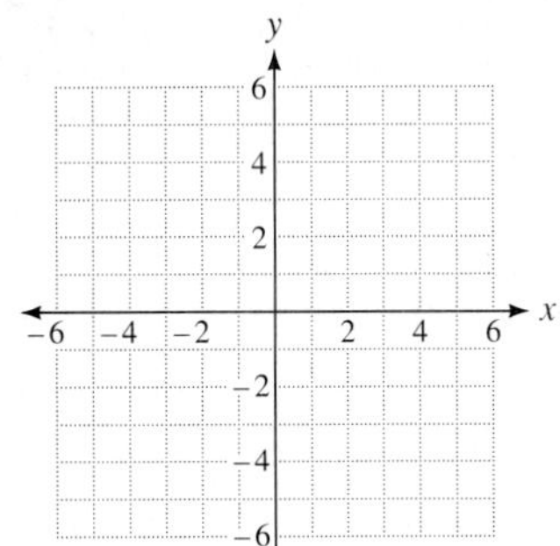

36. $(x - 1)^2 + (y - 3)^2 = 9$

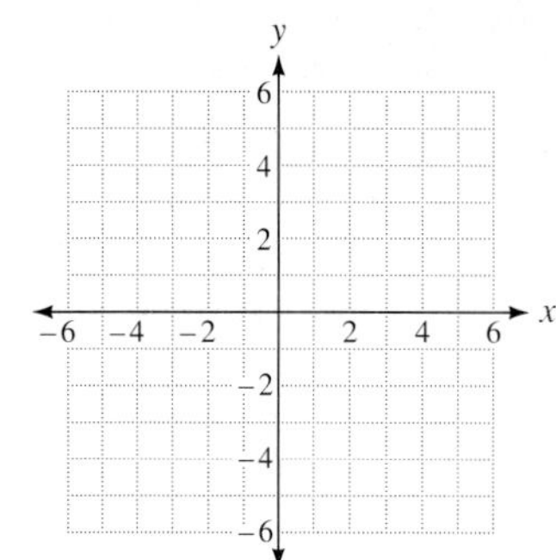

37. $xy = 3$

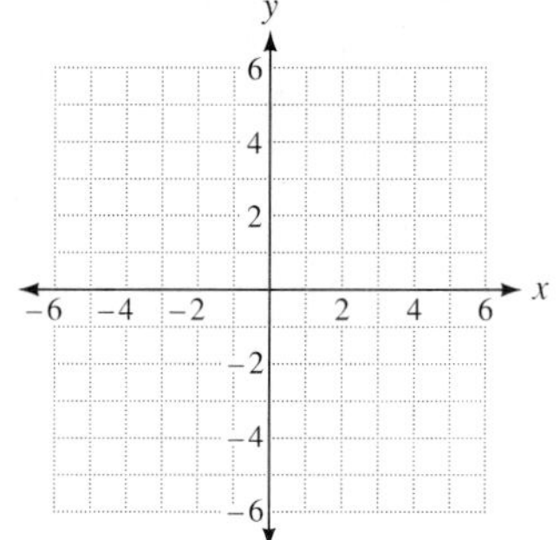

Identify each as the equation of a parabola, circle, ellipse, or hyperbola.

38. $x^2 - y^2 = 4$

39. $x^2 + y^2 - 8x + 2y - 3 = 0$

40. $y = \dfrac{16}{x}$

41. $x = -y^2 + 4y$

42. $4x^2 - 8y^2 = 32$

43. $x^2 + y^2 = 4$

44. Solve.

$$xy + 2x^2 = -8$$
$$x + y = 5$$

45. Write in standard form and give the center and radius of the circle with equation $x^2 + y^2 + 8x + 2y - 3 = 0$.

ANSWERS: 1.

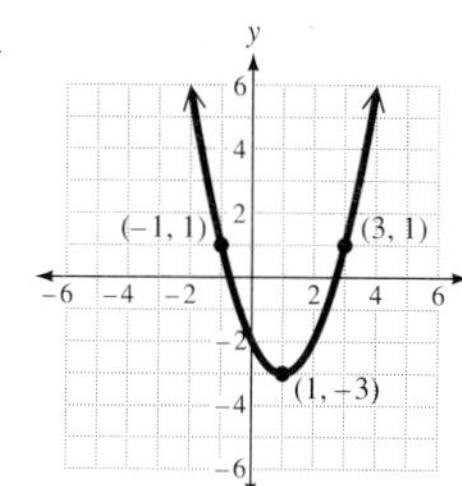

2.

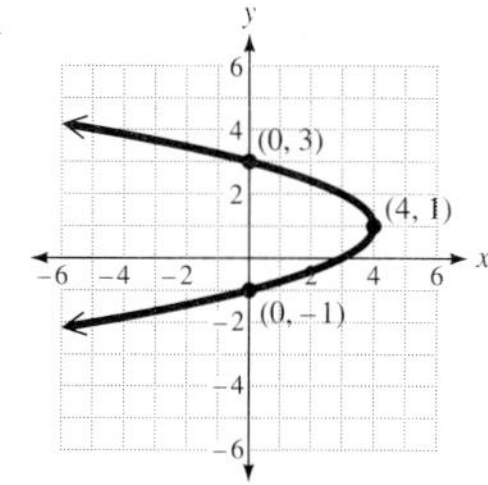

3.

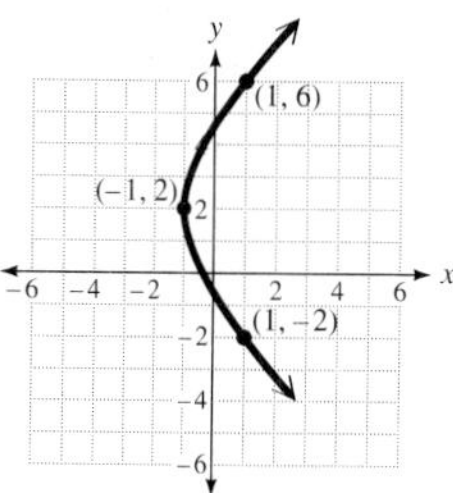

4. $(x + 5)^2 + (y - 2)^2 = 4^2$, $x^2 + y^2 + 10x - 4y + 13 = 0$ 5. $(x - 2)^2 + (y - 6)^2 = 3^2$, $x^2 + y^2 - 4x - 12y + 31 = 0$ 6. $\left(x - \frac{1}{2}\right)^2 + \left(y - \frac{1}{2}\right)^2 = 2^2$, $2x^2 + 2y^2 - 2x - 2y - 7 = 0$ 7. $x^2 + (y - 5)^2 = 1^2$, $x^2 + y^2 - 10y + 24 = 0$ 8. $(x - 2)^2 + (y + 8)^2 = 7^2$; $(2, -8)$; $r = 7$ 9. $\left(x + \frac{1}{2}\right)^2 + (y - 3)^2 = 1^2$; $\left(-\frac{1}{2}, 3\right)$; $r = 1$

10.

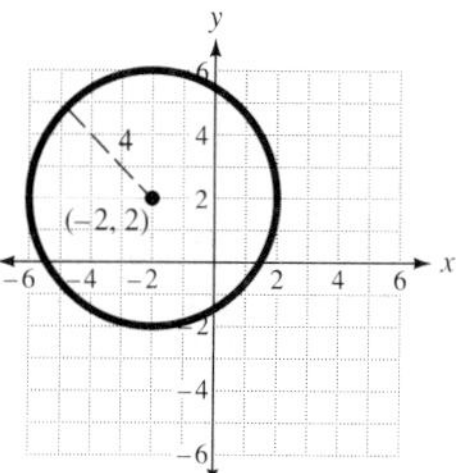

11.

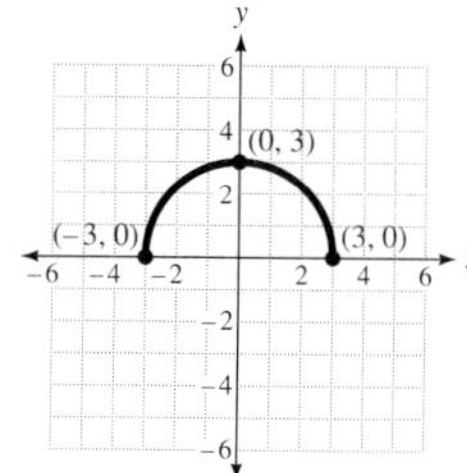

12.

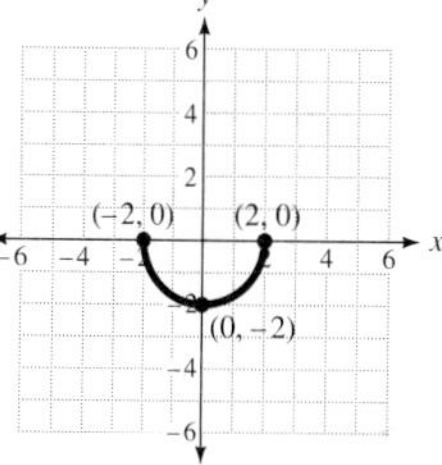

13. $y = \sqrt{100 - x^2}$; 10 ft 14. $\frac{x^2}{1} + \frac{y^2}{25} = 1$ 15. $\frac{x^2}{16} + \frac{y^2}{9} = 1$ 16. $\frac{y^2}{9} - \frac{x^2}{25} = 1$ 17. $\frac{x^2}{4} - \frac{y^2}{49} = 1$

18.

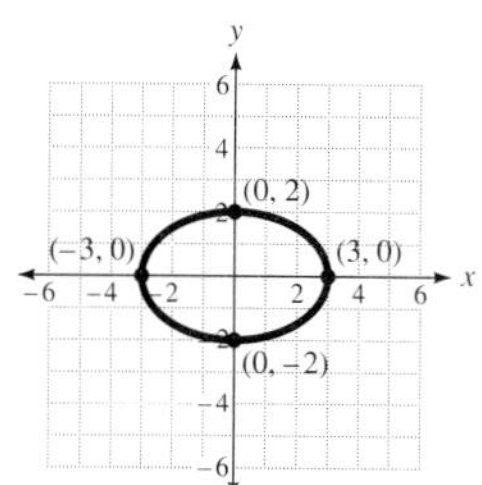

19.

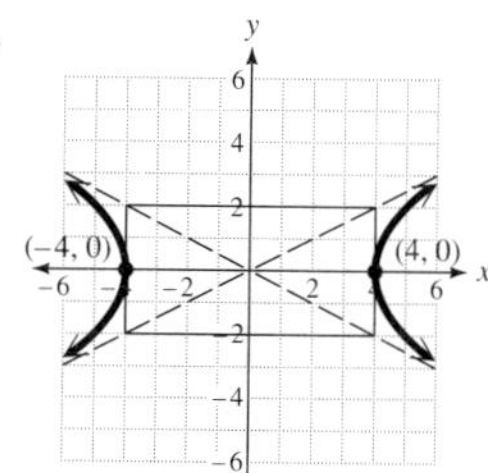

20.

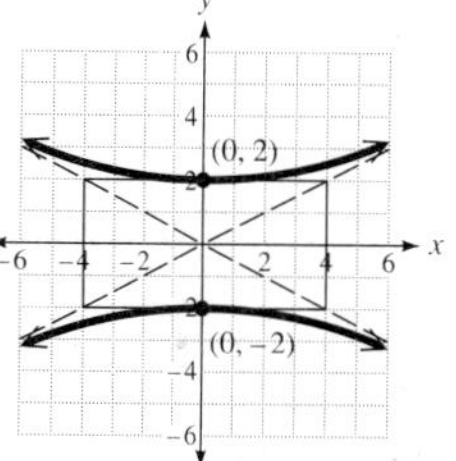

21. circle 22. parabola 23. circle 24. ellipse 25. hyperbola 26. parabola 27. parabola 28. hyperbola 29. ellipse 30. $\left(\frac{1}{3}, -\frac{5}{3}\right)$, $(1, -1)$ 31. $(2, 4)$, $(-2, -4)$, $(4, 2)$, $(-4, -2)$ 32. $(1, 0)$, $(-1, 0)$ 33. $(4, 2\sqrt{3})$, $(4, -2\sqrt{3})$ 34. 3 ft; 4 ft

35.

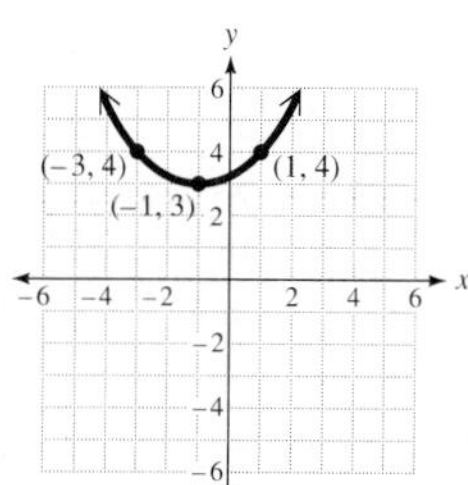

36.

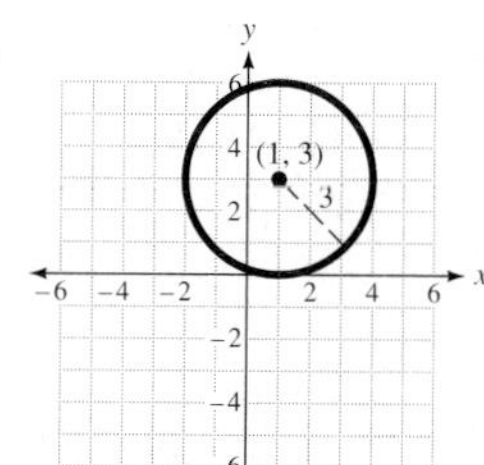

37.

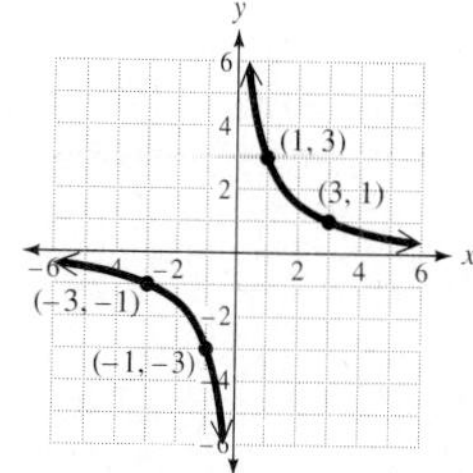

38. hyperbola 39. circle 40. hyperbola 41. parabola 42. hyperbola 43. circle 44. no real solutions 45. $(x + 4)^2 + (y + 1)^2 = 20$; $(-4, -1)$; $r = 2\sqrt{5}$

CHAPTER 11 TEST

Graph.

1. $y = -x^2 + 2$

1.

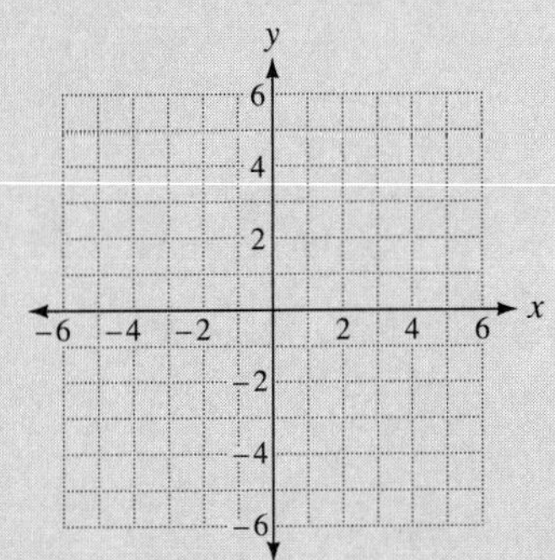

2. $x = y^2 + 4y + 3$

2.

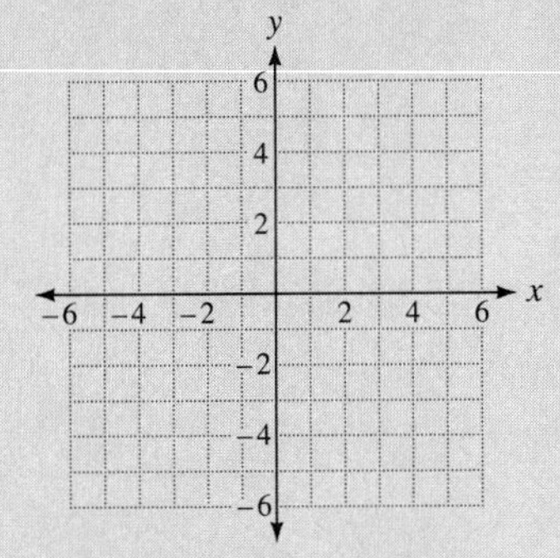

3. $x^2 + y^2 = 16$

3.

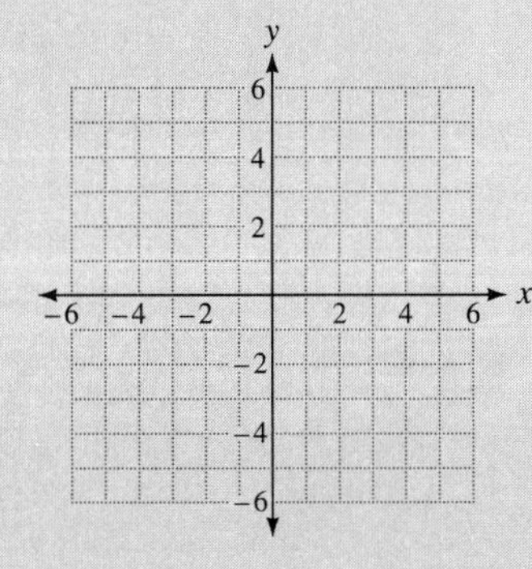

4. $(x - 2)^2 + (y + 1)^2 = 9$

4.

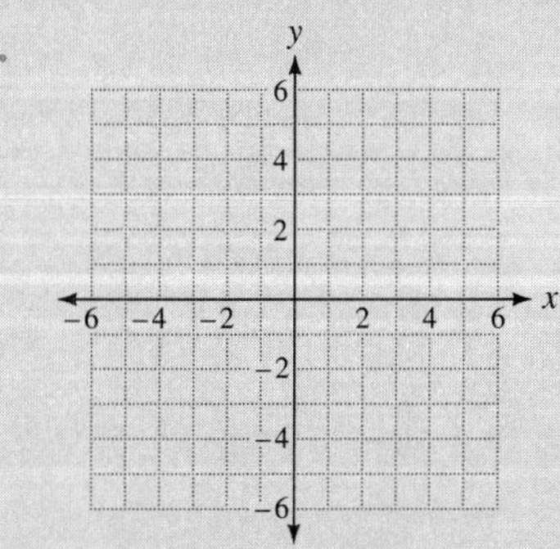

5. Find the standard and general forms of the equation of a circle centered at $(-2, -1)$ with radius $r = 5$.

5. ______________

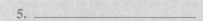

6. Find the standard form of the equation of the circle with general equation $x^2 + y^2 + 14x - 12y + 60 = 0$. What are the center and radius of the circle?

6. ______________

Graph.

7. $\dfrac{x^2}{16} + \dfrac{y^2}{4} = 1$

7.

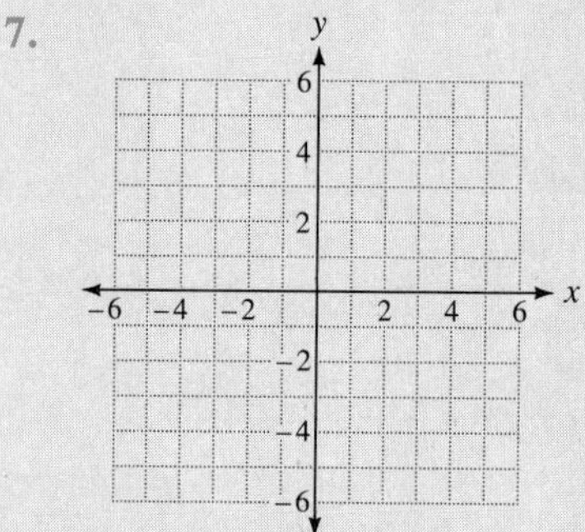

8. $\dfrac{y^2}{4} - \dfrac{x^2}{9} = 1$

8.

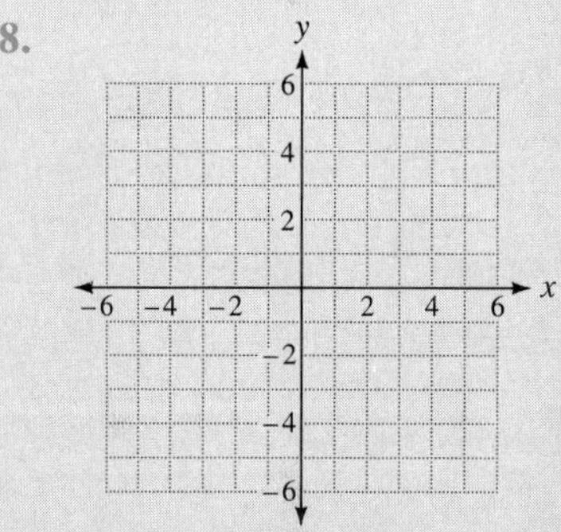

Identify each as the equation of a parabola, circle, ellipse, or hyperbola.

9. $\dfrac{x^2}{4} - \dfrac{y^2}{9} = 1$

9. ______________

10. $x^2 + (y - 1)^2 = 9$

10. ______________

11. $y = x^2 - 3x + 5$

11. ______________

12. $4x^2 + y^2 = 4$

12. ______________

Solve.

13. $3x^2 - y^2 = -13$
$2x + y = -1$

13. ______________

14. At what points do the circle $x^2 + y^2 = 9$ and the parabola $x^2 = 8y$ intersect?

14. ______________

CHAPTER 12

Sequences and Series

12.1 BASIC DEFINITIONS

STUDENT GUIDEPOSTS

1. Infinite Sequence
2. Finite Sequence
3. Series
4. Sigma Summation Notation

1 INFINITE SEQUENCE

Informally, we think of a sequence as a collection of numbers arranged in a particular order. There is a first number, a second number, a third number, and so forth; each number in the sequence corresponds to a natural number. This suggests the following formal definition.

Infinite Sequence

An **infinite sequence** is a function with domain the set of natural numbers $N = \{1, 2, 3, \ldots\}$.

For example, consider the function a defined by

$$a(n) = n^2 \quad (n = 1, 2, 3, \ldots).$$

Instead of the usual functional notation $a(n)$, for sequences we usually write

$$a_n = n^2.$$

That is, a letter with a subscript, such as a_n, represents numbers in the range of a sequence. The sequence defined by $a_n = n^2$ gives the following terms.

$$\begin{aligned} a_1 &= 1^2 = 1 \\ a_2 &= 2^2 = 4 \\ a_3 &= 3^2 = 9 \\ a_4 &= 4^2 = 16 \\ &\vdots \end{aligned}$$

A sequence is frequently defined by giving its range. The sequence above can be written

$$1, 4, 9, 16, \ldots, n^2, \ldots.$$

Each number in the range of a sequence is a **term** of the sequence, with a_n (in our example n^2) the ***n*th term** or **general term** of the sequence. The formula for the nth term generates the terms of a sequence by repeated substitution of counting numbers for n.

EXAMPLE 1 Finding Terms of an Infinite Sequence

Write the first four terms of the infinite sequence with general term $a_n = 2n - 1$.

$$\begin{array}{cccc} 1 & 2 & 3 & 4 \\ a_1 = 2(1) - 1 & a_2 = 2(2) - 1 & a_3 = 2(3) - 1 & a_4 = 2(4) - 1 \\ = 1 & = 3 & = 5 & = 7 \end{array}$$

The first four terms are 1, 3, 5, and 7, and the sequence is

$$1, 3, 5, 7, \ldots, 2n - 1, \ldots.$$

Practice Exercise 1

Write the first five terms of the sequence with general term $x_n = 3n + 5$.

Answer: $x_1 = 8$, $x_2 = 11$, $x_3 = 14$, $x_4 = 17$, $x_5 = 20$

2 FINITE SEQUENCE

Some sequences we will consider are finite rather than infinite.

Finite Sequence

A **finite sequence** with m terms is a function with domain the set of natural numbers $\{1, 2, \ldots, m\}$.

For example,

$$2, 4, 6, 8, 10$$

is a finite sequence with 5 terms where $a_n = 2n$, for $n = 1, 2, 3, 4, 5$. In contrast,

$$2, 4, 6, 8, 10, \ldots$$

is an infinite sequence where $a_n = 2n$, for $n = 1, 2, 3, \ldots$. The three dots after the last-mentioned term distinguish an infinite sequence from a finite sequence.

EXAMPLE 2 Finding the Terms of a Finite Sequence

A finite sequence has four terms, and the formula for the nth term is

$$x_n = (-1)^n \frac{1}{2^{n-1}}.$$

What is the sequence?

When $n = 1$: $\quad x_n = x_1 = (-1)^1 \frac{1}{2^{1-1}} = (-1)\frac{1}{2^0} = -1.$

When $n = 2$: $\quad x_n = x_2 = (-1)^2 \frac{1}{2^{2-1}} = (-1)^2 \frac{1}{2^1} = \frac{1}{2}.$

When $n = 3$: $\quad x_n = x_3 = (-1)^3 \frac{1}{2^{3-1}} = (-1)^3 \frac{1}{2^2} = -\frac{1}{4}.$

Practice Exercise 2

Find the finite sequence with three terms having nth term

$$a_n = (-1)^{n-1} \frac{1}{3^n}.$$

When $n = 4$: $\quad x_n = x_4 = (-1)^4 \frac{1}{2^{4-1}} = (-1)^4 \frac{1}{2^3} = \frac{1}{8}.$

The sequence is

$$-1, \frac{1}{2}, -\frac{1}{4}, \frac{1}{8}.$$

Answer: $a_1 = \frac{1}{3}$; $a_2 = -\frac{1}{9}$; $a_3 = \frac{1}{27}$; the sequence is $\frac{1}{3}, -\frac{1}{9}, \frac{1}{27}$.

In the examples above we were given a formula for the general or nth term of the sequence and asked to calculate several terms. If the first few terms of the sequence are given, we may be able to find a formula for the nth term.

EXAMPLE 3 Finding the General Term

Find a formula for a_n given the first few terms of the sequence.

(a) 2, 3, 4, 5, . . .

Here each term is one larger than the corresponding natural number, n. We have

$$\begin{aligned} a_1 &= 2 = 1 + 1 \\ a_2 &= 3 = 2 + 1 \\ a_3 &= 4 = 3 + 1 \\ a_4 &= 5 = 4 + 1. \end{aligned}$$

Hence, $a_n = n + 1$.

(b) $-3, 6, -9, 12, \ldots$

When the terms alternate in sign, a factor of $(-1)^n$ or $(-1)^{n+1}$ should be included in the general term. Here we have

$$\begin{aligned} a_1 &= -3 = 3 \cdot 1 \cdot (-1)^1 \\ a_2 &= 6 = 3 \cdot 2 \cdot (-1)^2 \\ a_3 &= -9 = 3 \cdot 3 \cdot (-1)^3 \\ a_4 &= 12 = 3 \cdot 4 \cdot (-1)^4. \end{aligned}$$

Thus, $a_n = 3n(-1)^n$ or $a_n = (-1)^n 3n$.

PRACTICE EXERCISE 3

Find a formula for a_n for each sequence.

(a) 4, 7, 10, 13, . . .

(b) $2, -4, 6, -8, \ldots$

Answers: (a) $a_n = 3n + 1$ (b) $a_n = (-1)^{n+1}2n$

3 SERIES

Associated with every sequence is a **series,** the indicated sum of the terms of the sequence. For example, associated with the sequence

$$2, 4, 6, 8, 10$$

is the series $\quad 2 + 4 + 6 + 8 + 10,$

and associated with the sequence

$$-1, \frac{1}{2}, -\frac{1}{4}, \frac{1}{8}$$

is the series $\quad (-1) + \left(\frac{1}{2}\right) + \left(-\frac{1}{4}\right) + \left(\frac{1}{8}\right).$

These are examples of **finite series,** or finite sums of numbers.

Associated with the sequence

$$1, 3, 5, 7, 9, \ldots$$

is the **infinite series**

$$1 + 3 + 5 + 7 + 9 + \ldots + (2n - 1) + \ldots.$$

The terms of a sequence are separated by commas, but the word *series* means that the terms are separated by plus signs and are added. Although every finite series "sums up" to a number, we will see that the same is not true for infinite series.

4 SIGMA SUMMATION NOTATION

The Greek letter Σ (sigma), called the **summation symbol,** is often used to abbreviate a series. For example, the series

$$2 + 4 + 6 + 8 + 10$$

which has general term $x_n = 2n$, can be written

$$\sum_{n=1}^{5} x_n \quad \text{or} \quad \sum_{n=1}^{5} 2n$$

and is read "the sum of the terms x_n or $2n$ as n varies over the counting numbers from 1 to 5." The letter n is the **index** on the summation, while 1 and 5 are the **lower** and **upper limits of summation,** respectively. In general, if

$$x_1, x_2, x_3, x_4, \ldots, x_n$$

is a sequence, its associated series is

$$\sum_{k=1}^{n} x_k = x_1 + x_2 + x_3 + x_4 + \ldots + x_n.$$

Note that any letter may be used for the index.

EXAMPLE 4 Using Sigma Summation Notation

Write out each series without using sigma summation notation.

(a) $\sum_{k=1}^{5} (k^2 + 1)$

When $k = 1$: $x_1 = (1)^2 + 1 = 1 + 1 = 2.$

When $k = 2$: $x_2 = (2)^2 + 1 = 4 + 1 = 5.$

When $k = 3$: $x_3 = (3)^2 + 1 = 9 + 1 = 10.$

When $k = 4$: $x_4 = (4)^2 + 1 = 16 + 1 = 17.$

When $k = 5$: $x_5 = (5)^2 + 1 = 25 + 1 = 26.$

Thus,

$$\sum_{k=1}^{5} (k^2 + 1) = 2 + 5 + 10 + 17 + 26 = 60.$$

Practice Exercise 4

Write out each series without using sigma summation notation.

(a) $\sum_{n=1}^{4} (n^3 - 5)$

(b) $\sum_{k=2}^{4} (-1)^k \sqrt{k+1}$

Observe that the lower limit need not always be 1.

When $k = 2$: $x_2 = (-1)^2 \sqrt{2+1} = (1)\sqrt{3} = \sqrt{3}$.

When $k = 3$: $x_3 = (-1)^3 \sqrt{3+1} = (-1)\sqrt{4} = -\sqrt{4} = -2$.

When $k = 4$: $x_4 = (-1)^4 \sqrt{4+1} = (1)\sqrt{5} = \sqrt{5}$.

Then $$\sum_{k=2}^{4} (-1)^k \sqrt{k+1} = \sqrt{3} - 2 + \sqrt{5}.$$

(b) $\sum_{j=3}^{6} (-1)^{j-1} \sqrt{j^2+7}$

Answers: (a) $-4 + 3 + 22 + 59 = 80$ (b) $4 - \sqrt{23} + 4\sqrt{2} - \sqrt{43}$

In Example 4(b) the lower limit of summation is 2 (not 1 as before). The lower limit can be any whole number less than or equal to the upper limit.

When a sequence is infinite, such as

$$\frac{1}{2}, \frac{1}{4}, \frac{1}{8}, \frac{1}{16}, \frac{1}{32}, \ldots, \frac{1}{2^n}, \ldots$$

the infinite series

$$\frac{1}{2} + \frac{1}{4} + \frac{1}{8} + \frac{1}{16} + \frac{1}{32} + \ldots + \frac{1}{2^n} + \ldots$$

is written in sigma notation as

$$\sum_{n=1}^{\infty} \frac{1}{2^n}$$

and is read "the sum of $\frac{1}{2^n}$ as n varies from 1 to infinity." The symbol ∞ does not represent a number. Rather, it means that the summation continues on without ending.

12.1 EXERCISES A

Answer true *or* false *in Exercises 1–4. If the statement is false, explain why.*

1. The series associated with a sequence is the indicated sum of the terms of a sequence.

2. When a sequence has a last term it is an infinite sequence.

3. The nth term of a sequence is also called the general term of the sequence.

4. In the sigma summation notation $\sum_{k=1}^{n} a_k$, k is the index of summation, 1 is the lower limit, and n is the upper limit of summation.

5. Write $a_2 + a_3 + a_4 + a_5 + a_6$ using sigma summation notation.

6. Write $\sum_{k=3}^{7} b_k$ without sigma summation notation.

In each of the following, the nth term of a sequence is given. Find the first four terms ($n = 1, 2, 3, 4$) and the seventh term ($n = 7$).

7 $x_n = \dfrac{(-1)^n}{n}$

8. $a_n = \dfrac{n+1}{n}$

9. $b_n = \left(-\dfrac{1}{2}\right)^n$

10. $x_n = n^2 - 2$

Find a formula for a_n given the first few terms of the sequence.

11. 5, 6, 7, 8, . . .

12. 4, 8, 12, 16, . . .

13. −2, −4, −6, −8, . . .

14. −1, 1, −1, 1, −1, 1, . . .

15. 1, 4, 9, 16, 25, . . .

16. $\dfrac{1}{2}, \dfrac{1}{4}, \dfrac{1}{8}, \dfrac{1}{16}, \ldots$

Which sequences are finite and which are infinite?

17. $2, 1, \dfrac{1}{2}, \dfrac{1}{4}, \ldots$

18. −1, 1, −1, 1, −1, 1, . . .

19. 1, 100, 10,000, 1,000,000

Find the associated series for each sequence.

20. $\dfrac{1}{2}, \dfrac{1}{3}, \dfrac{1}{4}, \dfrac{1}{5}, \ldots$

21. 5, 5, 5, 5, . . .

22. −1000, 2000, −3000

Rewrite each series using sigma summation notation.

23. $2 + 5 + 8 + 11 + \ldots + (3n - 1)$

24. $1 + \sqrt{2} + \sqrt{3} + 2 + \ldots + \sqrt{n}$

25. $\dfrac{1}{2} + \dfrac{1}{3} + \dfrac{1}{4} + \ldots + \dfrac{1}{n+1} + \ldots$

26. $-1 + \dfrac{1}{4} - \dfrac{1}{9} + \dfrac{1}{16} + \ldots + (-1)^n \dfrac{1}{n^2} + \ldots$

Rewrite each series without sigma summation notation.

27. $\sum_{k=1}^{6} a_k$

28. $\sum_{i=2}^{8} x_i$

29. $\sum_{k=1}^{5} (2k + 5)$

30. $\sum_{n=1}^{4} \dfrac{1}{n}$

31 $\sum_{k=4}^{5} (k^2 + 1)$

32. $\sum_{n=1}^{\infty} (-1)^n 2^n$

*The **arithmetic mean** of a collection of data (numbers),* $x_1, x_2, x_3, \ldots, x_n$, *is given by*

$$\bar{x} = \frac{1}{n} \sum_{k=1}^{n} x_k.$$

Use this formula to find $\bar{x}$ *for the data given in Exercises 33–34.*

33. 2, 5, 8, 11, 14

34. 1, 4, 9, 16

FOR REVIEW

Exercises 35–36 involving literal equations review material from Section 2.4 to help you prepare for the next section.

35. Solve $a_n = a_1 + (n - 1)d$ for d and use the result to find d when $a_n = -7$, $a_1 = 5$, and $n = 7$.

36. Solve $S_n = \frac{n}{2}[a_1 + a_n]$ for a_1 and use the result to find a_1 when $S_n = 116$, $n = 8$, and $a_n = 25$.

ANSWERS: 1. true 2. false; (finite) 3. true 4. true 5. $\sum_{k=2}^{6} a_k$ 6. $b_3 + b_4 + b_5 + b_6 + b_7$ 7. $x_1 = -1$, $x_2 = \frac{1}{2}$, $x_3 = -\frac{1}{3}$, $x_4 = \frac{1}{4}$, $x_7 = -\frac{1}{7}$ 8. $a_1 = 2$, $a_2 = \frac{3}{2}$, $a_3 = \frac{4}{3}$, $a_4 = \frac{5}{4}$, $a_7 = \frac{8}{7}$ 9. $b_1 = -\frac{1}{2}$, $b_2 = \frac{1}{4}$, $b_3 = -\frac{1}{8}$, $b_4 = \frac{1}{16}$, $b_7 = -\frac{1}{128}$ 10. $x_1 = -1$, $x_2 = 2$, $x_3 = 7$, $x_4 = 14$, $x_7 = 47$ 11. $a_n = n + 4$ 12. $a_n = 4n$ 13. $a_n = -2n$ 14. $a_n = (-1)^n$ 15. $a_n = n^2$ 16. $a_n = \frac{1}{2^n}$ 17. infinite 18. infinite 19. finite 20. $\frac{1}{2} + \frac{1}{3} + \frac{1}{4} + \frac{1}{5} + \ldots$ 21. $5 + 5 + 5 + 5 + \ldots$ 22. $-1000 + 2000 - 3000$ 23. $\sum_{k=1}^{n} (3k - 1)$ 24. $\sum_{k=1}^{n} \sqrt{k}$ 25. $\sum_{n=1}^{\infty} \frac{1}{n+1}$ 26. $\sum_{n=1}^{\infty} (-1)^n \frac{1}{n^2}$ 27. $a_1 + a_2 + a_3 + a_4 + a_5 + a_6$ 28. $x_2 + x_3 + x_4 + x_5 + x_6 + x_7 + x_8$ 29. $7 + 9 + 11 + 13 + 15$ 30. $1 + \frac{1}{2} + \frac{1}{3} + \frac{1}{4}$ 31. $17 + 26$ 32. $-2 + 4 - 8 + 16 - 32 + \ldots + (-1)^n 2^n + \ldots$ 33. 8 34. 7.5 35. $d = \frac{a_n - a_1}{n - 1}$; -2 36. $a_1 = \frac{2S_n - na_n}{n}$; 4

12.1 EXERCISES B

Answer true *or* false *in Exercises 1–4. If the statement is false, explain why.*

1. Sequences that do not have a last term but continue indefinitely are finite sequences.

2. The general term of a sequence is also called the nth term of the sequence.

3. The Greek letter Σ is the summation symbol.

4. In the sigma summation notation $\sum_{j=3}^{m} b_j$, j is the index of summation, 3 is the lower limit, and m is the upper limit of summation.

5. Write $d_5 + d_6 + d_7 + d_8$ using sigma summation notation.

6. Write $\sum_{m=4}^{7} a_m$ without sigma summation notation.

In each of the following, the nth term of a sequence is given. Find the first four terms ($n = 1, 2, 3, 4$) *and the seventh term* ($n = 7$).

7. $x_n = (-1)^n(n + 1)$

8. $a_n = \dfrac{n + 2}{n + 3}$

9. $b_n = (-2)^n$

10. $x_n = 3n^2$

Find a formula for a_n given the first few terms of the sequence.

11. $-1, 0, 1, 2, 3, \ldots$

12. $5, 10, 15, 20, \ldots$

13. $-1, -2, -3, -4, \ldots$

14. $-1, 2, -3, 4, -5, 6, \ldots$

15. $-2, 4, -8, 16, -32, \ldots$

16. $\frac{1}{2}, \frac{2}{3}, \frac{3}{4}, \frac{4}{5}, \ldots$

Which sequences are finite and which are infinite?

17. $\frac{1}{2}, \frac{1}{3}, \frac{1}{4}, \frac{1}{5}, \ldots$

18. $5, 5, 5, 5, \ldots$

19. $-1000, 2000, -3000$

Find the associated series for each sequence.

20. $2, 1, \frac{1}{2}, \frac{1}{4}, \ldots$

21. $-1, 1, -1, 1, \ldots$

22. $1, 100, 10{,}000, 100{,}000$

Rewrite each series using sigma summation notation.

23. $3 + 5 + 7 + 9 + \ldots + (2n + 1)$

24. $1 + 4 + 9 + 16 + \ldots + n^2$

25. $\frac{2}{3} + \frac{3}{4} + \frac{4}{5} + \ldots + \frac{n+1}{n+2} + \ldots$

26. $1 - \sqrt[3]{2} + \sqrt[3]{3} - \sqrt[3]{4} + \ldots + (-1)^{n+1}\sqrt[3]{n} + \ldots$

Rewrite each series without sigma summation notation.

27. $\sum_{k=1}^{4} a_k$

28. $\sum_{j=4}^{10} x_j$

29. $\sum_{k=1}^{4} (2k + 7)$

30. $\sum_{j=1}^{3} \frac{1}{j+1}$

31. $\sum_{n=5}^{6} (n^2 - 1)$

32. $\sum_{j=1}^{\infty} (-1)^j 3^j$

*The **arithmetic mean** of a collection of data (numbers), $x_1, x_2, x_3, \ldots x_n$, is given by*

$$\overline{x} = \frac{1}{n}\sum_{k=1}^{n} x_k.$$

Use this formula to find $\overline{x}$ for the data given in Exercises 33–34.

33. $4, 7, 12, 14, 13$

34. $2, 5, 11, 20$

FOR REVIEW

Exercises 35–36 involving literal equations review material from Section 2.4 to help you prepare for the next section.

35. Solve $a_n = a_1 + (n - 1)d$ for n and use the results to find n when $a_n = 33$, $a_1 = -2$, and $d = 5$.

36. Solve $S_n = \frac{n}{2}[a_1 + a_n]$ for n and use the result to find n when $S_n = 51$, $a_1 = -4$, and $a_n = 21$.

12.1 EXERCISES C

Find a formula for the nth term of the sequence.

1. $\frac{1}{3}, \frac{2}{4}, \frac{3}{5}, \frac{4}{6}, \ldots$

2. $-1, \sqrt{2}, -\sqrt{3}, 2, -\sqrt{5}, \ldots$

3. $\frac{1}{2}, \frac{4}{3}, \frac{9}{4}, \frac{16}{5}, \ldots$

$\left[\text{Answer: } a_n = \frac{n^2}{n+1}\right]$

4. $9, -27, 81, -243, \ldots$

12.2 ARITHMETIC SEQUENCES

STUDENT GUIDEPOSTS

1. Arithmetic Sequence
2. Formula for the nth Term
3. Sum of the First n Terms

1 ARITHMETIC SEQUENCE

In the sequence

$$2, 5, 8, 11, 14, \ldots$$

each term (after the first) can be obtained by adding 3 to the term immediately preceding it. That is,

the second term = the first term plus 3 $\quad 5 = 2 + 3$

the third term = the second term plus 3 $\quad 8 = 5 + 3$

and so forth. A sequence like this has a special name.

Arithmetic Sequence

An **arithmetic sequence** (sometimes called **arithmetic progression**) is a sequence in which every term after the first is the sum of the preceding term and a fixed number called the **common difference** of the sequence.

Notice that the common difference of the sequence above is 3. We use the following general notation.

a_1 for the first term

a_n for the nth term

d for the common difference

n for the number of terms from a_1 to a_n, inclusive

S_n for the sum of the first n terms

For the arithmetic sequence

$$1, 6, 11, 16, \ldots,$$

$a_1 = 1$ and $d = 5$ (each term after the first is found by adding 5 to the preceding term). Thus,

$$a_2 = a_1 + d = 1 + 5 = 6,$$
$$a_3 = a_2 + d = 6 + 5 = 11,$$

and so forth. If we take any term and subtract the preceding term ($6 - 1 = 5$, $11 - 6 = 5$, $16 - 11 = 5$, etc.) the difference is always 5. This is why d is called the *common difference* of the sequence.

2 FORMULA FOR THE *n*TH TERM

A general formula for calculating any particular term of an arithmetic sequence is a useful tool. Suppose we calculate several terms of an arbitrary arithmetic sequence.

$$\begin{aligned}
\text{1st term} &= a_1 = a_1 + 0d \\
\text{2nd term} &= a_2 = a_1 + d = a_1 + 1d \\
\text{3rd term} &= a_3 = a_2 + d = (a_1 + d) + d = a_1 + 2d \\
\text{4th term} &= a_4 = a_3 + d = (a_1 + 2d) + d = a_1 + 3d \\
\text{5th term} &= a_5 = a_4 + d = (a_1 + 3d) + d = a_1 + 4d
\end{aligned}$$

In each case, the nth term (5th for example) is the first term plus $(n - 1)$ (for example, $5 - 1 = 4$) times d. This gives us

$$n\text{th term} = a_n = a_1 + (n - 1)d$$

which is the formula for the general term of an arithmetic sequence.

EXAMPLE 1 FINDING TERMS OF AN ARITHMETIC SEQUENCE

Find the 5th and 11th terms of the arithmetic sequence with first term 3 and common difference 4.

We are given that $a_1 = 3$ and $d = 4$.

$$\begin{aligned}
a_5 &= a_1 + (5 - 1)d \\
&= 3 + (4)(4) = 19 \\
a_{11} &= a_1 + (11 - 1)d \\
&= 3 + (10)(4) = 43
\end{aligned}$$

Notice that the n in a_n is the same as the n in $(n - 1)$.

PRACTICE EXERCISE 1

Find the 4th and 10th terms of the arithmetic sequence with first term -2 and common difference 7.

Answer: $a_4 = 19$ and $a_{10} = 61$

3 SUM OF THE FIRST *n* TERMS

The sum of the first n terms in an arithmetic sequence can also be calculated by a formula. Let S_n denote the sum of the first n terms of an arithmetic sequence. Then

$$\begin{aligned}
S_n &= a_1 + \quad a_2 \quad + \quad a_3 \quad + \quad a_4 \quad + \ldots + \quad a_n \\
&= a_1 + (a_1 + d) + (a_1 + 2d) + (a_1 + 3d) + \ldots + [a_1 + (n - 1)d].
\end{aligned}$$

Reversing the order of addition, we obtain

$$\begin{aligned}
S_n &= \quad a_n \quad + \quad a_{n-1} \quad + \quad a_{n-2} \quad + \ldots + a_1 \\
&= [a_1 + (n - 1)d] + [a_1 + (n - 2)d] + [a_1 + (n - 3)d] + \ldots + a_1.
\end{aligned}$$

Add corresponding terms in both representations of S_n.

$$
\begin{aligned}
S_n &= a_1 + a_1 + d + a_1 + 2d + \dots + a_1 + (n-1)d \\
S_n &= a_1 + (n-1)d + a_1 + (n-2)d + a_1 + (n-3)d + \dots + a_1 \\
\hline
2S_n &= 2a_1 + (n-1)d + 2a_1 + (n-1)d + 2a_1 + (n-1)d + \dots + 2a_1 + (n-1)d \\
2S_n &= n[2a_1 + (n-1)d] \quad \text{There are } n \text{ terms of the form } 2a_1 + (n-1)d
\end{aligned}
$$

$$S_n = \frac{n}{2}[2a_1 + (n-1)d]$$

EXAMPLE 2 FINDING THE SUM OF TERMS

Find the 9th term and the sum of the first 9 terms of the arithmetic sequence with $a_1 = -2$ and $d = 5$.

We need to calculate a_9 and S_9.

$$
\begin{aligned}
a_9 &= a_1 + (9-1)d = -2 + (8)5 \\
&= -2 + 40 = 38 \\
S_9 &= \frac{9}{2}[2a_1 + (9-1)d] = \frac{9}{2}[2(-2) + (8)(5)] \\
&= \frac{9}{2}[-4 + 40] = \frac{9}{2}[36] = 162
\end{aligned}
$$

PRACTICE EXERCISE 2

Find the 8th term and the sum of the first 8 terms of the arithmetic sequence with $a_1 = 13$ and $d = -4$.

Answer: $a_8 = -15$ and $S_8 = -8$

An alternate form for S_n, which would have been useful in Example 2 when we had already calculated a_9, is easily derived from

$$S_n = \frac{n}{2}[2a_1 + (n-1)d].$$

By writing $2a_1$ as $a_1 + a_1$ and observing that $a_1 + (n-1)d = a_n$, we have

$$S_n = \frac{n}{2}[a_1 + \underbrace{a_1 + (n-1)d}_{a_n}]$$

$$S_n = \frac{n}{2}[a_1 + a_n].$$

Thus, the sum of the first n terms is n times the average of the first and the nth terms. In Example 2 we had already calculated $a_9 = 38$. When we know the nth term, it is easier to substitute in the second formula.

$$
\begin{aligned}
S_9 &= \frac{9}{2}[a_1 + a_n] = \frac{9}{2}[-2 + 38] \\
&= \frac{9}{2}[36] = 162
\end{aligned}
$$

CAUTION

In the example above, do not substitute n (which is 9) for a_n (which is 38). That is, $n \neq a_n$. This is a common mistake.

Let us summarize the important concepts and formulas relative to arithmetic sequences.

Formulas for Arithmetic Sequences

A sequence in which each term after the first, a_1, is obtained by adding a fixed number d, the **common difference,** to the preceding term is an **arithmetic sequence.** The **general** or ***n*th term** of an arithmetic sequence is given by

$$a_n = a_1 + (n - 1)d.$$

The sum of the first n terms of an arithmetic sequence is given by

$$S_n = \frac{n}{2}[a_1 + a_n] \quad \text{or} \quad S_n = \frac{n}{2}[2a_1 + (n - 1)d].$$

As we see in the following examples, there are many applications of arithmetic sequences.

EXAMPLE 3 Number Problem

Find the sum of the even integers from 2 through 100.

Since there are fifty even integers from 2 through 100, we must calculate

$$S_{50} = 2 + 4 + 6 + 8 + \ldots + 100.$$

Since $a_1 = 2$, $d = 2$, $n = 50$, and $a_{50} = 100$,

$$S_{50} = \frac{50}{2}[a_1 + a_{50}] = 25[2 + 100] = 25(102) = 2550.$$

Practice Exercise 3

Find the sum of the odd integers from 1 through 99.

Answer: 2500

EXAMPLE 4 Consumer Problem

A new car costs \$10,000. Assume that it depreciates 24% the first year, 20% the second year, 16% the third year, and continues in the same manner for 6 years. If all depreciations apply to the original cost, what is the value of the car in 6 years?

We calculate the sum of the depreciations with $a_1 = 24$, $a_2 = 20$, $a_3 = 16, \ldots$, and $d = -4$.

$$\begin{aligned} S_6 &= \frac{6}{2}[2a_1 + (6 - 1)d] \\ &= \frac{6}{2}[(2)(24) + (5)(-4)] \\ &= 3[48 - 20] = 3[28] = 84 \end{aligned}$$

The total percentage of depreciation is 84%, so the total depreciation is

$$84\% \text{ of } 10{,}000 = (0.84)(10{,}000) = 8400.$$

Thus, the total value of the car in 6 years is

$$\$10{,}000 - \$8400 = \$1600.$$

Practice Exercise 4

A lot costs \$15,000, and appreciates 10% the first year, 12% the second year, 14% the third year, and continues in the same manner for 8 years. If all appreciations apply to the initial value, what is the value of the lot in 8 years?

Answer: \$35,400

EXAMPLE 5 COUNTING PROBLEM

A theater has 50 rows with 20 seats in the first row, 22 in the second, 24 in the third, and so forth. How many seats are in the theater?

We have $a_1 = 20$, $d = 2$, and $n = 50$, and we must calculate S_{50}.

$$S_{50} = \frac{50}{2}[2a_1 + (50 - 1)d] = 25[(2)(20) + (49)(2)]$$

$$= 25[40 + 98] = 25[138] = 3450$$

PRACTICE EXERCISE 5

A field is in the shape of a trapezoid with 40 rows of rose plants. If there are 15 roses in the first row, 18 in the second, 21 in the third, and so on, how many plants are in the field?

Answer: 2940 plants

12.2 EXERCISES A

Answer true *or* false. *If the statement is false, explain why.*

1. In an arithmetic sequence, each term after the first is obtained by adding a fixed number called the common difference to the preceding term.

2. The formula for calculating the *n*th term of an arithmetic sequence is $S_n = \frac{n}{2}[a_1 + a_n]$.

Decide whether each sequence is arithmetic. If it is, give the common difference.

3. 3, 6, 9, 12, . . .

4. 2, −2, 2, −2, . . .

5. 1, 1, 1, 1, . . .

6. −3, −1, 1, 3, . . .

Write the first five terms of the arithmetic sequence with the given a_1 and d.

7. $a_1 = 8$, $d = -3$

8. $a_1 = 5$, $d = 7$

9. $a_1 = 0$, $d = -5$

With the given information about the arithmetic sequence, find the indicated unknowns.

10. $a_1 = 6$, $d = 3$; a_5, S_5

11. $a_1 = 18$, $d = -3$; a_{15}, S_{15}

12. $a_1 = 5$, $a_2 = 9$; d, a_8

13 $a_2 = 3$, $a_4 = 7$, d, a_1

14. $a_1 = -12$, $d = 3$; a_5, S_{20}

15. $a_1 = -20$, $a_{16} = -80$; d, S_{16}

Solve.

16. Find the sum of the even integers from 2 through 200.

17. Find the sum of the multiples of 3 from 3 through 45.

18. A new car costs \$9000. Assume that it depreciates 21% the first year, 18% the second year, 15% the third, and continues in the same manner for 5 years. If all depreciations apply to the original cost, what is the value of the car in 5 years?

19. An auditorium has 40 rows with 30 seats in the first row, 33 in the second row, 36 in the third row, and so forth. How many seats are in the auditorium?

FOR REVIEW

20. Find x_1, x_2, x_3, and x_5 for a general sequence for which $x_n = 3 + n^2$.

21. Rewrite $\sum_{n=1}^{4} \frac{2}{3n+1}$ without sigma summation notation.

Exercises 22–25 will help you prepare for the next section.

22. Solve $S_n = \frac{a_1 - ra_n}{1 - r}$ for a_1 and use the result to find a_1 when $S_n = -242$, $r = 3$, and $a_n = -162$.

23. Find the value of a_1 if $a_n = \frac{1}{27}$, $r = \frac{1}{3}$, and $n = 5$, in the formula $a_n = a_1 r^{n-1}$.

24. Find the value of r if $a_n = 96$, $a_1 = 3$, and $n = 6$, in the formula $a_n = a_1 r^{n-1}$.

25. Find the value of n if $a_n = 320$, $a_1 = 5$, and $r = 2$, in the formula $a_n = a_1 r^{n-1}$.

ANSWERS: 1. true 2. false (this is the formula for the sum of the first n terms) 3. yes; $d = 3$ 4. no 5. yes; $d = 0$ 6. yes; $d = 2$ 7. 8, 5, 2, −1, −4 8. 5, 12, 19, 26, 33 9. 0, −5, −10, −15, −20 10. $a_5 = 18$, $S_5 = 60$ 11. $a_{15} = -24$, $S_{15} = -45$ 12. $d = 4$, $a_8 = 33$ 13. $d = 2$, $a_1 = 1$ 14. $a_5 = 0$, $S_{20} = 330$ 15. $d = -4$, $S_{16} = -800$ 16. 10,100 17. 360 18. \$2250 19. 3540 seats 20. $x_1 = 4$, $x_2 = 7$, $x_3 = 12$, $x_5 = 28$ 21. $\frac{2}{4} + \frac{2}{7} + \frac{2}{10} + \frac{2}{3}$ 22. $a_1 = S_n - S_n r + ra_n$; −2 23. 3 24. 2 25. 7

12.2 EXERCISES B

Answer true *or* false. *If the statement is false, explain why.*

1. In an arithmetic sequence, a_1 symbolizes the first term, a_n symbolizes the nth term, d symbolizes the common difference, and S_n symbolizes the sum of the first n terms.

2. The formula for calculating the sum of the first n terms of an arithmetic sequence is $a_n = a_1 + (n - 1)d$.

Decide whether each sequence is arithmetic. If it is, give the common difference.

3. 3, 7, 11, 15, . . .

4. 5, 0, 5, 0, 5, . . .

5. −6, −6, −6, −6, . . .

6. 1, 4, 8, 13, . . .

Write the first five terms of the arithmetic sequence with the given a_1 and d.

7. $a_1 = -3$, $d = -4$

8. $a_1 = 9$, $d = -5$

9. $a_1 = -4$, $d = 4$

With the given information about the arithmetic sequence, find the indicated unknowns.

10. $a_1 = -3$, $d = 4$; a_{10}, S_{10}

11. $a_1 = -7$, $d = 2$; a_{20}, S_{20}

12. $a_1 = 7$, $a_3 = 15$; d, a_9

13. $a_1 = 10$, $a_4 = -5$; d, S_{14}

14. $a_1 = 50$, $a_{10} = 95$; d, S_{10}

15. $a_5 = 22$, $a_6 = 25$; d, a_1

Solve.

16. Find the sum of the multiples of 5 from 5 through 350.

17. Find the sum of the odd integers from 1 through 199.

18. A new house costs $80,000 and increases in value 20% the first year, 22% the second, 24% the third, and continues in the same manner for 7 years. If all appreciations apply to the original cost, what will be the value of the house in 7 years?

19. A collection of coins is arranged in the shape of a triangle with 30 coins in the first row, 29 in the second, and so forth down to 1 coin. How many coins are in the collection?

FOR REVIEW

20. Find x_1, x_2, x_3, and x_5 for a general sequence for which $x_n = \frac{n-1}{n+1}$.

21. Rewrite $\sum_{k=1}^{\infty} (-1)^k a_k$ without sigma summation notation.

Exercises 22–25 will help you prepare for the next section.

22. Solve $S_n = \frac{a_1 - ra_n}{1 - r}$ for a_1 and use the result to find a_1 when $S_n = 86$, $r = -2$, and $a_n = 128$.

23. Find the value of a_1 if $a_n = \frac{1}{2}$, $r = \frac{1}{2}$, and $n = 6$, in the formula $a_n = a_1 r^{n-1}$.

24. Find the value of r if $a_n = 1$, $a_1 = 27$, and $n = 4$, in the formula $a_n = a_1 r^{n-1}$.

25. Find the value of n if $a_n = 320$, $a_1 = 10$, and $r = 2$, in the formula $a_n = a_1 r^{n-1}$.

12.2 EXERCISES C

Solve.

1. The population of Kingston is decreasing by 500 inhabitants each year. If its population at the beginning of 1980 was 20,135, what was its population at the beginning of 1990?
[Answer: 15,135]

2. A rock is dropped from the top of a tall cliff and falls 16 ft during the first second, 48 ft during the second, 80 ft during the third, and so on. How many feet does the rock fall during the twelfth second?

Use the formula for S_n to evaluate each series in Exercises 3–6.

3. $\sum_{k=1}^{5} (2k + 3)$
[Answer: 45]

4. $\sum_{k=1}^{6} (3k - 1)$

5. $\sum_{n=1}^{10} (2n + 5)$

6. $\sum_{n=1}^{12} (3n + 2)$

The terms between a_1 and a_n of an arithmetic sequence are called the ***arithmetic means*** *of a_1 and a_n. Thus the means between a_1 and a_5 are a_2, a_3, and a_4. Use this to insert the indicated number of arithmetic means between the given two numbers in Exercises 7–10.*

7. Two between 5 and 11
[Answer: 7, 9]

8. Three between 50 and 70

9. Four between −6 and 9

10. Five between 13 and −11

12.3 GEOMETRIC SEQUENCES

STUDENT GUIDEPOSTS

1. Geometric Sequence
2. Formula for the nth Term
3. Sum of the First n Terms

1 GEOMETRIC SEQUENCE

Consider the sequence

$$2, 6, 18, 54, 162, \ldots$$

in which each term (after the first) can be found by multiplying the preceding term by 3. That is,

the second term = the first term times 3 $\quad 6 = 2 \cdot 3$

the third term = the second term times 3 $\quad 18 = 6 \cdot 3$

and so forth. A sequence like this is given a special name.

Geometric Sequence

A **geometric sequence** (sometimes called **geometric progression**) is a sequence in which every term after the first is the product of the preceding term and a fixed number called the **common ratio** of the sequence.

Notice that the common ratio of the sequence given above is 3. We use the following general notation.

a_1 for the first term

a_n for the nth term

r for the common ratio

n for the number of terms from a_1 to a_n, inclusive

S_n for the sum of the first n terms

For the geometric sequence

$$1, \frac{1}{2}, \frac{1}{4}, \frac{1}{8}, \frac{1}{16}, \ldots$$

$a_1 = 1$ and $r = \frac{1}{2}$ (to obtain each succeeding term multiply the preceding one by $\frac{1}{2}$).

Thus,

$$a_2 = a_1 \cdot r = 1 \cdot \frac{1}{2} = \frac{1}{2},$$

$$a_3 = a_2 \cdot r = \frac{1}{2} \cdot \frac{1}{2} = \frac{1}{4},$$

$$a_4 = a_3 \cdot r = \frac{1}{4} \cdot \frac{1}{2} = \frac{1}{8},$$

and so on. Notice that if we divide any term by the preceding one ($\frac{1}{2} \div 1 = \frac{1}{2}$, $\frac{1}{4} \div \frac{1}{2} = \frac{1}{2}$, $\frac{1}{8} \div \frac{1}{4} = \frac{1}{2}$, etc.), the quotient or ratio is always $\frac{1}{2}$. This is why we call r the *common ratio* of the sequence.

2 FORMULA FOR THE *n*TH TERM

As with arithmetic sequences, there is a formula for calculating the *n*th term of a geometric sequence. Let us calculate several terms of an arbitrary geometry sequence.

$$\begin{aligned}
\text{1st term} &= a_1 = a_1r^0\\
\text{2nd term} &= a_2 = a_1 \cdot r = a_1r^1\\
\text{3rd term} &= a_3 = a_2 \cdot r = (a_1r) \cdot r = a_1r^2\\
\text{4th term} &= a_4 = a_3 \cdot r = (a_1r^2) \cdot r = a_1r^3\\
\text{5th term} &= a_5 = a_4 \cdot r = (a_1r^3) \cdot r = a_1r^4
\end{aligned}$$

In each case, the *n*th term (5th for example) is the first term times r raised to the $(n - 1)$ power $(5 - 1 = 4)$. Thus, we have

$$n\text{th term} = a_n = a_1r^{n-1}.$$

EXAMPLE 1 FINDING TERMS OF A GEOMETRIC SEQUENCE

Find the 7th and 10th terms of the following geometric sequence.

$$9,\ 3,\ 1,\ \frac{1}{3},\ \frac{1}{9},\ \ldots$$

We have $a_1 = 9$ and $r = \frac{1}{3}$.

$$a_7 = a_1r^{7-1} = 9\left(\frac{1}{3}\right)^6 = 9\left(\frac{1}{9}\right)\left(\frac{1}{3^4}\right) = \frac{1}{3^4} = \frac{1}{81}$$

$$a_{10} = a_1r^{10-1} = 9\left(\frac{1}{3}\right)^9 = 9\left(\frac{1}{9}\right)\left(\frac{1}{3^7}\right) = \frac{1}{3^7} = \frac{1}{2187}$$

Notice that the n in a_n is the same as the n in the exponent $n - 1$.

PRACTICE EXERCISE 1

Find the 6th and 8th terms of the geometric sequence $\frac{1}{9}, \frac{1}{3}, 1, 3, 9, \ldots$.

Answer: $a_6 = 27$ and $a_8 = 243$

3 SUM OF THE FIRST *n* TERMS

We now derive a formula for calculating the sum of the first n terms of a geometric sequence. Let S_n denote the sum of the first n terms. The series is

$$S_n = a_1 + a_1r + a_1r^2 + a_1r^3 + \ldots + a_1r^{n-2} + a_1r^{n-1}.$$

Multiply S_n by r.

$$rS_n = a_1r + a_1r^2 + a_1r^3 + \ldots + a_1r^{n-2} + a_1r^{n-1} + a_1r^n$$

Subtract rS_n from S_n.

$$\begin{aligned}
S_n &= a_1 + a_1r + a_1r^2 + a_1r^3 + \ldots + a_1r^{n-2} + a_1r^{n-1}\\
-rS_n &= \quad - a_1r - a_1r^2 - a_1r^3 - \ldots - a_1r^{n-2} - a_1r^{n-1} - a_1r^n\\
\hline
S_n - rS_n &= a_1 \qquad - a_1r^n\\
S_n(1 - r) &= a_1 - a_1r^n
\end{aligned}$$

$$S_n = \frac{a_1 - a_1r^n}{1 - r}$$

EXAMPLE 2 Finding the Sum of Terms

Find the 8th term and the sum of the first 8 terms of the geometric sequence

$$-2,\ 1,\ -\frac{1}{2},\ \frac{1}{4},\ \ldots .$$

We have $a_1 = -2$ and $r = -\frac{1}{2}$. Since r is negative, the signs of the terms alternate. We must calculate $a_8(n = 8)$ and S_8.

$$a_8 = a_1 r^{8-1} = (-2)\left(-\frac{1}{2}\right)^7 = (-2)\left(-\frac{1}{2}\right)\left(-\frac{1}{2}\right)^6 = \frac{1}{2^6} = \frac{1}{64}$$

$$S_8 = \frac{a_1 - a_1 r^8}{1 - r} = \frac{(-2) - (-2)\left(-\frac{1}{2}\right)^8}{1 - \left(-\frac{1}{2}\right)}$$

$$= \frac{(-2) - (-2)\left(-\frac{1}{2}\right)\left(-\frac{1}{2}\right)^7}{\frac{3}{2}}$$

$$= \frac{(-2) - \left(-\frac{1}{2^7}\right)}{\frac{3}{2}} = \frac{(-2) + \frac{1}{128}}{\frac{3}{2}}$$

$$= \frac{-256 + 1}{128} \cdot \frac{2}{3} = -\frac{255}{128} \cdot \frac{2}{3}$$

$$= -\frac{85}{64} \approx -1.33$$

Practice Exercise 2

Find the 7th term and the sum of the first 7 terms of the geometric sequence $\frac{1}{256}, -\frac{1}{64}, \frac{1}{16}, -\frac{1}{4}, \ldots .$

Answer: $a_7 = 16$ and $S_7 = \frac{3277}{256} \approx 12.80$

An alternate form for S_n can be derived from

$$S_n = \frac{a_1 - a_1 r^n}{1 - r}$$

by writing $a_1 r^n = r(a_1 r^{n-1}) = r a_n$.

$$S_n = \frac{a_1 - r a_n}{1 - r}$$

In Example 2 we calculated $a_8 = \frac{1}{64}$. It might have been easier to obtain S_8 by substituting a_8 in the formula above.

$$S_8 = \frac{a_1 - r a_8}{1 - r} = \frac{(-2) - \left(-\frac{1}{2}\right)\left(\frac{1}{64}\right)}{1 - \left(-\frac{1}{2}\right)}$$

$$= \frac{-2 + \frac{1}{128}}{\frac{3}{2}} = -\frac{255}{128} \cdot \frac{2}{3} = -\frac{85}{64}$$

Here is a summary of important ideas and formulas about geometric sequences.

Formulas for Geometric Sequences

A sequence in which each term after the first, a_1, is obtained by multiplying the preceding term by a fixed number r, the **common ratio,** is a **geometric sequence.** The **general** or ***n*th term** of a geometric sequence is given by

$$a_n = a_1 r^{n-1}.$$

The sum of the first n terms of a geometric sequence is given by

$$S_n = \frac{a_1 - ra_n}{1 - r} \quad \text{or} \quad S_n = \frac{a_1 - a_1 r^n}{1 - r}.$$

EXAMPLE 3 FINDING A GEOMETRIC SEQUENCE

Find a_1 and r for the geometric sequence that has $a_2 = 10$ and $a_5 = 80$.

$$80 = a_5 = a_1 r^{5-1} = a_1 r^4$$
$$10 = a_2 = a_1 r^{2-1} = a_1 r$$

Divide these terms.

$$\frac{80}{10} = \frac{a_5}{a_2} = \frac{a_1 r^4}{a_1 r} = r^3$$
$$8 = r^3$$
$$2 = r$$

Then

$$10 = a_2 = a_1 r = a_1 2$$
$$10 = a_1 2$$
$$5 = a_1.$$

Thus, $a_1 = 5$ and $r = 2$.

PRACTICE EXERCISE 3

Find a_1 and r for the geometric sequence that has $a_3 = -1$ and $a_6 = \frac{1}{27}$.

Answer: $a_1 = -9$ and $r = -\frac{1}{3}$

The next examples illustrate the many applications of geometric sequences.

EXAMPLE 4 CONSUMER PROBLEM

A new car costing \$8000 depreciates 20% of its value each year. How much is the car worth at the end of 5 years?

To solve a word problem involving geometric sequences (arithmetic sequences also), it is wise to write out the first few terms. Once this is done, a_1, r, and n are much easier to identify.

At beginning of 1st year	*At beginning of 2nd year*	*At beginning of 3rd year*	*At beginning of 4th year*
$a_1 = 8000$	$a_2 = (0.80)(8000)$ $= 6400$	$a_3 = (0.80)^2(8000)$ $= (0.80)(6400)$ $= 5120$	$a_4 = (0.80)^3(8000)$ $= (0.80)(5120)$ $= 4096$

PRACTICE EXERCISE 4

A motorhome costing \$30,000 depreciates 15% of its value each year. How much will the motorhome be worth at the end of 6 years?

Notice that to obtain the next term in the sequence (the value of the car each succeeding year), the preceding term is multiplied by 0.80 (80%). Thus, $a_1 = 8000$ and $r = 0.8$. To obtain the value of the car at the end of 5 years we need to find a_6 (the value at the end of the fifth year is equal to the value at the beginning of the sixth year). Thus, $n = 6$ and

$$a_6 = a_1 r^{6-1} = (8000)(0.80)^5 \approx 2621.$$

The value of the car after five years is approximately \$2621. Compare this example with Example 4 of the previous section. There the rate of depreciation was computed on the original amount as opposed to the current amount here.

Answer: approximately \$11,314

CAUTION

Often a problem such as Example 4 is solved incorrectly by assuming that $n = 5$ ("at end of five years"). However, when we write out a few terms, it is clear that n must be 6. Similar remarks apply to interest (compounded annually) problems and population growth problems.

EXAMPLE 5 Compound Interest Problem

A woman borrows \$1000.00 at 12% interest compounded annually. If she pays off the loan in full at the end of 3 years, how much does she pay?

At beginning of 1st year	*At beginning of 2nd year (or end of 1st year)*	*At beginning of 3rd year (or end of 2nd year)*
$a_1 = 1000$	$a_2 = 1000 + 0.12(1000)$	$a_3 = (1.12)(1000) + (0.12)(1.12)(1000)$
	$= [1 + 0.12](1000)$	$= [1 + 0.12](1.12)(1000)$
	$= (1.12)(1000)$	$= (1.12)^2(1000)$

To obtain the next term of the sequence, we multiply the preceding term by 1.12. To find the amount that would have to be repaid at the end of 3 years, we calculate a_4 (4 not 3). We have $a_1 = 1000$, $r = 1.12$, and

$$a_4 = a_1 r^{4-1} = (1000)(1.12)^3 \approx (1000)(1.40493) = 1404.93.$$

At the end of 3 years the woman would have to pay back \$1404.93.

Practice Exercise 5

A house appreciates 12% of its value each year. If it is now worth \$60,000, what will be its value at the end of 4 years?

Answer: approximately \$94,411

12.3 EXERCISES A

Write true *or* false *for each of the following. If the statement is false, explain why.*

1. In a geometric sequence, each term after the first is obtained by multiplying a fixed number called the common ratio by the preceding term.

2. The formula for calculating the nth term of a geometric sequence is $S_n = \frac{a_1 - ra_n}{1 - r}$.

Decide whether each sequence is geometric. If it is, give the common ratio.

3. 5, 15, 45, 135, . . .

4. −1, 1, −1, 1, −1, 1, . . .

5. $4, 2, 1, \frac{1}{2}, \ldots$

6. $-3, 1, -\frac{1}{3}, \frac{1}{9}, \ldots$

Write the first five terms of the geometric sequence with the given a_1 and r.

7. $a_1 = \frac{1}{8}, r = -2$

8. $a_1 = -15, r = \frac{1}{3}$

9. $a_1 = 32, r = \frac{1}{4}$

With the given information about the geometric sequence, find the indicated unknowns.

10. $a_1 = 128, r = -\frac{1}{2}; a_8, S_8$

11. $a_1 = -\frac{1}{9}, a_6 = -27; r, S_6$

12. $a_2 = 6, r = \frac{1}{3}; a_1, a_5$

13 $a_2 = \frac{1}{3}, a_5 = -9; r, a_1$

14. $a_1 = 2, r = 4; a_4, S_4$

15. $a_1 = 15, r = -\frac{1}{5}; a_6, S_3$

Solve.

16. A new car costing \$9000 depreciates 30% of its value each year. How much is the car worth at the end of 5 years?

17. A woman borrows \$1000 at 12% interest compounded annually. If she pays off the loan in full at the end of 4 years, how much does she pay?

18. Burford was offered a job for the month of June (30 days), and was told he would be paid 1¢ at the end of the first day, 2¢ at the end of the second day, 4¢ at the end of the third day, and so forth, doubling each previous day's salary. However, Burford refused the job thinking that the pay was inferior. Would you take the job? Why?

19. A ball dropped from a height of 20 ft always rebounds 80% of its previous fall. How high would the ball rebound on the fifth bounce? Give answer to the nearest tenth of a foot.

FOR REVIEW

20. Write the first 5 terms of the arithmetic sequence having $a_1 = 7$ and $d = -2$.

21. Find d and S_8 for the arithmetic sequence having $a_1 = -18$ and $a_8 = 38$.

Solve.

22. The population of a town is increasing by 100 each year. If its present population is 1500, what will be its population in 5 years?

23. What is the sum of the odd integers from 1 through 49?

Exercises 24–26 will help you prepare for the next section.

24. Simplify the complex fraction $\dfrac{\frac{2}{3}}{1-\frac{1}{6}}$.

25. Solve $S = \frac{a_1}{1-r}$ for r and use the result to find r when $S = 45$ and $a_1 = 36$.

26. Consider the geometric sequence $1, \frac{1}{3}, \frac{1}{9}, \frac{1}{27}, \ldots$. Use your calculator to approximate S_2, S_4, S_6, and S_8. Do you see anything special about these sums?

ANSWERS: 1. true 2. false (this is the formula for the sum of the first n terms) 3. yes; $r = 3$ 4. yes; $r = -1$ 5. yes; $r = \frac{1}{2}$ 6. yes; $r = -\frac{1}{3}$ 7. $\frac{1}{8}, -\frac{1}{4}, \frac{1}{2}, -1, 2$ 8. $-15, -5, -\frac{5}{3}, -\frac{5}{9}, -\frac{5}{27}$ 9. $32, 8, 2, \frac{1}{2}, \frac{1}{8}$ 10. $a_8 = -1$; $S_8 = 85$ 11. $r = 3$; $S_6 = -\frac{364}{9}$ 12. $a_1 = 18$; $a_5 = \frac{2}{9}$ 13. $r = -3$; $a_1 = -\frac{1}{9}$ 14. $a_4 = 128$; $S_4 = 170$ 15. $a_6 = -\frac{3}{625}$; $S_3 = \frac{63}{5}$ 16. approximately \$1500 17. \$1573.52 18. Your answer should be *yes*, since you would be working for approximately \$10,700,000 for the month. 19. 6.6 ft 20. 7, 5, 3, 1, −1 21. $d = 8$; $S_8 = 80$ 22. 2000 people 23. 625 24. $\frac{4}{5}$ 25. $r = \frac{S - a_1}{S}$; $\frac{1}{5}$ 26. 1.33, 1.48, 1.498, 1.4998; the sums appear to be approaching 1.5.

12.3 EXERCISES B

Write true *or* false *for each of the following. If the statement is false, explain why.*

1. In a geometric sequence, a_1 symbolizes the first term, a_n symbolizes the nth term, r symbolizes the common ratio, and S_n symbolizes the sum of the first n terms.

2. The formula for calculating the sum of the first n terms of a geometric sequence is $a_n = a_1 r^{n-1}$.

Decide whether each sequence is geometric. If it is, give the common ratio.

3. 7, 14, 28, 56, . . .

4. 1, 1, 1, 1, . . .

5. 6, 3, 5, 7, . . .

6. $4, -1, \frac{1}{4}, -\frac{1}{16}, \ldots$

Write the first five terms of the geometric sequence with the given a_1 and r.

7. $a_1 = \frac{5}{6}$, $r = -6$

8. $a_1 = 24$, $r = \frac{1}{8}$

9. $a_1 = -10$, $r = -\frac{1}{5}$

With the given information about the geometric sequence, find the indicated unknowns.

10. $a_1 = 3$, $r = 4$; a_3, S_3

11. $a_1 = \frac{1}{8}$, $a_8 = 16$; r, S_8

12. $a_3 = 10$, $a_6 = \frac{1}{100}$; r, a_1

13. $a_4 = 16$, $r = -2$; a_1, a_6

14. $a_1 = -3$, $r = -\frac{1}{3}$; a_5, S_5

15. $a_1 = -\frac{1}{5}$, $r = 5$; a_5, S_4

Solve.

16. A house costing \$70,000 increases at a rate of 30% of its value each year. How much is it worth at the end of 4 years?

17. Harry has a savings account which pays 8% interest compounded annually. If he started with \$2000, how much will he have at the end of 5 years?

18. The population of a town is increasing 10% each year. If its present population is 1500, what will be its population in 5 years?

19. A bacteria culture in a lab doubles in number every four hours. If there were 1000 bacteria present at 12:00 noon on Friday, how many bacteria would there be 20 hours later?

FOR REVIEW

20. Write the first five terms of the arithmetic sequence having $a_1 = -\frac{1}{2}$ and $d = \frac{1}{2}$.

21. Find n and S_n for the arithmetic sequence having $a_1 = 4$, $a_n = -11$, and $d = -3$.

Solve.

22. A child puts 1¢ in her bank on the first day of June, 2¢ on the second day of June, 3¢ on the third day, and so forth. How much money will be in her bank at the end of the month?

23. A collection of dimes is arranged in a triangular array with 10 coins in the base row, 9 in the next, 8 in the next, and so forth. Find the value of the collection.

Exercises 24–26 will help you prepare for the next section.

24. Simplify the complex fraction $\dfrac{\frac{1}{4}}{1 - \frac{1}{3}}$.

25. Solve $S = \frac{a_1}{1 - r}$ for r and use the result to find r when $S = 32$ and $a_1 = 16$.

26. Consider the geometric sequence $2, 1, \frac{1}{2}, \frac{1}{4}, \ldots$. Use your calculator to approximate S_2, S_4, S_6, and S_8. Do you see anything special about these sums?

12.3 EXERCISES C

Solve.

1. The terms between a_1 and a_n of a geometric sequence are called *geometric means* of a_1 and a_n. Insert two geometric means between 5 and -40.

2. A ball is dropped from a height of 18 ft. If on each rebound, it rises to a height $\frac{2}{3}$ the distance from which it fell, how far (up and down) will the ball have traveled when it hits the ground for the sixth time? [Answer: approximately 80.5 ft]

3. A pendulum is released and swings through an arc measuring 15 inches. Each swing thereafter it travels a distance 0.85 times the length of the previous pass. Approximately how far will the pendulum swing on the eighth pass?

4. A radioactive dye is injected into a system in a medical test. After one hour, 60% of the dye remains. At the end of two hours 60% of the remaining dye remains, and so on. If one unit of dye is injected, approximately what percent will remain after 12 hours?

Use the formula for S_n to evaluate each series in Exercises 5–8.

5. $\sum_{k=1}^{6} \left(\frac{2}{3}\right)^k$

6. $\sum_{k=1}^{5} \left(\frac{1}{4}\right)^k$

7. $\sum_{n=1}^{5} \left(-\frac{2}{5}\right)^n$

8. $\sum_{n=1}^{6} \left(-\frac{1}{3}\right)^n$

12.4 INFINITE GEOMETRIC SEQUENCES

STUDENT GUIDEPOSTS

1. Infinite Geometric Sequence
2. Sum of an Infinite Geometric Sequence
3. Converting Repeating Decimals to Fractions

1 INFINITE GEOMETRIC SEQUENCE

We have formulas for the sum of the first n terms of a geometric sequence. In some instances, it is possible to find the sum of all the terms of a geometric sequence even if there are infinitely many. To do this, we need to generalize "finding a sum." Recall the story of the man who lives 1 mile from town and plans to walk from his home to town by walking one-half the remaining distance each day until he arrives. Perhaps this story can help us to see the significance of an infinite geometric series.

Distance walked first day: $a_1 = \frac{1}{2} = \left(\frac{1}{2}\right)^1$

Distance walked second day: $a_2 = \frac{1}{4} = \left(\frac{1}{2}\right)^2$

Distance walked third day: $a_3 = \frac{1}{8} = \left(\frac{1}{2}\right)^3$

$\vdots$

Distance walked nth day: $a_n = \left(\frac{1}{2}\right)^n$

The distances walked each day form the following geometric sequence.

$$\frac{1}{2}, \frac{1}{4}, \frac{1}{8}, \ldots, \left(\frac{1}{2}\right)^n, \ldots \qquad \left(r = \frac{1}{2}, a_1 = \frac{1}{2}\right)$$

Suppose that we calculate the sum of the first n terms of this sequence, for $n =$ 1, 2, 3, 4, 5, 6, 7.

$$S_1 = \frac{1}{2}$$

$$S_2 = \frac{1}{2} + \frac{1}{4} = \frac{3}{4}$$

$$S_3 = \frac{1}{2} + \frac{1}{4} + \frac{1}{8} = \frac{7}{8}$$

$$S_4 = \frac{1}{2} + \frac{1}{4} + \frac{1}{8} + \frac{1}{16} = \frac{15}{16}$$

$$S_5 = \frac{1}{2} + \frac{1}{4} + \frac{1}{8} + \frac{1}{16} + \frac{1}{32} = \frac{31}{32}$$

$$S_6 = \frac{1}{2} + \frac{1}{4} + \frac{1}{8} + \frac{1}{16} + \frac{1}{32} + \frac{1}{64} = \frac{63}{64}$$

$$S_7 = \frac{1}{2} + \frac{1}{4} + \frac{1}{8} + \frac{1}{16} + \frac{1}{32} + \frac{1}{64} + \frac{1}{128} = \frac{127}{128}$$

When will S_n equal 1? That is, when will the man reach town? He will never actually reach town since on any given day, he walks only half the remaining distance. However, if we assume that the man continues to walk in the prescribed manner, the total distance that he has walked gets closer and closer to 1 (the sequence of sums $\frac{1}{2}, \frac{3}{4}, \frac{7}{8}, \frac{15}{16}, \frac{31}{32}, \frac{63}{64}, \frac{127}{128}, \ldots$ gets closer and closer to 1). In a sense, then, we could say that the total distance walked is 1 mile (or that the sum of the geometric sequence $\frac{1}{2} + \frac{1}{4} + \frac{1}{8} + \frac{1}{16} + \ldots$ is 1). It is in this sense that we use the term *sum* when applying it to an infinite geometric sequence. That is, if an infinite geometric sequence

$$a_1, a_2, a_3, a_4, \ldots$$

has

$$S_1, S_2, S_3, S_4, \ldots$$

as its **sequence of partial sums** (sequence of sums of the first n terms), and if the S_n's approach some fixed number S as n gets larger and larger, we call S the **sum of the geometric sequence,** and

$$S = a_1 + a_2 + a_3 + a_4 + \ldots.$$

2 SUM OF AN INFINITE GEOMETRIC SEQUENCE

Some infinite geometric sequences have sums and others do not. For example,

$$1 + 3 + 9 + 27 + 81 + \ldots \qquad (r = 3, a_1 = 1)$$

does not have a sum since the S_n's

$$S_1 = 1, \quad S_2 = 4, \quad S_3 = 13, \quad S_4 = 40, \quad S_5 = 121, \ldots$$

get larger and do not approach some fixed number S. On the other hand,

$$1 + \frac{1}{3} + \frac{1}{9} + \frac{1}{27} + \frac{1}{81} + \ldots \qquad \left(r = \frac{1}{3}, a_1 = 1\right)$$

does have a sum, since the S_n's

$$S_1 = 1, \quad S_2 = 1\frac{1}{3}, \quad S_3 = 1\frac{4}{9}, \quad S_4 = 1\frac{13}{27}, \quad S_5 = 1\frac{40}{81}, \ldots$$

are approaching $1\frac{1}{2} = \frac{3}{2}$. In the first example, $r = 3$, while in the second, $r = \frac{1}{3}$.

When the common ratio, r, in an infinite geometric sequence is such that $|r| < 1$, then the sequence has a sum in the sense described above. Considering the formula for the sum of the first n terms in a geometric sequence,

$$S_n = \frac{a_1 - a_1 r^n}{1 - r},$$

if $|r| < 1$, then r^n gets smaller as n gets larger. Thus, $a_1 r^n$ gets smaller and S_n approaches

$$\frac{a_1}{1 - r}.$$

Sum of an Infinite Geometric Sequence

An infinite geometric sequence with first term a_1 and common ratio r satisfying $|r| < 1$ has a sum given by

$$S = \frac{a_1}{1 - r}.$$

EXAMPLE 1 The Sum of an Infinite Geometric Sequence

Find the sum of each infinite geometric sequence.

(a) $\frac{1}{2}, \frac{1}{4}, \frac{1}{8}, \frac{1}{16}, \frac{1}{32}, \ldots$

Since $r = \frac{1}{2}$ and $|r| = \left|\frac{1}{2}\right| = \frac{1}{2} < 1$, the sum exists and is given by

$$S = \frac{a_1}{1 - r} = \frac{\frac{1}{2}}{1 - \frac{1}{2}} = \frac{\frac{1}{2}}{\frac{1}{2}} = 1.$$

(Compare with the discussion earlier.)

(b) $\frac{1}{64}, \frac{1}{32}, \frac{1}{16}, \frac{1}{8}, \ldots$

Since $r = 2$ and $|r| = |2| = 2 > 1$, the sum does not exist. This becomes clear when the next few terms $\left(\frac{1}{4}, \frac{1}{2}, 1, 2, 4, 8, 16\right)$ are computed.

Practice Exercise 1

Find the sum of each infinite geometric sequence.

(a) $1, \frac{1}{3}, \frac{1}{9}, \frac{1}{27}, \ldots$

(b) $\frac{1}{27}, -\frac{1}{9}, \frac{1}{3}, -1, \ldots$

Answers: (a) $\frac{3}{2}$ (b) The sum does not exist since $r = -3$, and $|-3| = 3 > 1$.

EXAMPLE 2 Finding a Sequence Given Its Sum

Is there an infinite geometric sequence with $a_1 = 2$ and $S = -4$?

Since $S = \frac{a_1}{1 - r}$, we have

$$-4 = \frac{2}{1 - r}$$

$$-4(1 - r) = 2 \quad \text{Multiply both sides by } 1 - r$$

$$-4 + 4r = 2 \quad \text{Distribute}$$

$$4r = 6$$

$$r = \frac{6}{4} = \frac{3}{2}.$$

But since $|r| = \left|\frac{3}{2}\right| = \frac{3}{2} > 1$, there cannot be an infinite geometric sequence with first term 2 and sum -4.

Practice Exercise 2

Is there an infinite geometric sequence with $a_1 = -4$ and $S = \frac{-16}{3}$?

Answer: Yes, $r = \frac{1}{4}$ and $|r| = \left|\frac{1}{4}\right| = \frac{1}{4} < 1$.

The following is an application of infinite geometric sequences.

EXAMPLE 3 DISTANCE PROBLEM

The tip of a pendulum moves back and forth in such a way that it sweeps out an arc 18 in long, and on each succeeding pass the length of the arc traveled is $\frac{8}{9}$ of the length of the preceding pass. What is the total distance traveled by the tip of the pendulum?

In reality, the pendulum will eventually stop because of friction. However, for simplicity we use an infinite geometric series to represent the total distance traveled.

$$18 + \frac{8}{9}(18) + \left(\frac{8}{9}\right)^2(18) + \left(\frac{8}{9}\right)^3(18) + \ldots$$

Then $a_1 = 18$, $r = \frac{8}{9}$ $\left(|r| = \left|\frac{8}{9}\right| = \frac{8}{9} < 1\right)$, so that

$$S = \frac{a_1}{1-r} = \frac{18}{1-\frac{8}{9}} = \frac{18}{\frac{1}{9}} = 18 \cdot 9 = 162.$$

The total distance traveled is approximately 162 in.

PRACTICE EXERCISE 3

A child on a swing sweeps out an arc 21 ft long, and on each succeeding pass the length of the arc traveled is $\frac{6}{7}$ of the length of the preceding pass. What is the total distance traveled by the child?

Answer: 147 ft

3 CONVERTING REPEATING DECIMALS TO FRACTIONS

The formula for the sum of an infinite geometric sequence is also used to convert repeating decimals to fractions.

EXAMPLE 4 CONVERTING A REPEATING DECIMAL TO A FRACTION

Convert $0.\overline{3}$ to a fraction.

Since $0.\overline{3} = 0.33333\ldots$

$$= 0.3 + 0.03 + 0.003 + 0.0003 + \ldots$$
$$= 0.3 + (0.1)(0.3) + (0.1)^2(0.3) + (0.1)^3(0.3) + \ldots,$$

we see that $0.\overline{3}$ is really an infinite geometric series with $a_1 = 0.3$ and $r = 0.1$. Then,

$$S = \frac{a_1}{1-r} = \frac{0.3}{1-0.1} = \frac{0.3}{0.9} = \frac{3}{9} = \frac{1}{3}.$$

Thus, $0.\overline{3} = \frac{1}{3}$ (which can be checked by division).

PRACTICE EXERCISE 4

Convert $0.\overline{27}$ to a fraction. Expand the decimal to see that $a_1 = 0.27$ and $r = 0.01$.

Answer: $\frac{3}{11}$ (Check this by dividing.)

12.4 EXERCISES A

Find the sum of each infinite geometric sequence.

1. $1, \frac{3}{4}, \frac{9}{16}, \frac{27}{64}, \ldots$

2. $\frac{1}{50}, \frac{1}{10}, \frac{1}{2}, \frac{5}{2}, \ldots$

3. $9, 1, \frac{1}{9}, \frac{1}{81}, \ldots$

4. $16, -4, 1, -\frac{1}{4}, \ldots$

5. $25, 15, 9, \frac{27}{5}, \ldots$

6. $-36, 6, -1, \frac{1}{6}, -\frac{1}{36}, \ldots$

7 $\sum_{k=1}^{\infty} \left(\frac{1}{4}\right)^k$

8. $\sum_{k=1}^{\infty} 5^{-k}$

9. $\sum_{k=1}^{\infty} \left(\frac{9}{5}\right)^k$

With the given information about the infinite geometric sequence, find the indicated unknowns.

10. $a_1 = 3, r = \frac{1}{2}; S, a_5$

11. $a_1 = 15, S = 30; r, a_4$

12. $a_1 = 16, S = 4; r, a_3$

13 $a_2 = 10, a_5 = \frac{2}{25}; a_1, S$

Convert each decimal to a fraction.

14. $0.\overline{4}$

15. $0.\overline{72}$

16. $0.\overline{312}$

17. $1.\overline{3}$

18. $2.\overline{15}$

19. $3.\overline{121}$

Solve.

20 The water which runs a waterwheel is decreased in such a way that the wheel makes 16 revolutions during the first minute, 4 revolutions during the second minute, 1 the third minute, $\frac{1}{4}$ the fourth, and so forth. How many revolutions will it take before it stops?

21. The tip of a pendulum moves back and forth in such a way that it sweeps out an arc 24 in long, and on each succeeding pass the length of the arc traveled is $\frac{7}{8}$ the length of the preceding pass. What is the total distance traveled by the tip of the pendulum?

FOR REVIEW

Solve.

22. Find a_7 and S_7 for the geometric sequence having $a_1 = \frac{1}{16}$ and $r = -4$.

23. Nancy borrows \$2000 at 11% interest compounded annually. If she pays off the loan in full at the end of 5 years, how much does she pay?

Exercises 24–29 review material from Chapter 5 to help you prepare for the next section. Multiply and simplify.

24. $(x + 1)^2$

25. $(x + 1)^3$

26. $(x + 1)^4$

27. $(x + 1)^5$

28. $(x + 1)^6$

29. $(x + 1)^7$

ANSWERS: 1. 4 2. no sum ($r = 5$) 3. $\frac{81}{8}$ 4. $\frac{64}{5}$ 5. $\frac{125}{2}$ 6. $-\frac{216}{7}$ 7. $\frac{1}{3}$ 8. $\frac{1}{4}$ 9. no sum $\left(r = \frac{9}{5}\right)$ 10. $S = 6$; $a_5 = \frac{3}{16}$ 11. $r = \frac{1}{2}$; $a_4 = \frac{15}{8}$ 12. No such infinite geometric sequence exists, since $r = -3$. 13. $a_1 = 50$; $S = \frac{125}{2}$ 14. $\frac{4}{9}$ 15. $\frac{8}{11}$ 16. $\frac{104}{333}$ 17. $\frac{4}{3}$ 18. $\frac{71}{33}$ 19. $\frac{3118}{999}$ 20. $21\frac{1}{3}$ revolutions 21. 192 inches 22. $a_7 = 256$; $S_7 = \frac{3277}{16}$ 23. \$3370.12 24. $x^2 + 2x + 1$ 25. $x^3 + 3x^2 + 3x + 1$ 26. $x^4 + 4x^3 + 6x^2 + 4x + 1$ 27. $x^5 + 5x^4 + 10x^3 + 10x^2 + 5x + 1$ 28. $x^6 + 6x^5 + 15x^4 + 20x^3 + 15x^2 + 6x + 1$ 29. $x^7 + 7x^6 + 21x^5 + 35x^4 + 35x^3 + 21x^2 + 7x + 1$

12.4 EXERCISES B

Find the sum of each infinite geometric sequence.

1. $24, 3, \frac{3}{8}, \frac{3}{64}, \ldots$

2. $10, -5, \frac{5}{2}, -\frac{5}{4}, \ldots$

3. $\frac{1}{16}, \frac{1}{4}, 1, 4, \ldots$

4. $-120, -20, -\frac{10}{3}, -\frac{5}{9}, \ldots$

5. $72, 60, 50, \frac{125}{3}, \ldots$

6. $-12, 9, -\frac{27}{4}, \frac{81}{16}, \ldots$

7. $\sum_{k=1}^{\infty} \left(\frac{1}{3}\right)^k$

8. $\sum_{k=1}^{\infty} 6^{-k}$

9. $\sum_{k=1}^{\infty} \left(-\frac{4}{5}\right)^k$

With the given information about each infinite geometric sequence, find the indicated unknowns.

10. $a_1 = 12$, $r = -\frac{1}{2}$; S, a_4

11. $a_1 = -8$, $S = 24$; r, a_5

12. $a_1 = -90$, $S = -70$; r, a_3

13. $a_3 = 12$, $a_6 = -\frac{32}{9}$; a_1, S

Convert each decimal to a fraction.

14. $0.\overline{9}$ **15.** $0.\overline{36}$ **16.** $0.\overline{561}$

17. $5.\overline{2}$ **18.** $4.\overline{81}$ **19.** $1.\overline{321}$

Solve.

20. A child on a swing traverses an arc of 21 ft. Each pass thereafter, the arc is $\frac{6}{7}$ the length of the previous arc. How far does he travel before coming to rest?

21. A rocket is fired upward in such a way that it travels 300 m the first second and each subsequent second it travels $\frac{2}{5}$ as far as it did the preceding second. How far will it go before it stops rising?

FOR REVIEW

Solve.

22. Find a_1 and a_6 for the geometric sequence having $a_3 = 24$ and $r = -\frac{1}{2}$.

23. The population of a town is decreasing 20% each year. If its present population is 250,000, what will be its population in 10 years?

Exercises 24–29 review material from Chapter 5 to help you prepare for the next section. Multiply and simplify.

24. $(y - 1)^2$ **25.** $(y - 1)^3$ **26.** $(y - 1)^4$

27. $(y - 1)^5$ **28.** $(y - 1)^6$ **29.** $(y - 1)^7$

12.4 EXERCISES C

Solve.

1. A ball dropped from a height of 15 ft always rebounds $\frac{3}{5}$ of the height of the previous fall. How far does it travel (up and down) before coming to rest? [Answer: 60 ft]

2. Charlotte received $5000 on the day of her birth, and $\frac{4}{5}$ as much on each birthday as on the previous one. Approximately how much will she receive in her lifetime (assuming she lives to a ripe old age)?

3. A square has area 64 in^2 (each side is 8 inches). A second square is constructed by connecting in order the midpoints of the sides of the first square, a third by connecting in order the midpoints of the sides of the second square, and so forth. Calculate the sum of the areas of all these squares.

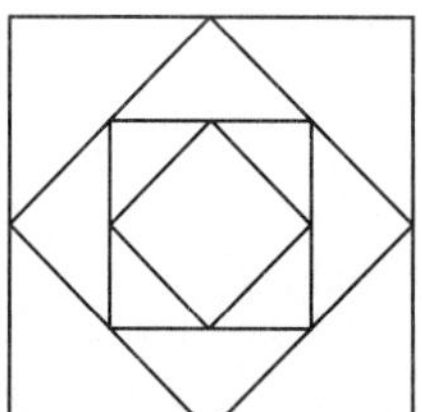

12.5 BINOMIAL EXPANSION

STUDENT GUIDEPOSTS

1 Binomial Expansions — 2 Pascal's Triangle

1 BINOMIAL EXPANSIONS

When a binomial of the form $a + b$ is raised to a power, the resulting polynomial can be thought of as a series. Suppose we expand several such powers and search for a pattern.

$$(a+b)^0 = 1$$
$$(a+b)^1 = a+b$$
$$(a+b)^2 = a^2+2ab+b^2$$
$$(a+b)^3 = a^3+3a^2b+3ab^2+b^3$$
$$(a+b)^4 = a^4+4a^3b+6a^2b^2+4ab^3+b^4$$
$$(a+b)^5 = a^5+5a^4b+10a^3b^2+10a^2b^3+5ab^4+b^5$$
$$(a+b)^6 = a^6+6a^5b+15a^4b^2+20a^3b^3+15a^2b^4+6ab^5+b^6$$

We might observe the following:

1. There are always $n + 1$ terms in the expansion.
2. The exponents on a start with n and decrease to 0.
3. The exponents on b start with 0 and increase to n.
4. The sum of the exponents in each term is always n.
5. If a and b are both positive, all terms are positive.
6. If a is positive and b is negative, the terms have alternating signs; those with odd powers of b are negative.
7. If a is negative and b is positive, the terms have alternating signs; those with odd powers of a are negative.
8. If a and b are both negative, all terms are positive if n is even and negative if n is odd.

2 PASCAL'S TRIANGLE

To discover the pattern of the numerical coefficients of each term, we write the coefficients in the same arrangement as in the preceding expansions.

Row 0	1
Row 1	1 1
Row 2	1 2 1
Row 3	1 3 3 1
Row 4	1 4 6 4 1
Row 5	1 5 10 10 5 1
Row 6	1 6 15 20 15 6 1

This triangular array forms what is known as **Pascal's triangle,** named for Blaise Pascal, a famous seventeenth-century mathematician, who was the first to use it. The row number corresponds to the exponent n in the expansion of $(a + b)^n$. The numbers in any row, other than the first and the last which are always 1, can be determined by adding the two numbers immediately above and to the left and right of it. For example, as indicated, 15 is $5 + 10$. Thus, Pascal's triangle gives us one way to determine the coefficients in the expansion of a given binomial.

EXAMPLE 1 Using Pascal's Triangle

Expand $(2x + y)^4$.

In this example $n = 4$, $a = 2x$, and $b = y$.

Row 4 of Pascal's triangle has the following numbers for the coefficients.

1 4 6 4 1

We thus substitute $a = 2x$ and $b = y$ into

$$a^4 + 4a^3b + 6a^2b^2 + 4ab^3 + b^4$$

to obtain

$$\begin{aligned}&(2x)^4 + 4(2x)^3y + 6(2x)^2y^2 + 4(2x)y^3 + y^4\\&= 16x^4 + 4(8x^3)y + 6(4x^2)y^2 + 4(2x)y^3 + y^4\\&= 16x^4 + 32x^3y + 24x^2y^2 + 8xy^3 + y^4.\end{aligned}$$

Practice Exercise 1

Expand $(z - 3)^5$.

We have $n = 5$, $a = z$, and $b = -3$ [since we are expanding $(z + (-3))^5$]. Note that when b is negative the signs alternate. Use Row 5 of Pascal's triangle to obtain the desired expansion.

Answer: $z^5 - 15z^4 + 90z^3 - 270z^2 + 405z - 243$

EXAMPLE 2 Expanding a Binomial

Expand $(a^2 - 2b)^6$.

Since we are expanding $(a^2 + (-2b))^6$, $n = 6$, $a = a^2$, and $b = -2b$. From Row 6 of Pascal's triangle, the coefficients are

1 6 15 20 15 6 1.

If we substitute a^2 for a and $-2b$ for b in

$$a^6 + 6a^5b + 15a^4b^2 + 20a^3b^3 + 15a^2b^4 + 6ab^5 + b^6$$

we obtain

$$\begin{aligned}&(a^2)^6 + 6(a^2)^5(-2b) + 15(a^2)^4(-2b)^2 + 20(a^2)^3(-2b)^3\\&+ 15(a^2)^2(-2b)^4 + 6(a^2)(-2b)^5 + (-2b)^6\\&= a^{12} - 12a^{10}b + 60a^8b^2 - 160a^6b^3 + 240a^4b^4 - 192a^2b^5 + 64b^6.\end{aligned}$$

Practice Exercise 2

Expand $(u^3 + 3v)^4$.

Answer: $u^{12} + 12u^9v + 54u^6v^2 + 108u^3v^3 + 81v^4$

The student who continues in mathematics will encounter the *binomial theorem,* which provides a more sophisticated technique for expanding binomials when n is very large and for determining only one particular term in an expansion.

12.5 EXERCISES A

1. Write out Row 8 of Pascal's triangle.

Expand each of the following binomials by using Pascal's triangle.

2. $(x + 3)^5$

3. $(x - 3)^5$

4. $(x - y)^6$

5. $(x - 2y)^4$

6. $(x^2 + y)^6$

7. $(x^2 - 3y)^4$

8. $(x + y)^8$

9. $(a - b)^8$

10 $(x^{-1} + x)^5$ Write without negative exponents.

FOR REVIEW

Find the sum of the given infinite geometric sequence.

11. $28, 21, \frac{63}{4}, \frac{189}{16}, \ldots$

12. $\sum_{k=1}^{\infty} \left(\frac{3}{5}\right)^k$

13. Convert $3.\overline{7}$ to a fraction.

14. A swing traverses an arc of 15 ft. Each pass thereafter the arc is $\frac{4}{9}$ the length of the previous pass. How far does it travel before coming to rest?

ANSWERS: 1. Row 8: 1 8 28 56 70 56 28 8 1 2. $x^5 + 15x^4 + 90x^3 + 270x^2 + 405x + 243$ 3. $x^5 - 15x^4 + 90x^3 - 270x^2 + 405x - 243$ 4. $x^6 - 6x^5y + 15x^4y^2 - 20x^3y^3 + 15x^2y^4 - 6xy^5 + y^6$ 5. $x^4 - 8x^3y + 24x^2y^2 - 32xy^3 + 16y^4$ 6. $x^{12} + 6x^{10}y + 15x^8y^2 + 20x^6y^3 + 15x^4y^4 + 6x^2y^5 + y^6$ 7. $x^8 - 12x^6y + 54x^4y^2 - 108x^2y^3 + 81y^4$ 8. $x^8 + 8x^7y + 28x^6y^2 + 56x^5y^3 + 70x^4y^4 + 56x^3y^5 + 28x^2y^6 + 8xy^7 + y^8$ 9. $a^8 - 8a^7b + 28a^6b^2 - 56a^5b^3 + 70a^4b^4 - 56a^3b^5 + 28a^2b^6 - 8ab^7 + b^8$ 10. $\frac{1}{x^5} + \frac{5}{x^3} + \frac{10}{x} + 10x + 5x^3 + x^5$ 11. 112 12. $\frac{3}{2}$ 13. $\frac{34}{9}$ 14. 27 ft

12.5 EXERCISES B

1. Write out row 9 of Pascal's triangle.

Expand each of the following binomials using Pascal's triangle.

2. $(3a - 1)^5$

3. $(2x + y)^4$

4. $(z - 3)^5$

5. $(a^2 + 2b)^6$

6. $(x^2 - 2)^4$

7. $(x^{-1} + x)^3$

8. $(x^3 - y^3)^4$

9. $(2x - 3y)^4$

10. $(1 + 1)^7$

FOR REVIEW

Find the sum of the given infinite geometric sequence.

11. $6, -4, \frac{8}{3}, -\frac{16}{9}, \ldots$

12. $\sum_{k=1}^{\infty} \left(-\frac{7}{9}\right)^k$

13. Convert $2.\overline{36}$ to a fraction.

14. A runner who started to tire ran 240 m in one minute, 180 m the second minute, 135 m the next minute, and so forth until he stopped. How far did he run before stopping?

12.5 EXERCISES C

Expand using Pascal's triangle.

1. $(3 - 2)^5$

2. $(1 + \sqrt{2})^3$

3. $(5 - 3)^6$

4. $(x^{-2} + y^{-2})^6$ Write the answer without negative exponents.

CHAPTER 12 REVIEW

KEY WORDS

12.1 An **infinite sequence** is a function with domain the set of natural numbers.

Each number in the range of a sequence is called a **term** of the sequence with a_n the ***n*th term** or **general term.**

A **finite sequence** with m terms is a function with domain $\{1, 2, 3, \ldots, m\}$.

A **series** is the indicated sum of the terms of a sequence.

12.2 An **arithmetic sequence** is a sequence in which each term after the first is the sum of the preceding term and a fixed number called the **common difference** of the sequence.

12.3 A **geometric sequence** is a sequence in which each term after the first is the product of the preceding term and a fixed number called the **common ratio** of the sequence.

12.4 The sequence of sums of the first n terms of a geometric sequence is called the **sequence of partial sums.**

If the sequence of partial sums of an infinite geometric sequence approaches some number, that number is called the **sum of the infinite geometric sequence.**

12.5 **Pascal's triangle** is a triangular array of numbers corresponding to the coefficients in a binomial expansion.

KEY CONCEPTS

12.1 The Greek letter Σ, the summation symbol, is used to abbreviate a series. For example,

$$a_1 + a_2 + a_3 = \sum_{n=1}^{3} a_n,$$

where n is the index on the summation, and 1 and 3 are the lower and upper limits of summation.

12.2 Arithmetic sequence: $a_n = a_1 + (n - 1)d$

$$S_n = \frac{n}{2}[2a_1 + (n-1)d] = \frac{n}{2}[a_1 + a_n]$$

Do not confuse n, the *number* of terms, with a_n, the nth term.

12.3 Geometric sequence: $a_n = a_1 r^{n-1}$

$$S_n = \frac{a_1 - ra_n}{1 - r} = \frac{a_1 - a_1 r^n}{1 - r}$$

12.4 The sum of an infinite geometric sequence is given by

$$S = \frac{a_1}{1 - r} \text{ if } |r| < 1.$$

12.5 The coefficients in the expansion of a binomial of the form $a + b$ raised to a power n can be found in the nth row of Pascal's triangle.

REVIEW EXERCISES

Part I

12.1 **1.** In the sigma summation notation $\sum_{n=1}^{m} a_n$, what is n called?

2. Write $b_4 + b_5 + b_6 + b_7$ using sigma summation notation.

3. Write $x_1 + x_2 + x_3 + \ldots$ using sigma summation notation.

In each exercise the nth term of a general sequence is given. Find the first 4 terms and the 7th term of each.

4. $x_n = n^2 - n$

5. $a_n = \dfrac{(-1)^n}{3n}$

6. Rewrite without sigma summation notation.

$$\sum_{k=0}^{3} \sqrt{k + 1}$$

7. Rewrite using the sigma summation notation.

$$\frac{1}{4} + \frac{4}{5} + \frac{9}{6} + \ldots + \frac{n^2}{n + 3} + \ldots$$

12.2 **8.** Write the first 5 terms of an arithmetic sequence with $a_1 = -11$ and $d = 4$, and find S_{10}.

9. A new car costs \$9500 and depreciates 25% the first year, 21% the second year, 17% the third, and so on for 5 years. If all depreciations apply to the original cost, what is the value of the car in 5 years?

12.3 **10.** Write the first 5 terms of a geometric sequence with $a_1 = \frac{1}{5}$ and $r = -5$, and find S_5.

11. A man borrows \$3000 at 8% interest compounded annually. If he pays off the loan in full at the end of 6 years, how much does he pay?

12.4 **12.** Find the sum of the infinite geometric sequence $18, -6, 2, -\frac{2}{3}, \ldots$.

13. A child on a swing traverses an arc of 20 ft. Each pass thereafter, he traverses an arc that is $\frac{16}{17}$ the length of the previous arc. How far does he travel before coming to rest?

Convert each decimal to a fraction and check by division.

14. $0.\overline{7}$

15. $6.\overline{12}$

12.5 *Expand each of the following binomials using Pascal's triangle.*

16. $(y + 2)^6$

17. $(a - a^{-1})^4$ Write without negative exponents.

Part II

18. A collection of nickels is arranged in a triangular array with 20 coins in the base row, 19 in the next, 18 in the next, and so forth. Find the value of the collection.

19. Find the sum of the first 8 terms of the geometric sequence with $a_1 = 81$ and $a_8 = \frac{1}{27}$.

20. Is there an infinite geometric sequence with $a_1 = -3$ and $S = 10$?

21. Find x_1, x_2, and x_5 if $x_n = \frac{(-1)^{n+1}}{3^n}$.

22. Find the 20th term and the sum of the first 20 terms of the sequence $-7, -4, -1, 2, \ldots$.

23. Mary plans to start an exercise program by walking 15 minutes each day for a week. Each week thereafter, she plans to increase the time walking each day by 5 minutes. How many weeks will it take for Mary to be walking an hour daily?

Use Pascal's triangle to expand each binomial.

24. $(3a - z)^5$:

25. $(y^{-1} + y)^4$

Find the formula for a_n given the first few terms of each general sequence.

26. $1, -4, 9, -16, 25, \ldots$

27. $0, \frac{1}{4}, \frac{2}{5}, \frac{3}{6}, \frac{4}{7}, \ldots$

Tell whether the following statements are true *or* false. *If the statement is* false, *explain why.*

28. The indicated sum of a sequence is called its associated series.

29. Each member of a sequence is a term of the sequence.

30. Sequences that do not have a last term but continue on indefinitely are called finite sequences.

31. The general term of a sequence is often called the nth term of the sequence.

32. In sigma notation $\sum_{n=1}^{m} a_n$, m is the lower limit of summation.

ANSWERS: **1.** index **2.** $\sum_{k=4}^{7} b_k$ **3.** $\sum_{k=1}^{\infty} x_k$ **4.** $x_1 = 0$; $x_2 = 2$; $x_3 = 6$; $x_4 = 12$; $x_7 = 42$ **5.** $a_1 = -\frac{1}{3}$; $a_2 = \frac{1}{6}$; $a_3 = -\frac{1}{9}$; $a_4 = \frac{1}{12}$; $a_7 = -\frac{1}{21}$ **6.** $1 + \sqrt{2} + \sqrt{3} + 2$ **7.** $\sum_{n=1}^{\infty} \frac{n^2}{n+3}$ **8.** $-11, -7, -3, 1, 5$; $S_{10} = 70$ **9.** \$1425 **10.** $\frac{1}{5}, -1, 5, -25, 125$; $S_5 = \frac{521}{5}$ **11.** \$4760.62 **12.** $\frac{27}{2}$ **13.** 340 ft **14.** $\frac{7}{9}$ **15.** $\frac{202}{33}$ **16.** $y^6 + 12y^5 + 60y^4 + 160y^3 + 240y^2 + 192y + 64$ **17.** $a^4 - 4a^2 + 6 - \frac{4}{a^2} + \frac{1}{a^4}$ **18.** \$10.50 **19.** $\frac{3280}{27}$ **20.** No such infinite sequence exists since r is $\frac{13}{10}$. **21.** $x_1 = \frac{1}{3}$, $x_2 = -\frac{1}{9}$, $x_5 = \frac{1}{243}$ **22.** $a_{20} = 50$; $S_{20} = 430$ **23.** 10 weeks **24.** $243a^5 - 405a^4z + 270a^3z^2 - 90a^2z^3 + 15az^4 - z^5$ **25.** $y^{-4} + 4y^{-2} + 6 + 4y^2 + y^4$ **26.** $a_n = (-1)^{n+1}n^2$ **27.** $a_n = \frac{n-1}{n+2}$ **28.** true **29.** true **30.** false; they are called infinite **31.** true **32.** false; m is the upper limit

CHAPTER 12 TEST

1. True or false: Each term after the first in an arithmetic sequence can be obtained from the preceding term by adding a fixed number called the common difference. 1. ______

2. Write $a_3 + a_4 + a_5 + a_6$ using sigma summation notation. 2. ______

3. The formula for finding the sum of an infinite geometric sequence only applies when the common ratio satisfies what inequality? 3. ______

4. Given $x_n = 5n - 3$, find the first two terms ($n = 1, 2$) and the 5th term. 4. ______

5. Find the first 3 terms ($n = 1, 2, 3$) and S_6 of an arithmetic sequence with $a_1 = 8$ and $d = -5$. 5. ______

6. A collection of dimes is arranged in a triangular array with 50 coins in the first row, 48 coins in the second row, 46 in the third row, and so forth (2 are in the last row). What is the value of the collection? 6. ______

7. Find the first 3 terms ($n = 1, 2, 3$) and S_5 of a geometric sequence with $a_1 = 3$ and $r = -2$. 7. ______

8. A truck costing \$12,000 depreciates 30% of its value each year. How much is the truck worth at the end of 5 years? 8. ______

9. The tip of a pendulum swings back and forth in such a way that it sweeps out an arc 24 inches long on the first pass. On each succeeding pass, the length of the arc traveled is $\frac{9}{10}$ of the length of the previous pass. What is the total distance traveled by the tip of the pendulum before it comes to rest? 9. ______

10. Convert $0.\overline{4}$ to a fraction. 10. ______

11. Expand $(a - 4x)^5$ using Pascal's triangle. 11. ______

12. Jeff is offered a job that pays a salary of \$23,500 the first year. If he receives a raise of \$1200 each year thereafter, how much will he be earning annually during his 15th year with the company? 12. ______

FINAL REVIEW EXERCISES

These exercises, organized by chapter, will help you review the course material. Since workspace is not provided, you will need to do the problems on separate sheets of paper.

CHAPTER 1

1. Write $\{x|x \text{ is a whole number less than } 2\}$ using the listing method.

2. Write $\{1, 3, 5\}$ using set-builder notation.

3. True or false: Every integer is a real number.

4. Simplify $-|-(-10)|$

Perform the indicated operations.

5. $\left(-\frac{3}{8}\right) + \frac{1}{4}$

6. $\left(-\frac{4}{5}\right) - \left(-\frac{3}{10}\right)$

7. $\left(-\frac{4}{3}\right)\left(\frac{3}{16}\right)$

8. $\left(-\frac{5}{8}\right) \div \left(-\frac{15}{2}\right)$

Name the property that justifies each statement.

9. $3(a + 4) = 3a + 3 \cdot 4$

10. $-(-8.2) = 8.2$

11. $7 + a = a + 7$

12. If $a = -5$ then $-5 = a$.

13. Factor. $-2a - 8x + 6$

14. Multiply. $-3(1 - x)$

15. Collect like terms.
$3a - 2z + a - z - 4a$

16. Remove parentheses and simplify.
$4[5y - (3y - 1)]$

Simplify and write without negative exponents.

17. $\frac{a^3b^{-7}}{a^{-2}b^5}$

18. $3^0x^{-8}x^{-2}x^7$

19. $(2y^{-2}z^4)^{-3}$

20. $\left(\frac{2xy^2}{y^{-3}}\right)\left(\frac{3x^{-2}y}{xy^2}\right)^{-1}$

Evaluate when $a = -2$.

21. $4a^2$

22. $(4a)^2$

23. $-4a^2$

24. $-4a^{-2}$

25. Write 42,100,000 in scientific notation.

26. Write 8.41×10^{-7} without using scientific notation.

CHAPTER 2

Solve.

27. $3 - 3(x - 1) = 4x - 1$

28. $z(z + 3) = z^2 - 9$

29. A wire 36 cm long is to be cut into two pieces in such a way that one piece is 8 cm longer than the other. How long is each piece?

30. Two cars leave the same city at 9:00 A.M., one heading north and the other heading south. If one is traveling 50 mph and the other 35 mph, at what time will they be 340 mi apart?

31. After a 6% raise, Paul's new salary is \$9752. What was his former salary?

32. The sum of two consecutive odd integers is 96. Find the two integers.

33. Find the measure of each angle of a triangle if the second is four times the first and the third is 5° more than twice the first.

34. After receiving a 20% discount on the selling price, Jeff Altman paid \$33.60, including 5% sales tax, for a pair of shoes. What was the price of the shoes before the discount?

Solve and graph. Give answer in interval notation when appropriate.

35. $3z - 5 \geq 4(z - 1)$

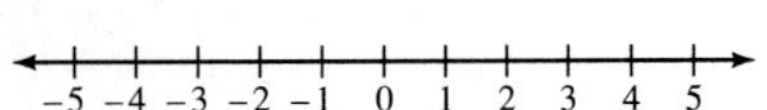

36. $y - 2 < -4$ or $y - 2 \geq 4$

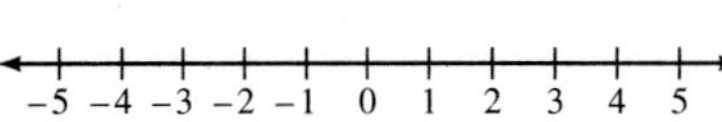

37. $-5 \leq 2x - 1 < 1$

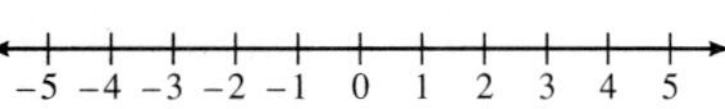

38. $|4 - 3x| = 7$

−5 −4 −3 −2 −1 0 1 2 3 4 5

39. $|2y + 1| > 5$

−5 −4 −3 −2 −1 0 1 2 3 4 5

40. $|2z - 3| \leq 7$

−5 −4 −3 −2 −1 0 1 2 3 4 5

41. Solve $P = q + 50r$ for r.

42. Solve $|x - 1| = |2x - 3|$.

CHAPTER 3

43. Give the x-intercept, y-intercept, slope, and graph of the line with equation $2x - 5y + 10 = 0$.

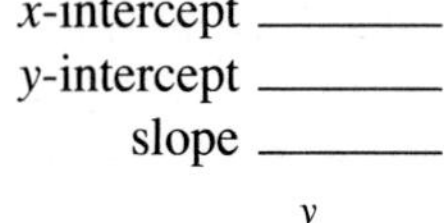

x-intercept ________
y-intercept ________
slope ________

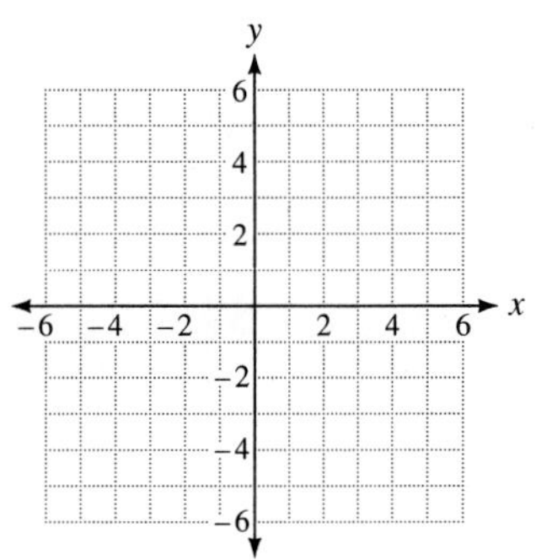

44. Graph using the slope method.

$$y = \frac{2}{3}x + 1$$

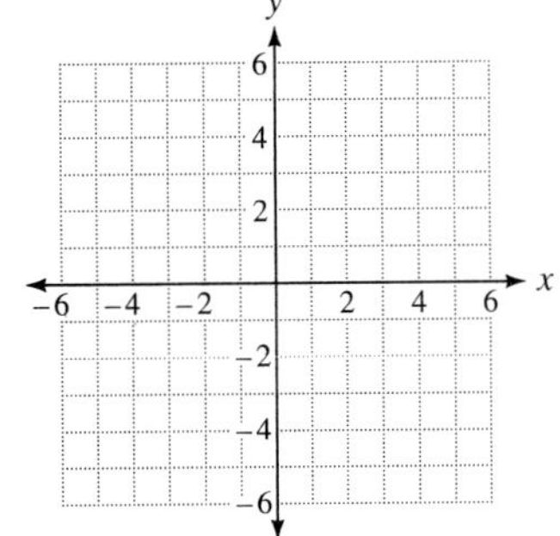

45. Find the general form of the equation of the line passing through the points $(4, -2)$ and $(-1, 3)$.

46. Write the equation, in general form, of the line containing the point $(2, 5)$ and perpendicular to $4x + 3y - 8 = 0$.

47. Is the following a function?

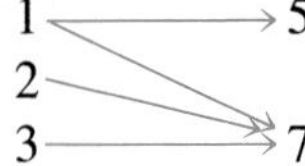

48. Is the following the graph of a function?

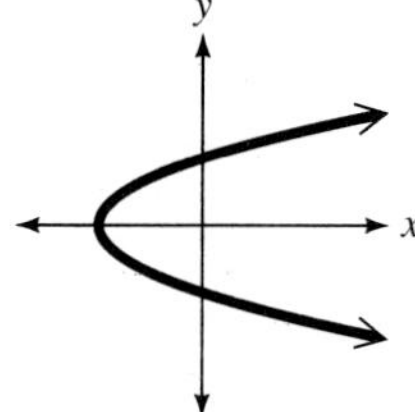

49. Is the relation $\{(1, 3), (2, 3), (3, 3)\}$ a function?

50. Specify the domain of the function defined by

$$y = \frac{1}{x + 5}.$$

51. Given the function $f(x) = x^2 + 3$, find the following.

(a) $f(-2)$
(b) $f(1)$
(c) $f(-a)$

52. Steve Caparella earns an annual salary of \$15,000 plus a 20% commission on all of his sales. Express his annual earnings as a function, f, of his total sales, x. What are his annual earnings when his yearly sales total is \$520,000?

CHAPTER 4

53. Tell whether $(3, -2)$ is a solution to the system.
$$\begin{aligned} x - 5y - 13 &= 0 \\ 2x + 3y &= 0 \end{aligned}$$

54. Without solving, tell the number of solutions to the system.
$$\begin{aligned} 4x - 3y &= 2 \\ -8x + 6y &= -2 \end{aligned}$$

Solve.

55. $$\begin{aligned} 3x + 5y &= -2 \\ x - 3y &= 4 \end{aligned}$$

56. $$\begin{aligned} x - 5y &= 2 \\ -3x + 15y &= -6 \end{aligned}$$

57. Tom Reese wishes to mix two kinds of candy, one worth 90¢/lb and the other worth \$1.50/lb. How many pounds of each must there be in 60 lb of the mix if it is to sell for \$1.30/lb?

58. A collection of dimes and quarters is worth \$3.75. If there are 27 coins in the collection, how many of each are there?

59. $$\begin{aligned} x - 3y + 2z &= 6 \\ 2x + 4y - z &= 2 \\ 4x - 2y + 3z &= 14 \end{aligned}$$

60. $$\begin{aligned} x - y - z &= 2 \\ 3x - y - 3z &= -4 \\ x + y - z &= 4 \end{aligned}$$

61. $$\begin{aligned} x + y - z &= 4 \\ 2x - y + 3z &= -3 \\ x + 2y - z &= 6 \end{aligned}$$

62. The average of a student's three scores is 79. The sum of the first and the second scores is 57 more than the third, and the first is 7 less than the second. Find the three scores.

Evaluate the determinants.

63. $\begin{vmatrix} 3 & 5 \\ -1 & 7 \end{vmatrix}$

64. $\begin{vmatrix} 1 & -2 & 3 \\ 0 & 5 & -1 \\ 2 & -1 & 4 \end{vmatrix}$

Solve for x only using Cramer's rule. [Hint: Use Exercises 63 and 64.]

65. $$\begin{aligned} 3x + 5y &= 1 \\ -x + 7y &= -9 \end{aligned}$$

66. $$\begin{aligned} x - 2y + 3z &= 7 \\ 5y - z &= -15 \\ 2x - y + 4z &= 5 \end{aligned}$$

Graph in a Cartesian coordinate system.

67. $4x + 2y > 8$

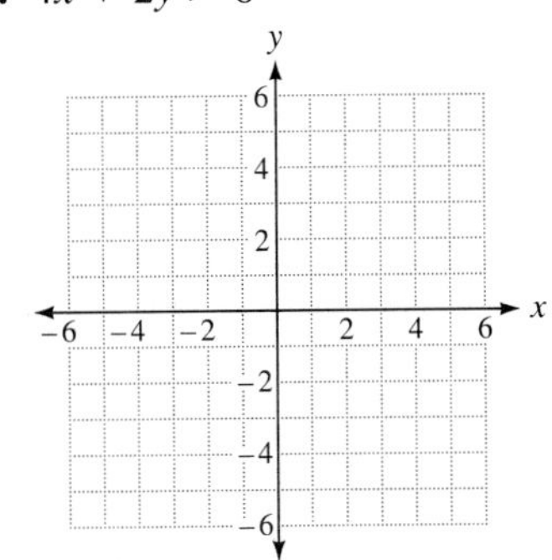

68. $5x - 2y \leq 10$
$x + 2 > 0$

69. $x \geq 0$
$y \geq 0$
$5x + 3y \leq 15$

CHAPTER 5

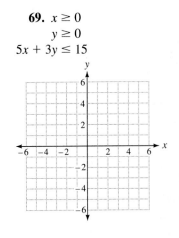

In Exercises 70–71, collect like terms and write in descending powers of x.

70. $3x - 5x^3 + x^2 - x^7 + 2$

71. $3x^2y^2 - 2x^4y + x^2y^2 + 3x^4y - xy$

72. Evaluate $2a^2b - 3ab^2$ when $a = -1$ and $b = 3$.

73. Add $3y^2 - 14y + y^3 - 5$ and $7 - y + 2y^3 - y^2$.

74. Subtract $-2xy + 3x^2y^2 - 9$ from $5x^2y^2 - 2xy + 4$.

75. Perform the indicated operations. $(3ab - 2a^2b^2) - (a^2b^2 - 5) + (1 - 2ab)$

Mutiply.

76. $-2xy^2(3x - 4y)$

77. $(2x + y)(x - 4y)$

78. $(a - 3b)^2$

79. $(3u - v)(3u + v)$

Factor.

80. $3x^2 - 12$

81. $a^2 + 10a + 25$

82. $4y^3 - 32$

83. $3x^2 - 8x + 4$

84. $3x^3 + 3$

85. $-6u^2 + 10uv + 4v^2$

Solve.

86. $2x(x - 3) = 0$

87. $x^2 - 1 = 3(x + 1)$

88. Twice the product of two consecutive positive odd integers is 126. Find the integers.

89. The length of a rectangle is 7 cm more than the width and the area is 44 cm^2. Find its dimensions.

The current I in amperes in an electrical circuit varies according to time t in seconds by the equation $I = 4t^2 - 72t + 180$. *Use this in Exercises 90–91.*

90. What is the initial amperage in the circuit?

91. In how many seconds will the amperage be 0 amperes?

CHAPTER 6

92. What values of the variable must be excluded?

$$\frac{3x + 1}{x^2 - 2x}$$

93. Reduce to lowest terms.

$$\frac{8x^3 + 1}{2x + 1}$$

Perform the indicated operations.

94. $\dfrac{x - y}{4x + 4y} \cdot \dfrac{x^2 - y^2}{x^2 - 2xy + y^2}$

95. $\dfrac{a^2 - 9}{a^2 - 8a + 16} \div \dfrac{3a - 9}{a - 4}$

96. $\dfrac{4}{x + y} + \dfrac{2x + y}{x^2 - y^2}$

97. $\dfrac{2}{a + b} - \dfrac{2a}{a^2 + 2ab + b^2}$

98. $\dfrac{4y}{y^2 - 1} - \dfrac{4}{y - 1} + \dfrac{3y}{1 - y}$

99. $\dfrac{x^2 - 9}{x^2 - x - 6} \div \dfrac{2x^2 + x - 15}{x^2 + 7x + 10} \cdot \dfrac{2x^2 - x - 10}{x^2 - 25}$

Solve.

100. $\dfrac{1}{y} = \dfrac{y - 2}{24}$

101. $\dfrac{a}{a + 3} - \dfrac{18}{a^2 - 9} = 1$

102. The speed of a river is 5 mph. If a boat travels 75 mi downstream in the same time that it takes to travel 45 mi upstream, find the speed of the boat in still water.

103. A small pipe can fill a reservoir in 5 days and a large pipe can fill it in 3 days. How long will it take to fill it if both pipes are turned on together?

104. If a car travels 675 mi on 30 gal of gas, how many miles can it travel on 46 gal of gas?

105. Suppose y varies jointly as x and the square root of z and inversely as the square of w. Find the equation of variation if $y = 2$, when $x = 3$, $z = 4$, and $w = 3$.

106. Solve $\dfrac{1}{a-b} = c$ for a.

107. Simplify. $\dfrac{\dfrac{a}{b} - \dfrac{b}{a}}{\dfrac{1}{a} + \dfrac{1}{b}}$

108. Divide. $\dfrac{9z^6 + 27z^5 - 3z^4}{-3z^4}$

109. Divide. Use synthetic division to check your work. $(x^4 - x^3 + x^2 - 3x + 2) \div (x - 1)$

CHAPTER 7

Simplify each radical in Exercises 110–113. For these exercises, do not assume that variables and algebraic expressions under even-indexed radicals necessarily represent nonnegative numbers.

110. $\sqrt{(-5x)^2}$

111. $\sqrt[3]{(-5x)^3}$

112. $-\sqrt{w^2}$

113. $\sqrt{(a+b)^2}$

Evaluate, assuming all algebraic expressions under even-indexed radicals are positive.

114. $\sqrt{(-11)^2}$

115. $\sqrt[3]{27a^3b^6}$

116. $\sqrt[4]{(-2)^4a^8}$

Simplify each radical and assume all variables under even-indexed radicals are positive.

117. $2\sqrt{20a^3b^2}$

118. $\sqrt{\dfrac{18}{z^2}}$

119. $\sqrt[3]{\dfrac{-125x^3y}{8y^4}}$

Multiply or divide. Rationalize all denominators.

120. $\sqrt{6xy}\sqrt{3x^3y^2}$

121. $\dfrac{\sqrt[3]{16a^3}}{\sqrt[3]{2b}}$

Add or subtract and simplify.

122. $\sqrt[3]{27a^5b^9} - 3b\sqrt[3]{125a^5b^6}$

123. $2x\sqrt{63x} + 5\sqrt{28x^3}$

Multiply and simplify.

124. $\sqrt{3}(\sqrt{12} - \sqrt{3})$

125. $(\sqrt{2} + 1)^2$

126. $(\sqrt{x} - 5)(\sqrt{x} + 5)$

127. $(\sqrt{2} - \sqrt{5})(2\sqrt{2} + \sqrt{5})$

128. Rationalize the denominator.

$$\frac{\sqrt{7}}{\sqrt{7} - 2}$$

129. Simplify.

$$\left(\frac{3a^{-1}b^{-1/2}}{a^3b}\right)^{-2}$$

Rationalize the numerators.

130. $\dfrac{\sqrt{14}}{2}$

131. $\dfrac{\sqrt[3]{x}}{7}$

132. $\dfrac{\sqrt{x} - 2}{3}$

Solve.

133. $\sqrt{y + 11} - 1 = \sqrt{y + 4}$

134. $\sqrt{2x + 1} - \sqrt{x + 6} = 0$

135. $x^{1/3}y^{1/3} = a$ for y

136. A hiker walked 9 mi east, then 12 mi north. How far was she then from the starting point?

137. Find the distance between the points $(-2, 6)$ and $(-3, -1)$.

138. Find the midpoint of the segment joining $(-4, -1)$ and $(6, 7)$.

Perform the indicated operations.

139. $(2 + 3i) + (8 - i)$

140. $(3 + i)(2 - 4i)$

141. $\dfrac{2 - i}{3 + 2i}$

142. i^{89}

143. $\sqrt{-25} - \sqrt{-4}$

144. $\dfrac{\sqrt{-9}}{\sqrt{-4}}$

CHAPTER 8

Solve.

145. $y^2 - 6y - 7 = 0$

146. $4x^2 = 8x$

147. $z^2 - 2z - 1 = 0$

148. $y^4 - 3y^2 + 2 = 0$

149. $(x - 3)^2 - 3(x - 3) - 4 = 0$

150. $(z - 3)(z + 3) = -13$

151. $x^2 + 2x + 4 = 0$

152. $\dfrac{1}{x - 1} + \dfrac{2}{x} = 1$

153. $\sqrt{x^2 + 2} - \sqrt{x + 1} = 0$

154. $\sqrt{3y + 1} - \sqrt{y - 4} = 3$

155. Cindy can complete a lab test in 3 hr less time than Burford. If they work together, they can do the test in 4 hr. To the nearest tenth of an hour, how long will it take each working alone to complete the test?

156. A boat sails downstream 96 miles and returns to the dock in a total time of 7 hours. Find the speed of the boat downstream, to the nearest tenth, if the speed downstream is 1 mph faster than the speed upstream.

157. Twice the product of two positive consecutive even integers is 240. Find the integers.

158. The number of square inches in the area of a square is the same as 5 more than the number of inches in its perimeter. Find the length of a side.

159. Solve by completing the square.
$x^2 - 5x + 2 = 0$

160. Solve for x. $\dfrac{n}{x - 2} + \dfrac{n}{x + 2} = 1$

161. Use the discriminant to determine the nature of the solutions of $2x^2 - 3x + 7 = 0$

162. Find a quadratic equation that has $3, -4$ as solutions.

163. Solve $2x^2 - 3x - 2 < 0$ and give the answer using interval notation.

164. Solve $\dfrac{x + 2}{x - 7} \geq 0$ and give the answer using interval notation.

CHAPTER 9

165. Sketch the graph of $h(x) = -x^2 + 6$ by sliding the graph of $f(x) = -x^2$ up or down.

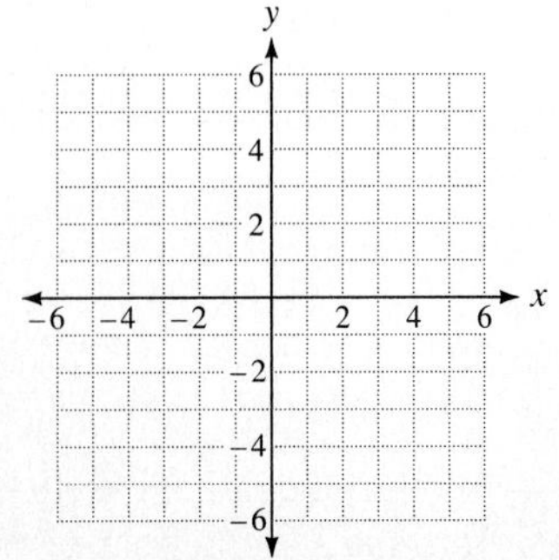

166. Find the x-intercepts, the vertex, and the graph of $f(x) = x^2 - 6x + 5$.

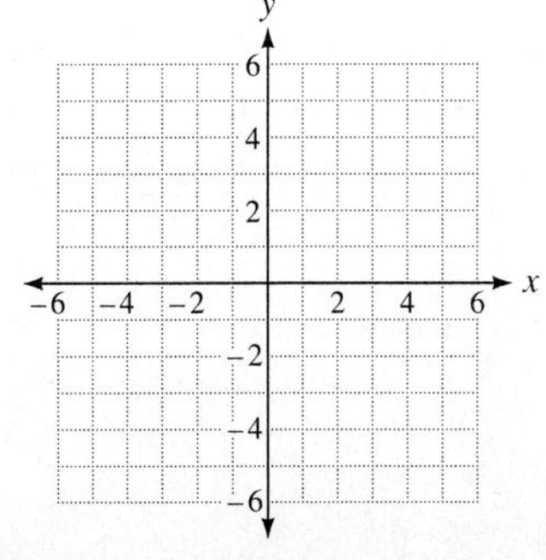

167. Without graphing, determine whether the graph of the quadratic function opens up or down and find the vertex and x-intercepts (if they exist).
$f(x) = -x^2 + 3x - 7$

168. What is the maximum area of a rectangular pasture enclosed by 240 yd of fencing? What are the dimensions of the pasture?

Find the domain of each function.

169. $f(x) = \sqrt{5 - x}$

170. $g(x) = \dfrac{9}{\sqrt{5 - x}}$

171. Assume that $f(x) = 1 - x$ and $g(x) = x^2 + 3$. Find each function value.

(a) $(f + g)(2)$ **(b)** $(f - g)(2)$ **(c)** $(fg)(2)$

(d) $\left(\dfrac{f}{g}\right)(2)$ **(e)** $f[g(2)]$ **(f)** $g[f(2)]$

Make a table of values, sketch the graph of each function, and tell whether the function is one-to-one.

172. $f(x) = x^4 - 6$

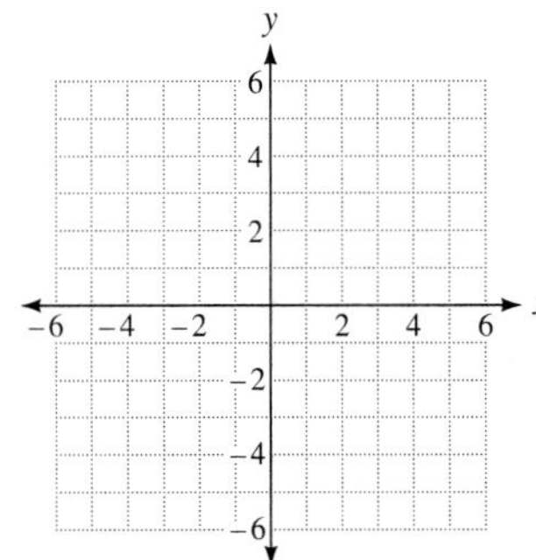

173. $g(x) = |x| - 4$

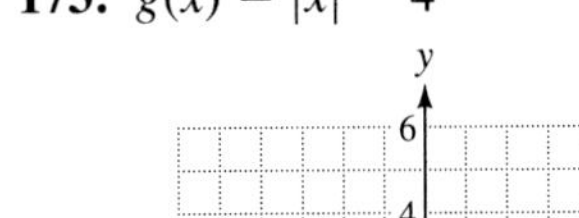

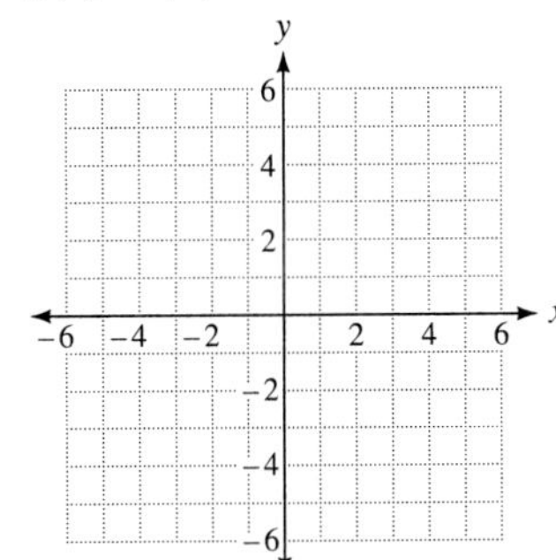

174. $h(x) = -\sqrt{x + 3}$

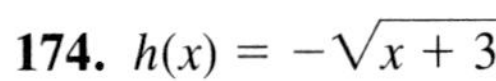

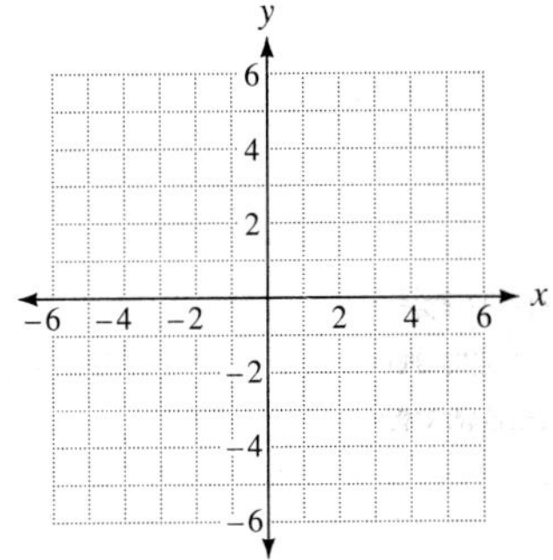

175. Consider the function $f(x) = \frac{1}{5}x - \frac{2}{5}$. Find the equation for the inverse of f and use it to find the following.

(a) $f(-1)$ **(b)** $f^{-1}(-1)$ **(c)** $f[f^{-1}(-1)]$ **(d)** $f^{-1}[f(-1)]$

CHAPTER 10

176. Convert $7^x = 3$ to logarithmic form.

177. Convert $\log_a 25 = 6$ to exponential form.

178. Express $\log_a \frac{x^2}{y\sqrt{z}}$ in terms of logarithms of x, y, and z.

179. Express $-3 \log_a x + \frac{1}{2} \log_a y$ as a single logarithm.

Write true *or* false *for each of the following. If the statement is false, change the right side to make it true.*

180. $\log_a \dfrac{u}{v} = \dfrac{\log_a u}{\log_a v}$

181. $\log_a a = 1$

Given that $\log_a 3 = 0.4771$ *and* $\log_a 5 = 0.6990$, *find the following logarithms.*

182. $\log_a 9$

183. $\log_a \dfrac{3}{5}$

184. $\log_a 15$

Find each logarithm.

185. $\log 0.000328$

186. $\ln 298.3$

187. $\log_5 491$

Find each antilogarithm, correct to three significant digits.

188. $\log x = 2.6255$

189. $\log x = -0.4135$

190. $\ln x = 1.3759$

Solve.

191. $\log x - \log (x + 9) = -1$

192. $3^{2x+6} = 27$

193. Use the sound equation.
$$y = 10 \log \frac{x}{x_0}$$
to find the loudness of a sound in decibels if the given sound has intensity 75,000 times the weakest sound detected.

194. If \$1000 is invested at an annual rate of 12% compounded quarterly, use $S = A(1 + i)^n$ to find the value of the account at the end of 6 years.

195. The population of a county is increasing according to $y = 100{,}000\, e^{0.13t}$. In how many years will the population double?

196. Graph $y = 2^x$ and $y = \log_2 x$ in the same coordinate system. What is the relationship between these two functions?

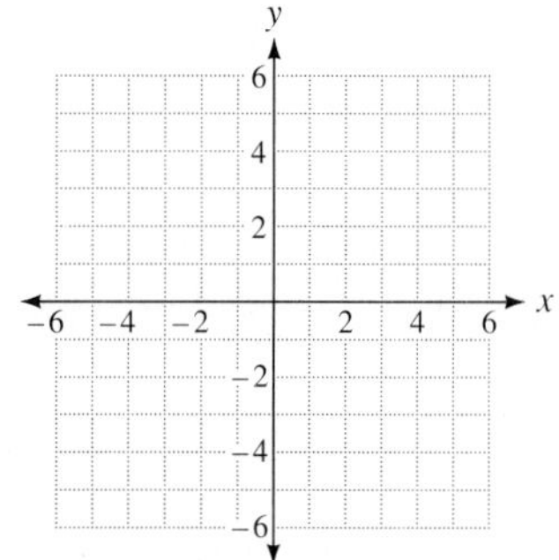

CHAPTER 11

197. Find the standard and general forms of the equation of a circle with center $(-4, 2)$ and radius 6.

198. Find the standard form of the equation of the following circle, and give the center and radius. $x^2 + y^2 - 2x + 8y + 13 = 0$

199. Give the standard form of the equation of the ellipse with intercepts $(2, 0)$, $(-2, 0)$, $(0, 5)$, and $(0, -5)$.

200. Give the standard form of the equation of the hyperbola that opens left and right and has $a = 3$ and $b = 9$.

Graph.

201. $y = -x^2 + 2x - 2$

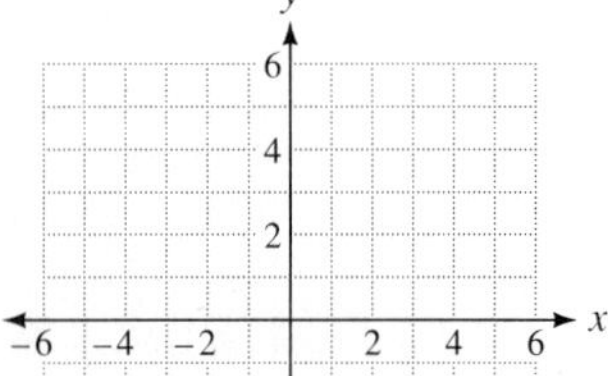

202. $x = y^2 - 4$

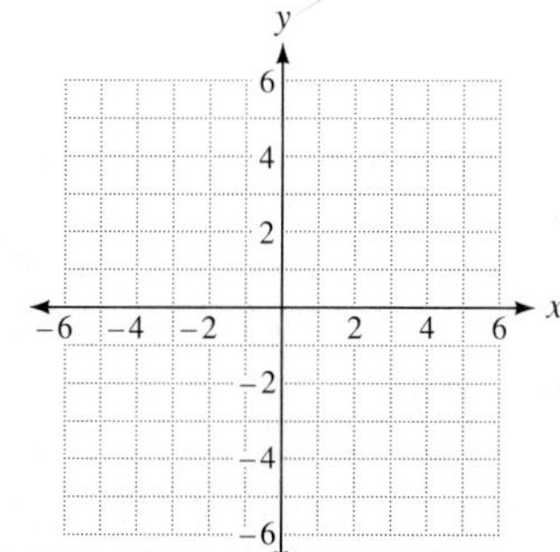

203. $(x + 1)^2 + (y - 3)^2 = 4$

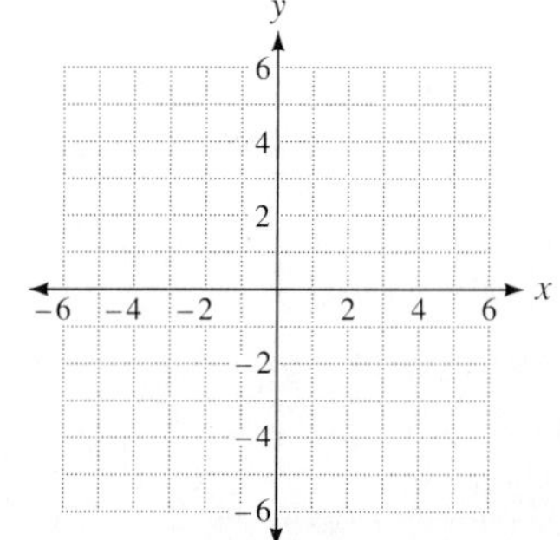

204. $\dfrac{x^2}{16} + \dfrac{y^2}{9} = 1$

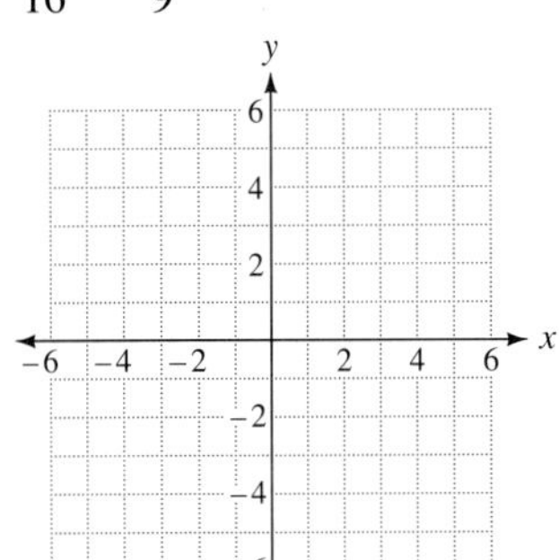

205. $\dfrac{x^2}{4} - \dfrac{y^2}{9} = 1$

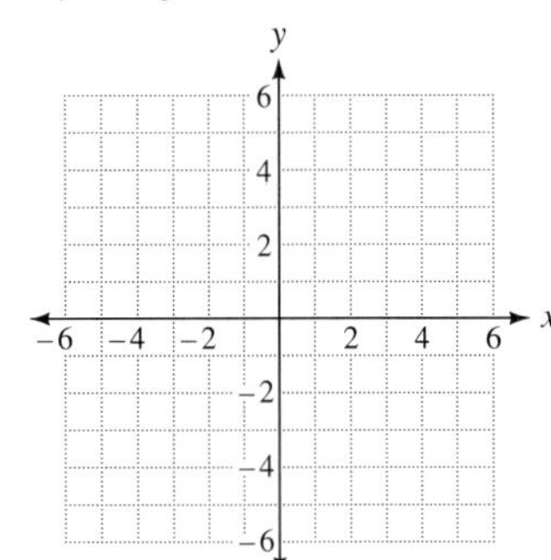

206. $xy = 6$

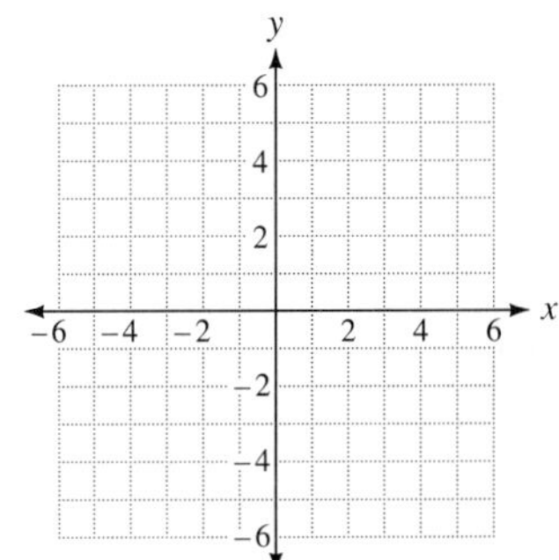

Solve the following systems.

207. $x^2 + 3y^2 = 19$
$x - y = 3$

208. $2x^2 + y^2 = 6$
$x^2 - y^2 = -3$

CHAPTER 12

209. Rewrite $b_1 + b_2 + b_3 + b_4$ using sigma summation notation.

210. Rewrite $\sum_{k=1}^{5} x_k$ without sigma summation notation.

211. The nth term of a general sequence is $x_n = 3^n + 1$; find the first three terms x_1, x_2, x_3.

212. An arithmetic sequence has $a_1 = 11$ and $d = -4$; find a_6 and S_6.

213. A geometric sequence has $a_1 = 32$ and $r = -\frac{1}{2}$; find a_5 and S_5.

214. An auditorium has 30 rows with 20 seats in the first row, 23 seats in the second row, 26 in the third row, and so forth. How many seats are in the auditorium?

215. A new car costing \$10,000 depreciates 25% of its value each year. How much is the car worth at the end of 6 years?

216. The tip of a pendulum moves back and forth in such a way that it sweeps out an arc 24 in long, and on each succeeding pass, the length of the arc traveled is $\frac{9}{10}$ the length of the preceding pass. What is the total distance traveled by the tip of the pendulum?

217. Convert $3.\overline{45}$ to a fraction.

218. Expand $(x + 3)^5$ using Pascal's triangle.

ANSWERS: **1.** $\{0, 1\}$ **2.** $\{x|x$ is an odd integer between 0 and 6$\}$ **3.** true **4.** -10 **5.** $-\frac{1}{8}$ **6.** $-\frac{1}{2}$ **7.** $-\frac{1}{4}$ **8.** $\frac{1}{12}$ **9.** distributive law **10.** double-negation property **11.** commutative law of addition **12.** symmetric law of equality **13.** $-2(a + 4x - 3)$ **14.** $-3 + 3x$ **15.** $-3z$ **16.** $8y + 4$ **17.** $\frac{a^5}{b^{12}}$ **18.** $\frac{1}{x^3}$ **19.** $\frac{y^6}{8z^{12}}$ **20.** $\frac{2x^4y^6}{3}$ **21.** 16 **22.** 64 **23.** -16 **24.** -1 **25.** 4.21×10^7 **26.** 0.000000841 **27.** 1 **28.** -3 **29.** 14 cm, 22 cm **30.** 1:00 P.M. **31.** \$9200 **32.** 47, 49 **33.** 25°, 100°, 55° **34.** \$40.00 **35.** $(-\infty, -1]$ **36.** $(-\infty, -2)$ or $[6, \infty)$

37. $[-2, 1)$ 38. $-1, \frac{11}{3}$

39. $(-\infty, -3)$ or $(2, \infty)$ 40. $[-2, 5]$ 41. $r = \frac{P-q}{50}$ 42. $2, \frac{4}{3}$

43. x-intercept $= (-5, 0)$, y-intercept $= (0, 2)$, slope $= \frac{2}{5}$

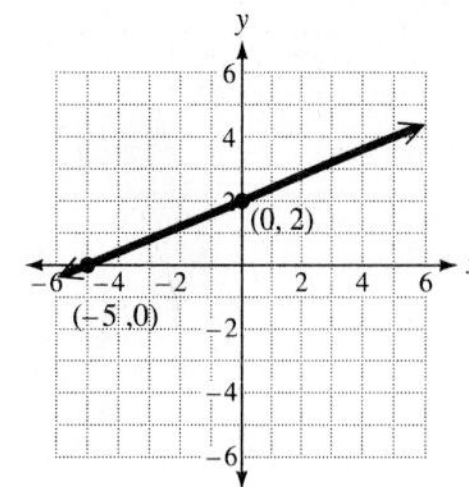

44.

45. $x + y - 2 = 0$
46. $3x - 4y + 14 = 0$
47. no 48. no 49. yes
50. all real numbers except -5
51. (a) 7 (b) 4 (c) $a^2 + 3$
52. $f(x) = (0.20)x + 15{,}000$; $f(520{,}000) = \$119{,}000$ 53. yes
54. no solutions; lines are parallel

55. $(1, -1)$ 56. infinitely many solutions of the form $\left(x, \frac{1}{5}x - \frac{2}{5}\right)$ for x any real number 57. 20 lb of 90¢ candy; 40 lbs of \$1.50 candy 58. 7 quarters, 20 dimes 59. infinitely many solutions; the system is dependent 60. no solution; the system is inconsistent 61. $(1, 2, -1)$ 62. 70, 77, 90 63. 26 64. -7 65. $A_x = 52$, $A = 26$, $x = 2$ 66. $A_x = -7$, $A = -7$, $x = 1$

67.

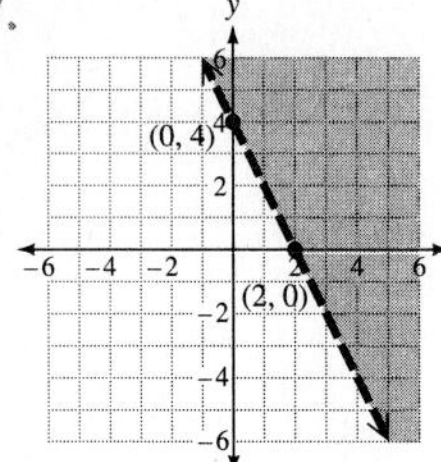

68.

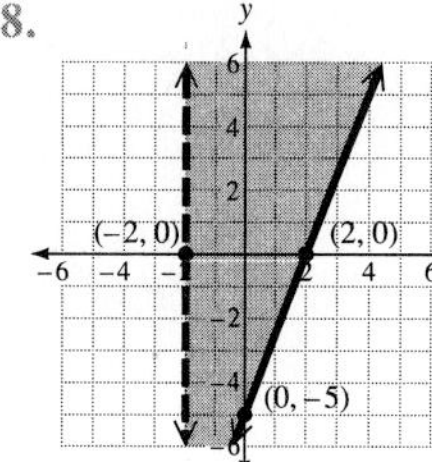

69.

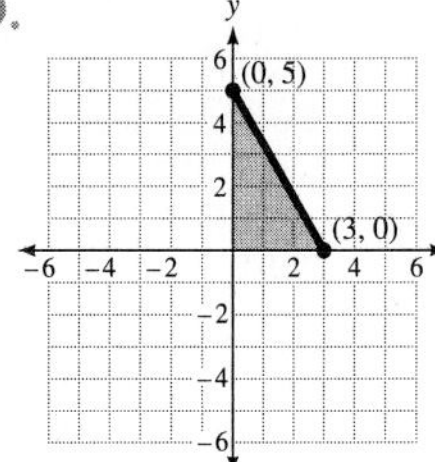

70. $-x^7 - 5x^3 + x^2 + 3x + 2$
71. $x^4y + 4x^2y^2 - xy$ 72. 33
73. $3y^3 + 2y^2 - 15y + 2$
74. $2x^2y^2 + 13$ 75. $-3a^2b^2 + ab + 6$ 76. $-6x^2y^2 + 8xy^3$
77. $2x^2 - 7xy - 4y^2$
78. $a^2 - 6ab + 9b^2$ 79. $9u^2 - v^2$
80. $3(x - 2)(x + 2)$

81. $(a + 5)(a + 5)$ 82. $4(y - 2)(y^2 + 2y + 4)$ 83. $(3x - 2)(x - 2)$ 84. $3(x + 1)(x^2 - x + 1)$ 85. $-2(u - 2v)(3u + v)$ 86. 0, 3 87. $-1, 4$ 88. 7 and 9 89. 4 cm by 11 cm 90. 180 amperes 91. first at 3 sec and again at 15 sec 92. 0, 2 93. $4x^2 - 2x + 1$ 94. $\frac{1}{4}$ 95. $\frac{a+3}{3(a-4)}$ 96. $\frac{3(2x-y)}{(x-y)(x+y)}$ 97. $\frac{2b}{(a+b)(a+b)}$ 98. $\frac{-3y^2-3y-4}{(y-1)(y+1)}$ 99. $\frac{x+2}{x-5}$ 100. $6, -4$ 101. no solution 102. 20 mph 103. $\frac{15}{8}$ days 104. 1035 mi 105. $y = \frac{3x\sqrt{z}}{w^2}$ 106. $a = \frac{1}{c} + b$ 107. $a - b$ 108. $-3z^2 - 9z + 1$ 109. $x^3 + x - 2$; 0 110. $5|x|$ 111. $-5x$ 112. $-|w|$ 113. $|a + b|$ 114. 11 115. $3ab^2$ 116. $2a^2$ 117. $4ab\sqrt{5a}$ 118. $\frac{3\sqrt{2}}{z}$ 119. $\frac{-5x}{2y}$ 120. $3x^2y\sqrt{2y}$ 121. $\frac{2a\sqrt[3]{b^2}}{b}$ 122. $-12ab^3\sqrt[3]{a^2}$ 123. $16x\sqrt{7x}$ 124. 3 125. $3 + 2\sqrt{2}$ 126. $x - 25$ 127. $-1 - \sqrt{10}$ 128. $\frac{7+2\sqrt{7}}{3}$ 129. $\frac{a^8b^3}{9}$ 130. $\frac{7}{\sqrt{14}}$ 131. $\frac{x}{7\sqrt[3]{x^2}}$ 132. $\frac{x-4}{3(\sqrt{x}+2)}$ 133. 5 134. 5 135. $\frac{a^3}{x}$ 136. 15 mi 137. $5\sqrt{2}$ 138. $(1, 3)$ 139. $10 + 2i$ 140. $10 - 10i$ 141. $\frac{4-7i}{13}$ 142. i 143. $3i$ 144. $\frac{3}{2}$ 145. $-1, 7$ 146. 0, 2 147. $1 \pm \sqrt{2}$ 148. $\pm\sqrt{2}, \pm 1$ 149. 7, 2 150. $\pm 2i$ 151. $-1 \pm i\sqrt{3}$ 152. $2 \pm \sqrt{2}$ 153. $\frac{1 \pm i\sqrt{3}}{2}$ 154. 5, 8 155. Cindy takes 6.8 hr; Burford takes 9.8 hr 156. 27.9 mph 157. 10, 12 158. 5 in 159. $\frac{5 \pm \sqrt{17}}{2}$ 160. $n \pm \sqrt{n^2 + 4}$ 161. two complex 162. $x^2 + x - 12 = 0$ 163. $\left(-\frac{1}{2}, 2\right)$ 164. $(-\infty, 2]$ or $(7, \infty)$ 165.

166. x-intercepts: $(1, 0)$ and $(5, 0)$ vertex: $(3, -4)$

167. down; the vertex is $\left(\frac{3}{2}, -\frac{19}{4}\right)$; there are no x-intercepts 168. 3600 yd^2; 60 yd by 60 yd 169. all real numbers $x \le 5$ 170. all real numbers $x < 5$ 171. (a) 6 (b) -8 (c) -7 (d) $-\frac{1}{7}$ (e) -6 (f) 4

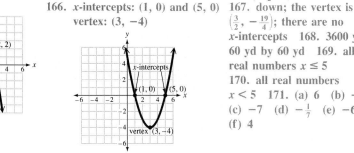

172. not one-to-one

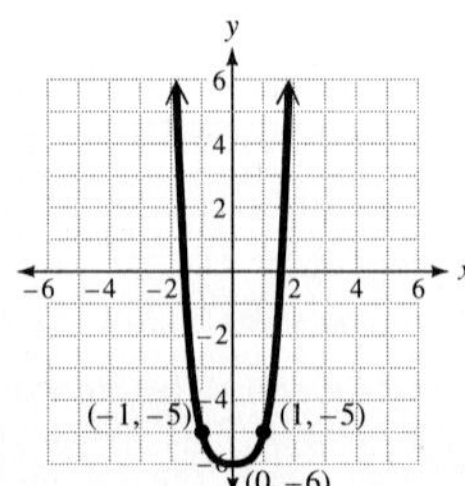

173. not one-to-one

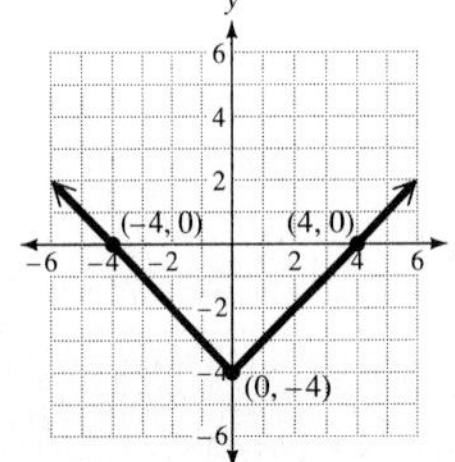

174. is one-to-one

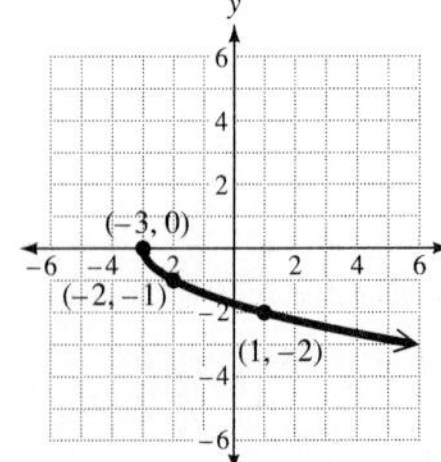

175. $f^{-1}(x) = 5x + 2$; (a) $-\frac{3}{5}$ (b) -3 (c) -1 (d) -1 176. $\log_7 3 = x$ 177. $a^6 = 25$ 178. $2 \log_a x - \log_a y - \frac{1}{2} \log_a z$ 179. $\log_a \frac{\sqrt{y}}{x^3}$ 180. false; $\log_a u - \log_a v$ 181. true 182. 0.9542 183. -0.2219 184. 1.1761 185. -3.4841

186. 5.6981 187. 3.8501 188. 422 189. 0.386 190. 3.96 191. 1 192. $-\frac{3}{2}$ 193. 49 decibels 194. \$2032.79

195. about 5.3 years 196. The functions are inverses.

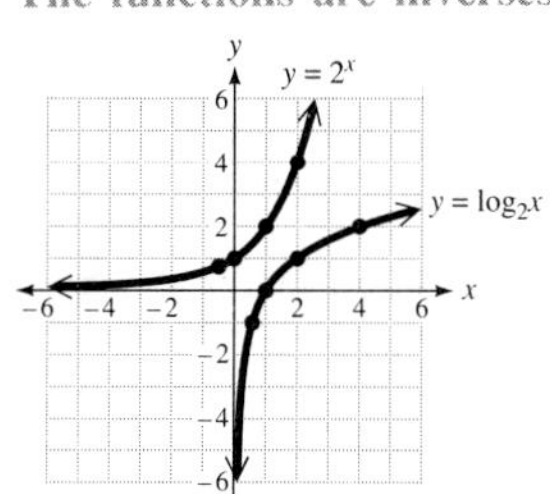

197. $(x + 4)^2 + (y - 2)^2 = 6^2$; $x^2 + y^2 + 8x - 4y - 16 = 0$ 198. $(x - 1)^2 + (y + 4)^2 = 2^2$; $(1, -4)$; $r = 2$ 199. $\frac{x^2}{4} + \frac{y^2}{25} = 1$ 200. $\frac{x^2}{9} - \frac{y^2}{81} = 1$

201.

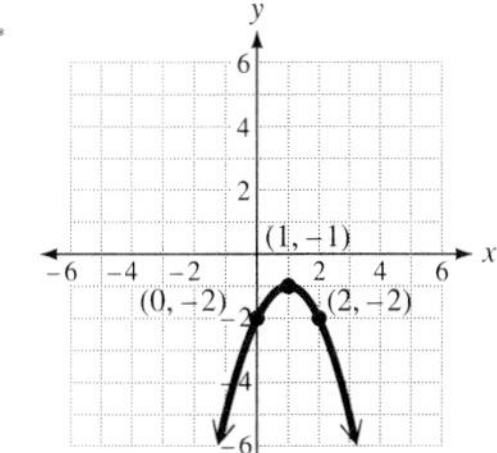

202.

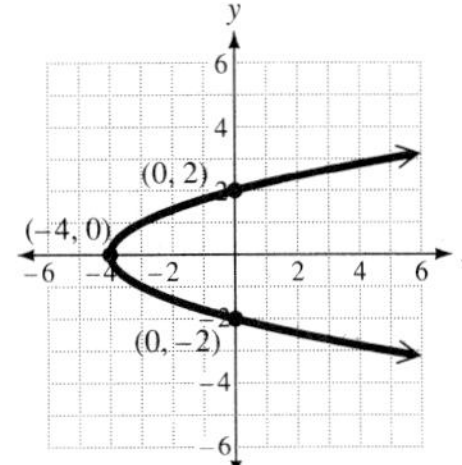

203.

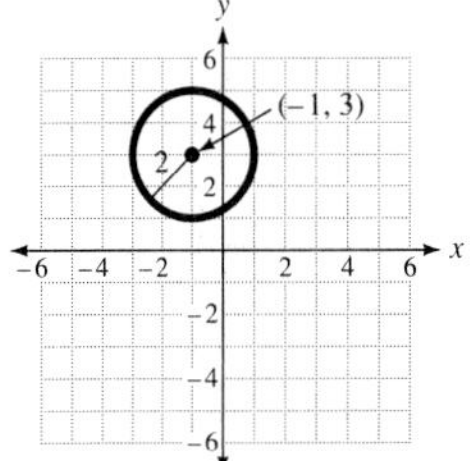

204.

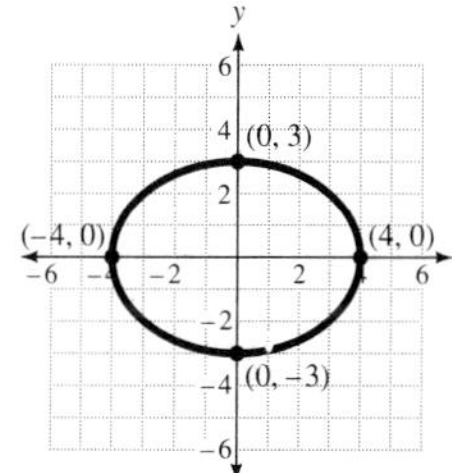

205.

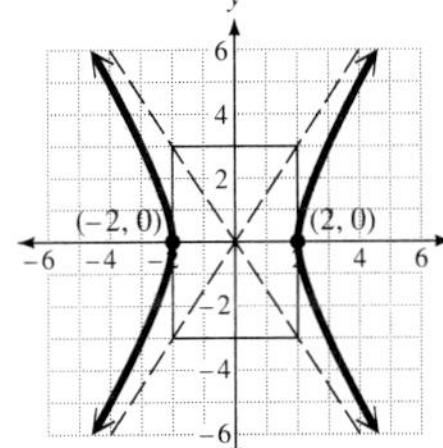

206.

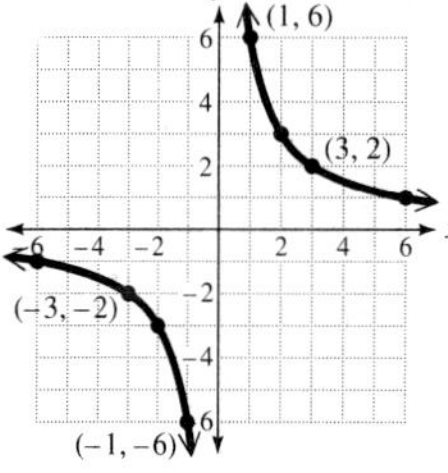

207. $(4, 1)$, $\left(\frac{1}{2}, -\frac{5}{2}\right)$ 208. $(1, 2)$, $(1, -2)$, $(-1, 2)$, $(-1, -2)$ 209. $\sum_{k=1}^{4} b_k$ 210. $x_1 + x_2 + x_3 + x_4 + x_5$ 211. 4, 10, 28 212. $a_6 = -9$, $S_6 = 6$ 213. $a_5 = 2$, $S_5 = 22$ 214. 1905 215. $1780 216. 240 in 217. $\frac{38}{11}$ 218. $x^5 + 15x^4 + 90x^3 + 270x^2 + 405x + 243$

ANSWERS TO CHAPTER TESTS

CHAPTER 1 TEST **1.** $\{0, 1, 2, 3, 4\}$ **2.** 0.875 **3.** true **4.** $<$ **5.** -4 **6.** $-5 \le x < 1$ **7.** -2 **8.** -16 **9.** 1 **10.** 3 **11.** 3 **12.** commutative law of addition **13.** double-negation property **14.** transitive law **15.** 6 **16.** $-\frac{1}{5}$ **17.** -15 **18.** $(a + 5 - c)x$ **19.** $-6 + 3c$ **20.** 18 **21.** 36 **22.** -9 **23.** 81 **24.** 27 **25.** 13 **26.** $x - 1$ **27.** $4a - 3$ **28.** 176°F **29.** x^4 **30.** $\frac{a^3}{b^{15}}$ **31.** $12x^9y^6$ **32.** 2.37×10^{-5}

CHAPTER 2 TEST **1.** -5 **2.** -6 **3.** 0 **4.** -7 **5.** -60 **6.** 14 in; 19 in **7.** 15°; 45°; 120° **8.** 1:00 P.M. **9.** 350 **10.** $15,600 **11.** 15% **12.** $a = \frac{W - c}{30}$ **13.** $w = \frac{uv + 2}{u + v}$ **14.** $x \le 2$ **15.** $y < -2$ or $y > 4$ **16.** $1 < z < 5$ **17.** $-1, \frac{11}{3}$ **18.** $[-2, 7]$ **19.** $-6, \frac{4}{3}$ **20.** $x > -6$

CHAPTER 3 TEST 1. $y = -\frac{3}{2}x + 6$ 2. (4, 0) 3. (0, 6) 4. $-\frac{3}{2}$ 5.

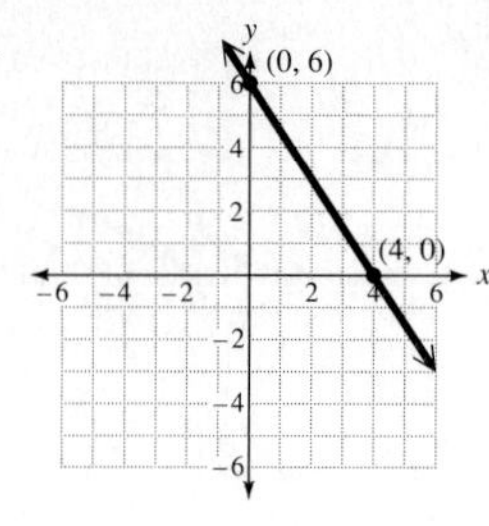

6. (10, 70); \$70 7. -2; $\frac{1}{2}$ 8. parallel 9. $3x + y + 2 = 0$
10. $2x + y - 5 = 0$ 11. $y = 8$ 12. yes 13. yes
14. no 15. -3 16. $3a$ 17. linear 18.

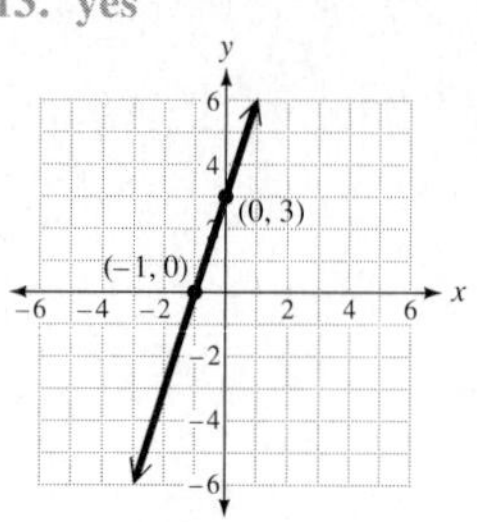

19. all real numbers except -3
20. $g(x) = 20x$

CHAPTER 4 TEST 1. no 2. infinitely many 3. (5, −1) 4. $(x, 2x - 3)$ x any real number 5. 6; 11 6. 18 lb of \$2.00/lb nuts and 30 lb of \$2.80/lb nuts 7. no solution 8. (1, 0, −2) 9. $x + y + z = 10{,}000$; $0.08x + 0.07y + 0.10z = 810$; $0.08x + 0.07y - 0.04z = 390$ 10. 16 11. -14 12. $x = 1$, $A_x = 11$, $A = 11$ 13. $x = 3$, $A_x = -9$, $A = -3$
14. 15.

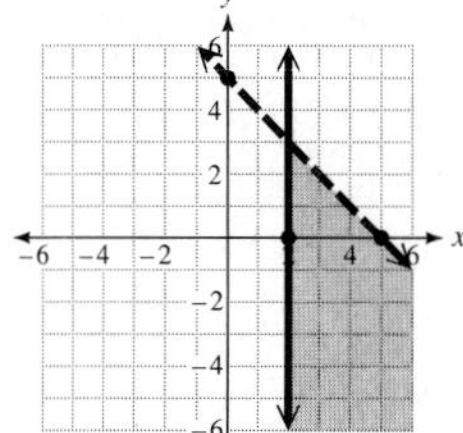

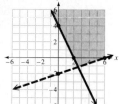

CHAPTER 5 TEST 1. $-5x^2y$ 2. 4 3. trinomial 4. 53 5. $3x^3 - 4$ 6. $-a^3b^2 + 4b^4$ 7. $5x^2y^3$ 8. $8ab + 1$
9. $-12u^3v + 8u^2v^2$ 10. $a^2 - 9b^2$ 11. $z^2 + 10z + 25$ 12. $z^2 - 10z + 25$ 13. $12x^2 + 7xy - 10y^2$ 14. $a^4 - b^4$
15. $-11xy^2(xy - 6 + 4x)$ 16. $(3u + v)(2u - v)$ 17. $2(x - 2)(x + 2)$ 18. $(a + 2b)(a^2 - 2ab + 4b^2)$ 19. $(x - 3y)^2$
20. $(b - y)(x - 2)$ 21. $(x + y - 2)(x + y - 3)$ 22. 7, −1 23. $-5, \frac{1}{2}$ 24. Mack is 15; Betty is 12 25. 30

CHAPTER 6 TEST 1. 1, −1 2. $x^2 + xy + y^2$ 3. $\frac{a+1}{a+3}$ 4. 1 5. $\frac{(x-2)^2}{x^2}$ 6. $\frac{1}{x-1}$ 7. $\frac{3b}{(a-b)^2}$ 8. $\frac{-2(2a+1)}{(a+2)(a+3)(a-2)}$
9. 1, −3 10. Jeff is 12; Sue is 10 11. $y = \frac{2 - wx}{w}$ 12. $y = \frac{2x\sqrt{z}}{w}$ 13. 65 minutes 14. 12 mph 15. $x - 3$
16. $-6a^3 + 4a + 1$ 17. $4a^2 - a + 3$ 18. $x^3 + x^2 - 3x - 1$; -2

CHAPTER 7 TEST 1. 7 2. -7 3. $4ab$ 4. $2x$ 5. $-2y$ 6. $5\sqrt{3}$ 7. $3\sqrt[3]{2}$ 8. $2y\sqrt[4]{y}$ 9. $\frac{3x\sqrt[4]{x^3}}{y^2}$ 10. $5\sqrt{2} - \sqrt{10}$
11. $4 + 2\sqrt{3}$ 12. $3ab\sqrt[4]{b^3}$ 13. $x + 2\sqrt{ax} - 3a$ 14. $-3\sqrt{2}$ 15. 0 16. 81 17. -1.5 18. $\frac{5y^2\sqrt{y}}{x}$ 19. $-\sqrt{2} - \sqrt{5}$
20. $\frac{1}{\sqrt{6} - 3}$ 21. 21 22. -4 23. (a) $2\sqrt{10}$ (b) (−1, 0) 24. 5 mi 25. $a = \frac{c^2}{b}$ 26. $3 + i$ 27. $-6 + 32i$ 28. $\frac{9 + 7i}{26}$
29. about 16 seats 30. $5|a|$

CHAPTER 8 TEST 1. $\frac{1}{2}, -5$ 2. $-1 \pm 2\sqrt{2}$ 3. $-1 \pm i\sqrt{2}$ 4. 3, −5 5. 1, −4 6. 8 cm, 6 cm 7. $\frac{3}{2}$ 8. 8
9. $c = \frac{\pm\sqrt{a+b}}{3}$ 10. 36 days 11. 7.5% 12. one time; 3 sec after it is launched 13. two complex
14. $5x^2 + 7x - 6 = 0$ 15. $x < -1$ or $x > 5$; $(-\infty, -1)$ or $(5, \infty)$ −5 −4 −3 −2 −1 0 1 2 3 4 5
16. $1 \le x < 4$; $[1, 4)$ −5 −4 −3 −2 −1 0 1 2 3 4 5

CHAPTER 9 TEST 1. Slide the graph of $f(x) = x$ up 3 units.

2.

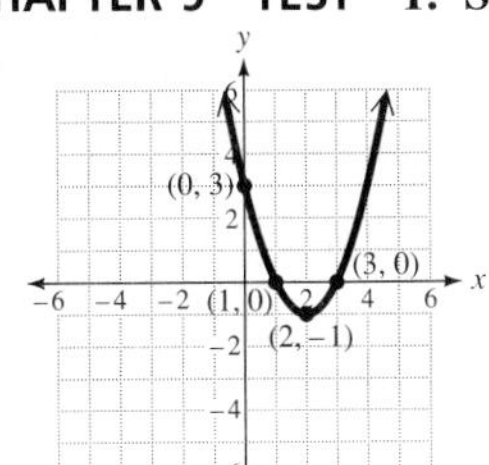

3. 144 4. all real numbers $x > -10$ 5. -16 6. 5 7. -10 8. 9 9. -13 10. -1 11. no 12. $f^{-1}(x) = 4x + 3$; (a) 11 (b) 2

13.

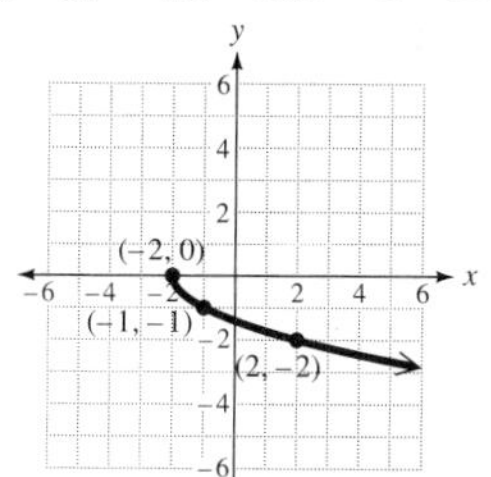

14.

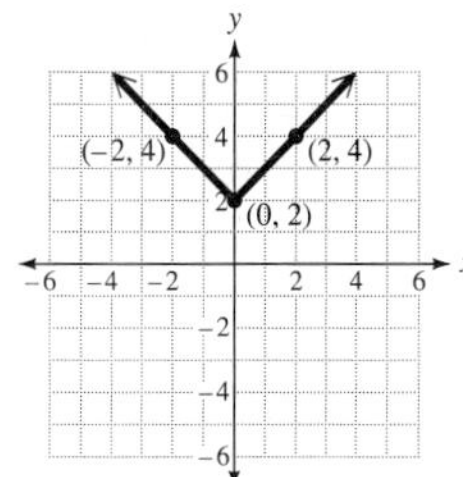

15.

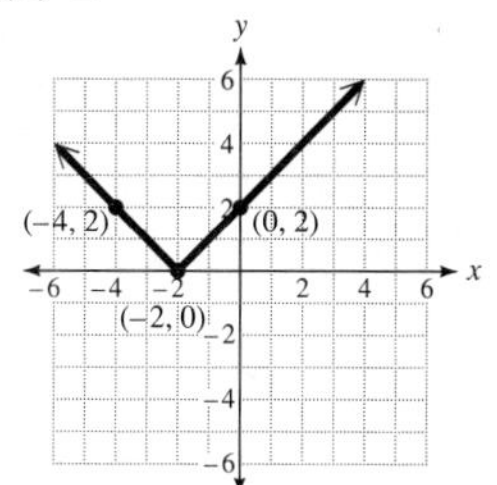

CHAPTER 10 TEST 1. $b^x = 5$ 2. $\log_u 7 = v$ 3. 4 4. -1 5. 3 6. 11 7. $2 \log_a x + 3 \log_a y - \log_a z$ 8. $\log_a \frac{z^2}{y^{1/3}}$ 9. false 10. true 11. 1 12. -0.2132 13. 1.8485 14. 1960 15. 0.647 16. 2, -2 17. $\frac{1}{9}$ 18. 50.8 decibels 19. \$8351.04 20. 15.4 years 21.

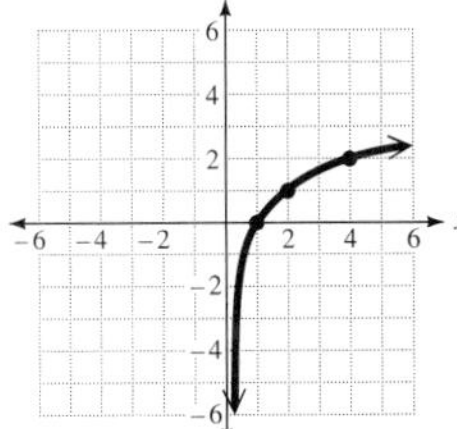

CHAPTER 11 TEST

1.

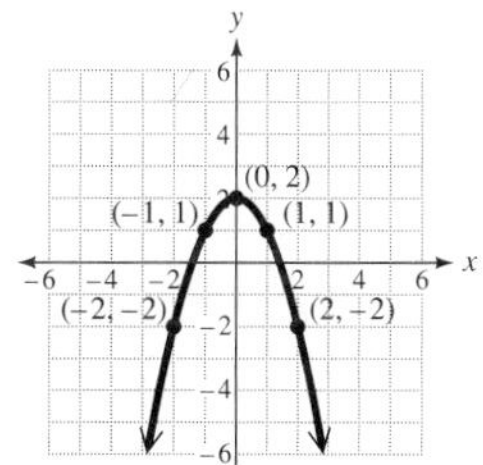

2.

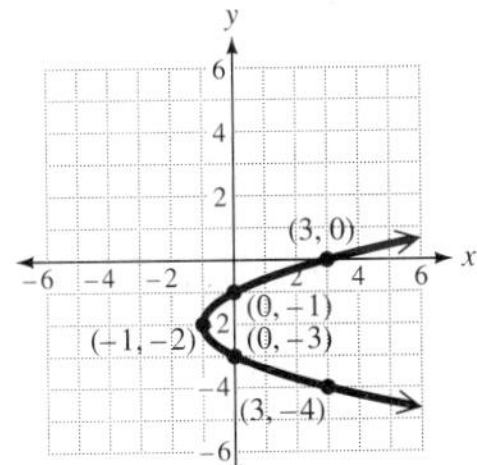

3.

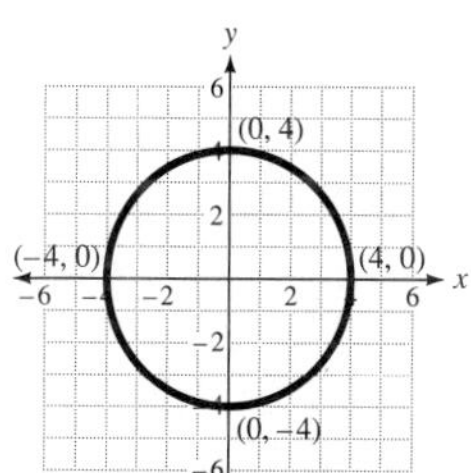

4.

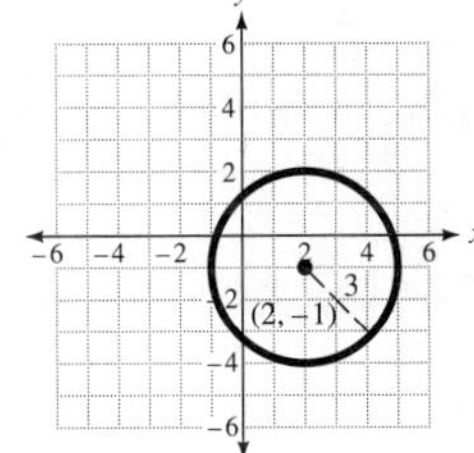

5. $(x + 2)^2 + (y + 1)^2 = 5^2$; $x^2 + y^2 + 4x + 2y - 20 = 0$ 6. $(x + 7)^2 + (y - 6)^2 = 5^2$; $(-7, 6)$; $r = 5$

7.

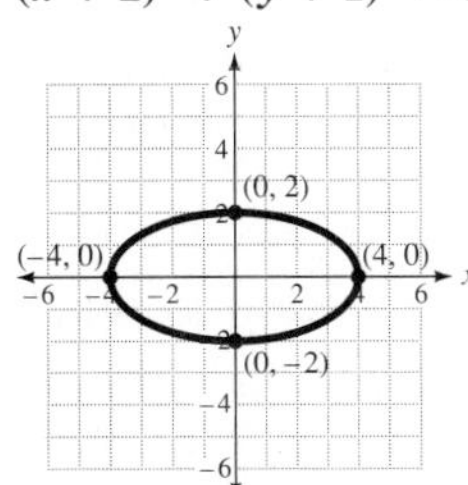

8.

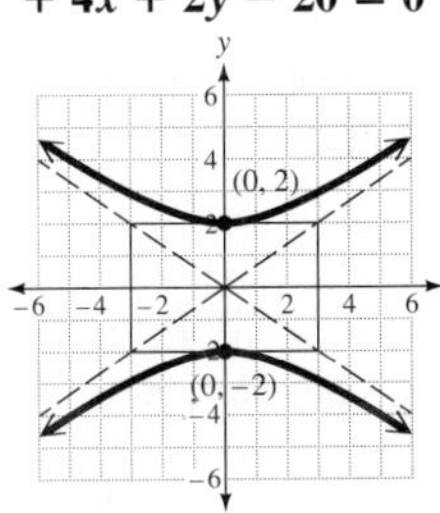

9. hyperbola 10. circle 11. parabola 12. ellipse 13. $(2, -5)$, $(-6, 11)$ 14. $(2\sqrt{2}, 1)$, $(-2\sqrt{2}, 1)$

CHAPTER 12 TEST 1. true 2. $\sum_{n=3}^{6} a_n$ 3. $|r| < 1$ 4. $x_1 = 2$, $x_2 = 7$, $x_5 = 22$ 5. $a_1 = 8$, $a_2 = 3$, $a_3 = -2$, $S_6 = -27$ 6. \$65.00 7. $a_1 = 3$, $a_2 = -6$, $a_3 = 12$, $S_5 = 33$ 8. \$2016.84 9. 240 in 10. $\frac{4}{9}$ 11. $a^5 - 20a^4x + 160a^3x^2 - 640a^2x^3 + 1280ax^4 - 1024x^5$ 12. \$40,300

APPENDIX

Logarithmic Tables and Interpolation

STUDENT GUIDEPOSTS

1. Characteristic and Mantissa
2. Finding Common Logarithms
3. Standard Form of a Logarithm
4. Finding Antilogarithms
5. Linear Interpolation
6. Logarithmic Computation

1 CHARACTERISTIC AND MANTISSA

In the past few years, scientific calculators have all but replaced logarithm tables. However, when a calculator is unavailable, a table of common logarithms can be used. A table such as the one at the end of this appendix is limited in accuracy to three significant digits (four if interpolation is used).

If n is any positive number, n can be written in scientific notation as

$$n = m \times 10^c,$$

where $1 \le m < 10$ and c is an integer. Using the product rule,

$$\log n = \log (m \times 10^c) = \log m + \log 10^c = \log m + c.$$

Log m, the **mantissa** of log n, is a decimal greater than or equal to 0 and less than 1. The **characteristic** c of log n is an integer. Since any positive number can be written in scientific notation, the table of common logarithms, with logarithms of numbers between 1.00 and 9.99 in increments of 0.01, gives us logarithms of any number with three significant digits. The left-hand column in the table shows numbers 1.0 through 9.9 while the numbers 0 through 9 head each of the other columns. To find log 1.23, for example, we look down the left column to 1.2, read across that row to the column headed 3, and find the decimal 0.0899.

n	0	1	2	3	4	5	6	7	8	9
1.0	.0000	.0043	.0086	.0128	.0170	.0212	.0253	.0294	.0334	.0374
1.1	.0414	.0453	.0492	.0531	.0569	.0607	.0645	.0682	.0719	.0755
1.2	.0792	.0828	.0864	.0899	.0934	.0969	.1004	.1038	.1072	.1106
1.3	.1139	.1173	.1206	.1239	.1271	.1303	.1335	.1367	.1399	.1430
1.4	.1461	.1492	.1523	.1553	.1584	.1614	.1644	.1673	.1703	.1732

2 FINDING COMMON LOGARITHMS

We now summarize the method for finding a common logarithm using the table.

To Find the Common Logarithm of a Given Number

1. Write the number in scientific notation.
2. Use the table of common logarithms to find the logarithm of the number between 1 and 10 (the mantissa of the logarithm of the given number).
3. The logarithm of the original number is the characteristic (the exponent on 10) plus the mantissa.

EXAMPLE 1 FINDING COMMON LOGARITHMS

(a) Find log 45,600.

$$\begin{aligned} \log 45{,}600 &= \log (4.56 \times 10^4) && \text{Scientific notation} \\ &= \log 4.56 + \log 10^4 && \log uv = \log u + \log v \\ &= \log 4.56 + 4 && \log 10^4 = 4 \\ &= 0.6590 + 4 && \text{From table of common logarithms} \\ &= 4.6590 \end{aligned}$$

(b) Find log 4.56.

$$\begin{aligned} \log 4.56 &= \log (4.56 \times 10^0) \\ &= \log 4.56 + \log 10^0 \\ &= 0.6590 + 0 \\ &= 0.6590 \end{aligned}$$

(c) Find log 0.00456.

$$\begin{aligned} \log 0.00456 &= \log (4.56 \times 10^{-3}) \\ &= \log 4.56 + \log 10^{-3} \\ &= 0.6590 + (-3) \\ &= 0.6590 - 3 && \textit{Not } -3.6590 \end{aligned}$$

PRACTICE EXERCISE 1

(a) Find log 283,000.

(b) Find log 2.83.

(c) Find log 0.0000283.

Answers: (a) 5.4518 (b) 0.4518 (c) 0.4518 − 5

3 STANDARD FORM OF A LOGARITHM

In Example 1(c), notice that we left the characteristic separate from the mantissa in what is called **standard form.** If we subtract, we obtain

$$0.6590 - 3 = -2.3410.$$

Although −2.3410 is a correct form for the logarithm of 0.00456, it does not directly display the characteristic (−3) and positive mantissa (0.6590). When tables are used, it is wise to leave the logarithm in standard form. (If we use a calculator to find logarithms, the result is displayed in the latter form. You should be able to convert from one form of a logarithm to the other. Changing from

$$0.6590 - 3 \quad \text{to} \quad -2.3410$$

is simply a matter of subtracting. Converting from

$$-2.3410 \quad \text{to} \quad 0.6590 - 3$$

is a bit more challenging. In this case we add and subtract 3 (any integer greater than the absolute value of −2.3410 will work, and 3 is the smallest such integer).

$$\begin{array}{r} 3.0000 - 3 \\ -2.3410 \\ \hline 0.6590 - 3 \end{array}$$

Suppose we add and subtract 10.

$$\begin{array}{r} 10.0000 - 10 \\ -2.3410 \\ \hline 7.6590 - 10 \end{array}$$

We clearly obtain the same result since

$$\begin{aligned} 7.6590 - 10 &= 7 + 0.6590 - 10 \\ &= 0.6590 + 7 - 10 = 0.6590 - 3. \end{aligned}$$

4 FINDING ANTILOGARITHMS

Up to now we have illustrated how to find a logarithm of a given number. Now we show how to find a number when its logarithm is given, that is, how to find an **antilogarithm (antilog).** To use the table, the given logarithm must be in standard form showing the positive mantissa and the integer characteristic.

EXAMPLE 2 FINDING COMMON ANTILOGARITHMS

(a) Find the antilog of 2.7832. (That is, find $10^{2.7832}$.)
We need to find n in the equation

$$\begin{aligned} \log n &= 2.7832 \\ &= 0.7832 + 2. \end{aligned}$$

The mantissa is 0.7832, and the characteristic is 2. In the table we find 0.7832 in the row headed by 6.0 and in the column headed by 7.

n	0	1	2	3	4	5	6	7	8	9
5.5	.7404	.7412	.7419	.7427	.7435	.7443	.7451	.7459	.7466	.7474
5.6	.7482	.7490	.7497	.7505	.7513	.7520	.7528	.7536	.7543	.7551
5.7	.7559	.7566	.7574	.7582	.7589	.7597	.7604	.7612	.7619	.7627
5.8	.7634	.7642	.7649	.7657	.7664	.7672	.7679	.7686	.7694	.7701
5.9	.7709	.7716	.7723	.7731	.7738	.7745	.7752	.7760	.7767	.7774
6.0	.7782	.7789	.7796	.7803	.7810	.7818	.7825	**.7832**	.7839	.7846
6.1	.7853	.7860	.7868	.7875	.7882	.7889	.7896	.7903	.7910	.7917
6.2	.7924	.7931	.7938	.7945	.7952	.7959	.7966	.7973	.7980	.7987
6.3	.7993	.8000	.8007	.8014	.8021	.8028	.8035	.8041	.8048	.8055
6.4	.8062	.8069	.8075	.8082	.8089	.8096	.8102	.8109	.8116	.8122

$$\begin{aligned} n &= 6.07 \times 10^2 \\ &= 607 \end{aligned}$$

Thus, $\quad 10^{2.7832} = 607.$

(b) Find the antilog of −2.2321.
We must first express

$$\log n = -2.2321$$

in standard form by adding and subtracting 3.

PRACTICE EXERCISE 2

(a) Find the antilog of 5.8075.

(b) Find the antilog of −1.5935

$$\begin{array}{r} 3.0000 - 3 \\ -2.2321 \\ \hline 0.7679 - 3 \end{array}$$

Thus, the mantissa is .7679 and the characteristic is -3. From the table,

$$\begin{aligned} n &= 5.86 \times 10^{-3} \\ &= 0.00586. \end{aligned}$$

Answers: (a) 642,000 (b) 0.0255

5 LINEAR INTERPOLATION

To find the common logarithm of a number with four significant digits, such as 5138, we can use a process called **linear interpolation.** The word *linear* appears because the process uses a straight line to approximate the portion of the graph of $y = \log x$ between two values in the table. Consider Figure A.1.

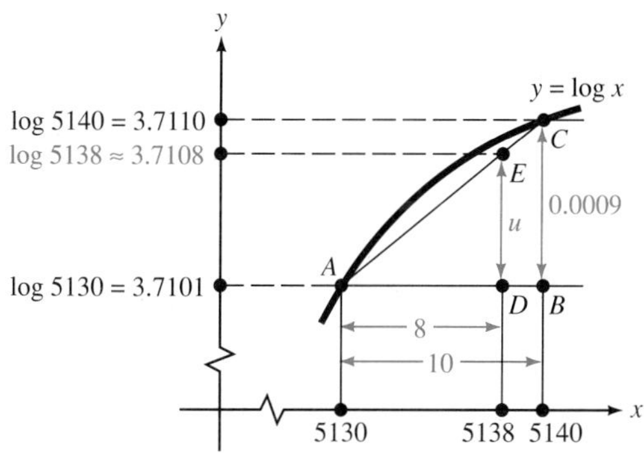

Figure A.1 Interpolation

Noting that triangles *ABC* and *ADE* are similar, we conclude that

$$\frac{8}{10} = \frac{u}{0.0009}.$$

Thus,
$$u = \frac{8}{10}(0.0009) \approx 0.0007.$$

To obtain the approximation for log 5138, add 0.0007 to log 5130.

$$\log 5138 \approx \log 5130 + 0.0007 = 3.7101 + 0.0007 = 3.7108.$$

EXAMPLE 3 Using Linear Interpolation (Logarithms)

Use linear interpolation to find log 0.006412. The characteristic is -3 so we concentrate on finding the mantissa for log 6.412.

		x	$\log x$		
0.010	0.002	6.410	0.8069	u	0.0006
		6.412	?		
		6.420	0.8075		

$$\frac{0.002}{0.010} = \frac{2}{10} = \frac{u}{0.0006}$$

Practice Exercise 3

Use linear interpolation to find log 75,620.

$$u = \frac{2}{10}(0.0006) \approx 0.0001$$

$$\begin{aligned}\log 6.412 &\approx \log 6.41 + 0.0001\\ &= 0.8069 + 0.0001\\ &= 0.8070\\ \log 0.006412 &= 0.8070 - 3 = -2.1930\end{aligned}$$

Answer: 4.8786

If a calculation requires finding the antilog of a logarithm whose mantissa is not in the table, linear interpolation can be used to approximate the four-digit answer.

EXAMPLE 4 USING LINEAR INTERPOLATION (ANTILOGS)

Find the antilog of 2.4705.

The characteristic 2 tells us where to place the decimal point in the final answer. Thus we can concentrate on 0.4705. In the table, 0.4705 is between log 2.95 = 0.4698 and log 2.96 = 0.4713.

		x	$\log x$		
0.010	u	2.950	0.4698	0.0007	0.0015
		?	0.4705		
		2.960	0.4713		

$$\frac{u}{0.010} = \frac{0.0007}{0.0015} = \frac{7}{15}$$

$$u = \frac{7}{15}(0.010) \approx 0.005$$

$$\log 2.955 \approx 0.4705$$

Thus, the antilog of 2.4705 is approximately

$$2.955 \times 10^2 = 295.5.$$

PRACTICE EXERCISE 4

Find the antilog of -1.6329.

Answer: 0.02329

6 LOGARITHMIC COMPUTATION

In the past, difficult and time-consuming calculations such as

$$\frac{(0.0325) \cdot (42.3)^2}{\sqrt[3]{1.07}}$$

were often done using logarithms. Products, quotients, and powers of decimals can be found by converting to logarithms and adding, subtracting, and multiplying the results. Since, for example, adding two decimals is easier than multiplying them, logarithmic calculation was a time-saving device. Today, however, the calculator has made logarithmic computation nearly obsolete. Nevertheless, in order to understand better the rules of logarithms and to learn more about logarithms in general, we briefly illustrate the techniques of logarithmic computation in the following examples.

EXAMPLE 5 Logarithmic Computation

Use logarithms to find N to three significant digits if $N = \dfrac{(1.06)^{20}}{1.35}$.

$$N = \frac{(1.06)^{20}}{1.35}$$

$$\log N = \log \frac{(1.06)^{20}}{1.35} \qquad \text{If } x = y\text{, then } \log x = \log y$$

$$= \log (1.06)^{20} - \log 1.35 \qquad \text{Quotient rule}$$

$$= 20 \log 1.06 - \log 1.35 \qquad \text{Power rule}$$

$$= 20(0.0253) - 0.1303$$

$$= 0.3757 \qquad \text{Remember that this is } \log N$$

N is the antilog of 0.3757.

$$N = 2.38 \times 10^0 \qquad \text{To three significant digits}$$

$$= 2.38$$

Practice Exercise 5

Use logarithms to find N to three significant digits if $N = \dfrac{(1.15)^{42}}{(1.02)^{15}}$.

$$N = \frac{(1.15)^{42}}{(1.02)^{15}}$$

$$\log N = \log \frac{(1.15)^{42}}{(1.02)^{15}}$$

$$= \log (1.15)^{42} - \log (1.02)^{15}$$

$$= 42 \log 1.15 - 15 \log 1.02$$

Answer: 263

EXAMPLE 6 Logarithmic Computation

Use logarithms to find N to three significant digits if $N = \sqrt[5]{21.5}\,(0.0349)^3$.

$$N = (21.5)^{1/5}(0.0349)^3 \qquad \text{Convert to fractional exponent}$$

$$\log N = \log [(21.5)^{1/5}(0.0349)^3] \qquad \text{Take log of both sides}$$

$$= \log (21.5)^{1/5} + \log (0.0349)^3 \qquad \text{Product rule}$$

$$= \frac{1}{5} \log 21.5 + 3 \log 0.0349 \qquad \text{Power rule}$$

$$= \frac{1}{5}(1.3324) + 3(0.5428 - 2) \qquad \text{Watch the characteristics}$$

$$= 0.2665 + 1.6284 - 6$$

$$= 1.8949 - 6$$

$$= 0.8949 - 5 \qquad \text{This is } \log N\text{, not } N$$

$$N = 7.85 \times 10^{-5} \qquad \text{Find the antilog: } 10^{0.8949} = 7.85$$

$$= 0.0000785$$

Practice Exercise 6

Use logarithms to find N to three significant digits if $N = \sqrt{3160}\,(0.00871)^6$.

Answer: 2.45×10^{-11}

These examples show the usefulness of logarithms for power calculations when a calculator with a $\boxed{y^x}$ key is not available. When performing logarithmic calculations, write the problem in detail, as above. If this is not done, we may find ourselves multiplying or dividing when we should be adding or subtracting, and we might not recognize that in the last step, an antilog must be found.

EXAMPLE 7 Logarithmic Computation

Use logarithms to find N correct to three significant digits if $N = \dfrac{\sqrt[3]{0.0943}}{251}$.

$$N = \frac{(0.0943)^{1/3}}{251}$$

$$\begin{aligned}\log N &= \log \frac{(0.0943)^{1/3}}{251}\\ &= \log (0.0943)^{1/3} - \log 251\\ &= \frac{1}{3}\log (0.0943) - \log 251\\ &= \frac{1}{3}(0.9745 - 2) - 2.3997\end{aligned}$$

If we divide $0.9745 - 2$ by 3, we no longer have an integer characteristic, since 2 is not evenly divisible by 3. Thus, we need to convert $0.9745 - 2$ to $1.9745 - 3$ and then divide by 3 to obtain

$$\log N = (0.6582 - 1) - (2.3997).$$

Next, we change the form of the characteristic in $0.6582 - 1$ to $9.6582 - 10$ in order to subtract 2.3997 and keep the mantissa positive.

$$\begin{array}{r} 9.6582 - 10\\ -2.3997\\ \hline 7.2585 - 10\end{array}$$

Then

$$\begin{aligned}\log N &= 7.2585 - 10\\ &= 0.2585 - 3\\ N &= 1.81 \times 10^{-3} \quad \text{Find the antilog: } 10^{0.2585} = 1.81\\ &= 0.00181.\end{aligned}$$

Practice Exercise 7

Use logarithms to find N correct to three significant digits if $N = \dfrac{(216)^{1/2}}{\sqrt[5]{0.00375}}$.

Answer: 44.9

The two difficulties we met in the preceding example are not as bothersome when a calculator is used to find logarithms. In such cases, the nonnegative mantissas necessary for using the table of common logarithms need not be kept.

EXERCISES A

1. The common logarithm of a number can be expressed in two parts: an integer, called the **(a)** ________________ and a decimal part between 0 and 1, called the **(b)** ________________.

2. Given that $\log 73.2 = 1.8645$, the characteristic is **(a)** ________________, and the mantissa is **(b)** ________________.

3. Given that $\log 0.0732 = 0.8645 - 2$, the characteristic is **(a)** ________________, and the mantissa is **(b)** ________________.

4. Given that $\log 7.32 = 0.8645$, the characteristic is **(a)** ________________, and the mantissa is **(b)** ________________.

5. When using the table of common logarithms to find the antilog of 3.9974, we find the number whose mantissa is **(a)** ________________ and multiply it by 10 to the **(b)** ________________ power.

Using the table of common logarithms, find the following.

6. log 0.00759 **7.** log 7590 **8.** log 0.376

9. log 5 **10.** log 0.5 **11.** log 500

Convert each logarithm to standard form, which displays the positive mantissa and the characteristic.

12. -5.1258 **13.** -2.0013 **14.** -0.3176

Using the table of common logarithms, find the antilogarithm of each number.

15. $0.6571 - 1$ **16.** 4.9754 **17.** $0.9090 - 3$

18. -1.4318 **19.** -0.9508 **20.** $9.8401 - 10$

Use the table and linear interpolation to approximate the logarithm of each number, correct to four decimal places.

21. 3.278 **22.** 6399 **23.** 437.6

24. 12,420 **25.** 0.003972 **26.** 0.05979

Use the table and linear interpolation to approximate the antilogarithm of each number, correct to four significant digits.

27. 1.6974 **28.** 2.2390 **29.** $0.5409 - 2$

30. $0.9738 - 1$ **31.** -3.1155 **32.** -2.3031

Use logarithmic computation to find N, correct to three significant digits.

33. $N = \dfrac{(1.07)^{30}}{5.13}$ **34.** $N = \sqrt[7]{32.8}\,(0.491)^3$ **35.** $N = (1.04)^{25}(3.78)$

$$\log N = \log \frac{(1.07)^{30}}{5.13}$$
$$= \log (1.07)^{30} - \log 5.13$$
$$= 30 \log 1.07 - \log 5.13$$
$$= 30(\qquad) - (\qquad)$$

$$\log N = \log [\sqrt[7]{32.8}\,(0.491)^3]$$
$$= \log (32.8)^{1/7} + \log (0.491)^3$$
$$= \frac{1}{7} \log (32.8) + 3 \log (0.491)$$
$$=$$

36. $N = [(23.5)(1.08)]^{1/5}$

37. $N = \dfrac{(30{,}700)^{1/3}}{(22.5)^3}$

38. $N = [\sqrt{1.03}\,(1.08)^3]^{2/3}$

39. $N = \dfrac{(0.00351)^{1/4}}{(4.12)^2}$

40. $N = \dfrac{(3.78)^2(0.135)^{1/2}}{7280}$

41. $N = \dfrac{(6810)(1.21)^{40}}{225}$

ANSWERS: 1. (a) characteristic (b) mantissa 2. (a) 1 (b) 0.8645 3. (a) −2 (b) 0.8645 4. (a) 0 (b) 0.8645 5. (a) 0.9974 (b) 3rd 6. 0.8802 − 3 or −2.1198 7. 3.8802 8. 0.5752 − 1 or −0.4248 9. 0.6990 10. 0.6990 − 1 or −0.3010 11. 2.6990 12. 0.8742 − 6 13. 0.9987 − 3 14. 0.6824 − 1 15. 0.454 16. 94,500 17. 0.00811 18. 0.0370 19. 0.112 20. 0.692 21. 0.5156 22. 3.8061 23. 2.6411 24. 4.0941 25. 0.5990 − 3 or −2.4010 26. 0.7766 − 2 or −1.2234 27. 49.82 28. 173.4 29. 0.03475 30. 0.9415 31. 0.0007665 32. 0.004976 33. 1.48 34. 0.195 35. 10.1 36. 1.91 37. 0.00275 38. 1.18 39. 0.0143 40. 0.000721 41. 62,000

EXERCISES B

1. When a logarithm is written so that the characteristic is separate from the mantissa, the logarithm is in ________________ form.

2. Given that log 25.7 = 1.4099, the characteristic is **(a)** ________________ and the mantissa is **(b)** ________________.

3. Given that log 0.0257 = 0.4099 − 2, the characteristic is **(a)** ________________ and the mantissa is **(b)** ________________.

4. Given that log 2.57 = 0.4099, the characteristic is **(a)** ________________ and the mantissa is **(b)** ________________.

5. When using the table of common logarithms to find the antilog of 4.0691, we find the number whose mantissa is **(a)** ________________ and multiply it by 10 to the **(b)** ________________ power.

Using the table of common logarithms, find the following.

6. log 0.00325

7. log 3250

8. log 0.211

9. log 9

10. log 0.9

11. log 900

Convert each logarithm to standard form, which displays the positive mantissa and the characteristic.

12. −6.2215

13. −3.0045

14. −0.2987

Using the table of common logarithms, find the antilogarithm of each number.

15. 0.0607 − 1

16. 5.2201

17. −2.3382

18. −1.3655

19. −0.3002

20. 9.2405 − 10

Use the table and linear interpolation to approximate the logarithm of each number, correct to four decimal places.

21. 7429

22. 6.157

23. 29,340

24. 249.7

25. 0.03995

26. 0.0004256

Use the table and linear interpolation to approximate the antilogarithm of each number, correct to four significant digits.

27. 2.4675

28. 1.2390

29. 0.9286 − 1

30. 3.6754

31. −3.1595

32. −2.8118

Use logarithmic computation to find N, correct to three significant digits.

33. $N = \dfrac{(1.12)^{25}}{62.5}$

34. $N = \sqrt[6]{526}\,(0.028)^4$

35. $N = (1.07)^{42}(5.69)$

36. $N = [(52.6)(1.52)]^{1/5}$

37. $N = \dfrac{(426{,}000)^{1/4}}{(0.0911)^2}$

38. $N = [\sqrt[3]{5.6}\,(7.22)^3]^{1/6}$

39. $N = \dfrac{(291{,}000)^{2/5}}{(0.0861)^3}$

40. $N = \dfrac{(7.99)^3(37.2)^{3/2}}{0.222}$

41. $N = \dfrac{(4090)(1.13)^5}{6630}$

EXERCISES C

Use logarithms to find N, correct to three significant digits.

1. $N = \dfrac{\sqrt[9]{(1.25)^5(0.00367)^2}}{\sqrt[4]{(0.0169)^3}}$

[Answer: 6.95]

2. $N = \sqrt[8]{[(1.08)^3(0.0569)^2]^7}$

TABLE OF COMMON LOGARITHMS

n	0	1	2	3	4	5	6	7	8	9
1.0	.0000	.0043	.0086	.0128	.0170	.0212	.0253	.0294	.0334	.0374
1.1	.0414	.0453	.0492	.0531	.0569	.0607	.0645	.0682	.0719	.0755
1.2	.0792	.0828	.0864	.0899	.0934	.0969	.1004	.1038	.1072	.1106
1.3	.1139	.1173	.1206	.1239	.1271	.1303	.1335	.1367	.1399	.1430
1.4	.1461	.1492	.1523	.1553	.1584	.1614	.1644	.1673	.1703	.1732
1.5	.1761	.1790	.1818	.1847	.1875	.1903	.1931	.1959	.1987	.2014
1.6	.2041	.2068	.2095	.2122	.2148	.2175	.2201	.2227	.2253	.2279
1.7	.2304	.2330	.2355	.2380	.2405	.2430	.2455	.2480	.2504	.2529
1.8	.2553	.2577	.2601	.2625	.2648	.2672	.2695	.2718	.2742	.2765
1.9	.2788	.2810	.2833	.2856	.2878	.2900	.2923	.2945	.2967	.2989
2.0	.3010	.3032	.3054	.3075	.3096	.3118	.3139	.3160	.3181	.3201
2.1	.3222	.3243	.3263	.3284	.3304	.3324	.3345	.3365	.3385	.3404
2.2	.3424	.3444	.3464	.3483	.3502	.3522	.3541	.3560	.3579	.3598
2.3	.3617	.3636	.3655	.3674	.3692	.3711	.3729	.3747	.3766	.3784
2.4	.3802	.3820	.3838	.3856	.3874	.3892	.3909	.3927	.3945	.3962
2.5	.3979	.3997	.4014	.4031	.4048	.4065	.4082	.4099	.4116	.4133
2.6	.4150	.4166	.4183	.4200	.4216	.4232	.4249	.4265	.4281	.4298
2.7	.4314	.4330	.4346	.4362	.4378	.4393	.4409	.4425	.4440	.4456
2.8	.4472	.4487	.4502	.4518	.4533	.4548	.4564	.4579	.4594	.4609
2.9	.4624	.4639	.4654	.4669	.4683	.4698	.4713	.4728	.4742	.4757
3.0	.4771	.4786	.4800	.4814	.4829	.4843	.4857	.4871	.4886	.4900
3.1	.4914	.4928	.4942	.4955	.4969	.4983	.4997	.5011	.5024	.5038
3.2	.5051	.5065	.5079	.5092	.5105	.5119	.5132	.5145	.5159	.5172
3.3	.5185	.5198	.5211	.5224	.5237	.5250	.5263	.5276	.5289	.5302
3.4	.5315	.5328	.5340	.5353	.5366	.5378	.5391	.5403	.5416	.5428
3.5	.5441	.5453	.5465	.5478	.5490	.5502	.5514	.5527	.5539	.5551
3.6	.5563	.5575	.5587	.5599	.5611	.5623	.5635	.5647	.5658	.5670
3.7	.5682	.5694	.5705	.5717	.5729	.5740	.5752	.5763	.5775	.5786
3.8	.5798	.5809	.5821	.5832	.5843	.5855	.5866	.5877	.5888	.5899
3.9	.5911	.5922	.5933	.5944	.5955	.5966	.5977	.5988	.5999	.6010
4.0	.6021	.6031	.6042	.6053	.6064	.6075	.6085	.6096	.6107	.6117
4.1	.6128	.6138	.6149	.6160	.6170	.6180	.6191	.6201	.6212	.6222
4.2	.6232	.6243	.6253	.6263	.6274	.6284	.6294	.6304	.6314	.6325
4.3	.6335	.6345	.6355	.6365	.6375	.6385	.6395	.6405	.6415	.6425
4.4	.6435	.6444	.6454	.6464	.6474	.6484	.6493	.6503	.6513	.6522
4.5	.6532	.6542	.6551	.6561	.6571	.6580	.6590	.6599	.6609	.6618
4.6	.6628	.6637	.6646	.6656	.6665	.6675	.6684	.6693	.6702	.6712
4.7	.6721	.6730	.6739	.6749	.6758	.6767	.6776	.6785	.6794	.6803
4.8	.6812	.6821	.6830	.6839	.6848	.6857	.6866	.6875	.6884	.6893
4.9	.6902	.6911	.6920	.6928	.6937	.6946	.6955	.6964	.6972	.6981
5.0	.6990	.6998	.7007	.7016	.7024	.7033	.7042	.7050	.7059	.7067
5.1	.7076	.7084	.7093	.7101	.7110	.7118	.7126	.7135	.7143	.7152
5.2	.7160	.7168	.7177	.7185	.7193	.7202	.7210	.7218	.7226	.7235
5.3	.7243	.7251	.7259	.7267	.7275	.7284	.7292	.7300	.7308	.7316
5.4	.7324	.7332	.7340	.7348	.7356	.7364	.7372	.7380	.7388	.7396
n	0	1	2	3	4	5	6	7	8	9

n	0	1	2	3	4	5	6	7	8	9
5.5	.7404	.7412	.7419	.7427	.7435	.7443	.7451	.7459	.7466	.7474
5.6	.7482	.7490	.7497	.7505	.7513	.7520	.7528	.7536	.7543	.7551
5.7	.7559	.7566	.7574	.7582	.7589	.7597	.7604	.7612	.7619	.7627
5.8	.7634	.7642	.7649	.7657	.7664	.7672	.7679	.7686	.7694	.7701
5.9	.7709	.7716	.7723	.7731	.7738	.7745	.7752	.7760	.7767	.7774
6.0	.7782	.7789	.7796	.7803	.7810	.7818	.7825	.7832	.7839	.7846
6.1	.7853	.7860	.7868	.7875	.7882	.7889	.7896	.7903	.7910	.7917
6.2	.7924	.7931	.7938	.7945	.7952	.7959	.7966	.7973	.7980	.7987
6.3	.7993	.8000	.8007	.8014	.8021	.8028	.8035	.8041	.8048	.8055
6.4	.8062	.8069	.8075	.8082	.8089	.8096	.8102	.8109	.8116	.8122
6.5	.8129	.8136	.8142	.8149	.8156	.8162	.8169	.8176	.8182	.8189
6.6	.8195	.8202	.8209	.8215	.8222	.8228	.8235	.8241	.8248	.8254
6.7	.8261	.8267	.8274	.8280	.8287	.8293	.8299	.8306	.8312	.8319
6.8	.8325	.8331	.8338	.8344	.8351	.8357	.8363	.8370	.8376	.8382
6.9	.8388	.8395	.8401	.8407	.8414	.8420	.8426	.8432	.8439	.8445
7.0	.8451	.8457	.8463	.8470	.8476	.8482	.8488	.8494	.8500	.8506
7.1	.8513	.8519	.8525	.8531	.8537	.8543	.8549	.8555	.8561	.8567
7.2	.8573	.8579	.8585	.8591	.8597	.8603	.8609	.8615	.8621	.8627
7.3	.8633	.8639	.8645	.8651	.8657	.8663	.8669	.8675	.8681	.8686
7.4	.8692	.8698	.8704	.8710	.8716	.8722	.8727	.8733	.8739	.8745
7.5	.8751	.8756	.8762	.8768	.8774	.8779	.8785	.8791	.8797	.8802
7.6	.8808	.8814	.8820	.8825	.8831	.8837	.8842	.8848	.8854	.8859
7.7	.8865	.8871	.8876	.8882	.8887	.8893	.8899	.8904	.8910	.8915
7.8	.8921	.8927	.8932	.8938	.8943	.8949	.8954	.8960	.8965	.8971
7.9	.8976	.8982	.8987	.8993	.8998	.9004	.9009	.9015	.9020	.9025
8.0	.9031	.9036	.9042	.9047	.9053	.9058	.9063	.9069	.9074	.9079
8.1	.9085	.9090	.9096	.9101	.9106	.9112	.9117	.9122	.9128	.9133
8.2	.9138	.9143	.9149	.9154	.9159	.9165	.9170	.9175	.9180	.9186
8.3	.9191	.9196	.9201	.9206	.9212	.9217	.9222	.9227	.9232	.9238
8.4	.9243	.9248	.9253	.9258	.9263	.9269	.9274	.9279	.9284	.9289
8.5	.9294	.9299	.9304	.9309	.9315	.9320	.9325	.9330	.9335	.9340
8.6	.9345	.9350	.9355	.9360	.9365	.9370	.9375	.9380	.9385	.9390
8.7	.9395	.9400	.9405	.9410	.9415	.9420	.9425	.9430	.9435	.9440
8.8	.9445	.9450	.9455	.9460	.9465	.9469	.9474	.9479	.9484	.9489
8.9	.9494	.9499	.9504	.9509	.9513	.9518	.9523	.9528	.9533	.9538
9.0	.9542	.9547	.9552	.9557	.9562	.9566	.9571	.9576	.9581	.9586
9.1	.9590	.9595	.9600	.9605	.9609	.9614	.9619	.9624	.9628	.9633
9.2	.9638	.9643	.9647	.9652	.9657	.9661	.9666	.9671	.9675	.9680
9.3	.9685	.9689	.9694	.9699	.9703	.9708	.9713	.9717	.9722	.9727
9.4	.9731	.9736	.9741	.9745	.9750	.9754	.9759	.9763	.9768	.9773
9.5	.9777	.9782	.9786	.9791	.9795	.9800	.9805	.9809	.9814	.9818
9.6	.9823	.9827	.9832	.9836	.9841	.9845	.9850	.9854	.9859	.9863
9.7	.9868	.9872	.9877	.9881	.9886	.9890	.9894	.9899	.9903	.9908
9.8	.9912	.9917	.9921	.9926	.9930	.9934	.9939	.9943	.9948	.9952
9.9	.9956	.9961	.9965	.9969	.9974	.9978	.9983	.9987	.9991	.9996
n	0	1	2	3	4	5	6	7	8	9

SOLUTIONS TO SELECTED EXERCISES

The exercises solved here are marked with a blue circle and white number in the A exercise sets.

1.4

41. $\dfrac{4 \cdot 5^2}{\sqrt{6^2 + 8^2}} = \dfrac{4 \cdot 25}{\sqrt{36 + 64}}$ Find all powers first

$= \dfrac{100}{\sqrt{100}} = \dfrac{100}{10} = 10$

42. $3 - (-2)\dfrac{4 - (-1)}{2 + (-3)} = 3 - (-2)\dfrac{5}{-1}$

$= 3 - (-2)(-5)$

$= 3 - 10 = -7$

1.5

15. $\dfrac{\sqrt{c^2 + 11}}{b} = \dfrac{\sqrt{(5)^2 + 11}}{-2}$ Substitute 5 for *c* and −2 for *b*

$= \dfrac{\sqrt{25 + 11}}{-2} = \dfrac{\sqrt{36}}{-2} = \dfrac{6}{-2} = -3$

25. $-x^2 = -(-3)^2$ Substitute −3 for *x*; use parentheses

$= -(9)$ Square before negating

$= -9$

26. $(-x)^2 = (-(-3))^2$ Substitute −3 for *x*

$= (3)^2$ Negate before squaring

$= 9$

Notice the difference between Exercises 25 and 26.

27. $2x^2 = 2(-3)^2$ Substitute −3 for *x*, use parentheses

$= 2(9)$ Square before multiplying by 2

$= 18$

34. $2w^2 = 2\left(-\frac{1}{2}\right)^2$ Substitute $-\frac{1}{2}$ for *w*

$= 2\left(\frac{1}{4}\right)$ Square first

$= \frac{1}{2}$

35. $(2w)^2 = \left(2\left(-\frac{1}{2}\right)\right)^2$

$= (-1)^2$ Multiply before squaring

$= 1$

65. $x - [3x - (1 - 2x)] = x - [3x - 1 + 2x]$

$= x - [5x - 1]$

$= x - 5x + 1$

$= -4x + 1$

1.6

32. $(3a)^{-1} = \dfrac{1}{(3a)^1} = \dfrac{1}{3a}$

33. $3a^{-1} = 3 \cdot \dfrac{1}{a} = \dfrac{3}{a}$

Compare Exercises 32 and 33. Notice that 3 is not raised to the −1 power in 33 while it is raised to the −1 power in 32.

34. $x^{-1} + a^{-1} = \dfrac{1}{x} + \dfrac{1}{a}$

35. $(x + a)^{-1} = \dfrac{1}{x + a}$

Notice that Exercises 34 and 35 are not the same. This is easy to see if you substitute values for *x* and *a*, say $x = 2$ and $a = 2$. Then 34 simplifies to $\frac{1}{2} + \frac{1}{2} = 1$, but 35 simplifies to $\frac{1}{2+2} = \frac{1}{4}$.

41. $\left(\dfrac{5^0x^{-2}}{3^{-1}y^{-2}}\right)^2 = \left(\dfrac{3x^{-2}}{y^{-2}}\right)^2$ $5^0 = 1$ and 3^{-1} moves to the numerator

$= \dfrac{3^2(x^{-2})^2}{(y^{-2})^2}$ Apply exponent of 2 to all factors

$= \dfrac{9x^{-4}}{y^{-4}}$ Use power rule

$= \dfrac{9y^4}{x^4}$ Switch powers from numerator to denominator and denominator to numerator, changing signs

44. $\left(\dfrac{x^4y^{-2}}{x^3y^{-4}}\right)^{-2}\left(\dfrac{x^{-1}y^{-3}}{x^{-4}y^5}\right)^{-3}$

Simplify inside parentheses first using the quotient rule.

$(x^{4-3}y^{-2-(-4)})^{-2}(x^{-1-(-4)}y^{-3-5})^{-3}$

$= (xy^2)^{-2}(x^3y^{-8})^{-3}$

$= x^{-2}(y^2)^{-2}(x^3)^{-3}(y^{-8})^{-3}$ $(ab)^n = a^nb^n$

$= x^{-2}y^{-4}x^{-9}y^{24}$ $(a^m)^n = a^{mn}$

$= x^{-11}y^{20}$ $a^ma^n = a^{m+n}$

$= \dfrac{y^{20}}{x^{11}}$ $a^{-n} = \frac{1}{a^n}$

45. $\dfrac{(x^{-2}y^{-1})^2(xy^{-5})^{-2}}{(3x^2y^{-1})^{-2}(2xy)^{-1}}$

$= \dfrac{(x^{-2})^2(y^{-1})^2x^{-2}(y^{-5})^{-2}}{3^{-2}(x^2)^{-2}(y^{-1})^{-2}2^{-1}x^{-1}y^{-1}}$ $(ab)^n = a^nb^n$

$= \dfrac{x^{-4}y^{-2}x^{-2}y^{10}}{3^{-2}x^{-4}y^2 2^{-1}x^{-1}y^{-1}}$ $(a^m)^n = a^{mn}$

$= \dfrac{x^{-6}y^8}{3^{-2}x^{-5}2^{-1}y}$ $a^ma^n = a^{m+n}$

$= 3^2 \cdot 2 \cdot x^{-6-(-5)}y^{8-1}$ $\frac{1}{a^{-n}} = a^n$ and $\frac{a^m}{a^n} = a^{m-n}$

$= 18x^{-1}y^7 = \dfrac{18y^7}{x}$

61. $(ab)^{-2} = ((-1)(-1))^{-2}$ Substitute -1 for a and for b

$= (1)^{-2} = \frac{1}{(1)^2} = \frac{1}{1} = 1$

62. $ab^{-2} = (-1)(-1)^{-2} = \frac{(-1)}{(-1)^2} = \frac{-1}{1} = -1$

Notice in Exercise 61 that both a and b are raised to the -2 power while in 62 only b is.

63. $a^0 - b^2 = (-1)^0 - (-1)^2$ Substitute -1 for a and -1 for b

$= 1 - 1$ $\quad (-1)^0 = 1$

$= 0$

2.1

34. $2.1y + 45.2 = 3.2 - 8.4y$

$21y + 452 = 32 - 84y$ Multiply through by 10 to clear decimals

$105y = -420$ Add $84y$ and subtract 452 from both sides

$y = -4$ Divide by 105

35. $3z + \frac{3}{2} + \frac{5}{2}z = \frac{1}{2}z + \frac{5}{2}z$

$6z + 3 + 5z = z + 5z$ Multiply through by 2

$11z + 3 = 6z$ Collect like terms

$5z = -3$ Subtract $6z$ and 3 from both sides

$z = -\frac{3}{5}$ Divide by 5

43. $2[x - (x + 1)] = -2$

$2[x - x - 1] = -2$ Clear parentheses

$2[-1] = -2$

$-2 = -2$

Since we obtain an identity, every real number is a solution.

47. $\frac{x+1}{3} - \frac{x+7}{2} = 1$

$6\left[\frac{x+1}{3} - \frac{x+7}{2}\right] = 6 \cdot 1$ Multiply both sides by 6 to clear fractions

$2(x + 1) - 3(x + 7) = 6$

$2x + 2 - 3x - 21 = 6$

$-x - 19 = 6$

$-x = 25$

$x = -25$

51. Substitute -1 for y and solve the result for k.

$$3y + 2 = k - 1$$
$$3(-1) + 2 = k - 1$$
$$-1 = k - 1$$
$$0 = k$$

When k is 0, the equation becomes

$$3y + 2 = -1,$$

which has solution -1.

53. First solve $5x - 1 = 9$.

$$5x = 10$$
$$x = 2$$

Substitute 2 for x in the first equation.

$$3x + 1 = m$$
$$3(2) + 1 = m$$
$$7 = m$$

When m is 7, the equation becomes

$$3x + 1 = 7,$$

which has the same solution as (is equivalent to) $5x - 1 = 9$.

2.2

32. Let x = the length of the pole,

$\frac{1}{6}x$ = the length of the pole in the sand,

25 = the length of the pole in the water,

$\frac{2}{3}x$ = the length of the pole in the air.

The equation to solve is

$$x = \frac{1}{6}x + 25 + \frac{2}{3}x.$$

$6x = 6\left[\frac{1}{6}x + 25 + \frac{2}{3}x\right]$ Multiply by 6 to clear fractions

$6x = x + 150 + 4x$

$6x = 5x + 150$

$x = 150$

The length of the pole is 150 ft.

33. Let x = the score Becky must make on the fourth test. To average four scores, we add them and divide by 4. Thus, we need to solve

$$\frac{96 + 78 + 91 + x}{4} = 90.$$

$\not{4}\left[\frac{96 + 78 + 91 + x}{\not{4}}\right] = 4 \cdot 90$ Multiply both sides by 4

$96 + 78 + 91 + x = 360$

$265 + x = 360$

$x = 95$

Becky must get at least a 95 on the fourth test.

34. Let x = the number of miles Charlene drove. We need to solve

$$2(12.50) + (0.10)x = 73.$$
$$25 + 0.10x = 73$$
$$0.10x = 48$$
$$x = 480$$

Charlene drove 480 miles.

41. Let x = rise in the water level. Since the volume of water is a rectangular solid of length 10 ft, width 8 ft, and height x ft, we know that the volume is

$$(10)(8)x = 80x \text{ ft}^3.$$

But the volume is also equal to the volume of the sphere, which is about 33.5 ft^3. Thus, we need to solve

$$80x = 33.5.$$
$$x = \frac{33.5}{80} \approx 0.4 \text{ ft}$$

2.3

24. Let x = amount of fuel added. The increase in fuel is $1182 - 985 = 197$ gal. The question asked is: 197 is what % of 985? This translates to:

$$197 = (x)(985).$$
$$\frac{197}{985} = x$$
$$0.2 = x$$

The percent increase is 20%.

25. Let x = the population in 1980. The increase in population in 1990 was 35% of x, or $0.35x$. Thus, the population in 1990 was the population in 1980 plus the increase. Then we need to solve

$$x + 0.35x = 567.$$
$$(1 + 0.35)x = 567$$
$$1.35x = 567$$
$$x = \frac{567}{1.35}$$
$$x = 420$$

The population in 1980 was 420.

27. Let x = the amount to invest. Over a five-year period, 11% simple interest is 55%. The equation to solve is

$$x + 0.55x = 11{,}470.$$
$$1.55x = 11{,}470$$
$$x = 7400$$

Paul must invest $7400.

28. Let x = amount invested in a savings account,
$10{,}000 - x$ = amount invested in a mutual fund,
$0.12x$ = earnings in the savings account,
$0.15(10{,}000 - x)$ = earnings in the mutual fund.
We need to solve

$$0.12x + 0.15(10{,}000 - x) = 1275.$$
$$0.12x + 1500 - 0.15x = 1275$$
$$-0.03x = -225$$
$$x = 7500$$
$$10{,}000 - x = 2500$$

Lucky Lucia invested $7500 in the savings account and $2500 in the mutual fund.

31. Let x = the number of miles Jeff drove. If y is the bill less sales tax, then $y + 0.04y = 199.68$. Thus $y(1.04) = 199.68$ so that $y = 192$. We must solve

$$(4)(13) + (0.08)x = 192.$$
$$0.08x = 140$$
$$x = 1750$$

Jeff drove a total of 1750 miles.

34. Let t = the time the woman walks,
$7 - t$ = the time she rides,
$4.5t$ = the distance she walks,
$2.5(7 - t)$ = the distance she rides.
Since the distance down and the distance back out are equal, we need to solve

$$4.5t = 2.5(7 - t).$$
$$4.5t = 17.5 - 2.5t$$
$$7t = 17.5$$
$$t = 2.5$$

She walks 2.5 hours, and the total length of the trip is $2(4.5)(2.5) = 22.5$ miles.

35. Let t = the time of travel,
$55t$ = the distance that one travels,
$50t$ = the distance that the second travels.
When they meet, the total of the two distances traveled is 315, so the equation to solve is

$$55t + 50t = 315.$$
$$105t = 315$$
$$t = 3$$

Since they traveled for 3 hr and left at 9:30 A.M., they will meet at 12:30 P.M.

38. Let x = rate for one press,
$x - 15$ = rate for second press,
$150x$ = number flyers printed by one,
$150(x - 15)$ = number flyers printed by other.

Then we need to solve

$$150x + 150(x - 15) = 14{,}850.$$
$$150x + 150x - 2250 = 14{,}850$$
$$300x = 12{,}600$$
$$x = \frac{12{,}600}{300} = 42$$
$$x - 15 = 42 - 15 = 27$$

One press prints 42 fliers/min and the other 27 fliers/min.

39. Let t = time for cutter to overtake the private boat,
$35t$ = distance cutter travels.
Since the private boat has traveled 10 nautical miles before the cutter starts, the distance the private boat travels can be written as

$$10 + 20t.$$

The two distances are equal so we solve

$$35t = 10 + 20t.$$
$$15t = 10$$
$$t = \frac{10}{15} = \frac{2}{3}$$

It takes $\frac{2}{3}$ hr or 40 minutes for the cutter to overtake the private boat.

2.4

14. $$A = \frac{1}{2}(b_1 + b_2)h$$
$$2A = (b_1 + b_2)h \quad \text{Multiply both sides by 2}$$
$$2A = b_1h + b_2h \quad \text{Remove parentheses}$$
$$2A - b_2h = b_1h \quad \text{Subtract } b_2h$$
$$\frac{2A - b_2h}{h} = b_1 \quad \text{Divide by } h$$

26. $$2(x - b) + 5x = 4c$$
$$2x - 2b + 5x = 4c$$
$$7x - 2b = 4c$$
$$7x = 4c + 2b$$
$$x = \frac{4c + 2b}{7}$$

2.5

11. $$\frac{x+3}{-2} < 17$$
$$(-2)\left[\frac{x+3}{-2}\right] > (-2)(17) \quad \text{Multiply both sides by } -2 \text{ and reverse inequality}$$
$$x + 3 > -34 \quad \text{Do not reverse inequality when subtracting 3}$$
$$x > -37$$

The solution is $(-37, \infty)$.

16. $$5(3 - x) - 10 \le 25$$
$$15 - 5x - 10 \le 25$$
$$-5x + 5 \le 25$$
$$-5x \le 20$$
$$x \ge -4 \quad \text{Divide both sides by } -5 \text{ and reverse inequality}$$

The solution is $[-4, \infty)$.

28. Let x = the unknown number,
$\frac{1}{3}x$ = one-third the number.

If $\frac{1}{3}x$ is no more than 15, we solve

$$\frac{1}{3}x \le 15.$$
$$x \le 45 \quad \text{Multiply by 3}$$

The solution is $(-\infty, 45]$.

29. Let s = Walt's score on the fourth test. The average of his four scores is

$$\frac{86 + 83 + 97 + s}{4} = \frac{266 + s}{4}.$$

Since the average must be 90 or better, we need to solve

$$\frac{266 + s}{4} \ge 90.$$
$$266 + s \ge (4)(90) \quad \text{Clear fraction}$$
$$266 + s \ge 360$$
$$s \ge 94$$

Walt's score must be 94 or better.

30. Since S must exceed (be greater than) C, and $S = 150n$ while $C = 125n + 350$, we need to solve

$$150n > 125n + 350.$$
$$25n > 350 \quad \text{Subtract } 125n \text{ from both sides}$$
$$n > \frac{350}{25} \quad \text{Divide both sides by 25}$$
$$n > 14$$

To make a profit, the owner must sell 15 or more sets.

2.6

15. Let s = Jake's score on the fourth test. We need to solve

$$80 \le \frac{69 + 82 + 79 + s}{4} \le 90.$$
$$320 \le 69 + 82 + 79 + s \le 360$$
$$320 \le 230 + x \le 360$$
$$90 \le x \le 130$$

But the maximum percent is 100%. Thus, Jake must score between 90% and 100% (he can't score more than 100%), inclusive.

2.7

14. We need to solve

$$\frac{1}{2}z + 1 = \frac{3}{4}z - 1 \quad \text{and} \quad \frac{1}{2}z + 1 = -\left(\frac{3}{4}z - 1\right).$$

$$4\left(\frac{1}{2}z + 1\right) = 4\left(\frac{3}{4}z - 1\right) \quad 4\left(\frac{1}{2}z + 1\right) = -4\left(\frac{3}{4}z - 1\right)$$

$$2z + 4 = 3z - 4 \qquad 2z + 4 = -3z + 4$$

$$8 = z \qquad 5z = 0$$

$$z = 0$$

The two solutions are 8 and 0.

19. Let x = the desired number,
$|x + 1|$ = absolute value of the sum of the number and 1.

We need to solve $|x + 1| = |x|$. Thus, we get

$$x + 1 = x \quad \text{and} \quad x + 1 = -x.$$

$$1 = 0 \qquad 1 = -2x$$

$$\text{no solution} \qquad -\frac{1}{2} = x$$

The number is $-\frac{1}{2}$.

35. $\left|\frac{x+1}{2}\right| > 3$ This translates into the following.

$$\frac{x+1}{2} > 3 \quad \text{or} \quad \frac{x+1}{2} < -3$$

$$x + 1 > 6 \quad \text{or} \quad x + 1 < -6$$

$$x > 5 \quad \text{or} \quad x < -7$$

The solution is $(-\infty, -7)$ or $(5, \infty)$.

41. Let x = the desired number.
We need to solve $|2x + 3| \geq 7$, which translates into the following.

$$2x + 3 \leq -7 \quad \text{or} \quad 2x + 3 \geq 7$$

$$2x \leq -10 \quad \text{or} \quad 2x \geq 4$$

$$x \leq -5 \quad \text{or} \quad x \geq 2$$

3.1

39. To complete (1,   ), substitute 1 for x in $y = 200x + 30$ and solve for y.

$$y = 200(1) + 30 = 230$$

Similarly, when $x = 5$, $y = 200(5) + 30 = 1030$, and when $x = 10$, $y = 200(10) + 30 = 2030$.
(a) The cost of producing 1 part is \$230, taken from the ordered pair (1,230). **(b)** The cost of producing 5 parts is \$1030. **(c)** The cost of producing 10 parts is \$2030.

3.3

17. The slope of l_1 is

$$m_1 = \frac{2 - (-1)}{0 - 1} = \frac{3}{-1} = -3.$$

The slope of l_2 is

$$m_2 = \frac{-1 - 0}{0 - 3} = \frac{-1}{-3} = \frac{1}{3}.$$

Since -3 and $\frac{1}{3}$ are negative reciprocals of each other, l_1 and l_2 are perpendicular.

3.4

13. First write $-3x - y + 4 = 0$ in slope-intercept form.

$$-y = 3x - 4$$

$$y = -3x + 4$$

The slope of this line is -3, so a line perpendicular to it has slope $\frac{1}{3}$. Use the point-slope form.

$$y - 4 = \frac{1}{3}(x - (-1))$$

$$3y - 12 = x + 1 \quad \text{Multiply both sides by 3}$$

$$-x + 3y - 13 = 0$$

$$x - 3y + 13 = 0 \quad \text{General form}$$

16. We are given x-intercept (3, 0) and slope $\frac{1}{2}$. Use the x-intercept like any other point in the point-slope form.

$$y - y_1 = m(x - x_1)$$

$$y - 0 = \frac{1}{2}(x - 3)$$

$$2y = x - 3 \quad \text{Multiply by 2}$$

$$x - 2y - 3 = 0 \quad \text{General form}$$

20. Since lines with undefined slope are vertical lines (parallel to the y-axis), the desired line must be parallel to the y-axis passing through (2, 3). Since all x-coordinates of points on this line are 2, the equation is

$$x = 2 \text{ or } x - 2 = 0.$$

22. Basically we are given two points, (10, 2800) and (25, 6125). Using the formula for m and the point-slope form,

$$y - 2800 = \frac{6125 - 2800}{25 - 10}(x - 10)$$

$$y - 2800 = \frac{3325}{15}(x - 10)$$

$$y - 2800 = \frac{665}{3}(x - 10).$$

Multiply each side of the equation by 3 to obtain

$$3y - 8400 = 665(x - 10)$$

$$3y - 8400 = 665x - 6650$$

$$665x - 3y + 1750 = 0. \quad \text{General form}$$

Use this equation to find y when $x = 31$.

3.6

23. Let x = his total sales,
$0.12x$ = his commission.
Thus, his annual earnings are found by adding \$15,000 to his commission.

$$f(x) = 15{,}000 + 0.12x$$

When x is 100,000, we find

$$\begin{aligned} f(100{,}000) &= 15{,}000 + (0.12)(100{,}000) \\ &= 15{,}000 + 12{,}000 \\ &= 27{,}000. \end{aligned}$$

His earnings are \$27,000.

24. We need to evaluate $f(x)$ when $x = 2$.

$$\begin{aligned} f(x) &= -16x^2 + 128x \\ f(2) &= -16(2)^2 + 128(2) \\ &= -16(4) + 256 \\ &= -64 + 256 = 192 \end{aligned}$$

The height of the object is 192 ft in 2 sec.

4.1

20. We graph each line using a scale which will allow us to read the intersection. The solution appears to be $(\frac{1}{2}, -1)$, which does check in both equations.

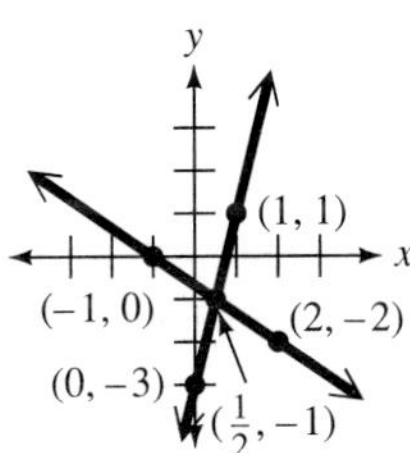

4.2

5. $5x - 5y = 10$
$x - y = 2$

Solve the second equation for y,

$$y = x - 2,$$

and substitute into the first.

$$\begin{aligned} 5x - 5(x - 2) &= 10 \\ 5x - 5x + 10 &= 10 \\ 10 &= 10 \end{aligned}$$

Since we obtain an identity, the system has infinitely many solutions. Since $y = x - 2$, any point on the common line of solutions can be written in the form $(x, x - 2)$, for any real number x.

10. $5x - 7y = 3$
$-10x + 14y = -1$

Multiply the second equation by $\frac{1}{2}$ and add.

$$\begin{aligned} 5x - 7y &= 3 \\ -5x + 7y &= -\tfrac{1}{2} \\ \hline 0 &= \tfrac{5}{2} \end{aligned}$$

Since this is a contradiction, there is no solution.

19. $\frac{3}{4}x - \frac{2}{3}y = 1$

$\frac{3}{8}x - \frac{1}{6}y = 1$

First, clear all fractions by multiplying the first equation by 12 and the second by 24.

$$\begin{aligned} 9x - 8y &= 12 \\ 9x - 4y &= 24 \\ \hline -4y &= -12 \quad \text{Subtract to eliminate } x \\ y &= 3 \end{aligned}$$

Substitute 3 for y in $9x - 8y = 12$.

$$\begin{aligned} 9x - 8(3) &= 12 \\ 9x - 24 &= 12 \\ 9x &= 36 \\ x &= 4 \end{aligned}$$

The solution is (4, 3).

20. $0.6x - 0.7y = 1.3$
$-1.2x + 1.4y = 7.3$

First, multiply both equations by 10 to clear the decimals.

$$\begin{aligned} 6x - 7y &= 13 \\ -12x + 14y &= 73 \\ 12x - 14y &= 26 \\ \hline 0 &= 99 \quad \text{Multiply first equation by 2 and add} \end{aligned}$$

Since we obtain a contradiction, the system has no solution.

21. $\frac{4}{x} - \frac{2}{y} = -2$

$\frac{2}{x} + \frac{1}{y} = 3$

The system can be written in the form

$$4\left(\frac{1}{x}\right) - 2\left(\frac{1}{y}\right) = -2$$

$$2\left(\frac{1}{x}\right) + \left(\frac{1}{y}\right) = 3.$$

Substitute u for $\frac{1}{x}$ and v for $\frac{1}{y}$ to obtain the following system.

$$4u - 2v = -2$$

$$2u + v = 3$$

The first equation can be simplified by multiplying both sides by $\frac{1}{2}$.

$$\begin{aligned} 2u - v &= -1 \\ 2u + v &= 3 \\ \hline 4u \quad &= 2 \quad \text{Add to eliminate } v \end{aligned}$$

$$u = \frac{2}{4}$$

$$u = \frac{1}{2}$$

Substitute $\frac{1}{2}$ for u in $2u + v = 3$.

$$2\left(\frac{1}{2}\right) + v = 3$$

$$1 + v = 3$$

$$v = 2$$

Since $u = \frac{1}{2}$, $\frac{1}{x} = \frac{1}{2}$ so $x = 2$. Also, since $v = 2$, $\frac{1}{y} = 2$ so $y = \frac{1}{2}$. The solution is $\left(2, \frac{1}{2}\right)$.

4.3

2. Let x = Jim's age,
y = Pete's age,
$x + 3$ = Jim's age in 3 years,
$y + 3$ = Pete's age in 3 years.

$$y = x - 7 \quad \text{Pete is 7 years younger than Jim}$$

$$x + 3 + y + 3 = 33 \quad \text{Sum of ages in 3 years is 33}$$

$$x + y = 27 \quad \text{Simplify}$$

$$x + x - 7 = 27 \quad \text{Substitute } y = x - 7$$

$$2x - 7 = 27$$

$$2x = 34$$

$$x = 17$$

Jim is 17 and Pete is $y = x - 7 = 17 - 7 = 10$.

4. Let x = number of hours at 30 mph,
y = number of hours at 40 mph,
$30x$ = distance traveled at 30 mph,
$40y$ = distance traveled at 40 mph.
The first equation is

$$30x + 40y = 230.$$

By traveling 10 mph *faster* throughout, she would have traveled at 40 mph for x hr and 50 mph for y hr. The second equation is

$$40x + 50y = 300.$$

First divide through both equations by 10 to obtain the following system.

$$\begin{aligned} 3x + 4y &= 23 \\ 4x + 5y &= 30 \end{aligned}$$

$$\begin{aligned} 12x + 16y &= 92 \quad \text{Multiply first by 4} \\ -12x - 15y &= -90 \quad \text{Multiply second by } -3 \\ \hline y &= 2 \quad \text{Add} \end{aligned}$$

Substitute 2 for y in $3x + 4y = 23$.

$$3x + 4(2) = 23$$

$$3x + 8 = 23$$

$$3x = 15$$

$$x = 5$$

She traveled for 5 hr at 30 mph and 2 hr at 40 mph.

5. Let x = cost of one book in dollars,
y = cost of one pen in dollars,
$4x$ = value of 4 books,
$6y$ = value of 6 pens,
$3x$ = value of 3 books,
$9y$ = value of 9 pens.

$$4x + 6y = 9 \quad \text{Cost of 4 books and 6 pens}$$

$$3x + 9y = 9 \quad \text{Cost of 3 books and 9 pens}$$

Divide the second equation by 3 to simplify, and solve for x.

$$x + 3y = 3 \quad \text{Divide by 3}$$

$$x = -3y + 3 \quad \text{Solve for } x$$

$$4x + 6y = 9 \quad \text{First equation}$$

$$4(-3y + 3) + 6y = 9 \quad \text{Substitute } x = -3y + 3$$

$$-12y + 12 + 6y = 9$$

$$-6y + 12 = 9$$

$$-6y = -3$$

$$y = \frac{1}{2} = 0.5$$

$$x = -3(0.5) + 3$$

$$= -1.5 + 3 = 1.5$$

One book costs \$1.50, and one pen costs \$0.50.

8. Let x = number of adults,
y = number of children.
The first equation is

$$x + y = 450.$$

Changing to cents, we obtain the second equation.

$$200x + 75y = 60{,}000$$

Solve the first equation for x,

$$x = 450 - y,$$

and substitute into the second.

$$200(450 - y) + 75y = 60{,}000$$

$$90{,}000 - 200y + 75y = 60{,}000$$

$$-125y = -30{,}000$$

$$y = 240$$

Then $x = 450 - y = 450 - 240 = 210$. Thus, there were 210 adults and 240 children at the play.

14. Let x = amount invested at 10%,
y = amount invested at 9%.
Since \$500 more was invested at 10% than 9%, one equation is

$$x = 500 + y.$$

$$0.10x = \text{interest earned on 10\% account,}$$

$$0.09y = \text{interest earned on 9\% account.}$$

Then the total interest earned, $0.10x + 0.09y$, was \$183 giving us the second equation,

$$0.10x + 0.09y = 183,$$

which can be simplified by multiplying both sides by 100.

$$10x + 9y = 18{,}300$$

Since the first equation is already solved for x, substitute this expression into the second.

$$10(500 + y) + 9y = 18{,}300$$
$$5000 + 10y + 9y = 18{,}300$$
$$19y = 13{,}300$$
$$y = 700$$

Then $x = 500 + y = 500 + 700 = 1200$. Thus, \$1200 was invested at 10%, and \$700 at 9%.

16. Let x = number of liters of 50% acid solution needed,
y = number of liters of 25% acid solution needed,
$0.50x$ = amount of acid in x liters of 50% acid,
$0.25y$ = amount of acid in y liters of 25% acid,
$0.40(10)$ = amount of acid in 10 liters of 40% acid.

$x + y = 10$ **Total amount of solution**

$0.50x + 0.25y = (0.40)(10)$ **Amount of acid**

$50x + 25y = 40(10)$ **Multiply by 100 to clear decimals**

$2x + y = 16$ **Divide by 25**

Now solve the system by subtracting to eliminate y.

$$\begin{aligned} x + y &= 10 \\ 2x + y &= 16 \\ \hline -x \quad &= -6 \end{aligned}$$

$$x = 6$$
$$6 + y = 10$$
$$y = 4$$

Thus, 6L of 50% and 4L of 25% acid are required.

18. Let x = dollars charged per day,
y = dollars charged per mile,
$3x$ = dollars charged for 3 days,
$316y$ = dollars charged for 316 miles,
$5x$ = dollars charged for 5 days,
$242y$ = dollars charged for 242 miles.

$3x + 316y = 143.12$ **Mr. Green's total bill**

$5x + 242y = 147.44$ **Mr. Brown's total bill**

Multiply the first equation by 5 and the second by -3 and add.

$$\begin{aligned} 15x + 1580y &= 715.60 \\ -15x - 726y &= -442.32 \\ \hline 854y &= 273.28 \end{aligned}$$

$$y = \frac{273.28}{854} = 0.32$$

$$5x + 242(0.32) = 147.44$$
$$5x + 77.44 = 147.44$$
$$5x = 70$$
$$x = 14$$

The charge is \$14 per day and \$0.32 per mile.

19. Let x = rate of houseboat in still water,
y = rate of current in river,
$x + y$ = rate traveling downstream,
$x - y$ = rate traveling upstream.

Using the formula $d = rt$, we have:

$30 = (x + y)3$ **Distance downstream**

$30 = (x - y)5.$ **Distance upstream**

Dividing the first equation by 3 and the second by 5 we have the following system.

$$x + y = 10$$
$$x - y = 6$$

Add to eliminate y.

$$2x = 16$$
$$x = 8$$

Substitute 8 for x in $x + y = 10$.

$$8 + y = 10$$
$$y = 2$$

The houseboat traveled 8 mph, and the river current was 2 mph.

4.4

5. (A) $3x + y - z = 4$
(B) $y - 2z = 5$
(C) $2x + z = -1$

If we use (A) and (B) and eliminate y, the result can be paired with (C) immediately.

(A) − (B) $3x + z = -1$
(C) $2x + z = -1$

$x \quad = 0$ **Subtract to eliminate z**

Substitute 0 for x in $3x + z = -1$.

$$0 + z = -1$$
$$z = -1$$

Substitute -1 for z in (B) to obtain y.

$$y - 2(-1) = 5$$
$$y + 2 = 5$$
$$y = 3$$

The solution is $(0, 3, -1)$.

12. (A) $0.3x - 2.1y + 1.2z = -3$
(B) $5.7x + 4.2y - 4.8z = -3$
(C) $1.1x - 6.9y + 3.6z = -19$

To eliminate decimals, multiply each equation by 10.

$$\begin{array}{rl} 10(A) & 3x - 21y + 12z = -30 \\ 10(B) & 57x + 42y - 48z = -30 \\ 10(C) & 11x - 69y + 36z = -190 \end{array}$$

$$\begin{array}{rl} 40(A) & 12x - 84y + 48z = -120 \\ 10(B) & \underline{57x + 42y - 48z = -\ 30} \\ (D) & 69x - 42y \qquad = -150 \end{array}$$

$$\begin{array}{rl} -30(A) & -9x + 63y - 36z = 90 \\ 10(C) & \underline{11x - 69y + 36z = -190} \\ (E) & 2x - 6y \qquad = -100 \end{array}$$

$$\begin{array}{rl} (D) & 69x - 42y = -150 \\ -7(E) & \underline{-14x + 42y = 700} \\ & 55x = 550 \quad \text{(D) − 7(E)} \\ & x = 10 \end{array}$$

$$\begin{array}{rl} (E) & 2(10) - 6y = -100 \\ & -6y = -120 \\ & y = 20 \end{array}$$

$$\begin{array}{rl} 10(A) & 3(10) - 21(20) + 12z = -30 \\ & 30 - 420 + 12z = -30 \\ & 12z = 360 \\ & z = 30 \end{array}$$

The solution is $(10, 20, 30)$.

18. Let x = amount invested in mutual fund,
y = amount invested in certificates,
z = amount invested in business.

The system to solve is:

$$x + \quad y + \quad z = 10{,}000$$
$$0.08x + 0.07y + 0.10z = 840$$
$$0.08x + 0.07y - 0.02z = 360.$$

Clearing the decimals and subtracting the third from the second gives

$$12z = 48{,}000$$
$$z = 4000.$$

Substituting 4000 for z in the first two equations will give a system in x and y which, when solved, gives $x = 2000$ and $y = 4000$.

21. Let x = price in cents per pound of oranges,
y = price in cents per pound of apples,
z = price in cents per pound of limes.

The average of three numbers is found by adding the numbers and dividing by 3.

$$\frac{x + y + z}{3} = 50 \quad \text{Average price is 50 cents per pound}$$

$$x + y + z = 150 \quad \text{Multiply by three}$$

$$6x + 5y + 2z = 600 \quad \text{Cost of 6 lb of oranges, 5 lb of apples, and 2 lb of limes}$$

$$y = z - 25 \quad \text{Cost of limes minus 25 is cost of apples}$$

The system to solve is the following.

$$\begin{array}{rl} (A) & x + \ y + \ z = 150 \\ (B) & 6x + 5y + 2z = 600 \\ (C) & y - \ z = -25 \end{array}$$

Multiply (A) by -6 and add to (B) to get

$$\begin{array}{rl} -6(A) & -6x - 6y - 6z = -900 \\ (B) & \underline{6x + 5y + 2z = \ 600} \\ (D) & -y - 4z = -300. \end{array}$$

$$\begin{array}{rl} (C) & y - \ z = -25 \\ (D) & \underline{-y - 4z = -300} \\ & -5z = -325 \quad \text{(C) + (D)} \\ & z = 65 \end{array}$$

$$\begin{array}{rl} (C) & y - 65 = -25 \\ & y = 40 \end{array}$$

$$\begin{array}{rl} (A) & x + 40 + 65 = 150 \\ & x = 45 \end{array}$$

Oranges are 45¢/lb, apples are 40¢/lb, and limes are 65¢/lb.

4.5

14. $3x + 2y = 7$
$-x + 4y = 0$

$$A = \begin{vmatrix} 3 & 2 \\ -1 & 4 \end{vmatrix} = (3)(4) - (-1)(2) = 12 + 2 = 14$$

$$A_x = \begin{vmatrix} 7 & 2 \\ 0 & 4 \end{vmatrix} = (7)(4) - (0)(2) = 28 \qquad A_y = \begin{vmatrix} 3 & 7 \\ -1 & 0 \end{vmatrix} = (3)(0) - (-1)(7) = 7$$

$$x = \frac{A_x}{A} = \frac{28}{14} = 2 \qquad y = \frac{A_y}{A} = \frac{7}{14} = \frac{1}{2}$$

18. $4x - 3y + 8z = 12$ $\qquad 4x - 3y + 8z = 12$
$2x - \frac{3}{2}y + 4z = 11$ $\qquad 4x - 3y + 8z = 22$ **Remove fractions**
$x \qquad - 5z = -10$ $\qquad x \qquad - 5z = -10$

$$A = \begin{vmatrix} 4 & -3 & 8 \\ 4 & -3 & 8 \\ 1 & 0 & -5 \end{vmatrix} = 4(15) - 4(15) + 1(0) = 0$$

$$A_x = \begin{vmatrix} 12 & -3 & 8 \\ 22 & -3 & 8 \\ -10 & 0 & -5 \end{vmatrix}$$
$$= 12(15) - 22(15) + (-10)(0) = -150$$

We do not need to calculate A_y and A_z. Since $A = 0$ and $A_x = -150 \neq 0$, there are no solutions to the system.

5.1

20. $\frac{1}{2}x^3 - \frac{1}{3}x - \frac{1}{4}x^3 - \frac{1}{9}x + 1$

$= \frac{1}{2}x^3 - \frac{1}{4}x^3 - \frac{1}{3}x - \frac{1}{9}x + 1$

$= \left(\frac{1}{2} - \frac{1}{4}\right)x^3 + \left(-\frac{1}{3} - \frac{1}{9}\right)x + 1$

$= \frac{1}{4}x^3 - \frac{4}{9}x + 1$

25. $3x^4y^8 - 2xy^9 - 4x^4y^8 + 5xy^9 - 4x^{10}$

$= 5xy^9 - 2xy^9 + 3x^4y^8 - 4x^4y^8 - 4x^{10}$

$= (5 - 2)xy^9 + (3 - 4)x^4y^8 - 4x^{10}$

$= 3xy^9 - x^4y^8 - 4x^{10}$

38. $(6a^3b^3 + 3a^2b - 4ab) - (-8a^3b^3 + 5a^2b + 10ab^2)$

$= 6a^3b^3 + 3a^2b - 4ab + 8a^3b^3 - 5a^2b - 10ab^2$

$= 6a^3b^3 + 8a^3b^3 + 3a^2b - 5a^2b - 10ab^2 - 4ab$

$= (6 + 8)a^3b^3 + (3 - 5)a^2b - 10ab^2 - 4ab$

$= 14a^3b^3 - 2a^2b - 10ab^2 - 4ab$

41. $\left(\frac{2}{3}z^2 - \frac{1}{5}z + 2\right) + \left(\frac{1}{2}z^2 + z - \frac{3}{4}\right)$

$= \frac{2}{3}z^2 + \frac{1}{2}z^2 - \frac{1}{5}z + z + 2 - \frac{3}{4}$

$= \left(\frac{2}{3} + \frac{1}{2}\right)z^2 + \left(-\frac{1}{5} + 1\right)z + \left(2 - \frac{3}{4}\right)$

$= \frac{7}{6}z^2 + \frac{4}{5}z + \frac{5}{4}$

43. $(7.3x^2y^2 - 8.7xy + 14.2) + (-6.7x^2y^2 + 2.3xy + 5.9)$

$= 7.3x^2y^2 - 6.7x^2y^2 - 8.7xy + 2.3xy + 14.2 + 5.9$

$= 0.6x^2y^2 - 6.4xy + 20.1$

44. $(6x^3y^3 + 5x^2y^2) + (11x^2y^2 - 4xy) - (2x^3y^3 - 2x^2y^2 + 7xy)$

$= 6x^3y^3 + 5x^2y^2 + 11x^2y^2 - 4xy - 2x^3y^3 + 2x^2y^2 - 7xy$

$= 6x^3y^3 - 2x^3y^3 + 5x^2y^2 + 11x^2y^2 + 2x^2y^2 - 4xy - 7xy$

$= (6 - 2)x^3y^3 + (5 + 11 + 2)x^2y^2 + (-4 - 7)xy$

$= 4x^3y^3 + 18x^2y^2 - 11xy$

46. $5a^2 - 4b(3a - [2b - b(a - 1) - (ab + b)])$

$= 5a^2 - 4b(3a - [2b - ba + b - ab - b])$

$= 5a^2 - 4b(3a - [2b - 2ab])$

$= 5a^2 - 4b(3a - 2b + 2ab)$

$= 5a^2 - 12ab + 8b^2 - 8ab^2$

51. Substitute 5 for u, -2 for v, and 3 for w in $3uvw - 4u^2v$.

$3(5)(-2)(3) - 4(5)^2(-2) = (15)(-6) - 4(25)(-2)$

$= -90 - 4(-50)$

$= -90 + 200 = 110$

53. The total cost of manufacturing and wholesale is $M(u) + W(u)$.

$M(u) + W(u) = (3u^2 - 4u + 4) + (u^2 - u + 3)$

$= 3u^2 - 4u + 4 + u^2 - u + 3$

$= 3u^2 + u^2 - 4u - u + 4 + 3$

$= (3 + 1)u^2 + (-4 - 1)u + (4 + 3)$

$= 4u^2 - 5u + 7$

56. $P(u) = I(u) - (M(u) + W(u))$

$= (5u^2 - 2u) - (4u^2 - 5u + 7)$

$= 5u^2 - 2u - 4u^2 + 5u - 7$

$= u^2 + 3u - 7$

5.2

28. First we multiply $(a - 5b)$ and $(2a + b)$, and then multiply the result by $2a - 3b$.

$$\begin{array}{r} a - 5b \\ 2a + b \\ \hline 2a^2 - 10ab \\ ab - 5b^2 \\ \hline 2a^2 - 9ab - 5b^2 \end{array} \qquad \begin{array}{r} 2a^2 - 9ab - 5b^2 \\ 2a - 3b \\ \hline 4a^3 - 18a^2b - 10ab^2 \\ -6a^2b + 27ab^2 + 15b^3 \\ \hline 4a^3 - 24a^2b + 17ab^2 + 15b^3 \end{array}$$

30. $(a - b)[3a^2 - (a + b)(a - 2b)]$

$= (a - b)[3a^2 - (a^2 - ab - 2b^2)]$

$= (a - b)[3a^2 - a^2 + ab + 2b^2]$

$= (a - b)(2a^2 + ab + 2b^2)$

$= 2a^3 + a^2b + 2ab^2 - 2a^2b - ab^2 - 2b^3$

$= 2a^3 + a^2b - 2a^2b + 2ab^2 - ab^2 - 2b^3$

$= 2a^3 - a^2b + ab^2 - 2b^3$

34. $F(t) = N(t)S(t)$

$= (t^2 - 2t - 5)(2t^2 + t - 6)$

$= 2t^4 - 3t^3 - 18t^2 + 7t + 30$

$$\begin{array}{r} t^2 - 2t - 5 \\ 2t^2 + t - 6 \\ \hline 2t^4 - 4t^3 - 10t^2 \\ t^3 - 2t^2 - 5t \\ -6t^2 + 12t + 30 \\ \hline 2t^4 - 3t^3 - 18t^2 + 7t + 30 \end{array}$$

48. $(0.5x + 0.1y)^2$

$= (0.5x)^2 + 2(0.5x)(0.1y) + (0.1y)^2$

$= 0.25x^2 + 0.1xy + 0.01y^2$

56. $[a - (b + c)][a + (b + c)]$

$= a^2 - (b + c)^2$

$= a^2 - (b^2 + 2bc + c^2)$

$= a^2 - b^2 - 2bc - c^2$

62. $S = A(1 + i)^2 = 1000(1 + 0.12)^2$

$= 1000(1.12)^2$

$= 1000(1.2544)$

$= \$1254.40$

$S = A + 2Ai + Ai^2$

$= 1000 + 2(1000)(0.12) + (1000)(0.12)^2$

$= 1000 + (2000)(0.12) + (1000)(0.0144)$

$= 1000 + 240 + 14.4$

$= \$1254.40$

5.3

13. $64xy^5z^2 - 128x^2y^3z^3 - 144xy^2z^4$

$= (16xy^2z^2)4y^3 - (16xy^2z^2)8xyz - (16xy^2z^2)9z^2$

$= 16xy^2z^2(4y^3 - 8xyz - 9z^2)$

14. Notice that $a + b$ is a common factor of the polynomial.

$3a(a + b) - 2b(a + b) = (3a - 2b)(a + b)$

17. $5x^2 + 2xy + 10x + 4y = (5x^2 + 2xy) + (10x + 4y)$

$= x(5x + 2y) + 2(5x + 2y)$

$= (x + 2)(5x + 2y)$

20. $2a^2b + 6ab - 3a - 9$

$= 2ab(a + 3) - 3(a + 3)$

$= (2ab - 3)(a + 3)$ Factor out (a + 3)

24. $7x^3y^3 + 21x^2y^2 - 10x^3y^2 - 30x^2y$

$= x^2y(7xy^2 + 21y - 10xy - 30)$

$= x^2y[(7xy^2 + 21y) - (10xy + 30)]$

$= x^2y[7y(xy + 3) - 10(xy + 3)]$

$= x^2y(7y - 10)(xy + 3)$

25. $E = 25c^2e^3 - 10c^3e^2$

$= 5c^2e^2(5e - 2c)$ Factor out $5c^2e^2$, the GCF

5.4

24. $-4w^2 - 34w - 70 = -2(2w^2 + 17w + 35)$
Factor $2w^2 + 17w + 35$ with $a = 2$, $b = 17$, and $c = 35$.

Factors of a	*Factors of c*
2, 1	35, 1 and 1, 35
	7, 5 and 5, 7

$-2(2w^2 + 17w + 35)$

$= -2(2w + \_\_)(w + \_\_)$

$\stackrel{?}{=} -2(2w + 5)(w + 7)$ Does not work

$\stackrel{?}{=} -2(2w + 7)(w + 5)$ This works

$-2(2w^2 + 17w + 35) = -2(2w + 7)(w + 5)$

31. $6x^2 + 23xy + 20y^2$

Factors of a	*Factors of c*
1, 6	20, 1 and 1, 20
2, 3	10, 2 and 2, 10
	5, 4 and 4, 5

$6x^2 + 23xy + 20y^2$

$= (\_\_x + \_\_y)(\_\_x + \_\_y)$

$\stackrel{?}{=} (x + 10y)(6x + 2y)$ Does not work

$\stackrel{?}{=} (2x + 5y)(3x + 4y)$ This works

$6x^2 + 23xy + 20y^2 = (2x + 5y)(3x + 4y)$

37. $(x + 2)^2 - 3(x + 2) + 2$
Substitute u for $x + 2$ and factor.

$$u^2 - 3u + 2 = (u - 2)(u - 1)$$

Then substitute $x + 2$ for u.

$(x + 2)^2 - 3(x + 2) + 2 = (x + 2 - 2)(x + 2 - 1)$

$= x(x + 1)$

40. $u^6 - 7u^3 + 12$
Substitute x for u^3 and factor.

$$x^2 - 7x + 12 = (x - 3)(x - 4)$$

Then substitute u^3 for x.

$$u^6 - 7u^3 + 12 = (u^3 - 3)(u^3 - 4)$$

45. $h = -16t^2 + 656t - 640$

First factor the right side.

$h = -16(t^2 - 41t + 40)$ Factor out −16

$= -16(t - 40)(t - 1)$ Factor trinomial

(a) When $t = 0.5$,

$h = -16(0.5 - 40)(0.5 - 1)$

$= -16(-39.5)(-0.5) = -316$

After 0.5 sec, the missile is 316 ft below the surface of the water.

(b) When $t = 1$,

$h = -16(1 - 40)(1 - 1)$

$= -16(-39)(0)$

$= 0.$

After 1 sec, the missile is at the water level.

(c) When $t = 20$,

$h = -16(20 - 40)(20 - 1)$

$= -16(-20)(19)$

$= 6080.$

After 20 sec, the missile is 6080 ft high.

5.5

14. $(x - y)^2 - 9 = u^2 - 9$ Let $u = x - y$

$= u^2 - 3^2$ $u = a$, $3 = b$, in $a^2 - b^2 = (a + b)(a - b)$

$= (u + 3)(u - 3)$

$= [(x - y) + 3][(x - y) - 3]$ $u = x - y$

$= (x - y + 3)(x - y - 3)$

23. $7u^3 + 56v^3 = 7(u^3 + 8v^3)$
$= 7(u^3 + (2v)^3)$

Substitute u for a and $2v$ for b in $(a^3 + b^3) = (a + b)(a^2 - ab + b^2)$.

$$7(u^3 + (2v)^3) = 7(u + 2v)(u^2 - 2uv + 4v^2)$$

37. $(x + y)^2 + 2(x + y) + 1$
First substitute u for $(x + y)$ and factor.

$$u^2 + 2u + 1 = (u + 1)^2$$

Then substitute $(x + y)$ for u.

$$(x + y)^2 + 2(x + y) + 1 = (x + y + 1)^2$$

39. $3uv + u - 3v^2 - v = (3uv + u) - (3v^2 + v)$
$= u(3v + 1) - v(3v + 1)$
$= (u - v)(3v + 1)$

44. $-18u^2 - 69uv - 60v^2$
$= -3(6u^2 + 23uv + 20v^2)$
$= -3(2u + 5v)(3u + 4v)$

Always factor out the GCF first.

48. $(u + v)^2 + 3(u + v)(x + y) + 2(x + y)^2$
$= w^2 + 3wz + 2z^2$ **Let $w = u + v$ and $z = x + y$**
$= (w + z)(w + 2z)$
$= [(u + v) + (x + y)][(u + v) + 2(x + y)]$ **$w = u + v$ and $z = x + y$**
$= (u + v + x + y)(u + v + 2x + 2y)$

5.6

16. $3u^2 = 5u$ **Do *not* divide by u**
$3u^2 - 5u = 0$
$u(3u - 5) = 0$
$u = 0$ or $3u - 5 = 0$
$3u = 5$
$u = \frac{5}{3}$

26. $y^3 - 2y^2 + y = 0$
$y(y^2 - 2y + 1) = 0$
$y(y - 1)(y - 1) = 0$
$y = 0$ or $y - 1 = 0$ or $y - 1 = 0$
$y = 1$ $y = 1$

34. Let x = length of a side,
x^2 = area of square,
$4x$ = perimeter of square.
The equation to solve is

$$x^2 = 12 + 4x.$$

$x^2 - 4x - 12 = 0$
$(x + 2)(x - 6) = 0$
$x = -2, 6$ **Zero-product rule**

We need to discard -2 since the length of a side cannot be negative. Thus, the side is 6 in in length.

37. Let n = the first even integer,
$n + 2$ = the next even integer.
The equation to solve is

$$n^2 + (n + 2)^2 = 100.$$

$n^2 + n^2 + 4n + 4 = 100$
$2n^2 + 4n - 96 = 0$
$n^2 + 2n - 48 = 0$ **Divide out 2**
$(n + 8)(n - 6) = 0$
$n = -8, 6$ **Zero-product rule**

Since the integers were to be positive, we discard -8 and obtain 6 and 8.

40. Let w = width of the box,
$w + 7$ = length of the box.

$V = 1728 = w(w + 7)(12)$
$144 = w(w + 7)$ **Divide by 12**
$144 = w^2 + 7w$
$0 = w^2 + 7w - 144$
$0 = (w + 16)(w - 9)$
$w + 16 = 0$ or $w - 9 = 0$
$w = -16$ $w = 9$

The width is 9 cm and the length is $9 + 7 = 16$ cm.

41. Let x = height of triangle,
$x - 8$ = base of triangle.
The area of a triangle is given by $A = \frac{1}{2}bh$, so we need to solve

$$42 = \frac{1}{2}(x - 8)(x).$$

$84 = x^2 - 8x$ **Multiply by 2**
$x^2 - 8x - 84 = 0$
$(x + 6)(x - 14) = 0$
$x = -6, 14$ **Zero-product rule**

We discard -6, so the height is 14 cm and the base is 6 cm.

49. We find n when $P(n)$ is 30.

$30 = 3n^2 - 2n - 10$
$0 = 3n^2 - 2n - 40$
$0 = (3n + 10)(n - 4)$
$n = -\frac{10}{3}, 4$

Since the number of suits sold cannot be negative (nor a fraction), 4 suits were sold.

6.1

31. $$\frac{xz + xw + yz + yw}{z^2 + zw} = \frac{x(z + w) + y(z + w)}{z(z + w)}$$
$$= \frac{(x + y)\cancel{(z + w)}}{z\cancel{(z + w)}} = \frac{x + y}{z}$$

34. $$\frac{9a^2 - 15a + 25}{27a^3 + 125} = \frac{9a^2 - 15a + 25}{(3a)^3 + 5^3}$$

$$= \frac{\cancel{9a^2 - 15a + 25}}{(3a + 5)(\cancel{9a^2 - 15a + 25})}$$

$$= \frac{1}{3a + 5}$$

44. We evaluate $P = \dfrac{A}{1 + 0.07\,t}$ when $A = 5000$ and $t = 6$. The steps needed to evaluate

$$\frac{5000}{1 + 0.07(6)}$$

on a calculator follow.

6 [×] 0.07 [+] 1 [=] [1/x] [×] 5000 [=] → 3521.1268

To the nearest cent, $P = \$3521.13$.

6.2

10. $$\frac{4 - a^2}{c^2 + 2cd + d^2} \cdot \frac{c^2 - d^2}{2 - a - a^2} \cdot \frac{c + d - ca - da}{2a - a^2}$$

$$= \frac{(2 - a)(2 + a)(c - d)(c + d)}{(c + d)(c + d)(2 + a)(1 - a)} \cdot$$

$$\frac{1(c + d) - a(c + d)}{a(2 - a)}$$

$$= \frac{\cancel{(2 - a)}\cancel{(2 + a)}(c - d)\cancel{(c + d)}\cancel{(1 - a)}\cancel{(c + d)}}{\cancel{(c + d)}\cancel{(c + d)}\cancel{(2 + a)}\cancel{(1 - a)}a\cancel{(2 - a)}}$$

$$= \frac{c - d}{a}$$

18. 0.27 is actually equal to $\frac{27}{100}$, so its reciprocal is $\frac{100}{27}$.

21. $$\frac{(u^2v^3)^2}{3uv} \div \frac{(uv^2)^3}{12u^3v}$$

$$= \frac{u^4v^6}{3uv} \cdot \frac{12u^3v}{u^3v^6}$$

$$= \frac{\overset{u^3}{\cancel{u^4}}\cancel{v^6} \cdot \overset{4}{\cancel{12}}\cancel{u^3}\cancel{v}}{\cancel{3}\cancel{u}\cancel{v}\cancel{u^3}\cancel{v^6}}$$

$$= 4u^3$$

28. $$\frac{3s^2 - 5st - 2t^2}{t^2 + st - 2s^2} \cdot \frac{t^2 + 4st - 5s^2}{12s^2 + 7st + t^2} \div \frac{5s^2 - 9st - 2t^2}{8s^2 - 2st - t^2}$$

$$= \frac{(3s + t)(s - 2t)(t + 5s)(t - s)}{(t + 2s)(t - s)(3s + t)(4s + t)} \cdot \frac{8s^2 - 2st - t^2}{5s^2 - 9st - 2t^2}$$

$$= \frac{\cancel{(3s + t)}\cancel{(s - 2t)}\cancel{(t + 5s)}\cancel{(t - s)}\cancel{(4s + t)}(2s - t)}{(t + 2s)\cancel{(t - s)}\cancel{(3s + t)}\cancel{(4s + t)}\cancel{(5s + t)}\cancel{(s - 2t)}}$$

$$= \frac{2s - t}{2s + t}$$

6.3

5. $$\frac{x}{8x - 6} - \frac{5}{2(3 - 4x)} = \frac{x}{2(4x - 3)} - \frac{5}{2(3 - 4x)}$$

$$= \frac{x}{2(4x - 3)} - \frac{(-1)5}{(-1)(2)(3 - 4x)}$$

$$= \frac{x}{2(4x - 3)} - \frac{-5}{2(4x - 3)}$$

$$= \frac{x}{2(4x - 3)} + \frac{5}{2(4x - 3)}$$ With practice, you can skip from the first step to this one

$$= \frac{x + 5}{2(4x - 3)}$$

15. $$\frac{z + 2}{z^2 + 4z + 4} \quad \text{and} \quad \frac{z - 1}{z^2 + z - 2}$$

$$\frac{\cancel{z + 2}}{(z + 2)(\cancel{z + 2})} \quad \text{and} \quad \frac{\cancel{z - 1}}{(\cancel{z - 1})(z + 2)}$$

$$\frac{1}{z + 2} \quad \text{and} \quad \frac{1}{z + 2}$$

Once the fractions have been reduced to lowest terms, the LCD is clearly $z + 2$.

17. $$\frac{x + 3}{5x^4 - 15x^3 - 50x^2} \quad \text{and} \quad \frac{x - 7}{10x^3 - 100x^2 + 250x}$$

$$\frac{x + 3}{5x^2(x^2 - 3x - 10)} \quad \text{and} \quad \frac{x - 7}{10x(x^2 - 10x + 25)}$$

$$\frac{x + 3}{5xx(x - 5)(x + 2)} \quad \text{and} \quad \frac{x - 7}{2 \cdot 5x(x - 5)(x - 5)}$$

Since the fractions cannot be reduced, we concentrate on the denominators as shown. The LCD must consist of one 2, one 5, two x's, two $(x - 5)$'s, and one $(x + 2)$. Thus,

$$\text{LCD} = 2 \cdot 5 \cdot x \cdot x(x - 5)(x - 5)(x + 2)$$
$$= 10x^2(x - 5)^2(x + 2).$$

33. $$a - b - \frac{-ab^2}{a^2 + ab} = a - b - \frac{-\cancel{a}b^2}{\cancel{a}(a + b)}$$

$$= \frac{(a - b)}{1} + \frac{b^2}{a + b}$$ The LCD $= a + b$

$$= \frac{(a - b)(a + b)}{(a + b)} + \frac{b^2}{a + b}$$

$$= \frac{a^2 - b^2}{a + b} + \frac{b^2}{a + b}$$

$$= \frac{a^2 - b^2 + b^2}{a + b}$$

$$= \frac{a^2}{a + b}$$

35. $\dfrac{2a-1}{a^2-4a+4}+\dfrac{2a+3}{4-a^2}$

$$=\frac{2a-1}{(a-2)(a-2)}+\frac{2a+3}{(2-a)(2+a)}$$

$$=\frac{2a-1}{(a-2)(a-2)}+\frac{(-1)(2a+3)}{(a-2)(a+2)}$$

The LCD = $(a-2)^2(a+2)$

$$=\frac{(2a-1)(a+2)}{(a-2)^2(a+2)}-\frac{(2a+3)(a-2)}{(a-2)^2(a+2)}$$

$$=\frac{(2a^2+3a-2)-(2a^2-a-6)}{(a-2)^2(a+2)}$$ **Use parentheses**

$$=\frac{2a^2+3a-2-2a^2+a+6}{(a-2)^2(a+2)}$$

$$=\frac{4a+4}{(a-2)^2(a+2)}=\frac{4(a+1)}{(a-2)^2(a+2)}$$

44. $\dfrac{x+y}{x^2+2xy+y^2}-\dfrac{x}{x^2-xy-2y^2}+\dfrac{y}{x^2-4xy-5y^2}$

$$=\frac{\cancel{(x+y)}}{\cancel{(x+y)}(x+y)}-\frac{x}{(x+y)(x-2y)}+\frac{y}{(x+y)(x-5y)}$$

$$=\frac{(x-2y)(x-5y)}{(x+y)(x-2y)(x-5y)}-\frac{x(x-5y)}{(x+y)(x-2y)(x-5y)}+\frac{y(x-2y)}{(x+y)(x-5y)(x-2y)}$$

$$=\frac{(x^2-7xy+10y^2)-(x^2-5xy)+(xy-2y^2)}{(x+y)(x-2y)(x-5y)}$$

$$=\frac{x^2-7xy+10y^2-x^2+5xy+xy-2y^2}{(x+y)(x-2y)(x-5y)}$$

$$=\frac{8y^2-xy}{(x+y)(x-2y)(x-5y)}=\frac{y(8y-x)}{(x+y)(x-2y)(x-5y)}$$

6.4

7. $\dfrac{x}{x+4}=\dfrac{4}{x-4}+\dfrac{x^2+16}{x^2-16}$

$$\frac{x}{x+4}=\frac{4}{x-4}+\frac{x^2+16}{(x+4)(x-4)}$$ **The LCD = $(x-4)(x+4)$**

$$(x-4)\cancel{(x+4)}\left[\frac{x}{\cancel{x+4}}\right]=(x-4)(x+4)\left[\frac{4}{x-4}+\frac{x^2+16}{(x-4)(x+4)}\right]$$

$$x(x-4)=\cancel{(x-4)}(x+4)\left[\frac{4}{\cancel{x-4}}\right]+\cancel{(x-4)}\cancel{(x+4)}\left[\frac{x^2+16}{\cancel{(x-4)}\cancel{(x+4)}}\right]$$

$$x^2-4x=4x+16+x^2+16$$

$$-4x=4x+32$$

$$-8x=32$$

$$x=-4$$

But since -4 gives zero in the denominator of $\frac{x}{x+4}$, it cannot be a solution. Thus, there are no solutions.

14. $$\frac{-1}{a^2-3a}=\frac{1}{a}+\frac{a}{a-3}$$

$$\frac{-1}{a(a-3)}=\frac{1}{a}+\frac{a}{a-3}$$

$$a(a-3)\left[\frac{-1}{a(a-3)}\right]=a(a-3)\left[\frac{1}{a}+\frac{a}{a-3}\right]$$

$$-1=(a-3)+a^2$$

$$-1=a-3+a^2$$

$$0=a^2+a-2$$

$$0=(a+2)(a-1)$$

$$a=-2,\ 1$$ **Zero-product rule**

Both -2 and 1 will check.

20. Let x = Lorraine's age,

$\frac{3}{4}x$ = Harriet's age.

The equation to solve is

$$x-\frac{3}{4}x=10.$$

$$4x-3x=40$$ **Multiply through by 4**

$$x=40$$

Lorraine is 40 and Harriet is 30.

23. Let x = the score to be made on the fourth test. To average the four scores, add them and divide by 4.

$$\frac{75+80+82+x}{4}=83$$

$$\cancel{4}\left[\frac{75+80+82+x}{\cancel{4}}\right]=4[83]$$

$$237+x=332$$

$$x=95$$

She needs a 95 to average 83 on the four tests.

30. $$a^{-1}=b+c^{-1}$$

$$\frac{1}{a}=b+\frac{1}{c}$$

$$\cancel{a}c\left[\frac{1}{\cancel{a}}\right]=ac[b]+a\cancel{c}\left[\frac{1}{\cancel{c}}\right]$$

$$c=abc+a$$

$$c-abc=a$$

$$c(1-ab)=a$$

$$c=\frac{a}{1-ab}$$

6.5

2. Let x = number of votes for loser. The equation to solve is

$$\frac{5}{3}=\frac{820}{x}.$$

$$5x = 3(820) \quad \textbf{Cross-product equation}$$
$$x = \frac{3(820)}{5} = 3(164) = 492$$

The loser received 492 votes.

7. Let x = the length of one piece,
$65 - x$ = the length of the second piece.

$$[x + (65 - x) = 65.]$$

With the ratio of the two pieces 8:5, we solve

$$\frac{x}{65 - x} = \frac{8}{5}.$$
$$5x = 8(65 - x)$$
$$5x = 520 - 8x$$
$$13x = 520$$
$$x = 40$$
$$65 - x = 65 - 40 = 25$$

The two pieces have length 40 ft and 25 ft.

13. $a = \frac{cu}{vw}$ is the equation describing the variation. Substitute 4 for a, 3 for u, 3 for v, and 4 for w to find the constant of variation, c.

$$4 = \frac{c(3)}{(3)(4)}$$
$$4 = \frac{3c}{12}$$
$$48 = 3c$$
$$16 = c$$

The equation of variation is $a = \frac{16u}{vw}$.

17. $A = \frac{c}{d^2}$ is the equation describing the variation. Substitute 64 for A and 9 for d to find the constant of variation, c.

$$64 = \frac{c}{9^2}$$
$$(64)(81) = c$$
$$5184 = c$$

Next, substitute 24 for d in $A = \frac{5184}{d^2}$ to find the desired attraction.

$$A = \frac{5184}{(24)^2} = \frac{5184}{576} = 9$$

The attraction is 9 lb.

20. $w = \frac{c}{d^2}$
Substitute 150 for w and 4000 for d to find c.

$$150 = \frac{c}{(4000)^2}, \quad \text{so} \quad (4000)^2(150) = c$$

Then substitute 5000 for d (1000 miles above the surface is 5000 miles from the center) and substitute $150(4000)^2$ for c.

$$w = \frac{(150)(4000)^2}{(5000)^2} = \frac{(150)(4)^2\cancel{(1000)^2}}{5^2\cancel{(1000)^2}} = 96$$

The object weighs 96 lb 1000 mi above the surface.

6.6

6. Let x = her rate going to Denver,
$x + 40$ = her rate returning to Phoenix.
Since the distance going and the distance returning are both 800, using $d = rt$, or $t = \frac{d}{r}$, we have

$$\frac{800}{x} = \text{time going to Denver,}$$
$$\frac{800}{x + 40} = \text{time returning to Phoenix.}$$

Since the total time of the trip is 9 hours, we solve $\frac{800}{x} + \frac{800}{x + 40} = 9$.

$$\cancel{x}(x + 40)\left[\frac{800}{\cancel{x}}\right] + x\cancel{(x + 40)}\left[\frac{800}{\cancel{(x + 40)}}\right]$$
$$= 9x(x + 40)$$
$$800x + 32{,}000 + 800x = 9x^2 + 360x$$
$$9x^2 - 1240x - 32{,}000 = 0$$
$$(x - 160)(9x + 200) = 0$$
$$x = 160 \text{ or } x = -\tfrac{200}{9}, \text{ which is meaningless}$$

Her rate going was 160 mph and her rate returning was 200 mph ($x + 40 = 160 + 40 = 200$).

8. Let x = number of hours they travel upstream,
$4 - x$ = number of hours they return downstream.
Since they go upstream the same distance that they return downstream, we equate distance using $d = rt$.
$6x$ = distance traveled upstream
$12(4 - x)$ = distance traveled downstream

$$6x = 12(4 - x)$$
$$6x = 48 - 12x$$
$$18x = 48$$
$$x = \frac{48}{18} = \frac{8}{3} = 2\frac{2}{3} \text{ hr}$$

Thus, they must turn back in $2\frac{2}{3}$ hr, or in 2 hr 40 min, which is at 11:40 A.M. At that time they will have traveled $\left(2\frac{2}{3}\right)(6) = \left(\frac{8}{3}\right)(6) = 16$ mi upstream.

15. Let t = time to fill tank when both pipes are open,
$\frac{1}{4}$ = amount filled by inlet in 1 hr,
$\frac{1}{10}$ = amount drained by outlet in 1 hr,
$\frac{1}{t}$ = amount filled in 1 hr when both are open.

We need to solve

$$\frac{1}{t} = \frac{1}{4} - \frac{1}{10}.$$

$$(20t)\frac{1}{t} = 20t\left(\frac{1}{4} - \frac{1}{10}\right)$$

$$20 = 5t - 2t$$

$$20 = 3t$$

$$\frac{20}{3} = t$$

It takes $\frac{20}{3}$ hr or $6\frac{2}{3}$ hr to fill the tank.

6.7

12. $$\frac{\dfrac{1}{x+y} - \dfrac{1}{x-y}}{\dfrac{-2}{x-y}} = \frac{\left[\dfrac{1}{x+y} - \dfrac{1}{x-y}\right](x+y)(x-y)}{\left[\dfrac{-2}{x-y}\right](x+y)(x-y)}$$

$$= \frac{(x-y)-(x+y)}{-2(x+y)} = \frac{x-y-x-y}{-2(x+y)}$$

$$= \frac{-\not{2}y}{-\not{2}(x+y)} = \frac{y}{x+y}$$

13. $$\frac{a+1}{a-1} = \frac{\dfrac{x+y}{y}+1}{\dfrac{x+y}{y}-1} = \frac{\left[\dfrac{x+y}{y}+1\right]y}{\left[\dfrac{x+y}{y}-1\right]y}$$

$$= \frac{x+y+y}{x+y-y} = \frac{x+2y}{x}$$

17. $$a - \frac{a}{1-\dfrac{a}{1-a}} = a - \frac{a}{\dfrac{1-a}{1-a} - \dfrac{a}{1-a}}$$

$$= a - \frac{a}{\dfrac{1-2a}{1-a}} = a - \left[\frac{a}{1}\cdot\frac{1-a}{1-2a}\right]$$

$$= a - \frac{a(1-a)}{1-2a} = \frac{a(1-2a)}{1-2a} - \frac{a-a^2}{1-2a}$$

$$= \frac{a-2a^2-a+a^2}{1-2a} = \frac{-a^2}{1-2a}$$

23. $$\frac{2x^{-1}+2y^{-1}}{(xy)^{-1}} = \frac{\dfrac{2}{x}+\dfrac{2}{y}}{\dfrac{1}{xy}} = \frac{\left[\dfrac{2}{x}+\dfrac{2}{y}\right]xy}{\left[\dfrac{1}{xy}\right]xy}$$

$$= \frac{2y+2x}{1} = 2(x+y)$$

25. Evaluate

$$f = \frac{d_1d_2}{d_2+d_1}$$

when $d_1 = 8.0$ and $d_2 = 14.0$.

$$f = \frac{(8.0)(14.0)}{14.0+8.0} = \frac{112}{22} \approx 5.1$$

The focal length is about 5.1 cm.

27. We need to simplify

$$\frac{2D}{\dfrac{D}{4}+\dfrac{D}{2}}.$$

First add the fractions in the denominator; they have LCD 4.

$$\frac{2D}{\dfrac{D}{4}+\dfrac{2D}{4}} = \frac{2D}{\dfrac{3D}{4}}$$

$$= \frac{4}{3D}(2D) \quad \textbf{The } D\textbf{'s divide out}$$

$$= \frac{8}{3} \approx 2.7$$

Marla's average rate of speed was about 2.7 mph.

6.8

3. $$\frac{-27a^5+9a^3-81a^2}{-3a}$$

$$= \frac{-27a^5}{-3a} + \frac{9a^3}{-3a} + \frac{-81a^2}{-3a}$$

$$= 9a^4 - 3a^2 + 27a$$

16.
$$\begin{array}{r}
x^2 + 4x + 5 \\
x^2 - x + 1\,\overline{)\,x^4 + 3x^3 + 2x^2 - x + 5} \\
\underline{x^4 - x^3 + x^2} \qquad\qquad \\
4x^3 + x^2 - x \qquad \\
\underline{4x^3 - 4x^2 + 4x} \qquad \\
5x^2 - 5x + 5 \\
\underline{5x^2 - 5x + 5} \\
0
\end{array}$$

Thus, the quotient is $x^2 + 4x + 5$.

20.
$$\begin{array}{r}
3x^2 + x - 2 \qquad\quad \\
2x^2 + 3\,\overline{)\,6x^4 + 2x^3 + 5x^2 - x + 1} \\
\underline{6x^4 \qquad + 9x^2} \qquad\quad\;\; \\
2x^3 - 4x^2 \qquad\quad\;\; \\
\underline{2x^3 \qquad + 3x} \qquad \\
-4x^2 - 4x \qquad \\
\underline{-4x^2 \qquad - 6} \\
-4x + 7
\end{array}$$

The quotient is $3x^2 + x - 2$ with remainder $-4x + 7$, or $3x^2 + x - 2 + \frac{-4x+7}{2x^2+3}$.

21. Divide $y^3 - y^2 + 2y + p$ by $y - 1$.

$$\begin{array}{r} y^2 + 2 \\ y - 1\overline{)y^3 - y^2 + 2y + p} \\ \underline{y^3 - y^2} \\ 0 + 2y + p \\ \underline{2y - 2} \\ p + 2 \end{array}$$

For $p + 2$ to be 0, we solve

$$p + 2 = 0$$
$$p = -2.$$

6.9

7. First write the dividend in descending powers of x.

$$(x^4 + x^3 - x - 3) \div (x + 1)$$

Write coefficients (don't forget $0x^2$) and begin.

$$\begin{array}{r|l} -1 & 1 + 1 + 0 - 1 - 3 \\ & -1 + 0 + 0 + 1 \\ \hline & 1 + 0 + 0 - 1 - 2 \\ & (x^3 + 0x^2 + 0x - 1) \end{array}$$

We obtain $x^3 - 1$ with a remainder of -2.

13.
$$\begin{array}{r|l} -1 & 3 + 0 + 0 + 1 + 0 - 5 \\ & -3 + 3 - 3 + 2 - 2 \\ \hline & 3 - 3 + 3 - 2 + 2 - 7 \end{array}$$

We obtain $3x^4 - 3x^3 + 3x^2 - 2x + 2$ with a remainder of -7.

7.1

27. $\sqrt[4]{(-2x)^4} = |-2x| = |-2||x| = 2|x|$

30. $-\sqrt[5]{(-2x)^5} = -(-2x) = 2x$

31. $\sqrt{(x + 7)^2} = |x + 7|$
Remember that even-indexed radicals require absolute value.

32. $\sqrt[3]{(x + 7)^3} = x + 7$
Odd-indexed radicals don't require absolute value.

49. $\sqrt{x^2 - 14x + 49} = \sqrt{(x - 7)^2}$ Factor the radicand
$= x - 7$ We assume $x - 7$ is nonnegative so $|x - 7| = x - 7$

59. To evaluate $\frac{\sqrt{1{,}250{,}000{,}000}}{0.00623}$, first write 1,250,000,000 in scientific notation as 1.25×10^9. The steps on your calculator are:

1.25 [EE] 9 [√] [÷] 0.00623 [=] → [5.675 06]

The display gives 5.675×10^6 which to three significant digits is $5.68 \times 10^6 = 5{,}680{,}000$.

66. Find d when $h = 1800$ using $d = 1.4\sqrt{h}$.

$d = 1.4\sqrt{1800} = 59.39697$ Using a calculator

The viewing distance is about 59.4 mi, which means that an object 50 mi away could be seen.

7.2

20.
$$\begin{aligned} \sqrt[3]{-27x^4y^5z^8} &= \sqrt[3]{(-3)^3x^3 \cdot x \cdot y^3 \cdot y^2 \cdot z^6 \cdot z^2} \\ &= \sqrt[3]{(-3)^3x^3y^3z^6xy^2z^2} \\ &= -3xyz^2\sqrt[3]{xy^2z^2} \end{aligned}$$

29.
$$\begin{aligned} \frac{3x}{y}\sqrt[5]{\frac{x^5y^6}{243}} &= \frac{3x}{y}\sqrt[5]{\frac{x^5y^5y}{3^5}} \\ &= \frac{3x}{y} \cdot \frac{xy}{3}\sqrt[5]{y} \\ &= \frac{\cancel{3}x^2\cancel{y}}{\cancel{3y}}\sqrt[5]{y} = x^2\sqrt[5]{y} \end{aligned}$$

34. If $\sqrt[3]{5x^2}$ is in a denominator, we need to multiply numerator and denominator by $\sqrt[3]{5^2x}$, for then

$$\begin{aligned} \sqrt[3]{5x^2}\,\sqrt[3]{5^2x} &= \sqrt[3]{5^3x^3} \\ &= 5x, \end{aligned}$$

and the radical in the denominator will be removed.

43.
$$\begin{aligned} \frac{7}{\sqrt[4]{y}} &= \frac{7\sqrt[4]{y^3}}{\sqrt[4]{y}\sqrt[4]{y^3}} \\ &= \frac{7\sqrt[4]{y^3}}{\sqrt[4]{y \cdot y^3}} = \frac{7\sqrt[4]{y^3}}{y} \end{aligned}$$

48.
$$\begin{aligned} \frac{\sqrt[3]{x^2z^4}}{\sqrt[3]{2y^2}} &= \frac{\sqrt[3]{x^2z^3z}}{\sqrt[3]{2y^2}} \\ &= \frac{z\sqrt[3]{x^2z}}{\sqrt[3]{2y^2}} \\ &= \frac{z\sqrt[3]{x^2z}\,\sqrt[3]{2^2y}}{\sqrt[3]{2y^2}\,\sqrt[3]{2^2y}} \\ &= \frac{z\sqrt[3]{4x^2yz}}{\sqrt[3]{2^3y^3}} = \frac{z\sqrt[3]{4x^2yz}}{2y} \end{aligned}$$

52.
$$\begin{aligned} \frac{\sqrt[3]{3}}{\sqrt[3]{x}} &= \frac{\sqrt[3]{3}\,\sqrt[3]{3^2}}{\sqrt[3]{x}\,\sqrt[3]{3^2}} \\ &= \frac{\sqrt[3]{3^3}}{\sqrt[3]{3^2x}} \\ &= \frac{3}{\sqrt[3]{9x}} \end{aligned}$$

7.3

5.
$$\begin{aligned} &\sqrt{5x^2y}\sqrt{35xy^2} \\ &= \sqrt{5x^2y \cdot 5 \cdot 7 \cdot xy^2} \\ &= \sqrt{5^2x^2y^2 \cdot 7xy} \\ &= 5xy\sqrt{7xy} \end{aligned}$$

8.
$$\begin{aligned} \sqrt{5xy}\,\sqrt{15x^2y}\,\sqrt{3xy^2} &= \sqrt{5xy \cdot 15x^2y \cdot 3xy^2} \\ &= \sqrt{5^2 \cdot 3^2 \cdot x^4 \cdot y^4} \\ &= 5 \cdot 3 \cdot x^2 \cdot y^2 = 15x^2y^2 \end{aligned}$$

18.
$$\begin{aligned} \sqrt[4]{9x^2y^5}\,\sqrt[4]{45x^3y^2} &= \sqrt[4]{9x^2y^5 \cdot 45x^3y^2} \\ &= \sqrt[4]{3^4x^4y^4 \cdot 5xy^3} \\ &= 3xy\sqrt[4]{5xy^3} \end{aligned}$$

27. $\dfrac{\sqrt{98x^3y^5}}{\sqrt{18x^2y^7}} = \sqrt{\dfrac{98x^3y^5}{18x^2y^7}}$

$= \sqrt{\dfrac{\not{2} \cdot 49 \cdot \not{x^2} \cdot x \cdot \not{y^5}}{\not{2} \cdot 9 \cdot \not{x^2} \cdot y^2 \cdot \not{y^5}}}$

$= \dfrac{\sqrt{49x}}{\sqrt{9y^2}} = \dfrac{7\sqrt{x}}{3y}$

46. Since the circles have radii 3.0, 4.0, and 5.0 cm, the sides of the triangle are

$$3.0 + 4.0 = 7.0 = a$$
$$4.0 + 5.0 = 9.0 = b$$
$$5.0 + 3.0 = 8.0 = c$$

We use the formula

$$r = \sqrt{\frac{(s-a)(s-b)(s-c)}{s}}$$

where $s = \dfrac{7.0 + 9.0 + 8.0}{2} = \dfrac{24.0}{2} = 12.0.$

Thus, $r = \sqrt{\dfrac{(12.0 - 7.0)(12.0 - 9.0)(12.0 - 8.0)}{12.0}}$

$= \sqrt{\dfrac{(5.0)(3.0)(4.0)}{12.0}}$

$= \sqrt{5.0} \approx 2.236068.$

The radius of the inscribed circle is about 2.2 cm.

7.4

9. $9\sqrt[3]{250} - 4\sqrt[3]{128} = 9\sqrt[3]{125 \cdot 2} - 4\sqrt[3]{64 \cdot 2}$

$= 9\sqrt[3]{125}\,\sqrt[3]{2} - 4\sqrt[3]{64}\,\sqrt[3]{2}$

$= 9 \cdot 5\sqrt[3]{2} - 4 \cdot 4\sqrt[3]{2}$

$= 45\sqrt[3]{2} - 16\sqrt[3]{2} = 29\sqrt[3]{2}$

14. $6\sqrt{48y^4} + 4\sqrt{12y^4} = 6\sqrt{16y^4 \cdot 3} + 4\sqrt{4y^4 \cdot 3}$

$= 6\sqrt{16y^4}\,\sqrt{3} + 4\sqrt{4y^4}\,\sqrt{3}$

$= 6(4y^2)\sqrt{3} + 4(2y^2)\sqrt{3}$

$= 24y^2\sqrt{3} + 8y^2\sqrt{3} = 32y^2\sqrt{3}$

21. $3\sqrt[5]{a^{10}b^{12}} - ab\sqrt[5]{a^5b^7}$

$= 3\sqrt[5]{a^{10}b^{10}b^2} - ab\sqrt[5]{a^5b^5b^2}$

$= 3\sqrt[5]{(a^2)^5(b^2)^5}\,\sqrt[5]{b^2} - ab\sqrt[5]{a^5b^5}\,\sqrt[5]{b^2}$

$= 3a^2b^2\sqrt[5]{b^2} - ab(ab)\sqrt[5]{b^2}$

$= 3a^2b^2\sqrt[5]{b^2} - a^2b^2\sqrt[5]{b^2} = 2a^2b^2\sqrt[5]{b^2}$

30. $\dfrac{\sqrt[3]{x^2y^2}}{xy} + \dfrac{1}{\sqrt[3]{xy}} = \dfrac{\sqrt[3]{x^2y^2}}{xy} + \dfrac{\sqrt[3]{x^2y^2}}{\sqrt[3]{xy}\,\sqrt[3]{x^2y^2}}$

$= \dfrac{\sqrt[3]{x^2y^2}}{xy} + \dfrac{\sqrt[3]{x^2y^2}}{xy}$

$= \dfrac{2\sqrt[3]{x^2y^2}}{xy}$

40. $\sqrt[4]{x^3}(\sqrt[4]{xy^3} + \sqrt[4]{xy^4}) = \sqrt[4]{x^3}\,\sqrt[4]{xy^3} + \sqrt[4]{x^3}\,\sqrt[4]{xy^4}$

$= \sqrt[4]{x^3xy^3} + \sqrt[4]{x^3xy^4}$

$= \sqrt[4]{x^4y^3} + \sqrt[4]{x^4y^4}$

$= x\sqrt[4]{y^3} + xy$

44. $(2\sqrt{5} - \sqrt{2})(3\sqrt{5} + 4\sqrt{2}) = 2\sqrt{5} \cdot 3\sqrt{5} +$
$2\sqrt{5} \cdot 4\sqrt{2} - \sqrt{2} \cdot 3\sqrt{5} - \sqrt{2} \cdot 4\sqrt{2}$

$= 2 \cdot 3 \cdot 5 + 2 \cdot 4\sqrt{10} - 3\sqrt{10} - 4 \cdot 2$

$= 30 + 8\sqrt{10} - 3\sqrt{10} - 8$

$= 22 + 5\sqrt{10}$

54. $\dfrac{\sqrt{2}}{2\sqrt{2} + \sqrt{3}} = \dfrac{\sqrt{2}(2\sqrt{2} - \sqrt{3})}{(2\sqrt{2} + \sqrt{3})(2\sqrt{2} - \sqrt{3})}$

$= \dfrac{\sqrt{2}(2\sqrt{2} - \sqrt{3})}{4(2) - 3}$

$= \dfrac{2 \cdot 2 - \sqrt{6}}{8 - 3} = \dfrac{4 - \sqrt{6}}{5}$

58. $\dfrac{4x - 25}{2\sqrt{x} - 5} = \dfrac{(4x - 25)(2\sqrt{x} + 5)}{(2\sqrt{x} - 5)(2\sqrt{x} + 5)}$

$= \dfrac{\not{(4x - 25)}(2\sqrt{x} + 5)}{\not{4x - 25}} = 2\sqrt{x} + 5$

66. $\dfrac{\sqrt{x + h} - \sqrt{x}}{h}$

$= \dfrac{(\sqrt{x + h} - \sqrt{x})(\sqrt{x + h} + \sqrt{x})}{h(\sqrt{x + h} + \sqrt{x})}$

$= \dfrac{(x + h) - x}{h(\sqrt{x + h} + \sqrt{x})}$

$= \dfrac{\not{h}}{\not{h}(\sqrt{x + h} + \sqrt{x})} = \dfrac{1}{\sqrt{x + h} + \sqrt{x}}$

68. $\dfrac{5 + \sqrt{125}}{10} = \dfrac{5 + \sqrt{25 \cdot 5}}{10}$

$= \dfrac{5 + 5\sqrt{5}}{10}$

$= \dfrac{\not{5}(1 + \sqrt{5})}{\not{5} \cdot 2} = \dfrac{1 + \sqrt{5}}{2}$

7.5

23. $\left(\dfrac{a^3}{b^{-6}}\right)^{1/3} = \dfrac{a^{(3)(1/3)}}{b^{(-6)(1/3)}} = \dfrac{(a^3)^{1/3}}{(b^{-6})^{1/3}} = \dfrac{a^{3(1/3)}}{b^{(-6)1/3}}$

$= \dfrac{a^1}{b^{-2}} = ab^2$

27. $\left(\dfrac{96a^3b^{-2}}{3a^{-2}b^8}\right)^{1/5} = \left(\dfrac{\not{3} \cdot 32a^{3-(-2)}}{\not{3}b^{8-(-2)}}\right)^{1/5}$

$= \left(\dfrac{32a^5}{b^{10}}\right)^{1/5}$

$= \dfrac{(32)^{1/5}(a^5)^{1/5}}{(b^{10})^{1/5}} = \dfrac{\sqrt[5]{32a^5}}{b^2} = \dfrac{2a}{b^2}$

34. $\sqrt[5]{\dfrac{a^{20}}{625b^{15}}} = \left(\dfrac{a^{20}}{625b^{15}}\right)^{1/5}$

$= \dfrac{(a^{20})^{1/5}}{(5^4)^{1/5}(b^{15})^{1/5}}$

$= \dfrac{a^4}{(5^4)^{1/5}b^3}$

$= \dfrac{a^4}{b^3\sqrt[5]{5^4}}$

$= \dfrac{a^4\sqrt[5]{5}}{b^3\sqrt[5]{5^4}\,\sqrt[5]{5}} = \dfrac{a^4\sqrt[5]{5}}{5b^3}$

36. $\dfrac{1}{\sqrt[3]{5^2}} = \dfrac{1}{5^{2/3}} = \dfrac{5^{1/3}}{5^{2/3}5^{1/3}} = \dfrac{5^{1/3}}{5} = \dfrac{\sqrt[3]{5}}{5}$

7.6

3. $8\sqrt{3x-1} + 7 = 0$
Since $8\sqrt{3x-1}$ is nonnegative for any x and 7 is positive, there is no value of x for which their sum is 0. Thus, there is no solution.

5. $\sqrt{x^2-5} + x - 5 = 0$

$\sqrt{x^2-5} = 5 - x$

$(\sqrt{x^2-5})^2 = (5-x)^2$

$x^2 - 5 = 25 - 10x + x^2$

$-5 = 25 - 10x$

$-30 = -10x$

$3 = x$

The solution is 3 since it checks in the original equation.

7. $\sqrt{y+6} + \sqrt{y+11} = 5$

$\sqrt{y+6} = 5 - \sqrt{y+11}$

$(\sqrt{y+6})^2 = (5 - \sqrt{y+11})^2$

$y + 6 = 25 - 10\sqrt{y+11} + y + 11$

$y + 6 = 36 - 10\sqrt{y+11} + y$

$6 = 36 - 10\sqrt{y+11}$

$-30 = -10\sqrt{y+11}$

$3 = \sqrt{y+11}$

$9 = y + 11$

$-2 = y$ **This checks**

10. $\dfrac{8}{\sqrt{a}} = 2$

$\dfrac{8\sqrt{a}}{\sqrt{a}} = 2\sqrt{a}$ **Multiply by $\sqrt{a}$**

$8 = 2\sqrt{a}$

$4 = \sqrt{a}$

$16 = a$ **Square both sides**

This does check.

15. $\sqrt{3x-2} - \sqrt{2x+5} = -1$

$\sqrt{3x-2} = \sqrt{2x+5} - 1$

$(\sqrt{3x-2})^2 = (\sqrt{2x+5} - 1)^2$

$3x - 2 = 2x + 5 - 2\sqrt{2x+5} + 1$

$x - 8 = -2\sqrt{2x+5}$

$(x-8)^2 = (-2\sqrt{2x+5})^2$

$x^2 - 16x + 64 = 4(2x+5)$

$x^2 - 16x + 64 = 8x + 20$

$x^2 - 24x + 44 = 0$

$(x-2)(x-22) = 0$

$x - 2 = 0$ or $x - 22 = 0$

$x = 2$ $\quad x = 22$

2 checks but 22 does not.

18. $\sqrt[4]{w+2} = \sqrt[4]{5w}$

$(\sqrt[4]{w+2})^4 = (\sqrt[4]{5w})^4$

$w + 2 = 5w$

$2 = 4w$

$\dfrac{1}{2} = w$ **This does check**

28. $a = \sqrt{1 + \dfrac{b}{x}}$ for x

$a^2 = \left(\sqrt{1 + \dfrac{b}{x}}\right)^2$

$a^2 = 1 + \dfrac{b}{x}$

$a^2 - 1 = \dfrac{b}{x}$

$x(a^2 - 1) = b$

$x = \dfrac{b}{a^2 - 1}$

30. $z^{1/3} + x = y$ for z

$z^{1/3} = y - x$

$(z^{1/3})^3 = (y-x)^3$

$z = (y-x)^3$

31. Let x = the number,
$2\sqrt{x}$ = twice the square root of the number,
$\sqrt{3x+9}$ = the square root of 9 more than 3 times the number.

$2\sqrt{x} = \sqrt{3x+9}$

$(2\sqrt{x})^2 = (\sqrt{3x+9})^2$

$4x = 3x + 9$

$x = 9$ **This does check**

34. $c = 100\sqrt[3]{n} + 1200$

$= 100\sqrt[3]{8} + 1200 = 100(2) + 1200$

$= 200 + 1200 = 1400$

The cost is \$1400.

43. We need to solve

$$1.5 = 2\pi\sqrt{\frac{L}{32}}.$$

$$(1.5)^2 = 4\pi^2\left(\frac{L}{32}\right) \quad \text{Square both sides}$$

$$\frac{(1.5)^2}{4\pi^2} = \frac{L}{32} \quad \text{Divide both sides by } 4\pi^2$$

$$\frac{32(1.5)^2}{4\pi^2} = L \quad \text{Multiply both sides by 32}$$

Using a calculator and 3.14 for π we obtain $L \approx 1.8256319$. To the nearest hundredth, the pendulum is 1.83 ft long.

7.7

4. Substitute 5 for b and $5\sqrt{3}$ for c into

$$a^2 + b^2 = c^2.$$
$$a^2 + (5)^2 = (5\sqrt{3})^2$$
$$a^2 + 25 = (25)(3)$$
$$a^2 + 25 = 75$$
$$a^2 = 50$$
$$a = \sqrt{50} \quad \text{Use positive root only}$$
$$a = \sqrt{25 \cdot 2} = 5\sqrt{2}$$

9. The diagonal of a square is the hypotenuse of the right triangle formed with the sides. If x is the length of a side,

$$x^2 + x^2 = (2)^2.$$
$$2x^2 = 4$$
$$x^2 = 2$$
$$x = \sqrt{2}$$

Thus, each side is $\sqrt{2}$ mi or about 1.4 mi.

14. By the Pythagorean theorem we have:

$$a^2 + a^2 = h^2.$$
$$2a^2 = h^2$$
$$\sqrt{2a^2} = h$$
$$a\sqrt{2} = h$$

27. The midpoint of the segment joining $(a, 3)$ and $(-2, b + 1)$ is

$$(\bar{x}, \bar{y}) = \left(\frac{a + (-2)}{2}, \frac{3 + (b + 1)}{2}\right)$$
$$= \left(\frac{a - 2}{2}, \frac{b + 4}{2}\right).$$

Since $(\bar{x}, \bar{y}) = (2, 1)$, $\bar{x} = 2$ and $\bar{y} = 1$ so that

$$\frac{a - 2}{2} = 2 \quad \text{and} \quad \frac{b + 4}{2} = 1.$$

$$a - 2 = 4 \qquad b + 4 = 2$$
$$a = 6 \qquad b = -2$$

Thus, $a = 6$ and $b = -2$.

7.8

10. $\sqrt{-4}\,\sqrt{-9} = \sqrt{(-1)(4)}\,\sqrt{(-1)(9)}$

$= \sqrt{-1}\,\sqrt{4}\,\sqrt{-1}\,\sqrt{9}$

$= i(2)i(3) = 6i^2 = -6$

13. $\dfrac{\sqrt{-4}}{\sqrt{-9}} = \dfrac{\sqrt{(-1)(4)}}{\sqrt{(-1)(9)}}$

$= \dfrac{\sqrt{-1}\,\sqrt{4}}{\sqrt{-1}\,\sqrt{9}} = \dfrac{i\sqrt{4}}{i\sqrt{9}} = \dfrac{2}{3}$

17. $(4x + 1) + 7yi = 5 - i$

$4x + 1 = 5$ and $7y = -1$

$4x = 4 \qquad y = -\dfrac{1}{7}$

$x = 1$

41. $\dfrac{8 - 7i}{5 + 4i} = \dfrac{(8 - 7i)(5 - 4i)}{(5 + 4i)(5 - 4i)}$

$= \dfrac{40 - 67i + 28i^2}{25 - 16i^2}$

$= \dfrac{40 - 67i - 28}{25 + 16} \quad i^2 = -1$

$= \dfrac{12 - 67i}{41}$

$= \dfrac{12}{41} - \dfrac{67}{41}i$

8.1

8. $(x - 1)(x + 1)(x - 2) = 0$

The zero-product rule applies to more than 2 factors so we solve

$x - 1 = 0$ or $x + 1 = 0$ or $x - 2 = 0$

$x = 1$ or $x = -1$ or $x = 2$

The three solutions are 1, −1, 2.

21.

$$5x^2 - 4x = 3x^2 + 9x + 7$$
$$2x^2 - 4x = 9x + 7 \quad \text{Subtract } 3x^2$$
$$2x^2 - 13x = 7 \quad \text{Subtract } 9x$$
$$2x^2 - 13x - 7 = 0 \quad \text{Subtract 7}$$
$$(2x + 1)(x - 7) = 0$$

$2x + 1 = 0$ or $x - 7 = 0$

$2x = -1 \qquad x = 7$

$x = -\dfrac{1}{2}$

The solutions are $-\frac{1}{2}$ and 7.

25. $5y(y + 1) = 5y^2 + 5y$

$5y^2 + 5y = 5y^2 + 5y \quad$ Clear parentheses

Since we obtain an identity, any real number is a solution.

33.
$$x^3 + 3x^2 + 2x = 0$$
$$x(x^2 + 3x + 2) = 0 \quad \text{Factor } x$$
$$x(x + 1)(x + 2) = 0$$
$$x = 0 \quad \text{or} \quad x + 1 = 0 \quad \text{or} \quad x + 2 = 0$$
$$x = -1 \qquad x = -2$$

The solutions are 0, -1, and -2.

38. $(2x - 1)^2 + 4 = 0$
$$(2x - 1)^2 = -4$$
$$2x - 1 = \pm\sqrt{-4} \quad \text{Take square root of both sides}$$
$$2x - 1 = \pm 2i \quad \sqrt{-4} = \sqrt{(4)(-1)} = \sqrt{4}\sqrt{-1} = 2i$$
$$2x = 1 \pm 2i$$
$$x = \frac{1 \pm 2i}{2} = \frac{1}{2} \pm i$$

46. Let w = the width of the rectangle, $w + 5$ = the length of the rectangle. Use the formula $A = lw$ and substitute.
$$(w + 5)w = 84$$
$$w^2 + 5w = 84$$
$$w^2 + 5w - 84 = 0$$
$$(w - 7)(w + 12) = 0$$
$$w - 7 = 0 \quad \text{or} \quad w + 12 = 0$$
$$w = 7 \qquad w = -12$$

Since the width cannot be negative, we discard -12. Thus, the width is 7 cm and the length is 12 cm ($w + 5 = 7 + 5 = 12$).

49. Since the supply must equal the demand, $S = D$, which gives us
$$p - 95 = \frac{500}{p}.$$
$$p(p - 95) = 500 \quad \text{Multiply both sides by } p$$
$$p^2 - 95p = 500 \quad \text{Distribute}$$
$$p^2 - 95p - 500 = 0 \quad \text{General form}$$
$$(p + 5)(p - 100) = 0 \quad \text{Factor}$$
$$p + 5 = 0 \quad \text{and} \quad p - 100 = 0 \quad \text{Zero-product rule}$$
$$p = -5 \qquad p = 100$$

Since -5 cannot be a price per gallon, we have 100¢/gal as the desired price.

8.2

7. $(3z - 1)^2 = \frac{4}{9}$
$$3z - 1 = \pm\sqrt{\frac{4}{9}} = \pm\frac{2}{3}$$
$$3z = 1 \pm \frac{2}{3}$$
$$3z = 1 + \frac{2}{3} \quad \text{or} \quad 3z = 1 - \frac{2}{3}$$
$$3z = \frac{5}{3} \quad \text{or} \quad 3z = \frac{1}{3}$$
$$z = \frac{5}{9} \quad \text{or} \quad z = \frac{1}{9}$$

The solutions are $\frac{5}{9}$ and $\frac{1}{9}$.

11. We must add $\left(\frac{1}{2} \cdot \frac{8}{5}\right)^2$ or $\left(\frac{8}{10}\right)^2 = \left(\frac{4}{5}\right)^2 = \frac{16}{25}$ to complete the square.

14. $2u^2 + 3u - 1 = 0$
$$u^2 + \frac{3}{2}u - \frac{1}{2} = 0 \quad \text{Divide through by 2}$$
$$u^2 + \frac{3}{2}u = \frac{1}{2}$$
$$u^2 + \frac{3}{2}u + \frac{9}{16} = \frac{1}{2} + \frac{9}{16} \quad \text{Add } \frac{9}{16} = \left[\frac{1}{2}\left(\frac{3}{2}\right)\right]^2 \text{ to both sides}$$
$$\left(u + \frac{3}{4}\right)^2 = \frac{17}{16}$$
$$u + \frac{3}{4} = \pm\sqrt{\frac{17}{16}} = \pm\frac{\sqrt{17}}{4}$$
$$u = -\frac{3}{4} \pm \frac{\sqrt{17}}{4}$$
$$u = \frac{-3 \pm \sqrt{17}}{4}$$

23. $\frac{1}{3}x^2 - \frac{4}{9}x - \frac{2}{3} = 0$
$$9\left(\frac{1}{3}x^2 - \frac{4}{9}x - \frac{2}{3}\right) = 9 \cdot 0 \quad \text{Multiply by 9 to clear fractions}$$
$$3x^2 - 4x - 6 = 0$$
$$x = \frac{-b \pm \sqrt{b^2 - 4ac}}{2a}$$
$$= \frac{-(-4) \pm \sqrt{(-4)^2 - 4(3)(-6)}}{2(3)}$$
$$= \frac{4 \pm \sqrt{16 + 72}}{6} = \frac{4 \pm \sqrt{88}}{6}$$
$$= \frac{4 \pm 2\sqrt{22}}{6} = \frac{\not{2}(2 \pm \sqrt{22})}{\not{2} \cdot 3} = \frac{2 \pm \sqrt{22}}{3}$$

31. $2x^2 - 3x + 4 = 0$

$a = 2$, $b = -3$, $c = 4$
$$x = \frac{-b \pm \sqrt{b^2 - 4ac}}{2a}$$
$$= \frac{-(-3) \pm \sqrt{(-3)^2 - 4(2)(4)}}{2(2)}$$
$$= \frac{3 \pm \sqrt{9 - 32}}{4} = \frac{3 \pm \sqrt{-23}}{4}$$
$$= \frac{3 \pm i\sqrt{23}}{4} \quad \sqrt{-23} = \sqrt{23(-1)} = \sqrt{23}\sqrt{-1} = i\sqrt{23}$$

32. $(2z - 1)(z - 2) - 11 = 2(z + 4) - 8$

$$2z^2 - 4z - z + 2 - 11 = 2z + 8 - 8$$
$$2z^2 - 5z - 9 = 2z$$
$$2z^2 - 7z - 9 = 0$$
$$(2z - 9)(z + 1) = 0$$
$$2z - 9 = 0 \quad \text{or} \quad z + 1 = 0$$
$$2z = 9 \qquad z = -1$$
$$z = \frac{9}{2}$$

The solutions are $\frac{9}{2}$ and -1.

34. Let $x =$ the length of one leg,
$x + 2 =$ the length of the second leg.
Use the Pythagorean theorem to obtain the desired equation.

$$x^2 + (x + 2)^2 = 11^2$$
$$x^2 + x^2 + 4x + 4 = 121$$
$$2x^2 + 4x - 117 = 0$$
$$x = \frac{-b \pm \sqrt{b^2 - 4ac}}{2a} = \frac{-4 \pm \sqrt{16 - 4(2)(-117)}}{4}$$
$$= \frac{-4 \pm \sqrt{952}}{4} = \frac{-4 \pm 2\sqrt{238}}{4}$$

Since the length of a leg cannot be negative, we discard $\frac{-4 - 2\sqrt{238}}{4}$. Use 15.4 for an approximation of $\sqrt{238}$.

$$\frac{-4 + 2(15.4)}{4} = 6.7$$

One leg is about 6.7 m and the other is 8.7 m.

37. Since the catcher is 6 ft down the line towards third base, he is actually 84 ft from third base. Let $x =$ distance from the catcher to second base. Then x is the hypotenuse of a right triangle with legs 90 ft and 84 ft. Use the Pythagorean theorem.

$$90^2 + 84^2 = x^2$$
$$15{,}156 = x^2$$
$$\sqrt{15{,}156} = x \quad \textbf{Use positive root only}$$
$$123.10971 \approx x$$

The ball must travel about 123.1 ft.

8.3

14.
$$[x - (2 + i)][x - (2 - i)] = 0$$
$$[x - 2 - i][x - 2 + i] = 0$$
$$x^2 - 2x + ix - 2x + 4 - 2i - ix + 2i - i^2 = 0$$
$$x^2 - 4x + 4 + 1 = 0 \quad \mathbf{i^2 = -1}$$
$$x^2 - 4x + 5 = 0$$

17. With $x_1 = 2 + 3i$ and $x_2 = 2 - 3i$, we have:

$$(x - x_1)(x - x_2) = 0$$
$$[x - (2 + 3i)][x - (2 - 3i)] = 0$$
$$x^2 - 2x - 3ix - 2x + 3ix + (2 + 3i)(2 - 3i) = 0$$
$$x^2 - 4x + (4 - 9i^2) = 0$$
$$x^2 - 4x + (4 + 9) = 0$$
$$x^2 - 4x + 13 = 0.$$

8.4

9. $x^{-2} - x^{-1} - 12 = 0$ **Let $u = x^{-1}$; then $u^2 = x^{-2}$**

$$u^2 - u - 12 = 0$$
$$(u + 3)(u - 4) = 0$$
$$u + 3 = 0 \quad \text{or} \quad u - 4 = 0$$
$$u = -3 \qquad u = 4$$

To complete the solution, we substitute $x^{-1} = \frac{1}{x}$ for u and solve for x.

$$\frac{1}{x} = -3 \quad \text{or} \quad \frac{1}{x} = 4$$
$$x = -\frac{1}{3} \qquad x = \frac{1}{4}$$

10. $(z - 7)^4 - 13(z - 7)^2 + 42 = 0$
Substitute u for $(z - 7)^2$.

$$u^2 - 13u + 42 = 0$$
$$(u - 6)(u - 7) = 0 \quad \textbf{Factor}$$
$$u - 6 = 0 \quad \text{or} \quad u - 7 = 0 \quad \textbf{Zero-product rule}$$
$$u = 6 \quad \text{or} \quad u = 7$$

Then substitute $(z - 7)^2$ for u and solve for z.

$$(z - 7)^2 = 6 \quad \text{or} \quad (z - 7)^2 = 7$$
$$z - 7 = \pm\sqrt{6} \qquad z - 7 = \pm\sqrt{7}$$
$$z = 7 \pm \sqrt{6} \qquad z = 7 \pm \sqrt{7}$$

There are four solutions, $7 \pm \sqrt{6}$ and $7 \pm \sqrt{7}$.

16. $\dfrac{x^2}{2x + 4} = 1$
Multiply both sides by the LCD $2x + 4$.

$$x^2 = 2x + 4$$
$$x^2 - 2x - 4 = 0$$
$$x = \frac{-(-2) \pm \sqrt{(-2)^2 - 4(1)(-4)}}{2(1)}$$
$$= \frac{2 \pm \sqrt{4 + 16}}{2}$$
$$= \frac{2 \pm \sqrt{20}}{2}$$
$$= \frac{2 \pm 2\sqrt{5}}{2} = 1 \pm \sqrt{5}$$

20. $\dfrac{-2}{v^2 - 2v} = \dfrac{1}{v} + \dfrac{3v}{v - 2}$

$\dfrac{-2}{v(v-2)} = \dfrac{1}{v} + \dfrac{3v}{v-2}$ LCD = $v(v - 2)$

$\cancel{v}(\cancel{v-2})\left[\dfrac{-2}{\cancel{v}(\cancel{v-2})}\right]$

$= \cancel{v}(v-2)\left[\dfrac{1}{\cancel{v}}\right] + v(\cancel{v-2})\left[\dfrac{3v}{\cancel{v-2}}\right]$

$$-2 = v - 2 + 3v^2$$
$$0 = 3v^2 + v$$
$$0 = v(3v + 1)$$
$$v = 0 \quad \text{or} \quad 3v + 1 = 0$$
$$3v = -1$$
$$v = -\frac{1}{3}$$

Since 0 is a meaningless replacement, the only solution is $-\frac{1}{3}$.

25. $\sqrt{x+4} - x + 8 = 0$

$(\sqrt{x+4})^2 = (x-8)^2$ Isolate radical and square

$$x + 4 = x^2 - 16x + 64$$
$$0 = x^2 - 17x + 60$$
$$0 = (x - 12)(x - 5)$$
$$x - 12 = 0 \quad \text{or} \quad x - 5 = 0$$
$$x = 12 \qquad\qquad x = 5$$

Check: $\sqrt{12 + 4} - 12 + 8 \stackrel{?}{=} 0$

$$\sqrt{16} - 12 + 8 \stackrel{?}{=} 0$$
$$0 = 0$$
$$\sqrt{5 + 4} - 5 + 8 \stackrel{?}{=} 0$$
$$\sqrt{9} - 5 + 8 \stackrel{?}{=} 0$$
$$6 \neq 0$$

The only solution is 12.

29. $\sqrt{2z + 5} - \sqrt{z + 2} = 1$

$$(\sqrt{2z+5})^2 = (\sqrt{z+2} + 1)^2$$
$$2z + 5 = z + 2 + 2\sqrt{z+2} + 1$$
$$(z + 2)^2 = (2\sqrt{z+2})^2$$
$$z^2 + 4z + 4 = 4(z + 2)$$
$$z^2 + 4z + 4 = 4z + 8$$
$$z^2 = 4$$
$$z = \pm\sqrt{4} = \pm 2$$

Since +2 and −2 check, they are both solutions.

31. $(x - 5)^{1/2} = x^{1/2} + 1$

$$(\sqrt{x-5})^2 = (\sqrt{x} + 1)^2$$
$$x - 5 = x + 2\sqrt{x} + 1$$
$$-6 = 2\sqrt{x}$$
$$-3 = \sqrt{x}$$

But since $\sqrt{x}$ represents the principal root of x, that is, the nonnegative root, it cannot equal −3. Thus, we conclude that the equation has no solution.

35. Let x = time for Wanda to do the job,
$x - 3$ = time for Maria to do the job,
4 = time to do the job together,
$\dfrac{1}{x}$ = amount Wanda does in 1 hour,
$\dfrac{1}{x-3}$ = amount Maria does in 1 hour,
$\dfrac{1}{4}$ = amount done together in 1 hour.

Then we must solve:

$$\frac{1}{x} + \frac{1}{x-3} = \frac{1}{4}.$$

$4x(x-3)\left[\dfrac{1}{x} + \dfrac{1}{x-3}\right] = 4x(x-3)\left[\dfrac{1}{4}\right]$ Multiply both sides by the LCD = $4x(x - 3)$

$4(x - 3) + 4x = x(x - 3)$ Distribute and simplify

$$4x - 12 + 4x = x^2 - 3x$$
$$0 = x^2 - 11x + 12$$
$$x = \frac{-b \pm \sqrt{b^2 - 4ac}}{2a}$$
$$= \frac{-(-11) \pm \sqrt{(-11)^2 - 4(1)(12)}}{2(1)}$$
$$= \frac{11 \pm \sqrt{121 - 48}}{2}$$
$$= \frac{11 \pm \sqrt{73}}{2}$$
$$\approx \frac{11 \pm 8.5}{2} = \begin{cases} \dfrac{11 + 8.5}{2} = 9.75 \\[2ex] \dfrac{11 - 8.5}{2} = 1.25 \end{cases}$$

But x cannot be 1.25 since $x - 3$ is then negative. Thus, to the nearest tenth, it would take Wanda 9.8 hr and Maria 6.8 hr to do the job alone.

42. Let x = rate of the airplane going,
$x + 25$ = rate of the airplane returning.
Then using $d = rt$ solved for t, $t = \frac{d}{r}$, we have:

$$\frac{2000}{x} = \text{time going,}$$
$$\frac{2000}{x + 25} = \text{time returning.}$$

Since the total time of the trip was 9.5 hr, we must solve:

$$\frac{2000}{x} + \frac{2000}{x + 25} = 9.5.$$

$$x(x+25)\left[\frac{2000}{x}+\frac{2000}{x+25}\right]=9.5x(x+25) \quad \text{Multiply by LCD}$$

$$2000(x+25)+2000x=9.5x(x+25)$$

$$2000x+50{,}000+2000x=9.5x^2+237.5x$$

$$0=9.5x^2-3762.5x-50{,}000$$

Clear the decimals by multiplying through by 10.

$$95x^2-37{,}625x-500{,}000=0$$

Divide through by 5.

$$19x^2-7525x-100{,}000=0$$

Use the quadratic formula.

$$x=\frac{7525\pm\sqrt{(-7525)^2-4(19)(-100{,}000)}}{2(19)}$$

$$=\frac{7525\pm\sqrt{64{,}225{,}625}}{38}$$

$$\approx\frac{7525\pm 8014.09}{38}$$

Since the negative solution is meaningless, $x\approx\frac{7525+8014.09}{38}\approx 408.92\approx 409$. Thus, the airplane traveled $x+25=409+25=434$ mph returning.

48. $\frac{1}{x}+\frac{n}{x+1}=1$ for x

$$\cancel{x}(x+1)\left[\frac{1}{\cancel{x}}\right]+x\cancel{(x+1)}\left[\frac{n}{\cancel{x+1}}\right]=x(x+1)[1]$$

Multiply by LCD = $x(x+1)$

$$x+1+nx=x^2+x$$

$$0=x^2-nx-1$$

$$x=\frac{-b\pm\sqrt{b^2-4ac}}{2a}$$

$$=\frac{-(-n)\pm\sqrt{(-n)^2-4(1)(-1)}}{2(1)}$$

$$=\frac{n\pm\sqrt{n^2+4}}{2}$$

8.5

4. $2x^2+x\geq 15$
$2x^2+x-15\geq 0$

First solve for the critical points.

$$2x^2+x-15=0$$

$$(2x-5)(x+3)=0$$

$$x=\frac{5}{2},\ -3$$

The critical points form the intervals $(-\infty,-3)$, $\left(-3,\frac{5}{2}\right)$, and $\left(\frac{5}{2},\infty\right)$. Use -4, 0, and 3 as test points.

$$-4:\ 2(-4)^2+(-4)-15=13>0$$

$$0:\ 2(0)^2+(0)-15=-15<0$$

$$3:\ 2(3)^2+(3)-15=6>0$$

Thus, the solution includes the intervals containing -4 and 3 (where $2x^2+x-15\geq 0$) together with $\frac{5}{2}$ and -3 since the inequality is $\geq$. The solution is

$$x\leq -3 \quad\text{or}\quad x\geq\frac{5}{2},$$

or, in interval notation, $(-\infty,-3]$ or $\left[\frac{5}{2},\infty\right)$.

9. $3x^2-2x\leq 2$
$3x^2-2x-2=0$

$$x=\frac{-(-2)\pm\sqrt{(-2)^2-4(3)(-2)}}{2(3)}=\frac{2\pm\sqrt{4+24}}{6}$$

$$=\frac{2\pm\sqrt{28}}{6}=\frac{2\pm 2\sqrt{7}}{6}=\frac{1\pm\sqrt{7}}{3}$$

Thus, the critical points are $\frac{1+\sqrt{7}}{3}$ (approximately 1.2) and $\frac{1-\sqrt{7}}{3}$ (approximately -0.5). Using test points -1, 0, and 2 we can see that the solution is all numbers between the critical points. That is, the solution is

$$\frac{1-\sqrt{7}}{3}\leq x\leq\frac{1+\sqrt{7}}{3}$$

or, in interval notation, $\left[\frac{1-\sqrt{7}}{3},\frac{1+\sqrt{7}}{3}\right]$

13. $x^2+9\leq 0$
First solve for the critical points.

$$x^2+9=0$$

$$x^2=-9$$

$$x=\pm 3i$$

Since the solutions are imaginary (not real), there are no critical points. Thus, the solution to the inequality will be all real numbers or no real numbers, depending on what happens at a single test point. Use 0 as the test point to get

$$0+9=9\leq 0,$$

which is false. Since 0 is not a solution, no real number is a solution. The inequality has no solution.

17. $$\frac{2x+1}{x-1}<3$$

$$\frac{2x+1}{x-1}-3<0$$

$$\frac{2x+1}{x-1}-\frac{3(x-1)}{x-1}<0$$

$$\frac{2x+1-3x+3}{x-1}<0$$

$$\frac{-x+4}{x-1}<0$$

The critical points are solutions to:

$$-x+4=0 \quad\text{and}\quad x-1=0$$

$$x=4 \qquad\qquad x=1.$$

The critical points separate the number line into $(-\infty, 1)$, $(1, 4)$, and $(4, \infty)$. Using a test point in each, we see that $\frac{-x+4}{x-1} < 0$ when x is in $(-\infty, 1)$ or $(4, \infty)$. The solution is

$$x < 1 \quad \text{or} \quad x > 4$$

or, in interval notation,

$$(-\infty, 1) \quad \text{or} \quad (4, \infty).$$

22. $\dfrac{(x-3)(x+2)}{x-1} > 0$

The critical points are solutions to:

$$\begin{array}{lll} x - 3 = 0 & \text{or} \quad x + 2 = 0 & \text{or} \quad x - 1 = 0 \\ x = 3 & \quad\;\; x = -2 & \quad\;\; x = 1 \end{array}$$

The critical points separate the number line into the intervals $(-\infty, -2)$, $(-2, 1)$, $(1, 3)$, and $(3, \infty)$. Choosing a test point in each we discover that $\frac{(x-3)(x+2)}{x-1} > 0$ when x is in $(-2, 1)$ or $(3, \infty)$ which gives the solution in interval notation. Alternatively, the solution could be given as $-2 < x < 1$ or $x > 3$.

9.1

5. $f(x) = x^2 - 4x + 3$

Then $a = 1$, $b = -4$, and $c = 3$.

Since $a = 1 > 0$, the parabola opens up. The x-coordinate of the vertex is

$$-\frac{b}{2a} = -\frac{-4}{2(1)} = \frac{4}{2} = 2.$$

The y-coordinate of the vertex is

$$\begin{aligned} f\left(-\frac{b}{2a}\right) &= f(2) = (2)^2 - 4(2) + 3 \\ &= 4 - 8 + 3 \\ &= -1. \end{aligned}$$

The vertex is $(2, -1)$. Since the vertex is below the x-axis and the parabola opens up, there will be x-intercepts found by solving

$$\begin{gathered} x^2 - 4x + 3 = 0. \\ (x-1)(x-3) = 0 \\ x - 1 = 0 \quad \text{and} \quad x - 3 = 0 \\ x = 1 \qquad\qquad x = 3 \end{gathered}$$

The x-intercepts are $(1, 0)$ and $(3, 0)$.

12. $g(x) = 3x^2 + 5x + 1$

Then $a = 3$, $b = 5$, and $c = 1$.

Since $a = 3 > 0$, the parabola opens up. The x-coordinate of the vertex is

$$-\frac{b}{2a} = -\frac{5}{2(3)} = -\frac{5}{6}.$$

The y-coordinate of the vertex is

$$\begin{aligned} f\left(-\frac{b}{2a}\right) = f\left(-\frac{5}{6}\right) &= 3\left(-\frac{5}{6}\right)^2 + 5\left(-\frac{5}{6}\right) + 1 \\ &= 3\left(\frac{25}{36}\right) - \frac{25}{6} + 1 \\ &= \frac{25}{12} - \frac{50}{12} + \frac{12}{12} \\ &= -\frac{13}{12} \end{aligned}$$

The vertex is $\left(-\frac{5}{6}, -\frac{13}{12}\right)$. Since the vertex is below the x-axis and the parabola opens up, there will be x-intercepts found by solving

$$3x^2 + 5x + 1 = 0.$$

Since this won't factor, we use the quadratic formula.

$$\begin{aligned} x = \frac{-b \pm \sqrt{b^2 - 4ac}}{2a} &= \frac{-5 \pm \sqrt{(5)^2 - 4(3)(1)}}{2(3)} \\ &= \frac{-5 \pm \sqrt{25 - 12}}{6} \\ &= \frac{-5 \pm \sqrt{13}}{6} \end{aligned}$$

The intercepts are $\left(\frac{-5+\sqrt{13}}{6}, 0\right)$ and $\left(\frac{-5-\sqrt{13}}{6}, 0\right)$.

21. $P = -x^2 + 80x - 1500$ must be maximized.

With $a = -1$ and $b = 80$, $-\frac{b}{2a} = -\frac{80}{2(-1)} = 40$.

Thus, the x-coordinate of the vertex is 40, and the y-coordinate (the value of P when x is 40) is $-(40)^2 + 80(40) - 1500 = 100$. The maximum profit ($P$) is \$100 when 40 (x) sandwiches are sold.

9.2

8. $f(x) = \sqrt{x^2 - 8x + 12}$

The domain of f is all real numbers x for which $x^2 - 8x + 12 \geq 0$. We need to solve the inequality. First solve the equation $x^2 - 8x + 12 = 0$ to obtain the critical points.

$$\begin{gathered} x^2 - 8x + 12 = 0 \\ (x-2)(x-6) = 0 \\ x - 2 = 0 \quad \text{and} \quad x - 6 = 0 \\ x = 2 \qquad\qquad x = 6 \end{gathered}$$

The critical points separate the number line into the intervals $(-\infty, 2)$, $(2, 6)$, and $(6, \infty)$. Using test points in each of these we find that $x^2 - 8x + 12 \geq 0$ when x is in $(-\infty, 2]$ or $[6, \infty)$. Thus, the domain of f is all real numbers $x \leq 2$ or $x \geq 6$.

17. Given $f(x) = 4 + 3x$ and $g(x) = x^2 - 3$, then

$$g(-1) = (-1)^2 - 3 = 1 - 3 = -2,$$

and $f[g(-1)] = f[-2] = 4 + 3(-2) = 4 - 6 = -2$.

9.3

16. $f(x) = 2x^3$
Then $y = 2x^3$, so we interchange x and y and solve for y.

$$x = 2y^3 \quad \text{Interchange } x \text{ and } y$$
$$\frac{x}{2} = y^3$$
$$\sqrt[3]{\frac{x}{2}} = y \quad \text{Take cube root of both sides}$$

Thus, $f^{-1}(x) = \sqrt[3]{\frac{x}{2}}$.

20. $g(x) = x^2 + 1,\ x \geq 0$
Then $y = x^2 + 1$, $x \geq 0$, so interchange x and y and solve for y.

$$x = y^2 + 1$$
$$x - 1 = y^2$$
$$\pm\sqrt{x - 1} = y \quad \text{Take square root of both sides}$$

But since $x \geq 0$ in the original function, when x and y are interchanged, $y \geq 0$ now, so we discard the negative root. Thus, $g^{-1}(x) = \sqrt{x - 1}$.

10.1

38. Given $N = (2000)2^t$, when $t = 0$, we have

$$N = (2000)2^0.$$
$$= (2000)(1) \quad 2^0 = 1$$
$$= 2000$$

Thus there are 2000 bacteria initially (at time $t = 0$). When $t = 7$,

$$N = (2000)2^7$$
$$= (2000)(128)$$
$$= 256{,}000.$$

In 7 hours there are 256,000 bacteria.

10.3

15. $\log_a xy^2\sqrt{z}$

$$= \log_a x + \log_a y^2 + \log_a \sqrt{z} \quad \text{Product rule}$$
$$= \log_a x + \log_a y^2 + \log_a z^{1/2} \quad \sqrt{z} = z^{1/2}$$
$$= \log_a x + 2\log_a y + \frac{1}{2}\log_a z \quad \text{Power rule}$$

17. $\frac{1}{2}\log_a y - \log_a x + 2\log_a z$

$$= \log_a y^{1/2} - \log_a x + \log_a z^2 \quad \text{Power rule}$$
$$= \log_a \sqrt{y} - \log_a x + \log_a z^2 \quad y^{1/2} = \sqrt{y}$$
$$= \log_a \frac{\sqrt{y}}{x} + \log_a z^2 \quad \text{Quotient rule}$$
$$= \log_a \frac{z^2\sqrt{y}}{x}$$

25. $\log_a 0.2 = \log_a \frac{2}{10}$

$$= \log_a \frac{1}{5}$$
$$= \log_a 1 - \log_a 5$$
$$= 0 - 0.6990$$
$$= -0.6990$$

10.4

34. We need to evaluate

$$p = 200 - 0.4e^{0.004x}$$

when $x = 100$.

$$p = 200 - 0.4e^{(0.004)(100)}$$
$$= 200 - 0.4e^{0.4}$$

Using a calculator, we evaluate $e^{0.4}$. The steps are:

0.4 [INV] [ln x] ⟶ [1.4918247]

Multiply this by 0.4.

[×] 0.4 ⟶ [0.5967299]

Change the sign and add 200.

[±] [+] 200 [=] ⟶ [199.40327]

The price is $199.40, to the nearest cent.

37. We need to evaluate

$$L = 8.8 + 2.2 \ln d$$

when $d = 15$.

$$L = 8.8 + 2.2 \ln 15$$

The steps on a calculator are:

15 [ln x] [×] 2.2 [+] 8.8 [=] ⟶ [14.75771]

To the nearest tenth, the limiting magnitude is 14.8.

10.5

5. $\log_2(x + 1) + \log_2(x - 1) = 3$

$$\log_2(x + 1)(x - 1) = 3 \quad \text{Product rule}$$
$$\log_2(x^2 - 1) = 3$$
$$x^2 - 1 = 2^3 \quad \text{Change to exponential equation}$$
$$x^2 - 1 = 8$$
$$x^2 - 9 = 0$$
$$(x - 3)(x + 3) = 0$$
$$x - 3 = 0 \quad \text{or} \quad x + 3 = 0$$
$$x = 3 \qquad\qquad x = -3$$

Check: $x = 3$ $\quad \log_2 (3 + 1) + \log_2 (3 - 1) \stackrel{?}{=} 3$

$$\log_2 4 + \log_2 2 \stackrel{?}{=} 3$$
$$2 + 1 \stackrel{?}{=} 3$$
$$3 = 3$$

$x = -3$ $\quad \log_2 (-3 + 1) + \log_2 (-3 - 1) \stackrel{?}{=} 3$

Since $\log_2 (-2)$ and $\log_2 (-4)$ are not defined, the only solution is 3.

16.
$$3^{x^2}9^x = \frac{1}{3}$$
$$3^{x^2}(3^2)^x = 3^{-1}$$
$$3^{x^2}3^{2x} = 3^{-1}$$
$$3^{x^2+2x} = 3^{-1}$$
$$x^2 + 2x = -1$$
$$x^2 + 2x + 1 = 0$$
$$(x + 1)^2 = 0$$
$$x + 1 = 0$$
$$x = -1$$

18. $2^{2x}3^{x+1} = 5$

Take the common log of both sides.

$$\log[2^{2x}3^{x+1}] = \log 5$$
$$\log 2^{2x} + \log 3^{x+1} = \log 5 \quad \text{Product rule}$$
$$2x \log 2 + (x + 1) \log 3 = \log 5 \quad \text{Power rule}$$
$$2x \log 2 + x \log 3 + \log 3 = \log 5$$
$$2x \log 2 + x \log 3 = \log 5 - \log 3$$
$$x(2 \log 2 + \log 3) = \log 5 - \log 3$$
$$x = \frac{\log 5 - \log 3}{2 \log 2 + \log 3}$$

This is the exact answer. We can use a calculator to get an approximate solution, $x \approx 0.206$.

20. If the intensity is *reduced* by 80% of I_0, then we are looking for the depth at which the intensity is 20% of I_0, which is $0.20I_0$. Thus, we need to solve

$$0.20I_0 = I_0e^{-0.008x}.$$
$$0.20 = e^{-0.008x} \quad \text{Divide both sides by } I_0$$
$$\ln 0.20 = \ln e^{-0.008x} \quad \text{Take natural log of both sides}$$
$$\ln 0.20 = -0.008x \quad \ln e^u = u$$
$$\frac{\ln 0.20}{-0.008} = x$$

Using a calculator we obtain $x \approx 201.17974$. To the nearest tenth, the depth at which the intensity is reduced by 80% is about 201.2 ft.

10.6

9. We need to find y when $x = 50{,}100x_0$ in $y = \log\frac{x}{x_0}$.

$$y = \log\frac{50{,}100x_0}{x_0}$$
$$= \log 50{,}100 \quad \text{Divide out } x_0$$
$$\approx 4.6998377$$

To the nearest tenth, the magnitude of the earthquake on the Richter scale is about 4.7.

12. Consider $y = 5000e^{0.02t}$. The initial population, when $t = 0$, is

$$y = 5000e^{(0.02)(0)} = 5000e^0 = 5000. \quad e^0 = 1$$

We want t when the population, y, is double this, which is 10,000.

$$10{,}000 = 5000e^{0.02t}$$
$$2 = e^{0.02t} \quad \text{Divide both sides by 5000}$$
$$\ln 2 = \ln e^{0.02t} \quad \text{Take natural log of both sides}$$
$$\ln 2 = 0.02t \quad \ln e^u = u$$
$$\frac{\ln 2}{0.02} = t$$
$$34.657359 \approx t \quad \text{Use a calculator}$$

Thus, the population will double in about 34.7 years.

13. Consider $y = 250e^{-0.5t}$. The initial amount is when $t = 0$.

$$y = 250e^{(-0.5)(0)} = 250e^0 = 250.$$

We want t when the amount y is 125 (half the initial amount).

$$125 = 250e^{-0.5t}$$
$$0.5 = e^{-0.5t}$$
$$\ln 0.5 = \ln e^{-0.5t}$$
$$\ln 0.5 = -0.5t$$
$$\frac{\ln 0.5}{-0.5} = t$$
$$1.3862944 \approx t$$

The half-life is about 1.39 years.

11.1

5. $y = x^2 - 5$ This can be written as

$$y = 1 \cdot (x - 0)^2 - 5.$$

The vertex is $(0, -5)$, and since $a = 1 > 0$, the parabola opens up.

9. $x = -2y^2 - 4y - 7$ We complete the square in y to write the equation in standard form.

$$x = -2(y^2 + 2y \qquad) - 7 \quad \text{Factor out } -2$$
$$= -2(y^2 + 2y + 1) - 7 + 2$$

Since 1 is added inside the parentheses to complete the square, and it is multiplied by the factor of -2, we have really added -2 to the expression which is then balanced out by adding $+2$ to the -7. We have

$$x = -2(y + 1)^2 - 5.$$

The vertex is $(-5, -1)$, and since $a = -2 < 0$, the parabola opens left.

11.2

9. $4x^2 + 4y^2 + 40x - 4y + 37 = 0$
We need to complete the square in both x and y.

$$4x^2 + 40x \qquad + 4y^2 - 4y \qquad = -37$$

But to complete the square we must have the coefficient of the squared terms equal to 1. Divide through by 4.

$$x^2 + 10x \qquad + y^2 - y \qquad = -\frac{37}{4}$$

Add $\left[\frac{1}{2}(10)\right]^2 = [5]^2 = 25$ and $\left[\frac{1}{2}(-1)\right]^2 = \left[-\frac{1}{2}\right]^2 = \frac{1}{4}$ to both sides.

$$x^2 + 10x + 25 + y^2 - y + \frac{1}{4} = -\frac{37}{4} + 25 + \frac{1}{4}$$

$$(x + 5)^2 + \left(y - \frac{1}{2}\right)^2 = 16 = 4^2$$

The radius is 4 and the center is $\left(-5, \frac{1}{2}\right)$.

11.4

4. $5x^2 + xy - y^2 = -1$
$y - 2x = 1$

Solve the second equation for y and substitute into the first.

$$y = 2x + 1$$
$$5x^2 + x(2x + 1) - (2x + 1)^2 = -1$$
$$5x^2 + 2x^2 + x - (4x^2 + 4x + 1) = -1$$
$$5x^2 + 2x^2 + x - 4x^2 - 4x - 1 = -1$$
$$3x^2 - 3x - 1 = -1$$
$$3x^2 - 3x = 0$$
$$3x(x - 1) = 0$$

$x = 0$ or $x - 1 = 0$
$x = 1$
$y = 2(0) + 1 = 1$ $\quad y = 2(1) + 1 = 3$

The solutions are $(0, 1)$ and $(1, 3)$.

9. $3xy - y^2 = -13$
$2xy + y^2 = -2$

$6xy - 2y^2 = -26$ **Multiply first equation by 2**
$-6xy - 3y^2 = 6$ **Multiply second equation by -3**

$$-5y^2 = -20$$
$$y^2 = 4$$
$$y = \pm 2$$

For $y = 2$, $3x(2) - (2)^2 = -13$
$$6x - 4 = -13$$
$$6x = -9$$
$$x = \frac{-9}{6} = -\frac{3}{2}.$$

For $y = -2$, $3x(-2) - (2)^2 = -13$
$$-6x - 4 = -16$$
$$-6x = -9$$
$$x = \frac{-9}{-6} = \frac{3}{2}.$$

The solutions are $\left(-\frac{3}{2}, 2\right)$ and $\left(\frac{3}{2}, -2\right)$.

13. Let x = one number,
y = the other number.

$x^2 + y^2 = 100$ **The sum of the squares is 100**
$x - y = 2$ **The difference of the numbers is 2**
$x = y + 2$ **Solve the second equation for x**

$$(y + 2)^2 + y^2 = 100$$
$$y^2 + 4y + 4 + y^2 = 100$$
$$2y^2 + 4y + 4 - 100 = 0$$
$$2y^2 + 4y - 96 = 0$$
$$y^2 + 2y - 48 = 0$$
$$(y - 6)(y + 8) = 0$$

$y - 6 = 0$ or $y + 8 = 0$
$y = 6$ $\qquad y = -8$
$x = 6 + 2 = 8$ $\qquad x = -8 + 2 = -6$

12.1

7. $x_n = \dfrac{(-1)^n}{n}$ To find x_1, substitute 1 for n.

$$x_1 = \frac{(-1)^1}{1} = \frac{-1}{1} = -1$$

To find x_2, substitute 2 for n.

$$x_2 = \frac{(-1)^2}{2} = \frac{1}{2}$$

Similarly, substitute 3 for n, 4 for n, and 7 for n.

$$x_3 = \frac{(-1)^3}{3} = -\frac{1}{3}$$
$$x_4 = \frac{(-1)^4}{4} = \frac{1}{4}$$
$$x_7 = \frac{(-1)^7}{7} = -\frac{1}{7}$$

31. $\sum_{k=4}^{5} (k^2 + 1) = (4^2 + 1) + (5^2 + 1)$
$= (16 + 1) + (25 + 1) = 17 + 26$

12.2

13. $a_2 = 3$, $a_4 = 7$, find d and a_1.
We know that $a_n = a_1 + (n-1)d$. Thus,

$$a_2 = 3 = a_1 + (2-1)d \quad \text{Note that } n = 2$$

and $a_4 = 7 = a_1 + (4-1)d$. Note that $n = 4$

Thus, we have the system

$$a_1 + d = 3$$
$$a_1 + 3d = 7.$$

Subtract to eliminate a_1.

$$-2d = -4$$
$$d = 2$$

Substitute $d = 2$ to find a_1.

$$a_1 + 2 = 3$$
$$a_1 = 1$$

12.3

13. $a_2 = \frac{1}{3}$, $a_5 = -9$; find r and a_1.
We know that $a_n = a_1r^{n-1}$. Thus,

$$a_2 = \frac{1}{3} = a_1r^1 \quad \text{and} \quad a_5 = -9 = a_1r^4.$$

Divide these two to obtain

$$\frac{a_2}{a_5} = \frac{\frac{1}{3}}{-9} = \frac{a_1r}{a_1r^4}.$$

$$-\frac{1}{27} = \frac{1}{r^3}$$

Thus, $r^3 = -27$ so that $r = -3$. Substitute $r = -3$ to find a_1.

$$\frac{1}{3} = a_1(-3)$$

$$-\frac{1}{9} = a_1$$

12.4

7. $\sum_{k=1}^{\infty} \left(\frac{1}{4}\right)^k = \left(\frac{1}{4}\right)^1 + \left(\frac{1}{4}\right)^2 + \left(\frac{1}{4}\right)^3 + \left(\frac{1}{4}\right)^4 + \cdots$
When we expand in this way, we see that the sequence of terms forms a geometric sequence with $a_1 = \frac{1}{4}$ and $r = \frac{1}{4}$. Since $|r| = \left|\frac{1}{4}\right| < 1$, the sum will exist.

$$S = \frac{a_1}{1-r} = \frac{\frac{1}{4}}{1-\frac{1}{4}} = \frac{\frac{1}{4}}{\frac{3}{4}} = \frac{1}{3}$$

13. $a_2 = 10$, $a_5 = \frac{2}{25}$, find a_1 and S.

$$a_2 = 10 = a_1r^{2-1} = a_1r \text{ and } a_5 = \frac{2}{25} = a_1r^{5-1} = a_1r^4$$

Divide these two to obtain

$$\frac{a_2}{a_5} = \frac{10}{\frac{2}{25}} = \frac{a_1r}{a_4r^4}.$$

$$125 = \frac{1}{r^3}$$

$$r^3 = \frac{1}{125} = \frac{1}{5^3} = \left(\frac{1}{5}\right)^3$$

Thus, $r = \frac{1}{5}$.

Then $a_2 = 10 = a_1r = a_1\left(\frac{1}{5}\right)$, so that

$$10 = \frac{a_1}{5}$$

$$50 = a_1.$$

Next we find S.

$$S = \frac{a_1}{1-r} = \frac{50}{1-\frac{1}{5}} = \frac{50}{\frac{4}{5}} = \frac{250}{4} = \frac{125}{2}$$

20. If we add up the number of revolutions,

$$16 + 4 + 1 + \frac{1}{4} + \cdots,$$

we recognize an infinite geometric sequence with $a_1 = 16$ and $r = \frac{1}{4}$. Thus,

$$S = \frac{a_1}{1-r} = \frac{16}{1-\frac{1}{4}} = \frac{16}{\frac{3}{4}} = \frac{64}{3} = 21\frac{1}{3}$$

revolutions.

12.5

10. $(x^{-1} + x)^5$
Since we are expanding $(x^{-1} + x)^5$, $n = 5$, $a = x^{-1}$, and $b = x$. From Row 5 of Pascal's triangle, the coefficients are

1 5 10 10 5 1.

Substitute x^{-1} for a and x for b in

$$a^5 + 5a^4b + 10a^3b^2 + 10a^2b^3 + 5ab^4 + b^5.$$

We obtain

$$(x^{-1})^5 + 5(x^{-1})^4x + 10(x^{-1})^3x^2 + 10(x^{-1})^2x^3 + 5(x^{-1})x^4 + x^5$$
$$= x^{-5} + 5x^{-4}x + 10x^{-3}x^2 + 10x^{-2}x^3 + 5x^{-1}x^4 + x^5$$
$$= x^{-5} + 5x^{-3} + 10x^{-1} + 10x + 5x^3 + x^5$$
$$= \frac{1}{x^5} + \frac{5}{x^3} + \frac{10}{x} + 10x + 5x^3 + x^5.$$

INDEX